UNIT CONVERSIONS (EQUIVALENTS)

Length

1 in = 2.54 cm
1 cm = 0.394 in
1 ft = 30.5 cm
1 m = 39.4 in = 3.28 ft
1 mi = 5280 ft = 1.61 km
1 km = 0.621 mi
1 nautical mile = 6080 ft = 1.85 km
1 fermi = 1 femtometer (fm) = 10^{-15} m
1 angstrom (Å) = 10^{-10} m
1 light-year = 9.46×10^{15} m
1 parsec = 3.26 light-years

Time

1 day = 8.64×10^4 s
1 year = 3.156×10^7 s

Speed

1 mi/h = 1.47 ft/s = 1.61 km/h = 0.447 m/s
1 km/h = 0.278 m/s = 0.621 mi/h
1 ft/s = 0.305 m/s = 0.682 mi/h
1 m/s = 3.28 ft/s = 3.60 km/h
1 knot = 1.151 mi/h = 0.5144 m/s

Angle

1 radian (rad) = 57.30° = 57°18′
1° = 0.01745 rad
1 rev/min (rpm) = 0.1047 rad/s

Mass

1 atomic mass unit (u) = 1.6606×10^{-27} kg
1 slug = 14.6 kg
1 kg = 0.0685 slug
[1 kg has a weight of 2.21 lb where g = 9.80 m/s².]

Force

1 lb = 4.45 N
1 N = 10^5 dyne = 0.225 lb

Energy and Work

1 J = 10^7 ergs = 0.738 ft · lb
1 ft · lb = 1.36 J = 1.29×10^{-3} Btu = 3.25×10^{-4} kcal
1 kcal = 4.18×10^3 J = 3.97 Btu
1 Btu = 252 cal = 778 ft · lb = 1054 J
1 eV = 1.602×10^{-19} J
1 kWh = 3.60×10^6 J = 860 kcal

Power

1 W = 1 J/s = 0.738 ft · lb/s
1 hp (U.S.) = 550 ft · lb/s = 746 W
1 hp (metric) = 750 W

Pressure

1 atm = 1.013 bar = 1.013×10^5 N/m²
**   = 14.7 lb/in² = 760 torr**
1 lb/in² = 6.90×10^3 N/m²
1 Pa = 1 N/m² = 1.45×10^{-4} lb/in²

SI DERIVED UNITS AND THEIR ABBREVIATIONS

Quantity	Unit	Abbreviation	In terms of base units[†]
Force	newton	**N**	$kg \cdot m/s^2$
Energy and work	joule	**J**	$kg \cdot m^2/s^2$
Power	watt	**W**	$kg \cdot m^2/s^3$
Pressure	pascal	**Pa**	$kg/(m \cdot s^2)$
Frequency	hertz	**Hz**	s^{-1}
Electric charge	coulomb	**C**	$A \cdot s$
Electric potential	volt	**V**	$kg \cdot m^2/(A \cdot s^3)$
Electric resistance	ohm	**Ω**	$kg \cdot m^2/(A^2 \cdot s^3)$
Capacitance	farad	**F**	$A^2 \cdot s^4/(kg \cdot m^2)$
Magnetic field strength	tesla	**T**	$kg/(A \cdot s^2)$
Magnetic flux	weber	**Wb**	$kg \cdot m^2/A \cdot s^2$
Inductance	henry	**H**	$kg \cdot m^2/(s^2 \cdot A^2)$

[†] kg = kilogram (mass), m = meter (length), s = second (time),
  A = ampere (electric current).

METRIC (SI) MULTIPLIERS

Prefix	Abbreviation	Value
Tera	**T**	10^{12}
Giga	**G**	10^9
Mega	**M**	10^6
Kilo	**k**	10^3
Hecto	**h**	10^2
Deka	**da**	10^1
Deci	**d**	10^{-1}
Centi	**c**	10^{-2}
Milli	**m**	10^{-3}
Micro	**μ**	10^{-6}
Nano	**n**	10^{-9}
Pico	**p**	10^{-12}
Femto	**f**	10^{-15}

2nd Edition

PHYSICS

PRINCIPLES WITH
APPLICATIONS

Douglas C. Giancoli

PRENTICE HALL, INC. Englewood Cliffs, New Jersey 07632

Library of Congress Cataloging in Publication Data

GIANCOLI, DOUGLAS C.
  Physics, principles with applications.

  Includes index.
  1. Physics. I. Title.
QC23.G399 1985   530   84-22250
ISBN 0-13-672627-5

Editorial/production supervision: Virginia Huebner
Interior design: Janet Schmid
Cover photograph by Kate Kline May/D. Giancoli.
Development editor: Raymond Mullaney
Manufacturing buyer: John Hall

Printed in the United States of America

10 9 8 7 6 5 4

ISBN 0-13-672627-5 01

Prentice-Hall International, Inc., *London*
Prentice-Hall of Australia Pty. Limited, *Sydney*
Editora Prentice-Hall do Brasil, Ltda., *Rio de Janeiro*
Prentice-Hall Canada Inc., *Toronto*
Prentice-Hall Hispanoamericana, S.A., *Mexico*
Prentice-Hall of India Private Limited, *New Delhi*
Prentice-Hall of Japan, Inc., *Tokyo*
Prentice-Hall of Southeast Asia Pte. Ltd., *Singapore*
Whitehall Books Limited, *Wellington, New Zealand*

For Andrea, Francesca, and Mary Teresa

CONTENTS

PREFACE

This revised introductory algebra-based physics textbook incorporates a number of changes and updatings (described below). But its basic intent and general outline remain the same. The book is intended to be readable, interesting, and accessible to students, and is meant to give them a thorough understanding of the basic concepts of physics and, by means of many interesting applications, to prepare them to use physics in their own lives and professions. It is particularly appropriate for introductory physics courses taken by students studying biology, (pre)medicine, architecture, technology, earth and environmental sciences, and other disciplines; it is also suitable for use in colleges that offer only one introductory physics course.

    This book offers, above all, an in-depth presentation of physics. I have tried to avoid the dogmatic approach of treating topics formally and abstractly first, and only later (if at all) coming down to earth and relating the material to the students' own experience. This approach may be appealing (it's elegant), but it can slow down the learning process for all but the best students. My approach is to recognize that physics is a description of reality and thus to start each topic with concrete observations and experiences that students can directly relate to. Only then are readers led into the more formal and abstract treatment of topics. Not only does this make the material more interesting and easier to understand, but it is closer to the way physics is actually practiced. Historically, we didn't start with the first law of thermodynamics, for example, and then derive all kinds of consequences from it; rather, the law was a generalization of all kinds of phenomena. In addition, I have endeavored to present the basic concepts of physics in their historical and philosophical context. I have also tried to avoid the problem of leaving certain topics "hanging", with students wondering, "Why did we study that?" Thus I have tried to indicate why each topic is important, and to bring each topic to completion. We study static forces in structures, for example, partly because real materials are elastic and also can fracture; so I have included the latter topics in the statics chapter.

    Another important feature of this book is the inclusion of a wide range of examples and applications from other fields: biology, medicine, architecture, technology, earth sciences, the environment, and daily living. These serve not only to enliven the book but also to show how physics is important in other disciplines and professions, and in everyday life. Some applications serve only as examples of physical principles. Others are treated in depth, with whole sections devoted to them (among these are the study of medical imaging systems, constructing arches and domes, and the effects of radiation). But applications do not dominate the text (this is, after all, a physics book!). They have been carefully chosen and integrated into the text so as not to interfere with the development of the physics but rather to illuminate it.

Much attention is given to problem solving. Explicit hints on how to attack problems (rewritten for this new edition) are given in several places early in the book, notably in Sections 1-9, 3-8, and 3-10 (whose title is "Notes on Problem Solving"). The last-named section is placed after the students have had some experience wrestling with problems and hopefully will then be motivated to read and pay attention to this section; it can, of course, be covered earlier if an instructor so desires. Some 250 examples are fully worked out in the text; these help students to fix ideas in their minds, to demonstrate interesting applications, and to help students develop problem-solving skills. Many examples are taken from everyday life and aim at being realistic. There are over 2250 end-of-chapter exercises, including more than 650 questions that require verbal answers based on an understanding of the concepts, and over 1600 problems involving mathematical calculation. The wide range of problems relates directly to the physics as well as the applications. They are arranged by sections, and they are graded according to difficulty: level I problems are simple, requiring an understanding of basic definitions and concepts and typically the use of a single equation; level II are normal problems, requiring more thought and often the combination of two different concepts; level III are the most difficult, often requiring synthesis of three or more concepts or perhaps dealing with more advanced material. The arrangement by section number means only that those problems depend on material up to and including that section; earlier sections and chapters are often relied upon, particularly in level II or III problems. The ranking of the problems by difficulty (I, II, III) is necessarily subjective and is intended only as a guide. Level II problems, particularly, are of a very wide range. I suggest that instructors assign a significant number of the level I and level II problems and reserve level III problems mainly to challenge superior students. Although most level I problems may seem easy, they help to build self-confidence—an important part of learning, especially in physics. Answers to odd-numbered problems are given at the back of the book. Throughout the text *Système international* (SI) units are used. Other metric and British units are defined for informational purposes.

This second edition of *Physics, Principles with Applications*, has been thoroughly revised to improve coverage, thoroughness, and clarity, as well as to include the latest developments in physics and its applications. The most significant improvements over the first edition involve some expansion and reorganization of mechanics, some reorganization of modern physics with the latest developments, and new applications. The principal changes are:

1. Kinematics is now divided into two chapters: Chapter 1 is devoted to one-dimensional motion, whereas Chapter 2 deals with kinematics in two (or three) dimensions, including projectile motion. This allows a more gradual introduction to physics, and also a more complete treatment of kinematics including more on relative velocity, vectors, and the use of Δ (to represent "change in").

2. There are now separate chapters on work and energy (Chapter 5) and on linear momentum (Chapter 6), with fuller and more careful treatments of each, particularly for potential energy, conservation of energy, collisions, and center of mass.

3. Rotational motion now has a chapter all to itself (Chapter 7), allowing a fuller treatment in one place of all aspects of rotation about an axis: angular

quantities, rotational kinematics and dynamics, torque, rotational kinetic energy, angular momentum and its conservation including its vector nature.

4. Circular motion and gravitation, now Chapter 4, is treated right after dynamics (before energy and momentum) since it deals with acceleration and force. (It could, instead, be treated just before Chapter 7 on rotational motion, if desired.)

5. The old chapter on ac circuits and electronics has been eliminated; ac circuits are now included in Chapter 20 with electromagnetic induction; semiconductors and their use in electronics are now included with modern physics in the new Chapter 28 on molecules and solids; vacuum tubes have been dropped except for the CRT, which is treated in Chapter 19.

6. Modern physics has been reorganized a little and brought up to date. Principal changes are: (a) a special chapter (28) on molecules and solids; (b) expansion of the effects and uses of radiation, addition of new imaging techniques (see below), NMR, and the new SI units for dosimetry; (c) new particles such as the τ lepton, W^+ and Z^0 particles, the bottom and top quarks; (d) quantum chromodynamics (QCD), electroweak theory, grand unified theories (GUT) and symmetry breaking.

7. Some parts of the book have been rewritten to improve clarity or pedagogy. Examples are more careful and consistent treatments of work, energy, potential energy, center of mass, Newton's third law, heat and thermal energy, electric potential, and sign conventions for lenses and mirrors.

8. There are new applications to biology and medicine, including the new imaging techniques using ultrasound (Chapter 14), computerized tomography (CT scanning) along with conventional X ray (Chapter 24), nuclear medicine, emission tomography (SPET and PET), and NMR imaging (Chapter 30).

9. New or expanded geophysical applications include examples involving continental plates (Sections 9-7 and 9-13), the tides (Chapter 4), gravity anomalies and mineral exploration (Chapter 4), the geophone or seismometer (Section 20-8).

10. Nearly all the figures have been redrawn to improve clarity and accuracy. As a general rule, real (tangible) objects are shown in color whereas our analysis (coordinate axes, graphs, field lines, arrows representing vectors, and so on) is shown in black.

11. The division of each chapter has been simplified so there are only main sections, and no subsections; many of the subsections of the first edition are now regular sections, and all section headings are in the table of contents.

12. Among other changes, this revision incorporates the new definition of the meter in terms of the speed of light. The worked-out examples include a number of more sophisticated ones so students can see how to treat more complex situations. Some of the old problems have been retained, sometimes with the numbers changed to protect the innocent (and foil the guilty), a few have been dropped, and many new problems (and questions) have been added.

To make room for the additional material in this new edition, it was necessary to eliminate some material that was in the first edition. Almost nothing of the basic physics was cut (except, perhaps, slightly, to improve clarity). To incorporate the expanded treatments of basic physics, a few of the more esoteric topics were dropped or shortened. And some of the applications which are less current, or for which adequate treatments can readily be found elsewhere, were shortened or dropped to make room for the latest applications, such as the various new forms of medical imaging, which are deeply based in physics and for which treatments at this level are not yet readily available.

Although there has been some reorganization of material within some areas, the general outline of this new edition retains the traditional order of topics: mechanics, including fluid mechanics (Chapters 1–9), kinetic theory and thermodynamics (Chapters 10–12), vibrations, waves, and sound (Chapters 13 and 14), electricity and magnetism (Chapters 15–21), light (Chapters 22–24), and modern physics (Chapters 25–31).

Nearly all topics customarily taught in introductory physics courses are included here. The tradition of beginning with mechanics is sensible, I believe, since it was developed first, historically, and since so much else in physics depends on it. Within mechanics, there are various ways to order the topics, and this book allows for considerable flexibility. Statics, for example, can be covered either before or after dynamics. I prefer to cover statics after dynamics partly because many students seem to have trouble with the concept of force. (They tend to associate force with motion, and it seems to help if they understand the nature of this connection before dealing with forces without motion.) This order also allows full development of the concept of torque before it is used in statics. Moreover statics is a special case of dynamics—we study statics so that we can prevent structures from becoming dynamic (falling down)—and that sense of being at the limit of dynamics is intuitively helpful. None-theless, statics (Chapter 8) can be covered earlier, if desired, before dynamics, after a brief introduction to vectors. Another option is the position of the chapters on light. I have placed them after electricity and magnetism and EM waves, as is typical. However, light could be treated immediately after the chapters on waves (Chapters 13 and 14), thereby keeping the various types of wave motion in one place. Another position choice involves special relativity (Chapter 25), which is located along with the other chapters on modern physics, after EM waves and light have been covered. Relativity could be covered, if desired, along with mechanics—say after Chapter 6—since it depends (except for the op-tional Section 25-2) mainly on material through Chapter 6.

The book contains more material than can be covered in most one-year courses. This was done to give instructors flexibility in choice of topics. The wide range of subjects also means that students can learn about many topics even though there is not class time for them. This aspect makes the book more valuable to students as a resource and as a reference book. Sections marked with a star (asterisk) are considered optional. These sections contain slightly more advanced physics material, often material not usually covered in typical courses, and/or interesting applications. They contain no material needed in later chapters (except, perhaps, in later optional sections). This does not imply that all non-starred sections must be covered; there still remains considerable flexibil-ity in the choice of material to suit the needs of students and instructors.

For use in a two-semester course, the book can be considered to be divided roughly in half between Chapters 14 and 15. The first half would then include mechanics, heat, and sound, with the second half containing electricity and magnetism, light, and modern physics. (The slightly longer second half has more optional sections.) As mentioned earlier, the order and choice of topics is flexible, so other course outlines are possible. For a brief or ''bare bones'' course, all the optional material could be dropped as well as major parts of Chapters 9, 14, 18, 21, 27, and 31, as well as selected parts of Chapters 6, 7, 8, 12, 20, 23, 24, and 30.

Mathematics can be an obstacle to student understanding. To avoid frightening students with an initial chapter on mathematics, I have instead used an appendix for review of algebra, geometry, accuracy and signifi-cant figures, exponents, powers of ten, and proportionality. Other appen-dixes cover order-of-magnitude estimating and dimensional analysis. Other important mathematical tools, such as addition of vectors and trig-onometry, are dealt with in the text where first needed. Difficult language too can hinder understanding; and to put students at their ease, I have tried to write in a relaxed, colloquial style, avoiding jargon. New or unusual terms are carefully defined when first used. It is necessary, I feel, to pay careful attention to detail, especially when deriving an important result. Whether it is a verbal discussion, or a mathematical one, I have aimed at including all steps in a derivation so that students don't get bogged down in details and then fail to understand the concept as a whole. I have tried to make clear which equations are general, and which are not, by explicitly stating the limitations of important equations in brackets next to the equation, such as

$$x = x_0 + v_0 t + \tfrac{1}{2} at^2. \quad \text{[constant acceleration]}$$

The new revision of this book has depended to a great extent on the hundreds of instructors who have used the book in class and were kind enough to send me their comments and suggestions for improvement. To all of them I owe a debt of thanks. I also wish to thank the professors who read through the text, either the first edition or the newly revised manuscript (or both), and offered valuable suggestions; these include John Anderson (University of Pittsburgh), Gene Barnes (CSU, Sacramento), Isaac Bass, Paul A. Bender (Washington State University), Joseph Boyle (Miami-Dade Community College), Peter Brancazio (Brooklyn College, CUNY), Warren Deshotels (Marquette), Laurent Hodges (Iowa State), Gordon Jones (Mississippi State), Michael Lieber (U. Arkansas), Robert Messina, David Mills (College of the Redwoods), Ed Nelson (U. Iowa), John Reading (Texas A & M), William Riley (Ohio State U.), D. Lee Rutledge (Oklahoma State U.), Paul Urone (CSU, Sacramento), Jearl Walker (Cleveland State U.), Gareth Williams (San Jose State U.), Peter Zimmerman (Louisiana State U.). I owe special thanks to Professors John Heilbron, Richard Marrus, and Howard Shugart for many helpful discussions and suggestions, and for their hospitality at the University of California, Berkeley, as well as to Professors Tito Arecchi, Director of the National Optics Institute, University of Florence, and Paolo Galluzzi and the staff of the Institute and Museum of the History of Science, Florence. I also wish to acknowledge the work of Joseph Boyle in preparing a study guide for this new edition and of John Reading who prepared the solutions manual with the assistance of Ed Gibson, Stewart Ryan, and J. Neal Huffaker. Finally, I wish to thank the many people at Prentice-Hall who worked on this project and made it possible, especially Ray Mullaney, Doug Humphrey, John Davis, Tim Moore, and Virginia Huebner. The responsibility for all errors lies, of course, with me. I welcome comments and corrections.

Douglas C. Giancoli

NOTES TO STUDENTS AND INSTRUCTORS ON THE FORMAT

1. Sections marked with a star (*) are considered optional (see the Preface).

2. The customary conventions are used: Symbols for physical quantities are italicized (such as *m* for mass), whereas units are not italicized (m for meter); boldface (**F**) is used for vectors, and this is discussed in the text.

3. Important terms are italicized where they are introduced, and the most important are in boldface (such as *coefficient of friction* and **acceleration**).

4. Few equations are valid in all situations. Where practical, the limitations of important equations are stated in square brackets next to the equation.

5. Worked-out Examples and their Solutions in the text are set off with a vertical colored line in the margin.

6. Each chapter ends with a Summary, giving a brief review of important concepts and terms (the most important ones are italicized here). The summaries are not intended to give an understanding of the material, which can only be had from a study of the chapter. Optional topics are not normally included in the summaries (except for Chapter 28 which is all optional).

7. Following the Summary in each chapter are sets of Questions that students should attempt to answer (to themselves at least) and Problems arranged according to section and difficulty (see the Preface). Questions and Problems that relate to optional sections are starred.

8. The appendixes contain useful background and reference material such as a mathematical review, discussions of order-of-magnitude estimating and dimensional analysis, and a table of isotopes with atomic masses and other data. Tables used frequently are located inside the front and back covers.

9. The extensive Index can be a useful tool. For example, it can be used to look up concepts or words whose meanings have been forgotten.

INTRODUCTION

Physics is the most basic of the sciences. It deals with the behavior and structure of matter. The field of physics is usually divided into the areas of motion, fluids, heat, sound, light, electricity and magnetism, and the modern topics of relativity, atomic structure, and nuclear physics. We will cover all these topics in this book, beginning with motion (or mechanics, as it is sometimes called). But before we begin on the physics itself, let us take a brief look at how this activity called "science," including physics, is actually practiced.

▌ The practice of science: science and creativity

The principal aim of all sciences, including physics, is generally considered to be the ordering of the complex appearances detected by our senses—that is, an ordering of what we often refer to as the "world around us." Many people think of science as a mechanical process of collecting facts and devising theories. This is not the case. Science is a creative activity that in many respects resembles other creative activities of the human mind.

Let's take some examples to see why this is true. One important aspect of science is *observation* of events. But observation requires imagination, for scientists can never include everything in a description of what they observe. Hence, scientists must make judgments about what is relevant in their observations. As an example, let us consider how two great minds, Aristotle (384–322 B.C.) and Galileo (1564–1642), interpreted motion along a horizontal surface. Aristotle noted that objects given an initial push along the ground (or on a table top) always slow down and stop. Consequently Aristotle believed that the natural state of a body is at rest. Galileo, in his reexamination of horizontal motion in the early 1600s, chose rather to study the idealized case of motion free from resistance. In fact, Galileo imagined that if friction could be eliminated, an object given an initial push along a horizontal surface would continue to move indefinitely without stopping. He concluded that for an object to be in motion was just as natural as to be at rest. By seeing something new in the same "facts," Galileo is often given credit for founding our modern view of motion (more details in Chapters 1, 2, and 3). This seeing of something new was surely inspired thinking.

Theories are never derived from observations—they are *created* to explain observations. They are inspirations that come from the minds of human beings. For example, the idea that matter is made up of atoms (the atomic theory) was certainly not arrived at because someone observed atoms. Rather, the idea sprang from a creative mind. The theory of relativity, the electromagnetic theory of light, and Newton's law of universal gravitation were likewise the result of inspiration.

The great theories of science may be compared, as creative achievements, with great works of art or literature. But how does science differ from these other creative activities? One important difference is that science requires *testing*[†] of its ideas or theories to see if predictions are borne out by experiment. Indeed, careful experimentation is a crucial part of physics.

Although the testing of theories can be considered to distinguish science from other creative fields, it should not be assumed that a theory is "proved" by testing. First of all, no measuring instrument is perfect, so precise confirmation cannot be possible. Furthermore, it is not possible to test a theory in every single possible circumstance. Hence a theory can never be absolutely verified. In fact theories themselves are generally not perfect—a theory rarely agrees with experiment exactly, within experimental error, in every single case in which it is tested. Indeed, the history of science tells us that theories come and go; that long-held theories are replaced by new ones. The process of one theory replacing another is an important subject in the philosophy of science today; we can discuss it here only briefly.

A new theory is accepted by scientists in some cases because its predictions are quantitatively in much better agreement with experiment than is the older theory. But in many cases, a new theory is accepted only if it explains a greater *range* of phenomena than does the older one. Copernicus's sun-centered theory of the universe, for example, was no more accurate than Ptolemy's earth-centered theory for predicting the motion of heavenly bodies. But Copernicus's theory had consequences that Ptolemy's did not: for example, it made possible a determination of the order and distance of the planets and predicted the moonlike phases in the appearances of Venus. A simpler (or no more complex) and richer theory, one which unifies and explains a greater variety of phenomena, is more useful and beautiful to a scientist. And this aspect, as well as quantitative agreement, plays a major role in the acceptance of a theory.

An important aspect of any theory is how well it can quantitatively predict phenomena; and from this point of view, a new theory may often seem to be only a minor advance over the old one. For example, Einstein's theory of relativity gives predictions that differ very little from the older theories of Galileo and Newton in nearly all everyday situations; its predictions are better mainly in the extreme case of very high speeds close to the speed of light. From this point of view, the theory of relativity might be considered as mere "fine-tuning" of the older theory. But quantitative prediction is not the only important outcome of a theory. Our view of the world is affected as well. As a result of Einstein's theory of relativity, for example, our concepts of space and time have been completely changed; and we have

---

[†] Some philosophers therefore emphasize that testing of a theory can be used only to *falsify* it, not to confirm it—and/or to put a limit on its range of validity.

come to see mass and energy as a single entity (via the famous equation $E = mc^2$). Indeed, our view of the world underwent a major change when relativity theory came to be accepted.

▌ Physics and its relation to other fields

For a long time science was more or less a united whole known as natural philosophy. Not until the last century or two did the distinctions between physics and chemistry and even the life sciences become prominent. Indeed, the sharp distinctions we now see between the arts and the sciences is itself but a few centuries old. It is no wonder then that the development of physics has both influenced, and been influenced by, other fields. For example, Leonardo da Vinci's notebooks contain the first references to the forces present within a structure, a subject we consider as physics today; but Leonardo was interested, at least in part, because of the relevance to architecture and building. Early work in electricity that led to the discovery of the electric battery and electric current was done by an eighteenth-century physiologist, Luigi Galvani (1737–1798); he noticed the twitching of frogs' legs in response to an electric spark and later that the muscles twitched when in contact with two dissimilar metals (see Chapter 17). At first this phenomenon was known as "animal electricity," but it shortly became clear that current electricity itself could exist in the absence of an animal. More recently, in the 1930s and 1940s, a number of scientists trained as physicists became interested in applying the ideas and techniques of physics to problems in microbiology. Among the most prominent were Max Delbrück (1906–1981) and Erwin Schrödinger (1887–1961). They hoped, among other things, that by studying biological organisms, some new unsuspected laws of physics might be uncovered. Alas, this hope has not been realized; but their efforts gave rise to the field we now call "molecular biology" and have resulted in a dramatic increase in our understanding of the genetics and structure of living beings.

One does not have to be a research scientist in, say, medicine or molecular biology to be able to use physics in his or her work. A zoologist, for example, may find it useful to know how prairie dogs and other animals can live underground without suffocating. A botanist may wonder how water can get to the tops of trees. A physical therapist will do a more effective job if aware of the principles of lever action, center of gravity, and the action of forces within the human body. A knowledge of the operating principles of optical and electronic equipment is helpful in a variety of fields. Life scientists and architects alike will be interested in the nature of heat loss and gain in human beings and the resulting comfort or discomfort. Architects themselves may never have to calculate, for example, the dimensions of the pipes in a heating system or the forces involved in a given structure to determine if it is stable; but they must know the principles behind these systems in order to make realistic designs and to communicate effectively with engineering consultants and other specialists. From the aesthetic or psychological point of view, too, architects must be aware of the forces involved in a structure—for instability, even if only apparent, can be discomforting to those who must live or work in the structure. Indeed, many of the features we admire in the architecture of the past three millennia were introduced not for their decorative effect but rather for practical purposes. For example, the cornices and pediments of windows protected the facades of buildings from the erosive effects of water. The base and capital of a column acted to diffuse the load from the column onto the masonry above and below. The development of the arch as a means to span a space and at the same time support a heavy load will be discussed in Chapter 8. There we will see that the pointed, or Gothic, arch was not originally a decorative device but a technological development of considerable importance.

The list of how physics relates to other fields is extensive. In the chapters that follow we will discuss many such applications in detail, as we carry out our principal aim of explaining basic physics.

▋ Models, theories, and laws

When scientists are trying to understand a particular set of phenomena, they often make use of a **model.** A model, in the scientists' sense, is a kind of analogy or mental image of the phenomena in terms of something we are familiar with. One example is the wave model of light. We cannot see waves of light as we can water waves; but it is valuable to think of light as if it were made up of waves because experiments on light indicate that it behaves in many respects as water waves do.

The purpose of a model is to give us a mental or visual picture—something to hold onto—when we cannot see what actually is happening. Models often give us a deeper understanding: the analogy to a known system (for instance, water waves in the above example) can suggest new experiments to perform and can provide ideas about what other related phenomena might occur.

No model is ever perfect, and scientists are constantly trying to refine their models or to think up new ones when old ones do not seem adequate. The atomic model of matter has gone through many refinements. At one time or another, atoms were imagined to be tiny spheres with hooks on them (to explain chemical bonding), or as tiny billiard balls continually bouncing against each other. More recently, the "planetary model" of the atom visualized the atom as a nucleus with electrons revolving around it, just as the planets revolve about the sun.

You may wonder what the difference is between a theory and a model. Sometimes the words are used interchangeably. Usually, however, a model is fairly simple and provides a structural similarity to the phenomena being studied, whereas a **theory** is broader, more detailed, and attempts to solve a set of problems often with mathematical precision. Often, as a model is developed and modified and corresponds more closely to experiment over a wide range of phenomena, it may come to be referred to as a theory. The atomic theory is an example, as is the wave theory of light.

Models can be very helpful, and they often lead to important theories; but it is important not to confuse a model, or a theory, with the real system or the phenomena themselves.

Scientists give the title **law** to certain concise but general statements about how nature behaves (that momentum is conserved, for example); sometimes the statement takes the form of a relationship or equation between quantities (such as Einstein's famous equation, $E = mc^2$).

To be called a law, a statement must be found experimentally valid over a wide range of observed phenomena; in a sense, the law brings a unity to many observations. For less general statements, the term **principle** is often used (such as Archimedes' principle). Where to draw the line between laws and principles is, of course, arbitrary, and there is not always complete consistency.

Scientific laws are different from political laws in that the latter are *prescriptive:* they tell us how we must behave. Scientific laws are *descriptive:* they do not say how nature *must* behave, but rather describe how nature *does* behave. As with theories, laws cannot be tested in the infinite variety of cases possible. So we cannot be sure that any law is absolutely true. We use the term law when its validity has been tested over a wide range of cases, and when any limitations and the range of validity are clearly understood. Even then, as new information comes in, certain laws may have to be modified or discarded.

Scientists normally do their work as if the accepted laws and theories were true; but they are obliged to keep an open mind in case new information should alter the validity of any given law or theory.

▌ Mathematics

The appendixes that follow the last chapter contain a brief summary of simple mathematical techniques that will be used in this book. You may find it useful to examine them now to review old topics or learn any new ones. You may also want to reread them later when you need those concepts. Other mathematical techniques, such as vectors and trigonometric functions, are treated in the text itself.

1 DESCRIBING MOTION: KINEMATICS IN ONE DIMENSION

The motion of objects—baseballs, automobiles, joggers, and even the sun and moon—is an obvious part of everyday life. Motion was undoubtedly the first aspect of the physical world to be thoroughly studied, and this study can be traced back to the ancient civilizations of Asia Minor. Although the ancients acquired significant insight into motion, it was not until comparatively recently, in the sixteenth and seventeenth centuries, that our modern understanding of motion was established. Many contributed to this understanding, but, as we shall soon see, two individuals stand out above the rest: Galileo Galilei (1564–1642) and Isaac Newton (1642–1727).

The study of the motion of objects, and the related concepts of force and energy, forms the field called *mechanics*. Mechanics is customarily divided into two parts: *kinematics*, which is the description of how objects move, and *dynamics*, which deals with why objects move as they do. This chapter and the next deal with kinematics.

We start by discussing objects that move without rotating. Such motion is called translational motion. In the present chapter we will mainly be concerned with describing an object that moves along a straight-line path, which we call one-dimensional, or *rectilinear*, motion. In Chapter 2 we will study how to describe motion in two (or three) dimensions.

1-1 ▌ Speed

"Speed" and "velocity" are words we often use interchangeably in everyday language. In physics we make a distinction between them (see Section 1-5). For now, we discuss only speed, which refers to how far an object travels in a given time interval. If a car travels 240 kilometers (km) in 3 hours, we say its average speed was 80 km/h. In general, the **average speed** of an object is defined as *the distance traveled divided by the time it takes to travel this distance.*

$$\text{Average speed} = \frac{\text{distance traveled}}{\text{time elapsed}}.$$

This definition can be written briefly as:

$$\boxed{\bar{v} = \frac{d}{t},}$$

(1-1)

where d stands for distance, t for the elapsed time, and v for the speed (v is the abbreviation for the word velocity). The bar ($^-$) over the v is the standard symbol meaning "average,"

> **EXAMPLE 1-1**   How far can a cyclist travel in 4.0 h if his average speed is 11.5 km/h?
>
> SOLUTION   Since we want to find the distance traveled, we rewrite the equation above as $d = \bar{v}t$. Since $\bar{v} = 11.5$ km/h and $t = 4.0$ h, $d = (11.5$ km/h)(4.0 h) $= 46$ km.

1-2 ∎ Reference frames and coordinate systems

When riding in a train you may observe a bird flying by overhead and remark that it looks as if it is moving at a speed of 20 km/h. But do you mean it is traveling 20 km/h with respect to the train, or with respect to the ground?

Every measurement must be made with respect to a **frame of reference.** For example, while on a train traveling at 80 km/h, you might notice a person walk past you toward the front of the train at a speed of, say, 5 km/h. Of course this is the person's speed with respect to the train. With respect to the ground that person is moving at 85 km/h. It is always important to specify the frame of reference when stating a speed. We almost always mean "with respect to the earth" without even thinking about it, but the reference frame should be specified whenever there might be confusion.

The values of other physical quantities also depend on the frame of reference. For example, there is no point in telling you that Yosemite National Park is 300 km away unless I specify 300 km from where. Distances are always measured in some frame of reference. Furthermore, when specifying the motion of an object, it is important to specify not only the speed but also the direction of motion. For example, if a friend leaves New York on a jet plane that travels at a speed of 1000 km/h, you would like to know in what direction this person is going—toward Washington, San Francisco, Paris, or wherever. Often we can specify a direction by using the cardinal points, north, east, south, and west, and by "up" and "down." This is not always convenient; so in physics we often draw a set of coordinate axes, as shown in Fig. 1-1, to represent a frame of reference. Objects positioned to the right of the origin of coordinates (O) on the x axis are usually said to have an x coordinate with a positive value, if to the left a negative value. The position along the y axis is usually considered positive when above and negative when below O, although the reverse convention can be used if convenient. Any point on the plane can then be specified by giving its x and y coordinates. In three dimensions, a z axis perpendicular to the x and y axes is also used.

FIGURE 1-1

Standard set of coordinate (Cartesian) axes.

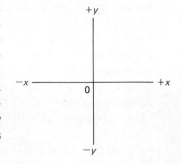

Although most measurements are made in reference frames fixed on the earth, it is important to recognize that reference frames other than the earth are perfectly legitimate. For example, scientific measurements are often made on moving ships and even on the moon.

1-3 ∎ Standards and units; the SI system

The measurement of any quantity is made relative to a particular standard or unit, and this unit must be specified along with the numerical value of the quantity. For example, we can measure length in such units as inches, feet, or miles, or in the metric system in centimeters, meters, or kilometers. To specify that the length of a particular object is 18.6 is meaningless. The unit *must* be given; for clearly, 18.6 meters is very different from 18.6 inches or 18.6 millimeters.

Until about 200 years ago, the units of measurement were not standardized, and that made scientific communication difficult. Different people used different units: cubits, leagues, hands; and even the length of the foot varied from place to place.

The first real international standard was the establishment of the standard *meter* (abbreviated m) by the French Academy of Sciences in the 1790s. In a spirit of rationality, the standard meter was originally chosen to be one ten-millionth of the distance from the earth's equator to either pole[†]; and a platinum rod to represent this length was made. In 1889, the meter was defined more precisely as the distance between two finely engraved marks on a particular bar of platinum-iridium alloy now kept near Paris at the International Bureau of Weights and Measures. By the end of the nineteenth century it became possible, through the work of the American physicist A. A. Michelson (see Section 23-9), to define the meter in terms of the wavelength of light; and in 1960 the meter was redefined to be 1,650,763.73 wavelengths of a particular orange light emitted by the gas krypton-86. In 1983, for reasons of greater precision, the meter was again redefined, this time in terms of the speed of light (whose best measured value, in terms of the older definition of the meter, was 299,792,458 m/s). The new definition reads: "The meter is the length of path traveled by light in vacuum during a time interval of 1/299,792,458 of a second."

British units (inch, foot, mile) are now defined in terms of the meter. The inch (in) is defined as precisely 2.54 centimeters (cm; 1 cm = 0.01 m). Other conversion factors are given in the table on the inside of the front cover of this book.

The standard unit of *time* is the *second* (s). For many years, the second was defined as 1/86,400 of a mean solar day. The standard second is now defined more precisely in terms of vibrations within a cesium atom. Specifically, one second is defined as the time required for 9,192,631,770 vibrations of cesium atoms when they are vibrating in a specific manner. There are, of course, 60 s in one minute (min) and 60 minutes in one hour (h).

[†] Modern measurements of the earth's circumference reveal that the intended length was off by about one-fiftieth of 1 percent.

The definitions of other standard units for other quantities will be defined as we encounter them in later chapters.

In the metric system, the larger and smaller units are defined in multiples of 10 from the standard unit, and this makes calculation particularly easy. Thus one centimeter is $\frac{1}{100}$ m, one kilometer (km) is 1000 m, and so on. The prefixes "centi-," "kilo-," and others are listed in Table 1-1 and can be applied not only to units of length, but to units of volume, mass, or any other metric unit. For example a centiliter (cL) is $\frac{1}{100}$ liter (L) and a kilogram (kg) is 1000 grams (g). The conversion factors between the various units of the British system (for example 12 inches in a foot) make it unwieldy for calculation. This is one reason why scientists, and nearly all countries of the world, have adopted the metric system.

When dealing with the laws and equations of physics it is very important to use a consistent set of units. To take a very simple example, suppose you want to know how far you can go in your car in 40 min at an average speed of 90 km/h. As we saw in Example 1-1, the distance, d, can be written as the product of the average speed, $\bar{v}$, and the time, t: $d = \bar{v}t$. But if you multiply 90 km/h by 40 min, you will get a ridiculous answer. Since the speed, v, is given in terms of hours, the time, t, must also be in hours, and in this case it is $\frac{2}{3}$ h; therefore $d = (90 \text{ km/h})(\frac{2}{3} \text{ h}) = 60$ km. (Note how the hour units cancel in this equation; the equals sign in an equation applies not only to the numerical values but also to the units.) When dealing with more complicated equations, the necessity of using a consistent set of units is even more crucial.

Several systems of units have been in use over the years. Today the most important by far is the **Système International** (French for International System), which is abbreviated SI. In SI units, the standard of length is the meter, that for time is the second, and that for mass is the kilogram. This system used to be called the MKS (meter-kilogram-second) system.

A second metric system is the cgs system in which the centimeter, gram, and second are the standard units of length, mass, and time, as abbreviated in the title.

The British engineering system takes as its standards the foot for length, the pound for force, and the second for time.

SI units are the principal ones used today in scientific work. We will therefore use SI units almost exclusively in this book, although we will give the cgs and British units for various quantities when introduced, as well as conversion factors.

Physical quantities can be divided into two categories: **base quantities** and **derived quantities.** The corresponding units for these quantities are called *base units* and *derived units.* Scientists, in the interest of simplicity, want the smallest number of base quantities possible that is consistent with a full description of the physical world. This number turns out to be seven, and for the SI are given in Table 1-2. All other quantities can be defined in terms of these seven base quantities.[†] An example is speed, which is defined (Section 1-1) as distance divided by time. Base quantities are not definable in terms

---

[†] The only exceptions are for angle (radians—see Chapter 7) and solid angle (steradian) for which general agreement has not been reached as to whether these are base or derived quantities.

TABLE 1-1
Metric (SI) prefixes (multipliers)

Prefix	Abbreviation	Value
Tera	T	10^{12}
Giga	G	10^{9}
Mega	M	10^{6}
Kilo	k	10^{3}
Hecto	h	10^{2}
Deka	da	10^{1}
Deci	d	10^{-1}
Centi	c	10^{-2}
Milli	m	10^{-3}
Micro	μ	10^{-6}
Nano	n	10^{-9}
Pico	p	10^{-12}
Femto	f	10^{-15}

TABLE 1-2
SI base quantities and units

Quantity	Unit	Unit Abbreviation
Length	meter	m
Time	second	s
Mass	kilogram	kg
Electric current	ampere	A
Temperature	kelvin	K
Amount of substance	mole	mol
Luminous intensity	candela	cd

of other quantities, which is why they are called base. What we can do is specify a rule (or set of rules) for measuring that base quantity; this is called an *operational definition*, and such definitions can be made for both base and derived quantities.

1-4 ∎ Changing units

It is often necessary or useful to change from one set of units to another. For example it is often preferable to specify the speed of a car in m/s rather than in km/h, such as when measuring braking distances since the time and distances involved are usually seconds and meters rather than hours and kilometers. Also, one sometimes must change units from metric to British units, or vice versa.

To determine what a speed of 80 km/h is in m/s, we proceed as follows. There are 1000 m in 1 km, and 3600 s in 1 h. Thus:

$$80 \text{ km/h} = \left(\frac{80 \text{ km}}{1 \text{ h}}\right)\left(\frac{1000 \text{ m}}{1 \text{ km}}\right)\left(\frac{1 \text{ h}}{3600 \text{ s}}\right)$$

$$= (80)\left(\frac{1000}{3600}\right)\frac{\text{m}}{\text{s}}$$

$$= (80)(0.278) \text{ m/s} = 22 \text{ m/s}.$$

Notice that in the first line we have multiplied our original number $(80 \frac{\text{km}}{\text{h}})$ by two conversion factors $(\frac{1000 \text{ m}}{1 \text{ km}} = 1$ and $\frac{1 \text{ h}}{3600 \text{ s}} = 1)$, both of which in essence are equal to one and thus do not change the equation. The units of hours and of kilometers cancel out so we obtain m/s. (Note also that we kept only two significant figures in the final answer, 22 m/s, since the least significant number in the product contained two significant figures; see Appendix A-2 for a discussion of significant figures.)

When changing units it is often a problem to figure out whether the conversion factor is to go in the numerator or in the denominator. The easiest way to be sure is to check if the units cancel out as they did above. We would have gotten a wrong result if we had written $(80 \frac{\text{km}}{\text{h}})(\frac{1000 \text{ m}}{1 \text{ km}})(\frac{3600 \text{ s}}{1 \text{ h}})$, and this would have been obvious since the hour units do not cancel.

In the last line in the calculation above we see the factor 0.278; this is the conversion factor between m/s and km/h. This can more quickly be obtained by noting that 1 km/h is the same as 1000 m in 3600 s. That is, $1 \frac{\text{km}}{\text{h}} = \frac{1000 \text{ m}}{3600 \text{ s}} = 0.278 \text{ m/s}$. Thus, any speed given in km/h can be changed to m/s by multiplying by the factor of 0.278.

Changes between metric and British units can be done in a similar way, and we leave these calculations as an exercise.

1-5 ∎ Average velocity; displacement

The terms velocity and speed are used interchangeably in ordinary language. However, in physics we make a distinction between the two. Most importantly, the term **velocity** is used to signify both the *magnitude* (numerical value) of how fast an object is moving and the *direction* in which it is moving. Speed, on the other hand, is a magnitude only. There is a second difference between speed and velocity: namely, the average velocity is defined in terms of "displacement," rather than total distance traveled; that is

$$\text{average velocity} = \frac{\text{displacement}}{\text{time elapsed}},$$

where *displacement* is defined as *the change in position of an object.* To see the distinction between total distance and displacement, imagine a person walking 50 m to the east and then turning around and walking back (west) a distance of 10 m. The total *distance* traveled is 60 m but the *displacement* is only 40 m since the person is now only 40 m from the starting point. Suppose the walk took 40 s; then the average speed was (60 m)/(40 s) = 1.5 m/s, but the magnitude of the average velocity in this case was only (40 m)/(40 s) = 1.0 m/s. This discrepancy between the magnitude of the velocity and the speed occurs in some cases, but only for the *average* values, and we rarely need be concerned with it. The magnitude of the instantaneous velocity (see below) and the instantaneous speed are always the same.

To discuss the motion of an object in general, suppose that at some moment in time, call it t_1, the object is on the x axis at point x_1 in the coordinate system shown in Fig. 1-2. At some later time, t_2, suppose it is at point x_2. The elapsed time is $t_2 - t_1$, and during this time interval the displacement of our object was $x_2 - x_1$. It is convenient to write

$$\Delta x = x_2 - x_1$$

where the symbol Δ (Greek letter delta) means "change in." Then Δx means the displacement or the "change in x." Similarly, we can write the elapsed time (change in time) as $\Delta t = t_2 - t_1$. Then the **average velocity** ($\bar{v}$), defined as *the displacement divided by the elapsed time,* can be written

$$\bar{v} = \frac{x_2 - x_1}{t_2 - t_1} = \frac{\Delta x}{\Delta t}. \tag{1-2}$$

Notice that if x_2 is less than x_1, the object is moving to the left, so $\Delta x = x_2 - x_1$ is less than zero. The sign of the displacement, and/or velocity, indicates the direction; the average velocity for an object moving to the right along the x axis is positive, whereas it is negative when the object moves to the left. Another way to state the definition of velocity is that it is the *rate of change of position in time.*

EXAMPLE 1-2   The position of a runner as a function of time is plotted as moving along the x axis of a coordinate system. During a 3.00-s time interval, the runner's position changes from $x_1 = 50.0$ m to $x_2 = 30.5$ m as shown in Fig. 1-3. What was the runner's average velocity?

SOLUTION   The displacement is $\Delta x = x_2 - x_1 = 30.5$ m $- 50.0$ m $= -19.5$ m. The time interval is $\Delta t = 3.00$ s. Therefore the average velocity is

$$\bar{v} = \frac{\Delta x}{\Delta t} = \frac{-19.5 \text{ m}}{3.00 \text{ s}} = -6.50 \text{ m/s}.$$

The displacement and average velocity are negative, so the direction the runner is moving is to the left along the x axis.

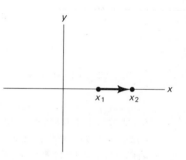

FIGURE 1-2

The arrow represents the displacement $x_2 - x_1$.

FIGURE 1-3

Object moves from $x = 50.0$ m to $x = 30.5$ m. The displacement is -19.5 m.

1-6 ▮ Instantaneous velocity

If you drive a car along a straight road for 150 km in 2.0 h, the magnitude of your average velocity is 75 km/h. It is unlikely, though, that you were moving at 75 km/h at every instant. To deal with this situation we need the concept of *instantaneous velocity*, which is the velocity at any instant of time (it is what a speedometer is supposed to indicate). More precisely, the **instantaneous velocity** at any moment is defined as *the average velocity over an indefinitely short time interval*. That is, Eq. 1-2 is to be evaluated in the limit of Δt being infinitesimally small, approaching zero. We can write the definition of instantaneous velocity, v, for one-dimensional motion as

$$v = \lim_{\Delta t \to 0} \frac{\Delta x}{\Delta t}. \tag{1-3}$$

The notation $\lim_{\Delta t \to 0}$ means the ratio $\Delta x/\Delta t$ is to be evaluated in the limit of Δt approaching zero. But note that we do not simply set $\Delta t = 0$ in this definition, for then Δx would also be zero, and we would have an undefined number. Rather, we are considering the ratio $\Delta x/\Delta t$ as a whole; as we let Δt approach zero, Δx approaches zero also; and the ratio $\Delta x/\Delta t$ approaches some definite value which is the instantaneous velocity at a given instant. Note also that for instantaneous velocity we use the symbol v whereas for average velocity we use $\bar{v}$, with a bar. In the rest of this book, when we use the term "velocity" it will refer to instantaneous velocity; when we want to speak of the average velocity, we will make this clear by including the word "average."

If an object moves at a uniform (or constant) velocity over a particular time interval, then its instantaneous velocity at any instant is the same as its average velocity (see Fig. 1-4a). But in most situations this will not be the case. For example, a car may start from rest, speed up to 50 km/h, remain at that velocity for a time, then slow down to 20 km/h in a traffic jam, and finally stop at its destination after traveling a total of 15 km in 30 min. This trip is plotted on the graph of Fig. 1-4b. Also shown on the graph is the average velocity (dashed line), which is $\bar{v} = \Delta x/\Delta t = 15$ km/0.50 h = 30 km/h.

Finally note that the instantaneous speed equals the magnitude of the instantaneous velocity under all circumstances; this is true because in the limit of $\Delta t \to 0$, the displacement also approaches zero, and so there is no way for the magnitude of this infinitesimal displacement to differ from the infinitesimal distance traveled.

(a)

(b)

FIGURE 1-4

Velocity of an object as a function of time: (a) at constant velocity, (b) with varying velocity.

1-7 ▮ Vectors and scalars

A quantity such as velocity, which has direction as well as a magnitude, is called a **vector.** Other quantities that are also vectors are displacement, force, and momentum. However, many quantities such as time, temperature, and energy, have no direction associated with them; they only have a magnitude;

that is, they are specified completely by giving a number (and units, if any). Such quantities are called **scalars.**

Drawing a diagram of a particular physical situation is very helpful in physics, and this is especially true when dealing with vectors. On a diagram, each vector is represented by an arrow. The arrow is always drawn so that it points in the direction of the vector it represents; the length of the arrow is drawn proportional to the magnitude of the vector. For example, in Fig. 1-5 arrows have been drawn representing the velocity of a car at various places as it rounds the curve. The magnitude of the velocity at each point can be read off this figure by measuring the length of the corresponding arrow.

When we write the symbol for a vector, we will always use boldface type.[†] Thus for velocity we write **v.** If we are only concerned with the magnitude we will simply write v. We will deal with vectors more fully in Chapter 2 where we discuss motion in two (or three) dimensions. For now, we deal mainly with motion along a line (rectilinear motion); vectors which point in one direction along this line will be assigned a positive value, and those pointing in the opposite direction will have a negative sign. In Example 1-2, for instance, the velocity was negative since it pointed to the left along the x axis (as did the displacement—see Fig. 1-3). A velocity or displacement vector pointing to the right (Fig. 1-2) would be considered positive.

1-8 ▌ Acceleration

An object whose velocity is changing is said to be accelerating. An automobile increasing in speed from rest to 80 km/h is accelerating. If one car can accomplish this change in speed in less time than another, it is said to undergo a greater acceleration. **Average acceleration** is defined as the **rate of change of velocity,** or the change in velocity divided by the time taken to make this change:

$$\text{Average acceleration} = \frac{\text{change of velocity}}{\text{time elapsed}}.$$

In symbols, the average acceleration, $\bar{\mathbf{a}}$, over a time interval $\Delta t = t_2 - t_1$ during which the velocity changes by $\Delta \mathbf{v} = \mathbf{v}_2 - \mathbf{v}_1$, is defined as

$$\bar{\mathbf{a}} = \frac{\mathbf{v}_2 - \mathbf{v}_1}{t_2 - t_1} = \frac{\Delta \mathbf{v}}{\Delta t}. \tag{1-4}$$

Equation 1-4 is a vector equation. For linear motion, a, v_1, v_2, and Δv can be considered as scalars so we need only use numerical values, with a plus or minus sign to indicate direction.

The **instantaneous acceleration, a,** can be defined in analogy to instantaneous velocity:

$$\mathbf{a} = \lim_{\Delta t \to 0} \frac{\Delta \mathbf{v}}{\Delta t}; \tag{1-5}$$

Scale: 1 cm = 90 km/h

FIGURE 1-5
Car traveling on a road. The arrows represent the velocity vector at each position.

[†] The symbol for a vector is usually indicated in handwritten work with an arrow over it, as $\vec{v}$ for velocity.

here $\Delta \mathbf{v}$ represents the infinitesimally small change in velocity during the infinitesimally short time interval Δt.

The concept of acceleration is often confused with velocity. To help make the distinction, let us take an example.

EXAMPLE 1-3 A car accelerates along a straight road from rest to 60 km/h in 5.0 s. What is the magnitude of its average acceleration?

SOLUTION From Eq. 1-4 we have

$$\bar{a} = \frac{60 \text{ km/h} - 0 \text{ km/h}}{5.0 \text{ s}} = 12 \text{ km/h/s}.$$

This is read as "twelve kilometers per hour per second" and means that, on the average, the velocity changed by 12 km/h each second. That is, assuming the acceleration was constant, during the first second the car's velocity increased from zero to 12 km/h; during the next second its velocity increased by another 12 km/h up to 24 km/h, and so on. (Of course if the instantaneous acceleration was not constant, these numbers would be different.) Note carefully that acceleration is the rate at which the velocity changes whereas velocity is the rate at which position changes. Also note that the calculated acceleration above contained two different time units: hours and seconds. We usually prefer to use only seconds; to do so we can change the 60 km/h to $(60 \text{ km/h})(0.278 \frac{\text{m/s}}{\text{km/h}}) = 17$ m/s (see Section 1-4). Then we get

$$\bar{a} = \frac{17 \text{ m/s} - 0.0 \text{ m/s}}{5.0 \text{ s}} = 3.4 \text{ m/s}^2,$$

where instead of m/s/s we have written the units as m/s^2 (meters per second squared); we can do this since

$$\frac{\text{m/s}}{\text{s}} = \frac{\text{m}}{\text{s} \cdot \text{s}} = \frac{\text{m}}{\text{s}^2}.$$

According to the above calculation, the velocity changed on the average by 3.4 m/s during each second, for a total change of 17 m/s over the 5.0 s.

EXAMPLE 1-4 An automobile is moving along a straight highway which we choose to be to the right along the positive x axis and the driver puts on the brakes. If the initial velocity is $v_1 = 15.0$ m/s and it takes 5.0 s to slow down to $v_2 = 5.0$ m/s, what was the car's acceleration?

SOLUTION Let us say that the initial time was $t_1 = 0$; then $t_2 = 5.0$ s. (Note that our choice of $t_1 = 0$ doesn't matter since only $\Delta t = t_2 - t_1$ appears in Eq. 1-4.) Then

$$\bar{a} = \frac{5.0 \text{ m/s} - 15.0 \text{ m/s}}{5.0 \text{ s}} = -2.0 \text{ m/s}^2.$$

The negative sign appears because the final velocity is less than the initial velocity and thus indicates that the speed is decreasing. In this case the direction of the acceleration is to the left—even though the velocity is always pointing to the right.

The definition of acceleration involves the change in *velocity*. Thus an acceleration results not only when the *magnitude* of the velocity changes but also if the *direction* changes. For example, a person riding in a car traveling at constant speed around a curve or a child riding on a merry-go-round, will both experience an acceleration because of a change in direction of the velocity. Such accelerations due to change in direction of velocity will be discussed later.

1-9 ▮ Uniformly accelerated motion

Many practical situations occur in which the acceleration is constant. In many other situations the variation in acceleration is sufficiently small that we are justified in assuming that it is constant. We now treat this situation of *uniformly accelerated motion:* that is, when the magnitude of the acceleration is constant and the motion is in a straight line. In this case, the instantaneous and average accelerations are equal.

To simplify our notation, let us take the initial time in any discussion to be zero: $t_1 = 0$. We can then let $t_2 = t$ be the elapsed time. The initial position (x_1) and initial velocity (v_1) of an object will now be represented by x_0 and v_0, and at time t the position and velocity will be called x and v (rather than x_2 and v_2). The average velocity during the time t will be (from Eq. 1-2)

$$\bar{v} = \frac{x - x_0}{t};$$

and the acceleration (which is assumed constant in time) will be (from Eq. 1-4)

$$a = \frac{v - v_0}{t}.$$

A common problem is to determine the velocity of an object after a certain time given its acceleration. We can solve such problems by solving for v in the last equation:

$$v = v_0 + at. \quad \text{[constant acceleration]} \quad (1\text{-}6)$$

For example, it may be known that the acceleration of a particular motorcycle is 4.0 m/s^2 and we wish to determine how fast it will be going after, say, 6.0 s. Assuming it starts from rest ($v_0 = 0$), after 6.0 s the velocity will be $v = at = (4.0 \text{ m/s}^2)(6.0 \text{ s}) = 24 \text{ m/s}$.

Next, let us see how to calculate the position of an object after a time t when it is undergoing constant acceleration. From the definition of average velocity (Eq. 1-2) $\bar{v} = (x - x_0)/t$, so

$$x = x_0 + \bar{v}t.$$

Because the velocity increases at a uniform rate, the average velocity, $\bar{v}$, will be midway between the initial and final velocities:

$$\bar{v} = \frac{v + v_0}{2}. \quad \text{[constant acceleration]} \quad (1\text{-}7)$$

(Be careful. This is not generally true if the acceleration is not constant. See, for example, problems 9 and 10.) We combine the last three equations and find

$$x = x_0 + \bar{v}t$$

$$= x_0 + \left(\frac{v + v_0}{2}\right)t$$

$$= x_0 + \left(\frac{v_0 + at + v_0}{2}\right)t$$

or

$$x = x_0 + v_0t + \tfrac{1}{2}at^2. \quad [\text{constant acceleration}] \quad (1\text{-}8)$$

Equations 1-6, 1-7, and 1-8 are three of the four most useful equations of uniformly accelerated motion. We now derive the fourth equation, which is useful in a situation when, for instance, the acceleration, position, and initial velocity are known, and the final velocity is desired but the time t is not known. To obtain the velocity v at time t in terms of v_0, a, x, and x_0, we begin as above:

$$x = x_0 + \bar{v}t = x_0 + \left(\frac{v + v_0}{2}\right)t$$

where we have substituted in Eq. 1-7. Next we solve Eq. 1-6 for t, obtaining $t = (v - v_0)/a$, and substituting this into the above equation we have

$$x = x_0 + \left(\frac{v + v_0}{2}\right)\left(\frac{v - v_0}{a}\right)$$

$$= x_0 + \frac{v^2 - v_0^2}{2a}.$$

We solve this for v^2 and obtain the desired relation:

$$v^2 = v_0^2 + 2a(x - x_0). \quad [\text{constant acceleration}] \quad (1\text{-}9)$$

We now have four equations relating the various quantities important in uniformly accelerated motion (that is, a is a constant). We collect them here in one place for further reference:

$$v = v_0 + at \qquad [a = \text{constant}] \quad (1\text{-}10a)$$

$$x = x_0 + v_0t + \tfrac{1}{2}at^2 \qquad [a = \text{constant}] \quad (1\text{-}10b)$$

$$v^2 = v_0^2 + 2a(x - x_0) \qquad [a = \text{constant}] \quad (1\text{-}10c)$$

$$\bar{v} = \frac{v + v_0}{2}. \qquad [a = \text{constant}] \quad (1\text{-}10d)$$

These equations are not valid unless a is a constant. They can be applied only during time periods when the acceleration is constant. In many cases we can set $x_0 = 0$, and this simplifies the above equations a bit. In the examples that follow, we assume $x_0 = 0$ unless stated otherwise.

⬜ **EXAMPLE 1-5** Suppose a planner is designing an airport for small planes. One kind of airplane that might use this airfield must reach a speed before takeoff of 200 km/h (55.6 m/s) and can accelerate at 12.0 m/s². If the runway is 100 m long, can this airplane reach the proper speed to take off?

SOLUTION We use Eq. 1-10c with $x_0 = 0$, $v_0 = 0$, $x = 100$ m, and $a = 12.0$ m/s². Then

$$v^2 = 0 + 2(12.0 \text{ m/s}^2)(100 \text{ m}) = 2400 \text{ m}^2/\text{s}^2$$
$$v = \sqrt{2400 \text{ m}^2/\text{s}^2} = 49.0 \text{ m/s}.$$

Unfortunately, this length runway is *not* sufficient. By solving Eq. 1-10c for $(x - x_0)$ you can determine how long a runway is needed for this plane.

The solving of problems, such as the examples we have already given, serves two purposes. First, solving problems is useful and practical in itself; second, solving problems makes you think about the ideas and concepts, and applying the concepts helps you to understand them. But knowing how to do a problem—even to begin it—may not always seem easy. First, and most important, is to read the problem through carefully; and more than once. Spend a moment thinking and trying to understand what physics principles might be involved. Up to this point in the book, we have been concerned mainly with the definitions of velocity and acceleration and the "kinematic equations for constant acceleration," Eqs. 1-10, that we derived from those definitions. We will discuss problem solving more in Chapter 3, after we have covered more physics. For now it is important to note that physics is *not* a collection of equations to be memorized. (In fact, instead of memorizing the very useful Eqs. 1-10, it might be better to understand how to derive them from the definitions of velocity and acceleration as we did above.) To simply search for an equation that might work can be very dangerous and can lead you to a wrong result. After careful readings of the problem, and letting it sink in, draw a picture of the situation and then follow this (rough) procedure. (1) Write down what quantities are "known" or "given," and then what you *want* to know. (2) Think about what principles, definitions, and/or equations relate the quantities involved; before using them, be sure their range of validity includes your problem (for example, Eqs. 1-10 are valid only when the acceleration is constant); if you find an applicable equation that involves only known quantities and one desired unknown, solve the equation algebraically for the unknown; in many instances several sequential calculations, and/or a combination of equations, may be needed. (3) Think carefully about the result you obtain: is it reasonable? Does it make sense according to your own intuition and experience? (4) A very important aspect of doing problems is keeping track of units. Note that an equals sign implies the units on each side must be the same just as the numbers must. If the units do not balance, a mistake has no doubt been made. This can serve as a check on your solutions (but it only tells you if you're wrong, not if you're right).

EXAMPLE 1-6 How long does it take a car to travel 30.0 m if it accelerates from rest at a rate of 2.00 m/s²?

SOLUTION First we make a table, choosing $x_0 = 0$:

Known	Wanted
$x_0 = 0$	t
$x = 30.0$ m	
$a = 2.00$ m/s²	
$v_0 = 0$	

Since a is constant, we can use Eqs. 1-10. Equation 1-10a is not helpful in this case since it contains v, an unknown, as well as the desired unknown t. Equation 1-10c is worse—it contains v but not t. Equation 1-10b is perfect since the only unknown quantity is t. Before we solve for t in this equation, we simplify it by setting $v_0 = 0$ first. Then

$$x = \tfrac{1}{2}at^2$$

$$t^2 = \frac{2x}{a} = \frac{2(30.0 \text{ m})}{2.00 \text{ m/s}^2} = 30.0 \text{ s}^2$$

$$t = \sqrt{30.0 \text{ s}^2} = 5.48 \text{ s}.$$

EXAMPLE 1-7 We now consider the stopping distances for a car which are important not only for traffic safety but are needed for traffic design problems. The problem divides itself naturally into two parts, in each of which the car can be considered to have constant acceleration, although the acceleration is different in the two parts: (1) the time between the decision to apply the brakes and their actual application (the "reaction time") during which we assume $a = 0$; and (2) the actual braking period when the vehicle decelerates ($a \neq 0$). We assume the car is moving to the right along the positive x axis with $x_0 = 0$. The stopping distance depends on the reaction time of the driver, the initial speed of the car (the final speed is zero), and the deceleration of the car. For a dry road, good brakes can decelerate a car at a rate of about 5 m/s² to 8 m/s². We make the calculation for an initial velocity of 100 km/h (28 m/s) and assume the acceleration of the car is -6.0 m/s² (the minus sign appears because the velocity is taken to be positive and its magnitude is decreasing). Reaction time for normal drivers varies from perhaps 0.3 s to about 1.0 s; we take it to be 0.50 s.

SOLUTION For the first part of the problem, the car travels at a constant speed of 28 m/s during the time the driver is reacting (0.50 s). Thus:

Known	Wanted
$t = 0.50$ s	x
$v_0 = 28$ m/s	
$v = 28$ m/s	
$a = 0$	
$x_0 = 0$	

To find x we can use Eq. 1-10b (note that Eq. 1-10c won't work because x is multiplied by a which is zero):

$$x = v_0 t + 0 = (28 \text{ m/s})(0.50 \text{ s}) = 14 \text{ m}.$$

Now for the second part during which the brakes are applied and the car brought to rest:

Known	Wanted
$v_0 = 28$ m/s	x
$v = 0$	
$a = -6.0$ m/s^2	

Equation 1-10a doesn't contain x; Eq. 1-10b contains x but also the unknown t. Equation 1-10c is what we want; we solve for x (after setting $x_0 = 0$):

$$v^2 - v_0^2 = 2ax$$

$$x = \frac{v^2 - v_0^2}{2a}$$

$$= \frac{0 - (28 \text{ m/s})^2}{2(-6.0 \text{ m/s}^2)} = \frac{-784 \text{ m}^2/\text{s}^2}{-12 \text{ m/s}^2}$$

$$= 65 \text{ m}.$$

The car traveled 14 m while the driver was reacting and another 65 m during the braking period before coming to a stop; the total distance traveled was then 79 m. Under wet or icy conditions, the value of a may be only one-third the value for a dry road since the brakes cannot be applied as hard without skidding, and hence stopping distances are much greater. Note also that stopping distance increases with the *square* of the speed, not just linearly with speed!

◻   EXAMPLE 1-8   A baseball pitcher throws a fastball with a speed of 30.0 m/s. Calculate approximately the average acceleration of the ball during the throwing motion. It is observed that in throwing the baseball, the pitcher accelerates the ball through a total distance of about 3.50 m from behind the ear to the point where it is released (Fig. 1-6).

SOLUTION   We want to find the acceleration a given that $x = 3.50$ m, $v_0 = 0$, and $v = 30.0$ m/s. We use Eq. 1-10c and solve for a:

$$a = \frac{v^2 - v_0^2}{2x}$$

$$= \frac{(30.0 \text{ m/s})^2 - (0 \text{ m/s})^2}{2(3.50 \text{ m})} = \frac{900 \text{ m}^2/\text{s}^2}{7.00 \text{ m}} = 129 \text{ m/s}^2.$$

This is a very large acceleration!

FIGURE 1-6

A baseball pitcher accelerates the ball over a distance of about 3.5 m.

The analysis of motion we have been discussing in this chapter is basically algebraic. It is sometimes helpful to use a graphical interpretation as well and this is done in Section 1-11 (optional).

1-10 ■ Falling bodies

One of the commonest examples of uniformly accelerated motion is that of an object allowed to fall vertically to the earth. That a falling body is accelerating may not be obvious at first. And beware in thinking, as was widely believed until the time of Galileo, that heavier bodies fall faster than lighter bodies and that the speed of fall is proportional to how heavy the object is.

Galileo's analysis made use of his new and creative technique of abstraction and simplification—that is, of imagining what would happen in idealized (simplified) cases. For free fall, he postulated that all bodies would fall with the *same constant acceleration* in the absence of air or other resistance. He showed that this postulate predicts that for an object falling from rest, the distance traveled will be proportional to the square of the time ($d \propto t^2$). This we can see from Eq. 1-10b; but Galileo was the first to derive this mathematical relation. In fact, one of Galileo's great contributions to science was to establish such mathematical relations and to insist on their importance; another great contribution of Galileo was the proposing of a theory with specific experimental consequences that could be quantitatively checked ($d \propto t^2$).

To support his claim that the speed of falling objects increases as they fall, Galileo made use of the following argument: a heavy stone dropped from a height of 2 m will drive a stake into the ground much farther than if the same stone is dropped from a height of only 10 cm; the stone must clearly be moving faster in the former case. As we saw, Galileo also claimed that *all* objects, light or heavy, fall with the *same* acceleration, at least in the absence of air. Now common sense may say that the ancients were perhaps closer to the truth; for if you hold a piece of paper horizontally in one hand and a heavier object, say, a baseball, in the other and release them at the same time (see Fig. 1-7a) surely the heavier object will reach the ground first. Repeat the experiment but this time crumple up the paper into a small wad (see Fig. 1-7b). You will find that the two objects reach the floor at nearly the same time.

Galileo was sure that air acts like a sort of friction on very light objects that have a large surface area; but in many ordinary circumstances this air resistance is negligible. In a chamber from which the air has been removed even light objects like a feather or a horizontally held piece of paper will fall with the same acceleration as any other object (see Fig. 1-8). Such a demonstration in vacuum was of course not possible in Galileo's time, which makes Galileo's achievement all the greater. Galileo is often called the "father of modern science," not only for the content of his science (astronomical discoveries, inertia, free fall), but also for his style or approach to science (idealization and simplification, mathematization of theory, prediction of testable consequences that must be checked by experiment).

Galileo's specific contribution to our understanding of the motion of falling objects can be summarized as follows:

■ **At a given location on the earth and in the absence of air resistance, all objects fall with the same uniform acceleration.**

We call this acceleration the *acceleration due to gravity* on the earth, and we

FIGURE 1-7

(a) A ball and a light piece of paper are dropped at the same time. (b) Repeated, with the paper wadded up.

FIGURE 1-8

A rock and a feather are dropped simultaneously (a) in air, (b) in a vacuum.

Air-filled tube (a)   Evacuated tube (b)

give it the symbol g. Its value is approximately:

$$g = 9.80 \text{ m/s}^2.$$

In British units g is about 32 ft/s^2. Actually, g varies slightly according to latitude and elevation (see Table 1-3) but these variations are so small that we will ignore them for most purposes (but see Section 4-6). The effects of air resistance are usually small and we will neglect them for the most part. However, air resistance will be noticeable even on a reasonably heavy object if the distance of fall is very large.[†]

When dealing with freely falling objects we can make use of Eqs. 1-10 where for a we use the numerical value of g given above. Also, since the motion is vertical we will substitute y in place of x and y_0 in place of x_0 (and we take $y_0 = 0$ unless otherwise specified). It is arbitrary whether we choose y to be positive in the upward direction or in the downward direction, but we must be consistent about it throughout a problem.

TABLE 1-3

Acceleration due to gravity at various locations on earth

Location	Elevation (ft)	g (m/s²)
New York	0	9.80
San Francisco	350	9.80
Denver	540	9.79
Pikes Peak	14,100	9.78
Equator	0	9.78
North Pole (calculated)	0	9.83

□ **EXAMPLE 1-9** Suppose that a ball is dropped from a tower 70.0 m high. How far will it have fallen after 1.00, 2.00, and 3.00 s? Assume y is positive downward.

SOLUTION Equation 1-10b is the appropriate one, with $v_0 = 0$ and $y_0 = 0$. Then, after 1.00 s,

$$y = \tfrac{1}{2}at^2 = \tfrac{1}{2}(9.80 \text{ m/s}^2)(1.00 \text{ s})^2 = 4.90 \text{ m}.$$

Similarly after 2.00 s, $y = 19.6$ m, and after 3.00 s, $y = 44.1$ m. See Fig. 1-9.

□ **EXAMPLE 1-10** A person throws a ball *upward* into the air with an initial velocity of 15.0 m/s. Calculate (*a*) how high it goes, and (*b*) how long the ball is in the air before it comes back to his hand. We are not concerned here with the throwing action, but only with the motion of the ball *after* it leaves the thrower's hand, Fig. 1-10.

SOLUTION Let us choose y to be positive in the upward direction and negative in the downward direction. (Note: this is a different convention from that used in the previous example.) Then the acceleration will have a negative sign (since it changes the velocity in a downward direction), $a = -g = -9.80 \text{ m/s}^2$.

(*a*) To determine the maximum height we calculate the position of the ball when its velocity equals zero ($v = 0$ at the highest point). At $t = 0$ we have $y_0 = 0$, $v_0 = 15.0$ m/s; at time t (*maximum height*), $v = 0$, $a = -9.80 \text{ m/s}^2$, and we wish to find y. We use Eq. 1-10c (replacing x with y) and solve for y:

$$v^2 = v_0^2 + 2ay$$

$$y = \frac{v^2 - v_0^2}{2a} = \frac{0 - (15.0 \text{ m/s})^2}{2(-9.80 \text{ m/s}^2)} = 11.5 \text{ m}.$$

[†] The speed of an object falling in air (or other fluid) does not increase indefinitely; if it falls far enough, it will reach a maximum velocity called the *terminal velocity*. The maximum velocity is reached when the force of air resistance (which increases with speed) reaches a magnitude equal to the force of gravity.

4.90 m
After 1.00 s

19.6 m

After 2.00 s

44.1 m

After 3.00 s

FIGURE 1-9

When an object is dropped from the top of a tower, it falls with progressively greater speed and covers greater distance with every successive second.

FIGURE 1-10

Object thrown into air leaves thrower's hand at A, reaches its maximum height at B, and returns to original height at C. (See Example 1-10.)

(*b*) We could do this section of the example in two parts by first calculating the time required for the ball to reach its highest point, and then calculating the time it takes to fall back down. However, it is simpler to consider the motion from *A* to *B* to *C* (Fig. 1-10) in one step and use Eq. 1-10b; we can do this because *y* (or *x*) represents position or displacement, and not the total distance traveled. Thus, at both points *A* and *C*, *y* = 0. We use Eq. 1-10b with *a* = −9.80 m/s² and find

$$y = v_0 t + \tfrac{1}{2}at^2$$
$$0 = (15.0 \text{ m/s})t + \tfrac{1}{2}(-9.80 \text{ m/s}^2)t^2.$$

There are two solutions[†]: *t* = 0 and *t* = (2)(15.0 m/s)/(9.80 m/s²) = 3.06 s. The first solution corresponds to the initial point, *A*, and the second to the return, *C*. Thus, the ball is in the air for 3.06 s.

This last example illustrates the errors of two common misconceptions: that acceleration and velocity are always in the same direction, and that an object thrown upward has zero acceleration at the highest point (*B* in Fig. 1-10). Both are wrong. In the latter case, the velocity is indeed zero at the highest point, but the acceleration is still that of gravity (*a* = −*g*); if the acceleration were zero, the ball's velocity would not change—it would not have slowed down to *v* = 0 and then started increasing in the downward direction. As to the first error mentioned above, velocity and acceleration are *not* necessarily in the same direction; when the ball in the above example is moving upward, its velocity is positive (upward), although the acceleration is negative (downward); in the downward flight (*B* to *C* in Fig. 1-10) both velocity and acceleration are negative (downward).

EXAMPLE 1-11 The jump of a flea can be analyzed using slow-motion photographs. The motion can be separated into two parts. The first is the "push off," lasting about 10⁻³ s, during which the flea's legs push against the ground and accelerate it to a speed of about 1.0 m/s. The second part is the flight of the flea into the air subject only to gravity (Fig. 1-11). Calculate: (*a*) the acceleration of the flea during push off expressed as a multiple of *g*, the acceleration of gravity; (*b*) the distance above the ground the flea reaches during push off; and (*c*) the height the flea should reach during the second period when its acceleration is that of gravity. Assume the jump is vertical.

SOLUTION Let us take the upward direction as positive, with $y_0 = 0$.
(*a*) During the push off

$$a = \frac{v - v_0}{t} = \frac{1.0 \text{ m/s} - 0}{10^{-3} \text{ s}} = 1000 \text{ m/s}^2.$$

The magnitude of the acceleration is (1000 m/s²)/(9.80 m/s²) ≈ 100 times the acceleration of gravity. Thus *a* ≈ 100 *g*.

FIGURE 1-11

The jump of a flea (Example 1-11).

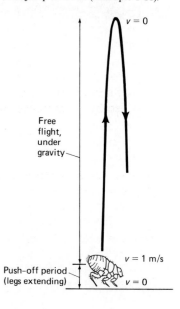

v = 0

Free flight, under gravity

v = 1 m/s

Push-off period (legs extending)

v = 0

[†] Quadratic equations (involving t^2) can be solved using the quadratic formula; see Appendix A-5.

(b) We use Eq. 1-10b where again we use y instead of x:

$$y = \tfrac{1}{2}at^2 = 0.5(10^3 \text{ m/s}^2)(10^{-3} \text{ s})^2$$

$$= 0.5 \times 10^{-3} \text{ m} = 0.05 \text{ cm}.$$

(c) Now the acceleration is $a = -g = -9.80 \text{ m/s}^2$ and the initial velocity for this second period is 1.0 m/s. We use Eq. 1-10c, with $v = 0$ (we are considering the highest point):

$$y = \frac{v^2 - v_0^2}{2a} = \frac{0 - (1.0 \text{ m/s})^2}{2(-9.80 \text{ m/s}^2)}$$

$$= 0.05 \text{ m} = 5 \text{ cm}.$$

Photographs indicate that the flea only jumps about two-thirds this high. Can you guess why our calculation is off?

The acceleration of an object, particularly rockets and fast airplanes, is often given as a multiple of $g = 9.80 \text{ m/s}^2$. For example, a diving plane undergoing 3.00 g's would have an acceleration of $(3.00)(9.80 \text{ m/s}^2) = 29.4 \text{ m/s}^2$.

*1-11 ∎ Graphical analysis of linear motion†

Figure 1-4 shows the graph of the velocity of a car versus time for two cases of linear motion: (a) constant velocity, and (b) a particular case where the magnitude of the velocity varied. It is also useful to graph (or "plot") the position x as a function of time. The time t is considered the independent variable and is measured along the horizontal axis (called the abscissa); and the position, x, the dependent variable, is measured along the vertical axis (ordinate). We assume that at $t = 0$ the position $x_0 = 0$. First we consider a car moving at a constant velocity of 40 km/h, which is equivalent to 11 m/s (see Section 1-4). From Eq. 1-10b, $x = vt$, and we see that x increases by 11 m every second. Thus, the position increases linearly in time, so the graph of x vs. t is a straight line as shown in Fig. 1-12. Each point on this straight line tells us the car's position at a particular time. For example, at $t = 3.0$ s, the position is 33 m, and at $t = 4.0$ s, $x = 44$ m; this is indicated by the dashed lines. The small triangle on the graph indicates the *slope* of the straight line which is defined as the change in the dependent variable (Δx) divided by the corresponding change in the independent variable (Δt):

$$\text{slope} = \frac{\Delta x}{\Delta t}.$$

We see, using the definition of velocity (Eq. 1-2), that the slope of the x vs. t graph is equal to the velocity. And, as can be seen from the little triangle on the graph, $\Delta x/\Delta t = (11 \text{ m})/(1.0 \text{ s}) = 11$ m/s, which is the given velocity.

The slope of the x vs. t graph is everywhere the same if the velocity is constant. But if the velocity changes, the slope of the x vs. t graph also varies.

† Certain sections and subsections of this book, such as this one, are considered *optional* at the discretion of the instructor. See the Preface for more details.

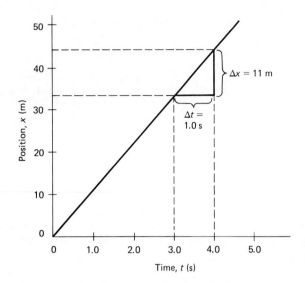

FIGURE 1-12

Graph of position vs. time for object moving at a uniform velocity of 11 m/s.

Consider, for example, a car that accelerates uniformly from rest to 15 m/s in 10 s, after which it remains at a constant velocity of 15 m/s for the next 10 s, decelerates to 5.0 m/s during the following 5.0 s, and then remains at this constant velocity; the velocity as a function of time is shown in the graph of Fig. 1-13a. Now, to construct the x vs. t graph, we can use Eq. 1-10b with constant acceleration for the intervals $t = 0$ to $t = 10$ s and $t = 20$ s to $t = 25$ s, and with constant velocity for the periods $t = 10$ s to $t = 20$ s and after $t = 25$ s. The result is the x vs. t graph of Fig. 1-13b.

From the origin to point A on the plot, the x vs. t graph is not a straight line, but a curve. The *slope* of the curve at any point is defined as the *slope of the tangent to the curve at that point*. For example, the tangent to the curve at the time $t = 5.0$ s is drawn on the graph (it is labeled "tangent"). A triangle is drawn with $\Delta t = 4.0$ s; Δx can be measured off the graph and is found to be 30 m. Thus the slope of the curve at $t = 5.0$ s, which equals the instantaneous velocity at that instant, is $v = \Delta x / \Delta t = 30$ m/4.0 s = 7.5 m/s. (This is the same result we get if we apply $v = at$ with $t = 5.0$ s and $a = [15$ m/s $- 0$ m/s]/$[10$ s] = 1.5 m/s^2.) In the region between A and B (Fig. 1-13) the x vs. t graph is a straight line and the slope can be measured using the triangle shown between $t = 15$ s and $t = 20$ s: $\Delta x / \Delta t = 75$ m/5.0 s = 15 m/s.

FIGURE 1-13

(a) Velocity vs. time and (b) displacement vs. time for an object with variable velocity. See text.

(a)

(b)

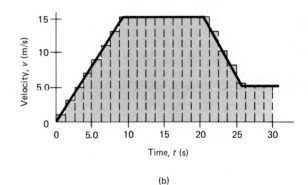

(a)

(b)

FIGURE 1-14

Determining displacement from v vs. t graph is done by calculating areas.

Suppose we were given the x vs. t graph of Fig. 1-13b; we could measure the slopes at a number of points and plot these slopes as a function of time. Since the slope equals the velocity we would thus have reconstructed the v vs. t graph! In other words, if we were given the graph of x vs. t, we could determine the velocity as a function of time using graphical methods, instead of using equations. This technique is particularly useful when the acceleration is not constant, for then Eqs. 1-10 cannot be used.

The reverse process is also possible. If we are given the v vs. t graph, we can determine the position, x, as a function of time. To do so, we use the following procedure which we apply to the v vs. t graph of Fig. 1-14a (which is the same as Fig. 1-13a). We first divide the time axis into many subintervals (in Fig. 1-14a, only six for simplicity) which are indicated in the figure by the dashed vertical lines. In each interval, a *horizontal* dashed line is drawn to indicate the average velocity during that time interval. For example, in the first interval, the velocity increases at a constant rate from zero to 7.5 m/s so $\bar{v} = 3.75$ m/s; and in the third interval the velocity is a constant 15 m/s, so $\bar{v} = 15$ m/s (no horizontal dashed line is shown since it coincides with the curve itself). The displacement (change in position) during any subinterval is, from the definition of average velocity, $\Delta x = \bar{v}\,\Delta t$. Thus the displacement during each subinterval equals the product of $\bar{v}$ and Δt, and this is just the *area of the rectangle*, shown in gray, for that interval. The total displacement after, say, 25 s, will be the sum of the first five rectangles.

If the velocity varies a great deal, it may be difficult to estimate $\bar{v}$ from the graph. To reduce this difficulty, more—but narrower—subintervals are used; that is, we make each Δt smaller, as in Fig. 1-14b. Ideally, we could let Δt approach zero; this leads to the subject of integral calculus, which we don't discuss here. The result, in any case, is that *the total displacement between any two times is equal to the area under the v vs. t graph between these two times*.

FIGURE 1-15

The gray area represents the displacement during the time interval $t = 2.0$ s to $t = 6.0$ s.

EXAMPLE 1-12 A space vehicle accelerates uniformly from 50 m/s at $t = 0$ to 150 m/s at $t = 10$ s. How far did it move between $t = 2.0$ s and $t = 6.0$ s?

SOLUTION The graph of v vs. t is shown in Fig. 1-15. We simply need to calculate the gray area shown, which is a trapezoid. The area will be the average of the heights (in units of velocity) times the width (which is 4.0 s). At $t = 2.0$ s, $v = 70$ m/s; and at $t = 6.0$ s, $v = 110$ m/s. Thus

$$\Delta x = \left(\frac{70 \text{ m/s} + 110 \text{ m/s}}{2}\right)(4.0 \text{ s}) = 360 \text{ m}.$$

In cases where the acceleration is not constant, the area can be obtained by counting squares on graph paper (see the problems).

▌ Summary[†]

Kinematics deals with the description of how objects move, as compared to *dynamics*, which deals with why they move as they do. The description of the motion of any object must always be given relative to some particular *reference frame*. To describe the motion of an object, which in this chapter we restrict to only one dimension (along a line), we use the concepts of displacement, velocity, and acceleration. The *average speed* is defined as the distance traveled divided by the elapsed time. *Displacement* refers to the change in position of the object. Velocity is the rate of change of position: the *average velocity* of an object over a particular time interval is defined as the displacement Δx divided by the time interval Δt: $\bar{v} = \Delta x / \Delta t$. The *instantaneous velocity*, whose magnitude is the same as the *instantaneous speed*, is the average velocity taken over an indefinitely short time.

[†] The summaries that appear at the end of each chapter in this book are intended to give a brief overview of the main ideas of each chapter. They *cannot* act to give an understanding of the material, which can only be had by a detailed reading of the chapter.

Acceleration is defined as the rate of change of velocity: the *average acceleration* over a time interval Δt is defined as $\bar{a} = \Delta v / \Delta t$ where Δv is the change of velocity during the time interval Δt; the *instantaneous acceleration* is defined as the average acceleration taken over an indefinitely short time interval.

If an object moves in a straight line with constant acceleration (*uniformly accelerated motion*), the velocity, v, and position, x, are related to the acceleration, a, the elapsed time, t, and the initial position and velocity, x_0 and v_0, by Eqs. 1-10:

$$v = v_0 + at, \qquad x = x_0 + v_0 t + \tfrac{1}{2}at^2,$$
$$v^2 = v_0^2 + 2a(x - x_0), \qquad \bar{v} = \frac{v + v_0}{2}.$$

Objects that move vertically near the surface of the earth, either falling or having been projected vertically up or down, move with the constant downward *acceleration due to gravity* of about $g = 9.80$ m/s^2, if air resistance can be ignored.

▌ Questions

1. If an accurate speedometer registers a constant value for a period of time, can you determine the average velocity over that period of time using only the speedometer? Explain.

2. Can the average velocity of a particle be zero over a given time interval if it is not zero over a shorter time interval? Explain.

3. Can an object have a varying velocity if its speed is constant? If yes, give examples.

4. Can an object have a varying speed if its velocity is constant? If yes, give examples.

5. Does a car speedometer measure speed, velocity, or both?

6. Can the velocity of an object be zero at the same instant its acceleration is not zero? Give an example.

7. If an object has a greater speed, does it necessarily have a greater acceleration? Explain, using examples.

8. Compare the acceleration of a motorcycle that accelerates from 80 km/h to 90 km/h with the acceleration of a bicycle that accelerates from rest to 10 km/h in the same time.

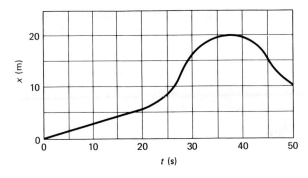

FIGURE 1-16

FIGURE 1-17

9. Can you conclude that a car is not accelerating if its speedometer indicates a steady 60 km/h?

10. Can an object have a northward velocity and a southward acceleration?

11. Can the velocity of an object be negative when its acceleration is positive? What about vice versa?

12. Give an example where both the velocity and acceleration are negative.

13. A rock is thrown vertically upward with speed v from the edge of a cliff. A second rock is thrown vertically downward with the same initial speed. Which rock has the greater speed when it reaches the bottom of the cliff? Ignore air resistance.

14. A ball, thrown vertically upward, returns to the thrower's hand. Which part of the journey requires the longer time, upward or downward? Answer for (a) no air resistance, (b) in the presence of air resistance. (Hint: the acceleration due to air resistance is always in a direction opposite to the motion.)

*15. Describe in words the motion plotted in Fig. 1-16.

*16. Describe in words the motion of the object graphed in Fig. 1-17.

▌ Problems

[*The problems at the end of each chapter are arranged by section and marked according to difficulty: (I) is easiest, (III) the most difficult, as discussed in the Preface.*]

SECTIONS 1-1 TO 1-7

1. (I) What must your average speed be in order to travel 880 km in 8.10 h?

2. (I) At an average speed of 11.8 km/h, how far will a bicyclist travel in 175 min?

3. (I) A bird can fly 30 km/h. How long does it take to fly 235 km?

4. (I) If you are driving 90 km/h and you look to the side for 2.0 s, how far do you travel during this inattentive period?

5. (I) 55 mph is how many (a) km/h, (b) m/s, (c) ft/s?

6. (I) Determine the conversion factor between (a) km/h and mi/h, (b) m/s and ft/s, (c) mi/h and m/s.

7. (II) A person jogs eight complete laps around a quarter-mile track in a total time of 10.5 min. Calculate (a) the average speed, and (b) the average velocity, in m/s.

8. (II) A rock thrown horizontally at a large bell 50 m away is heard to hit the bell 4.5 s later. If the speed of sound is 330 m/s, what was the speed of the rock? (Disregard the effect of gravity.)

9. (II) An airplane travels 1800 km at a speed of 1000 km/h. It then encounters a headwind that slows it to 850 km/h for the next 2300 km. What was the average speed of the plane for this trip? (Hint: think carefully before using Eq. 1-10d.)   243 km/h

10. (II) A race car driver must average 200 km/h for four laps to qualify for a race. Because of engine trouble, the car averages only 170 km/h over the first two laps. What average speed must be maintained for the last two laps? (Hint: think carefully before using Eq. 1-10d.)

11. (II) Calculate the carrying capacity (number of cars passing a given point per hour) on a freeway with three lanes (in one direction) using the following assumptions: the average speed is 90 km/h, the average length of a car is 6.0 m, and the average distance between cars should be 65 m.

12. (II) A car traveling 90 km/h is 100 m behind a truck traveling 50 km/h. How long will it take the car to reach the truck?   9 s

SECTION 1-8

13. (I) At high speeds, a particular automobile is capable of an acceleration of about 0.50 m/s². At this rate how long does it take to accelerate from 90 km/h to 100 km/h?

14. (I) An advertisement claims that a sportscar can accel-

erate from rest to 100 km/h in 7.6s. What is its acceleration in m/s^2?

15. (II) A sportscar is advertised to be able to stop, from a speed of 100 km/h, within 45 m. What is its acceleration in m/s^2? How many g's is this ($g = 9.80$ m/s^2)?

16. (III) The position of a racing car, which starts from rest at $t = 0$ and moves in a straight line has been measured as a function of time as given in the following table. Estimate (a) its velocity and (b) its acceleration as a function of time. Display each in a table and on a graph.

t (s)	0	0.25	0.50	0.75	1.00	1.50	2.00
x (m)	0	0.11	0.46	1.06	1.94	4.62	8.55

t (s)	2.50	3.00	3.50	4.00	4.50	5.00	5.50	6.00
x (m)	13.79	20.36	28.31	37.65	48.37	60.30	73.26	87.16

SECTION 1-9

17. (I) The principal kinematic equations, 1-10a through 1-10d, become particularly simple if the initial speed is zero. Write down the equations for this special case. (Also put $x_0 = 0$.)

18. (I) A car accelerates from 40 km/h up to 80 km/h in 5.0 s. What was its acceleration in m/s^2, and how far did it travel in this time? Assume constant acceleration.

19. (I) An eagle accelerates from 15 m/s to 20 m/s over a distance of 50 m. What was its average acceleration?

20. (II) A car decelerates from a speed of 30.0 m/s to rest in 6.00 s. How far did it travel in that time? 90 m

21. (II) A 100-m-long train accelerates uniformly from rest. If the front of the train passes a railway worker 150 m down the track at a speed of 25 m/s, what will be the speed of the last car as it passes the worker?

22. (II) A car traveling 80 km/h decelerates at a constant 1.5 m/s^2. Calculate (a) the distance it goes before it stops, (b) the time it takes to stop, and (c) the distance it travels during the first and third seconds.

23. (II) Make up a table of stopping distances for an automobile with an initial speed of 80 km/h and human reaction time of 1.0 s: (a) for an acceleration $a = -4.0$ m/s^2; (b) for $a = -8.0$ m/s^2.

24. (II) Repeat problem 23 using a reaction time of 0.40 s.

25. (III) Show that the equation for the stopping distance of a car is $d_s = v_0 t_R - v_0^2/(2a)$ where v_0 is the initial speed of the car, t_R is the driver's reaction time, and a is the constant acceleration (and is negative).

26. (III) In putting, the force with which a golfer strikes a ball is determined so that the ball will stop within some small distance of the cup, say 1.0 m long or short, even if the putt is missed. Accomplishing this from an uphill lie (that is, putting downhill) is more difficult than from a downhill lie. To see why, assume that on a particular green the ball decelerates constantly at

3.0 m/s^2 going downhill and constantly at 4.0 m/s^2 going uphill. Suppose we have an uphill lie 7.0 m from the cup. Calculate the allowable range of initial velocities we may impart to the ball so that it stops in the range 1.0 m short to 1.0 m long of the cup. Do the same for a downhill lie 7.0 m from the cup. What in your results suggests that the downhill putt is more difficult?

27. (III) A woman driving her car at 50 km/h approaches an intersection when the traffic light turns yellow. She knows that the yellow light lasts only 2.0 s before turning to red, and she is 30 m away from the near side of the intersection. Should she try to stop, or should she make a run for it? The intersection is 12 m wide, and her car's maximum deceleration is -6.0 m/s^2. Also her car takes 7.0 s to accelerate from 50 km/h to 70 km/h. Ignore the length of her car and her reaction time.

28. (III) In the design of a rapid transit system, it is necessary to balance out the average speed of a train against the distance between stops. The more stops there are, the slower the train's average speed. To get an idea of this problem, calculate the time it takes a train to make a 36-km trip in two situations: (a) The stations at which the trains must stop are 0.80 km apart; (b) the stations are 3.0 km apart. Assume that at each station the train accelerates at a rate of 1.5 m/s^2 until it reaches 80 km/h, then stays at this speed until it puts on its brakes to arrive at the next station at which time it decelerates at -3.0 m/s^2. Assume it stops at each station for 20 s.

29. (III) For the design of a rapid transit system as discussed in the previous problem, derive a general formula for the average speed of a train. Specify the symbols used for all quantities involved such as the acceleration, deceleration, maximum velocity, distance between stations, and time stopped at each station.

30. (III) A runner hopes to complete the 5000-m run in less than 13.0 min. After exactly 11.0 min, there are still 800 m to go. The runner must accelerate at 0.20 m/s^2 for how many seconds in order to achieve the desired time?

SECTION 1-10 (Neglect Air Resistance)

31. (I) Calculate the acceleration of the baseball in Example 1-8 in "g's."

32. (I) A stone is dropped from the top of a cliff. It is seen to hit the ground below after 5.5 s. How high is the cliff? 148 m

33. (I) A baseball is thrown vertically into the air with a speed of 24.7 m/s. (a) How high does it go? (b) How long does it take to return to the ground?

34. (I) (a) How long does it take a brick to reach the ground if dropped from a height of 80.0 m? (b) What will be its velocity just before it reaches the ground?

35. (I) A trout jumps a waterfall 2.5 m high. With what minimum speed did it leave the water below to reach the top?

28

36. (II) The flea in Example 1-11 actually reached a height of only 3.5 cm because of air resistance. Using this height, and the initial speed of 1.0 m/s, calculate the actual deceleration rate of the flea, assuming it to be constant.

37. (II) A stone is dropped from the roof of a high building. A second stone is dropped 1.00 s later. How far apart are the stones when the second one has reached a speed of 23.0 m/s?

38. (II) A kangeroo jumps to a vertical height of 2.8 m. How long was it in the air before returning to earth?

39. (II) A helicopter is ascending vertically with a speed of 8.00 m/s; at a height of 120 m above the earth, a package is dropped from a window. How much time does it take for the package to reach the ground?

40. (II) A person who is properly constrained by a shoulder harness has a good chance of surviving a car collision if the deceleration does not exceed 30 g's. Assuming uniform deceleration at this rate, calculate the distance over which the front end of the car must be designed to collapse if a crash occurs at 100 km/h.

41. (II) The acceleration due to gravity on the moon is about one-sixth what it is on earth. If an object is thrown vertically upward on the moon, how many times higher will it go as compared to earth assuming the same initial velocity.

42. (II) Draw a graph of (a) the speed and (b) the distance fallen, as a function of time, for a body falling under the influence of gravity for $t = 0$ to $t = 5.00$ s.

43. (II) If air resistance is neglected, show that a ball thrown vertically upward with a speed v_0 will have the same speed, v_0, when it comes back down to the starting point.

44. (II) A falling stone takes 0.30 s to pass a window 2.4 m high. From what height above the top of the window did the stone fall?

45. (II) A stone is thrown vertically upward with a speed of 18.0 m/s. (a) How fast is it moving when it reaches a height of 16.0 m? (b) How long is required to reach this height? (c) Why are there two answers to (b)?

46. (III) A rock is dropped from a seacliff and the sound of it striking the ocean is heard 3.0 s later. If the speed of sound is 330 m/s, how high is the cliff?

47. (III) Pelicans tuck their wings and free fall straight down when diving for fish. Suppose a pelican starts its dive from a height of 20 m and cannot change its path once committed. If it takes a fish 0.10 s to perform evasive action, at what minimum height must it spot the pelican to escape? Assume the fish is at the surface of the water.

48. (III) Suppose you adjust your garden hose nozzle for a hard stream of water. You point the nozzle vertically upward at a height of 1.5 m above the ground. When you quickly move the nozzle away from the vertical, you hear the water striking the ground next to you for 2.0 s. What is the water speed as it leaves the nozzle?

* SECTION 1-11

*49. (I) The position of a rabbit along a straight tunnel as a function of time is plotted in Fig. 1-16. What is its instantaneous velocity (a) at $t = 10.0$ s and (b) at $t = 30.0$ s? What is its average velocity (c) between $t = 0$ and $t = 5.0$ s, (d) between $t = 25.0$ s and $t = 30.0$ s, and (e) between $t = 40.0$ s and $t = 50.0$ s?

*50. (I) In Fig. 1-16, (a) during what time periods, if any, is the rabbit's velocity constant? (b) At what time is its velocity the greatest? (c) At what time, if any, is the velocity zero? (d) Does the rabbit run in one direction or in both along its tunnel during the time shown?

*51. (I) Figure 1-17 shows the velocity of a train as a function of time. (a) At what time was its velocity greatest? (b) During what periods, if any, was the velocity constant? (c) During what periods, if any, was the acceleration constant? (d) When was the magnitude of the acceleration greatest?

*52. (II) In Fig. 1-17, estimate the distance the train traveled during (a) the first minute, (b) the second minute.

*53. (II) Construct the v vs. t graph for the object whose displacement as a function of time is given by Fig. 1-16.

*54. (II) Construct the x vs. t graph for the object whose velocity as a function of time is given by Fig. 1-17.

*55. (II) (a) Suppose at times $t_0 = 0$, $t_1 = 1.0$ s, $t_2 = 2.0$ s, $t_3 = 3.0$ s, an object is at the positions $x_0 = 0$, $x_1 = 2.0$ m, $x_2 = 4.0$ m, $x_3 = 6.0$ m; plot these points on an x vs. t graph. Compute the average velocity over the time intervals $t_1 - t_0$, $t_2 - t_1$, $t_3 - t_2$. (b) Do the same supposing the object is at positions $x_0 = 10.0$ m, $x_1 = 8.0$ m, $x_2 = 6.0$ m, $x_3 = 4.0$ m for the respective times. (c) Do the same assuming the object was at $x_0 = 0$, $x_1 = 3.0$ m, $x_2 = 6.0$ m, $x_3 = 9.0$ m at times $t_0 = -2.0$ s, $t_1 = -1.0$ s, $t_2 = 0.0$ s, $t_3 = 1.0$ s.

*56. (III) Suppose an object's position as a function of time is given by the following table:

t_i	x_i
0.0 s	0.000 m
1.0 s	0.333 m
2.0 s	1.333 m
2.1 s	1.470 m
2.2 s	1.610 m
2.5 s	2.083 m
3.0 s	3.000 m

Plot these points on a graph. Compute the average velocity $\bar{v}$ for the time intervals $t_{3.0} - t_{2.0}$, $t_{2.5} - t_{2.0}$, $t_{2.2} - t_{2.0}$, and $t_{2.1} - t_{2.0}$. To what value of the instantaneous velocity at $t_{2.0}$ do these numbers seem to be converging? On the graph, draw a long straight line through each pair points; for example, draw the first long straight line through the points ($t_{2.0} = 2.0$ s, $x_{2.0} = 1.333$ m) and ($t_{3.0} = 3.0$ s, $x_{3.0} = 3.0$ m). The slope of each line equals the average velocity over the corresponding time interval. Note the slope to which the lines are converging as $t \rightarrow t_{2.0}$. This slope equals the instantaneous velocity at $t_{2.0}$.

2 KINEMATICS IN TWO DIMENSIONS; VECTORS

In Chapter 1 we dealt with motion along a straight line. We now consider the description of the motion of objects that move in paths in two (or three) dimensions. To do so we first need to discuss vectors and how they are added.

As we discussed in Section 1-7, a *vector* quantity is one that has both magnitude and direction. Examples are displacement, velocity, acceleration, and force. *Scalar* quantities, on the other hand, have magnitude only. They are specified completely by giving a number (plus units, if any). Examples are temperature and time.

2-1 ∎ Addition of vectors — graphical methods

Because vectors are quantities that have direction as well as magnitude, they must be added in a special way. In this chapter we will deal mainly with displacement vectors (for which we now use the symbol **D**) and velocity vectors (**v**). But the results will apply for other vectors that we encounter later.

We use simple arithmetic for adding scalars, such as time. Simple arithmetic can also be used for adding vectors if they are in the same direction. For example, if a person walks 8 km east one day, and 6 km east the next day, the person will be 8 km + 6 km = 14 km from the point of origin. We say that the *net* or *resultant* displacement is 14 km to the east. If, on the other hand, the person walks 8 km east on the first day, and 6 km west (in the reverse direction) on the second day, then he will be 2 km from the origin after the 2 days so his resultant displacement is 2 km to the east. In this case, the resultant displacement is obtained by subtraction: 8 km − 6 km = 2 km.

But simple arithmetic cannot be used if the two vectors are not along the same line. For example, suppose a person walks 10.0 km east and then walks 5.0 km north. This motion can be represented on a graph in which the positive y axis points north and the positive x axis points east, Fig. 2-1. On this graph we draw an arrow, labeled $\mathbf{D}_1$, to represent the displacement vector of the 10.0-km displacement to the east; and a second arrow, $\mathbf{D}_2$, to represent the 5.0-km displacement to the north. Both vectors are drawn to scale.

FIGURE 2-1
A person walks 10.0 km east and then 5.0 km north. These two displacements are represented by the vectors $\mathbf{D}_1$ and $\mathbf{D}_2$, which are shown as arrows. The resultant displacement vector, $\mathbf{D}_R$, which is the sum of $\mathbf{D}_1$ and $\mathbf{D}_2$, is also shown. Measurement on the graph with ruler and protractor shows that $\mathbf{D}_R$ has magnitude 11.2 km and it is at an angle $\theta = 27°$ north of east.

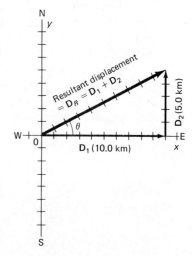

After taking this walk, the person is now 10.0 km east and 5.0 km north of the point of origin. The *resultant displacement* is represented by the arrow labeled $\mathbf{D_R}$ on the diagram. If you use a ruler and a protractor; you will measure on this diagram that the person is 11.2 km from the origin at an angle of 27° north of east. In other words, the resultant displacement vector has a magnitude of 11.2 km and makes an angle (θ) with the x axis of 27°. The magnitude (length) of $\mathbf{D_R}$ can also be obtained using the theorem *of* Pythagoras since D_1, D_2, and D_R form a right triangle with D_R as the hypotenuse. Thus $D_R = \sqrt{D_1^2 + D_2^2} = \sqrt{(10.0 \text{ km})^2 + (5.0 \text{ km})^2} = \sqrt{125 \text{ km}^2} = 11.2$ km. You can use this theorem, of course, only when the vectors are *perpendicular* to each other.

The resultant displacement vector, $\mathbf{D_R}$, is the sum of the vectors $\mathbf{D_1}$ and $\mathbf{D_2}$. That is, $\mathbf{D_R} = \mathbf{D_1} + \mathbf{D_2}$. This is a *vector* equation. An important feature of adding two vectors that are not along the same line is that the magnitude of the resultant vector is not equal to the sum of the magnitudes of the two separate vectors ($D_R \neq D_1 + D_2$).

Figure 2-1 illustrates the general rules for adding two vectors together, no matter what angles they make, to get their sum. Specifically, the rules are as follows: (1) on a diagram, draw one of the vectors—call it $\mathbf{V_1}$—to scale; (2) next draw the second vector, $\mathbf{V_2}$, to scale, placing its tail at the tip of the first vector and being sure its direction is correct; (3) the arrow drawn from the tail of the first vector to the tip of the second represents the *sum*, or **resultant,** of the two vectors. The length of the resultant can be measured and compared to the scale. Angles can be measured with a protractor. This method is known as the *tail-to-tip method of adding vectors.* It is, in fact, the definition of how to add vectors.

Note that it is not important in which order the vectors are taken. For example, a displacement of 5.0 km north, to which is added a displacement of 10.0 km east, yields a resultant of 11.2 km and angle $\theta = 27°$ (Fig. 2-2)—the same as when they were added in reverse order (Fig. 2-1). That is

$$\mathbf{V_1} + \mathbf{V_2} = \mathbf{V_2} + \mathbf{V_1}.$$

The tail-to-tip method can be extended to three or more vectors. An example is shown in Fig. 2-3; the three vectors could represent displacements (northeast, south, west) or perhaps three forces. Check for yourself that you get the same resultant no matter in which order you add the three vectors.

A second way to add two vectors is the *parallelogram method;* it is equivalent to the tail-to-tip method. In this method, the two vectors are drawn from a common origin and a parallelogram is constructed using these two vectors as adjacent sides. The resultant is the diagonal drawn from the

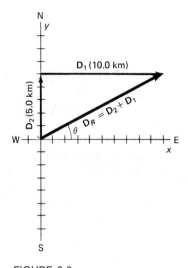

FIGURE 2-2

If the order of the vectors is reversed, the resultant is the same; see Fig. 2-1.

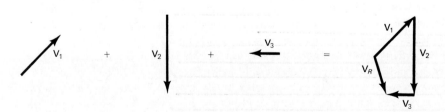

FIGURE 2-3

The resultant of three vectors, $\mathbf{V_R} = \mathbf{V_1} + \mathbf{V_2} + \mathbf{V_3}$.

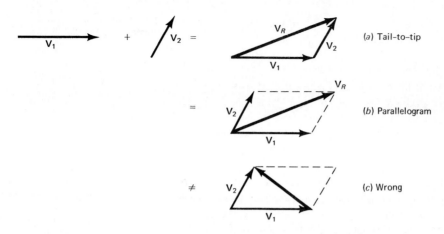

(a) Tail-to-tip

(b) Parallelogram

(c) Wrong

FIGURE 2-4

Vector addition by two different methods, (a) and (b); (c) is incorrect.

common origin. An example is shown in Fig. 2-4b. In Fig. 2-4a the tail-to-tip method is shown, and it is clear that both methods yield the same result. It is a common error to draw the sum vector as the diagonal running between the tips of the two vectors, as in Fig. 2-4c; this is incorrect: it does not represent the sum of the two vectors (in fact, it represents their difference—see below).

Vectors, when considered as mathematical quantities, can be moved about parallel to themselves as long as their direction and magnitude are not changed. We used this above when adding vectors. (Physically, the position of a vector can be important; for example, the vector representing a force on a body must be placed at the point where the force acts if we are to understand the resulting movement of that body correctly.)

FIGURE 2-5

(a) The negative of a vector is a vector with the same length but opposite direction; (b) subtracting two vectors: $\mathbf{A} - \mathbf{B}$.

2-2 ∎ Subtraction of vectors and multiplication of a vector by a scalar

Given a vector $\mathbf{V}$, we define the *negative* of this vector $(-\mathbf{V})$ to be a vector with the same magnitude but opposite direction to $\mathbf{V}$, Fig. 2-5a. We can use this to define the subtraction of one vector from another: the difference between two vectors, $\mathbf{A} - \mathbf{B}$, is defined as

$$\mathbf{A} - \mathbf{B} = \mathbf{A} + (-\mathbf{B});$$

that is, the difference between two vectors is equal to the sum of the first plus the negative of the second. Thus our rules for addition of vectors can be applied as shown in Fig. 2-5b using the tail-to-tip method.

A vector $\mathbf{V}$ can be multiplied by a scalar c. We define this product so that $c\mathbf{V}$ has the same direction as $\mathbf{V}$ and has magnitude cV. That is, multiplication of a vector by a positive scalar c changes the magnitude of the vector by a factor c but doesn't alter the direction. If c is a negative scalar, the magnitude of the product $c\mathbf{V}$ is still cV (without the minus sign) but the direction is precisely opposite to that of $\mathbf{V}$.

2-3 ▮ Analytic method for adding vectors; components

Adding vectors graphically using a ruler and protractor is not very accurate, and is not useful for vectors in three dimensions. We discuss now a more precise method for adding vectors.

Consider first a vector **V** which lies in a plane. It can be expressed as the sum of two other vectors, called the **components** of the original vector. The components are usually chosen to be along two perpendicular directions. The process of finding the components is known as _resolving the vector into its components._ An example is shown in Fig. 2-6; the vector **V** could be a displacement vector that points at an angle $\theta = 30°$ north of east, where we have chosen the x axis to be to the east and the y axis north. This vector is resolved into its x and y components by drawing lines from the tip (A) of our given vector that are perpendicular to the x and y axes (lines AB and AC). Then the lines OB and OC represent the x and y components of **V** respectively; these _vector components_ are written $\mathbf{V}_x$ and $\mathbf{V}_y$. The magnitudes of $\mathbf{V}_x$ and $\mathbf{V}_y$, namely V_x and V_y, are referred to simply as the components, and thus are numbers (with units) that are $+$ or $-$ depending on whether they point along the positive or negative x or y axis. As can be seen in Fig. 2-6, $\mathbf{V}_x + \mathbf{V}_y = \mathbf{V}$ by the parallelogram method of adding vectors.

Space is made up of three dimensions, and sometimes it is necessary to resolve a vector into components along three mutually perpendicular directions. The components are then called $\mathbf{V}_x$, $\mathbf{V}_y$, and $\mathbf{V}_z$. Although resolution of a vector in three dimensions is merely an extension of the above technique, we will mainly be concerned with situations in which the vectors are in a plane and two components are all that are necessary.

In order to add vectors using the method of components, we need to use the trigonometric functions sine, cosine, and tangent, which we now discuss.

Given any angle, θ, as in Fig. 2-7a, a right triangle can be constructed by drawing a line perpendicular to either of its sides, as in Fig. 2-7b. The longest side of a right triangle, opposite the right angle, is called the hypotenuse, which we label c. The side opposite the angle θ is labeled o, and the side adjacent is labeled a; we let c, o, and a represent the lengths of these sides, respectively. We now define the three trigonometric functions, sine, cosine, and tangent (abbreviated sin, cos, tan) in terms of the right triangle, as follows:

$$\left.\begin{array}{l} \sin \theta = \dfrac{\text{side opposite}}{\text{hypotenuse}} = \dfrac{o}{c} \\[2mm] \cos \theta = \dfrac{\text{side adjacent}}{\text{hypotenuse}} = \dfrac{a}{c} \\[2mm] \tan \theta = \dfrac{\text{side opposite}}{\text{side adjacent}} = \dfrac{o}{a}. \end{array}\right\} \qquad (2\text{-}1)$$

Now it is an interesting fact that if we make the triangle bigger, but keep the same angles, then the ratio of the length of one side to the other, or of one side to the hypotenuse, remains the same; that is, in Fig. 2-7c we have

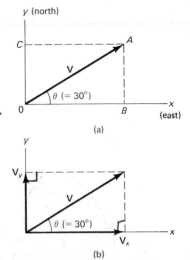

(a)

(b)

FIGURE 2-6

Resolving a vector **V** into its components along an arbitrarily chosen set of x and y axes.

FIGURE 2-7

Starting with an angle θ (a) we can construct right triangles of different sizes, (b) and (c), but the ratio of the lengths of the sides does not depend on the size of the triangle.

(a)

(b)

(c)

$a/c = a'/c'$; $o/c = o'/c'$; and $o/a = o'/a'$. Thus the values of sin, cos, and tan do not depend on how big the triangle is. They depend only on the size of the angle. The values of sin, cos, and tan for different angles have been computed precisely, and a table of them is located on the inside back cover of this book.

The use of trigonometric functions for finding the components of a vector is illustrated in Fig. 2-8 where it is seen that a vector and its two components can be thought of as making up a right triangle. If we multiply the definition of $\sin\theta = V_y/V$ by V on both sides, we get

$$V_y = V \sin\theta; \qquad (2\text{-}2a)$$

similarly we obtain

$$V_x = V \cos\theta. \qquad (2\text{-}2b)$$

Note that θ is chosen (by convention) to be the angle that the vector makes with the positive x axis.

Using Eqs. 2-2 we can calculate V_x and V_y for the vector illustrated in Fig. 2-6. Suppose **V** represents a displacement of 500 m in a direction 30° north of east. Then $V = 500$ m. From the trigonometric tables, $\sin 30° = 0.500$ and $\cos 30° = 0.866$. Then $V_x = V \cos\theta = (500 \text{ m})(0.866) = 433$ m (east) and $V_y = V \sin\theta = (500 \text{ m})(0.500) = 250$ m (north).

Note that there are two ways to specify a vector in a given coordinate system: we can give its components, V_x and V_y; or we can give its magnitude V and the angle θ it makes with the x axis. We can shift from one description to the other using Eqs. 2-2, or by using the theorem of Pythagoras[†] and the definition of tangent

$$V = \sqrt{V_x^2 + V_y^2} \qquad (2\text{-}3a)$$

$$\tan\theta = \frac{V_y}{V_x} \qquad (2\text{-}3b)$$

as can be seen in Fig. 2-8.

We can now discuss how to add vectors analytically by using components. The first step is to resolve each vector into its components. Next we can show (see Fig. 2-9) that the addition of two vectors $\mathbf{V}_1$ and $\mathbf{V}_2$ to give a resultant, $\mathbf{V} = \mathbf{V}_1 + \mathbf{V}_2$, implies that

$$\begin{aligned} V_x &= V_{1x} + V_{2x} \\ V_y &= V_{1y} + V_{2y} \end{aligned} \qquad (2\text{-}4)$$

A careful examination of Fig. 2-9 will verify these equations for the x and y components. If the magnitude and direction of the resultant vector are desired, they can be obtained using Eqs. 2-4.

The choice of coordinate axes is, of course, always arbitrary. You can often reduce the work involved in adding vectors analytically by a good choice of axes—for example, by choosing one of the axes to be in the same direction as one of the vectors so that vector will have only one nonzero component.

[†] In three dimensions, the theorem of Pythagoras becomes $V^2 = \sqrt{V_x^2 + V_y^2 + V_z^2}$, where V_z is the component along the third, or z, axis.

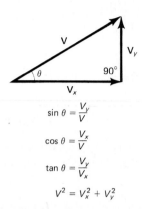

$$\sin\theta = \frac{V_y}{V}$$

$$\cos\theta = \frac{V_x}{V}$$

$$\tan\theta = \frac{V_y}{V_x}$$

$$V^2 = V_x^2 + V_y^2$$

FIGURE 2-8

Finding the components of a vector using trigonometric functions.

FIGURE 2-9

The components of $\mathbf{V} = \mathbf{V}_1 + \mathbf{V}_2$ are $V_x = V_{1x} + V_{2x}$ and $V_y = V_{1y} + V_{2y}$.

EXAMPLE 2-1 An explorer walks 22.0 km in a northerly direction, and then walks in a direction 60° south of east for 47.0 km (Fig. 2-10a). What is his resultant displacement from the origin?

SOLUTION We choose the positive x axis to be east and the positive y axis north. We resolve each displacement vector into its components (Fig. 2-10b). Since $\mathbf{D}_1$ has magnitude 22.0 km and points north, it has only a y component

$$D_{1x} = 0, \qquad D_{1y} = 22.0 \text{ km,}$$

whereas $\mathbf{D}_2$ has both x and y components:

$$D_{2x} = (47.0 \text{ km})(\cos 60°) = (47.0 \text{ km})(0.500) = 23.5 \text{ km}$$
$$D_{2y} = -(47.0 \text{ km})(\sin 60°) = -(47.0 \text{ km})(0.866) = -40.7 \text{ km.}$$

Notice that D_{2y} is negative because this vector component points along the negative y axis. The resultant vector ($\mathbf{D}$) has components:

$$D_x = D_{1x} + D_{2x} = 0 + 23.5 \text{ km} = 23.5 \text{ km}$$
$$D_y = D_{1y} + D_{2y} = 22.0 \text{ km} - 40.7 \text{ km} = -18.7 \text{ km.}$$

This specifies the resultant vector completely: $D_x = 23.5$ km, $D_y = -18.7$ km. We can also specify it by giving its magnitude and angle:

$$D = \sqrt{D_x^2 + D_y^2} = \sqrt{(23.5 \text{ km})^2 + (-18.7 \text{ km})^2} = 30.0 \text{ km}$$

$$\tan \theta = \frac{D_y}{D_x} = -\frac{18.7 \text{ km}}{23.5 \text{ km}} = -0.796$$

so $\theta = 38.5°$ south of east. The negative sign for the tangent results because θ is below the x axis, Fig. 2-10c.

Even though adding vectors analytically is generally more precise than graphical methods, it is important to draw at least rough vector diagrams for given situations since they give you something tangible to look at when analyzing a problem, and they can provide a rough check on results.

2-4 ∎ Relative velocity

A very useful example of adding vectors involves determining *relative velocity*. As a simple example consider two trains approaching one another, each with a speed of 80 km/h with respect to earth; then the speed of one train relative to the other is 160 km/h. That is, to an observer on one train, the other train seems to be approaching at 160 km/h. Similarly, when one car traveling 90 km/h passes a second car traveling 75 km/h, the first car has a speed relative to the second car of 90 km/h − 75 km/h = 15 km/h.

When the velocities are along the same line, simple addition or subtraction is sufficient to obtain the relative velocity. But if they are not along the same line, we must make use of vector addition. But we note, as mentioned in Section 1-2, that when specifying a velocity, it is important to specify what the reference frame is.

FIGURE 2-10
Example 2-1.

(a)

(b)

(c)

Before considering numerical examples, let us extend the definition of velocity given in Chapter 1 (for one-dimensional motion) to the case of two (or three) dimensions. For simplicity, suppose an object starts at the origin of some coordinate system at time $t = 0$. If after a time t its position is given by the displacement vector $\mathbf{D}$, then the average velocity $\bar{\mathbf{v}}$ during that time interval is defined as

$$\bar{\mathbf{v}} = \mathbf{D}/t. \tag{2-5}$$

This is the same definition we used in Chapter 1: average velocity is displacement divided by time; the only difference now is that this is a relation between displacement and velocity vectors. If we want to consider the more general case of an object being at a position given by $\mathbf{D}_1$ at some initial time t_1, and at a point given by $\mathbf{D}_2$ at some later time t_2, then the average velocity is

$$\bar{\mathbf{v}} = (\mathbf{D}_2 - \mathbf{D}_1)/(t_2 - t_1) = \Delta\mathbf{D}/\Delta t.$$

Usually we can take $t_1 = 0$ and $\mathbf{D}_1 = 0$ (at $t = 0$ the object is at the origin) and then Eq. 2-5 will suffice. The instantaneous velocity can be defined, as in Chapter 1, by considering the limit of the ratio $\Delta\mathbf{D}/\Delta t$ over an infinitesimally short time interval.

EXAMPLE 2-2 An airplane whose airspeed is 200 km/h heads due north. But a 100 km/h northeast wind suddenly begins to blow. What is the resulting velocity of the plane with respect to the ground?

SOLUTION The two velocity vectors are shown resolved into components in Fig. 2-11a. They are drawn with a common origin for convenience; $\mathbf{v}_1$ represents the velocity of the airplane with respect to the air and $\mathbf{v}_2$ the velocity of the wind with respect to the ground. The resultant velocity, $\mathbf{v}_R$, is the velocity of the plane with respect to the ground and is the vector sum of $\mathbf{v}_1$ and $\mathbf{v}_2$. Since $\mathbf{v}_1$ is along the y-axis, it has only a y-component:

$$v_{1x} = 0 \text{ km/h}$$

$$v_{1y} = v_1 = 200 \text{ km/h}.$$

The components of $\mathbf{v}_2$ are

$$v_{2x} = -v_2 \cos 45° = -(100 \text{ km/h})(0.707) = -70.7 \text{ km/h}$$

$$v_{2y} = -v_2 \sin 45° = -(100 \text{ km/h})(0.707) = -70.7 \text{ km/h}.$$

Both v_{2x} and v_{2y} are negative because their directions are, respectively, along the negative x and y axes. The components of the resultant velocity are

$$v_{Rx} = 0 \text{ km/h} - 70.7 \text{ km/h} = -70.7 \text{ km/h},$$

$$v_{Ry} = 200 \text{ km/h} - 70.7 \text{ km/h} = 129 \text{ km/h}.$$

We find the magnitude of the resultant velocity using the Pythagorean theorem:

$$v_R = \sqrt{v_{Rx}^2 + v_{Ry}^2} = 147 \text{ km/h}.$$

FIGURE 2-11
Example 2-2.

(a)

(b)

To find the angle θ that $\mathbf{v_R}$ makes with the x-axis (Fig. 2-11b) we use

$$\tan \theta = \frac{v_{Ry}}{v_{Rx}} = \frac{129 \text{ km/h}}{-70.7 \text{ km/h}} = -1.82.$$

(The negative sign results because θ is with respect to the negative x axis, which we already know from the diagram.[†]) Using trig tables we find that $\tan 61° = 1.804$ and $\tan 62° = 1.881$. So our angle θ is about 61°. A calculator with an INV TAN or "$\tan^{-1}$" key gives $\theta = -61.2°$.

When determining relative velocity, it is easy to make a mistake by adding or subtracting the wrong velocities. It is useful, therefore, to use a careful labeling process that makes things clear. Each velocity is labeled by two subscripts: the first refers to the object, the second to the reference frame in which it has this velocity. For example, suppose a boat is headed directly across a stream as shown in Fig. 2-12. We let $\mathbf{v_{BW}}$ be the velocity of the **B**oat with respect to the **W**ater. Similarly, $\mathbf{v_{WS}}$ is the velocity of the **W**ater with respect to the **S**hore (this is the river current) and $\mathbf{v_{BS}}$ is the resultant velocity of the **B**oat with respect to the **S**hore (it includes the effect of the current pulling the boat downstream). Note that $\mathbf{v_{BW}}$ is what the boat's motor produces (against the water), whereas $\mathbf{v_{BS}}$ is $\mathbf{v_{BW}}$ plus the effect of the current. Therefore the velocity of the boat relative to the shore is

$$\mathbf{v_{BS}} = \mathbf{v_{BW}} + \mathbf{v_{WS}}. \tag{2-6a}$$

By writing the subscripts via the convention above, we see that the inner subscripts on the right-hand side of Eq. 2-6a are the same (the two W's), and the outer subscripts on right of Eq. 2-6a (the B and the S) are the same as the two subscripts for the sum vector on the left, $\mathbf{v_{BS}}$. By following this convention, one can usually write down the correct equation relating velocities in different reference frames.[‡] Our derivation of Eq. 2-6a is valid in general and can be extended to three or more velocities. For example, if a fisherman on the boat walks with a velocity $\mathbf{v_{FB}}$ relative to the boat, his velocity relative to the shore is $\mathbf{v_{FS}} = \mathbf{v_{FB}} + \mathbf{v_{BW}} + \mathbf{v_{WS}}$. The equations involving relative velocity will be correct when adjacent inner subscripts are identical and when the outermost ones correspond exactly to the two on the velocity on the left of the equation. But this works only with plus signs (on the right), not minus signs.

It is often useful to remember that for any two objects or reference frames, A and B, the velocity of A relative to B has the same magnitude but opposite direction as the velocity of B relative to A:

$$\mathbf{v}_{BA} = -\mathbf{v}_{AB}. \tag{2-6b}$$

For example, if a train is traveling 100 km/h relative to earth in a certain direction, objects on earth (such as trees) appear to an observer on the train to be traveling 100 km/h in the opposite direction.

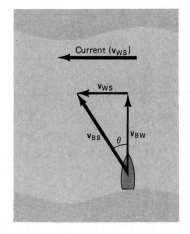

FIGURE 2-12
A boat heading directly across a river whose current moves at 1.20 m/s.

[†] The tangent is negative when either the x or the y component, but not both, is negative; that is, when θ is in the second quadrant ($90° < \theta < 180°$) or fourth quadrant ($270° < \theta < 360°$, or $0° > \theta > -90°$). These two possibilities occurred in this example (2-2) and the previous one (2-1).

[‡] We thus would know by inspection that (for example) the equation $\mathbf{v_{BW}} = \mathbf{v_{BS}} + \mathbf{v_{WS}}$ is wrong.

EXAMPLE 2-3 A boat can travel 1.85 m/s in still water. (*a*) If the boat heads directly across a stream whose current is 1.20 m/s, what is the velocity (magnitude and direction) of the boat relative to the shore? (*b*) What will be the position of the boat, relative to its point of orgin, after 3.00 s?

SOLUTION (*a*) As shown in Fig. 2-12, the boat is pulled downstream by the current. As discussed above, the boat's velocity with respect to the shore, $\mathbf{v}_{BS}$, is the sum of its velocity with respect to the water, $\mathbf{v}_{BW}$, plus the velocity of the water with respect to the shore, $\mathbf{v}_{WS}$:

$$\mathbf{v}_{BS} = \mathbf{v}_{BW} + \mathbf{v}_{WS}.$$

We could measure θ and v_{BS} (the magnitude of $\mathbf{v}_{BS}$) from the diagram, and roughly obtain $v_{BS} \approx 2.2$ m/s, and $\theta \approx 33°$. Or, since $\mathbf{v}_{BW}$ is perpendicular to $\mathbf{v}_{WS}$, we can get v_{BS} using the theorem of Pythagoras, $v_{BS} = \sqrt{v_{BW}^2 + v_{WS}^2} = \sqrt{(1.85 \text{ m/s})^2 + (1.20 \text{ m/s})^2} = 2.21$ m/s; and $\tan\theta = v_{WS}/v_{BW} = (1.20 \text{ m/s})/(1.85 \text{ m/s}) = 0.6486$[†], so $\theta = 33.0°$. (*b*) To find the position after 3.00 s, we determine the displacement vector $\mathbf{D}$ using the definition of velocity (see Eq. 2-5): $\mathbf{D} = \mathbf{v}t$, where t is the elapsed time. Since t is a scalar, it does not affect directions; so the direction of $\mathbf{D}$ is the same as that for $\mathbf{v}$. The magnitude of $\mathbf{D}$ in our case is $D = v_{BS}t = (2.21 \text{ m/s})(3.00 \text{ s}) = 6.63$ m. So the boat, after 3.00 s, is 6.63 m from its point of origin in a direction 33.0° downstream from directly across the river.

EXAMPLE 2-4 Suppose the boat of the previous example must cross the stream to a point on the other bank directly opposite its starting point; at what upstream angle must it point?

SOLUTION In this case, $\mathbf{v}_{BW}$ must be perpendicular to the bank since the boat has to arrive at a point directly opposite its starting point. To accomplish this, the boat has to head at an angle upstream, so we draw the rough diagram of Fig. 2-13. We want to find the angle θ. We have the same vector equation as we had in Example 2-3,

$$\mathbf{v}_{BS} = \mathbf{v}_{BW} + \mathbf{v}_{WS},$$

but the vectors $\mathbf{v}_{BS}$ and $\mathbf{v}_{BW}$ have different directions now than they did in Example 2-3. The magnitudes of $\mathbf{v}_{BW}$ and $\mathbf{v}_{WS}$ remain the same: $v_{BW} = 1.85$ m/s and $v_{WS} = 1.20$ m/s. We have a right triangle (see Fig. 2-13) and we know v_{BW} and v_{WS}; they are related to the sine of the angle θ:

$$\sin\theta = \frac{v_{WS}}{v_{BW}} = \frac{1.20 \text{ m/s}}{1.85 \text{ m/s}} = 0.6486.$$

Thus $\theta = 40.4°$, so the boat must head upstream at a 40.4° angle. (Notice that we didn't use $\tan\theta$ to get θ since we would first have had to determine v_{BS}.)

FIGURE 2-13
Boat must head upstream at an angle θ if it is to move directly across the river.

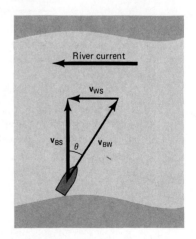

† For this value of the tangent we have kept 4 significant figures even though all the numbers in this example are given to 3 significant figures. It is usually a good idea that one extra significant figure be kept within a calculation when convenient. See discussion of significant figures in Appendix A-2.

EXAMPLE 2-5 Two aircraft on a collision course approach each other at a 90° angle with the same speed of 400 km/h (111 m/s), Fig. 2-14a. What is the relative velocity of one with respect to the other? (That is, determine the velocity of aircraft 1 as seen by aircraft 2.

SOLUTION Figure 2-14a shows the situation in a reference frame fixed to the earth. But we want to view the situation from a reference frame in which aircraft 2 is at rest, and this is shown in Fig. 2-14b. In this reference frame (the world as seen by the pilot of aircraft 2) the earth moves toward aircraft 2 with velocity $\mathbf{v}_{E2}$ (speed of 400 km/h) which is of course equal and opposite to $\mathbf{v}_{2E}$, the velocity of aircraft 2 with respect to the earth (Eq. 2-6b):

$$\mathbf{v}_{E2} = -\mathbf{v}_{2E}.$$

Thus the velocity of aircraft 1 as seen by aircraft 2 is

$$\mathbf{v}_{12} = \mathbf{v}_{1E} + \mathbf{v}_{E2}$$

or (since $\mathbf{v}_{E2} = -\mathbf{v}_{2E}$)

$$\mathbf{v}_{12} = \mathbf{v}_{1E} - \mathbf{v}_{2E}.$$

That is, the velocity of aircraft 1 as seen by aircraft 2 is the difference of their velocities, $\mathbf{v}_{1E} - \mathbf{v}_{2E}$, both measured relative to the earth. See Fig. 2-14c. Since the magnitudes of $\mathbf{v}_{1E}$, $\mathbf{v}_{2E}$, and $\mathbf{v}_{E2}$ are equal, we see (Fig. 2-14b) that $\mathbf{v}_{12}$ points at a 45° angle toward aircraft 2; the speed is

$$v_{12} = \sqrt{(111 \text{ m/s})^2 + (111 \text{ m/s})^2} = 157 \text{ m/s (565 km/h)}.$$

(a)

(b)

(c)

FIGURE 2-14
Example 2-5.

2-5 ∎ Projectile motion

So far in this chapter, our discussion of motion in a plane (two dimensions) has dealt only with displacement and velocity. But objects often have acceleration, so now we examine how to deal with accelerated motion. We do so by discussing a very important and useful special case, that of projectile motion.

Projectile motion refers to the motion of an object projected into the air near the surface of the earth and subjected to the downward acceleration of gravity. A thrown baseball, a jumping grasshopper, a speeding bullet, and an athlete doing the high jump, are all examples of projectile motion.

Although air resistance is often important, in many cases its effect can be ignored, and we will ignore it in the following analysis. We will not be concerned now with the process by which the object is thrown or projected; we consider only its motion after it has been projected and is moving freely through the air under the action of gravity. Thus the acceleration of the object[†] is that of gravity, **g**, acting downward with magnitude $g = 9.80 \text{ m/s}^2$.

---

† We restrict ourselves to objects whose distance traveled and maximum height above the earth are small compared to the earth's radius (6.4×10^6 m) so that **g** can be considered constant.

FIGURE 2-15
Projectile motion.

It was Galileo who first accurately described projectile motion; he showed that it could be understood by analyzing the horizontal and vertical components of the motion separately. This was an innovative analysis, not done in this way by anyone prior to Galileo. (It was also idealized in that it did not take into account air resistance.) For convenience assume that the motion begins at time $t = 0$ at the origin of an xy coordinate system (so $x_0 = y_0 = 0$).

First let us look at a (tiny) ball rolling off the end of a table with an initial velocity v_{x0} in the horizontal (x) direction; see Fig. 2-15. The velocity vector **v** at each instant points in the direction of the motion at that instant and is always tangent to the path. Following Galileo's ideas, we treat the horizontal and vertical components of the velocity v_x and v_y separately and we can apply the kinematic equations (1-10a through 1-10d) to each. Once the ball leaves the table (at $t = 0$) it experiences a vertically downward acceleration equal to g, the acceleration of gravity. Thus v_y is continually increasing in the downward direction (until the ball hits the ground). From Eq. 1-10a we can write $v_y = gt$ since the initial velocity in the vertical direction is zero. The vertical displacement, y, is given by $y = \frac{1}{2}gt^2$. Since there is no acceleration in the horizontal direction, the horizontal component of velocity, v_x, remains constant, equal to its initial value v_{x0}, and thus has the same magnitude at each point on the path. The two components, $\mathbf{v}_x$ and $\mathbf{v}_y$, can be added vectorially to obtain the velocity **v** for each point on the path, as shown in Fig. 2-15.

One result of this analysis, which Galileo himself predicted, is that an object projected horizontally will reach the ground in the same time as an object dropped vertically. This is because the vertical motions are the same in both cases. Figure 2-16 is a multiple-exposure photograph of an experiment that confirms this.

FIGURE 2-16

Multiple exposure photograph showing positions of 2 balls at (the same) equal intervals. One ball was dropped from rest at the same time the other was projected horizontally outward. The vertical motion for each ball is the same. Reproduced from PSSC Physics, second edition, 1965; D. C. Heath & Company with Education Development Center, Newton, MA.

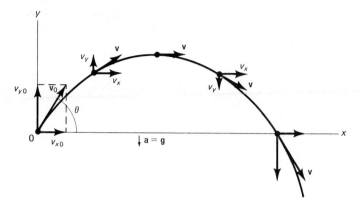

FIGURE 2-17

Motion of a projectile fired with initial velocity v_0 at angle θ_0 to the horizontal.

If an object is projected at an angle, as in Fig. 2-17, the analysis is essentially the same except that now there is an initial vertical component of velocity, v_{y0}. Because of the downward acceleration of gravity, v_y continually decreases until the object reaches the highest point on its path, at which point $v_y = 0$. Then v_y starts to increase in the downward direction, as shown.

It can be shown that the path followed by a projectile is a parabola.[†]

In the following examples of projectile motion we use the kinematic equations (1-10a through 1-10d) separately for the vertical and horizontal components of the motion; we use x or y for the respective displacement. See Table 2-1 which assumes y is positive upward so $a_y = -g = -9.80 \text{ m/s}^2$. Note, however, that if the y axis is taken positive downward, $a_y = g = 9.80 \text{ m/s}^2$.

TABLE 2-1

Kinematic equations for projectile motion (y positive upward)

Horizontal motion		Vertical Motion
$v_x = v_{x0}$	(Eq. 1-10a)	$v_y = v_{y0} - gt$
$x = v_{x0}t$	(Eq. 1-10b)	$y = v_{y0}t - \frac{1}{2}gt^2$
	(Eq. 1-10c)	$v_y^2 = v_{y0}^2 - 2gy$

☐ **EXAMPLE 2-6** A rock is thrown horizontally from a 100.0-m-high cliff. It strikes level ground 90.0 m from the base of the cliff. At what speed was it thrown? Take the y direction to be positive downward (which means we can't use Table 2-1 unless we change the sign of g).

SOLUTION First we find how long it takes to reach the ground below. We use Eq. 1-10b for the vertical (y) direction (which we take as posi-

[†] This is actually an approximation, valid if the acceleration is constant in magnitude and direction (that is, $\mathbf{a} = \mathbf{g}$ = constant). If the projectile path is long (that is, not small compared to the earth's radius), the variation in acceleration of gravity becomes important; the projectile's path in this more general case is an ellipse (or part of an ellipse) if air resistance can be ignored (see Section 4-10).

tive downward) with $v_{y0} = 0$ and $a = g = 9.80$ m/s². Thus

$$y = \tfrac{1}{2}gt^2,$$

$$t = \sqrt{\frac{2y}{g}} = \sqrt{\frac{200.0 \text{ m}}{9.80 \text{ m/s}^2}} = 4.52 \text{ s.}$$

To calculate the initial velocity, v_{x0}, we again use Eq. 1-10b but this time for the horizontal (x) direction, so $a = 0$. Thus

$$x = v_{x0}t$$

$$v_{x0} = \frac{x}{t} = \frac{90.0 \text{ m}}{4.52 \text{ s}} = 19.9 \text{ m/s.}$$

EXAMPLE 2-7 A football is kicked at an angle to the horizontal of $\theta = 37.0°$ with a velocity of magnitude 20.0 m/s as in Fig. 2-17. Calculate (a) the maximum height, (b) the time of travel before it hits the ground, and (c) how far away it hits the ground. Assume for simplicity that the ball leaves the foot at ground level.

SOLUTION We take the y direction as positive upward. The components of the initial velocity are

$$v_{x0} = v_0 \cos 37.0° = (20.0 \text{ m/s})(0.799) = 16.0 \text{ m/s}$$

$$v_{y0} = v_0 \sin 37.0° = (20.0 \text{ m/s})(0.602) = 12.0 \text{ m/s.}$$

(a) The maximum height is attained where $v_y = 0$, and this occurs (Eq. 1-10a, Table 2-1, with $g = 9.80$ m/s²) at time $t = v_{y0}/g = (12.0 \text{ m/s})/(9.80 \text{ m/s}^2) = 1.22$ s. From Eq. 1-10b we have

$$y = v_{y0}t - \tfrac{1}{2}gt^2$$
$$= (12.0 \text{ m/s})(1.22 \text{ s}) - \tfrac{1}{2}(9.80 \text{ m/s}^2)(1.22 \text{ s})^2$$
$$= 7.35 \text{ m.}$$

Alternatively, we could have used Eq. 1-10c solved for y, and found

$$y = \frac{v_{y0}^2 - v_y^2}{2g} = \frac{(12.0 \text{ m/s})^2 - (0 \text{ m/s})^2}{2(9.80 \text{ m/s}^2)} = 7.35 \text{ m.}$$

(b) To find the time it takes for the ball to return to the ground we use Eq. 1-10b and set $y = 0$ (ground level)

$$y = v_{y0}t - \tfrac{1}{2}gt^2$$
$$0 = (12.0 \text{ m/s})t - \tfrac{1}{2}(9.80 \text{ m/s}^2)t^2,$$

and therefore

$$t = \frac{2(12.0 \text{ m/s})}{(9.80 \text{ m/s}^2)} = 2.45 \text{ s.}$$

(The solution $t = 0$ also satisfies the equation but this corresponds to the initial point when y is also zero.)

(c) The total distance traveled in the x direction is found by applying Eq. 1-10b with $a = 0$, $v_{x0} = 16.0$ m/s

$$x = v_{x0}t = (16.0 \text{ m/s})(2.45 \text{ s}) = 39.2 \text{ m.}$$

▮ Summary

A quantity that has both a magnitude and a direction is called a *vector*. A quantity that has only a magnitude is called a *scalar*.

Addition of vectors can be done graphically by placing the tail of each successive arrow (representing each vector) at the tip of the previous one. The sum or *resultant vector* is the arrow drawn from the tail of the first to the tip of the last. Two vectors can also be added using the parallelogram method. Vectors can be added more accurately using the analytical method of adding their *components* along chosen axes with the aid of trigonometric functions.

The velocity of an object relative to one frame of reference can be found by vector addition if its velocity relative to a second frame of reference, and the relative velocity of the two reference frames, are known.

Projectile motion—that of an object moving through the air—can be analyzed as two separate motions in the horizontal and vertical directions. If air resistance can be ignored, the horizontal motion is at constant velocity whereas the vertical motion is uniformly accelerated and is the same as for a body falling vertically under the action of gravity.

▮ Questions

1. Does the odometer of a car measure a scalar or a vector quantity? What about the speedometer?
2. Can the displacement vector for a particle moving in two dimensions ever be longer than the length of path traveled by the particle over the same time interval? Can it ever be less? Discuss.
3. During practice, a baseball player hits a very high fly ball, and then runs in a straight line and catches it. Which had the greater displacement, the player or the ball?
4. If $\mathbf{V} = \mathbf{V}_1 + \mathbf{V}_2$, is V necessarily greater than V_1 and/or V_2? Discuss.
5. Two vectors have length $V_1 = 3.5$ *km* and $V_2 = 4.0$ km. What are the maximum and minimum magnitudes of their vector sum?
6. Can two vectors, of unequal magnitude, add up to give the zero vector? Can *three* unequal vectors?
7. Can the magnitude of a vector ever (*a*) equal, or (*b*) be less than, one of its components?

8. Can a vector of magnitude zero have a nonzero component?
9. One car travels due east at 40 km/h, and a second car travels north at 40 km/h. Are their velocities equal? Explain.
10. What factors are important for an athlete doing the broad jump? What about the high jump?
11. A projectile has the least speed at what point in its path?
12. A child wishes to determine the speed a slingshot imparts to a rock. How can this be done using only a meter stick?
13. Two rowers who can row at the same speed in still water set off across a river at the same time. One heads straight across and is pulled downstream somewhat by the current. The other one heads upstream at an angle so as to arrive at a point opposite the starting point. Which rower reaches the opposite side first?

▮ Problems

SECTIONS 2-1 TO 2-3

1. (I) A car is driven 30 km west and then 80 km southwest. What is the displacement of the car from the point of origin (magnitude and direction)?
2. (I) A delivery truck travels 18 blocks north, 16 blocks east, and 10 blocks south. What is its final displacement from the origin? Assume the blocks are equal length.
3. (I) The three vectors in Fig. 2-3 can be added in six different orders ($\mathbf{V}_1 + \mathbf{V}_2 + \mathbf{V}_3$, $\mathbf{V}_1 + \mathbf{V}_3 + \mathbf{V}_2$, etc.).

Show on a diagram that the same resultant is obtained no matter what the order.

4. (I) Show that the vector labeled "wrong" in Fig. 2-4c is actually the difference of the two vectors. Is it $\mathbf{V}_2 - \mathbf{V}_1$ or $\mathbf{V}_1 - \mathbf{V}_2$?
5. (I) Figure 2-18 shows two vectors, $\mathbf{A}$ and $\mathbf{B}$ whose magnitudes are $A = 6.3$ units and $B = 3.5$ units. Determine $\mathbf{C}$ if (*a*) $\mathbf{C} = \mathbf{A} + \mathbf{B}$, (*b*) $\mathbf{C} = \mathbf{A} - \mathbf{B}$, (*c*) $\mathbf{C} = \mathbf{B} - \mathbf{A}$. Give magnitude and direction for each.

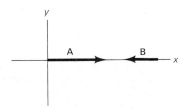

FIGURE 2-18

6.3 m
5.5°
E of S

6. (II) Graphically determine the resultant of the following three vector displacements: (1) 10 m, 30° north of east; (2) 6 m, 37° east of north; (3) 12 m, 30° west of south.

7. (II) Two vectors $\mathbf{V}_1$ and $\mathbf{V}_2$ add to a resultant $\mathbf{V} = \mathbf{V}_1 + \mathbf{V}_2$. Describe $\mathbf{V}_1$ and $\mathbf{V}_2$ if (a) $V = V_1 + V_2$, (b) $V^2 = V_1^2 + V_2^2$, (c) $V_1 + V_2 = V_1 - V_2$.

8. (II) Three vectors are shown in Fig. 2-19; their magnitudes are given in arbitrary units. Determine the sum of the three vectors. Give resultant in terms of (a) components, (b) magnitude and angle with x axis.

9. (II) Determine the vector $\mathbf{A} - \mathbf{C}$ given the vectors $\mathbf{A}$ and $\mathbf{C}$ in Fig. 2-19.

10. (II) Given the vectors $\mathbf{A}$ and $\mathbf{B}$ shown in Fig. 2-19 determine $\mathbf{B} - \mathbf{A}$.

11. (II) For the vectors given in Fig. 2-19, determine (a) $\mathbf{A} - \mathbf{B} + \mathbf{C}$, (b) $\mathbf{A} + \mathbf{B} - \mathbf{C}$.

1.8 m/s²
1 2.9s

12. (II) (a) A skier is accelerating down a 30.0° hill at 3.60 m/s². What is the vertical component of her acceleration? (b) How long will it take her to reach the bottom of the hill, assuming she starts from rest and accelerates uniformly, if the elevation change is 150 m?

13. (II) The components of a vector $\mathbf{V}$ are often written (V_x, V_y, V_z). What are the components and length of a vector which is the sum of the two vectors $\mathbf{V}_1$ and $\mathbf{V}_2$ whose components are (6, 0, 2) and (1, 4, 3)?

14. (II) Given the vectors $\mathbf{V}_1$ and $\mathbf{V}_2$ from problem 13 determine a third vector, $\mathbf{V}_3$, such that (a) $\mathbf{V}_1 + \mathbf{V}_2 + \mathbf{V}_3 = 0$. (b) $\mathbf{V}_1 - \mathbf{V}_2 + \mathbf{V}_3 = 0$.

FIGURE 2-19

15. (II) A deliveryman travels 30 m north, 25 m east, 12 m south, and then takes an elevator 36 m up into a building. What is his final displacement from his starting point? (Give components and also magnitude.)

16. (II) What is the y component of a vector in the xy plane whose magnitude is 36.5 and whose x component is 25.4? What is the direction of this vector?

17. (II) The summit of a mountain, 2150 m above a camp, is measured on a map to be 4750 m horizontally from the camp in a direction 28.2° west of north. What are the components of the displacement vector from camp to summit. What is its length? Choose the x axis east, y axis north, and z axis up.

SECTION 2-4

18. (I) A dog walks at a speed of 1.8 m/s along the deck toward the front of a boat which is traveling at 7.6 m/s with respect to the water. What is the velocity of the dog with respect to the water? What if the dog were walking toward the back of the boat?

19. (I) An airplane is traveling 1000 km/h in a direction 37° east of north. (a) Find the components of the velocity vector in the northerly and easterly directions. (b) How far north and how far east has the plane traveled after 2.0 h?

20. (II) A vacationer walks 4.0 km/h directly across a cruise ship whose speed relative to the earth is 11.2 km/h. What is the speed of the vacationer with respect to the earth?

21. (II) Raindrops make an angle θ with the vertical when viewed through a moving train window. If the speed of the train is v_t, what is the speed of the raindrops in the reference frame of the earth in which they are assumed to fall vertically.

22. (II) A boat can travel 2.60 m/s in still water. (a) If the boat heads directly across a stream whose current is 0.90 m/s, what is the velocity (magnitude and direction) of the boat relative to the shore? (b) What will be the position of the boat, relative to its point of origin, after 4.0 s?

3.8 min

23. (II) Two trains approach each other on parallel tracks. Each has a speed of 80 km/h with respect to the earth. If they are initially 10 km apart, how long will it be before they pass each other?

24. (II) An airplane is heading due north at a speed of 300 km/h. If a wind begins blowing from the southwest at a speed of 50 km/h (average), calculate: (a) the velocity (magnitude and direction) of the plane, and (b) how far off course it will be after 30 min if the pilot takes no corrective action.

25. (II) Determine the speed of the boat with respect to the shore in Example 2-4.

26. (II) A motorboat whose speed in still water is 8.25 km/h must aim upstream at an angle of 25.5° (with respect to a line perpendicular to the shore) in order to travel directly across the stream. (a) What is the speed of the current? (b) What is the resultant speed of the boat with respect to the shore?

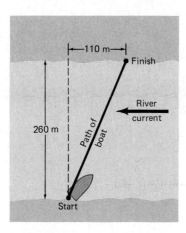

FIGURE 2-20

27. (II) A ferryboat, whose speed in still water is 2.85 m/s, must cross a 260-m-wide river and arrive at a point 110 m upstream from where it starts (Fig. 2-20). To do so, the pilot must head the boat at a 45° upstream angle. What is the speed of the river's current?

28. (II) A helicopter heads due south with an air speed of 50 km/h. The pilot observes, however, that they have covered 30 km in the previous 45 min in a southwesterly direction. What is the wind speed and direction?

29. (II) A swimmer is capable of swimming 1.80 m/s in still water. (a) If she aims her body directly across a 200-m wide river whose current is 0.80 m/s, how far downstream (from a point opposite her starting point) will she land? (b) How long will it take her to reach the other side? (c) At what upstream angle must the swimmer aim if she is to arrive at a point directly across the stream?

30. (II) An airplane, whose air speed is 380 km/h, is supposed to fly in a straight path 23.0° N of E. But a steady 100 km/h wind is blowing from the north. In what direction should the plane head?

31. (II) An automobile traveling 90.0 km/h overtakes a 1.00-km long train traveling in the same direction on a track parallel to the road. If the train's speed is 66.0 km/h, how long does it take the car to pass it and how far will the car have traveled in this time? What are the results if the car and train are traveling in opposite directions?

32. (III) A motorcycle traveling 90.0 km/h approaches a car traveling in the same direction at 80.0 km/h. When the motorcycle is 50 m behind the car, the rider pushes down on the accelerator and passes the car 10.0 s later. What was the acceleration of the motorcycle?

33. (III) An unmarked police car traveling a constant 80 km/h is passed by a speeder traveling 100 km/h. Precisely 1.0 s after the speeder passes, the policeman steps on the accelerator; if the police car's acceleration is 2.0 m/s², how much time passes before the police car overtakes the speeder (assumed moving at constant speed)?

34. (III) Assume in the previous problem that the speeder's speed is not known. If the police car accelerates uniformly as given above for 6.0 s, what was the speeder's speed?

35. (III) The speed of a boat in still water is v. The boat is to make a round trip in a river whose current travels at speed u. Derive a formula for the time needed to make a round trip of total distance D if the boat makes the round trip by moving (a) upstream and back downstream, (b) directly across the river and back. We must assume u < v; why?

SECTION 2-5 (Neglect air resistance)

36. (I) A diver running 3.6 m/s dives out horizontally from the edge of a vertical cliff and reaches the water below 2.0 s later. How high was the cliff and how far from its base did the diver hit the water?

37. (I) A tiger leaps horizontally from a 15-m high rock with a speed of 7.0 m/s. How far from the base of the rock will it land?   12 m

38. (II) A ball thrown horizontally at 18 m/s from the roof of a building lands 24 m from the base of the building. How high is the building?

39. (II) Show that the speed with which a projectile leaves the ground is equal to its speed just before it strikes the ground at the end of its journey, assuming the firing level equals the landing level.

40. (II) An airplane traveling a constant 150 km/h is to drop supplies to flood victims isolated on a patch of land 200 m below. The supplies should be dropped how many seconds before the plane is directly overhead?

41. (II) A football is kicked with a speed of 21.0 m/s at an angle of 37° to the horizontal. How much later does it hit the ground? Ignore air resistance.

42. (II) A hunter aims directly at a target (on the same level) 220 m away. If the bullet leaves the gun at a speed of 550 m/s, by how much will it miss the target?

43. (II) An athlete throws the shotput (mass = 7.3 kg) with an initial speed of 14 m/s at a 40° angle to the horizontal. Calculate the distance traveled. The shot leaves the shotputter's hand at a height of 2.2 m above the ground.

44. (II) Show that the time required for a projectile to reach its highest point is equal to the time for it to return to its original height.

45. (II) An Olympic long jumper is capable of jumping 8.0 m. Assuming his horizontal speed is 9.0 m/s (top sprinting speed is slightly over 10 m/s) as he leaves the ground, how long was he in the air and how high did he go? Assume that he lands standing upright, that is, the same way he left the ground.

46. (II) A projectile is fired into the air with an initial speed of 40.0 m/s. Plot on graph paper its trajectory for initial projection angles of θ = 15°, 30°, 45°, 60°, 75°, 90°. Plot at least 10 points for each curve.

47. (III) A hunter aims his gun directly at a monkey hanging from a high tree branch some distance away. At the

instant the gun is shot the monkey drops from the branch, hoping to avoid the bullet. Show analytically that the monkey made the wrong move. Ignore air resistance.

48. (III) An athlete executing a long jump leaves the ground at a 30° angle and travels 8.90 m. What was the takeoff speed?

49. (III) Determine how much further a person can jump on the moon as compared to the earth if the takeoff speed and angle are the same. The acceleration due to gravity on the moon is one-sixth what it is on earth.

50. (III) (a) Show that the range R of a projectile, which is defined as the horizontal distance traveled when the final point is at the same level as the initial point, is given by the equation $R = (v_0^2 \sin 2\theta)/g$. [Hint: use the trigonometry identity $\sin 2\theta = 2 \sin \theta \cos \theta$.] Ignore air resistance. (b) Show that the maximum range is obtained when $\theta = 45°$. (c) When performing the long jump, athletes gain speed by running before making the jump. Give a practical reason why the longest jumps are made at angles somewhat less than 45°.

51. (III) Prove that, in the absence of air resistance, the path of an object projected horizontally is a parabola.

52. (III) Police agents flying a constant 200 km/h horizontally in a low-flying airplane wish to drop an explosive onto a master criminal's automobile traveling 130 km/h on a level highway 78.0 m below. At what angle (with the horizontal) should the car be in their sights when the bomb is released?

53. (III) A ball is thrown horizontally from the top of a cliff with initial speed v_0. At any moment its direction of motion makes an angle θ to the horizontal. Derive a formula for θ as a function of time.

54. (III) A basketball leaves a player's hands at a height of 2.1 m above the floor. The basket is 2.6 m above the floor. The player likes to shoot the ball at a 35° angle. If the shot is made from a horizontal distance of 12.0 m and must be accurate to ±0.22 m (horizontally), what is the range of initial speeds allowed to make the basket?

3 MOTION AND FORCE: DYNAMICS

We have discussed how motion is described in terms of velocity and acceleration. Now we deal with the question of *why* objects move as they do: what makes an object at rest begin to move? What causes a body to accelerate or decelerate? What is involved when an object moves in a circle? We might answer that in each case a force is required. In this chapter we will investigate the connection between force and motion. Before we delve into this subject of *dynamics*, we first discuss the concept of force in a qualitative way.

3-1 ∎ Force

Intuitively, we can define **force** as any kind of a push or a pull. When you push a grocery cart, you are exerting a force on it. When children pull a wagon, they are exerting a force on the wagon. When a motor lifts an elevator, or a hammer hits a nail, or the wind blows the leaves of a tree, a force is being exerted. We say that an object falls because of the *force of gravity*. Forces do not always give rise to motion. For example, you may push very hard on a heavy desk or sofa and it may not move.

Whether or not an object moves when a force is exerted on it, the object does change shape. This is obvious when you squeeze a balloon, or push on a mattress. You can also see the slight deformation of the metal when you push on the side of a refrigerator or the fender of a car. There is always some deformation when a force is exerted, although it may take delicate instruments to detect it for a very rigid object like a heavy steel plate.

One way to quantitatively measure the magnitude (or strength) of a force is to make use of a spring scale (Fig. 3-1). Normally such a spring scale is used to find the weight of an object; by weight we mean the force of gravity acting on the body (Section 3-7). The spring scale, once calibrated[†], can be

---

[†] A spring scale is calibrated by hanging from it a series of identical objects of equal weight or mass, say 1 pound or 1 kilogram. The positions of the pointer when 0, 1, 2, 3, . . . units of weight or mass are hung from it are marked. Although the amount the spring stretches is very nearly proportional to the amount of weight hung from it, as long as it isn't stretched too far, we don't have to make use of this fact in this calibration method. We only assume that the pointer rests at the same position when the same weight (force) acts. This could serve as an operational definition (Section 1-3) of force.

FIGURE 3-1

A spring scale used to measure a force.

used to measure other kinds of forces as well, such as the pulling force shown in Fig. 3-1.

A force has direction as well as magnitude and is therefore a vector. We can represent any force on a diagram by an arrow, just as we did with velocity. The direction of the arrow is of course in the direction of the push or pull, and its length is drawn proportional to the strength or magnitude of the force. Although the definition of force as a push or a pull is adequate for the moment, a more precise definition is possible as discussed in Sections 3-4 and 3-5.

3-2 ▌ Newton's first law of motion

What is the exact connection between force and motion? Aristotle (384–322 B.C.) believed that a force was required to keep an object moving along a horizontal plane. He would argue that to make a book move across the table, you would have to exert a force on it continuously. To Aristotle, the natural state of a body was at rest, and a force was believed necessary to keep a body in motion. Furthermore, Aristotle argued, the greater the force on the body, the greater its speed.

Some 2000 years later, Galileo (1564–1642), skeptical about these Aristotelian views just as he was of those on falling bodies, came to a radically different conclusion. Galileo maintained that it is just as natural for an object to be in horizontal motion with a constant speed as it is to be at rest. To understand Galileo's idea, consider the following observations involving motion along a horizontal plane (where the effects of gravity do not enter). It will take a certain amount of force to push an object with a rough surface along a table top at constant speed. To push an equally heavy object with a very smooth surface across the table at the same speed will require less force. Finally, if a layer of oil or other lubricant is placed between the surface of the object and the table, then almost no force is required to move the object. (These observations may well be obvious to you, but if not, you should do these simple experiments for yourself). Notice that in each step the force required was less and less. As the next step we can extend the data to a situation in which the object does not rub against the table at all—or there is a perfect lubricant between them—and theorize that once started, the object would move across the table at constant speed with *no* force applied. A steel ball bearing rolling on a hard horizontal surface approaches this situation closely.

It was Galileo's genius to imagine an idealized world—in this case, one where there is no friction—and to see it could produce a more useful view of the real world. It was this idealization that led him to his remarkable conclusion that if no force is applied to a moving object, it will continue to move with constant speed in a straight line. An object slows down only if a force is exerted on it. Galileo thus interpreted friction as a force akin to ordinary pushes and pulls.

To push an object across a table at constant speed requires a force from your hand only to balance out the force of friction; the pushing force is equal in magnitude to the friction force but they are in opposite directions, so the *net* force on the object is zero, Fig. 3-2. This is consistent with Galileo's viewpoint, for the object moves with constant speed when no net force is exerted on it.

The difference between Aristotle's view and Galileo's is not simply one of right or wrong. Aristotle's view was not really wrong, for our everyday experience indicates that moving objects do tend to come to a stop if not continually pushed. The real difference lies in the fact that Aristotle's view about the "natural state" of a body was essentially a final statement—no further development was possible. Galileo's analysis, on the other hand, could be extended to explain a great many more phenomena. By making the creative leap of imagining the experimentally unattainable situation of no friction and by interpreting friction as a force, Galileo was able to reach his conclusion that an object will continue moving with constant velocity if no force acts to change this motion.

Upon this foundation, Isaac Newton (1642–1727) built his great theory of motion. Newton's analysis of motion is summarized in his famous "three laws of motion." In his great work, the *Principia* (published in 1687, it contains nearly all his work on motion), Newton readily acknowledged his debt to Galileo. In fact, **Newton's first law of motion** is very close to Galileo's conclusions[†]:

FIGURE 3-2

F represents the force applied by the person and **F**$_{fr}$ represents the force of friction.

> ■ **Every body continues in its state of rest or of uniform speed in a straight line unless it is compelled to change that state by forces acting on it.**

The tendency of a body to maintain its state of rest or of uniform motion in a straight line is called **inertia**. As a result, Newton's first law is often called the **law of inertia.**

3-3 ▌ Mass

Newton's second law (which we discuss in the next section) makes use of the concept of *mass*. Newton himself used the term mass as a synonym for *quantity of matter*. This intuitive notion of the mass of a body is not very

---

† It is not clear in Galileo's works if he adopted "linear" inertia, or "spherical" inertia—that is, natural motion continuing along a spherical surface such as on the surface of the earth. Newton, and Descartes before him adopted the principle of inertia for motion along a straight line.

precise because the concept "quantity of matter" is itself not well defined. A more precise definition is that **mass** *is a measure of the inertia of a body*. The more inertia a body has, the harder it is to change its state of motion; it is harder to start it moving from rest, or to stop it when it is moving, or to change its motion sideways out of a straight line path. A piano or a truck has much more inertia than a baseball, and it is much harder to change its state. It therefore has much more mass. The definition of mass as a measure of the inertia of a body is completely compatible with the notion of mass as "quantity of matter."

The terms *mass* and *weight* are often confused with one another, but it is important to distinguish between them. Mass is a property of a body itself. Weight on the other hand is a force, the force of gravity acting on a body. To see the difference, suppose we take an object to the moon; it will weigh only about one-sixth as much as it did on earth since the force of gravity is weaker. But its mass will be the same. It will have the same amount of matter and it will have just as much inertia—for in the absence of friction, it will be just as hard to start it moving or to stop it once it is moving.

3-4 ∎ Newton's second law of motion

Newton's first law states that if no net force is acting on a body it remains at rest, or if moving it continues moving with constant speed in a straight line. But what happens if a force is exerted on a body? Newton perceived that the velocity will change. A net force exerted on an object may make its speed increase or if it is in a direction opposite to the motion, it will reduce the speed. If the net force acts sideways on a moving object, the direction as well as the magnitude of the velocity changes. Thus, *a net force gives rise to acceleration*.

What precisely is the relationship between acceleration and force? Ordinary experience can answer this question. Consider the force required to push a roller skate or cart whose friction is minimal. (If there is friction, consider the net force, which is the force you exert minus the force of friction.) Now if you push with a gentle but constant force for a certain period of time, you will make the skate or cart accelerate from rest up to some speed, say 3 km/h. If you push twice as hard, you will find that it will reach 3 km/h in half the time; that is, its acceleration will have been twice as great. If you double the force, the acceleration doubles; if you triple the force, the acceleration is tripled, and so on. Thus, the acceleration of a body is directly proportional to the net applied force. But the acceleration depends on the mass of the object as well. If you push an empty grocery cart with the same force as you push one that is filled with groceries, you will find that the latter accelerates more slowly. The greater the mass, the less the acceleration for a given net force. It is found, as Newton argued, that the acceleration of a body is inversely proportional to its mass. These relationships are found to hold in general and can be summarized as follows:

∎ **The acceleration of an object is directly proportional to the net force acting on it and is inversely proportional to its mass. The direction of the acceleration is in the direction of the applied net force.**

This is **Newton's second law of motion.** In symbols, we can write it as:

$$\mathbf{a} \propto \frac{\mathbf{F}}{m},$$

where $\mathbf{a}$ stands for acceleration, m for the mass, and $\mathbf{F}$ for the *net force*. By **net force** we mean the vector sum of all the forces acting on the body. To change a proportion into an equation we merely need to insert a constant of proportionality (see Appendix A-1). The choice of a constant is arbitrary in this case since we are relating quantities with different units; we can therefore choose the unit of force or of mass so that the proportionality constant equals one. Then $\mathbf{a} = \mathbf{F}/m$. Rearranging, we have the familiar statement of Newton's second law in the form of an equation:

$$\mathbf{F} = m\mathbf{a}. \tag{3-1}$$

Newton's second law relates the description of motion to the cause of motion, force. It is one of the most fundamental and useful relationships in all of physics. From Newton's second law we can make a more precise definition of *force* as *an action capable of accelerating an object* (see the next section).

To use Newton's second law, we must discuss the units for mass and for force. The proper unit of mass in the SI (metric) system of units (see Section 1-3) is the kilogram (kg). The actual standard is a particular platinum-iridium cylinder whose mass is by definition one *kilogram.*[†] It is kept near Paris at the International Bureau of Weights and Measures.

The unit of force is chosen so that the proportionality constant in Newton's second law, $\mathbf{F} \propto m\mathbf{a}$, will be one, and thus $\mathbf{F} = m\mathbf{a}$. The SI unit of force (with mass in kilograms) is called the *newton* (N). One newton, then, is the force required to impart an acceleration of one m/s^2 to a mass of one kilogram. Thus $1\,N = 1\,kg \cdot m/s^2$.

In cgs units, the unit of mass is the gram (g) as mentioned earlier. The unit of force is the *dyne*, which is defined as the force required to impart an acceleration of $1\,cm/s^2$ to a mass of $1\,g$; thus $1\,dyne = 1\,g \cdot cm/s^2$. It is easy to show that $1\,dyne = 10^{-5}\,N$. (Do not confuse the abbreviation g for gram with the symbol g for the acceleration due to gravity, which is always italicized.)

In the English system, the unit of force is the pound, which is defined as the weight (which is a force) of a body with a mass of 0.454 kg at a particular place on the earth where the acceleration due to gravity is $g = 32.1734\,ft/s^2$. The unit of mass is the *slug*, defined as a mass that will undergo an acceleration of $1\,ft/s^2$ when a force of 1 lb is applied to it. Thus, $1\,lb = 1\,slug \cdot ft/s^2$. It is easy to show that $1\,lb = 4.45\,N$. The table inside this book's front cover contains additional conversion factors.

It is very important that only one set of units be used in a given calculation or problem. SI is preferred, and is used almost exclusively in this book. If the force is given in newtons and the mass in grams, then before attempting to solve for the acceleration in SI units, the mass must be changed to kilograms. For example, if the force is given as 2.0 N and the mass is 500 g, we change the latter to 0.50 kg and the acceleration will then automatically

---

[†] Other masses can be measured by comparing them on an equal arm balance.

come out in m/s^2 when Newton's second law is used:

3 Motion and Force: Dynamics

$$a = \frac{F}{m} = \frac{2.0 \text{ N}}{0.50 \text{ kg}} = 4.0 \text{ m/s}^2.$$

EXAMPLE 3-1 Calculate the average force that the pitcher in Example 1-8 exerts on the baseball, whose mass is 0.145 kg.

SOLUTION We calculated the acceleration to be 129 m/s^2, and the mass of the ball is 0.145 kg. Then

$$F = ma$$
$$= (0.145 \text{ kg})(129 \text{ m/s}^2) = 18.7 \text{ N}.$$

EXAMPLE 3-2 According to a simplified model of a mammalian heart, at each pulse approximately 20 g of blood is accelerated from 0.25 m/s to 0.35 m/s during a period of 0.10 s. What is the magnitude of the average force exerted by the heart muscle?

SOLUTION First we find the average acceleration a:

$$a = \frac{\Delta v}{\Delta t} = \frac{0.35 \text{ m/s} - 0.25 \text{ m/s}}{0.10 \text{ s}} = 1.0 \text{ m/s}^2.$$

Then the force required is: $F = ma = (0.020 \text{ kg})(1.0 \text{ m/s}^2) = 2.0 \times 10^{-2}$ N. Since we used SI units throughout, the force automatically comes out in newtons.

3-5 ∎ Laws and equations

When is an equation a law and when is it just an equation? To answer this, let us compare Newton's second law, $\mathbf{F} = m\mathbf{a}$, with the kinematic equations, Eqs. 1-10a, b, c, and d. The kinematic equations, as useful as they are, do not tell us anything startling about nature itself, since they were derived from the definitions of velocity and acceleration and the assumption of constant acceleration. Newton's second law, on the other hand, is based on experiment and tells us something about how things behave in nature. Given the simple definitions of mass, force, and acceleration, there is no obvious relation between them until one actually looks at natural objects and sees how they behave. Because there is actual physical content in the relation $\mathbf{F} = m\mathbf{a}$, and because it summarizes so much of experience, it is called a law.

However, in this particular case there is a problem. It is believed by many scientists (this is called the "positivist" or "operationalist" school of philosophy) that the intuitive definition of force as a push or a pull is not a good one because it is too vague. They claim that force must be defined more carefully, and that the only way to do this is via Newton's second law itself. Thus $\mathbf{F} = m\mathbf{a}$ becomes the definition of force. In other words, the force applied to an object is defined by measurement of the object's acceleration.

Whether you accept the positivist point of view of $\mathbf{F} = ma$ as a definition of force, or whether you wish to think of this relation as a law, has no bearing on the use of this relation in practical situations.

3-6 ▌ Newton's third law of motion

Newton's second law of motion describes quantitatively how forces affect motion. But where, we may ask, do forces come from? Observations suggest that a force applied to any object is always applied *by another object*. A horse pulls a wagon, a person pushes a grocery cart, a hammer pushes on a nail, a magnet attracts an iron nail. In each of these examples, one object exerts the force and the other body feels it: for example the hammer exerts the force, and the nail feels it.

But Newton realized things cannot be so one-sided. True, the hammer exerts a force on the nail. But the nail evidently exerts a force back on the hammer as well, for the hammer's speed is rapidly reduced to zero upon contact; only a strong force could cause such a rapid deceleration. Thus, said Newton, the two bodies must be treated on an equal basis. The hammer exerts a force on the nail, and the nail exerts a force back on the hammer. This led Newton to his **third law of motion:**

> ▌ **Whenever one object exerts a force on a second object, the second exerts an equal and opposite force on the first.**

This law is sometimes paraphrased as "to every action there is an equal and opposite reaction." To avoid confusion, it is very important to remember that the "action" force and the "reaction" force are acting on *different* objects.

As evidence for the validity of this third law look at your hand when you push against a grocery cart or against the edge of a desk (Fig. 3-3). Your hand's shape is distorted, clear evidence that a force is being exerted on it. You can *see* the edge of the desk pressing into your hand. You can even *feel* the desk exerting a force on your hand; it hurts. The harder you push against the desk, the harder the desk pushes back on your hand.

Consider next the ice skater in Fig. 3-4. Since there is very little friction between her skates and the ice, she will move freely if a force is exerted on her. She pushes against the wall, and then *she* starts moving backwards. Clearly there had to be a force exerted on her to make her move. The force she exerts on the wall cannot make *her* move, for that force pushes on the wall and can only affect the wall. Something had to exert a force on her to make her move, and that force could only have been exerted by the wall. The force with which the wall pushes on her is equal and opposite to the force she exerts on the wall.

When a person throws a package out of a boat the boat moves in the opposite direction. The person exerts a force on the package; the package exerts an equal and opposite force back on the person, and this force propels the boat backward slightly. Rockets work on the same principle. A common misconception is that rockets accelerate because the gases rushing out the back of the engine push against the ground or the atmosphere. Actually, a

Force exerted on hand by desk.

Force exerted on desk by hand.

FIGURE 3-3

If you push your hand against the edge of a table, the table edge pushes back against you.

FIGURE 3-4

When an ice skater pushes against a wall, the wall pushes back and forces her to move outward.

rocket accelerates because it exerts a strong force on the gases, expelling them. The gases exert an equal and opposite force *on the rocket* and it is this force that propels the rocket forward. Thus a space vehicle can maneuver in empty space just by firing its rockets in the direction opposite to that in which it wants to accelerate.

Let us consider how we walk. A person begins walking by pushing with the foot against the ground. The ground then exerts an equal and opposite force back on the person (Fig. 3-5); it is this force *on* the person that moves him or her forward. In a similar way, a bird flies forward by exerting a force on the air, but it is the air pushing back on the bird's wings that propels the bird forward. And an automobile moves forward because of the force exerted on it by the ground which is the reaction to the force exerted on the ground by the wheels.

From the above examples it is clear that it is quite important to remember *on* what object a given force is exerted and *by* what object that force is exerted. The point is that a force influences the motion of an object only when it is applied *on* that object. A force exerted *by* a body does not influence that body; it only influences the other body *on* which it is exerted. Thus, to avoid confusion, the two prepositions *on* and *by* must always be used—and used with care.

We tend to associate forces with active bodies such as humans, animals, engines, or a moving object like a hammer. It is often difficult to see how an inanimate object at rest, such as a wall or a desk, can exert a force. The explanation lies in the fact that every material, no matter how hard, is elastic, at least to some degree. No one can deny that a stretched rubber band can exert a force on a wad of paper and send it flying across the room. Although other materials may not stretch as easily as rubber, they do stretch when a force is applied to them. And just as a stretched rubber band exerts a force, so does a stretched (or compressed) wall or desk.

Force exerted by person on the ground

Force exerted by ground on the person

FIGURE 3-5

We can walk forward because the ground pushes forward on our feet when we push backward against the ground.

3-7 ■ Weight: the force of gravity

Galileo claimed that objects dropped near the surface of the earth would fall with the same acceleration, g, if air resistance could be neglected. The force that gives rise to this acceleration is called the force of gravity. We now apply Newton's second law to the gravitational force and for the acceleration, **a**, we use the acceleration of gravity, **g**. Thus the force of gravity, $\mathbf{F}_g$, on a body, which is also commonly called its **weight** (with symbol **w** meaning the same as $\mathbf{F}_g$) can be written as:

$$\mathbf{F}_g = \mathbf{w} = m\mathbf{g}, \qquad (3\text{-}2)$$

and the direction of this force is down toward the center of the earth.

In SI units, $g = 9.80 \text{ m/s}^2$, so the weight, w, of a 1.00-kg mass is $1.00 \text{ kg} \times 9.80 \text{ m/s}^2 = 9.80 \text{ N}$. This value of course is valid only at the surface of the earth. On the moon, on other planets, or in space, the weight of a given mass will be different. For example, on the moon, g is about 1.7 m/s^2, about one-sixth of what it is on earth. Thus 1.0 kg weighs only 1.7 N. Also,

(a)

(b)

FIGURE 3-6

(a) The net force on an object at rest is zero according to Newton's second law. Therefore the downward force of gravity ($\mathbf{F}_g$) on an object must be balanced by an upward force (the normal force $\mathbf{F}_N$) exerted by the table in this case. (b) $\mathbf{F}_N'$ is the force exerted on the table by the statue and is the reaction force to $\mathbf{F}_N$ as per Newton's third law. The reaction to $\mathbf{F}_g$ is not shown.

the weight of an object varies very slightly at different places on the earth's surface since g varies slightly, as we saw in Chapter 1. Although we will not have occasion to use British units, we note that for practical purposes a 1-kg mass has a weight of about 2.2 lb on the surface of the earth. (On the moon, 1 kg weighs about 0.4 lb.)

The force of gravity acts on an object when it is falling. When an object is at rest on the earth, the gravitational force does not disappear, as we know if we weigh it on a spring scale. The same force, given by Eq. 3-2, continues to act. Why then doesn't the object move? From Newton's second law, the net force on an object at rest is zero; there must be another force on the object to balance the gravitational force. For an object resting on a table, the table exerts this upward force; see Fig. 3-6a. The table is compressed slightly beneath the object and due to its elasticity it pushes up on the object as shown. The force exerted by the table is often called a *contact force*, since it occurs when two objects are in contact. (The force of your hand pushing on a cart is also a contact force.) When a contact force acts normal to the common surface of contact ("normal" means perpendicular), it is usually referred to as the **normal force**; hence it is labeled $\mathbf{F}_N$ in the diagram.

Let us examine the forces involved in Fig. 3-6 in more detail, making clear what we can understand from Newton's second law, and what from Newton's third law. The two forces shown in Fig. 3-6a are both acting on the statue, which remains at rest, so the vector sum of these two forces must be zero. We know this from Newton's second law: if the acceleration of an object is zero, the net force on it must be zero. Hence $\mathbf{F}_g$ and $\mathbf{F}_N$ must be of equal magnitude and opposite direction; but they are *not* the equal and opposite forces spoken of in Newton's third law. This is a critical point that can cause great confusion. The action and reaction forces of Newton's third law act on *different bodies.* (The two shown in Fig. 3-6a act on the same body.) For each of the forces shown in Fig. 3-6a, we can ask, "What is the reaction force?" The upward force, $\mathbf{F}_N$, on the statue is exerted by the table. The reaction to this force is a force exerted by the statue on the table. It is shown in Fig. 3-6b, where it is labeled $\mathbf{F}_N'$. This force, $\mathbf{F}_N'$, exerted on the table by the statue, is the reaction force to $\mathbf{F}_N$ as called for by Newton's third law. (We could equally well say the reverse: the force $\mathbf{F}_N$ on the statue exerted by the table is the reaction to the force $\mathbf{F}_N'$ exerted on the table by the statue.) Now, what

FIGURE 3-7

(a)Two forces, $\mathbf{F}_1$ and $\mathbf{F}_2$, act on an object. (b) The sum, or resultant, of $\mathbf{F}_1$ and $\mathbf{F}_2$ is $\mathbf{F}_R$.

$F_2 = 100$ N

$F_1 = 100$ N

(a)

$F_R = F_1 + F_2$

F_2

$45°$

F_1

(b)

about the other force on the statue, the force of gravity $\mathbf{F}_g$? Can you guess what the reaction is to this force?[†]

3-8 ∎ Applications of Newton's laws: vector forces

Newton's second law tells us that the acceleration of an object is proportional to the *net* force on the object. The **net force** is the *vector sum* of all forces acting on the object. Force vectors must be added according to the rules we developed for vectors in Chapter 2. This is illustrated in Fig. 3-7 where two equal magnitude forces are shown acting on an object at right angles to each other. Intuitively we can see that the object will move at a 45° angle and thus the net force is at a 45° angle. This is just what the rules of vector addition give. The Pythagorean theorem tells us that the resultant force must have magnitude $F_R = \sqrt{(100 \text{ N})^2 + (100 \text{ N})^2} = 141 \text{ N}$. That this is the correct answer is not intuitively obvious, although clearly the net force will be less than 200 N (since the two persons are pulling at cross purposes to some extent), and it is certainly greater than zero. Carefully done experiments show that two 100-N forces acting at 90° to one another have the same effect as one force of magnitude 141 N acting at 45° in agreement with our rules for vector addition.

□   EXAMPLE 3-3   Calculate the sum of the two vectors acting on the ship shown in Fig. 3-8a.

SOLUTION   These two vectors are shown resolved in Fig. 3-8b. The components of $\mathbf{F}_1$ are:

$$F_{1x} = F_1 \cos 45.0° = (40.0 \text{ N})(0.707) = 28.3 \text{ N},$$
$$F_{1y} = F_1 \sin 45.0° = (40.0 \text{ N})(0.707) = 28.3 \text{ N}.$$

The components of $\mathbf{F}_2$ are:

$$F_{2x} = F_2 \cos 37.0° = (30.0 \text{ N})(0.799) = 24.0 \text{ N},$$
$$F_{2y} = -F_2 \sin 37.0° = -(30.0 \text{ N})(0.602) = -18.1 \text{ N}$$

F_{2y} is negative since it points along the negative y axis. The components of the resultant force are (see Fig. 3-8c):

$$F_{Rx} = 28.3 \text{ N} + 24.0 \text{ N} = 52.3 \text{ N}$$
$$F_{Ry} = 28.3 \text{ N} - 18.1 \text{ N} = 10.2 \text{ N}.$$

To find the magnitude of the resultant force, we use the Pythagorean theorem:

$$F_R = \sqrt{F_{Rx}^2 + F_{Ry}^2} = \sqrt{(52.3)^2 + (10.2)^2} \text{ N} = 53.3 \text{ N}.$$

[†] This is a hard question since we don't discuss the force of gravity in detail until Chapter 4. In fact, it is a force exerted on the earth by the statue, which is a gravitational force (like $\mathbf{F}_g$) and can be considered to act at the earth's center as we shall discuss in Chapter 4; we won't be concerned with it in this chapter.

FIGURE 3-8

Two force vectors act on a boat (Example 3-3).

$F_1 = 40.0 \text{ N}$

$45.0°$
$37.0°$

$F_2 = 30.0 \text{ N}$

(a)

(b)

(c)

The only remaining question is the angle θ which the net force $\mathbf{F}_R$ makes with the x axis. We use:

$$\tan \theta = \frac{F_{Ry}}{F_{Rx}} = \frac{10.2 \text{ N}}{52.3 \text{ N}} = 0.195,$$

which corresponds to an angle of $11.0°$.

□ **EXAMPLE 3-4** Calculate the force required to accelerate the cart in Fig. 3-9a from rest to 0.50 m/s in 2.0 s. The mass of the cart is 20 kg.

SOLUTION   If we ignore friction, there are three forces on the cart as shown in Fig. 3-9b: the forward pushing force exerted by the person, $\mathbf{F}_p$; the downward force of gravity, $\mathbf{F}_g$; and the upward force exerted by the floor, $\mathbf{F}_N$ (which is the reaction to the force of the cart pushing down on the floor). The vertical forces, $\mathbf{F}_g$ and $\mathbf{F}_N$, must add up to zero; if they didn't the cart would accelerate vertically. Thus the net force on the cart is simply $\mathbf{F}_p$. To calculate how large F_p must be, we first calculate the acceleration required: $a = (0.50 \text{ m/s} - 0)/2.0 \text{ s} = 0.25 \text{ m/s}^2$. Then the magnitude of the force must be $F_p = ma = (20 \text{ kg})(0.25 \text{ m/s}^2) = 5.0 \text{ N}$.

This example illustrates several important aspects in the use of Newton's laws. First we draw a sketch of the situation. Next, if there is only one body involved, we show all the forces acting on that body; if several bodies are involved, we make a diagram of each body separately, showing all the forces acting *on that body;* such a diagram is called a **free body diagram.** An example is Fig. 3-10b, which is discussed below in Example 3-5. Newton's second law involves vectors, and it is usually important to resolve vectors into components. An x and a y axis should be chosen in a way that simplifies the calculation. Then Newton's second law can be applied to the x and y components separately. That is, the x component of the total force will be related to the x component of the acceleration: $F_x = ma_x$; and similarly for the y direction.

□ **EXAMPLE 3-5** Figure 3-10a shows a 10-kg package being pulled along the floor with a force of 40 N. The force is applied at an angle of $30°$. We wish to calculate (*a*) the acceleration of the package and (*b*) the magnitude of the upward force F_N exerted by the floor on the package. Assume friction can be neglected.

SOLUTION   Figure 3-10b shows the free body diagram of the package. With the y axis vertical, and x axis horizontal, the pull of 40 N has components:

$$F_x = (40 \text{ N})(\cos 30°) = (40 \text{ N})(0.866) = 35 \text{ N},$$
$$F_y = (40 \text{ N})(\sin 30°) = (40 \text{ N})(0.500) = 20 \text{ N}.$$

(*a*) In the horizontal direction:

$$ma_x = F_x$$
$$a_x = (35 \text{ N})/(10 \text{ kg}) = 3.5 \text{ m/s}^2.$$

(a)

(b)

FIGURE 3-9
Forces on a cart (Example 3-4).

FIGURE 3-10
Example 3-5; (b) is the free body diagram.

(a)

(b)

The acceleration of the package is thus 3.5 m/s^2.

(b) In the vertical direction we have:

$$ma_y = F_N - F_g + F_y,$$

where F_g is preceded by a negative sign since it acts in the negative y direction. Now $F_g = mg = (10 \text{ kg})(9.8 \text{ m/s}^2) = 98 \text{ N}$ and $F_y = 20 \text{ N}$. Furthermore, we know $a_y = 0$ since the package does not move vertically. Then

$$0 = F_N - 98 \text{ N} + 20 \text{ N}$$

and so

$$F_N = 78 \text{ N}.$$

Notice that in this case F_N is less than F_g; the ground does not push against the full weight of the package since part of the pull exerted by the person is in the upward direction.

We next take an example of projectile motion (see Section 2-5).

◻ **EXAMPLE 3-6** A locust jumps by extending its hind legs against the ground, Fig. 3-11a. Experiments indicate that a 3.0 g locust exerts an average force of about 0.45 N against the ground, at an angle of 57°; during this acceleration period the locust body moves a distance of about 4.0 cm. Calculate (a) the net force **F** acting on the locust, (b) its acceleration, (c) its takeoff velocity, and finally (d) the distance you expect it to travel during the jump, ignoring air resistance.

SOLUTION The first three parts deal with the jumping action; the last part deals with projectile motion per se. (a) Figure 3-11b shows the two forces acting on the locust: $\mathbf{F}_{gr}$, the 0.45 N force exerted by the ground ("reaction" to locust pushing against the ground), and the weight of the locust of magnitude $mg = (0.0030 \text{ kg})(9.8 \text{ m/s}^2) = 0.030 \text{ N}$. Using the component method of adding vectors, the net force **F** on the locust has components

$$F_x = (0.45 \text{ N})(\cos 57°) = 0.25 \text{ N},$$
$$F_y = (0.45 \text{ N})(\sin 57°) - 0.030 \text{ N} = 0.35 \text{ N}.$$

Then

$$F = \sqrt{F_x^2 + F_y^2} = \sqrt{(0.25 \text{ N})^2 + (0.35 \text{ N})^2} = 0.43 \text{ N}.$$

To find the angle θ, we note that

$$\frac{F_y}{F_x} = \frac{F \sin \theta}{F \cos \theta} = \tan \theta.$$

Substituting values for F_y and F_x we find $\tan \theta = (0.35 \text{ N})/(0.25 \text{ N}) = 1.40$, so $\theta = 54°$. The net force is 0.43 N and the angle at which it acts, which will be the takeoff angle, is 54°.

(b) The acceleration leading to takeoff is

$$a = \frac{F}{m} = \frac{0.43 \text{ N}}{0.0030 \text{ kg}} = 140 \text{ m/s}^2.$$

(a)

(b)

FIGURE 3-11
Locust (Example 3-6).

(c) The body of the locust accelerates at 140 m/s^2 over a distance of 0.040 m. Therefore its speed at the end of this period, at which point the legs leave the ground and the locust becomes airborne, is (using Eq. 1-10c with D being the distance moved along the direction of the acceleration):

$$v^2 = v_0^2 + 2aD$$
$$= 0 + (2)(140 \text{ m/s}^2)(0.040 \text{ m}) = 11 \text{ m}^2/\text{s}^2$$
$$v = 3.4 \text{ m/s}.$$

(d) In this part we are dealing with projectile motion. We assume there is no air resistance (a poor assumption, as we shall see). The locust leaves the ground with a speed of 3.4 m/s at an angle of 54°. We proceed as in Example 2-7. First we find the time to travel the full distance (assuming the flight is over level ground):

$$y = v_{y0}t - \tfrac{1}{2}gt^2$$
$$0 = (3.4 \text{ m/s})(\sin 54°)t - (\tfrac{1}{2})(9.8 \text{ m/s}^2)t^2$$
$$t = \frac{2.8 \text{ m/s}}{4.9 \text{ m/s}^2} = 0.57 \text{ s}.$$

Then the distance $x = v_{x0}t = (v_0 \cos 54°)t = (3.4 \text{ m/s})(0.59)(0.57 \text{ s}) = 1.1$ m. In fact, no locust jumps this far (at least if it doesn't use its wings, which we have assumed). Air resistance on this light object is quite considerable and limits its range to about a third of the above value, and its actual path is not a parabola.

3-9 ▎ Applications involving friction, inclines

So far we have ignored friction, but it must be taken into account in most practical situations. Friction exists between two solid surfaces because even the smoothest looking surface is quite rough on a microscopic scale. Even when a wheel rolls across a surface, there is still some friction, called *rolling friction*, although it is generally much less than when one body slides across the surface of another body. We will mainly be concerned with sliding friction in this section, and it is usually called *kinetic friction* (kinetic is Greek for "moving").

When a body is in motion, the force of kinetic friction always acts to oppose the sliding of the two surfaces past each other. Its magnitude depends on the nature of the two sliding surfaces. For a given pair of surfaces, it is proportional to the *normal force* between the two surfaces, which is the force that either object exerts on the other, perpendicular to their common surface of contact.[†] It does not depend appreciably on the total surface area of contact; that is, the force of friction on a brick is the same whether it is being slid on its flat face or on its end. We can write the proportionality as an equation

---

[†] Note that both the normal force and the friction force are forces exerted by one surface on the other; one is perpendicular to the contact surfaces, and the other is parallel.

by inserting a constant of proportionality, μ_k (Greek letter "mu"): 3 Motion and Force: Dynamics

$$F_{fr} = \mu_k F_N. \qquad \text{[kinetic friction]}$$

This is an approximate (but reasonably accurate and useful) relation. It is not a law; it is a relation between the magnitude of the friction force, F_{fr}, which acts parallel to the two surfaces, and the magnitude of the normal force, F_N, which acts perpendicular to the surfaces. It is not a vector equation since the two forces are perpendicular to one another. The term μ_k is called the *coefficient of kinetic friction* and its value depends on the two surfaces; measured values for a variety of surfaces are given in Table 3-1. These are only approximate however, since μ depends on whether the surfaces are wet or dry, on how much they have been sanded or rubbed and if any burrs remain.

What we have been discussing up to now is *kinetic friction*, when one body slides over another. There is also *static friction*, which refers to a force that resists any attempt to start a body moving. Suppose an object such as a desk is resting on a horizontal floor. If no horizontal force is exerted on the desk, there also is no friction force. But suppose, now, you try to push the desk, but it doesn't move; you are exerting a horizontal force, but the desk isn't moving, so there must be another force on the desk keeping it from moving ($\mathbf{F}_{net} = 0$). This is the force of *static friction* exerted by the floor on the desk. If you push with a greater force without moving the desk, the force of static friction also has increased. If you push hard enough, the desk will finally start to move. At this point, you have exceeded the maximum force of static friction which is given by $F_{fr} = \mu_s F_N$, where μ_s is the *coefficient of static friction* (Table 3-1). Since the force of static friction varies from zero to this maximum value, we can write

$$F_{fr} \leq \mu_s F_N. \qquad \text{[static friction]}$$

You may have noticed that it is often easier to keep a heavy object (like a desk) moving than it is to start it moving in the first place. This is a reflection of the fact that μ_s is almost always greater than μ_k. (It can never be less. Why?)

TABLE 3-1
Coefficients of friction[†]

Surfaces	Coefficient of Static Friction, μ_s	Coefficient of Kinetic Friction, μ_k
Wood on wood	0.4	0.2
Wood (waxed) on wet snow	0.14	0.1
Ice on ice	0.1	0.03
Metal on metal (lubricated)	0.15	0.07
Steel on steel (unlubricated)	0.6	0.3
Rubber on solids	1–4	1
Lubricated ball bearings	<0.01	<0.01
Synovial joints (in human limbs)	0.01	0.01

[†] Values are approximate and are intended only as a guide.

EXAMPLE 3-7   Recalculate the acceleration of the package in Example 3-5 assuming a coefficient of kinetic friction of 0.30. The free body diagram is shown in Fig. 3-12.

SOLUTION   The force of kinetic friction acts in the negative x direction, opposite to the direction of motion and is parallel to the surface of contact. The calculation for the y direction is just the same as before, Example 3-5. Thus the force with which the floor pushes up on the package, the normal force, is still 78 N. For the x direction we have a new term, the force of friction whose magnitude equals $\mu_k F_N = (0.30)(78\ \text{N}) = 23\ \text{N}$. Thus

$$ma_x = F_x - \mu_k F_N = 35\ \text{N} - 23\ \text{N} = 12\ \text{N}$$

$$a_x = \frac{12\ \text{N}}{10\ \text{kg}} = 1.2\ \text{m/s}^2.$$

FIGURE 3-12

Free body diagram for Example 3-7.

EXAMPLE 3-8   In Fig. 3-13a two objects are connected by a rope running over a pulley. The coefficient of kinetic friction between object I and the table is 0.20. (We ignore the mass of the rope and pulley and any friction in the pulley, which means we can assume that a force applied to one end of the rope will have the same magnitude at the other end). We wish to find the acceleration of the system, which will be the same for both objects assuming the rope doesn't stretch.

SOLUTION   Free body diagrams are shown for each body in Fig. 3-13b and c. Body I does not move vertically, so the normal force just balances the weight, $F_N = m_I g = (5.0\ \text{kg})(9.8\ \text{m/s}^2) = 49\ \text{N}$. In the horizontal direction there are two forces on body I: F_R, the pull of the rope (whose value we don't know), and the force of friction $= \mu_k F_N = (0.20)(49\ \text{N}) = 9.8\ \text{N}$. The horizontal acceleration is what we wish to find, and using Newton's second law, $F = ma$, we have (taking the positive direction to the right):

$$F_R - F_{fr} = m_I a.$$

Next consider block II. The force of gravity $F_g = m_{II} g = 19.6\ \text{N}$ pulls downward; and the rope pulls upward with a force F_R (which is the tension in the rope); this is the same force the rope exerts on body I since if the rope pulls on body I with a force F_R, then by Newton's third law, body I pulls back on the rope with the same force F_R; this force F_R is transmitted all

(a) (b) (c)

FIGURE 3-13

Example 3-8.

61

along the rope to body II, just as if a person were pulling on the end of it.[†]
Now we can write Newton's second law for body II (taking the downward
direction as positive):

$$m_{II}g - F_R = m_{II}a$$

We have two unknowns, a and F_R, but we now also have two equations.
We solve the first for F_R:

$$F_R = F_{fr} + m_I a,$$

and substitute this into the second equation:

$$m_{II}g - F_{fr} - m_I a = m_{II}a.$$

Now we solve for a and put in numerical values:

$$a = \frac{m_{II}g - F_{fr}}{m_I + m_{II}} = \frac{19.6 \text{ N} - 9.8 \text{ N}}{5.0 \text{ kg} + 2.0 \text{ kg}} = 1.4 \text{ m/s}^2.$$

If we wish, we can calculate F_R using the first equation:

$$F_R = F_{fr} + m_I a = 9.8 \text{ N} + (5.0 \text{ kg})(1.4 \text{ m/s}^2) = 17 \text{ N}.$$

FIGURE 3-14
Example 3-9.

(a)

(b)

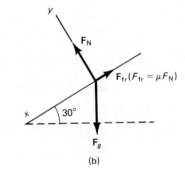

(c)

We now discuss some examples of objects moving on an incline such as a
hill or a ramp. Such situations commonly arise; they are interesting because
gravity is the accelerating force yet the acceleration is not vertical.

EXAMPLE 3-9 The skier in Fig. 3-14a has just begun descending the
30° slope. Assuming the coefficient of kinetic friction is 0.10, calculate (a)
her acceleration and (b) the speed she will reach after 6.0 s.

SOLUTION (a) The free body diagram in Fig. 3-14b indicates all the
forces acting on the skier: her weight ($\mathbf{F}_g = m\mathbf{g}$) downward, and the two
forces exerted on her skis by the snow—the normal force perpendicular to
the snow's surface, and the friction force parallel to the surface. For convenience, we choose the x axis parallel to the snow surface with positive
direction downhill, and the y axis perpendicular to the surface. Then we
only have to resolve one vector into components, the weight. The components of the weight are shown as dashed lines in Fig. 3-14c; they are given

This angle is θ, the
same as the slope angle,
because the right sides of
these two angles are perpendicular
to each other and the left
sides are perpendicular to
each other.

[†] Precisely true only if the pulley is massless and frictionless, as assumed here. If the rope
had had a mass m_R that could not be ignored, then the force at the two ends of the rope would
not be the same. The rope would pull on block I with force F_{RI} which is different from the force
F_{RII} it exerts as the other end on block II. By Newton's third law, block II pulls back on the
rope with a force F_{RII} and block I pulls on the rope with a force F_{RI}. The net force on the rope
(considered now as a third "free body") is $F_{RII} - F_{RI}$ and this must equal the mass of the rope
times its acceleration:

$$F_{RII} - F_{RI} = m_R a.$$

Here we see clearly that if $m_R = 0$, the force at either end is the same. If $m_R \neq 0$, the two
equations above become

$$F_{RI} - F_{fr} = m_I a, \quad \text{and} \quad m_{II}g - F_{RII} = m_{II}a,$$

so we get three equations in three unknowns (a, F_{RI}, and F_{RII}) which can be solved.

by (notice which angle is θ and the reason given on the diagram):

$$F_{gx} = mg \sin \theta,$$
$$F_{gy} = mg \cos \theta.$$

First we apply Newton's second law to the y direction:

$$F_N - mg \cos \theta = ma_y = 0,$$

since there is no motion in the y direction. Thus

$$F_N = mg \cos \theta.$$

In the x direction, $F_x = ma_x$ yields:

$$mg \sin \theta - \mu_k F_N = ma_x$$
$$mg \sin \theta - \mu_k mg \cos \theta = ma_x.$$

There is an m in each term, so they can be canceled out. Thus:

$$a_x = g \sin 30° - \mu_k g \cos 30°$$
$$= [0.50 - (0.10)(0.866)]g$$
$$= 0.41 \, g.$$

The acceleration is 0.41 times the acceleration of gravity, which in numbers is $a = (0.41)(9.8 \text{ m/s}^2) = 4.0 \text{ m/s}^2$. It is interesting that the mass canceled out and so the acceleration doesn't depend on the mass. The fact that such a cancellation sometimes occurs, and thus saves calculation, is one advantage of working with the algebraic equations and putting in the numbers only at the end.

(b) The speed after 6.0 s is found by using Eq. 1-10a:

$$v = v_0 + at = 0 + (4.0 \text{ m/s}^2)(6.0 \text{ s}) = 24 \text{ m/s},$$

where we assumed a start from rest.

EXAMPLE 3-10   Again consider a skier descending a hill, but this time we don't know the coefficient of kinetic friction between the skis and the snow and it is this we want to determine. This can be done by observing the skier descend slopes of different angles, and noticing at what angle she descends at constant speed. The $F = ma$ equations for the x and y components will be the same as above except that $a_x = 0$ in this case, and the angle θ will not necessarily be 30°. Thus:

$$F_N - mg \cos \theta = ma_y = 0$$
$$mg \sin \theta - \mu_k F_N = ma_x = 0.$$

From the first equation we have $F_N = mg \cos \theta$; we substitute this into the second equation:

$$mg \sin \theta - \mu_k (mg \cos \theta) = 0.$$

Now we solve for μ_k:

$$\mu_k = \frac{mg \sin \theta}{mg \cos \theta} = \frac{\sin \theta}{\cos \theta}.$$

From the definitions of sine and cosine, Eqs. 2-1 (see Fig. 2-7).

$$\frac{\sin \theta}{\cos \theta} = \frac{o/c}{a/c} = \frac{o}{a} = \tan \theta.$$

Thus

$$\mu_k = \tan \theta$$

where θ is the angle at which the skier moves at constant speed. For example if $\theta = 5°$, then $\mu_k = \tan 5° = 0.09$.

EXAMPLE 3-11   In the design of a supermarket, there are to be several ramps connecting different parts of the store. Customers will have to push grocery carts up the ramps and it is obviously desired that this not be too difficult. An engineer has done a survey and found that almost no one complains if the force required is no more than 20 N. Ignoring friction, at what maximum angle θ should the ramps be built assuming a full 20-kg grocery cart.

SOLUTION   As shown in Fig. 3-15 the person's force on the cart must just balance the x component of the weight. For the case of the maximum force of 20 N, we have

$$F_{gx} = F_g \sin \theta = 20 \text{ N}.$$

Since $F_g = mg = (20 \text{ kg})(9.8 \text{ m/s}^2) = 200 \text{ N}$, we have $\sin \theta = 20/200 = 0.10$; so $\theta = 5.7°$. In a real situation friction would have to be taken into account, particularly friction in the wheels of the carts when they get old, and this would probably be done by experiment.

FIGURE 3-15
Example 3-11.

Friction can be a hindrance. It slows down moving objects and causes heating and binding of moving parts in machinery. Friction can be reduced by using lubricants such as oil. More effective in reducing friction between two surfaces is to maintain a layer of air or other gas between them. Devices using this concept, which is not practical for most situations, include air tracks and air tables (or games) in which the layer of air is maintained by forcing air through many tiny holes. Another technique to maintain the air layer is to suspend objects in air using magnetic fields; this is finding applications for certain types of trains. On the other hand, friction can be helpful. Our ability to walk depends on friction between the soles of our shoes (or feet) and the ground. (Does walking involve static friction or kinetic friction?) The movement of a car, and also its stability, depend on friction. When friction is low, such as on ice, safe walking or driving become difficult.

3-10 ▮ Notes on problem solving

Problem solving is important not only because it is useful and practical in itself, but also because it makes you think about the ideas and concepts, and by applying the concepts you come to understand them better. Solving problems, except in the simplest cases, cannot generally be done by a set procedure. Instead, it often requires creativity, for each problem is different.

Nonetheless, we can outline a general approach that will be of some help in solving problems. But before attacking any problem, it is very important to know the definitions, terminology, and the basic principles and laws that apply.

1. Read written problems carefully. A common error is to leave out a word or two when reading, which can completely change the sense of a problem.

2. Draw a picture or diagram of the situation. (This is probably the most overlooked, yet the most crucial, part of solving a problem.) Use arrows to represent vectors such as velocity or force; use a separate diagram for different types of vectors: say, one for force and one for velocity, if both types of vector are involved. Make sure to include all forces on a given body, and make clear what forces act on what body (or you can make an error in determining the *net force* on a particular body). A diagram showing all the forces acting on a given body (and only on that body) is called a *free body diagram* as discussed in Section 3-8.

3. Determine what the unknowns are—that is, what you are trying to determine.

4. Decide what you need in order to find the unknowns. (*a*) It may help to see if there is one or more relationships (or equations) that relate the unknowns to the knowns; but be sure the relationship(s) is applicable in the given case. Beware of formulas that are not general but apply only in a specific case. (It is dangerous for this reason to thumb through a chapter looking for an equation that will work.) It is very important to know the limitations of each formula or relationship—when it is valid and when not. In this book, the more general equations have been given numbers, but even these can have a limited range of validity. The unnumbered equations are often valid only for very specific cases. (*b*) Also it is helpful to determine what information is relevant and what is irrelevant in a given situation or problem.

5. Try to solve the problem roughly, to see if it is doable (to check if enough information has been given) and reasonable. Use your intuition, and make rough calculations—see "Order of Magnitude Estimating" in Appendix B. A rough calculation, or a reasonable guess about what the range of final answers might be, is a very useful tool. And a rough calculation can be checked against the final answer to catch errors in calculation (such as in the powers of 10).

6. Solve the problem, which may include algebraic manipulation of equations and/or numerical calculations. Be sure to keep track of units, for this can serve as a check.

7. Again consider if your answer is reasonable. The use of dimensional analysis, described in Appendix C, can also serve as a check for many problems.

▍Summary

Newton's *three laws of motion* are the basic classical laws describing motion. Newton's *first law* states that if the net force on an object is zero, an object originally at rest remains at rest, and an object in motion remains in motion in a straight line with constant velocity. The tendency of a body to resist a change in its motion is called *inertia*. *Mass* is a measure of the inertia of a body; *weight* refers to the force of gravity on a body, and is equal to the product of the body's mass m and the acceleration of gravity g ($F_g = mg$).

Newton's *second law of motion* states that the acceleration of a body is directly proportional to the net force acting on it, and inversely proportional to its mass ($\mathbf{F} = m\mathbf{a}$). Newton's second law is one of the most important and fundamental laws in classical physics. *Force*, which is a vector, can be considered as a push or pull; or, from Newton's second law, force can be defined as an action capable of giving rise to acceleration. *Net* force on an object refers to the vector sum of all forces acting on it.

Newton's *third law of motion* states that whenever one body exerts a force on a second body, the second body always exerts a force on the first body which is equal in magnitude but opposite in direction.

A consistent set of units must be used when making calculations. The SI is usually used for scientific work.

When two bodies slide over one another, the force of friction which each exerts on the other can be written approximately as $F_{fr} = \mu F_N$ where F_N is the normal force (the force each body exerts on the other perpendicular to their contact surface), and μ is the coefficient of kinetic friction if there is relative motion of the bodies; if the bodies are at rest, μ is the coefficient of static friction and F_{fr} is the maximum friction force before motion starts.

▌ Questions

1. Compare the effort (or force) needed to lift a 10-kg object when you are on the moon as compared to on earth. Compare the force needed to throw a 2-kg object horizontally with a given speed when on the moon as compared to on earth.

2. A stone hangs by a fine thread from the ceiling and a section of the same thread dangles from the bottom of the stone. If a person gives a sharp pull on the dangling thread, where is the thread likely to break; below the stone or above it? What if the person gives a slow and steady pull? Explain your answers.

3. Why does a child in a wagon seem to fall backward when you give the wagon a sharp pull? Does the child "really," go backward? Explain.

4. Whiplash sometimes results from an automobile accident when the victim's car is struck violently from the rear. Explain why the head of the victim seems to be thrown backward in this situation. Is it really?

5. When a golf ball is dropped to the pavement, it bounces back up. (*a*) Is a force needed to make it bounce back up? (*b*) If so, what exerts the force?

6. Examine, in the light of Newton's first and second laws, the motion of your leg during one stride while walking.

7. A person wearing a cast on an arm or a leg experiences extra fatigue. Explain this on the basis of Newton's first and second laws.

8. If the acceleration of a body is zero, are no forces acting on it?

9. Why do you push harder on the pedals of a bicycle when first starting out than when moving at constant speed?

10. Only one force acts on an object. Can the object have zero acceleration? Can it have zero velocity?

11. When you are running and want to stop quickly, you must decelerate quickly. (*a*) What is the origin of the force that causes you to stop? (*b*) Estimate (using your own experience) the maximum rate of deceleration of a person running at top speed to come to rest.

12. A certain horse says to himself, "There's no use in trying to pull that cart. No matter how hard I pull on the cart, it will pull back on me with exactly the same force. So I can never make it go." Explain what is wrong with the horse's reasoning. Free body diagrams, showing all forces involved, will help.

13. Why, when you walk on a log floating in water, does the log move in the opposite direction?

14. Why does it hurt your foot when you kick a football?

15. When you stand still on the ground, how large a force does the ground exert on you? Why doesn't this force make you rise up into the air?

16. The force of gravity on a 2-kg rock is twice as great as that on a 1-kg rock. Why then doesn't the heavier rock fall faster?

17. A person exerts an upward force of 40 N to hold a bag of groceries. Describe the "reaction" force (Newton's third law) by stating: (a) its magnitude, (b) its direction, (c) *on* what body it is exerted, and (d) *by* what body it is exerted.

18. According to Newton's third law, each team in a tug-of-war pulls with equal force on the other team. What, then, determines which team will win?

19. Can a coefficient of friction exceed 1.0?

20. Devise a method for measuring the coefficient of friction between two surfaces using an inclined plane.

▌ Problems

SECTIONS 3-4 TO 3-7

1. (I) How much tension must a rope withstand if it is used to accelerate a 1700-kg car at 0.50 m/s²? Ignore friction.

2. (I) What force is needed to accelerate a bicycle of mass 80 kg (including a rider) at a rate of 1.25 m/s²?

3. (I) A net force of 41.6 N accelerates an object at 4.20 m/s². What is the mass of the object?

4. (I) What is your mass in kg, and your weight in newtons?

5. (I) What average force is required to stop a 1000-kg car in 6.0 s if it is traveling at 90 km/h?

6. (I) How much force is required to accelerate a 4.0-g object at 10,000 "g's" (say, in a centrifuge)?

7. (I) What is the weight of a 70-kg astronaut (*a*) on earth, (*b*) on the moon ($g = 1.7$ m/s^2), (*c*) on Venus ($g = 8.7$ m/s^2), (*d*) in outer space traveling with constant velocity.

8. (II) What average force is needed to accelerate a 6.50-g bullet from rest to 500 m/s over a distance of 0.700 m along the barrel of a rifle?

9. (II) A 0.10-g spider is descending on a strand which supports it with a force of 5.6×10^{-4} N. What is the acceleration of the spider? Ignore air resistance.

10. (II) A person has a reasonable chance of surviving an automobile crash if the deceleration is no more than 30 "g's." Calculate the force on a 70-kg person accelerating at this rate. What distance is traveled if brought to rest at this rate from 80 km/h?

11. (II) A 0.145-kg baseball traveling 35.0 m/s strikes the catcher's mitt which, in bringing the ball to rest, recoils backward 11.0 cm. What was the average force applied by the ball on the glove?

12. (II) What is the average force exerted by a shotputter on a 7.0-kg shot if the shot is moved through a distance of 2.9 m and is released with a speed of 13 m/s?

13. (II) An elevator (mass 4250 kg) is to be designed so that the maximum acceleration is 0.0500 g. What are the maximum and minimum forces the motor should exert on the supporting cable?

14. (II) A 40-kg child wants to escape from a third story window to avoid punishment. Unfortunately, a makeshift rope made of sheets can support a mass of only 30 kg. How can the child use this "rope" to escape? Give quantitative answer.

15. (II) What is the acceleration of a falling 65-kg skydiver if air resistance exerts a force of 250 N?

16. (II) The cable supporting a 1650-kg elevator has a maximum strength of 22,500 N. What maximum upward acceleration can it give the elevator without breaking.

17. (II) A typical home-run ball will leave the bat at a speed of 80 m/s after a contact time with the bat of perhaps 5.0×10^{-4} s. If the baseball has a mass of 0.14 kg (and initially was at rest) what was the force (assume constant) exerted by the bat on the ball? Is this force large enough to lift an average person?

18. (III) An exceptional standing jump would raise a person 0.80 m off the ground. To do this, what force must a 70-kg person exert against the ground? Assume the person lowers himself 0.20 m prior to jumping.

19. (III) A person jumps from a tower 5.0-m high. When he strikes the ground below, he bends his knees so that his torso decelerates over an approximate distance of 0.70 m. If the mass of his torso (excluding legs) is 50 kg,

find: (*a*) his velocity just before his feet strike the ground, and (*b*) the average force exerted on his torso by his legs during the deceleration.

SECTIONS 3-8 and 3-9

20. (I) For a new building, an architect has designed the rafters so that each exerts a force of 2000 N on the vertical studs at an angle of 40° to the horizontal. Calculate the horizontal and vertical components of this force.

21. (I) A 500-N force acts in a northwesterly direction. A second 500-N force must be exerted in what direction so that the resultant of the two vectors points westward?

22. (I) A force of 280 N is required to start a 24.0-kg box moving across a concrete floor. What is the coefficient of static friction between the box and the floor?

23. (I) If the coefficient of kinetic friction between a 25-kg crate and the floor is 0.45, how much force is required to move the crate at a steady speed across the floor? How much force is required if μ_k is zero?

24. (I) A man pushes a 15-kg lawnmower at constant speed with a force of 90 N directed along the handle, which is at an angle of 30° to the horizontal. Calculate: (*a*) the horizontal retarding force on the mower, and (*b*) the normal force on the mower.

25. (II) In the problem above, what average force must the man exert on the lawnmower to accelerate it from rest to 5.0 km/h in 3.0 s? Assume same retarding force.

26. (II) Suppose you are standing on a train accelerating at 0.50 g. What minimum coefficient of friction must exist between your feet and the floor if you are not to slide?

27. (II) The 100-m dash can be run by the best sprinters in 10.0 s. (*a*) What is the horizontal component of force exerted on the sprinter's feet by the ground during acceleration, which we assume is done uniformly over the first 10.0 m? (*b*) What is the average speed of the sprinter over the last 90 m of the race? Assume the sprinter's mass is 70 kg.

28. (II) A roller coaster reaches the top of the steepest hill with a speed of 5.0 km/h. It then descends the hill which is at an average angle of 45° and is 50-m long. What will its speed be when it reaches the bottom? Neglect friction.

29. (II) Referring to Fig. 3-13 how large a mass would body I have to have to prevent any motion from occurring? Assume $\mu_s = 0.20$.

30. (II) A wet bar of soap slides freely down a ramp 2.0 m long inclined at 6.8°. How long does it take to reach the bottom? Neglect friction.

31. (II) A box is given a push so that it slides across the floor. How far will it go, given that the coefficient of kinetic friction is 0.30 and the push imparts an initial speed of 3.0 m/s?

32. (II) What is the maximum acceleration a car can undergo on level ground if the static coefficient of friction between the tires and the ground is 0.65?

33. (II) (a) Show that the minimum stopping distance for an automobile traveling at speed v is equal to $(v^2)/(2\mu g)$ where μ is the coefficient of static friction between the tires and the road, and g is the acceleration of gravity. (b) What is this distance for a 1500-kg car traveling 90 km/h if $\mu = 0.85$?

34. (II) At the instant a race began a 60-kg sprinter was found to exert a force of 950 N on the starting block at a 20° angle with respect to the ground. (a) What was the horizontal acceleration of the sprinter? (b) If the force was exerted for 0.32 s, with what speed did the sprinter leave the starting block?

35. (II) One paint bucket weighing 20.0 N is hanging by a massless rope from another paint bucket (with a handle on the bottom) also weighing 20.0 N, and the two are being pulled upward with an acceleration of 1.65 m/s² by another massless rope attached to the upper bucket. Calculate the tension in each rope.

36. (II) A 5000-kg helicopter accelerates upward at 0.550 m/s² while lifting a 1500-kg car. (a) What is the lift force exerted by the air on the propellers? (b) What is the tension in the cable (ignore its mass) that connects car to helicopter?

37. (II) An 18.0-kg box is released on a 33.0° incline and accelerates down the incline at 0.300 m/s². Find the friction force impeding its motion. How large is the coefficient of friction?

38. (II) A 1000-kg car pulls a 450-kg trailer. The car exerts a horizontal force of 3.5×10^3 N against the ground in order to accelerate. What force does the car exert on the trailer? Assume a (rolling) friction coefficient of 0.45.

39. (II) A car can decelerate at -5.5 m/s² without skidding when coming to rest on a level road. What would its deceleration be if the road inclines 15° uphill? Assume the same friction force.

40. (II) The apparatus shown in Fig. 3-16 in which two masses are suspended over a pulley, is known as Atwood's machine. Assuming the pulley is massless and frictionless, and that the cord is massless and doesn't stretch, calculate: (a) the acceleration of the system, and (b) the tension in the cord.

41. (II) A train locomotive pulling three identical 8850-kg

cars accelerates at 0.225 m/s². With what force does the first car pull on the second? Ignore friction.

42. (II) Suppose the locomotive of problem 41 has a mass of 23,500 kg, and that the friction force on each car is 1100 N. What must be the force that the locomotive exerts against the rails?

43. (II) A car starts rolling down a 1-in-4 hill (1-in-4 means that for each 4 m traveled, the elevation change is 1 m). How fast is it going when it reaches the bottom after traveling 50 m? Ignore friction.

44. (II) Repeat problem 43 assuming an effective coefficient of friction equal to 0.10.

45. (II) Two crates, of mass 80 kg and 120 kg, are in contact and rest on a horizontal surface. A 700-N horizontal force is exerted on the 80-kg crate toward the second crate. If the coefficient of kinetic friction is 0.25, calculate: (a) the acceleration of the system, and (b) the force that each crate exerts on the other. (c) Repeat part (b) assuming the 700 N force is exerted on the 120-kg crate (in the opposite direction, naturally).

46. (II) A flatbed truck is carrying a 2800-kg crate of heavy machinery. If the coefficient of static friction between the crate and the bed of the truck is 0.55, what is the maximum rate the driver can decelerate when coming to a stop in order to avoid crushing the cab with the crate?

47. (III) If a bicyclist of mass 75 kg (including the bicycle) can coast down a 5.6° hill at a steady speed of 7.0 km/h, how much (average) force must be applied to climb the hill at the same speed?

48. (III) The two masses shown in Fig. 3-16 are each initially 1.60 m above the ground and the pulley is 4.0 m above the ground. What maximum height does the 1.2-kg object reach after the system is released from rest?

49. (III) What is the acceleration of the system shown in Fig. 3-17 if the kinetic coefficient of friction is 0.15?

50. (III) A bicyclist can coast down a 4.0° hill at 6.0 km/h. If the force of friction is proportional to the speed v so that $F_{fr} = cv$, calculate: (a) the value of the constant c, and (b) the average force that must be applied in order to descend the hill at 20 km/h. The mass of the cyclist plus bicycle is 80 kg.

51. (III) A motorcyclist is coasting with the engine off at a steady speed of 12 m/s but enters a sandy stretch where the coefficient of friction is 0.80. Will the cyclist emerge from the sandy stretch without having to start the engine if the sand lasts for 15 m? If so, what will be the speed upon emerging?

52. (III) A city planner is working on the redesign of a hilly portion of a city. An important consideration is how steep the roads can be so that even small cars can get up the hills without slowing down. It is given that a particular small car, with a mass of 1000 kg, can accelerate on the level from rest to 14 m/s (50 km/h) in 8.0 s. Using this figure, calculate the maximum steepness of a hill.

FIGURE 3-16   FIGURE 3-17

4 CIRCULAR MOTION; GRAVITATION

An object moves in a straight line if the net force on it acts in the direction of motion, or is zero. If the net force acts at an angle to the direction of motion at any moment, then the object moves in a curve. We have already discussed one example of this in projectile motion. Another important situation is that of an object moving in a circle, such as a ball at the end of a string revolving around one's head, or the nearly circular motion of the moon about the earth. These are examples of circular motion.

In this chapter we study circular motion of an object, and how Newton's laws of motion apply to it. We will also discuss how Newton conceived of another great law by applying the concepts of circular motion to the motion of the moon and the planets. This new law was his law of universal gravitation, which was the capstone of Newton's analysis of the physical world. Indeed, Newtonian mechanics with its three laws of motion and the law of universal gravitation, was accepted for centuries (until the early 1900s) as the way the universe works.

4-1 ▌ Uniform circular motion

An object that moves in a circle at constant speed v is said to undergo *uniform circular motion*. The magnitude of the velocity remains constant in this case; but the direction of the velocity is continually changing as the object moves around the circle, Fig. 4-1. Since acceleration is defined as the rate of change of velocity, a change in direction of velocity constitutes an acceleration just as does a change in magnitude. Thus, an object undergoing uniform circular motion is accelerating. We now investigate this acceleration quantitatively.

Acceleration is defined as

$$\mathbf{a} = \frac{\mathbf{v} - \mathbf{v}_0}{\Delta t} = \frac{\Delta \mathbf{v}}{\Delta t},$$

where $\Delta \mathbf{v}$ is the change in velocity during the short time interval Δt. We will eventually consider the situation when Δt approaches zero and thus obtain the instantaneous acceleration; but for purposes of making a clear drawing, Fig. 4-2, we consider a nonzero time interval. During the time Δt the particle in Fig. 4-2a moves from point A to point B, covering a small distance Δl which

FIGURE 4-1

A particle moving in a circle, showing how the velocity changes. Note that at each point, the instantaneous velocity is in a direction tangent to the circular path.

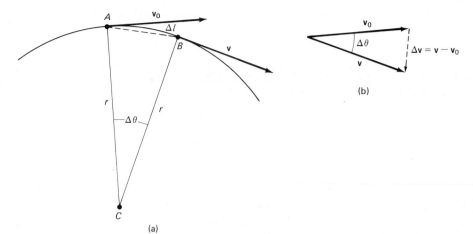

FIGURE 4-2

Determining the change in velocity, $\Delta\mathbf{v}$, for a particle moving in a circle.

subtends a small angle $\Delta\theta$. The change in the velocity vector is $\mathbf{v} - \mathbf{v}_0 = \Delta\mathbf{v}$. If we transfer $\mathbf{v}_0$ to the right side of this equation, we obtain $\mathbf{v} = \mathbf{v}_0 + \Delta\mathbf{v}$. Thus $\Delta\mathbf{v}$ added to $\mathbf{v}_0$ is equal to $\mathbf{v}$, so $\Delta\mathbf{v}$ is the vector shown dashed in Fig. 4-2b. In this diagram we notice that when Δt is very small (approaching zero), and therefore Δl and $\Delta\theta$ are also very small, $\mathbf{v}$ will be almost parallel to $\mathbf{v}_0$ and $\Delta\mathbf{v}$ will be essentially perpendicular to them. Thus $\Delta\mathbf{v}$ points toward the center of the circle. Since $\mathbf{a}$, by its definition above, is in the same direction as $\Delta\mathbf{v}$, it too must point toward the center of the circle. Therefore, this acceleration is called **centripetal acceleration** ("center-seeking" acceleration), and henceforth we denote it by $\mathbf{a}_c$.

Now that we have determined the direction of the acceleration, we next determine the magnitude of the centripetal acceleration, a_c. The vectors $\mathbf{v}$, $\mathbf{v}_0$ and $\Delta\mathbf{v}$ in Fig. 4-2b form a triangle that is geometrically similar[†] to triangle ABC in Fig. 4-2a. This relies on the fact that the angle between $\mathbf{v}_0$ and $\mathbf{v}$ is equal to $\Delta\theta$, defined as the angle between CA and CB in Fig. 4-2a; this is true because CB is perpendicular to $\mathbf{v}$, and CA is perpendicular to $\mathbf{v}_0$. Thus we can write

$$\frac{\Delta v}{v} \approx \frac{\Delta l}{r}.$$

This is an exact equality when Δt approaches zero, for then the arc length Δl equals the chord length AB. Since we want to find the instantaneous acceleration, which is the case for which Δt approaches zero, we write this as an equality and solve for Δv:

$$\Delta v = \frac{v}{r}\,\Delta l.$$

To get the centripetal acceleration, a_c, we divide Δv by Δt:

$$a_c = \frac{\Delta v}{\Delta t} = \frac{v}{r}\frac{\Delta l}{\Delta t},$$

[†] Appendix A contains a review of geometry.

70

and since $\Delta l/\Delta t$ is the linear speed, v, of the object:

$$a_c = \frac{v^2}{r}. \tag{4-1}$$

To summarize, an object moving in a circle of radius r with constant speed v has an acceleration whose direction is toward the center of the circle and whose magnitude is $a_c = v^2/r$. It is not surprising that this acceleration depends on v and r. For the greater the speed v, the faster the velocity changes direction; and the larger the radius, the less rapidly the velocity changes direction.

The acceleration vector points toward the center of the circle, as discussed above; but the velocity vector always points in the direction of motion, which is tangential to the circle; thus the velocity and acceleration vectors are perpendicular to each other at every point for uniform circular motion; see Fig. 4-3. This is another example that illustrates the common mistake of believing that acceleration and velocity are always in the same direction; we mentioned this in Section 1-10; another example is for projectile motion (Section 2-5), where the acceleration $\mathbf{a} = \mathbf{g}$ is always downward but the velocity vector can have various directions (Figs. 2-15 and 2-17). There is, in fact, *no* fixed general relationship between the directions of $\mathbf{a}$ and $\mathbf{v}$.

FIGURE 4-3

For uniform circular motion, $\mathbf{a}$ is always perpendicular to $\mathbf{v}$.

◻ EXAMPLE 4-1   A 150-g ball at the end of a string is swinging in a horizontal circle of radius 1.15 m, as in Fig. 4-1. The ball makes exactly 2.00 revolutions in a second. What is its centripetal acceleration?

SOLUTION   First we must determine the speed of the ball v. If the ball makes two complete revolutions per second, then the ball travels in a complete circle in 0.500 s. The distance traveled in this time is the circumference of the circle, $2\pi r$, where r is the radius of the circle. Therefore the ball has speed

$$v = \frac{2\pi r}{t} = \frac{2(3.14)(1.15 \text{ m})}{0.500 \text{ s}} = 14.4 \text{ m/s}.$$

The centripetal acceleration is

$$a_c = \frac{v^2}{r} = \frac{(14.4 \text{ m/s})^2}{1.15 \text{ m}} = 180 \text{ m/s}^2.$$

◻ EXAMPLE 4-2   The moon's nearly circular orbit about the earth has a radius of about 385,000 km and a period of 27.3 days. Determine the acceleration of the moon toward the earth.

SOLUTION   In orbit around the earth, the moon travels a distance $2\pi r$ where $r = 3.85 \times 10^8$ m is the radius of its circular path. The speed of the moon in its orbit about the earth is then

$$v = \frac{2\pi r}{T} = \frac{(6.28)(3.85 \times 10^8 \text{ m})}{(27.3 \text{ d})(24.0 \text{ h/d})(3600 \text{ s/h})} = 1.02 \times 10^3 \text{ m/s}.$$

Therefore

$$a_c = \frac{v^2}{r} = \frac{(1.02 \times 10^3 \text{ m/s})^2}{(3.85 \times 10^8 \text{ m})} = 0.00273 \text{ m/s}^2 = 2.73 \times 10^{-3} \text{ m/s}^2,$$

or about $2.78 \times 10^{-4}\,g$, where $g = 9.80 \text{ m/s}^2$ is the acceleration of gravity at the earth's surface.[†]

An object moving in a circle, such as a ball on the end of a string, must have a force applied to it to keep it in that circle; that is, a force is necessary to give it a centripetal acceleration. The magnitude of the required force can be calculated using Newton's second law, $F = ma$, where we use the value of the centripetal acceleration, $a_c = v^2/r$, and F must be the total (or net) force:

$$F = ma_c = m\frac{v^2}{r}.$$

Since a_c is directed toward the center of the circle, the force too must be directed toward the center of the circle. That a force is necessary can be seen intuitively from the fact that if no force were exerted on the object, it would not move in a circle but in a straight line, as Newton's first law tells us. To pull an object out of its "natural" straight-line path, a force to the side is necessary; for uniform circular motion, this sideways force must act toward the circle's center (see Fig. 4-4). This force is sometimes called a centripetal force. But be aware that "centripetal force" does not indicate some new kind of force. The term merely means that the force is directed toward the circle's center. The force must be applied by some object. For example, when a person swings a ball on the end of string in a circle, the person (via the string) exerts the force on the ball.

There is a common misconception that an object moving in a circle has an outward force acting on it, a so-called centrifugal ("center-fleeing") force. Consider, for example, a person swinging a ball on the end of a string around his head (Fig. 4-5). If you have ever done this yourself, you know that you feel a force pulling outward on your hand. The misconception arises when this pull is interpreted as an outward "centrifugal" force pulling on the ball which is transmitted along the string to the hand. But this is not what is happening at all. To keep the ball moving in a circle, the person pulls inwardly on the ball. The ball, then, exerts an equal and opposite force on the hand (Newton's third law), and *this* is the force your hand feels. The force *on the ball* is the one exerted *inwardly* on it by the person. For even more convincing evidence that a centrifugal force does not act on the ball, consider what happens when you let go of the string. If a centrifugal force were acting, the ball would fly outward, as shown in Fig. 4-6a. But it doesn't; it flies off tangentially (Fig. 4-6b), in the direction of the velocity it had at the moment it was released since the inward force no longer acts.

An example of centripetal acceleration occurs when a fast-moving automobile rounds a curve. In such a situation you may feel that you are thrust outward. This is not some mysterious centrifugal force pulling on you.

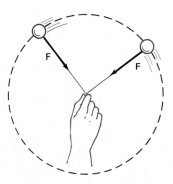

FIGURE 4-4

A force is required to keep an object moving in a circle. If the speed is constant, the force is directed toward the center of the circle.

FIGURE 4-5

Swinging a ball on the end of a string. Your hand pulls in on the ball, and the ball pulls outward on your hand (Newton's third law).

Force on hand exerted by ball   Force on ball exerted by hand

[†] Note: this acceleration $a = 2.78 \times 10^{-4}\,g$ is *not* the acceleration of gravity of objects at the moon's surface due to the moon's gravity. Rather it is the acceleration due to the earth's gravity of any object (such as the moon) which is 385,000 km from the earth.

What is happening is that you tend to move in a straight line whereas the car curves "in front of you," so to speak. To make you go in the curved path, the back of the seat or the door of the car exerts a force on you (Fig. 4-7a). The car itself must have an inward force on it if it is to move in a curve. On a flat road this force is supplied by friction between the tires and the pavement. (It is static friction as long as the tires are not slipping.) If the friction force is not great enough, as under icy conditions, sufficient force cannot be applied and the car will skid out of a circular path into a more nearly straight path.

☐ EXAMPLE 4-3 A 1000-kg car rounds a curve on a flat road of radius 50 m at a speed of 50 km/h (14 m/s). Will the car make the turn if (a) the pavement is dry and the coefficient of static friction is 0.60, (b) the pavement is icy and $\mu = 0.20$?

SOLUTION First we calculate the net force F required to keep the car moving in a circle around the curve:

$$F = m\frac{v^2}{r} = \frac{(1000 \text{ kg})(14 \text{ m/s})^2}{(50 \text{ m})} = 3900 \text{ N}.$$

Figure 4-7b shows the free-body diagram for the car. The normal force, F_N, on the car is equal to the weight since the road is flat: $F_N = mg = (1000 \text{ kg})(9.8 \text{ m/s}^2) = 9800 \text{ N}$. For (a), $\mu = 0.60$ and the maximum friction force attainable is

$$F_{fr} = \mu F_N = (0.60)(9800 \text{ N}) = 5900 \text{ N}.$$

Since a force of only 3900 N is needed, and that is how much will in fact be exerted by the road as a static friction force, the car can make the turn. But in (b) the maximum friction force possible is

$$F_{fr} = \mu F_N = (0.20)(9800 \text{ N}) = 2000 \text{ N}.$$

The car will skid.

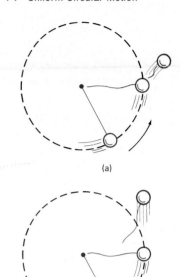

FIGURE 4-6

If centrifugal force existed, the ball would fly off as in (a) when released. In fact, it flies off as in (b).

FIGURE 4-7

(a) The road exerts an inward force on a car to make it move in a circle; and the car exerts an inward force on the passenger. (b) Forces on the car (front view).

(a)

(b)

The situation is worse if the wheels lock when the brakes are applied. When the tires are rolling, the bottom of the tire is at rest (for that moment) against the road, so static friction exists. But if the wheels lock, the tires slide and the friction force, which is now kinetic friction, is less. When the road is wet or icy, locking of the wheels occurs with less force on the brake pedal since there is less resistance from road friction.

The banking of curves can reduce the chance of skidding because the normal force of the road will have a component toward the center of the circle (Fig. 4-8), thus reducing the need for friction. For a given angle of banking, there will be one speed for which no friction at all is required. This will be the case when the horizontal component of the normal force, $F_N \sin \theta$, is just equal to the force required to give a vehicle its centripetal acceleration; that is

$$F_N \sin \theta = m \frac{v^2}{r}.$$

The banking angle of a road, θ, is chosen so that this condition holds for a particular speed, called the "design speed."

FIGURE 4-8
Normal force on a car rounding a banked curve, resolved into its horizontal and vertical components. Note that the centripetal acceleration is horizontal (and not parallel to the sloping road); why?

☐ EXAMPLE 4-4 (a) For a car traveling with speed v around a curve of radius r, determine a formula for the angle at which a road should be banked so that no friction is required. (b) What is this angle for a curve of radius 50 m at a design speed of 50 km/h?

SOLUTION (a) From the equation above, $F_N \sin \theta = mv^2/r$. In the vertical direction, the forces are $F_N \cos \theta$ upward (Fig. 4-8) and the weight of the car (mg) downward. Since there is no vertical motion, this component of the acceleration is zero so Newton's second law tells us

$$F_N \cos \theta - mg = 0.$$

Therefore

$$F_N = \frac{mg}{\cos \theta},$$

and substituting this into the equation above for the horizontal motion we have

$$\frac{mv^2}{r} = F_N \sin \theta$$

$$= mg \frac{\sin \theta}{\cos \theta} = mg \tan \theta,$$

or

$$\tan \theta = \frac{v^2}{rg}.$$

(b) For $r = 50$ m and $v = 50$ km/h (or 14 m/s), $\tan \theta = (14 \text{ m/s})^2/(50 \text{ m}) \cdot (9.8 \text{ m/s}^2) = 0.40$, so $\theta = 22°$.

*4-2 ∎ Nonuniform circular motion

Circular motion at constant speed occurs when the force on an object is exerted toward the center of the circle. If the force is not directed toward the center but is at an angle as shown in Fig. 4-9, the force has two components. The component directed toward the center of the circle, F_c, gives rise to the centripetal acceleration a_c and keeps the object moving in a circle. The component F_T, tangent to the circle, acts to increase (or decrease) the speed, and thus gives rise to a component of the acceleration tangent to the circle, a_T. When an F_T acts, the speed of the object is changed.

In summary, a_T is related to the change in the *magnitude* of the velocity of the object, whereas a_c is related to the change in *direction* of the velocity.

When you first start revolving a ball on the end of a string around your head, you must give it tangential acceleration. You do this by pulling on the string with your hand displaced from the center of the circle. In athletics, a hammer thrower accelerates the hammer tangentially in a similar way so that it reaches a high speed before release.

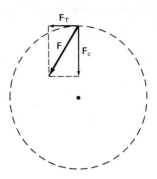

FIGURE 4-9

The speed of a particle moving in a circle changes if the force on it has a tangential component.

*4-3 ∎ Centrifugation

A useful device that very nicely illuminates the dynamical aspects of circular motion is the centrifuge, or the very high-speed ultracentrifuge. It is used to sediment materials quickly or to separate materials with slightly different characteristics. Test tubes or other containers are held in the centrifuge rotor which is accelerated to very high rotational speeds. This is shown in Fig. 4-10 where one test tube is shown in two different positions as the rotor turns. The small circle represents a small particle, perhaps a cell component or a macromolecule, in a fluid-filled test tube. When the tube is at position A and the rotor is turning, the particle has a tendency to move in a straight line in the direction of the dashed arrow. The fluid resists the motion of the particles, thus exerting the centripetal force necessary to keep the particles moving in a circle. Usually the resistance of the fluid does not quite equal the required

FIGURE 4-10

Rotating test tube in a centrifuge.

centripetal force and the particles eventually reach the bottom of the tube, and then the bottom of the tube exerts the force necessary to keep the particles moving in a circle. In fact the bottom of the tube must exert a force on the whole tube of fluid, keeping it moving in a circle. If the tube is not strong enough to exert this force, it will break. The details of the forces involved will be discussed in Chapter 9.

The kinds of materials placed in a centrifuge are those which do not sediment or separate quickly under the action of gravity. The point of a centrifuge is to provide an "effective gravity" much larger than normal gravity because of the high rotational speeds, and the particles move down the tube more rapidly. The effect on the rotating tube, and its contents, is just the same as if the tube were at rest in a vertical position with a much increased value of g.

EXAMPLE 4-5   An ultracentrifuge rotor rotates at 50,000 rpm (revolutions per minute). The top of a test-tube (perpendicular to the rotation axis) is 6.00 cm, and the bottom of the tube is 10.0 cm, from the axis of rotation. (a) Calculate the centripetal acceleration in "g's." (b) If the contents of the tube have a total mass of 12.0 g, what force must the bottom of the tube withstand?

SOLUTION   (a) At the top of the tube, a particle revolves in a circle of circumference $2\pi r = 2(3.14)(0.0600 \text{ m}) = 0.377$ m. It makes 5.00×10^4 such revolutions each minute, and so its speed is

$$v = \frac{(0.377 \text{ m/rev})(5.00 \times 10^4 \text{ rev/min})}{60.0 \text{ s/min}} = 3.14 \times 10^2 \text{ m/s}.$$

The centripetal acceleration is then

$$a_c = \frac{v^2}{r} = \frac{(3.14 \times 10^2 \text{ m/s})^2}{(0.0600 \text{ m})} = 1.64 \times 10^6 \text{ m/s}^2$$

which, dividing by $g = 9.80 \text{ m/s}^2$, is 1.67×10^5 g's. At the bottom of the tube ($r = 0.100$ m), the speed is $v = (2\pi r)(5.00 \times 10^4)/(60.0 \text{ s}) = (0.628)$ $(5.00 \times 10^4)/(60.0 \text{ s}) = 5.24 \times 10^2$ m/s and $a_c = v^2/r = 2.74 \times 10^6 \text{ m/s}^2$ or 2.79×10^5 g's.

(b) Since the accelerationn varies with distance from the axis, we estimate the force using the average acceleration $= (1.64 \times 10^6 \text{ m/s}^2 + 2.74 \times 10^6 \text{ m/s}^2)/2 = 2.19 \times 10^6 \text{ m/s}^2$.   Then $F = ma = (0.0120 \text{ kg})(2.19 \times 10^6 \text{ m/s}^2) = 2.63 \times 10^4$ N, which is equivalent to the weight of a 2680-kig mass ($m = F/g = 2.63 \times 10^4 \text{ N}/9.80 \text{ m/s}^2 = 2.68 \times 10^3$ kg) or almost 3 tons.

*4-4 ∎ Rotating frames of reference

We have so far examined rotational motion from the outside, as an observer fixed on the earth. Sometimes it is convenient to place ourselves (in theory, if not physically) into the rotating system. Let us, for example, examine the motion of objects from the point of view, or frame of reference, of a person seated on a rotating platform such as a merry-go-round. It looks to him as if the rest of the world is going around *him*. But let us focus attention on what he observes when he places a tennis ball on the floor of the merry-go-round, which we assume is frictionless. If he puts the ball down gently, without giving it any push, he will observe that it accelerates from rest and moves outward in a path similar to that shown in Fig. 4-11a. According to Newton's first law, an object initially at rest should stay at rest if no force acts on it. But, according to the observer on the merry-go-round, the ball starts moving even though there is no force applied to it. To an observer on the earth, this is all very clear; for to this person the ball has an initial velocity when it is released (because the platform is moving), and it simply continues moving in a straight-line path as shown in Fig. 4-11b, in accordance with Newton's first law.

But what shall we do about the frame of reference of the person on the merry-go-round? Clearly, Newton's first law, the law of inertia, does not hold in this rotating frame of reference. (For this reason, such a frame is called a *noninertial* reference frame. An *inertial* reference frame is one in which the law of inertia—Newton's first law—does hold, and so do Newton's second and third laws.) Newton's second law does not hold either in a rotating frame of reference. For instance in the situation described above, there is no force on the ball; yet, with respect to the merry-go-round, it accelerates.

Because Newton's laws do not hold when observations are made with respect to a rotating frame of reference, calculation of motion can be difficult. However, we can still apply Newton's laws in such a reference frame if we make use of a trick. We write down the equation $F = ma$ as if a force equal to mv^2/r were acting on the object[†] in addition to any other forces that may be acting. This extra force, which can be designated as "centrifugal force" since it acts outwards, is called a "pseudoforce" ("pseudo" means "false"). It is a pseudoforce because there is no other object that exerts this force; besides, when viewed from the earth or other inertial reference frames, the force disappears entirely. We have made it up, created it, so that we can make calculations using the relation $F = ma$. Thus the observer in Fig. 4-11a can calculate the motion of the ball by assuming that a force equal to mv^2/r acts on it.

We can examine the motion of a particle in a centrifuge from the frame of reference of the rotor. In this frame of reference, the particles move in a more-or-less straight path down the tube. (From the reference frame of the earth, the particles go round and round.) The acceleration of a particle with respect to the rotor can then be calculated using $F = ma$ if we include a pseudoforce, "F," equal to mv^2/r in addition to the resisting force F_R exerted by the liquid on the particle (Fig. 4-12).

The earth itself is rotating on its axis. Thus, strictly speaking, Newton's laws are not valid on the earth. However, the effect is usually so small that it can be ignored, although it does affect the movement of large air masses and ocean currents.

[†] This is slightly simplified. Careful analysis shows that other "pseudoforces" may also be present as a result of the rotation; one such is the Coriolis force. These are, however, often small enough to be ignored.

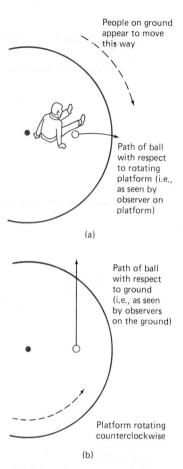

People on ground appear to move this way

Path of ball with respect to rotating platform (i.e., as seen by observer on platform)

(a)

Path of ball with respect to ground (i.e., as seen by observers on the ground)

Platform rotating counterclockwise

(b)

FIGURE 4-11

Path of a ball released on a rotating merry-go-round (a) in the reference frame of the merry-go-round and (b) in the reference frame of the earth.

FIGURE 4-12

The forces on a particle in a rotating centrifuge, in the reference frame of the centrifuge.

$"F" = mv^2/r$

F_R

Because of the earth's rotation, the material of the earth is concentrated slightly at the equator. The earth is thus not a perfect sphere but is slightly fatter at the equator than it is at the poles.

4-5 ∎ Newton's law of universal gravitation

Besides developing the three laws of motion, Sir Isaac Newton also examined the motion of the heavenly bodies—the planets and the moon. In particular, he wondered about the nature of the force that must act to keep the moon in its nearly circular orbit around the earth.

Newton was also thinking about the apparently unrelated problem of gravity. Since falling bodies accelerate, Newton had concluded that they must have a force exerted on them, a force we call the force of gravity. But what, we may ask, *exerts* this force of gravity—for, as we have seen, whenever a body has a force exerted *on* it, that force is exerted *by* some other body. Every object on the surface of the earth feels this force of gravity, and no matter where the object is, the force is directed toward the center of the earth. Newton concluded that it must be the earth itself that exerts the gravitational force on objects at its surface.

According to an early account, Newton was sitting in his garden and noticed an apple drop from a tree. He is said to have been struck with a sudden inspiration: If the effect of gravity acts at the tops of trees, and even at the tops of mountains, then perhaps it acts all the way to the moon! Whether this story is true or not,[†] it does capture something of Newton's reasoning and inspiration. With this idea that it is terrestrial gravity that holds the moon in its orbit, Newton developed his great theory of gravitation, with considerable help and encouragement from Robert Hooke (1635–1703).

Newton set about determining the magnitude of the gravitational effect that the earth exerts on the moon as compared to the gravitational effect on objects at the earth's surface. At the surface of the earth, the force of gravity accelerates objects at 9.80 m/s^2. But what is the centripetal acceleration of the moon? Since the moon moves with nearly uniform circular motion, the acceleration can be calculated from $a_c = v^2/r$; we already did this calculation in Example 4-2 and found that $a_c = 0.00273 \text{ m/s}^2$. In terms of the acceleration of gravity at the earth's surface, g, this is equivalent to

$$a_c \approx \frac{1}{3600} g.$$

That is, the acceleration of the moon toward the earth is about $\frac{1}{3600}$ as great as the acceleration of objects at the earth's surface. Now the moon is 385,000 km from the earth, which is about 60 times the earth's radius of

[†] Many historians of science today believe that Newton may have invented this story to push back the date of discovery to the 1660s, when in fact his correspondence and notebooks indicate he really didn't come to the law of universal gravitation until about 1685. The idea that there might be a connection between terrestrial gravity and the force on the moon, and even the $1/r^2$ dependence, had already been suggested by others, including Hooke; nonetheless, we still give most of the credit to Newton for the gravitational law since he proposed it in its fullness and gave very strong arguments to support it.

6380 km. That is, the moon is 60 times farther from the earth's center than are objects at the earth's surface. But $60 \times 60 = 3600$—again that number 3600! Newton concluded that the gravitational force exerted by the earth on any object decreases with the square of its distance, r, from the earth's center:

$$\text{force of gravity} \propto \frac{1}{r^2}.$$

The moon, being 60 earth radii away, feels a gravitational force only $\frac{1}{60^2} = \frac{1}{3600}$ times as strong as it would if it were at the earth's surface. Any object placed 385,000 km from the earth would experience the same acceleration due to the earth's gravity as the moon experiences: 0.00273 m/s².

Newton realized that the force of gravity on an object depends not only on distance but also on the object's mass. In fact, it is directly proportional to its mass, as we have seen. According to Newton's third law, when the earth exerts its gravitational force on an object, such as the moon, that object exerts an equal and opposite force on the earth (Fig. 4-13). Because of this symmetry, Newton reasoned, the magnitude of the force of gravity must be proportional to *both* the masses. Thus

$$F \propto \frac{m_e m_o}{r^2}$$

where m_e is the mass of the earth, m_o the mass of the other object, and r the distance from the earth's center to the center of the object.

Newton went a step further in his analysis of gravity. He had determined that the force required to hold the different planets in their orbits around the sun diminished as the inverse square of their distance from the sun. This led him to believe that it was also the gravitational force that acted between the sun and each of the planets to keep them in their orbits. And if gravity acts between these objects, why not between all objects? Thus he proposed his famous **law of universal gravitation,** which we can state as follows:

■ **Every particle in the universe attracts every other particle with a force that is proportional to the product of their masses and inversely proportional to the square of the distance between them. This force acts along the line joining the two particles.**

The magnitude of this force can be written as:

$$F = G \frac{m_1 m_2}{r^2}, \qquad (4\text{-}2)$$

where m_1 and m_2 are the masses of the two particles, r is the distance between them, and G is a universal constant which must be measured experimentally and has the same numerical value for all objects.

The value of G must be very small since we are not aware of any force existing between ordinary sized objects, such as between two baseballs. The force between two ordinary objects was first measured over 100 years after Newton's publication of his law, by Henry Cavendish in 1798. To detect and measure the incredibly small force, he used an apparatus like that shown

FIGURE 4-13
The gravitational force one body exerts on a second body is directed toward the first body and is equal and opposite to the force exerted by the second body on the first.

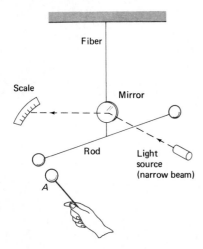

Fiber

Scale

Mirror

Rod

Light source
(narrow beam)

A

FIGURE 4-14

Schematic diagram of Cavendish's apparatus. Two spheres are attached to a light horizontal rod, which is suspended at its center by a thin fiber. When the sphere labeled A is brought close to one of the suspended spheres, the gravitational force causes the latter to move, and this twists the fiber slightly. The tiny movement is magnified by use of a narrow light beam directed at a mirror, which is mounted on the fiber, and the beam reflects onto a scale. Previous determination of what force twisted the fiber a given amount allows one to determine the magnitude of the gravitational force between two objects.

in Fig. 4-14. Cavendish not only confirmed Newton's hypothesis that two bodies attract one another and that Eq. 4-2 accurately describes this force, but because he could measure F, m_1, m_2, and r accurately, he was then able to determine the value of the constant G as well. The accepted value today is

$$G = 6.67 \times 10^{-11} \text{ N·m}^2/\text{kg}^2.$$

[Strictly speaking, Eq. 4-2 gives the magnitude of the gravitational force that one particle exerts on a second particle that is a distance r away. For an extended object (that is, not a point), we must consider how to measure the distance r. You might think that r would be the distance between the centers of the objects; this is not necessarily true, and to do a calculation correctly, each extended body must be considered as a collection of tiny particles and the total force is the sum of the forces due to all the particles. The sum over all these particles is often best done using integral calculus which Newton himself invented. Newton showed, however, that for two uniform spheres, Eq. 4-2 gives the correct force where r is the distance between their centers. Also, when extended bodies are small compared to the distance between them (as for the earth–sun system), little inaccuracy results from considering them as point particles.]

FIGURE 4-15

Orientation of sun (s), earth (e), and moon (m) for Example 5-7 (not to scale).

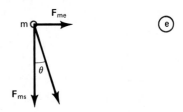

□ EXAMPLE 4-6 What is the force of gravity acting on a spacecraft when it orbits two earth radii from the earth's center (that is, 6400 km above the earth's surface) as compared to when it is on the earth's surface?

SOLUTION The spacecraft is twice as far from the earth's center as when at the surface of the earth; therefore, since the force of gravity decreases as the square of the distance (and $\frac{1}{2^2} = \frac{1}{4}$) the force of gravity on it will be only one-fourth as great.

□ EXAMPLE 4-7 Determine the net force on the moon ($m_m = 7.36 \times 10^{22}$ kg) due to the gravitational attraction of both the earth ($m_e = 5.98 \times 10^{24}$ kg) and the sun ($m_s = 1.99 \times 10^{30}$ kg) assuming they are at right angles to each other (Fig. 4-15).

SOLUTION   The earth is 3.85×10^5 km from the moon so F_{me} (force on the moon due to the earth) is

$$F_{me} = \frac{(6.67 \times 10^{-11}\ \text{N·m}^2/\text{kg}^2)(7.36 \times 10^{22}\ \text{kg})(5.98 \times 10^{24}\ \text{kg})}{(3.85 \times 10^8\ \text{m})^2}$$

$$= 1.98 \times 10^{20}\ \text{N}.$$

The sun is 1.50×10^8 km from the earth, so F_{ms} (force on the moon due to the sun) is

$$F_{ms} = \frac{(6.67 \times 10^{-11}\ \text{N·m}^2/\text{kg}^2)(7.36 \times 10^{22}\ \text{kg})(1.99 \times 10^{30}\ \text{kg})}{(1.50 \times 10^{11}\ \text{m})^2}$$

$$= 4.35 \times 10^{20}\ \text{N}.$$

Since the two forces act at right angles in the case we are considering (Fig. 4-15), the total force is

$$F = \sqrt{(1.98)^2 + (4.35)^2} \times 10^{20}\ \text{N} = 4.79 \times 10^{20}\ \text{N}$$

which acts at an angle $\theta = \tan^{-1}(1.98/4.35) = 24.5°$.

The law of universal gravitation should not be confused with Newton's second law of motion, $\mathbf{F} = m\mathbf{a}$. The former describes a particular force, gravity, and how its strength varies with the distance and masses involved. Newton's second law, on the other hand, relates the force on a body—it can be any force—to the mass and acceleration of that body.

4-6 ▌ Gravity near the earth's surface; geophysical applications

When Eq. 4-2 is applied to the gravitational force between the earth and an object at its surface, m_1 becomes the mass of the earth, m_e; and r becomes the distance from the earth's center,[†] namely the radius of the earth, r_e. This force of gravity due to the earth is the weight of the object, which we have been writing as mg. Thus

$$mg = G\frac{mm_e}{r_e^2}.$$

Hence

$$g = G\frac{m_e}{r_e^2}. \tag{4-3}$$

Thus the acceleration of gravity at the surface of the earth, g, is determined by m_e and r_e. (Be sure not to confuse G with g; they are very different quantities, but are related by Eq. 4-3.)

[†] That the distance is measured from the earth's center does not imply that the force of gravity somehow emanates from that one point. Rather, all parts of the earth attract gravitationally but the net effect is a force acting toward the earth's center.

Until G was measured, the mass of the earth was not known. But once G was measured, Eq. 4-3 could be used to calculate the earth's mass, and Cavendish was the first to do so. Since $g = 9.80$ m/s² and the radius of the earth $r_e = 6.38 \times 10^6$ m, then from Eq. 4-3 we obtain

$$m_e = \frac{gr_e^2}{G} = \frac{(9.80 \text{ m/s}^2)(6.38 \times 10^6 \text{ m})^2}{6.67 \times 10^{-11} \text{ N·m}^2/\text{kg}^2} = 5.98 \times 10^{24} \text{ kg}$$

for the mass of the earth.

When dealing with the weight of objects at the surface of the earth, we can continue to use simply mg. If we wish to calculate the force of gravity on an object some distance from the earth, or the force due to some other heavenly body such as that exerted by the moon or a planet, we can calculate the effective value of g from Eq. 4-3, replacing r_e (and m_e) by the appropriate distance (and mass), or we can use Eq. 4-2 directly.

EXAMPLE 4-8   What is the effective value of g at a height of 1000 km above the earth's surface? That is, what is the acceleration due to gravity of objects allowed to fall freely at this altitude?

SOLUTION   We use Eq. 4-3, with r_e replaced by $r = 7380$ km $= 7.38 \times 10^6$ m:

$$g = \frac{(6.67 \times 10^{-11} \text{ N·m}^2/\text{kg}^2)(5.98 \times 10^{24} \text{ kg})}{(7.38 \times 10^6 \text{ m})^2}$$

$$= 7.32 \text{ m/s}^2.$$

In Chapter 1 we saw that the value of g varied slightly from place to place. From Eq. 4-3 we might expect g to be slightly less on the tops of mountains, for example, than at sea level since r_e, the distance from the earth's center, is slightly greater there. Indeed, this is what is found experimentally. However, Eq. 4-3 does not give precise values for g at various locations since the earth is not a perfect sphere. It not only has mountains and valleys, and bulges at the equator, but its mass is not distributed precisely uniformly (its rotation also has an effect on the measurement of g).

The value of g can vary locally on the earth's surface because of the presence of irregularities and rocks of different densities. Such variations in g, known as "gravity anomalies," are very small—on the order of 1 part per 10^6 or 10^7 in the value of g. But they can be measured ("gravimeters" today can detect variations in g to 1 part in 10^8). Geophysicists use such measurements as part of their investigations into the structure of the earth's crust. Such measurements also are used in mineral and oil exploration. Mineral deposits, for example, often have a greater density than surrounding material; because of the greater mass in a given volume, g can have a slightly greater value on top of such a deposit than at its flanks. "Salt domes," under which petroleum is often found, have a lower than average density; searches for a slight reduction in the value of g in certain locales have lead to the discovery of oil.

4-7 ∎ Satellites and "weightlessness"

Artificial satellites circling the earth are now commonplace. A satellite is put into orbit by accelerating it to a sufficiently high tangential speed with the use of rockets, as shown in Fig. 4-16. If the speed is too high, the spacecraft will not be confined by the earth's gravity and will escape, never to return. If too low, it will fall back to earth. Satellites are usually put into circular, or nearly circular, orbits. This requires the least take-off speed. It is sometimes asked: "What keeps a satellite up?" The answer is its speed and the fact that the earth is round. If a satellite stopped moving, it would of course fall down to earth. At the high speed a satellite has, it would fly off into space if the gravitational force of the earth did not act on it—see Fig. 4-17. And if the earth were flat (and sufficiently long) the satellite would be pulled down until it hit the earth, but the earth is round, and so the satellite, though "falling" under the action of gravity, "falls" around the earth as shown if its speed is sufficiently high.

For satellites that move (at least approximately) with uniform circular motion, the acceleration is v^2/r. The force that gives a satellite this acceleration is the force of gravity, and since a satellite may be at a considerable distance from the earth, we must use Eq. 4-2 for the force acting on it. When we apply Newton's second law, $F = ma$, we find

$$G\frac{mm_e}{r^2} = m\frac{v^2}{r}.$$ (4-4)

This equation relates the distance of the satellite from the earth's center, r, to its speed, v. Note that only one force, gravity, is acting on the satellite.

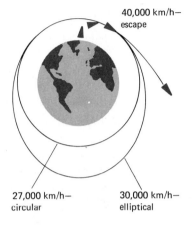

40,000 km/h—
escape

27,000 km/h—
circular

30,000 km/h—
elliptical

FIGURE 4-16
Artificial satellites.

FIGURE 4-17
A moving satellite "falls" out of a straight-line path, toward the earth.

□ EXAMPLE 4-9   Calculate the speed required for a satellite moving in a circular orbit 200 km above the earth's surface.

SOLUTION   We solve Eq. 4-4 for v;

$$v = \sqrt{\frac{Gm_e}{r}}.$$

The value of r is 6380 km + 200 km = 6580 km, or 6.58×10^6 m. Thus

$$v = \sqrt{\frac{(6.67 \times 10^{-11} \text{ N·m}^2/\text{kg}^2)(5.98 \times 10^{24} \text{ kg})}{6.58 \times 10^6 \text{ m}}} = 7.79 \times 10^3 \text{ m/s}.$$

This is approximately 27,000 km/h. Note that the speed required will decrease with increasing height.

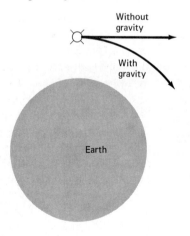

Without
gravity

With
gravity

Earth

People and other objects in a satellite circling the earth are said to experience apparent weightlessness. Before tackling the case of a satellite, however, let us first look at the simpler case of a falling elevator. In Fig. 4-18a we see an elevator at rest. Let us consider an object such as a bag hanging from a spring scale as shown. The scale indicates the downward force exerted on it by the bag. This force, exerted *on* the scale, is just equal and opposite to the force exerted *by* the scale upwards on the bag. We call this force **w**.

Since the mass, m, is not accelerating, we apply $F = ma$ to the bag and obtain:

$$w - mg = 0,$$

where mg is the weight of the bag; thus $w = mg$, and since the scale indicates the force w exerted on it by the bag, it registers a force equal to the weight of the bag as we expect. If, now, the elevator has an acceleration, a, then applying $F = ma$ to the bag we have $w - mg = ma$. Solving for w we obtain

$$w = mg + ma.$$

We have chosen the positive direction up. Thus, if the acceleration is up, a is positive and the scale, which measures w, will read more than mg. If the elevator accelerates downward, a will be negative and w will be less than mg. For example, if the elevator's acceleration is $-\frac{1}{2}g$, then we find $w = mg - \frac{1}{2}mg = \frac{1}{2}mg$. That is, the scale reads one-half the actual weight. If the elevator is in *free fall* (for example if the cables break) then $a = -g$ and $w = mg - mg = 0$. The scale reads zero! See Fig. 4-18b. The bag seems weightless. If the person in the elevator let go of a pencil, say, it would not fall to the floor. True, the pencil would fall with acceleration g. But so do the floor of the elevator and the person. The pencil would hover right in front of the person. This is called *apparent weightlessness* because, in fact, gravity is still acting on the object. The objects seem weightless only because the elevator is accelerating at $-g$.

The "weightlessness" experienced by people in a satellite orbiting close to the earth is the same apparent weightlessness experienced in a freely falling elevator. It may seem strange, at first, to think of a satellite as freely falling. But a satellite is indeed falling toward the earth, as was shown in Fig. 4-17; the force of gravity causes it to "fall" out of its natural straight-line path. The acceleration of the satellite must be the acceleration due to gravity, since the only force acting on it is gravity. (We made use of this fact in obtaining Eq. 4-4; the fact that we wrote the acceleration as v^2/r does not alter this fact.) Thus, although the force of gravity acts on objects within the satellite, the objects experience an apparent weightlessness because they, and the satellite, are accelerating as in free fall.

When a spacecraft is out in space far from the earth and other attracting bodies such as the moon, the force of gravity due to these bodies will be quite small because of the large distances involved. Persons in such a spacecraft will experience real weightlessness, although the effects will be the same as for the apparent weightlessness described above.

The effects on human beings of weightlessness (whether real or apparent makes no difference) are interesting. In ordinary circumstances, for example, a person can become quite tired holding out his arms horizontally. But for a person experiencing weightlessness, no effort is needed. The arms will just "float" there, since there is no sensation of weight. This effect has many applications in athletics; during a jump or a dive, while on a trampoline, and even between strides while running, a person is experiencing apparent weightlessness or free fall, although only for a short time. During these brief periods, limbs can be moved much more easily, since inertia only, not gravity, needs to be overcome. The loss of control because of lack of contact with the ground is compensated for by the increased mobility. Prolonged weightlessness in space, however, can have deleterious effects on

(a)

(b)

FIGURE 4-18

An object in an elevator at rest exerts a force on a spring scale equal to its weight (a); but in a freely falling elevator (b), it experiences "weightlessness".

FIGURE 4-19

Tides on the earth due to movement of water acted on by the gravitational force exerted by the moon.

health. Red blood cells diminish, blood collects in the thorax, bones lose calcium and become brittle, and muscles lose their tone. These effects are at present being carefully studied.

*4-8 ▌ Earth's tides

The reason for the earth's tides, and why high and low tides occur at regular intervals, was a subject of interest to thinkers for thousands of years. But it was only in the seventeenth century that a clear explanation was given—by Newton. Newton argued that the gravitational attraction of the moon causes the tides. This force has only a slight effect on the solid land masses (and it is detectable), but the great masses of water that make up the oceans are free to move. Because the gravitational force decreases with the square of the distance, the moon pulls harder on that part of the earth closest to it. The water in the oceans is pulled toward the moon, resulting in a high tide on the side of the earth facing the moon as shown in Fig. 4-19. Simultaneously a high tide occurs on the opposite side of the earth. Why this second high tide? Because the pull of the moon on the far side of the earth is less (since it is farther from the moon) than the pull on the central (nearer) parts of the earth. (We might visualize this as the earth being pulled away from the water on the far side, just as the water on the side of the earth near the moon is pulled away from the earth a bit.) The ocean water thus tends to collect on the sides closest and farthest from the moon, and is taken from the regions between, where there will be low tides (see Fig. 4-19). Since the moon moves relative to the earth so that it returns to the same position overhead after about 25 hours, there are two high and two low tides at any point every 25 hours[†] (approximately two of each per day). Because the high tides stay more or less in line with the moon, it is as if the solid earth moved beneath

---

[†] According to an amusing historical anecdote (probably apocryphal), Napoleon (or was it Philip of Spain?) was supposedly told by his admirals that he had nothing to fear from the British navy because it was run by superstitious people who believed the tides had something to do with the motion of the moon!

the tidal bulges. High tides do not necessarily correspond precisely to the time at which the moon is directly overhead since other forces are also acting; among the complicating factors are the rotation of the earth, friction between the solid earth and the moving water, and the gravitational pull of the sun. These factors also affect the height of the tides. For example, the highest and lowest tides occur when the sun and moon are lined up so both are pulling in the same direction (called "spring tides"); when the sun and moon are at right angles, the tides are smallest ("neap tides")—see Fig. 4-20.

FIGURE 4-20

Tides are greatest when sun and moon are aligned as in (a) and (b). Tides are least when sun and moon are at right angles (c).

*4-9 ∎ Gravitational versus inertial mass

We have dealt with two aspects of mass. In Chapter 3, we defined mass as a measure of the inertia of a body. Newton's second law relates the force acting on a body to its acceleration and its *inertial mass*, as we call it. Then in this chapter we have dealt with mass as a property related to the gravitational force—that is, mass as a quantity that determines the strength of the gravitational force between two bodies. This we call the *gravitational mass*.

Now it is not at all obvious that the inertial mass of a body should be equal to its gravitational mass. (The force of gravity might have depended on a completely different property of a body, just as the electrical force depends on a property called electric charge, as we shall see later.) Newton's and Cavendish's experiments indicated that these two types of mass are equal for a body, and modern experiments confirm it to a precision of about 1 part in 10^{11}.

If the gravitational and inertial masses were not equal, Galileo's conclusion that all bodies fall to earth with the same acceleration in the absence of air resistance would not be valid. For suppose a heavier body had the same inertia as a lighter body (and by lighter we mean the force of gravity on it is less). Then the heavier body would accelerate faster than the lighter one since $a = F/m$; that is, F would be greater for the heavy body but m would be the same as for the lighter body. Because the inertial mass is equal to the gravitational mass, the heavier body—in spite of the greater force on it—has greater inertia than the lighter one in exact proportion, so its acceleration is the same.

*4-10 ∎ Kepler's laws and Newton's synthesis

More than a half century before Newton proposed his three laws of motion and his law of universal gravitation, the German astronomer Johannes Kepler (1571–1630) had written a number of astronomical works in which we can find a detailed description of the motion of the planets about the sun. Kepler's work resulted in part from the many years he spent examining data collected by Tycho Brahe (1546—1601) on the positions of the planets in their motion through the heavens. Tucked away among these writings were three findings which we now refer to as **Kepler's laws of planetary motion.** These are summarized as follows, with additional explanation in Fig. 4-21.

Kepler's first law: The path of each planet about the sun is an ellipse with the sun at one focus (Fig. 4-21a).

Kepler's second law: Each planet moves so that an imaginary line drawn from the sun to the planet sweeps out equal areas in equal periods of time (Fig. 4-21b).

Kepler's third law: The ratio of the squares of the periods of any two planets revolving about the sun is equal to the ratio of the cubes of their average distances from the sun. That is, if T_1 and T_2 represent the periods (time needed for one revolution about the sun), and r_1 and r_2 represent their average distances from the sun, then

$$\left(\frac{T_1}{T_2}\right)^2 = \left(\frac{r_1}{r_2}\right)^3.$$

(Present-day data are given in Table 4-1; see last column.)

Newton was able to show that Kepler's laws could be derived mathematically from his law of universal gravitation and his laws of motion; and he showed that for any reasonable form for the gravitational force law, only one that depends on the inverse square of the distance is fully consistent with all three of Kepler's laws. He thus used Kepler's laws as evidence in favor of his law of universal gravitation, Eq. 4-2.

Kepler's third law is the easiest to derive, and we do it here for the special case of a circular orbit. (Most of the planetary orbits are fairly close to a circle, which is a special case of an ellipse.) First we write down Newton's second law of motion, $F = ma$; then for F we substitute the law of universal gravitation, Eq. 4-2, and for a the centripetal acceleration, v^2/r:

$$F = ma$$

$$G\frac{m_1 M}{r^2} = m_1 \frac{v_1^2}{r_1}.$$

Here m_1 is the mass of a particular planet, r_1 its average distance from the sun, and v_1 its average speed in orbit; M is the mass of the sun, since it is the gravitational attraction of the sun that keeps each planet in its orbit.

(a)

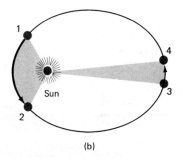

(b)

FIGURE 4-21

(a) Kepler's first law. An ellipse is a closed curve such that the sum of the distances from any point P on the curve to two fixed points (called the foci, F_1 and F_2) remains constant. That is, the sum of the distances $F_1P + F_2P$ is the same for all points on the curve. A circle is a special case of an ellipse when the two foci coincide, at the center of the circle. (b) Kepler's second law. The two shaded regions have equal areas. It takes the planet the same time to move from point 1 to point 2 as it takes it to move from point 3 to point 4. Planets move fastest in that part of their orbit where they are closest to the sun.

TABLE 4-1
Planetary data applied to Kepler's third law

Planet	Mean Distance from Sun, r (10^6 km)	Period, T (Earth Years)	$\dfrac{r^3/T^2}{\left(\dfrac{\text{km}^3}{\text{yr}^2} \times 10^{24}\right)}$
Mercury	57.9	0.241	3.34
Venus	108.2	0.615	3.35
Earth	149.6	1.0	3.35
Mars	227.9	1.88	3.35
Jupiter	778.3	11.86	3.35
Saturn	1427	29.5	3.34
Uranus	2870	84.0	3.35
Neptune	4497	165	3.34
Pluto	5900	248	3.33

Now the period T_1 of the planet is the time required for one complete orbit, a distance equal to $2\pi r_1$, the circumference of a circle; thus

$$v_1 = \frac{2\pi r_1}{T_1}.$$

We substitute this formula for v_1 into the equation above:

$$G\frac{m_1 M}{r_1^2} = m_1 \frac{4\pi^2 r_1}{T_1^2}.$$

We rearrange this to get

$$\frac{T_1^2}{r_1^3} = \frac{4\pi^2}{GM}.$$

We derived this for planet 1 (say Mars); the same derivation would apply for a second planet (say Saturn):

$$\frac{T_2^2}{r_2^3} = \frac{4\pi^2}{GM},$$

where T_2 and r_2 are the period and orbit radius for the second planet. Since the right sides of the two previous equations are equal, we have $T_1^2/r_1^3 = T_2^2/r_2^3$ or, rearranging,

$$\frac{T_1^2}{T_2^2} = \frac{r_1^3}{r_2^3},$$

which is Kepler's third law.

Accurate measurements on the orbits of the planets indicated that they did not precisely follow Kepler's laws. For example, slight deviations from perfectly elliptical orbits were observed. Newton was aware that this is to be expected from the law of universal gravitation ("every particle in the universe attracts every other particle...") because each planet exerts a gravitational force on the other planets. Since the mass of the sun is much greater than that of any planet, the force on one planet due to any other planet will be small in comparison to the force on it due to the sun. (The derivation of perfectly elliptical orbits ignores the forces due to other planets.) But because of this small force, each planetary orbit should depart from a perfect ellipse,[†] especially when a second planet is fairly close to it. Such deviations, or *perturbations*, as they are called, from perfect ellipses are indeed observed. In fact, Newton's recognition of perturbations in the orbit of Saturn was a hint that helped him formulate the law of universal gravitation, that all bodies can attract gravitationally. Observation of other perturbations later led to the discovery of Neptune and Pluto. Deviations in the orbit of Uranus, for example, could not be accounted for by perturbations due to the other known planets; careful calculation in the nineteenth century indicated that these deviations could be accounted for if there were another planet farther out in the solar system. The position of this planet was predicted from the devia-

---

[†] Newton said thusly that the "planets neither move exactly in ellipses nor revolve twice in the same orbit." He also pointed out, for example, that the sun is not precisely at one focus of the ellipse; even in the ideal case of only two bodies (sun and one planet), the two would revolve about their common center of gravity.

tions in the orbit of Uranus, and telescopes focused on that region of the sky quickly found it; the new planet was called Neptune. Similar but much smaller perturbations of Neptune's orbit led to the discovery of Pluto in 1930.

The development by Newton of the law of universal gravitation and the three laws of motion was a major achievement in Western thought. For with these laws, Newton was able to describe the motion of objects on earth and in the heavens. The motions of heavenly bodies and bodies on earth were seen to follow the same laws (something not previously recognized generally, although Galileo and Descartes had argued in its favor). For this reason (and also because Newton integrated the results of earlier workers into his system), we sometimes speak of Newton's "synthesis."

Newton's work was so encompassing that it constituted a theory of the universe, and influenced philosophy and other fields. The laws formulated by Newton are referred to as *causal laws*. By *causality* we mean the idea that one occurrence can cause another. We have repeatedly observed, for example, that when a rock strikes a window, the window almost immediately breaks. We infer that the rock *caused* the window to break. This idea of "cause and effect" took on more forceful meaning with Newton's laws. For the motion—or rather the acceleration—of any object was seen to be *caused* by the net force acting on it. As a result the universe came to be pictured by many scientists and philosophers as a big machine whose parts move in a predictable and predetermined way—according to natural laws. However, this *deterministic* view of the universe had to be modified by scientists in the twentieth century, as we shall see in Chapters 26 and 27.

4-11 ∎ Types of forces in nature

We have already discussed the fact that Newton's law of universal gravitation, Eq. 4-2, describes how a particular type of force—gravity—depends on the distance between, and masses of, the objects involved. Newton's second law, $\mathbf{F} = m\mathbf{a}$, tells how a body will accelerate due to any type of force $\mathbf{F}$. But what are the types of forces that occur in nature?

Physicists today recognize only four different forces in nature.[†] These are: (1) the gravitational force; (2) the electromagnetic force (we shall see later that electric and magnetic forces are intimately related), (3) the strong nuclear force, and (4) the weak nuclear force. We have discussed the gravitational force in detail. The nature of the electromagnetic force will be discussed in detail in later chapters. The strong and weak nuclear forces operate at the level of the atomic nucleus, and although they manifest themselves in such phenomena as radioactivity and nuclear energy, they are much less obvious in our daily lives.

But what about such forces as ordinary pushes and pulls, and friction? What category do they fit in? We call these types of force *contact forces* since they occur when one object is in contact with another object which exerts the force. According to modern quantum theory, these forces are due to the

---

[†] Physicists are now working on theories that would unify these four forces—that is, to consider some or all of these forces as different manifestations of the same basic force. This is discussed in the final chapter of this book.

electromagnetic force. For example, the force your fingers exert on a pencil is the result of electrical repulsion between the outer electrons of the atoms of your fingers and those of the pencil.

▌ Summary

An object moving in a circle of radius r with constant speed v is said to be in *uniform circular motion;* it has a *centripetal acceleration* a_c toward the center of the circle of magnitude

$$a_c = \frac{v^2}{r},$$

because the velocity vector is continually changing in direction. A force acting toward the center is thus needed to keep a particle revolving in a circle.

Newton's *law of universal gravitation* states that every particle in the universe attracts every other par-

ticle with a force proportional to the product of their masses and inversely proportional to the square of the distance between them:

$$F = G \frac{m_1 m_2}{r^2}.$$

The direction of this force is along the line joining the two particles. It is this gravitational force that keeps the moon revolving around the earth and the planets revolving around the sun.

▌ Questions

1. Sometimes it is said that water is removed from clothes in a spin dryer by centrifugal force throwing the water outward. Is this correct?

2. Will the acceleration of a car be the same if it travels around a sharp curve at 60 km/h as when it travels around a gentle curve at the same speed? Explain.

3. Suppose a car moves at constant speed along a mountain road. At what places does it exert the greatest and least forces on the road: (*a*) at the top of a hill, (*b*) at a dip between two hills, (*c*) on a level stretch near the bottom of a hill?

4. Describe all the forces acting on a child riding a horse on a merry-go-round.

5. Does an apple exert a gravitational force on the earth? If so, how large a force? Consider an apple (*a*) attached to a tree, (*b*) falling.

6. If the earth's mass were suddenly to double, in what ways would the moon's orbit change?

7. Describe how careful measurements of the variation in g in the vicinity of an ore deposit might be used to estimate the amount of ore present.

8. When will your apparent weight be the greatest, as measured by a scale in a moving elevator? When the elevator (*a*) accelerates downward, (*b*) accelerates upward, (*c*) is in free fall, (*d*) moves upward at constant speed. In which case would your weight be the least? When would it be the same as when you are on the ground?

9. If you were in a satellite orbiting the earth, how might you cope with walking, drinking, or putting a pair of scissors on a table?

10. An antenna loosens and becomes detached from a satellite in a circular orbit around the earth. Describe the antenna's motion subsequently. If it will land on earth, describe where; if not, describe how it could be made to land on earth.

11. The sun is directly below us at midnight, in line with the earth's center. Are we then heavier at midnight, due to the sun's gravitational force on us, than we are at noon? Explain.

12. Astronauts who spend long periods in outer space could be adversely affected by weightlessness. One way to simulate gravity is to shape the spaceship like a bicycle wheel which rotates about an axis just like a wheel, with the astronauts walking on the inside of the "tire." Explain how this simulates gravity. Consider (*a*) how objects fall, (*b*) the force we feel on our feet, and (*c*) any other aspects of gravity you can think of.

13. People sometimes ask, "What keeps a satellite up in its orbit around the earth?" How would you respond?

14. Explain why a person running experiences "free fall" or "apparent weightlessness" between steps.

*15. The sun's gravitational pull on the earth is much larger than the moon's. Yet the moon is mainly responsible for the tides. Explain.

*16. The earth moves faster in its orbit around the sun in winter than in summer. Is it closer to the sun in summer or in winter? Does this affect the seasons? Explain.

∎ Problems

SECTION 4-1

1. (I) A child moves with a speed of 1.80 m/s when 12.4 m from the center of a merry-go-round. Calculate (a) the centripetal acceleration of the child, and (b) the net force exerted on the child (mass = 25.0 kg).

2. (I) A jet plane traveling 1800 km/h (500 m/s) pulls out of a dive by moving in an arc of radius 3.00 km. What is the plane's acceleration in g's?

3. (I) Calculate the centripetal acceleration of the earth in its orbit around the sun and the net force exerted on it. What exerts this force on the earth? Assume that the earth's orbit is a circle of radius 1.50×10^{11} m.

4. (I) A horizontal force of 26.0 N is applied to a 0.60-kg stone to keep it rotating in a horizontal circle of radius 0.40 m. Calculate its speed.

5. (I) What is the maximum speed with which a 1300-kg car can round a turn of radius 95 m on a flat road if the coefficient of friction between tires and road is 0.55? Is this result independent of the mass of the car?

6. (I) How large must the coefficient of friction be between the tires and the road if a car is to round a level curve of radius 62 m at a speed of 55 km/h?

7. (II) Tarzan plans to cross a gorge by swinging in an arc from a hanging vine. If his arms are capable of exerting a force of 1500 N on the rope, what is the maximum speed he can tolerate at the lowest point of his swing? His mass is 85 kg; the vine is 4.0 m long.

8. (II) Is it possible to whirl a bucket of water fast enough in a vertical circle so the water won't fall out? If so, what is the minimum speed?

9. (II) A coin is placed 18.0 cm from the axis of a rotating turntable of variable speed. When the speed of the turntable is slowly increased, the coin remains fixed on the turntable until a rate of 58 rpm is reached, at which point the coin slides off. What is the coefficient of static friction between the coin and the turntable?

10. (II) Because the earth rotates once per day, the effective acceleration of gravity at the equator is slightly less than it would be if the earth didn't rotate. Estimate the magnitude of this effect. What fraction of g is this?

11. (II) Use dimensional analysis (see Appendix C) to obtain the form for the centripetal acceleration, $a_c = v^2/r$.

12. (II) What minimum speed must a roller coaster be traveling when upside down at the top of a circle (Fig. 4-22) if the passengers are not to fall out? Assume a radius of curvature of 8.0 m.

13. (II) A ball on the end of a string is revolving at a uniform rate in a vertical circle or radius 96.5 cm as shown in Fig. 4-23. If its speed is 3.15 m/s and its mass is 0.335 kg, calculate the tension in the string when the ball is (a) at the top of its path, and (b) at the bottom of its path.

14. (II) A ball of mass m is revolved in a vertical circle at the end of a cord of length L. What is the minimum speed v needed at the top of the circle if the cord is to remain taut?

15. (II) A projected space station consists of a circular tube which is set rotating about its center (like a tubular

FIGURE 4-22

FIGURE 4-23

bicycle tire). The circle formed by the tube has a diameter of 1.6 km. (a) On which part of the inside of the tube will people be able to walk? (b) What must be the rotation speed (revolutions per day) if an effect equal to gravity at the surface of the earth (1 g) is to be felt?

16. (III) If a curve with a radius of 60 m is properly banked for a car traveling 60 km/h, what must be the coefficient of static friction for a car not to skid when traveling at 90 km/h?

17. (III) A 1200-kg car rounds a curve of radius 65 m banked at an angle of 14°. If the car is traveling at 80 km/h, will a friction force be required? If so, how much and in what direction?

18. (III) If a plant is allowed to grow from seed on a rotating platform, it will grow at an angle, pointing inward. Calculate what this angle will be (put yourself in the rotating frame) in terms of g, r, and v.

*SECTION 4-4

*19. (III) In a rotating frame of reference, Newton's first and second laws remain valid if we assume that a "pseudoforce" equal to mv^2/r is acting. What effect does this assumption have on the validity of Newton's third law?

SECTIONS 4-5 AND 4-6

20. (I) Calculate the force of gravity between two bowling balls each of which has a mass of 8.0 kg, when they are 0.50 m apart (center to center).

21. (I) Calculate the force of gravity on a spacecraft 12,800 km above the earth's surface if its mass is 700 kg.

22. (I) Calculate the acceleration due to gravity on the moon. The moon's radius is about 1.7×10^6 m and its mass is 7.4×10^{22} kg.

23. (I) A hypothetical planet has a radius 1.6 times that of the earth, but has the same mass. What is the acceleration due to gravity near its surface?

24. (I) A hypothetical planet has a mass of 2.2 times that of earth, but the same radius; what is g near its surface?

25. (II) Suppose a hypothetical planet has a radius 20.0 times that of earth and a mass 100 times that of earth. What is g near the surface?

26. (II) What is the distance from the earth's center to a point outside the earth where the gravitational acceleration due to the earth is $\frac{1}{25}$ of its value at the earth's surface?

27. (II) At the surface of a certain planet the gravitational acceleration g has a magnitude of 2.0 m/s^2. A 4.0-kg brass ball is transported to this planet. Give: (a) the mass of the brass ball on the earth and on the planet; and (b) the weight of the brass ball on the earth and on the planet.

28. (II) Suppose the mass of the earth were doubled, but it kept the same density and spherical shape. How would the weight of objects at the earth's surface change?

29. (II) Calculate the effective value of g, the acceleration of gravity, (a) 3200 m, (b) 3200 km, above the earth's surface.

30. (II) Determine the mass of the sun using the known value for the period of the earth and its distance from the sun.

31. (II) How far above the earth's surface will the acceleration of gravity be half what it is on the surface?

32. (II) Compare the force of gravity exerted by the sun on the moon with the force exerted by the earth on the moon. The sun's mass is about 3.0×10^5 that of the earth and is about 1.5×10^8 km from the moon whereas the earth is only 3.8×10^5 km away.

33. (II) Derive a formula for the mass of a planet in terms of its radius, r, the acceleration due to gravity at its surface, g_p, and the gravitational constant, G.

34. (II) At what distance from the earth will a spacecraft traveling directly from the earth to the moon experience zero net force because the earth and moon pull with equal and opposite forces?

35. (III) Four 8.0-kg spheres are located at the corners of a square of side 0.50 m. Calculate the magnitude and direction of the gravitational force on one sphere due to the other three.

36. (III) How long would a day be if the earth were rotating so fast that objects at the equator would be weightless?

SECTION 4-7

37. (I) A 15.0-kg monkey hangs from a cord suspended from the ceiling of an elevator. The cord can withstand a tension of 185 N, and breaks as the elevator accelerates upwards. What was the elevator's minimum acceleration (magnitude and direction)?

38. (I) Calculate the speed of a satellite moving in a stable circular orbit about the earth at a height of 3200 km.

39. (II) One of the moons of Jupiter discovered by Galileo has a rotational period of 1.44×10^6 s and it is 1.9×10^9 m from Jupiter on the average. From these data, determine the mass of Jupiter.

40. (II) What will a spring scale read for the apparent weight of a 55.0-kg woman in an elevator that moves (a) with constant speed upward of 5.0 m/s, (b) with constant speed downward of 5.0 m/s, (c) with upward acceleration of 0.33 g, (d) with downward acceleration 0.33 g, (e) in free fall?

41. (II) A 60.0 kg person carries a 125-N suitcase onto an elevator. As the elevator descends, the apparent weight of the suitcase felt by the person is only 100 N. Calculate (a) the acceleration of the elevator, (b) the force exerted by the floor on the person, and (c) the net force on the person.

42. (II) At what height above the earth's surface must a satellite be placed if it is to remain over the same geographical point on the equator of the earth?

43. (II) What is the apparent weight of a 65-kg astronaut 4200 km from the center of the earth's moon in a space

vehicle, (a) moving at constant velocity, (b) accelerating toward the moon at 3.6 m/s². State "direction" in each case.

44. (III) (a) Show that the mass of the sun is given by the formula $M_S = (4\pi^2 r^3)/(GT^2)$, where T is the period of any planet (the time to make one revolution about the sun), r is the radius of its orbit (assumed to be circular), and G is the gravitational constant. (b) Calculate the mass of the sun using the earth as the planet.

45. (III) Two stars maintain a constant distance apart of 8.0×10^{10} m and rotate about a point midway between them at a rate of one revolution every 12.6 yr. (a) Why don't the two stars crash into one another due to the gravitational force between them? (b) What must be the mass of each star (take the two masses to be the same)?

*SECTION 4-8

***46.** (II) (a) Calculate the force of the sun ($m = 2 \times 10^{30}$ kg) and that of the moon ($m = 7.4 \times 10^{22}$ kg) on the earth. (b) Explain why the moon has the greater influence on the earth's tides. The sun is 1.5×10^{11} m from earth.

*SECTION 4-10

***47.** (I) Use Kepler's laws and the period of the moon (27.4 days) to determine the period of an artifical satellite orbiting near the earth's surface.

***48.** (I) The asteroid Icarus, though only a few hundred meters across, orbits the sun like the other planets. Its period is about 410 days. What is its average distance from the sun?

***49.** (I) Venus is an average distance of 1.08×10^8 km from the sun. Estimate the length of the Venusian year using the fact that the earth is 1.49×10^8 km from the sun on the average.

***50.** (II) Use Kepler's third law to determine at what height above the earth an artificial satellite must orbit if it is to remain at the same place relative to earth?

***51.** (III) (a) Use Kepler's second law to show that the ratio of the speeds of a planet at its near and far points from the sun is equal to the inverse ratio of the near and far distances: $v_N/v_F = d_F/d_N$. (b) Given that the earth's distance from the sun varies from 1.47 to 1.52×10^{11} m, determine the minimum and maximum velocities of the earth in its orbit around the sun.

5 WORK AND ENERGY

Until now we have been studying the motion of a particle in terms of Newton's three laws of motion. In this analysis, *force* has played a central role as the quantity determining the motion. In this chapter and the next, we discuss an alternative analysis of the motion of a particle in terms of the quantities *energy* and *momentum*. The importance of these quantities is that they are *conserved;* that is, in quite general circumstances they remain constant. The fact that such quantities are conserved gives us not only a deeper insight into the nature of the world but also gives us another way to attack practical problems.

The conservation laws of energy and momentum are especially valuable in dealing with systems of many objects in which a detailed consideration of the forces involved would be difficult.

This chapter is devoted to the very important concept of *energy* and the closely related concept of *work*, which are scalar quantities and have no direction associated with them. Since these two quantities are scalars, they are often easier to work with than are vector forces. Energy derives its importance from two sources. First, it is a conserved quantity; and second, energy is a concept that is useful not only in the study of motion, but in all areas of physics and other sciences as well. But before discussing energy itself, we first examine the concept of work.

5-1 ∎ Work done by a constant force

The word *work* has a variety of meanings in everyday language. But in physics, work is given a very specific meaning to describe what is accomplished by the action of a force when it makes an object move through a distance. Specifically, the **work** done on a particle by a constant force (constant in both magnitude and direction) is defined to be *the product of the magnitude of the displacement times the component of the force in the direction of the displacement*. In equation form we can write this as

$$W = Fd \cos \theta \qquad (5\text{-}1)$$

where F is the constant force, d is the net displacement of the particle, and θ is the angle between the directions of the force and the net displacement.

(Note that $F \cos \theta$ is the component of **F** parallel to **d**.) To make this clear, let us start by considering the case where the motion and the force are in the same direction so $\cos \theta = 1$ and $W = Fd$. For example, if you push a loaded grocery cart a distance of 50 m by exerting a horizontal force of 30 N on the cart, you do 30 N × 50 m = 1500 N·m of work on the cart.

As this example shows, in SI units work is measured in newton-meters; for convenience, a special name is given to this unit, the *joule* (J): 1 J = 1 N·m. In the cgs system the unit of work is called the *erg* and is defined as 1 erg = 1 dyne·cm. In British units, work is measured in foot-pounds. It is easy to show that 1 J = 10^7 erg = 0.7376 ft·lb.

A force can be exerted on an object and yet do no work on that object. For example, if you hold a heavy bag of groceries in your hands at rest, you do no work on it; you may become tired (and indeed energy is being expended by your muscles), but because the bag is not moved through a distance (the displacement is zero) the work $W = 0$. You also do no work on the bag of groceries if you carry it as you walk horizontally across the floor as shown in Fig. 5-1. No horizontal force is required to move the package at a constant velocity. However, you do exert an upward force **F** on the package equal to its weight. But this upward force is perpendicular to the horizontal motion of the package and thus has nothing to do with that motion. Indeed the force, since it is vertical, has no component in the direction of the (horizontal) displacement. So, using our definition of work above, the work done by the upward force **F** on the bag must be zero. (Note that $\theta = 90°$ in this case so $\cos \theta = 0$.) Thus when the force is perpendicular to the motion, that force does no work on the object.

Consider now a boy pulling a wagon by exerting a force **F** at an angle θ to the horizontal as shown in Fig. 5-2. If he pulls the wagon a distance d along the ground, the work done is readily found using Eq. 5-1. If the force is 20 N, the angle 30°, and the wagon is pulled 100 m, the total work done is $W = Fd \cos \theta = (20\ \text{N})(100\ \text{m})(0.866) = 1700$ J.

In our definition of work above, Eq. 5-1, we stated that the work done by a force **F** on an object undergoing a displacement **d** was the product of the displacement and the component of the force in the direction of the displacement. If we use the symbol $F_{\parallel}$ to represent the component of **F** parallel to **d**, then

$$F_{\parallel} = F \cos \theta.$$

Then the definition of work (Eq. 5-1) can be written as

$$W = F_{\parallel} d.$$

There is another, equivalent, way to determine the work that is often convenient. Let us use $d_{\parallel}$ to represent the component of **d** that is parallel to **F**.

FIGURE 5-1
Work = 0 in this case since **F** is perpendicular to displacement.

FIGURE 5-2
Work done by a force **F** acting at an angle θ to the ground is $Fd \cos \theta$.

Then

$$d_{||} = d \cos \theta.$$

Then, putting this into Eq. 5-1, we can write

$$W = Fd_{||}.$$

That is, the work done by a force **F** on an object can be considered as the product of the force F times the component of displacement in the direction of **F**. In sum, we can write

$$W = F(d \cos \theta) = Fd_{||} \qquad (5\text{-}2a)$$

and

$$W = (F \cos \theta)d = F_{||}d. \qquad (5\text{-}2b)$$

The $\cos \theta$ factor can go with either the F or the d. We can choose to use Eq. 5-2a, 5-2b, or 5-1 in calculations according to which is most convenient.

□   **EXAMPLE 5-1**   Calculate the work done against gravity to carry a 15.0-kg backpack up a hill of height $h = 10.0$ m as shown in Fig. 5-3.

  SOLUTION   Neglecting any acceleration that might occur, the person exerts a constant upward force **F** on the pack of magnitude equal to its weight: $(15.0 \text{ kg})(9.80 \text{ m/s}^2) = 147$ N. We can use Eq. 5-2a by setting $d_{||} = h$, the vertical height of the hill. Then the work done is

$$W = Fh = (147 \text{ N})(10.0 \text{ m}) = 1470 \text{ J}.$$

Note that the work done depends only on the change in elevation and not on the steepness of the hill. The same work would be done to lift the pack vertically the same height h. To find the total work done by the person against gravity to climb the hill, we would proceed in the same way using the weight of the person plus the pack.

FIGURE 5-3
Example 5-1.

  When dealing with work, as with force, it is necessary to specify whether you are talking about work done *by* a specific object, or done *on* a specific object. It is also important to specify if the work done is due to one particular force, or due to the total *net force* on the object. The total (or net) work done on an object is the algebraic sum of the work done by each force; this total work is, of course, the work done by the resultant net force. For example, when a person gently lifts a 5.0-kg package from the ground up to a table 1.0 m high, there are two forces exerted on the package: the person pulling upward, and gravity (its weight $= mg = 49$ N) acting downward. If the package is lifted gently at constant speed, the upward force exerted by the person is equal to the weight, and the work done *by the person* is $W = (49 \text{ N})(1.0 \text{ m}) = 49$ J. If we take the upward direction as positive, then the work done *by gravity* will be $W = (-49 \text{ N})(1.0 \text{ m}) = -49$ J. (Note that when the force is in the direction opposite to the displacement, the work is negative; and in this case $\theta = 180°$ in Eq. 5-1.) Thus the *net work* done on the package is $49 \text{ J} + (-49 \text{ J}) = 0$. This is consistent with the fact that since the net force on the package is zero, the net work on it must be zero. This of course does not contradict the fact that the person did do work on the package equal to 49 J.

*5-2 ∎ Work done by a varying force

If the force acting on an object is constant, the work done by that force can be calculated using Eq. 5-1 or 5-2. But in many cases the force varies in magnitude or direction during a process. For example, as a rocket moves away from earth, work is done to overcome the force of gravity which varies as the inverse square of the distance from the earth's center; and the force exerted by a spring increases with the amount of stretch.

The work done by a varying force can be determined graphically. The procedure is like that for determining displacement when the velocity is known as a function of time (Section 1-11). To determine the work done by a variable force, we plot $F \cos \theta$ as a function of distance d, as in Fig. 5-4a. We divide the distance into small segments Δd. For each segment, we indicate the average of $F \cos \theta$ by a horizontal dashed line. Then the work done during each segment is $\Delta W = (F \cos \theta)(\Delta d)$, which is the area of a rectangle (Δd) wide and $(F \cos \theta)$ high. The total work done to move the object a total distance $d = d_b - d_a$ is the sum of the areas of the rectangles (5 in the case shown in Fig. 5-4a). Usually the average value of $(F \cos \theta)$ for each segment must be estimated, and a reasonable approximation of the work done can then be made. If we subdivide the distance into many more segments, Δd can be made smaller and our estimate more accurate. In the limit as Δd approaches zero, the total area of the many narrow rectangles approaches the area under the curve, Fig. 5-4b. That is, *the work done by a variable force in moving an object between two points is equal to the area under the* ($F \cos \theta$) *vs. d curve between those two points.*

FIGURE 5-4

Work done by a force F can be calculated (a) by taking sum of areas of the rectangles; (b) area under the curve of $F \cos \theta$ vs. d.

(a)

(b)

5-3 ∎ Kinetic energy and the work–energy theorem

Energy is one of the most important concepts in science. Yet we cannot give a simple but accurate and general definition of energy in only a few words. Energy of various types can be defined fairly simply; and in this chapter we will define translational kinetic energy and (mechanical) potential energy. In later chapters we will define other types of energy, such as that related to heat (Chapters 11 and 12). The crucial aspect of all the types of energy is that they can be defined consistently with one another (with the units of work) and in such a way that the sum of all types, the *total energy*, is the same after any process occurs as it was before: that is, the quantity "energy" can be defined so that it is a conserved quantity. But more on this later.

For the purposes of this chapter, we could define energy in the usual way as "the ability to do work." This simple definition is not very precise, nor is it really valid for all types of energy.[†] But it can be used for mechanical energy, which we discuss in the present chapter; and it can serve as a motivation for defining energy in purely mechanical systems since the connection between work and energy is fundamental. We now define and discuss one of the basic types of energy: kinetic energy.

A moving object can do work on another object it strikes. A flying cannonball does work on a brick wall it knocks down; a moving hammer does work on a nail it strikes. In either case, a moving object exerts a force on a second object and moves it through a distance. An object in motion has the ability to do work and thus can be said to have energy. The energy of motion is called **kinetic energy,** from the Greek word *kinetikos*, meaning "motion."

In order to obtain a quantitative definition for kinetic energy, let us calculate how much work a moving object actually can do in a particular case of motion in one dimension. For now we assume the object is a particle, or in any case is undergoing only translational motion. Suppose the moving object has mass m and velocity v, and it strikes a second object on which it does work and then comes to rest. To be concrete, we might think of a hammer striking a nail. Let us assume that, upon striking the nail, the hammer exerts a constant force $\mathbf{F}$ on the nail over a distance d parallel to $\mathbf{F}$, Fig. 5-5, until the hammer comes to rest. This force thus does an amount of work on the nail, $W = Fd$. By Newton's third law, the nail exerts a force equal to $-F$ on the hammer. Let us assume that $-F$ represents the net force on the hammer (imagine the person's hand as merely supporting the hammer after it has reached speed v). Then, by Newton's second law, the force that the nail exerts on the hammer decelerates it from its initial speed v to rest: $-F = ma$. The acceleration a of the hammer (which is constant since we assumed F is constant during this process) can be related to the initial speed v and the distance d using Eq. 1-10c; since the final speed is

FIGURE 5-5

A hammer does work when striking a nail.

[†] Energy associated with heat is often not available to do work, as we will discuss in Chapter 12.

zero, $v^2 = -2ad$. Combining these equations, we find the work done by the hammer is

$$W = Fd = -mad = m\left(\frac{v^2}{2d}\right)d = \tfrac{1}{2}mv^2.$$

Thus, a body of mass m moving with a velocity v can do an amount of work equal to $\tfrac{1}{2}mv^2$ when it is brought to rest. We therefore *define* the quantity $\tfrac{1}{2}mv^2$ as the **translational kinetic energy** (KE) of the body:

$$\text{KE} = \tfrac{1}{2}mv^2. \tag{5-3}$$

Note that whereas the translational KE is directly proportional to the mass of the object, it is proportional to the *square* of the velocity. Thus, if the mass is doubled, the KE is doubled; but if the velocity is doubled, the object has four times as much KE and is therefore capable of doing four times as much work.

We have seen that a moving object can do work. The opposite is true as well: work must be done on an object to give it KE. To find the precise relationship, we reverse the above argument. Suppose an object of mass m is moving with an initial speed v_0, and to accelerate it (uniformly) to a speed v a net force F is exerted on it parallel to its motion over a distance d; then the net work done on it is $W = Fd$. Using Newton's second law, $F = ma$, and Eq. 1-10c, $v^2 = v_0^2 + 2ad$, we find

$$W = Fd$$
$$= mad$$
$$= m\left(\frac{v^2 - v_0^2}{2d}\right)d$$

or

$$W = \tfrac{1}{2}mv^2 - \tfrac{1}{2}mv_0^2. \tag{5-4}$$
$$W = \Delta\text{KE}$$

That is

■ **The net work done on an object is equal to its change in kinetic energy.**

This is known as the **work-energy theorem.** Notice, however, that since we made use of Newton's second law, $F = ma$, F must be the *net* force—that is, the sum of all forces acting on the object. Thus the work-energy theorem is valid only if W is the *net work* done on the object.

The connection between work and kinetic energy thus operates in both directions. First, if work is done on an object, its kinetic energy increases. Second, if an object has kinetic energy, it can do work on something else; and if it does, its own KE decreases.

Because of the direct connection between work and kinetic energy, energy must be measured in the same units as work: joules in SI units, ergs in the cgs, and ft·lbs in the British system.

⊡ EXAMPLE 5-2   A 145-g baseball is thrown with a speed of 25 m/s. (*a*) What is its KE? (*b*) How much work was done to reach this speed starting from rest?

SOLUTION   (*a*) KE $= \frac{1}{2}mv^2 = \frac{1}{2}(0.145 \text{ kg})(25 \text{ m/s})^2 = 45$ J. (*b*) Since the initial KE was zero the work done is just equal to the final KE, 45 J.

⊡ EXAMPLE 5-3   How much work is required to accelerate a 1000-kg car from 20 m/s to 30 m/s?

SOLUTION   The work needed is equal to the increase in kinetic energy:

$$W = \frac{1}{2}mv^2 - \frac{1}{2}mv_0^2$$
$$= \frac{1}{2}(1000 \text{ kg})(30 \text{ m/s})^2 - \frac{1}{2}(1000 \text{ kg})(20 \text{ m/s})^2$$
$$= 2.5 \times 10^5 \text{ J.}$$

5-4 ▌ Potential energy

An object can be said to have energy not only by virtue of its motion, but also by virtue of its position or configuration (shape). This is called **potential energy** (PE).

A wound-up watch spring, for example, has potential energy. For, as it unwinds, it does work to move the watch hands around. The watch spring acquired its potential energy because work was done *on* it by the person winding the watch.

Perhaps the most common example of potential energy is *gravitational potential energy*. A heavy brick held high in the air has potential energy because of its position. It has the ability to do work, for if it is released it will fall to the ground and can do work on, say, a stake, driving it into the ground. Let us determine quantitatively the gravitational PE of an object near the surface of the earth. In order to lift an object of mass *m* vertically, a force at least equal to its weight, *mg*, must be exerted on it, say, by a person's hand. In order to lift it without acceleration to a height *y* above the ground, Fig. 5-6, a person must do work equal to the product of the force, *mg*, and the vertical distance *y*; that is, $W = mgy$. If we instead allow the object to fall freely under the action of gravity and drive a stake into the ground, the object can do an amount of work equal to *mgy* on the stake, a fact which can be verified by using Eqs. 1-10 as we did in Section 5-3 for kinetic energy. Thus, to raise an object of mass *m* to a height *y requires* an amount of work equal to *mgy*; and once at height *y* the object has the *ability* to do an amount of work equal to *mgy*. Therefore we define the *gravitational potential energy* of a body as the product of its weight *mg* and its height *y*:

$$\text{gravitational PE} = mgy. \tag{5-5a}$$

Notice that the gravitational PE depends on the *vertical height* of the object above some reference level, in this case the ground. As we saw earlier

FIGURE 5-6

An object of mass *m* at height *h* above the ground can do an amount of work $W = mgh$ when it falls.

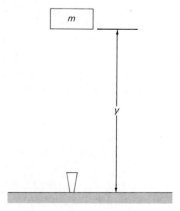

in this chapter, the work required to lift an object a height y does not depend on the path taken—that is, on whether one lifts it vertically or goes up a hill as in Fig. 5-3. Similarly, the work it can do after descending does not depend on whether the object falls vertically or by some other path (as down a frictionless incline), but depends only on the vertical height y.

In some situations you may wonder from what point to measure the height y. The gravitational PE of a book held high above a table, for example, depends on whether we measure y from the top of the table, from the floor, or from some other reference point.[†] What is physically important in any situation is the *change* in potential energy because that is what is related to the work done. We can thus choose to measure y from any reference point that is convenient, but we must be consistent throughout any given problem or situation. The change in PE between any two points does not depend on this choice.

FIGURE 5-7

Example 5-4.

◻   EXAMPLE 5-4   A 1000-kg car moves from point A, Fig. 5-7, to point B and then point C. (a) What is its PE at B and C relative to point A? (b) What is its change in potential energy when it goes from B to C? (c) Repeat parts (a) and (b) but take the reference point ($y = 0$) to be at point C.

SOLUTION   (a) Let us measure the heights from point A, so initially the car has zero PE. At point B, where $y = 10$ m,

$$\text{PE}_B = mgy = (1000 \text{ kg})(9.8 \text{ m/s}^2)(10 \text{ m}) = 1.0 \times 10^5 \text{ J}.$$

At point C, $y = -15$ m since C is below A. Therefore,

$$\text{PE}_C = mgy = (1000 \text{ kg})(9.8 \text{ m/s}^2)(-15 \text{ m}) = -1.5 \times 10^5 \text{ J}.$$

(b) In going from B to C, the potential energy change is:

$$\text{PE}_C - \text{PE}_B = (-1.5 \times 10^5 \text{ J}) - (1.0 \times 10^5 \text{ J}) = -2.5 \times 10^5 \text{ J}.$$

That is, the car loses 2.5×10^5 J.

(c) In this case, the car has a potential energy initially (at A) equal to $(1000 \text{ kg})(9.8 \text{ m/s}^2)(15 \text{ m}) = 1.5 \times 10^5$ J since $y = 15$ m. At B, its PE is 2.5×10^5 J, and at C it is zero. But the change in PE going from B to C (or from A to B) gives the same result as in part (b).

In general, if the object is initially a height y_1 above the reference point, then the change in potential energy when it moves to a different height y_2 is:

$$\text{PE}_2 - \text{PE}_1 = mgy_2 - mgy_1. \qquad [\text{gravitational PE}] \quad (5\text{-}5b)$$

As we already saw, the change in potential energy when an object of mass m moves from height y_1 to a height y_2 is equal to the work required by an

---

[†] A similar thing is true of kinetic energy: the KE of an object depends on the choice of reference frame; for example, a person sitting on a bus moving at speed v has zero KE when measured in the reference frame of the bus, but has KE $= \frac{1}{2}mv^2$ in a reference frame fixed to the ground. However, the *change* in PE does not depend on the reference frame, whereas for KE, even the change in KE depends on the reference frame.

external force to move the object from height y_1 to height y_2. The change in gravitational PE is also equal to the work done by gravity on the mass if it falls back from y_2 to y_1.

There are other kinds of PE besides gravitational. In general, we define the *change in potential energy associated with a particular force when an object is moved from one point to another to be the work required by some other (external) force to move the object from the first point to the second without accelerating it.* Equivalently, we can define the change in PE to be the work that would be done by the force itself in moving the object from the second point back to the first.

We consider now one other type of PE, that associated with elastic materials. This includes a great variety of practical applications. To take a simple example, consider a spring such as that in a watch, or better yet the simple coil spring shown in Fig. 5-8. The spring has potential energy when compressed (or stretched); for when it is released it can do work on a ball as shown. To hold a spring either stretched or compressed an amount x beyond its normal (unstretched) length requires a force that is directly proportional to x; that is, $F = kx$, where k is a constant, called the spring constant, and is a measure of the stiffness of the particular spring. This *spring equation,* sometimes referred to as *Hooke's law,*[†] is accurate for springs as long as x is not too great. In order to calculate the potential energy of a compressed spring, we merely need to calculate the work required to compress it (Fig. 5-8b), or the work it does when released (Fig. 5-8c). In either case, the work done is $W = Fx$ where x is the amount it is compressed from its normal length. The force F varies over this distance, becoming greater the more the spring is compressed. We want the average force, and since F varies linearly from zero at the uncompressed position to kx when fully compressed, the average force must be $\frac{1}{2}kx$. The work done is then $W = (\frac{1}{2}kx)(x) = \frac{1}{2}kx^2$. Hence the *elastic potential energy* is proportional to the square of the amount of compression

$$\text{elastic PE} = \tfrac{1}{2}kx^2. \tag{5-6}$$

(a)

(b)

(c)

FIGURE 5-8

A spring (a) can store energy (elastic PE) when compressed (b), which can be used to do work when released (c).

If a spring is *stretched* a distance x beyond its normal length, it too has potential energy given by this equation. Thus x can be either the amount compressed or amount stretched from the normal position.

In each of the above examples of potential energy—from a brick held at a height y, to a compressed spring—an object has the capacity or *potential* to do work even though it is not yet actually doing it. That is why we use the term "potential" energy. From these examples we can also see that energy can be *stored*, for later use, in the form of potential energy. It is also worth noting that although there is a single, universal formula for the kinetic energy of a particle, $\frac{1}{2}mv^2$, there is no single formula for potential energy; instead, the mathematical form of the PE depends on the force involved.

[†] The term "law" applied to this relation is not really appropriate since, first of all, it is only an approximation, and secondly because it refers only to a limited set of phenomena. Most physicists prefer to reserve the word "law" to those relations which are deeper and more encompassing and precise, such as Newton's laws of motion or the law of conservation of energy (Section 5-7).

Potential energy is always associated with a force, and that force is always exerted on one body by another body (the earth exerts a gravitational force on a falling stone; a compressed spring exerts a force on a ball; and so on). Thus potential energy is not something a body "has" by itself, but rather it is associated with the interaction of two (or more) bodies.

5-5 ∎ Conservative forces

The work done against gravity in moving an object from one point to another does not depend on the path taken. For example, it takes the same work ($= mgy$) to lift an object of mass m vertically a certain height as to carry it up a hill of the same height as in Fig. 5-3. (See Example 5-1). Forces such as gravity for which the work done does not depend on the path taken but depends only on the initial and final positions are called *conservative forces*.[†] Friction, on the other hand, is not a conservative force since the work done in, say, pushing a crate across a floor from one point to another depends on whether the path taken is straight, or is curved or zigzag. (In the two latter cases more work is required since the distance is greater and, unlike the gravitational force, the friction force is always directed precisely opposite to the direction of motion—so the work done depends on the particular path taken.)

An important fact is that *potential energy can be defined only for a conservative force*. Thus, although potential energy is always associated with a force, we can't write PE for just any force—such as friction which is a non-conservative force.

We can now extend the **work-energy theorem** (discussed earlier in Section 5-3) to include potential energy. Suppose several forces act on a particle, some of which are conservative, and we can write a potential energy function for these conservative forces. Then *the work, W', done by all the other forces acting on a particle is equal to the total change in kinetic and potential energy of the particle*:

$$W' = \Delta\text{KE} + \Delta\text{PE}. \qquad (5\text{-}7)$$

It must be emphasized that all the forces acting on a body must be included in Eq. 5-7, either in the potential energy term on the right, or in the work term, W', on the left (but not in both!). It is arbitrary whether a given conservative force is considered as doing work (and therefore included in W') or as involved in a change in potential energy, whereas nonconservative forces (such as friction) must be included in the work term, W'. We can thus bring the PE term over to the left side so that all forces are treated as doing work. In this case the work done is the *net work* since it includes the sum of all the forces. Hence we have

$$\text{net work done} = \Delta\text{KE}$$

which the original form of the work-energy theorem discussed earlier.

[†] You should be able to convince yourself that the elastic force of, say, a spring (or other elastic material) in which $F = kx$ is also a conservative force.

5-6 ∎ Other forms of energy; energy transformations

Besides the KE and PE of ordinary objects, other forms of energy can be defined as well. These include electric energy, nuclear energy, thermal energy, and the chemical energy stored in food and fuels. With the advent of the atomic theory, these other forms of energy have come to be considered as kinetic or potential energy at the atomic or molecular level. For example, according to the atomic theory, thermal energy is interpreted as the kinetic energy of rapidly moving molecules—when an object is heated, the molecules that make up the object move faster. On the other hand, the energy stored in food and fuel such as gasoline can be regarded as potential energy stored by virtue of the relative positions of the atoms within a molecule. For this energy to be used to do work, it must be released, usually through a chemical reaction. This is something like a compressed spring which, when released, can do work. Enzymes in our bodies allow the release of energy stored in food molecules. The violent spark of a spark plug in an automobile allows the mixture of gas and air to react chemically, releasing the stored energy and doing work against the piston to propel the car forward. Electric, magnetic, and nuclear energy can also be considered examples of kinetic and potential (or stored) energy. We will deal with these other forms of energy in detail in later chapters.

Energy can be transformed from one form to another, and we have already encountered several examples of this. A stone held high in the air has potential energy; as it falls, it loses potential energy since its height above the ground decreases (Eq. 5-5); at the same time it gains in kinetic energy since its velocity is increasing. Potential energy is being transformed into kinetic energy.

Often the transformation of energy involves a transfer of energy from one body to another. The PE stored in the spring of Fig. 5-8b is transformed into KE of the ball, Fig. 5-8c. Water at the top of a dam has potential energy which is transformed into kinetic energy as the water falls; at the base of the dam, the kinetic energy of the water can be transferred to turbine blades and further transformed to electric energy as we shall see in a later chapter. The potential energy stored in a bent bow can be transformed into kinetic energy of the arrow.

In each of these examples, the transfer of energy is accompanied by the performance of work. The spring of Fig. 5-8 does work on the ball; water does work on the turbine blades; the bow does work on the arrow. This observation gives us a further insight into the relation between work and energy: *Work is done whenever energy is transferred from one object to another.*[†] Another example is when a person throws a ball or pushes a grocery cart. The work done is a manifestation of energy being transferred from the person (ultimately derived from the chemical energy of food) to the ball or cart.

[†] If the objects are at different temperatures, heat can flow between them instead, or in addition. See Chapters 11 and 12.

5-7 ▊ The law of conservation of energy

Whenever energy is transformed, it is found that no energy is gained or lost in the process. As an example, let us consider a stone that is allowed to fall toward the ground (Fig. 5-9) without air resistance. Before it is dropped, it has potential energy equal to mgy (we take the ground as the reference level). As it falls, its PE decreases but its KE increases. Just before hitting the ground, it has only kinetic energy. In fact, its KE at the bottom is exactly equal to the PE it had at the top. We can prove this by using Eq. 1-10c; since $v_0 = 0$ and $a = g$, we have $v^2 = 2gy$ where y is the distance fallen. Thus, the KE after falling a distance y is

$$\tfrac{1}{2}mv^2 = \tfrac{1}{2}m(2gy)$$
$$= mgy.$$

In other words, the original PE, equal to mgy, has been transformed entirely into KE, with the total amount neither increasing nor decreasing.

This example is an illustration of one of the most important principles in physics, the **law of conservation of energy:**

▊ **The total energy is neither increased nor decreased in any process. Energy can be transformed from one form to another, and transferred from one body to another, but the total amount remains constant.**

For mechanical systems involving conservative forces, this law can be derived from Newton's laws and thus is equivalent to them. But in its full generality, the validity of the law of conservation of energy rests on experimental observation, and was not proposed until the middle of the nineteenth century. Even though Newton's laws are found to fail in the submicroscopic world of the atom, the law of conservation of energy has been found to hold in every experimental situation so far tested.

We need a brief aside here to distinguish the use of the word "conservation" in everyday life from its specific use in physics. In everyday usage, conservation means "saving" or "using wisely"—such as when it is said that we must "conserve energy" (what is really meant is that we must conserve *fuel*). In physics the word conservation refers to a quantity that remains strictly *constant*. To be clear, we can say that energy *is* conserved (we don't have to try to conserve it), but we should try to conserve our fuel resources so that we don't use them up too quickly.

Also, it is important to recognize that energy is not a material entity. It is not substance or matter of any kind. Energy is a concept, an abstraction. The concept of energy was invented by human minds. Some physicists would say that energy is a mathematical function. This, of course, doesn't lessen its importance. That we can define a quantity which can take various forms (such as $\tfrac{1}{2}mv^2$ or mgy) but which always remains constant is very remarkable—and very useful, as we shall see.

Let us now consider some simple illustrations of the law of conservation of energy. For the moment we consider mechanical systems in which friction (and other nonconservative forces) can be neglected. In this case the law of conservation of energy can be stated in the following way:

$$KE + PE = \text{constant}. \qquad (5-8)$$

That is, the sum of the kinetic plus potential energies of an object or system of objects, which is called the *total mechanical energy*, remains constant. For example, in the case of the rock falling under gravity, Fig. 5-9, the rock initially has only potential energy. As it falls, its PE decreases, but its KE increases to compensate so that the sum of the two remains constant. At any point along the path, the total mechanical energy is given by $\frac{1}{2}mv^2 + mgy$ where y is its height above the ground at the point and v is its velocity at that point. If we let the subscript 1 represent the rock at one point along its path (for example, the initial point), and 2 represent it at some other point, then we can write

total energy at point 1 = total energy at point 2

or (see Eq. 5-8)

$$\tfrac{1}{2}mv_1^2 + mgy_1 = \tfrac{1}{2}mv_2^2 + mgy_2. \qquad (5-9)$$

To see the practical value of this relation, suppose the original height of the stone in Fig. 5-9 is $y_1 = 3.0$ m, and we wish to calculate its speed when it has fallen to 1.0 m above the ground. Then, since $v_1 = 0$ (the moment of release), $y_2 = 1.0$ m, and $g = 9.8$ m/s^2, Eq. 5-9 tells us

$$0 + (m)(9.8 \text{ m/s}^2)(3.0 \text{ m}) = \tfrac{1}{2}mv_2^2 + (m)(9.8 \text{ m/s}^2)(1.0 \text{ m}).$$

The m's cancel out, and solving for v_2^2, we find $v_2^2 = 2[(9.8 \text{ m/s}^2)(3.0 \text{ m}) - (9.8 \text{ m/s}^2)(1.0 \text{ m})] = 39.2 \text{ m}^2/\text{s}^2$, and $v^2 = \sqrt{39.2}$ m/s $= 6.3$ m/s.

Equation 5-9 can be applied to any object moving without friction under the action of gravity. For example, Fig. 5-10 shows a car starting from rest at the top of a hill, and coasting without friction to the bottom and up the hill on the other side. Initially, the car has only PE. As it coasts down the hill, it loses PE and gains in KE, but the sum of the two remains constant. At the bottom of the hill, it has its maximum KE and as it climbs up the other side, the KE changes back to PE. When the car comes to rest again, all of its energy will be PE. Since PE is proportional to height and because energy is conserved, the car comes to rest at a height equal to its original height. If the two hills are the same height, the car just barely reaches the top of the second hill when it stops. If the second hill is lower than the first, not all of the car's KE will be transformed to PE and the car continues over the top and down the other side. If the second hill is higher, the car will only reach a height on it equal to its original height on the first hill. This is true no matter how steep the hill is, since PE depends only on the vertical height.

FIGURE 5-9

The stone's PE changes to KE as it falls.

FIGURE 5-10

Car coasting down a hill illustrates conservation of energy.

□ EXAMPLE 5-5 Assuming the height of the hill in Fig. 5-10 is 40 m, calculate (a) the velocity of the car at the bottom of the hill and (b) at what height it will have half this speed.

SOLUTION (a) We use Eq. 5-9 with $v_1 = 0$, $y_1 = 40$ m, and $y_2 = 0$. Then

$$0 + (m)(9.8 \text{ m/s}^2)(40 \text{ m}) = \tfrac{1}{2}mv_2^2 + 0.$$

The m's cancel out and we find $v_2 = 28$ m/s.

FIGURE 5-11

Transformation of energy during a
pole vault.

(b) We use the same equation but now $v_2 = 14$ m/s and y_2 is unknown:

$$0 + (m)(9.8 \text{ m/s}^2)(40 \text{ m}) = \tfrac{1}{2}(m)(14 \text{ m/s})^2 + (m)(9.8 \text{ m/s}^2)(y_2).$$

We cancel the m's and solve for y_2 and find $y_2 = 30$ m. That is, the car has a speed of 14 m/s when it is 30 *vertical* meters above the lowest point, both when descending the left-hand hill and when ascending the right-hand hill.

There are many interesting examples of the conservation of energy in athletics, one of which is the pole vault illustrated in Fig. 5-11. In terms of energy, the sequence of events is as follows: KE of the running athlete is transformed into elastic PE of the bending pole, and as he leaves the ground, into gravitational PE; then, as he reaches the top and the pole straightens out again, it has all been transformed to gravitational PE. (We ignored his low speed as he travels over the bar.) The pole does not supply any energy, but it acts as a very convenient device to *store* energy and thus aid in the transformation of KE into gravitational PE, which is the net result. The energy required to pass over the bar depends on how high the center of mass[†] (cm) of the vaulter must be raised. By bending their bodies, pole vaulters can keep their cm so low that it actually passes slightly beneath the bar, thus enabling them to cross over a higher bar than would otherwise be possible.

☐   EXAMPLE 5-6   Calculate the kinetic energy and the velocity required for a 70-kg pole vaulter to pass over a 5.0-m-high bar. Assume the vaulter's center of mass is initially 0.90 m off the ground and reaches its maximum height at the level of the bar itself.

SOLUTION   We equate the total energy just before the vaulter places the end of the pole into the ground (and the pole begins to bend and store PE) with his total energy as he passes over the bar (we ignore the small amount of KE at this point). His body must be raised a height $y = 5.0$ m $- 0.9$ m $= 4.1$ m. Thus

$$\tfrac{1}{2}mv^2 + 0 = 0 + mgy$$

[†] The center of mass (cm) of a body is that point at which the entire mass of the body can be considered as concentrated for purposes of describing its translational motion—that is, so we can consider it as a particle. (This is discussed in Chapter 6.)

and

$$\text{KE} = \tfrac{1}{2}mv^2 = mgy$$

$$= (70 \text{ kg})(9.8 \text{ m/s}^2)(4.1 \text{ m}) = 2.8 \times 10^3 \text{ J}.$$

The velocity $v = \sqrt{\dfrac{2\text{KE}}{m}} = \sqrt{\dfrac{2(2800 \text{ J})}{70 \text{ kg}}} = 8.9 \text{ m/s}.$

Notice that the kinematic equations for constant acceleration, Eqs. 1-10, could have been used to treat the simple case of a falling rock (discussed immmediately after Eq. 5-9) but they could not be used to solve either Example 5-5 or 5-6 since the acceleration is not constant in these cases. Here we have seen the great power of using the conservation of energy principle. Even though this principle is equivalent to Newton's second law, $\mathbf{F} = m\mathbf{a}$, and can be derived from it[†], we have used the conservation of energy to solve problems without having to deal with force and acceleration which vary in a complicated way for these cases.

As another example of the conservation of mechanical energy let us consider a mass m connected to a spring whose own mass can be neglected and whose stiffness constant is k. The mass m has speed v at any moment and the potential energy of the system is $\tfrac{1}{2}kx^2$ where x is the displacement of the spring from its unstretched length. If neither friction nor any other force is acting, the conservation of energy principle tells us that

$$\tfrac{1}{2}mv_1^2 + \tfrac{1}{2}kx_1^2 = \tfrac{1}{2}mv_2^2 + \tfrac{1}{2}kx_2^2$$

where the subscripts 1 and 2 refer to the velocity and displacement at two different times.

☐ **EXAMPLE 5-7** A ball of mass $m = 2.6$ kg, starting from rest, falls a vertical distance $h = 55$ cm before striking a vertical coiled spring, which it compresses (see Fig. 5-12). If the spring has stiffness constant $k = 72$ N/m, what is the maximum compression of the spring? Measure all distances from the point where the ball first touches the uncompressed spring. Ignore the mass of the spring.

SOLUTION Since the motion is vertical, we use y instead of x (y positive upward). Let the maximum compression of the spring be called Y. Then initially, the total energy of the system is

$$E_1 = mgh.$$

When the spring has been compressed to its maximum, the total energy is

$$E_2 = \tfrac{1}{2}kY^2 - mgY.$$

The first term on the right is the elastic PE, the second term is the gravitational PE since the ball fell a distance h before striking the spring ($y = 0$ at the top of the unstretched spring) and then drops a further distance Y as the spring compresses ($y_2 = -Y$ where $Y > 0$). The gravitational PE is $-mgY$ at this point. At both points 1 and 2 the kinetic energy is zero, although at

FIGURE 5-12
Example 5-7.

[†] This is done via the work-energy theorem. (See Sections 5-3 and 5-5.)

intermediate points it is nonzero since the ball is moving. Because energy is conserved, $E_1 = E_2$, so

$$mgh = \tfrac{1}{2}kY^2 - mgY$$

or

$$\tfrac{1}{2}kY^2 - mgY - mgh = 0.$$

We use the quadratic formula (Appendix A-5) to find Y:

$$Y = \frac{+mg \pm \sqrt{m^2g^2 + 2mghk}}{k} = 1.1 \text{ m}.$$

We used the root with the plus sign since this gives $Y > 0$, as assumed. The root with the negative sign yields $Y = -0.36$ m which would correspond to the ball and spring, connected to each other, having sprung back up a distance of 0.36 m above the unstretched ($y = 0$) length.

This example illustrates the fact that when dealing with potential energy we must consider a system, and not a single particle, since potential energy does not exist for an isolated body: potential energy is associated with a force, and a force on one object is always exerted by some other object. The potential energy is thus a property of the system as a whole. For the system of a mass m on the end of a spring the potential energy increases by $\tfrac{1}{2}kx^2$ when the spring is stretched a distance x from equilibrium, and this is related to the force $F = -kx$ exerted by the spring on the mass. Likewise, when a particle is raised to a height y above the earth's surface, the potential energy change is mgy; the system here is the particle plus the earth.

We have neglected friction up to now, but in many situations it cannot be ignored. In a real situation the car in Fig. 5-10 for example, will not in fact reach the same height on the second hill as it had on the first hill because of friction. In this and in other natural processes the sum of the kinetic and potential energies does not remain constant but decreases. (Because of this, frictional forces are called *dissipative forces*.) Historically this fact hindered the formulation of a comprehensive conservation of energy law until well into the nineteenth century. It was not until then that heat, which is always produced when there is friction retarding the motion (try rubbing your hands together), was recognized as energy. Quantitative studies by nineteenth century scientists (this is discussed in Chapters 11 and 12) demonstrated that if heat is interpreted as energy (properly called *thermal energy*) then the total energy is conserved in any process. For example, if the car in Fig. 5-10 is subject to frictional forces, then the initial total energy of the car will be equal to the kinetic plus potential energy of the car at any subsequent point along its path plus the amount of thermal energy produced in the process. The thermal energy produced by a friction force F_{fr} is equal to the work done against this force: $W = F_{fr}d$, where d is the distance over which the force acts. Equation 5-9 can then be rewritten to take into account friction:

$$\tfrac{1}{2}mv_1^2 + mgy_1 = \tfrac{1}{2}mv_2^2 + mgy_2 + F_{fr}d$$

where d is the actual distance along the path traveled by the object in going from point 1 to point 2.

EXAMPLE 5-8   The car in Example 5-5 is found to reach a vertical height of only 25 m on the second hill before coming to a stop. It traveled a total distance of 400 m. Calculate the average friction force on the car. Assume the car has a mass of 1000 kg.

SOLUTION   We take point one to be the instant when the car started coasting and point two when it stopped. Then $v_1 = 0$, $y_1 = 40$ m, $v_2 = 0$, $y_2 = 25$ m, and $d = 400$ m. Thus

$$0 + (1000 \text{ kg})(9.8 \text{ m/s}^2)(40 \text{ m})$$
$$= 0 + (1000 \text{ kg})(9.8 \text{ m/s}^2)(25 \text{ m}) + F_{\text{fr}}(400 \text{ m}).$$

We solve this for F_{fr} to find:

$$F_{\text{fr}} = 370 \text{ N}.$$

Another example of the transformation of kinetic energy into thermal energy occurs when an object, such as the rock of Fig. 5-9, strikes the ground. Both the rock and the ground will be slightly warmer as a result of their collision. A more apparent example of this transformation of kinetic energy into heat can be observed by vigorously striking a nail several times with a hammer and then gently touching the nail with your finger.

When other forms of energy are involved, such as chemical or electrical energy, the total amount of energy is always found to be conserved. Hence the law of conservation of energy is believed to be universally valid.

5-8 ∎ Power

Power is defined as the rate at which work is done or the rate at which energy is transformed:

$$\text{power} = \frac{\text{work}}{\text{time}} = \frac{\text{energy transformed}}{\text{time}}.$$

The power of a horse refers to how much work it can do per unit time; the power rating of an engine refers to how much chemical or electrical energy can be transformed into mechanical energy per unit time. In SI units, power is measured in joules per second and this is given a special name, the *watt* (W): 1 W = 1 J/s. We are most familiar with the watt for measuring the rate at which an electric light bulb or heater changes electric energy into light or heat energy, but it is used for other types of energy transformation as well. In the British system, the unit of work is the ft·lb/s. For practical purposes a larger unit is often used, the *horsepower*. One (British) horsepower[†] (hp) is defined as 550 ft·lb/s, which equals 746 W. In the metric system, 1 hp is

---

[†] The unit was first chosen by James Watt (1736–1819) who needed a way to specify the power of his newly developed steam engines. He found by experiment that a good horse can work all day at an average rate of 360 ft·lb/s. So as not to be accused of exaggeration in the sale of his steam engines, he multiplied this by 1.5 when he defined the hp.

defined as 750 W. The metric and British horsepower units differ by less than one percent.

To see the distinction between energy and power, consider the following example. A person is limited in the work he can do not only by the total energy required, but also by the rate this energy is transformed; that is, by power. For example a person may be able to walk a long distance or climb many flights of stairs before having to stop because he has used up so much energy. On the other hand, if a person runs very quickly upstairs, he may fall exhausted after only a flight or two. He is limited in this case by power, the rate at which his body can transform chemical energy into mechanical energy.

EXAMPLE 5-9   A 70-kg man runs up a long flight of stairs in 4.0 s. The vertical height of the stairs is 4.5 m. Calculate his power output in watts and horsepower.

SOLUTION   The work done is against gravity. Hence the power P is

$$P = \frac{mgy}{t} = \frac{(70 \text{ kg})(9.8 \text{ m/s}^2)(4.5 \text{ m})}{4.0 \text{ s}} = 770 \text{ W}.$$

Since there are 750 W in 1 hp, the man is doing work at a rate of just over 1 hp. It is worth noting that a human cannot do work at this rate for very long.

Automobiles do work to overcome the force of friction (and air resistance), to climb hills, and to accelerate. A car is limited by the rate it can do work, which is why automobile engines are rated in horsepower. A car needs power most when it is climbing hills and when accelerating (as for passing). In the next example we will calculate how much power is needed in these situations for a car of reasonable size. Even when a car travels on the level, it needs some power for it must do work to overcome friction and air resistance; these forces depend on the conditions but are usually in the region of 400–1000 N.

It is often convenient to write the power in terms of the net force F applied to an object and its speed v. This is readily done since $P = W/t$ and $W = Fd$ where d is the distance traveled. Then

$$P = \frac{W}{t} = \frac{Fd}{t}$$

$$P = Fv \tag{5-10}$$

where $v = d/t$ is speed of the object.

EXAMPLE 5-10   Calculate the power required of a 1400-kg car under the following circumstances: (a) the car climbs a 10° hill at a steady 80 km/h and (b) the car accelerates from 90 to 110 km/h in 6.0 s to pass another car on a level road. Assume the force of friction on the car is 700 N in both parts of the problem. See Fig. 5-13.

FIGURE 5-13

Calculation of power needed for a car to climb a hill (Example 5-10).

SOLUTION   (*a*) To move at a steady speed up the hill, the car must exert a force equal to the sum of the friction force, 700 N, and the component of gravity parallel to the hill, $mg \sin 10° = (1400 \text{ kg})(9.8 \text{ m/s}^2)(0.174) = 2400$ N. Since $v = 80 \text{ km/h} = 22 \text{ m/s}$,

$$P = Fv$$
$$= (2400 \text{ N} + 700 \text{ N})(22 \text{ m/s}) = 6.8 \times 10^4 \text{ W}$$
$$= 91 \text{ hp}.$$

(*b*) The car accelerates from 25.0 m/s to 30.6 m/s (90 to 110 km/h). Thus the car must exert a force that overcomes the 700 N friction force plus that required to give it the acceleration $a = (30.6 \text{ m/s} - 25.0 \text{ m/s})/6.0 \text{ s} = 0.93 \text{ m/s}^2$. Since the mass of the car is 1400 kg, the force required for acceleration is $F = ma = (1400 \text{ kg})(0.93 \text{ m/s}^2) = 1300$ N, and the total force required is then 2000 N. Since the force is constant, the power required is greatest when the velocity v is greatest (Eq. 5–10). Thus the maximum needed power is

$$P = (2000 \text{ N})(30.6 \text{ m/s})$$
$$= 6.12 \times 10^4 \text{ W}$$
$$= 82 \text{ hp}.$$

Even taking into account the fact that only 60 to 80 percent of the engine's power reaches the wheels, it is clear from these calculations that an engine of 100 to 150 hp is quite adequate[†] from a practical point of view even in a reasonably heavy car. In view of fuel requirements and environmental damage, larger engines are becoming a luxury that society is questioning.

∎ Summary

Work is done on an object by a force when the force moves the object through a distance, *d*. If the direction of a constant force makes an angle θ with the direction of motion, the work done by this force is $W = Fd \cos \theta$.

Energy can be defined as the ability to do work. Both work and energy are measured in joules (1 J = 1 N·m) in SI units. *Kinetic energy* (KE) is energy of motion; a body of mass *m* and speed *v* has translational KE equal to $\frac{1}{2}mv^2$. An object can have *potential energy* (PE) by virtue of its position or shape. Examples are gravitational PE (equal to *mgy* where *y* is the height of the object of mass *m* above an arbitrary reference point), elastic PE (such as in a compressed spring), as well as chemical, electrical, and nuclear energy. The change in potential energy of an object when it changes position is defined as the work needed to take it from one position to the other. The *work-energy theorem* states that the *net* work done on a body (by the *net* force) equals the change in kinetic energy of that body.

The law of *conservation of energy* states that energy can be transformed from one type to another, but the total energy remains constant. It is valid even when friction is present since the heat generated by friction can be considered a form of energy.

Power is defined as the rate at which work is done, or the rate at which energy is transformed. The SI unit of power is the watt (1 W = 1 J/s).

[†] But note that horsepower ratings given by manufacturers are often higher than the real available horsepower.

▌Questions

1. In what ways is the word "work" as used in everyday language the same as defined in physics? In what ways is it different?

2. Can a centripetal force ever do work? Explain.

3. Can the normal force on an object ever do work? Explain.

4. You have two springs that are identical except that spring 1 is stiffer than spring 2 $(k_1 > k_2)$. On which spring is more work done (a) if they are stretched using the same force, (b) if they are stretched the same distance?

5. Can kinetic energy ever be negative? Explain.

6. Does the net work done on a particle depend on the choice of reference frame? How does this affect the work-energy theorem?

7. By approximately how much does your gravitational potential energy change when you jump as high as you can?

8. Describe the energy transformations when a child hops around on a pogo stick.

9. Describe the energy transformations that take place when a skier starts skiing down a hill, but after a time is brought to rest by striking a snowdrift.

10. The net force acting on a particle is conservative and increases its KE by 300 J. What is the change in (a) the potential energy, (b) the total energy, of the particle?

11. Does the following statement make sense? "Fiberglass poles led to higher pole vaults because the additional potential energy of bending was converted to gravitational potential energy." Explain.

12. An inclined plane has a height h. A body of mass m is released from rest at the top. Does its velocity at the bottom depend on the angle of the plane if (a) there is no friction, (b) there is friction?

13. Why is it tiring to push hard against a solid wall even though no work is done?

14. Analyze the motion of a simple swinging pendulum in terms of energy, (a) ignoring friction, and (b) taking it into account. Explain why a grandfather clock has to be wound up.

15. When a "superball" is dropped, can it rebound to a greater height than its original height?

16. When an experienced hiker comes to a log across his path, he prefers to step over it rather than surmounting it and jumping down on the other side. Why?

17. A rock thrown with a certain speed from the top of a cliff will enter the water below with a speed that is the same whether the rock is thrown horizontally or at any angle. Explain.

18. Suppose you lift a suitcase from the floor to a table. Does the work you do on the suitcase depend on (a) whether you lift it straight up or along a more complicated path, (b) the time it takes, (c) the height of the table, (d) the weight of the suitcase?

19. Repeat the previous question for the *power* needed rather than the work.

20. Why is it easier to climb a mountain via a zigzag trail rather than to climb straight up?

▌Problems

SECTION 5-1

1. (I) A 50-kg woman climbs a flight of stairs 6.0-m high. How much work is required?

2. (I) A 550-N crate rests on the floor. How much work is required to move it at constant speed (a) 2.0 m along the floor against a friction force of 150 N, and (b) 2.0 m vertically?

3. (I) How much work did a horse do that pulled a 200-kg wagon 80 km without acceleration along a level road if the effective coefficient of friction was 0.060?

4. (I) How far must a 200-kg pile driver fall if it is to be capable of doing 13,000 J of work?

5. (II) What is the minimum work needed to push a 1000-kg car 45.0 m up a 12.5° incline? (a) Ignore friction. (b) Assume the effective coefficient of friction is 0.30.

6. (II) Eight bricks, each 6.0 cm thick with mass 1.5 kg lie flat on a table. How much work is required to stack them one on top of another?

7. (II) In pedaling a bicycle, a particular cyclist exerts a downward force of 90 N during each stroke. If the diameter of the circle traced by each pedal is 36 cm, calculate how much work is done in each stroke.

8. (II) Calculate the net work done on a helicopter of mass M that is accelerating upward at $0.10\,g$ for a distance h.

9. (III) A 300-kg piano slides at constant speed 4.5 m down a 25° incline; it is kept from accelerating by a man who is pushing back on it. The effective coefficient of friction is 0.39. Calculate (a) the net work done on the piano, (b) the work done by the man on the piano, (c) the work done by gravity on the piano.

10. (III) Assume a cyclist of weight w can exert a force on the pedals equal to $0.90w$ on the average. If the pedals rotate in a circle of radius 18 cm, the wheels have a radius of 34 cm, and the front and back sprockets on which the chain runs have 52 and 17 points respectively, determine the maximum steepness of hill the

cyclist can climb. Assume the mass of the bike is 12 kg and of the rider is 60 kg. Ignore friction.

*SECTION 5-2

*11. (II) In Fig. 5-4a, assume the distance axis is linear and that $d_a = 5.0$ m and $d_b = 32.0$ m. Estimate the work done by this force in moving a 20-kg object from d_a to d_b.

*12. (II) A spring has $k = 74$ N/m. Use a graph to determine the work needed to stretch it from $x = 3.0$ cm to $x = 5.5$ cm where $x = 0$ refers to its unstretched length.

*13. (II) The force exerted on a particle increases linearly from zero at $x = 0$ to 24.0 N at $x = 3.0$ m. It remains constant at 24.0 N from $x = 3.0$ m to $x = 8.0$ m and then decreases linearly to zero at $x = 11.0$ m. Determine the work done to move the particle from $x = 0$ to $x = 11.0$ m graphically by determining the area under the F vs. x graph.

*14. (III) A 1000-kg space vehicle falls vertically from a height of 2000 km above the earth's surface. Determine how much work is done by the force of gravity in bringing the vehicle to the earth's surface. (First construct an F vs. r graph, where r is the distance from the earth's center; then determine the work graphically.)

SECTION 5-3

15. (I) What is the initial KE of a 0.50-mg flea that leaves the ground at a speed of 100 cm/s? (1 mg = 10^{-3} gram.)

16. (I) A carbon atom of mass 1.99×10^{-26} kg has 4.64×10^{-19} J of kinetic energy. How fast is it moving?

17. (I) How much work must be done to stop a 1000-kg car traveling at 100 km/h?

18. (II) A 1250-kg car rolling on a horizontal surface has speed $v = 40$ km/h when it strikes a horizontal coiled spring and is brought to rest in a distance of 2.5 m. What is the spring constant of the spring? Ignore nonconservative forces such as friction.

19. (II) A baseball ($m = 140$ g) traveling 30 m/s moves a fielder's glove backward 35 cm when the ball is caught. What was the average force exerted by the ball on the glove?

20. (II) If the speed of a car is doubled, by what factor will its minimum braking distance be increased, assuming all else is the same? Ignore the driver's reaction time.

21. (III) One car has twice the mass of a second car, but only half as much KE. When both cars increase their speed by 3.0 m/s, they then have the same KE. What were the original speeds of the two cars?

22. (III) A 130-kg load is lifted 30 m vertically by a single cable with an acceleration $a = 0.15\,g$. Determine (a) the tension in the cable, (b) the net work done on the load, (c) the work done by the cable on the load, (d) the work done by gravity on the load, (e) the final speed of the load assuming it started from rest.

23. (III) An elevator cable breaks when a 750-kg elevator is 25 m above a huge spring ($k = 4.0 \times 10^4$ N/m) at the bottom of the shaft. Calculate (a) the work done by

gravity on the elevator before it hits the spring, (b) the speed of the elevator just before striking the spring, (c) the amount the spring compresses (note that work is done by both the spring and gravity in this part).

SECTIONS 5-4 AND 5-5

24. (I) A spring has a spring constant, k, of 320 N/m. How much must this spring be compressed to store 50 J?

25. (I) A 4.2-kg monkey swings from one branch to another 1.7 m higher. What is the change in its potential energy?

26. (II) A 1.80-m tall person lifts a 230-g book so it is 2.15 m off the ground. What is the potential energy of the book relative to (a) the ground, (b) the top of the person's head? (c) How is the work done by the person related to the answers in parts (a) and (b)?

27. (II) A 65-kg hiker starts at an elevation of 1500 m and climbs to the top of a 2600-m peak. (a) What is the hiker's change in potential energy? (b) What is the minimum work required of the hiker? (c) Can the actual work done be more than this? Explain.

28. (II) (a) A spring of spring constant k is initially compressed a distance x_0 from its normal length. What is the change in potential energy if it is now compressed by an amount x from equilibrium? (b) The spring is now *stretched* a distance x_0 from the normal length. What is the change in potential energy as compared to when it is compressed by an amount x_0 from normal length?

SECTION 5-7

29. (I) Tarzan is running at top speed (8.0 m/s) and grabs a vine hanging vertically from a tall tree in the jungle. How high can he swing upward? Does the length of the vine (or rope) affect your answer?

30. (I) An object slides down a frictionless 33.0° incline whose vertical height is 14.0 cm. How fast is it going when it reaches the bottom?

31. (II) In the high jump, the KE of an athlete is transformed into gravitational potential energy without the aid of a pole. With what minimum speed must the athlete leave the ground in order to lift his center of mass 2.20 m and cross the bar with a speed of 0.80 m/s?

32. (II) A projectile is fired at an upward angle of 45.0° from the top of a 380-m cliff with a speed of 500 m/s. What will be its speed when it strikes the ground at the base of the cliff? (Use conservation of energy.)

33. (II) A spring whose spring constant is 850 N/m is compressed 0.40 m. What speed can it give to a 0.30-kg ball when released? Ignore mass of the spring.

34. (II) A mass m is attached to the end of a spring of spring constant k, Fig. 5-14. The mass is given an initial displacement x_0, after which it oscillates back and forth. Write down a formula for the total mechanical energy (ignore friction and mass of the spring) in terms of position x and speed v.

FIGURE 5-14

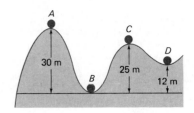

FIGURE 5-16

35. (II) A 24-kg child descends a slide 5.0-m high and reaches the bottom with a speed of 2.8 m/s. How much thermal energy due to friction was generated in this process?

36. (II) Two railroad cars, each of mass 4200 kg and traveling 80 km/h, collide head-on and come to rest. How much thermal energy is produced in this collision?

37. (II) A 0.25-kg pine cone falls from a branch 20 m above the ground. (*a*) With what speed would it hit the ground if air resistance could be ignored? (*b*) If it actually hits the ground with a speed of 9.0 m/s, what was the average force of air resistance exerted on it?

38. (II) A vertical spring (ignore its mass) whose spring constant is 850 N/m stands on a table and is compressed 0.400 m. (*a*) What speed can it give to a 0.300-kg ball when released? (*b*) How high above its original position (spring compressed) will the ball fly?

39. (II) A ball is attached to a horizontal cord of length L whose other end is fixed, Fig. 5-15. (*a*) If the ball is released, what will be its speed at the lowest point of its path? (*b*) A peg is located a distance h directly below the point of attachment of the cord. If $h = 0.75L$, what will be the speed of the ball when it reaches the top of its circular path about the peg?

40. (II) A skier traveling 12.6 m/s reaches the foot of a steady upward 20° incline and glides 11.4 m up (along) the incline before coming to rest. What was the average coefficient of friction?

41. (II) A roller coaster is shown in Fig. 5-16. Assuming no friction, calculate the speed at points B, C, D, assuming it has a speed of 2.10 m/s at point A.

42. (II) The roller coaster in Fig. 5-16 passes point A with

FIGURE 5-15

a speed of 1.10 m/s. If the average force of friction is equal to one-fifth of its weight, with what speed will it reach point B? The distance traveled is 67.0 m.

43. (III) A cyclist intends to cycle up a 12.0° hill 100 m high. Assuming the mass of bicycle plus person is 78.0 kg, (*a*) calculate how much work must be done against gravity; (*b*) if each complete revolution of the pedals moves the bike 5.10 m along its path, calculate the average force that must be exerted on the pedals tangent to their circular path. Neglect friction and other losses. The pedals turn in a circle of diameter 36.0 cm.

44. (III) A small mass m slides without friction along the looped apparatus shown in Fig. 5-17. If the object is to remain on the track, even at the top of the circle (whose radius is r), from what minimum height h must it be released?

45. (III) An engineer is designing a spring to be placed at the bottom of an elevator shaft; if the elevator cable should break at a height h above the top of the spring, calculate the required value of the spring constant, k, so that passengers undergo an acceleration of no more than $10\,g$ when brought to rest. Let M be the total mass of the elevator and passengers.

46. (III) A 200-g wood block is firmly attached to a horizontal spring, Fig. 5-14. The block can slide along a table where the coefficient of friction is 0.40. A force of 10 N compresses the spring 18 cm. If the spring is released from this position, how far beyond its equilibrium position will it stretch on its first swing?

SECTION 5-8

47. (I) If a car generates 15 hp when traveling at a steady 80 km/h, what must be the average force exerted on the car due to friction and air resistance?

48. (I) How long will it take a 1500-W motor to lift a 400-kg piano to a sixth-story window 32.0 m above?

49. (I) Show that a British horsepower is equal to 746 W. What is the percent difference between this unit and the metric hp?

FIGURE 5-17

50. (I) Electric energy is often stated in kilowatt-hours. Show that the kilowatt-hour (kWh) is a unit of energy and is equal to 3.60×10^6 J.

51. (II) A 1200-kg car slows down from 90 km/h to 70 km/h in about 5.0 s on the level when it is in neutral. Approximately what power (watts and hp) is needed to keep the car traveling at a constant 80 km/h?

52. (II) A shotputter accelerates a 7.3-kg shot from rest to 15 m/s. If this motion takes 2.0 s, what average power was developed?

53. (II) How much work can a 2.0-hp motor do in 1.0 h?

54. (II) A 70-kg hiker climbs to the top of a 4200-m high mountain. The climb is made in 4.0 h starting at an elevation of 3200 m. Calculate (a) the work done against gravity, (b) the average power output in watts and in horsepower, and (c) assuming the body is 15 percent efficient, what rate of energy input was required.

55. (II) How fast must a cyclist climb a 12.5° hill to maintain a power output of 0.23 hp? Ignore friction and assume the mass of cyclist plus bicycle is 85 kg.

56. (II) A pump is to lift 4.10 kg of water per minute through a height of 2.85 m. What output rating (watts) should the pump motor have?

57. (II) A bicyclist coasts down a 6.6° hill at a steady speed of 8.0 km/h. Assuming a total mass of 75-kg (bicycle plus rider), what must be the cyclist's power output to climb the same hill at the same speed?

58. (II) A 1000-kg car has a maximum power output of 100 hp. How steep a hill can it climb at a constant speed of 60 km/h if the frictional forces add up to 500 N?

59. (III) Water flows over a dam at the rate of 800 kg/s and falls vertically 150 m before striking the turbine blades. Calculate (a) the velocity of the water just before striking the turbine blades, and (b) the rate at which mechanical energy is transferred to the turbine blades. Assume the water loses 80 percent of its speed when striking the blades, and that 12 percent of the initial energy is transformed to thermal energy.

60. (III) A bicyclist of mass 80 kg (including the bicycle) can coast down a 3.4° hill at a steady speed of 6.0 km/h. Pumping hard, the cyclist can descend the hill at a speed of 30 km/h. Using the same power, at what speed can the cyclist climb the same hill? Assume the force of friction is directly proportional to the speed v; that is, $F_{fr} = bv$, where b is a constant.

6 LINEAR MOMENTUM

The law of conservation of energy, which we discussed in the previous chapter, is one of several great conservation laws in physics. Among the other quantities found to be conserved are linear momentum, angular momentum, and electric charge. We will eventually discuss all of these since the conservation laws are among the most important in all of science. In this chapter we discuss linear momentum (sometimes simply called "momentum") and its conservation; we will also make use of the laws of conservation of linear momentum and of energy to analyze collisions. Indeed, the law of conservation of momentum is particularly useful when dealing with two or more bodies that interact with each other, such as in collisions. Our focus up to now has been mainly on the motion of a single particle. But in this chapter we will deal with systems[†] of two or more particles (and with extended bodies which can be considered as collections of particles). An important idea for this study is that of center of mass, which we discuss later in the chapter.

6-1 ∎ Momentum and its relation to force

The **linear momentum** (or "momentum" for short) of a body is defined as the product of its mass and its velocity. Momentum is usually represented by the symbol **p**. If we let m represent the mass of a body and **v** represent its velocity, then its momentum, **p**, is

$$\mathbf{p} = m\mathbf{v}. \tag{6-1}$$

Since velocity is a vector, momentum is a vector. The direction of the momentum is the direction of the velocity; and the magnitude of the momentum, p, at any moment is equal to the numerical product of the mass m times the instantaneous velocity v. Since **v** depends on the reference frame, this frame must be specified. The unit of momentum is just that of mass × velocity, which in SI units is kg·m/s.

Everyday usage of the term *momentum* is in accord with the definition above. For according to Eq. 6-1, a fast-moving car has more momentum than a slow-moving car of the same mass; and a heavy truck has more

---

[†] A *system* is any body or group of bodies we choose to deal with.

momentum than a small car moving with the same speed. The more momentum an object has, the harder it is to stop it, and the greater effect it will have if it is brought to rest by impact or collision. A football player is more likely to be stunned if tackled by a heavy opponent running at top speed than by a lighter or slower-moving tackler. A heavy fast-moving truck can do more damage than a slow-moving small car.

A force is required to change the momentum of an object whether it is to increase the momentum, decrease it (such as to bring a moving object to rest), or to change its direction. Newton in fact originally stated his second law in terms of momentum (although he called the product mv the "quantity of motion"). Newton's statement of the second law, translated into modern language, is as follows: "the rate of change of momentum of a body is proportional to the net force applied to it." For a constant force we can write this as:

$$\mathbf{F} = \frac{\Delta \mathbf{p}}{\Delta t}, \qquad\qquad (6\text{-}2)$$

where $\mathbf{F}$ is the net force applied to the object and $\Delta \mathbf{p}$ is the resulting momentum change that occurs during the time interval Δt. The familiar form of the second law, $\mathbf{F} = m\mathbf{a}$, can readily be derived from Eq. 6-2, for the case of constant acceleration, using Eq. 1-10a now written in vector form:

$$\mathbf{v} = \mathbf{v}_0 + \mathbf{a}t;$$

to prove this, we assume an object of mass m is accelerated uniformly (although the result is valid even if it is not uniform) from a velocity $\mathbf{v}_0$ to velocity $\mathbf{v}$ in a time interval Δt. We start with Eq. 6-2, with t replacing Δt:

$$\mathbf{F} = \frac{\Delta \mathbf{p}}{t} = \frac{m\mathbf{v} - m\mathbf{v}_0}{t} = \frac{m}{t}(\mathbf{v} - \mathbf{v}_0).$$

We then substitute in Eq. 1-10a and obtain the desired result:

$$\mathbf{F} = \frac{m}{t}(\mathbf{a}t)$$

$$\mathbf{F} = m\mathbf{a}.$$

Newton's statement, Eq. 6-2, is actually more general than the more familiar one since it includes the situation in which the mass may change. This is important in certain situations such as rockets which lose mass as they burn fuel. Another interesting example is the following.

◻   EXAMPLE 6-1   Water leaves a hose at a rate of 5.0 kg/s with a speed of 50 m/s. It strikes a wall, which stops it. (That is, we ignore any splashing back.) What is the force exerted by the water on the wall?

SOLUTION   In each second, water with a momentum of (5.0 kg) (50 m/s) = 250 kg·m/s is brought to rest. The magnitude of the force (assumed constant) required to change the momentum by this amount is

$$F = \frac{\Delta p}{\Delta t} = \frac{0 - 250 \text{ kg·m/s}}{1.0 \text{ s}} = -250 \text{ N}.$$

The minus sign indicates that the force on the water is opposite to its original velocity. The wall exerts a force of 250 N to stop the water, so by Newton's third law, the water exerts a force of 250 N on the wall.

6-2 ■ Conservation of momentum

The real importance of the concept of momentum lies in the fact that, under certain circumstances, it is a conserved quantity. In the midseventeenth century, shortly before Newton's time, it had been observed that the sum of the momenta of two colliding objects remains constant. Consider, for example, the head-on collision of two billiard balls shown in Fig. 6-1. Although the momentum of each of the two balls changes as a result of the collision, the *sum* of their momenta is found to be the same before as after the collision. If $m_1\mathbf{v}_1$ is the momentum of ball number one and $m_2\mathbf{v}_2$ the momentum of ball two, both measured before the collision, then the total momentum of the two balls before the collision is $m_1\mathbf{v}_1 + m_2\mathbf{v}_2$. After the collision, the balls each have a different velocity and momentum which we will designate by a "prime" on the velocity: $m_1\mathbf{v}_1'$ and $m_2\mathbf{v}_2'$. The total momentum after the collision is $m_1\mathbf{v}_1' + m_2\mathbf{v}_2'$. No matter what the velocities and masses involved are, it is found that the total momentum before the collision is the same as afterwards whether the collision is head-on or not:

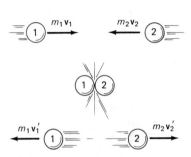

FIGURE 6-1

Momentum is conserved in collision of two balls.

$$\text{momentum before} = \text{momentum after}$$

$$m_1\mathbf{v}_1 + m_2\mathbf{v}_2 = m_1\mathbf{v}_1' + m_2\mathbf{v}_2'. \tag{6-3}$$

That is, the total momenta of the two balls is conserved: it stays constant.

Although the conservation of momentum principle was discovered experimentally, it is closely connected to Newton's laws of motion and can in fact be derived from them. We will do the derivation for the one-dimensional case illustrated in Fig. 6-1. We use Newton's second law as expressed in Eq. 6-2 and rewrite it by multiplying both sides by Δt (since the motion is in one dimension, we don't need vector notation):

$$F\,\Delta t = \Delta p. \tag{6-4}$$

We apply this to ball 1 in Fig. 6-1 letting v_1 be its velocity before collision and v_1' its velocity after the collision:

$$F\,\Delta t = m_1 v_1' - m_1 v_1.$$

In this relation, F is the force ball 2 exerts on ball 1, and Δt is the time the two balls are in contact during collision. When we apply Eq. 6-4 to ball 2, we note that by Newton's third law the force on ball 2 due to ball 1 is $-F$ so we have

$$-F\,\Delta t = m_2 v_2' - m_2 v_2.$$

Combining these last two equations we have

$$m_1 v_1' - m_1 v_1 = -(m_2 v_2' - m_2 v_2)$$

which tells us that any momentum lost by one ball is gained by the other. Thus the total momentum remains constant. This equation is readily re-arranged to give Eq. 6-3.

This derivation can be extended to include any number of interacting bodies. To show this in a simple way, we let **p** in Eq. 6-2 represent the total momentum of a system (that is, the vector sum of the momenta of all objects in the system). Then, if the net force **F** on the system is zero, $\Delta\mathbf{p} = 0$ so the total momentum doesn't change. Thus the general statement of the **law of conservation of momentum** is:

■ **The total momentum of an isolated system of bodies remains constant.**

By a *system* we simply mean a set of objects that interact with each other. An *isolated* system is one in which the only forces present are those between the objects of the system; that is, there is no net external force. If external forces do act—that is, forces exerted by objects outside the system—then the momentum may not be conserved. However, if the system is redefined so as to include the other objects, then the conservation of momentum principle will apply. For example, if we take as our system a falling rock, it does not conserve momentum since an external force, the force of gravity exerted by the earth, is acting on it. However, if we include the earth in the system, the total momentum of rock plus earth is conserved. (This of course means that the earth comes up to meet the ball; however, since the earth's mass is so great, its upward velocity is very tiny.)

> **EXAMPLE 6-2**   A 10,000-kg railroad car traveling at a speed of 24.0 m/s strikes an identical car at rest. If the cars lock together as a result of the collision, what is their common speed afterward?
>
> SOLUTION   The initial total momentum is
>
> $$m_1 v_1 + m_2 v_2 = (10{,}000 \text{ kg})(24.0 \text{ m/s}) + (10{,}000 \text{ kg})(0 \text{ m/s})$$
> $$= 2.40 \times 10^5 \text{ kg·m/s}.$$
>
> After the collision, the total momentum will be the same but it will be shared by both cars; since the two cars become attached, they will have the same velocity, call it v'. Then:
>
> $$(m_1 + m_2)v' = 2.40 \times 10^5 \text{ kg·m/s}$$
> $$v' = \frac{2.40 \times 10^5 \text{ kg·m/s}}{2.00 \times 10^4 \text{ kg}} = 12 \text{ m/s}.$$

The law of conservation of momentum is particularly useful when we are dealing with fairly simple systems such as collisions and certain types of ex-plosions. For example rocket propulsion, which we saw in Chapter 3 can be understood on the basis of action and reaction, can also be explained on the basis of the conservation of momentum: before a rocket is fired, the total mo-mentum of rocket plus fuel is zero; as the fuel burns, the total momentum re-mains unchanged, but the backward momentum of the expelled gases is just balanced by the forward momentum of the rocket itself. Similar examples are the recoil of a gun and the throwing of a package from a boat.

□   **EXAMPLE 6-3**   Calculate the recoil velocity of a 4.0-kg rifle which shoots a 0.050-kg bullet at a speed of 280 m/s.

SOLUTION   The total momentum of the system is conserved. We let the subscripts B represent the bullet and R the rifle; the final velocities are indicated by primes. Then

$$m_B v_B + m_R v_R = m_B v_B' + m_R v_R'$$

$$0 + 0 = (0.050 \text{ kg})(280 \text{ m/s}) + (4.0 \text{ kg})(v_R')$$

$$v_R' = -\frac{(0.050 \text{ kg})(280 \text{ m/s})}{(4.0 \text{ kg})}$$

$$= -3.5 \text{ m/s}.$$

Since the rifle has a much larger mass, its velocity is much less than that of the bullet. The minus sign indicates that the velocity (and momentum) of the rifle is in the opposite direction to that of the bullet. Notice that it is the *vector sum* of the momenta that is conserved.

6-3 ▎ Collisions and impulse

Conservation of momentum is a very useful tool for dealing with collision processes, as we already saw in the examples of the previous section. Collisions are a common occurrence in everyday life: a tennis racket striking a tennis ball, a baseball bat or golf club striking a ball, two billiard balls colliding, one railroad car striking another, a hammer hitting a nail. At the atomic and subatomic level, scientists learn about the structure of atoms and nuclei, and about the nature of the forces involved, by careful study of collisions between atoms and nuclei.

In a collision of two ordinary objects, both objects are deformed, often considerably, because of the strong forces involved, Fig. 6-2. When the collision occurs, the force normally jumps from zero at the moment of contact to a very large value within a very short time, and then abruptly returns to zero again. A graph of the magnitude of the force one object exerts on the other during a collision, as a function of time, is typically like that shown in Fig. 6-3. The time interval Δt is usually very distinct and usually very small.

From Newton's second law, Eq. 6-2, the *net* force on an object is equal to the rate of change of its momentum:

$$\mathbf{F} = \frac{\Delta \mathbf{p}}{\Delta t}.$$

This equation applies, of course, to *each* of the objects in a collision. If we multiply both sides of this equation by the time interval Δt we obtain

$$\mathbf{F} \, \Delta t = \Delta \mathbf{p}. \tag{6-5}$$

The quantity on the left, the product of the force **F** times the time Δt over which the force acts, is called the **impulse.** We see that the total change in momentum is equal to the impulse. The concept of impulse is of most help when dealing with forces that act over a short time, such as when a bat hits a baseball. The force is generally not constant and often its variation in time

FIGURE 6-2

Tennis racket striking a ball. Note deformation of both ball and racket due to the large force each exerts on the other. (Photograph by Russ Kinne, Photo Researchers, Inc.)

FIGURE 6-3

Force as a function of time during a typical collision.

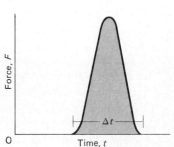

is like that graphed in Fig. 6-3. It is often sufficient to approximate this by an average force $\bar{F}$ acting over a time Δt, as indicated by the dashed line in Fig. 6-4.

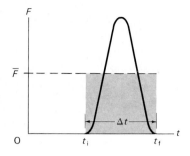

FIGURE 6-4

The average force $\bar{F}$ over Δt gives the same impulse ($\bar{F} \Delta t$) as the actual force.

□ EXAMPLE 6-4 (a) Calculate the impulse suffered when a 70-kg person lands on firm ground after jumping from a height of 5.0 m. Then estimate the average force exerted on the person's feet by the ground if the landing is (b) stiff-legged, and (c) with bent legs. In the former case, assume the body moves 1.0 cm during impact and in the second case, when the legs are bent, about 50 cm.

SOLUTION (a) Although we don't know F and thus can't calculate the impulse $F \Delta t$ directly, we can use the fact that the impulse equals the change in momentum of the object. After a fall of 5.0 m, the velocity of the person will be (using the appropriate kinematics equation for constant acceleration, Eq. 1-10c, in the vertical (y) direction, with $a = g = 9.8$ m/s^2, and initial velocity $v_0 = 0$):

$$v = \sqrt{2a(y - y_0)} = \sqrt{2(9.8 \text{ m/s}^2)(5.0 \text{ m})} = 9.9 \text{ m/s}.$$

After striking the ground, the momentum is quickly brought to zero, Fig. 6-5. The impulse on the person is then

$$\bar{F} \Delta t = \Delta p = p - p_0$$
$$= 0 - (70 \text{ kg})(9.9 \text{ m/s}) = -690 \text{ N·s}.$$

The negative sign tells us the force must be opposed to the original momentum—that is, it acts upward.

FIGURE 6-5

Period during which impulse acts (Example 6-4).

(b) In coming to rest, the body decelerates from 9.9 m/s to zero in a distance $d = 1.0 \times 10^{-2}$ m. The average speed during this period was $(9.9 \text{ m/s} + 0 \text{ m/s})/2 = 5.0$ m/s, so the time of the collision was $\Delta t = d/\bar{v} = (1.0 \times 10^{-2} \text{ m})/(5.0 \text{ m/s}) = 2.0 \times 10^{-3}$ s. Since the impulse $\bar{F} \Delta t = 690$ N·s and $\Delta t = 2.0 \times 10^{-3}$ s, the average net force $\bar{F}$ was

$$\bar{F} = \frac{690 \text{ N·s}}{2.0 \times 10^{-3} \text{ s}} = 3.5 \times 10^5 \text{ N}.$$

The force $\bar{F}$ is the *net* force on the person (we calculated it from Newton's second law), which is the sum of the upwards force exerted by the earth, F_e, and the downward force of gravity, $mg = (70 \text{ kg})(9.8 \text{ m/s}^2) = 6.9 \times 10^2$ N. Thus the earth exerts an average force of 3.5×10^5 N + 6.9×10^2 N = 3.5×10^5 N upwards on the person's feet.

(c) This is just like part (b) except $d = 0.50$ m so $\Delta t = (0.50 \text{ m})/(5.0 \text{ m/s}) = 0.10$ s and

$$\bar{F} = \frac{690 \text{ N·s}}{0.10 \text{ s}} = 6.9 \times 10^3 \text{ N}.$$

The upward force on the person's feet is, as in part (b), $\bar{F} + mg = 6.9 \times 10^3$ N + 0.69×10^3 N = 7.6×10^3 N. Clearly the force on the feet and legs is much less when the knees are bent. In fact the ultimate strength of the leg bone (see Chapter 8, Table 8-2) is not high enough to support the force calculated in part (b), so the leg would undoubtedly break in such a landing, whereas it probably would not in part (c).

6-4 ∎ Conservation of energy and momentum in collisions

During most collisions, we usually don't know how the collision force varies as a function of time. Nonetheless, we can still determine some of the details of the motion after a collision, given the initial motion, by making use of the conservation laws for momentum and energy. We saw in Section 6-2 that in the collision of two objects such as billiard balls the total momentum is conserved. If the two objects are very hard and elastic and no heat is produced in the collision, then kinetic energy is conserved as well. By this we mean that the sum of the kinetic energies of the two objects is the same after the collision as before. Of course for the brief moment during which the two objects are in contact, some (or all) of the energy is stored momentarily in the form of elastic potential energy; but if we compare the total kinetic energy before the collision with the total after the collision, they are found to be the same. Such a collision is called an **elastic collision.** If we use the subscripts 1 and 2 to represent the two objects, we can write the equation for conservation of total kinetic energy as

$$\tfrac{1}{2}m_1v_1^2 + \tfrac{1}{2}m_2v_2^2 = \tfrac{1}{2}m_1v_1'^2 + \tfrac{1}{2}m_2v_2'^2 \qquad \text{[elastic collision]} \qquad (6\text{-}6)$$

where primed quantities (') mean after the collision and unprimed mean before the collision.

Although at the atomic level the collisions of atoms and molecules are often elastic, in the "macroscopic" world of ordinary objects an elastic collision is an ideal that is never quite reached since at least a little thermal energy (and perhaps sound and other forms of energy) is always produced during a collision. The collision of two hard elastic balls, such as billiard balls, however, is very close to perfectly elastic and we often treat it as such. Even when the KE is not conserved, the *total* energy is, of course, conserved.

Collisions in which kinetic energy is not conserved are said to be *inelastic.* The kinetic energy that is lost is changed into other forms of energy, usually thermal energy, so that the total energy is conserved. In this case we can write that $KE_1 + KE_2 = KE_1' + KE_2' +$ thermal and other forms of energy.

6-5 ∎ Elastic collisions in one dimension

We now apply the conservation laws of momentum and kinetic energy to an elastic collision between two particles that is head-on, so all the motion is along a line. Let us assume both particles are moving initially, with velocities v_1 and v_2, along the x axis, Fig. 6-6a. After the collision, their velocities are v_1' and v_2', Fig. 6-6b. For any $v > 0$, the particle is moving to the right (increasing x); whereas for $v < 0$, the particle is moving to the left, toward decreasing values of x.

From conservation of momentum we have

$$m_1v_1 + m_2v_2 = m_1v_1' + m_2v_2'$$

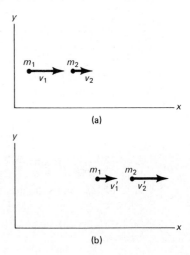

FIGURE 6-6

Two particles, of mass m_1 and m_2, (a) before collision, (b) after collision.

and because the collision is assumed to be elastic, kinetic energy is also conserved:

$$\tfrac{1}{2}m_1v_1^2 + \tfrac{1}{2}m_2v_2^2 = \tfrac{1}{2}m_1v_1'^2 + \tfrac{1}{2}m_2v_2'^2.$$

We have two equations; therefore we can solve for two unknowns. If we know the masses and initial velocities, then we can solve these two equations for the velocities after the collision, v_1' and v_2'. Let us take an example.

EXAMPLE 6-5   A particle of mass m moving with speed v collides head on with a second particle of equal mass at rest ($v_2 = 0$). What are the speeds of the two particles after the collision, assuming it is elastic?

SOLUTION   Since $v_1 = v$ and $v_2 = 0$, conservation of momentum and kinetic energy give

$$mv = mv_1' + mv_2'$$
$$\tfrac{1}{2}mv^2 = \tfrac{1}{2}mv_1'^2 + \tfrac{1}{2}mv_2'^2.$$

Since m is the same in all terms, we can divide it out, so these two equations become

$$v = v_1' + v_2'$$
$$v^2 = v_1'^2 + v_2'^2.$$

From the first equation, $v_1' = v - v_2'$, and we substitute this into the second equation:

$$v^2 = (v - v_2')^2 + v_2'^2$$
$$= v^2 - 2v_2'v + 2v_2'^2.$$

Or, after simplifying

$$0 = v_2'(v - v_2').$$

There are two solutions: $v_2' = 0$ and $v_2' = v$. The first solution $v_2' = 0$, corresponds to no collision at all; particle 2 is still at rest. The second solution

$$v_2' = v$$

is the interesting one (there is a collision) and when we substitute this into the momentum equation, $v = v_1' + v_2'$, we find

$$v_1' = v - v_2' = v - v = 0.$$

In summary, before the collision we have

$$v_1 = v, \qquad v_2 = 0$$

and after the collision

$$v_1 = 0 \qquad v_2 = v.$$

That is, particle 1 is brought to rest whereas particle 2 acquires the original velocity of particle 1. This fact is often observed by billiard players. It is valid only if the two objects have equal masses.

*6-6 ∎ Elastic collisions in two or three dimensions

Conservation of momentum and energy can also be applied to collisions in two or three dimensions, and in this case the vector nature of momentum is important. One common type of non-head-on collision is one for which one particle (called the "projectile") strikes a second particle initially at rest (the "target" particle). This is the common situation in games such as billiards and for experiments in atomic and nuclear physics (the projectiles, from radioactive decay or accelerated by a high-energy accelerator, strike a stationary target nucleus).

Figure 6-7 shows particle 1 (the projectile, m_1) heading along the x axis toward particle 2 (the target, m_2) which is initially at rest. If these are, say, billiard balls, m_1 strikes m_2 and they go off at the angles θ'_1 and θ'_2, which are measured relative to m_1's initial direction (the x axis). If they are electrically charged particles, or nuclear particles, they may begin to deflect even before they touch because of the force (electric or nuclear) between them. (You might think, for example, of two magnets oriented so they repel each other: when one moves toward the other, the second moves away before the first one touches it.)

Let us now apply the laws of conservation of momentum and kinetic energy for an elastic collision like that of Fig. 6-7. From conservation of kinetic energy, since $v_2 = 0$, we have

$$\tfrac{1}{2}m_1v_1^2 = \tfrac{1}{2}m_1v_1'^2 + \tfrac{1}{2}m_2v_2'^2. \qquad (6\text{-}7a)$$

We choose the xy plane to be the plane in which lie the initial and final momenta. Since momentum is a vector, and is conserved, its components in the x and y directions remain constant. In the x direction

$$m_1v_1 = m_1v_1' \cos\theta'_1 + m_2v_2' \cos\theta'_2. \qquad (6\text{-}7b)$$

Since there is no motion in the y direction initially, the y component of the total momentum is zero:

$$0 = m_1v_1' \sin\theta'_1 + m_2v_2' \sin\theta'_2. \qquad (6\text{-}7c)$$

We have three independent equations. This means we can solve for at most three unknowns. If we are given m_1, m_2, v_1 (and v_2, if not zero), we cannot

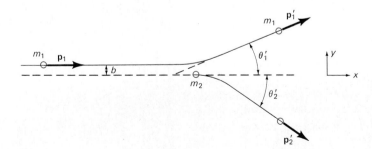

FIGURE 6-7

Particle 1, the projectile, collides with particle 2, the target. They move off, after the collision, with momenta $\mathbf{p}'_1$ and $\mathbf{p}'_2$ at angles θ'_1 and θ'_2.

uniquely predict the final variables v_1', v_2', θ_1', and θ_2' since there are four of them; θ_2', for example, can be anything. However, if we measure one of these variables, say θ_1', then the other three variables (v_1', v_2', and θ_2') are uniquely determined and we can calculate them using the above three equations.

EXAMPLE 6-6   A proton traveling with speed 8.2×10^5 m/s collides elastically with a stationary proton in a hydrogen target. One of the protons is observed to be scattered at a $60°$ angle. At what angle will the second proton be observed, and what will be the velocities of the two protons after the collision?

SOLUTION   Since $m_1 = m_2$, Eqs. 6-7a, b, and c become

$$v_1^2 = v_1'^2 + v_2'^2$$
$$v_1 = v_1' \cos \theta_1' + v_2' \cos \theta_2'$$
$$0 = v_1' \sin \theta_1' + v_2' \sin \theta_2'$$

where $v_1 = 8.2 \times 10^5$ m/s and $\theta_1' = 60°$ are known. In the second and third equations, we take the v_1' terms to the left side and square both sides of these two equations:

$$v_1^2 - 2v_1v_1' \cos \theta_1' + v_1'^2 \cos^2 \theta_1' = v_2'^2 \cos^2 \theta_2'$$
$$v_1'^2 \sin^2 \theta_1' = v_2'^2 \sin^2 \theta_2'.$$

We add these two equations and use the fact that $\sin^2 \theta + \cos^2 \theta = 1$ to get:

$$v_1^2 - 2v_1v_1' \cos \theta_1' + v_1'^2 = v_2'^2.$$

Into this equation we substitute, from the first equation above, $v_2'^2 = v_1^2 - v_1'^2$, and obtain

$$2v_1'^2 = 2v_1v_1' \cos \theta_1'$$

or

$$v_1' = v_1 \cos \theta_1' = (8.2 \times 10^5 \text{ m/s})(\cos 60°) = 4.1 \times 10^5 \text{ m/s}.$$

To obtain v_2', we use the first equation (conservation of KE):

$$v_2' = \sqrt{v_1^2 - v_1'^2} = \sqrt{(8.2 \times 10^5 \text{ m/s})^2 - (4.1 \times 10^5 \text{ m/s})^2} = 7.1 \times 10^5 \text{ m/s}.$$

Finally, from the third equation:

$$\sin \theta_2' = -\frac{v_1'}{v_2'} \sin \theta_1' = -\left(\frac{4.1 \times 10^5 \text{ m/s}}{7.1 \times 10^5 \text{ m/s}}\right)(0.866) = -0.50,$$

so $\theta_2' = -30°$. (The minus sign merely means particle 2 moves at an angle below the x axis if particle 1 is above the axis (as in Fig. 6-7)). An example of such a collision is shown in the bubble chamber photo of Fig. 6-8. Notice that the two trajectories are at right angles to each other after the collision. This can be shown to be true in general for non-head-on collisions of two particles of equal mass, one of which was at rest initially (see the problems).

FIGURE 6-8

Photo of a proton collision in a hydrogen bubble chamber (a device that makes visible the paths of elementary particles). The many lines represent incoming protons which can strike the protons of the hydrogen in the chamber. (Courtesy of Brookhaven National Laboratory.)

*6-7 ∎ Inelastic collisions

Collisions in which kinetic energy is not conserved are called *inelastic collisions*. Some of the initial kinetic energy in such collisions is transformed into other types of energy such as thermal or potential energy, so the total final KE is less then the total initial KE; the inverse can also happen when potential energy (such as chemical or nuclear) is released, and then the total final KE can be greater than the initial KE. If two objects stick together as a result of a collision, the collision is said to be *completely inelastic*. Two colliding balls of putty that stick together, or two railroad cars that couple together when they collide (Example 6-2) are examples. The kinetic energy is not necessarily all transformed to other forms of energy in an inelastic collision. In Example 6-2 for instance, we saw that when a traveling railroad car collided with a stationary one, the coupled cars traveled off with some KE. Even though KE is not conserved in inelastic collisions, the total energy is conserved, and the total vector momentum is also always conserved.

EXAMPLE 6-7   For the totally inelastic collision we considered in Example 6-2 calculate how much of the initial kinetic energy is transferred to thermal or other forms of energy.

SOLUTION   Initially, the total kinetic energy is $\frac{1}{2}m_1v_1^2 = \frac{1}{2}(10,000 \text{ kg})(24.0 \text{ m/s})^2 = 2.88 \times 10^6$ J. After the collision, the total kinetic energy is $\frac{1}{2}(20,000 \text{ kg})(12.0 \text{ m/s})^2 = 1.44 \times 10^6$ J. Hence the energy transformed to other forms is 2.88×10^6 J $- 1.44 \times 10^6$ J $= 1.44 \times 10^6$ J.

6-8 ∎ Center of mass

Until now we have been mainly concerned with the motion of single particles; when we have dealt with an extended body (that is, a body that has size) we have assumed that it approximated a point particle or that it underwent only translational motion. Real "extended" bodies, however, can undergo rotational and other types of motion as well. For example, the diver in Fig. 6-9a undergoes only translational motion (all parts of the body follow the same path), whereas the diver in Fig. 6-9b undergoes both translational and rotational motion. We will refer to motion that is not pure translation as *general motion*.

Observations of the motion of bodies indicates that even if a body rotates, or there are several bodies that move relative to one another, there is one point that moves in the same path that a particle would if subjected to the same net force. This point is called the **center of mass** (cm). The general motion of an extended body (or system of bodies) can be considered as the sum of the translational motion of its cm plus rotational, vibrational, or other type of motion about its cm.

As an example, consider the motion of the center of mass of the diver in Fig. 6-9a: the cm follows a parabolic path even when the diver rotates as

FIGURE 6-9

Motion of the diver is pure translation in (a), but is translation plus rotation in (b).

(a)

(b)

shown in Fig. 6-9b. This is the same parabolic path that a projected particle follows when acted on only by the force of gravity (that is, projectile motion). Other points in the rotating diver's body follow more complicated paths.

We will show in Section 6-9 (optional) that this important property of the cm follows from Newton's laws if the cm is defined in the following way. We can consider any extended body as being made up of many tiny particles. But first we consider a system made up of only two particles, of mass m_1 and m_2; we choose a coordinate system so that both particles lie on the x axis at positions x_1 and x_2, Fig. 6-10. The center of mass of this system is defined to be at the position, x_{cm}, given by

$$x_{cm} = \frac{m_1 x_1 + m_2 x_2}{m_1 + m_2} = \frac{m_1 x_1 + m_2 x_2}{M}, \qquad (6\text{-}8a)$$

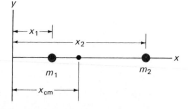

where $M = m_1 + m_2$ is the total mass of the system. The center of mass lies on the line joining m_1 and m_2. If the two masses are equal ($m_1 = m_2 = m$), x_{cm} is midway between them, since then $x_{cm} = m(x_1 + x_2)/2m = (x_1 + x_2)/2$. If one mass is greater than the other, say $m_1 > m_2$, then the cm is closer to the larger mass. If there are more than two particles along a line, there will be additional terms in Eq. 6-8a, as the following example shows.

FIGURE 6-10

The center of mass of a two-particle system lies on the line joining the two masses.

EXAMPLE 6-8   Three particles of equal mass m lie along the x axis at positions $x_1 = 1.0$ m, $x_2 = 5.0$ m, and $x_3 = 6.0$ m. Find the position of the cm.

SOLUTION   We use Eq. 6-8a with a third term:

$$x_{cm} = \frac{m x_1 + m x_2 + m x_3}{m + m + m} = \frac{m(x_1 + x_2 + x_3)}{3m}$$

$$= \frac{(1.0 \text{ m} + 5.0 \text{ m} + 6.0 \text{ m})}{3} = \frac{12.0 \text{ m}}{3} = 4.0 \text{ m}.$$

TABLE 6-1

Center of mass of parts of typical male human body[†]

Distance of Hinge Points (%)	Hinge points (●)(Joints)		Center of Mass (×) (% Height Above Floor)		Percent Mass
91.2	Base of skull on spine		Head	93.5	6.9
81.2	Shoulder joint		Trunk and neck	71.1	46.1
			Upper arms	71.7	6.6
		elbow 62.2	Lower arms	55.3	4.2
		wrist 46.2	Hands	43.1	1.7
52.1	Hip		Upper legs (thighs)	42.5	21.5
28.5	Knee				
			Lower legs	18.2	9.6
4.0	Ankle		Feet	1.8	3.4
				58.0	100.0

[†] From *Bioastronautics Data Book*, NASA, Washington, DC.

If the particles are spread out in two or three dimensions, then we need to specify not only the x coordinate of the cm (x_{cm}) but also the y and z co-ordinates, which will be given by formulas just like Eq. 6-8a. For example, for two particles of mass m_1 and m_2 whose y coordinates are y_1 and y_2, the y coordinate of this cm, y_{cm}, would be:

$$y_{cm} = \frac{m_1 y_1 + m_2 y_2}{m_1 + m_2} = \frac{m_1 y_1 + m_2 y_2}{M}. \qquad (6\text{-}8b)$$

For more particles, there would be more terms in this formula.

If we have a group of extended bodies, each of whose cm is known, we can find the cm of the group using Eqs. 6-8. As an example we consider the human body. In Table 6-1 the cm and hinge points (joints) for the different components of a "representative" person are shown. Of course, there are wide variations among people, so these data represent only a rough average. Note that the numbers represent a percentage of the total height, which is regarded as 100 units.

□ EXAMPLE 6-9 Calculate the cm of a whole leg when (*a*) stretched out and (*b*) bent at 90°, as shown in Fig. 6-11.

SOLUTION (*a*) We will use the percentage units; that is, the person has a mass of 100 units and a height of 100 units. We measure the distance from the hip joint using Table 6-1 and obtain the numbers shown on the diagram. Using Eq. 6-8a we obtain

$$x_{cm} = \frac{(21.5)(9.6) + (9.6)(33.9) + (3.4)(50.3)}{21.5 + 9.6 + 3.4}$$

$$= 20.4 \text{ units.}$$

Thus, the center of mass of the leg and foot is 20.4 units below the hip joint or $52.1 - 20.4 = 31.7$ units above the base of the foot. (*Note:* If the person is 170 cm tall, this is $31.7 \times 170 \text{ cm}/100 = 54$ cm.) (*b*) This is a two-dimensional problem. We use an xy coordinate system as shown in Fig. 6-11b. First we calculate how far to the right of the hip joint the cm lies:

$$x_{cm} = \frac{(21.5)(9.6) + (9.6)(23.6) + (3.4)(23.6)}{21.5 + 9.6 + 3.4} = 14.9 \text{ units.}$$

Next, we calculate the distance, y_{cm}, of the cm above the floor:

$$y_{cm} = \frac{(3.4)(1.8) + (9.6)(18.2) + (21.5)(28.5)}{21.5 + 9.6 + 3.4} = 23.1 \text{ units.}$$

Thus the cm is located 23.1 units above the floor and 14.9 units to the right of the hip joint.

Note in this last example that the cm can actually lie *outside* the body. Another example is a doughnut whose cm is at the center of the hole.

Knowing the cm of the body when it is in various positions is of great use in studying body mechanics. One simple example from athletics is shown in Fig. 6-12. If high jumpers can get in the position shown, their cm would actually pass below the bar, which means that for a particular takeoff speed, they could clear a higher bar. This is indeed what they try to do.

(a)

(b)

FIGURE 6-11

Example 6-9, finding the cm of a leg in two different positions (⊗ represents the calculated cm).

FIGURE 6-12

The cg of a high jumper may actually pass beneath the bar.

A concept similar to *center of mass* is **center of gravity** (cg). The cg of a body is that point at which the force of gravity can be considered to act. Of course, the force of gravity actually acts on all the different parts or particles of a body, but for purposes of determining the motion of a body as a whole, we can assume that the entire weight of the body (which is the sum of the weights of all its parts) acts at the cg. Strictly speaking, there is a conceptual difference between the center of gravity and the center of mass, but for practical purposes they are generally the same point.[†]

It is often easier to determine the cm or cg of an extended body experimentally rather than analytically. To do so, we make use of the fact that if a body is suspended from any point, it will swing (Fig. 6-13) unless its cm lies on a vertical line directly below the point from which it is suspended. If the object is two dimensional or has a plane of symmetry, it need only be hung from two different pivot points and the respective vertical lines drawn; then the center of mass will be at the intersection of the two lines, as in Fig. 6-14. If the object doesn't have a plane of symmetry, the cm with respect to the third dimension is found by suspending the object from at least three points that are not in a plane. For symmetrically shaped bodies such as uniform cylinders (wheels), spheres and rectangular solids, the cm is located at the geometric center of the body.

FIGURE 6-13
The force of gravity, considered to act at the cg, causes the body to rotate about the pivot point, unless the cg is on a vertical line directly below the pivot, in which case the body can stay at rest.

*6-9 ▌ Center of mass and translational motion

As mentioned in the previous section, a major reason for the importance of the concept of center of mass is that the motion of the cm for a system of particles (or an extended body) is directly related to the net force acting on the system as a whole. We now show this, taking the simple case of one-dimensional motion and only three particles, but the extension to more bodies and three dimensions follows the same lines.

Suppose the three particles lie on the x axis and have masses m_1, m_2, m_3 and positions x_1, x_2, x_3. From Eq. 6-8a for the cm, we can write

$$Mx_{cm} = m_1 x_1 + m_2 x_2 + m_3 x_3$$

where $M = m_1 + m_2 + m_3$ is the total mass of the system. If these particles are in motion, say with velocities v_1, v_2, v_3, respectively, along the x axis, then in a short time Δt they each will have traveled a distance:

$$\Delta x_1 = x_1' - x_1 = v_1 \Delta t$$
$$\Delta x_2 = x_2' - x_2 = v_2 \Delta t$$
$$\Delta x_3 = x_3' - x_3 = v_3 \Delta t$$

where x_1', x_2', and x_3' represent their new positions after a time Δt. The position of the new cm is given by

$$Mx_{cm}' = m_1 x_1' + m_2 x_2' + m_3 x_3'.$$

FIGURE 6-14
Finding the cm of an object.

[†] There would be a difference between the two only if a body were large enough so that the acceleration due to gravity would be different at different parts of the body.

If we subtract the two cm equations, we get

$$M \, \Delta x_{cm} = m_1 \, \Delta x_1 + m_2 \, \Delta x_2 + m_3 \, \Delta x_3.$$

During the time Δt, the cm will have moved a distance

$$\Delta x_{cm} = x'_{cm} - x_{cm} = v_{cm} \, \Delta t$$

where v_{cm} is the velocity of the cm. Into the equation just above, we substitute the relations for all the Δx's:

$$M v_{cm} \, \Delta t = m_1 v_1 \, \Delta t + m_2 v_2 \, \Delta t + m_3 v_3 \, \Delta t.$$

We divide out Δt and get

$$M v_{cm} = m_1 v_1 + m_2 v_2 + m_3 v_3. \tag{6-9}$$

Since $m_1 v_1 + m_2 v_2 + m_3 v_3$ is the sum of the momenta of the particles of the system, it represents the *total momentum* of the system. Thus we see from Eq. 6-9 that *the total linear momentum of a system of particles is equal to the product of the total mass* M *and the velocity of the center of mass of the system.* Or, *the linear momentum of an extended body is the product of the body's mass and the velocity of its cm.*

If there are forces acting on the particles, then the particles may be accelerating. In a short time Δt, each particle's velocity will change by an amount

$$\Delta v_1 = a_1 \, \Delta t, \quad \Delta v_2 = a_2 \, \Delta t, \quad \Delta v_3 = a_3 \, \Delta t.$$

If we now use the same reasoning as we did to derive Eq. 6-9, we obtain

$$M a_{cm} = m_1 a_1 + m_2 a_2 + m_3 a_3.$$

According to Newton's second law, $m_1 a_1 = F_1$, $m_2 a_2 = F_2$, and $m_3 a_3 = F_3$, where F_1, F_2, and F_3 are the net forces on the three particles. Thus we get for the system as a whole:

$$M a_{cm} = F_1 + F_2 + F_3 = F_{net}. \tag{6-10}$$

That is, *the sum of all the forces acting on the system is equal to the total mass of the system times the acceleration of its center of mass.* This is **Newton's second law** for a system of particles, and it also applies for an extended body (which can be thought of as a collection of particles). Thus we conclude that the *center of mass of a system of particles (or of an extended body of total mass* M) *moves like a single particle of mass* M *which is acted on by the same net external force.* That is, the system moves as if all its mass were concentrated at the cm and all the external forces acted at that point. We can thus treat the translational notion of any body or system of bodies as the motion of a particle. This theorem clearly simplifies our analysis of the motion of complex systems or extended bodies. Although the motion of various parts of the system may be complicated, we may often be satisfied with knowing the motion of the cm. This theorem also allows us to solve certain types of problems very easily, as illustrated by the following example.

⬜ EXAMPLE 6-10 A rocket is fired into the air as shown in Fig. 6-15. At the moment it reaches its highest point, a horizontal distance D from its starting point, it separates into two parts of equal mass. Part I falls vertically to earth. Where does part II land? Assume **g** = constant.

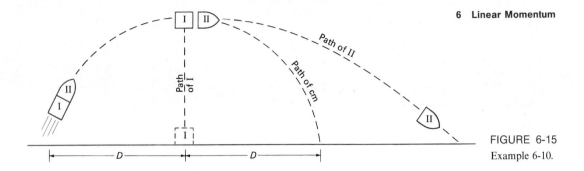

FIGURE 6-15
Example 6-10.

SOLUTION   The path of the cm of the system continues to follow the parabolic trajectory of a projectile acted on by a constant gravitational force. The cm will thus arrive at a point $2D$ from the origin. Since the masses of I and II are equal, the cm must be midway between them. Therefore II lands a distance $3D$ from the origin. (If part I had been given a kick up or down, instead of merely falling, the answer would have been somewhat more complicated.)

❚ Summary

The *momentum*, **p**, of a body is defined as its mass times its velocity: $\mathbf{p} = m\mathbf{v}$. In terms of momentum, Newton's second law can be written $\mathbf{F} = \Delta\mathbf{p}/\Delta t$: the rate of change of momentum equals the net applied force.

Momentum is a conserved quantity: the law of *conservation of momentum* states that the total momentum of an isolated system of objects remains constant. An isolated system is one on which the net external force is zero.

The law of conservation of momentum is very useful in dealing with *collisions*. In a collision two (or more) bodies interact with each other for a very short time and the force between them during this time is very large compared to any other forces acting. The *impulse* of a force on a body is defined as $\mathbf{F}\,\Delta t$ where

$\mathbf{F}$ is the average force acting during the (usually short) time Δt. The impulse is equal to the change in momentum of the body: $\mathbf{F}\,\Delta t = \Delta\mathbf{p}$. Momentum is conserved in any collision. The total energy is also conserved, but this may not be useful unless the only type of energy transformation involves kinetic energy; in this case, kinetic energy is conserved and the collision is called an *elastic collision*. If kinetic energy is not conserved, the collision is called *inelastic*.

The *center of mass* (cm) of a body (or group of bodies) is that point at which the net force can be considered to act for purposes of determining the motion of the body as a whole. The complete motion of a body can be described as the translational motion of its center of mass plus rotation (or other internal motion) about its center of mass.

❚ Questions

1. We claim that momentum is conserved. Yet most moving objects eventually slow down and stop. Explain.

2. When a person jumps from a tree to the ground, what happens to the momentum of the person upon striking the ground?

3. Explain, on the basis of conservation of momentum, how a fish propels itself forward by swishing its tail back and forth.

4. Why, when you release an inflated balloon, does it fly across the room?

5. It is said that in ancient times a rich man with a bag of gold coins was frozen to death stranded on the surface of a frozen lake. Because the ice was frictionless, he could not push himself to shore. What could he have done to save himself had he not been so miserly?

6. How can a rocket change direction when it is far out in space and is essentially in a vacuum?

7. According to Eq. 6-5 the shorter the impact time of an impulse, the greater the force must be for the same momentum change and hence the greater the deformation of the object on which the force acts. Explain on this basis the value of "air bags" which are intended to inflate during an automobile collision and reduce the possibility of fracture or death.

8. Is it possible for a body to receive a larger impulse from a small force than from a large force?

9. A light body and a heavy body have the same kinetic energy. Which has the greater momentum?

10. Is it possible for an object to have momentum without having energy? Can it have energy but no momentum? Explain.

11. In a collision between two cars, which would you expect to be more damaging to the occupants: if the cars collide and remain together, or if the two rebound backward? Explain.

12. A superball is dropped from a height h onto a hard steel plate (fixed to the earth), from which it rebounds at very nearly its original speed. (a) Is the momentum of the ball conserved during any part of this process? (b) If we consider the ball and earth as our system, during what parts of the process is momentum conserved? (c) Answer part (b) for a piece of putty that falls and sticks to the steel plate.

13. If a falling ball were to make a perfectly elastic collision with the floor, would it rebound to its original height? Explain.

FIGURE 6-16
A spool of thread.

*14. Use conservation of (vector) momentum to show that a two-particle collision occurs in a plane if (a) one particle is initially at rest; (b) the two particles have momenta in the same or opposite directions.

15. Why do you tend to lean backward when carrying a heavy load in your arms?

16. Why is the cm of a 1-m length of pipe at its midpoint whereas this is not true for your arm or leg?

17. Show on a diagram how your cm shifts when you change from a lying position to a sitting position.

18. Describe an analytic way of determining the cm of any triangular-shaped, thin uniform plate.

19. Explain why a uniform rectangular brick can be placed so that slightly less than half its length can be suspended over the edge of a table, but no more.

*20. A rocket following a parabolic path through the air suddenly explodes into many pieces. What can you say about the motion of this system of pieces?

*21. Analyze the motion of the spool in Fig. 6-16 if you pull the thread (a) directly upward, (b) horizontally. Consider the effect on the motion of how hard you pull.

*22. If only an external force can change the momentum of the center of mass of an object, how can the internal force of the engine accelerate a car?

▮ Problems

SECTIONS 6-1 and 6-2

1. (I) What is the momentum of an 18.0-g sparrow flying with a speed of 15.0 m/s.

2. (I) The air in a 200 km/h wind strikes a 30.0 m × 20.0 m face of a building at a rate of 5.45×10^4 kg/s. Calculate the net force on the building, assuming the air is brought to rest on impact.

3. (I) Calculate the force exerted on a rocket given that the propelling gases are expelled at a rate of 1000 kg/s with a speed of 60,000 m/s (at takeoff).

4. (I) A child throws a 5.40-kg package horizontally from a boat with a speed of 10.0 m/s. Calculate the resulting velocity of the boat assuming it was initially at rest. The mass of the child is 20.0 kg and that of the boat is 80.0 kg.

5. (II) A 15,000-kg railroad car travels alone on a level frictionless track with a constant speed of 23.0 m/s. A 5000-kg additional load is dropped onto the car. What then will be its speed?

6. (II) A 140-kg tackler moving at 3.0 m/s meets head on (and tackles) a 90-kg halfback moving at 7.5 m/s. What will be their mutual speed immediately after the collision?

7. (II) A 44-g bullet strikes and becomes embedded in a 1.54-kg block of wood placed on a horizontal surface just in front of the gun. If the coefficient of kinetic friction between the block and the surface is 0.28, and the impact drives the block a distance of 18.0 m before it comes to rest, what was the muzzle speed of the bullet?

8. (II) A gun is fired vertically into a 3.40-kg block of wood at rest directly above it. If the bullet has a mass

of 62.0 g and a speed of 340 m/s, how high will the block rise into the air?

9. (II) An atomic nucleus at rest decays radioactively into an alpha particle and a smaller nucleus. What will be the speed of this recoiling nucleus if the speed of the alpha particle is 6.2×10^5 m/s? Assume the nucleus has a mass 57 times greater than that of the alpha particle.

10. (II) A particle of mass m traveling with speed v_0 along the x axis suddenly shoots out one-third its mass parallel to the y axis with speed $2v_0$. Give the components of the velocity of the remainder of the particle.

11. (II) The decay of a neutron into a proton, an electron, and a neutrino is an example of a three-particle decay process. Use the vector nature of momentum to show that if the neutron is initially at rest, the velocity vectors of the three must be coplanar (that is, all in the same plane). The result is not true for numbers greater than three.

12. (II) A 940-kg two-stage rocket is traveling at a speed of 6.20×10^3 m/s with respect to earth when a predesigned explosion separates the rocket into two sections of equal mass that then move with a relative speed of 2.45×10^3 m/s along the original line of motion. (a) What is the speed and direction of each section after the explosion? (b) How much energy was supplied by the explosion? (Hint: what is the change in KE as a result of the explosion?)

13. (III) A 4200-kg rocket is traveling in outer space with a velocity of 150 m/s toward the sun. It wishes to alter its course by 30.0°, and can do this by shooting its rockets briefly in a direction perpendicular to its original motion. If the rocket gases are expelled at a speed of 2700 m/s, what mass of gas must be expelled?

SECTION 6-3

14. (I) A 90-kg fullback is traveling 5.0 m/s and is stopped by a tackler in 1 s. Calculate (a) the original momentum of the fullback, (b) the impulse imparted to the tackler, and (c) the average force exerted on the tackler.

15. (I) The tennis ball may leave the racket of a top player on the serve with a speed of 65.0 m/s. If the ball's mass is 0.0600 kg and is in contact with the racket for 0.0300 s, what is the average force on the ball? Would this force be large enough to lift an average size person?

16. (I) A 0.145-kg baseball pitched at 35.0 m/s is hit on a horizontal line drive straight back at the pitcher at 50.0 m/s. If the contact time between bat and ball is 5.00×10^{-4} s, calculate the force (assumed to be constant) between the ball and bat.

17. (II) A tennis ball of mass m and speed v strikes a wall at a 45° angle and rebounds with the same speed at 45°. What is the impulse given the wall?

18. (II) A 150-kg astronaut (including space suit) acquires a speed of 2.35 m/s by pushing off with his legs from a 2200-kg space capsule. (a) What is the change in speed of the space capsule? (b) If the push lasts 0.200 s, what was the average force exerted by each on the other? As the reference frame, use the position of the capsule before the push.

19. (II) Water strikes the turbine blades of a generator so that its rebounding velocity is reversed in direction but undiminished in magnitude. If the flow rate is 60.0 kg/s and the water speed is 16.5 m/s, what is the average force on the blades?

20. (II) Suppose the force acting on a tennis ball (mass 0.060 kg) as a function of time is given by the graph of Fig. 6-17. Use graphical methods to (a) estimate the

FIGURE 6-17

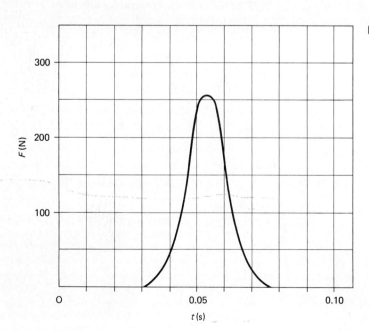

total impulse given the ball; (b) estimate the speed of the ball after being struck, assuming the ball is being served so it is nearly at rest initially.

21. (III) The force on a bullet is given by the formula $F = 480 - 1.6 \times 10^5 t$ over the time interval $t = 0$ to $t = 3.0 \times 10^{-3}$ s; in this formula t is in seconds and F is in newtons. (a) Plot a graph of F vs. t for $t = 0$ to $t = 3.0$ ms. (b) Estimate, using graphical methods, the impulse given the bullet. (c) If the bullet achieves a speed of 320 m/s as a result of this impulse, given to it in the barrel of a gun, what must its mass be?

22. (III) From what maximum height can a 60-kg person jump without breaking the lower leg bone whose cross-sectional area is 3.0×10^{-4} m²? Ignore air resistance and assume the cm of the person moves a distance of 0.60 m from the standing to the seated position (that is, in breaking the fall). Assume the breaking strength (force per unit area) of bone is 170×10^6 N/m².

SECTIONS 6-4 and 6-5

23. (II) A ball of mass 0.440 kg moving with a speed of 8.10 m/s collides head-on with a 0.220-kg ball at rest. If the collision is perfectly elastic, what will be the speeds and directions of the two balls after the collision?

24. (II) Two billiard balls of equal mass undergo a perfectly elastic head-on collision. If the speed of one ball was initially 2.00 m/s, and of the other 3.00 m/s in the opposite direction, what will be their speeds after the collision?

25. (II) A 0.160 kg ball makes an elastic head-on collision with a second ball initially at rest. The second ball moves off with half the original speed of the first ball. (a) What is the mass of the second ball? (b) What fraction of the original kinetic energy gets transferred to the second ball?

26. (II) A ball of mass m makes a head-on elastic collision with a second ball (at rest) and rebounds with a speed equal to one-third its original speed. What is the mass of the second ball?

27. (II) Suppose a heavy particle (mass m_1) has an elastic head-on collision with a very light particle of mass m_2 initially at rest. Show that if $m_1 \gg m_2$, the velocity of the projectile (m_1) is practically unchanged, whereas the target particle (m_2) acquires a velocity $v'_2 = 2v_1$.

28. (II) Suppose a light particle (m_1) with velocity v_1 strikes head-on a heavy particle (m_2) at rest. If $m_1 \ll m_2$, show that the projectile (m_1) rebounds in the opposite direction with virtually the same speed (that is, $v'_1 \approx -v_1$) and that the heavy target (m_2) acquires very little velocity, assuming the collision is elastic.

29. (III) A block of mass $m = 2.00$ kg slides down a 30.0° incline which is 3.60 m high. At the bottom, it strikes a block of mass $M = 6.00$ kg which is at rest on a horizontal surface, Fig. 6-18. If the collision is elastic, and friction can be ignored, determine (a) the speeds of the two blocks after the collision and (b) how far back up the incline the smaller mass will go.

FIGURE 6-18

*SECTION 6-6

*30. (II) A radioactive nucleus at rest decays into a second nucleus, an electron, and a neutrino. The electron and neutrino are emitted at right angles and have momenta of 8.6×10^{-23} kg·m/s and 6.2×10^{-23} kg·m/s, respectively. What is the magnitude and direction of the momentum of the recoiling nucleus?

*31. (II) A billiard ball of mass $M_A = 0.400$ kg moving with a speed $v_A = 2.00$ m/s strikes a second ball, initially at rest, of mass $m_B = 0.400$ kg. As a result of the collision, the first ball is deflected off at an angle of 30.0° with a speed $v'_A = 1.20$ m/s. (a) Taking the x axis to be the original direction of motion of ball A, write down the equations expressing the conservation of momentum for the components in the x and y directions separately. (b) Solve these equations for the speed, v'_B, and angle θ, of ball B. Assume the collision is elastic.

*32. (II) A particle of mass m traveling with speed v collides elastically with a target particle of mass $2m$ (initially at rest) and is scattered at 90°. (a) At what angle does the target particle move after the collision? (b) What are the final speeds of the two particles? (c) What fraction of the initial KE is transferred to the target particle?

*33. (III) Two billiard balls of equal mass move at right angles and meet at the origin of an xy coordinate system. One is moving upward along the y axis at 3.00 m/s, the other is moving to the right along the x axis with speed 4.80 m/s. After the collision (assumed elastic), the second ball is moving along the positive y axis. What is the final direction of the first ball, and what are their two speeds?

*34. (III) A billiard ball of mass $M_A = 0.40$ kg strikes a second ball, initially at rest, of mass $M_B = 0.60$ kg. As a result of this elastic collision, ball A is deflected at an angle of 30° and ball B at 53°. What is the ratio of their speeds after the collision?

*SECTION 6-7

*35. (II) A meteor whose mass was about 10^8 kg struck the earth ($m = 6.0 \times 10^{24}$ kg) with a speed of about 15 km/s and came to rest in the earth. (a) What was the earth's recoil speed? (b) What fraction of the meteor's kinetic energy was transformed to KE of the earth? (c) By how much did the earth's KE change as a result of this collision?

*36. (II) An eagle ($m_1 = 6.6$ kg) moving with speed $v_1 = 12.4$ m/s is on a collision course with a second eagle ($m_2 = 8.3$ kg) moving at $v_2 = 9.1$ m/s in a direction at right angles to the first. After they collide, they hold

FIGURE 6-19
Ballistic pendulum.

onto one another. In what direction, and with what speed, are they moving after the collision?

*37. (II) An explosion breaks an object into two pieces, one of which has 1.5 times the mass of the other. If 4500 J were released in the explosion, how much kinetic energy did each piece acquire?

*38. (III) The *ballistic pendulum* is a device used to measure the speed of a projectile, such as a bullet. The projectile, of mass m, is fired into a large block (of wood or other material) of mass M which is suspended like a pendulum. (Usually M is somewhat greater than m.) As a result of the collision, the center of mass of the pendulum (and embedded projectile) swings up to a maximum height h, Fig. 6-19. Determine the relationship between the initial speed of the projectile, v, and the height h.

*39. (III) After a completely inelastic collision between two objects of equal mass, each having initial speed v, the two move off with speed $v/3$. What was the angle between their initial directions?

SECTION 6-8

40. (I) Determine the cm of an outstretched arm using Table 6-1.

41. (I) Assume that your proportions are the same as those in Table 6-1, and calculate the mass of one of your legs.

42. (I) An empty 1500-kg car has its cm 3.10 m from the front of the car. How far from the front of the car will

the cm be when two people sit in the front seat 2.60 m from the front of the car and three in the back seat 3.85 m from the front? Assume that each person has a mass of 65 kg.

43. (I) The distance between a carbon atom ($m = 12$ u) and an oxygen atom ($m = 16$ u) in the CO molecule is 1.13×10^{-10} m. How far from the oxygen atom is the center of mass of the molecule?

44. (II) Use Table 6-1 to calculate the position of the cm of an arm bent at a right angle. Assume that the person is 170-cm tall.

45. (II) Calculate how far below the torso's median line the cm will be when a jumper is in a position such that his arms and legs are hanging vertically, and the trunk and head are horizontal. Will this be outside the body? Use Table 6-1.

46. (II) Three spheres, of radius r_0, $2r_0$, and $3r_0$, are placed next to one another (in contact) with their centers along a straight line and the $r = 2r_0$ sphere in the center. Where is the cm of this system? Assume the spheres are made of the same uniform material.

47. (III) A square uniform raft, 30 m by 30 m of mass 7400 kg, is used as a ferryboat carrying cars northward across a river. (a) If three cars, each of mass 1000 kg, occupy the NE, SE, and SW corners, determine the cm of the loaded ferryboat. (b) If the car in the SW corner accelerates northward at 1.80 m/s² relative to the raft, where will the cm be after 4.0 s?

*SECTION 6-9

*48. (II) The masses of the earth and moon are 5.98×10^{24} kg and 7.36×10^{22} kg, respectively, and they are separated by about 3.80×10^8 m. (a) Where is the cm of this system located? (b) What can you say about the motion of the earth-moon system about the sun, and of the earth and moon separately about the sun?

*49. (II) (a) Suppose that in Example 6-10 (Fig. 6-15) $\dot{m}_{II} = 3m_I$, where then would m_{II} land? (b) What if $m_I = 3m_{II}$?

*50. (II) A 50-kg girl and a 70-kg boy stand 8.0 m apart on frictionless ice. (a) How far from the girl is their cm? (b) If they hold on to the two ends of a rope, and the boy pulls on the rope so that he moves 2.2 m, how far from the girl will he be now? (c) How far will the boy have moved when he collides with the girl?

7 ROTATIONAL MOTION

We have, until now, been concerned with translational motion. In this chapter and the next, we will deal with rotational motion. We will mainly be concerned with *rigid bodies;* by a rigid body we mean a body that has a definite shape that doesn't change, so that the particles that make it up stay in fixed positions relative to one another. Of course any real body is capable of vibrating or deforming when a force is exerted on it; these effects are often very small, so the concept of an ideal rigid body is very useful as a good approximation.

The motion of a rigid body (as mentioned in Chapter 6) can be analyzed as the translational motion of its center of mass, plus rotational motion about its center of mass. We have already discussed translational motion in detail; so now, in this chapter, we focus our attention on purely rotational motion. By *purely rotational motion* we mean that all points in the body move in circles, such as the point P in the rotating wheel of Fig. 7-1, and that the centers of these circles all lie on a line called the **axis of rotation.**

7-1 ∎ Angular quantities

To describe rotational motion, we make use of angular quantities, such as angular velocity and angular acceleration. These are defined in analogy to the corresponding quantities in linear motion.

Every point in a body rotating about a fixed axis moves in a circle whose center is on the axis; the radius of that circle is r, the perpendicular

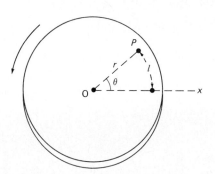

FIGURE 7-1

Looking down on a wheel that is rotating counterclockwise about an axis (perpendicular to the page), through the wheel's center at 0.

distance of that point from the axis of rotation. A line drawn perpendicular from the axis to any point sweeps out the same angle θ in the same time. To indicate the position of the body, or how far it has rotated, we specify the angle θ of some particular line in the body with respect to some reference line, such as the x axis (see Fig. 7-1). A point in the body (such as P in Fig. 7-1) moves through an angle θ when it travels the distance l measured along the circumference of its circular path. Angles are commonly measured in degrees, but the mathematics of circular motion is much simpler if we use the *radian* for angular measure. One *radian* (rad) is defined as the angle subtended by an arc whose length is equal to the radius. (For example, in Fig. 7-1, point P is a distance r from the axis of rotation and it has moved a distance l along the arc of a circle; if $l = r$, then θ is exactly equal to 1 rad.) In general, any angle θ is given by

$$\theta = \frac{l}{r} \tag{7-1}$$

where r is the radius of the circle, and l is the arc length subtended by the angle θ; θ itself will then be in radians. Radians can be related to degrees in the following way. In a complete circle there are $360°$ and this of course must correspond to an arc length equal to the circumference of the circle, $l = 2\pi r$; thus $\theta = l/r = 2\pi r/r = 2\pi$ rad in a complete circle, so

$$360° = 2\pi \text{ rad.}$$

One radian is then $360°/2\pi = 360°/6.28 = 57.3°$.

□ **EXAMPLE 7-1** A particular bird's eye can just distinguish objects which subtend an angle no smaller than about 3×10^{-4} rad (how many degrees is this?). How small an object can the bird just distinguish when flying at a height of 100 m (Fig. 7-2a)?

SOLUTION From Eq. 7-1, $l = r\theta$. Strictly speaking, l is the arc length, but for small angles the linear distance subtended is approximately the same.[†] Since $r = 100$ m and $\theta = 3 \times 10^{-4}$ rad we find

$$l = (100 \text{ m})(3 \times 10^{-4} \text{ rad}) = 3 \times 10^{-2} \text{ m} = 3 \text{ cm.}$$

Had the angle been given in degrees, we would first have had to change it to radians to make this calculation.

Note in this example that we used the fact that the radian is dimensionless (has no units) since it is the ratio of two lengths.

Angular velocity is defined in analogy with ordinary linear velocity. Instead of distance traveled, we use the angular distance θ. Thus the **average angular velocity** (denoted by ω, the Greek lowercase letter omega) is defined as

$$\bar{\omega} = \frac{\theta}{t}, \tag{7-2a}$$

[†] Even for an angle as large as $15°$, the error in making this estimate is only 1 percent, but for larger angles the error increases rapidly.

FIGURE 7-2

(a) Example 7-1; (b) for small angles, arc length s and the linear distance (the chord) are nearly equal.

Linear distance

Arc length

(a)   (b)

where θ is the angle through which the body has rotated in the time t. We define the **instantaneous angular velocity** as the small angle, $\Delta\theta$, through which the body turns in the very short time interval Δt:

$$\omega = \frac{\Delta\theta}{\Delta t} \quad [\Delta t \text{ very small, approaching zero}]. \qquad (7\text{-}2b)$$

Angular velocity is generally specified in radians per second. Note that all points in the body rotate with the same angular velocity. This follows since every position in the body moves through the same angle in the same time interval.

Angular acceleration, in analogy to ordinary linear acceleration, is defined as the change in angular velocity divided by the time required to make this change. The **average angular acceleration** (denoted by α, the Greek lowercase letter alpha) is defined as

$$\bar{\alpha} = \frac{\omega - \omega_0}{t}, \qquad (7\text{-}3a)$$

where ω_0 is the angular velocity initially and ω is the angular velocity after the time t has passed. **Instantaneous angular acceleration** is defined in the usual way:

$$\alpha = \frac{\Delta\omega}{\Delta t} \quad [\Delta t \text{ very small, approaching zero}]. \qquad (7\text{-}3b)$$

Since ω is the same for all points of a rotating body, Eq. 7-3 tells us that α also will be the same for all points. With ω measured in radians per second and t in seconds, α will be expressed as radians per second squared (rad/s^2).

Each particle or point of a rotating rigid body has, at any moment, a linear velocity v and linear acceleration a. We can relate these linear quantities, v and a, of each particle to the angular quantities, ω and α, of the rotating body as a whole. Consider a particle located a distance r from the axis of rotation. If the body rotates with angular velocity ω, any particle will have a linear velocity whose direction is tangent to its circular path; the magnitude of its linear velocity, v, is $v = \Delta l/\Delta t$. From Eq. 7-1, a change in rotation angle $\Delta\theta$ is related to the linear distance traveled by $\Delta l = r\,\Delta\theta$. Hence

$$v = \frac{\Delta l}{\Delta t} = r\frac{\Delta\theta}{\Delta t}$$

$$v = r\omega. \qquad (7\text{-}4)$$

Thus, although ω is the same for every point in the rotating body at any instant, the linear velocity v is greater for points farther from the axis. Note that Eq. 7-4 is valid both instantaneously and on the average.

We can use Eq. 7-4 to show that the angular acceleration α is related to the tangential linear acceleration a_T of a particle by

$$a_\text{T} = \frac{\Delta v}{\Delta t} = r\frac{\Delta\omega}{\Delta t}$$

$$a_\text{T} = r\alpha. \qquad (7\text{-}5)$$

In this equation r is the radius of the circle in which the particle is moving, and the subscript T in a_T stands for "tangential" since the acceleration considered here is along the circle (that is, tangent to it).

The total linear acceleration of a particle is

$$\mathbf{a} = \mathbf{a}_T + \mathbf{a}_c$$

where the radial component, $\mathbf{a}_c$, the "centripetal acceleration," points toward the center of the particle's circular path; as we saw in Chapter 4, $a_c = v^2/r$, and we rewrite this in terms of ω using Eq. 7-4:

$$a_c = \frac{v^2}{r} = \omega^2 r. \tag{7-6}$$

Equations 7-4, 7-5, and 7-6 relate the angular quantities describing the rotation of a body to the linear quantities for each particle of the body.

It is sometimes useful to relate the angular velocity ω to the frequency of rotation, f. By frequency, we mean the number of complete revolutions (rev) per second. One revolution (of, say, a wheel) corresponds to an angle of 2π radians, and thus 1 rev/s = 2π rad/s; hence, in general the frequency f is related to the angular velocity ω by

$$f = \frac{\omega}{2\pi}$$

or

$$\omega = 2\pi f.$$

□   EXAMPLE 7-2   What is the linear speed of a point on the edge of a steadily rotating 33-rpm (revolutions per minute) phonograph record whose diameter is 30 cm?

SOLUTION   First we find the angular velocity in radians per second: the frequency $f = 33$ rev/min = 33 rev/60 s = 0.55 rev/s; then $\omega = 2\pi f = (6.28)(0.55 \text{ rev/s}) = 3.5$ rad/s. The radius r is 0.15 m, so the speed v at the edge is

$$v = r\omega = (0.15 \text{ m})(3.5 \text{ rad/s}) = 0.52 \text{ m/s}.$$

□   EXAMPLE 7-3   What is the magnitude of the acceleration of a speck of dust on the edge of the record in Example 7-2?

SOLUTION   From Example 7-2, $\omega = 3.5$ rad/s and $v = 0.52$ m/s. Since $\omega = $ constant, $\alpha = 0$ so $a_T = 0$. From Eq. 7-6,

$$a_c = \omega^2 r = (3.5 \text{ rad/s})^2 (0.15 \text{ m}) = 1.8 \text{ m/s}^2;$$

or $a_c = v^2/r = (0.52 \text{ m/s})^2/(0.15 \text{ m}) = 1.8 \text{ m/s}^2$ which is the same result.

□   EXAMPLE 7-4   A centrifuge rotor is accelerated from rest to 20,000 rpm in 5.0 min. What is its average angular acceleration?

SOLUTION

$$\omega = (20{,}000 \text{ rev/min})\left(\frac{2\pi \text{ rad/rev}}{60 \text{ s/min}}\right) = 2100 \text{ rad/s};$$

then, since the initial angular velocity is 0, and the final angular velocity is 2100 rad/s, we have

$$\bar{\alpha} = \frac{2100 \text{ rad/s} - 0}{300 \text{ s}} = 7.0 \text{ rad/s}^2.$$

7-2 ∎ Kinematic equations for uniformly accelerated rotational motion

In Chapter 1 we derived the important equations that relate acceleration, velocity, and distance for the situation of uniform linear acceleration. Those equations were derived from the definitions of linear velocity and acceleration assuming constant acceleration. The definitions of angular velocity and angular acceleration are the same as for their linear counterparts, except that θ has replaced the linear displacement x, ω has replaced v, and α has replaced a. Therefore, the angular equations for constant angular acceleration will be analogous to Eqs. 1-10 with x replaced by θ, v by ω, and a by α, and they can be derived in exactly the same way. We summarize them here, opposite their linear equivalents:

Angular	Linear		
$\omega = \omega_0 + \alpha t$	$v = v_0 + at$	[constant α, a]	(7-7a)
$\theta = \omega_0 t + \frac{1}{2}\alpha t^2$	$x = v_0 t + \frac{1}{2}at^2$	[constant α, a]	(7-7b)
$\omega^2 = \omega_0^2 + 2\alpha\theta$	$v^2 = v_0^2 + 2ax$	[constant α, a]	(7-7c)
$\bar{\omega} = \dfrac{\omega + \omega_0}{2}$	$\bar{v} = \dfrac{v + v_0}{2}$	[constant α, a]	(7-7d)

Note that ω_0 represents the angular velocity at $t = 0$, whereas θ and ω represent the angular position and velocity at time t. Since the angular acceleration is constant, $\alpha = \bar{\alpha}$. These equations are valid of course also for constant angular velocity, for which case $\alpha = 0$ and we have $\omega = \omega_0$, $\theta = \omega_0 t$, and $\bar{\omega} = \omega$.

☐ **EXAMPLE 7-5** Through how many turns has the centrifuge rotor of Example 7-4 turned during its acceleration period? Assume constant angular acceleration.

SOLUTION We know that $\omega_0 = 0$, $\omega = 2100$ rad/s, $\alpha = \bar{\alpha} = 7.0$ rad/s^2, and $t = 300$ s. We could use either Eq. 7-7b or 7-7c; the former gives

$$\theta = 0 + \tfrac{1}{2}(7.0 \text{ rad/s}^2)(300 \text{ s})^2 = 3.2 \times 10^5 \text{ rad}.$$

To find the total number of revolutions we divide by 2π and obtain 5.0×10^4 revolutions. (To decide whether to multiply or to divide by 2π, it helps to remember that there are more radians than revolutions since 2π rad = 1 rev.)

7-3 ▌ Torque

Up to now in this chapter, we have discussed rotational kinematics—the description of rotational motion in terms of angle, angular velocity, and angular acceleration. Now we discuss the dynamics, or causes, of rotational motion. Just as we found analogies between linear rotational motion for the description of motion, so rotational equivalents for dynamics exist as well. For example the rotational equivalent of Newton's first law states that a freely rotating body will continue to rotate with constant angular velocity as long as no forces (or, as we shall see shortly, no torques) act to change that motion. More difficult is the question of a rotational equivalent for Newton's second law; that is, what gives rise to angular acceleration? To make an object start rotating about an axis clearly requires a force; but the direction of this force, and where it is applied, are also important. Take, for example, an ordinary situation such as the door in Fig. 7-3. If you apply a force $\mathbf{F}_1$ to the door as shown, you will find that the greater F_1 is the more quickly the door opens. But now if you apply the same magnitude force at a point closer to the hinge, say $\mathbf{F}_2$ in Fig. 7-3, you will find that the door will not open so quickly. The effect of the force is less.

So the angular acceleration of the door is proportional not only to the magnitude of the force; but, assuming for the moment that only this one force acts (we ignore friction in the hinges, and so on), it is also proportional to the perpendicular distance from the axis of rotation to the line along which the force acts. Thus, if the distance r_1 in Fig. 7-3b is three times larger than r_2 then the angular acceleration of the door will be three times as great, assuming of course that the magnitudes of the forces are the same. To say it another way, if $r_1 = 3r_2$ then F_2 must be three times as large as F_1 to give the same angular acceleration. The distances r_1 and r_2 are called the *lever arms* or *moment arms* of the respective forces. The angular acceleration, then, will be proportional to the product of the force times the lever arm. This product is called the *moment* of the force, or **torque,** and is abbreviated τ (Greek lowercase letter tau). The angular acceleration α of an object is found to be directly proportional to the net applied torque, τ:

$$\alpha \propto \tau.$$

Thus we see that it is torque that gives rise to angular acceleration. This is the rotational analog of Newton's second law for linear motion, $a \propto F$. (In

FIGURE 7-3

Applying the same force with different lever arms, r.

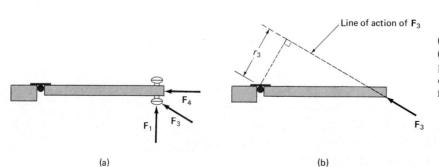

FIGURE 7-4

(a) Forces acting at different angles. (b) The lever arm is defined as the perpendicular distance from the axis of rotation to the line of action of the force.

Section 7-4 we will see what factor is needed to make this proportionality an equation.)

We define the *lever arm* as the *perpendicular* distance of the axis of rotation from the line of action of the force (we mean the distance which is perpendicular to both the axis of rotation and to an imaginary line drawn along the direction of the force); we do this to take into account the effect of forces acting at an angle. It is clear that a force applied at an angle, such as $\mathbf{F}_3$ in Fig. 7-4, will be less effective than the same magnitude force applied straight on, such as $\mathbf{F}_1$ (Fig. 7-4a). And if you push on the end of the door so that the force is directed at the hinge (the axis of rotation), as indicated by $\mathbf{F}_4$, the door will not move at all.

The lever arm for a force such as $\mathbf{F}_3$ is found by drawing a line along the direction of $\mathbf{F}_3$ (this is the "line of action" of $\mathbf{F}_3$), and then drawing another line from the axis of rotation perpendicular to the first line (and also perpendicular to the rotation axis). The length of this second line is the lever arm for $\mathbf{F}_3$ and is labeled r_3 in Fig. 7-4b.

The torque associated with $\mathbf{F}_3$ is then $r_3 F_3$. This short lever arm and the corresponding smaller torque associated with $\mathbf{F}_3$ is consistent with the observed fact that $\mathbf{F}_3$ is less effective in opening the door than is $\mathbf{F}_1$. When the lever arm is defined in this way, experiment shows that the relation $\alpha \propto \tau$ is valid in general. Notice in Fig. 7-4a that the line action of the force $\mathbf{F}_4$ passes through the hinge and hence its lever arm is zero. Consequently zero torque is associated with $\mathbf{F}_4$ and it gives rise to no angular acceleration, in accord with everyday experience.

In general, then, we can write the torque about a given axis as

$$\tau = r_\perp F \tag{7-8a}$$

where $r_\perp$ is the lever arm and the perpendicular symbol ($\perp$) reminds us that we must use the distance from the axis of rotation that is perpendicular to the line of action of the force (Fig. 7-5a). An alternate but equivalent way of determining the torque associated with a force is to resolve the force into components parallel and perpendicular to a line joining the point of application of the force to the axis as shown in Fig. 7-5b. Then the torque will be equal to $F_\perp$ times the distance r from the axis to the point of application of the force:

$$\tau = r F_\perp. \tag{7-8b}$$

That this gives the same result as Eq. 7-8a can be seen from the fact that $F_\perp = F \sin \theta$ and $r_\perp = r \sin \theta$; so

$$\tau = r F \sin \theta \tag{7-8c}$$

in either case. We can use any of Eqs. 7-8 to calculate the torque, whichever is easiest.

Since torque is a force times a distance, it is measured in units of N·m in SI units, dyne · cm in the cgs system, and lb · ft in the English system.

> **EXAMPLE 7-6**   What is the torque applied by the biceps on the lower arm in Figs. 7-6a and b. The axis of rotation is through the elbow joint and the muscle is inserted 5.0 cm from the elbow joint.

FIGURE 7-5
Torque $= r_\perp F = r F_\perp$.

FIGURE 7-6
Example 7-6.

(a)

(b)

SOLUTION   (a) $F = 700$ N, $r_\perp = 0.050$ m, so $\tau = (0.050 \text{ m}) (700 \text{ N}) = 35$ N·m. (b) Because the arm is at a 45° angle, the lever arm is shorter (Fig. 7-6b): $r_\perp = (0.050 \text{ m}) (\sin 45°)$; F is still 700 N, so

$$\tau = (0.050 \text{ m}) (0.71) (700 \text{ N}) = 25 \text{ N} \cdot \text{m}.$$

7-4 ▌ Rotational dynamics; torque and rotational inertia

We have seen that the angular acceleration α of a rotating body is proportional to the torque τ applied to it:

$$\alpha \propto \tau.$$

This corresponds to Newton's second law for translational motion, $a \propto F$, where torque has taken the place of force and correspondingly the angular acceleration α takes the place of the linear acceleration a. In the linear case, the acceleration is not only proportional to the net force but is also inversely proportional to the inertia of the body, which we call its mass m; thus we could write $a = F/m$. But what plays the role of mass for the rotational case? That is what we now set out to determine. At the same time we will see that the relation $\alpha \propto \tau$ follows directly from Newton's second law, $F = ma$.

We first consider a very simple case: a particle of mass m rotating in a circle of radius r at the end of a string or rod whose mass we can ignore (Fig. 7-7). The torque which gives rise to its angular acceleration is $\tau = rF$. If we make use of Newton's second law for linear quantities, $F = ma$, and Eq. 7-5 relating the angular acceleration to the tangential linear acceleration, $a_T = r\alpha$, we have

$$F = ma$$
$$= mr\alpha.$$

When we multiply both sides by r we find that the torque $\tau = rF$ is given by

$$\tau = mr^2\alpha. \qquad \text{[single particle]} \quad (7\text{-}9)$$

Here at last we have a direct relation between the angular acceleration and the applied torque τ. The quantity mr^2 represents the *rotational inertia* of the particle and is called its *moment of inertia*.

Now let us consider a rotating rigid body, such as a wheel, which we can think of as consisting of many particles located at various distances from the axis of rotation. We can apply Eq. 7-9 to each particle of the body and then sum over all the particles. The sum of the various torques is just the total torque, which we now call τ, so we obtain:

$$\tau = \left(\sum mr^2\right)\alpha \qquad (7\text{-}10)$$

where we factored out the α since it is the same for all the particles of the body. The $\sum$ (Greek capital letter sigma) means "the sum of." The sum, $\sum mr^2$, represents the sum of the masses of each particle in the body multiplied by the square of the distance of that particle from the axis of rotation.

FIGURE 7-7
A mass m rotating in a circle of radius r about a fixed point.

If we give each particle a number (1, 2, 3, ...), then $\sum mr^2 = m_1 r_1^2 + m_2 r_2^2 + m_3 r_3^2 + \cdots$. This quantity is called the **moment of inertia** of the body, I:

$$I = \sum mr^2. \tag{7-11}$$

Combining Eqs. 7-10 and 7-11, we can write

$$\tau = I\alpha. \tag{7-12}$$

This is the rotational equivalent of Newton's second law. It is valid for the rotation of a rigid body about a fixed axis. (It can be shown that Eq. 7-12 is valid also when the body is translating with acceleration, but only if I and α are calculated about the center of mass of the body, and the rotation axis through the cm doesn't change direction.)

We see that the moment of inertia, I, which is a measure of the rotational inertia of a body, plays the same role for rotational motion that mass does for translational motion. As can be seen from Eq. 7-11, the rotational inertia of an object depends not only on its mass, but also on how that mass is distributed. For example, a large diameter cylinder will have greater rotational inertia than one of equal mass but smaller diameter (and therefore longer), Fig. 7-8. The former will be harder to start rotating, and harder to stop. When the mass is concentrated further from the axis of rotation, the rotational inertia is greater. For rotational motion, the mass of a body *cannot* be considered as concentrated at its center of mass.

Whenever Eq. 7-12 is used, it must be remembered to give α in rad/s^2 and other quantities in a consistent set of units. The moment of inertia, I, has units of kg·m^2 in SI units.

FIGURE 7-8

A large diameter wheel has greater rotational inertia than one of smaller diameter but equal mass.

☐ **EXAMPLE 7-7** Two particles, of mass 5.0 kg and 7.0 kg, are mounted 4.0 m apart on a light rod (whose mass can be ignored) as shown in Fig. 7-9. Calculate the moment of inertia of the system (*a*) when rotated about an axis passing halfway between the masses, Fig. 7-9a, and (*b*) when the system rotates about an axis located 0.50 m to the left of the 5.0 kg mass (Fig. 7-9b).

SOLUTION (*a*) Both particles are the same distance, 2.0 m, from the axis of rotation. Thus $I = \sum mr^2 = (5.0 \text{ kg})(2.0 \text{ m})^2 + (7.0 \text{ kg})(2.0 \text{ m})^2 = 48$ kg·m^2.

(*b*) The 5.0 kg mass is now 0.50 m from the axis and the 7.0 kg mass is 4.50 m from the axis. Then

$$I = \sum mr^2 = (5.0 \text{ kg})(0.50 \text{ m})^2 + (7.0 \text{ kg})(4.5 \text{ m})^2$$
$$= 1.3 \text{ kg·m}^2 + 142 \text{ kg·m}^2$$
$$= 143 \text{ kg·m}^2.$$

FIGURE 7-9

Calculating moment of inertia (Example 7-7).

The above example illustrates two important points. First, the moment of inertia of a given system is different for different axes of rotation. Second, part (*b*) illustrates the fact that mass close to the axis of rotation contributes little to the total moment of inertia; in this example, the 5.0 kg object contributed less than 1 percent to the total.

For most ordinary bodies, the mass is distributed continuously and calculation of the moment of inertia, $\sum mr^2$, can be difficult. Expressions can be worked out, however, for the moments of inertia of regularly shaped

Object	Location of axis		Moment of inertia	Radius of gyration
Thin ring of radius R	Through center		MR^2	R
Cylinder of radius R	Through center		$\frac{1}{2}MR^2$	$\frac{R}{\sqrt{2}}$
Sphere of radius R	Through center		$\frac{2}{5}MR^2$	$\sqrt{\frac{2}{5}}R$
Long rod of length L	Through center		$\frac{1}{12}ML^2$	$\frac{L}{\sqrt{12}}$
	Through end		$\frac{1}{3}ML^2$	$\frac{L}{\sqrt{3}}$

FIGURE 7-10

Moments of inertia for various objects of uniform composition.

bodies in terms of their dimensions. Fig. 7-10 gives these expressions for a number of solids rotated about the axes specified. The only one for which the result is obvious is that for the thin hoop rotated about an axis passing through its center perpendicular to the plane of the hoop. For this object, all the mass is concentrated at the same distance from the axis, R. Thus $\sum mr^2 = (\sum m)R^2 = MR^2$, where M is the total mass of the hoop.

When discussing moments of inertia, it is often convenient to work with the *radius of gyration, k*, which is a sort of average radius. In particular, the radius of gyration of an object is defined so that if all the mass of the object were concentrated at this distance from the axis, it would have the same moment of inertia as the original object. For example (see Fig. 7-10), the radius of gyration of a cylinder is $(1/\sqrt{2})R \approx 0.71R$; this means that a solid cylinder of radius 100 cm has the same moment of inertia as an equal-mass thin hoop of radius 71 cm. The moment of inertia of any object can be written in terms of its radius of gyration as

$$I = Mk^2.$$

For unusual or irregularly shaped objects it is common to specify the radius of gyration.

FIGURE 7-11

Example 7-8.

EXAMPLE 7-8   A 15.0-N force (represented by **T**) is applied to a cord wrapped around a wheel of mass $M = 4.00$ kg, radius $R = 33.0$ cm, and radius of gyration $k = 30.0$ cm, Fig. 7-11. If there is a frictional torque $\tau_{\mathrm{fr}} = 1.10$ N·m at the hub, what is the acceleration of the wheel?

SOLUTION   First we calculate I:

$$I = Mk^2 = (4.00 \text{ kg})(0.300 \text{ m})^2 = 0.360 \text{ kg·m}^2.$$

There are two torques involved, that due to the 15.0-N force of magnitude $(0.330 \text{ m})(15.0 \text{ N}) = 4.95$ N·m, and the opposing friction torque of

1.10 N·m. Thus we use Eq. 7-12 and obtain

$$\alpha = \frac{\tau}{I} = \frac{TR - \tau_{\mathrm{fr}}}{I} = \frac{(4.95 \ \mathrm{N \cdot m}) - (1.10 \ \mathrm{N \cdot m})}{0.360 \ \mathrm{kg \cdot m^2}} = 10.7 \ \mathrm{rad/s^2}.$$

EXAMPLE 7-9   Suppose in Fig. 7-11 that instead of a constant 15.0-N force being exerted on the cord hanging from the edge of the wheel, a block of weight 15.0 N (mass $m = 1.53$ kg) hangs from the cord, which we assume not to stretch. (a) Calculate the angular acceleration α of the wheel and the linear acceleration a of the mass m. (b) Determine the angular velocity ω of the wheel and the linear velocity v of the mass m at $t = 3.00$ s if the wheel starts from rest at $t = 0$.

SOLUTION   (a) Let T be the tension in the cord. Then a force T acts at the edge of the wheel and we have (see Example 7-8) for the wheel:

$$\alpha = \frac{\tau}{I} = \frac{TR - \tau_{\mathrm{fr}}}{I}.$$

Two forces act on the block of mass m: the force of gravity, mg, acts downward, and the tension of the cord T pulls upward. So by $F = ma$ we have

$$ma = mg - T.$$

Note that the tension T, which is the force exerted on the edge of the wheel, is *not* in general equal to the weight of the block ($= mg = 15.0$ N); this is because the block is accelerating, so (by the last equation above) $T = mg - ma$. To obtain α, we eliminate T between the above two equations and use Eq. 7-5

$$a = R\alpha$$

which is valid since the tangential acceleration of a point on the edge of the wheel is the same as the acceleration of the block if the cord doesn't stretch or slip. Thus, since $T = mg - ma = mg - mR\alpha$, we have

$$\alpha = \frac{(mg - mR\alpha)R - \tau_{\mathrm{fr}}}{I};$$

or (since α appears on both sides above)

$$\alpha = \frac{mgR - \tau_{\mathrm{fr}}}{I + mR^2}$$

$$= \frac{(15.0 \ \mathrm{N})(0.330 \ \mathrm{m}) - 1.10 \ \mathrm{N \cdot m}}{0.360 \ \mathrm{kg \cdot m^2} + (1.53 \ \mathrm{kg})(0.330 \ \mathrm{m})^2} = 7.31 \ \mathrm{rad/s^2}.$$

The angular acceleration is somewhat less in this case than in Example 7-8; this reflects the fact that T is somewhat less than the weight of the block, mg. The linear acceleration of the block is

$$a = R\alpha = (0.330 \ \mathrm{m})(7.31 \ \mathrm{rad/s^2}) = 2.41 \ \mathrm{m/s^2}.$$

(b) Since the angular acceleration is constant,

$$\omega = \omega_0 + \alpha t$$

$$= 0 + (7.31 \ \mathrm{rad/s^2})(3.00 \ \mathrm{s}) = 21.9 \ \mathrm{rad/s};$$

and the velocity of the block is the same as that for a point on the wheel's edge:

$$v = R\omega = (0.330 \text{ m})(21.9 \text{ rad/s}) = 7.23 \text{ m/s}.$$

This can also be obtained using the linear equation $v = v_0 + at = 0 + (2.41 \text{ m/s}^2)(3.00 \text{ s}) = 7.23 \text{ m/s}.$

7-5 ▌ Rotational kinetic energy

The quantity $\frac{1}{2}mv^2$ is the kinetic energy of a body undergoing translational motion. A body rotating about an axis is said to have *rotational kinetic energy* and by analogy we would expect this to be given by the formula $\frac{1}{2}I\omega^2$ where I is the moment of inertia of the body and ω is its angular velocity. We can indeed prove that this is true. Consider any rotating object as made up of many tiny particles, each of mass m. If we let r represent the distance of any one particle from the axis of rotation, then its linear velocity is $v = r\omega$. The total kinetic energy of the whole body will be the sum of the KEs of all its particles:

$$\begin{aligned} \text{KE} &= \sum \left(\tfrac{1}{2}mv^2\right) \\ &= \sum \left(\tfrac{1}{2}mr^2\omega^2\right) \\ &= \tfrac{1}{2}\left(\sum mr^2\right)\omega^2 \end{aligned}$$

where we have factored out the $\frac{1}{2}$ and the ω^2 since they are the same for every particle. Since $\sum mr^2 = I$, the moment of inertia, we see that the kinetic energy of a rotating object, as expected, is:

$$\text{rotational KE} = \tfrac{1}{2}I\omega^2. \tag{7-13}$$

An object that rotates while its cm undergoes translational motion will have both translational and rotational KE. Eq. 7-13 gives the rotational KE if the rotation axis is fixed; or, if the object is moving (such as a wheel rolling down a hill), it is still valid as long as the rotation axis is fixed in direction.[†] Then the total kinetic energy is

$$\text{KE} = \tfrac{1}{2}mv_{\text{cm}}{}^2 + \tfrac{1}{2}I_{\text{cm}}\omega^2$$

where v_{cm} is the linear velocity of the cm, I_{cm} is the moment of inertia about an axis through the cm, ω is the angular velocity about this axis, and M is the total mass of the body.

FIGURE 7-12

Sphere rolling down a hill has both translational and rotational kinetic energy, Example 7-10.

☐ **EXAMPLE 7-10** What will be the speed of a solid sphere of mass M and radius R when it reaches the bottom of an incline if it starts from rest at a vertical height H and rolls without slipping? See Fig. 7-12. Ignore losses due to retarding forces.

SOLUTION We use the law of conservation of energy where we must now include rotational kinetic energy. The total energy at any point a verti-

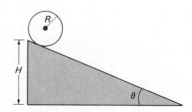

[†] If the axis is not fixed in space, the rotational KE can take on a more complicated form.

cal distance y above the base of the incline is

$$\tfrac{1}{2}Mv^2 + \tfrac{1}{2}I_{cm}\omega^2 + Mgy.$$

We equate the total energy at the top ($y = H$ and $v = \omega = 0$) to the total energy at the bottom ($y = 0$):

$$0 + 0 + MgH = \tfrac{1}{2}Mv^2 + \tfrac{1}{2}I_{cm}\omega^2 + 0.$$

From Fig. 7-10, the moment of inertia of a sphere about an axis through its cm is $I_{cm} = \tfrac{2}{5}MR^2$. Since the sphere rolls without slipping, the speed, v, of the center of mass with respect to the point of contact (which is momentarily at rest at any instant) is equal to the speed of a point on the edge relative to the center; we therefore have $\omega = v/R$. Hence

$$\tfrac{1}{2}Mv^2 + \tfrac{1}{2}(\tfrac{2}{5}MR^2)\left(\frac{v^2}{R^2}\right) = MgH.$$

Canceling the M's and R's, we obtain

$$(\tfrac{1}{2} + \tfrac{1}{5})v^2 = gH$$

or

$$v = \sqrt{\tfrac{10}{7}gH}.$$

This can be compared to the velocity of an object that slides down the same plane without rotating and without friction, in which case $v = \sqrt{2gH}$, which is greater.

If there were no friction between the sphere and plane in this example, the sphere would have slid instead of rolled. Friction must be present to make the sphere roll. We did not need to take it into account in the energy equation because it is static friction and does no work.

7-6 ▌ Angular momentum and its conservation

Throughout this chapter we have seen that by using the appropriate angular variables, the kinematic and dynamic equations for rotational motion are analogous to those for ordinary linear motion. In the last section (Section 7-5) we saw, for example, that rotational kinetic energy can be written as $\tfrac{1}{2}I\omega^2$, which is analogous to the translational KE $= \tfrac{1}{2}mv^2$. In like manner, the linear momentum, $p = mv$, has a rotational analog. It is called **angular momentum**, and for a body rotating about a fixed axis it is defined as

$$L = I\omega \tag{7-14}$$

where I is the moment of inertia and ω is the angular velocity.

We saw in Chapter 6 (Section 6-1) that Newton's second law can be written not only as $F = ma$, but also more generally in terms of momentum (Eq. 6-2), $F = \Delta p/\Delta t$. In a similar way, the rotational equivalent of Newton's second law, which we saw in Eq. 7-12 can be written as $\tau = I\alpha$, can also be written in terms of angular momentum:

$$\tau = \frac{\Delta L}{\Delta t} \tag{7-15}$$

where τ is the net torque acting to rotate the body, and ΔL is the change in angular momentum in the time Δt. Eq. 7-12, $\tau = I\alpha$, is a special case of Eq. 7-15 when the moment of inertia is constant. This can be seen as follows. If a body has angular velocity ω_0 at time $t = 0$, and angular velocity ω at a time Δt later, then its angular acceleration (Eq. 7-3) is

$$\alpha = \frac{\Delta\omega}{\Delta t} = \frac{\omega - \omega_0}{\Delta t}.$$

Then from Eq. 7-15 we have

$$\tau = \frac{\Delta L}{\Delta t} = \frac{I\omega - I\omega_0}{\Delta t} = \frac{I(\omega - \omega_0)}{\Delta t} = I\frac{\Delta\omega}{\Delta t}$$

$$\tau = I\alpha,$$

which is Eq. 7-12.

Angular momentum is an important concept in physics because under certain conditions it is a conserved quantity. We can see from Eq. 7-15 that if the net torque τ on a body is zero, then $\Delta L/\Delta t$ equals zero. That is L does not change. This, then, is the **law of conservation of angular momentum** for a rotating body:

> ■ The total angular momentum of a rotating body remains constant if the net torque acting on it is zero.

The law of conservation of momentum is one of the great conservation laws of physics.

When there is zero net torque acting on a body, and the body is rotating about a fixed axis or about an axis through its cm such that its direction doesn't change, we can write

$$I\omega = I_0\omega_0 = \text{constant};$$

I_0 and ω_0 are the moment of inertia and angular velocity about that axis at some initial time ($t = 0$), and I and ω are their values at some other time. The parts of the body may alter their positions relative to one another, so that I changes; ω changes as well and the product $I\omega$ remains constant.

Many interesting everyday phenomena can be understood on the basis of conservation of angular momentum. For example, a skater doing a spin on ice rotates at a relatively low speed when her arms are outstretched; but when she brings her arms in close to her body she suddenly spins much faster. Remembering the definition of moment of inertia as $I = \sum mr^2$ it is clear that when she pulls her arms in closer to the axis of rotation her moment of inertia is reduced. Since the angular momentum $I\omega$ remains constant (we ignore the small torque due to friction), if I decreases then the angular velocity ω must increase. If she reduces her moment of inertia by a factor of two, she will then rotate with twice the angular velocity.

A similar example is the diver shown in Fig. 7-13. The push as he leaves the board gives him an initial angular momentum about his cm. When he curls himself into the tuck position he rotates quickly one or more times; he then stretches out again, increasing his moment of inertia, which reduces the angular velocity to a small value, and then he enters the water. The change in moment of inertia from the straight position to the tuck position can be a factor of as much as $3\frac{1}{2}$.

FIGURE 7-13

Diver rotates faster when arms and legs are tucked in than when outstretched. Angular momentum is conserved.

FIGURE 7-14

Example 7-11.

EXAMPLE 7-11   A mass m attached to the end of a string revolves in a circle on a frictionless table top. The other end of the string passes through a hole in the table, Fig. 7-14. Initially the ball rotates with a speed $v_1 = 2.4$ m/s in a circle of radius $r_1 = 0.80$ m. The string is then pulled slowly through the hole so the radius is reduced to $r_2 = 0.48$ m. What is the speed, v_2, of the mass now?

SOLUTION   The force exerted by the string on the mass m does not alter its angular momentum about the axis of rotation, since the force is exerted toward the axis; so $\tau = 0$. Hence, from conservation of angular momentum:

$$I_1\omega_1 = I_2\omega_2;$$

since the moment of inertia of a single particle is $I = mr^2$, we have

$$mr_1^2\omega_1 = mr_2^2\omega_2.$$

Then, since $v = r\omega$, we can write:

$$v_2 = r_2\omega_2 = r_2\omega_1\left(\frac{r_1^2}{r_2^2}\right) = r_2\frac{v_1}{r_1}\left(\frac{r_1^2}{r_2^2}\right) = v_1\frac{r_1}{r_2}$$

$$= (2.4 \text{ m/s})\left(\frac{0.80 \text{ m}}{0.48 \text{ m}}\right) = 4.0 \text{ m/s}.$$

FIGURE 7-15

(a) Rotating wheel.
(b) Right-hand rule for obtaining direction of ω.

*7-7 ∎ Vector nature of angular quantities

Just as Newton's second law, $\mathbf{F} = \Delta\mathbf{p}/\Delta t$ or $\mathbf{F} = m\mathbf{a}$, is a vector equation, so too is the rotational equivalent which we have written only with magnitudes in Eqs. 7-12 ($\tau = I\alpha$) and 7-15 ($\tau = \Delta L/\Delta t$). That is, we can write

$$\boldsymbol{\tau} = I\boldsymbol{\alpha} \qquad (7\text{-}16\text{a})$$

or

$$\boldsymbol{\tau} = \frac{\Delta\mathbf{L}}{\Delta t} = \frac{\Delta(I\boldsymbol{\omega})}{\Delta t} \qquad (7\text{-}16\text{b})$$

where $\boldsymbol{\tau}$, $\boldsymbol{\alpha}$, $\mathbf{L}$, and $\boldsymbol{\omega}$ are vectors. But in what directions do these vectors point? In fact, we have to *define* the directions for rotational quantities, and we take first the angular velocity, $\boldsymbol{\omega}$.

Consider the rotating wheel shown in Fig. 7-15a. The linear velocity of different particles of the wheel point in all different directions. The only

(a)

(b)

unique direction in space associated with the rotation is along the axis of rotation, perpendicular to the actual motion. We therefore choose the axis of rotation to be the direction of the angular velocity vector, **ω**. Actually, there is still an ambiguity since **ω** could point in either direction along the axis of rotation (up or down in Fig. 7-15a). The convention we use, called the *right-hand rule*, is the following: when the fingers of the right hand are curled around the rotation axis and point in the direction of the rotation, then the thumb points in the direction of **ω**. This is shown in Fig. 7-15b. Note that **ω** points in the direction a right-handed screw would move when turned in the direction of rotation. Thus, if the rotation of the wheel in Fig. 7-15a is counterclockwise, the direction of **ω** is upward as shown in Fig. 7-15b; if the wheel rotates clockwise, then **ω** points in the opposite direction, downward. Note that no part of the rotating body moves in the direction of **ω**.

If the axis of rotation is fixed, then **ω** can change only in magnitude. Thus $\boldsymbol{\alpha} = \Delta\boldsymbol{\omega}/\Delta t$ must also point along the axis of rotation. If the rotation is clockwise as in Fig. 7-15a and if the magnitude ω is increasing, **α** points upward, but if ω is decreasing (the wheel is slowing down), **α** points downward. If the rotation is clockwise, **α** will point downward if ω is increasing, and points upward if ω is decreasing.

Since **ω** always points along the axis of rotation, if the axis of rotation changes direction, **ω** changes direction. In this case **α** will not point along the axis of rotation. We will see examples of this later in this chapter. But for motion about a fixed axis, **ω** and **α** are both along the rotation axis.

The vector nature of torque can be expressed in a similar way. Suppose, for example, the wheel of Fig. 7-15 is given an angular acceleration by a force F applied at the edge of the wheel (and in the plane of the wheel) as shown in Fig. 7-16. The position vector **r** is drawn from a point 0 on the rotation axis to the point where the force **F** acts. The torque vector can be defined in terms of the cross product of two vectors.

In general, the **vector** or **cross product** of two vectors, **A** and **B**, is defined as another vector $\mathbf{C} = \mathbf{A} \times \mathbf{B}$ *whose magnitude is*

$$C = |\mathbf{A} \times \mathbf{B}| = AB \sin \theta$$

where θ is the angle ($< 180°$) between **A** *and* **B**, *and whose direction is perpendicular to both* **A** *and* **B** *in the sense of the right-hand rule*, Fig. 7-17. According to the right-hand rule, as shown in the figure, you orient your hand so your fingers point along **A**, and when you bend your fingers they point along **B**. When your hand is correctly oriented in this way, your thumb will point along the direction of $\mathbf{C} = \mathbf{A} \times \mathbf{B}$.

The **torque vector** is defined as the cross product of **r** and **F**:

$$\boldsymbol{\tau} = \mathbf{r} \times \mathbf{F}.$$

Notice that this definition is consistent with what we have said above about torque: first, according to the definition of the cross product above, the magnitude of the torque is $\tau = rF \sin \theta$, and this is just what we already obtained in Eq. 7-8c. Secondly, the direction of **τ** is given by the right-hand rule; when we apply this to the vectors **r** and **F** of Fig. 7-16, we see that **τ** will point outward (toward the reader) along the axis of rotation. This is the same direction that **α** points, as we discussed above relative to Fig. 7-15. This

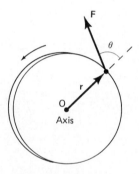

FIGURE 7-16

The torque due to **F** starts the wheel rotating counterclockwise so ω and α point out of the page.

FIGURE 7-17

The vector $\mathbf{C} = \mathbf{A} \times \mathbf{B}$ is perpendicular to the plane containing **A** and **B**; its direction is given by the right-hand rule.

checks Eq. 7-16a, above,

$$\tau = I\alpha,$$

as a vector equation: the directions are the same on the two sides of the equation, as are the magnitudes.

*7-8 ▌ Vector angular momentum; a rotating wheel

Angular momentum, like linear momentum, is a vector quantity. For a symmetrical body (such as a wheel, cylinder, hoop, sphere) rotating about a symmetry axis,[†] we can write the vector angular momentum as

$$\mathbf{L} = I\omega; \qquad (7\text{-}17)$$

the angular velocity vector ω (and therefore also $\mathbf{L}$) points along the axis of rotation in the direction given by the right-hand rule (Fig. 7-15b).

The vector nature of angular momentum can be used to explain a number of interesting (and sometimes surprising) phenomena. For example, consider a person standing at rest on a circular platform capable of rotating friction-free about an axis through its center (that is, a simplified merry-go-round). If the person now starts to walk along the edge of the platform, the platform starts rotating in the opposite direction. Why? Well, this is an example of the conservation of angular momentum. If the person starts walking counterclockwise, the person's angular momentum will be pointed upward along the axis of rotation (remember how we defined the direction of ω using the right-hand rule). The magnitude of the person's angular momentum will be $L = I\omega = (mr^2)(v/r)$ where v is the person's speed (relative to earth, not the platform), r is his distance from the rotation axis, m is his mass, and his moment of inertia is mr^2 if we consider him a particle (mass concentrated at one point). The platform rotates in the opposite direction, so its angular momentum points downward. If the initial total angular momentum was zero (person and platform at rest), it will remain zero after the person starts walking—that is, the upward angular momentum of the person just balances the oppositely directed angular momentum of the platform, so the total vector angular momentum is still zero. Even though the person exerts a force (and torque) on the platform, and vice versa, these are internal torques (internal to the system consisting of platform plus person); there are no external torques (assuming friction-free bearings of the platform) so from Eq. 7-16b the angular momentum remains constant.

Next we consider an experiment that is easily done, and gives a surprising result which illustrates the vector nature of the relation

$$\tau = \frac{\Delta \mathbf{L}}{\Delta t}.$$

Suppose you are holding a bicycle wheel by a handle connected to its axle as in Fig. 7-18a. The wheel is spinning rapidly so its angular momentum $\mathbf{L}$

[†] If the rotation axis is not a symmetry axis of the body, then Eq. 7-17 is valid for the component of $\mathbf{L}$ along the rotation axis. There may or may not be other components, but we won't consider these more complicated cases here.

FIGURE 7-18

When you try to tilt a rotating bicycle wheel vertically, it swerves to the side.

(a)

(b)

points horizontally as shown. Now you suddenly try to tilt the axle upward as shown by the dashed line in Fig. 7-18a (so the cm moves vertically). You expect the wheel to go up, but it unexpectedly swerves to the right! To explain this bizarre effect—you may need to do it to believe it—we only need to use the relation $\tau = \Delta\mathbf{L}/\Delta t$. In the short time Δt, you exert a torque (about an axis through your wrist) that points along the x axis perpendicular to $\mathbf{L}$. Thus the change in $\mathbf{L}$ is

$$\Delta\mathbf{L} \approx \tau \,\Delta t$$

so $\Delta\mathbf{L}$ must also point (approximately) along the x axis, since τ does (Fig. 7-18b). Thus the new angular momentum, $\mathbf{L} + \Delta\mathbf{L}$, points to the right of $\mathbf{L}$. Since the angular momentum is directed along the axle of the wheel, we see that the axle, which now is along $\mathbf{L} + \Delta\mathbf{L}$, must move sideways, to the right, which is what we observe.

Another example of the conservation of angular momentum is the gyroscope, used by mariners and as the basis for guidance systems of aircraft. The rapidly spinning wheel is mounted on a complicated set of bearings (Fig. 7-19) so that even when the mount moves, no net torque acts to change the direction of the angular momentum. Thus the axis of the wheel remains pointed in the same direction in space.

FIGURE 7-19

A gyroscope. From D. C. Giancoli, *Ideas of Physics*, 2nd ed. Reproduced by permission from Harcourt Brace Jovanovich, Inc.

❚ Summary

When a rigid body rotates about a fixed axis, each point of the body moves in a circular path. Lines drawn perpendicularly from the rotation axis to various points in the body all sweep out the same angle θ in any given time interval. Angles are conveniently measured in *radians* where one radian is the angle subtended by an arc whose length is equal to the radius.

Angular velocity, ω, is defined as the rate of change of angular position: $\omega = \Delta\theta/\Delta t$. All parts of a rigid body rotating about a fixed axis have the same angular velocity at any instant. *Angular acceleration*, α, is defined as the rate of change of angular velocity: $\alpha = \Delta\omega/\Delta t$. Angular velocity is related to *frequency* f by $\omega = 2\pi f$. *Angular acceleration*, α, is the rate at which angular velocity changes. The linear velocity v of a point fixed at a distance r from the axis of rotation is $v = r\omega$, and its tangential linear acceleration is $a_T = r\alpha$. The equations describing uniformly accelerated rotational motion (α = constant) have the same form as for uniformly accelerated linear motion:

$$\omega = \omega_0 + \alpha t; \qquad \theta = \omega_0 t + \tfrac{1}{2}\alpha t^2;$$

$$\omega^2 = \omega_0^2 + 2\alpha\theta; \qquad \bar{\omega} = \frac{\omega + \omega_0}{2}.$$

The dynamics of rotation is analogous to the dynamics of linear motion; force is replaced by *torque*, τ, which is defined as the product of force times lever arm (perpendicular distance from the line of action of the force to the axis of rotation); mass is replaced by *moment of inertia*, I, which depends not only on the mass of the body but also on how the mass is distributed about the axis of rotation; and linear acceleration is replaced by angular acceleration. The rotational equivalent of Newton's second law is then

$$\tau = I\alpha.$$

The *rotational kinetic energy* of a body rotating about a fixed axis with angular velocity ω is

$$\text{KE} = \tfrac{1}{2}I\omega^2.$$

For a body both translating and rotating, the total kinetic energy is the sum of the translational KE of the body's cm plus the rotational KE of the body about its cm:

$$\text{KE} = \tfrac{1}{2}Mv_{\text{cm}}^2 + \tfrac{1}{2}I_{\text{cm}}\omega^2$$

as long as the rotation axis is fixed in direction.

The *angular momentum*, L, of a body about a fixed rotation axis is given by

$$L = I\omega.$$

Newton's second law, in terms of angular momentum, becomes

$$\tau = \frac{\Delta L}{\Delta t}.$$

If the net torque on the body is zero, $\Delta L/\Delta t = 0$, so $L = $ constant. This is the *law of conservation of angular momentum* for a rotating body.

▌ Questions

1. You are standing a known distance from the Statue of Liberty. Describe how you could determine its height using only a meter stick, and without moving from your place.

2. A bicycle odometer (which measures distance traveled) is attached near the wheel hub and is designed for 27-in wheels. What happens if you use it on a bicycle with 24-in wheels?

3. Suppose a record turntable rotates at constant angular velocity. Does a point on the rim have radial and/or tangential acceleration? If the turntable accelerates uniformly, does the point have radial and/or tangential acceleration? For which cases would the magnitude of either (linear) acceleration change?

4. If the angular quantities θ, ω, and α were specified in terms of degrees rather than radians, how would Eqs. 7-7 for uniformly accelerated rotational motion have to be altered?

5. Can a small force exert a greater torque than a larger force? Explain.

6. If a force **F** acts on a body such that its lever arm is zero, does it have any effect on the body's motion?

7. Why is it more difficult to do a situp with your hands behind your head than when they are outstretched in front of you? A diagram may help you to answer this.

8. Expert bicyclists use very lightweight "sew-up" (tubular) tires. They claim that reducing the mass of the tires is far more significant than an equal reduction in mass elsewhere on the bicycle. Explain why this is true.

9. Mammals that depend on being able to run fast have slender lower legs with flesh and muscle concentrated high, near the shoulder. On the basis of rotational dynamics, explain why this distribution of mass is advantageous.

10. Why do tightrope walkers carry a long, narrow beam?

11. If the net force on a system is zero, is the net torque also zero? If the net torque on a system is zero, is the net force zero?

12. A stick stands vertically on its end on a frictionless surface. Describe its motion when it is tipped slightly to one side and falls.

13. A sphere, a cylinder, and a hoop, all of the same diameter, roll down an inclined plane. Which arrives at the bottom first? Which last?

14. Two spheres stand at rest at the top of an incline. One has twice the radius and twice the mass of the other. Which reaches the bottom of the incline first? Which has the greater speed there? Which has the greater total kinetic energy at the bottom?

15. A cyclist rides over the top of a hill. Is the bicycle's motion rotational, translational, or a combination of both?

16. If there were a great migration of people toward the equator, how would this affect the length of the day?

17. We claim that momentum and angular momentum are conserved. Yet most moving or rotating bodies eventually slow down and stop. Explain.

*18. In what direction is the earth's angular velocity for its daily rotation on its axis?

*19. The angular velocity of a wheel rotating on a horizontal axle points west. In what direction is the linear velocity of a point on the top of the wheel? If the angular acceleration points east, describe the tangential linear acceleration of this point. Is the angular speed increasing or decreasing?

*20. When a motorcyclist leaves the ground on a jump, if the throttle is left on (so the rear wheel spins) why does the front of the cycle rise up?

*21. A quarterback leaps into the air to throw a forward pass. As he throws the ball, the upper part of his body rotates; if you look quickly you will notice that his hips and legs rotate in the opposite direction. Explain.

*22. Look at the face of a clock with a second hand. In what direction is the angular momentum of the second hand?

*23. On the basis of the law of conservation of angular momentum, discuss why a helicopter must have more than one rotor (or propeller). Discuss one or more ways the second propeller can operate in order to keep the body stable.

*24. Describe the torque needed if the person in Fig. 7-18 is to tilt the axle of the rotating wheel directly upward with no swerving to the side.

*25. Suppose you are standing on a turntable that can freely rotate. When you hold a rotating bicycle wheel over your head, with its axis vertical, you are at rest. If you now move the wheel so its axis is horizontal, what happens to you? What happens if you then point the axis of the wheel downward?

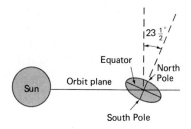

FIGURE 7-20

*26. *Precession of the equinoxes.* The axis of the earth precesses with a period of about 25,000 years. This is much like the precession of a top. Explain how the earth's equatorial bulge gives rise to a torque exerted by the sun on the earth; see Fig. 7-20 which is drawn for the winter solstice (December 21) with the equatorial bulge exaggerated. About what axis would you expect the earth's rotation axis to precess as a result of this torque? Does the torque exist 3 months later? Explain.

▌ Problems

SECTION 7-1

1. (I) What are the following angles in radians: (a) 30°, (b) 90°, and (c) 420°?

2. (I) The Eiffel Tower is 300 m tall. When you are standing at a certain place in Paris, it subtends an angle of 5°. How far are you, then, from the Eiffel Tower?

3. (I) A laser beam is directed at the moon, 380,000 km from earth. The beam diverges at an angle of 1.8×10^{-5} rad. How large a spot will it make on the moon?

4. (I) A bicycle with 68-cm-diameter tires travels 2.0 km. How many revolutions do the wheels make?

5. (I) A 20-cm-diameter grinding wheel rotates at 2000 rpm. Calculate its angular velocity in rad/s.

6. (I) What is the linear speed of a point on the edge of the grinding wheel in problem 5?

7. (I) A 70-cm-diameter wheel rotating at 1200 rpm is brought to rest in 15 s. Calculate its angular acceleration.

8. (I) A 33-rpm phonograph record reaches its rated speed 2.8 s after turning it on. What was the angular acceleration?

9. (I) A record player turntable of radius R_1 is turned by a circular rubber roller of radius R_2 in contact with it at their outer edges. What is the ratio of their angular velocities, ω_1/ω_2?

10. (II) Estimate the angle subtended by the moon using a ruler and your finger or other object to just blot out the moon. Describe your measurement and the result obtained and then use it to estimate the diameter of the moon. The moon is about 380,000 km from the earth.

11. (II) Calculate the angular velocity of the earth (a) in its orbit around the sun, and (b) about its axis.

12. (II) What is the linear speed of a point (a) on the equator, (b) at a latitude of 50°N, due to the earth's rotation? (See previous problem.)

13. (II) A large spool of rope lies on the ground with the end of the rope lying on the top edge of the spool. A person grabs the end of the rope and walks a distance L holding onto it. The spool rolls behind him without slipping. What length of rope unwinds from the spool? How far does the spool's cm move?

14. (II) A 40-cm-diameter wheel accelerates uniformly from 80 rpm to 300 rpm in 3.6 s. Determine (a) its angular acceleration and (b) the radial and tangential components of the linear acceleration of a point on the edge of the wheel 2.0 s after it started accelerating. Assume the rotation axis is fixed.

SECTION 7-2

15. (I) An automobile engine slows down from 4500 rpm to 1000 rpm in 6.5 s. Calculate (a) its angular acceleration, assumed uniform, and (b) the total number of revolutions the engine makes in this time.

16. (I) A centrifuge accelerates from rest to 10,000 rpm in 420 s. Through how many revolutions did it turn in this time?

17. (I) A phonograph turntable reaches its rated speed of 33 rpm after making 1.5 revolutions. What was its angular acceleration?

18. (II) Two rubber wheels are mounted next to one another so their circular edges touch. Wheel 1, of radius $R_1 = 3.0$ cm, accelerates at a rate 0.88 rad/s² and drives the second wheel, of radius $R_2 = 5.0$ cm, by contact (without slipping). (a) Starting from rest, how long does it take the second wheel to reach an angular speed of 33 rpm? (b) What was the angular acceleration of wheel 2?

19. (II) Starting from the definitions of ω and α, derive Eqs. 7-7 assuming constant angular acceleration.

20. (II) A cyclist accelerates from rest at a rate of 0.80 m/s². How fast will a point on the rim of the tire (diameter = 68 cm) at the top be moving after 4.0 s? (Hint: at any moment, the lowest point on the tire is in contact with the ground and hence is at rest.)

21. (II) A 40-cm-diameter wheel accelerates uniformly from 80 rpm to 300 rpm in 3.6 s. How far will a point on the edge of the wheel have traveled in this time?

156

22. (II) The tires of a car make 55 revolutions as the car reduces its speed uniformly from 80 km/h to 55 km/h. The tires have a diameter of 1.0 m. (*a*) What was the angular acceleration? (*b*) If the car continues to decelerate at this rate, how much more time is required for it to stop?

23. (III) A wheel, starting from rest, undergoes uniform angular acceleration α about its fixed axle. (*a*) Write the components of the linear acceleration, a_T and a_c, for a point P which is a distance r from the axle in terms of α, r and time t. (*b*) Let ϕ be the angle between the linear acceleration vector, **a**, and the line drawn between P and the axis. Express ϕ in terms of the total number of revolutions of the wheel, N.

SECTIONS 7-3 and 7-4

24. (I) A person exerts a force of 18 N on the end of a door 84 cm wide. What is the magnitude of the torque if the force is exerted (*a*) perpendicular to the door, and (*b*) at a 60° angle to the face of the door?

25. (I) What is the maximum torque exerted by a 60-kg person riding a bike if the person puts all the weight on each pedal when climbing a hill? The pedals rotate in a circle of radius 18 cm.

26. (I) Calculate the moment of inertia of a 12.0-kg sphere of radius 0.205 m when the axis of rotation is through its center.

27. (I) Calculate the moment of inertia of a 66.7-cm-diameter bicycle wheel. The rim and tire have a combined mass of 1.13 kg. The mass of the hub can be ignored. (Why?)

28. (II) A small 12.4-kg ball on the end of a light rod is rotated in a horizontal circle of radius 2.20 m. Calculate: (*a*) the moment of inertia of the system about the axis of rotation, and (*b*) the torque needed to keep the ball rotating at constant angular velocity if air resistance exerts a force of 0.0200 N on the ball.

29. (II) A grinding wheel is a uniform cylinder of radius 8.25 cm and mass 0.880 kg. Calculate: (*a*) its moment of inertia, and (*b*) the torque needed to accelerate it from rest to 1200 rpm in 4.00 s if a frictional torque of 0.0145 N·m is also acting.

30. (II) What is the radius of gyration of a 13.6-kg wheel that accelerates from rest to 600 rpm in 10.0 s when a net torque of 3.25 N·m acts on it?

31. (II) A 1.84-m-diameter sphere can be rotated about an axis through its center by a torque of 12.3 N·m which accelerates it uniformly from rest through a total of 180 rev in 15.0 s. What is the mass of the sphere?

32. (II) A merry-go-round accelerates from rest to 3.0 rad/s in 34 s. Assuming that the merry-go-round is a uniform disc of radius 8.0 m and mass 31,000 kg, calculate the net torque required to accelerate it.

33. (II) A centrifuge rotor rotating at 10,000 rpm is shut off and is eventually brought to rest by a frictional torque of 1.20 N·m. If the mass of the rotor is 4.37 kg and its radius of gyration is 0.0710 m, through how

FIGURE 7-21

many revolutions will the rotor turn before coming to rest and how long will it take?

34. (II) The forearm in Fig. 7-21 accelerates a 3.5-kg ball at 8.0 m/s² by means of the triceps muscle as shown. Calculate: (*a*) the torque needed, and (*b*) the force that must be exerted by the triceps muscle. Ignore the mass of the arm.

35. (III) An *Atwood machine* consists of two masses, m_1 and m_2, which are connected by a massless inelastic cord that passes over a pulley, Fig. 7-22. If the pulley has radius R_0 and moment of inertia I about its axle, determine the acceleration of the masses m_1 and m_2, and compare to the situation where the moment of inertia of the pulley is ignored.

36. (III) Assume that a 1.00-kg ball is thrown solely by the action of the forearm, which rotates about the elbow joint under the action of the triceps muscle, Fig. 7-21. The ball is accelerated from rest to 10.0 m/s in 0.220 s, at which point it is released. Calculate: (*a*) the angular acceleration of the arm, and (*b*) the force required of the triceps muscle. Assume that the forearm has a mass of 3.2 kg and rotates like a uniform rod about an axis at its end.

FIGURE 7-22
Atwood's machine.

FIGURE 7-23

37. (III) A hammer thrower accelerates the hammer (mass = 7.30 kg) from rest within four full turns (revolutions) and releases it at a speed of 27.2 m/s. Assuming a uniform rate of increase in angular velocity and a radius of 2.10 m, calculate: (a) the angular acceleration, (b) the (linear) tangential acceleration, (c) the centripetal acceleration just before release, (d) the net force being exerted on the hammer by the athlete just before release, and (e) the angle of this force with respect to the radius of the circular motion.

38. (III) A bicyclist traveling with speed $v = 5.2$ m/s on a flat road is making a turn with a radius $r = 3.8$ m. The forces acting on the cyclist and cycle are the normal force ($\mathbf{F_N}$) and friction force ($\mathbf{F_{fr}}$) exerted by the road on the tires, and $m\mathbf{g}$, the total weight of the cyclist and cycle. (a) Explain carefully why the angle θ the bicycle makes with the vertical (Fig. 7-23) must be given by $\tan \theta = F_{fr}/F_N$ if the cyclist is to maintain balance. (b) Calculate θ for the values given. (Hint: consider the "circular" translational motion of the bicycle and rider.) (c) If the coefficient of static friction between tires and road is $\mu_s = 0.65$, what is the minimum turning radius?

SECTION 7-5

39. (I) A centrifuge rotor has a moment of inertia of 4.00×10^{-2} kg·m². How much energy is required to bring it from rest to 10,000 rpm?

40. (II) Calculate the translational speed of a cyclinder when it reaches the foot of an incline 18.0 m high. Assume it starts from rest and rolls without slipping.

41. (II) A merry-go-round has a mass of 1560 kg and a radius of gyration of 18.5 m. How much work is required to accelerate it from rest to a rotation rate of one revolution in 7.10 s?

42. (II) (a) Show that the work done by a force F on a rotating body is equal to $W = \tau \Delta\theta$ where τ is the torque due to F (assumed constant) and $\Delta\theta$ is the angle through which the body is rotated. (b) Show that the rate of work done, or power P, is $P = \tau\omega$ where ω is the angular velocity at any instant. (c) An automobile engine develops a torque of 280 N·m at 4,000 rpm. What is the horsepower of the engine?

43. (II) One possibility for a low-pollution automobile is for it to use energy stored in a heavy rotating flywheel. Suppose such a car has a total mass of 1400 kg, uses a 1.50-m diameter uniform cylindrical flywheel of mass 240 kg, and should be able to travel 300 km without needing a flywheel "spinup." (a) Make reasonable assumptions (average frictional retarding force = 500 N, twenty acceleration periods from rest to 90 km/h, equal uphill and downhill—assuming during downhill, energy can be put back into the flywheel), and show that the total energy needed to be stored in the flywheel is about 1.6×10^8 J. (b) What is the angular velocity of the flywheel when it has a full "energy charge"? (c) About how long would it take a 150 hp motor to give the flywheel a full energy charge before a trip?

44. (II) A hollow cylinder (hoop) is rolling on a horizontal surface at speed $v = 3.4$ m/s when it reaches a 20° incline. (a) How far up the incline will it go? (b) How long will it be on the incline before it arrives back at the bottom?

45. (III) Two masses, $m_1 = 22.5$ kg and $m_2 = 33.2$ kg, are connected by a rope that hangs over a pulley (as in Fig. 7-22). The pulley is a uniform cylinder of radius 0.300 m and mass 8.80 kg. Initially m_1 is on the ground and m_2 rests 2.50 m above the ground. If the system is now released, use conservation of energy to determine the speed of m_2 just before it strikes the ground. Assume the pulley is frictionless.

46. (III) A 5.10-m-long pole is balanced vertically on its tip. What will be the speed of the tip of the pole just before it hits the ground? Assume the lower end of the pole does not slip.

SECTION 7-6

47. (I) What is the angular momentum of a 200-g ball rotating on the end of a string in a circle of radius 1.00 m at an angular speed of 9.45 rad/s?

48. (I) A person stands, hands at the side, on a platform that is rotating at a rate of 1.20 rev/s. If the person now raises her arms to a horizontal position, the speed of rotation decreases to 0.80 rev/s. (a) Why does this occur? (b) By what factor has the moment of inertia of the person changed?

49. (I) A diver (such as that shown in Fig. 7-13) can reduce his moment of inertia by a factor of about 3.5 when changing from the straight position to the tuck position. If he makes two rotations in 1.5 s when in the tuck position, what is his angular speed (rev/s) when in the straight position?

50. (II) (a) What is the angular momentum of a 2.13-kg uniform cylindrical grinding wheel of radius 12.5 cm when rotating at 1500 rpm? (b) How much torque is required to stop it in 7.00 s?

51. (II) The moon orbits the earth so that the same side always faces the earth. Determine the ratio of its spin angular momentum (about its own axis) and its orbital angular momentum. (In the latter case, treat the moon as a particle orbiting the earth.)

52. (II) A toy train can run on a horizontal circular track of radius R mounted near the edge of a solid circular board of uniform thickness with mass M that is free to rotate, without friction, about a vertical axis through its center. If the system starts from rest, what mass should the train have so that its speed relative to the ground is always equal to that of a point on the track (but opposite direction) no matter how fast it travels relative to the tracks?

53. (II) A 4.5-m-diameter merry-go-round is rotating freely with an angular velocity of 0.70 rad/s; its total moment of inertia is 1750 kg·m². Four people standing on the ground, each of 65-kg mass, suddenly step onto the edge of the merry-go-round. What will be the angular velocity of the merry-go-round now? What if the people were on it initially, and then jumped off?

54. (III) Suppose a 60-kg person stands at the edge of a 6.0-m-diameter merry-go-round turntable which is mounted on frictionless bearings and has a moment of inertia of 1800 kg·m². The turntable is at rest initially, but when the person begins running at a speed of 4.2 m/s (with respect to the turntable) around its edge, the turntable begins to rotate in the opposite direction. Calculate the angular velocity of the turntable.

*SECTION 7-8

*55. (II) A person whose moment of inertia is I_p stands at the center of a platform that can freely rotate without friction (but initially is at rest) and holds a spinning bicycle wheel with axis horizontal as in Fig. 7-18. The wheel has moment of inertia I_w and angular velocity, ω_w. What will be the angular velocity ω of the platform if the person moves the axis of the wheel so it points (a) vertically upward, (b) at a 60° angle to the vertical, (c) vertically downward. (d) What will ω be if the person reaches up and stops the wheel in part (a)?

8 BODIES IN EQUILIBRIUM; ELASTICITY AND FRACTURE

8-1 ∎ Statics—the study of forces in equilibrium

In this chapter we will study a special case of motion—when the net force and torque on an object or system of objects are zero. In this case either the object or system will be at rest, or the center of mass of the object or system will be moving at constant velocity. We will be concerned mainly with the first case, when the object or objects are all at rest. Now you may think that the study of objects at rest is not very interesting since the objects will have neither velocity nor acceleration; and the net force and the net torque will be zero. But it is just this last aspect that makes this subject of *statics* interesting: true, the *net* force on an object at rest is zero, but this does not imply that no forces act on it. In fact it is virtually impossible to find a body on which no forces act at all. Objects within our experience have at least one force acting on them (gravity) and if they are at rest then there must be other forces acting on them as well so that the total net force is zero. An object at rest on a table, for example, has two forces acting on it, the downward force of gravity and the force the table exerts upward on it, Fig. 8-1. Since the net force is zero, the upward force exerted by the table must be equal to the force of gravity acting downward. (Do not confuse this with the equal and opposite forces of Newton's third law which act on different bodies; here both forces act on the same body.) Such a body is said to be in **equilibrium** (Latin for 'equal forces") under the action of these two forces.[†]

The subject of statics is concerned with the calculation of the forces acting on bodies which are in equilibrium. The techniques for doing this find application in a wide range of fields. A knowledge of the forces in muscles and joints is of great value in medicine and physical therapy, and is also valuable in the study of athletic activity. Architects and engineers must be able to calculate the forces on the structural members of a building, for any material will break or buckle if too much force is applied.

Let us take a simple example of the addition of forces applied to orthodonture.

FIGURE 8-1

The book is in equilibrium; the net force on it is zero.

[†] A body in equilibrium at rest is said to be in *static equilibrium* whereas a body moving uniformly at constant velocity is in *dynamic equilibrium;* we consider mainly the former case.

☐   **EXAMPLE 8-1**   The wire band shown in Fig. 8-2 has a tension T of 2.0 N along it. It therefore exerts a force of 2.0 N on the tooth (to which it is attached) in the two directions shown. Calculate the resultant force on the tooth due to the wire.

SOLUTION   Since the two forces are equal, their sum will be directed along the line that bisects the angle between them, which we have labeled the y axis. The x components of the two forces add up to zero. The y component of each force is $(2.0 \text{ N})(\cos 70°) = 0.68$ N and adding the two together we get a total force of 1.36 N. Note that if the wire is firmly attached to the tooth, the tension to the right, say, can be made larger than to the left and the net force would correspondingly be directed more toward the right.

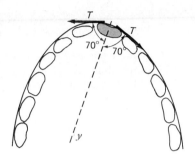

FIGURE 8-2

Forces on a tooth, Example 8-1.

Although it is intended that the tooth in this example should move, it moves very slowly and is essentially in equilibrium. The net force on it is therefore zero. The gums, then, must exert a force of 1.36 N on the tooth in a direction opposite to that exerted by the wire. In a similar way the force exerted on the leg by the traction apparatus shown in Fig. 8-3 can be calculated to be 320 N. Notice that if the pulleys are essentially frictionless, there is a tension of 20 kg $(9.8 \text{ m/s}^2) = 200$ N all along the cord; thus there are two 200-N forces acting at $37°$ angles on the central pulley and on the leg and the resultant is 320 N acting to the right. Now the leg is in equilibrium so there must be another 320-N force acting on the leg to keep it at rest. What exerts this force?

8-2 ∎ The conditions for equilibrium

For a body to be at rest, the sum of the forces acting on it must add up to zero. Since force is a vector, the components of the net force must each be zero. Hence, if the forces on the object act in a plane, a condition for equilibrium is that

$$\sum F_x = 0, \qquad \sum F_y = 0. \qquad (8\text{-}1)$$

If the forces act in three dimensions (a situation we will not consider) then the third component must be zero as well, $\sum F_z = 0$. We must remember that

FIGURE 8-3

Traction apparatus exerts force on a leg.

if a particular force component points along the negative x or y axis, it must have a negative sign. Equation 8-1 is called the *first condition* for equilibrium.

□ **EXAMPLE 8-2**   Calculate the tension in the two cords used to support the 200-kg chandelier shown in Fig. 8-4.

SOLUTION   The three forces, F_1, F_2, and the 200-kg weight of the chandelier, act at the point where the three cords join. We choose this junction point (it could be a knot) as the object for which we write $\sum F_x = 0$, $\sum F_y = 0$. (Note that we don't bother considering the chandelier itself since only two forces act on it, gravity downward and the equal but opposite force exerted upward by the cord, both of which equal $mg = 1960$ N.) There are only two unknowns and we can solve for them using Eq. 8-1. We first resolve F_1 into its horizontal (x) and vertical (y) components. Although we don't know the value of F_1 we can write $F_{1x} = F_1 \cos 60°$ and $F_{1y} = F_1 \sin 60°$. F_2 has only an x component. In the vertical direction we have only the weight of the chandelier $= (200$ kg$)(g)$ acting downward and the vertical component of F_1 upward. Since $\sum F_y = 0$, we have:

$$\sum F_y = F_1 \sin 60° - (200 \text{ kg})(g) = 0$$

so

$$F_1 = \frac{(200 \text{ kg})(g)}{\sin 60°} = (230 \text{ kg})(g) = 2260 \text{ N}.$$

In the horizontal direction,

$$\sum F_x = F_2 - F_1 \cos 60° = 0.$$

Thus

$$F_2 = F_1 \cos 60° = (230 \text{ kg})(g)(0.500) = (115 \text{ kg})(g) = 1130 \text{ N}.$$

The magnitudes of F_1 and F_2 determine the strength of cord or wire that must be used. In this case, the wire must be able to hold at least 230 kg.

FIGURE 8-4

Example 8-2.

Although Eq. 8-1 must be true if an object is to be in equilibrium, it is not a sufficient condition. Figure 8-5 shows an object on which the net force is zero. Although the two forces labeled F add up to give zero net force on the object, they do give rise to a torque which will rotate the object. Referring to Eq. 7-12, we see that if a body is to remain at rest, the net torque applied to it must be zero. Thus we have the *second condition* for equilibrium: that the sum of the torques acting on a body must be zero:

$$\sum \tau = 0. \tag{8-2}$$

This will assure that the angular acceleration, α, about any axis will be zero; if the body is not rotating initially ($\omega = 0$) it will not start rotating. Equations 8-1 and 8-2 are the only requirements for a body to be in equilibrium.

We will consider cases in which the forces all act in a plane[†] (we call it the xy plane). In this case the torque is calculated about an axis that is per-

FIGURE 8-5

Although the net force on it is zero, the pencil will move (rotate). A pair of equal forces acting in opposite directions but at different points on a body (as shown here) is referred to as a *couple*.

[†] If the forces are not in a plane, the situation is more complicated; there will be three force equations plus three torque equations (torque calculated about three mutually perpendicular axes). We won't usually be concerned with such complicated situations.

pendicular to the xy plane. The choice of this axis is arbitrary. Since $\tau = I\alpha$ for a rigid body about any fixed axis (Eq. 7-12), $\tau = 0$ about any fixed axis if the body is in equilibrium ($\alpha = 0$): therefore we can choose any axis that makes our calculation easier.

This subject of statics is important because it allows us to calculate certain forces on (or within) a structure when some of the forces on it are already known. There is no single technique for attacking such statics problems, but the following procedure is helpful. (1) Choose one body at a time for consideration, and on a diagram show all the forces acting on the body and the points at which these forces act; this is called a "free-body diagram." (2) Choose a convenient coordinate system and resolve the forces into their components. (3) Using letters to represent unknowns, write down equations for $\sum F_x = 0$, $\sum F_y = 0$, and $\sum \tau = 0$, and then solve these equations. Undoubtedly the hardest step is (1): *all* the forces *on* the body must be included, but the forces exerted *by* this body on other objects must *not* be included.

It is useful to refer back to Example 8-2 to see how this procedure was followed. Notice that we did not bother to consider the forces on the chandelier (which are gravity downward and the rope pulling upward with an equal force). Since we wanted to know the tension in the two cords, we chose the junction point (or knot) of the cords as our object. Since the object is essentially a point, there would be no torques, so the torque equation was not used. Of course, one does not have to use all three equations if they are not needed.

In many situations we must make use of the torque equation (8-2). Just as the components of a force are positive or negative depending on their direction, so too are torques. If a torque that tends to rotate the object counterclockwise is considered positive then a torque that tends to rotate it clockwise must be considered negative. Another way of stating the equilibrium conditions is that the sum of all clockwise torques is equal to the sum of all counterclockwise torques; and for the forces: the sum of the upward forces is equal to the sum of the downward forces, and the sum of horizontal forces to the left is equal to the sum of horizontal forces to the right.

One of the forces that acts on bodies is the force of gravity. Our analysis in this chapter is greatly simplified if we use the concept of center of gravity (cg); as we discussed in Section 6-8, we can consider the force of gravity on the body as a whole as acting at its cg. For uniform symmetrically shaped bodies, the cg is at the geometric center. For more complicated bodies, the cg can be determined as discussed in Section 6-8.

EXAMPLE 8-3   A uniform 1200-kg beam (beneath a floor, let us say) supports 15,000 kg of machinery, Fig. 8-6. Calculate the force on each of the supports.

SOLUTION   We analyze the forces on the beam, since the force it exerts on each support is equal and opposite to the force exerted by that support on the beam. We call these forces $\mathbf{F}_1$ and $\mathbf{F}_2$ in Fig. 8-6. The weight of the beam itself acts at its center of gravity, 10.0 m from either end. Since it doesn't matter which point we choose as the axis for writing the torque equation, we can choose one that is convenient. If we calculate the torques about the point of application of $\mathbf{F}_1$, then $\mathbf{F}_1$ will not enter (its lever arm

FIGURE 8-6

A 1200-kg beam supports a 15,000-kg machine (Example 8-3).

will be zero) and we will have an equation in only one unknown, F_2. Thus $\sum \tau = 0$ gives

$$-(10.0 \text{ m})(1200 \text{ kg})(g) - (15.0 \text{ m})(15{,}000 \text{ kg})(g) + (20.0 \text{ m})F_2 = 0.$$

Solving for F_2 we find $F_2 = (12{,}000 \text{ kg})(g) = 118{,}000 \text{ N}$. To find F_1 we use $\sum F_y = 0$:

$$\sum F_y = F_1 - (1200 \text{ kg})(g) - (15{,}000 \text{ kg})(g) + F_2 = 0.$$

Putting in $F_2 = (12{,}000 \text{ kg})(g)$ we find $F_1 = (4200 \text{ kg})(g) = 41{,}200 \text{ N}$.

Figure 8-7 shows a beam which extends beyond its support. Such a beam is called a *cantilever*. The forces acting on the beam in this figure are those due to the supports, F_1 and F_2, and the force of gravity, which acts at the cg, 5.0 m to the right of the right-hand support. If you follow the procedure of the last example to calculate F_1 and F_2 assuming they point upward as shown in the figure (see the problems) you will find that F_1 comes out negative; if the beam has a mass of 1200 kg, then $\mathbf{F}_2 = 15{,}000 \text{ N}$ and $\mathbf{F}_1 = -3000 \text{ N}$. Whenever an unknown force comes out negative, it merely means that the force actually points in the opposite direction from what you assumed. Thus in Fig. 8-7, $\mathbf{F}_1$ actually points downward. With a little reflection it should become obvious that the left-hand support must indeed pull downward on the beam if it is to be in equilibrium (and not push upward as $\mathbf{F}_2$ does).

FIGURE 8-7

A cantilever.

As another example, we calculate the stability of a building under the action of wind. It is obviously an important consideration in the design of high buildings to determine if the wind will knock them over. In Fig. 8-8, the forces on the building are shown. The weight $\mathbf{w}$ acts downward at the cg (assumed to be at the center of the building). $\mathbf{F}_a$ represents the force of the air (wind), which, if we assume it acts uniformly over the whole face of the building, can be considered as a single force acting at the center of that face of the building. Finally, $\mathbf{F}_E$ represents the force of the earth on the building which counterbalances the first two forces to keep the building in equilibrium. Normally the upward and sideways force of the earth ($\mathbf{F}_E$) acts all along the base of the building. But we are interested in the worst possible situation, when the building has started to tip slightly and $\mathbf{F}_E$ acts at the last point of support, the corner of the building. The components of $\mathbf{F}_E$ can be calculated from the force equations, but what we are really interested in is whether the building will reach this rather dire situation of starting to tip over. To determine this, we must calculate torques about the potential pivot point.

FIGURE 8-8

Force on a building subjected to wind (F_a) and gravity (w); F_E is the force on the building due to the earth in the situation when the building is just about to tip.

☐   EXAMPLE 8-4   A 50-story building is being planned. It is to be 200 m high with a base 40 m by 70 m. Its total mass will be about 1.5×10^7 kg and its weight therefore about 1.5×10^8 N. Will this building tip over in a 200-km/h wind[†] which exerts a force of 1450 N/m² over the 70-m-wide face?

[†] The force (in N) exerted per square meter on a flat surface by a wind of speed v (in km/h) is given roughly by $0.035 \, v^2$, assuming the wind blows perpendicular to the surface.

SOLUTION   To get F_a we must multiply the force per area by the total area: $F_a = (1450 \text{ N/m}^2)(70 \text{ m})(200 \text{ m}) = 2.0 \times 10^7$ N. We calculate the torque about the potential pivot point, the lower rear edge of the building (where F_E acts in Fig. 8-8). The wind, as we saw above, can be considered to act at the midpoint of the face, which is 100 m above the ground. The torque it produces, tending to overturn the building, is equal to (100 m) · $(2.0 \times 10^7 \text{ N}) = 2.0 \times 10^9$ N·m. The torque tending to keep the building upright is the force of gravity acting through a lever arm equal to half the width of the building, or 20 m; this torque equals $(20 \text{ m})(1.5 \times 10^8 \text{ N}) = 3.0 \times 10^9$ N·m. The building will not tip over, but it is close; a safety factor of two or more would be better.

*8-3 ∎ Applications to muscles and joints

The techniques for calculating forces on bodies in equilibrium can readily be applied to the human (or animal) body. This is of great use in studying the forces on muscles, bones, and joints for organisms in motion or at rest. Generally a muscle is attached, via tendons, to two different bones (see Fig. 8-9). The points of attachment are called *insertions*. The two bones are flexibly connected at a *joint*, such as those at the elbow, knee, and ankle. A muscle exerts a pull when its fibers contract under stimulation by a nerve, but it cannot exert a push. A muscle that tends to bring two limbs closer together, such as the biceps muscle in the upper arm (Fig. 8-9) are called *flexor* muscles; those that act to extend a limb outward, such as the triceps, are called *extensor* muscles. The flexor muscle in the upper arm is used when lifting an object in your hand; the extensor muscle is used when throwing a baseball.

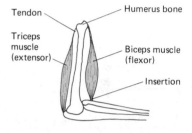

FIGURE 8-9

Diagram showing the biceps (flexor) and triceps (extensor) muscles in human arm.

FIGURE 8-10

Example 8-5.

EXAMPLE 8-5   How much force must the biceps muscle exert when a 5.0-kg mass is held in the hand (a) with the arm outstretched as in Fig. 8-10a. and (b) when the arm is at a 45° angle as in Fig. 8-10b? Assume the mass of forearm and hand together is 2.0 kg and their cg is as shown.

SOLUTION   (a) The forces acting on the forearm are shown in Fig. 8-10a and include the upward force F_M exerted by the muscle and a force F_J exerted at the joint by the bone in the upper arm. We wish to find F_M, which is done most easily by using the torque equation, calculated about the joint so that F_J does not enter:

$$(0.050 \text{ m})(F_M) - (0.15 \text{ m})(2.0 \text{ kg})(g) - (0.35 \text{ m})(5.0 \text{ kg})(g) = 0.$$

We solve this for F_M and find $F_M = (41 \text{ kg})(g) = 400$ N. (b) The lever arm for the muscle is reduced to $(0.050 \text{ m}) \sin 45° = 0.035$ m. But the lever arms for the downward forces are reduced by the same ratio; so the same result will be obtained, $F_M = 400$ N.

Note in the example that the force required of the muscle is quite large compared to the weight of the object lifted. Indeed, the muscles and joints of the body are generally subjected to quite large forces.

The point of insertion of a muscle varies from person to person. A slight increase in the point of insertion of the biceps muscle from 5.0 cm to 5.5 cm can be a considerable advantage for lifting and throwing. Indeed, athletes are often found to have muscle insertions farther from the joint than the average person, and if this applies to one muscle it usually applies to all.

As another example of the large forces acting within the human body we consider the muscles used to support the trunk when a person bends forward, Fig. 8-11a. The lowest vertebra on the spinal column (fifth lumbar vertebra) acts as a fulcrum for this bending position. The "erector spinae" muscles that support the trunk act at an effective angle of about 12° to the axis of the spine. Figure 8-11b is a simplified schematic drawing showing the forces on the upper body; we assume the trunk makes an angle of 30° with the horizontal. The force exerted by the muscles is represented by $\mathbf{F_M}$; the force exerted on the base of the spine at the lowest vertebra is $\mathbf{F_V}$; and $\mathbf{w_1}$, $\mathbf{w_2}$, $\mathbf{w_3}$ represent the weights of the head, freely hanging arms, and trunk, respectively. The values shown are approximations taken from Table 6-1; the distances (in cm) refer to a person 180 cm tall, but are approximately in the same ratio of 1:2:4 for an average person of any height and the answer below is then independent of the height of the person.

Erector spinae muscles

Fifth lumbar vertebra

(a)

◻ **EXAMPLE 8-6** Calculate the magnitude and direction of the force $\mathbf{F_V}$ acting on the fifth lumbar vertebra for the example shown in Fig. 8-11b.

SOLUTION   First we calculate F_M using the torque equation (Eq. 8-2) about the base of the spine; the torques are calculated using $\tau = rF_\perp = r(F \sin \theta)$ where θ is 12° for $\mathbf{F_M}$ and 60° for $\mathbf{w_1}$, $\mathbf{w_2}$, and $\mathbf{w_3}$:

$$(0.48\ \text{m})(\sin 12°)(F_M) - (0.72\ \text{m})(\sin 60°)(w_1)$$
$$-(0.48\ \text{m})(\sin 60°)(w_2) - (0.36\ \text{m})(\sin 60°)(w_3) = 0.$$

Putting in the values for w_1, w_2, w_3 given in the figure, we find $F_M = 2.2w$, where w is the total weight of the body. To get the components of F_V we use the x and y components of the force equation:

$$\sum F_y = F_{V_y} - F_M \sin 18° - w_1 - w_2 - w_3 = 0$$
$$F_{V_y} = 1.3w.$$
$$\sum F_x = F_{V_x} - F_M \cos 18° = 0$$
$$F_{V_x} = 2.1w.$$

Then $F_V = \sqrt{F_{V_x}^2 + F_{V_y}^2} = 2.5w$. The angle θ that F_V makes with the horizontal is given by $\tan \theta = F_{V_y}/F_{V_x} = 0.62$, so $\theta = 32°$.

$w_1 = 0.07w$
(head)
$w_2 = 0.12w$
(arms)
$w_3 = 0.46w$ w = total weight
(trunk)   of person

(b)

FIGURE 8-11

(a) Person bending over. (b) Forces required of back muscles ($\mathbf{F_M}$) and in the vertebrae ($\mathbf{F_V}$) when a person bends over.

The force on the lowest vertebra is thus $2\frac{1}{2}$ times the body weight! This force is transmitted from the "sacral" bone at the base of the spine, through the fluid-filled and somewhat flexible *intervertebral disc*. The discs at the base of the spine are clearly being compressed under very large forces.

If the person in Fig. 8-11 has a mass of 90 kg, and is holding 20 kg in his hands (this increases w_2 to $0.33w$) then F_V is increased to nearly five times the person's weight ($5w$)! (For this 200-lb person, the force on the disc would be 1000 lb!) With such strong forces acting, it is little wonder that so many people suffer from lower back pain at one time or another in their lives.

*8-4 ∎ Simple machines: levers and pulleys

Simple machines, such as levers, pulleys, gears, and the wheel and axle, are designed to reduce the force needed to lift a heavy load. They can be analyzed using the principles we have discussed in this chapter. Consider, for example, the lever shown in Fig. 8-12. If the stone is lifted slowly, it is essentially in equilibrium, and using the torque rule ($\sum \tau = 0$), with the torques calculated about the pivot point, we see that a force of only $(50 \text{ kg})(g) = 500$ N is needed to lift the stone whose weight is $(200 \text{ kg})(g) = 2000$ N. In a sense, the force has been multiplied by a factor of 4, and we say the lever gives us a *mechanical advantage* of 4. From the geometry of the situation as shown in Fig. 8-12b, the 500-N force must be exerted through a distance of 2.0 m to raise the stone 0.50 m. Thus the work done by the person is $(500 \text{ N})(2.0 \text{ m}) = 1000$ J, and the work done on the stone is $(2000 \text{ N})(0.50 \text{ m}) = 1000$ J, also. In other words, the input work equals the output work (this is essentially conservation of energy).

This example has been idealized, for in practice it would be found that the force necessary to lift the 2000-N stone would be different from 500 N. This is due to two factors common to all machines: friction and the weight of the machine itself (for example, in Fig. 8-12 we would need to include the weight of the lever in calculating the torques). We therefore define an "ideal" and an "actual" mechanical advantage. The *ideal mechanical advantage* (IMA) is defined as the (output force)/(input force) for the ideal situation, which means the output work = input work, or $F_i' d_i = F_o' d_o$; the subscript "o" means output and "i" means input and the primes remind us we are talking of the force in the ideal case. From this relation we have $F_o'/F_i' = d_i/d_o$. Thus the ideal mechanical advantage is also the ratio of the distance d_i moved by the input force to the distance d_o moved by the load (the output):

$$\text{IMA} = \frac{d_i}{d_o}.$$

The *actual mechanical advantage* (AMA) is the ratio of the actual output

4.0 m

200 kg

1.0 m

(a)

2.0 m

0.50 m

(b)

FIGURE 8-12
A simple lever.

force to actual input force for the particular machine

$$\text{AMA} = \frac{F_o}{F_i}.$$

Finally, the efficiency of any machine, be it simple or complex, is defined as (work output)/(work input). This can be written in terms of the IMA and AMA:

$$\text{efficiency} = \frac{F_o d_o}{F_i d_i}$$

$$= \left(\frac{F_o}{F_i}\right)\Big/\left(\frac{d_i}{d_o}\right) = \frac{\text{AMA}}{\text{IMA}}.$$

For example, suppose it actually took a 550-N force to lift the 2000 N stone in Fig. 8-12. The IMA = 2.0 m/0.50 m = 4.0, whereas the AMA = 2000 N/ 550 N = 3.6. The efficiency is then (3.6)/(4.0) = 0.90 or 90 percent.

Pulleys are often used to mechanical advantage. The pulley system shown in Fig. 8-13 has an IMA of 2.0. This is readily seen by noting that there are two ropes supporting the weight and in the ideal case each need carry only half the weight. This can also be seen from the fact, see Fig. 8-13, that $d_o = d_i/2$. For more complicated systems, such as those shown in Fig. 8-14, similar considerations will reveal that the IMA in general is equal to the number of ropes pulling up on the load. The AMA will always be less than the IMA; hence the work output will be less than the work input.

8-5 ∎ Stability and balance

A body in static equilibrium, if left undisturbed, will undergo no translational or rotational acceleration since the sum of all the forces and the sum of all the torques acting on it are zero. However, if the object is displaced slightly, three different outcomes are possible: (1) the object may return to its original position, in which case it is said to be in **stable equilibrium;** (2) the object will move even farther from its original position, in which case it is said to be in **unstable equilibrium;** or (3) it remains in its new position, in which case it is said to be in **neutral equilibrium.** Consider the following examples. A ball suspended freely from a string is in stable equilibrium for if it is displaced to one side it will quickly return to its original position (Fig. 8-15a). On the other hand, a pencil standing on its point is in unstable equilibrium. If its cg is directly over its tip (Fig. 8-15b), the net force and net torque on it will be zero. But if it is displaced ever so slightly—say by a slight vibration or tiny air current—there will be a torque on it and it will continue to fall in the direction of the original displacement. Finally, an example of an object in neutral equilibrium is a sphere resting on a horizontal table top. If it is placed slightly to one side, it will remain in its new position.

In most situations, such as in the design of structures and in working with the human body, we are interested in maintaining stable equilibrium or

FIGURE 8-13
Simple pulley.

FIGURE 8-14
Various pulley systems.

| IMA = 3 | IMA = 3 | IMA = 4 |

balance, as we sometimes say. In general, an object whose cg is below its point of support, such as a ball on a string, will be in stable equilibrium. If the cg is above the base of support we have a more complicated situation. Consider a block standing on its end (Fig. 8-16a). If it is tipped slightly, it will return to its original position due to the torque on it as shown in Fig. 8-16b. But if it is tipped too far, Fig. 8-16c, it will fall over. The critical point is reached when the cg is no longer above the base of support. In general, *a body whose cg is above its base of support will be stable if a vertical line projected downward from its cg falls within the base of support.* This is a result of the fact that the upward force on the object (which balances out gravity) can only be exerted within the area of contact, so that if the force of gravity acts beyond this area, a net torque will act to topple the object. Stability in this case is relative. A brick lying on its widest face is more stable than a brick standing on its end, for it will take more of an effort to tip it over. In the extreme case of the pencil in Fig. 8-15b, the base is practically a point and the slightest disturbance will topple it. In general, the larger the base and the lower the cg, the more stable the object.

In this sense, humans are much less stable than four-legged mammals, which not only have a much larger base of support because of their four legs, but have a lower cg. The human species has had to develop special apparatus, such as certain very strong muscles, in order to deal with the problem of keeping a person upright and at the same time stable. Because of this upright position, humans suffer from numerous ailments such as low back pain due to the large forces involved as we saw in Example 8-6. When walking and performing other kinds of movement a person continually shifts the body so its cg is over the feet, although in the normal adult this requires no conscious thought. Even as simple a movement as bending over requires moving the hips backward so that the cg remains over the feet, and this repositioning is done without thinking about it. To illustrate this, position yourself with your heels and back to a wall and try to touch your toes. Persons carrying heavy loads automatically adjust their posture so that the cg of the total mass is over their feet, as shown in Fig. 8-17. However, to people who suffer from muscle weakness in the trunk or legs, or who have suffered damage to the spinal cord, even walking can be difficult. A conscious effort is required to balance the body over the feet and the slightest miscalculation can lead to a fall.

FIGURE 8-15

(a) Stable equilibrium; (b) unstable equilibrium.

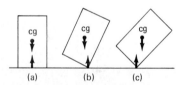

FIGURE 8-16

Equilibrium of a block resting on a surface.

FIGURE 8-17

Humans adjust their posture when carrying loads to achieve stability.

8-6 ▌ Elasticity; stress and strain

In the first part of this chapter we studied how to calculate the forces on objects in equilibrium. In this section we study the effects of these forces, for any object changes shape under the action of applied forces. If the forces are great enough, the object will break or *fracture,* and we discuss this in Section 8-7.

If a force is exerted on an object, such as the vertically suspended metal bar shown in Fig. 8-18, the length of the object changes. If the amount of elongation, ΔL, is small compared to the length of the object, experiment shows that ΔL is proportional to the weight or force exerted on the object;

this was first noted by Robert Hooke (1635–1703). This proportionality can be written as an equation and is sometimes called *Hooke's law*[†]:

$$F = k\,\Delta L. \tag{8-3}$$

Here F represents the force (or weight) pulling on the object, ΔL is the increase in length, and k is a proportionality constant. Equation 8-3 is found to be valid for almost any solid material from iron to bone, but it is valid only up to a point. For if the force is too great, the object stretches excessively and eventually breaks. Figure 8-19 shows a typical graph of elongation versus applied force. Up to a point called the *elastic limit*, the object will return to its original length if the applied force is removed. This is called the *elastic region*. If the object is stretched beyond the elastic limit, it will be permanently deformed. For most common materials Eq. 8-3 is a good approximation almost up to the elastic limit and the graph is a straight line. Beyond this point, the graph deviates from a straight line and no simple relationship exists between F and ΔL. If the object is stretched much beyond the elastic limit, it will break. The maximum force that can be applied without breaking is called the *ultimate strength* of the material.

The amount of elongation of an object, such as the bar shown in Fig. 8-18, depends not only on the force applied to it, but also on the material from which it is made and on its dimensions. That is, the constant k in Eq. 8-3 can be written in terms of these factors. If we compare bars made of the same material but of different lengths and cross-sectional areas, it is found that for the same applied force the amount of stretch (again assumed small compared to the total length) is proportional to the original length and inversely proportional to the cross-sectional area. That is, the longer the object, the more it elongates for a given force; and the fatter it is, the less it elongates. These experimental findings can be combined with Eq. 8-3 to yield the relation

$$\Delta L = \frac{1}{E}\frac{F}{A}L_0, \tag{8-4}$$

where L_0 is the original length of the object, A is the cross-sectional area, and ΔL is the change in length due to the applied force F; E is a constant of proportionality[‡] known as the *elastic modulus*, or *Young's modulus*, and its value depends only on the material. The value of Young's modulus for various materials is given in Table 8-1. Because E is a property only of the material and is independent of the object's size or shape, Eq. 8-4 is far more useful for practical calculation than Eq. 8-3. From Eq. 8-4 we see that the change in length of an object is directly proportional to the product of the object's length L_0 and the force per unit area F/A applied to it. It is general practice to define the force per unit area, as the **stress:**

$$\text{stress} = \frac{\text{force}}{\text{area}} = \frac{F}{A}.$$

Also, the **strain** is defined to be the ratio of the change in length to the

[†] The term "law" applied to this relation is not really appropriate—see footnote on this point in Section 5-4.

[‡] The fact that E is in the denominator so that $1/E$ is the actual proportionality constant is merely a convention.

FIGURE 8-18

Hooke's law: $\Delta L \propto$ applied force.

FIGURE 8-19

Applied force vs. elongation for a typical solid.

TABLE 8-1
Elastic moduli

Material	Elastic modulus, $E\,(N/m^2)$	Shear modulus, $G\,(N/m^2)$	Bulk modulus, $B\,(N/m^2)$
Solids			
Iron, cast	100×10^9	40×10^9	90×10^9
Steel	200×10^9	80×10^9	140×10^9
Brass	100×10^9	35×10^9	80×10^9
Aluminum	70×10^9	25×10^9	70×10^9
Concrete	20×10^9		
Brick	14×10^9		
Marble	50×10^9	20×10^9	70×10^9
Granite	45×10^9	20×10^9	45×10^9
Wood (pine)			
(parallel to grain)	10×10^9		
(perpendicular to grain)	1×10^9		
Nylon	5×10^9		
Bone (limb)	15×10^9	80×10^9	
Liquids			
Water			2.0×10^9
Alcohol (ethyl)			1.0×10^9
Mercury			2.5×10^9

original length:

$$\text{strain} = \frac{\text{change in length}}{\text{original length}} = \frac{\Delta L}{L_0}.$$

Strain is thus the fractional change in length of the object and is a measure of how much the bar has been deformed. Equation 8-4 can be rewritten as

$$\frac{F}{A} = E\,\frac{\Delta L}{L_0} \qquad (8\text{-}5)$$

or

$$E = \frac{F/A}{\Delta L/L_0} = \frac{\text{stress}}{\text{strain}}.$$

Thus we see that the strain is directly proportional to the stress.

☐ EXAMPLE 8-7 A 1.60-m-long steel piano wire has a diameter of 0.20 cm. How great is the tension in the wire if it stretches 0.30 cm when tightened?

SOLUTION We solve for F in Eq. 8-4 and note that the area $A = \pi r^2 = (3.14)(0.0010\ \text{m})^2 = 3.1 \times 10^{-6}\ \text{m}^2$. Then

$$F = E\,\frac{\Delta L}{L_0}\,A$$

$$= (2.0 \times 10^{11}\ \text{N/m}^2)\left(\frac{0.0030\ \text{m}}{1.60\ \text{m}}\right)(3.1 \times 10^{-6}\ \text{m}^2) = 1200\ \text{N},$$

where we obtained the value for E from Table 8-1.

The bar shown in Fig. 8-18, is said to be under *tension* or *tensile stress*. For not only is there a force pulling down on the bar at its lower end, but since the bar is in equilibrium we know that the support at the top is exerting an equal upward force[†] on the bar at its upper end, Fig. 8-20a. In fact, this tensile stress exists throughout the material. Consider for example the lower half of a suspended bar as shown in Fig. 8-20b. This lower half is in equilibrium so there must be an upward force on it to balance the downward force at its lower end. What exerts this upward force? It must be the upper part of the bar. Thus we see that external forces applied to an object give rise to internal forces, or stress, within the material itself.

Strain or deformation due to tensile stress is but one type of stress to which materials can be subjected. There are two other common types of stress: compressive and shear. *Compressive* stress is the exact opposite of tensile stress. Instead of being stretched, the material is compressed: the forces act inwardly on the body. Any sort of column that supports a weight, such as the columns of a Greek temple (Fig. 8-21) is subjected to a compressive stress. Equations 8-4 and 8-5 apply equally well to compression and tension, and the values for E are usually the same.

Figure 8-22 compares tensile and compressive stresses as well as the third type, shear stress. An object under *shear* stress has equal and opposite forces applied *across* its opposite faces. An example is a book or brick firmly attached to a table top with a force exerted parallel to the top surface; the table exerts an equal and opposite force along the bottom surface. Although the dimensions of the object do not change significantly, the shape of the object does change as shown in the figure. An equation similar to 8-4 can be applied to calculate shear strain:

$$\Delta L = \frac{1}{G}\frac{F}{A}L_0 \qquad (8\text{-}6)$$

but ΔL, L, and A must be reinterpreted as indicated in Fig. 8-22c. Note that A is the area of the surface *parallel* to the applied force (and not perpendicular as for tension and compression), and ΔL is *perpendicular* to L_0. The constant of proportionality, G, is called the *shear modulus* and is generally one-half to one-third the value of the elastic modulus, E (see Table 8-1).

The rectangular object undergoing shear in Fig. 8-22c would not actually be in equilibrium under the forces shown, for a net torque would exist. If the object is in fact in equilibrium, there must be two more forces acting on it which balance out this torque. One acts vertically upward on the right, and the other acts vertically downward on the left as shown in Fig. 8-23. This is generally true of shear forces. If the object is a brick or book lying on a table, these two additional forces can be exerted by the table and whatever exerts the other horizontal force.

If an object is subjected to a pressure on all sides, its volume will be compressed. A common situation is a body submerged in a fluid; for in this case, the fluid exerts a pressure on the object in all directions, as we shall see in Chapter 9. Pressure is defined as force per area and thus is the equivalent of "stress." For this situation the change in volume, ΔV, is found to be proportional to the original volume, V_0, and to the increase in the pressure,

[†] If we ignore the weight of the bar.

FIGURE 8-20

Stress exists *within* the material.

FIGURE 8-21

A Greek temple (in Paestum, Italy). (D. Giancoli).

Tension

(a)

Compression

(b)

Shear

(c)

FIGURE 8-22

The three types of stress.

ΔP. We thus obtain a relation of the same form as Eqs. 8-4 and 8-5 but with a proportionality constant called the *bulk modulus*, B:

$$\frac{\Delta V}{V_0} = -\frac{1}{B}\,\Delta P \qquad (8\text{-}7)$$

or

$$B = -\frac{\Delta P}{\Delta V/V_0}.$$

The minus sign is included to indicate that the volume *decreases* with an increase in pressure. Values for the bulk modulus are given in Table 8-1. Since liquids and gases do not have a fixed shape, only the bulk modulus applies to them.

FIGURE 8-23

Balance of forces and torques for shear stress.

8-7 ∎ Fracture

If the stress on a solid object is too great, the object fractures or breaks, Fig. 8-24. In Table 8-2 are listed the ultimate tensile strength, compressive strength, and shear strength for a variety of materials. These give the maximum force per unit area that an object can withstand under each of these three types of stress. They are, however, representative values only and the actual value for a given specimen can differ considerably. It is therefore necessary to maintain a safety factor of from 3 to perhaps 10 or more—that is, the actual stresses on a structure should not exceed one-tenth to one-third of the values given in the table. You may encounter tables of the "allowable stresses" in which an appropriate safety factor has already been included.

FIGURE 8-24

Fracture as a result of the three types of stress.

Tension

Compression

Shear

␣ EXAMPLE 8-8 (*a*) What minimum cross-sectional area should the supports have to support the beam of Example 8-3 (Fig. 8-6) assuming the supports are made of concrete and a safety factor of 6 is required? (*b*) How much will the chosen supports compress under the given load?

TABLE 8-2

8 Bodies in Equilibrium; Elasticity
and Fracture

Ultimate strengths of materials (force/area)

Material	Tensile strength (N/m^2)	Compressive strength (N/m^2)	Shear strength (N/m^2)
Iron, cast	170×10^6	550×10^6	170×10^6
Steel	500×10^6	500×10^6	250×10^6
Brass	250×10^6	250×10^6	200×10^6
Aluminum	200×10^6	200×10^6	200×10^6
Concrete	2×10^6	20×10^6	2×10^6
Brick		35×10^6	
Marble		80×10^6	
Granite		170×10^6	
Wood (pine)			
(parallel to grain)	40×10^6	35×10^6	5×10^6
(perpendicular to grain)		10×10^6	
Nylon	500×10^6		
Bone (limb)	130×10^6	170×10^6	

SOLUTION (a) The right-hand support receives the larger force, 1.18×10^5 N. It is clearly under compression and from Table 8-2 we see that the ultimate compressive strength of concrete is 2.0×10^7 N/m². Using a safety factor of 6, the maximum allowable stress is 3.3×10^6 N/m². Since $F/A = 3.3 \times 10^6$ N/m² and $F = 1.18 \times 10^5$ N we can solve for A and find

$$A = \frac{1.18 \times 10^5 \text{ N}}{3.3 \times 10^6 \text{ N/m}^2} = 3.6 \times 10^{-2} \text{ m}^2 \text{ or } 360 \text{ cm}^2.$$

A support 18 cm × 20 cm will be adequate.

(b) We solve for

$$\frac{\Delta L}{L_0} = \frac{1}{E}\frac{F}{A} = \left(\frac{1}{2.0 \times 10^{10} \text{ N/m}^2}\right)(3.3 \times 10^6 \text{ N/m}^2) = 1.7 \times 10^{-4}.$$

Thus, if the support has a length $L_0 = 5.0$ m, $\Delta L = 0.85 \times 10^{-3}$ m, or about 1 mm. This calculation was for the right-hand support. If the left-hand support is made of the same cross-sectional area, it will compress less and this should be taken into account.

As can be seen in Table 8-2, concrete (like stone and brick) is reasonably strong under compression but extremely weak under tension. Thus concrete can be used as vertical columns placed under compression but is of little value as a beam since it cannot withstand the tensile forces that arise (see Fig. 8-25). Reinforced concrete, in which iron rods are embedded in the concrete, is much stronger in terms of stability; but the concrete on the lower edge of a loaded beam still tends to crack since concrete is weak under tension. This problem is solved with **prestressed concrete** which also contains iron rods or a wire mesh, but during the pouring of the concrete, the rods or wire are held under tension. After the concrete dries, the tension on the iron is released and this puts the concrete under compression. The amount of compressive stress is carefully predetermined so that when the design loads are applied to the beam, they reduce the compression on the lower edge but never put the concrete into tension.

FIGURE 8-25
A beam sags, at least a little (but is exaggerated here), even under its own weight. The beam thus changes shape so that the upper portion is compressed, and the lower portion is elongated (under tension). Shearing stress also occurs within the beam.

FIGURE 8-26
Round arches in the Roman Forum. The one in the background is the Arch of Titus. (D. Giancoli).

*8-8 ■ Spanning a space: arches and domes

There are a great many areas where the arts and humanities overlap the sciences and this is especially clear in architecture. Many of the features we admire in the architecture of the past were introduced not just for their decorative effect but often for practical purposes. One example is the development of methods to span a space, from the simple beam to arches and domes.

It might be said that the first important architectural invention was the post-and-beam (or post-and-lintel) construction in which two upright posts support a horizontal beam. It is still used today; but until steel was introduced in the nineteenth century, the length of a beam was quite limited because the strongest building materials were then stone and brick. Hence the width of a span was limited by the size of available stones. But equally important, all of these materials—though strong under compression—are very weak under tension and shear; all these stresses occur in a beam as shown in Fig. 8-25. The minimal space that could be spanned using stone is shown by the closely spaced columns of the great Greek temples (Fig. 8-21).

The introduction of the semicircular arch by the Romans (Fig. 8-26), aside from its aesthetic appeal, was a tremendous technological innovation. It had been preceded by the so-called "triangle arch" and the "corbeled arch," but these were only slight improvements over the post-and-beam (see Fig. 8-27). The advantage of the "true" or semicircular arch is that, if well designed, its wedgeshaped stones experience stress which is mainly compressive even when supporting a large load such as the wall and roof of a cathedral. Because the stones are forced to squeeze against each other, they are thus mainly under compression (see Fig. 8-28). Note however, that the arch transfers horizontal as well as vertical forces to the supports. A round arch consisting of many well-shaped stones could span a very wide space. However, considerable buttressing on the sides was needed to support the horizontal components of the forces. The buttressing was frequently provided by heavy walls leading off to the side.

The pointed arch came into use in Europe about A.D. 1100 and soon became the hallmark of the great Gothic cathedrals. It too was an important technical innovation, and it is likely that this was at least part of the reason it was originally introduced: it was first used to support heavy loads such as the tower of a cathedral and as the central arch across the nave; the lesser arches in these "transitional" buildings often remained round. Apparently the builders realized that because of the steepness of the pointed arch, the forces due to the weight above could be brought down more nearly vertically, so less horizontal buttressing would be needed. The pointed arch reduced the load on the walls, so there could be more openness and light. The smaller amount of buttressing needed was provided on the outside by graceful flying buttresses (Fig. 8-29).

The technical innovation of the pointed arch was achieved not through calculation but through experience and intuition. The beginnings of the science of statics and building appeared during the Renaissance and was advanced by Galileo; but it was not until somewhat later that detailed calculations, such as those presented earlier in this chapter, came into use. To

Post-and-beam

Corbeled arch

Triangle arch

Semicircular arch

Pointed arch (Gothic)

Tudor arch

FIGURE 8-27

Various means to span a space.

FIGURE 8-28

Stones in a round (or "true") arch are mainly under compression.

FIGURE 8-29

A flying buttress (on the cathedral of Notre Dame, in Paris). (D. Giancoli).

make an accurate analysis of a stone arch is quite difficult in practice. But if we make some simplifying assumptions, we can show why the horizontal component of the force at the base is less for a pointed arch than for a round one. Figure 8-30 shows a round arch and a pointed arch, each with an 8.0 m span. The height of the round arch is thus 4.0 m whereas that of the pointed arch is larger and has been chosen to be 8.0 m. Each arch supports a weight of 12.0×10^4 N ($= 12,000$ kg $\times g$) which for simplicity we have divided into two parts (each 6.0×10^4 N) acting on the two halves of each arch as shown. To be in equilibrium, each of the supports must exert an upward force of 6.0×10^4 N; they also exert a horizontal force, F_H, at the base of the arch, and it is this we want to calculate. We focus only on the right half of each arch; we set equal to zero the toal torque calculated about the apex of the arch due to the forces exerted on that half arch, as if there were a hinge at the apex. For the round arch, the torque equation is:

$$(4.0 \text{ m})(6.0 \times 10^4 \text{ N}) - (2.0 \text{ m})(6.0 \times 10^4 \text{ N}) - (4.0 \text{ m})(F_H) = 0.$$

Thus $F_H = 3.0 \times 10^4$ N. For the pointed arch the torque equation is

$$(4.0 \text{ m})(6.0 \times 10^4 \text{ N}) - (2.0 \text{ m})(6.0 \times 10^4 \text{ N}) - (8.0 \text{ m})(F_H) = 0.$$

Solving, we find $F_H = 1.5 \times 10^4$ N—only half as much! From this calculation we can see that the horizontal buttressing force required for a pointed arch is less because the arch is higher, and there is therefore a longer lever arm for this force. Indeed, the steeper the arch, the less the horizontal component of the force needs to be, and hence the more nearly vertical is the force exerted at the base of the arch.

The further development of the arch was one of decline; for the subsequent flattened arches, such as the Tudor arch (Fig. 8-27), were structurally weaker than the simple pointed arch. However, with the coming of advanced methods of calculation in the nineteenth and twentieth centuries it became possible to calculate the best shape of arch for a given load condition. For example, if the load is uniform across the span, it can be shown that the stresses within the arch will be purely compressive if the arch has a parabolic shape.

Whereas an arch spans a two-dimensional space, a dome—which is basically an arch rotated about a vertical axis—spans a three-dimensional

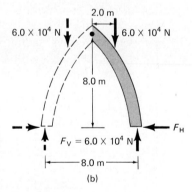

FIGURE 8-30

Forces in a round arch (a) compared with those in a pointed arch (b).

space. The Romans built the first large domes; their shape was hemispherical and some still stand, such as that of the Pantheon in Rome (Fig. 8-31). By the time of the Renaissance, the technique for constructing large domes seems to have been lost. Indeed, the dome of the Pantheon was a source of wonder to Renaissance architects. (Even today we do not know precisely how it was built). The problem came to the fore in fifteenth-century Florence with the designing of a new cathedral which was to have a dome 43 m in diameter to rival that of the Pantheon. A dome, like an arch, is not stable until all the stones are in place. It had been the custom to support a dome during construction with a wooden framework. But no trees big enough or strong enough could be found to support the 43 m space required for the cathedral in Florence. In 1418, after the cathedral was finished except for the dome, a competition for the design of the dome was held and was won by Brunelleschi. One problem that had to be dealt with was that the dome was to rest on a drum that had been completed with no external abutments; and there was no place to put any. Hence the dome must exert the minimum of horizontal force. Brunelleschi solved this by designing a pointed dome (Fig. 8-32), since a pointed dome, like a pointed arch, exerts a smaller side thrust against its base.

The other major problem was how to support the dome during construction. Instead of using a wooden framework, Brunelleschi built the dome in horizontal layers. Each layer was bonded to the previous one which held it in place until the last stone of the circle was placed. Each closed ring was then strong enough to support the next layer. It was an amazing feat.

To end this section we will consider the forces necessary to support a modern dome, that of the Small Sports Palace in Rome, Fig. 8-33. A dome, like an arch, is statically most stable when under compression. The 36 buttresses which support the 1.2×10^6 kg dome are positioned at a 38° angle and connect smoothly with the dome.

EXAMPLE 8-9 Calculate the components of the force, F_V and F_H, which each buttress exerts on the dome so that the force acts compressively, that is, at a 38° angle (Fig. 8-34).

SOLUTION The vertical load on *each* buttress is 1/36 of the total weight. Thus

$$F_V = \frac{(1.2 \times 10^6 \text{ kg})(9.8 \text{ m/s}^2)}{36} = 3.4 \times 10^5 \text{ N}.$$

FIGURE 8-31

The Pantheon in Rome, built in the first century. (Italian Government Travel Office).

FIGURE 8-32

Brunelleschi's dome for the cathedral of Florence, Italy. (Italian Government Travel Office).

FIGURE 8-33

The dome of the Small Sports Palace in Rome, built for the 1960 Olympics. (Italian Government Travel Office).

177

The force must act at a 38° angle at the base of the dome in order to be purely compressive. Thus

$$\tan 38° = \frac{F_V}{F_H} = \frac{340{,}000 \text{ N}}{F_H}$$

$$F_H = \frac{340{,}000 \text{ N}}{\tan 38°} = 430{,}000 \text{ N}.$$

In order that each of the buttresses be able to exert this 430,000-N horizontal force, a prestressed concrete tension ring surrounds the base of the buttresses beneath the ground.

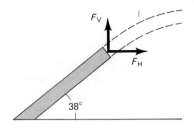

FIGURE 8-34
Example 8-9.

▮ Summary

A body at rest, or one in uniform motion at constant velocity, is said to be in *equilibrium*. The determination of the forces within a structure at rest is the field called *statics*. The two necessary conditions for a body to be in equilibrium are (1) the vector sum of all the forces on it must be zero, and (2) the sum of all the torques (calculated about any arbitrary point as axis) must also be zero. It is important when doing statics problems to apply the equilibrium conditions to only one body at a time.

A body in static equilibrium is said to be in (*a*) *stable*, (*b*) *unstable*, or (*c*) *neutral equilibrium*, depending on whether a slight displacement leads to (*a*) a return to the original position, (*b*) further movement, or (*c*) rest in the new position. An object in stable equilibrium is also said to be in *balance*.

Hooke's law applies to many elastic solids and states that the change in length of an object is proportional to the applied force. If the force is too great, the object will exceed its elastic limit, which means it will no longer return to its original shape when the distorting force is removed. If the force is even greater, the ultimate strength of the material can be exceeded and the object fractures. The force per unit area acting on a body is called the *stress*, and the resulting fractional change in length is called the *strain*. The stress on a body is present within the body and can be of three types: *compression*, *tension*, and *shear*. The ratio of stress to strain is called the elastic modulus of the material; Young's modulus applies for compression and tension, and the shear modulus for shear; bulk modulus applies to an object whose volume changes as a result of pressure on all sides. All three moduli are constants for a given material when distorted within its elastic region.

▮ Questions

1. Describe several situations where a body is not in equilibrium, even though the net force on it is zero.
2. A bear sling, Fig. 8-35, is used in some national parks for placing backpackers' food out of the reach of bears. Explain why the force needed to pull the backpack up increases as the backpack gets higher and higher.
3. A ladder, leaning against a wall, makes a 60° angle with the ground. When is it more likely to slip: when a person stands near the top or near the bottom?
4. Explain why touching the toes while seated on the floor with outstretched legs produces less stress on the lower spinal column than when touching the toes from a standing position. Use a diagram.
5. An earth retaining wall is shown in Fig. 8-36a. The earth, particularly when wet, can exert a significant force *F* on the wall. (*a*) What force produces the torque to keep the wall upright? (*b*) Explain why the retaining wall in Fig. 8-36b would be much less likely to overturn.
6. Explain why a rectangular brick can be placed so that slightly less than half its length can be suspended over the edge of a table, but no more.

FIGURE 8-35

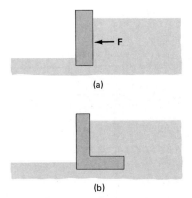

(a)

(b)

FIGURE 8-36

FIGURE 8-37

The dots indicate the cg of each brick. The fractions $\frac{1}{4}$ and $\frac{1}{2}$ indicate what portion of each brick is hanging beyond its support.

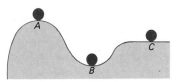

FIGURE 8-38

7. Which of the configurations of brick, (a) or (b) of Fig. 8-37, is the more likely to be stable? Why?

8. Name the type of equilibrium for each position of the ball in Fig. 8-38.

9. In what state of equilibrium is a cube (a) when resting on its face, (b) when on its edge?

10. Why do you tend to lean backward when carrying a heavy load in your arms?

11. What purpose does a walking stick serve when you are hiking in rough country? Be specific.

12. Place yourself facing the edge of an open door. Position your feet astride the door with your nose and abdomen touching the door's edge. Try to rise on your tiptoes. Why can't this be done?

13. Why is it not possible to sit upright in a chair and rise to one's feet without first leaning forward?

14. Why is it more difficult to do situps when your knees are bent than when your legs are stretched out?

15. Examine how a pair of scissors or shears cuts through a piece of cardboard. Is the name "shears" justified?

16. Materials such as ordinary concrete and stone are very weak under tension or shear. Would it be wise to use such a material for either of the supports of the cantilever shown in Fig. 8-7? If so, which one(s)?

Problems

SECTIONS 8-1 AND 8-2

1. (I) What should be the tension in the wire if the net force exerted on the tooth in Fig. 8-2 is to be 0.50 N? Assume that the angle between the two forces is 160° rather than the 140° in the figure.

2. (I) Calculate the torque about the front support of a diving board exerted by a 60-kg person 3.0 m from that support.

3. (I) Calculate the mass m needed in order to suspend the leg shown in Fig. 8-39. Assume the leg has a mass

FIGURE 8-39

FIGURE 8-40

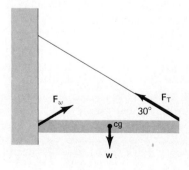

FIGURE 8-42

of 12.0 kg, and its cg is 35.0 cm from the hip joint; the sling is 77.5 cm from the hip joint.

4. (I) Two cords support a chandelier in the manner shown in Fig. 8-4 except the upper wire makes an angle of 45° with the ceiling. If the cords can sustain a force of 1000 N without breaking, what is the maximum weight chandelier that can be supported?

5. (II) Assume in Example 8-1 that the net force as calculated is 10° to the left of where it should point, if the tooth is to move correctly. If the tension to the left is 2.0 N, what should the tension to the right be to make the net force act in the correct direction?

6. (II) Find the tension in the two wires shown in Fig. 8-40.

7. (II) Calculate F_1 and F_2 for the uniform cantilever shown in Fig. 8-7 whose mass is 1200 kg.

8. (II) Calculate the net force required to topple a 60-kg rectangular retaining wall 3.0-m high and 1.2-m thick. Assume the toppling force acts 1.5 m above the base, and that the force exerted by the ground to resist toppling acts at the corner of the base (as in Fig. 8-8).

9. (II) Calculate the forces F_1 and F_2 that the supports exert on the diving board of Fig. 8-41 when a 50-kg person stands at its tip. Ignore the weight of the board.

10. (II) Repeat the problem above taking into account the board's mass of 40 kg. Assume the board's cg is at its center.

11. (II) Calculate the mass m required in Fig. 8-39 to support the leg using the result of Example 6-9 and the values given in Table 6-1 assuming an 88.5-kg person

170 cm tall. The leg pivots about the hip joint and the support acts at the ankle joint.

12. (II) Calculate F_1 and F_2 for the beam shown in Fig. 8-42. Assume it is uniform and has a mass of 250 kg.

13. (II) Calculate the tension F_T in the wire which supports the 20-kg beam shown in Fig. 8-43 and the force $\mathbf{F_w}$ exerted by the wall on the beam (give magnitude and direction).

14. (II) A 150-cm-tall person lies on a light (massless) board which is supported by two scales, one under the feet and one beneath the top of the head. The two scales read, respectively, 29.4 kg and 32.8 kg. Where is the center of gravity of this person?

15. (II) A tightly stretched "high wire" is 50 m long. It sags 3.5 m when a 60-kg tightrope walker stands at its center. What is the tension in the wire? Is it possible to increase the tension in the wire so that there is no sag?

16. (II) The two trees in Fig. 8-35 are 12.5 m apart. Calculate the magnitude of the force $\mathbf{F}$ a backpacker must exert to hold the 15-kg backpack so the rope sags at its midpoint by (a) 2.0 m, (b) 0.20 m.

17. (II) A uniform 9.5-m-long ladder of mass 18.0 kg leans against a smooth wall (so the force exerted by the wall, $\mathbf{F_w}$, is perpendicular to the wall). The ladder makes a 20° angle with the vertical wall; and the ground is rough. (a) Calculate the components of the force exerted by the ground on the ladder at its base, and (b) determine what the coefficient of friction at the base of the ladder must be if the ladder is not to slip when a 75-kg person stands three-fourths of the way up the ladder.

18. (II) If the coefficient of friction between the ladder and the ground in the situation described in the preceding

FIGURE 8-41

FIGURE 8-43

FIGURE 8-45

FIGURE 8-46

FIGURE 8-44

problem is 0.30, how far up the ladder can the person climb before the ladder starts to slip?

19. (III) A person wants to push a lamp (mass 9.6 kg) across the floor. Assuming the person pushes at a height of 60 cm above the ground and the coefficient of friction is 0.20, determine (a) whether the lamp will slide or tip over (Fig. 8-44). (b) Calculate the maximum height above the floor at which the person can push the lamp so it slides rather than tips.

20. (III) A home mechanic wants to raise the 220-kg engine out of a car. The plan is to stretch a rope vertically from the engine to a branch of a tree 25 m above. The rope passes over a firmly placed pulley 1.0 m above the engine so that when the mechanic climbs halfway up the tree and pulls horizontally on the rope at its midpoint (12 m below the branch), the engine rises out of the car. How much force must the mechanic exert to raise the engine 0.50 m? Assume the rope doesn't stretch.

21. (III) A 30-kg round table is supported by three legs placed equal distances apart on the edge. What minimum mass, placed on the table's edge, will cause the table to overturn?

22. (III) Two guy wires run from the top of a pole 2.4 m tall that supports a volleyball net. The two wires are anchored to the ground 2.0 m apart and each 2.0 m from the pole. The tension in each wire is 85 N. What is the tension in the net, assumed horizontal and attached at the top of the pole?

*SECTION 8-3

*23. (I) If the point of insertion of the biceps muscle into the lower arm shown in Fig. 8-10a were 6.0 cm, how much mass could the person hold with a muscle exertion of 400 N?

*24. (I) Approximately what force, F_M, must the extensor muscle in the upper arm exert on the lower arm to hold a 7.3 kg shotput (Fig. 8-45)? Assume the lower arm has

a mass of 2.8 kg and its cg is 12 cm from the pivot point.

*25. (II) Calculate the force required of the "deltoid" muscle, F_M, to hold up the outstretched arm shown in Fig. 8-46. The total mass of the arm is 2.8 kg.

*26. (II) Suppose the hand in the above problem holds a 10-kg mass. What force, F_M, is required of the deltoid muscle assuming the mass is 50 cm from the shoulder joint?

*27. (II) Calculate the magnitude of the force F_J exerted by the shoulder on the upper arm at the joint for problems 25 and 26.

*28. (III) Calculate the magnitude of the force at the base of the spine, F_V in Fig. 8-11b, if a mass of 20 kg is held in the person's hands. Assume the person has a mass of 70 kg.

*29. (III) When walking, a person momentarily puts all the weight on one foot. The cg of the body lies over the supporting foot. Figure 8-47 shows the supporting leg and the forces on it. Calculate the force exerted by the "hip abductor" muscles, F_M, and the x and y components of the force F_J acting on the joint. Take the whole leg as the object under discussion.

*30. (III) (a) Using the information in problem 29, calculate F_M and F_J, assuming the person carries a 20-kg suitcase in each hand. Assume the cg of each suitcase lies directly below the edge of the hipbone. (b) Calculate F_M and F_J assuming the person carries a suitcase only in the hand on the side opposite the supporting leg. [Hint: first calculate the common cg of the person plus suitcase; this point will be above the foot, thus shifting the horizontal measurements from those shown in Fig. 8-47.]

$w_1 = 90$ N (wt. of leg)
$w = 700$ N (wt. of whole body)

FIGURE 8-47

*SECTION 8-4

*31. (I) One kind of car jack is essentially a lever with a ratchet to hold the lifted object in place while the lever is returned for a new thrust. If the lever is pivoted 2.0 cm from the ratchet and the handle is 35.0 cm long, what is the IMA?

*32. (II) Calculate the force required in Fig. 8-12 taking into account the lever's mass of 12 kg. Assume the cg of the lever is at its center.

*33. (II) Some car jacks work on the principle of a screw. Determine the IMA of a jack in terms of the pitch (distance between threads) P, radius r_1 of the screw, and the turning radius r_2 of the jack handle. Ignore any other effects, such as that due to gears or levers in the jack.

*34. (II) Design a pulley system which can lift 80 kg with a force of $(18$ kg $\times g)$ assuming an efficiency of 75 percent.

*35. (II) A 38-cm long crank is attached to a spindle with a 2.5-cm radius. Assuming an efficiency of 80 percent, how much weight can be lifted on a cord wrapped around the spindle if the person can exert a steady 50-N force on the crank handle?

*36. (II) An inclined plane (such as used by the Egyptians to build the pyramids) can be considered to be a simple machine since a heavy weight can be raised by the application of a smaller force. (a) Calculate the IMA of a plane inclined at an angle of 15° to the horizontal.

(b) If the coefficient of friction for a particular object on this plane is 0.40, what is the AMA?

SECTION 8-5

37. (II) The Leaning Tower of Pisa is 55 m tall and about 7.0 m in diameter. The top is 4.5 m off-center. Is the tower in stable equilibrium? If so, how much further can it lean before it becomes unstable? Assume the tower is of uniform composition.

38. (III) Four bricks are to be stacked at the edge of a table, each brick overhanging the one below it, so that the top brick extends as far as possible beyond the edge of the table. (a) To achieve this, show that the bricks must extend no more than (starting at the top) 1/2, 1/4, 1/6, and 1/8 of their length beyond the one below. (b) Is the top brick completely beyond the base? (c) Determine a general formula for the maximum total distance spanned by n bricks if they are to remain stable. (d) A builder wants to construct a corbeled arch (Figure 8-27) based on the principle of stability discussed in (a) and (c) above. What minimum number of bricks, each 0.30 m long, is needed if the arch is to span 1.0 m?

SECTION 8-6

39. (I) A marble column of cross-sectional area 2.0 m² supports a mass of 20,000 kg. (a) What is the stress within the column? (b) What is the strain?

40. (I) By how much is the column in the previous problem shortened if it is 10-m high?

41. (I) If the supports in Example 8-8 were 10 m high, how much would they shorten?

42. (II) How much force is required to stretch a steel piano wire of 0.050 cm diameter by 0.015 percent?

43. (II) If the two wires in Fig. 8-40 are made of steel wire 1.0 mm in diameter, what is the percentage stretch of each because of the load?

44. (II) A 15-cm-long animal tendon was found to stretch 3.1 mm by a force of 12.4 N. The tendon was approximately round with an average diameter of 8.5 mm. Calculate the elastic modulus of this tendon.

45. (II) How much pressure is needed to compress the volume of an iron block by 0.10 percent? Express answer in N/m², and compare it to atmospheric pressure $(1.0 \times 10^5$ N/m²).

46. (II) One liter of alcohol (1000 cm³) in a flexible container is carried to the bottom of the sea where the pressure is 2.4×10^6 N/m². What will its volume be now?

47. (III) A scallop forces open its shell with an elastic material called abductin whose elastic modulus is about 2.0×10^6 N/m². If this piece of abductin is 3.0 mm thick and has a cross-sectional area of 0.50 cm², how much potential energy does it store when compressed 1.0 mm?

48. (III) A pole projects horizontally from the front wall of a shop. A 4.5-kg sign hangs from the pole at a point 2.4 m from the wall. (a) What is the torque due to this

sign calculated about the point where the pole meets the wall? (b) If the pole is not to fall off, there must be another torque exerted to balance it. What exerts this torque? Use a diagram to show how this torque must act. (c) Discuss whether compression, tension, and/or shear play a role in part (b).

SECTION 8-7

49. (I) The femur bone in the leg has an average effective cross section of about 3.0 cm² ($=3.0 \times 10^{-4}$ m²). How much compressive force can it withstand before breaking?

50. (II) If a compressive force of 3.0×10^4 N is exerted on the end of a 20-cm-long bone of cross-sectional area 3.6 cm², (a) will the bone break, and (b) if not, by how much does it shorten?

51. (II) What is the minimum cross-sectional area required of a vertical steel wire from which is suspended a 250-kg chandelier? Assume a safety factor of 4.

52. (II) Assume the supports of the cantilever shown in Fig. 8-7 (mass = 7200 kg) are made of wood. Calculate the minimum cross-sectional area required of each, assuming a safety factor of 8.

53. (II) An iron bolt is used to connect two iron plates together. The bolt must withstand shear forces up to about 2000 N. Calculate the minimum diameter for the bolt, based on a safety factor of 4.

54. (II) In Example 6-4 in Chapter 6 we calculated the impulse and average force on the leg of a person who jumps 5.0 m down to the ground. If the legs are not bent upon landing, so the body moves a distance d of only about $d = 1.0$ cm during collision, determine (a) the stress in the tibia bone (area $= 3.0 \times 10^{-4}$ m²) and (b) whether or not the bone will break. (c) Repeat for a bent-knees landing ($d = 50.0$ cm).

55. (II) The roof of a 7.5 m × 10.0 m room in a school is to support a uniformly distributed load of 10,000 kg (133 kg/m²) in addition to its own mass of 4100 kg. The roof is to be supported by "2 × 4s" (actually about 4.0 × 9.0 cm) along the 10.0-m sides. How many supports are required on each side and how far apart must they be? Consider only compression and assume a safety factor of 15.

56. (III) There is a maximum height of a uniform vertical column made of any material that can support itself without buckling, and it is independent of the cross-sectional area (why?). Calculate this height for (a) steel (density 7.8×10^3 kg/m³), (b) granite (density 2.7×10^3 kg/m³).

57. (III) From what height must a 1.2-kg rectangular brick 15.0 cm × 6.0 cm × 4.0 cm be dropped above a rigid steel floor in order to break the brick? Assume the brick strikes the floor directly on its largest face, and that the compression of the brick is much greater than that of the steel (that is, ignore compression of the steel). State other simplifying assumptions that may be necessary.

58. (III) A steel cable is to support an elevator whose total (loaded) mass is not to exceed 2200 kg. If the maximum acceleration of the elevator is 1.0 m/s², calculate the diameter of cable required. Assume a safety factor of 5.0.

*SECTION 8-8

*59. (II) How high must a pointed arch be if it is to span a space 8.0 m wide and exert one-third the horizontal force at its base that a round arch would?

*60. (II) The subterranean tension ring that exerts the balancing horizontal force on the abutments in Fig. 8-33 is 36 sided so each segment makes a 10° angle with the adjacent one, Fig. 8-48. Calculate the tension F that must exist in each segment so that the required force of 4.3×10^5 N can be exerted at each corner (Example 8-9).

FIGURE 8-48

430,000 N

9 FLUIDS

The three common states, or *phases*, of matter are solid, liquid, and gas. We can distinguish these three phases as follows. A *solid* maintains a fixed shape and a fixed size; even if a large force is applied to a solid, it does not readily change its shape or volume. A *liquid* does not maintain a fixed shape—it takes on the shape of its container—but like a solid it is not readily compressible and its volume can be changed significantly only by a very large force. A *gas* has neither a fixed shape nor a fixed volume—it will expand to fill its container. For example when air is pumped into an automobile tire, the air does not all run to the bottom of the tire as a liquid would; it fills the whole volume of the tire. Since liquids and gases do not maintain a fixed shape, they each have the ability to flow; they are thus often referred to collectively as **fluids.**

The division of matter into three states is not always simple. How, for example, should butter be classified? We should also mention that a fourth state of matter can be distinguished. This state is called the *plasma*, and it occurs only at very high temperatures, as we will discuss in Chapter 30. Some scientists believe that so-called colloids (suspensions of tiny particles in a liquid) should also be considered a separate state of matter. However, for our present purposes we will mainly be interested in the three ordinary states of matter. In Chapter 8 we discussed some of the properties of solid materials. In this chapter we will discuss the properties of fluids.

9-1 ∎ Density and specific gravity

It is sometimes said that iron is "heavier" than wood. This cannot really be true since a large log clearly weighs more than an iron nail. What we should say is that iron is more *dense* than wood.

The **density,** ρ, of a substance is defined as its mass per unit volume:

$$\rho = \frac{m}{V}, \tag{9-1}$$

where m is the mass of an amount of the substance whose volume is V. Den-

TABLE 9-1

Densities of substances†

Substance	Mass density, ρ (kg/m³)	Substance	Mass density, ρ(kg/m³)
Solids		Liquids	
Aluminum	2.70×10^3	Water (4°C)	1.00×10^3
Iron and steel	7.8×10^3	Blood, plasma	1.03×10^3
Copper	8.9×10^3	Blood, whole	1.05×10^3
Lead	11.3×10^3	Seawater	1.025×10^3
Gold	19.3×10^3	Mercury	13.6×10^3
Concrete	2.3×10^3	Alcohol, ethyl	0.79×10^3
Granite	2.7×10^3	Gasoline	0.68×10^3
Wood	$0.3–0.9 \times 10^3$	Gases†	
Glass	2.6×10^3	Air	1.29
Ice	0.917×10^3	Helium	0.179
Bone	1.7×10^3	Carbon dioxide	1.98
		Water (steam)(100°C)	0.598

† Densities are given at 0°C and 1 atm pressure unless otherwise specified.

sity is a characteristic property of a substance; objects made of a given substance, say pure iron, can have any size or mass, but the density will be the same for each.

The SI unit for density is kg/m³. Sometimes densities are given in g/cm³. Note that since $1 \text{ kg/m}^3 = 1000 \text{ g}/(100 \text{ cm})^3 = 10^{-3} \text{ g/cm}^3$, a density given in g/cm³ must be multiplied by 1000 to give the result in kg/m³. Thus the density of aluminum is $\rho = 2.70 \text{ g/cm}^3$, which is equal to 2700 kg/m³. The densities of a variety of materials are listed in Table 9-1. Temperature affects the density of substances (although the effect is slight for liquids and solids), and hence the temperature is specified in Table 9-1.

☐   **EXAMPLE 9-1**   What is the mass of a lead sphere of radius 0.500 m?

SOLUTION   The volume of the sphere is

$$V = \tfrac{4}{3}\pi r^3 = \tfrac{4}{3}(3.14)(0.500 \text{ m})^3 = 0.523 \text{ m}^3.$$

From Table 9-1, the density of lead is $\rho = 11,300 \text{ kg/m}^3$, so we have from Eq. 9-1

$$m = \rho V = (11,300 \text{ kg/m}^3)(0.523 \text{ m}^3) = 5910 \text{ kg}.$$

The **specific gravity** of a substance is defined as the ratio of the density of that substance to the density of water at 4.0°C. Specific gravity (abbreviated SG) is a pure number, without dimensions or units. Since the density of water is $1.00 \text{ g/cm}^3 = 1.00 \times 10^3 \text{ kg/m}^3$, the specific gravity of any substance will be precisely equal numerically to its density specified in g/cm³, or 10^{-3} times its density specified in kg/m³. For example (see Table 9-1) the specific gravity of lead is 11.3 and of alcohol 0.79.

Pressure is defined as force per unit area, where the force F is understood to be acting perpendicular to the surface area A:

$$\text{pressure} = P = \frac{F}{A}. \qquad (9\text{-}2)$$

The SI unit of pressure is N/m^2. This unit has the official name pascal (Pa): $1 \ Pa = 1 \ N/m^2$; however, for simplicity, we will often use N/m^2. Other units sometimes used are dynes/cm^2, lb/in^2 (sometimes abbreviated "psi"), and kg/cm^2 (as if kilograms were a force: that is, $1 \ kg/cm^2 = 9.8 \ N/cm^2 = 9.8 \times 10^4 \ N/m^2$). The last two are often used on tire gauges. We will meet several other units shortly.

As an example of calculating pressure, a 60-kg person whose two feet cover an area of 500 cm^2 exerts a pressure of $F/A = mg/A = (60 \ kg) \cdot (9.8 \ m/s^2)/(0.050 \ m^2) = 12 \times 10^3 \ N/m^2$ on the ground. If the person stands on one foot, the force is the same but the area will be half, so the pressure will be twice as much: $24 \times 10^3 \ N/m^2$. The concept of pressure is particularly useful in dealing with fluids, which is why we introduced it in this chapter.

It is an experimental fact that *a fluid exerts a pressure in all directions.* This is well known to swimmers and divers who feel the water pressure on all parts of their bodies. At a particular point in a fluid at rest, the pressure is the same in all directions. This is illustrated in Fig. 9-1; consider a tiny cube of the fluid which is so small that we can ignore the force of gravity on it. Then the pressure on one side of it must equal the pressure on the opposite side. If this weren't true, the net force on this cube would not be zero, and it would move until the pressure did become equal; if the fluid is not flowing, then the pressures must be equal.

Another important property of a fluid at rest is that the force due to fluid pressure always acts *perpendicularly* to any surface it is in contact with. If there were a component of the force parallel to the surface as shown in Fig. 9-2, then according to Newton's third law, the surface would exert a force back on the fluid that also would have a component parallel to the surface; this component would cause the fluid to flow, in contradiction to our assumption that the fluid is at rest.

Let us now calculate quantitatively how the pressure in a liquid of uniform density varies with depth. Consider a point which is at a depth h below the surface of the liquid (that is, the surface is a height h above this point) as shown in Fig. 9-3. The pressure due to the liquid at this depth h is due to the weight of the column of liquid above it. Thus the force acting on the area is $F = mg = \rho A h g$ where Ah is the volume of the column, ρ is the density of the liquid (assumed to be constant), and g is the acceleration of gravity. The pressure, P, is then

$$P = \frac{F}{A} = \frac{\rho A h g}{A}$$

$$P = \rho g h. \qquad (9\text{-}3a)$$

Thus the pressure is directly proportional to the density of the liquid, and to the depth within the liquid. In general, the *pressure at equal depths within a*

FIGURE 9-1

Pressure is the same in every direction in a fluid at a given depth; if it weren't the fluid would be in motion.

FIGURE 9-2

The force a non-moving fluid exerts on a solid surface is perpendicular to the surface; that is, $F_{\parallel} = 0$.

FIGURE 9-3

Calculating the pressure at a depth h in a liquid.

uniform liquid is the same. (Equation 9-3a gives the pressure due to the liquid itself. If an external pressure is exerted at the surface of the liquid, this must be taken into account as we will discuss in Section 9-4.)

Equation 9-3a is extremely useful. It is valid for fluids whose density is constant and does not change with depth—that is, if the fluid is *incompressible*. This is usually a good approximation for liquids (although at great depths in the ocean the density of water is increased substantially by compression due to the great weight of water above). Gases, on the other hand, are very compressible and density can vary significantly with depth. If the density varies only slightly, Eq. 9-3a can be used to determine the difference in pressure ΔP at different heights with ρ being the average density:

$$\Delta P = \rho g\, \Delta h. \qquad (9\text{-}3b)$$

☐   EXAMPLE 9-2   The surface of the water in a storage tank is 30 m above a water faucet in the kitchen of a house. Calculate the water pressure at the faucet.

SOLUTION   Atmospheric pressure acts both at the surface of the water in the storage tank, and on the water leaving the faucet. The pressure difference between the inside and outside of the faucet will be

$$\Delta P = \rho g h = (1.0 \times 10^3 \text{ kg/m}^3)(9.8 \text{ m/s}^2)(30 \text{ m})$$
$$= 2.9 \times 10^5 \text{ N/m}^2.$$

The height h is sometimes called the *pressure head*. In this example, the head of the water is 30 m.

9-3 ∎ Atmospheric pressure and gauge pressure

The pressure of the earth's atmosphere, as in any fluid, decreases with decreased depth (or increased height). But the earth's atmosphere is somewhat complicated because not only does the density of air vary greatly with altitude, but there is no distinct top surface to the atmosphere from which h (in Eq. 9-3a) can be measured. We can, however, calculate the approximate difference in pressure between two altitudes using Eq. 9-3b.

The pressure of the air at a given place varies slightly according to the weather. At sea level, the pressure of the atmosphere on the average is 1.013×10^5 N/m^2 (or 14.7 lb/in^2). This value is used to define another unit of pressure in common use, the *atmosphere* (abbreviated atm):

$$1 \text{ atm} = 1.013 \times 10^5 \text{ N/m}^2 = 101.3 \text{ kPa}.$$

Another unit of pressure sometimes used (in meteorology and on weather maps) is the *bar*, which is defined as 1 bar = 1.00×10^5 N/m^2; thus standard atmospheric pressure is slightly more than 1 bar.

The pressure due to the weight of the atmosphere is exerted on all objects immersed in this great sea of air, including our bodies. How does a human body withstand the enormous pressure on its surface? The answer lies in the fact that living cells maintain an internal pressure that just balances the external pressure. Similarly, the pressure inside a balloon balances the outside pressure of the atmosphere. An automobile tire, because of its rigidity, can maintain pressures much greater than the external pressure.

One must be careful, however, when determining the pressure in a tire, for tire gauges, and most other pressure gauges as well, register the pressure over and above atmospheric pressure. This is called **gauge pressure.** Thus, to get the absolute pressure P, one must add the atmospheric pressure, P_a, to the gauge pressure, P_G:

$$P = P_a + P_G.$$

For example, if a tire gauge registers 220 kPa, the actual pressure within the tire is 220 kPa + 100 kPa = 320 kPa. This is equivalent to about 3.2 atm (2.2 atm gauge pressure).

9-4 ∎ Pascal's principle

The earth's atmosphere exerts a pressure on all objects with which it is in contact, including other fluids. Atmospheric pressure acting on a fluid is transmitted throughout that fluid. For instance, according to Eq. 9-3a, the water pressure at a depth of 100 m below the surface of a lake is $P = \rho g h = (1000 \text{ kg/m}^3)(9.8 \text{ m/s}^2)(100 \text{ m}) = 9.8 \times 10^5 \text{ N/m}^2$ or 9.7 atm. However, the total pressure at this point is due to the pressure of water plus the pressure of the air above it. Hence the total pressure (if the lake is near sea level) is 9.7 atm + 1.0 atm = 10.7 atm. This is just one example of a general principle attributed to the French philosopher and scientist Blaise Pascal (1623–1662). **Pascal's principle** states that *pressure applied to a confined fluid increases the pressure throughout by the same amount.*

FIGURE 9-4

Applications of Pascal's principle: (a) hydraulic brakes in a car; (b) hydraulic lift.

A number of practical devices make use of Pascal's principle. Two examples, hydraulic brakes in an automobile and the hydraulic lift, are illustrated in Fig. 9-4. In the case of a hydraulic lift, a small force can be used to exert a large force by making the area of one piston (the output) larger than the area of the other (the input). This is another example of a simple machine (Chapter 8) and relies on the fact that the pressures on the input and output cylinders are the same at equal heights. (If the difference in height of the two pistons is not too great this result is not much altered.) Thus, if the input quantities are represented by the subscript "i" and the output by "o," we have

$$P_o = P_i$$

$$\frac{F_o}{A_o} = \frac{F_i}{A_i},$$

or, finally

$$\frac{F_o}{F_i} = \frac{A_o}{A_i}.$$

F_o/F_i is the "ideal mechanical advantage" of the hydraulic press, and is equal to the ratio of the areas. For example, if the area of the output piston is 20 times that of the input cylinder, the force is multiplied by a factor of 20; thus a force of 200 lb could lift a 4000-lb car.

A number of living organisms make use of hydraulic pressure. A sea anemone can achieve a variety of shapes by the action of muscles on its sea-water-filled body cavity. Its body is sometimes referred to as a "hydrostatic skeleton." Earthworms move forward by successive contractions of circular muscles along the body axis acting on the hydrostatic skeleton. The legs of spiders have flexor muscles but possess no extensor muscles; their legs are extended by fluid driven into them under pressure.

9-5 ▌ Measurement of pressure; gauges and the barometer

Many devices have been invented to measure pressure, some of which are shown in Fig. 9-5. The simplest is the open-tube *manometer* (Fig. 9-5a) which is a U-shaped tube partially filled with a liquid, usually mercury or water. The pressure P being measured is related to the difference in height of the two levels of the liquid by the relation (see Eq. 9-3b)

$$P = P_0 + \rho g h,$$

where P_0 is atmospheric pressure (acting on the top of the fluid in the left-hand tube) and ρ is the density of the liquid. Note that the quantity $\rho g h$ is the "gauge pressure"—the amount by which P exceeds atmospheric pressure. If the liquid in the left-hand column were lower than that in the right-hand column, this would indicate that P was less than atmospheric pressure (and h would be negative).

Instead of calculating the product $\rho g h$, it is common to merely specify the height h. In fact pressures are sometimes specified as so many "millimeters of mercury" (mm Hg), and sometimes as so many "mm of water."

FIGURE 9-5

Pressure gauges: (a) open-tube manometer; (b) Bourdon gauge; (c) aneroid gauge.

(a) Open–tube manometer

(b) Bourdon gauge

(c) Aneroid gauge (mainly used for air pressure and then is called an aneroid barometer)

TABLE 9-2   **9   Fluids**

Conversion factors between different units of pressure

In Terms of $1\ Pa = 1\ N/m^2$	Related to 1 atm
1 atm $= 1.013 \times 10^5\ N/m^2$	1 atm $= 1.013 \times 10^5\ N/m^2$
$= 1.013 \times 10^5\ Pa = 101.3\ kPa$	
1 bar $= 1.000 \times 10^5\ N/m^2$	$= 1.013$ bar
1 dyne/cm^2 $= 0.1\ N/m^2$	$= 1.013 \times 10^6$ dyne/cm^2
1 kg/cm^2 $= 9.85 \times 10^4\ N/m^2$	$= 1.03$ kg/cm^2
1 lb/in^2 $= 6.90 \times 10^3\ N/m^2$	$= 14.7$ lb/in^2
1 lb/ft^2 $= 47.9\ N/m^2$	$= 2.12 \times 10^3$ lb/ft^2
1 cm Hg $= 1.33 \times 10^3\ N/m^2$	$= 76$ cm Hg
1 mm Hg $= 133\ N/m^2$	$= 760$ mm Hg
1 torr $= 133\ N/m^2$	$= 760$ torr
1 mm H$_2$O (4°C) $= 9.81\ N/m^2$	$= 1.03 \times 10^4$ mm H$_2$O (4°C)

The unit mm Hg is equivalent to a pressure of 133 N/m², since 1.00 mm = 1.00×10^{-3} m and

$$\rho g h = (13.6 \times 10^3\ kg/m^3)(9.80\ m/s^2)(1.00 \times 10^{-3}\ m)$$

$$= 1.33 \times 10^2\ N/m^2.$$

The unit mm Hg is also called the "torr" in honor of Evangelista Torricelli (1608–1647), who invented the barometer (see below). Conversion factors among the various units of pressure (an incredible nuisance!) are given in Table 9-2. It is important that only N/m² = Pa, the proper SI unit, be used in calculations involving other quantities specified in SI units.

Other types of pressure gauge are the Bourdon gauge (Fig. 9-5b) in which increased pressure tends to straighten out the delicate tube which is attached to a pointer, and the aneroid gauge (Fig. 9-5c) in which the pointer is connected to the flexible ends of an evacuated thin metal chamber. In a more sophisticated type of gauge, the pressure to be measured is applied to a thin metal diaphragm and the resulting distortion of the diaphragm is detected electrically with the use of a so-called pressure transducer (see Chapter 18).

Atmospheric pressure is often measured by a modified kind of mercury manometer with the end closed. This is called a mercury *barometer*, Fig. 9-6. The glass tube is filled with mercury and then inverted into the bowl of mercury. If the tube is long enough, the level of the mercury will drop, leaving a vacuum at the top of the tube, since atmospheric pressure can support a column of mercury only about 76 cm high (exactly 76.0 cm at normal atmospheric pressure). That is, a column of mercury 76 cm high exerts the same pressure as the atmosphere: from the formula $P = \rho g h$, with $\rho = 13.6 \times 10^3\ kg/m^3$ for mercury and $h = 76.0$ cm we have

$$P = (13.6 \times 10^3\ kg/m^3)(9.80\ m/s^2)(0.760\ m)$$

$$= 1.013 \times 10^5\ N/m^2$$

$$= 1.00\ atm.$$

Household barometers are usually of the aneroid type, Fig. 9-5c.

A calculation similar to that above will show that atmospheric pressure can maintain a column of water 10.3 m high in a tube whose top is under vacuum. It was a source of wonder and frustration a few centuries ago that

FIGURE 9-6

A mercury barometer, when the air pressure is 76 cm Hg.

$P = 0$

76 cm

$P = 1$ atm

no matter how good a vacuum pump was, it could not lift water more than about 10 m. It was, for example, a great practical difficulty in trying to pump water out of deep mine shafts which required multiple stages for depths greater than 10 m. This problem, which concerned Galileo, was first understood by Torricelli. The point is that a pump does not really suck water up a tube. It is air pressure that pushes the water up the tube if the other end is under a vacuum, just as it is air pressure that pushes (or keeps) the mercury 76 cm high in a barometer.

*9-6 ∎ Pumps; the heart and blood pressure

Pumps can be classified into categories according to their function. A *vacuum pump* is designed to reduce the pressure (usually of air) in a given vessel. A *force pump*, on the other hand, is a pump which is intended to increase the pressure—for example, to lift a liquid (such as water from a well) or to push a fluid through a pipe. Figure 9-7 illustrates the principle behind a simple "reciprocating" pump. It could be a vacuum pump, in which case the "intake" is connected to the vessel to be evacuated; a similar mechanism is used in some force pumps, and in this case the fluid is forced under increased pressure through the outlet. Other kinds of pumps are illustrated in Fig. 9-8.

FIGURE 9-7

An example of one kind of pump. As the diagram indicates, the intake valve opens and air (or fluid that is being pumped) fills the empty space when the cylinder moves to the left. When the cylinder moves to the right (not shown), the outlet valve opens and fluid is forced out.

FIGURE 9-8

(a) Centrifugal pump: the rotating blades force fluids through the outlet pipe; this kind of pump is used in vacuum cleaners and as a water pump in automobiles. (b) Rotary oil pump, used to obtain vacuum as low as 10^{-4} mm Hg: gas (usually air) from the vessel to be evacuated diffuses into the space G via the intake pipe I; the rotating off-center cylinder C traps the gas in G and pushes it out the exhaust valve E, in the meantime allowing more gas to diffuse into G for the next cycle. The sliding valve V is kept in contact with C by a spring S and this prevents the exhaust gas from returning to G. (c) Diffusion pump, used to obtain vacuums as low as 10^{-8} mm Hg: air molecules from the vessel to be evacuated diffuse into the jet, where the rapidly moving jet of oil sweeps the molecules away. The "forepump" is a mechanical pump, such as the rotary type (b), and acts as a first stage in reducing the pressure.

(a)

(b)

(c)

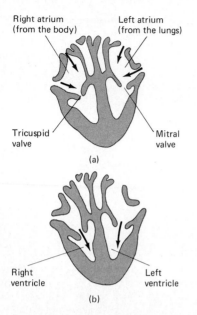

Right atrium (from the body)   Left atrium (from the lungs)

Tricuspid valve   Mitral valve

(a)

Right ventricle   Left ventricle

(b)

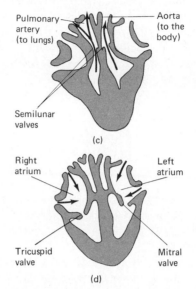

Pulmonary artery (to lungs)   Aorta (to the body)

Semilunar valves

(c)

Right atrium   Left atrium

Tricuspid valve   Mitral valve

(d)

FIGURE 9-9

(a) In the diastole phase, the heart relaxes between beats. Blood moves into the heart; both atria are filled rapidly. (b) When the atria contract, the systole or pumping phase begins. The contraction pushes the blood through the mitral and tricuspid valves into the ventricles. (c) The contraction of the ventricles forces the blood through the semilunar valves into the pulmonary artery (which leads to the lungs) and to the aorta (the body's largest artery), which leads to the arteries serving all the body (Figure 9-26). (d) When the heart relaxes, the semilunar valves close; blood fills the atria, beginning the cycle again.

The centrifugal pump, or any force pump, can be used as a *circulating pump*—that is, to circulate a fluid around a closed path, such as the cooling water or lubricating oil in an automobile.

The heart of a human (and of other animals as well) is essentially a circulating pump. The action of a human heart is shown in Fig. 9-9. There are actually two separate paths for blood flow. The longer path takes blood to the parts of the body, via the arteries, bringing oxygen to body tissues and picking up carbon dioxide which it carries back to the heart via veins. This blood is then pumped into the lungs (the second path) where the carbon dioxide is released and oxygen is taken up. The oxygen-laden blood is returned to the heart where it is again pumped to the tissues of the body.

Blood pressure is measured using either a mercury-filled manometer or one of the other types of gauge mentioned above; it is usually calibrated in mm Hg. The gauge is attached to a closed air-filled jacket which is wrapped around the upper arm at the level of the heart, Fig. 9-10. Two values of blood pressure are measured: the maximum pressure when the heart is pumping, which is called *systolic pressure;* and the pressure when the heart is in the resting part of the cycle, which is called *diastolic pressure*. Initially, the air pressure in the jacket is increased high above the systolic pressure by means of a hand pump, and this compresses the main (brachial) artery in the arm and cuts off the flow of blood. The air pressure is then slowly reduced to the point where blood again begins to flow into the arm; it is detected by listening with a stethoscope to the characteristic tapping sound[†] of the blood returning to the forearm. At this point, systolic pressure is just equal to the air pressure in the jacket which can be read off the gauge. The air pressure is subsequently reduced further and the tapping sound disappears when blood at low pressure can enter the artery. At this point, the gauge indi-

FIGURE 9-10

Device for measuring blood pressure.

Jacket

Hand pump

Gauge

[†] When the blood flows through the constriction caused by the tight jacket, its velocity is high and the flow is turbulent (see Section 9-13); it is the turbulence that causes the tapping sound.

cates the diastolic pressure. Normal systolic pressure is around 120 mm Hg whereas normal diastolic pressure is around 80 mm Hg.

9-7 ❙ Buoyancy and Archimedes' principle

Objects submerged in a fluid appear to weigh less than they do when outside the fluid. For example, a large rock that you would have difficulty lifting off the ground can often be easily lifted from the bottom of a stream; when the rock breaks through the surface of the water it suddenly seems to be much heavier. Many objects, such as wood, float on the surface of the water. These are two examples of *buoyancy*. In each example, the force of gravity is acting downward; but in addition, an upward buoyant force is exerted by the liquid.

The buoyant force arises from the fact that the pressure in a fluid increases with depth. Thus the upward pressure on the bottom surface of a submerged object is greater than the downward pressure on its top surface. To see the effect of this, consider a cylinder of height h whose top and bottom ends have an area A and which is completely submerged in a fluid of density ρ_f as shown in Fig. 9-11. The fluid exerts a pressure $P_1 = \rho_f g h_1$ against the top surface of the cylinder; the force due to this pressure on top of the cylinder is $F_1 = P_1 A = \rho_f g h_1 A$, and it is directed downward. Similarly, the fluid exerts an upward force on the bottom of the cylinder equal to $F_2 = P_2 A = \rho_f g h_2 A$. The net force due to the fluid pressure, which is the *buoyant force* $\mathbf{F_B}$, acts upward and has the magnitude

$$F_B = F_2 - F_1$$
$$= \rho_f g A(h_2 - h_1)$$
$$= \rho_f g A h$$
$$= \rho_f g V$$

where $V = Ah$ is the volume of the cylinder. Since ρ_f is the density of the fluid, the product $\rho_f g V = m_f g$ is the weight of fluid which takes up a volume equal to the volume of the cylinder. Thus the buoyant force on the cylinder is equal to the weight of fluid displaced[†] by the cylinder. This result is valid no matter what the shape of the object. Its discovery is credited to Archimedes (287?–212 B.C.), and it is called **Archimedes' principle:** *The buoyant force on a body immersed in a fluid is equal to the weight of the fluid displaced by that object.*

We can derive Archimedes' principle in general by the following simple but elegant argument. The irregularly shaped object D shown in Fig. 9-12a is acted on by the force of gravity (its weight, w) and the buoyant force, F_B. (If there is no other force acting on the object, such as a hand pulling up, the object shown will move downward since $w > F_B$.) We wish to determine F_B and to do so we next consider a body of (the same) fluid (D' in Fig. 9-12b) with the same shape and size as the original object, and located at the same

---

[†] By "fluid displaced" we mean a volume of fluid equal to the volume of the object, or that part of the object submerged if it floats (the fluid that used to be where the object was). If the object is placed in a glass initially filled to the brim with water, the water that flows over the top represents the water displaced by the object.

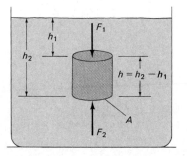

FIGURE 9-11
Determination of buoyant force.

FIGURE 9-12
Archimedes' principle.

(a)

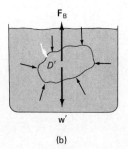

(b)

depth; you might think of this body of fluid as being separated from the rest of the fluid by an imaginary transparent membrane. The buoyant force F_B on this body of fluid will be exactly the same as that on the original object since the surrounding fluid, which exerts F_B, is in exactly the same configuration. Now the body of fluid D' is in equilibrium (the fluid as a whole is at rest); therefore, $F_B = w'$, where w' is the weight of the body of fluid. Hence the buoyant force F_B is equal to the weight of the body of fluid whose volume is equal to the volume of the original submerged object, which is Archimedes' principle.

EXAMPLE 9-3   A 70-kg rock lies at the bottom of a lake. Its volume is 3.0×10^4 cm^3. How much force is needed to lift it?

SOLUTION   The buoyant force on the rock due to the water is equal to the weight of 3.0×10^{-2} m^3 of water:

$$F_B = m_{H_2O}g = \rho_{H_2O}gV$$
$$= (1.0 \times 10^3 \text{ kg/m}^3)(9.8 \text{ m/s}^2)(3.0 \times 10^{-2} \text{ m}^3)$$
$$= 2.9 \times 10^2 \text{ N}.$$

The weight of the rock is $mg = (70 \text{ kg})(9.8 \text{ m/s}^2) = 6.9 \times 10^2$ N. Hence the force needed to lift it is 690 N − 290 N = 400 N. It is as if the rock had a mass of only $(400 \text{ N})/(9.8 \text{ m/s}^2) = 41$ kg.

Archimedes is said to have discovered his principle in his bath while thinking how he might determine whether the king's new crown was pure gold or a fake. Gold has a specific gravity of 19.3, somewhat higher than most metals, but a determination of specific gravity or density is not readily done directly since the volume of an irregularly shaped object is not easily determined. However, if the object is weighed in air ($= w$) and also "weighed" while it is under water ($= w'$), the density can be determined using Archimedes' principle as the following example shows.

EXAMPLE 9-4   A 14.7-kg crown has an apparent weight when submerged in water corresponding to 13.4 kg. Is it gold?

SOLUTION   The apparent weight of the submerged object, w', equals its actual weight w minus the buoyant force F_B:

$$w' = w - F_B = \rho_o gV - \rho_f gV$$

where V is the volume of the object, ρ_o its density, and ρ_f the density of the fluid (water in this case). Then we can write

$$\frac{w}{w - w'} = \frac{\rho_o gV}{\rho_f gV} = \frac{\rho_o}{\rho_f}.$$

[Thus $w/(w - w')$ is equal to the specific gravity of the object if the fluid in which it is submerged is water.] For the crown we have

$$\frac{\rho_o}{\rho_{H_2O}} = \frac{w}{w - w'} = \frac{14.7 \text{ kg}}{1.3 \text{ kg}} = 11.3.$$

This corresponds to a density of 11,300 kg/m^3. The crown seems to be made of lead!

Archimedes' principle applies equally well to objects that float such as wood. In general, *an object floats on a fluid if its density is less than that of the fluid.* For example, a log whose specific gravity is 0.60 and whose volume is 2.0 m³ will have a mass of 1200 kg. If the log is fully submerged, it will displace a mass of water $m = \rho V = (1000 \text{ kg/m}^3)(2.0 \text{ m}^3) = 2000 \text{ kg}$. Hence the buoyant force on it will be greater than its weight and it will float to the top. It will come to equilibrium when it displaces 1200 kg of water, which means that 1.2 m³ or 0.60 of its volume will be submerged. In general, the fraction of the object submerged is given by the ratio of the object's density to that of the fluid.

EXAMPLE 9-5   A hydrometer is a simple instrument used to measure specific gravity. A particular hydrometer (Fig. 9-13) consists of a glass tube, weighted at the bottom, which is 25.0 cm long, 2.00 cm² in cross-sectional area, and has a mass of 45.0 g. How far from the end should the 1.000 mark be placed?

SOLUTION   The hydrometer has a density

$$\rho = \frac{m}{V} = \frac{45.0 \text{ g}}{(2.00 \text{ cm}^2)(25.0 \text{ cm})} = 0.900 \text{ g/cm}^3.$$

Thus, when placed in water, it will come to equilibrium when 0.900 of its volume is submerged. Since it is of uniform cross section, $(0.900) \cdot (25.0 \text{ cm}) = 22.5 \text{ cm}$ of its length will be submerged. Since the specific gravity of water is defined to be 1.000, the mark should be placed 22.5 cm from the end.

FIGURE 9-13

A hydrometer.

Archimedes' principle is useful also in geology. According to the modern theory of plate tectonics (and continental drift), the continents can be considered to be floating on a "sea" of slightly deformable rock around it (mantle rock). Some interesting calculations can be done using very simple models, which we consider here and in the problems.

EXAMPLE 9-6   A simple model considers a continent as a block (density = 2800 kg/m³) floating in the mantle rock around it (density = 3300 kg/m³), Fig. 9-14. Assuming the continent is 35 km thick (average thickness of earth's crust), estimate the height of the continent above the surrounding rock.

SOLUTION   As discussed above, the fraction of continent submerged is equal to the ratio of the densities: $(2800 \text{ kg/m}^3)/(3300 \text{ kg/m}^3) = 0.85$. Hence 0.15 is above the surrounding rock, which is $(0.15)(35 \text{ km}) = 5.3 \text{ km}$.

FIGURE 9-14

Model of a continent "floating" in earth's crust.

This result, 5.3 km, represents a rough estimate of the average depth of the ocean bottom relative to the continents. (It is rough since it ignores such things as the weight of the oceans.)

Air is a fluid and it too exerts a buoyant force. Ordinary objects weigh less in air than they do if weighed in a vacuum. Because the density of air is so small, the effect for ordinary solids is slight. There are objects, however, that float in air—helium balloons, for example.

◻ **EXAMPLE 9-7** What volume of helium is needed if a balloon is to lift a load of 800 kg (including the weight of the empty balloon)?

SOLUTION The buoyant force on the helium, F_B, which is equal to the weight of displaced air, must at least be equal to the weight of the helium plus the load:

$$F_B = (m_{He} + 800 \text{ kg})g$$

where g is the acceleration due to gravity. This equation can be written in terms of density:

$$\rho_{air} V g = (\rho_{He} V + 800 \text{ kg})g$$

Solving now for V we find

$$V = \frac{800 \text{ kg}}{\rho_{air} - \rho_{He}} = \frac{800 \text{ kg}}{(1.29 \text{ kg/m}^3 - 0.18 \text{ kg/m}^3)}$$

$$= 720 \text{ m}^3.$$

*9-8 ∎ Surface tension and capillarity

Up to now in this chapter we have mainly been studying what happens beneath the surface of a liquid (or gas). But the actual surface of a liquid also behaves in an interesting way. A number of common observations suggest that the surface of a liquid acts like a stretched membrane under tension. For example, a drop of water on the end of a dripping faucet, or hanging from a thin branch in the early morning dew, forms into a nearly spherical shape as if it were a tiny balloon filled with water; and a steel needle can be made to float on the surface of water even though it is denser than the water. The surface of a liquid acts like it is under tension, and this tension, acting parallel to the surface, arises from the attractive forces between the molecules. This effect is called *surface tension*. More specifically a quantity called the *surface tension*, γ (the Greek letter gamma), is defined as the force F per unit length L that acts across any line in a surface, tending to pull the surface open:

$$\gamma = \frac{F}{L}. \tag{9-4}$$

To understand this, consider the U-shaped apparatus shown in Fig. 9-15 which encloses a thin film of liquid; because of surface tension, a force

FIGURE 9-15

U-shaped wire apparatus holding film of liquid to measure surface tension ($\gamma = F/2l$).

(a) Top view

(b) Edge view (magnified)

F is required to pull the movable wire and thus increase the surface area of the liquid. The liquid contained by the wire apparatus is a thin film having both a top and a bottom surface; hence the length of the surface being increased is $2l$. Thus the surface tension $\gamma = F/2l$. A delicate apparatus of this type can be used to measure the surface tension of various liquids. The surface tension of water is 0.072 N/m at 20°C. Table 9-3 gives the values for other liquids; note that temperature has a considerable effect on the surface tension.

We can see how surface tension arises by examining the process from the molecular point of view. The molecules of a liquid exert attractive forces on each other; these attractive forces are shown acting, in Fig. 9-16, on a molecule deep within the liquid and on a second molecule at the surface. The molecule inside the liquid is in equilibrium due to the forces of other molecules acting in all directions. The molecule at the surface is also normally in equilibrium (the liquid is at rest); this is true even though the forces on a molecule at the surface can only be exerted by molecules below it (or at an equal height). Hence there is a net attractive force downward, which tends to compress the surface layer slightly—but only to the point where this downward force is balanced by an upward (repulsive) force due to close contact (or collision with) the molecules below.[†] This compression of the surface means that, in essence, the liquid tries to minimize its surface area. This is why water tends to form spherical droplets, for a sphere represents the minimum surface area for a given volume.

In order to increase the surface area of a liquid, a force is required and work must be done to bring molecules from the interior to the surface (Fig. 9-17). This work increases the potential energy of the molecules and is sometimes called *surface energy*. The greater the surface area, the greater the surface energy.

The amount of work needed to increase the surface area can be calculated from Fig. 9-15 and Eq. 9-4:

$$W = F\,\Delta x$$
$$= \gamma L\,\Delta x$$
$$= \gamma\,\Delta A$$

where Δx is the change in distance and ΔA is the total increase in area (at both surfaces in Fig. 9-15) and so we can write

$$\gamma = \frac{W}{\Delta A}.$$

Thus, the surface tension γ is not only equal to the force per unit length; it is also equal to the work done per unit increase in surface area. Hence, γ can be specified in N/m or J/m^2.

Because of surface tension, insects can walk on water; and objects more dense than water, such as a steel needle, can actually float on the surface.

[†] The air molecules above exert a force, but this is a small effect since the air molecules are so far apart. The surface tension therefore does depend on the material above the surface but the effect is generally small if the latter is a dilute gas. But because of this small effect, surface tensions are specified for the boundary between two materials. If the second material is not mentioned, it is assumed to be air at atmospheric pressure.

TABLE 9-3
Surface tension (N/m) of some substances

Substance	γ(N/m)
Mercury (20°C)	0.44
Blood, whole (37°C)	0.058
Blood, plasma (37°C)	0.073
Alcohol, ethyl (20°C)	0.023
Water (0°C)	0.076
(20°C)	0.072
(100°C)	0.059

FIGURE 9-16

Molecular theory of surface tension, showing attractive forces (only) on a molecule at the surface, and on one deep inside the liquid.

FIGURE 9-17

Forces on a molecule being brought to the surface because of an increase in the surface area. The molecules at the surface must exert a strong force on the molecule being pulled up, and they in turn feel a strong force back (Newton's third law). This is the surface tension force.

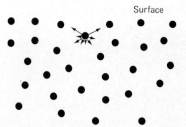

Figure 9-18a shows how the surface tension can support the weight w of an object. Actually, w is the "effective weight" of the object—its true weight less the buoyant force—since the object sinks slightly into the fluid. If the object is spherical in shape, which is close for the base of an insect's leg (see Fig. 9-18b), the surface tension acts at all points along a circle of radius r. Only the vertical component, $\gamma \cos \theta$, acts to balance w. Hence the net upward force due to surface tension is $2\pi r\gamma \cos \theta$.

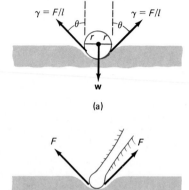

(a)

□ EXAMPLE 9-8   The base of an insect's leg is approximately spherical in shape with a radius of about 2.0×10^{-5} m. The 0.0030 g mass of the insect is supported equally by the six legs. Calculate the angle θ (see Fig. 9-18). Assume the water temperature is 20°C.

SOLUTION   Since

$$2\pi r\gamma \cos \theta = w$$

where w is one-sixth the weight of the insect (since it has six legs) we have

$$(6.28)(2.0 \times 10^{-5}\text{ m})(0.072\text{ N/m}) \cos \theta = \tfrac{1}{6}(3.0 \times 10^{-6}\text{ kg})(9.8\text{ m/s}^2)$$

$$\cos \theta = \frac{0.49}{0.90} = 0.54$$

$$\theta = 57°.$$

Notice that if $\cos \theta$ were greater than 1, this would indicate that the surface tension would not be great enough to support the weight.

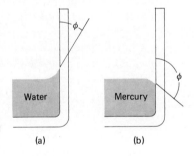

(b)

FIGURE 9-18
Surface tension acting on (a) a sphere, (b) an insect leg.

Calculations such as that above are not always accurate since the radius r of the surface depression is not precisely equal to the radius of the object. However, approximate estimates can be made as to whether an object will or will not remain on the surface.

Soaps and detergents have the effect of lowering the surface tension of water. This is desirable for washing and cleaning since the high surface tension of pure water prevents it from penetrating easily between the fibers of material and into tiny crevices. Substances which reduce the surface tension of a liquid are called *surfactants*.

Surface tension plays a role in another interesting phenomenon, capillarity. It is a common observation that water in a glass container rises up slightly where it touches the glass, Fig. 9-19a. The water is said to "wet" the glass. Mercury, on the other hand, is depressed when it touches the glass, Fig. 9-19b; the mercury does not wet the glass. Whether or not a liquid wets a solid surface is determined by the relative strengths of the cohesive forces between the molecules of the liquid compared to the adhesive forces between the molecules of the liquid and those of the container. (*Cohesion* refers to the force between molecules of the same type and *adhesion* to the force between molecules of different types.) Water wets glass because the water molecules are more strongly attracted to the glass molecules than they are to other water molecules. The opposite is true for mercury: the cohesive forces are stronger than the adhesive forces. The angle that the tangent to the liquid surface makes with the solid surface is called the *angle of contact* ϕ. Its value

FIGURE 9-19
Water (a) "wets" the surface of glass, whereas mercury (b) does not.

TABLE 9-4

Contact angle of some substances

Substance	Contact Angle
Water–glass	0°
Organic liquids (most)–glass	0°
Mercury–glass	140°
Water–parafin	107°
Kerosene–glass	26°

depends on the relative strengths of the cohesive and adhesive forces (see Fig. 9-19 and Table 9-4). When ϕ is less than 90°, the liquid wets the solid; if ϕ is greater than 90°, the liquid does not wet the solid.

In tubes having very small diameters, liquids are observed to rise or fall relative to the level of the surrounding liquid. This phenomenon is called *capillarity*, and such thin tubes are called *capillaries*. Whether the liquid rises or falls (Fig. 9-20) depends on the relative strengths of the adhesive and co-hesive forces. Thus water rises in a glass tube whereas mercury falls. The actual amount of rise (or fall) depends on the surface tension—since it is this which keeps the liquid surface from breaking apart—as well as on the con-tact angle ϕ and the radius r of the tube. To calculate h, the height of rise, we refer to Fig. 9-21. The surface tension γ acts at an angle ϕ all around a circle of radius r. The magnitude of the vertical force F due to surface tension is, by Eq. 9-4, $F = (\gamma \cos \phi)(L)$. Since $L = 2\pi r$, we have $F = 2\pi r \gamma \cos \phi$. This force is just balanced by the weight of the liquid below, which is approximately a cylinder of height h and volume $V = \pi r^2 h$. Hence

$$2\pi r \gamma \cos \phi = mg$$
$$= \rho V g$$
$$= \rho \pi r^2 h g$$

where ρ is the density of the liquid. Solving for h we find

$$h = \frac{2\gamma \cos \phi}{\rho g r}. \tag{9-5}$$

For many liquids, such as water in glass, ϕ is close to zero, and since $\cos 0° = 1$, Eq. 9-5 reduces to the simpler form $h = 2\gamma/\rho g r$.

Equation 9-5 is also valid when the liquid drops, as in the case of mer-cury in a glass tube. In this situation, the contact angle ϕ is greater than 90°, and $\cos \phi$ will be negative; this makes h negative which corresponds to a de-crease in level. Notice from Eq. 9-5 that the thinner the tube the greater will be the rise (or fall) of the liquid.

EXAMPLE 9-9   If the xylem (tiny tubes which carry nutrients up-ward) of a plant has a radius of 0.0010 cm, calculate how high the surface tension can be expected to pull a column of water. Assume $\phi = 0°$.

SOLUTION   Using $\gamma = .072$ N/m in Eq. 9-5 we find

$$h = \frac{(2)(0.072 \text{ N/m})(1.0)}{(1.0 \times 10^3 \text{ kg/m}^3)(9.8 \text{ m/s}^2)(1.0 \times 10^{-5} \text{ m})}$$
$$= 1.5 \text{ m}.$$

One example of how capillary action is important in practical situa-tions is in regard to buildings. Water on the exterior (say from rain) can be brought to the interior by capillary action along tiny cracks or joints. To pre-vent this, whenever materials are joined together such that a narrow channel exists—say at a window—it is general practice to widen the channel at one point as shown in Fig. 9-22.

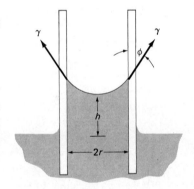

(a)
Glass tube in water

(b)
Glass tube in mercury

FIGURE 9-20

Capillarity.

FIGURE 9-21

A liquid rises a height h in a capillary tube given by $h = 2\gamma \cos \phi/\rho g r$ (see text).

FIGURE 9-22

A window frame and/or the sill have a widened channel to prevent inward seepage of moisture due to capillarity.

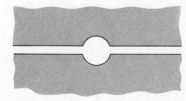

*9-9 ∎ Negative pressure and the cohesion of water; the rise of fluids in trees

Normally, a fluid exerts an outward pressure on its container and the container exerts an inward pressure on the fluid. Pressures can range from zero up to very large positive values. But can fluid pressure be negative? The answer, strange as it may seem, is yes; under certain circumstances a negative pressure can exist in a liquid, but not in a gas.

Negative pressures have been obtained, although it is difficult, using the apparatus shown in Fig. 9-23. The closed tube is initially filled with liquid and then the reservoir on the right is pumped down with a vacuum pump. According to Eq. 9-3 the pressure difference between points A and B is $\rho g h$ where ρ is the density of the liquid:

$$P_B - P_A = \rho g h.$$

When the pump has reduced the pressure to near zero, $P_B = 0$ since point B is at the same level as the surface of the liquid in the reservoir. Hence the pressure at A must be negative:

$$P_A = -\rho g h.$$

FIGURE 9-23
Producing a negative pressure.

Normally we would expect the liquid to fall out of the tube when the pump reduces the pressure. But if the tube is scrupulously clean and the liquid is free from impurities the liquid may remain in the tube. Negative pressures reaching -400 atmospheres have been achieved using this technique. The liquid under negative pressure is unstable, however, and the smallest disturbance will cause the liquid to shatter into droplets, and it falls.

A liquid under negative pressure actually pulls inward on its container; the liquid acts as if it were all under tension (not just its surface) much like a rope pulled at the two ends. How can this occur? It is due to the cohesive forces between the liquid molecules as well as adhesive forces between the liquid and the container. The cohesive forces between water molecules are quite strong—the tensile strength of pure water is about 30×10^6 N/m^2. These are the same forces that hold the water molecules together as ice at low temperatures. Liquid water does not normally sustain a tension force as a solid does because the slightest intrusion of an impurity (such as air) allows the water to flow and take on another shape; it can then fall under gravity.

It has long been wondered how water can rise to the top of tall trees such as redwoods which reach heights of over 100 m. Water, containing mineral nutrients from the soil, is carried upward by tiny capillaries (the xylem). The radius of these capillaries varies from about 0.01 mm to 0.3 mm. As we saw in Example 9-9, even the smallest sized capillaries can raise water by only about 1.5 m. On the other hand, atmospheric pressure can raise water only about 10 m even when the top end is under vacuum (Section 9-5). Neither of these effects account for the rise of water in tall trees. Today it is generally accepted that although the above effects may help, it is the cohesive forces between water molecules acting under tension at negative pressure that are responsible, particularly for very tall trees. This theory was first proposed in the last century, but until recently there was much opposition to it. Within the last few years actual measurements of the pressure within the xylem of trees have been made; they were indeed found to be negative, often as low as −25 atm at the top. When water evaporates from leaves, other water molecules move in to take their place. Through cohesive forces, the water below is pulled upward.

(a)

(b)

FIGURE 9-24

(a) Streamline or laminar flow; (b) turbulent flow.

9-10 ∎ Fluids in motion; flow rate and the equation of continuity

We now turn from the study of fluids at rest to the more complex subject of fluids in motion. Many aspects of fluid motion are still not completely understood; nonetheless, with certain simplifying assumptions, a good comprehension of this subject can be obtained.

To begin with, we usually distinguish two types of fluid flow. If the flow is smooth such that neighboring layers of the fluid slide by each other smoothly, the flow is said to be **laminar** or **streamline flow.** This kind of flow is characterized by the fact that each particle of the fluid follows a smooth path, called a streamline, and these paths do not cross over one another, Fig. 9-24a. Above a certain speed, which depends on a number of factors as we shall see later, the flow becomes turbulent. **Turbulent flow** is characterized by flow in small whirlpool-like circles called eddy currents or eddies, Fig. 9-24b. Eddies absorb a great deal of energy and although a certain amount of internal friction, called **viscosity,** is present during streamline flow, it is much greater when the flow is turbulent. A few tiny drops of ink or food coloring dropped into a moving liquid can quickly reveal whether the flow is streamline or turbulent.

Let us consider the steady laminar flow of a fluid through an enclosed tube or pipe as shown in Fig. 9-25. First we determine how the speed of the fluid changes with the size of the tube. The flow rate is defined as the mass m of fluid that passes a given point per unit time t: m/t. In Fig. 9-25, the volume of fluid passing point 1 in a time t is just $A_1 l_1$, where l_1 is the distance the fluid moves in time t and A_1 is the cross-sectional area of the tube at that point. Since the velocity[†] of fluid passing point 1 is $v_1 = l_1/t$, the flow rate m/t past point 1 is (where V refers to volume)

FIGURE 9-25

Fluid flow through a pipe of varying diameter.

$$\frac{m}{t} = \frac{\rho_1 V_1}{t} = \frac{\rho_1 A_1 l_1}{t} = \rho_1 A_1 v_1.$$

Similarly, at point 2, the flow rate is $\rho_2 A_2 v_2$. Since no fluid flows in or out the sides, the flow rate at points 1 and 2 must be equal. Thus

$$\rho_1 A_1 v_1 = \rho_2 A_2 v_2.$$

[†] If there were no viscosity, the velocity would be the same across a cross section of the tube. Real fluids have viscosity and this internal friction causes different layers of the fluid to flow at different speeds. In this case v_1 and v_2 represent the average speeds at each cross section.

This is called the **equation of continuity.** If the fluid is incompressible, which is an excellent approximation for liquids under most circumstances (and often for gases as well), then $\rho_1 = \rho_2$ and the equation of continuity becomes

$$A_1 v_1 = A_2 v_2. \qquad (9\text{-}6)$$

Notice that the product Av represents the *volume* rate of flow (volume of fluid passing a given point per second), since $V/t = Al/t = Av$. Equation 9-6 tells us that where the cross-sectional area is large the velocity is small, and where the area is small the velocity is high. That this makes sense can be seen by looking at a river; a river flows slowly and languidly through a meadow where it is broad, but speeds up to torrential speed when passing through a narrow gorge.

Equation 9-6 can be applied to the flow of blood in the body. Blood flows from the heart into the aorta, from which it passes into the major arteries; these branch into the small arteries (arterioles), which in turn branch into myriads of tiny capillaries; the blood returns to the heart via the veins, Fig. 9-26.

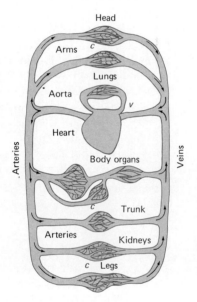

□   EXAMPLE 9-10   The radius of the aorta is about 1.0 cm and the blood flowing through it has a speed of about 30 cm/s. Calculate the average speed of the blood in the capillaries using the fact that although each capillary has a diameter of about 8×10^{-4} cm, there are literally billions of them so that their total cross section is about 2000 cm^2.

SOLUTION   The speed of blood in the capillaries is

$$v_2 = \frac{v_1 A_1}{A_2} = \frac{(0.30 \text{ m/s})(3.14)(0.010 \text{ m})^2}{(2 \times 10^{-1} \text{ m}^2)}$$
$$= 5 \times 10^{-4} \text{ m/s}$$

or 0.5 mm/s.

Another example that makes use of the equation of continuity and the argument leading up to it is the following.

□   EXAMPLE 9-11   How large must a heating duct be if air moving 3.0 m/s along it can replenish the air in a room of 300 m^3 volume every 15 min? Assume the air's density remains constant.

SOLUTION   We consider the room as a large section of duct for purposes of applying Eq. 9-6 (call it point 2). Reasoning in the same way we did to obtain Eq. 9-6, we see that $A_2 v_2 = A_2 l_2/t = V_2/t$ where V_2 is the volume of the room. Then $A_1 v_1 = A_2 v_2 = V_2/t$ and

$$A_1 = \frac{V_2}{v_1 t} = \frac{300 \text{ m}^3}{(3.0 \text{ m/s})(900 \text{ s})} = 0.11 \text{ m}^2.$$

If the duct has a circular cross section, $A = \pi r^2$ and we find that the radius must be 0.19 m or 19 cm.

v—valves
c—capillaries

FIGURE 9-26
Human circulatory system.

9-11 ∎ Bernoulli's equation

Have you ever wondered how air can circulate in a prairie dog's burrow, why smoke goes up a chimney, or why a car's convertible top bulges upward at high speeds? These are examples of a principle worked out by Daniel Bernoulli (1700–1782) in the early eighteenth century. In essence, *Bernoulli's principle* states that *where the velocity of a fluid is high, the pressure is low, and where the velocity is low, the pressure is high.* For example, if the pressures at points 1 and 2 in Fig. 9-25 are measured, it will be found that the pressure is lower at point 2, where the velocity is higher, than it is at point 1, where the velocity is lower. At first glance, this might seem strange; you might expect that the higher speed at point 2 would imply a greater pressure. But this cannot be the case; for if the pressure at point 2 were higher than at 1, this higher pressure would slow the fluid down, whereas in fact it has speeded up in going from point 1 to point 2. Thus the pressure at 2 must be less than at 1 which will allow the fluid to accelerate.

Bernoulli developed an equation that expresses this principle quantitatively. To derive Bernoulli's equation, we assume the flow is steady and laminar, the fluid is incompressible, and the viscosity is small enough to be ignored. To be general, we assume the fluid is flowing in a tube of nonuniform cross section that varies in height above some reference level, Fig. 9-27. We will consider the amount of fluid shown in color and calculate the work done to move it from the position shown in (a) to that shown in (b). In this process, fluid at point 1 flows a distance Δl_1 and forces the fluid at the point 2 to move a distance Δl_2. The fluid to the left of point 1 exerts a pressure P_1 on the fluid and does an amount of work $W_1 = F_1 \Delta l_1 = P_1 A_1 \Delta l_1$. At point 2, the work done is $W_2 = -P_2 A_2 \Delta l_2$; the negative sign is present because the force exerted on the fluid is opposite to the motion (thus the fluid shown in color does work on the fluid to the right of point 2). Work is also done on the fluid by the force of gravity; since the net effect of the process shown in Fig. 9-27 is to move a mass m of volume $A_1 \Delta l_1$ ($= A_2 \Delta l_2$, since the fluid is

(a)

(b)

FIGURE 9-27

Fluid flow: for derivation of Bernoulli's equation.

incompressible) from point 1 to point 2, the work done by gravity is

$$W_3 = -mg(y_2 - y_1),$$

where y_1 and y_2 are heights of the center of the tube above some (arbitrary) reference level. Notice that in the case shown in Fig. 9-27 this term is negative since the motion is uphill against the force of gravity. The net work W done on the fluid is thus:

$$W = W_1 + W_2 + W_3$$
$$W = P_1 A_1 \Delta l_1 - P_2 A_2 \Delta l_2 - mgy_2 + mgy_1.$$

According to the work-energy theorem (Section 5-3), the net work done on a system is equal to its change in kinetic energy. Thus

$$\tfrac{1}{2}mv_2^2 - \tfrac{1}{2}mv_1^2 = P_1 A_1 \Delta l_1 - P_2 A_2 \Delta l_2 - mgy_2 + mgy_1.$$

The mass m has volume $A_1 \Delta l_1 = A_2 \Delta l_2$ and thus we can substitute $m = \rho A_1 \Delta l_1 = \rho A_2 \Delta l_2$ and obtain (after dividing through by $A_1 \Delta l_1 = A_2 \Delta l_2$ and rearranging):

$$P_1 + \tfrac{1}{2}\rho v_1^2 + \rho g y_1 = P_2 + \tfrac{1}{2}\rho v_2^2 + \rho g y_2. \tag{9-7}$$

This is **Bernoulli's equation.** Since points 1 and 2 can be any two points along a tube of flow, Bernoulli's equation can be written:

$$P + \tfrac{1}{2}\rho v^2 + \rho g y = \text{constant}$$

at every point in the fluid.

Bernoulli's equation can be applied to a great many situations. One example is to calculate the velocity, v_1, of a liquid flowing out of a spigot at the bottom of a reservoir, Fig. 9-28. We choose point 2 in Eq. 9-7 to be the top surface of the liquid; and assuming the diameter of the reservoir is large compared to that of the spigot, v_2 will be almost zero. The pressure at points 1 (the spigot) and 2 are each equal to atmospheric pressure so $P_1 = P_2$. Then Bernoulli's equation becomes

$$\tfrac{1}{2}\rho v_1^2 + \rho g y_1 = \rho g y_2$$

or

$$v_1 = \sqrt{2g(y_2 - y_1)}. \tag{9-8}$$

FIGURE 9-28
Torricelli's theorem.

This result is called *Torricelli's theorem.* Although it is seen to be a special case of Bernoulli's equation, it was discovered a century before Bernoulli by Evangelista Torricelli; hence its name. Notice that the liquid leaves the spigot with the same speed that a freely falling object would attain falling the same height. This should not be too surprising since the derivation of Bernoulli's equation relies on the conservation of energy.

Another special case of Bernoulli's equation arises when the fluid is flowing but there is no appreciable change in height; that is, $y_1 = y_2$. Then Eq. 9-7 becomes

$$P_1 + \tfrac{1}{2}\rho v_1^2 = P_2 + \tfrac{1}{2}\rho v_2^2. \tag{9-9}$$

This tells us quantitatively that where the speed is high the pressure is low, and vice versa. It explains many everyday phenomena, some of which are illustrated in Fig. 9-29. The pressure in the air blown at high speed across

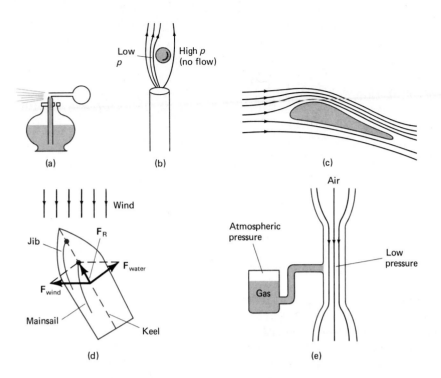

FIGURE 9-29
Examples of Bernoulli's principle.

the top of the vertical tube of a perfume atomizer (Fig. 9-29a) is less than the normal air pressure acting on the surface of the liquid in the bowl; thus perfume is pushed up the tube because of the reduced pressure at the top. A Ping-Pong ball can be made to float above a blowing jet of air (some vacuum cleaners can blow air), Fig. 9-29b; if the ball begins to leave the jet of air, the higher pressure outside the jet pushes the ball back in.

Airplane wings and other airfoils are designed to deflect the air so that although streamline flow is largely maintained, the streamlines are crowded together above the wing, Fig. 9-29c. Just as the flow lines are crowded together in a pipe constriction where the velocity is high, so the crowded streamlines above the wing indicate the air speed is greater than below the wing. Hence the air pressure above the wing is less than that below and there is thus a net upward force; this is called *dynamic lift*. Actually, Bernoulli's principle is only one aspect of the lift on a wing. Wings are usually tilted slightly upward so that air striking the bottom surface is deflected downward; the change in momentum of the rebounding air molecules results in an additional upward force on the wing. Turbulence also plays an important role; for example, if the wing is tilted upward too much, turbulence sets in behind the wing with a consequent loss of lift.

A sailboat can move against the wind, Fig. 9-29d, and the Bernoulli effect aids in this considerably if the sails are arranged so the air velocity increases in the narrow constriction between the two sails. (The normal pressure behind the mainsail is larger than the reduced pressure in front of it and this pushes the boat forward.) When going against the wind, the mainsail is set at an angle approximately midway between the wind direction and the boat's axis (keel line) as shown. The force of the wind on the sail (momentum change of wind bouncing off the sail), plus the Bernoulli effect, acts nearly perpendicular to the sail ($\mathbf{F}_{wind}$). This would tend to make the boat move sideways but the keel beneath prevents this—for the water exerts a force ($\mathbf{F}_{water}$) on the keel nearly perpendicular to it. The resultant of these two forces ($\mathbf{F}_R$) is almost directly forward as shown.

A *Venturi tube* is essentially a pipe with a narrow constriction (the throat). One example of a Venturi tube is the barrel of a carburetor in a car, Fig. 9-29e. The flowing air speeds up as it passes this constriction (Eq. 9-6) and so the pressure is lower. Because of the reduced pressure, gasoline under atmospheric pressure in the carburetor reservoir is forced into the air stream and mixes with the air before entering the cylinders.

The Venturi tube is also the basis of the *Venturi meter* which is used to measure the flow speed of fluids, Fig. 9-30. Venturi meters can be used to measure the flow velocities of gases and liquids and have even been designed to measure blood velocity in arteries. (See the problems.)

FIGURE 9-30

Venturi meter.

Why does smoke go up a chimney? It's partly because hot air rises (it's less dense and therefore buoyant). But Bernoulli's principle also plays a role. Because wind blows across the top of a chimney, the pressure is less there than inside the house. Hence, air and smoke are pushed up the chimney. Even on an apparently still night there is usually enough ambient air flow at the top of a chimney to allow upward flow of smoke.

If gophers, prairie dogs, rabbits, and other animals that live underground are to avoid suffocation, the air must circulate in their burrows. The burrows are always made to have at least two entrances (Fig. 9-31). The speed of air flow across different holes will usually be slightly different. This results in a slight pressure difference which forces a flow of air through the burrow a la Bernoulli. The flow of air is enhanced if one hole is higher than the other (and this is often done by animals) since wind speed tends to increase with height.

FIGURE 9-31

Bernoulli's principle is responsible for air flow in underground burrows.

☐ EXAMPLE 9-12   Water circulates throughout a house in a hot water heating system. If the water is pumped at a speed of 0.50 m/s through a 2.0-cm-diameter pipe in the basement under a pressure of 3.0 atm, what will be the flow speed and pressure in a 1.3-cm-diameter pipe on the second floor 5.0 m above?

SOLUTION   We first calculate the velocity v_2 using the equation of continuity, Eq. 9-6, noting that the areas are proportional to the diameters squared ($A = \pi D^2/4$):

$$v_2 = \frac{v_1 A_1}{A_2} = (0.50 \text{ m/s}) \frac{(0.020 \text{ m})^2}{(0.013 \text{ m})^2} = 1.2 \text{ m/s}.$$

To find the pressure we use Bernoulli's equation:

$$P_2 = P_1 + \rho g(y_1 - y_2) + \tfrac{1}{2}\rho(v_1^2 - v_2^2)$$
$$= (3.0 \times 10^5 \text{ N/m}^2) + (1.0 \times 10^3 \text{ kg/m}^3)(9.8 \text{ m/s}^2)(-5.0 \text{ m})$$
$$+ \tfrac{1}{2}(1.0 \times 10^3 \text{ kg/m}^3)[(0.50 \text{ m/s})^2 - (1.2 \text{ m/s})^2]$$
$$= 3.0 \times 10^5 \text{ N/m}^2 - 4.9 \times 10^4 \text{ N/m}^2 - 6.0 \times 10^2 \text{ N/m}^2$$
$$= 2.5 \times 10^5 \text{ N/m}^2.$$

Notice that the velocity term contributes very little in this case.

Bernoulli's equation ignores the effects of friction (viscosity) and the compressibility of the fluid. The energy that is transformed to internal (or potential) energy due to compression and to thermal energy by friction can be taken into account by adding terms to the right side of Eq. 9-7. These terms are difficult to calculate theoretically and are normally determined empirically. We will not pursue it here, but merely note that it does not significantly alter the explanations for the phenomena described above.

*9-12 ∎ Viscosity

As already mentioned, real fluids have a certain amount of internal friction which is called **viscosity.** It exists in both liquids and gases and is essentially a frictional force between different layers of fluid as they move past one another. In liquids it is due to the cohesive forces between the molecules; in gases it arises from collisions between the molecules.

Different fluids possess different amounts of viscosity: syrup is more viscous than water; grease is more viscous than engine oil; liquids in general are much more viscous than gases. The viscosity of different fluids can be expressed quantitatively by a *coefficient of viscosity*, η (the Greek lowercase letter eta), which is defined in the following way. A thin layer of fluid is placed between two flat plates. One plate is stationary and the other is made to move, Fig. 9-32. The fluid directly in contact with either plate is held to the surface by the adhesive force between the molecules of the liquid and those of the plate. Thus the upper surface of the fluid moves with the same speed v as the upper plate, whereas the fluid in contact with the stationary plate remains stationary. The stationary layer of fluid retards the flow of the layer just above it; this layer retards the flow of the next layer and so on. Thus the velocity varies linearly from 0 to v as shown. The increase in velocity divided by the distance over which this change is made—equal to v/l—is called the *velocity gradient*. To move the upper plate requires a force, a fact you can verify by moving a flat plate across a puddle of syrup on a table. For a given fluid, it is found that the force required, F, is proportional to the area of either plate, A, and to the speed, v, and is inversely proportional to the separation, l, of the plates: $F \propto vA/l$. For different fluids, the more viscous the fluid, the greater is the required force. Hence the proportionality constant for this equation is defined as the *coefficient of viscosity, η:*

$$F = \eta A \frac{v}{l}.\tag{9-10}$$

Solving for η, we find $\eta = Fl/vA$. Thus the SI units of η are N·s/m² = Pa·s (pascal·second); in the cgs system the unit is dyne·s/cm² and this unit is

FIGURE 9-32

Determination of viscosity.

TABLE 9-5

9   Fluids

Coefficient of viscosity for various fluids

Fluid	Temperature (°C)	Coefficient of Viscosity, η (Pa·s)[†]
Water	0	1.8×10^{-3}
	20	1.0×10^{-3}
	100	0.3×10^{-3}
Whole blood	37	$\approx 4 \times 10^{-3}$
Blood plasma	37	$\approx 1.5 \times 10^{-3}$
Ethyl alcohol	20	1.2×10^{-3}
Engine oil (SAE 10)	30	200×10^{-3}
Glycerine	20	1500×10^{-3}
Air	20	0.018×10^{-3}
Hydrogen	0	0.009×10^{-3}
Water vapor	100	0.013×10^{-3}

[†] 1 Pa·s = 10 P = 1×10^3 cP

called a *poise* (P). Viscosities are often stated in *centipoise* (cP), which is one-hundredth of a poise. Table 9-5 lists the coefficient of viscosity for various fluids; the temperature is also specified, since it has a strong effect—the viscosity of liquids such as motor oil, for example, decreases rapidly as temperature increases.

*9-13 ▮ Flow in tubes: Poiseuille's equation, blood flow, Reynolds number

If a fluid had no viscosity, it could flow through a level pipe without a force being applied. Because of viscosity, a pressure difference between the ends of a tube is necessary for the steady flow of any real fluid, be it water or oil in a pipe, or blood in the circulatory system of a human.

The rate of flow of a fluid in a round tube depends on the viscosity of the fluid, the pressure difference, and the dimensions of the tube. The French scientist J. L. Poiseuille (1799–1869), who was interested in the physics of blood circulation (and after whom the "poise" is named), determined how these variables affect the flow rate of an incompressible fluid undergoing laminar flow in a cylindrical tube. His result, known as *Poiseuille's equation*, is as follows:

$$Q = \frac{\pi r^4 (P_1 - P_2)}{8\eta L} \tag{9-11}$$

where r is the inside radius of the tube, L is its length, $P_1 - P_2$ is the pressure difference between the ends, η is the viscosity, and Q is the volume rate of flow (volume of fluid flowing past a given point per unit time). The derivation of Eq. 9-11, which is long and requires the use of calculus, can be found in more advanced textbooks. Equation 9-11 applies to laminar flow. There is no such simple mathematical relation if the flow is turbulent.

⬜ **EXAMPLE 9-13** Engine oil (assume SAE 10—Table 9-5) passes through a fine 1.80-mm-diameter tube in a prototype engine. The tube is 5.5 cm long. What pressure difference is needed to maintain a flow rate of 5.6 mL/min?

SOLUTION The flow rate in SI units is $Q = 5.6 \times 10^{-6}$ m³/60 s = 9.3×10^{-8} m³/s. We solve for $P_1 - P_2$ in Eq. 9-11 and put all terms in SI units:

$$P_1 - P_2 = \frac{8\eta LQ}{\pi r^4}$$

$$= \frac{8(2.0 \times 10^{-1} \text{ N·s/m}^2)(5.5 \times 10^{-2} \text{ m})(9.3 \times 10^{-8} \text{ m}^3/\text{s})}{3.14(0.90 \times 10^{-3} \text{ m})^4}$$

$$= 4.0 \times 10^3 \text{ N/m}^2$$

or about 0.040 atm.

Poiseuille's equation tells us that the flow rate Q is directly proportional to the "pressure gradient," $(P_1 - P_2)/L$, and it is inversely proportional to the viscosity of the fluid. This is just what we might expect. It may be surprising, however, that Q also depends on the *fourth* power of the tube's radius. This means that for the same pressure gradient, if the tube radius is doubled the flow rate is increased by a factor of sixteen! Thus the rate of flow, or alternately the pressure required to maintain a given flow rate, is greatly affected by only a small change in tube radius.

An interesting example of this r^4 dependence is *blood flow* in the human body. Poiseuille's equation is valid only for the streamline flow of an incompressible fluid with constant viscosity η, so it cannot be precisely accurate for blood whose flow is not without turbulence; blood also contains corpuscles (whose diameter is almost equal to that of a capillary), and η depends to a certain extent on the flow speed v. Nonetheless Poiseuille's equation does give a reasonable first approximation. The body controls the flow of blood by means of tiny bands of muscle surrounding the arteries. Contraction of these muscles reduces the diameter of an artery and, due to the r^4 in Eq. 9-11, the flow rate is greatly reduced for only a small change in radius. Very small actions by these muscles can thus control precisely the flow of blood to different parts of the body. Another aspect is that the radius of arteries is reduced as a result of arteriosclerosis (hardening of the arteries) and by cholesterol buildup; when this happens, the pressure gradient must be increased to maintain the same flow rate. If the radius is reduced by half, the heart would have to increase the pressure by a factor of about 16 in order to maintain the same blood-flow rate. The heart must work much harder under these conditions but usually cannot maintain the original flow rate. Thus, high blood pressure is an indication both that the heart is working harder and that blood-flow rate is reduced.

If the flow velocity is large, the flow through a tube will become turbulent and Poiseuille's equation will no longer hold. When the flow is turbulent, the flow rate Q for a given pressure difference will be less than for laminar flow as given in Eq. 9-11. This is a result of the fact that friction forces are much greater when turbulence is present.

The onset of turbulence is often abrupt and can be characterized approximately by the so-called *Reynolds number*, *Re*:

$$Re = \frac{2\bar{v}r\rho}{\eta}, \qquad (9\text{-}12)$$

where $\bar{v}$ is the average speed of the fluid, ρ is its density, η is its viscosity, and r is the radius of the tube in which the fluid is flowing. Experiments show that the flow is laminar if *Re* has a value less than about 2000 but is turbulent if *Re* exceeds this value.

EXAMPLE 9-14   The average speed of blood in the aorta ($r = 1.0$ cm) during the resting part of the heart's cycle is about 30 cm/s. Is the flow laminar or turbulent?

SOLUTION   To answer this we calculate the Reynolds number using the values of ρ and η from Tables 9-1 and 9-5:

$$Re = \frac{(2)(0.30 \text{ m/s})(0.010 \text{ m})(1.05 \times 10^3 \text{ kg/m}^3)}{(4.0 \times 10^{-3} \text{ N}\cdot\text{s/m}^2)} = 1600.$$

The flow will probably be laminar, but it is close to turbulence.

Notice in this example that since $1 \text{ N} = 1 \text{ kg}\cdot\text{m/s}^2$, *Re* has no units. Hence we see that Reynolds number is always a *dimensionless* quantity; its value is the same in any consistent set of units. Next we take a geophysical example.

EXAMPLE 9-15   According to the plate tectonic model, the plates supporting the earth's continents move very slowly on the hot deformable rock below. Show that this flow is laminar using the following data: speed $v = 50$ mm/yr, density and viscosity of deformable rock below $\rho = 3200 \text{ kg/m}^3$ and $\eta = 4 \times 10^{19}$ Pa·s (notice the magnitude!), with thickness ≈ 200 km.

SOLUTION   This is not exactly fluid flow in a tube, but it is similar (one edge of tube moves with the fluid); also the tube is not round, but Eq. 9-12 can be used for a rough guide, and we set the 200-km thickness equal to $2r$. Then

$$Re = \frac{2\bar{v}r\rho}{\eta}$$

$$= \frac{(5.0 \times 10^{-2} \text{ m/3.14} \times 10^7 \text{ s})(2.0 \times 10^5 \text{ m})(3200 \text{ kg/m}^3)}{4 \times 10^{19} \text{ Pa} \cdot \text{s}} \ll 1.$$

The flow is laminar, due to the low speed of plate movement and the enormous viscosity of the underlying rock.

*9-14 ▌ Object moving in a fluid; sedimentation and drag

In the previous section we saw how viscosity (and other factors) affect the flow of a fluid through a tube. In this section we will examine a slightly different situation, that of an object moving relative to a fluid. It could be an obstacle that obstructs the flow of a fluid, such as a large rock in a river, or it could be an object moving in a fluid, such as a glider or car moving through air, a fish in water, or a molecule sedimenting in a centrifuge.

When an object moves relative to a fluid, the fluid exerts a force on the object. This force, which is referred to as a *drag force*, is due to the viscosity of the fluid and also, at high speeds, to turbulence behind the object.

To characterize the motion of an object relative to a fluid, it is useful to define another *Reynolds number*

$$Re' = \frac{vL\rho}{\eta}$$

where ρ and η are the density and viscosity of the fluid, v is the object's velocity relative to the fluid, and L is a characteristic length of the object. This Reynolds number must be clearly distinguished from the one used for fluid flow in a tube (although the form is similar) since the phenomena are different. When the Reynolds number for our present case is less than about 1,[†] the flow around an object is essentially laminar and it is found experimentally that the viscous force F_v is directly proportional to the speed of the object:

$$F_v = kv. \tag{9-13}$$

The magnitude of k depends on the size and shape of the object and on the viscosity of the fluid. For a sphere of radius r it has been calculated to be

$$k = 6\pi r \eta.$$

Thus the viscous force on a small sphere, when the flow is laminar, is given by an equation known as Stokes's equation:

$$F_v = 6\pi r \eta v. \qquad \text{[for a sphere]}$$

For larger Reynolds numbers (usually above a value between about 1 and 10), there will be turbulence behind the body known as the wake (see Fig. 9-24b), and the drag force will be larger than that given by Stokes's equation for a sphere. For more streamlined objects, however, there will be less turbulence and hence less drag. When turbulence is present, experiments show that the drag force increases as the square of the speed, $F_v \propto v^2$. The increase with speed is thus much more rapid than in the case of strictly laminar flow. When the Reynolds number approaches a value around 10^6, the drag force increases abruptly. For above this value, turbulence exists not only behind the object but also in the layer of fluid lying next to the body (called the *boundary layer*) all along its sides.

Sedimentation refers to small objects falling in a fluid—examples are tiny particles of rock or minerals sedimenting under the sea, and red blood cells sedimenting in the fluid plasma in a laboratory.

An object of mass m falling through a fluid under the action of gravity has several forces on it as shown in Fig. 9-33: the force of gravity, mg; F_B, the

---

[†] An object 1 mm long moving at a speed of 1 mm/s through water has a Reynolds number equal to 1. So does an object 2 mm long traveling 7 mm/s in air. This situation thus applies mainly to fairly small objects such as raindrops, pollen grains, and molecules in a centrifuge.

FIGURE 9-33

Forces on a small object falling through a fluid.

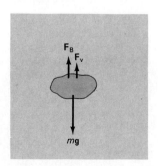

211

buoyant force of the fluid; and F_v the viscous force. By Newton's second law, the net force is equal to the mass times the acceleration of the object:

$$mg - F_B - F_v = ma.$$

The buoyant force F_B is equal to the weight of fluid displaced; that is, $F_B = \rho_f V g$, where ρ_f is the density of the fluid, V is the volume of the object (and hence the volume displaced), and g is the acceleration due to gravity. We can also write $mg = \rho_o V g$ where ρ_o is the density of the object. Using Eq. 9-13 we can write the above equation as:

$$(\rho_o - \rho_f)Vg - kv = ma.$$

The first term is the effective weight of the object in the fluid. As the object increases in speed, the viscous force increases until its magnitude equals the effective weight of the object. At this point the acceleration is zero and the speed increases no further. This maximum speed, v_T, is called the *terminal velocity* or *sedimentation velocity* and is obtained from the equation above by setting $a = 0$:

$$v_T = \frac{(\rho_o - \rho_f)Vg}{k}.$$

The sedimentation velocity of tiny objects such as macromolecules and other constituents of cells is extremely small. It can be increased using a centrifuge (Chapter 4) since the effect of a centrifuge is to exert a force on a particle as if the acceleration of gravity were increased to a value of $\omega^2 r$ (see Eq. 7-6); here, ω is the angular velocity of the rotor and r is the distance of the object from the axis of rotation. Thus, we can use the equation above for a centrifuge if we substitute the effective acceleration of gravity, equal to $\omega^2 r$, for g:

$$v_T = \frac{(\rho_o - \rho_f)V\omega^2 r}{k}. \tag{9-14}$$

▌ Summary

The three common phases of matter are solid, liquid, and gas. Liquids and gases are collectively called *fluids*, meaning they have the ability to flow. The *density* of a material is defined as its mass per unit volume. *Specific gravity* is the ratio of the density of the material to the density of water.

Pressure is defined as force per unit area. The pressure at a depth h in a liquid is given by $\rho g h$ where ρ is the density of the liquid and g is the acceleration due to gravity. In addition, if an external pressure is applied to a confined fluid this pressure is transmitted throughout the fluid; this is known as *Pascal's principle*. Pressure is measured using a manometer or other type of gauge. A barometer is used to measure atmospheric pressure. Standard atmospheric pressure (average at sea level) is 1.013×10^5 N/m^2. *Gauge pressure* is the total pressure less atmospheric pressure.

Archimedes' principle states that an object submerged wholly or partially in a fluid is buoyed up by a force equal to the weight of fluid it displaces. This principle is used in a method to determine specific gravity, and explains why objects whose density is less than that of a fluid will float in that fluid.

Fluid flow rate is the mass or volume of fluid that passes a given point per unit time. The equation of continuity states that for an incompressible fluid flowing in an enclosed tube, the product of the velocity of flow and the cross-sectional area of the tube remains constant: $Av = $ constant. *Bernoulli's equation* tells us that where the velocity of a fluid is high the pressure in it is low, and where the velocity is low the pressure is high. Bernoulli's principle explains many common phenomena.

Fluid flow can be characterized either as *streamline* (sometimes called *laminar*), in which the layers of fluid move smoothly and regularly along paths called streamlines, or as *turbulent*, in which case the flow is not smooth and regular but is characterized by irregularly shaped whirlpools. *Viscosity* refers to friction within a fluid that prevents the fluid from flowing freely and is essentially a frictional force between different layers of fluid as they move past one another.

▌ Questions

1. Which of the following are fluids at room temperature: gold, mercury, air, glass, alcohol, carbon dioxide?

2. What is the specific gravity of (*a*) gold, (*b*) ice, (*c*) air?

3. If one material has a higher density than another, does this mean the molecules of the first must be heavier than those of the second? Explain.

4. A small amount of water is boiled in a one gallon gasoline can. The can is removed from the heat and the lid put on. Shortly thereafter, the can collapses. Explain.

5. It is often said that "water seeks its own level." Explain.

6. When measuring blood pressure, why must the jacket be held at the level of the heart?

7. Consider what happens when you push both a pin and a stick against your skin with the same force. Decide what determines whether your skin suffers a cut—the net force applied to it or the pressure.

8. Explain how the tube in Fig. 9-34, known as a siphon, can transfer liquid from one container to a lower one even though the liquid must flow uphill for part of its journey. (Note that the tube must be filled with liquid to start with.) Why doesn't the liquid in each side of the tube flow back into its container?

9. Devise a method for determining the mass of one of your legs, using a swimming pool.

10. (*a*) Show that the buoyant force on a partially submerged object acts at the center of gravity of the fluid before it is displaced. This point is called the *center of buoyancy*. (*b*) For a ship to be stable, should its center of buoyancy be above, below, or at the same point as its center of gravity? Explain.

FIGURE 9-34

A siphon.

11. Will an empty balloon have precisely the same apparent weight on a scale as one that is filled with air? Explain.

12. Does the buoyant force on a diving bell deep beneath the ocean have precisely the same value as when the bell is just beneath the surface? Explain.

13. Will a helium balloon rise indefinitely in the air? Explain.

14. A barge filled to overflowing with sand approaches a low bridge over the river and cannot quite pass under it. Should sand be added to or removed from the barge?

15. It is harder to pull the plug out of the drain of a bathtub when the tub is full of water than when it is empty. Is this a contradiction of Archimedes' principle? Explain.

16. An ice cube floats in a glass of water filled to the brim. As the ice melts, will the glass overflow?

*17. A duck can float in water because it preens its feathers to apply a layer of grease. Explain how the increased surface tension allows the duck to float.

18. Roofs of houses are sometimes pushed off during a tornado or hurricane. Explain, using Bernoulli's principle.

19. Why does the canvas top of a convertible bulge out when the car is traveling at high speed?

20. A tall Styrofoam cup is filled with water. Two holes are punched in the cup near the bottom, and water begins rushing out. If the cup is dropped so it falls freely, will the water continue to flow from the holes? Explain.

21. Why does a sailboat need a keel? In a small sailboat the keel is removed when the boat is anchored. Why?

22. Blood corpuscles tend to flow in the center of blood vessels. Explain.

23. Children are told to avoid standing too close to a rapidly moving train because they might get sucked under it. Is this possible? Explain.

24. Hummingbirds expend 20 times as much energy to hover in front of a flower as they do in normal flight. Explain.

25. Why do airplanes normally take off into the wind?

26. What forces act to accelerate the fluid in Fig. 9-27 when it moves from region 1 to region 2?

27. Why does the stream of water from a faucet become narrower as it falls?

*28. Identical steel ball bearings are dropped into tubs of water at 10°C and 40°C. In which tub will the ball bearing reach the bottom more quickly?

▮ Problems

SECTION 9-1

1. (I) The approximate volume of the granite monolith known as El Capitan in Yosemite National Park is about 10^8 m^3. What is its approximate mass?

2. (I) What is the approximate mass of air in a living room 6.6 m $\times$ 4.4 m $\times$ 2.9 m?

3. (I) If 5.0 L of antifreeze solution (SG = 0.80) is added to 4.0 L of water to make a 9.0 L mixture, what is the SG of the mixture?

4. (II) A bottle has a mass of 31.20 g when empty and 98.44 g when filled with water. When filled with another fluid the mass is 88.78 g. What is the SG of the fluid?

SECTIONS 9-2 TO 9-5

5. (I) The arm of a record player exerts a force of (1.0 g) $\times$ g on a record. If the diameter of the stylus is 0.0013 cm ($= 0.5$ mil $= 0.5 \times 10^{-3}$ in), calculate the pressure on the record groove in N/m^2 and in atmospheres.

6. (I) A typical value for systolic blood pressure is 120 mm Hg. Convert this to (a) torr, (b) N/m^2, (c) atm, (d) lb/in^2.

7. (I) What is the difference in blood pressure between the top of the head and bottom of the feet of a 1.60-m-tall person standing vertically?

8. (I) What is the total force and the absolute pressure on the bottom of a swimming pool 8.0 m by 15.0 m whose uniform depth is 2.0 m? What will be the pressure against the *side* of the pool near the bottom?

9. (I) (a) Calculate the total force of the atmosphere acting on the top of a table which measures 2.0 m $\times$ 1.2 m. (b) What is the total force acting upward on the underside of the table?

10. (II) Intravenous infusions are often made under gravity as shown in Fig. 9-35. Assuming the fluid has a density of 1.00 g/cm^3, at what height h should the bottle be placed so the liquid pressure is (a) 60 mm Hg, (b) 600 mm H$_2$O? (c) If the blood pressure is 18 mm Hg above atmospheric pressure, how high should the bottle be placed so that the fluid just barely enters the vein?

11. (II) The gauge pressure in each of the four tires of an 1800-kg automobile is 220 kPa. How much area of each tire is in contact with the ground?

12. (II) A 3.0 N force is applied to the plunger of a hypodermic needle. If the diameter of the plunger is 1.0 cm and that of the needle 0.20 mm, (a) with what force does the fluid leave the needle? (b) What force on the plunger would be needed to push fluid into a vein where the gauge pressure is 18 mm Hg? Answer for the instant just before the fluid starts to move.

13. (II) What is the approximate difference in air pressure between the top and the bottom of the World Trade

FIGURE 9-35

Center Building in New York City which is 410 m tall and is located at sea level? Express as a fraction of atmospheric pressure at sea level.

14. (II) When you drive up into the mountains, or descend rapidly from the mountains, your ears "pop," which means that the pressure behind the ear drum is being equalized to that outside. If this did not happen, what would be the approximate force on an eardrum of area 0.50 cm^2 if a change in altitude of 1000 m takes place?

15. (II) What is the total force on a rectangular-shaped dam 60 m high and 110 m wide if the water is filled to the top?

16. (II) One arm of a U-shaped tube (open at both ends) contains water and the other alcohol. If the two fluids meet exactly at the bottom of the U, and the alcohol is at a height of 18.0 cm, at what height will the water be?

17. (II) The maximum gauge pressure in a hydraulic lift is 18 atm. What is the largest size vehicle (kg) it can lift if the diameter of the output line is 20 cm?

18. (II) Determine the minimum gauge pressure needed in the water pipe leading into a building if water is to come out of a faucet on the twelfth floor 30 m above.

19. (II) Show that the work done when a pressure P acts to move a volume of fluid ΔV is $W = P \Delta V$.

20. (II) The contraction of the left ventricle (chamber) of the heart pumps blood to the body. Assuming that the inner surface of the left ventricle has an area of 85 cm^2 and the maximum pressure in the blood is 120 mm Hg, estimate the force exerted by the ventricle at maximum pressure.

21. (II) Estimate the total mass of the earth's atmosphere using the known value of atmospheric pressure at sea level.

22. (III) During each heartbeat, approximately 70 cm³ of blood is pushed from the heart at an average pressure of 105 mm Hg. Calculate the power output of the heart, in watts, assuming 60 beats per minute.

23. (III) In working out his principle, Pascal showed dramatically how force can be multiplied with fluid pressure. He placed a long tube of 0.30 cm radius vertically into a 20-cm-radius wine barrel. He found that, when the barrel was filled with water and the tube filled to a height of 12 m, the barrel burst. Calculate (a) the mass of fluid in the tube, (b) the net force on the lid of the barrel.

24. (III) Estimate the density of the water 10.0 km deep in the sea. (See Section 8-6 and Table 8-1.) By what fraction does it differ from the density at the surface?

SECTION 9-7

25. (I) Determine the specific gravity of a fish weighing 820 g in air and 52.0 g in pure water.

26. (I) The hydrometer of Example 9-5 sinks to a depth of 22.3 cm when placed in a fermenting vat. What is the density of the brewing liquid?

27. (I) A geologist finds that a moon rock whose mass is 7.20 kg has an apparent of mass of 5.88 kg when submerged in water. What is the density of the rock?

28. (I) What fraction of a piece of iron will be submerged when it floats in mercury?

29. (II) A 0.40-kg piece of wood floats in water but is found to sink in alcohol (SG = 0.79) in which it has an apparent mass of 0.020 kg. What is the SG of the wood?

30. (II) A 142.0-g cuttlefish is found to have an effective mass of only 0.55 g when submerged in pure water. (a) Calculate its density. (b) How much would it weigh in seawater (SG = 1.025)?

31. (II) A 78-kg person has an effective mass of 54 kg (because of buoyancy) when standing in water that comes up to the hips. Calculate the mass of each leg. Assume the body has SG = 1.00.

32. (II) A freighter has a horizontal cross-sectional area of 3800 m² at the water line. When loaded, the ship lowers 6.5 m. What is the mass of the load?

33. (II) A small animal is found to remain suspended in a mixture of 18 percent (by weight) alcohol and 82 percent water. What is the density of the animal?

34. (II) A row boat has a volume of 1.5 m³ and a mass of 35 kg. How many people, of mass 70 kg each, can the boat hold without sinking?

35. (II) Archimedes' principle can be used not only to determine the SG of a solid using a known liquid (Example 9-4); the reverse can be done as well. (a) As an example, a 2.00-kg aluminium ball has an apparent mass of 1.40 kg when submerged in a particular liquid; calculate the density of the liquid. (b) Derive a simple formula for determining the density of a liquid using this procedure.

36. (II) Calculate the true mass (in vacuum) of a piece of aluminium whose apparent mass is 3.0000 kg when weighed in air.

37. (II) The specific gravity of ice is 0.92, whereas that for seawater is 1.03. What fraction of an iceberg is above the surface of the water?

38. (III) A 1.24-kg piece of wood (SG = 0.50) floats on water. What minimum mass of lead, hung from it by a string, will cause it to sink?

39. (III) If an object floats in water, its density can be determined by tying a "sinker" on it so that both the object and weight sink. Show that the specific gravity is given by $w/(w_1 - w_2)$ where w is the weight of the object alone in air, w_1 is the apparent weight when a "sinker" is tied to it and the sinker only is submerged, and w_2 is the apparent weight when both the object and the sinker are submerged.

40. (III) A bucket of water is accelerated upward at 3.5 g. What is the buoyant force on a 1.0-kg granite rock (SG = 2.7) submerged in the water? Will the rock float? Why or why not?

*SECTION 9-8

*41. (I) If the force F needed to move the wire in Fig. 9-15 is 4.8×10^{-3} N, calculate the surface tension γ of the enclosed fluid. Assume $l = 0.060$ m.

*42. (I) How high will water rise in a glass tube 0.10 mm in radius?

*43. (II) If the base of an insect's leg has a radius of about 3.0×10^{-5} m and its mass is 0.015 g, would you expect the six-legged insect to remain on top of the water?

*44. (II) The surface tension of a liquid can be determined by measuring the force F needed to just lift a circular platinum ring of radius r from the surface of the liquid. (a) Find a formula for γ in terms of F and r. (b) At 30°C, if $F = 8.40 \times 10^{-3}$ N and $r = 2.8$ cm calculate γ for the tested liquid.

*45. (II) A pencil 1.0 cm in diameter is held vertically in a glass of water. The water wets the pencil so that the contact angle is 0°. Calculate the magnitude and direction of the net force on the pencil due to surface tension.

*46. (II) A glass tube 1.0 mm in diameter is placed in mercury. Where will the level of the mercury be in the tube relative to the rest of the liquid?

*47. (II) A small pool of water on a table is broken into 100 droplets. By what factor does the surface energy change, assuming the original pool is flat and of depth h and the droplets are hemispheres of radius h.

*48. (III) How high will water rise due to capillarity between two flat plates of glass 0.10 mm apart when placed vertically in water?

*49. (III) Show that inside a soap bubble there must be a pressure ΔP in excess of that outside equal to $\Delta P = 4\gamma/r$ where r is the radius of the bubble and γ is the surface tension. (Hint: think of the bubble as two hemispheres in contact with each other; and remember that there are two surfaces to the bubble. Note that this result applies to any kind of membrane where $T = 2\gamma$ is the tension per unit length in that membrane.)

SECTIONS 9-10 and 9-11

50. (I) Using the data of Example 9-10 calculate the average speed of blood flow in the major arteries of the body which have a total cross-sectional area of about 2.0 cm^2.

51. (I) If wind blows at 25 m/s over your house, what is the net force on the roof if its area is 250 m^2?

52. (I) A 15-cm-radius air duct is used to replenish the air of a room 10 m $\times$ 6.0 m $\times$ 3.0 m every 10 min. How fast does the air flow in the duct?

53. (II) What gauge pressure in the water mains is necessary if a firehose is to spray water to a height of 25 m?

54. (II) How high should the pressure head be if water is to come from a faucet at a speed of 8.0 m/s? Ignore viscosity.

55. (II) What is volume rate of flow of water from a 1.0 cm diameter faucet if the pressure head is 10 m?

56. (II) Show that the power needed to drive a fluid through a pipe is equal to the volume rate of flow, Q, times the pressure difference, $P_1 - P_2$.

57. (II) What is the lift (in newtons) due to Bernoulli's principle on a wing of area 50 m^2 if the air passes over the top and bottom surfaces at speeds of 320 m/s and 290 m/s respectively?

58. (II) Water at a pressure of 3.3 atm at street level flows into an office building at a speed of 0.50 m/s through a pipe 5.0 cm in diameter. The pipes taper down to 2.5 cm diameter by the top floor, 25 m above. Calculate the flow velocity and the pressure in such a pipe on the top floor. Ignore viscosity. Pressures are gauge pressures.

59. (II) A 4-cylinder automobile has a displacement of 2000 cm^3; this is approximately how much air is brought into the four cylinders per revolution. If the engine runs at a speed of 1500 rpm, and the venturi of the carburetor has a radius of 1.25 cm, (a) what is the speed of air flow through the venturi; (b) what is the pressure (in atmospheres) in the venturi?

60. (III) (a) Show that the flow velocity measured by a Venturi meter is given by the relation $v_1 = A_2\sqrt{2(P_1 - P_2)/\rho(A_1^2 - A_2^2)}$. See Fig. 9-30. (b) A Venturi tube is measuring the flow of water; it has a main diameter of 3.0 cm tapering down to a throat diameter of 1.0 cm; if the pressure difference is measured to be 18 mm Hg, what is the velocity of the water?

*SECTION 9-13

*61. (I) A gardener feels it is taking him too long to water a garden with a $\frac{3}{8}$-in-diameter hose. By what factor will his time be cut if he uses a $\frac{5}{8}$-in-diameter hose? Assume nothing else is changed.

*62. (I) During heavy exercise, the flow speed of blood increases by perhaps a factor of two. Referring to Example 9-14, calculate the Reynolds number and determine what type of flow you would expect in the aorta.

*63. (II) Blood from an animal is placed in a bottle 1.50 m above a 3.8-cm-long needle of inside diameter 0.40 mm from which it flows at a rate of 4.1 cm^3 per minute. What is the viscosity of this blood?

*64. (II) What must be the pressure difference between the two ends of a 2.0-km section of pipe 40 cm in diameter if it is to transport oil ($\rho = 950$ kg/m^3, $\eta = 0.20$ Pa·s) at a rate of 400 cm^3/s?

*65. (II) What diameter must a 30-m-long air duct have if the pressure of a ventilation and heating system is to replenish the air in a room 10 m $\times$ 18 m $\times$ 4.0 m every 10 min? Assume the pump can exert a gauge pressure of 0.64×10^{-3} atm.

*66. (II) Calculate the pressure drop per cm along the aorta using the data of Example 9-14 and Table 9-5.

*67. (III) What is the (approximate) maximum flow rate, Q, of water in a 10-cm diameter pipe if turbulence is to be avoided?

*68. (III) A patient is to be given a blood transfusion. The blood is to flow through a tube from a raised bottle to a needle inserted in the vein (Fig. 9-35). The inside diameter of the 4.0-cm-long needle is 0.45 mm and the required flow rate is 4.0 cm^3 of blood per minute. How high should the bottle be placed above the needle? Obtain ρ and η from the tables. Assume the blood pressure is 20 torr above atmospheric pressure.

*69. (III) Water shoots 14.6 m high from a 1.0-cm-diameter pipe in a fountain. What must be the pressure at the pump, which is 4.2 m below the nozzle (in the ground)? Take into account viscosity, but ignore air resistance. Note any simplifying assumptions you make.

*SECTION 9-14

*70. (III) Calculate the magnitude and direction of the terminal velocity of a 1.0 mm radius air bubble rising in oil of viscosity 0.20 Pa·s and SG 0.90.

*71. (III) If an object requires 30 min to sediment in an ultracentrifuge rotating at 30,000 rpm at an average distance from the axis of 8.0 cm, how long would it take to sediment under gravity in the same tube standing vertically in the lab?

*72. (III) (a) Show that the terminal velocity of a small sphere of density ρ_0 falling through a fluid of density ρ_F and viscosity η is

$$v_T = \frac{2}{9}\frac{(\rho_0 - \rho_F)r^2g}{\eta}.$$

(b) What is the terminal velocity of a spherical raindrop of radius $r = 0.020$ cm falling in air?

10 TEMPERATURE AND KINETIC THEORY

Much of this chapter will be devoted to an investigation of the theory that matter is made up of atoms and that these atoms are in continuous random motion. This theory is called the *kinetic theory*. ("Kinetic," you may recall, is Greek for "moving.")

In order to investigate this atomic view of matter, we must also discuss the concept of temperature as well as the experimentally measured properties of gases; for these will serve as a foundation for testing the kinetic theory.

10-1 ∎ Atoms

The idea that matter is made up of atoms dates back to the ancient Greeks. According to the Greek philosopher Democritus, if a given substance—say a piece of iron—were cut into smaller and smaller bits, eventually a smallest piece of that substance would be obtained which could not be divided further. This smallest piece was called an **atom,** which in Greek means "indivisible."[†] The only real alternative to the atomic theory of matter was the idea that matter is continuous and can be subdivided indefinitely.

Today the atomic theory is generally accepted by scientists. The experimental evidence in its favor, however, came mainly in the eighteenth, nineteenth, and twentieth centuries, and much of it was obtained from the analysis of chemical reactions. A crucial piece of evidence was the *law of definite proportions*, which is a summation of experimental results collected during the half century prior to 1800; it states that when two or more elements[‡] combine to form a compound, they always do so in the same proportions by weight. For example, table salt is always formed from 23 parts sodium and 35 parts chlorine; and water is formed from one part hydrogen

---

[†] Today, of course, we don't consider the atom as indivisible, but rather as consisting of a nucleus (containing protons and neutrons) and electrons. See Chapters 26–29.

[‡] An *element* is a substance, such as gold, iron, or copper, which cannot be broken down into simpler substances by chemical means. *Compounds* are substances made up of elements, such as carbon dioxide and water. The smallest piece of an element is an atom, the smallest piece of a compound is a molecule. Molecules are made up of atoms; a molecule of water is made up of two atoms of hydrogen and one of oxygen; its chemical formula is H_2O.

and eight parts oxygen. A continuous theory of matter could hardly account for the law of definite proportions but, as John Dalton (1766–1844) pointed out, the atomic theory could: the weight proportions of each element required to form a compound corresponded to the relative weights of the combining atoms; one atom of sodium (Na), for example, could combine with one atom of chlorine (Cl) to form one molecule of salt (NaCl), and one atom of sodium would have a mass 23/35 times as large as one of chlorine. By measuring the relative amounts of each element needed to form a large variety of compounds, experimenters established the relative weights of atoms. Hydrogen, the lightest atom, was arbitrarily assigned the relative weight of 1; on this scale, carbon was about 12, oxygen 16, sodium 23, and so on.[†]

Today, we speak of the relative masses of atoms and molecules—what we call the **atomic mass** or **molecular mass,** respectively[‡]—and these are based on assigning an ordinary carbon atom, ^{12}C, the value of exactly 12.0000 atomic mass units (u). The atomic mass of hydrogen is then 1.0078 u, and the values for other atoms are as listed in Appendix D.

Another important piece of evidence for the atomic theory is the so-called *Brownian movement*, named after the biologist Robert Brown, who is credited with its discovery in 1827. While he was observing tiny pollen grains suspended in water under his microscope, Brown noticed that the tiny grains moved about in tortuous paths even though the water appeared to be perfectly still. The atomic theory easily explains Brownian movement if the further reasonable assumption is made that the atoms of any substance are continually in motion. Then Brown's tiny pollen grains are jostled about by the vigorous barrage of rapidly moving molecules of water.

In 1905, Albert Einstein examined Brownian movement from a theoretical point of view and was able to calculate from the experimental data the approximate size and mass of atoms and molecules. His calculations showed that the diameter of a typical atom is about 10^{-10} m.

At the start of Chapter 9 we distinguished the three common states of matter—solid, liquid, gas—based on *macroscopic*, or "large-scale," properties. Now let us see how these three phases of matter differ from the atomic, or *microscopic*, point of view. Clearly, atoms and molecules must exert attractive forces on each other. For how else could a brick or a piece of aluminum stay together in one piece? These forces are of an electrical nature (more on this in later chapters). On the other hand, if the molecules come too close together, the force becomes repulsive (electric repulsion between their outer electrons); they thus maintain a minimum distance from each other. In a solid material, the attractive forces are strong enough that the atoms or molecules are held in more or less fixed positions, often in an array known as a crystal lattice, as shown in Fig. 10-1a. The atoms or molecules in a solid are in motion—they vibrate about their nearly fixed positions. In a liquid,

FIGURE 10-1

Atomic arrangements in (a) a crystalline solid, (b) a liquid, (c) a gas.

(a)

(b)

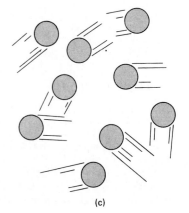

(c)

[†] It was not quite so simple, however. For example, from the various compounds oxygen formed, its relative weight was judged to be 16; but this was inconsistent with the weight ratio in water of oxygen to hydrogen (only 8 to 1). This difficulty was explained by assuming two H atoms combine with one O atom to form a water molecule. To make a self-consistent scheme, many other molecules also had to be judged as containing more than one atom of a given type.

[‡] The terms *atomic weight* and *molecular weight* are popularly used for these quantities, but properly speaking we are comparing masses.

the atoms or molecules are moving more rapidly, or the forces between them are weaker, so that they are sufficiently free to roll over one another as in Fig. 10-1b. In a gas, the forces are so weak, or the speeds so high, that the molecules do not even stay close together. They move rapidly every which way, Fig. 10-1c, filling any container and occasionally colliding with one another. On the average, the speeds are sufficiently high in a gas that when two molecules collide, the force of attraction is not strong enough to keep them close together and they fly off in new directions.

10-2 ▌ Temperature

In everyday life, **temperature** refers to how hot or cold an object is. A hot oven is said to have a high temperature whereas a cold tray of ice is said to have a low temperature.

Many properties of matter change with temperature. For example, most materials expand when heated. An iron beam is longer when hot than when cold; concrete roads and sidewalks expand and contract slightly according to temperature, which is why compressible spacers are placed at regular intervals. The electrical resistance of matter changes with temperature (see Chapter 17). So too does the color radiated by objects, at least at high temperatures: you may have noticed that the heating element of an electric stove glows with a red color when hot; at higher temperatures, solids such as iron glow orange or even white; the white light from an ordinary incandescent light bulb comes from an extremely hot tungsten wire.

An instrument designed to measure temperature is called a **thermometer.** There are many kinds of thermometers, but they all have in common the fact that they depend on some property of matter that changes with temperature. Most common thermometers rely on the expansion of a material with an increase in temperature. The first thermometer, invented by Galileo, made use of the expansion of a gas. Common thermometers today consist of a hollow glass tube filled with mercury or with alcohol colored with a red dye (Fig. 10-2a). The liquid expands more than the glass when the temperature is increased, so the liquid level rises in the tube. Although metals also expand with temperature, the change in length of, say, a metal rod is generally too small to measure accurately for ordinary changes in temperature. However, a useful thermometer can be made by bonding together two dissimilar metals whose rate of expansion is different (Fig. 10-2b). When the temperature is increased, the different amounts of expansion cause the bimetallic strip to bend. Often the bimetallic strip is in the form of a coil, one end of which is fixed and the other is attached to a pointer. This kind of thermometer (Fig. 10-2c) is used for ordinary air thermometers, oven thermometers, and in automobiles as the automatic choke.

In order to measure temperature quantitatively, some sort of numerical scale must be defined. The most common scale today is the *Celsius* scale, sometimes called the *centigrade* scale. In the United States, the *Fahrenheit* scale is also common. The most important scale in scientific work is the absolute, or Kelvin, scale, and it will be discussed later in this chapter.

FIGURE 10-2

(a) Mercury or alcohol in glass thermometer; (b) bimetallic strip; (c) bimetallic-strip thermometer.

Tube

Bulb (acts as a reservoir)

(a)

(b)

(c)

One way to define a temperature scale is to assign arbitrary values to two readily reproducible temperatures. For both the Celsius and Fahrenheit scales these two fixed points are chosen to be the freezing point and the boiling point[†] of water, both taken at atmospheric pressure. On the Celsius scale, the freezing point of water is chosen to be 0°C ("zero degrees Celsius") and the boiling point 100°C. On the Fahrenheit scale, the freezing point is defined as 32°F and the boiling point 212°F. A practical thermometer is calibrated by placing it in carefully prepared environments at each of the two temperatures and marking the position of the mercury or pointer. For a Celsius scale, the distance between the two marks is then divided into one hundred equal intervals separated by small marks representing each degree between 0°C and 100°C (hence the name "centigrade scale" meaning "hundred steps"). For a Fahrenheit scale, the two points are labeled 32°F and 212°F and the distance between them is divided into 180 equal intervals. For temperatures below the freezing point of water and above the boiling point of water the scales can be extended using the same equally spaced intervals. However, ordinary thermometers can be used only over a limited temperature range because of their own limitations—for example, the mercury in a mercury-in-glass thermometer solidifies at some point, below which the thermometer will be useless; it is also rendered useless above temperatures where the fluid vaporizes. For very low or very high temperatures, specialized thermometers are required, some of which we will mention later.

Every temperature on the Celsius scale corresponds to a particular temperature on the Fahrenheit scale, Fig. 10-3. It is easy to convert from one to the other if you remember that 0°C corresponds to 32°F and that a range of 100° on the Celsius scale corresponds to a range of 180° on the Fahrenheit scale. Thus, one Fahrenheit degree (1 F°) corresponds to $100/180 = \frac{5}{9}$ of a Celsius degree (1 C°). That is, $1 \text{ F}° = \frac{5}{9} \text{ C}°$. (Notice that when we refer to a specific temperature, we say "degrees Celsius," as in 20°C; but when we merely refer to a change in temperature or a temperature interval, we say "Celsius degrees," as in "1 C°.")

FIGURE 10-3
Celsius and Fahrenheit scales compared.

□   EXAMPLE 10-1   Normal body temperature is 98.6°F. What is this on the Celsius scale?

SOLUTION   First we note that 98.6°F is $98.6 - 32.0 = 66.6$ F° above the freezing point of water. Since each F° is equal to $\frac{5}{9}$ C°, this corresponds to $66.6 \times \frac{5}{9} = 37.0$ Celsius degrees above the freezing point; as the freezing point is 0°C, the temperature is 37.0°C.

Different materials do not expand in quite the same way over a wide temperature range. Consequently, if we calibrate different kinds of thermometers exactly as described above, they will not usually agree precisely. Because of

[†] The freezing point of a substance is defined as that temperature at which the solid and liquid phases coexist in equilibrium—that is, without the liquid changing into the solid or vice versa. Experimentally, this is found to occur at only one definite temperature, for a given pressure. Similarly, the boiling point is defined as that temperature at which the liquid and gas coexist in equilibrium. Since these points vary with pressure, the pressure must be specified (usually it is 1 atm).

how we calibrated them, they will agree at 0°C and at 100°C. But because of different expansion properties, they may not agree at intermediate temperatures (remember we arbitrarily divided the thermometer scale into 100 equal divisions between 0°C and 100°C). Thus a carefully calibrated mercury-in-glass thermometer might register a temperature of 52.0°C whereas a carefully calibrated thermometer of another type might read 52.6°C.

Because of this discrepancy, some standard kind of thermometer must be chosen so that these intermediate temperatures can be precisely defined. The chosen standard for this purpose is the so-called *constant-volume gas thermometer*. As shown in the simplified diagram of Fig. 10-4, this thermometer consists of a bulb filled with a dilute gas connected by a thin tube to a mercury manometer. The volume of the gas is kept constant by raising or lowering the right-hand tube of the manometer so that the mercury in the left tube coincides with the reference mark. An increase in temperature causes a proportional increase in pressure in the bulb; so the tube must be lifted higher to keep the gas volume constant. The height of the mercury in the right-hand column is then a measure of the temperature. This thermometer can be calibrated and gives the same results for all gases in the limit of the pressure approaching zero. The resulting scale is defined as the standard temperature scale.

FIGURE 10-4
A constant-volume gas thermometer.

10-3 ▮ Thermal expansion

Most substances expand when heated and contract when cooled. However, the amount of expansion or contraction varies, depending on the material.

Experiments indicate that the change in length ΔL of almost all solids is, to a very good approximation, directly proportional to the change in temperature ΔT. As might be expected, the change in length is also proportional to the original length of the object, L_0. That is, for the same temperature change, a 4-m-long iron rod will increase twice as much in length as a 2-m-long iron rod. We can write this proportionality as an equation:

$$\Delta L = \alpha L_0 \, \Delta T, \qquad (10\text{-}1)$$

where α, the proportionality constant, is called the *coefficient of linear expansion* for the particular material and has units of $(\text{C}°)^{-1}$. The values of α for various materials[†] at 20°C are listed in Table 10-1. It should be noted that α does vary slightly with temperature (which is why thermometers made of different materials do not agree precisely). However, if the temperature range is not too great, the variation can usually be ignored.

▯   EXAMPLE 10-2   A steel girder is 200 m long at 20°C. If the extremes of temperature to which it might be exposed are −30°C to +40°C, how much will it contract and expand?

[†] For certain crystalline materials, α may be different for the three spatial directions. We won't be concerned with this.

TABLE 10-1

Coefficients of expansion at 20°

Material	Coefficient of Linear Expansion, α $(C°)^{-1}$	Coefficient of Volume Expansion, β $(C°)^{-1}$
Solids		
Aluminum	25×10^{-6}	75×10^{-6}
Brass	19×10^{-6}	56×10^{-6}
Iron or steel	12×10^{-6}	35×10^{-6}
Lead	29×10^{-6}	87×10^{-6}
Glass (Pyrex)	3×10^{-6}	9×10^{-6}
Glass (ordinary)	9×10^{-6}	27×10^{-6}
Quartz	0.4×10^{-6}	1×10^{-6}
Concrete and brick	$\approx 12 \times 10^{-6}$	$\approx 36 \times 10^{-6}$
Marble	$1.4–3.5 \times 10^{-6}$	$4–10 \times 10^{-6}$
Liquids		
Gasoline		950×10^{-6}
Mercury		180×10^{-6}
Ethyl alcohol		1100×10^{-6}
Glycerin		500×10^{-6}
Water		210×10^{-6}
Gases		
Air (and most other gases at atmospheric pressure)		3400×10^{-6}

SOLUTION   From Table 10-1, we find that $\alpha = 12 \times 10^{-6}(C°)^{-1}$. The increase in length when it is at 40°C will be

$$\Delta L = (12 \times 10^{-6}/C°)(200 \text{ m})(40°C - 20°C),$$
$$= 4.8 \times 10^{-2} \text{ m},$$

or 4.8 cm. When the temperature decreases to $-30°C$, $\Delta T = -50 \text{ C°}$; so the change in length is

$$\Delta L = (12 \times 10^{-6}/C°)(200 \text{ m})(-50 \text{ C°}) = -12.0 \times 10^{-2} \text{ m},$$

or a decrease in length of 12 cm.

The change in volume of a material which undergoes a temperature change is given by a relation which is similar to Eq. 10-1, namely

$$\Delta V = \beta V_0 \Delta T, \tag{10-2}$$

where ΔT is the change in temperature, V_0 is the original volume, ΔV is the change in volume, and β is the *coefficient of volume expansion*. The units of β are $(C°)^{-1}$. The value of β for a number of materials is given in Table 10-1. Notice that for solids β is normally equal to approximately 3α; but note that this is not true for solids that are not isotropic (isotropic means having the same properties in all directions). Notice also that linear expansion has no meaning for liquids and gases since they do not have fixed shapes.

Equations 10-1 and 10-2 are accurate only if ΔL or ΔV is small compared to L_0 or V_0. This is of particular concern for liquids and even more so for gases because of the large values of β. Furthermore, the variation of

β with temperature is also quite large for gases. Therefore a more convenient way of dealing with gases is needed and this is discussed starting in Section 10-5.

□   EXAMPLE 10-3   The 70-L steel gas tank of a car is filled to the top with gasoline at 20°C. The car is then left to sit in the sun and the tank reaches a temperature of 50°C. How much gasoline do you expect to overflow from the tank?

SOLUTION   The gasoline expands by

$$\Delta V = \beta V_0 \Delta T = (950 \times 10^{-6} \text{ C}^{\circ-1})(70 \text{ L})(30 \text{ C}^{\circ}) = 2.0 \text{ L}.$$

The tank also expands. We can think of it as a shell that undergoes volume expansion ($\beta \approx 3\alpha = 36 \times 10^{-6} \text{ C}^{\circ-1}$); but if it were solid, the exterior surface layer (the shell) would expand just the same. Thus the tank increases in volume by

$$\Delta V = (36 \times 10^{-6} \text{ C}^{\circ-1})(70 \text{ L})(30 \text{ C}^{\circ}) = 0.075 \text{ L},$$

so the tank expansion has little effect. If this full tank were left in the sun, a couple of liters of gas could spill out into the road.

Most substances expand more or less uniformly with an increase in temperature. Water, however, does not follow the usual pattern. If water at 0°C is heated, it actually *decreases* in volume until it reaches 4°C; above 4°C water behaves normally and expands as the temperature is increased, Fig. 10-5. Water thus has its greatest density at 4°C. It is for this reason that water in a lake freezes first at its top surface; when the water cools below 4°C, the less dense cooler water rises (or remains) at the top whereas

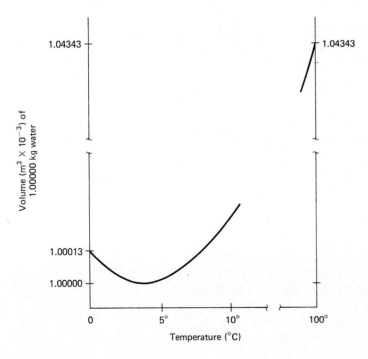

FIGURE 10-5
Volume ($m^3 \times 10^{-3}$) of 1.00000 kg water as a function of temperature (note the break in each axis).

223

the denser 4°C water remains below; the cooler water at the top freezes first since it drops below 0°C first.

10-4 ∎ Thermal stresses

In some situations, the ends of a rod or slab of material are rigidly fixed, which prevents expansion or contraction. If the temperature should change, large compressive or tensile stresses will be set up; these are sometimes referred to as *thermal stresses*. The magnitude of these stresses can be calculated using the concept of elastic modulus developed in Chapter 8. To calculate the internal stress, we can think of this process as occurring in two steps. The rod expands (or contracts) by an amount ΔL given by Eq. 10-1, and then a force is applied to compress (or expand) the material back to its original length. The force F required is given by Eq. 8-4

$$\Delta L = \frac{1}{E}\frac{F}{A}L_0$$

where E is Young's modulus for the material. To calculate the internal stress, F/A, we then set ΔL in Eq. 10-1 equal to ΔL in the equation above and find

$$\alpha L_0 \Delta T = \frac{1}{E}\frac{F}{A}L_0.$$

EXAMPLE 10-4   Blocks of concrete 10 m long are placed end to end with no space in between them to allow for expansion. If the blocks were placed at a temperature of 10°C, what will be the force of compression when the temperature reaches 40°C? The contact area between each block is 0.20 m². Will fracture occur?

SOLUTION   We solve for F in the equation above and use the value of E found in Table 8-1;

$$F = \alpha \Delta T E A$$
$$= (12 \times 10^{-6}/\text{C}°)(30 \text{ C}°)(20 \times 10^9 \text{ N/m}^2)(0.20 \text{ m}^2)$$
$$= 1.4 \times 10^6 \text{ N}.$$

The stress, F/A, is $(1.4 \times 10^6 \text{ N})/(0.20 \text{ m}^2) = 7.0 \times 10^6 \text{ N/m}^2$. This is not far from the ultimate strength of concrete (Table 8-2) under compression and exceeds it for tension and shear. Hence, assuming the concrete is not perfectly aligned, part of the force will act in shear and fracture is likely.

10-5 ∎ The gas laws and absolute temperature

The use of Eq. 10-2 to describe the expansion of a gas is not very useful, partly because the expansion can be so great, and partly because gases generally expand to fill whatever container they are in. Indeed, Eq. 10-2 is meaningful only if the pressure is kept constant. The volume of a gas depends very much on the pressure as well as on the temperature. It is therefore valuable to determine a relation between the volume, the pressure, the temperature, and the mass of a gas. Such a relation is called an **equation of state.**[†] (By the word *state* we mean the physical condition of the system.)

If the state of a system is changed we will always wait until the pressure and temperature have reached the same values throughout. We thus consider only *equilibrium states* of a system—when the variables that describe it (such as temperature and pressure) are the same throughout the system and are not changing in time. We also note that the results of this section are accurate only for gases that are not too dense (the pressure is not too high, on the order of 1 atm or less) and not close to their liquefaction (boiling) point.

For a given quantity of gas it is found experimentally that to a good approximation *the volume of a gas is inversely proportional to the pressure applied to it when the temperature is kept constant.* That is,

$$V \propto \frac{1}{P}. \qquad \text{[constant } T]$$

For example, if the pressure on a gas is doubled, the volume is reduced to half its original volume. This relation is known as *Boyle's law*, after Robert Boyle (1627–1691), who first stated it on the basis of his own experiments. Boyle's law can also be written:

$$PV = \text{constant.} \qquad \text{[constant } T]$$

That is, at constant temperature, if either the pressure or volume of the gas is allowed to vary, the other variable also changes so that the product PV remains constant.

Temperature also affects the volume of a gas, but a quantitative relationship between V and T was not found until more than a century after Boyle's work. The Frenchman Jacques Charles (1746–1823) found that when the pressure is not too high and is kept constant, the volume of a gas increases with temperature at a nearly constant rate as shown in Fig. 10-6a. However, all gases liquefy at low temperatures (for example oxygen liquefies at $-183°C$) and so the graph cannot be extended below the liquefaction point. Nonetheless, the graph is essentially a straight line and if projected to lower temperatures as shown by the dashed line, it crosses the axis at about $-273°C$.

This kind of graph can be drawn for any gas, and the straight line always projects back to $-273°C$ at zero volume. This seems to imply that if a gas could be cooled to $-273°C$ it would have zero volume, and at lower temperatures a negative volume—which makes no sense, of course. This led to the suggestion that perhaps $-273°C$ was the lowest temperature possible, and many other more recent experiments indicate that this is so. This temperature is referred to as the *absolute zero* of temperature. Its value has been measured to be about $-273.15°C$.

[†] An equation of state can also be sought for solids and liquids since for them too the volume depends on the mass, temperature, and external pressure, although the temperature and pressure have much less effect than on gases. However, the situation with solids and liquids is much more complicated.

(a)

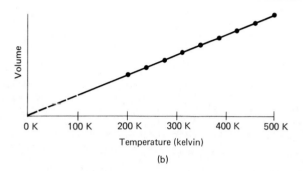

(b)

FIGURE 10-6

Volume of a gas as a function of (a) Celsius temperature, (b) Kelvin temperature, when the pressure is kept constant.

Absolute zero forms the basis of a temperature scale known as the **absolute,** or **Kelvin, scale,** and it is used extensively in scientific work. On this scale the temperature is specified as degrees Kelvin or, preferably, simply as kelvins (K) without the degree sign. The intervals are the same as for the Celsius scale, but the zero on this scale (0 K) is chosen as absolute zero itself. Thus the freezing point of water (0°C) is 273.15 K and the boiling point 373.15 K. Indeed, any temperature on the Celsius scale can be changed to kelvins by adding 273.15 to it:

$$T(\text{K}) = T(°\text{C}) + 273.15.$$

Now let us look at Fig. 10-6b, where we see that the graph of the volume of a gas versus absolute temperature is essentially a straight line. Thus, to a good approximation, *the volume of a given amount of gas is directly proportional to the absolute temperature when the pressure is kept constant.* This is known as *Charles's law*, and can be written

$$V \propto T. \qquad \text{[constant } P]$$

A third gas law, known as *Gay-Lussac's law*, after Joseph Gay-Lussac (1778–1850), states that *at constant volume, the pressure of a gas is directly proportional to the absolute temperature:*

$$P \propto T. \qquad \text{[constant } V]$$

A familiar example of this is the fact that a closed jar or aerosol can thrown into a fire will explode due to the increase in gas pressure inside.

The laws of Boyle, Charles, and Gay-Lussac are not really laws in the sense that we use this term today (precise, deep, wide-ranging validity). They are really only approximations that are accurate for real gases only as long as the pressure and density of the gas are not too high, and the gas is not too close to condensation. Unfortunately, the term law applied to these three relationships has become traditional, so we have stuck with this usage.

10-6 ▌ The ideal gas law

The gas laws of Boyle, Charles, and Gay-Lussac were obtained by using a technique that is very useful in science: namely, to hold one or more variables constant in order to see clearly the effects of changing only one of the variables. These laws can now be combined into a single more general relation between the pressure, volume, and temperature of a fixed quantity of gas:

$$PV \propto T.$$

This relation indicates how any of the quantities P, V, or T will vary when the other two quantities change. This relation reduces to Boyle's, Charles's, or Gay-Lussac's law when either the temperature, the pressure, or the volume, respectively, is held constant.

Finally, we must incorporate the effect of the amount (or mass) of gas that is present. Anyone who has blown up a balloon knows that the more air forced into the balloon, the bigger it gets. Indeed, careful experiment shows that at constant temperature and pressure, the volume V of an enclosed gas increases in direct proportion to the mass m of gas present. Hence we can write

$$PV \propto mT.$$

This proportion relates the significant variables for gases. It can be made into an equation by inserting a constant of proportionality. Experiment shows that this constant has a different value for different gases. However, the constant of proportionality turns out to be the same for all gases if instead of the mass m we use the number of moles; one **mole** (abbreviated mol) is defined as the amount of substance which contains as many atoms or molecules as there are in 0.012 kg of carbon 12 (whose atomic mass is exactly 12 u). A simpler but equivalent definition is: 1 mol of a pure substance contains that number of grams numerically equal to the molecular mass of the substance. For example, the molecular mass of hydrogen gas (H_2) is 2.0 u (since each molecule contains two atoms of hydrogen and each atom has an atomic mass of 1.0 u); thus 1 mol of H_2 is equal to 0.0020 kg. Similarly, 1 mol of neon gas has a mass of 0.010 kg, and 1 mol of CO_2 has a mass of $[12 + (2 \times 16)] \times 10^{-3}$ kg $= 0.044$ kg. The mole is the official unit in the SI system. Sometimes the kilomole (kmol) is used, which is that number of kilograms numerically equal to the molecular mass of the substance.

Now we write the above proportion as an equation:

$$PV = nRT, \tag{10-3}$$

where n represents the number of moles and R is the constant of proportionality. R is called the *universal gas constant* because its value is found experimentally to be the same for all gases; its value, in several sets of units (only the first is the proper SI unit), is:

$$R = 8.314 \text{ J/(mol·K)} \qquad \text{[SI units]}$$
$$= 0.0821 \text{ (L·atm)/(mol·K)}$$
$$= 1.99 \text{ calories/(mol·K)}.$$

Equation 10-3 is called the **ideal gas law,** or the **equation of state for an ideal gas.** We use the term "ideal" because real gases do not follow Eq. 10-3 precisely, particularly at high pressures or when the gas is near the liquefaction point. However, at pressures on the order of 1 atm or less, and when T is not close to the boiling point of the gas, Eq. 10-3 is quite accurate.

The ideal gas law is an extremely useful tool, and we now consider some examples. We will often refer to "standard conditions" or "standard temperature and pressure" (STP) which means $T = 273 \text{ K}(0°C)$ and $P = 1.00 \text{ atm} = 1.01 \times 10^5 \text{ N/m}^2$.

> **EXAMPLE 10-5**   Determine the volume of 1 mol of any gas at STP, assuming it behaves like an ideal gas.
>
> SOLUTION   We solve for V in Eq. 10-3:
>
> $$V = \frac{nRT}{P} = \frac{(1.00 \text{ mol})(8.31 \text{ J/mol·K})(273 \text{ K})}{(1.01 \times 10^5 \text{ N/m}^2)}$$
> $$= 22.4 \times 10^{-3} \text{ m}^3.$$
>
> Since 1 L is $1000 \text{ cm}^3 = 1 \times 10^{-3} \text{ m}^3$, 1 mol of any gas has a volume of 22.4 L at STP.

The value for the volume of 1 mol of gas at STP (22.4 L) is worth remembering for it sometimes makes calculation simpler as the following example shows.

> **EXAMPLE 10-6**   A flexible container of oxygen (O_2, molecular mass = 32 u) at STP has a volume of 10.0 m^3. What is the mass of gas enclosed?
>
> SOLUTION   Since 1 mol occupies a volume of $22.4 \times 10^{-3} \text{ m}^3$ the 10.0 m^3 of oxygen corresponds to $(10.0)/(22.4 \times 10^{-3}) = 446 \text{ mol}$. Since 1 mol has a mass of 0.0320 kg, the mass of oxygen is $(446 \text{ mol})(0.0320 \text{ kg/mol}) = 14.3 \text{ kg}$.

Frequently, volume is specified in liters and pressure in atmospheres. Rather than convert these to SI units, we can instead use the value of R given above as 0.0821 L·atm/mol·K. In many situations it is not necessary to use the values of R at all. For example, many problems involve a change in the pressure, temperature, and volume of a fixed amount of gas. In this case, $PV/T = nR = \text{constant}$, since n and R remain constant. If we now let P_1, V_1, and T_1 represent the appropriate variables initially, and P_2, V_2, T_2 represent the variables after the change is made, then we can write

$$\frac{P_1 V_1}{T_1} = \frac{P_2 V_2}{T_2}.$$

If we know any five of the quantities in this equation, we can solve for the sixth.

⬜ **EXAMPLE 10-7** An automobile tire is filled to a gauge pressure of 200 kPa at 10°C. After driving 100 km the temperature within the tire rises to 40°C. What is the pressure within the tire now?

SOLUTION   Since the volume remains essentially constant, $V_1 = V_2$ and therefore

$$\frac{P_1}{T_1} = \frac{P_2}{T_2}.$$

This is, incidentally, a statement of Gay-Lussac's law. Since the pressure given is the gauge pressure, we must add atmospheric pressure = 101 kPa to get the absolute pressure $P_1 = 301$ kPa. Then

$$P_2 = \frac{P_1}{T_1} T_2 = \frac{(3.01 \times 10^5 \text{ Pa})(313 \text{ K})}{(283 \text{ K})} = 333 \text{ kPa}.$$

Subtracting out atmospheric pressure we find the resulting gauge pressure to be 232 kPa, which is a 15 percent higher increase. This example illustrates why car manuals suggest checking the pressure when the tires are cold.

10-7 ▮ Ideal gas law in terms of molecules: Avogadro's number

The fact that the gas constant, R, has the same value for all gases is a remarkable reflection of simplicity in nature. This fact was first recognized, although in a slightly different form, by the Italian scientist Amedeo Avogadro (1776–1856). Avogadro stated that *equal volumes of gas at the same pressure and temperature contain equal numbers of molecules*. This is sometimes called *Avogadro's hypothesis*. That this is consistent with R being the same for all gases can be seen as follows. First of all from Eq. 10-3 we see that for the same number of moles n, and the same pressure and temperature, the volume will be the same for all gases as long as R is the same. And secondly the number of molecules in 1 mol is the same for all gases; this follows directly from the definition of a mole.[†] Thus Avogadro's hypothesis is equivalent to R being the same for all gases.

The number of molecules in a mole is known as **Avogadro's number,** N_A. Although Avogadro conceived the notion, he was not able to actually determine the value of N_A. Indeed, precise measurements had to await the twentieth century. A number of methods have been devised to measure N_A

[†] For example, the molecular mass of H_2 gas is 2.0 atomic mass units (u) while that of O_2 gas is 32.0 u. Thus 1 mol of H_2 has a mass of 0.0020 kg and 1 mol of O_2 gas, 0.032 kg. Now the number of molecules in a mole is equal to the total mass M of a mole divided by the mass m of one molecule; since this ratio (M/m) is the same for all gases by definition of the mole, a mole of any gas must contain the same number of molecules.

and the accepted value is

$$N_A = 6.02 \times 10^{23} \quad \text{[molecules/mol]}.$$

Since the total number of molecules N in a gas is equal to the number per mole times the number of moles ($N = nN_A$), the ideal gas law, Eq. 10-3, can be written in terms of the number of molecules present:

$$PV = nRT = \frac{N}{N_A} RT,$$

or

$$PV = NkT, \tag{10-4}$$

where $k = R/N_A$ is called *Boltzmann's constant* and has the value $k = (8.314 \text{ J/mol·K})/(6.02 \times 10^{23}/\text{mol}) = 1.38 \times 10^{-23}$ J/K.

☐ **EXAMPLE 10-8** Use Avogadro's number to determine the mass of a hydrogen atom.

SOLUTION One mole of hydrogen (atomic mass = 1.008) has a mass of 1.008×10^{-3} kg and contains 6.02×10^{23} atoms. Thus one atom has a mass

$$m = \frac{1.008 \times 10^{-3} \text{ kg}}{6.02 \times 10^{23}} = 1.67 \times 10^{-27} \text{ kg}.$$

Historically, the reverse process was one method used to obtain N_A—that is, from the measured mass of the hydrogen atom.

10-8 ∎ Kinetic theory and the molecular interpretation of temperature

The concept that matter is made up of atoms which are in continual random motion is called the **kinetic theory.** We now investigate the properties of a gas from the point of view of kinetic theory and in particular we will calculate the pressure in a gas in terms of its molecular properties. We will also arrive at an important relation between the average kinetic energy of molecules in a gas and the absolute temperature.

We make the following assumptions about the molecules in a gas. These assumptions reflect a simple view of a gas but nonetheless correspond well to the essential features of real gases that are at low pressures and far from the liquefaction point. Under these circumstances gases follow the ideal gas law closely, and indeed we shall refer to such a gas as an *ideal gas.* The assumptions, which represent the basic postulates of the kinetic theory, are: (1) There are a large number of molecules, N, each of mass m, moving in random directions with a variety of speeds. This assumption is in accord with our observation that a gas fills its container and, in the case of air on the earth, is kept from escaping only by the force of gravity. (2) The molecules are, on the average, far apart from one another; that is, their average separation is much greater than the diameter of each molecule. (3) The molecules are assumed to obey the laws of classical mechanics, and are presumed to interact with one another only when they collide. Although molecules exert weak attractive forces on each other between collisions, the potential energy associated with these forces is small compared to the kinetic energy, and we ignore it for now. (4) Collisions with another molecule or the wall of the vessel are assumed to be perfectly elastic, like the collisions of perfectly elastic billiard balls (Chapter 6).

We can see immediately how this kinetic view of a gas can explain Boyle's law: the pressure exerted on a wall of a container of gas is due to the constant bombardment of molecules; if the volume is reduced by (say) half, the molecules are closer together and twice as many will be striking a given area of the wall per second. Hence we expect the pressure to be twice as great, which is Boyle's law.

Now let us calculate quantitatively the pressure in a gas based on kinetic theory. For purposes of argument, we imagine that the molecules are contained in a rectangular vessel whose ends have area A and whose length is l as shown in Fig. 10-7. The pressure exerted by the gas on the walls of its container is, according to our model, due to the collisions of the molecules with the walls. Let us focus our attention on the wall, of area A, at the left end of the container and examine what happens when one molecule strikes this wall, as shown in Fig. 10-8. This molecule exerts a force on the wall and the wall exerts an equal and opposite force back on the molecule. The magnitude of this force, according to Newton's second law, is equal to the molecule's rate of change of momentum (Section 6-1). Assuming the collision is elastic, only the x component of the molecule's momentum changes, and it changes from $-mv_x$ (it is moving in the negative x direction) to $+mv_x$. Thus the change in momentum, $\Delta(mv)$, which is the final momentum minus the initial momentum, is

$$\Delta(mv) = mv_x - (-mv_x) = 2mv_x$$

for one collision. This molecule will make many collisions with the wall, each separated by a time Δt, which is the time it takes the molecule to travel across the box and back again, a distance equal to $2l$. Thus $2l = v_x \Delta t$ or $\Delta t = 2l/v_x$. The time Δt between collisions is very small so the number of collisions per second is very large. Thus the average force—averaged over many collisions—will be equal to the force exerted during one collision divided by the time between collisions (Newton's second law):

$$F = \frac{\Delta(mv)}{\Delta t}$$

$$= \frac{2mv_x}{2l/v_x} = \frac{mv_x^2}{l}. \qquad \text{[due to one molecule]}$$

Of course the actual force due to one molecule is intermittent, but because a huge number of molecules are striking the wall per second, the force is pretty steady. To calculate the force due to all the molecules in the box, we have to add the contributions of each. Thus the net force on the wall must be:

$$F = \frac{m}{l}(v_{x1}^2 + v_{x2}^2 + \cdots + v_{xN}^2),$$

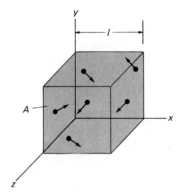

FIGURE 10-7

Molecules of a gas moving about in a cubical container.

FIGURE 10-8

Arrows indicate the momentum of one molecule as it rebounds from the end wall.

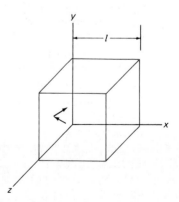

where v_{x1} means v_x for molecule number 1 (we arbitrarily assign each molecule a number) and the sum extends over the total of N molecules. Now the average value of the square of the x component of velocity is

$$\overline{v_x^2} = \frac{v_{x1}^2 + v_{x2}^2 + \cdots + v_{xN}^2}{N}. \tag{10-5}$$

Thus we can write the force as

$$F = \frac{m}{l} N \overline{v_x^2}.$$

We know that the square of any vector is equal to the sum of the squares of its components (theorem of Pythagoras). Thus $v^2 = v_x^2 + v_y^2 + v_z^2$ for any velocity v. Taking averages we obtain

$$\overline{v^2} = \overline{v_x^2} + \overline{v_y^2} + \overline{v_z^2}.$$

Since the velocities of the molecules in our gas are assumed to be random, there is no preference to one direction or another; hence

$$\overline{v_x^2} = \overline{v_y^2} = \overline{v_z^2}.$$

Combining this relation to the one just above, we get

$$\overline{v^2} = 3\overline{v_x^2}.$$

We substitute this into the equation for the force F:

$$F = \frac{m}{l} N \frac{\overline{v^2}}{3}.$$

The pressure on the wall is then

$$P = \frac{F}{A} = \frac{1}{3} \frac{Nm\overline{v^2}}{Al}$$

or

$$P = \frac{1}{3} \frac{Nm\overline{v^2}}{V}, \tag{10-6}$$

where $V = lA$ is the volume of the container. This is the result we were seeking, the pressure in a gas expressed in terms of molecular properties.

Equation 10-6 can be rewritten in a clearer form by multiplying both sides by V and slightly rearranging the right side:

$$PV = \tfrac{2}{3}N(\tfrac{1}{2}m\overline{v^2}). \tag{10-7}$$

The quantity $\tfrac{1}{2}m\overline{v^2}$ is the average kinetic energy ($\overline{\mathrm{KE}}$) of the molecules in the gas. If we compare Eq. 10-7 with Eq. 10-4, the ideal gas law, we see that the two agree if

$$\tfrac{2}{3}(\tfrac{1}{2}m\overline{v^2}) = kT,$$

or,

$$\tfrac{1}{2}m\overline{v^2} = \tfrac{3}{2}kT. \tag{10-8}$$

This equation tells us that the *average translational kinetic energy of molecules in a gas is directly proportional to the absolute temperature.* The higher the temperature, according to kinetic theory, the faster the molecules are moving on the average. This relation is one of the triumphs of the kinetic theory.

☐ **EXAMPLE 10-9** What is the average translational kinetic energy of molecules in a gas at 37°C?

SOLUTION We use Eq. 10-8 and change 37°C to 310 K:

$$\overline{KE} = \tfrac{3}{2}kT$$
$$= \tfrac{3}{2}(1.38 \times 10^{-23} \text{ J/K})(310 \text{ K}) = 6.42 \times 10^{-21} \text{ J}.$$

Equation 10-8 holds not only for gases, but also applies reasonably accurately to liquids and solids. Thus the result of Example 10-9 would apply to molecules within living cells at body temperature (37°C).

We can also use Eq. 10-8 to calculate how fast molecules are moving on the average. Notice that the average in Eqs. 10-5 through 10-8 is over the *square* of the velocity. The square root of v^2 is called the "root mean square" velocity, v_{rms} (since we are taking the square *root* of the *mean* of the *square* of the velocity):

$$v_{rms} = \sqrt{\overline{v^2}} = \sqrt{\frac{3kT}{m}}. \tag{10-9}$$

The *mean speed*, $\bar{v}$, is the average of the magnitudes of the speeds themselves; $\bar{v}$ is generally not equal to v_{rms}. To see the difference between the mean speed and the rms speed, we take the following example.

☐ **EXAMPLE 10-10** Eight particles have the following speeds, given in m/s: 1.0, 6.0, 4.0, 2.0, 6.0, 3.0, 2.0, 5.0. Calculate (a) the mean speed, and (b) the rms speed.

SOLUTION (a) The mean speed is

$$\bar{v} = \frac{1.0 + 6.0 + 4.0 + 2.0 + 6.0 + 3.0 + 2.0 + 5.0}{8} \text{ m/s} = 3.6 \text{ m/s}.$$

(b) The rms speed is

$$v_{rms}$$
$$= \sqrt{\frac{(1.0)^2 + (6.0)^2 + (4.0)^2 + (2.0)^2 + (6.0)^2 + (3.0)^2 + (2.0)^2 + (5.0)^2}{8}} \text{ m/s}$$
$$= 4.1 \text{ m/s}.$$

We see in this example that $\bar{v}$ and v_{rms} are not necessarily equal. In fact, for an ideal gas they differ by about 8 percent. We usually won't be concerned with this small difference however.

☐ **EXAMPLE 10-11** What is the rms speed of air molecules at room temperature (20°C)?

SOLUTION We must apply Eq. 10-9 to oxygen and nitrogen separately since they have different masses. Using the result of Example 10-8, the masses of one molecule of O_2 (molecular mass = 32 u) and N_2 (molecular mass = 28 u) are:

$$m(O_2) = (32)(1.67 \times 10^{-27} \text{ kg}) = 5.3 \times 10^{-26} \text{ kg}$$
$$m(N_2) = (28)(1.67 \times 10^{-27} \text{ kg}) = 4.7 \times 10^{-26} \text{ kg}.$$

Thus, for oxygen

$$v_{\text{rms}} = \sqrt{\frac{3kT}{m}} = \sqrt{\frac{(3)(1.38 \times 10^{-23} \text{ J/K})(293 \text{ K})}{(5.3 \times 10^{-26} \text{ kg})}}$$
$$= 480 \text{ m/s},$$

and for nitrogen the result is $v_{\text{rms}} = 510$ m/s. (This is over 1500 km/h or 1000 mi/h.)

10-9 ∎ Distribution of molecular speeds

The molecules in a gas are assumed to be in random motion, which means that many molecules have speeds less than the average speed and others have speeds greater than the average. In 1859, James Clerk Maxwell (1831–1879) derived, on the basis of kinetic theory, that the speeds of molecules in a gas are distributed according to the graph shown in Fig. 10-9. This is known as the *Maxwell distribution of speeds*.[†] The speeds vary from zero up to many times the average speed, but as can be seen from the graph, most molecules have speeds that are not far from the average. Less than one percent of the molecules exceed four times v_{rms}. Experiments to determine the distribution were first performed in the 1920s. They confirmed with considerable accuracy the Maxwell distribution and the direct proportion between average kinetic energy and absolute temperature, Eq. 10-8.

Figure 10-10 shows the distribution for two different temperatures; just as v_{rms} increases with temperature, so the whole distribution curve shifts to the right at higher temperatures. This figure illustrates how kinetic theory

[†] Mathematically, the distribution is given by $\Delta N = Cv^2 \exp(-\frac{1}{2}mv^2/kT)\Delta v$ where ΔN is the number of molecules with speed between v and $v + \Delta v$, C is a constant, and exp means the expression in parenthesis is an exponent on the natural number $e = 2.718 \ldots$.

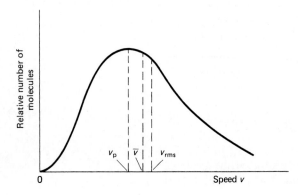

FIGURE 10-9

Distribution of speeds of molecules in an ideal gas. Note that $\bar{v}$ and v_{rms} are not at the peak of the curve (that speed is called the "most probable speed", v_p). This is because the curve is skewed to the right: it is not symmetrical.

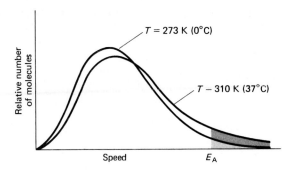

T = 273 K (0°C)

T – 310 K (37°C)

Relative number
of molecules

Speed E_A

FIGURE 10-10

Distribution of molecular speeds for two different temperatures.

can be used to explain the fact that many chemical reactions take place more rapidly as the temperature increases, a fact that is very important in life processes. Two molecules will chemically react only if their kinetic energy is great enough so that when they collide they penetrate into one another somewhat. This minimum energy is called the *activation energy*, E_A, and it has a specific value for each chemical reaction. The molecular speed corresponding to a kinetic energy of E_A for a particular reaction is indicated in Fig. 10-10. The number of molecules with energy greater than this value is given by the area under the curve to the right of E_A. In Fig. 10-10, the respective areas (which are proportional to the number of molecules) for two different temperatures are shown by gray shadings. It is clear that the number of molecules that have kinetic energies in excess of E_A increases greatly for only a small increase in temperature. The rate at which a chemical reaction occurs is proportional to the number of molecules with energy greater than E_A, and thus we see why reaction rates increase rapidly with increased temperature.

*10-10 ∎ Real gases and changes of phase

The ideal gas law, as mentioned before, is an accurate description of the behavior of a gas as long as the pressure is not too high, and as long as the temperature is far from the liquefaction point. But what happens to real gases when these two criteria are not satisfied?

To answer this question, let us look at a graph of pressure plotted against volume for a given amount of gas. On such a "*PV* diagram," Fig. 10-11, each point represents an equilibrium state of the given substance. The various curves show how the pressure varies as the volume is changed at constant temperature for several different values of the temperature. The dashed curve A' represents the behavior of a gas as predicted by the ideal gas law; that is, $PV =$ constant. The solid curve A represents the behavior of a real gas at the same temperature. Notice that at high pressure, the volume in a real gas is less than that predicted by the ideal gas law. At lower temperatures, curves B and C in Fig. 10-11, the behavior deviates even more from the curves predicted by the ideal gas law (for example, B'), and the deviation is greater the closer the gas is to liquefying.

To explain this, we note that at higher pressures we expect the molecules to be closer together; and, particularly at lower temperatures, the

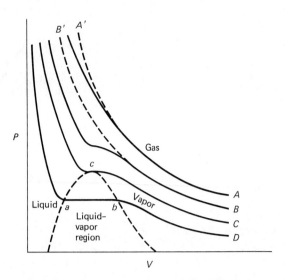

FIGURE 10-11

A *PV* diagram for a real substance.

potential energy associated with the attractive forces between the molecules that we ignored before is no longer negligible compared to the now reduced kinetic energy of the molecules. These forces tend to pull the molecules closer together so that at a given pressure, the volume is less than expected from the ideal gas law. At still lower temperatures, these forces cause lique-faction and the molecules become very close together.

Curve *D* represents the situation when liquefaction occurs. At low pressure on curve *D* (on the right in Fig. 10-11), the substance is a gas and occupies a large volume. As the pressure is increased, the volume decreases until point *b* is reached. Beyond *b*, the volume decreases with no change in pressure; the substance is gradually changing from the gas to the liquid phase. At point *a*, all of the substance has changed to liquid. Further in-crease in pressure reduces the volume only slightly—liquids are nearly incompressible—so the curve is very steep as shown. The area within the tongue-shaped dashed line represents the region where the gas and liquid phases exist together in equilibrium.

Curve *C* in Fig. 10-11 represents the behavior of the substance at its *critical temperature;* and the point *c* (the one point where this curve is hori-zontal) is called the **critical point.** At temperatures less than the critical tem-perature (and this is the definition of the term), a gas will change to the liquid phase if sufficient pressure is applied. Above the critical temperature, no amount of pressure can cause a gas to change phase and become a liquid—what happens instead is that the gas becomes denser and denser as the pres-sure is increased and gradually it acquires properties resembling a liquid, but no liquid surface forms. The critical temperatures for various gases are given in Table 10-2. Scientists tried for many years to liquefy oxygen without success. Only after the discovery of the behavior of substances associated with the critical point was it realized that oxygen can be liquefied only if first cooled below its critical temperature of −118°C.

Often a distinction is made between the terms "gas" and "vapor": a substance below its critical temperature in the gaseous state is called a *vapor;* when above the critical temperature, it is called a *gas;* this is indicated in Fig. 10-11.

TABLE 10-2

Critical temperatures and pressures

Substance	Critical Temperature		Critical Pressure (atm)
	°C	K	
Water	374	647	218
Carbon dioxide	31	304	72.8
Oxygen	−118	155	50
Nitrogen	−147	126	33.5
Hydrogen	−239.9	33.3	12.8
Helium	−267.9	5.3	2.3

The behavior of a substance can be diagrammed not only on a PV diagram but also on a PT diagram. A PT diagram, often called a *phase diagram*, is particularly convenient for comparing the different phases of a substance. Figure 10-12 is the phase diagram for water. The curve labeled l-v represents those points where the liquid and vapor phases are in equilibrium—it is thus a graph of the boiling point versus pressure. Note that the curve correctly shows that at a pressure of 1 atm the boiling point is 100°C and that the boiling point is lowered for a decreased pressure. The curve s-l represents points where solid and liquid exist in equilibrium and thus is a graph of the freezing point versus pressure. At 1 atm, the freezing point of water is of course 0°C as shown. Notice also in Fig. 10-12 that at a pressure of 1 atm, the substance is in the liquid phase if the temperature is between 0°C and 100°C but is in the solid or vapor phase if the temperature is below 0°C or above 100°C. The curve labeled s-v is the *sublimation point* versus pressure curve. *Sublimation* refers to the process whereby at low pressures (in the case of water, at pressures less than 0.0060 atm) a solid changes directly into the vapor phase without passing through the liquid phase. Carbon dioxide, for example, which in the solid phase is called dry ice, sublimates even at atmospheric pressure.

The intersection of the three curves is the **triple point.** The triple point represents a unique temperature and pressure (see Table 10-3) and it is only

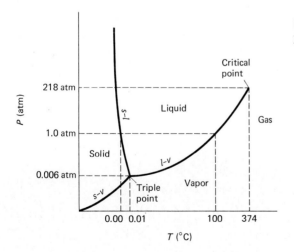

FIGURE 10-12

Phase diagram for water (note that scales are not linear).

TABLE 10-3
Triple-point data

Substance	Temperature (K)	Pressure	
		N/m²	atm
Water	273.16 (0.01°C)	6.10×10^2	6.03×10^{-3}
Carbon dioxide	216.6	5.16×10^5	5.10
Ammonia	195.40	6.06×10^3	6.00×10^{-2}
Nitrogen	63.2	1.25×10^4	1.24×10^{-1}
Oxygen	54.4	1.52×10^2	1.50×10^{-3}
Hydrogen	13.8	7.03×10^3	6.95×10^{-2}

at this point that the three phases can exist together in equilibrium. Because the triple point corresponds to a unique value of temperature and pressure, it is precisely reproducible and is often used as a point of reference. For example, the standard of temperature is usually specified as exactly 273.16 K at the triple point of water, rather than 273.15 K at the freezing point of water at 1 atm.

Notice that the *s-l* curve for water slopes upward to the left. This is true only of substances that *expand* upon freezing; for at a higher pressure, a lower temperature is needed to cause the liquid to freeze. More commonly, substances contract upon freezing and the *s-l* curve slopes upward to the right, as shown for CO_2 in Fig. 10-13.

The phase transitions we have been discussing are the common ones. Some substances, however, can exist in several forms in the solid phase. A transition from one of these phases to another occurs at a particular temperature and pressure, just like ordinary phase changes. For example, ice has been observed in at least eight different modifications at very high pressure. Ordinary helium is a unique substance in that it has two distinct liquid phases, called helium I and II. They exist only at temperatures within a few degrees of absolute zero. Helium II exhibits very unusual properties referred

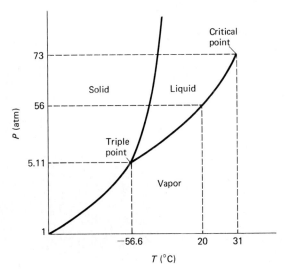

FIGURE 10-13
Phase diagram for carbon dioxide.

to as *superfluidity*. It has extremely low viscosity and exhibits strange properties such as actually climbing up the sides of an open container and over the top. Some materials, like glass, certain resins, and sulfur, are called *amorphous solids*. They do not have a crystal structure like most solids and they do not have a definite melting point. When heated, they gradually soften and there is no phase transition. Amorphous solids are thus often considered to be extremely viscous liquids rather than true solids.

*10-11 ▌ Vapor pressure and humidity

If a glass of water is left out overnight, the water level will have dropped by morning. We say the water has evaporated, meaning that some of the water has changed to the vapor or gas phase.

This process of *evaporation* can be explained on the basis of kinetic theory. The molecules in a liquid move past one another with a variety of speeds that follow, approximately, the Maxwell distribution. There are strong attractive forces between these molecules, which is what keeps them close together in the liquid phase. A molecule in the upper regions of the liquid may, because of its speed, leave the liquid momentarily. But just as a rock thrown into the air returns to the earth, so the attractive forces of the other molecules can pull the vagabond molecule back to the liquid surface— that is, if its velocity is not too large. If the molecule has a high enough velocity, it will escape from the liquid entirely, like a rocket escaping the earth, and become part of the gas phase. Only those molecules that have kinetic energy above a particular value can escape to the gas phase. We have already seen that kinetic theory predicts that the number of molecules with kinetic energy above a particular value (such as E_A in Fig. 10-10) increases with temperature, and this is in accord with the well-known observation that the evaporation rate is greater at higher temperatures.

Since it is the fastest molecules that escape from the surface, the average speed of those remaining is less. When the average speed is less, the absolute temperature is less. Thus kinetic theory predicts that evaporation is a cooling process. You have no doubt noticed this when you stepped out of a warm shower and felt cold as the water on your body began to evaporate; and after working up a sweat on a hot day even a slight breeze makes you feel cool.

Air normally contains water vapor (water in the gas phase) and it comes mainly from evaporation. To look at this process in a little more detail, consider a closed container that is partially filled with water (it could just as well be any other liquid), and from which the air has been removed (Fig. 10-14). The fastest moving molecules quickly evaporate into the space above. As they move about, some of these molecules will strike the liquid surface and again become part of the liquid phase (this is called *condensation*). The number of molecules in the vapor increases for a time, until a point is reached where the number returning to the liquid equals the number leaving in the same time interval. Equilibrium then exists and the space is said to be *saturated*. The pressure of the vapor when it is saturated is called the *saturated vapor pressure* (or sometimes simply the vapor pressure).

FIGURE 10-14

Vapor appears above a liquid in a closed container.

The saturated vapor pressure does not depend on the volume of the container. If the volume above the liquid were reduced, the density of molecules in the vapor phase would be increased. More molecules would then be striking the liquid surface per second. There would be a net flow of molecules back to the liquid phase until equilibrium was again reached and this would occur at the same value of the saturated vapor pressure.

The saturated vapor pressure of any substance depends on the temperature. At higher temperatures, a greater number of molecules have sufficient kinetic energy to break from the liquid surface into the vapor phase. Hence equilibrium will be reached at a higher pressure. The saturated vapor pressure of water at various temperatures is given in Table 10-4. Notice that even solids—for example, ice—have a measurable saturated vapor pressure.

In everyday situations, evaporation from a liquid takes place into the air above it rather than into a vacuum. This does not materially alter the discussion above relating to Fig. 10-14. Equilibrium will still be reached when there are sufficient molecules in the gas phase that the number re-entering the liquid equals the number leaving. This number is not affected by the presence of air although collisions with air molecules may lengthen the time needed to reach equilibrium. Thus equilibrium occurs at the same value of the saturated vapor pressure as if air weren't there.

Of course, if the container is large or is not closed, all the liquid may evaporate before saturation is reached. And if the container is not sealed—as, for example, a room in your house—it is not likely that the air will become saturated with water vapor; unless of course, it is raining outside.

The saturated vapor pressure of a liquid increases with temperature. When the temperature is raised to the point where the saturated vapor pressure at that temperature equals the external pressure, *boiling* occurs. As the boiling point is approached, tiny bubbles tend to form in the liquid, which indicates a change from the liquid to the gas phase. However, if the vapor pressure inside the bubbles is less than the external pressure, the bubbles immediately are crushed. As the temperature is increased, the saturated vapor pressure inside a bubble eventually becomes equal to or exceeds the external air pressure. The bubble will then not collapse but will increase in size and rise to the surface. Boiling has then begun. A liquid boils when its saturated

TABLE 10-4
Saturated vapor pressure of water

Temperature (°C)	Saturated Vapor Pressure		Temperature (°C)	Saturated Vapor Pressure	
	torr (=mmHg)	Pa (=N/m²)		torr (=mmHg)	Pa (=N/m²)
−50	0.030	4.0	40	55.3	7.37×10^3
−10	1.95	2.60×10^2	50	92.5	1.23×10^4
0	4.58	6.11×10^2	60	149	1.99×10^4
5	6.54	8.72×10^2	70	234	3.12×10^4
10	9.21	1.23×10^3	80	355	4.73×10^4
15	12.8	1.71×10^3	90	526	7.01×10^4
20	17.5	2.33×10^3	100	760	1.01×10^5
25	23.8	3.17×10^3	120	1489	1.99×10^5
30	31.8	4.24×10^3	150	3570	4.76×10^5

vapor pressure equals the external pressure. This occurs for water under 1 atm (760 torr) of pressure at 100°C as can be seen from Table 10-4.

The boiling point of a liquid clearly depends on the external pressure. At high elevations, the boiling point of water is somewhat less than at sea level since the air pressure is less. For example, on the summit of Mt. Everest (8850 m) the air pressure is about one-third of what it is at sea level, and from Table 10-4 we can see that water will boil at about 70°C. Cooking food by boiling takes longer at high elevations, since the temperature is less. Pressure cookers, however, reduce cooking time, since they build up a pressure as high as 2 atm.

Vapor pressure of water affects the weather; when we refer to dry or humid weather, we are referring to the water vapor content of air. In a gas such as air which is a mixture of several types of gases, the total pressure is the sum of the *partial pressures* of each gas present. (By partial pressure we mean the pressure each gas would extert if it alone occupied the whole volume.) The partial pressure of water in the air can be as low as zero, and can vary up to a maximum equal to the saturated vapor pressure of water at the given temperature. Thus at 20°C the partial pressure cannot exceed 17.5 torr (see Table 10-4). The *relative humidity* is defined as the ratio of the partial pressure to the saturated vapor pressure at a given temperature. It is usually expressed as a percentage:

$$\text{Relative humidity} = \frac{\text{partial pressure of } H_2O}{\text{saturated vapor pressure of } H_2O} \times 100.$$

Thus, when the humidity is close to 100%, the air holds nearly all the water vapor it can.

EXAMPLE 10-12   On a particular hot day, the temperature is 30°C and the partial pressure of water vapor in the air is 21.0 torr. What is the relative humidity?

SOLUTION   From Table 10-4 the saturated vapor pressure of water at 30°C is 31.8 torr. Hence the relative humidity is

$$\frac{21.0 \text{ torr}}{31.8 \text{ torr}} \times 100 = 66.0\%.$$

Humans are sensitive to humidity. A relative humidity of 40–50 percent is generally optimum for both health and comfort. High humidity, particularly on a hot day, reduces the evaporation of moisture from the skin, which is one of the body's vital mechanisms for regulating body temperature. Very low humidity, on the other hand, can have a drying effect on the skin and mucous membranes.

Proper humidity must be maintained to prevent damage to paintings, tape recordings, and a wide variety of sensitive objects. Thus the design of heating and air-conditioning systems for buildings must take into account not only heating and cooling but the control of relative humidity as well.

Air is saturated with water vapor when the partial pressure of water in the air is equal to the saturated vapor pressure at that temperature. If the partial pressure of water exceeds the saturated vapor pressure, the air is said to be *supersaturated*. This situation can occur when a temperature decrease occurs. For example, suppose the temperature is 30°C and the partial pressure of water is 21 torr, which is a humidity of 66 percent as we saw above. Suppose now, the temperature falls to, say, 20°C, such as might happen at nightfall. From Table 10-4 we see that the saturated vapor pressure of water at 20°C is 17.5 torr. Hence the relative humidity would be greater than 100 percent and the supersaturated air cannot hold this much water. The excess water condenses and appears as *dew*; this process is also responsible for the formation of *fog, clouds*, and rain.

When air containing a given amount of water is cooled, a temperature is reached where the partial pressure of water equals the saturated vapor pressure. This is called the *dew point*. Measurement of the dew point is the most accurate means of determining the relative humidity. One method uses a polished metal surface in contact with air which is gradually cooled down. The temperature at which moisture begins to appear on the surface is the dew point, and the partial pressure of water can then be obtained from saturated vapor pressure tables. If, for example, on a given day the temperature is 20°C and the dew point is 5°C, then the partial pressure of water (Table 10-4) in the original air was 6.54 torr whereas its saturated vapor pressure was 17.5 torr; hence the relative humidity was 6.54/17.5 = 37.4 percent.

A more convenient but less accurate method for measuring relative humidity is the so-called wet bulb–dry bulb technique which makes use of two thermometers. One thermometer bulb is fitted with a snug cloth jacket that is soaking wet. The apparatus is usually swung in the air, and the lower the humidity, the more evaporation takes place from the wet bulb; this causes its temperature reading to be less. A comparison of the temperature readings on the wet bulb thermometer and the dry (ordinary) thermometer can then be checked against special tables that have been compiled to obtain the relative humidity.

*10-12 ▮ Diffusion

If you carefully place a drop of food coloring in a glass of water, you will find that the color spreads throughout the water. The process may take several hours (assuming you don't shake the glass), but eventually the color will become uniform. This mixing occurs because of the random movement of the molecules, and is called *diffusion*. Diffusion occurs in gases too. Common examples include perfume or smoke (or the odor of something burned on the stove) diffusing in air, although convection often plays a greater role in spreading the odor than does diffusion. In each case the diffusing substance moves from a region where its concentration is high to one where its concentration is low.

Diffusion can be readily understood on the basis of kinetic theory and the random motion of molecules. Consider a tube of cross-sectional area A containing molecules in a higher concentration on the left than on the right, Fig. 10-15. We assume the molecules are in random motion. Yet there will be a net flow of molecules to the right. To see why this is true, let us consider the small section of tube of length Δx as shown. Molecules from both regions

FIGURE 10-15

Diffusion occurs from a region of high concentration to one of lower concentration (only one type of molecule is shown).

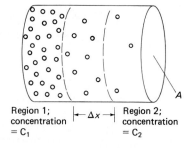

Region 1; concentration = c_1 $\leftarrow \Delta x \rightarrow$ Region 2; concentration = c_2

1 and 2 cross into this central section as a result of their random motion. The more molecules there are in a region, the more will strike a given area or cross a boundary. Since there is a greater concentration of molecules in region 1 than in region 2, more molecules cross into the central section from region 1 than from region 2. There is, then, a net flow of molecules from left to right, from high concentration toward low concentration. The flow stops only when the concentrations become equal.

You might expect that the greater the difference in concentration, the greater the flow rate. This is indeed the case. In 1855, the physiologist Adolf Fick (1829–1901) determined experimentally that the rate of diffusion (J) is directly proportional to the change in concentration per unit distance $(C_1 - C_2)/\Delta x$ (which is called the *concentration gradient*), and to the cross-sectional area A (see Fig. 10-15):

$$J = DA \frac{C_1 - C_2}{\Delta x}. \qquad (10\text{-}10)$$

D is a constant of proportionality called the *diffusion constant*. Equation 10-10 is known as the *diffusion equation*, or *Fick's law*. If the concentrations are given in mol/m^3, then J is the number of moles passing a given point per second; if the concentrations are given in kg/m^3, then J is the mass movement per second (kg/s). The length Δx, of course, is given in meters.

Equation 10-10 applies not only to the simple situation of a gas diffusing as shown in Fig. 10-15, which is called *self-diffusion*. The same considerations apply for a gas diffusing in a second gas (perfume vapor in air), or for a substance dissolved in a liquid; these are the more common situations. The rate of diffusion will be slower, particularly in liquids, because there will be collisions with the other molecules. Thus, the diffusion constant D will depend on the properties of the substances involved, and also on the temperature and the external pressure. The values of D for a variety of substances are given in Table 10-5.

EXAMPLE 10-13   To get an idea of the time required for diffusion, estimate how long it might take for ammonia (NH_3) to be detected 10 cm from a bottle after it is opened, assuming only diffusion.

SOLUTION   This will be an order of magnitude calculation. The rate of diffusion J can be set equal to the number of molecules N diffusing

TABLE 10-5

Diffusion constants, D (20°C, 1 atm)

Diffusing Molecules	Medium	D(m^2/s)
H_2	Air	6.3×10^{-5}
O_2	Air	1.8×10^{-5}
O_2	Water	100×10^{-11}
Blood hemoglobin	Water	6.9×10^{-11}
Glycine (an amino acid)	Water	95×10^{-11}
DNA (molecular mass 6×10^6 u)	Water	0.13×10^{-11}

across area A in a time t: $J = N/t$. We solve for t:

$$t = \frac{N}{J}$$

and then use Eq. 10-10:

$$t = \frac{N}{AD}\frac{\Delta x}{\Delta C}.$$

The average concentration (midway between bottle and nose) can be approximated by $\bar{C} \approx N/V$ where V is the volume over which the molecules move; V is roughly of the order of $V \approx A\,\Delta x$ where Δx is 10 cm, so we substitute $N = \bar{C}A\,\Delta x$ into the above equation:

$$t \approx \frac{(\bar{C}A\,\Delta x)\Delta x}{AD\,\Delta C} = \frac{\bar{C}}{\Delta C}\frac{(\Delta x)^2}{D}.$$

The concentration of ammonia is high near the bottle and low near the detecting nose, so $\bar{C} \approx \Delta C/2$ or $(\bar{C}/\Delta C) \approx \frac{1}{2}$. Since NH_3 has a size somewhere between H_2 and O_2, from Table 10-5 we can choose $D \approx 4 \times 10^{-5}$ m^2/s. Then

$$t \approx \frac{1}{2}\frac{(0.10 \text{ m})^2}{(4 \times 10^{-5} \text{ m}^2/\text{s})} \approx 100 \text{ s},$$

or about a minute or two. This seems a bit long from experience, probably because air currents (convection) usually help it along.

Diffusion is extremely important for living organisms. Within cells, for example, molecules produced in certain chemical reactions must diffuse through water to other areas where they take part in other reactions.

Gas diffusion is important too. Plants require carbon dioxide for photosynthesis. The CO_2 diffuses into leaves from the outside through tiny openings called stomata. As CO_2 is utilized by the cells, its concentration drops below that in the air outside, and more diffuses inward as described by Fick's law. Water vapor and oxygen produced by the cells diffuses outward into the air.

Animals also exchange oxygen and CO_2 with the environment. Oxygen is required for energy producing reactions and must diffuse into cells; CO_2 is produced as an end product of many metabolic reactions and must diffuse out of cells. Because of the slowness of diffusion over longer distances, all but the smallest animals have developed complex respiratory and circulatory systems. In humans, oxygen is taken into the lungs where it diffuses across lung tissue and into the blood which distributes it to cells throughout the body. The blood also carries carbon dioxide produced by the cells back to the lungs where it diffuses outward.

❚ Summary

The atomic theory of matter postulates that all matter is made up of tiny entities called *atoms* which are typically 10^{-10} m in diameter. *Atomic* and *molecular masses* are specified on a scale where ordinary carbon

(^{12}C) is arbitrarily given the value 12.0000 u (atomic mass units). The distinction between solids, liquids, and gases can be attributed to the strength of the attractive forces between the atoms or molecules and to their average speed.

Temperature is a measure of how hot or cold a body is. *Thermometers* are used to measure temperature on the Celsius (°C), Fahrenheit (°F), and Kelvin (K) scales. Two standard points on each scale are the freezing point of water (0°C, 32°F, 273.15 K) and the boiling point of water (100°C, 212°F, 373.15 K). A change in temperature of one kelvin equals a change of one Celsius degree or 9/5 Fahrenheit degrees.

The change in length, ΔL, of a solid when its temperature changes by an amount ΔT is directly proportional to the temperature change and to its original length L_0. That is, $\Delta L = \alpha L_0 \Delta T$ where α is the coefficient of linear expansion. The change in volume of most solids, liquids, and gases is proportional to the temperature change and to the original volume V_0: $\Delta V = \beta V_0 \Delta T$; the coefficient of volume expansion, β, is approximately equal to 3α for solids. Water is unusual because, unlike most materials whose volume increases with temperature, its volume actually decreases as the temperature increases in the range from 0°C to 4°C.

The *ideal gas law*, or *equation of state for an ideal gas*, relates the pressure, P, volume, V, and temperature, T (in kelvins), of n moles of gas by the equation

$$PV = nRT$$

where $R = 8.314\ \text{J/mol·K}$ for all gases. Real gases obey the ideal gas law quite accurately if they are not at too high a pressure or near their liquefaction point. One *mole* of a substance is defined as the number of grams which is numerically equal to its atomic or molecular mass. *Avogadro's number*, $N_A = 6.02 \times 10^{23}$, is the number of atoms or molecules in 1 mol of any pure substance. The ideal gas law can be written in terms of the number of molecules, N, in the gas as

$$PV = NkT$$

where $k = R/N_A = 1.38 \times 10^{-23}\ \text{J/K}$ is Boltzmann's constant. According to the kinetic theory of gases, which is based on the idea that a gas is made up of molecules that are moving rapidly and at random, the average kinetic energy of the molecules is proportional to the Kelvin temperature T:

$$\overline{KE} = \tfrac{3}{2}kT$$

where k is Boltzmann's constant. At any moment, there exists a wide distribution of molecular speeds within a gas.

▮ Questions

1. Name several properties of materials that could be exploited to make a thermometer.
2. Figure 10-16 shows a diagram of a typical *thermostat* used to control a furnace (or other heating or cooling system). The bimetallic strip consists of two strips of different metals bonded together. Explain why this strip bends when the temperature changes, and how this controls the furnace.

FIGURE 10-16

A typical thermostat.

Wires to heater

Rotate this cam to adjust for desired temperature

Contact

3. In the relation $\Delta L = \alpha L_0 \Delta T$, should L_0 be the initial length, the final length, or does it matter?
4. Why is it sometimes easier to remove the lid from a tightly closed jar after warming it under hot running water?
5. Long steam pipes often have a section in the shape of a U. Why?
6. A circular ring is heated from 20°C to 80°C. Will the hole in the ring become larger or smaller?
7. A flat, uniform cylinder of lead floats in mercury at 0°C. Will the lead float higher or lower when the temperature is raised?
8. In view of the fact that fireplaces become hot, why should chimneys not be used for support of a house or building?
9. Explain why it is advisable to add water to an overheated automobile engine only slowly, and only with the engine running.

10. The principal virtue of Pyrex glass is that its coefficient of linear expansion is much smaller than that for ordinary glass (Table 10-1). Explain why this gives rise to the increased heat resistance of Pyrex.

11. Why doesn't the size of different molecules enter into the gas laws?

12. When a gas is rapidly compressed, say, by pushing down a piston, its temperature increases. When a gas expands against a piston, it cools. Explain these changes in temperature using the kinetic theory, in particular noting what happens to the momentum of molecules when they strike the moving piston.

13. In Section 10-8 we assumed the gas molecules made perfectly elastic collisions with the walls of the container. This assumption is not necessary as long as the walls are at the same temperature as the gas. Why?

14. Explain in words how Charles's law follows from kinetic theory and the relation between average kinetic energy and the absolute temperature.

15. Explain in words how Gay-Lussac's law follows from kinetic theory.

16. As you go higher in the earth's atmosphere, the ratio of N_2 molecules to O_2 molecules increases. Why?

17. Escape velocity for the earth refers to the minimum speed an object must have to leave the earth and never return. The escape velocity for the moon is about one-tenth what it is for the earth since the moon is smaller. Explain, then, why the moon has practically no atmosphere.

18. A thermal gradient precipitator is a device for removing particulate matter from polluted air. It consists, basi- cally of two solid surfaces, one hot and one cold, close to each other. Particles in air that is allowed to pass between the two surfaces tend to collect on the cold surface and thus are removed. Explain. (This effect can be noticed on the wall behind a heating radiator, particularly if the wall is an outside wall and therefore quite cool.)

19. If a container of gas is at rest, the average velocity of molecules must be zero. Yet the average speed is not zero. Explain.

*20. How do a gas and a vapor differ?

*21. Liquid CO_2 can exist under what conditions? Be specific. Can it exist as a liquid at normal room temperature?

*22. Why doesn't Dry Ice last long at room temperature?

*23. We saw that the vapor pressure of a liquid (say, water) does not depend on the external pressure. Yet the temperature of boiling does depend on the external pressure. Is there a contradiction? Explain.

*24. Alcohol evaporates more quickly than water at room temperature. What can you infer about the molecular properties of one relative to the other?

*25. Explain why a hot humid day is far more uncomfortable than a hot dry day at the same temperature.

*26. Is it possible to boil water at room temperature (20°C) without heating it? Explain.

*27. Why must we breathe? That is, why can't we simply rely on diffusion of oxygen into our lungs?

▌ Problems

SECTION 10-1

1. (I) What mass of gold has the same number of atoms as 1.0 kg of iron?

SECTION 10-2

2. (I) (a) "Room temperature" is often taken to be 68°F; what is this on the Celsius scale? (b) The temperature of the filament in a light bulb is about 1800°C; what is this on the Fahrenheit scale?

3. (I) The original Celsius temperature scale [due to Anders Celsius (1701–1744)] defined the freezing point of water as 100° and the boiling point as 0°. What temperature on this scale corresponds to 25°C?

4. (I) In an alcohol-in-glass thermometer, the alcohol column has length 12.45 cm at 0.0°C and length 21.30 cm at 100.0°C. What is the temperature if the column has length (a) 15.10 cm, and (b) 22.95 cm.

5. (II) At what temperature will the Fahrenheit and Centigrade scales yield the same numerical value?

SECTION 10-3

6. (I) A concrete highway is built of slabs 22 m long. How wide should the expansion cracks between the slabs be to prevent buckling if the range of temperature is −20°C to +50°C?

7. (I) A precise steel tape measure has been calibrated at 20°C. At 40°C, (a) will it read high or low, and (b) what will be the percentage error?

8. (II) To make a secure fit, rivets that are larger than the rivet hole are often used and the rivet is cooled (usually in dry ice) before it is placed in the hole. A steel rivet 2.385 cm in diameter is to be placed in a hole 2.382 cm in diameter. To what temperature must the rivet be cooled if it is to fit in the hole (at 20°C)?

9. (II) If the density of mercury is 13.59×10^3 kg/m³ at 20°C, what will its density be at 65°C?

10. (II) A uniform rectangular plate of length l and width w has coefficient of linear expansion α. Show that, if we neglect very small quantities, the change in area of

the plate due to a temperature change ΔT is $\Delta A = 2\alpha l w \Delta T$.

11. (II) An iron sphere is 28.0 cm in diameter. What will be its change in volume if it is heated from 20°C to 200°C?

12. (II) An iron plug is to be placed in a ring made of brass. At room temperature the diameter of the plug is 9.114 cm and that of the inside of the ring is 9.097 cm. They must both be brought to what common temperature in order to fit?

13. (II) If a fluid is contained in a long narrow vessel so it can expand in essentially one direction only, show that the effective coefficient of linear expansion α is approximately equal to the coefficient of volume expansion β.

14. (II) (a) Show that the change in the density ρ of a substance when the temperature changes by ΔT is given by $\Delta\rho = -\beta\rho\,\Delta T$. (b) What is the fractional change in density of a lead sphere whose temperature decreases from 30°C to −30°C?

15. (II) An ordinary glass is filled to the brim with 288.3 mL of water at 10°C. If the temperature is now increased by 30°C, about how much water (if any) will spill from the glass?

16. (III) Show that for an isotropic solid, $\beta = 3\alpha$ as long as the amount of expansion is small. β and α are the coefficients of volume and linear expansion, respectively.

17. (III) An iron cube floats in a bowl of liquid mercury at 0°C. (a) If the temperature is raised to 30°C, will the cube float higher or lower in the mercury? (b) By what percent will the fraction of volume submerged change?

18. (III) A 12.4-kg solid iron cylindrical wheel of radius 0.45 m is rotating about its axle in frictionless bearings with angular velocity $\omega = 32.8$ rad/s. If its temperature is now raised from 20°C to 80°C, what is the fractional change in ω?

19. (III) (a) The tube of a mercury thermometer has an inside diameter of 0.120 mm. The bulb has a volume of 0.250 cm³. How far will the thread of mercury move when the temperature changes from 10.0°C to 20.0°C? Take into account expansion of the glass (Pyrex). (b) Determine a formula for the length of the mercury column in terms of relevant variables.

SECTION 10-4

20. (I) At what temperature will the ultimate compressive strength of concrete be exceeded for the blocks discussed in Example 10-4?

21. (II) (a) A horizontal steel I-beam of cross-sectional area 0.016 m² is rigidly connected to two vertical steel girders. If the beam was installed when the temperature was 25°C, what stress is developed in the beam when the temperature drops to −14°C? (b) Is the ultimate strength of the steel exceeded? (c) What stress is developed if the beam is concrete and has a cross-sectional area of 0.13 m²? Will it fracture?

22. (III) A wine barrel of diameter 122.860 cm at 20°C is to be enclosed by an iron band. The circular band has an inside diameter of 122.848 cm at 20°C. It is 8.7 cm wide and 0.55 cm thick. (a) To what temperature must the band be heated so that it will fit over the barrel? (b) What will be the tension in the band when it cools to 20°C?

SECTION 10-5

23. (I) What are the following temperatures on the Kelvin scale: (a) 37°C, (b) 80°F, (c) −196°C?

24. (I) Absolute zero is what temperature on the Fahrenheit scale?

25. (II) Typical temperatures in the interior of the earth and sun are about 4×10^3°C and 1.5×10^7°C, respectively. (a) What are these temperatures in kelvins? (b) What percent error is made in each case if a person forgets to change °C to K?

SECTION 10-6

26. (I) If 5.00 m³ of a gas initially at STP is placed under a pressure of 4.0 atm, the temperature of the gas rises to 25°C. What is the volume?

27. (I) The pressure in a helium gas cylinder is initially 30 atmospheres. After many balloons have been blown up, the pressure has decreased to 6 atm. What fraction of the original gas remains in the cylinder?

28. (II) Write the ideal gas law in terms of the density of the gas.

29. (II) Calculate the density of oxygen at STP using the ideal gas law.

30. (II) A storage tank contains 25.7 kg of nitrogen (N_2) at an absolute pressure of 2.60 atm. What will the pressure be if the nitrogen is replaced by an equal mass of CO_2?

31. (II) A tank contains 28.0 kg of O_2 gas at a gauge pressure of 6.80 atm. If the oxygen is replaced by helium, how many kg of the latter will be needed to produce a gauge pressure of 8.25 atm?

32. (II) If 8.10 mol of helium gas is at 20°C and a gauge pressure of 0.190 atm, calculate (a) the volume of the helium gas under these conditions and (b) the temperature if the gas is compressed to half the volume at a gauge pressure of 2.10 atm.

33. (II) A house has a volume of 600 m³. (a) What is the total mass of air inside the house at 0°C? (b) If the temperature rises to 25°C, what mass of air enters or leaves the house?

34. (II) What is the pressure inside a 22.0-L container holding 24.5 kg of argon gas at 20°C?

35. (II) A tire is filled with air at 15°C to a gauge pressure of 190 kPa. If the tire reaches a temperature of 40°C, what fraction of the original air must be removed if the original pressure of 190 kPa is to be maintained?

36. (II) If 60.0 L of oxygen at 11.0°C and an absolute pressure of 1.88 atm are compressed to 38.6 L and at

the same time the temperature is raised to 78.0°C, what will the new pressure be?

37. (II) If a skin diver fills his lungs to full capacity of 5.5 L when 12 m below the surface, to what volume would his lungs expand if he quickly rose to the surface? Is this advisable?

38. (II) Determine a formula for the coefficient of volume expansion, β, for an ideal gas at constant pressure in terms of P, V, T, and/or n.

39. (III) An air bubble at the bottom of a lake 16 m deep has a volume of 1.10 cm^3. If the temperature at the bottom is 5.5°C and at the top it is 17.0°C, what is the volume of the bubble just before it reaches the surface?

40. (III) Compare the value for the density of water vapor at 100°C and 1 atm (Table 9-1) with the value predicted from the ideal gas law. Why would you expect a difference?

SECTION 10-7

41. (I) Calculate the number of molecules/m^3 in an ideal gas at STP.

42. (I) How many moles of water are there in 1.000 L? How many molecules?

43. (II) The lowest pressure attainable using the best available vacuum techniques is about 10^{-12} N/m^2. At such a pressure, how many molecules are there per cm^3 at 0°C?

44. (II) From the known value of atmospheric pressure at the surface of the earth, estimate the total number of air molecules in the earth's atmosphere.

45. (II) A cubic box of volume 8.0×10^{-3} m^3 is filled with air at atmospheric pressure at 20°C. The box is closed and heated to 150°C. What is the net force on each side of the box?

SECTION 10-8

46. (I) (a) What is the average kinetic energy of an oxygen molecule at STP? (b) What is the total translational kinetic energy of 1 mol of O$_2$ molecules at 20°C?

47. (I) Calculate the rms speed of helium atoms near the surface of the sun at a temperature of about 6000°K.

48. (I) Twelve molecules have the following speeds, given in arbitrary units: 5, 2, 2, 6, 0, 4, 1, 3, 5, 1, 7, 3. Calculate (a) the mean speed and (b) the rms speed.

49. (I) A gas is at 20°C. To what temperature must it be raised to double the rms speed of its molecules?

50. (II) Show that for a mixture of two gases at the same temperature, the ratio of their rms speeds is equal to the inverse ratio of the square roots of their molecular masses.

51. (II) What is the rms speed of nitrogen molecules contained in an 8.0-m^3 volume at 2.1 atm if the total amount of nitrogen is 1300 mol?

52. (II) (a) Calculate the approximate rms speed of an amino acid whose molecular mass is 89 u in a living cell at 37°C. (b) What would be the rms speed of a protein of molecular mass 50,000 u at 37°C?

53. (II) In outer space the density of matter is about one atom per cm^3, mainly hydrogen atoms, and the temperature is about 3.4 K. Calculate the average speed of these hydrogen atoms, and the pressure (in atmospheres).

54. (II) Calculate (a) the rms speed of an oxygen molecule at 0°C, and (b) determine how many times per second it would move back and forth across a 5.0-m-long room on the average, assuming it made very few collisions with other molecules.

55. (II) Show that the pressure P in a gas can be written $P = \frac{1}{3}\rho v^2$ where ρ is the density of the gas and v is the rms speed of the molecules.

56. (III) What is the average distance between oxygen molecules at STP?

57. (III) The two isotopes of uranium, ^{235}U and ^{238}U (the superscripts refer to their atomic mass), can be separated by a gas diffusion process by combining them with fluorine to make the gaseous compound UF$_6$. Calculate the ratio of the rms speeds of these molecules for the two isotopes.

58. (III) Calculate approximately the total translational kinetic energy of all the molecules in an *E. coil* bacterium of mass 2.0×10^{-15} kg at 37°C. Assume 70 percent of the cell, by weight, is water, and the other molecules have an average molecular weight on the order of 10^5.

*SECTION 10-10

*59. (I) (a) At atmospheric pressure, in what phases can CO$_2$ exist? (b) For what range of pressures and temperatures can CO$_2$ be a liquid. Refer to Fig. 10-13.

*60. (I) CO$_2$ exists in what phase when the pressure is 30 atm and the temperature is 30°C? (Fig. 10-13.)

*61. (I) Water is in which phase when the temperature is 90°C and the pressure is 0.01 atm?

*SECTION 10-11

*62. (I) What is the dew point (approximately) if the humidity is 40 percent on a day when the temperature is 30°C.

*63. (I) If the air pressure at a particular place in the mountains is 0.77 atm, at what temperature does water boil?

*64. (I) What is the air pressure at a place where water boils at 80°C?

*65. (I) What is the partial pressure of water on a day when the temperature is 25°C and the relative humidity is 65 percent?

*66. (II) What is the approximate pressure inside a pressure cooker if the water is boiling at a temperature of 120°C? Assume no air escaped during the heating process which started at 20°C.

*67. (II) If the humidity in a room of volume 600 m^3 at 25°C is 90 percent, what mass of water must be removed to reduce the humidity to 50 percent?

*68. (II) If the relative humidity in a room 3.0 m by 6.5 m by 8.8 m is 60 percent, what mass of water can still evaporate from an open pan if the temperature is 20°C?

*69. (II) An *autoclave* is a device used to sterilize laboratory instruments. It is essentially a high pressure steam boiler and operates on the same principle as a pressure cooker. However, because hot steam under pressure is more effective in killing microorganisms than moist air at the same temperature and pressure, the air is removed and replaced by steam. Typically, the gauge pressure inside the autoclave is 1.0 atm; what is the temperature of the steam? Assume the steam is in equilibrium with boiling water.

*70. (III) Air that is at its dew point of 5°C is drawn into a building where it is heated to 25°C. What will be the relative humidity at this temperature? Assume constant pressure of 1.0 atm.

*71. (III) On a day when the temperature is 30°C, the wet bulb thermometer of a humidity tester falls to 10°C. What is the relative humidity?

*SECTION 10-12

*72. (II) What is the time needed for a glycine molecule (see Table 10-5) to diffuse a distance of 10^{-3} cm in water at 20°C if its concentration varies over that distance from 1.0 mol/m^3 to half this value. Compare this "speed" to its average thermal speed. The molecular mass of glycine is about 75 u.

*73. (II) Oxygen diffuses from the surface of insects to the interior through tiny tubes called tracheae. An average trachea is about 2 mm long and has cross-sectional area of 2×10^{-9} m^2. Assuming the concentration of oxygen inside is half what it is outside in the atmosphere, calculate (a) the diffusion rate J and (b) estimate the average time for a molecule to diffuse in. Assume the diffusion constant is 1×10^{-5} m^2/s.

*74. (III) (a) Derive *Graham's law*, which states that "the rate of diffusion of gas molecules is inversely proportional to the square root of the molecular mass." (b) Which would diffuse faster, N_2 gas or O_2 gas, and by how much (percentage)?

11 HEAT

When a pot of cold water is placed on a hot burner of a stove, the temperature of the water increases. We say that heat flows from the hot burner to the cold water. When two objects at different temperatures are put in contact, heat spontaneously flows from the hotter one to the colder one. The spontaneous flow of heat is in the direction tending to equalize the temperature. If the two objects are kept in contact long enough for their temperatures to become equal, the two bodies are said to be in thermal equilibrium, and there is no further heat flow between them. For example, when the mercury in a fever thermometer is still rising, heat is flowing from the patient's mouth to the thermometer; when the mercury stops, the thermometer is then in equilibrium with the person's mouth, and they are at the same temperature.

11-1 ▌ Heat as energy transfer

We use the term "heat" in everyday life as if we knew what we meant. But, in fact, the term is often used inconsistently; so it is important for us to clearly define heat and clarify the phenomena and concepts related to heat.

It is common to speak of the flow of heat—heat flows from a stove burner to a pot of coffee, from the sun to the earth, from a person's mouth into a fever thermometer. It flows spontaneously from an object at higher temperature to one at lower temperature. Indeed, an eighteenth-century model of heat pictured heat flow as movement of a fluid substance called *caloric*. However, the caloric fluid was never able to be detected, and in the nineteenth century, it was found that the various phenomena associated with heat could be described consistently without the need to use the fluid model. The new model viewed heat as being akin to work and energy as we will discuss in a moment. First we note that a common unit for heat, still in use today, is named after caloric. It is called the *calorie* (cal) and is defined as *the amount of heat necessary to raise the temperature of 1 gram of water by 1 Celsius degree, from* 14.5°C *to* 15.5°C. This particular temperature range is specified since the heat required is very slightly different at different temperatures. (The difference is less than 1 percent over the range 0 to 100°C and can be ignored for most purposes.) More often used than the calorie is the *kilocalorie* (kcal), which is 1000 calories. Thus 1 *kcal is the heat needed to raise* 1 *kg of water by* 1 *C°, from* 14.5°C *to* 15.5°C. Sometimes a kilocalorie is called a Calorie (with a capital C), and it is by this unit that the energy value of food is specified. (In the British system of units, heat is measured in British thermal units (Btu). One Btu is defined as the heat needed to raise the temperature of 1 lb of water by 1 F°. It can be shown (Problem 4) that 1 Btu = 0.252 kcal.)

The idea that heat is related to energy was pursued by a number of scientists in the 1800s, particularly by an English brewer, James Prescott Joule (1818–1889). Joule performed a number of experiments that were crucial in establishing our present-day view that heat, like work, represents a transfer of energy. One of Joule's experiments is shown (simplified) in Fig. 11-1. The falling weight causes the paddle wheel to turn; the friction between the water and the paddle wheel causes the temperature of the water to rise slightly (barely measurable, in fact, by Joule). Of course the same temperature rise could also be obtained by heating the water on a hot stove. In this and a great many other experiments (some involving electrical energy) Joule found that a given amount of work was always equivalent to a particular amount of heat. Quantitatively, 4.18 joules (J) of work was found to be equivalent to 1 calorie (cal) of heat. This is known as the *mechanical equivalent of heat:*

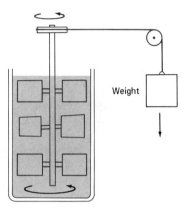

$$4.18 \text{ J} = 1 \text{ cal}$$

$$4.18 \times 10^3 \text{ J} = 1 \text{ kcal}.$$

FIGURE 11-1

Joule's experiment on the mechanical equivalent of heat.

As a result of these and other experiments, scientists came to interpret heat not as a substance, and not even as a form of energy, but rather as a transfer of energy: when heat flows from a hot object to a cooler one, it is energy that is being transferred from the hot to the cold object. Thus, **heat** refers to *energy that is transferred from one body to another because of a difference in temperature.* In SI units, the unit for heat, as for any form of energy, is the joule. Nonetheless, calories and kcal are still sometimes used. Today the calorie is *defined* in terms of the joule (rather than in terms of the properties of water, as given above).

The development of kinetic theory fully supported, and indeed nicely explains, the idea of heat as a transfer of energy. Let us examine the process of heating a pot of water on a hot stove. According to kinetic theory, the average kinetic energy of molecules increases with temperature; thus the molecules of the stove burner have much more kinetic energy on the average than those of the cold water or the pot. When the high-KE stove molecules collide with the lower-KE molecules of the pot, some of their kinetic energy is transferred to the pot molecules, just as a fast-moving billiard ball can transfer some of its kinetic energy to a ball it collides with. The molecules of the pot gain in kinetic energy (those of the stove lose energy). The now-higher-KE pot molecules, in turn, transfer some of their kinetic energy, by collision, to the lower-KE water molecules. The temperature of the water and the pot consequently increases. Thus we see how heat is a transfer of energy.

11-2 ∎ Distinction between temperature, heat, and internal energy

We introduce the concept of internal energy now since it will help clarify ideas about heat. The sum total of all the energy of all the molecules in an object is called its **thermal energy** or **internal energy.** (We will use the two terms interchangeably.) Occasionally the term "heat content" of a body is used for this purpose; however, this is not a good term to use for it can be confused with heat itself. Heat, as we have seen, is not the energy a body contains, but rather refers to the amount of energy transferred from one body to another at a different temperature.

Using the kinetic theory, we can make a clear distinction between temperature, heat, and internal energy. Temperature is a measure of the *average* kinetic energy of individual molecules. Thermal or internal energy refers to the *total* energy of all the molecules in the object. (Thus two equal-mass hot ingots of iron may have the same temperature, but two of them have twice as much thermal energy as one does.) Heat, finally, refers to a *transfer* of energy (such as thermal energy) from one object to another because of a difference in temperature.

Notice that the direction of spontaneous heat flow between two objects depends on their temperatures, not on how much internal energy they each have. Thus, if 50 g of water at 30°C is placed in contact (or mixed) with 200 g of water at 25°C, heat flows *from* the water at 30°C *to* the water at 25°C even though the internal energy of the 25°C water is much greater because there is so much more of it.

11-3 ∎ Internal energy of an ideal gas

Let us calculate the internal energy of n moles of an ideal monatomic (one atom per molecule) gas. The internal energy, U, is the sum of the translational kinetic energies of all the atoms. This sum is just equal to the average kinetic energy per molecule times the total number of molecules, N:

$$U = N(\tfrac{1}{2}m\overline{v^2}).$$

From Eq. 10-8 we then have

$$U = \tfrac{3}{2}NkT$$

or

$$U = \tfrac{3}{2}nRT \quad [\text{monatomic ideal gas}] \quad (11\text{-}1)$$

where n is the number of moles. Thus, the internal energy of an ideal gas depends only on temperature and the number of moles of gas.

If the gas molecules contain more than one atom, then the rotational and vibrational energy of the molecules (Fig. 11-2) must also be taken into account. The internal energy will be greater at a given temperature than for a monatomic gas, but it will still be a function only of temperature.

FIGURE 11-2

Molecules can have (a) rotational and (b) vibrational energy.

(a)

(b)

The internal energy of real gases also depends mainly on temperature, but where they deviate from ideal gas behavior, it also depends somewhat on pressure and volume; we usually won't worry about this.

The internal energy of liquids and solids is quite complicated, for it includes electrical potential energy associated with the forces (or "chemical" bonds) between atoms and molecules.

11-4 ∎ Specific heat; calorimetry

The amount of heat Q required to change the temperature of a system is found to be proportional to the mass m of the system and to the temperature change ΔT, a fact known as early as the eighteenth century. This is expressed in the equation

$$Q = mc\,\Delta T \qquad\qquad (11\text{-}2)$$

where c is a quantity characteristic of the material called *specific heat*. For water at 15°C and a constant pressure of 1 atm, $c = 1.00 \text{ kcal/kg·C}°$ or $4.18 \times 10^3 \text{ J/kg·C}°$; this corresponds to the fact that it takes 1 kcal of heat to raise the temperature of 1 kg of water by 1 C°. Table 11-1 gives the values of specific heat for other substances at 20°C. The values of c depend to some extent on temperature (as well as slightly on pressure), but for temperature changes that are not too great, c can often be considered constant.

TABLE 11-1

Specific Heats (20°C and constant, 1 atm, pressure)

Substance	Specific Heat c_p	
	kcal/kg · C°	J/kg · C°
Aluminum	0.22	900
Copper	0.093	390
Glass	0.20	840
Iron or steel	0.11	450
Lead	0.031	130
Marble	0.21	860
Silver	0.056	230
Wood	0.4	1700
Alcohol (ethyl)	0.58	2400
Mercury	0.033	140
Water		
Ice $(-5°C)$	0.50	2100
Liquid (15°C)	1.00	4180
Steam (110°C)	0.48	2010
Human body (average)	0.83	3470
Protein	0.4	1700

EXAMPLE 11-1   How much heat is required to raise the temperature of 20 kg of iron from 10°C to 90°C?

SOLUTION   From Table 11-1, the specific heat of iron is 450 J/kg·C°. The change in temperature is (90°C − 10°C) = 80°C. Thus

$$Q = mc\,\Delta T = (20 \text{ kg})(450 \text{ J/kg·C°})(80 \text{ C°}) = 7.2 \times 10^5 \text{ J} = 720 \text{ kJ}.$$

(Note: 1 kJ = 10^3 J = 0.239 kcal since 1 kcal = 4.18×10^3 J.)

If the iron had been cooled from 90°C to 10°C, 7.2×10^5 J of heat would have flowed out of the iron. In other words, Eq. 11-2 is valid for heat flow either in or out, with a corresponding increase or decrease in temperature. If 20 kg of water, rather than iron, had been heated from 10°C to 90°C in the above example, the heat required would have been almost 10 times as much (specific heat = 4180 J/kg·C° vs. 450 J/kg·C°). Water has one of the highest specific heats of all substances, which makes it an ideal substance for hot-water space-heating systems and other uses that require a minimal drop in temperature.

When different parts of an isolated system are at different temperatures, heat can flow from the part at higher temperature to the part at lower temperature. If the system is completely isolated, no energy can flow into or out of it; so, from the conservation energy, the heat lost by one part of the system is equal to the heat gained by the other part:

$$\text{heat lost} = \text{heat gained.}$$

Let us take an example.

EXAMPLE 11-2   If 200 cm^3 of tea at 95°C is poured into a 150-g glass cup initially at 25°C, what will be the final temperature of the mixture when equilibrium is reached, assuming that no heat flows to the surroundings?

SOLUTION   Since tea is mainly water, its specific heat is 4180 J/kg·C° and its mass m is its density times its volume: $m = \rho V = (1.0 \times 10^3 \text{ kg/m}^3) \cdot (200 \times 10^{-6} \text{ m}^3) = 0.20$ kg. We set

$$\text{heat lost by tea} = \text{heat gained by cup}$$

$$m_{tea}c_{tea}\,(95°C - T) = m_{cup}c_{cup}\,(T - 25°C)$$

where T is the as yet unknown final temperature. Putting in numbers and using Table 11-1, we solve for T, and find

$$(0.20 \text{ kg})(4180 \text{ J/kg·C°})(95°C - T) = (0.15 \text{ kg})(840 \text{ J/kg·C°})(T - 25°C)$$

$$79{,}400 - 836T = 126T - 6300$$

$$T = 89°C$$

The tea drops in temperature by 6 C° by coming into equilibrium with the cup.

FIGURE 11-3

Simple water calorimeter.

The exchange of energy as discussed in this example is the basis for a technique known as *calorimetry*, which is the quantitative measurement of heat exchange. To make such measurements, a *calorimeter* is used; a simple water calorimeter is shown in Fig. 11-3. It is very important that the calorimeter be well insulated so that only a minimal amount of heat is exchanged with the outside. One important use of the calorimeter is in the determination of specific heats of substances. In the technique known as the "method of mixtures," a sample of the substance is heated to a high temperature, which is accurately measured, and then quickly placed in the cool water of the calorimeter. The heat lost by the sample will be gained by the water and the calorimeter; by measuring the final temperature of the mixture, the specific heat can be calculated as illustrated in the following example.

EXAMPLE 11-3 We wish to determine the specific heat of a new alloy. A 0.150-kg sample of the alloy is heated to 540°C. It is then quickly placed in 400 g of water at 10.0°C which is contained in a 200-g aluminum calorimeter cup. (We do not need to know the mass of the insulating jacket since we assume it does not change significantly in temperature.) The final temperature of the mixture is 30.5°C. Calculate the specific heat of the alloy.

SOLUTION We use the fact that the heat lost equals the heat gained; that is

$$\left(\begin{array}{c}\text{heat lost}\\\text{by sample}\end{array}\right) = \left(\begin{array}{c}\text{heat gained}\\\text{by water}\end{array}\right) + \left(\begin{array}{c}\text{heat gained by}\\\text{calorimeter cup}\end{array}\right)$$

$$m_s c_s \Delta T_s = m_w c_w \Delta T_w + m_{cal} c_{cal} \Delta T_{cal}.$$

Putting in values and using Table 11-1 this equation becomes

$$(0.150 \text{ kg})(c_s)(540°C - 30.5°C) = (0.40 \text{ kg})(4180 \text{ J/kg·C}°)(30.5°C - 10.0°C)$$
$$+ (0.20 \text{ kg})(900 \text{ J/kg·C}°)(30.5°C - 10.0°C)$$

$$76.4 c_s = (34,300 + 3700) \text{ J/kg·C}°$$

$$c_s = 500 \text{ J/kg·C}°$$

In making this calculation, we have ignored any heat transferred to the thermometer and the stirrer (which is needed to quicken the heat transfer process and thus reduce heat loss to the outside). These can be taken into account by adding additional terms to the right side of the above equation and will result in a slight correction to the value of c_s (see the problems). It should be noted that the quantity $m_{cal} c_{cal}$ is often called the *water equivalent* of the calorimeter—that is, $m_{cal} c_{cal}$ is numerically equal to the mass of water (in kilograms) that would absorb the same amount of heat.

A *bomb calorimeter* is used to measure the heat released when a substance burns. Important applications are the burning of foods to determine their Calorie content, and burning of seeds and other substances to determine their "energy content," or heat of combustion. A carefully weighed sample of the substance, together with an excess amount of oxygen at high pressure, is placed in a sealed container (the "bomb"). The bomb is placed in the water of the calorimeter and a fine wire passing into the bomb is then heated briefly, which causes the mixture to ignite.

EXAMPLE 11-4 Determine the Calorie content of 100 g of a certain kind of cake from the following measurements. A 10-g sample of the cake is allowed to dry before putting it in the bomb. The aluminum bomb has a mass of 0.615 kg and is placed in 2.00 kg of water contained in an aluminum calorimeter cup of mass 0.524 kg. The initial temperature of the mixture is 15.0°C and its temperature after ignition is 36.0°C.

SOLUTION   In this case the heat Q released in the burning is absorbed by the entire mixture:

$$
\begin{aligned}
Q &= (m_w c_w + m_{cal} c_{cal} + m_{bomb} c_{bomb}) \Delta T \\
&= [(2.00 \text{ kg})(1.0 \text{ kcal/kg} \cdot \text{C}°) + (0.524 \text{ kg})(0.22 \text{ kcal/kg} \cdot \text{C}°) \\
&\quad + (0.615 \text{ kg})(0.22 \text{ kcal/kg} \cdot \text{C}°)][36.0°\text{C} - 15.0°\text{C}] \\
&= 47 \text{ kcal}
\end{aligned}
$$

Since 47 kcal is released in the burning of 10 g of cake, a 100-g portion would contain 470 "Calories."

11-5 ▌ Latent heat

When a material changes phase from solid to liquid, or liquid to gas (see also Section 10-10), a certain amount of energy is involved in this *change of phase*. For example, let us trace what happens when 1.0 kg of water is heated at a steady rate from, say, $-20°$C, where it is ice, until it has all changed to water and then to steam above 100°C, all at 1 atm pressure. As shown in the graph of Fig. 11-4, as heat is added to the ice, its temperature rises at a rate of about 2 C°/kcal of heat added (since for ice $c \approx 0.50$ kcal/kg·C°). However, when 0°C is reached, the temperature stops increasing even though heat is still being added. Instead, as heat is added, the ice is gradually observed to change to water in the liquid state without any change in temperature. After about 40 kcal have been added at 0°C, half the ice remains and half has changed to water. After about 80 kcal, or 330 kJ, has been added, all the ice has changed to water, still at 0°C. Further addition of heat causes the temperature to again increase (now at a rate of 1 C°/kcal). When 100°C is reached, the temperature again remains constant as the heat added changes the water to vapor (steam). It requires about 540 kcal (2260 kJ) to change the 1.0 kg of

FIGURE 11-4

Heat added to bring 1.0 kg of ice at $-20°$C to steam.

water completely to steam, after which the curve rises again indicating that the temperature of the steam now rises as heat is added.

The heat required to change 1.0 kg of a substance from the solid to the liquid state is called the *heat of fusion;* it is denoted by l_F. The heat of fusion of water is 79.7 kcal/kg or, in proper SI units, 333 kJ/kg. The heat required to change a substance from the liquid to the vapor phase is called the *heat of vaporization*, l_V, and for water it is 539 kcal/kg or 2260 kJ/kg. Other substances follow graphs similar to Fig. 11-4, although the melting-point and boiling-point temperatures are different, as are the specific heats and heats of fusion and vaporization. Values for the heats of fusion and vaporization, which are also called the *latent heats*, are given in Table 11-2 for a number of substances.

The heats of vaporization and fusion also refer to the amount of heat released by a substance when it changes from a gas to a liquid, or from a liquid to a solid. Thus, steam releases 2260 kJ/kg when it changes to water, and water releases 333 kJ/kg when it becomes ice.

Of course, the heat involved in a change of phase depends not only on the latent heat but also on the total mass of the substance. That is

$$Q = ml$$

where l is the latent heat of the particular process, m is the mass of substance, and Q is the heat required or given off. For example, when 5.00 kg of water freezes at 0°C, (5.00 kg) × (3.33 × 10⁵ J) = 1.67 × 10⁶ J of energy is released.

Calorimetry sometimes involves a change of state as the following examples show. Indeed, latent heats are often measured using calorimetry.

EXAMPLE 11-5 How much energy does a refrigerator have to remove from 1.5 kg of water at 20°C to make ice at −12°C?

SOLUTION Heat must flow out to reduce the water from 20°C to 0°C, to change it to ice, and then to lower the ice from 0°C to −12°C:

$$Q = mc_{water}(20°C - 0°C) + ml_F + mc_{ice}[0° - (-12°C)]$$
$$= (1.5 \text{ kg})(4180 \text{ J/kg·C°})(20 \text{ C°}) + (1.5 \text{ kg})(3.33 \times 10^5 \text{ J/kg})$$
$$+ (1.5 \text{ kg})(2100 \text{ J/kg·C°})(12 \text{ C°})$$
$$= 6.6 \times 10^5 \text{ J} = 660 \text{ kJ.}$$

TABLE 11-2
Latent heats (at 1 atm)

Substance	Melting Point (°C)	Heat of Fusion kcal/kg[†]	Heat of Fusion J/kg	Boiling Point (°C)	Heat of Vaporization kcal/kg[†]	Heat of Vaporization J/kg
Oxygen	−218.8	3.3	0.14 × 10⁵	−183	51	2.1 × 10⁵
Ethyl alcohol	−114	25	1.04 × 10⁵	78	204	8.5 × 10⁵
Water	0	79.7	3.33 × 10⁵	100	539	22.6 × 10⁵
Lead	327	5.9	0.25 × 10⁵	1750	208	8.7 × 10⁵
Silver	961	21	0.88 × 10⁵	2193	558	23 × 10⁵
Tungsten	3410	44	1.84 × 10⁵	5900	1150	48 × 10⁵

[†] Numerical values in kcal/kg are the same in cal/g

EXAMPLE 11-6   A 0.50-kg chunk of ice at $-10°C$ is placed in 3.0 kg of water at 20°C. At what temperature and in what phase will the final mixture be?

SOLUTION   In this situation, before we can write down an equation, we must first check to see if the final mixture will be ice, a mixture of ice and water at 0°C, or all water. To bring the 3.0 kg of water at 20°C down to 0°C would require an energy release of $mc\,\Delta T = (3.0\text{ kg})(4180\text{ J/kg·C°})(20\text{ C°}) = 250$ kJ. To raise the ice from $-10°C$ to 0°C would require $(0.50\text{ kg})(2100\text{ J/kg·C°})(10\text{ C°}) = 10.5$ kJ and to change the ice to water at 0°C will require $(0.50\text{ kg})(333\text{ kJ/kg}) = 167$ kJ for a total of 177 kJ. This is not enough energy to bring the 3.0 kg of water at 20°C down to 0°C, so we know that the mixture must end up all water, somewhere between 0°C and 20°C. Now we can determine the final temperature by writing

$$\begin{pmatrix} \text{heat to raise} \\ \text{ice from} \\ -10°C\text{ to }0°C \end{pmatrix} + \begin{pmatrix} \text{heat to} \\ \text{change ice} \\ \text{to water} \end{pmatrix} + \begin{pmatrix} \text{heat to raise} \\ \text{the 0.50 kg water} \\ \text{from }0°C\text{ to }T \end{pmatrix}$$

$$= \begin{pmatrix} \text{heat lost by 3.0 kg} \\ \text{of water cooling} \\ \text{from }20°C\text{ to }T \end{pmatrix};$$

then

$$10.5\text{ kJ} + 167\text{ kJ} + (0.50\text{ kg})(4180\text{ J/kg·C°})(T)$$

$$= (3.0\text{ kg})(4180\text{ J/kg·C°})(20°C - T)$$

or

$$14{,}600\,T = 73{,}800$$

$$T = 5.1°C.$$

EXAMPLE 11-7   The specific heat of mercury is 0.033 kcal/kg·C°. When 1.0 kg of solid mercury at its melting point of $-39°C$ is placed in a 0.50-kg aluminum calorimeter filled with 1.2 kg of water at 20.0°C, the final temperature of the mixture is found to be 16.5°C. What is the heat of fusion of mercury in kcal/kg?

SOLUTION   The heat gained by the mercury (Hg) equals the heat lost by the water and calorimeter:

$$m_{Hg}l_{Hg} + m_{Hg}c_{Hg}[16.5°C - (-39°C)]$$

$$= m_w c_w(20°C - 16.5°C) + m_{Al}c_{Al}(20.0°C - 16.5°C)$$

$$(1.0\text{ kg})(l_{Hg}) + (1.0\text{ kg})(0.033\text{ kcal/kg·C°})(55.5\text{ C°})$$

$$= (1.2\text{ kg})(1.0\text{ kcal/kg·C°})(3.5\text{ C°})$$

$$+ (0.50\text{ kg})(0.22\text{ kcal/kg·C°})(3.5\text{ C°})$$

$$l_{Hg} = (4.2 + 0.4 - 1.8)\text{ kcal/kg} = 2.8\text{ kcal/kg}.$$

The latent heat to change a liquid to a gas is needed not only at the boiling point. Water can change from the liquid to the gas phase even at room temperature. This process is called *evaporation* (see also Section 10-11). The value of the heat of vaporization increases slightly with a decrease in temperature: at 20°C, for example, it is 2450 kJ/kg (585 kcal/kg) compared to 2260 kJ/kg (539 kcal/kg) at 100°C. When water evaporates, it cools, since the energy required (the latent heat of vaporization) comes from the water itself; so its internal energy, and therefore its temperature, must drop.[†]

Evaporation of water from the skin is one of the most important methods the body uses to control its temperature; when the temperature of the blood rises slightly above normal, the hypothalamus gland detects this temperature increase and sends a signal to the sweat glands to increase their production. The energy required to vaporize this water comes from the body, and hence the body cools.

We can make use of kinetic theory to see why energy is needed to melt or vaporize a substance. At the melting point, the latent heat of fusion does not increase the kinetic energy (and the temperature) of the molecules in the solid but instead is used to overcome the potential energy associated with the forces between the molecules. That is, work must be done against these attractive forces to break the molecules loose from their relatively fixed positions in the solid so they can freely roll over one another in the liquid phase. Similarly, energy is required for molecules held close together in the liquid phase to escape into the gaseous phase. This process is a more violent reorganization of the molecules than is melting (the average distance between the molecules is greatly increased) and hence the heat of vaporization is generally much greater than the heat of fusion for a given substance.

11-6 ▌ Heat transfer: conduction

Heat is transferred from one place or body to another in three different ways: by *conduction*, *convection*, and *radiation*. We now discuss each of these in turn; but in practical situations, any two or all three may be operating at the same time. We start with conduction.

When a metal poker is put in a hot fire, or a silver spoon is placed in a hot bowl of soup, the exposed end of the poker or spoon soon becomes hot as well, even though it is not directly in contact with the source of heat. We say that heat has been conducted from the hot end to the cold end.

Heat **conduction** in many materials can be visualized as the result of molecular collisions. As one end of the object is heated, the molecules there move faster and faster. As they collide with their slower moving neighbors, they transfer some of their energy to these molecules whose speeds thus increase. These in turn transfer some of their energy by collision with molecules still farther down the object. Thus the energy of thermal motion is transferred by molecular collision down the object. In metals, according to

---

[†] According to kinetic theory, evaporation is a cooling process because it is the fastest-moving molecules that escape from the surface (Section 10-11); hence the average speed of the remaining molecules is less, so by Eq. 10-8 the temperature is less.

TABLE 11-3

11 Heat

Thermal conductivities

Substance	Thermal Conductivity, k	
	kcal/s·m·C°	J/s·m·C°
Silver	10×10^{-2}	420
Copper	9.2×10^{-2}	380
Aluminum	5.0×10^{-2}	200
Steel	1.1×10^{-2}	40
Glass (typical)	2.0×10^{-4}	0.84
Brick and concrete	2.0×10^{-4}	0.84
Water	1.4×10^{-4}	0.56
Human tissue (without blood)	0.5×10^{-4}	0.2
Asbestos	0.4×10^{-4}	0.16
Wood	$0.2-0.4 \times 10^{-4}$	0.08–0.16
Cork and glass wool	0.1×10^{-4}	0.042
Down	$.06 \times 10^{-4}$	0.025
Air	$.055 \times 10^{-4}$	0.023

modern theory, it is collisions between free electrons[†] within the metal that are visualized as being mainly responsible for conduction.

Heat conduction takes place only if there is a difference in temperature. Indeed, it is found experimentally that the rate of heat flow through a substance is proportional to the difference in temperature between its ends. The rate of heat flow also depends on the size and shape of the object, and to investigate this quantitatively, let us consider the heat flow through a uniform object as illustrated in Fig. 11-5. It is found experimentally that the heat flow ΔQ per time interval Δt is given by the relation

$$\frac{\Delta Q}{\Delta t} = kA \frac{T_2 - T_1}{l} \tag{11-3}$$

where A is the cross-sectional area of the object, l is the distance between the two ends, which are at temperatures T_1 and T_2, and k is a proportionality constant called the *thermal conductivity*, which is characteristic of the material. From Eq. 11-3 we see that the rate of heat flow[‡] is directly proportional to the cross-sectional area and to the temperature gradient $(T_1 - T_2)/l$.

The thermal conductivities, k, for a variety of substances are given in Table 11-3. Substances for which k is large conduct heat rapidly and are said to be *good conductors*. Most metals fall in this category, although there is a wide range even among them as you may observe by holding the ends of a silver spoon and a stainless steel spoon immersed in the same hot cup

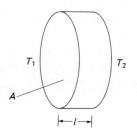

FIGURE 11-5

Heat conduction.

[†] As we will discuss later (Chapter 15 and especially Chapter 28), in metals some of the electrons of the atoms are quite free to move about.

[‡] This is quite similar to the relations describing diffusion (Chapter 10) and the flow of fluids through a pipe (Chapter 9). In those cases the flow of matter was found to be proportional to the pressure gradient $(P_1 - P_2)/l$. This close similarity is one reason we speak of the flow of heat; yet we must keep in mind that no substance is flowing in this case—it is energy that is being transferred.

of soup. Substances for which k is small, such as asbestos and down, are poor conductors of heat and are therefore good *insulators*. The relative magnitude of k can explain simple phenomena such as why a tile floor is much colder on the feet than a rug-covered floor at the same temperature: tile is a better conductor of heat than the rug; heat transferred from your foot to the rug is not conducted away rapidly, so the rug quickly heats up on its surface to the temperature of your foot; but the tile conducts the heat away rapidly and thus can take more heat from your foot, so its surface temperature drops.

EXAMPLE 11-8 A major source of heat loss from a house is through the windows. Calculate the rate of heat flow through a glass window 2.0 m × 1.5 m in area and 3.2 mm thick if the temperatures at the inner and outer surfaces are 15.0°C and 14.0°C, respectively.

SOLUTION Since $A = (2.0 \text{ m})(1.5 \text{ m}) = 3.0 \text{ m}^2$, $l = 3.2 \times 10^{-3}$ m, and using Table 11-3 to get k, we have from Eq. 11-3

$$\frac{\Delta Q}{\Delta t} = kA \frac{T_2 - T_1}{l} = \frac{(0.84 \text{ J/s} \cdot \text{m} \cdot \text{C°})(3.0 \text{ m}^2)(15.0°\text{C} - 14.0°\text{C})}{(3.2 \times 10^{-3} \text{ m})}$$

$$= 790 \text{ W}.$$

This is equivalent to $(790 \text{ J/s})/(4.18 \times 10^3 \text{ J/kcal}) = 0.19 \text{ kcal/s}$, or $(0.19 \text{ kcal/s}) \cdot (3600 \text{ s/h}) = 680 \text{ kcal/h}$.

You might notice that 15°C is not very warm for the living room of a house. The room itself may indeed be much warmer, and the outside might be colder than 14°C. But the temperatures of 15°C and 14°C were specified as those at the window surfaces and there is usually a considerable drop in temperature of the air in the vicinity of the window both on the inside and the outside. That is, the layer of air on either side of the window acts as an insulator and normally the major part of the temperature drop between the inside and outside of the house takes place across the air layer. (If there is a heavy wind, the air outside a window will constantly be replaced with cold air; the temperature gradient across the glass will be greater and there will be a much greater rate of heat loss.) Thus, increasing the width of the air layer (such as using two panes of glass separated by an air gap) is far more helpful than simply increasing the glass thickness.

The insulating properties of clothing come from the insulating properties of air. Without clothes, our bodies would heat the air in contact with the skin and soon become reasonably comfortable since air is a very good insulator; but since air moves—there are breezes and drafts, and people themselves move about—the warm air would be replaced by cold air, thus increasing the heat loss from the body. Clothes keep us warm by holding air so it cannot move readily. It is not the cloth that insulates us, but the air that the cloth traps. Down is a very good insulator because even a small amount of it fluffs up and traps a great amount of air. On this basis, can you see one reason why drapes in front of a window reduce heat loss from a house?

Although liquids and gases are generally not very good conductors of heat, they can transfer heat quite rapidly by convection. **Convection** is the process whereby heat is transferred by the mass movement of molecules from one place to another. Whereas conduction involves molecules (and/or electrons) moving only over small distances and colliding, convection involves the movement of molecules over large distances.

A forced-air furnace, in which air is heated and then blown by a fan into a room, is an example of *forced convection. Natural convection* occurs as well, and one familiar example is that hot air rises. For instance the air above a radiator (or other type of heater) expands as it is heated and hence its density decreases; because its density is less, it rises. Warm or cold ocean currents, such as the Gulf Stream, represent natural convection on a large scale. Wind is another example of convection, and weather in general is a result of convective air currents.

When a pot of water is heated (Fig. 11-6), convection currents are set up as the heated water at the bottom of the pot rises because of its reduced density and is replaced by cooler water from above. This principle is used in many heating systems, such as the hot-water-radiator system shown in Fig. 11-7. Cold water is heated in the furnace and as its temperature rises, it expands and rises as shown. This causes the water to circulate in the system; hot water than enters the radiators, heat is transferred by conduction to the air, and the cooled water returns to the furnace. Thus the water circulates because of convection; however, pumps are sometimes used to

FIGURE 11-6

Convection currents in a pot of water being heated on a stove.

FIGURE 11-7

Convection plays a role in heating a house.

facilitate circulation. The air in the room also becomes uniformly heated as a result of convection. The air heated by the radiator rises and is replaced by cooler air, resulting in convective air currents, as shown.

Other types of furnace also depend on convection. Hot-air furnaces with registers (openings) near the floor often do not have fans but depend on natural convection, which can be appreciable. In other systems a fan is used. In either case it is important that cold air can return to the furnace so that convective currents circulate throughout the room if the room is to be uniformly heated.

Another example of convection and its effects is given in the following excerpt from François Matthes's "The Winds of Yosemite Valley"[†]:

> It happens to be so ordained in nature that the sun shall heat the ground more rapidly than the air. And so it comes that every slope or hillside basking in the morning sun soon becomes itself a source of heat. It gradually warms the air immediately over it, and the latter, becoming lighter, begins to rise. But not vertically upward, for above it is still the cool air pressing down. Up along the warm slope it ascends, much as shown by the arrows in the accompanying diagram (Fig. 11-8a). Few visitors to the valley but will remember toiling up some neverending zigzags on a hot and breathless day, with the sun on their backs and their own dust floating upward with them in an exasperating, choking cloud. Perhaps they thought it was simply their misfortune that the dust should happen to rise on that particular day. It always does on a sun-warmed slope.
>
> But again, memories may arise of another occasion when, on coming down a certain trail the dust ever descended with the travelers, wafting down upon them from zigzag to zigzag as if with malicious pleasure. That, however, undoubtedly happened on the shady side of the valley. For there the conditions are exactly reversed. When the sun leaves a slope the latter begins at once to lose its heat by radiation, and in a short time is colder than the air. The layer next to the ground then gradually chills by contact, and, becoming heavier as it condenses, begins to creep down along the slope (Fig. 11-8b). There is, thus, normally a warm updraft on a sunlit slope and a cold downdraft on a shaded slope—and that rule one may depend on almost any day in a windless region like the Yosemite. Indeed, one might readily take advantage of it and plan his trips so as to have a dust-free journey.

The human body produces a great deal of thermal energy; of the food energy transformed within the body, at best 20 percent is used to do work, so over 80 percent appears as thermal energy. During light activity, for example, if this thermal energy were not dissipated, the body temperature would rise about 3 C° per hour. Clearly, the heat generated by the body must be transferred to the outside. Is the heat transferred by conduction? The temperature of the skin in a comfortable environment is 33 to 35°C, whereas the interior of the body is at 37°C. A simple calculation (see the Problems) shows that, because of this small temperature difference, plus the low thermal conductivity of tissue, conduction is responsible for very little of the heat that must be dissipated. Instead, the heat is carried to the surface by the blood. In addition to all its other important responsibilities, blood

FIGURE 11-8

Convection on a hiking trail:
(a) upward movement of air in the morning because it is heated;
(b) downward movement in evening because it is cooled.

[†] Reprinted from the *Science Club Bulletin*, June, 1911, pp. 91–92.

acts as a convective fluid to transfer heat to just beneath the surface of the skin. It is then conducted (over a very short distance) to the surface. Once at the surface, the heat is transferred to the environment by convection, evaporation, and radiation (see the next section).

11-8 ∎ Heat transfer: radiation

Convection and conduction require the presence of matter. Yet all life on earth depends on the transfer of energy from the sun, and this energy is transferred to the earth over empty (or nearly empty) space. This form of energy transfer is heat—since the sun's temperature is much higher (6000 K) than earth's—and is referred to as **radiation.** The warmth we receive from a fire is mainly radiant energy (most of the air heated by a fire rises by convection up the chimney and does not reach us).

As we shall see in later chapters, radiation consists essentially of electromagnetic waves. Suffice it to say for now that radiation from the sun consists of visible light plus many other wavelengths that the eye is not sensitive to; this includes infrared (IR) radiation, which is mainly responsible for heating the earth.

The rate at which an object radiates energy has been found to be proportional to the fourth power of the Kelvin temperature, T. That is, a body at 2000 K as compared to one at 1000 K radiates energy at a rate $2^4 = 16$ times as much. The rate of radiation is also proportional to the area A of the emitting object, so the rate at which energy leaves the object, $\Delta Q / \Delta t$, is

$$\frac{\Delta Q}{\Delta t} = e\sigma A T^4; \qquad (11\text{-}4)$$

here σ is a universal constant called the *Stefan-Boltzmann constant* which has the value $\sigma = 5.67 \times 10^{-8}$ W/m$^2 \cdot$K^4; and e, called the *emissivity*, is a number between 0 and 1 that is characteristic of the material. Very black surfaces, such as charcoal, have emissivity close to 1 whereas shiny surfaces have e close to zero and thus emit correspondingly less radiation. The value of e depends somewhat on the temperature of the body. Skin has a surprisingly high emissivity, about 0.6 for light skin to 0.8 for dark skin.

Not only do shiny surfaces emit less radiation, but they absorb little of the radiation that falls upon them (most is reflected). Black and very dark objects, on the other hand, absorb nearly all the radiation that falls on them—which is why light-colored clothing is preferable to dark clothing on a hot day. Thus *a good absorber is also a good emitter.*

Any object not only emits energy by radiation, but it also absorbs energy radiated by other bodies. If an object of emissivity e and area A is at a temperature T_1, it radiates energy at a rate $e\sigma A T_1^4$. If the object is surrounded by an environment at temperature T_2 and high emissivity (≈ 1), then the rate the surroundings radiate energy is proportional to T_2^4 and thus the rate that energy is absorbed by the object is proportional to T_2^4. The *net* rate of radiant heat flow from the object is given by the equation

$$\frac{\Delta Q}{\Delta t} = e\sigma A(T_1^4 - T_2^4), \qquad (11-5)$$

where A is the surface area of the object, T_1 its temperature and e its emissivity (at temperature T_1), and T_2 is the temperature of the surroundings. Notice in this equation that the rate of heat absorption by an object was taken to be $e\sigma A T_2^4$; that is, the proportionality constant is the same for both emission and absorption. This must be true to correspond with the experimental fact that equilibrium between object and surroundings is reached when they come to the same temperature; that is, $\Delta Q/\Delta t$ must equal zero when $T_1 = T_2$, so the coefficients of emission and absorption terms must be the same. This confirms the idea that a good emitter is a good absorber.

Because both the object and its surroundings radiate energy, there is a net transfer of energy from one to the other unless everything is at the same temperature. From Eq. 11-5 it is clear that if $T_1 > T_2$, the net flow of heat is from the body to the surroundings, so the body cools; but if $T_1 < T_2$, the net heat flow is from the surroundings into the body and its temperature rises. If different parts of the surroundings are at different temperatures, Eq. 11-5 becomes more complicated.

☐   EXAMPLE 11-9   A person is sitting unclothed in a room whose dark walls are at a temperature of 15°C. Calculate the rate of heat loss by radiation assuming a skin temperature of 34°C and $e = 0.70$. Take the surface area of the body not in contact with the chair to be 1.5 m².

SOLUTION   From Eq. 11-5 we have

$$\frac{\Delta Q}{\Delta t} = e\sigma A(T_1^4 - T_2^4)$$

$$= (0.70)(5.67 \times 10^{-8} \text{ J/s·m}^2\text{·K}^4)(1.5 \text{ m}^2)[(307 \text{ K})^4 - (288 \text{ K})^4]$$

$$= 120 \text{ W}.$$

We shall see in Chapter 12 that a resting person produces heat internally at a lower rate than the heat loss by radiation as calculated in this example. Because of this, and because other forms of heat loss are also present, the person's temperature would drop, causing considerable discomfort. The body responds to excessive heat loss by increasing its metabolic rate (Section 12-3) and shivering is one method by which the body increases its metabolism. Naturally, clothes help a great deal. Example 11-9 illustrates the fact that a person may be uncomfortable even if the temperature of the air were, say 25°C, which is quite a warm room. If the walls or floor are cold, radiation to them occurs no matter how warm the air is. Indeed, it is estimated that radiation accounts for about 50 percent of the heat loss from a sedentary person in a normal room. Rooms are most comfortable when the walls and floor are warm and the air not so warm. Floors and walls can be heated by means of hot-water conduits or electric heating elements. Although such first-rate heating systems are not common today, it is interesting to note that 2000 years ago the Romans, even in houses in the remote province of

(a)

FIGURE 11-9
Radiant energy striking a body at an angle θ.

(b)

(c)

Great Britain, made use of hot water and steam conduits in the floor to heat their houses.

Heating of an object by radiation from the sun cannot be calculated using Eq. 11-5 since this equation assumes a uniform temperature, T_2, of the environment surrounding the object, whereas the sun is essentially a point source; hence it must be treated as an additional source of energy. Heating by the sun is calculated using the fact that about 1350 J of energy strikes the atmosphere of the earth from the sun per second per meter squared of area at right angles to the sun's rays. This number, 1350 W/m², is called the *solar constant*. The atmosphere may absorb as much as 70 percent of this energy before it reaches the ground, depending on the cloud cover. On a clear day, about 1000 W/m² reaches the earth's surface; an object of emissivity e with area A facing the sun absorbs heat at a rate, in watts, of about $1000eA \cos \theta$ where θ is the angle between the sun's rays and a line perpendicular to the area A (Fig. 11-9). That is, $A \cos \theta$ is the "effective" area, at right angles to the sun's rays. The explanations for the seasons, the polar ice caps, and why the sun heats the earth more at midday than at sunrise or sunset are also related to this $\cos \theta$ factor.

(d)

EXAMPLE 11-10   What is the rate of energy absorption from the sun by a person lying flat on the beach on a clear day if the sun makes a 30° angle with the vertical? Assume that $e = 0.70$, the area of the body exposed to the sun is 0.80 m², and that 1000 W/m² reaches the earth's surface.

SOLUTION   Since cos 30° = 0.866, we have

$$\frac{\Delta Q}{\Delta t} = (1000 \text{ W/m}^2)eA \cos \theta$$

$$= (1000 \text{ W/m}^2)(0.70)(0.80 \text{ m}^2)(0.866)$$

$$= 490 \text{ W}.$$

FIGURE 11-10
Thermograms of 25-year-old woman smoker illustrating temperature decrease (due to impaired blood circulation) associated with smoking in the healthy individual. (a) Presmoking thermogram shows normal warm temperature level (white) of the hand; (b) 1 minute and (c) 15 minutes after smoking a single filtered cigarette reveals a 6°C drop in temperature of fingers. (d) After 30 minutes, temperature begins to return to normal. (Courtesy of American Cancer Society.)

Notice that if a person wears light-colored clothing, e is much smaller, so the energy absorbed is less.

An interesting application of thermal radiation to diagnostic medicine is *thermography*. A special instrument, the thermograph, scans the body, measuring the intensity of radiation from many points and forming a picture that resembles an X ray (Fig. 11-10). Areas where metabolic activity is high, such as in tumors, can often be detected on a thermogram as a result of their higher temperature and thereby increased radiation.

▌ Summary

Thermal energy, or *internal energy*, U, refers to the total energy of all the molecules in a body. *Heat* refers to the transfer of energy from one body to another because of a difference of temperature. Heat is thus measured in energy units, such as joules; heat and thermal energy are also sometimes specified in calories or kilocalories, where 1 cal = 4.18 J is the amount of heat needed to raise the temperature of 1 g of water by 1 C°.

The *specific heat*, c, of a substance is defined as the energy (or heat) required to change the temperature of unit mass of substance by 1 degree; as an equation, $Q = mc\,\Delta T$, where Q is the heat absorbed or given off, ΔT the temperature rise or decline, and m the mass of the substance. When heat flows within an isolated system, the heat gained by one part of the system is equal to the heat lost by the other part of the system; this is the basis of calorimetry, which is the quantitative measurement of heat exchange.

An exchange of energy occurs, without a change in temperature, whenever a substance changes phase; this happens because the potential energy of the molecules changes as a result of the changes in the relative positions of the molecules. The *heat of fusion* is the heat required to melt 1 kg of a solid into the liquid phase; it is also equal to the heat given off when the substance changes from liquid to solid. The *heat of vaporization* is the energy required to change 1 kg of a substance from the liquid to the vapor phase; it is also the energy given off when the substance changes from vapor to liquid.

Heat is transferred from one place (or body) to another in three different ways. In *conduction*, energy is transferred from higher-KE molecules or electrons to lower-KE neighbors when they collide. *Convection* is the transfer of energy by the mass movement of molecules over considerable distances. *Radiation*, which does not require the presence of matter, is energy transfer by electromagnetic waves, such as from the sun. All bodies radiate energy in an amount that is proportional to their surface area and to the fourth power of their Kelvin temperature; the energy radiated (or absorbed) also depends on the nature of the surface (dark and absorbing versus brightly reflecting), which is characterized by the emissivity.

▌ Questions

1. What happens to the work done when a jar of orange juice is vigorously shaken?
2. Is heat really involved in the Joule experiment of Fig. 11-1? Explain.
3. When a hot object warms a cooler object, does temperature flow between them? Are the temperature changes of the two objects equal?
4. In warmer areas where tropical plants grow but the temperature may drop below freezing a few times in the winter, the destruction of sensitive plants due to freezing can be reduced by watering them in the evening. Explain.
5. The specific heat of water is quite large. Explain why this fact makes water particularly good for heating systems (that is, hot-water radiators).
6. Why does water in a canteen stay cooler if the cloth jacket surrounding the canteen is kept moist?
7. Explain why burns caused by steam on the skin are often so severe.
8. Explain, using the concepts of latent heat and internal energy, why water cools (its temperature drops) when it evaporates.
9. Will potatoes cook faster if the water is boiling faster?
10. Does an ordinary electric fan cool the air? Why or why not? If not, why use it?
11. The temperature very high in the earth's atmosphere can be as high as 700°C. Yet an animal there would freeze to death rather than boil. Explain.
12. Why do Bedouins wear several layers of clothing in their desert environment even when the temperature reaches 50°C (122°F) or higher?
13. Down sleeping bags and parkas are often specified as so many inches or centimeters of *loft*, the actual thickness of the garment when it is fluffed up. Explain.
14. Sea breezes are often encountered on sunny days at the shore of a large body of water. Explain in light of the fact that the temperature of the land rises more rapidly than that of the nearby water.
15. The floor of a house on a foundation under which the air can flow is often cooler than a floor which rests directly on the ground (such as a concrete slab foundation). Explain.
16. The earth cools off at night much more quickly when the weather is clear than when cloudy. Why?

ermos bottle silvered, and why
between its two walls?

erature readings are always taken
er in the shade.

red clothes more comfortable in hot
dark clothes?

by in an incubator can be dangerously
en the air temperature in the incubator
ain.

ern Hemisphere the amount of heat re-
eat a room where the windows face north
is much gher than that required where the windows
face south. Explain.

22. Suppose you are designing one of the following (choose one): house, concert hall, medical office building. List as many sources of heat as you can think of. Estimate the heat produced by each.

23. Heat loss occurs through windows by the following processes: (1) ventilation around edges, (2) through the frame, particularly if it is metal, (3) through the glass panes, (4) radiation. (a) For the first three, what is (are) the mechanism(s): conduction, convection, or radiation? (b) Heavy curtains reduce which of these heat losses? Explain in detail.

24. A piece of wood lying in the sun absorbs more heat than a piece of shiny metal. Yet the wood feels less hot than the metal when you pick it up. Explain.

▌ Problems

SECTION 11-1

1. (I) How much heat (joules) is required to raise the temperature of 2.0 kg of water from 0°C to 20°C?

2. (I) To what temperature will 9500 J of work raise 2.0 kg of water initially at 10.0°C?

3. (I) How much work must a person do to offset eating a 400-Cal piece of cake?

4. (II) A British thermal unit (Btu) is a unit of heat in the British system of units. One Btu is defined as the heat needed to raise 1 pound of water by 1 F°. Show that 1 Btu = 0.252 kcal = 1055 J.

5. (II) A water heater can generate 8500 kcal/h. How much water can it heat from 10°C to 60°C per hour?

6. (II) How many kilocalories of heat are generated when the brakes are used to bring a 1400-kg car to rest from a speed of 80 km/h?

7. (II) If coal gives off 7000 kcal/kg when it is burned, how much coal will be needed to heat a house that requires 4.2×10^7 kcal for the whole winter? Assume that an additional 30 percent of the heat is lost up the chimney.

SECTION 11-4

8. (I) An automobile cooling system holds 12 L of water. How much heat does it absorb if its temperature rises from 20°C to 70°C?

9. (I) How many kilograms of copper will experience the same temperature rise as 10 kg of water when the same amount of heat is absorbed?

10. (I) What is the specific heat capacity of a metal substance if 135 kJ of heat is needed to raise 4.5 kg of the metal from 20°C to 42°C?

11. (I) What is the water equivalent of 0.288 kg of glass?

12. (I) What is the specific heat capacity of water in Btu/lb·F°? (See problem 4.)

13. (II) The *heat capacity*, C, of an object is defined as the amount of heat needed to raise its temperature by 1 C°.

Thus, to raise the temperature by ΔT requires heat Q given by

$$Q = C \Delta T.$$

(a) Write the heat capacity C in terms of the specific heat, c, of the material. (b) What is the heat capacity (at constant pressure) of 1 kg of water? (c) Of 35 kg of water?

14. (II) A 30-g lead bullet traveling 200 m/s is brought to rest when it penetrates a 1.0-kg block of wood. What is the temperature rise of the block and bullet (assumed to have reached thermal equilibrium) as a result of this collision?

15. (II) A large 300-kg marble rock falls vertically a height of 200 m before striking the ground. What will be the temperature rise of the rock if 50 percent of the heat generated remains in the rock?

16. (II) A 45-g glass thermometer reads 19.0°C before it is placed in 220 mL of water. When the water and thermometer come to equilibrium, the thermometer reads 38.5°C. What was the original temperature of the water?

17. (II) The 0.50-kg head of a hammer has a speed of 5.0 m/s just before it strikes a nail and is brought to rest. Estimate the temperature rise of a 15-g iron nail generated by ten such hammer blows done in quick succession. Assume the nail absorbs all the "heat."

18. (II) What will be the equilibrium temperature when a 200-g block of copper at 210°C is placed in a 180-g aluminum calorimeter cup containing 800 g of water at 11.0°C?

19. (II) When a 290-g piece of iron at 180°C is placed in a 100-g aluminum calorimeter cup containing 250-g of glycerin at 10°C, the final temperature is observed to be 38°C. What is the specific heat of glycerin?

20. (II) During light activity, a 70-kg person may generate 200 kcal/h. Assuming that 20 percent of this goes into useful work and the other 80 percent is converted

to heat, calculate the temperature rise of the body after 1.00h if none of this heat were transferred to the environment.

21. (II) How long does it take a 500-W coffee pot to bring to a boil 0.45 L of water initially at 10°C? Assume that the part of the pot which is heated with the water is made of 400 g of aluminum, and that no water boils away.

22. (II) When 220 g of a substance is heated to 330°C and then plunged into a 90-g aluminum calorimeter cup containing 150 g of water at 11.5°C, the final temperature, as registered by a 17-g glass thermometer, is 33.8°C. What is the specific heat of the substance?

23. (II) A 1.50-cm-diameter lead rod absorbs 275 kJ of heat. By how much does its length change? What would happen if the rod were only 2.0 cm long?

SECTION 11-5

24. (I) During exercise, a person may give off 180 kcal of heat in 30 min by evaporation of water from the skin. How much water has been lost?

25. (I) How much heat is needed to melt 13.00 kg of silver that is initially at 22.0°C?

26. (I) If 1.70×10^5 J of energy is supplied to a flask of oxygen at -183°C, how much oxygen will evaporate?

27. (II) An iron boiler of mass 128 kg contains 780 kg of water at 20°C. A heater supplies energy at the rate of 46,000 kJ/h. How long does it take for the water (a) to reach the boiling point and (b) to all have changed to steam?

28. (II) The specific heat of mercury is 138 J/kg·°C. Determine the latent heat of fusion of mercury using the following calorimeter data: 1.00 kg of solid Hg at its melting point of -39.0°C is placed in a 0.620-kg aluminum calorimeter with 0.400 kg of water at 12.80°C; the resulting equilibrium temperature is 5.06°C.

29. (II) How much water would have to evaporate from the skin per minute to take away all the heat generated by the basal metabolism (60 kcal/h) of a 65-kg person?

30. (II) What will be the final result when equal amounts of ice at 0°C and steam at 100°C are mixed together?

31. (II) A 55.0-kg ice skater moving at 8.5 m/s glides to a stop. Assuming the ice is at 0°C and that 50 percent of the heat generated by friction is absorbed by the ice, how much ice melts?

32. (III) A 25-g lead bullet traveling at 400 m/s passes through a thin iron wall and emerges at a speed of 250 m/s. If the bullet absorbs 50 percent of the heat generated, (a) what will be the temperature rise of the bullet? (b) If the ambient temperature is 20°C, will any of the bullet melt, and if so, how much?

SECTIONS 11-6 TO 11-8

33. (I) Calculate the rate of heat flow by conduction in Example 11-8 assuming there are strong gusty winds and that the external temperature is -5°C.

34. (I) (a) How much power is radiated by a tungsten sphere (emissivity $e = 0.35$) of radius 10 cm at a temperature of 20°C? (b) If the sphere is enclosed in a room whose walls are kept at -5°C, what is the *net* flow rate of energy out of the sphere?

35. (I) Over what distance must there be heat flow by conduction from the blood capillaries beneath the skin to the surface if the temperature difference is 0.50°C? Assume 200 W must be transferred through the whole body's surface area of 1.5 m².

36. (I) Approximately how much radiation does a person, of total area 1.5 m², absorb per hour from the sun when it makes a 40° angle to the vertical on a clear day and the person lies flat? Assume $e = 0.80$.

37. (II) Estimate the rate that heat can be conducted from the interior of the body to the surface. Assume that the thickness of tissue is 4.0 cm, that the skin is at 34°C and the interior at 37°C, and that the surface area is 1.5 m². Compare this to the measured value of about 230 W that must be dissipated by a person working lightly. This clearly shows the necessity of convective cooling by the blood.

38. (II) A 100-W light bulb generates 95 W of heat which is dissipated through a glass bulb that has a radius of 3.0 cm and is 1.0 mm thick. What is the difference in temperature between the inner and outer surfaces of the glass?

39. (II) A mountain climber wears down clothing 3.8 cm thick whose total surface area is 1.8 m². The temperature at the surface of the clothing is -10°C and at the skin is 34°C. Determine the rate of heat flow by conduction through the clothing (a) assuming it is dry and k, the thermal conductivity, is that of down, and (b) assuming the clothing is wet, so that k is that of water and the jacket has matted down to 0.50 cm thickness.

40. (II) Write an equation for the total rate of heat flow through the wall of a house if the wall consists of material with thermal conductivity k_1, total area A_1, and thickness l_1, and of windows with thermal conductivity k_2, area A_2, and thickness l_2. The temperature difference is ΔT.

41. (II) *Newton's law of cooling* states that for small temperature differences, if a body at a temperature T_1 is in surroundings at a temperature T_2, the body cools at a rate given by:

$$\frac{\Delta Q}{\Delta t} = K(T_1 - T_2)$$

where K is a constant. It includes the effects of conduction, convection, and radiation. That this linear relationship should hold is obvious if only conduction is considered. Show that it is also approximately true for radiation by showing that Eq. 11-5 reduces to

$$\Delta Q/\Delta t = 4\sigma e A T_2^3(T_1 - T_2) = \text{constant} \times (T_1 - T_2)$$

if $(T_1 - T_2)$ is small.

42. (II) A *thermal transmission coefficient*, U, defined by the equation:

$$\frac{\Delta Q}{\Delta t} = AU(T_2 - T_1),$$

is often used in practical work. Note that U is the rate of heat flow per unit area per degree of temperature. (a) How is U related to the thermal conductivity coefficient k? (b) Is U a characteristic property of a material? If not, what else must be known about the material before U can be determined? (c) Write an equation describing the total heat loss by conduction from a room in terms of the values of U for the different materials; assume the windows are of total area A_1, and have thermal transmission coefficient U_1. The walls have a total area A_2 and consist of a layer of brick (U_2) separated by an air gap (U_3) from a layer of wood (U_4). (d) What is the thermal transmission coefficient U for a brick wall 12 cm thick and 8.0 m wide by 2.5 m high?

43. (III) A house has well-insulated walls 22.0 cm thick (assume conductivity of air) and area 350 m², a roof of wood 5.5 cm thick and area 280 m², and uncovered windows 0.65 cm thick and total area 28 m². (a) Assuming that the heat loss is only by conduction, calculate the rate at which heat must be supplied to this house to maintain its temperature at 20°C if the outside temperature is −5°C. (b) If the house is initially at 10°C, estimate how much heat must be supplied to raise the temperature to 20°C within 30 min. Assume that only the air needs to be heated and that its volume is 700 m³. (c) If natural gas costs $0.055 per kilogram and its heat of combustion is 5.4×10^7 J/kg, how much is the monthly cost to maintain the house as in part (a) above for 12 h each day assuming 90 percent of the heat produced is used to heat the house? Take the specific heat of air to be 0.17 kcal/kg·C°.

44. (III) A double-glazed window is one with two panes of glass separated by an air space, Fig. 11-11. (a) Show that the rate of heat flow by conduction is given by:

$$\frac{\Delta Q}{\Delta t} = \frac{A(T_2 - T_1)}{l_1/k_1 + l_2/k_2 + l_3/k_3},$$

where k_1, k_2, and k_3 are the thermal conductivities for glass, air, and glass, respectively. (b) Generalize this expression for any number of materials placed next to one another.

45. (III) Approximately how long should it take 20 kg of

FIGURE 11-11

ice at 0°C to melt when it is placed in a carefully sealed Styrofoam "icebox" of dimensions 30 cm × 20 cm × 50 cm whose walls are 1.5 cm thick. Assume that the conductivity of Styrofoam is equal to that of air and that the outside temperature is 25°C.

46. (III) A leaf of area 40 cm² and mass 4.5×10^{-4} kg directly faces the sun on a clear day. The leaf has an emissivity of 0.85, and a specific heat of 0.80 kcal/kg·K. (a) Estimate the rate of rise of the leaf's temperature. (b) Calculate the temperature the leaf would reach if it lost all its heat by radiation (the surroundings are at 20°C). (c) In what other ways can the heat be dissipated by the leaf?

47. (III) Use the result of part (a) in the previous problem and take into account radiation from the leaf to calculate how much water must be transpired (evaporated) by the leaf per hour to maintain a temperature of 35°C.

48. (III) A house thermostat is normally set to 22°C but at night it is turned down to 12°C for 8 h. Estimate how much more heat would be needed (state as a percentage of daily usage) if the thermostat were not turned down at night. Assume that the outside temperature averages 0°C for the 8 h at night and 8°C for the remainder of the day, and that the heat loss from the house is proportional to the difference in temperature inside and out. To obtain an estimate from the data, you will have to make other simplifying assumptions; state what these are.

12 THE FIRST AND SECOND LAWS OF THERMODYNAMICS

Thermodynamics is the name we give to the study of processes in which energy is transferred as heat and as work.

In Chapter 5 we saw that work is done when energy is transferred from one body to another by mechanical means. In Chapter 11 we saw that heat is also a transfer of energy from one body to a second body which is at a lower temperature. Thus heat is much like work. To distinguish them, *heat* is defined as a *transfer of energy due to a difference in temperature*, whereas work results from a transfer of energy that is not because of a temperature difference.

In discussing thermodynamics we shall often refer to a particular system. A *system* is any object or set of objects that we wish to consider. Everything else in the universe is called the "environment." There are several kinds of systems. A *closed system* is one whose mass is constant. In an *open system*, mass may enter or leave. Many systems we study in physics are closed systems; but many systems, including plants and animals, are open systems since they exchange materials with the environment (food, oxygen, waste products). A closed system is said to be *isolated* if no energy in any form passes across its boundaries; otherwise it is not isolated.

12-1 ∎ The first law of thermodynamics

In Section 11-2, we defined the internal energy of a system as the sum total of all the energy of the molecules of the system. We would expect that the internal energy of a system would be increased either by doing work on the system, or by adding heat to it; and the internal energy would be decreased if heat flows out of the system or if work is done by the system on something else.

Thus, it is reasonable to propose an important law: the change in internal energy of a closed system, ΔU, will be given by

$$\Delta U = Q - W \qquad (12\text{-}1)$$

where Q is the heat *added* to the system and W is the work done *by* the system. If the work W done by the system is negative, which is equivalent to saying that positive work is done on the system by the environment, then U will increase. (Of couse we could have defined W as the work done on the system, in which case there would be a plus sign in Eq. 12-1; but it is conventional to define W and Q as we have done.) Similarly, if heat leaves the system, Q is negative. Equation 12-1 is known as the **first law of thermodynamics.** It is one of the great laws of physics, and its validity rests on experiment (such as Joule's) in which no exceptions have been seen. Since Q and W represent energy transferred into or out of the system, the internal energy changes accordingly; thus the first law of thermodynamics is a statement of the *law of conservation of energy*. It is worth noting that the conservation of energy law was not formulated until the nineteenth century, for it depended on the interpretation of heat as a transfer of energy.

Equation 12-1 applies to a closed system. It also applies to an open system if we take into account the change in internal energy due to the increase or decrease in the amount of matter. For an isolated system, $W = Q = 0$ and $\Delta U = 0$.

Although the kilocalorie is still sometimes used as the unit of heat, it is becoming more common to express heat in the SI unit of energy, the joule: $1 \text{ kcal} = 4.18 \times 10^3 \text{ J}$ or $1 \text{ J} = 2.39 \times 10^{-4} \text{ kcal}$.

A given system, in a particular state, can be said to have a certain amount of internal energy U. This cannot be said of heat or work. A system in a given state does not "have" a certain amount of heat or work. Rather, when work is done on a system (such as compressing a gas), or when heat is added or removed from a system, the state of the system *changes*. Thus work and heat are involved in *thermodynamic processes* which can change the system from one state to another; they are not characteristic of the state itself, as are pressure, volume, temperature, mass, and internal energy.

12-2 ∎ First law of thermodynamics applied to some simple systems

Let us analyze some simple processes in the light of the first law. First we consider an idealized process that is carried out at constant temperature. Such a process is called an **isothermal** process (from the Greek meaning "same temperature"). If the system is an ideal gas, the process follows the curve AB on the PV diagram shown in Fig. 12-1. This is the curve for $PV = $ constant, which is Boyle's law (see Chapter 10). We assume that the gas is enclosed in a container fitted with a movable piston. The gas is assumed to be in contact with a *heat reservoir* (a body whose mass is so large that, ideally, its temperature does not change significantly when heat is exchanged with our system), and the process is done very slowly to make certain that all of the gas stays at the same constant temperature. If the gas is initially at point A in Fig. 12-1, an amount of heat Q must be added to bring the system to point B. If the temperature is to remain constant, the gas must expand and do an amount of work W on the environment (it exerts a force through a distance on the piston). Since the temperature is kept constant, the internal energy does not change (see Eq. 11-1). Hence, by the first law

FIGURE 12-1

PV diagram for an ideal gas undergoing an isothermal process.

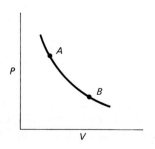

of thermodynamics (Eq. 12-1), $W = Q$: the work done by the gas equals the heat put into it.

An **adiabatic** process is one in which no heat is allowed to flow into or out of the system: $Q = 0$. This situation can occur if the system is extremely well insulated, or the process happens so quickly that heat has no time to flow in or out. The expansion of gases in an internal combustion engine is one example of a process that is very nearly adiabatic. An adiabatic expansion of an ideal gas follows a curve like that labeled AC in Fig. 12-2. Since $Q = 0$, we have from Eq. 12-1 that $\Delta U = -W$. That is, the internal energy decreases; hence the temperature decreases as well. This should be evident in Fig. 12-2, where the product $PV(=nRT)$ is less at point C than at point B. In an adiabatic compression, work is done on the gas; hence the internal energy increases and the temperatures rises. In a diesel engine, air is compressed adiabatically by a factor of 15 or more; the temperature rise is so great that, when the fuel is injected, the mixture ignites.

Isothermal and adiabatic processes are just two possible processes that can occur. In these, and in all other processes, the first law of thermodynamics holds.

It is often valuable to calculate the work done in a process. If the pressure is kept constant during a process (such a process is called *isobaric*, meaning "same pressure"), the work done is easily calculated. For example, if the gas in Fig. 12-3 expands slowly against the piston, the work done by the gas to raise the piston is the force F times the distance d. But the force is just the pressure P of the gas times the area A of the piston, $F = PA$. Thus

$$W = Fd$$
$$= PAd$$

or

$$W = P\,\Delta V, \tag{12-2}$$

where $\Delta V = Ad$ is the change in volume of the gas. This equation also holds if the gas is compressed, in which case ΔV is negative; W is then negative, which indicates that positive work is done *on* the gas. Equation 12-2 is also valid for liquids and solids, as long as the pressure is constant during the process.

Suppose that a gas, instead of being taken isothermally from point A to point B, is taken by way of the path ADB as shown in Fig. 12-4. In going from A to D, the gas does no work since the volume does not change; but in going from D to B, the gas does work equal to $P_B(V_B - V_A)$. This is the total work done in the process.

If the pressure varies during a process, such as for the isothermal process AB in Fig. 12-1, Eq. 12-2 cannot be used directly. A rough estimate can be obtained, however, by using an "average" value for P in Eq. 12-2. More accurately, the work done is equal to the area under the PV curve. This is obvious when the pressure is constant, for as can be seen from Fig. 12-5a, the shaded area is just $P_B(V_B - V_A)$ and this is the work done. Similarly, the work done during an isothermal process is equal to the shaded area shown in Fig. 12-5b. The calculation of work done in this case can be carried out using calculus.

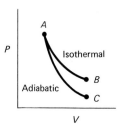

FIGURE 12-2

PV diagram for adiabatic (AC) and isothermal (AB) processes on an ideal gas.

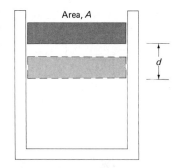

FIGURE 12-3

Work is done on the piston when the gas expands, moving the piston a distance d.

FIGURE 12-4

PV diagram for several processes (see the text).

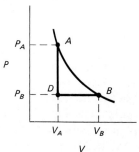

□ **EXAMPLE 12-1** An ideal gas is compressed at a constant pressure of 2.0 atm from 10.0 L to 2.0 L. (In this process some heat flows out and the temperature drops.) Heat is then added to the gas, holding the volume constant, and the pressure and temperature are allowed to rise until the temperature reaches its original value. This total process is shown as *BDA* in Fig. 12-4. Calculate (a) the total work done by the gas in the process and (b) the total heat flow into the gas.

SOLUTION (a) Work is done only in the first part of the compression (*BD*):

$$W = P \Delta V = (2.0 \times 10^5 \text{ N/m}^2)(2.0 \times 10^{-3} \text{ m}^3 - 10.0 \times 10^{-3} \text{ m}^3)$$
$$= -1.6 \times 10^3 \text{ J}.$$

(b) Since the temperatures at the beginning and at the end of the process are the same, there is no change in internal energy, $\Delta U = 0$. From the first law of thermodynamics we have

$$Q = W$$
$$= -1.60 \times 10^3 \text{ J}.$$

Since Q is negative, we know that 1600 J of heat flow out of the gas. Note that this is for the whole process, *BDA*.

(a)

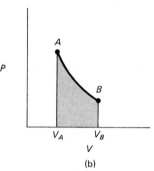
(b)

FIGURE 12-5

Work done by a gas is equal to the area under the *PV* curve.

*12-3 ∎ Human metabolism and the first law

Human beings and other animals do work. Work is done when a person walks or runs, or lifts a heavy object. Work requires energy. Energy is also needed for growth—to make new cells, and to replace old cells that have died. A great many energy-transforming processes occur within an organism, and they are referred to as *metabolism*.

We can apply the first law of thermodynamics

$$\Delta U = Q - W$$

to an organism, say the human body. Work W is done by the body, and this would result in a decrease in the body's internal energy (and temperature), which must be replenished. The body's internal energy is not maintained by a flow of heat Q into the body, however. Normally, the body is at a higher temperature than its surroundings, so heat usually flows *out* of the body. Even on a very hot day when heat is absorbed, the body has no way of utilizing this heat to support its vital processes. What then is the source of energy? It is the internal energy (chemical potential energy) stored in foods. Now in a closed system the internal energy changes only as a result of heat flow or work done; in an open system, such as an animal, internal energy itself can flow into or out of the system. When we eat food, we are bringing internal energy into our bodies directly, which thus increases the total internal energy U in our bodies. This energy eventually goes into work and heat flow from the body according to the first law.

The metabolic rate is the rate at which internal energy is transformed within the body. It is usually specified in kcal/h or in watts. Typical metabolic rates for a variety of human activities are given in Table 12-1 for an "average" 65-kg adult.

TABLE 12-1
Metabolic rate for 65-kg human being

Activity	Approximate Metabolic Rate (Watts)
Sleeping	70
Sitting upright	115
Light activity (eating, dressing, household chores)	230
Moderate work (tennis, walking)	480
Running (15 km/h)	1150
Bicycling (race)	1250

□ **EXAMPLE 12-2** How much energy is transformed in 24 h by a 65-kg person who spends 8.0 h sleeping, 1.0 h at moderate physical labor, 4.0 h in light activity, and 11.0 h working at a desk or relaxing?

SOLUTION Table 12-1 gives the metabolic rate in watts (J/s). Since there are 3600 s in an hour, the total energy transformed is

$$\left[\begin{array}{l} (8.0 \text{ h})(70 \text{ J/s}) + (1.0 \text{ h})(480 \text{ J/s}) \\ + (4.0 \text{ h})(230 \text{ J/s}) + (11.0 \text{ h})(115 \text{ J/s}) \end{array} \right] [3600 \text{ s/h}] = 1.16 \times 10^7 \text{ J}.$$

Since 4.18×10^3 J $= 1$ kcal, this is equivalent to 2800 kcal; so a food intake of 2800 cal would compensate for this energy output.

12-4 ∎ The second law of thermodynamics—introduction

The first law of thermodynamics states that energy is conserved. There are, however, many processes we can imagine that conserve energy but are not observed to occur in nature. For example, when a hot object is placed in contact with a cold object, heat flows from the hotter one to the colder one, never spontaneously the reverse. If heat were to leave the colder and pass to the hotter one, energy would still be conserved. Yet it doesn't happen. As a second example, consider what happens when you drop a rock and it hits the ground. The initial potential energy of the rock changes to kinetic as the rock falls and when the rock hits the ground this energy in turn is transformed into internal energy of the rock and the ground in the vicinity of the impact; the molecules move faster and the temperature rises slightly. But have you seen the reverse happen—a rock at rest on the ground suddenly rise up in the air because the thermal energy of molecules is transformed into kinetic energy of the rock as a whole? Energy would be conserved in this process, yet we never see it happen.

There are many other examples of processes that occur in nature but whose reverse does not. Here are two more. If you put a layer of salt in a jar and cover it with a layer of pepper, when you shake it you get a thorough mixture; no matter how long you shake it, the mixture is unlikely to separate into two layers again. Coffee cups and glasses break spontaneously if you drop them; but they don't go back together spontaneously.

The first law of thermodynamics would not be violated if any of these processes occurred in reverse. To explain this lack of reversibility, scientists in the latter half of the nineteenth century came to formulate a new principle known as the **second law of thermodynamics.** This law is a statement about which processes occur in nature and which do not. It can be stated in a variety of ways, all of which are equivalent. One statement, due to R. J. E. Clausius (1822–1888), is that *heat flows naturally from a hot object to a cold object; heat will not flow spontaneously from a cold object to a hot object.* Since this statement applies to one particular process, it is not obvious how it applies to other processes. A more general statement is needed that will include other possible processes in a more obvious way.

The development of a general statement of the second law was based partly on the study of heat engines. A **heat engine** is any device that changes thermal energy into mechanical work. We now examine heat engines, both from a practical point of view and to show their importance in developing the second law of thermodynamics.

12-5 ∎ Heat engines and refrigerators

It is easy to produce thermal energy by doing work—for example, by simply rubbing your hands together briskly, or indeed in any frictional process. But to get work from thermal energy is more difficult, and the invention of a practical device to do this came only about 1700 with the development of the steam engine.

The basic idea behind any heat engine is that mechanical energy can be obtained from thermal energy only when heat is allowed to flow from a high temperature to a low temperature; in the process some of the heat can then be transformed to mechanical work, as diagrammed in Fig. 12-6. The high and low temperatures, T_H and T_L, are called the *operating temperatures* of the engine. We will be interested only in engines that run in a repeating *cycle* (that is, the system returns repeatedly to its starting point) and thus can run continuously.

The operation of two practical engines, the steam engine and the internal combustion engine (used in most automobiles), is illustrated in Figs. 12-7 and 12-8. Steam engines are of two main types. In the so-called reciprocating type, Fig. 12-7a, the heated steam passes through the intake valve and expands against a piston, forcing it to move; as the piston returns to its original position, it forces the gases out the exhaust valve. In a steam turbine, Fig. 12-7b, everything is essentially the same, except that the reciprocating piston is replaced by a rotating turbine that resembles a paddlewheel with many sets of blades. Most of our electricity today is generated using steam turbines.[†] In a steam engine, the high temperature is obtained by burning coal, oil, or other fuel to heat the steam. In an internal combustion engine, the high temperature is achieved by burning the gasoline–air mixture in the cylinder itself (ignited by the spark plug), Fig. 12-8.

To see why a *temperature difference* is required to run an engine, let us examine the steam engine. In the reciprocating engine, for example (Fig.

[†] Even nuclear power plants utilize steam turbines; the nuclear fuel—uranium—merely serves as fuel to heat the steam.

FIGURE 12-6

Schematic diagram of a heat engine.

Steam

Boiler

Water

Intake valve (open)

Exhaust valve (closed)

Pump

Condenser

(a) Reciprocating type

High-pressure steam, from boiler

Low-pressure steam, exhausted to condenser

(b) Turbine (boiler, condenser not shown)

FIGURE 12-7

Steam engines.

Gas–air mixture from carburetor

Both valves closed

Both valves closed

Both valves closed

To exhaust pipe

Intake valve (open)

Exhaust valve (closed)

Cylinder

Spark plug (firing)

Rings

Exhaust valve (open)

Piston

Crankshaft

Connecting rod

(a) Intake   (b) Compression   (c) Ignition   (d) Expansion (power stroke)   (e) Exhaust

FIGURE 12-8

Four-cycle internal combustion engine: (a) the gasoline–air mixture flows into the cylinder as the piston moves down; (b) the piston moves upward and compresses the gas; (c) firing of the spark plug ignites the gasoline–air mixture, raising it to a high temperature; (d) the gases, now at high temperature and pressure, expand against the piston in this, the power stroke; (e) the burned gases are pushed out to the exhaust pipe; the intake valve then opens, and the whole cycle repeats.

12-7a), suppose there were no condenser or pump, and that the steam was at the same temperature throughout the system. This would mean that the pressure of the gas being exhausted would be the same as that on intake. Thus, although work would be done by the gas *on* the piston when it expands, an equal amount of work would have to be done *by* the piston to force the steam out the exhaust; hence no net work would be done. In a real engine, the exhausted gas is cooled to a lower temperature and condensed so that the exhaust pressure is less than the intake pressure. Thus, although the piston must do work on the gas to expel it on the exhaust stroke, it is less than the work done by the gas on the piston during the intake. So a net amount of work can be obtained—but only if there is a difference of temperature. Similarly, in the gas turbine if the gas were not cooled, the pressure on each side of the blades would be the same; by cooling the gas on the exhaust side, the pressure on the front side of the blade is greater and hence the turbine turns.

The efficiency, e, of any heat engine can be defined as the ratio of the work it does, W, to the heat input at the high temperature, Q_H (Fig. 12-6):

$$e = \frac{W}{Q_H}.$$

This is a very practical definition since W is the output (what you get from the engine) whereas Q_H is what you put in and pay for in burned fuel. Since energy is conserved, the heat input Q_H must equal the work done plus the heat that flows out at the low temperature (Q_L):

$$Q_H = W + Q_L.$$

Thus $W = Q_H - Q_L$ and the efficiency of an engine is

$$e = \frac{W}{Q_H} = \frac{Q_H - Q_L}{Q_H}$$

$$= 1 - \frac{Q_L}{Q_H}. \tag{12-3}$$

It is clear from Eq. 12-3 that the efficiency will be greater if Q_L can be made small.

To see how to increase efficiency, the French scientist Sadi Carnot (1796–1832) examined the characteristics of an ideal engine (now called a Carnot engine). Each of the processes of heat addition and exhaust, of gas expansion or compression, were considered to be done *reversibly*. That is, each of the processes (say, during expansion of the gases against a piston) was done so slowly that the process could be considered a series of equilibrium states, and the whole process could be done in reverse with no change in the magnitude of work done or heat exchanged. A real process, on the other hand, would occur more quickly; there would be turbulence present in the gas, friction would be present, and so on; because of these factors, a real process cannot be done precisely in reverse—the turbulence would be different and the heat lost to friction would not reverse itself. Thus real processes are called *irreversible*. Carnot found, in essence, that for an idealized reversible engine the efficiency could be written in terms of the operating temperatures of the engine, T_H and T_L (where $T_H > T_L$), specified in degrees Kelvin:

$$e = \frac{T_H - T_L}{T_H} = 1 - \frac{T_L}{T_H}. \tag{12-4}$$

FIGURE 12-9

Schematic diagram of a hypothetical perfect heat engine in which all the heat input is used to do work. Such a perfect heat engine is not possible to construct.

Real engines cannot have an efficiency even this high because of losses due to friction and the like. Real engines that are well designed reach 60 to 80 percent of the Carnot efficiency. (Note that to give the efficiency in percent, Eqs. 12-3 and 12-4 must be multiplied by 100 percent.)

EXAMPLE 12-3   A steam engine operates between 500°C and 270°C. What is the maximum possible efficiency of this engine?

SOLUTION   We must first change the temperature to kelvins. Thus $T_H = 773$ K and $T_L = 543$ K, and

$$e = 1 - \frac{543}{773} = 0.30.$$

Thus the maximum (or Carnot) efficiency is 30 percent. Realistically, an engine might attain 0.70 of this value or 21 percent. Note in this example that the exhaust temperature is still rather high, 270°C. Steam engines are often arranged in series so that the exhaust of one engine is used as intake by a second or third engine.

It is quite clear from Eq. 12-4 that at normal temperatures, a 100 percent efficient engine is not possible. Only if the exhaust temperature, T_L, were at absolute zero could 100 percent efficiency be obtained; but this is a practical (as well as theoretical) impossibility.[†] Thus it is possible to state that *no device is possible whose sole effect is to transform a given amount of heat completely into work.* That is, there can be no perfect (100 percent efficient) heat engine such as the one diagrammed in Fig. 12-9. This statement is another way of expressing the *second law of thermodynamics*. If the second law were not true, so that a perfect engine could be built, some rather remarkable things could happen; for example, if the engine of a ship did not need a low-temperature reservoir to exhaust heat into, the ship could sail across the ocean using the vast resources of the internal energy of the ocean water; indeed, we would have no fuel problems at all!

The operating principle of a **refrigerator,** or other **heat pump** (such as one to produce a flow of heat into or out of a house—the latter is called an air conditioner) is just the reverse of a heat engine. As diagrammed in Fig. 12-10, by doing work W, heat is taken from a low-temperature region, T_L (inside a refrigerator, say), and a greater amount of heat is exhausted at a high temperature, T_H (the room). You can often feel this heat blowing out beneath a refrigerator. The work W is usually done by a compressor motor which compresses a fluid, as illustrated in Fig. 12-11. A perfect refrigerator—one in which no work is required to take heat from the low-temperature region to the high-temperature region—is not possible. This is another way to state the *second law of thermodynamics*. In fact this was implicit in our first

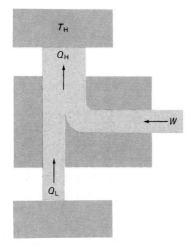

FIGURE 12-10

Schematic diagram of the operation of a refrigerator or air conditioner.

[†] It seems, from careful experimentation, that absolute zero is unattainable. This result is known as the *third law of thermodynamics*.

Refrigeration box
Low-pressure vapor
High-pressure vapor
Condenser
B
Sensor
Cool fluid
Compressor
Hot fluid
Low pressure
High pressure
Expansion valve

FIGURE 12-11

Typical refrigeration system.

statement of the second law of thermodynamics in Section 12-4: that heat does not flow spontaneously from a cold object to a hot object. To accomplish such a task, work must be done. Thus, *there can be no perfect refrigerator.*

12-6 ∎ Entropy and the second law of thermodynamics

We have seen several aspects of the second law of thermodynamics, but we have not yet arrived at a general statement of it. Indeed, it was not until the latter half of the nineteenth century that the *second law of thermodynamics* was finally stated in a general way, namely in terms of a quantity called **entropy,** introduced by Clausius in the 1860s. Entropy, unlike heat, is a function of the state of a system. (In the next section, we will see that entropy can be interpreted as a measure of the order or disorder of a system.) When we deal with entropy—as with potential energy—it is the *change* in entropy during a process that is important, not the absolute amount. According to Clausius, the change in entropy S of a system when an amount of heat Q is *added* to it by a reversible process at constant temperature is given by

$$\Delta S = \frac{Q}{T}, \tag{12-5}$$

where T is the Kelvin temperature. If the temperature varies during a process, a summation of the heat flow over the changing temperature can be calculated using calculus or a computer. However, if the temperature change is not too great, a reasonable approximation can be made using the average value of the temperature, as indicated in the next example.

EXAMPLE 12-4   A sample of 50.0 kg of water at 20.0°C is mixed with 50.0 kg of water at 24.0°C. Calculate the change in entropy.

SOLUTION   The final temperature of the mixture will be 22.0°C, since we started with equal amounts of water. A quantity of heat $Q = mc\,\Delta T = (50.0 \text{ kg})(1.00 \text{ kcal/kg·K})(2.0 \text{ C}°) = 100 \text{ kcal}$ flows out of the hot water as it cools down from 24°C to 22°C, and into the cold water as it warms from 20°C to 22°C. The total change in entropy, ΔS, will be the sum of the changes in entropy of the hot water, ΔS_H, and that of the cold water, ΔS_C:

$$\Delta S = \Delta S_H + \Delta S_C.$$

We now make use of Eq. 12-5; for the hot water we use an average temperature of 23.0°C (296 K) and for the cold water an average temperature of 21.0°C (294 K). Thus

$$\Delta S_H = -\frac{100 \text{ kcal}}{296 \text{ K}} = -0.338 \text{ kcal/K}$$

and

$$\Delta S_C = \frac{100 \text{ kcal}}{294 \text{ K}} = 0.340 \text{ kcal/K}.$$

Note that S_H decreased since heat flowed out of the hot water. But S_C increased by a greater amount, and the total change in entropy was

$$\Delta S = \Delta S_H + \Delta S_C = -0.338 \text{ kcal/K} + 0.340 \text{ kcal/K}$$
$$= +0.002 \text{ kcal/K}.$$

In this example we see that although the entropy of one part of the system decreased, the entropy of the other part increased by a greater amount so that the net change in entropy of the whole system is positive. This result, which we have calculated for a specific case in the above example, has been found to hold in all other cases tested. That is, the total entropy is found to increase in all natural processes. The second law of thermodynamics can be stated in terms of entropy as: *The entropy of an isolated system never decreases. It can only stay the same or increase.* Thus, in any natural process the change in entropy ΔS is greater than zero:

$$\Delta S > 0. \tag{12-6}$$

If the system is not isolated, then the change in entropy of the system, ΔS_s, plus the change in entropy of the environment, ΔS_{env}, must be greater than or equal to zero:

$$\Delta S = \Delta S_s + \Delta S_{env} > 0. \tag{12-7}$$

This, then, is the *general statement of the second law of thermodynamics:*

■ **The total entropy of any system plus that of its environment increases as a result of any natural process.**

Although the entropy of one part of the universe may decrease in any process (see Example 12-4), the entropy of some other part of the universe always increases by a greater amount, so the total entropy always increases.

Now that we finally have a quantitative general statement of the second law of thermodynamics, we can see that it is an unusual law. It differs considerably from other laws of physics, which are typically equalities (such as $F = ma$) or conservation laws (such as for energy and momentum). The second law introduces a new quantity, the entropy, S, but does not tell us it is conserved. Quite the opposite. Entropy is not conserved; it always increases in time.

12-7 ■ Order to disorder

The concept of entropy, as we have discussed it so far, may seem rather abstract. To get a feel for the concept of entropy we can relate it to the concepts of *order* and *disorder*. In fact, the entropy of a system can be considered a *measure of the disorder of the system.* Then the second law of thermodynamics can be stated simply as:

■ **Natural processes tend to move toward a state of greater disorder.**

Exactly what we mean by disorder may not always be clear; so to make it clear, we now consider a few examples. Some of these will show us how this very general statement of the second law actually applies beyond what we usually consider as thermodynamics.

Let us first look at the simple processes mentioned in Section 12-4. A jar containing separate layers of salt and pepper is more orderly than when the salt and pepper are all mixed up. Shaking a jar containing separate layers results in a mixture, and no amount of shaking brings the layers back again. The natural process is from a state of relative order (layers) to one of relative disorder (a mixture), not the reverse. That is, disorder increases. Similarly, a solid coffee cup is a more "orderly" object than the pieces of a broken cup. Cups break when they fall, but they do not spontaneously mend themselves. Again, the normal course of events is an increase of disorder.

When a hot object is put in contact with a cold object, heat flows from the high temperature to the low until the two objects reach the same intermediate temperature. At the beginning of the process we can distinguish two classes of molecules: those with a high average kinetic energy and those with a low average kinetic energy. After the process, all the molecules are in one class with the same average kinetic energy, and we no longer have the more orderly arrangement of molecules in two classes. Order has gone to disorder. To see this more clearly, note that the separate hot and cold objects could serve as the hot- and cold-temperature regions of a heat engine and thus could be used to obtain useful work. But once the two objects are put in contact and reach the same temperature, no work can be obtained. Disorder has increased.

This example illustrates the general concept that an increase in entropy corresponds to an increase in disorder. (We discuss this connection quantitatively in Section 12-9). In general, we associate disorder with uniformity, with randomness: salt and pepper in layers is more orderly than a random mixture; a neat stack of numbered pages is more orderly than pages strewn randomly about on the floor.

The remaining example of those we discussed earlier was that of a stone falling to the ground, its kinetic energy being transformed to thermal energy; and we noted that the reverse never happens: a stone never rises into the air of its own accord. This is another example of order changing to disorder. For although thermal energy is associated with the disorderly random motion of molecules, the molecules in the falling stone all have the same velocity downward in addition to their own random velocities. Thus the more orderly kinetic energy of the stone is changed to disordered thermal energy when it strikes the ground. Disorder increases in this process, as it does in all processes that occur in nature.

These last two examples of order going to disorder illustrate another very important aspect of the second law of thermodynamics—that thermal energy can be considered as a lower and less useful form of energy than mechanical energy. Mechanical energy (such as the PE of a rock before it falls, or the output of a heat engine when there is a difference in temperature) can be used to do work. But after it is changed to thermal energy, the possibility of doing work is lost—the energy is no longer available to do work. Thus we can consider mechanical and other forms of energy as more ordered than thermal energy is. In natural processes, energy is in a sense **degraded,** going from more orderly forms to less orderly forms, eventually ending up as thermal energy. This illustrates another important aspect of the second law of thermodynamics—that *in any natural process, some energy becomes unavailable to do useful work.*

*12-8 ▮ Evolution and growth, "time's arrow," "heat death"

An interesting example of the increase in entropy relates to biological evolution and to growth of organisms. Clearly, a human being is a highly ordered organism. The process of evolution from the early macromolecules and simple forms of life to *Homo sapiens* is a process of increasing order. So is the development of an individual from a single cell to a grown person. Do these processes violate the second law of thermodynamics? No, they do not. In the processes of evolution and growth, and even during the mature life of an individual, waste products are eliminated. These small molecules that remain as a result of metabolism are simple molecules without much order. Thus they represent relatively great disorder or entropy. Indeed, the total entropy of the molecules cast aside by organisms during the processes of evolution and growth is greater than the decrease in entropy associated with the order of the growing individual or evolving species.

Another aspect of the second law of thermodynamics is that it tells us in which *direction* processes go. If you were to see a film being run backward, you would undoubtedly be able to tell that it *was* run backward. For you would see odd occurrences, such as a broken coffee cup rising from the floor and reassembling on a table, or a torn balloon suddenly becoming whole again and filled with air. We know these things don't happen in real life; they are processes in which order increases—or entropy decreases. They violate the second law of thermodynamics. When watching a movie (or imagining that time could go backward), we are tipped off to a reversal of time by observing whether entropy is increasing or decreasing. Hence, entropy has been called "time's arrow"; for it tells in which direction time is going.

A prediction of the second law of thermodynamics is that as time goes on, the universe will approach a state of maximum disorder. Matter will become a uniform mixture, heat will have flowed from high-temperature regions to low-temperature regions until the whole universe is at one temperature. No work can then be done. All the energy of the universe will have become degraded to thermal energy. All change will cease. This, the so-called *heat death* of the universe, has been much discussed by philosophers. This final state seems an inevitable consequence of the second law of thermodynamics, although very far in the future. Yet it is based on the assumption that the universe is finite, which cosmologists are not really sure of. Furthermore, there is some question as to whether the second law of thermodynamics, as we know it, actually applies in the vast reaches of the universe. The answers are not yet in.

*12-9 ∎ Statistical interpretation of entropy and the second law

The ideas of entropy and disorder are made clearer with the use of a statistical or probabilistic analysis of the molecular state of a system. This statistical approach, which was first applied toward the end of the nineteenth century by Ludwig Boltzmann (1844–1906), makes a distinction between the "macrostate" and the "microstate" of a system. The *microstate* of a system would be specified when the position and velocity of every particle (or molecule) is given. The *macrostate* of a system is specified by giving the macroscopic properties of the system—the temperature, pressure, number of moles, and so on. In reality, we can know only the macrostate of a system. There are generally far too many molecules in a system to be able to know the velocity and position of every one at a given moment. Nonetheless, it is important to recognize that a great many different microstates can correspond to the *same* macrostate.

Let us take a simple example. Suppose you repeatedly shake four coins in your hand and drop them on the table. The number of heads and the number of tails that appear on a given throw is the macrostate of this system. Specifying each coin as being a head or a tail is the microstate of the system. In the following table we see the number of microstates that correspond to each macrostate:

Macrostate	Possible Microstates (H = heads, T = tails)	Number of Microstates
4 heads	H H H H	1
3 heads, 1 tail	H H H T, H H T H, H T H H, T H H H	4
2 heads, 2 tails	H H T T, H T H T, T H H T, H T T H, T H T H, T T H H	6
1 head, 3 tails	T T T H, T T H T, T H T T, H T T T	4
4 tails	T T T T	1

A basic assumption behind the statistical approach is that *each microstate is equally probable*. Thus the number of microstates that give the same macrostate corresponds to the relative probability of that macrostate occurring. The macrostate of two heads and two tails is the most probable one in our case of tossing four coins; out of the total of 16 possible microstates, six correspond to two heads and two tails, so the probability of throwing two heads and two tails is 6 out of 16, or 38 percent. The probability of throwing one head and three tails is 4 out of 16, or 25 percent; the probability of four heads is only 1 in 16, or 6 percent. Of course if you threw the coins 16 times, you might not find that two heads and two tails appears exactly 6 times, or four tails exactly once. These are only probabilities or averages. But if you made 1600 throws, very nearly 38 percent of them would be two heads and two tails. The greater the number of tries, the closer are the percentages to the calculated probabilities.

If we consider tossing more coins, say 100, the relative probability of throwing all heads (or all tails) is greatly reduced. There is only one microstate corresponding to all heads. For 99 heads and 1 tail, there are 100 microstates since each of the coins could be the one tail. The relative probabilities for other macrostates are given in Table 12-2. There are a total of about 10^{30} microstates possible.[†] Thus the relative probability of finding all heads is 1 in 10^{30}, an incredibly unlikely event! The probability of obtaining 50 heads and 50 tails (see Table 12-2) is $1.0 \times 10^{29}/10^{30} = 0.10$ or 10 percent. The probability of obtaining between 45 and 55 heads is 90 percent.

Thus we see that as the number of coins increases, the probability of obtaining an orderly arrangement (all heads or all tails) becomes extremely unlikely. The least orderly arrangement (half heads, half tails) is the most probable and the probability of being within a certain percentage (say 5 percent) of the most probable greatly increases as the number of coins increases. These same ideas can be applied to the molecules of a system. For example the most probable state of a gas (say the air in a room) is one in which the molecules take up the whole space and move about randomly; this corresponds to the Maxwellian distribution, Fig. 12-12a (and see Chapter 10). On the other hand, the very orderly arrangement of all the molecules located in one corner of the room and all moving with the same velocity (Fig. 12-12b) is extremely unlikely.

From these examples, it is clear that probability is directly related to disorder and hence to entropy. That is, the most probable state is the one with greatest entropy or greatest disorder and randomness. Boltzmann showed that, consistent with Clausius's definition ($\Delta S = Q/T$), the entropy of a system in a given (macro) state can be written[‡]

$$S = 2.3\, k \log W, \qquad (12\text{-}8)$$

where k is Boltzmann's constant ($k = 1.38 \times 10^{-23}$ J/K) and W is the number of microstates corresponding to the given macrostate; that is, W is proportional to the probability of occurrence of that state.

In terms of probability, the second law of thermodynamics—which tells us that entropy increases in any process—reduces to the statement that those processes occur which are most probable. The second law thus becomes a trivial statement. However, there is an additional element now. The second law in terms of probability does not *forbid* a decrease in entropy. Rather, it says the probability is extremely low. It is not impossible that salt

[†] Each coin has two possibilities, heads or tails. Then the possible number of microstates is $2 \times 2 \times 2 \times \ldots = 2^{100}$, which, using a calculator, gives 1.27×10^{30}.

[‡] The factor 2.3 comes from the fact that this equation is usually written $S = k \ln W$. Here the "natural logarithm," ln, to the base e, where $e = 2.718\ldots$, is used instead of the usual base 10. Since $\log e = \log 2.718 = 2.3$, $\ln W = 2.3 \log_{10} W$.

TABLE 12-2

Probabilities of various macrostates for 100 coin tosses

Macrostate		Number of Microstates, W
Heads	Tails	
100	0	1
99	1	1.0×10^2
90	10	1.7×10^{13}
80	20	5.4×10^{20}
60	40	1.4×10^{28}
55	45	6.1×10^{28}
50	50	1.0×10^{29}
45	55	6.1×10^{28}
40	60	1.4×10^{28}
20	80	5.4×10^{20}
10	90	1.7×10^{13}
1	99	1.0×10^2
0	100	1

(a)   (b)

FIGURE 12-12

(a) Most probable distribution of molecular speeds in a gas (Maxwellian, or random); (b) orderly, but highly unlikely, distribution of speeds in which all molecules have nearly the same speed.

and pepper should separate spontaneously into layers, or that a broken tea cup should mend itself. It is even possible that a lake should freeze over on a hot summer day (that is, heat flow out of the cold lake into the warmer surroundings). But the probability for such events occurring is extremely small. In our coin examples above, we saw that increasing the number of coins from 4 to 100 reduced drastically the probability of large deviations from the average, or most probable, arrangement. In ordinary systems we are dealing with incredibly large numbers of molecules: in 1 mol alone there are 6×10^{23} molecules. Hence the probability of deviation far from the average is incredibly tiny. For example, it has been calculated that the probability that a stone resting on the ground should transform 1 cal of thermal energy into mechanical energy and rise up into the air is much less likely than the probability that a group of monkeys typing randomly would by chance produce the complete works of Shakespeare.

*12-10 ▌ Energy resources; thermal pollution

When we speak in everyday life of energy usage, we are speaking of the transformation of energy from one form to another form that we want to use. For space heating of homes and buildings, the direct burning of fuels such as gas, oil, or coal releases energy stored as potential energy of the molecules. For many transformations, however, a *heat engine* is required, such as to run cars, aircraft, and other vehicles and, very importantly, to generate electricity.

We will discuss various means for producing electricity in a moment. But first let us look at two types of pollution associated with heat engines: air pollution and thermal pollution. Air pollution can result from burning of any fossil fuels (coal, oil, gas) such as in industrial furnaces for smelting and electric generating plants. The internal combustion engines of automobiles are especially polluting because the burning takes place so quickly that complete combustion does not take place and more noxious gases are thus produced. To help reduce air pollution, special devices are used (for example, to reburn the exhaust gases).

Another type of environmental pollution is **thermal pollution.** Every heat engine, from automobiles to power plants, exhausts heat to the environment (Q_L in Fig. 12-6). Most electricity-producing power plants today make use of a heat engine to transform thermal energy into electricity, and the exhaust heat is generally absorbed by a coolant such as water. If the engine is run efficiently (at best, 30 to 40 percent today), the temperature T_L (see Eq. 12-4) must be kept as low as possible. Hence a great deal of water must flow as coolant through a power plant. The water is usually obtained from a nearby river or lake, or from the ocean. As a result of the transfer of heat to the water, its temperature rises. This can cause great damage to aquatic life in the vicinity, in large part because the warmed water holds less dissolved oxygen. The lack of oxygen can adversely affect fish and other organisms and at the same time may encourage excessive growth of other (perhaps alien) organisms, such as algae, thus disrupting the ecology of an area. Another way of exhausting waste heat is to discharge it into the atmosphere by means of large cooling towers. Unfortunately this method may also have an environmental effect, for the heated air can alter the weather of a region. Although careful controls may eventually reduce air pollution to an acceptable level, thermal pollution cannot be avoided. All we can do, in light of the second law of thermodynamics, is use less energy and try to build more efficient engines. One hope is that eventually we can develop an engine that can transform nuclear energy, say, directly into electricity without using a heat engine. This would greatly improve efficiency and reduce pollution, but such an engine is not yet a real possibility.

We now discuss some of the practical aspects of electric power generation, and various means of doing so. Most of the electricity produced in the United States at the present time makes use of a heat engine coupled with an electric generator; there are a few other possibilities, and we will also discuss them in this section. Electric generators are devices that transform mechanical energy, usually rotational kinetic energy of turbines containing many blades (Fig. 12-13), into electric energy. (Details of how electric generators work will be discussed in Chapter 20.) The various means to turn the turbine will now be discussed, along with some of the advantages and disadvantages of each.

Fossil-fuel steam plants. At such a plant, coal, oil, or natural gas is burned to boil water and produce high-pressure steam that turns the turbine. The basic principles of this sort of steam engine were covered in Section 12-5. The advantages of such plants are that we know how to build them and they are not too expensive to run (yet). The disadvantages are that the products of combustion create air pollution; like all heat engines, their efficiency is limited (30 to 40 percent is typical); the waste heat produces thermal pollution; the extraction of the raw materials can be devastating to the land, especially in coal strip mining and oil-shale recovery; accidents are possible, such as oil spills in the sea; and there may not be a great deal of fossil fuel left—estimates range from a supply lasting a few centuries to only a few decades.

Nuclear power. Two nuclear processes release energy: fission and fusion. In fission, the nuclei of uranium or plutonium atoms are made to split ("fission") with the release of energy. In fusion, energy is released when small nuclei, such as those of hydrogen, combine ("fuse"); these processes are described in greater detail in Chapter 30. The fission process is used in all present nuclear power plants, since fusion has not yet been controlled. Nuclear energy is used to heat steam just as fossil fuels are. A nuclear power plant is thus essentially a steam engine using uranium as its fuel; it suffers from the low efficiency characteristic of all heat engines and the accompanying thermal pollution. Although nuclear power plants produce practically no air pollution (unless an accident occurs), they do present problems: radioactive substances are produced that are difficult to dispose of; a serious accident could result in the release of radioactive material into the air; there is the possibility of diversion of nuclear material by terrorists; and the fuel supply is limited. However, the energy produced per kilogram of fuel is very large, and extraction is less damaging to the land than for fossil fuels. The fusion process, on the other hand, seems to have fewer disadvantages and has the advantage of a vast supply of fuel—the hydrogen in the water molecules (H_2O) in the oceans. Unfortunately, this process cannot yet be controlled sufficiently, but holds promise for the future. (See Chapter 30.)

FIGURE 12-13

Mechanical or heat energy is transformed to electric energy with a turbine and generator.

Geothermal power. Both fossil-fuel and nuclear plants heat water to steam for the steam turbine. Natural steam can be obtained from the earth itself. In many places, water beneath the ground is in contact with the hot interior of the earth and is raised to high temperature and pressure. It comes to the surface as hot springs, geysers, or steam vents. We can use not only natural vents but can drill down to trapped steam beds. Already a large geothermal plant operates at the Geysers in northern California. A similar plant has been successfully operating for over 60 years in Italy, and a number of others are functioning in various areas around the world. There are a limited number of sites where high temperature and pressure exist naturally. Another possibility is to drill two parallel wells down to hot dry rock in contact with the earth's interior and to pass cold water under pressure down one well, the heated water (or steam) returning up the other well. Geothermal energy appears to be clean in that it produces little air pollution, although there is some (nonsteam) gas emission; there is also the thermal pollution of the spent hot water; the mineral content of the water (often high) may not only be environmentally polluting but can be corrosive to the parts of the apparatus itself. Nonetheless, it is a reasonably inexpensive means of production, and may hold some promise.

Tropical seas. There is a difference in temperature of perhaps 20 C° between the surface and the depths of the sea in the tropics; although not yet tried, a number of designs have been suggested in which this temperature difference could be used to drive a heat engine. (Remember that a heat engine must exhaust heat at a lower temperature than that of the heat input.) The working fluid, of course, could not be steam; rather, it would have to be a substance with a lower boiling point that could drive the turbines. Before this method is tried, many difficulties need to be overcome. The efficiency (Eq. 12-4) would be low because of the small temperature difference (at best, 7 percent); losses due to heat transfer must be kept very low, so the efficiency does not drop to zero. The ocean environment causes problems, such as corrosion and fouling by biological organisms caught in the intake water. On a large scale, the ocean currents of the world could be affected. Yet there would be little problem with air pollution or radioactive disposal, so the advantages, if done on a limited scale, might outweigh disadvantages.

Hydroelectric power plants. Hydroelectric plants use falling water instead of steam to turn the turbines. They are usually located at the base of a dam. Hydroelectric power plants produce practically no air or water pollution; furthermore, they are nearly 100 percent efficient, since very little waste heat is produced. However, there are not many good locations remaining in which to build dams, and the resulting reservoirs inundate land that may be fertile or of great scenic beauty. Another form of water power is the following.

Tidal power. The earth's tides have been put to work at a plant in France (the details are given in problem 39). As shown in Fig. 12-14, a basin behind a dam is filled at high tide and the water is released at low tide to drive turbines. At the next high tide the reservoir is filled again and the in-rushing water also turns turbines. Good sites for tidal power (where there is a large difference between high and low tides) are not plentiful and would require large dams across natural or artificial bays. The rather abrupt changes of water level could have an effect on wildlife, but otherwise tidal power

FIGURE 12-14
Tidal power plant. Turbines are located inside dam.

High tide

Low tide: Water is beginning to flow out of basin to ocean, driving turbines.

Water level equalized.

High tide: Water is allowed to flow back into basin, driving turbines.

Water level equalized.

would seem to have a minimal environmental effect. Unfortunately, reasonable estimates of available sites indicate that tidal power could at best produce only a small fraction of the world's energy needs. Nonetheless, since our future energy needs may have to come from a variety of small sources, rather than mainly one, as is the case now (fossil fuels), tidal power could become more important.

Wind power. Windmills were once a very practical device. Their resurrection—on a much grander scale—is a possibility as a means of turning a generator to produce electricity. One specific proposal projects some 300,000 of them, each over 800 ft high with blades 50 ft in diameter, dispersed throughout the midwestern United States where winds are strong and steady. Such an array might produce a good portion of U.S. energy needs. Of course, such a vast project might be something of an eyesore; it also might affect the weather and noise could be a problem. But wind power, on a modest scale, offers possibilities.

Solar power. Many kinds of solar energy are already in use: fossil fuels are the remains of plant life that grew by photosynthesis of light from the sun; hydroelectric power depends on the sun to evaporate water that later comes down as rain; and wind power depends on convection currents produced by the sun heating the atmosphere. But now let us see what we can do with the sun's rays directly. The sun's rays could be concentrated by large mirrors or lenses onto a small surface in order to produce temperatures high enough to produce steam to drive a turbine. Such a system could be used for a home generating system, although a backup system would probably be needed for cloudy days. On a larger scale, large areas of land would be needed to collect sufficient sunlight. The principal disadvantages are that the sun does not shine brightly every day, and not at all at night; thus a safe storage method would be needed. Also, large areas of land are needed for the collectors and concentrators. Although the ubiquitous thermal pollution would exist and the climate might be affected, there would be essentially no air or water pollution and no radioactivity, and the technology would not be too difficult. Another direct user of sunlight is the *solar cell*—or more correctly *photovoltaic cell*—which was developed by the U.S. space program to directly convert sunlight into electricity. Unfortunately, they are very expensive, costing many times per unit of power as ordinary means. However, if the cost should decrease, they would be very desirable, since thermal (and other) pollution would be very low (note that no heat engine is involved). They might be placed on roofs for home use. On a large scale they would again require a large land area since the sun's energy is not very concentrated. Another possibility is to place solar cells and concentrators in orbit around the earth. The sun's radiation would be greater there, before the atmosphere absorbs some of it, and the satellites would spend a minimal time in the earth's shadow. The electricity produced might be fed to a microwave generator and transmitted to receivers on earth.

It should be clear that all forms of energy production have undesirable side effects. Some are worse than others, and not all problems can be anticipated. New forms of energy production will be needed in the future as old reserves of fuel are used up. It is possible that many different methods—including those we have just discussed—will have to be used in the future to satisfy the needs of our energy-consuming society. The problems posed by any one form of energy production might not loom as large since there would be diverse fuel sources available to us. Finally, it is clear that conserving our limited fuel supplies, by avoiding wasteful use of energy, should be of great concern to our society.

▌ Summary

The *first law of thermodynamics* states that the change in internal energy ΔU of a system is equal to the heat added to the system, Q, minus the work, W, done by the system: $\Delta U = Q - W$. This is a statement of the conservation of energy, and is found to hold for all types of processes. Two simple thermodynamic processes are *isothermal*, which is a process carried out at constant temperature; and *adiabatic*, a process in which no heat is exchanged. The work W done by a gas at constant pressure, P, is given by $W = P \Delta V$, where ΔV is the change in volume of the gas.

A *heat engine* is a device for changing thermal energy, by means of heat flow, into useful work. The efficiency of a heat engine is defined as the ratio of the work W done by the engine to the heat input Q_H. Because of conservation of energy, the work output equals $Q_H - Q_L$, where Q_L is the heat exhausted to the environment; hence the efficiency $e = W/Q_H = 1 - Q_L/Q_H$. The upper limit on the efficiency can be written in terms of the upper and lower operating temperatures (in kelvins) of the engine, T_H and T_L; $e = 1 - T_L/T_H$. The operation of refrigerators and air conditioners is the reverse of that of a heat engine: work is done to extract heat from a cold region and exhaust it to a region at a higher temperature.

The *second law of thermodynamics* can be stated in several equivalent ways: (1) heat flows spontaneously from a hot object to a cold one, but not the reverse; (2) there can be no 100 percent efficient heat engine— that is, one that can change a given amount of heat completely into work; and (3) natural processes tend to move toward a state of greater disorder or greater entropy. This last is the most general statement and can be restated as: the total entropy, S, of any system plus that of its environment increases as a result of any natural process: $\Delta S > 0$. Entropy, is a quantitative measure of the disorder of a system. As time goes on, energy is degraded to less useful forms—that is, it is less available to do useful work.

▌ Questions

1. What happens to the internal energy of water vapor in the air that condenses on the outside of a cold glass of water? Is work done or heat exchanged? Explain.

2. Use the conservation of energy to explain why the temperature of a gas increases when it is compressed—say, by pushing down on a cylinder—whereas the temperature decreases when the gas expands.

3. In Fig. 12-2, will more work be done in the isothermal process AB or in the adiabatic process AC? In which process will there be a greater change in internal energy? In which will there be a greater flow of heat?

4. In an isothermal process, 3700 J of work is done by an ideal gas. Is this enough information to tell how much heat has been added to the system? If so, how much?

5. For the processes illustrated in Fig. 12-4, is more work or less work done in the isothermal process AB or in the process ADB?

6. Is it possible for the temperature of a system to remain constant even though heat flows into or out of it? If so, give examples.

7. Discuss how the first law of thermodynamics can apply to metabolism in humans. In particular, note that a person does work W, but very little heat Q is added to the body (rather, it tends to flow out); why then doesn't the internal energy drop drastically in time?

8. Is it possible to cool down a room on a hot summer day by leaving the refrigerator door open?

9. Can mechanical energy ever be transformed completely into heat or internal energy? Can the reverse happen? In each case, if your answer is no, explain why not; if yes, give examples.

10. The oceans contain a tremendous amount of thermal energy. Why, in general, is it not possible to put this energy to useful work?

11. A gas is allowed to expand (*a*) adiabatically and (*b*) iso-

thermally. In each process, does the entropy increase, decrease, or stay the same?

12. Give three examples, other than those mentioned in this chapter, of naturally occurring processes in which order goes to disorder. Discuss the observability of the reverse process.

13. Which do you think has the greater entropy, 1 kg of solid iron or 1 kg of liquid iron? Why?

14. What happens if you remove the lid of a bottle containing chlorine gas? Does the reverse process ever happen? Why or why not?

15. Think up several processes (other than those already mentioned) that would obey the first law of thermo-dynamics, but, if they actually occurred, would violate the second law.

16. Suppose you collect a lot of papers strewn all over the floor and put them in a neat stack; does this violate the second law of thermodynamics? Explain.

17. The first law of thermodynamics is sometimes whimsically stated as, "You can't get something for nothing," and the second law as, "You can't even break even." Explain how these statements could be equivalent to the formal statements.

18. Give three examples of naturally occurring processes that illustrate the degradation of usable energy into internal energy.

▮ Problems

SECTIONS 12-1 AND 12-2

1. (I) One liter of air is cooled at constant pressure until its volume is halved, and then it is allowed to expand isothermally back to its original volume. Draw the process on a PV diagram.

2. (I) In Example 12-1, if the heat lost from the gas in the process BD is 2.18×10^3 J, what is the change in internal energy of the gas?

3. (II) When 539 kcal of heat is added to 1.00 kg of water at 100°C, it is completely changed to steam at 100°C. Calculate (a) the work done in this process (the pressure is 1 atm), and (b) the change in internal energy of the water. One kilogram of steam occupies 1.67 m³ at 100°C and 1 atm.

4. (II) An ideal gas was slowly compressed at constant temperature to one-half its original volume. In the process, 80 kcal of heat was given off. (a) How much work was done (in joules)? (b) What was the change in internal energy of the gas? (Hint: see Eq. 11-1.)

5. (II) An ideal gas is allowed to expand adiabatically to twice its volume. In doing so, the gas does 850 J of work. (a) How much heat flowed into the gas? (b) What is the change in internal energy of the gas? (c) Did its temperature rise or fall?

6. (II) When 3.50×10^4 J of heat is added to a gas enclosed in a cylinder fitted with a light frictionless piston maintained at atmospheric pressure, the volume is observed to increase from 2.0 m³ to 3.2 m³. Calculate (a) the work done by the gas, and (b) the change in internal energy of the gas. (c) Graph this process on a PV diagram.

7. (II) An ideal gas expands at a constant pressure of 7.0 atm from 280 mL to 630 mL. Heat then flows out of the gas, at constant volume, and the pressure and temperature are allowed to drop until the temperature reaches its original value. Calculate (a) the total work done by the gas in the process, and (b) the total heat flow into the gas.

8. (II) When a gas is taken from a to c along the curved path in Fig. 12-15, the work done by the gas is $W = -35$ J and the heat added to the gas is $Q = -63$ J. Along path abc, the work done is $W = -48$ J. (a) What is Q for path abc? (b) If $P_c = \frac{1}{2}P_b$, what is W for path cda? (c) What is Q for path cda? (d) What is $U_a - U_c$? (e) If $U_d - U_c = 5$ J, what is Q for path da?

9. (III) A vertical, steel I-beam at the base of a building is 6.0 m tall, has a mass of 300 kg, and supports a load of 3.0×10^5 N. If the beam's temperature decreases by 4.0°C, calculate the change in its internal energy using the facts that for steel c_p is 0.11 kcal/kg·C° and the coefficient of linear expansion is 11×10^{-6} C°$^{-1}$.

*SECTION 12-3

*10. (I) How much energy would the person of Example 12-2 transform if 2.0 h of sleeping were used instead to read?

*11. (I) What is the average power output of a person who spends 10 h each day sleeping, 4 h at light activity, and 10 h watching television or loafing?

*12. (II) A person decides to lose weight by sleeping $\frac{1}{2}$ h less per day, using the time for light activity. How

FIGURE 12-15

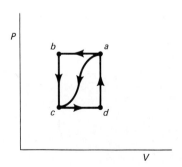

much weight (or mass) can this person expect to lose in 1 year, assuming no change in food intake? Use the fact that 1 kg of fat stores about 9000 kcal of energy.

SECTION 12-5

13. (I) What is the maximum efficiency of a heat engine whose operating temperatures are 480°C and 318°C?

14. (I) A heat engine produces 9500 J of heat while performing 2500 J of useful work. What is the efficiency of this engine?

15. (I) The exhaust temperature of a heat engine is 270°C. What must be the high temperature if the Carnot efficiency is to be 35 percent?

16. (I) It has been suggested that a heat engine could be developed that made use of the fact that the temperature several hundred meters beneath the surface of the ocean is several degrees cooler than the temperature at the surface. In the tropics, the temperatures may be 5°C and 25°C, respectively. What is the maximum efficiency such an engine could have? Why might such an engine be feasible in spite of the low efficiency? Can you imagine any adverse environmental effects that might occur?

17. (II) An engine that operates at half its theoretical (Carnot) efficiency operates between 545°C and 310°C while producing work at the rate of 1000 kW. How much heat is discharged per hour?

18. (II) A Carnot engine performs work at the rate of 500 kW while using 960 kcal of heat per second. If the temperature of the heat source is 620°C, at what temperature is the waste heat exhausted?

19. (II) A heat engine utilizes a heat source at 550°C and has an ideal (Carnot) efficiency of 30 percent. To increase the efficiency to 40 percent, what must be the temperature of the heat source?

20. (II) The burning of gasoline in a car releases about 3.0×10^4 kcal/gal. If a car averages 35 km/gal when driving 90 km/h, which requires 25 hp, what is the efficiency of the engine under those conditions?

21. (III) At a steam power plant, steam engines work in pairs, the output of heat from one being the approximate heat input of the second. The operating temperatures of the first are 700°C and 440°C, and of the second 430°C and 310°C. If the heat of combustion of coal is 2.8×10^7 J/kg, at what rate must coal be burned if the plant is to put out 500 MW of power? Assume the efficiency of the engines is 65 percent of the ideal (Carnot) efficiency.

22. (III) Water is used to cool the power plant in problem 21. If the water temperature is allowed to increase by no more than 7.0 C°, estimate how much water must pass through the plant per hour.

23. (III) A 40-percent efficient power plant puts out 850 MW (megawatts) of work (electrical energy). Cooling towers are used to take away the exhaust heat. If the air temperature is allowed to rise 7.5 C°, what volume of air (km³) is heated per day? Will the local climate

be heated significantly? If the heated air were to form a layer 200 m thick, how large an area would it cover for 24 h of operation? (The heat capacity of air is about 7.0 cal/mol·C° at constant pressure.)

24. (III) A "perfect" refrigerator absorbs heat from the freezer compartment at a temperature of −20°C and exhausts it into the room at 25°C. (a) How much work must be done by the refrigerator to change 0.50 kg of water at 25°C into ice at −20°C? (Hint: think of the refrigerator as a heat engine, and then reverse the process.) (b) If the compressor motor does work at the rate of 200 W, what minimum time is needed to take 0.50 kg of 25°C water and freeze it at 0°C?

SECTION 12-6

25. (I) One kilogram of water is heated from 0°C to 100°C. Calculate approximately the change in entropy of the water.

26. (I) What is the change in entropy of 2.0 kg of water at 100°C when it is vaporized to steam at 100°C?

27. (I) What is the change in entropy of 10 kg of water at 0°C when it is frozen to ice at 0°C?

28. (II) If the water in problem 27 were frozen by being in contact with a great deal of ice at −10°C, what would be the total change in entropy of the process?

29. (II) An aluminum rod conducts 180 cal/s from a heat source maintained at 250°C to a large body of water at 27°C. Calculate the rate entropy increases per unit time in this process.

30. (II) One kilogram of water at 20°C is mixed with 1 kg of water at 80°C in a well-insulated container. Calculate (approximately) the net change in entropy of the system.

31. (II) Two 1400-kg cars are traveling 30 km/h in opposite directions when they collide and are brought to rest. Estimate the change in entropy of the universe as a result of this collision? Assume $T = 20°C$.

*SECTION 12-9

*32. (II) Calculate the probabilities, when you throw two dice, of obtaining (a) a 2, and (b) a 7.

*33. (II) Suppose that you repeatedly shake six coins in your hand and drop them on the table. Construct a table showing the number of microstates that correspond to each macrostate. What is the probability of obtaining (a) three heads and three tails, and (b) six heads?

*34. (III) (a) Suppose you have four coins, all with tails up. You now rearrange them so two heads and two tails are up. What was the change in entropy of the coins? (b) Suppose your system is the 100 coins of Table 12-2; what is the change in entropy of the coins if they are mixed randomly initially, 50 heads and 50 tails, and you arrange them so all 100 are heads? (c) Compare these entropy changes to ordinary thermodynamic entropy changes.

*35. (I) Solar cells can produce about 40 W of electricity per square meter of surface area if directly facing the sun. How large an area is needed to supply the needs of a house that requires 100 kWh/day? Would this fit on the roof of an average house? (Assume that the sun shines 12 h/day.)

*36. (II) Water falls 80 m over a dam at a rate of 15,000 kg/s. How many megawatts of electric power could be produced by a power plant using this energy?

*37. (II) One way of storing energy for use during peak demand periods is to pump water to a high reservoir when the demand is low, and then release it to drive turbines when needed. Suppose that water is pumped to a lake 100 m above the turbines at a rate of 1.0×10^6 kg/s for 8.0 h during the night. (a) How much energy (kWh) is needed to do this each night? (b) If all this energy is released during a 16-h day, what is the average power output? Assume that the process is 80 percent efficient.

*38. (II) Suppose a power plant delivers energy at 1000 MW using steam turbines. The steam goes into the turbines superheated at 520 K and deposits its unused heat in river water at 290 K. Assume that the turbine operates as an ideal Carnot engine. If the river flow rate is 40 m³/s, calculate the temperature increase of the river water downstream from the power plant.

*39. (II) The basin of the tidal power plant at the mouth of the Rance River in France covers an area of 23 km². The average difference in water height between high and low tide is 8.5 m. Estimate how much work the falling water can do on the turbines per day, assuming there are two high and two low tides per day. Assume the basin is flat, and neglect any KE of water before and after its fall.

13 VIBRATIONS AND WAVES

Many objects vibrate or oscillate—an object on the end of a spring, a tuning fork, the balance wheel of a watch, a pendulum, a plastic ruler held firmly over the edge of a table and gently struck, the strings of a guitar or piano. Spiders detect prey by the vibrations of their webs, cars oscillate up and down when they hit a bump, buildings and bridges vibrate when heavy trucks pass or the wind is fierce. Indeed, because most solids are elastic (see Chapter 8), most material objects vibrate (at least briefly) when given an impulse. Electrical oscillations occur in radio and television sets. And at the atomic level, atoms vibrate within a molecule, and the atoms of a solid vibrate about their relatively fixed positions.

Vibrations and wave motion are intimately related subjects. Waves—whether ocean waves, waves on a string, earthquake waves, or sound waves in air—have as their source a vibration. In the case of sound, not only is the source a vibrating object, but so is the detector—the ear drum or the membrane of a microphone. Indeed, the medium through which a wave travels itself vibrates (such as air for sound waves). In the second half of this chapter, after we discuss vibrations, we will discuss simple waves such as those on water and on a string. In Chapter 14 we will study sound waves, and in later chapters we will encounter other forms of wave motion, including light.

Vibrations and wave motion are not really "new" phenomena, however, because for mechanical systems they are fully described on the basis of Newtonian mechanics.

13-1 ▌ Simple harmonic motion

When we speak of a **vibration** or an **oscillation** we mean the motion of an object that repeats itself, back and forth, over the same path. That is, the motion is *periodic*. The simplest form of periodic motion is represented by an object oscillating on the end of a coil spring. Because many other types of vibrational motion closely resemble this system, we will look at it in detail. We assume that the mass of the spring can be ignored, and that the spring is mounted horizontally as shown in Fig. 13-1a, so that the object of mass m slides without friction on the horizontal surface. Any spring has a natural length at which it exerts no force on the mass m, and this is called

FIGURE 13-1

Mass vibrating at the end of a spring.

$x = 0$

(a)

(b)

(c)

294

the *equilibrium position*. If the mass is moved either to the left, which compresses the spring, or to the right, which stretches it, the spring exerts a force on the mass which acts in the direction of returning it to the equilibrium position; hence it is called a "restoring force." The magnitude of the restoring force F is found to be directly proportional to the distance x the spring has been stretched or compressed, Fig. 13-1b and c:

$$F = -kx. \qquad (13\text{-}1)$$

Equation 13-1 is accurate as long as the spring is not compressed to the point where the coils touch, or stretched beyond the elastic region (see Fig. 8-19). The minus sign indicates that the restoring force is always in the direction opposite to the displacement x. For example, if we choose the positive direction to the right in Fig. 13-1, x is positive when the spring is stretched, but the direction of the restoring force is to the left. Notice that we have taken $x = 0$ at the equilibrium position. If the spring is compressed, x is negative but the force F acts toward the right (Fig. 13-1c). The proportionality constant k in Eq. 13-1 is called the "spring constant." Notice that in order to stretch the spring a distance x one has to exert an (external) force on the spring at least equal to $F = +kx$. The greater the value of k, the greater the force needed to stretch a spring a given distance. That is, the stiffer the spring, the greater the spring constant k.

Let us examine what happens when the spring is initially stretched a distance $x = A$ as shown in Fig. 13-2a and then released. The spring exerts a force on the mass that pulls it toward the equilibrium position; but because the mass has been accelerated by the force, it passes the equilibrium position with considerable speed. Notice that as the mass reaches the equilibrium position, the force on it decreases to zero, but its speed at this point is a maximum, Fig. 13-2b. As it moves farther to the left, the force on it acts to slow it down, and it stops momentarily at $x = -A$, Fig. 13-2c. It then begins moving back in the opposite direction, Fig. 13-2d, until it reaches the original starting point, $x = A$, Fig. 13-2e. It then repeats the motion.

To discuss vibrational motion, we need to define a few terms. The distance x of the mass from the equilibrium point at any moment is called the **displacement**. The maximum displacement—the greatest distance from the equilibrium point—is called the **amplitude, A.** One **cycle** refers to the complete to-and-fro motion from some initial point back to that same point, say from $x = A$ to $x = -A$ back to $x = A$. The **period, T,** is defined as the time required for one complete cycle. Finally, the **frequency, f,** is the number of complete cycles per second. Frequency is usually specified in hertz (Hz) where 1 Hz = 1 cycle per second. It is evident that

$$f = \frac{1}{T} \quad \text{and} \quad T = \frac{1}{f}. \qquad (13\text{-}2)$$

For example, if the frequency is 5 cycles per second then each cycle takes $\frac{1}{5}$ s.

The oscillation of a spring hung vertically is essentially the same as that of a horizontal spring. Because of the force of gravity, the length of the vertical spring at equilibrium will be longer than when it is horizontal. If x is measured from the new equilibrium position, Eq. 13-1 can be used directly with the same value of k. (See the problems.)

FIGURE 13-2

Force on, and velocity of, mass at different positions of its oscillation.

(a)

(b)

(c)

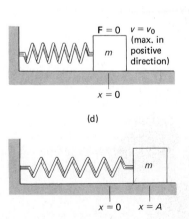

(d)

(e)

Any vibrating system for which the restoring force is directly proportional to the negative of the displacement (as in Eq. 13-1, $F = -kx$) is said to exhibit **simple harmonic motion** (SHM). Such a system is often called a **simple harmonic oscillator** (SHO). We saw in Section 8-6 that most solid materials stretch or compress according to Eq. 13-1 as long as the displacement is not too great. Because of this, many natural vibrations are simple harmonic or close to it.

☐ **EXAMPLE 13-1** When a family of four people with a total mass of 200 kg step into their 1200-kg car, the car's springs compress 3.0 cm. (a) What is the spring constant of the car's springs (assuming they act as a single spring)? (b) How far will the car descend if loaded with 300 kg?

SOLUTION (a) The added force of $(200 \text{ kg})(9.8 \text{ m/s}^2) = 1960 \text{ N}$ causes the springs to compress 3.0×10^{-2} m. Therefore, by Eq. 13-1, we have

$$k = \frac{F}{x} = \frac{1960 \text{ N}}{3.0 \times 10^{-2} \text{ m}} = 6.5 \times 10^4 \text{ N/m}.$$

(b) If the car is loaded with 300 kg, $x = F/k = (300 \text{ kg})(9.8 \text{ m/s}^2)/(6.5 \times 10^4 \text{ N/m}) = 4.5 \times 10^{-2}$ m, or 4.5 cm. We could have obtained this answer without solving for k: since x is proportional to F, if 200 kg compresses the spring 3.0 cm, then 1.5 times the force will compress the spring 1.5 times as much, or 4.5 cm.

FIGURE 13-3

Energy changes from kinetic energy to potential energy and back again as the spring oscillates.

(a)

(b)

(c)

(d)

13-2 ▌ Energy in the simple harmonic oscillator

To stretch or compress a spring, work has to be done; hence potential energy is stored in a stretched or compressed spring. Indeed, we have already seen in Section 5-4 that the potential energy is given by

$$\text{PE} = \tfrac{1}{2}kx^2.$$

Thus, since the total mechanical energy E is the sum of the kinetic and potential energies, we have

$$E = \tfrac{1}{2}mv^2 + \tfrac{1}{2}kx^2, \tag{13-3}$$

where v is the velocity of the mass m when it is a distance x from the equilibrium position. As long as there is no friction, the total mechanical energy E remains constant. As the mass oscillates back and forth, the energy continuously changes from potential energy to kinetic energy, and back again (Fig. 13-3). At the extreme points, $x = A$ and $x = -A$, all the energy is potential energy (and is the same whether the spring is compressed or stretched to the full amplitude), so, since $v = 0$ at these points:

$$E = \tfrac{1}{2}m(0)^2 + \tfrac{1}{2}kA^2 = \tfrac{1}{2}kA^2. \tag{13-4a}$$

Thus, the *total mechanical energy of a SHO is proportional to the square of the amplitude.* At the equilibrium point, $x = 0$, all the energy is kinetic:

$$E = \tfrac{1}{2}mv_0^2 + \tfrac{1}{2}k(0)^2 = \tfrac{1}{2}mv_0^2, \qquad (13\text{-}4b)$$

where v_0 represents the maximum velocity during the motion (which occurs at $x = 0$). At intermediate points the energy is part kinetic and part potential. By combining Eq. 13-4a with Eq. 13-3, we can find a useful equation for the velocity as a function of the position x:

$$\tfrac{1}{2}mv^2 + \tfrac{1}{2}kx^2 = \tfrac{1}{2}kA^2.$$

Solving for v^2, we have

$$v^2 = \frac{k}{m}(A^2 - x^2) = \frac{k}{m}A^2\left(1 - \frac{x^2}{A^2}\right).$$

Now, from Eqs. 13-4a and 13-4b, we have $\tfrac{1}{2}mv_0^2 = \tfrac{1}{2}kA^2$, so $v_0^2 = (k/m)A^2$. Inserting this in the equation above and taking the square root, we have

$$v = v_0\sqrt{1 - x^2/A^2}. \qquad (13\text{-}5)$$

This gives the velocity of the object at any position x.

EXAMPLE 13-2 A spring stretches 0.150 m when a 0.30-kg mass is hung from it. The spring is then stretched an additional 0.100 m from this equilibrium point and released. Determine (*a*) the spring constant k, (*b*) the maximum velocity v_0, (*c*) the velocity v when the mass is 0.050 m from equilibrium, and (*d*) the maximum acceleration of the mass.

SOLUTION (*a*) Since the spring stretches 0.150 m when 0.30 kg is hung from it, we find k from Eq. 13-1 to be

$$k = \frac{F}{x} = \frac{mg}{x} = \frac{(0.30\text{ kg})(9.8\text{ m/s}^2)}{0.150\text{ m}} = 19.6\text{ N/m}.$$

(*b*) The maximum velocity v_0 is attained as the mass passes through the equilibrium point and all the energy is kinetic. By conservation of energy, $\tfrac{1}{2}mv_0^2 = \tfrac{1}{2}kA^2$. Solving for v_0, we have

$$v_0 = A\sqrt{\frac{k}{m}} = (0.100\text{ m})\sqrt{\frac{19.6\text{ N/m}}{0.30\text{ kg}}} = 0.81\text{ m/s}.$$

(*c*) We use Eq. 13-5 and find that

$$v = v_0\sqrt{1 - x^2/A^2}$$

$$= (0.81\text{ m/s})\sqrt{1 - \frac{(0.050\text{ m})^2}{(0.100\text{ m})^2}} = 0.70\text{ m/s}.$$

(*d*) By Newton's second law, $F = ma$. So the maximum acceleration occurs where the force is greatest—that is, when $x = A = 0.100$ m. Thus

$$a = \frac{kA}{m} = \frac{(19.6\text{ N/m})(0.100\text{ m})}{0.30\text{ kg}} = 6.5\text{ m/s}^2.$$

13-3 ∎ The reference circle: the period and sinusoidal nature of SHM

The period of a simple harmonic oscillator is found to depend on the stiffness of the spring and also on the mass m that is oscillating; but—strange as it may seem—the period does not depend on the amplitude; you can find this out for yourself by using a watch and counting 10 or 20 cycles of a spring for a small amplitude and then for a large amplitude.

We can derive a formula for the period of SHM, and this can be done by comparing SHM to an object rotating in a circle. From this same "reference circle" we can obtain a second useful result—a formula for the position of an oscillating mass as a function of time.

Consider now a mass m rotating in a circle of radius A with speed v_0 on top of a table as shown in Fig. 13-4. As viewed from above, the motion is a circle. But a person who looks at the motion from the edge of the table, sees an oscillatory motion back and forth, and this corresponds precisely to SHM, as we shall now see. What the person sees, and what we are interested in, is the projection of the circulatory motion onto the x axis (Fig. 13-4b). To see that this motion is analogous to SHM, let us calculate the x component of the velocity v_0, which is labeled v in Fig. 13-4. The two triangles shown are similar, so

$$\frac{v}{v_0} = \frac{\sqrt{A^2 - x^2}}{A}$$

or

$$v = v_0 \sqrt{1 - \frac{x^2}{A^2}}.$$

This is exactly the equation for the speed of a mass oscillating with SHM (Eq. 13-5). Thus the projection on the x axis of a ball rotating in a circle has the same motion as a mass at the end of a spring. The period of the SHO is equal to that of the rotating object making one complete revolution; the time for one revolution is equal to the circumference of the circle (distance) divided by the speed

$$T = \frac{2\pi A}{v_0}.$$

From Eqs. 13-4a and b, we have $\frac{1}{2}kA^2 = \frac{1}{2}mv_0^2$, so $A/v_0 = \sqrt{m/k}$. Thus

$$T = 2\pi \sqrt{\frac{m}{k}}. \tag{13-6}$$

This is the formula we were looking for. The period depends on the mass m and the spring constant k, but not on the amplitude. We see from Eq. 13-6 that the greater the mass, the longer the period; and the stiffer the spring, the shorter the period. This makes sense since a greater mass means more inertia and therefore slower response (or acceleration); and larger k means greater force and therefore quicker response. Notice that Eq. 13-6 is not a direct proportion: the period varies as the square root of m/k. For example, the mass must be quadrupled to double the period. Equation 13-6 is fully

FIGURE 13-4

Analysis of simple harmonic motion as a side view (b) of circular motion (a).

(a)

(b)

in accord with experiment and is valid not only for a spring, but for all kinds of SHM that obey Eq. 13-1.

□   EXAMPLE 13-3   What are the period and frequency of the spring in Example 13-2?

SOLUTION   From Eq. 13-6,

$$T = 2\pi \sqrt{\frac{m}{k}} = 6.28 \sqrt{\frac{0.30 \text{ kg}}{19.6 \text{ N/m}}} = 0.78 \text{ s.}$$

The frequency $f = 1/T = 1.28$ Hz (Eq. 13-2).

□   EXAMPLE 13-4   A small cockroach of mass 0.30 g is caught in a spider's web. The web vibrates predominately with a frequency of 15 Hz. (a) Estimate the value of the spring constant k for the web. (b) At what frequency would you expect the web to vibrate if an insect of mass 0.10 g were trapped?

SOLUTION   (a) Using Eq. 13-6 and $f = 1/T$, we find

$$f = \frac{1}{2\pi} \sqrt{\frac{k}{m}}.$$

We solve for k and find that

$$k = (2\pi f)^2 m$$
$$= (6.28 \times 15 \text{ s}^{-1})^2 (3.0 \times 10^{-4} \text{ kg}) = 2.7 \text{ N/m.}$$

(b) We could substitute $m = 1.0 \times 10^{-4}$ kg in the above equation for f; but instead, we notice that the frequency decreases with the square root of the mass. Since the new mass is one-third the first mass, the frequency increases by a factor of $\sqrt{3}$. That is, $f = (15 \text{ Hz})(\sqrt{3}) = 26$ Hz.

We now use the reference circle to find the position of a mass undergoing SHM as a function of time. From Fig. 13-4 we see that $\cos \theta = x/A$, so the projection of the ball's position on the x axis is

$$x = A \cos \theta.$$

Since the ball is rotating with angular velocity ω, we can write $\theta = \omega t$ (see Chapter 7). Thus

$$x = A \cos \omega t. \tag{13-7a}$$

Furthermore, since the angular velocity ω (specified in radians per second) can be written as $\omega = 2\pi f$, where f is the frequency, we also can write

$$x = A \cos 2\pi f t, \tag{13-7b}$$

or in terms of the period T:

$$x = A \cos \frac{2\pi t}{T}. \tag{13-7c}$$

Notice in Eq. 13-7c that when $t = T$ (that is, after a time equal to the period), we have the cosine of 2π, which is the same as the cosine of zero. This is just a reflection of the fact that the motion repeats itself after a time T.

FIGURE 13-5

Sinusoidal nature of SHM as a function of time; in this case, $x = A \cos (2\pi t/T)$.

As we have seen, the projection on the x axis of the rotating object corresponds precisely to the motion of a simple harmonic oscillator. Thus, Eqs. 13-7 give the position of an object oscillating under simple harmonic motion. Since the cosine function varies between 1 and -1, x varies between A and $-A$, as it must. If a pen is attached to a vibrating mass as shown in Fig. 13-5 and a sheet of paper is moved at a steady rate beneath it, a curve will be drawn that accurately follows Eqs. 13-7.

Other equations for SHM are also possible depending on the initial conditions. For example, if at $t = 0$ the oscillations are begun by giving the mass a push when it is at the equilibrium position, the equation would be

$$x = A \sin \frac{2\pi t}{T}.$$

This curve has precisely the same shape as the cosine curve shown in Fig. 13-5, except it will be shifted by a quarter cycle so that it starts out at $x = 0$ instead of at $x = A$.

Either of the curves, sine or cosine, are referred to as being **sinusoidal.** Thus simple harmonic motion is said to be sinusoidal since the position varies as a sinusoidal function of time. It is easily shown that the velocity and acceleration of SHM are also sinusoidal. This is particularly easy to show for the acceleration, a. From Newton's second law, $F = ma$, the acceleration of the mass m is $a = F/m$ where the force exerted by the spring on the mass is $F = -kx$. Hence

$$a = \frac{F}{m} = \frac{-kx}{m} = -\left(\frac{kA}{m}\right) \sin \frac{2\pi t}{T}.$$

Since the acceleration of a SHO is *not* constant, the equations of uniformly accelerating motion, Eqs. 1-10, do *not* apply to SHM.

⬜ EXAMPLE 13-5 (*a*) What is the equation describing the motion of a spring that is stretched 20 cm from equilibrium and then released, and whose period is 1.5 s? (*b*) What will be its displacement after 1.8 s?

SOLUTION (*a*) The amplitude $A = 0.20$ m, and $2\pi/T = 6.28/(1.5 \text{ s}) = 4.2 \text{ s}^{-1}$. Thus

$$x = 0.20 \cos 4.2t \text{ m}.$$

(b) At $t = 1.80$ s, $x = 0.20 \cos 7.56$. The cosine repeats itself after $2\pi = 6.28$ rad so 7.56 rad gives the same result as $7.56 - 6.28 = 1.28$ rad; thus $x = 0.20 \cos 1.28$. From the tables or a calculator, $\cos 1.28$ is 0.29, so $x = (0.20)(0.29) = 0.058$ m, or 5.8 cm.

13-4 ▮ The simple pendulum

A simple pendulum consists of a small object (the pendulum "bob") suspended from the end of a light cord, Fig. 13-6. We assume that the cord doesn't stretch and that its mass can be ignored relative to that of the bob. A simple pendulum moving back and forth with negligible friction resembles simple harmonic motion: it oscillates along the arc of a circle with equal amplitude on either side of its equilibrium point (where it hangs vertically) and as it passes through the equilibrium point it has its maximum speed. But is it really undergoing SHM? That is, is the restoring force proportional to its displacement? Let us find out.

The displacement of the pendulum along the arc, x, is given by $x = L\theta$, where θ is the angle the string makes with the vertical and L is the length of the string (Fig. 13-6). Thus, if the restoring force is proportional to x or to θ, the motion will be simple harmonic. The restoring force is the component of the weight, mg, tangent to the arc:

$$F = -mg \sin \theta.$$

Since F is proportional to the sine of θ and not to θ itself, the motion is *not* SHM. However, if θ is small, then $\sin \theta$ is very nearly equal to θ if the latter is specified in radians. This can be seen by looking at the trigonometry table inside the back cover or by noting that in Fig. 13-6 the arc length $x (= L\theta)$ is nearly the same length as the chord $(= L \sin \theta)$ indicated by the dashed line if θ is small. For angles less than 15°, the difference between θ and $\sin \theta$ is less than 1 percent. Thus, to a very good approximation for small angles,

$$F \approx -mg\theta.$$

Using the fact that $x = L\theta$, we have

$$F \approx -\frac{mg}{L}x.$$

Thus, for small displacements, the motion is essentially simple harmonic, with an effective force constant of $k = mg/L$ (see Eq. 13-1). The period of a simple pendulum can be found using Eq. 13-6, where for k we substitute mg/L:

$$T = 2\pi \sqrt{\frac{m}{mg/L}}$$

$$= 2\pi \sqrt{\frac{L}{g}} \qquad (\theta \text{ small}). \tag{13-8}$$

A surprising result is that the period does not depend on the mass of the pendulum bob! You may have noticed this if you pushed a small child and a large one on the same swing.

FIGURE 13-6
Simple pendulum.

We saw in Section 13-3 that the period of any SHM, including a pendulum, does not depend on the amplitude. Galileo is said to have first noted this fact while watching a swinging lamp in the cathedral at Pisa; this discovery led to the pendulum clock,[†] the first really precise timepiece, which became the standard for centuries.

Because a pendulum does not undergo precisely SHM, the period does depend slightly on the amplitude, mainly for large amplitudes.

The accuracy of a pendulum clock would be affected, after many swings, by the decrease in amplitude due to friction; but the mainspring in a pendulum clock (or the falling weight in a grandfather clock) supplies energy to compensate for the friction and to maintain the amplitude constant, so that the timing remains accurate.

The pendulum finds use in geology, for geologists are interested in surface irregularities of the earth and frequently need to measure the acceleration of gravity at a given location very accurately. They often use a carefully designed pendulum to do this as illustrated in the next example.

EXAMPLE 13-6   A geologist's simple pendulum, whose length is 37.10 cm, has a frequency of 0.8190 Hz at a particular location on the earth. What is the acceleration of gravity at this location?

SOLUTION   From Eq. 13-8, we have

$$f = \frac{1}{T} = \frac{1}{2\pi}\sqrt{\frac{g}{L}}.$$

Solving for g, we obtain

$$g = (2\pi f)^2 L$$
$$= (6.283 \times 0.8190 \text{ s}^{-1})^2 (0.3710 \text{ m})$$
$$= 9.824 \text{ m/s}^2.$$

13-5 ∎ Damped harmonic motion

The amplitude of any real oscillating spring or swinging pendulum slowly decreases in time until the oscillations stop altogether. Fig. 13-7 shows a typical graph of the displacement as a function of time. This is called **damped**

[†] It is not known for sure whether or not Galileo actually built a pendulum clock; Christiaan Huygens (1629–1695), more than 10 years after Galileo's death, did build one.

FIGURE 13-7
Damped harmonic motion.

harmonic motion. The damping[†] is generally due to the resistance of air and to internal friction within the oscillating system. The energy that is thus dissipated to thermal energy is reflected in a decreased amplitude of oscillation.

Since natural oscillating systems are damped in general, why do we even talk about (undamped) simple harmonic motion? The answer is that SHM is much easier to deal with mathematically. And if the damping is not large, the oscillations can be thought of as simple harmonic motion on which is superposed the damping—that is, the decrease in amplitude represented by the dashed curves in Fig. 13-7. Although frictional damping does alter the frequency of vibration, the effect is usually small unless the damping is large; thus Eq. 13-6 can still be used in most cases.

Sometimes the damping is so large that the motion no longer resembles simple harmonic motion. Three common cases of heavily damped systems are shown in Fig. 13-8. Curve C represents the situation when the damping is so large that it takes a long time to reach equilibrium; the system is *over-damped*. Curve A represents an *underdamped* situation in which the system makes several swings before coming to rest. Curve B represents *critical damping;* in this case equilibrium is reached the quickest. These terms all derive from the use of practical damped systems such as door-closing mechanisms and shock absorbers in a car. These are usually designed to give critical damping; but as they wear out, underdamping occurs: a door slams and a car bounces up and down several times every time it hits a bump. Needles on electronic instruments (voltmeters, ammeters, level indicators on tape recorders) are usually critically damped or slightly underdamped. If they were very underdamped, they would swing back and forth excessively before arriving at the correct value; and if overdamped they would take too long to reach equilibrium, so rapid changes in the signal (say, recording level) would not be detected.

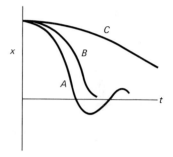

FIGURE 13-8

Underdamped (A), critically damped (B), and overdamped (C) motion.

13-6 ∎ Forced vibrations; resonance

When a vibrating system is set into motion, it vibrates at its natural frequency. In Sections 13-3 and 13-4 we developed formulas that relate the natural frequency (or period) to the properties of the system for elastic objects (like springs) and pendulums.

However, a system is often not left to merely oscillate on its own, but may have an external force applied to it which itself oscillates at a particular frequency. For example, we might pull the mass on the spring of Fig. 13-1 back and forth at a frequency f. The mass thus vibrates at the frequency f of the external force, even if this frequency is different from the **natural frequency** of the spring, which we will now denote by f_0 where (see Eq. 13-6)

$$f_0 = \frac{1}{2\pi}\sqrt{\frac{k}{m}}.$$

This is an example of **forced vibration.** The amplitude of vibration, and hence the energy transferred to the vibrating system, is found to depend on

[†] To "damp" means to diminish, restrain, or extinguish, as to "dampen one's spirits."

the difference between f and f_0. It is a maximum when the frequency of the external force equals the natural frequency of the system; that is, when $f = f_0$. The energy transferred to the system (proportional to the amplitude squared) is plotted in Fig. 13-9 as a function of the external frequency f. Curve A represents light damping and curve B heavy damping. The amplitude becomes very large when $f = f_0$, particularly when the damping (friction) is small. This is known as **resonance,** and the natural vibrating frequency of a system is often called its **resonant frequency.**

A simple illustration of resonance is pushing a child on a swing. A swing, like any pendulum, has a natural frequency of oscillation. If you were to close your eyes and push on the swing at a random frequency, the swing would bounce around and reach no great amplitude. But if you push with a frequency equal to the natural frequency of the swing, the amplitude increases greatly. This clearly illustrates the fact that at resonance, relatively little effort is required to obtain a large amplitude.

The great tenor Enrico Caruso was said to be able to break a goblet by singing a note of just the right frequency at full voice. This is an example of resonance, for the sound waves emitted by the voice act as a forced vibration on the glass. At resonance, the resulting vibration of the goblet may be large enough in amplitude that the glass exceeds its elastic limit and breaks.

Since material objects are, in general, elastic, resonance is an important phenomenon in a variety of situations. It is particularly important in building, although the effects are not always to be foreseen. For example, it has been reported that a railway bridge collapsed because a nick in one of the wheels of a passing train set up a resonant vibration in the bridge. Indeed, marching soldiers break step when crossing a bridge to avoid the possibility of a similar catastrophe. And the famous collapse of the Tacoma Narrows bridge (Fig. 13-10) in 1940 was due in part to resonance of the bridge.

We will meet important examples of resonance later in this chapter and in succeeding chapters. We will also see that vibrating objects often have not one, but many resonant frequencies.

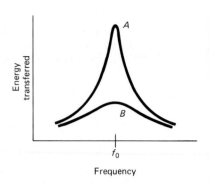

FIGURE 13-9

Resonance for lightly damped (A) and heavily damped (B) systems.

FIGURE 13-10

Large amplitude oscillation of the Tacoma Narrows Bridge due to heavy gusty winds just prior to collapse (November 7, 1940). (Wide World Photo.)

13-7 ▌ Wave motion

When you throw a stone into a lake or pool of water, circular waves form and move outward, Fig. 13-11. Waves will also travel along a cord (or a "slinky") that is stretched out straight on a table if you vibrate one end back and forth as shown in Fig. 13-12. Water waves, and waves on a cord, are two common examples of wave motion. We will meet other kinds of wave motion later, but for now we will concentrate on these simple "mechanical" waves.

If you have ever watched ocean waves moving toward shore, you may have wondered if the waves were carrying water into the beach. This is, in fact, not the case.[†] Water waves move with a recognizable velocity. But each particle of the water itself merely oscillates about an equilibrium point. This is clearly demonstrated by observing leaves on a pond as waves move by. The leaves (or a cork) are not carried forward by the waves, but simply oscillate about an equilibrium point because this is the motion of the water itself. Similarly, the wave on the rope of Fig. 13-12 moves to the right, but each piece of the rope only vibrates to and fro. This is a general feature of waves: waves can move over large distances but the medium (the water or the rope) itself has only a limited movement. Thus, although a wave is not matter, the wave pattern can travel in matter. A wave consists of oscillations that move without carrying matter with them.

Waves carry energy from one place to another. Energy is given to a water wave, for example, by a rock thrown into the water, or by wind far out at sea. The energy is transported by waves to the shore. If you have been under an ocean wave when it breaks, you know the energy it carries. The oscillating hand in Fig. 13-12 transfers energy to the rope, which is then transported down the rope and can be transferred to an object at the other end. All forms of wave motion transport energy.

Let us look a little more closely at how a wave is formed and how it comes to "travel". We first look at a single wave bump or *pulse*. A single pulse can be formed on a rope by a quick up-and-down motion of the hand, Fig. 13-13. The hand pulls up on one end of the rope and because the end piece is attached to adjacent pieces, these also feel an upward force and they, too, begin to move upward. As each succeeding piece of rope moves upward, the wave crest moves outward along the rope. Meanwhile, the end piece of rope has been returned to its original position by the hand, and as each succeeding piece of rope reaches its peak position, it, too, is pulled back down

[†] Do not be confused by the "breaking" of ocean waves, which occurs when the wave interacts with the bottom in shallow water and hence is no longer a simple wave.

(a)

(b)

(c)

(d)

FIGURE 13-11
Water waves spreading outward from a source. (D. C. Giancoli.)

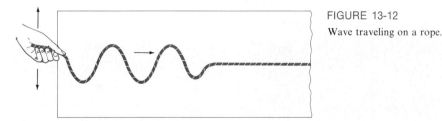

FIGURE 13-12
Wave traveling on a rope.

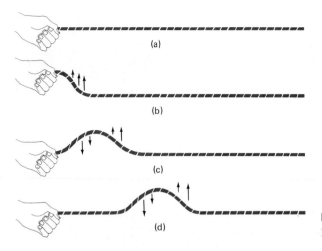

FIGURE 13-13

Motion of a wave pulse. Arrows indicate velocity of rope particles.

again. Thus the source of a traveling wave pulse is a disturbance, and cohesive forces between adjacent pieces of rope cause the pulse to travel outward. Waves in other media are created and propagate outward in a similar fashion.

A *continuous* or *periodic wave*, such as that shown in Fig. 13-12, has as its source a disturbance that is continuous and oscillating; that is, the source is a *vibration* or *oscillation*. In Fig. 13-12 a hand oscillates one end of the rope. Water waves may be produced by any vibrating object placed at the surface such as your hand; or the water itself is made to vibrate when wind blows across it or a rock is thrown into it. A vibrating tuning fork or drum membrane gives rise to sound waves in air; and we will see later that oscillating electric charges give rise to light waves. Indeed, almost any vibrating object sends out waves.

The source of any wave, then, is a vibration. And it is the *vibration* that propagates outward and thus constitutes the wave. If the source vibrates sinusoidally in SHM, then the wave itself—if the medium is perfectly elastic—will have a sinusoidal shape both in space and in time. That is, if you take a picture of the wave spread throughout space at a given instant of time, the wave will have the shape of a sine or cosine function. On the other hand, if you look at the motion of the medium at one place over a long period of time—for example, if you look between two closely spaced posts of a pier or out of a ship's porthole as water waves pass by—the up and down motion of that small segment of water will be simple harmonic motion—the water moves up and down sinusoidally in time.

Some of the important quantities used to describe a periodic sinusoidal wave are shown in Fig. 13-14. The high points on a wave are called crests, the low points troughs. The **amplitude** is the maximum height of a crest, or depth of a trough, relative to the normal (or equilibrium) level; the total swing from a crest to a trough is twice the amplitude. The distance between two successive crests is called the **wavelength,** λ (the Greek letter lambda). The wavelength is also equal to the distance between *any* two successive identical points on the wave. The **frequency,** f is the number of crests—or complete cycles—that pass a given point per unit time. The period, T, of course, is just $1/f$.

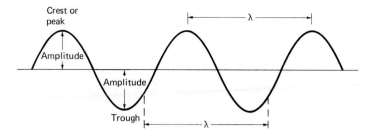

Crest or peak

Amplitude

Amplitude

Trough

λ

λ

FIGURE 13-14

Characteristics of a continuous wave.

The **wave velocity,** v, is the velocity at which wave crests appear to move. (The wave velocity must be distinguished from the velocity of a particle of the medium itself. For example, for a wave traveling along a string as in Fig. 13-12, the wave velocity is to the right, along the string, whereas the velocity of a particle of the string is perpendicular to it.) Since a wave crest travels a distance of one wavelength, λ, in one period, T, the wave velocity $v = \lambda/T$ or

$$v = \lambda f. \tag{13-9}$$

For example, suppose a wave has a wavelength of 5 m and a frequency of 3 Hz. Since 3 crests pass a given point per second, and the crests are 5 m apart, the first crest (or any other part of the wave) must travel a distance of 15 m during the 1 s; so its speed is 15 m/s.

The velocity of a wave depends on the properties of the medium in which it travels. The velocity of a wave on a stretched string, for example, depends on the tension in the string, F_T, and on the string's mass per unit length, m/L; for waves of small amplitude, the relationship is

$$v = \sqrt{\frac{F_T}{m/L}}. \tag{13-10}$$

This formula, makes sense qualitatively on the basis of Newtonian mechanics: that is, we expect the tension to be in the numerator and the mass per unit length in the denominator because when the tension is greater we expect the velocity to be greater since each segment of string is in better contact with its neighbor; and the greater the mass per unit length, the more inertia the string has and the more slowly the wave would be expected to propagate.

   EXAMPLE 13-7 A wave whose wavelength is 0.50 m is traveling down a 300-m-long wire whose total mass is 30 kg. If the wire is under a tension of 4000 N, what is the velocity and frequency of this wave?

   SOLUTION From Eq. 13-10, the velocity is

$$v = \sqrt{\frac{4000 \text{ N}}{(30 \text{ kg})/(300 \text{ m})}} = 200 \text{ m/s}.$$

The frequency then is

$$f = \frac{v}{\lambda} = \frac{200 \text{ m/s}}{0.50 \text{ m}} = 400 \text{ Hz}.$$

13-8 ▪ Types of waves

We saw earlier that although waves may travel over long distances, the particles of the medium vibrate only over a limited region of space. When a wave travels down a rope, say from left to right, the particles of the rope vibrate up and down in a direction transverse (or perpendicular) to the motion of the wave itself. Such a wave is called a **transverse wave.** There exists another type of wave known as a **longitudinal wave.** In a longitudinal wave, the vibration of the particles of the medium is along the *same* direction as the motion of the wave. Longitudinal waves are readily formed on a stretched spring or "slinky" by alternately compressing and expanding one end. This is shown in Fig. 13-15b, and can be compared to the transverse wave in Fig. 13-15a. A series of compressions and expansions propagate along the spring. The *compressions* are those areas where the coils are momentarily close together. *Expansions* (sometimes called *rarefactions*) are regions where the coils are momentarily far apart. Compressions and expansions correspond to the crests and troughs of a transverse wave.

An important example of a longitudinal wave is a sound wave in air. A vibrating drum head, for example, alternately compresses and rarefies the air and produces a longitudinal wave that travels outward in the air as shown in Fig. 13-16.

As in the case of transverse waves, each section of the medium in which a longitudinal wave passes oscillates over a very small distance whereas the wave itself can travel large distances. Wavelength, frequency, and wave velocity all have meaning for a longitudinal wave. The wavelength is the distance between successive compressions (or between successive expansions), and frequency is the number of compressions that pass a given point per second. The wave velocity is the velocity with which each compression appears to move and is equal to the product of wavelength and frequency (Eq. 13-9).

A longitudinal wave can be represented graphically by plotting the density of air molecules (or coils of a slinky) versus position as shown in Fig. 13-17. We will often use such a graphical representation because it is much easier to illustrate what is happening. Note that the graph looks much like a transverse wave.

Wavelength

(a)

Compression   Expansion

Wavelength

(b)

FIGURE 13-15
(a) Transverse wave; (b) longitudinal wave.

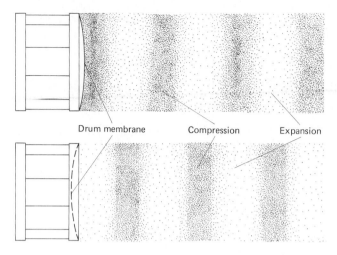

Drum membrane   Compression   Expansion

FIGURE 13-16

Production of a sound wave which is longitudinal.

The velocity of a longitudinal wave has a form similar to that for a transverse wave on a string (Eq. 13-10); that is

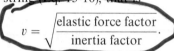

$$v = \sqrt{\frac{\text{elastic force factor}}{\text{inertia factor}}}.$$

In particular, for a longitudinal wave traveling down a long solid rod,

$$v = \sqrt{\frac{E}{\rho}} \qquad \text{(13-11a)}$$

where E is the elastic modulus (Chapter 8) of the material and ρ is its density. For a longitudinal wave traveling in a liquid or gas,

$$v = \sqrt{\frac{B}{\rho}} \qquad \text{(13-11b)}$$

where B is the bulk modulus and ρ the density.

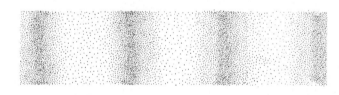

FIGURE 13-17

(a) A longitudinal wave with (b) its graphical representation.

(a)

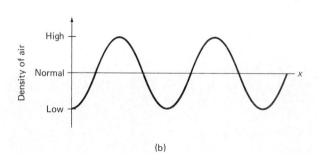

(b)

309

EXAMPLE 13-8   You can often hear an approaching train by putting your ear to the track. How long does it take for the wave to travel down the steel track if the train is 1.0 km away?

SOLUTION   Referring to Tables 8-1 and 9-1 for the elastic modulus and density of steel, respectively, we have

$$v = \sqrt{\frac{2.0 \times 10^{11} \text{ N/m}^2}{7.8 \times 10^3 \text{ kg/m}^3}} = 5.1 \times 10^3 \text{ m/s}.$$

Then the time t = distance/velocity = $(1.0 \times 10^3 \text{ m})/(5.1 \times 10^3 \text{ m/s}) = 0.20$ seconds.

Both transverse and longitudinal waves are produced when a disturbance known as an earthquake occurs. The transverse waves that travel through the body of the earth are called S waves and the longitudinal waves are called P waves. Both longitudinal and transverse waves can travel through a solid since the atoms or molecules can vibrate about their relatively fixed positions in any direction. But in a fluid, only longitudinal waves can propagate; this is because any transverse motion would experience no restoring force since a fluid can flow. This fact was used by geophysicists to infer that the earth's outer core is molten: longitudinal waves are detected diametrically across the earth, but not transverse waves; the only explanation for this was that the core of the earth must be liquid.

Besides these two types of waves that can pass through the body of the earth (or other substance), there can also be *surface waves* that travel along the boundary between two materials. A wave on water is actually a surface wave that moves on the boundary between water and air. The motion of each particle of water at the surface is elliptical (Fig. 13-18) if the wavelength is less than the depth, so it is a combination of transverse and longitudinal motions. Below the surface, there is also transverse plus longitudinal wave motion, in an elliptical shape as shown; at the bottom, the motion is only longitudinal. Surface waves are also set up on the earth when an earthquake occurs. The waves that travel along the surface are mainly responsible for the damage caused by earthquakes.

FIGURE 13-18

A water wave is an example of a *surface wave*, which is a combination of transverse and longitudinal wave motion.

*13-9 ▮ Energy transmitted by waves

Waves transmit energy from one place to another. As waves travel through a medium, the energy is transmitted as vibrational energy from particle to particle of the medium. For a sinusoidal wave of frequency f, the particles move in SHM as a wave passes, so each particle has an energy $E = \frac{1}{2}kx_0^2$, where x_0 is the amplitude of its motion, either transversely or longitudinally. (See Eq. 13-4a; we have replaced A by x_0.) Using Eq. 13-6, we can write k in terms of the frequency, $k = 4\pi^2 m/T^2 = 4\pi^2 mf^2$, so that

$$E = 2\pi^2 mf^2 x_0^2.$$

The mass $m = \rho V$, where ρ is the density of the medium and V its volume. Also, the volume $V = Al$, where A is the cross-sectional area through which the wave travels, and we can write l as the distance the wave travels in a time t as $l = vt$; here v is the speed of the wave. Thus $m = \rho V = \rho Al = \rho Avt$ and

$$E = 2\pi^2 \rho Avtf^2 x_0^2. \qquad (13\text{-}12)$$

From this equation we have the important result that the *energy transported by a wave is proportional to the square of the amplitude*. The average *rate* of energy transferred is the power $\bar{P}$:

$$\bar{P} = \frac{E}{t} = 2\pi^2 \rho Avf^2 x_0^2. \qquad (13\text{-}13)$$

Finally, the **intensity** I of a wave is defined as the power transferred across unit area perpendicular to the direction of energy flow:

$$I = \frac{\bar{P}}{A} = 2\pi^2 v\rho f^2 x_0^2. \qquad (13\text{-}14)$$

If the waves flow out from the source in all directions, it is a three-dimensional wave. Examples are sound traveling in the open air, earthquake waves, and light waves. If the medium is isotropic (same in all directions), the wave has a *spherical shape* (Fig. 13-19). As the wave moves outward, it is spread over a larger and larger area since the surface area of a sphere of radius r is $4\pi r^2$. Because energy is conserved, we can see from Eq. 13-12 or 13-13 that as the area A increases, the amplitude x_0 must decrease; that is, at two different distances from the source, r_1 and r_2 (see Fig. 13-19), $A_1 x_{01}^2 = A_2 x_{02}^2$, where x_{01} and x_{02} are the amplitudes of the wave at r_1 and r_2, respectively. Since $A_1 = 4\pi r_1^2$ and $A_2 = 4\pi r_2^2$, we have $(x_{01}^2 r_1^2) = (x_{02}^2 r_2^2)$, or

$$\frac{x_{02}}{x_{01}} = \frac{r_1}{r_2}.$$

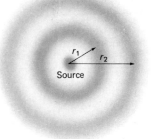

FIGURE 13-19

Wave traveling outward from source has spherical shape; two different crests (or compressions) are shown, of radius r_1 and r_2.

Thus the amplitude decreases inversely as the distance from the source. When the wave is twice as far from the source, the amplitude is half as large, and so on.

The intensity I also decreases with distance. Since I is proportional to x_0^2 (Eq. 13-14), I must decrease as the *square* of the distance from the source. This inverse-square law applies to sound and light and other types of waves. Another way to view this is to consider two points r_1 and r_2 at the same time. If the power output remains constant, the intensity at r_1 is $I_1 = \bar{P}/4\pi r_1^2$ and $I_2 = \bar{P}/4\pi r_2^2$. Thus

$$\frac{I_2}{I_1} = \frac{r_1^2}{r_2^2}. \qquad (13\text{-}15)$$

EXAMPLE 13-9   If the intensity of an earthquake P wave 100 km from the source is 1.0×10^6 W/m^2, what is the intensity 400 km from the source?

SOLUTION   The intensity decreases as the square of the distance from the source. Therefore, at 400 km, the intensity will be $(\frac{1}{4})^2 = \frac{1}{16}$ what it is at 100 km, or 6.2×10^4 W/m^2. Alternatively, Eq. 13-15 could be used: $I_2 = I_1 r_1^2 / r_2^2 = (1.0 \times 10^6 \text{ W/m}^2)(100 \text{ km})^2 / (400 \text{ km})^2 = 6.2 \times 10^4$ W/m^2.

The situation is different for a one-dimensional wave, such as a transverse wave on a string or a longitudinal wave pulse traveling down a uniform metal rod. The area A remains constant so the amplitude x_0 also remains constant; thus the amplitude and the intensity do not decrease with distance.

In practice, frictional damping is generally present and some of the energy is transformed into thermal energy. Thus the amplitude and intensity of a one-dimensional wave decrease with distance from the source, and for a three-dimensional wave the decrease will be greater than that discussed above, although the effect is often small.

13-10 ▮ Behavior of waves: reflection, refraction, interference, and diffraction

Reflection. When a wave strikes an obstacle, or comes to the end of the medium it is traveling in, at least a part of the wave is reflected. You have probably seen water waves reflect off a rock or the side of a swimming pool. And you may have heard a shout reflected from a distant cliff—which we call an "echo." A wave pulse traveling down a rope (or slinky) is reflected as shown in Fig. 13-20. You can observe this for yourself and see that the reflected pulse is inverted as in part (a) if the end of the rope is fixed, and returns right side up if the end is free as in part (b). In the case of two-dimensional waves (for example, waves on the surface of water) or three-dimensional waves, we are concerned with **wave fronts,** by which we mean the whole length of a wave crest. A line drawn in the direction of motion, perpendicular to the wave front, is called a *ray*. As shown in Fig. 13-21, the angle that the incoming or *incident wave* makes with the reflecting surface is equal to the angle made by the reflected wave. That is, *the angle of reflection equals the angle of incidence;* the "angle of incidence" is defined as the

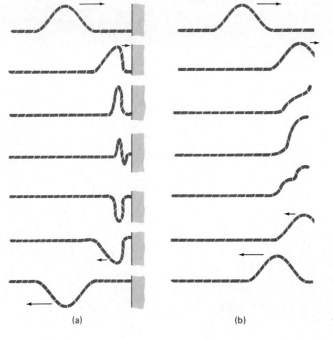

FIGURE 13-20

Reflection of a wave pulse on a rope when the end of the rope is (a) fixed and (b) free.

(a)   (b)

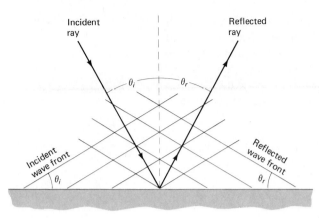

FIGURE 13-21

Law of reflection.

angle the incident ray makes with the perpendicular to the reflecting surface (or the wave front makes with a tangent to the surface) and the "angle of reflection" is the corresponding angle for the reflected wave.

When the wave pulse in Fig. 13-20a reaches the wall, not all of the energy is reflected. Some of it is absorbed by the wall. Part of the absorbed energy is transformed into thermal energy, and part continues to propagate through the material of the wall. This is more clearly illustrated by considering a pulse that travels down a rope which consists of a light section and a heavy section as shown in Fig. 13-22. When the wave reaches the boundary between the two sections, part of the pulse is reflected and part is transmitted as shown. The heavier the second section, the less is transmitted; and when the second section is a wall or rigid support, very little is transmitted.

Refraction. When any wave strikes a boundary, some of the energy is reflected and some is transmitted or absorbed. When a two- or three-dimensional wave traveling in one medium crosses a boundary into a medium where its velocity is different, the transmitted wave may move in a different direction than the incident wave, as shown in Fig. 13-23. This phenomenon

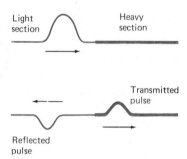

FIGURE 13-22

When a wave pulse reaches a discontinuity, part is reflected and part transmitted.

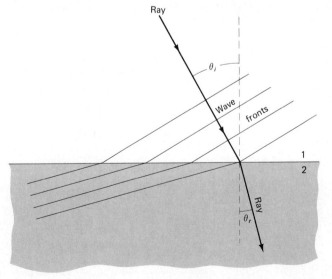

FIGURE 13-23

Refraction of waves passing a boundary.

313

is known as **refraction.** One example is a water wave; the velocity decreases in shallow water and the waves refract, Fig. 13-24. In Fig. 13-23, the velocity of the wave in medium 2 is less than in medium 1. In this case, the direction of the wave bends so it moves more nearly perpendicular to the boundary. That is, the angle of refraction, θ_r, is less than the angle of incidence, θ_i. To see why this is so, and to help us get a quantitative relation between θ_r and θ_i, let us think of each wave front as a row of soldiers; the soldiers are marching from firm ground (medium 1) into mud (medium 2) and hence are slowed down. The soldiers that reach the mud first are slowed down first and the row bends as shown in Fig. 13-25a. Let us consider the wave front (or row of soldiers) labeled A in Fig. 13-25b. In the same time t that A_1 moves a distance $l_1 = v_1 t$, we see that A_2 moves a distance $l_2 = v_2 t$. The two triangles shown have the side labeled a in common. Thus

$$\sin \theta_1 = \frac{l_1}{a} = \frac{v_1 t}{a}$$

and

$$\sin \theta_2 = \frac{l_2}{a} = \frac{v_2 t}{a}.$$

Dividing these two equations, we find that

$$\frac{\sin \theta_2}{\sin \theta_1} = \frac{v_2}{v_1}. \tag{13-16}$$

Since θ_1 is the angle of incidence (θ_i), and θ_2 is the angle of refraction (θ_r), Eq. 13-16 gives the quantitative relation between the two. Of course, if the wave were going in the opposite direction, the argument would not be changed; only θ_1 and θ_2 would change roles: θ_1 would be the angle of refraction and θ_2 the angle of incidence. Clearly then, if the wave travels into a medium where it can move faster, it will bend in the opposite way, $\theta_r > \theta_i$. We see from Eq. 13-16 that if the velocity increases, the angle increases and vice versa.

Earthquake waves refract within the earth as they travel through rock of different densities (and therefore the velocity is different) just as water

FIGURE 13-24

Water waves refracting. (Courtesy of U.S. Air Force, Cambridge Research Laboratories.)

FIGURE 13-25

Soldier analogy (a), to derive law of refraction for waves (b).

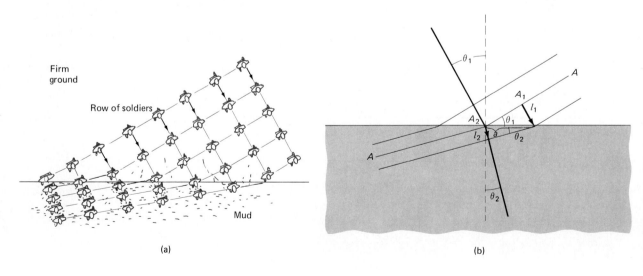

(a)

(b)

waves do. Light waves refract as well, and when we discuss light we shall find Eq. 13-16 very useful.

☐   **EXAMPLE 13-10**   An earthquake *P* wave passes across a boundary in rock where its velocity increases from 6.5 km/s to 8.0 km/s. If it strikes this boundary at 30°, what is the angle of refraction?

SOLUTION   Since sin 30° = 0.50, Eq. 13-16 yields

$$\sin \theta_2 = \frac{(8.0 \text{ m/s})}{(6.5 \text{ m/s})}(0.50) = 0.62,$$

so $\theta_2 = 38°$.

Interference. **Interference** refers to what happens when two waves pass through the same region of space at the same time. Consider for example the two wave pulses on a string traveling toward each other as shown in Fig. 13-26; in part (a) the two pulses have the same amplitude but one is a crest and the other a trough, and in part (b) they are both crests. In both cases, the waves meet and pass right on by each other. However, in the region where they overlap, the resultant displacement is the *algebraic sum of their separate displacements* (a crest is considered positive and a trough negative). This is called the **principle of superposition.** In Fig. 13-26a the two waves oppose one another as they pass by and the result is called **destructive interference.** In Fig. 13-26b, the resultant displacement is greater than that of either pulse and the result is called **constructive interference.** When two rocks are thrown into a pond simultaneously, the two sets of circular waves interfere with one another as shown in Fig. 13-27. In some areas of overlap, crests of one wave meet crests of the other (and troughs meet troughs); this is constructive interference and the water oscillates up and down with greater amplitude than either wave separately. In other areas, destructive interference occurs where the water actually does not move at all—this is where crests of one wave meet troughs of the other, and vice versa. In the first case, we say the two waves are *in phase* (by *phase* we mean the relative

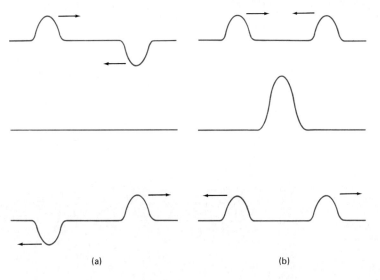

FIGURE 13-26

Two wave pulses pass each other. Where they overlap, interference occurs: (a) destructive; (b) constructive.

(a)   (b)

(a)

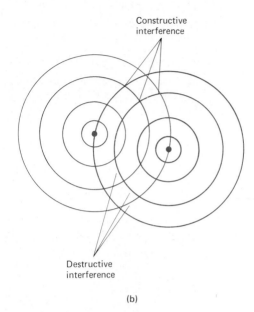

Constructive interference

Destructive interference

(b)

FIGURE 13-27

Interference of water waves. (D. Giancoli).

positions of the crests), whereas in destructive interference the two waves are *out of phase*. Of course, the relative phases of the two waves in most areas is intermediate between these two extremes, and this results in *partially destructive* interference. All three of these situations are shown in Fig. 13-28, where the amplitudes are plotted versus time at a given point in space.

Diffraction. Waves exhibit another important characteristic known as **diffraction.** This refers to the fact that waves spread as they travel, and when they encounter an obstacle they bend around it somewhat and pass into the region behind as shown in Fig. 13-29 for water waves. The amount of diffraction depends on the wavelength of the wave and on the size of the obstacle. This is shown in Fig. 13-30. If the wavelength is much larger than the object, such as the grass blades of Fig. 13-30a, the wave bends around them almost as if they were not there. For larger objects—parts (b) and (c)—there is more of a "shadow" region behind the obstacle. But notice in part (d), where the obstacle is the same as in part (c) but the wavelength is longer,

FIGURE 13-28

Two waves interfere: (a) constructively, (b) destructively, (c) partially destructively.

(a)

(b)

(c)

FIGURE 13-29

Wave diffraction. The waves come from the upper right. Note how they bend around and behind the headland. Note also how they bend around behind the rock outcrop from both sides, as diagrammed in Fig. 13-30b. (The headland and rock outcrop can be considered to form a "slit" through which the waves pass and spread out—compare to Fig. 23-2c for light.) (D. Giancoli)

that there is more diffraction into the shadow region. As a rule of thumb, only if the wavelength is less than the size of the object will there be a significant shadow region. It is worth noting that this rule applies to *reflection* from an obstacle as well. Very little of a wave is reflected unless the wavelength is less than the size of the obstacle. The fact that waves can bend around obstacles, and thus can carry energy to areas behind obstacles, is in clear distinction to energy carried by material particles. A clear example is the following: if you are standing around a corner on one side of a building, you can't be hit by a baseball thrown from the other side, but you can hear a shout or other sound because the sound waves diffract around the edges. Both interference and diffraction occur only for energy carried by waves and not for energy carried by material particles. This distinction was important for understanding the nature of light, and of matter itself, as we shall see in later chapters.

13-11 ▍ Standing waves; resonance

If you shake one end of a rope (or slinky) and the other end is kept fixed, a continuous wave will travel down to the fixed end and be reflected back. As you continue to vibrate the rope, there will be waves traveling in both directions, and the wave traveling down the rope will interfere with the reflected wave coming back. Usually there will be quite a jumble. But if you vibrate the rope at just the right frequency, these two waves will interfere in such a

(a) Water waves passing blades of grass

(b) Stick in water

(c) Short-wavelength waves passing log

(d) Long-wavelength waves passing log

FIGURE 13-30

Water waves passing objects of various sizes. Note that the larger the wavelength compared to the size of the object, the more diffraction there is into the "shadow region".

way that a large amplitude **standing wave** will be produced, Fig. 13-31. It is called a "standing wave" because it doesn't appear to be moving. The points of destructive interference, called **nodes,** and of constructive interference, called **antinodes,** remain in fixed positions. Standing waves occur at more than one frequency. The lowest frequency of vibration that produces a standing wave gives rise to the pattern shown in Fig. 13-31a. The standing waves shown in parts b and c are produced at precisely twice and three times the lowest frequency, assuming the tension in the rope is the same. The rope can also vibrate with four loops at four times the lowest frequency, and so on.

The frequencies at which standing waves are produced are the **natural frequencies** or **resonant frequencies** of the rope, and the different standing wave patterns shown in Fig. 13-31 are different "resonant modes of vibration." For although a standing wave is the result of the interference of two waves traveling in opposite directions, it is also an example of a vibrating object at resonance (Section 13-6). When a standing wave exists on a rope, the rope is vibrating in place; and at the frequencies at which resonance occurs, little effort is required to achieve a large amplitude. Standing waves then represent the same phenomenon as the resonance of a vibrating spring or pendulum which we discussed earlier. The only difference is that a spring or pendulum has only one resonant frequency, whereas the rope has an infinite number of resonant frequencies, each of which is a whole number multiple of the lowest resonant frequency.

Now let us consider a string stretched between two supports that is plucked like a guitar or violin string, Fig. 13-32a. Waves of a great variety of frequencies will travel in both directions along the string, will be reflected at the ends, and will travel back in the opposite direction. Most of these

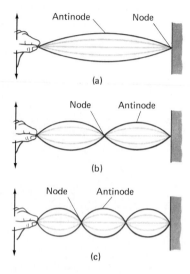

FIGURE 13-31

Standing waves corresponding to three resonant frequencies.

FIGURE 13-32

(a) A string is plucked. (b) Only standing waves corresponding to resonant frequencies persist for long.

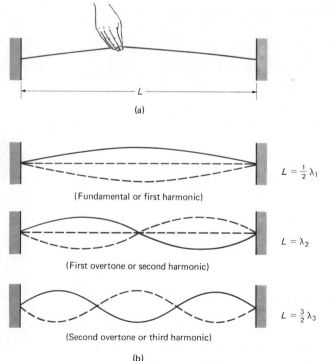

$L = \frac{1}{2}\lambda_1$

(Fundamental or first harmonic)

$L = \lambda_2$

(First overtone or second harmonic)

$L = \frac{3}{2}\lambda_3$

(Second overtone or third harmonic)

(b)

waves interfere in a random way with each other and quickly die away. However, those waves that correspond to the resonant frequencies of the string will persist. The ends of the string, since they are fixed, will be nodes. There may be other nodes as well. Some of the possible resonant modes of vibration (standing waves) are shown in Fig. 13-32b. Generally, the motion will be a combination of these different resonant modes; but only those frequencies that correspond to a resonant frequency will be present.

To determine the resonant frequencies, we first note that the wavelengths of the standing waves bear a simple relationship to the length L of the string. The lowest frequency, called the **fundamental frequency,** or **first harmonic,**[†] corresponds to one antinode (or loop); and as can be seen in Fig. 13-32b, the whole length corresponds to one-half wavelength. Thus $L = \frac{1}{2}\lambda_1$, where λ_1 stands for the wavelength of the fundamental. The next mode has two loops and is called the **second harmonic;** the length of the string L corresponds to one complete wavelength: $L = \lambda_2$. For the third and fourth harmonics, $L = \frac{3}{2}\lambda_3$ and $L = 2\lambda_4$, respectively, and so on. In general, we can write

$$L = \frac{n\lambda_n}{2}, \qquad \text{where} \quad n = 1, 2, 3, \ldots .$$

The integer n labels the number of the harmonic; $n = 1$ for the fundamental, $n = 2$ corresponds to the second harmonic, and so on. (The second harmonic is also called the first *overtone*, the third harmonic, the second overtone, and so on.) We solve for λ_n and find:

$$\lambda_n = \frac{2L}{n}, \qquad n = 1, 2, 3, \ldots . \qquad (13\text{-}17)$$

In order to find the frequency f of each vibration we use Eq. 13-9, $f = v/\lambda$.

Because a standing wave is equivalent to two traveling waves moving in opposite directions, the concept of velocity still makes sense and is given by Eq. 13-10 in terms of the tension F_T in the string and its mass per unit length $m/L = v = \sqrt{F_T/(m/L)}$.

☐ EXAMPLE 13-11   A piano string is 1.10 m long and has a mass of 9.00 g. (*a*) How much tension must the string be under if it is to vibrate at a fundamental frequency of 131 Hz? (*b*) What are the frequencies of the first four harmonics?

SOLUTION   (*a*) The wavelength of the fundamental is $\lambda = 2L = 2.20$ m (Eq. 13-17). The velocity is then $v = \lambda f = (2.20 \text{ m})(131 \text{ s}^{-1}) = 288$ m/s. Then, from Eq. 13-10 we have

$$F_T = \frac{m}{L} v^2$$

$$= \left(\frac{9.00 \times 10^{-3} \text{ kg}}{1.10 \text{ m}}\right)(288 \text{ m/s})^2 = 679 \text{ N}.$$

(*b*) The frequencies of the second, third, and fourth harmonics are 2, 3, and 4 times the fundamental frequency: 262, 393, and 524 Hz.

[†] The term "harmonic" comes from music, because such integral multiples of frequencies "harmonize."

A standing wave does appear to be standing still (and a traveling wave appears to move). The term "standing" wave is also meaningful from the point of view of energy. Since the string is at rest at the nodes, no energy flows past these points. Hence the energy "stands' in place in the string and is not transmitted down the string.

Standing waves are produced not only on strings, but on any object that is set into vibration. Even when a rock or a piece of wood is struck with a hammer, standing waves are set up that correspond to the natural resonant frequencies of that object. In general, the resonant frequencies depend on the dimensions of the object, just as for a string they depend on its length. For example a small object does not have as low resonant frequencies as a large object made of the same material. All musical instruments depend on standing waves to produce their musical sounds, from string instruments to wind instruments (in which a column of air vibrates as a standing wave) to drums and other percussion instruments. We will look at this in more detail in the next chapter.

❚ Summary

A vibrating object undergoes *simple harmonic motion* (SHM) if the restoring force is proportional to the displacement, $F = -kx$. The force constant k is the ratio of restoring force to displacement, $k = F/x$. The maximum displacement is called the *amplitude*. The *period* T is the time required for one complete cycle (back and forth) and the *frequency* f is the number of cycles per second; they are related by $f = 1/T$. The period of vibration for a mass m on the end of a spring is given by $T = 2\pi\sqrt{m/k}$. SHM is *sinusoidal*, which means that the displacement as a function of time follows a sine or cosine curve.

During SHM, the total energy $E = \frac{1}{2}mv^2 + \frac{1}{2}kx^2$ is continually changing from potential to kinetic and back again.

A *simple pendulum* of length L approximates SHM if its amplitude is small and friction can be ignored; its period is then given by $T = 2\pi\sqrt{L/g}$ where g is the acceleration of gravity.

When friction is present (for all real springs and pendulums), the motion is said to be *damped*. The maximum displacement decreases in time and the energy is eventually all transformed to heat.

If an oscillating force is applied to a system capable of vibrating, the amplitude of vibration can be very large if the frequency of the applied force is near the *natural* (or *resonant*) frequency of the oscillator; this is called *resonance*.

Vibrating objects act as sources of waves that travel outward from the source. Waves on water and on a string are examples. The wave may be a *pulse* (a single crest) or it may be continuous (many crests and troughs). The *wavelength* of a continuous wave is the distance between two successive crests. The *frequency* is the number of full wavelengths (or crests) that pass a given point per unit time. The *wave velocity* (how fast a crest moves) is equal to the product of wavelength and frequency, $v = \lambda f$. The *amplitude* of a wave is the maximum height of a crest, or depth of a trough, relative to the normal (or equilibrium) level.

In a *transverse wave*, the oscillations are perpendicular to the direction in which the wave travels; an example is a wave on a string. In a *longitudinal wave*, the oscillations are along (parallel to) the line of travel; sound is an example.

Waves reflect off objects in their path. When the *wave front* (of a two- or three-dimensional wave) strikes an object, the angle of reflection is equal to the angle of incidence. When a wave strikes a boundary between two materials in which it can travel, part of the wave is reflected and part is transmitted. The transmitted wave front may undergo *refraction* or bending. When two waves pass through the same region of space at the same time, they *interfere*. The resultant displacement at any point and time is the sum of their separate displacements; this can result in *constructive interference, destructive interference,* or something in between depending on the amplitudes and relative phases of the waves. Waves also undergo *diffraction,* meaning they can bend around an obstacle, so there is some wave motion behind the obstacle. The smaller the wavelength relative to the size of the object, the less the diffraction.

Waves traveling on a string (or other medium) of fixed length interfere with waves that have reflected off the end and are traveling back in the opposite direction. At certain frequencies, *standing waves* can be produced in which the waves seem to be standing still rather than traveling. The string (or other medium) is vibrating as a whole. This is a resonance phenomenon and the frequencies at which standing waves occur are called *resonant frequencies.* The points of destructive interference (no vibration) are called *nodes;* points of constructive interference (maximum amplitude of vibration) are called *antinodes.*

▌Questions

1. Give some examples of everyday vibrating objects. Which follow SHM, at least approximately?

2. Is the motion of a piston in an automobile engine simple harmonic? Explain. (See Fig. 12-12.)

3. Is the acceleration of a simple harmonic oscillator ever zero? If so, where?

4. Real springs have mass. How will the true period and frequency differ from those given by the equations for a mass oscillating on the end of an idealized massless spring?

5. How could you double the maximum speed of a SHO?

6. If you double the amplitude of a SHO, how does this change the frequency, maximum velocity, maximum acceleration, total mechanical energy?

7. A 10-kg fish is attached to the hook of a vertical spring scale, and is then released. Describe the scale reading as a function of time.

8. If a pendulum clock is accurate at sea level, will it gain or lose time when taken to the mountains?

9. The fact that harmonic motion is damped is said to be another example of the second law of thermodynamics. Explain.

10. Why can you make water slosh back and forth in a pan only if you shake the pan at a certain frequency?

11. A tuning fork of natural frequency 264 Hz sits on a table at the front of a room. At the back of the room, two tuning forks, one of natural frequency 260 Hz and one of 420 Hz, are initially silent, but when the tuning fork at the front of the room is set into vibration, the 260-Hz fork spontaneously begins to vibrate, but the 420-Hz fork does not. Explain.

12. Give several everyday examples of resonance.

13. Is a rattle in a car ever a resonance phenomenon? Explain.

14. Over the years, buildings have been able to be built out of lighter and lighter materials. How has this affected the natural vibration frequencies of buildings and the problems of resonance due to passing trucks, airplanes, or natural sources of vibration?

15. Is the frequency of a simple periodic wave equal to the frequency of its source? Why or why not?

16. Explain the difference between the speed of a transverse wave traveling down a rope and the speed of a tiny piece of the rope.

17. The speed of sound in most solids is somewhat greater than in air, yet the density of solids is much greater ($10^3 - 10^4$ times). Explain.

18. Give two reasons why circular water waves decrease in amplitude as they travel away from the source.

*19. Two linear waves have the same amplitude and otherwise are identical, except one has half the wavelength of the other. Which transmits more energy? How much more energy does it transfer?

*20. The intensity of a sound in real-life situations does not decrease precisely with the square of the distance from the source as we might expect from Eq. 13-15. Why not?

21. When a sinusoidal wave crosses the boundary between two sections of rope as in Fig. 13-22, the frequency does not change (although the wavelength and velocity do change). Explain why.

22. If we knew that energy was being transmitted from one place to another, how might we determine whether the energy was being carried by particles (material bodies) or by waves?

23. AM radio signals can usually be heard behind a hill, but FM often cannot. That is, AM signals bend more than FM. Explain. (Radio signals, as we shall see, are carried by electromagnetic waves whose wavelength for AM is typically 200 to 600 m and for FM about 3 m.)

24. If a string is vibrating in three segments, are there any places one can touch it with a knife blade without disturbing the motion?

25. When a standing wave exists on a string, the vibrations of incident and reflected waves cancel at the nodes. Does this mean that energy was destroyed?

Problems

SECTIONS 13-1 TO 13-3

1. (I) A piece of rubber is 45 cm long when a weight of 8.0 N hangs from it and is 58 cm long when a weight of 12.5 N hangs from it. What is the "spring" constant of this piece of rubber?

2. (I) If a particle undergoes SHM with amplitude A, what is the total distance it travels in one period?

3. (I) When an 80-kg person climbs into an 1100-kg car, the car's springs compress vertically by 1.2 cm. What will be the frequency of vibration when the car hits a bump? (Ignore damping.)

4. (I) A spring vibrates at 8.0 Hz, when a weight of 0.82 kg is hung from it. What is the spring constant?

5. (I) A fisherman's scale stretches 2.5 cm when a 2.1-kg fish hangs from it. What is the spring constant and what will be the frequency of vibration if the fish is pulled down and released so that it vibrates up and down?

6. (I) Construct a table indicating the position of the mass in Fig. 13-2 at the times $t = 0, \frac{1}{4}T, \frac{1}{2}T, \frac{3}{4}T, T, \frac{5}{4}T$, where T is the period of oscillation. On a graph of x versus t, plot these six points. Now connect these points with a smooth curve. Does your curve, based on these simple considerations, resemble that of a sine or cosine wave (Fig. 13-5)?

7. (II) A 0.350-kg mass at the end of a spring vibrates 2.0 times per second with an amplitude of 0.12 m. Determine (a) the velocity when it passes the equilibrium point, (b) the velocity when it is 0.10 m from equilibrium, (c) the total energy of the system, and (d) the equation describing the motion of the mass, assuming that at $t = 0$, x was a maximum.

8. (II) A spring vibrates with a frequency of 2.4 Hz when a weight of 0.60 kg is hung from it. What will its frequency be if only 0.30 kg hangs from it?

9. (II) A mass m at the end of a spring vibrates with a frequency of 0.62 Hz; when an additional 700-g mass is added to m, the frequency is 0.48 Hz. What is the value of m?

10. (II) A block of mass m is supported by two identical parallel vertical springs, each with spring constant k. What will be the frequency of vibration?

11. (II) A 1.0-kg mass vibrates according to the equation $x = 0.42 \cos 7.40t$ where x is in meters and t in seconds. Determine (a) the amplitude, (b) the frequency, (c) the total energy, (d) the kinetic energy and potential energy when $x = 0.16$ m.

12. (II) A 300-kg wooden raft floats on a lake. When a 75-kg man stands on the raft, it sinks deeper into the water by 5.0 cm. When the man steps off, the raft vibrates briefly. (a) What is the frequency of vibration? (b) What is the total energy of vibration (ignoring damping)?

13. (II) A mass of 2.70 kg stretches a vertical spring 0.325 m. If the spring is stretched an additional 0.110 m and released, how long does it take to reach the (new) equilibrium position again?

14. (II) At $t = 0$, a 650-g mass at rest on the end of a horizontal spring ($k = 64$ N/m) is struck by a hammer which gives it an initial speed of 1.16 m/s. Determine (a) the period and frequency of the motion, (b) the amplitude, (c) the maximum acceleration, (d) the position as a function of time, (e) the total energy.

15. (II) It takes a force of 60 N to compress the spring of a popgun 0.10 m to "load" a 0.200-kg ball. With what speed will the ball leave the gun?

16. (II) A mass on the end of a spring is stretched a distance x_0 from equilibrium and released. At what distance from equilibrium will it have (a) velocity equal to half its maximum velocity and (b) acceleration equal to half its maximum acceleration?

17. (II) Show that if a simple harmonic oscillator at rest at $x = 0$ is given a push at $t = 0$, the position x is given by $x = x_0 \sin\left[(2\pi/T)t\right]$.

18. (II) A spring of force constant 250 N/m vibrates with an amplitude of 8.50 cm when 0.330 kg hangs from it. (a) What is the equation describing this motion as a function of time? Assume that the mass passes through the equilibrium point at $t = 0$. (b) At what times will the spring have its maximum and minimum extensions?

19. (II) A 0.0120-kg bullet strikes a 0.300-kg block attached to a fixed horizontal spring whose spring constant is 5.20×10^3 N/m and sets it into vibration with an amplitude of 12.4 cm. What was the speed of the bullet before impact if the two objects move together after impact?

20. (II) A 70-kg person jumps from a window to a fire net 15 m below, which stretches the net 1.2 m. Assume that the net behaves like a simple spring, and calculate how much it would stretch if the same person were lying in it. How much would it stretch if the person jumped from 30 m?

21. (II) If one vibration has 10 times the energy of a second, but their frequencies and masses are the same, how do their amplitudes compare?

22. (III) The water in a U-shaped tube is displaced an amount Δx from equilibrium. (The level in one side is $2\Delta x$ above the level in the other side.) If friction is neglected, will the water oscillate harmonically? Determine a formula for the equivalent of the spring constant k. Does k depend on the density of the liquid, the cross section of the tube, or the length of the water column?

23. (III) If a mass m hangs from a vertical spring, show that the equilibrium length of the spring is a distance $l = mg/k$ longer than when the spring is horizontal, as in Fig. 13-1. Show also that $F = -kx$ holds for a vertical spring, where x is the displacement from the (vertical position) equilibrium point.

24. (III) In some diatomic molecules, the force each atom exerts on the other can be approximated by $F = -C/r^2 + D/r^3$ where C and D are positive constants. (a) Graph F versus r from $r = 0$ to $r = 2D/C$. (b) Show that equilibrium occurs at $r = r_0 = D/C$. (c) Let $\Delta r = r - r_0$ be a small displacement from equilibrium, where $\Delta r \ll r_0$. Show that for such small displacements the motion is approximately simple harmonic, and (d) determine the force constant. (e) What is the period of such motion? (Hint: assume one atom is kept at rest.)

25. (III) Use the reference circle to show that the velocity and acceleration of an object undergoing SHM are given by:

$$v = -v_0 \sin \omega t, \qquad a = -a_0 \cos \omega t,$$

where $v_0 = x_0\sqrt{k/m}$ and $a_0 = kx_0/m$ refer to the maximum speed and acceleration.

26. (III) A mass m is gently placed on the end of a freely hanging spring. The mass then falls 30 cm before it stops and begins to rise. What is the frequency of the motion?

27. (III) A rectangular block of wood floats in a calm lake. Show that, if friction is ignored, when the block is pushed gently down into the water, it will then vibrate with SHM. Also, determine an equation for the force constant.

SECTION 13-4

28. (I) A pendulum makes 24 vibrations in 30 s. What is its (a) period, and (b) frequency?

29. (I) How long must a simple pendulum be if it is to make exactly one complete vibration per second?

30. (II) What is the period of a simple pendulum 60 cm long (a) on the earth, and (b) when it is in a freely falling elevator.

31. (II) A pendulum oscillates at a frequency of 2.0 Hz. At $t = 0$ it is released starting at an angle of $20°$. Ignoring friction, what will be the position (angle) of the pendulum at (a) $t = 0.25$ s, (b) $t = 1.60$ s, and (c) $t = 500$ s?

32. (II) The length of a simple pendulum is 0.25 m and it is released at an angle of $20°$ to the vertical. (a) With what frequency does it vibrate? (b) What is the pendulum bob's speed when it passes through the lowest point of the swing? (Assume SHM.)

33. (II) Derive a formula for the maximum speed v_0 of a simple pendulum bob in terms of g, the length L, and the angle of swing θ_0.

34. (II) A simple pendulum oscillates with frequency f. What is its frequency if it accelerates at $\frac{1}{2}g$ (a) upward, (b) downward?

35. (III) A pendulum vibrates with an amplitude of $10°$. What fraction of the time does it spend between $+5°$ and $-5°$? Assume SHM.

36. (III) The pendulum of a clock consists of a small heavy bob at the end of a brass rod. The clock keeps accurate time at $15°C$, at which point the pendulum has a period of 0.4220 s. When the temperature is $35°C$, will the clock be fast or slow? By how much will it be in error after 1 day at $35°C$?

SECTIONS 13-7 AND 13-8

37. (I) A fisherman notices that wave crests pass the bow of his anchored boat every 6.0 s. He measures the distance between two crests to be 20 m. How fast are the waves traveling?

38. (I) A sound wave in air has a frequency of 262 Hz and travels with a speed of 330 m/s. How far apart are the wave crests (compressions)?

39. (I) AM radio signals have frequencies between 550 kHz and 1600 kHz (kilohertz) and travel with a speed of 3.0×10^8 m/s. What are the wavelengths of these signals? On FM the frequencies range from 88 MHz to 108 MHz (megahertz) and travel at the same speed; what are their wavelengths?

40. (I) Calculate the speed of longitudinal waves in (a) water, (b) granite.

41. (I) Two solid rods have the same bulk modulus, but one is twice as dense as the other. In which rod will the speed of longitudinal waves be greater, and by what factor?

42. (II) A rope of mass 0.75 kg is stretched between two supports 30 m apart. If the tension in the rope is 1200 N, how long will it take a pulse to travel from one support to the other?

43. (II) The ripples in a certain groove 12 cm from the center of a 33-rpm phonograph record have a wavelength of 2.4 mm. What will be the frequency of the sound emitted?

44. (II) Determine the wavelength of a 10,000-Hz sound wave traveling along an iron rod. (Hint: see Table 8-1.)

45. (II) A sailor strikes the side of his ship just below the surface of the sea. He hears the echo of the wave reflected from the ocean floor directly below 1.5 s later. How deep is the ocean at this point?

46. (II) S and P waves from an earthquake travel at different speeds, and this fact helps in the determination of the earthquake "focus" (where the disturbance took place). (a) Assuming typical speeds of 9.0 km/s and 5.0 km/s for P and S waves, respectively, how far away did the earthquake occur if a particular seismic station detects the arrival of these two types of waves 2.0 min apart? (b) Is one seismic station sufficient to determine the position of the focus? Explain.

*SECTION 13-9

***47. (I)** Two earthquake waves have the same frequency as they travel through the same portion of the earth, but one is carrying twice the energy. What is the ratio of the amplitudes of the two waves?

***48. (II)** Leaves on the surface of a pond are observed to move up and down a total vertical distance of 0.20 m as a wave passes. (a) What is the amplitude of the wave? (b) If the amplitude were changed to 0.15 m, by what factor would the energy in the wave change?

FIGURE 13-33

*49. (II) Compare (a) the intensities and (b) the amplitudes of an earthquake wave as it passes two points 10 km and 20 km from the source.

*50. (II) The intensity of a particular earthquake wave is measured to be 1.0×10^6 J/m²·s at a distance of 100 km from the source. (a) What was the intensity when it passed a point only 1.0 km from the source? (b) What was the rate energy passed through an area of 5.0 m² at 1.0 km?

*51. (III) Show that the amplitude x_0 of circular water waves decreases as the square root of the distance r from the source: $x_0 \propto 1/\sqrt{r}$. Ignore damping.

SECTION 13-10

52. (I) The two pulses shown in Fig. 13-33 are moving toward each other. (a) Sketch the shape of the string at the moment they directly overlap. (b) Sketch the shape of the string a few moments later. (c) In Fig. 13-26a, at the moment the pulses pass each other, the string is straight. What has happened to the energy at this moment?

53. (I) An earthquake P wave traveling 13.9 km/s strikes a boundary within the earth between two kinds of material. If it approaches the boundary at an incident angle of 42° and the angle of refraction is 23°, what is the speed in the second medium?

54. (I) Water waves approach a shelf where the velocity changes from 2.8 m/s to 1.8 m/s. If the incident waves make a 25° angle with the shelf, what will be the angle of refraction?

55. (II) A longitudinal earthquake wave strikes a boundary between two types of rock at a 10° angle. As it crosses the boundary, the specific gravity of the rock changes from 3.8 to 4.7. Assuming that the elastic modulus is the same for both types of rock, determine the angle of refraction.

56. (II) Consider a sine wave traveling down the stretched two-part cord of Fig. 13-22. Determine a formula (a) for the ratio of the speeds of the wave in the two sections, v_2/v_1, and (b) for the ratio of the wavelengths in the two sections. (The frequency is the same in both

sections. Why?) (c) Is the wavelength greater in the heavier cord or the lighter?

57. (II) It is found for any type of wave, say an earthquake wave, that if it reaches a boundary beyond which its speed is increased, there is a maximum incident angle if there is to be a transmitted refracted wave. This maximum incident angle θ_{iM} corresponds to an angle of refraction equal to 90°. If $\theta_i > \theta_{iM}$, all the wave is reflected at the boundary and none is refracted (because this would correspond to $\sin \theta_r > 1$, where θ_r is the angle of refraction, which is impossible); this is referred to as *total internal reflection*. (a) Find a formula for θ_{iM} using Eq. 13-16. (b) At what angles of incidence will there be only reflection and no transmission for an earthquake P wave traveling 6.5 km/s when it reaches a different kind of rock where its speed is 8.2 km/s?

SECTION 13-11

58. (I) A violin string vibrates at 196 Hz when unfingered. At what frequency will it vibrate if it is fingered one-fourth of the way down from the end?

59. (I) If a violin string vibrates at 440 Hz as its fundamental frequency, what are the frequencies of the first four harmonics?

60. (I) A particular string resonates in four loops at a frequency of 220 Hz. Name at least three other frequencies at which it will resonate.

61. (I) The velocity of waves on a string is 400 m/s. If the frequency of standing waves is 380 Hz, how far apart are the nodes?

62. (II) If two successive overtones of a vibrating string are 400 Hz and 450 Hz, what is the frequency of the fundamental?

63. (II) When you slosh the water back and forth in a tub at just the right frequency, the water alternately rises and falls at each end. Suppose the frequency to produce such a standing wave in a 50-cm wide tub is 0.85 Hz. What is the speed of the water wave?

64. (II) A guitar string is 90 cm long and has a mass of 3.6 g. From the bridge to the support post ($=L$) is 60 cm and the string is under a tension of 520 N. What are the frequencies of the fundamental and first two overtones?

65. (II) A particular violin string plays at a frequency of 294 Hz. If the tension is increased 10 percent, what will the frequency be now?

66. (II) Show that the frequency of standing waves on a string of length L and linear density μ, which is stretched to a tension F_T, is given by

$$f = \frac{n}{2L}\sqrt{\frac{F_T}{\mu}}$$

where n is an integer.

14 SOUND

Sound is associated with our sense of hearing and, therefore, with the physiology of our ears and the psychology of our brain which interprets the sensations that reach our ears. The term *sound* also refers to the physical sensation that stimulates our ears: namely, longitudinal waves.

We can distinguish three aspects of any sound. First, there must be a *source* for a sound; and as with any wave, the source of a sound wave is a vibrating object. Second, the energy is transferred from the source in the form of longitudinal sound *waves*. And third, the sound is *detected* by an ear or an instrument. We will discuss sources and detectors of sound later in this chapter, as well as some important applications to fields such as medicine, but now we look at some aspects of sound waves themselves.

14-1 ∎ Characteristics of sound

We already saw in Chapter 13, Fig. 13-16, how a vibrating drumhead produces a sound wave in air. Indeed, we usually think of sound waves traveling in the air, for normally it is the vibrations of the air that force our eardrums to vibrate. But sound waves can also travel in other materials. Two stones struck together under water can be heard by a swimmer beneath the surface, for the vibrations are carried to the ear by the water. When you put your ear flat against the ground, you can hear an approaching train or truck. In this case the ground does not actually touch your cardrum but the longitudinal wave transmitted by the ground is called a sound wave just the same. Indeed, longitudinal waves traveling in any material medium are often referred to as sound waves. Clearly sound cannot travel in the absence of matter. For example, a bell ringing inside an evacuated jar cannot be heard.

The **speed of sound** is different in different materials. In air at 0°C and 1 atm, sound travels at a speed of 331 m/s. We saw in Eq. 13-11 that the speed depends on the elastic modulus and the density of the material. Thus for helium, whose density is much less than that of air but whose elastic modulus is not greatly different, the speed is about three times as large as in air. In liquids and solids, which are much less compressible and therefore have much greater elastic moduli, the speed is larger still. The speed of sound in various materials is given in Table 14-1. The values depend somewhat on temperature, but this is significant mainly for gases. For example, in air, the

TABLE 14-1

Speed of sound in various materials at 20°C

Material	Speed (m/s)
Air	343
Air (0°C)	331
Helium	1005
Hydrogen	1300
Water	1440
Sea water	1560
Iron and steel	≈5000
Glass	≈4500
Aluminum	≈5100
Hard wood	≈4000

speed increases approximately 0.60 m/s for each Celsius degree increase in temperature:

$$v \approx (331 + 0.60\ T)\ \text{m/s},$$

where T is the temperature in °C. Unless stated otherwise, we will assume in this chapter that $T = 20°C$, so that $v = [331 + (0.60)(20)]$ m/s $= 343$ m/s.

Two aspects of any sound are immediately evident to a human listener. These are "loudness" and "pitch," and each refers to a sensation in the consciousness of the listener. But to each of these subjective sensations there corresponds a physically measurable quantity. **Loudness** is related to the energy in the sound wave and we shall discuss it in the next section.

The **pitch** of a sound refers to whether it is high, like the sound of a piccolo or violin, or low, like the sound of a bass drum or string bass. The physical quantity that determines pitch is the frequency, a fact that was first noted by Galileo. The lower the frequency, the lower the pitch, and the higher the frequency, the higher the pitch.[†] The human ear responds to frequencies in the range from about 20 Hz to about 20,000 Hz. This is called the **audible range.** These limits vary somewhat from one individual to another. One general trend is that as people age, they are less able to hear the high frequencies, so that the high-frequency limit may be 10,000 Hz or less.

Sound waves whose frequencies are outside the audible range may reach the ear, but we are not generally aware of them. Frequencies above 20,000 Hz are called *ultrasonic* (do not confuse with "supersonic," which means a speed faster than the speed of sound). Many animals can hear ultrasonic frequencies; dogs, for example, can hear sounds as high as 50,000 Hz and bats can detect frequencies as high as 100,000 Hz. Ultrasonic waves have a number of applications in medicine and other fields, which we will discuss later in this chapter.

Sound waves whose frequencies are below the audible range (that is, less than 20 Hz) are called *infrasonic.* Sources of infrasonic waves are earthquakes, thunder, volcanoes, and waves produced by vibrating heavy machinery. This last source can be particularly troublesome to workers, for infrasonic waves—even though inaudible—can cause damage to the human body. These low-frequency waves act in a resonant fashion, causing considerable motion and irritation of internal organs of the body.

14-2 ▌ Intensity of sound

Like pitch, **loudness** is a sensation in the consciousness of a human being. It, too, is related to a physically measurable quantity, the **intensity** of the wave. Intensity is defined as the energy transported by a wave per unit time across unit area. Since energy per unit time is power, intensity has units of power per unit area, or watts/meter² (W/m²). The intensity depends on the amplitude of the wave (actually it is proportional to the square of the amplitude, as was shown in Section 13-9).

The human ear can detect sounds with an intensity as low as 10^{-12} W/m² and as high as 1 W/m² (and even higher, although above this it is

---

[†] Although pitch is determined mainly by frequency, it also depends to a slight extent on loudness; for example, a very loud sound may seem slightly lower in pitch than a quiet sound of the same frequency.

painful). This is an incredibly wide range of intensity, spanning a factor of 10^{12} from lowest to highest. Presumably because of this wide range, what we perceive as loudness is not directly proportional to the intensity. True, the greater the intensity, the louder the sound. But to produce a sound that sounds about twice as loud requires a sound wave that has about 10 times the intensity. This is true to a first approximation at any sound level. For example, a sound wave of intensity 10^{-9} W/m^2 sounds to an average human being as if it is about twice as loud as one whose intensity is 10^{-10} W/m^2; and an intensity of 10^{-2} W/m^2 sounds about twice as loud as 10^{-3} W/m^2 and four times as loud as 10^{-4} W/m^2.

Because of this relationship between the subjective sensation of loudness and the physically measurable quantity intensity, it is usual to specify sound intensity levels using a logarithmic scale. The unit on this scale is a *bel*, or much more commonly, the *decibel* (dB), which is $\frac{1}{10}$ bel (1 dB = 0.1 bel). The **intensity level,** β, of any sound is defined in terms of its intensity, I, as follows:

$$\beta \text{ (in dB)} = 10 \log \frac{I}{I_0} \qquad (14\text{-}1)$$

where I_0 is the intensity of some reference level and the logarithm is to the base 10. I_0 is usually taken as the minimum intensity audible to an average person, the "threshold of hearing," which is $I_0 = 1.0 \times 10^{-12}$ W/m^2. The intensity level of a sound, for example, whose intensity is $I = 1.0 \times 10^{-10}$ W/m^2 will be (logarithms are discussed in Appendix A)

$$\beta = 10 \log \frac{10^{-10}}{10^{-12}} = 10 \log 100 = 20 \text{ dB},$$

since log 100 is equal to 2.0. Notice that the intensity level at the threshold of hearing is 0 dB; that is, $\beta = 10 \log (10^{-12}/10^{-12}) = 10 \log 1 = 0$. Notice, too, that an increase in intensity by a factor of 10 corresponds to a level increase of 10 dB. An increase in intensity by a factor of 100 corresponds to a level increase of 20 dB. Thus a 50-dB sound is 100 times more intense than a 30-dB sound.

The intensities and intensity levels for a number of common sounds are listed in Table 14-2.

TABLE 14-2
Intensity of various sounds

Source of the Sound	Intensity Level (dB)	Intensity (W/m^2)
Jet plane at 30 m	140	100
Threshold of pain	120	1
Loud indoor rock concert	120	1
Siren at 30 m	100	1×10^{-2}
Auto interior, moving at 90 km/h	75	3×10^{-5}
Busy street traffic	70	1×10^{-5}
Ordinary conversation, at 50 cm	65	3×10^{-6}
Quiet radio	40	1×10^{-8}
Whisper	20	1×10^{-10}
Rustle of leaves	10	1×10^{-11}
Threshold of hearing	0	1×10^{-12}

EXAMPLE 14-1   A high-quality loudspeaker is advertised to repro-duce, at full volume, frequencies from 30 Hz to 18,000 Hz with uniform intensity ± 3 dB. That is, over this frequency range, the intensity level does not vary by more than 3 dB from the center point. By what factor does the intensity change for the maximum intensity-level change of 3 dB?

SOLUTION   Let us call the average intensity I_1 and the average level β_1. Then the maximum intensity, I_2, corresponds to a level $\beta_2 = \beta_1 + 3$ dB. Thus

$$\beta_2 - \beta_1 = 10 \log \frac{I_2}{I_0} - 10 \log \frac{I_1}{I_0}$$

$$3 \text{ dB} = 10 \left(\log \frac{I_2}{I_0} - \log \frac{I_1}{I_0} \right)$$

$$= 10 \log \frac{I_2}{I_1}$$

since $(\log a - \log b) = \log a/b$. Then

$$\log \frac{I_2}{I_1} = 0.30.$$

Using a calculator we calculate 10^x with $x = 0.30$, or from a log table we find the logarithm of what number equals 0.30. The result is 2.0, so

$$\frac{I_2}{I_1} = 2.0$$

or I_2 is twice as intense as I_1.

It is worth noting that a sound level difference of 3 dB (which corre-sponds to a doubled intensity as we just saw) corresponds to only a very small change in the subjective sensation of apparent loudness. Indeed, the average human can distinguish a difference in level of only about 1 dB.

*14-3 ▮ Intensity related to amplitude and pressure amplitude

The intensity I is proportional to the square of the wave amplitude, x_0, as discussed in Section 13-9. Indeed, using Eq. 13-14 we can relate the ampli-tude quantitatively to the intensity I or level β, as the following example shows.

EXAMPLE 14-2   Calculate the maximum displacement of air mole-cules for a sound at the threshold of hearing, having a frequency of 1000 Hz.

SOLUTION   We use Eq. 13-14 and solve for x_0:

$$x_0 = \frac{1}{\pi f} \sqrt{\frac{I}{2\rho v}}$$

$$= \frac{1}{(3.14)(1.0 \times 10^3 \text{ s}^{-1})} \sqrt{\frac{1.0 \times 10^{-12} \text{ W/m}^2}{(2)(1.29 \text{ kg/m}^3)(343 \text{ m/s})}}$$

where we have taken the density of air to be 1.29 kg/m^3 and the speed of sound in air (assumed 20°C) as 343 m/s. When we do the arithmetic, we find that $x_0 = 1.1 \times 10^{-11}$ m.

The result of this example illustrates just how sensitive the human ear is. For it can detect displacements of air molecules which are actually less than the diameter of atoms (about 10^{-10} m)!

We have so far described sound waves in terms of the vibration of the molecules of the medium. But they can also be viewed from the point of view of pressure. Indeed, longitudinal waves are often referred to as **pressure waves.** Referring to Fig. 13-16, it should be clear that the pressure of the air in a compression is higher than normal, whereas in a rarefaction (or expansion) the pressure is less than normal. Figure 14-1 shows a graphical representation of a sound wave in air both in terms of displacement (part a) and pressure (part b). The pressure in a 1000-Hz sound wave at the threshold of hearing differs from atmospheric by at most $3 \times 10^{-5} \text{ N/m}^2$ ($= 3 \times 10^{-10}$ atm). This is called the *pressure amplitude*. Just as the intensity is proportional to the square of the displacement amplitude, x_0, it is also proportional to the square of the pressure amplitude. Indeed, it can be shown that the intensity is related to the pressure amplitude P_0 by the formula

$$I = \frac{P_0^2}{2\rho v},$$

where ρ is the density of the medium and v the velocity of sound in that medium.

Normally, the loudness or intensity of a sound decreases as you get farther from the source of the sound. In interior rooms, this effect is reduced because of reflections from the walls. However, if a source is in the open so that sound can radiate out freely in all directions, the intensity decreases as the inverse square of the distance,

$$I \propto \frac{1}{r^2},$$

as we saw in Eq. 13-15. Of course, if there is significant reflection from structures or the ground, the situation will be more complicated.

(a) Displacement

(b) Pressure

FIGURE 14-1

Representation of a sound wave in terms of (a) displacement; (b) pressure.

EXAMPLE 14-3   The intensity level of the sound from a jet plane at a distance of 30 m is 140 dB. What is the intensity level at 300 m? (Ignore reflections from the ground.)

SOLUTION   The intensity I at 30 m is found from Eq. 14-1: 140 dB = $10 \log I/(10^{-12} \text{ W/m}^2)$. Reversing the log equation to solve for I we have: $10^{14} = I/(10^{-12} \text{ W/m}^2)$, so $I = 10^2 \text{ W/m}^2$. At 300 m, ten times as far away, the intensity will be $(\frac{1}{10})^2 = 1/100$ as much, or 1 W/m^2. Hence, the intensity level $\beta = 10 \log (1/10^{-12}) = 120$ dB. Even at 300 m, the sound is at the threshold of pain. This is why workers at airports wear ear covers to protect their ears from damage.

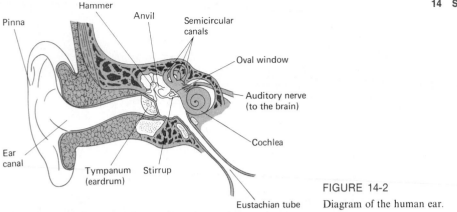

Pinna
Hammer
Anvil
Semicircular canals
Oval window
Auditory nerve (to the brain)
Cochlea
Ear canal
Tympanum (eardrum)
Stirrup
Eustachian tube

FIGURE 14-2

Diagram of the human ear.

*14-4 ▮ The ear and its response; loudness

The human ear, as we have seen, is a remarkably sensitive detector of sound. Mechanical detectors of sound, namely microphones, can barely match the ear in detecting low-intensity sounds.

The function of the ear is to efficiently transform the vibrational energy of sound waves into electrical signals which are carried to the brain by way of nerves. A microphone performs a similar task. Sound waves striking the diaphragm of a microphone set it into vibration. A "transducer" transforms this vibration into an electrical signal of the same frequency, which can then be amplified and sent to a loudspeaker or tape recorder. We shall discuss the operation of transducers when we study electricity and magnetism in later chapters. Here we shall discuss the structure and response of the ear.

Figure 14-2 is a diagram of the human ear. The ear is conveniently divided into three main divisions; the outer ear, middle ear, and inner ear. In the outer ear, sound waves from the outside travel down the ear canal to the ear drum (the tympanum), which vibrates in response to the impinging waves. The middle ear consists of three small bones known as the hammer, anvil, and stirrup, which transfer the vibrations of the eardrum to the inner ear at the oval window. This delicate system of levers offers a mechanical advantage of about two. The amplification of the pressure in the wave is much greater than this, however, since the area of the eardrum is about 20 times the area of the oval window. Thus the pressure is amplified by a factor of about 40. The inner ear consists of the semicircular canals, which are important for controlling balance, and the liquid-filled cochlea, where the vibrational energy of sound waves is transformed into electrical energy and sent to the brain. Figure 14-3 is a diagrammatic representation of the cochlea. The sound vibration travels from the oval window down the vestibular canal and back up the tympanic canal. Because of the viscosity of the liquid, considerable damping occurs, but any remaining energy is dissipated at the round window at the end of the tympanic canal. Between these two canals is a third, known as the cochlear duct. On the membrane separating the cochlear duct from the tympanic canal (the basilar membrane) is the "organ of Corti," which contains some 30,000 nerve endings. As a pressure wave passes along the tympanic canal, it causes ripples in the basilar mem-

FIGURE 14-3

Diagram of cochlea.

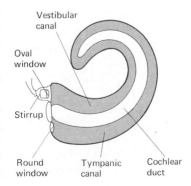

Vestibular canal
Oval window
Stirrup
Round window
Tympanic canal
Cochlear duct

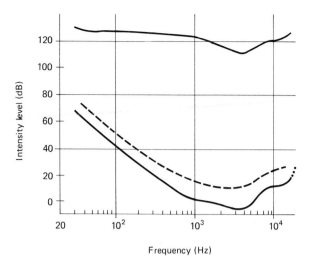

FIGURE 14-4

Sensitivity of the human ear as a function of frequency (see text). Note that the frequency scale is "logarithmic", in order to cover a wide range of frequencies.

brane and the attached organ of Corti; this is where the energy is transformed into electrical impulses and sent to the brain by way of the auditory nerve. The basilar membrane is under tension, but is less taut and becomes thicker as it goes from the middle ear to the apex of the cochlea. From our earlier considerations we might expect that the thicker, less taut end would be more sensitive to low frequencies and the tighter, thinner end more sensitive to the higher frequencies. Careful experiments indicate that this is indeed true, and this fact is important for our sensing of pitch.

The ear is not equally sensitive to all frequencies. The lowest curve in Fig. 14-4 represents the intensity level, as a function of frequency, in the softest sound that is just audible. As can be seen, the ear is most sensitive to sounds of frequency between 2000 and 3000 Hz. Whereas a 1000-Hz sound is audible at a level of 0 dB, a 100-Hz sound must be at least 30 dB to be heard. This lowest (solid) curve represents a very good ear; only about 1 percent of the population, mostly young people, have such a low "threshold of hearing." The middle (dashed) curve represents a more typical curve; 50 percent of the population have a threshold of hearing equal to or better than this. The top curve represents the "threshold of feeling or pain." Sounds above this level can actually be felt and cause pain. As can be seen, it does not vary much with frequency.

Our subjective sensation of loudness obviously depends not only on the intensity but also on frequency. For example, as seen in Fig. 14-4, an average person will detect a 30-dB sound at 1000 Hz as reasonably loud but a 30-dB sound at 50 Hz would not be heard at all.

To obtain the same loudness for sounds of different frequencies requires different intensities. Studies averaged over large numbers of people have produced the curves shown in Fig. 14-5. On this graph, each curve represents sounds that seemed to be equally loud. The number labeling each curve represents the *loudness level* (the units are called *phons*), which is numerically equal to the intensity level at 1000 Hz for that curve. For example, the curve labeled 40 represents sounds that are heard to have the same loudness as a 1000-Hz sound with an intensity level of 40 dB. From this curve we see that a 100-Hz tone must have an intensity of about 62 dB to sound as loud (to an average person) as a 1000-Hz tone of only 40 dB.

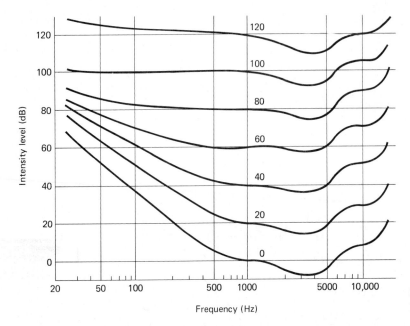

FIGURE 14-5

Loudness—level curves (see text).

It is quite clear that at lower intensity levels, our ears are less sensitive to the high and low frequencies relative to middle frequencies. The "loudness" control on stereo and hi-fi systems is intended to compensate for this. As the volume is turned down, the loudness control boosts the high and low frequencies relative to the middle frequencies so that the sound will be more uniform. Many listeners, however, find the sound more pleasing or natural without the loudness control.

14-5 ▌ Sources of sound: vibrating strings and air columns

The source of any sound is a vibrating object. Almost any object can vibrate and hence be a source of sound. We now discuss some simple sources of sound, particularly musical instruments. In musical instruments, the source is set into vibration by striking, plucking, bowing, or blowing. Standing waves are produced and the object vibrates at its natural resonant frequencies. A drum has a stretched membrane that vibrates. Xylophones and marimbas have metal or wood bars that can be set into vibration. Bells, cymbals, and gongs also make use of a vibrating metal. The most widely used instruments make use of vibrating strings, such as the violin, guitar, and piano, or make use of vibrating columns of air, such as the flute, trumpet, and pipe organ. (We have already seen that the pitch of a pure sound is determined by the frequency. Typical frequencies for musical notes on the so-called "equally-tempered chromatic scale" are given, simply for reference, in Table 14-3 for the octave beginning with middle C.)

We saw in Chapter 13, Fig. 13-32, how standing waves are established on a string. This is the basis for all stringed instruments. The pitch is normally determined by the lowest resonant frequency, the *fundamental*. which

TABLE 14-3

Equally tempered chromatic scale[†]

Note	Frequency (Hz)
C	262
C# or D♭	277
D	294
D# or E♭	311
E	330
F	349
F# or G♭	370
G	392
G# or A♭	415
A	440
A# or A♭	466
B	494
C′	524

[†] Only one octave is included.

corresponds to nodes occurring only at the ends. The wavelength of the fundamental is equal to twice the length of the string; therefore, the fundamental frequency $f = v/\lambda = v/2L$ where v is the velocity of the wave on the string. When a finger is placed on the string of, say, a guitar or violin, the effective length of the string is shortened; so its pitch is higher since the wavelength of the fundamental is shorter (Fig. 14-6). The strings on a guitar or violin are all the same length. They sound at a different pitch because the strings have different mass per unit length, m/L, which affects the velocity as seen in Eq. 13-10, $v = \sqrt{F_T/(m/L)}$. (The tension may also be different; adjusting the tension is the means for tuning the instrument.) Thus the velocity on a heavier string is less and the frequency will be less for the same wavelength. In pianos and harps, the strings are each of different length; for the lower notes they are not only longer but heavier as well, and the reason why is illustrated in the following example.

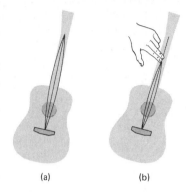

(a)   (b)

FIGURE 14-6

The wavelength of a fingered string (b) is shorter than that of an unfingered string (a). Hence, the frequency of the fingered string is higher. Only the simplest standing wave, the fundamental, is shown.

□ EXAMPLE 14-4 The highest key on a piano corresponds to a frequency about 150 times that of the lowest key. If the string for the highest note is 5.0 cm long, how long would the string for the lowest note have to be if it had the same mass per unit length and was under the same tension?

SOLUTION The velocity would be the same on each string so the frequency is inversely proportional to the length L of the string ($f = v/\lambda = v/2L$). Thus

$$\frac{L_1}{L_h} = \frac{f_h}{f_1}$$

where the subscripts l and h refer to the lowest and highest notes, respectively. Thus $L_1 = L_h(f_h/f_1) = (5.0 \text{ cm})(150) = 750$ cm, or 7.5 m. This would be ridiculously long for a piano. The longer lower strings are made heavier partly to avoid this, so that even on grand pianos the strings are no longer than about 3 m.

Stringed instruments would not be very loud if they relied on their vibrating strings to produce the sound waves since the strings are simply too thin to compress and expand much air. Stringed instruments therefore make use of a kind of mechanical amplifier known as a *sounding board* (piano) or *sounding box* (guitar, violin) which acts to amplify the sound by putting a greater surface area in contact with the air. When the strings are set into vibration, the sounding board or box is set into vibration as well. Since it has much greater area in contact with the air, it can produce a much stronger sound wave; it thus acts to amplify the sound. On an electric guitar the sounding box is not so important since the vibrations of the strings are amplified electrically.

Instruments such as woodwinds, the brasses, and the pipe organ produce sound from the vibrations of standing waves in a column of air within a tube or pipe. Standing waves can occur in the air of any cavity, but the frequencies present are complicated for any but very simple shapes such as a long, narrow tube. Fortunately, and for good reason, this is the situation for most wind instruments. In some instruments, a vibrating reed or the vibrating lip of the player helps to set up vibrations of the air column. In others a stream of air is directed against one edge of the opening or mouth-piece, leading to turbulence which sets up the vibrations. Because of the disturbance, whatever its source, the air within the tube vibrates with a variety of frequencies; but only certain frequencies persist, which correspond to standing waves.

For a string fixed at both ends, we saw that the standing waves have nodes (no movement) at the two ends, and one or more antinodes (large amplitude of vibration) in between; a node also separates each antinode. The lowest-frequency standing wave, the *fundamental*, corresponds to a single antinode. The higher-frequency standing waves are called overtones or harmonics; specifically, the first harmonic is the fundamental, the second harmonic has twice the frequency of the fundamental,[†] and so on; see Fig. 13-32.

The situation is similar for a column of air, but we must remember that it is now air itself that is vibrating. Thus the air at the closed end of a tube must be a (displacement) node since the air is not free to move there, whereas at the open end of a tube there will be an antinode since the air can move freely. The air within the tube vibrates in the form of longitudinal standing waves. The possible modes of vibration for a tube open at both ends (called an *open tube*) and for one that is open at one end but closed at the other (called a *closed tube*) are shown graphically in Fig. 14-7. The graphs represent the displacement amplitude of the vibrating air within the tube. The antinodes do not occur precisely at the open ends of tubes; their position depends on the diameter of the tube, but if this is small compared to the length, which is the usual case, the antinode occurs very close to the end as shown, and we assume this is the case in what follows. (Wavelength and other factors can also affect the position of the antinode.)

Let us look first at the open pipe, Fig. 14-7a. An open pipe has displacement antinodes at both ends. Notice that there must be at least one node within an open pipe if there is to be a standing wave at all. This corresponds to the fundamental frequency of the tube; since the distance between two successive nodes, or between two successive antinodes, is $\frac{1}{2}\lambda$, there is one-half a wavelength within the length of the tube in this case: $L = \frac{1}{2}\lambda$, so the fundamental frequency is $f_1 = v/\lambda = v/2L$, where v is the velocity of sound in air. The standing wave with two nodes is the first overtone or second harmonic and has half the wavelength ($L = \lambda$) and twice the frequency. Indeed, the frequency of each overtone is an integral multiple of the fundamental frequency. This is just what is found for a string.

For a closed tube, Fig. 14-7b, there is always a displacement node at the closed end and an antinode at the open end. Since the distance between a node and the nearest antinode is $\frac{1}{4}\lambda$, we see that the fundamental in this case corresponds to only one-fourth a wavelength within the length of the tube: $L = \lambda/4$. The fundamental frequency is thus $f_1 = v/4L$, or half what it is for an open pipe of the same length. There is another difference, for as we can see from Fig. 14-7b only the odd harmonics are present in a closed pipe: the overtones have frequencies equal to 3, 5, 7, ... times the fundamental

---

[†] When the resonant frequencies above the fundamental (that is, the overtones) are integral multiples of the fundamental, they are called harmonics. But if the overtones are not integral multiples of the fundamental, as is the case for a vibrating drum head, for example, they are not harmonics.

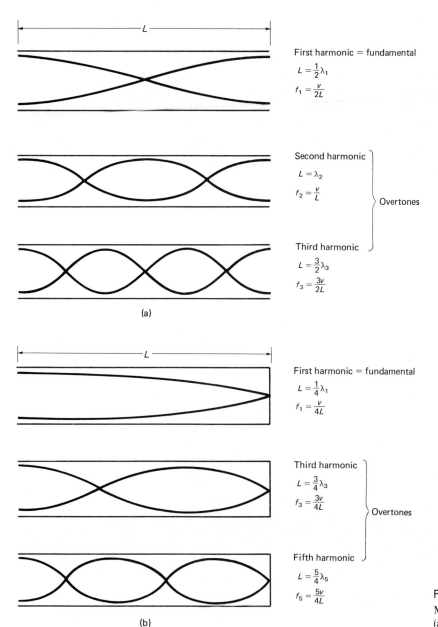

First harmonic = fundamental

$L = \frac{1}{2}\lambda_1$

$f_1 = \frac{v}{2L}$

Second harmonic

$L = \lambda_2$

$f_2 = \frac{v}{L}$

Third harmonic

$L = \frac{3}{2}\lambda_3$

$f_3 = \frac{3v}{2L}$

Overtones

(a)

First harmonic = fundamental

$L = \frac{1}{4}\lambda_1$

$f_1 = \frac{v}{4L}$

Third harmonic

$L = \frac{3}{4}\lambda_3$

$f_3 = \frac{3v}{4L}$

Fifth harmonic

$L = \frac{5}{4}\lambda_5$

$f_5 = \frac{5v}{4L}$

Overtones

(b)

FIGURE 14-7

Modes of vibration (standing waves) for (a) an open tube; (b) a closed tube.

frequency. There is no way for a wave with 2, 4, ... times the fundamental to have a node at one end and an antinode at the other, and thus they cannot exist as standing waves in a closed tube.

Organs make use of both open and closed pipes. Notes of different pitch are sounded using different pipes which vary in length from a few centimeters to 5 m or more. Other musical instruments act either like a closed tube or an open tube. A flute for example is an open tube for it is open not only where you blow into it, but also at the opposite end as well. The different notes on a flute and many other instruments are obtained by shortening the length of the tube—that is, by uncovering holes along its length. In a trumpet, on the other hand, the pushing down of the valves opens additional lengths of tube. In all these instruments, the longer the length of the vibrating air column, the lower the pitch.

The diagrams of Fig. 14-7 represent the *displacement* of the standing waves. The amplitude of the *pressure*, on the other hand, will be 90° out of phase with the displacement, just as for a traveling wave. Thus there is a pressure node at an open end of a tube (which makes sense since it is open to the atmosphere) and a pressure antinode at a closed end of a tube.

EXAMPLE 14-5 What will be the fundamental frequencies and first three overtones for a 26-cm long organ pipe at 20°C if it is (*a*) open and (*b*) closed?

SOLUTION At 20°C, the speed of sound in air is 343 m/s (Section 14-1). (*a*) For the open pipe, the fundamental frequency is:

$$f_1 = \frac{v}{2L} = \frac{343 \text{ m/s}}{2(0.26 \text{ m})} = 660 \text{ Hz}.$$

The overtones, which include all harmonics, are 1320 Hz, 1980 Hz, 2640 Hz, and so on.

(*b*) Referring to Fig. 14-7 we see that

$$f_1 = \frac{v}{4L} = \frac{343 \text{ m/s}}{4(0.26 \text{ m})} = 330 \text{ Hz}.$$

But only the odd harmonics will be present, so the first three overtones will be 990 Hz, 1650 Hz, and 2310 Hz.

EXAMPLE 14-6 A flute is designed to play middle C (264 Hz) as the fundamental frequency when all the holes are covered. Approximately how long should the distance be from the mouthpiece to end of the flute. (Note: this is only approximate since the antinode does not occur precisely at the mouthpiece.) Assume the temperature is 20°C.

SOLUTION The speed of sound in air at 20°C is 343 m/s. Then, from Fig. 14-7 the fundamental frequency f_1 is related to the length of the vibrating air column by $f = v/2L$. Solving for L we find

$$L = \frac{v}{2f} = \frac{343 \text{ m/s}}{(2)(264 \text{ s}^{-1})} = 0.650 \text{ m}.$$

EXAMPLE 14-7 If the temperature is only 10°C, what will be the frequency of the note played when all the openings are covered in the flute of Example 14-6?

SOLUTION The length L is still 65.0 cm. But now the velocity of sound is less since it changes by 0.60 m/s per each C°. For a drop of 10 C°, the velocity decreases by 6 m/s to 337 m/s. The frequency will be

$$f = \frac{v}{2L} = \frac{337 \text{ m/s}}{2(0.650 \text{ m})} = 259 \text{ Hz}.$$

This example illustrates why players of wind instruments take time to "warm up" their instruments so they will be in tune. The effect of temperature on stringed instruments is much smaller.

*14-6 ∎ Quality of sound

Whenever we hear a sound, particularly a musical sound, we are aware of its loudness, its pitch, and also of a third aspect called "quality." For example, when a piano and then a flute play a note of the same loudness and the same pitch (say middle C), there is a clear difference in the overall sound. We would never mistake a piano for a flute. This is what is meant by the **quality**[†] of a sound; for musical instruments the terms *timbre* or *tone color* are also used.

Just as loudness and pitch can be related to physically measurable quantities, so too can quality. The quality of a sound depends on the presence of overtones—their number and their relative amplitudes. Generally, when a note is played on a musical instrument, the fundamental as well as overtones are present simultaneously. Figure 14-8 illustrates how the superposition of three wave forms, in this case the fundamental and first two

[†] Note that quality in this sense does not refer to the goodness or badness of a sound or to the craftmanship that went into building the instrument.

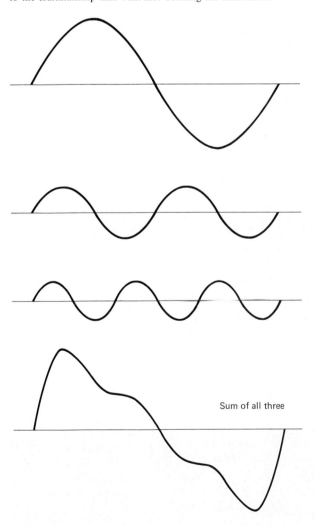

FIGURE 14-8
The amplitudes of the fundamental and first two overtones are added at each point to get the "sum", or composite waveform.

Sum of all three

overtones (with particular amplitudes), would combine to give a composite *wave form*. Of course more than two overtones are usually present.

The relative amplitudes of the various overtones are different for different musical instruments, and this is what gives each instrument its characteristic quality or timbre. A graph showing the relative amplitudes of the harmonics produced by an instrument is called a "sound spectrum"; several typical examples for different instruments are shown in Fig. 14-9. Normally, the fundamental has the greatest amplitude and its frequency is what is heard as the pitch.

The manner in which an instrument is played strongly influences the sound quality. Plucking a violin string, for example, makes a very different sound than pulling a bow across it. The sound spectrum at the very start (or end) of a note (as when a hammer strikes a piano string) can be very different from the subsequent sustained tone. This too affects the subjective tone quality of an instrument.

An ordinary sound, like that made by striking two stones together, is a noise that has a certain quality, but a clear pitch is not discernible. A noise such as this is a mixture of many frequencies which bear little relation to one another. If a sound spectrum were made of this noise, it would not show discrete lines like those of Fig. 14-9; instead it would show a continuous, or nearly continuous, spectrum of frequencies. Such a sound we call "noise" in comparison with the more harmonious sounds which contain frequencies that are simple multiples of the fundamental.

FIGURE 14-9

Sound spectra for several instruments.

FIGURE 14-10

Sound waves from two loudspeakers interfere.

14-7 ∎ Interference of sound waves; beats

We saw in Section 13-10 that when two waves simultaneously pass through the same region of space, they interfere with one another. Since this is true of any kind of wave, we should expect that interference will occur with sound waves, and indeed it does.

As a simple example, consider two large loudspeakers, *A* and *B*, a distance *d* apart on the stage of an auditorium as shown in Fig. 14-10. Let us assume the two speakers are emitting sound waves of the same single frequency and that they are in phase: that is, when one speaker is forming a compression, so is the other. The curved lines in the diagram represent the crests of sound waves from each speaker. Of course we must remember that for a sound wave, a crest is a compression in the air whereas a trough—which falls between two crests—is a rarefaction. A person or detector at a point such as *C*, which is the same distance from each speaker, will experience a loud sound because the interference will be constructive. On the other hand, at a point such as *D* in the diagram, little if any sound will be heard because destructive interference occurs—compressions of one wave meet rarefactions of the other and vice versa (see Fig. 13-27 and related discussion on water waves in Section 13-10). An analysis of this situation is clearer if we graphically represent the wave forms as in Fig. 14-11. There it can be seen that a point *C* constructive interference occurs since both waves simultaneously have crests or simultaneously have troughs. In Fig. 14-11b we see the situation for point *D*. The wave from speaker *B* must travel a greater

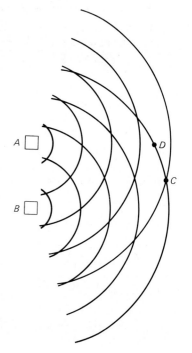

distance than the wave from A. Thus the wave from B lags behind that from A. In this diagram, point E is chosen so that the distance ED is equal to AD. Thus we see that if the distance BE is equal to precisely one-half the wavelength of the sound, the two waves will be exactly out of phase when they reach D, and destructive interference occurs. This then is the criterion for determining at what points destructive interference occurs: destructive interference occurs at any point whose distance from one speaker is greater than its distance from the other speaker by exactly one-half wavelength. Notice that if this extra distance (BE in the diagram) is equal to a whole wavelength (or 2, 3, ... wavelengths) then the two waves will be in phase and *constructive interference* occurs. If the distance BE equals $\frac{1}{2}$, $1\frac{1}{2}$, $2\frac{1}{2}$, ... wavelengths, *destructive interference* occurs.

If a loudspeaker emits a whole range of frequencies, not all wavelengths will destructively interfere at any one point such as D; rather only specific wavelengths will destructively interfere completely, according to the criteria above.

EXAMPLE 14-8   The two loudspeakers in Fig. 14-10 are 1.00 m apart. A person stands 4.00 m from one speaker. How far must this person be from the second speaker in order to detect destructive interference when the speakers emit an 1150-Hz sound? Assume the temperature is 20°C.

SOLUTION   The wavelength of this sound is

$$\lambda = \frac{v}{f} = \frac{343 \text{ m/s}}{1150 \text{ Hz}} = 0.30 \text{ m}.$$

For destructive interference to occur, the person must be one-half wavelength further from one loudspeaker than the other, or 0.15 m. Thus the person must be 4.15 m (or 3.85 m) from the second speaker. Notice in this example that if the speakers were less than 0.15 m apart, there would be no point that was 0.15 m farther from one speaker than the other and there would be no point where destructive interference would occur.

An interesting and important example of interference occurs in the phenomenon known as **beats.** This is the phenomenon that occurs if two sources of sound—say, two tuning forks—are close in frequency but not exactly the same. Sound waves from the two sources interfere with each other and the sound level alternately rises and falls; the regularly spaced intensity changes are called beats and often sound eerie.

To see how beats arise, consider two equal-amplitude[†] sound waves of frequency $f_1 = 50$ Hz and $f_2 = 55$ Hz, respectively. In 1.00 s, the first source makes 50 vibrations whereas the second makes 55. We now examine the waves at one point in space equidistant from the two sources. The wave forms for each wave as a function of time are shown on the first two lines of Fig. 14-12; the third line shows the sum of the two waves. At time $t = 0$ the two waves are shown to be in phase and interfere constructively. Because

---

[†] Beats will be heard even if the amplitudes are not equal, as long as the difference is not great.

(a)

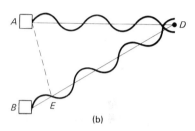

(b)

FIGURE 14-11

Sound waves from loudspeakers A and B (see Fig. 14-10) constructively interfere at C and destructively interfere at D.

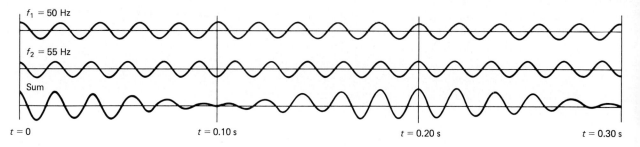

$f_1 = 50$ Hz

$f_2 = 55$ Hz

Sum

$t = 0$ $t = 0.10$ s $t = 0.20$ s $t = 0.30$ s

FIGURE 14-12

Beats occur as a result of the superposition of two sound waves of slightly different frequency.

the two waves vibrate at different rates, at time $t = 0.10$ s they are completely out of phase and destructive interference occurs as shown in the figure. At $t = 0.20$ s they are again in phase and the resultant amplitude again large. Thus the resultant amplitude is large every 0.20 s and in between it drops drastically. This rising and falling of the intensity is what is heard as beats. In this case the beats are 0.20 s apart. That is, the *beat frequency* is five per second or 5 Hz. This result, that the beat frequency equals the difference in frequency of the two waves, is valid in general.

The phenomenon of beats can occur with any kind of wave and is a very sensitive method for comparing frequencies. For example, to tune a piano, a piano tuner listens for beats produced between his standard tuning fork and that of a particular string on the piano and knows it is in tune when the beats disappear; he also listens for beats between the different piano strings. The members of an orchestra can tune up by listening for beats between their instruments and that of a standard tone (usually A above middle C at 440 Hz) produced by a piano or an oboe.

14-8 ∎ Doppler effect

You may have noticed that the pitch of the siren on a speeding firetruck drops abruptly as it passes you. Or you may have noticed the change in pitch of a blaring horn on a fast moving car as it passes by. The pitch of the sound from the engine of a race car changes as it passes an observer. When a source of sound is moving toward an observer, the pitch is higher than when the source is at rest; and when the source is traveling away from the observer, the pitch is lower. This phenomenon is known as the **Doppler effect** and occurs for all types of waves. Let us now see why it occurs, and calculate the change in frequency for sound waves.

To be concrete, consider the siren of a firetruck at rest which is emitting sound of a particular frequency in all directions as shown in Fig. 14-13a. If the firetruck is moving, the siren emits sound at the same frequency. But the sound waves it emits forward are closer together than normal as shown in Fig. 14-13b. This is because the fire engine, as it moves, is "catching up" with the previously emitted wavefronts. Thus an observer on the sidewalk will detect more wave crests passing per second so the frequency is higher. The wavefronts emitted behind the truck are, on the other hand, farther apart than normal because the truck is speeding away from them. Hence, fewer

(a) At rest

(b) Firetruck moving

FIGURE 14-13

(a) Both observers on the sidewalk hear the same frequency from the firetruck at rest. (b) Doppler effect: observer toward whom the firetruck moves hears a higher-frequency sound, and observer behind the firetruck hears a lower frequency.

wave crests per second pass by an observer behind the truck and the pitch is lower.

To calculate the change in frequency, we make use of Fig. 14-14, and we assume the air (or other medium) is at rest in our reference frame. In Fig. 14-14a, the source of the sound (say a siren) is at rest; two successive wave crests are shown, the second of which is just in the process of being emitted. The distance between these crests is λ, the wavelength. If the frequency of the source is f, then the time between emissions of wave crests is

$$T = \frac{1}{f}.$$

In Fig. 14-14b the source is moving with a velocity v_s. In a time T (as just defined) the first wave crest has moved a distance $d = vT$ where v is the velocity of the sound wave in air (which is of course the same whether the source is moving or not). In this same time, the source has moved a distance $d_s = v_s T$. Then the distance between successive wave crests, which is the new wavelength λ', is

$$\lambda' = d - d_s = vT - v_s T$$
$$= (v - v_s)T$$
$$= (v - v_s)\frac{1}{f}$$

since $T = 1/f$. The frequency f' of the wave is

$$f' = \frac{v}{\lambda'}$$
$$= \left(\frac{v}{v - v_s}\right)f$$

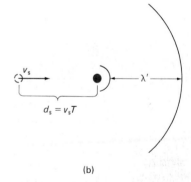

(a)

(b)

FIGURE 14-14

Determination of frequency change in the Doppler effect (see text).

or

$$f' = \frac{1}{\left(1 - \dfrac{v_s}{v}\right)}f. \qquad \begin{bmatrix} \text{source moving toward} \\ \text{stationary observer} \end{bmatrix} \quad (14\text{-}2a)$$

Because the denominator is less than one, $f' > f$. For example, if a source emits a sound of frequency 400 Hz when at rest, then when the source moves toward a fixed observer with a speed of 30 m/s, the observer hears a frequency (at 20°C) of

$$f' = \frac{400 \text{ Hz}}{\left(1 - \dfrac{30 \text{ m/s}}{343 \text{ m/s}}\right)} = 438 \text{ Hz}.$$

For a source which is moving *away* from the observer at a speed v_s, the new wavelength will be

$$\lambda' = d + d_s,$$

and the frequency f' will then be

$$f' = \frac{1}{\left(1 + \dfrac{v_s}{v}\right)} f. \qquad \begin{bmatrix} \text{source moving away from} \\ \text{stationary observer} \end{bmatrix} \quad (14\text{-}2b)$$

In this case, an observer behind the fire truck traveling at 30 m/s and whose siren vibrates at 400 Hz would hear a frequency of about 368 Hz.

The Doppler effect also occurs when the source is at rest and the observer is in motion. If the observer is traveling toward the source, the pitch is higher; and if the observer is traveling away from the source, the pitch is lower. Quantitatively, the change in frequency is slightly different than for the case of a moving source. In this case the distance between wave crests, the wavelength λ, is not changed. But the velocity of the crests with respect to the observer is changed. If the observer is moving toward the source, the speed of the waves relative to the observer is $v' = v + v_o$ where v is the velocity of the sound in the air (we assume the air is still) and v_o is the velocity of the observer. Hence, the new frequency is

$$f' = \frac{v'}{\lambda} = \frac{v + v_o}{\lambda}$$

or, since $\lambda = v/f$

$$f' = \left(1 + \frac{v_o}{v}\right) f. \qquad \begin{bmatrix} \text{observer moving toward} \\ \text{stationary source} \end{bmatrix} \quad (14\text{-}3a)$$

If the observer is moving away from the source, the relative velocity is $v' = v - v_o$ and

$$f' = \left(1 - \frac{v_o}{v}\right) f. \qquad \begin{bmatrix} \text{observer moving away} \\ \text{from stationary source} \end{bmatrix} \quad (14\text{-}3b)$$

When a sound wave is reflected from a moving obstacle, the frequency of the reflected wave will, because of the Doppler effect, be different from that of the incident wave. This is illustrated in the following example.

□ EXAMPLE 14-9 A 5000-Hz sound wave is directed toward an object moving 3.50 m/s toward the (stationary) source. What is the frequency of the reflected wave?

SOLUTION There are actually two Doppler shifts in this situation. First, the object acts like a moving observer and "detects" a sound wave

of frequency

$$f' = \left(1 + \frac{v_o}{v}\right)f = \left(1 + \frac{3.50 \text{ m/s}}{343 \text{ m/s}}\right)(5000 \text{ Hz}) = 5051 \text{ Hz}.$$

Secondly, the object acts like a moving source in reemitting (reflecting) the sound, so the reflected frequency is

$$f'' = \frac{1}{(1 - v_s/v)}f' = \frac{5051 \text{ Hz}}{\left(1 - \frac{3.50 \text{ m/s}}{343 \text{ m/s}}\right)} = 5103 \text{ Hz}.$$

Thus the frequency shifts by 103 Hz.

The incident wave and the reflected wave, when mixed together (say, electronically), interfere with one another and beats are produced. The beat frequency is equal to the difference in the two frequencies, and in the above example would be 103 Hz. This Doppler technique is used in a variety of medical applications, usually with ultrasonic waves in the megahertz frequency range. For example, ultrasonic waves reflected from red blood cells can be used to determine the velocity of blood flow. Similarly, the technique can be used to detect the movement of the chest of a young fetus and also to monitor its heartbeat.

The Doppler effect occurs for other types of waves as well. Light (and other types of electromagnetic waves) exhibits the Doppler effect; although the formulas for the frequency shift are not identical to Eqs. 14-2 and 14-3, the effect is similar. Important applications are to astronomy where, for example, the velocities of distant galaxies can be determined from the Doppler shift. Light from such galaxies is shifted toward lower frequencies (this is called the "red shift" since red has the lowest frequency of visible light), indicating that the galaxies are moving away from us. The greater the frequency shift, the greater the velocity of recession. It is found that the farther the galaxies are from us, the faster they move away. This observation is the basis for the idea that the universe is expanding.

On a less lofty plane, police radar speed traps make use of the Doppler effect (also of electromagnetic waves) to measure the speeds of approaching cars.

*14-9 ∎ Shock waves and the sonic boom

An object such as an airplane traveling faster than the speed of sound is said to have a *supersonic speed*. Such a speed is often given as a *Mach*[†] *number* which is defined as the ratio of the object's speed to that of sound in the medium at that location. For example, a plane traveling 900 m/s high in the atmosphere where the speed of sound is only 300 m/s has a speed of Mach 3.

When a source of sound moves at subsonic speeds, the pitch of the sound is altered as we have seen (the Doppler effect); see also Fig. 14-15a

[†] After the Austrian physicist Ernst Mach (1838–1916).

FIGURE 14-15

Sound waves emitted by an object at rest (a) or moving (b, c, d). If the object's velocity is less than the velocity of sound, the Doppler effect occurs (b); if its velocity is greater than the velocity of sound, a shock wave is produced (d).

(a) $v_{obj} = 0$

(b) $v_{obj} < v_{snd}$

(c) $v_{obj} = v_{snd}$

(d) $v_{obj} > v_{snd}$

FIGURE 14-16

Bow wave produced by a ship. (Courtesy of Naval Photographic Center.)

and b. But if a source of sound moves faster than the speed of sound, a more dramatic effect known as a *shock wave* occurs. In this case the source is actually "outrunning" the waves it produces. As shown in Fig. 14-15c, when the source is traveling at the speed of sound, the wavefronts it emits in the forward direction "pile up" directly in front of it. When the object moves at a supersonic speed, the wavefronts pile up on one another along the sides as shown in Fig. 14-15d. The different wave crests overlap one another and form a single very large crest which is the shock wave. Behind this very large crest there is usually a very large trough. A shock wave is essentially due to the constructive interference of a large number of wavefronts. A shock wave in air is analogous to the bow wave of a boat traveling faster than the speed of the water waves it produces, Fig. 14-16.

When an airplane travels at supersonic speeds, the noise it makes and its disturbance of the air form into a shock wave containing a tremendous amount of sound energy. When the shock wave passes a listener, it is heard as a loud "sonic boom." A sonic boom lasts only a fraction of a second, but the energy it contains is often sufficient to break windows and cause other damage. It can be psychologically unnerving as well. Actually the sonic boom from a supersonic aircraft is a double boom since a shock wave forms at both the front and the rear of the aircraft (Fig. 14-17).

When an aircraft approaches the speed of sound, it encounters a barrier of sound waves in front of it (see Fig. 14-15c). In order to exceed the speed of sound, extra thrust is needed to pass through this "sound barrier." This is called "breaking the sound barrier." Once a supersonic speed is attained, this barrier no longer impedes the motion. It is sometimes erroneously thought that a sonic boom is produced only at the moment an aircraft is breaking through the sound barrier. Actually, a shock wave follows

FIGURE 14-17

The (double) sonic boom has already been heard by the person on the right; it is just being heard by the person in the center; and it will shortly be heard by the person on the left.

the aircraft at all times it is traveling at supersonic speeds. A series of observers on the ground will each hear a loud "boom" as the shock wave passes, Fig. 14-17. The shock wave consists of a cone whose apex is at the aircraft. The angle of this cone, θ, (see Fig. 14-15d) is given by

$$\sin \theta = \frac{v_{snd}}{v_{obj}}, \qquad (14\text{-}4)$$

where v_{obj} is the velocity of the object (the aircraft) and v_{snd} is the velocity of sound in the medium (the proof is left as a problem).

*14-10 ∎ Applications; ultrasound and medical imaging

The reflection of sound is used in many applications to determine distance. The *sonar* or pulse-echo technique is used to locate underwater objects. (It is also used in medical applications as we shall see shortly.) A transmitter sends out a sound pulse through the water, and a detector receives its reflection, or echo, a short time later. This time is carefully measured, and from it the distance to the reflecting object can be determined since the speed of sound in water is known. The depths of the sea and the location of reefs, sunken ships, submarines, or schools of fish can be determined in this way. The interior structure of the earth is studied in a similar way by detecting reflections of waves traveling through the earth whose source was a deliberate explosion (called "soundings"). An analysis of waves reflected from various structures and boundaries within the earth reveals characteristic patterns that are also useful in the exploration for oil and minerals.

Sonar generally makes use of *ultrasonic* frequencies: that is, waves whose frequencies are above 20 kHz, beyond the range of human detection. For sonar, the frequencies are typically in the range 20 kHz to 100 kHz. One reason for using ultrasound, other than the fact that they are inaudible, is that for shorter wavelengths, there is less diffraction, so the beam spreads less and smaller objects can be detected. For as we saw in Chapter 13, particularly in Fig. 13-30, an obstacle intercepts and reflects a portion of a wave significantly only if the wavelength is less than the size of the object. Indeed, the smallest-sized objects that can be detected are on the order of the wavelength of the wave used. With the higher frequencies of ultrasound, the wavelength is smaller, so smaller objects can be detected.

In medicine, ultrasonic waves are used both in diagnosis and in treatment. Treatment involves destruction of unwanted tissue or objects in the body (such as tumors or kidney stones) using ultrasonic waves of very high intensity (as high as 10^7 W/m²) that are focused on the undesired material. This is simply destruction through deposit of large amounts of energy in a small area. Ultrasound can also be used in physical therapy to provide local heating of injured muscles.

The diagnostic use of ultrasound in medicine is a more complicated and very interesting application of physical principles. A *pulse-echo technique* is used, much like sonar. A high-frequency sound pulse is directed into the body, and its reflection from boundaries or interfaces between organs and other structures and lesions in the body are then detected. Using this technique, tumors and other abnormal growths, or pockets of fluid, can be distinguished; the action of heart valves and the development of a fetus can be examined; and information about various organs of the body, such as the brain, heart, liver, and kidneys, can be obtained. Although ultrasound does not replace X rays, for certain kinds of diagnosis it is more helpful; some kinds of tissue or fluid are not detected in X-ray photographs, but ultrasound waves are reflected from the boundaries. It is also possible to have "real-time" ultrasound images, as if one were watching a movie of a section of the interior of the body. Furthermore, at the low levels used for diagnosis ($< 3 \times 10^4$ W/m^2) no adverse affects have been reported; so ultrasound is considered a noninvasive method for probing the body.

The frequencies used in ultrasonic diagnosis are in the range 1 to 10 MHz (1 megahertz $= 10^6$ Hz). The speed of sound waves in the tissues of the human body averages about 1540 m/s (close to that for water); so the wavelength of a 1-MHz wave is about $\lambda = v/f = (1540 \text{ m/s})/(10^6 \text{ s}^{-1}) \approx 1.5 \times 10^{-3}$ m $= 1.5$ mm, and this sets the limit to the smallest-sized objects that can be detected. Higher frequencies mean shorter wavelengths, and thus in principle visualization of finer detail. However, the higher the frequency, the more the wave is absorbed by the body, and reflections from deeper inside the body are lost.

The technique works as follows. A brief pulse of ultrasound is emitted by a transducer, which is a device that transforms an electrical pulse into a mechanical vibration that produces the sound wave pulse. (This is what a loudspeaker does—we will discuss how transducers work in Chapters 18 and 19.) Part of the pulse is reflected at various interface surfaces in the body, and most (usually) continue on (see Fig. 13-22 for the case of a pulse on a rope). The detection of the reflected pulses is done by the same transducer, which transforms the sound pulses into electrical pulses (as a microphone does), and these pulses can then be displayed on the screen of a cathode-ray tube (the "picture tube" of an oscilloscope, display terminal, or TV monitor). As an example, consider a sound pulse passing through the abdomen as shown in Fig. 14-18a. At various interfaces in the body, part of the pulse is reflected. The time elapsed from when the pulse is emitted and when its reflection (echo) is received is proportional to the distance to the reflecting surface. For example, if the distance from transducer to the vertebra is 25 cm, the pulse travels a round trip distance of 2×25 cm $= 0.50$ m, and the time taken is $t = d/v = 0.50$ m/(1540 m/s) $= 3.2 \times 10^{-4}$ s $= 320$ μs. A pulse reflected from an object only 10 cm from the transducer would be received after only 130 μs. Figure 14-18b plots the reflected pulses as a function of the time they are received by the transducer for the situation shown in part (a).

The strength of a reflected pulse depends mainly on the differences in density of the two materials on either side of the interface. It also depends on the sound velocity in each material, but this normally has little effect since in most tissue the velocity is within a few percent of the average 1540 m/s. [Exceptions are bone (4000 m/s) and air (340 m/s); at interfaces involving bone or the lungs, most of the sound pulse is reflected, so ultrasound cannot be used as a probe beyond such interfaces.[†]]

[†] Also note that the transducer is placed in contact with a person's body, or the body and transducer are immersed in water, so as to avoid the high reflectivity at the first air-body interface which would greatly reduce the amount of sound waves passing into the body.

FIGURE 14-18

(a) Ultrasound pulse passes through abdomen, reflecting from surfaces in its path. (b) Reflected pulses plotted as a function of time (A-scan) when received by transducer; time is proportional to distance of travel. The vertical dashed lines simply point out which reflected pulse goes with which surface. (c) B–mode display for the same echoes: brightness of each dot is related to signal strength.

FIGURE 14-19

An A-scan of an eye. (From A. Macovski, *Medical Imaging Systems*, copyright 1983 by Prentice-Hall, Inc., Englewood Cliffs, NJ.)

A trace such as that shown in Fig. 14-18b can be displayed directly (such as on the screen of a cathode-ray-tube). An example is shown in Fig. 14-19. Such a display is called an *A-scan* or A-mode scan. Far more common today is a *B-scan*, which can be used to form two-dimensional images of a cross section of the body. In a B-mode scan, each echo is represented as a point whose position is given by the time delay and whose brightness depends on the strength of the echo. Figure 14-18c shows the B-scan for the echoes produced in part (a). A two-dimensional image can be formed from a series of B-scans. The transducer is moved and at each position it sends out a pulse and receives echoes as shown in Fig. 14-20. Each B-scan trace can be plotted, spaced appropriately one below the other, to form an

(a)

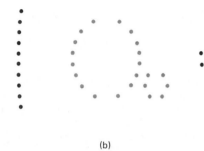

(b)

FIGURE 14-20

(a) Ten traces are made across the abdomen by moving the transducer, or by using an array of transducers. (b) The echoes are plotted in the B-scan mode to produce the image. More closely spaced traces would give a more detailed image.

image as shown in Fig. 14-20b. The display can be done on the screen of a cathode-ray tube. Only ten lines are shown in Fig. 14-20, so the image is crude. More lines would give a more precise image.[†] A photograph of an ultrasound image is shown in Fig. 14-21.

A faster scan can be obtained by using an array of transducers.[‡] Since the echo reflections do not simply travel back along the same line (as suggested in Fig. 14-18a) but actually spread out in other directions, all the transducers in an array could not be pulsed at once. The time between the pulse emitted by one transducer and the pulse emitted by the adjacent one must be greater than the time for the farthest echo to return; as we saw earlier, the echo from a surface 25 cm from a transducer takes about 320 μs. Typical spacing between pulses is on the order of 1000 μs or 1 ms, so transducers can be pulsed sequentially at a rate of 1000 per second. This is done electronically and is much faster than moving a single transducer. This speed allows a series of pictures to be taken per second, providing real-time "moving pictures" of moving objects such as the heart. There is a compromise between the number of lines per picture, and the number of frames per second. For example, at a pulse rate of 1000 per second, if each image has 100 lines, there can be 10 frames per second; with images of 50 lines, there can be 20 frames per second.

FIGURE 14-21

Ultrasound image of a 14-week-old human fetus. (Courtesy of A. Macovski, Stanford University Medical Center).

[†] *Radar* used for aircraft involves a similar pulse-echo techique except that it uses electromagnetic (EM) waves which, like light, travel with a speed of 3×10^8 m/s. This works for the large distances in the sky, but this high speed precludes the use of EM wave pulse-echo techniques in the body since the picosecond times involved $[\approx (0.50 \text{ m})/(3 \times 10^8 \text{ m/s}) \approx 3 \times 10^{-10} \text{ s} = 300 \text{ ps}]$ can not at present be separated electronically. Other techniques using EM waves will be discussed later.

[‡] Another fast-scanning technique is to rotate a single transducer about a point so it "looks" in a series of different directions through the body. This produces a "sector scan."

Ultrasound imaging has been an important advance in medicine. It, along with other types of medical imaging which we will discuss in later chapters (see Chapters 24 and 30), has in many cases replaced exploratory surgery and other risky, painful, and/or costly procedures. There is no evidence of toxic effects with ultrasound imaging, as there is with X-ray imaging. It is thus considered noninvasive. Nonetheless it is not a technique that replaces all others. Beam spreading limits the sharpness of images. And the way sound reflects from matter is not the same as for light or X-rays, so different kinds of information can be obtained from different imaging techniques. A new mode of ultrasound imaging is now being developed based on *transmitted* waves (instead of reflected) making use of computer assisted tomography techniques (normally done with X rays—see Chapter 24). Another technique, which we already discussed in Section 14-8, employs the Doppler shift of ultrasound echoes to measure velocities within the human body, such as blood flow and fetal heartbeat.

▮ Summary

Sound travels as a longitudinal wave in air and other materials. In air, the speed of sound increases with temperature; at 20°C it is about 343 m/s.

The *pitch* of a sound is determined by the frequency; the higher the frequency, the higher the pitch. The *audible range* of frequencies is roughly 20 to 20,000 Hz (1 Hz = 1 cycle per second). The *loudness* or *intensity* of a sound is related to the amplitude of the wave. Because the human ear can detect sound intensities from 10^{-12} W/m^2 to over 1 W/m^2, intensity levels are specified on a logarithmic scale. The *intensity level* β, specified in decibels, is defined in terms of intensity I as $\beta = 10 \log (I/I_0)$, where the reference intensity I_0 is usually taken to be 10^{-12} W/m^2. An increase in intensity by a factor of 100, for example, corresponds to a level increase of 20 dB.

Musical instruments are simple sources of sound in which standing waves are produced. The strings of a stringed instrument may vibrate as a whole with nodes only at the ends; the frequency at which this occurs is called the *fundamental*. The string can also vibrate at higher frequencies, called overtones or *harmonics*, in which there are one or more additional nodes. The frequency of each harmonic is a whole-number multiple of the fundamental. In wind instruments, standing waves are set up in the column of air within the tube. The vibrating air in an open tube (open at both ends) has antinodes at both ends; the fundamental frequency corresponds to a wavelength equal to twice the tube length. The harmonics have frequencies that are 2, 3, 4, . . . times the fundamental frequency. For a closed tube (closed at one end) the fundamental corresponds to a wavelength four times the length of the tube; only the odd harmonics are present, equal to 1, 3, 5, 7, . . . times the fundamental frequency.

Sound waves from different sources can interfere with each other. If two sounds are at slightly different frequencies, *beats* can be heard at a frequency equal to the difference in frequency of the two sources.

The *Doppler effect* refers to the change in pitch of a sound due to the motion either of the source or of the listener. If they are approaching each other, the pitch is higher; if they are moving apart, the pitch is lower.

▮ Questions

1. What is the evidence that sound travels as a wave?
2. What is the evidence that sound is a form of energy?
3. Country folk have a rule of thumb that the time delay between seeing lightning and hearing the thunder is an indication of how far away the lightning hit, and in particular, that each 5 seconds corresponds to 1 mile. Explain. What would be the rule for kilometers?
4. When boating on a lake or river at night, one can often hear clearly the voices or radios of people quite a distance away on the shore. Yet this rarely happens in the

daytime. This phenomenon can be explained by considering the refraction of sound due to different layers of air having different densities (because of differences in temperature). Draw a diagram and explain this phenomenon and determine if the layer of air nearest the water surface is denser than the one above at nighttime or in the daytime.

5. When a sound wave passes from air into water, do you expect the frequency or wavelength to change?

6. What evidence can you give that the speed of sound in air does not depend significantly on frequency?

7. The voice of a person who has inhaled helium sounds very high pitched. Why?

8. Explain why the larger area of the eardrum, as compared to that of the oval window, leads to an amplification of the pressure.

9. What is the reason that catgut strings on some musical instruments are wrapped with fine wire?

10. Whistle through your lips and describe how the pitch of the whistle is controlled.

11. Explain how a tube might be used as a filter to reduce the amplitude of sounds in various frequency ranges. (An example is a car muffler.)

12. How will the air temperature in a room affect the pitch of organ pipes?

13. Noise control is an important goal today. One mode of attack is to reduce the area of vibration of noisy machinery, for example by keeping it as small as possible or isolating it (acoustically) from the floor and walls. A second method is to make the surface out of a thicker material. Explain how each of these can reduce the noise level.

14. Why are the frets on a guitar spaced closer together as you move down the fingerboard toward the bridge?

15. Standing waves can be said to be due to "interference in space," whereas beats can be said to be due to "interference in time." Explain.

16. Suppose a source of sound moves at right angles to the line of sight of a listener at rest in still air. Will there be a Doppler effect? Explain.

17. If a wind is blowing, will this alter the frequency of the sound heard by a person at rest with respect to the source? Is the wavelength or velocity changed?

18. A sonic boom sounds much like an explosion. Explain the similarity between the two.

▌ Problems

(Unless stated otherwise, assume $T = 20°C$ and $v_{sound} = 343$ m/s.)

SECTION 14-1

1. (I) Ultrasonic waves with frequencies as high as 250,000 Hz are emitted by dolphins. What would be the wavelength of such a wave (a) in water and (b) in air?

2. (I) A hiker determines the length of a lake by listening for the echo of her shout reflected by a cliff at the far end of the lake. She hears the echo 1.20 s after shouting. How long is the lake?

3. (II) A stone is dropped from the top of a cliff. The splash it makes when striking the water below is heard 4.0 s later. How high is the cliff?

4. (II) A person sees a heavy stone strike the concrete pavement. A moment later two sounds are heard from the impact: one travels in the air and the other in the concrete, and they are 1.2 s apart. How far away did the impact occur?

SECTION 14-2

5. (I) What is the intensity level of a sound whose intensity is 1.0×10^{-8} W/m²?

6. (I) What is the intensity of a sound whose intensity level is 50 dB?

7. (II) A stereo tape recorder is said to have a signal-to-noise ratio of 62 dB. What is the ratio of intensities of the signal and the background noise?

8. (II) Human beings can detect a difference in sound intensity level of 1.0 dB. What is the ratio of the amplitudes of two sounds whose levels differ by this amount?

9. (II) If two firecrackers produce an intensity level of 95 dB at a certain place, what will be the intensity level if only one is exploded?

10. (II) A single mosquito 10 m from a person makes a sound close to the threshold of human hearing (0 dB). What will be the intensity level of 1000 such mosquitoes?

11. (II) A 75-dB sound wave strikes an eardrum whose area is 5.0×10^{-5} m². How much energy is absorbed by the eardrum per second?

12. (II) (a) Estimate the power output of sound from a person speaking in normal conversation. Use Table 14-2. Assume the sound spreads roughly uniformly over a hemisphere in front of the mouth. (b) How many people would produce a total sound output of 100 W of ordinary conversation?

13. (II) A stereo amplifier is rated at 40 W output at 1000 Hz. The output drops by 2 dB at 20 Hz. What is the power output at 20 Hz?

14. (II) What is the resultant intensity level when an 80-dB sound and an 85-dB sound are heard simultaneously?

*SECTION 14-3

*15. (II) Two sound waves have equal displacement amplitudes, but one has twice the frequency of the other.

(a) Which has the greater pressure amplitude and by what factor is it greater? (b) What is the ratio of their intensities?

*16. (II) If the amplitude of a sound wave is tripled, (a) by what factor will the intensity increase? (b) By how many dB will the intensity level increase?

*17. (II) Calculate the maximum displacement of air molecules when a 120-Hz sound wave passes whose intensity is at the threshold of pain (120 dB).

*18. (II) What would be the intensity level (in dB) of a sound wave in air that corresponds to a displacement amplitude of vibrating air molecules of 1.2 mm at 80 Hz?

*19. (II) In audio and communications systems, the *gain*, β, in decibels is defined as:

$$\beta = 10 \log \frac{P_{out}}{P_{in}},$$

where P_{in} is the power input to the system and P_{out} is the power output. A particular stereo amplifier puts out 35 W of power for an input of 1 mW. What is its gain in dB?

*20. (III) The intensity level 12.0 m from a loudspeaker, placed in the open, is 100 dB. What is the acoustic power output (W) of the speaker?

*21. (III) A jet plane emits 2.0×10^5 J of sound energy per second. (a) What is the intensity level 40 m away? Air absorbs sound at a rate of about 7 dB/km; calculate what the intensity level will be (b) 1 km and (c) 5 km away from this jet plane, taking into account air absorption.

*SECTION 14-4

*22. (I) What is the lowest intensity level that can be heard by an average ear (middle curve of Fig. 14-4)?

*23. (I) An 8000-Hz tone must have what intensity level to seem as loud as an 80-Hz tone that has a 60-dB intensity level? (See Fig. 14-5.)

*24. (I) What are the lowest and highest frequencies that the average ear (middle curve in Fig. 14-4) can hear when the intensity level is 25 dB?

SECTION 14-5

25. (I) The G string on a violin has a fundamental frequency of 196 Hz. The length of the vibrating portion is 32 cm and has a mass of 0.50 g. Under what tension must the string be placed?

26. (I) An unfingered guitar string is 0.70 m long and is tuned to play E above middle C (330 Hz). How far from the end of this string must the finger be placed to play A above middle C (440 Hz)?

27. (I) How far from the end of the flute in Example 14-6 should the hole be that must be uncovered to play D above middle C at 294 Hz?

28. (I) Determine the length of a closed organ pipe that emits middle C (264 Hz) when the temperature is 15°C.

29. (I) An organ pipe is 80 cm long. What are the fundamental and first three audible overtones (a) if the pipe is closed at one end and (b) if it is open at both ends?

30. (II) An organ is in tune at 20°C. By what fraction will the frequency be off at 0°C?

31. (II) Calculate the resonant frequency of the column of air in the outer ear of a human being, which is about 2.5 cm long. Does this correspond to a region of high sensitivity of the ear? Explain.

32. (II) Each string on a violin is tuned to a frequency $1\frac{1}{2}$ times that of its neighbor. If all the strings are to be placed under the same tension, what must be the mass per unit length of each string relative to that of the lowest string?

33. (II) The A string of a violin is 32 cm long between fixed points with a fundamental frequency of 440 Hz and a linear density of 5.0×10^{-4} kg/m. (a) What are the wave speed and tension in the string? (b) What is the length of the tube of a simple wind instrument (say an organ pipe) closed at one end whose fundamental is also 440 Hz if the speed of sound is 331 m/s in air? (c) What is the frequency of the first overtone of each instrument?

34. (II) (a) At $T = 20°C$, how long must an open organ pipe be if it is to have a fundamental frequency of 264 Hz? (b) If this pipe was filled with helium, what would its fundamental frequency be?

35. (II) A tuning fork is set into vibration above a vertical open tube filled with water. The water level is allowed to drop slowly; as it does so the air in the tube above the water level is heard to resonate with the tuning fork when the distance from the tube opening to the water level is 0.125 m and again at 0.395 m. What is the frequency of the tuning fork?

36. (II) How many overtones are present within the audible range for a 100-cm-long organ pipe at 20°C (a) if it is open and (b) if it is closed?

37. (III) An 80-cm-long guitar string of mass 1.50 g is placed near a tube open at one end, and also 80 cm long. How much tension should be in the string so that its fourth harmonic has the same frequency as the third harmonic of the tube?

SECTION 14-7

38. (I) Two horns emitting sounds of frequency 665 and 671 Hz, respectively, produce beats of what frequency?

39. (I) Two flutes are slightly out of tune when they play middle C so that beats are heard every 2 s. By how much do the two flutes differ in frequency?

40. (I) What will be the "beat frequency" if middle C (262 Hz) and C# (277 Hz) are played together? Will this be audible? What if each is played two octaves lower (each frequency reduced by a factor of 4)?

41. (II) Two violin strings are tuned to the same frequency, 440 Hz. The tension in one string is then decreased by 2 percent. What will be the frequency of beats heard when the two strings are played together?

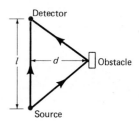

FIGURE 14-22

42. (II) A person hears a pure tone coming from two sources that seems to be in the 500–1000 Hz range. The sound is loudest at points equidistant from the two sources. In order to determine exactly what the frequency is, the person moves about and finds that the sound level is minimal at a point 0.22 m farther from one source than the other. What is the frequency of the sound?

43. (II) Two piano strings are supposed to be vibrating at 132 Hz but a piano tuner hears one beat every 2.0 s when they are played together. (a) If one is vibrating at 132 Hz, what must be the frequency of the other (is there only one answer)? (b) By how much (in percent) must the tension be increased or decreased to bring them in tune?

44. (II) How many beats will be heard if two identical flutes each try to play middle C (262 Hz), but one is at 0.0°C and the other at 20.0°C?

45. (II) Two loudspeakers are 2.5 m apart. A person stands 3.0 m from one speaker and 3.5 m from the other. (a) What is the lowest frequency at which destructive interference will occur at this point? (b) Calculate two other frequencies that also result in destructive interference at this point (give the next two highest).

46. (III) Show that the two speakers in Fig. 14-10 must be separated by at least a distance d equal to one-half the wavelength λ of sound if there is to be any place where complete destructive interference occurs. The speakers are in phase.

47. (III) A source emits sound of wavelength 2.80 m and 3.10 m in air. (a) How many beats per second will be heard (assume $T = 20°C$)? (b) How far apart in space are the regions of maximum intensity?

48. (III) A source of sound waves (wavelength λ) is a distance l from a detector. Sound reaches the detector directly, and also by reflecting off an obstacle as shown in Fig. 14-22. The obstacle is equidistant from source and detector. When the obstacle is a distance d to the right of the line of sight between source and detector, as shown, the two waves arrive in phase. How much farther to the right must the obstacle be moved if the two waves are to be out of phase by 180°, so destructive interference occurs? (Assume $\lambda \ll l, d$.)

SECTION 14-8

49. (I) A factory whistle emits a sound at 900 Hz. What frequency will be heard by an observer in an auto-mobile traveling at 70 km/h (a) away from the source, and (b) toward the source?

50. (I) The predominant frequency of a certain police car's siren is 1800 Hz when at rest. What is the detected frequency if the car is (a) moving toward an observer at 30 m/s, and (b) if moving away at the same speed?

51. (II) Sound waves of frequency 1.0×10^6 Hz are directed at the chest of a fetus and travel with a speed in the body of 1.5×10^3 m/s. What will be the expected shift in frequency if the chest of a normal fetus moves at a maximum speed of 0.10 m/s?

52. (II) Derive a general formula for the changed frequency f' due to the Doppler effect when both the source and the observer are moving.

53. (II) Two trains emit whistles of the same frequency, 380 Hz. If one train is at rest and the other is traveling at 90 km/h away from an observer at rest, what will the observer detect as the beat frequency?

54. (III) If the velocity of blood flow in the aorta is normally about 0.30 m/s, what beat frequency would you expect if 4.00 MHz ultrasound waves were directed along the flow and reflected from the red blood cells? Assume that the waves travel with a speed of 1.54×10^3 m/s.

55. (III) The Doppler effect using ultrasonic waves of frequency 2.25×10^6 Hz is used to monitor the heartbeat of a fetus. A (maximum) beat frequency of 600 Hz is observed. Assuming that the speed of sound in tissue is 1.54×10^3 m/s, calculate the maximum velocity of the surface of the beating heart.

56. (III) In problem 55, the beat frequency is found to appear and then disappear 180 times per minute which reflects the fact that the heart is beating and its surface changes speed. What is the heartbeat rate?

*SECTION 14-9

*57. (I) How fast is an object moving on the earth if its speed is specified as Mach 0.5?

*58. (II) A boat traveling 9.2 m/s makes a bow wave at an angle of 18° to its direction of motion. What is the speed of the water waves?

*59. (II) Show that the angle θ a sonic boom makes with the path of a supersonic object is given by Eq. 14-4.

*60. (II) An airplane travels at Mach 2.8 where the speed of sound is 310 m/s. (a) What is the angle the shock wave makes with the direction of the airplane's motion? (b) If the plane is flying at a height of 6500 m, how long after it is directly overhead will a person on the ground hear the shock wave?

*61. (III) The excess pressure P in a sonic boom decreases with the altitude h of an aircraft approximately as

$$P \propto \frac{1}{h^{3/4}}.$$

By what factor does the pressure change if a plane flies twice as high? Compare to an inverse-square law.

15 ELECTRIC CHARGE AND ELECTRIC FIELD

The word "electricity" may evoke an image of complex modern technology: computers, lights, motors, electric power. But the electric force plays an even deeper role in our lives, since according to atomic theory, the forces that act between atoms and molecules to hold them together to form liquids and solids are electrical forces. Similarly the electric force is responsible for the metabolic processes that occur within our bodies. Even ordinary pushes and pulls are the result of the electric force between the molecules of your hand and those of the object being pushed or pulled. Indeed, most of the forces we have dealt with so far, such as elastic forces and the normal force acting on a body, are now considered to be electric forces acting at the atomic level. This does not include gravity, however, which is a separate force.[†]

The earliest studies on electricity date back to the ancients, but it has been only in the past two centuries that electricity was studied in detail. We will discuss the development of ideas about electricity, including practical devices, as well as the relation to magnetism, in the next seven chapters.

15-1 ∎ Static electricity; electric charge and its conservation

The word *electricity* comes from the Greek word *elektron*, which means "amber." Amber is petrified tree resin, and the ancients knew that if you rub an amber rod with a piece of cloth, the amber attracts small pieces of leaves or dust. A piece of hard rubber, a glass rod, or a plastic ruler rubbed with a cloth will also display this "amber effect," or *static electricity* as we call it today. You can readily pick up small pieces of paper with a plastic ruler that you've just vigorously rubbed with even a paper towel, Fig. 15-1. You have probably experienced static electricity when combing your hair or upon taking a synthetic blouse or shirt from a clothes dryer. And you may have

FIGURE 15-1

Rub a plastic ruler (a) and bring it close to some tiny pieces of paper (b).

(a)

(b)

felt a shock when you touched a metal door knob after sliding across a car seat or walking across a nylon carpet. In each case, an object becomes "charged" due to a rubbing process and is said to possess an **electric charge.**

Is all electric charge the same, or is it possible that there is more than one type? In fact, there are two types of electric charge as the following simple experiments show. A plastic ruler is suspended by a thread and rubbed vigorously with a cloth to charge it; when a second ruler, which has also been charged in the same way, is brought close to the first, it is found that the one ruler *repels* the other. This is shown in Fig. 15-2a. Similarly, if a rubbed glass rod is brought close to a second charged glass rod, again a repulsive force is seen to act, Fig. 15-2b. However, if the charged glass rod is brought close to the charged plastic ruler, it is found that they *attract* each other, Fig. 15-2c. The charge on the glass must therefore be different than that on the plastic. Indeed, it is found experimentally that all charged objects fall into one of two categories. Either they are attracted to the plastic and repelled by the glass, just as glass is; or they are repelled by the plastic and attracted to the glass, just as the plastic ruler is. Thus there seems to be two, and only two, types of electric charge. Each type of charge repels the same type but attracts the opposite type. That is: *unlike charges attract; like charges repel.*

These two types of charge were referred to as *positive* and *negative* by the American statesman, philosopher, and scientist Benjamin Franklin (1706–1790). The choice of which name went with which type of charge was of course arbitrary. Franklin's choice sets the charge on the rubbed glass rod to be positive charge, so the charge on a rubbed plastic ruler (or amber) is called negative charge; we still follow this convention today.

Franklin's theory of electric charge was actually a "single-fluid" theory that viewed a positive charge as an excess of the electric fluid beyond an object's normal content of electricity, and a negative charge as a deficiency. Franklin argued that whenever a certain amount of charge is produced on one body in a process, an equal amount of the opposite type of charge is produced on another body. The names positive and negative are to be taken algebraically, so that during any process the net change in the amount of charge produced is zero. For example, when a plastic ruler is rubbed with a paper towel, the plastic acquires a negative charge and the towel an equal amount of positive charge; the charge is separated, but the sum of the two is zero. This is an example of a law that is now well established: the **law of conservation of electric charge,** which states that

❚ the net amount of electric charge produced in any process is zero.

No violations have ever been found, and this conservation law is as firmly established as those for energy and momentum.

15-2 ❚ Electric charge in the atom

Only within the past century has it become clear that electric charge has its origin within the atom itself. In later chapters we will discuss atomic structure and the ideas that led to our present view of the atom in more detail; but it will help our understanding of electricity if we discuss it briefly now.

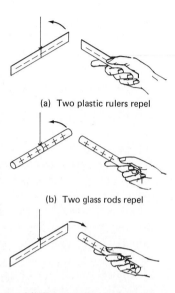

(a) Two plastic rulers repel

(b) Two glass rods repel

(c) Glass rod attracts plastic ruler

FIGURE 15-2

Unlike charges attract, whereas like charges repel one another.

Today's view, slightly simplified, shows the atom as having a heavy, positively charged nucleus surrounded by one or more negatively charged electrons (Fig. 15-3). In its normal state, the positive and negative charges within the atom are equal, and the atom is electrically neutral. Sometimes, however, an atom may lose one or more of its electrons, or may gain extra electrons. In this case the atom will have a net positive or negative charge, and is called an *ion*.

The nuclei in a solid material can vibrate but they tend to remain close to fixed positions, whereas some of the electrons move quite freely. The charging of an object by rubbing is explained by the transfer of electrons or ions from one material to the other. When a plastic ruler becomes negatively charged by rubbing with a paper towel, the transfer of charged particles from the one to the other leaves the towel with a positive charge equal in magnitude to the negative charge acquired by the plastic.

Normally when objects are charged by rubbing, they hold their charge only for a limited time and eventually return to the neutral state. Where does the charge go? In some cases it is neutralized by charged ions in the air (formed by cosmic rays, et cetera). Often more importantly, the charge can "leak off" onto water molecules in the air. This is because water molecules are *polar*—that is, even though they are neutral their charge is not distributed uniformly, Fig. 15-4. Thus the extra electrons on, say, a charged plastic ruler can "leak off" into the air because they are attracted to the positive end of water molecules. A positively charged object, on the other hand, can be neutralized by transfer of loosely held electrons from water molecules in the air. On dry days, static electricity is much more noticeable since the air contains fewer water molecules to allow leakage. On humid or rainy days, it is difficult to make any object hold its charge for long.

FIGURE 15-3
Simple view of the atom.

FIGURE 15-4
Diagram of a water molecule. Because it has opposite charges on different ends, it is called a "polar" molecule.

15-3 ∎ Insulators and conductors

Suppose we have two metal spheres, one highly charged and the other electrically neutral. If we now place an iron nail so that it touches both the spheres, it is found that the previously uncharged sphere quickly becomes charged. If, however, we had connected the two spheres together with a wooden rod or a piece of rubber, the uncharged ball would not have become noticeably charged. Materials like the iron nail are said to be **conductors** of electricity, whereas wood and rubber are called **nonconductors** or **insulators.**

Metals are generally good conductors whereas most other materials are insulators (although even insulators conduct electricity very slightly). It is interesting that nearly all natural materials fall into one or the other of these two quite distinct categories. There are a few materials, however, (notably silicon, germanium, and carbon) that fall into an intermediate (but distinct) category known as *semiconductors.*

From the atomic point of view the electrons in an insulating material are bound very tightly to the nuclei, whereas in a conductor many of them are bound very loosely and can move about freely within the material. When a positively charged object is brought close to or touches a conductor, the free electrons move quickly toward this positive charge. On the other hand, the free electrons move swiftly away from a negative charge that is brought close. In a semiconductor, there are very few free electrons, and in an insulator, almost none.

15-4 ▌ Induced charge; the electroscope

Suppose a positively charged metal object is brought close to a second (neutral) metal object. If the two touch, the free electrons in the neutral one are attracted to the positively charged object and some will pass over to it. Since the second object is now missing some of its negative electrons, it will have a net positive charge. This process is called "charging by conduction."

Now suppose a positively charged object is brought close to a neutral metal rod, but does not touch it. Although the electrons of the metal rod do not leave the rod, they still move within the metal toward the charged object; this leaves a positive charge at the opposite end, Fig. 15-5. A charge is said to have been *induced* at the two ends of the metal rod. Of course no charge has been created; it has merely been *separated;* the net charge on the metal rod is still zero. However, if the metal were divided in two pieces, we could have two charged objects, one charged positively and one charged negatively.

Another way to induce a net charge on a metal object is to connect it with a conducting wire to the ground (or a conducting pipe leading into the ground) as shown in Fig. 15-6a (⏚ means "ground"). The object is then said to be "grounded" or "earthed." Now the earth, since it is so large and can conduct, can easily accept or give up electrons; hence it acts like a reservoir for charge. If a charged object—let's say negative this time—is brought up close to the metal, free electrons in the metal are repelled and many of them move down the wire into the earth, Fig. 15-6b. This leaves the metal positively charged. If the wire is now cut, the metal will have a positive induced charge on it. If the wire were cut after the negative object is moved away, the electrons would all have moved back into the metal and it would be neutral.

An *electroscope* (or simple *electrometer*) is a device that can be used for detecting charge. As shown in Fig. 15-7, inside of a case are two movable leaves, often made of gold. (Sometimes only one leaf is movable.) The leaves are connected by a conductor to a metal ball on the outside of the case, but is insulated from the case itself. If a charged object is brought close to the knob, a separation of charge is induced on it, Fig. 15-8a; the two leaves become charged and repel each other as shown. If, instead, the knob is charged by conduction the whole apparatus becomes charged as shown in Fig. 15-8b. In either case, the greater the amount of charge, the greater the separation of the leaves.

Note however, that you cannot tell the sign of the charge in this way since a negative charge will cause the leaves to separate just as much as an equal magnitude positive charge—in either case the two leaves repel each other. An electroscope can, however, be used to determine the sign of the charge if it is first charged by conduction, say negatively, as in Fig. 15-9a.

Neutral metal rod

(a)

Metal rod still neutral, but with a separation of charge

(b)

FIGURE 15-5
Charging by induction.

(a)

(b)

FIGURE 15-6
Inducing a charge on an object connected to ground.

FIGURE 15-7
Electroscope.

Insulator

Metal

Gold leaves

Glass

Now if a negative object is brought close, as in Fig. 15-9b, electrons are induced to move further down into the leaves and they separate further. On the other hand, if a positive object is brought close, the electrons are induced to flow upward, leaving the leaves less negative and their separation is reduced, Fig. 15-9c.

The electroscope was much used in the early days of electricity. The same principle, aided by some electronics, is used in much more sensitive modern *electrometers*.

(a)   (b)

FIGURE 15-8
Electroscope charged (a) by induction, (b) by conduction.

15-5 ▌ Coulomb's law

We have seen that an electric charge exerts a force on other electric charges. But how does the magnitude of the charges and other factors affect the magnitude of this force? To answer this, the French physicist Charles Coulomb (1736–1806) investigated electric forces in the 1780s using a torsion balance (Fig. 15-10) much like that used by Cavendish for his studies of the gravitational force (Section 4-5).

Although precise instruments for the measurement of electric charge were not available in Coulomb's time, he was able to prepare small spheres with different amounts of charge in which the *ratio* of the charges was known. He reasoned that if a charged conducting sphere is placed in contact with an identical uncharged sphere, the charge on the first would then be shared equally by the two of them because of symmetry. He thus had a way to produce charges equal to $\frac{1}{2}$, $\frac{1}{4}$, and so on, of the original charge. Although he had some difficulty with induced charges, Coulomb was able to argue that the force one tiny charged object exerted on a second tiny charged object is directly proportional to the charge on each of them. That is, if the charge on either of the objects was doubled, the force was doubled; and if the charge on both of the objects was doubled the force increased to four times the original value. This was the case when the distance between the two spheres remained the same. If the distance between them was allowed to change, he found that the force decreased with the square of the distance between them. That is, if the distance was doubled, the force fell to only one-fourth of its original value. Thus, Coulomb concluded, the force one tiny charged object (ideally a *point charge*—one which, like an idealized point particle, has no spatial extent) exerts on a second one is proportional to the product of the amount of charge on one, Q_1, times the amount of charge on the other, Q_2, and inversely proportional to the square of the distance r between them; that is,

$$F = k \frac{Q_1 Q_2}{r^2} \qquad (15\text{-}1)$$

where k is a proportionality constant. This is known as **Coulomb's law.** Its validity rests on careful experiments that are much more sophisticated than Coulomb's difficult-to-perform experiment. The exponent 2 has been shown to be accurate to 1 part in 10^{16}.

Since we are dealing here with a new quantity (electric charge) we could choose its unit so that the proportionality constant k in Eq. 15-1 would be

(a)   (b)   (c)

FIGURE 15-9
A previously charged electroscope can be used to determine the sign of a given charge.

FIGURE 15-10
Schematic diagram of Coulomb's apparatus. It is similar to Cavendish's which was used to measure the gravitational force. When a charged sphere is placed close to the one on the suspended bar, the bar rotates slightly. The suspending fiber resists the twisting motion and the angle of twist is proportional to the force applied. By use of this apparatus, Coulomb was able to investigate how the electric force varied as a function of the magnitude of the charges and of the distance between them.

one. Indeed, such a system of units was once common.[†] However, the most widely used unit now is the *coulomb* (C) which is the SI unit. The precise definition of the coulomb today is in terms of electric current and magnetic field and will be discussed later (Section 19-13). In SI units, k has the value

$$k = 8.988 \times 10^9 \text{ N} \cdot \text{m}^2/\text{C}^2 \approx 9.0 \times 10^9 \text{ N} \cdot \text{m}^2/\text{C}^2.$$

Thus, 1 C is that amount of charge which, if it exists on each of two point objects placed 1 m apart, will result in each object exerting a force of about $(9.0 \times 10^9 \text{ N} \cdot \text{m}^2/\text{C}^2)(1.0 \text{ C})(1.0 \text{ C})/(1.0 \text{ m})^2 = 9.0 \times 10^9$ N on the other.

Charges produced by rubbing ordinary objects (such as a comb or plastic ruler) are typically a microcoulomb ($1 \mu\text{C} = 10^{-6}$ C) or less. The magnitude of the charge on one electron, on the other hand, is measured to be about 1.602×10^{-19} C (and its sign is negative). This is the smallest known charge;[‡] and because of its fundamental nature it is given the symbol e and is often referred to as the *elementary charge:*

$$e = 1.602 \times 10^{-19} \text{ C}.$$

Since an object cannot gain or lose a fraction of an electron, the net charge on any object must be an integral multiple of this charge. Electric charge is thus said to be *quantized* (existing only in discrete amounts); because e is so small, however, we normally don't notice this discreteness in macroscopic charges (1 μC requires about 10^{13} electrons) which thus seem continuous.

Equation 15-1 gives the magnitude of the force that either object exerts on the other, when the magnitudes of the charges Q_1 and Q_2 are given. The direction of this force is along the line joining the two objects. If the two charges have the same sign, the force on either object is directed away from the other; if the two objects have opposite charges, the force on one is directed toward the other, Fig. 15-11. If the signs of the two charges are included when using Eq. 15-1, F will be negative for attraction and positive for repulsion. Notice that the force one charge exerts on the second is equal but opposite to that exerted by the second on the first; this is in accord with Newton's third law.

It should be recognized that Eq. 15-1 applies to objects whose size is much smaller than the distance between them. Ideally, it is precise for point charges. For finite-sized objects, it is not always clear what value to use for r, particularly since the charge may not be distributed uniformly on the objects. If the two objects are spheres and the charge is known to be distributed uniformly on each, then r is the distance between their centers. It is important to keep in mind that Eq. 15-1 gives the force on a charge due to only *one* other charge. If several (or many) charges are present, the net force on any one of them will be the vector sum of the forces due to each of the others.

FIGURE 15-11

Direction of force depends on whether the charges have (a) same sign or (b) opposite sign.

[†] This is a cgs system of units, and the unit of electric charge is called the *electrostatic unit* (esu) or the statcoulomb. One esu is defined as that charge, on each of two point objects 1 cm apart, that gives rise to a force of 1 dyne.

[‡] Elementary-particle physicists have theorized the existence of smaller particles, called quarks, that would have a smaller charge equal to $\frac{1}{3}$ or $\frac{2}{3}e$. They have not been detected directly, experimentally, and theory indicates that free quarks may not be detectable (see Chapter 31).

EXAMPLE 15-1 What is the magnitude of the electric force on the electron ($Q_1 = -e$) of a hydrogen atom exerted by the single proton ($Q_2 = +e$) that is its nucleus, when the electron "orbits" the proton at its average distance of 0.53×10^{-10} m.

SOLUTION We use Eq. 15-1 with $Q_2 = +1.6 \times 10^{-19}$ C, $Q_1 = -Q_2$, and $r = 0.53 \times 10^{-10}$ m:

$$F = \frac{(9.0 \times 10^9 \text{ N·m}^2/\text{C}^2)(+1.6 \times 10^{-19} \text{ C})(-1.6 \times 10^{-19} \text{ C})}{(0.53 \times 10^{-10} \text{ m})^2}$$

$$= -8.2 \times 10^{-8} \text{ N}.$$

The minus sign means the force on the electron is toward the proton.

EXAMPLE 15-2 Calculate the net electric force on particle 3 in Fig. 15-12a (the -4.0 μC on the right) due to the other two charges.

SOLUTION The net force on particle 3 will be the sum of the force $\mathbf{F}_{31}$ exerted by particle 1 and the force $\mathbf{F}_{32}$ exerted by particle 2. The magnitudes of these two forces are

$$F_{31} = \frac{(9.0 \times 10^9 \text{ N·m}^2/\text{C}^2)(-4.0 \times 10^{-6} \text{ C})(-3.0 \times 10^{-6} \text{ C})}{(0.50 \text{ m})^2} = 0.43 \text{ N},$$

$$F_{32} = \frac{(9.0 \times 10^9 \text{ N·m}^2/\text{C}^2)(5.0 \times 10^{-6} \text{ C})(-4.0 \times 10^{-6} \text{ C})}{(0.20 \text{ m})^2} = -4.5 \text{ N}.$$

$\mathbf{F}_{31}$ is repulsive and $\mathbf{F}_{32}$ is attractive, so the direction of the forces is as shown in Fig. 15-12b. The net force on particle 3 is then

$$F = F_{32} + F_{31} = -4.5 \text{ N} + 0.4 \text{ N} = -4.1 \text{ N}.$$

The magnitude is 4.1 N, and it points to the left. (Notice that the charge in the middle (Q_2) in no way blocks the effect of the other one on the end (Q_1); it does exert its own force, of course.)

EXAMPLE 15-3 Calculate the force on charge Q_3 shown in Fig. 15-13a due to the charges Q_1 and Q_2.

FIGURE 15-12

Diagram for Example 15-2.

FIGURE 15-13

Determining the forces for Example 15-3.

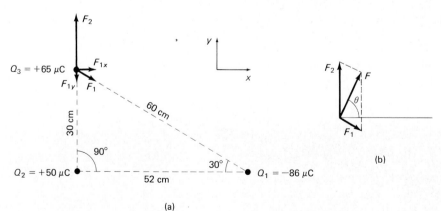

(a)

SOLUTION The forces $\mathbf{F}_1$ and $\mathbf{F}_2$ have the directions shown in the diagram since Q_1 exerts an attractive force and Q_2 a repulsive force. The magnitudes of $\mathbf{F}_1$ and $\mathbf{F}_2$ are (ignoring signs since we know the directions)

$$F_1 = \frac{(9.0 \times 10^9 \text{ N·m}^2/\text{C}^2)(6.5 \times 10^{-5} \text{ C})(8.6 \times 10^{-5} \text{ C})}{(0.60 \text{ m})^2} = 140 \text{ N},$$

$$F_2 = \frac{(9.0 \times 10^9 \text{ N·m}^2/\text{C}^2)(6.5 \times 10^{-5} \text{ C})(5.0 \times 10^{-5} \text{ C})}{(0.30 \text{ m})^2} = 330 \text{ N}.$$

We resolve $\mathbf{F}_1$ into its components along the x and y axes as shown:

$$F_{1x} = F_1 \cos 30° = 120 \text{ N},$$
$$F_{1y} = -F_1 \sin 30° = -70 \text{ N}.$$

The force $\mathbf{F}_2$ has only a y component. So the net force $\mathbf{F}$ on Q_3 has components $F_x = F_{1x} = 120$ N, and $F_y = F_2 + F_{1y} = 330 \text{ N} - 70 \text{ N} = 260$ N. Thus the magnitude of the net force $F = \sqrt{F_x^2 + F_y^2} = \sqrt{(120 \text{ N})^2 + (260 \text{ N})^2} = 290$ N; and it acts at an angle θ (see Fig. 15-13b) given by $\tan \theta = F_y/F_x = 260 \text{ N}/120 \text{ N} = 2.2$; so $\theta = 65°$.

The constant k in Eq. 15-1 is usually written in terms of another constant, ε_0, called the *permittivity of free space*; it is related to k by: $k = 1/4\pi\varepsilon_0$. Coulomb's law can then be written

$$F = \frac{1}{4\pi\varepsilon_0} \frac{Q_1Q_2}{r^2},$$

where

$$\varepsilon_0 = \frac{1}{4\pi k} = 8.85 \times 10^{-12} \text{ C}^2/\text{N·m}^2.$$

Coulomb's law describes the force between two charges when they are at rest. Additional forces come into play when charges are in motion, and these will be discussed in later chapters. In this chapter we discuss only charges at rest, and this is called *electrostatics*.

The value of the constant k given above, and that just given for ε_0, corresponds to the situation when the two charges are in vacuum or in air. In other materials, the force between two charges is often weaker. To express this fact, a second constant, K, called the *dielectric constant* of the material, is inserted into Coulomb's law:

$$F = \frac{k}{K} \frac{Q_1Q_2}{r^2}$$
$$= \frac{1}{4\pi\varepsilon_0 K} \frac{Q_1Q_2}{r^2}.$$

We can also define $\varepsilon = K\varepsilon_0$, where ε is the *permittivity of the material*. For vacuum, $K = 1$ and $\varepsilon = \varepsilon_0$. For air and other gases, K is very close to 1 and is generally taken to be so. However, for liquids and solids it can be quite large. For water, $K = 80$. This large value has a profound effect on materials, in that it allows many to be readily dissolved. Consider, for example, or-

FIGURE 15-14
(a) Sodium chloride crystal; (b) sodium chloride dissolving in water.

(a)

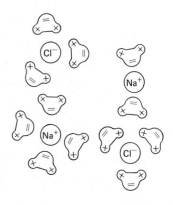

(b)

dinary table salt, NaCl (sodium chloride), which arranges itself in a crystal structure as shown in Fig. 15-14a; the attractive forces between Na^+ ions and Cl^- ions hold the crystal together. But when salt is placed in water, the water molecules that come between a Na^+ ion and a Cl^- ion cause the force between them to be reduced by a factor of 80. Because the force is now so weak, the ions can move about the liquid freely (Fig. 15-14b). The high dielectric constant of water thus accounts for its ability to dissolve many substances. The reason behind the large value of K for water is the fact that it is a polar molecule and can surround an ion as shown in the figure.

15-6 ▌ The electric field

Many common forces might be referred to as "contact forces." That is, you exert a force on an object by coming into contact with it; for example, you push or pull on a box, a broom, or a stalled car. Similarly, a tennis racket exerts a force on a tennis ball when it makes contact; or a ball exerts a force on a window when it makes contact.

On the other hand, both the gravitational force and the electrical force act over a distance; that is, there is a force even when the two objects are not in contact. The idea of a force *acting at a distance* was a difficult one for early thinkers. Newton himself felt uneasy with this idea when he published his law of universal gravitation. The conceptual difficulties can be overcome with the idea of the *field*, developed by the British scientist Michael Faraday (1791–1867). In the electrical case, according to Faraday, an *electric field* extends outward from every charge and permeates all of space (Fig. 15-15). When a second charge is placed near the first charge, it feels a force because of the electric field that is there (say, at point P in Fig. 15-15). The electric field at the location of the second charge is considered to interact directly with this charge to produce the force. It must be emphasized, however, that a field is *not* a kind of matter. It is, rather, a concept[†]—and a very useful one.

We can investigate the electric field surrounding a charge or group of charges by measuring the force on a small positive **test charge.** By a test charge we mean a charge so small that the force it exerts does not significantly alter the distribution of the other charges, the ones that cause the field being measured. The force on a tiny positive test charge q placed at various locations in the vicinity of a single positive charge Q would be as shown in Fig. 15-16. The force at b is less than at a because the distance is greater (Coulomb's law); and the force at c is smaller still. In each case, the force is directed radially outward from Q. The electric field is defined in terms of the force on such a positive test charge. In particular, the **electric field, E,** at any point in space is defined as the force **F** exerted on a tiny positive test charge at that point divided by the magnitude of the test charge q:

$$E = \frac{F}{q}. \tag{15-2}$$

[†] Whether the electric field is "real," and really exists, is a philosophical, even metaphysical, question. In physics it is a very useful idea, in fact a great invention of the human mind.

FIGURE 15-15

An electric field surrounds every charge. P is an arbitrary point.

FIGURE 15-16

Force exerted by charge $+Q$ on a small test charge, q, placed at points a, b, and c.

(Ideally, **E** is defined as the limit of **F**/q as q is taken smaller and smaller, approaching zero.) From this definition (Eq. 15-2) we see that the direction of the electric field at any point in space is defined as the direction of the force on a positive test charge at that point. And the magnitude of the electric field is the *force per unit charge*. Thus **E** is measured in units of newtons per coulomb (N/C).

The reason for defining **E** as **F**/q is so that **E** does not depend on the magnitude of the test charge q. This means that **E** describes only the effect of the charges creating the electric field at that point.

The electric field at any point in space can be measured, based on the definition, Eq. 15-2. For simple situations involving one or several point charges, we can calculate what **E** will be. For example the electric field at a distance r from a single point charge Q would have magnitude

$$E = k \frac{qQ}{r^2} \frac{1}{q}$$

$$= k \frac{Q}{r^2}. \qquad \text{[single point charge]} \quad (15\text{-}3)$$

This relation for the electric field due to a single point charge is also (in addition to Eq. 15-1) referred to as Coulomb's law. Notice that E is independent of q—that is, it depends only on the charge Q which produces the field, and not on the value of the test charge q.

☐ **EXAMPLE 15-4** Calculate the magnitude and direction of the electric field at a point P which is 30 cm to the right of a point charge $Q = -3.0 \times 10^{-4}$ C.

SOLUTION The magnitude of the electric field is

$$E = k \frac{Q}{r^2} = \frac{(9.0 \times 10^9 \text{ N·m}^2/\text{C}^2)(3.0 \times 10^{-4} \text{ C})}{(0.30 \text{ m})^2} = 3.0 \times 10^7 \text{ N/C}.$$

The direction of the electric field is *toward* the charge Q as shown in Fig. 15-17a since we defined the direction as that of the force on a positive test charge. If Q had been positive, the electric field would have pointed away, as in Fig. 15-17b.

(a)

(b)

FIGURE 15-17

Electric field at point P due to (a) a negative charge Q and (b) a positive charge Q (Example 15-4).

If the field is due to more than one charge, the individual fields due to each charge are added vectorially to get the total field.

☐ **EXAMPLE 15-5** Calculate the electric fields at points A and B in Fig. 15-18.

SOLUTION (*a*) The calculation is much like that of Example 15-3, but now we are dealing with electric fields. The electric field at A is the vector sum of the fields $\mathbf{E}_{A1}$ due to Q_1, and $\mathbf{E}_{A2}$ due to Q_2: $E_{A1} = (9.0 \times 10^9 \text{N·m}^2/\text{C}^2)(50 \times 10^{-6} \text{ C})/(0.60 \text{ m})^2 = 1.25 \times 10^6$ N/C, and similarly $E_{A2} = 5.0 \times 10^6$ N/C. The directions are as shown, so the total electric field at A, $\mathbf{E}_A$, has components $E_{Ax} = E_{A1} \cos 30° = 1.1 \times 10^6$ N/C and $E_{Ay} = E_{A2} - E_{A1} \sin 30° = 4.4 \times 10^6$ N/C. Thus the magnitude of $\mathbf{E}_A$ is

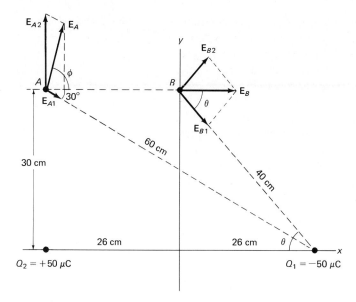

FIGURE 15-18

Calculation of the electric field at points A and B for Example 15-5.

$E_A = \sqrt{(1.1)^2 + (4.4)^2} \times 10^6$ N/C $= 4.5 \times 10^6$ N/C. It will act at an angle ϕ (see diagram) where $\tan \phi = 4.4/1.1 = 4.0$; so $\phi = 76°$.

(b) Since B is equidistant (40 cm by Pythagorean theorem) from the two equal charges, the magnitudes of E_{B1} and E_{B2} are the same; that is, $E_{B1} = E_{B2} = (9.0 \times 10^9$ N·m^2/C$^2)(50 \times 10^{-6}$ N)$/(0.40$ m$)^2 = 2.8 \times 10^6$ N/C. Also, because of the symmetry, the y components are equal and opposite. Hence the total field E_B is horizontal and equals $E_{B1} \cos \theta + E_{B2} \cos \theta = 2E_{B1} \cos \theta$; from the diagram, $\cos \theta = 26$ cm$/40$ cm $= 0.65$. Then $E_B = 2(2.8 \times 10^6$ N/C$)(0.65) = 3.6 \times 10^6$ N/C, and is along the x direction.

15-7 ∎ Lines of force

Since the electric field is a vector, it is sometimes referred to as a *vector field*. We could indicate the electric field with arrows at various points in a given situation, such as at a, b, and c in Fig. 15-16; the directions of $\mathbf{E}_a$, $\mathbf{E}_b$, and $\mathbf{E}_c$ would be the same as that of the forces already shown, but the lengths (magnitudes) would be different (since we divide by q); however, the relative lengths of $\mathbf{E}_a$, $\mathbf{E}_b$, and $\mathbf{E}_c$ would be the same as for the forces since we divide by the same q each time. However, to indicate the electric field in such a way at many points would result in many arrows, which would appear confusing. To avoid this, we use another technique, that of lines of force.

In order to visualize the electric field, we draw a series of lines to indicate the direction of the electric field at various points in space. These **electric field lines,** or **lines of force,** are drawn so that they indicate the direction of the force due to the given field on a positive test charge. The lines of force due to a single positive charge are shown in Fig. 15-19a and for a single negative charge in Fig. 15-19b. In part (a) the lines point radially outward

FIGURE 15-19

Electric field lines near (a) a single positive point charge and (b) a single negative point charge.

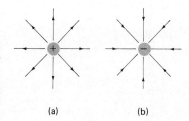

(a)   (b)

from the charge, and in part (b) they point radially inward toward the charge; these are the directions of the force that would be exerted on a positive test charge in each case. Only a few representative lines have been shown. One could just as well draw lines in between those shown since the electric field exists there as well. However, we can always draw the lines so the number of lines starting on a positive charge, or ending on a negative charge, is proportional to the magnitude of the charge. Notice that near the charge, where the force is greatest, the lines are closer together. This is a general property of electric field lines; *the closer the lines are together, the stronger the electric field in that region.* In fact the lines can always be drawn so that the number crossing unit area perpendicular to **E** is proportional to the magnitude of the electric field.

Figure 15-20a shows the electric field lines surrounding two charges of opposite sign. The electric field lines are curved in this case and they are directed from the positive charge to the negative charge. The direction of the field at any point is directed tangentially as shown by the faint arrow at point *P*. To satisfy yourself that this is the correct pattern for the electric field lines, you can make a few calculations such as those done in Example 15-5 for just this case (see Fig. 15-18). Figures 15-20b and c show the electric field lines surrounding two positive charges, and between two oppositely charged parallel plates. Notice that the electric field lines between the two plates are parallel and equally spaced, except near the edges. Thus, in the central region, the electric field has the same magnitude at all points and we can write

$$E = \text{constant.} \quad [\text{between two closely spaced parallel plates}] \quad (15\text{-}4)$$

Although the field fringes near the edges (the lines curve), we can often ignore this, particularly if the separation of the plates is small compared to their size. This should be compared to the field of a single point charge where the field decreases as the square of the distance, Eq. 15-3.

We summarize the properties of lines of force as follows:

1. The lines of force indicate the direction of the electric field; the field points in the direction tangent to the field lines at any point.

2. The lines are drawn so that the magnitude of the electric field, *E*, is proportional to the number of lines crossing unit area perpendicular to the lines. The closer the lines, the stronger the field.

3. Electric field lines start on positive charges and end on negative charges; and the number starting or ending is proportional to the magnitude of the charge.

Another way of visualizing the electric field lines is that each represents the path that would be followed by a tiny test charge placed on it. (This is strictly true only if the test charge has no inertia, or moves slowly because of, say, a frictional force.) Note that no two field lines can cross. (If they did, it would mean the electric field would point in two different directions at the same point, which doesn't make sense.)

The field concept can also be applied to the gravitational force. Thus we can say that a *gravitational field* exists for every object that has mass. One object attracts another by means of the gravitational field. The earth, for example, can be said to possess a gravitational field (Fig. 15-21), which is responsible for the force on objects that we drop. The *gravitational field*

FIGURE 15-20

Electric field lines for three arrangements of charges.

(a)

(b)

(c)

intensity is defined as the *force per unit mass*. The magnitude of the earth's gravitational field intensity at any point is then (GM_e/r^2), where M_e is the mass of the earth, r the distance of the point from the earth's center, and G the gravitational constant (Chapter 4); at the earth's surface, r is simply the radius of the earth and the gravitational field intensity is simply equal to g, the acceleration due to gravity (since $F/m = mg/m = g$). In outer space the gravitational field intensity can be calculated at any point as a sum of terms due to earth, sun, moon, or other bodies that contribute significantly.

FIGURE 15-21
The earth's gravitational field.

15-8 ∎ Electric fields and conductors

The electric field inside a good conductor is zero in the static situation—that is, when the charges are at rest. If there were an electric field within a conductor, there would be a force on its free electrons since $\mathbf{F} = q\mathbf{E}$; the electrons would move until they reached positions where the electric field, and therefore the force on them, did become zero.

This reasoning has some interesting consequences. For one, the charge on a conductor distributes itself on the outer surface. For a negatively charged conductor you can imagine that the negative charges repel one another and race to the surface to get as far away from one another as possible. Another consequence is the following. Suppose that a positive charge Q is placed at the center of a cavity within an uncharged isolated metal conductor whose shape is a spherical shell, Fig. 15-22. Because there can be no field within the metal, the lines leaving the positive charge must end on negative charges on the inner surface of the metal. Thus an equal amount of negative charge, $-Q$, is induced on the inner surface of the spherical shell; and a positive charge, $+Q$, of the same magnitude must then exist on the outer surface (since the shell is neutral). Thus, although no field exists in the metal itself, an electric field exists outside of it as shown in Fig. 15-22 as if the metal were not even there.

A related property of electric fields and conductors is that, when no charges are moving, the electric field is always perpendicular to the surface of a conductor. If there were a component of $\mathbf{E}$ parallel to the surface, electrons at the surface would move along the surface in response to this force, until they reached positions where no force was exerted on them; that is, until the electric field was perpendicular to the surface.

These properties pertain only to conductors. In a nonconductor, which does not have free electrons, an electric field can exist; and the electric field does not necessarily make an angle of 90° to its surface.

FIGURE 15-22

A charge placed inside a spherical shell. Charges are induced on the conductor surfaces. The electric field exists even beyond the shell but not within the conductor itself.

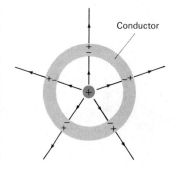

*15-9 ∎ Electric forces in molecular biology: DNA structure and replication

The study of the structure and functioning of a living cell at the molecular level is known as molecular biology. It is an important area for application of physics. Since the interior of a cell is fluid (mostly water), we can imagine it as a vast sea of molecules continually in motion, colliding into one another with various amounts of kinetic energy. These molecules interact with one another in various ways—chemical reactions (making and breaking of covalent bonds) and more brief interactions or unions that occur because of electrostatic attraction between molecules.

The many activities that occur within the cell, and which help to differentiate living matter from nonliving, are now considered to be the result of random ("thermal") molecular motion plus the ordering effect of the electrostatic force. We now use these ideas to analyze some cellular processes involving macromolecules (large molecules) such as DNA and proteins. The picture we present here has not been seen "in action." Rather, it is a model of what happens based on accepted physical theories and a great variety of experimental results.

The genetic information that is passed on from generation to generation in all living objects is contained in the chromosomes, which are made up of genes. Each gene contains the information needed to produce a particular type of protein molecule. A protein consists of one or more chains of small molecules called amino acids; it is used either as part of a structure (cell wall or muscle fiber, for example) or as an enzyme to catalyze a chemical reaction needed for the growth or survival of the organism. The genetic information contained in a gene is built into the principal molecule of a chromosome, the DNA (deoxyribonucleic acid). A DNA molecule consists of a long chain of many small molecules known as nucleotide bases. There are only four types of bases: adenine (A), cytosine (C), guanine (G), and thymine (T). These are arranged along the molecule according to a code, the "genetic code," which is "translated" into the amino acids that form the protein molecule.

A chromosome actually consists of two long DNA molecules wrapped about one another in the shape of a "double helix." As shown in Fig. 15-23, the two strands are held together by electrostatic forces—that is, by the attraction of positive charges to negative charges. We see in part (a), that an A (adenine) on one strand is always opposite a T on the other strand; similarly, a G is always opposite a C. This happens because the shape of the four molecules A, T, C, and G is such that a T fits closely only into an A, and a G into a C, and only in the case of this close proximity is the electrostatic force great enough to hold them together even for a short time (Fig. 15-23b). The electrostatic force between A and T, and between C and G, arises because some electrons in each of these molecules spend more time orbiting one atom than another; for example, the electron normally on the H atom of adenine spends some of its time orbiting the adjacent N atom, so the N has a net negative charge and the H a positive charge (upper part of Fig. 15-23b). This H^+ atom is then attracted to the O^- atom of the thymine.

How does the arrangement shown in Fig. 15-23 come about? It occurs when the chromosome replicates (duplicates) itself just before cell division; indeed, the arrangement of A opposite T and G opposite C ensures that the genetic information is passed on accurately to the next generation. The process of replication is shown in a simplified form in Fig. 15-24. The two strands of the DNA chain become unwound at one end, leaving the charged parts of the bases exposed. To see how the correct order of bases occurs, let us focus our attention on the G molecule indicated by the arrow on the

(a)

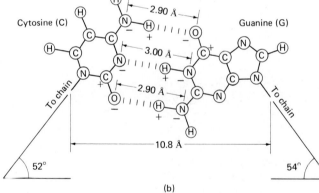

(b)

FIGURE 15-23

(a) Section of a DNA double helix. (b) "Close-up" view of the helix, showing how A and T attract each other and how G and C attract each other through electrostatic forces; this is how the double helix is held together. The vertical bars (||||) are used to indicate the electrostatic attraction (or "weak bond"). Note that there are two weak bonds between A and T, and three between C and G. The distance unit is the angstrom (1 Å = 10^{-10} m). (After J. D. Watson, *Molecular Biology of the Gene*, 3rd ed., W. A. Benjamin, Inc., Menlo Park, Calif., 1976.)

bottom strand in the figure. There are many unattached nucleotide bases of all four kinds bouncing around in the cellular fluid. The only one of the four bases that will experience attraction to our G, if it bounces close to it, will be a C. The charges on the other three bases are not arranged so they can get close to those on the G; they will thus have no significant force exerted on them—remember that the force decreases rapidly with distance. The electrostatic forces will often hold the C opposite our G long enough so that an enzyme can attach the C to the growing end of the new chain.

Thus we see that electrostatic forces not only hold the two chains together; they also operate to select the bases in the proper order during

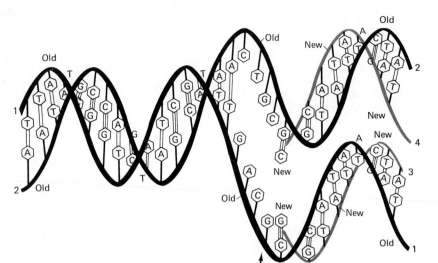

FIGURE 15-24
Replication of DNA. (After J. D. Watson, *Molecular Biology of the Gene*, 3rd ed., W. A. Benjamin, Inc., Menlo Park, Calif., 1976.)

replication so the genetic information is passed on accurately to the next generation. Note in Fig. 15-24 that the new number 4 strand has the same order of bases as the old number 1 strand; and the new number 3 strand is the same as the old number 2; so the two new helixes, 1–3 and 2–4, are identical to the original 1–2 helix. The error rate—say a T being incorporated in a new chain opposite a G—is less than 1 in 10^8. Such an error would correspond to a spontaneous mutation and a possible change in some characteristic of the organism.

*15-10 ∎ Protein synthesis and structure

Protein molecules play two very important roles in the cell: as structural units of the cell (cell walls, membranes, organelles); and as enzymes which catalyze the many chemical reactions that occur within the cell.

Protein molecules are long chains of single small molecules known as amino acids. There are 20 different amino acids of importance. These chains are generally linear, without branches, although several chains may be linked together. The order of the different amino acids determines the structure and role of each protein molecule.

The mechanism by which protein molecules are synthesized is illustrated in Fig. 15-25. It begins at the DNA double helix, for each gene on a chromosome contains the information for one protein. The ordering of the four bases, A, C, G, and T, provides the "code" for the order of amino acids in the protein. First the DNA double helix unwinds and a new molecule called messenger-RNA (m-RNA) is synthesized; m-RNA is a chain molecule containing four different bases, just like those of DNA except that thymine (T) is replaced by a very similar molecule called uracil (U). In the diagram a C has just been added to the growing m-RNA chain in much the same way that DNA replicates; and an A, attracted to the T on the DNA chain, will

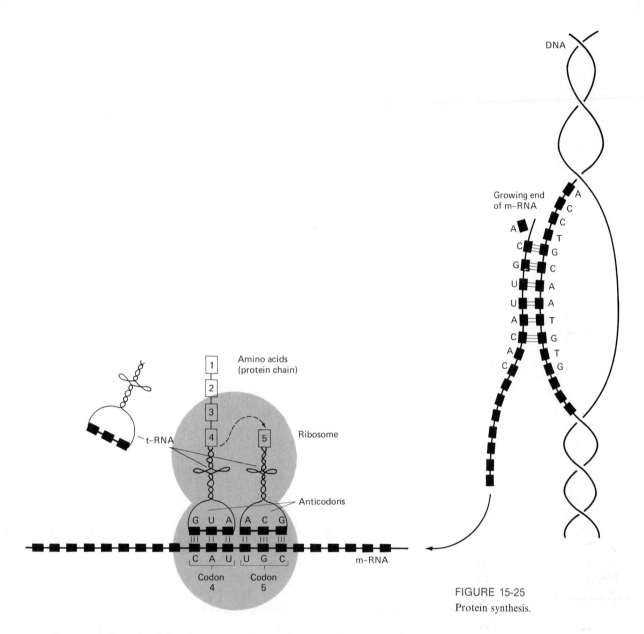

FIGURE 15-25
Protein synthesis.

soon be attached to the C by an enzyme. The order of the bases, and thus the genetic information, is preserved in the m-RNA. The m-RNA is buffeted about by all the moving molecules in the cell (remember kinetic theory) until it gets close to a tiny organelle known as a ribosome, to which it becomes attached by electrostatic attraction, Fig. 15-25. Also held by the electrostatic force to the ribosome are one or two "transfer-RNA" (t-RNA) molecules. These t-RNA molecules "translate" the genetic code of nucleotide bases into amino acids in the following way. There are different t-RNA molecules for each combination of three bases. At one end of the t-RNA molecule is the "anticodon," which is a set of three nucleotide bases. If all three bases match the three (the "codon") on that part of the m-RNA attached to the ribosome, the anticodon is attracted to the m-RNA and becomes attached briefly to the ribosome. Attached to the other end of a t-RNA molecule is the appropriate amino acid; for example, the CAU codon codes for the amino acid histidine. There are two attachment sites on the ribosome; these hold two t-RNA molecules while enzymes link the two amino acids together to lengthen the amino acid chain. As each amino acid becomes added, the old t-RNA molecule—having lost its amino acid—is removed, probably by a random collision with some molecule in the cellular fluid. A new one then becomes attached as the ribosome moves along the m-RNA molecule.

This process of protein synthesis is often presented as if it occurred in clockwork fashion—as if each molecule knew its role and went to its assigned place, like bees in a hive. But this is not the case. The forces of attraction between the electric charges of the molecules are rather weak; hence they become significant only when the molecules can come close together and several weak "bonds" can be made. Indeed, if the shapes are not just right, there is almost no attraction, which is why there are few mistakes. The point is that there are many molecules in the cell, all jostling about. Many different bases approach the end of the growing m-RNA chain, but only the A (see Fig. 15-25) will be held long enough to the T on the DNA to be enzymatically attached to the m-RNA chain. And many different t-RNA molecules will approach an open spot on the ribosome but only that one which has the proper anticodon will remain opposite the codon long enough for its amino acid to become attached to the growing protein chain. Thus, out of the random motion of the molecules, the electrostatic force between opposite electric charges acts to bring order out of chaos.

The fact that the attraction is weak is very important. If it were strong, collisions with other molecules would not allow a t-RNA molecule to be released from the ribosome, or the m-RNA to be released from the DNA. If they were not temporary encounters, metabolism would grind to a halt.

As amino acid is added to amino acid, the protein molecule grows in length until the last one is added and the protein is complete. Even as it is being made, this chain is being buffeted about in the cellular sea—we might think of a wiggling worm. But a protein molecule has charged polar groups along its length; and as it takes on various shapes, the electric forces of attraction between different parts of the molecule will eventually lead to a particular configuration which is quite stable. Each type of protein has its own peculiar shape depending on the location of charged atoms; in the last analysis, this depends on the order of the amino acids.

Amino acids all have the same basic structure except for a side group, which we designate R (Fig. 15-26a); R can be H, CH_3, or many other combinations of atoms. In many proteins, particularly in structural proteins, electrical attraction occurs between the regular parts of the amino acids and produces a repeating structure. A common example is the "α-helix structure," shown schematically in Fig. 15-26b. This structure is held together by the forces of attraction between the charged C—O and N—H groups.

Not all proteins have an α-helix structure. Often attraction between the R groups dominates. The usual situation is that a protein molecule has a twisted irregular shape, although it may have short sections of α-helix as well.

The *action* of proteins as enzymes, or as structureal units, also involves electrical forces of attraction between unlike electric charges, and so do many

FIGURE 15-26
(a) Amino acid. (b) α-Helix structure.

(a)

(b)

other cellular processes, such as the action of antibodies on antigens. The "lock-and-key" mechanism used to explain some of these processes involves molecules whose shapes are complementary, so they fit closely enough that electrostatic forces can hold them together.

▌ Summary

There are two kinds of *electric charge*, positive and negative. These designations are to be taken algebraically—that is, any charge is plus or minus so many coulombs (C), in SI units. Electric charge is *conserved:* if a certain amount of one type of charge is produced in a process, an equal amount of the opposite type is also produced on the same body or on a different body; thus the *net* charge produced is zero. According to the atomic theory, electric charge originates in the atom, which consists of a positively charged nucleus surrounded by negatively charged electrons. Each electron has a charge $-e = -1.6 \times 10^{-19}$ C. Conductors are those materials in which many electrons are relatively free to move, whereas electric insulators are those in which very few electrons are free to move. An object is negatively charged when it has an excess of electrons, and positively charged when it has less than its normal amount of electrons. An object can become charged by rubbing, (in which electrons are transferred from one material to another), by conduction (which is transfer of charge from one charged object to another by touching), or by induction (the separation of charge within an object because of the close approach of another charged object but without touching).

Electric charges exert a force on each other. If two charges are of opposite types, one positive and one negative, they each exert an attractive force on the other. If the two charges are the same type, each repels the other. The magnitude of the force one point charge exerts on another is proportional to the product of their charges and inversely proportional to the square of the distance between them:

$$F = k \frac{Q_1 Q_2}{r^2};$$

this is *Coulomb's law*.

An *electric field* is imagined to exist in space due to any charge or group of charges. The force on another charged object is then conceived to be due to the electric field present at its location. The *electric field*, **E**, at any point in space due to one or more charges is defined as the force per unit charge that would act on a test charge q placed at that point: $\mathbf{E} = \mathbf{F}/q$. Electric fields are represented by *lines of force* which start on positive charges and end on negative charges. Their direction indicates the direction the force would be on a tiny positive test charge placed at a point; the lines can be drawn so that the number per unit area is proportional to the magnitude of E. The static electric field (that is, no charges moving) inside a good conductor is zero, and the electric field lines just outside a charged conductor are perpendicular to its surface.

▌ Questions

1. If you charge a pocket comb by rubbing with a silk scarf, how can you determine if the comb is positively or negatively charged?

2. Why does a shirt or blouse taken from a clothes dryer sometimes cling to your body?

3. Why does a phonograph record attract dust just after it has been wiped clean?

4. Explain why fog or rain droplets tend to form around ions or electrons in the air.

5. Can you guess why trucks carrying flammable fluids drag a chain along the ground? (Hint: Have you ever experienced a slight shock when getting out of your car?)

6. A positively charged rod is brought close to a neutral piece of paper, which it attracts. Draw a diagram showing the separation of charge and explain why attraction occurs.

7. Why does a plastic ruler that has been rubbed with a cloth have the ability to pick up small pieces of paper? Why is this difficult to do on a humid day?

8. Contrast the *net charge* on a conductor to the "free charges" in the conductor.

9. Figures 15-5 and 15-6 show how a charged rod placed near an uncharged metal object can attract (or repel) electrons. There are a great many electrons in the metal, yet only some of them move as shown. Why not all of them?

10. When an electroscope is charged, the two leaves repel each other and remain at an angle. What balances the electric force of repulsion so that the leaves don't separate further?

11. The form of Coulomb's law is very similar to that for Newton's law of universal gravitation. What are the differences between these two laws? Compare also gravitational mass and electric charge.

12. We are not normally aware of the gravitational or electrical force between two ordinary objects. What is the reason in each case? Give an example where we are aware of each one and why.

13. Is the electric force a conservative force? Why or why not? (See Chapter 5.)

14. Is there an attraction between the two objects in Fig. 15-6b? Why or why not?

15. What experimental observations mentioned in the text rule out the possibility that the numerator in Coulomb's law contains the quantity $(Q_1 + Q_2)$ rather than the product $Q_1 Q_2$?

16. When a charged ruler attracts small pieces of paper, sometimes a piece jumps quickly away after touching the ruler. Explain.

17. Explain why we use *small* test charges when measuring electric fields.

18. When determining an electric field, must we use a *positive* test charge, or would a negative one do as well? Explain.

19. Draw the electric field lines surrounding two negative electric charges a distance l apart.

20. Assume that the two opposite charges in Fig. 15-20a are 12.0 cm apart. Consider the magnitude of the electric field 2.5 cm from the positive charge. On which side of this charge—top, bottom, left, or right—is the electric field the strongest? The weakest?

21. Why can lines of force never cross?

▌ Problems

SECTION 15-5

1. (I) How many electrons make up a charge of 100 μC?

2. (I) Two charged bodies exert a force of 480 mN on each other. What will be the force if they are moved so they are only one-eighth as far apart?

3. (I) What is the magnitude of the electric force of attraction between an iron nucleus ($q = +26e$) and its innermost electron if the distance between them is 1.0×10^{-12} m?

4. (I) How far apart must two electrons be if the force between them is to be 1.0×10^{-12} N?

5. (I) What is the magnitude of the force a 10-μC charge exerts on a 3.0-mC charge 2.0 m away? (1μC $= 10^{-6}$ C, 1 mC $= 10^{-3}$ C.)

6. (II) What is the total charge on all the electrons in 1.0 kg of H_2O?

7. (II) How close must two electrons be if the electric force between them is equal to the weight of either at the earth's surface?

8. (II) Three positive particles of charges 3.0 μC are located at the corners of an equilateral triangle of side 10 cm. Calculate the magnitude and direction of the net force on each particle.

9. (II) Particles of charge $+86$, $+48$, and -90μC are placed in a line. The center one is 1.5 m from each of the others. Calculate the net force on each due to the other two.

10. (II) A charge of 0.0500 C is placed at each corner of a square 15.0 m on a side. Determine the magnitude and direction of the force on each charge.

11. (II) Repeat problem 10 for the case when two of the positive charges, on opposite corners, are replaced by negative charges of the same magnitude.

12. (II) A 3.0-g copper penny has a positive charge of 0.55 mC. What fraction of its electrons has it lost?

13. (III) In one model of the hydrogen atom the electron revolves in a circular orbit around the proton with a speed of 1.1×10^6 m/s. What is the radius of the electron's orbit?

14. (III) A large electroscope is made with "leaves" that are 60 cm long wires with 28-g balls at the ends. When charged, nearly all the charge resides on the balls. If the wires each make a 30° angle with the vertical, what total charge must have been applied to the electroscope?

15. (III) Two charges, $-Q_0$ and $-3Q_0$, are a distance l apart. These two charges are free to move but do not because there is a third charge nearby. What must be the charge and placement of the third charge for the first two to be in equilibrium?

16. (III) A 3.0- and a -1.3-μC charge are placed 18 cm apart. Where can a third charge be placed so that it experiences no net force?

17. (III) Two nonconducting spheres have a total charge of 850 μC. When placed 1.30 m apart, the force each exerts on the other is 28.5 N and is repulsive. What is the charge on each? What if the force were attractive?

SECTION 15-6

18. (I) A force of 2.4 N is exerted on a -1.8-μC charge in a downward direction. What is the magnitude and direction of the electric field at this point?

FIGURE 15-27

FIGURE 15-28

19. (I) An electron in an electric field experiences a force of 8.0×10^{-16} N. What is the magnitude and direction of the electric field at this point?

20. (I) What is the magnitude of the force on an electron in an electric field of 600 N/C?

21. (I) What is the magnitude and direction of the electric field 12.0 m directly above a 13.0×10^{-6} C charge?

22. (II) What is the magnitude and direction of the electric field at a point midway between a -20-μC and a $+60$-μC charge 40 cm apart?

23. (II) A proton ($m = 1.67 \times 10^{-27}$ kg) is suspended at rest in a uniform electric field $\mathbf{E}$. Take into account gravity and determine $\mathbf{E}$.

24. (II) You are given two unknown point charges, Q_1 and Q_2. At a point on the line joining them, one-third of the way from Q_1 to Q_2, the electric field is zero. What can you say about these two charges?

25. (II) Use Coulomb's law to determine the electric field at A and B in Fig. 15-27 due to the two positive charges ($Q = 4.0$ μC) shown. Calculate magnitude and direction. Is your result consistent with Fig. 15-20b?

26. (II) Calculate the electric field at the center of a square 25 cm on a side if one corner is occupied by a $+33.0$-μC charge and the other three are occupied by -21.0-μC charges.

27. (II) Calculate the electric field at one corner of a square 80 cm on a side if the other three corners are occupied by 18.2×10^{-4} C charges.

28. (II) Calculate the magnitude of the electric field at the center of a square with 10 cm sides if the corners, taken in rotation, have charges of 1.0 μC, 2.0 μC, 3.0 μC, and 4.0 μC.

29. (II) Measurements indicate that there is an electric field surrounding the earth. Its magnitude is about 150 N/C at the earth's surface and points inward toward the earth's center. What is the magnitude of the electric charge on the earth? Is it positive or negative? (Hint: the electric field due to a uniformly charged sphere is the same as if all the charge were concentrated at its center.)

30. (II) What is the acceleration of an electron in a 2200-N/C electric field?

31. (II) What is the electric field strength at a point in space where a proton ($m = 1.67 \times 10^{-27}$ kg) experiences an acceleration of 7.6×10^4 m/s²?

32. (II) A spaceship makes a trip from earth to the moon, 380,000 km away. At what point in the trip will the gravitational field be zero? The mass of the moon is about $\frac{1}{81}$ that of earth.

33. (III) A water droplet of radius 0.020 mm remains stationary in the air. If the electric field of the earth is 150 N/C, how many excess electron charges must the water droplet have?

34. (III) An electron (mass $m = 9.1 \times 10^{-31}$ kg) is accelerated in the uniform field $\mathbf{E}$ ($E = 2.0 \times 10^4$ N/C) between two parallel charged plates. The separation of the plates is 1.5 cm. The electron is accelerated from rest near the negative plate and passes through a tiny hole in the positive plate, Fig. 15-28. (a) With what speed does it leave the hole? (b) Show that the gravitational force can be ignored.

*SECTION 15-9

*35. (III) The two strands of the helix-shaped DNA molecule are held together by electrostatic forces as shown in Fig. 15-23. Assume that the charge indicated on H and N atoms is 0.2e, and on the indicated C and O atoms is 0.4e, that atoms on each molecule are separated by 1.0×10^{-10} m, and that all relevant angles are 120°. Estimate the net force between (a) a thymine and an adenine; and (b) between a cytosine and a guanine. (c) Estimate the total force for a DNA molecule containing 10^5 pairs of such molecules.

*SECTION 15-10

*36. (II) Hair consists of protein molecules that have an α-helix structure. If hair grows 15 cm/yr, what is the rate of synthesis of the protein molecules? Specify in terms of number of turns of the α-helix per second (see Fig. 15-26b).

16 ELECTRIC POTENTIAL AND ELECTRIC ENERGY

We saw in Chapter 5 that the concept of energy was extremely valuable in dealing with mechanical problems. For one thing energy is a conserved quantity and thus an important aspect of nature. Furthermore we saw that many problems could be solved using the energy concept even though a detailed knowledge of the forces involved was not possible, or when a calculation involving Newton's laws would have been too difficult.

The energy point of view can be used in electricity, and it is especially useful. It not only extends the law of conservation of energy, but it gives us another way to view electrical phenomena and is a tool in solving problems more easily, in many cases, than by using forces and electric fields.

16-1 ▌ Electric potential and potential difference

Energy will be conserved if we can define electric potential energy as for other types of potential energy. That is, we define the change in electric potential energy, $PE_b - PE_a$, when a charge q moves from some point a to a second point b as the work required of an external force[†] to move the charge ("against" the electric field) from a to b without accelerating it. Equivalently, as we saw in Chapter 5, $PE_b - PE_a$ is equal to the work the electric field itself would do to move the charge back from b to a. For example, consider the electric field between two equally but oppositely charged parallel plates which are large compared to their separation so the field will be uniform over most of the region, Fig. 16-1. Now consider a positive point charge q placed at point b very near the positive plate as shown. The electric force on the charge will tend to move the charge toward the negative plate and will do work on the charge to move it from point b to point a (near the negative plate). In the process the charged particle will be accelerated and its kinetic energy increased; and the potential energy will be decreased by an equal amount, which is equal to the work done by the electric force in

[†] Note that if positive external work is done on the charge to move it from point a to point b, the potential energy increases ($PE_b - PE_a$ is positive); this is consistent with our earlier discussion of work and energy in Chapter 5, and also with the first law of thermodynamics (Section 12-1).

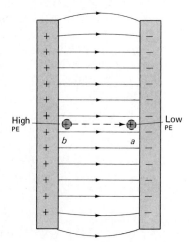

FIGURE 16-1

Work is done by the electric field in moving the positive charge from position a to position b.

moving the particle from b to a. In accord with the conservation of energy, electric potential energy is transformed into kinetic energy and the total energy is conserved. On the other hand, if an external force moves the positive charge (without accelerating it) from point a to point b against the electric field, this external work done on the particle increases its potential energy. The positive charge q has its greatest potential energy at point b, near the positive plate,[†] so $PE_b - PE_a > 0$. The reverse is true for a negative charge: its potential energy is greatest near the negative plate.

We defined the electric field (Chapter 15) as the force per unit charge; similarly, it is useful to define the **electric potential** (or simply the **potential** when "electric" is understood) as the *potential energy per unit charge*. Electric potential is given the symbol V; so we define the electric potential, V_a, at some point a, as

$$V_a = \frac{PE_a}{q},$$

where PE_a is the potential energy of a charge q placed at point a. As we discussed in Chapter 5, only differences in potential energy are physically measurable. Hence only the **difference in potential,** or the **potential difference,** between two points a and b (such as between a and b in Fig. 16-1) is measurable. Since the difference in potential energy, $PE_b - PE_a$, is equal to the work, W_{ba}, done by an external force (against the electric field) to move the charge from point a to point b, we have that the potential difference V_{ba} is

$$V_{ba} = V_b - V_a = \frac{W_{ba}}{q}.$$

The unit of electric potential, and of potential difference, is joules/coulomb and is given a special name, the *volt*, in honor of Alessandro Volta (1745–1827; he is best known for having invented the electric battery); the volt is abbreviated V, so $1\ V = 1 J/C$. Note from our definition that the positive plate in Fig. 16-1 is at a higher potential than the negative plate. Thus a positively charged object moves naturally from a high potential to a low potential; a negative charge does the reverse. Potential difference, since it is measured in volts, is often referred to as **voltage.**

If we wish to speak of the potential, V_a, at some point a, we must be aware that V_a depends on where the potential is chosen to be zero; and the zero point for electric potential in a given situation, just as for potential energy, can be chosen arbitrarily since only differences in potential energy can be measured. Often the ground, or a conductor connected directly to the ground, is taken as zero potential, and other potentials are given with respect to ground. (That is, a point where the voltage is 50 V is one where the difference of potential between it and ground is 50 V.) In other cases, as we shall see, we may choose the potential to be zero at infinity (see Section 16-5).

Since the electric potential is defined as the potential energy per unit charge, then the change in potential energy of a charge q when moved between two points a and b is

$$\Delta PE = PE_b - PE_a = qV_{ba}. \tag{16-1}$$

[†] At this point it has its greatest ability to do work (on some other object or system).

That is, if an object with charge q moves through a potential difference V_{ba} its potential energy changes by an amount qV_{ba}. For example, if the potential difference between the two plates in Fig. 16-1 is 6 V, then if a 1-C charge is moved (say by an external force) from a to b, it will gain $(1\ C)(6\ V) = 6\ J$ of electric potential energy. (And it will lose 6 J of electric PE as it moves from b to a). Similarly, a 2-C charge will gain 12 J, and so on. Thus, electric potential is a measure of how much energy an electric charge can acquire in a given situation. And, since energy is the ability to do work, the electric potential is also a measure of how much work a given charge can do. The exact amount depends both on the potential and on the charge.

To better understand electric potential, let's make a comparison to the gravitational case when a rock falls from the top of a cliff. The greater the height of a cliff, the more potential energy the rock has and the more kinetic energy it will have when it reaches the bottom. The actual amount of kinetic energy it will acquire, and the amount of work it can do, depends both on the height of the cliff and the mass of the rock. Similarly in the electrical case: the potential energy change, or the work that can be done, depends both on the potential difference (corresponding to the height of the cliff) and on the charge (corresponding to mass), Eq. 16-1.

Practical sources of electrical energy such as batteries and electric generators are intended to maintain a particular potential difference; the actual amount of energy used or (better) transformed depends on how much charge flows. For example, consider an automobile headlight connected to a 12.0-V battery. The amount of energy transformed (into light, and of course thermal energy) is proportional to how much charge flows, which in turn depends on how long the light is on. If over a given period 5.0 C of charge flows through the light, the total energy transformed is $(5.0\ C)(12.0\ V) = 60\ J$. If the headlight is left on twice as long, 10.0 C of charge will flow and the energy transformed is $(10.0\ C)(12.0\ V) = 120\ J$.

□ **EXAMPLE 16-1** An electron in the picture tube of a TV set is accelerated from rest through a potential difference $V_{ba} = 5000\ V$. (a) What is the change in potential energy of the electron? (b) What is the speed of the electron as a result of this acceleration?

SOLUTION (a) The charge on an electron is $e = -1.6 \times 10^{-19}\ C$. Therefore its change in potential energy is equal to

$$\Delta PE = qV_{ba} = (-1.6 \times 10^{-19}\ C)(+5000\ V)$$
$$= -8.0 \times 10^{-16}\ J.$$

The minus sign in the result indicates that the PE decreases. (The potential difference, V_{ba}, has a positive sign since the final potential is higher than the initial potential; that is, negative electrons are attracted from a negative electrode to a positive one.)

(b) The potential energy lost by the electron becomes kinetic energy. From conservation of energy (see Eq. 5-8), $\Delta KE + \Delta PE = 0$, so

$$\Delta KE = -\Delta PE$$
$$\tfrac{1}{2}mv^2 - 0 = -qV_{ba}$$

where the initial KE = 0 since we assume the electron started from rest. We solve for v and put in the mass of the electron $m = 9.1 \times 10^{-31}$ kg:

$$v = \sqrt{-\frac{2qV_{ba}}{m}}$$

$$= \sqrt{\frac{-2(-1.6 \times 10^{-19} \text{ C})(5000 \text{ V})}{9.1 \times 10^{-31} \text{ kg}}}$$

$$= 4.2 \times 10^7 \text{ m/s}.$$

16-2 ▌ Relation between electric potential and electric field

The effects of any charge distribution can be described either in terms of electric field or in terms of electric potential. Furthermore, there is an intimate connection between the potential and the field. Let us examine this relation for the case of a uniform electric field, such as that between the plates of Fig. 16-1 whose difference of potential is V_{ba}. The work done by the electric field to move a positive charge q from b to a is

$$W = qV_{ba}.$$

We can also write the work done as the force times distance and use the fact that the force on q is $F = qE$ where E is the uniform electric field between the plates. Thus

$$W = Fd = qEd$$

where d is the distance (parallel to the field lines) between points a and b. We now set these two expressions for W equal and find $qV_{ba} = qEd$, or

$$V_{ba} = Ed. \qquad [E \text{ uniform}] \quad (16\text{-}2)$$

If we solve for E, we find that $E = V_{ba}/d$. From this equation we can see that the units for electric field can be written as volts per meter (V/m) as well as newtons per coulomb (N/C). These are equivalent in general, since 1 N/C = 1 N·m/C·m = 1 J/C·m = 1 V/m.

▢   EXAMPLE 16-2   Two parallel plates are charged to a voltage of 50 V. If the separation between the plates is 0.050 m, calculate the electric field between them.

SOLUTION   We have from Eq. 16-2,

$$E = \frac{V}{d} = \frac{50 \text{ V}}{0.050 \text{ m}} = 1000 \text{ V/m}.$$

(In a region where E is not uniform, the connection between E and V takes on a slightly different form than Eq. 16-2. In general, it is possible to show that the electric field in a given direction at any point in space is equal to the *rate at which the electric potential changes over distance in that direction.* Actually, if we take into account direction, this gives the negative of the electric field. For example, the x component of the electric field is given by $E_x = -\Delta V/\Delta x$, where ΔV is the change in potential over the very short distant Δx. Note that this relation resembles Eq. 16-2 except that the distance Δx must be very small—so small that V does not change appreciably over this distance. Similar relations apply for the y and z components of **E**. Another way of stating the relation is this: if we plot V on a graph versus x, the slope of the graph at any point equals the magnitude of the x component of the electric field at that point. And we must insert a minus sign if we want the direction to come out right.)

16-3 ▮ Equipotential lines

The electric potential can be represented diagrammatically by drawing **equipotential lines** or, in three dimensions, **equipotential surfaces.** An equipotential surface is one on which all points are at the same potential. That is, the potential difference between any two points on the surface is zero, and no work is required to move a charge from one point to the other. An *equipotential surface must be perpendicular to the electric field* at any point. If this were not so—that is, if there were a component of **E** parallel to the surface—it would require work to move the charge along the surface against this component of **E**; and this would contradict the idea that it is an *equi*potential surface.

The fact that the electric field lines and equipotential surfaces are mutually perpendicular helps us locate the equipotentials when the electric field lines are known. In Fig. 16-2, a few of the equipotential lines are drawn (dashed lines) for the field between two parallel plates at a potential difference of 20 V. (The lines are only a part of the equipotential surface that extends into and out of the paper.) The negative plate is arbitrarily chosen to

FIGURE 16-2

Equipotential lines (the dashed lines) between two charged parallel plates; note that they are perpendicular to the electric field lines (shown solid).

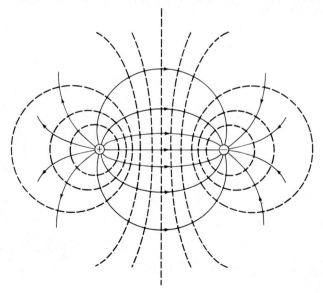

FIGURE 16-3

Equipotential lines (the dashed lines) and electric field lines (the solid lines) for two oppositely charged particles.

be zero volts and the potential of each equipotential line is indicated. The equipotential lines for the case of two equal but oppositely charged particles are shown in Fig. 16-3 as dashed lines.

We saw in Section 15-8 that there can be no electric field within a conductor in the static case, for otherwise the free electrons would feel a force and would move. Indeed *a conductor must be entirely at the same potential in the static case*, and the surface of a conductor is then an equipotential surface. (If it weren't, the free electrons at the surface would move since whenever there is a potential difference between two points, work can be done on charged particles to move them.) This is fully consistent with the fact, discussed earlier, that the electric field at the surface of a conductor must be perpendicular to the surface.

16-4 ▌ The electron volt, a unit of energy

The joule is a very large unit for dealing with energies of electrons, atoms, or molecules, whether in atomic and nuclear physics or in chemistry and molecular biology (see Example 16-1). For this purpose, the *electron volt* (eV) is used. One electron volt is defined as the energy acquired by an electron as a result of moving through a potential difference of 1 V. Since the charge on an electron is 1.6×10^{-19} C, and since the change in potential energy equals qV, 1 eV is equal to $(1.6 \times 10^{-19}$ C$)\cdot(1.0$ V$) = 1.6 \times 10^{-19}$ J; thus

$$1 \text{ eV} = 1.6 \times 10^{-19} \text{ J}.$$

An electron that accelerates through a potential difference of 1000 V will lose 1000 eV of potential energy and will thus gain 1000 eV or 1 keV (kilo-electron volt) of kinetic energy. On the other hand, if a particle has a charge equal to twice the charge on the electron ($= 2e = 3.2 \times 10^{-19}$ C), when it moves through a potential difference of 1000 V its energy will change by 2000 eV.

Although the electron volt is handy for *stating* the energies of molecules and elementary particles, it is not a proper SI unit; for calculations it should be converted to joules using the conversion factor given above. In Example 16-1, the electron acquired a kinetic energy of 8.0×10^{-16} J. We normally would quote this energy as 5000 eV ($= 8.0 \times 10^{-16}$ J$/1.6 \times 10^{-19}$ J/eV); but in determining its speed in SI units we had to use the KE in J.

16-5 ▌ Electric potential due to single point charges

The electric potential at a distance r from a single point charge Q can be derived from the expression for its electric field (Eq. 15-3) using calculus. The potential in this case is usually taken to be zero at infinity (∞), where the electric field is also zero. The result is

$$V = k\frac{Q}{r}. \quad \text{[single point charge]} \quad (16\text{-}3)$$

16-8 ∎ Dielectrics

In most capacitors there is an insulating sheet (such as paper or plastic) called a *dielectric* between the plates. This serves several purposes. First, dielectrics break down less readily than air, so higher voltages can be applied without charge passing across the gap. Furthermore, a dielectric allows the plates to be placed closer together to increase the capacitance without fear of them touching.[†] Finally, the dielectric itself increases the capacitance by a factor K known as the *dielectric constant*. Thus, for a parallel-plate capacitor

$$C = K\varepsilon_0 \frac{A}{d}. \qquad \text{[parallel-plate capacitor]} \quad (16\text{-}7)$$

This can also be written

$$C = \varepsilon \frac{A}{d},$$

where

$$\varepsilon = K\varepsilon_0$$

is the *permittivity of the material*. The values of the dielectric constant for various materials are given in Table 16-2. This is the same dielectric constant (and permittivity) discussed in Section 15-5; there we saw that the force between two charges is reduced by a factor equal to the dielectric constant. This is basically what happens when a dielectric is placed between the two plates of a capacitor. Let us see why, from the molecular point of view, the capacitance should increase.

Consider a capacitor whose plates are separated by an air gap. This capacitor has a charge $+Q$ on one plate and $-Q$ on the other (Fig. 16-7a). The capacitor is isolated (not connected to a battery). The potential difference between the plates, V_0, is given by Eq. 16-5: $Q = C_0 V_0$; the subscripts ($_0$) refer to the situation when only air is between the plates. Now we insert a dielectric between the plates (Fig. 16-7b). The molecules of the dielectric may be *polar*; that is, although the molecules are neutral, the electrons may not be evenly distributed so that one part of the molecule is positive and another negative. Because of the electric field between the plates, the molecules will tend to become oriented as shown. Even if the molecules are not polar, the electric field between the plates will induce a separation of charge in the molecules. Although the electrons do not leave the molecules, they will move slightly within the molecules toward the positive plate. So the situation is still as illustrated in Fig. 16-7b. The net effect in either case is as if there were a net negative charge on the outer edge of the dielectric facing the positive plate, and a net positive charge on the opposite side as shown in Fig. 16-7c.

Now imagine a positive test charge within the dielectric. The force that it feels is reduced by a factor K, the dielectric constant (Section 15-5). This is reflected by the fact that some of the electric field lines actually do not pass through the dielectric but end (and restart) on the charges induced on the

TABLE 16-2
Dielectric constants (at 20°C)

Material	Dielectric Constant K
Vacuum	1.0000
Air (1 atm)	1.0006
Paraffin	2.2
Rubber, hard	2.8
Vinyl (plastic)	2.8–4.5
Paper	3–7
Quartz	4.3
Glass	4–7
Porcelain	6–8
Mica	7
Ethyl alcohol	24
Water	80

[†] As we saw in Example 16-6, a simple parallel-plate capacitor of ordinary dimensions has a capacitance of only a few picofarads. A capacitance of 1 μF or more would require huge plates if dielectrics were not used.

surface of the dielectric (Fig. 16-7c). Because the force on our test charge is reduced by a factor K, the work needed to move it from one plate to the other is reduced by a factor K. (We assume that the dielectric fills all the space between the plates.) The voltage, which is the work done per unit charge, must therefore also have decreased by the factor K. That is, the voltage between the plates is now

$$V = \frac{V_0}{K}.$$

Now the charge Q on the plates has not changed, because they are isolated. So we have

$$Q = CV,$$

where C is the capacitance when the dielectric is present. When we combine this with the relation, $V = V_0/K$, we obtain

$$C = \frac{Q}{V} = \frac{Q}{V_0/K} = \frac{QK}{V_0} = KC_0,$$

since $C_0 = Q/V_0$. Thus we see, from an atomic point of view, why the capacitance is increased by the factor K.

16-9 ∎ Storage of electric energy

A charged capacitor stores electric energy. The energy stored in a capacitor will be equal to the work done to charge it up. The net effect of charging a capacitor is to remove charge from one plate and add it to the other plate. This is what a battery does when it is connected to a capacitor. When some charge is on each plate, it requires work to add more charge of the same sign. The more charge already on a plate, the more work is required to add more. The work needed to add a small amount of charge Δq, when a potential difference V is across the plates, is $\Delta W = \Delta q V$. Initially, when the capacitor is uncharged, it requires no work to move the first bit of charge over. By the end of the charging process, however, the work needed to add a charge Δq will be much greater because the voltage across the capacitor, which is proportional to the charge on the plates ($V = Q/C$), is now large. The voltage increases during this process from zero to its final value, V_f, at the end; then the total work done, W, will be equivalent to moving all the charge Q at once across a voltage equal to the average during the whole process; the average voltage is $V_f/2$, so

$$W = Q\frac{V_f}{2}.$$

Thus we can say that the energy stored in a capacitor is

$$\text{energy} = \tfrac{1}{2}QV,$$

where V is the potential difference between the plates (we have dropped the subscript) and Q is the charge on each plate. Since $Q = CV$, we can also write

$$\text{energy} = \tfrac{1}{2}QV = \tfrac{1}{2}CV^2 = \tfrac{1}{2}\frac{Q^2}{C}. \tag{16-8}$$

FIGURE 16-7

Molecular view of the effects of a dielectric.

(a)

(b)

(c)

(d)

☐ EXAMPLE 16-7 A 12-V battery is connected to a 20-μF capacitor which uses a paper dielectric. How much electric energy can be stored in the capacitor?

SOLUTION We do not need the dielectric constant since the value $C = 20 \times 10^{-6}$ F includes its effect. From Eq. 16-8, we have

$$\text{energy} = \tfrac{1}{2}CV^2 = \tfrac{1}{2}(20 \times 10^{-6} \text{ F})(12 \text{ V})^2 = 1.4 \times 10^{-3} \text{ J.}$$

Energy is not a substance and does not have a definite location. Nonetheless, it is often useful to think of it as being stored in the electric field between the plates. As an example let us calculate the energy stored in a parallel-plate capacitor in terms of the electric field.

We saw in Eq. 16-2 that the electric field E between two parallel plates is uniform (approximately) and is related to the potential difference by $V = Ed$, where d is the separation. Also, Eq. 16-6 tells us that $C = \varepsilon_0 A/d$. Thus

$$\text{energy} = \tfrac{1}{2}CV^2 = \frac{1}{2}\left(\frac{\varepsilon_0 A}{d}\right)(E^2 d^2)$$

$$= \tfrac{1}{2}\varepsilon_0 E^2 A d.$$

The quantity Ad is simply the volume between the plates in which the electric field E exists. If we divide both sides by the volume, we obtain an expression for the energy per unit volume or **energy density:**

$$\text{energy density} = \tfrac{1}{2}\varepsilon_0 E^2. \tag{16-9}$$

The energy stored per unit volume is proportional to the square of the electric field in that region. If a dielectric is present, ε_0 is replaced by ε. We derived Eq. 16-9 for the special case of a capacitor. But it can be shown to be true for any region of space where there is an electric field.

*16-10 ∎ The electrocardiogram

Each time the heart beats, changes in electrical potential occur on its surface that can be detected using metal contacts, called "electrodes," which are attached to the skin. The changes in potential are small, on the order of millivolts (mV), and must be amplified; they are displayed either with a chart recorder on paper, or on a cathode-ray-tube oscilloscope screen (discussed in Chapter 19). The record of the potential changes for a given person's heart is called an electrocardiogram (EKG or ECG); an example is shown in Fig. 16-8. The instrument itself is called an electrocardiograph. We are not so interested now in the electronics, but in the source of these potential changes and their relation to heart activity.

Muscle cells and nerve cells are similar in that both have an electric dipole layer across the cell wall; that is, in the normal situation there is a net positive charge on the exterior surface and a net negative charge on the interior surface as shown in Fig. 16-9a. The amount of charge depends on the size of the cell, but is approximately 10^{-3} C/m^2 of surface. For a cell whose surface area is 10^{-5} m^2, the total charge on either surface is thus

FIGURE 16-8

Typical ECG. Two heart beats are shown.

FIGURE 16-9

Heart muscle cell showing (a) charge dipole layer in resting state; (b) depolarization of cell progressing as muscle begins to contract; (c) potential V at points P and P' as a function of time.

$\pm 10^{-8}$ C. Just before the contraction of heart muscles, changes occur in the cell wall so that positive ions on the exterior of the cell are able to pass through the wall and neutralize those on the inside, or even make the inside surface slightly positive compared to the exterior, as shown in Fig. 16-9b. This depolarization, as it is called, starts at one end of the cell and progresses toward the opposite end, as indicated by the arrow in part (b), until the whole muscle is depolarized. The muscle then slowly repolarizes to its original state (Fig. 16-9a). This whole process requires less than a second.

The equal but opposite charges on the two sides of a cell wall can be considered as a set of electric dipoles (Section 16-6). When the cell is in the resting state (Fig. 16-9a) the potential at any point P outside the cell (see the figure) is zero; this is due to the fact that on opposite sides of the cell the individual dipoles are oriented in opposite directions, so the net effect of all of them is to give zero potential (the $\cos \theta$ factor in Eq. 16-4 is important since it takes on values of -1 to $+1$). The potential at P' will also be zero, so the potential difference between electrodes at P and P' will be zero. When the cell begins to depolarize, however, the potential at P will decrease while that at P' will increase. This is because there is no longer symmetry, for there are more dipoles oriented with the positive charge on the right than vice versa. Point P is nearer the negative ends of the dipoles; so the potential there becomes negative (V is due to a sum of terms in which the θ of Fig. 16-4 is $\approx 180°$, and therefore the $\cos \theta$ term in Eq. 16-4 is -1). Point P', however, is closer to the positive ends of the dipoles, so V will be positive (these terms have $\theta \approx 0°$, so $\cos \theta = +1$ predominantly). As the depolarization is completed, the potentials at P and P' return to zero. This is shown in Fig. 16-9c.

In the heart, the path of depolarization is complicated. Furthermore, after depolarization the muscles repolarize to the resting state (Fig. 16-9a). Thus the potential difference as a function of time is quite complicated (Fig. 16-8).

It is standard procedure to divide a typical electrocardiogram into regions corresponding to the various deflections (or "waves" as they are called), as shown in Fig. 16-8. Each of the deflections corresponds to the activity of a particular part of the heart beat (Fig. 9-9). The P wave corresponds to contraction of the atria. The QRS group corresponds to contraction of the ventricles; this group has three main phases because the depolarization follows a complicated path from left to right, and toward the front, then downward to the left and toward the rear. The T wave corresponds to recovery (repolarization) of the heart in preparation for the next cycle.

Electrocardiograms make use of three basic electrodes, one placed on either side of the heart on the hands and one on the left foot. Sometimes six additional electrodes are placed at other locations. The measurement of so many potential differences provides additional information (some of it redundant), since the heart is a three-dimensional object and depolarization takes place in all three dimensions. A complete electrocardiogram may include as many as 12 graphs.

The ECG is a powerful tool in identifying heart defects. For example, the right side of the heart enlarges if the right ventricle must push against an abnormally large load (as when blood vessels become hardened or clogged). This problem is readily observed on an ECG, since the S wave becomes very large (negatively). *Infarcts*, which are dead regions of the heart muscle that result from a heart attack, are also detected on an ECG because they reflect the depolarization wave.

The interpretation of an ECG depends to a great extent on experience obtained with many patients rather than on theoretical understanding. A good deal of scientific research remains to be done.

▌ Summary

The *electric potential* at any point in space is defined as the electric potential energy per unit charge. The *electric potential difference* between any two points is defined as the work required of an external force to move a 1-C electric charge between the two points. Potential difference is measured in volts (1 V = 1 J/C) and is sometimes referred to as *voltage*. The change in PE of a charge q when it moves through a potential difference V_{ba} is qV_{ba}.

The potential difference V between two points where a uniform electric field E exists is given by $V = Ed$, where d is the distance between the two points. An *equipotential line* or surface is all at the same potential, and is perpendicular to the electric field at all points. The electric potential due to a single point charge Q, relative to zero potential at infinity, is given by $V = kQ/r$.

A *capacitor* is a device used to store charge and consists of two nontouching conductors. The two conductors generally hold equal and opposite charges, Q, and the ratio of this charge to the potential difference V between the conductors is called the *capacitance*,

C; so $Q = CV$. The capacitance of a parallel-plate capacitor is proportional to the area of each plate and inversely proportional to their separation. The space between the conductors contains a nonconducting material such as air, paper, or plastic; these materials are referred to as dielectrics, and the capacitance is proportional to a property of dielectrics called the dielectric constant, K (nearly equal to 1 for air).

A charged capacitor stores an amount of electric energy given by $\frac{1}{2}QV = \frac{1}{2}CV^2 = \frac{1}{2}Q^2/C$. This energy can be thought of as stored in the electric field between the plates. The energy stored in any electric field E has a density (energy per unit volume) of $\frac{1}{2}\varepsilon_0 E^2$.

▋ Questions

1. If two points are at the same potential, does this mean that no work is done in moving a test charge from one point to the other? Does this imply that no force must be exerted?

2. Is it possible for a plastic comb to acquire a charge of 0.5 V?

3. Draw in a few equipotential lines in Fig. 15-20b.

4. Can two equipotential lines cross? Explain.

5. Is there a point between two equal positive charges where the electric field is zero? Where the electric potential is zero? Explain.

6. An electron is accelerated by a potential difference of, say, 100 V. How much greater would its final speed be if it were accelerated with four times as much voltage?

7. If a negative charge is initially at rest in an electric field, will it move toward a region of higher potential or lower potential? What about a positive charge? How does the potential energy of the charge change in each case?

8. State clearly the difference between (a) electric potential and electric field, (b) electric potential and electric potential energy.

9. If the potential at a point is zero, must the electric field also be zero? Give an example.

10. What can you say about the electric field in a region of space that has the same potential throughout?

11. How does the earth's gravitational field change with distance? What about its gravitational potential?

12. A satellite orbits the earth along an equipotential line. What shape must the orbit be?

13. How does the energy in a capacitor change if (a) the potential difference is doubled. (b) the charge on each plate is doubled, (c) the separation of the plates is doubled, as the capacitor remains connected to a battery?

14. How does the energy stored in a capacitor change when a dielectric is inserted if (a) the capacitor is isolated so Q doesn't change, (b) the capacitor remains connected to a battery so V doesn't change?

15. We have seen that the capacitance C depends on the size, shape, and position of the two conductors, as well as on the dielectric constant K. What then did we mean when we said that C is a constant in Eq. 16-5?

▋ Problems

SECTIONS 16-1 TO 16-4

1. (I) The electric field between two parallel plates connected to a 45-V battery is 600 V/m. How far apart are the plates?

2. (I) How strong is the electric field between two parallel plates 5.0 mm apart if the potential difference between them is 110 V?

3. (I) An electric field of 800 V/m is desired between two parallel plates 6.0 cm apart. How large a voltage should be applied?

4. (I) How much work is needed to move a $-3.0\text{-}\mu C$ charge from ground to a point whose potential is $+60$ V?

5. (I) How much kinetic energy will an electron gain (joules) if it falls through a potential difference of 800 V?

6. (I) What potential difference is needed to give a helium nucleus ($Q = 3.2 \times 10^{-19}$ C) 18.0 keV of KE?

7. (I) An electron acquires 6.4×10^{-16} J of kinetic energy when it is accelerated by an electric field from plate A to plate B. What is the potential difference between the plates, and which plate is at the higher potential?

8. (II) A lightning flash transfers 30 C of charge to earth through a potential difference of 3.5×10^7 V. (a) How much energy is dissipated? (b) How much water at 0°C could be brought to boiling?

FIGURE 16-10

FIGURE 16-11

FIGURE 16-12

9. (II) The work done by an external force to move a $-2.0\text{-}\mu C$ charge from point a to point b is 8.0×10^{-4} J. If the charge was started from rest and had 1.0×10^{-4} J of kinetic energy when it reached point b, what must be the potential difference between a and b?

10. (II) What is the speed of a 350-eV electron?

11. (II) What is the KE in electron volts of an oxygen molecule at STP (0°C, 1 atm)?

12. (II) In a television picture tube, electrons are accelerated by thousands of volts through a vacuum. If a television set were laid on its back, would electrons be able to move upward against the force of gravity? What potential difference, acting over a distance of 20 cm, would be needed to balance the downward force of gravity so that an electron would remain stationary? Assume that the electric field is uniform.

13. (III) An electron is accelerated horizontally from rest in a television picture tube by a potential difference of 10,000 V. It then passes between two horizontal plates 6.0 cm long and 1.0 cm apart which have a potential difference of 200 V (Fig. 16-10). At what angle θ will the electron be traveling after it passes between the plates?

SECTION 16-5

14. (I) What is the electric potential 11.0 cm from a 4.25-μC point change?

15. (I) What is the electric potential 2.0×10^{-10} m from a proton (charge $+e$)?

16. (I) What is the potential energy of an electron that is 0.53×10^{-10} m from a proton?

17. (II) A $+20\text{-}\mu C$ charge is placed 60 cm from an identical $+20\text{-}\mu C$ charge. How much work would be required to move a $+0.20\text{-}\mu C$ test charge from a point midway between them to a point 10 cm closer to either of the charges?

18. (II) How much voltage must be used to accelerate a proton so that it has sufficient energy to penetrate an iron nucleus? An iron nucleus has a charge of $+26e$ and its radius is about 4.0×10^{-15} m. Assume the potential is that for a point charge.

19. (III) A 30-μC and a $-2.0\text{-}\mu C$ charge are placed 2.0 cm apart. At what point along the line joining them is (a) the electric field zero, and (b) the potential zero?

20. (III) How much work must be done to bring three electrons from a great distance apart to within 1.0×10^{-10} m from one another?

21. (III) Consider point a which is 85 cm north of a $-45 \ \mu C$ point charge and point b which is 60 cm west

of the charge. Determine (a) $V_{ba} = V_b - V_a$, (b) $\mathbf{E}_b - \mathbf{E}_a$ (magnitude and direction).

*SECTION 16-6

*22. (II) An electron and a proton are 0.53×10^{-10} cm apart. (a) What is their dipole moment if they are at rest? (b) What is the average dipole moment if the electron revolves about the proton in a circular orbit?

*23. (II) Calculate the electric potential due to a dipole whose dipole moment is 4.8×10^{-30} C·m at a point 1.0×10^{-9} m away if this point is (a) along the axis of the dipole nearer the positive charge; (b) 45° above the axis but nearer the positive charge; (c) 45° above the axis but nearer the negative charge.

*24. (III) The dipole moment, considered as a vector, points from the negative to the positive charge. The water molecule, Fig. 16-11, has a dipole moment $\mathbf{p}$ which can be considered as the vector sum of the two dipole moments $\mathbf{p}_1$ and $\mathbf{p}_2$ as shown. The distance between each H and the O is about 0.96×10^{-10} m; the lines joining the center of the O atom with each H atom make an angle of 104° as shown, and the net dipole moment has been measured to be $p = 6.1 \times 10^{-30}$ C·m. Determine the charge q on each H atom.

*25. (III) Show that if two dipoles with dipole moments p_1 and p_2 are in line with one another (Fig. 16-12) the potential energy of one in the presence of the other (their "interaction energy") is given by

$$\text{PE} = -\frac{2kp_1p_2}{r^3},$$

where r is the distance between the two dipoles. Assume that r is much greater than the length of either dipole.

SECTIONS 16-7 AND 16-8

26. (I) How much charge flows from a 12.0-V battery when it is connected to a 2.50-μF capacitor?

27. (I) The two plates of a capacitor hold $+1500\ \mu C$ and $-1500\ \mu C$ of charge, respectively, when the potential difference is 388 V. What is the capacitance?

28. (I) A 1.0-F capacitor is desired. What area must the plates have if they are to be separated by a 5.0-mm air gap?

29. (I) What is the capacitance of two square parallel plates 6.0 cm on a side that are separated by 2.0 mm of paraffin?

30. (II) An electric field of 16.0×10^6 V/m is desired between two parallel plates each of area 110 cm^2 and separated by 2.00 cm of air. What charge must be on each plate?

31. (II) How strong is the electric field between the plates of a 20-μF air-gap capacitor if they are 2.0 mm apart and each has a charge of $300\ \mu C$?

32. (II) It takes 6.0 J of energy to move a 2.0-mC charge from one plate of a 60-μF capacitor to the other. How much charge is on each plate?

33. (II) A 3000-pF air-gap capacitor is connected to a 12-V battery. If a piece of mica is placed between the plates, how much charge will then flow from the battery?

34. (II) Dry air will break down if the electric field exceeds 3.0×10^6 V/m. What amount of charge can be placed on a capacitor if the area of each plate is 10 cm^2?

35. (III) A capacitor C_1 carries a charge Q_0. It is then connected directly to a second, uncharged, capacitor C_2. What charge will each carry now? What will be the potential difference across each?

36. (III) A 2.5-μF capacitor is charged by a 35-V battery and then is disconnected from the battery. When it is then connected to a second (initially uncharged) capacitor, C_2, the voltage on the first drops to 16 V. What is the value of C_2?

37. (III) A 4.0-μF capacitor is charged to 200 V and a 10.0-μF capacitor is charged to 150 V. The positive plates are now connected to each other and the negative plates are connected to each other. What will be the potential difference across each and the charge on each?

SECTION 16-9

38. (I) 200 V is applied to a 3600-pF capacitor. How much energy is stored?

39. (I) There is an electric field near the earth's surface whose intensity is about 150 V/m. How much energy is stored per cubic meter in this field?

40. (II) How much energy is stored by the electric field between two square plates, 10 cm on a side, separated by a 3.0-mm air gap? The charges on the plates are equal and opposite and of magnitude $300\ \mu C$. What if the gap were filled with mica?

41. (II) A parallel-plate capacitor has a fixed charge Q. The separation of the plates is then doubled. By what factor does the energy stored in the electric field change?

42. (III) A 2.0-μF capacitor is charged by a 12-V battery. It is disconnected from the battery and then connected to an uncharged 5.0-μF capacitor. Determine the total stored energy (a) before the two capacitors are connected and (b) after they are connected. (c) What is the change in energy? (d) Is energy conserved? Explain why.

17 ELECTRIC CURRENTS

Until the year 1800, the technical development of electricity consisted mainly of producing a static charge by friction. In the preceding century a number of machines had been built that could produce rather large potential by frictional means; one type of such apparatus is shown in Fig. 17-1. Large sparks could be produced by these machines, but they had little practical value.

In nature itself there were grander displays of electricity such as lightning and "St. Elmo's fire," which is a glow that appeared around the yardarms of ships during storms. That these phenomena were electrical in origin was not recognized until the eighteenth century. For example, it was only in 1752 that Franklin, in his famous kite experiment, showed that lightning was an electric discharge—a giant electric spark.

Finally, in 1800, an event of great practical importance occurred: Alessandro Volta (1745–1827) invented the electric battery, and with it produced the first steady flow of electric charge—that is, a steady electric current. This discovery opened a new era, which transformed our civilization, for today's electrical technology is based on electric current.

FIGURE 17-1
Early electrostatic generator. (Courtesy of John Heilbron)

17-1 ▌ The electric battery

The events that led to the discovery of the battery are interesting; for not only was this an important discovery, but it also gave rise to a famous scientific debate between Volta and Luigi Galvani (1737–1798), eventually involving many others in the scientific world.

In the 1780s Galvani, a professor at the University of Bologna (thought to be the world's oldest university still in existence), carried out a long series of experiments on the contraction of frog's leg muscle through electricity produced by a static-electricity machine. In the course of these investigations, Galvani found, much to his surprise, that contraction of the muscle could be produced by other means as well: when a brass hook was pressed into the frog's spinal cord and then hung from an iron railing that also touched the frog, the leg muscles again would contract. With further investigation, Galvani found that this strange but important phenomenon occurred for other pairs of metals as well.

What was the source of this unusual phenomenon? Galvani believed that the source of the electric charge was in the frog muscle or nerve itself, and that the wire merely transmitted the charge to the proper points; and when he published his work in 1791, he termed it "animal electricity." Many wondered, including Galvani himself, if he had discovered the long-sought "life-force."

Volta, at the University of Pavia 200 km away, was at first skeptical of Galvani's results; but at the urging of his colleagues he soon confirmed and extended those experiments. But Volta doubted Galvani's idea of animal electricity. Instead he came to believe that the source of the electricity was not in the animal itself but rather in the *contact between the two metals*. Volta made public his views and soon had many followers, although others still sided with Galvani.

Volta was both a strong theoretician and a careful and skillful experimenter. He soon realized that a moist conductor, such as a frog muscle or moisture at the contact point of the two dissimilar metals, was necessary in the circuit if it was to be effective. He also saw that the contracting frog muscle was a sensitive instrument for detecting electric "tension" or "electromotive force" (his words for what we now call potential), in fact more sensitive than the best available electrometers that he and others had developed. Most important, he recognized that a decisive answer to Galvani could be given only if the sensitive frog leg was replaced by an inorganic detector; that is, to cement his view that it was the contact of two dissimilar metals that caused the frog muscle to contract, he would have to connect the two dissimilar metals directly to an electrometer and observe a separation of the leaves representing a potential difference. This proved difficult since his most sensitive[†] electrometers were much less sensitive than the frog muscle. But the eventual success of this experiment vindicated Volta's theory.

Volta's research found that certain combinations of metals produced a greater effect than others, and, using his measurements, he listed them in order of effectiveness. (This "electrochemical series" is still used by chemists today.) And he found that carbon could be used in place of one of the metals.

Volta then conceived what is his greatest contribution to science. Between a disc of zinc and one of silver, he placed a piece of cloth or paper soaked in salt solution or dilute acid and piled a "battery" of such couplings, one on top of another, as shown in Fig. 17-2; this "pile" or "battery" produced a much increased potential difference. Indeed, when strips of metal connected to the two ends of the pile were brought close, a spark was produced. Volta had designed and built the first electric battery. A second design, known as the "crown of cups" is also shown in Fig. 17-2. Volta made public this great discovery in 1800.

The potential produced by Volta's battery was still weak compared to that produced by the best friction machines of the time, although it could produce considerable charge. (The electrostatic machines were high-potential, low-charge devices.) But it had a great advantage: it was "self-renewing"—it could produce a flow of electric charge continuously for a

---

† Volta's most sensitive electrometer measured about 40 V per degree (of leaf separation). Nonetheless, he was finally able to estimate the potential differences produced by dissimilar metals in contact: for a silver-zinc contact he got about 0.7 V, remarkably close to today's value of 0.78 V.

(a)   (b)

relatively long period of time. It was not long before even more powerful batteries were constructed.

After Volta's discovery of the electric battery, it was eventually recognized that a battery produces electricity by transforming chemical energy into electrical energy. Today a great variety of electric cells and batteries are available, from flashlight batteries (sometimes called "dry cells") to the storage battery of a car. The simplest batteries contain two plates or rods made of dissimilar metals (one can be carbon) called *electrodes*. The electrodes are immersed in a solution, such as a dilute acid, called the *electrolyte*. (In a dry cell, the electrolyte is absorbed in a powdery paste.) Such a device is properly called an *electric cell*, and several cells connected together is a battery. The chemical reactions involved in most electric cells are quite complicated. A detailed description of different electric cells can be found in chemistry textbooks. Here we describe how one very simple cell works, emphasizing the physical aspects.

The simple cell shown in Fig. 17-3 uses dilute sulfuric acid as the electrolyte. One of the electrodes is made of carbon, the other of zinc. That part of each electrode remaining outside the solution is called the *terminal*, and connections to wires and circuits are made here. The acid attacks the zinc electrode and tends to dissolve it. But each zinc atom leaves two electrons behind, so it enters the solution as a positive ion. The zinc electrode thus acquires a negative charge. As more zinc ions enter solution, the electrolyte can momentarily become positively charged. Because of this, and through other chemical reactions, electrons are pulled off the carbon electrode. Thus the carbon electrode becomes positively charged. The positive electrode is called the *anode*, the negative electrode, the *cathode*. Because there is an opposite charge on the two electrodes, there is a potential difference between the two terminals. In a cell whose terminals are not connected, only a small amount of the zinc is dissolved; for as the zinc electrode becomes increasingly negative, any new positive zinc ions produced are attracted back to the electrode. Thus a particular potential difference or voltage is maintained between the two terminals. If charge is allowed to flow between the terminals, say through a wire (or a light bulb), then more zinc can be dissolved. After a time, one or the other electrode is used up and the cell becomes "dead."

The voltage that exists between the terminals of a battery depends on what the electrodes are made of and their relative ability to be dissolved or give up electrons. The potential difference when no charge flows to an ex-

FIGURE 17-2
Two types of voltaic battery: (a) a pile; (b) "crown of cups." Z stands for zinc and A for silver (*argentum* in Latin). Taken from Volta's original publication. (Burndy Library, Norwalk, CT)

FIGURE 17-3
Simple electric cell.

ternal circuit is called the **electromotive force,** or **emf,** of the battery. The symbol $\mathscr{E}$ is often used for emf (don't confuse it with E for electric field). The emf of typical cells is 1.0 to 2.0 V. When charge is allowed to flow from a battery to an external circuit, the voltage at the terminals drops below the emf due to "internal resistance" (this is discussed in Section 18-2). When two or more cells are connected so that the positive terminal of one is connected to the negative terminal of the next, they are said to be connected in *series* and their voltages add up. Thus the voltage between the ends of two flash-light batteries so connected is 3.0 V, while the six 2-V cells of an automobile storage battery give 12 V.

Any device, such as a battery, that transforms one type of energy (mechanical, chemical, light, thermal) into electrical energy so that a potential difference can be maintained and charge can be supplied to an external circuit is called a **source,** or **seat, of emf.** Other sources of emf are electric generators, photocells, thermocouples, and so on (more on them later).

17-2 ▌ Electric current

When a continuous conducting path, such as a wire, is connected to the terminals of a battery, we have an electric **circuit,** Fig. 17-4a. On any diagram of a circuit as in Fig. 17-4b, we will represent a battery by the symbol "—┤├—"; the longer line on this symbol represents the positive terminal and the shorter line the negative terminal. When such a circuit is formed, charge can flow through the circuit from one terminal of the battery to the other. A flow of charge such as this is called an **electric current.**

More precisely, the electric current in a wire is defined as the net amount of charge that passes through it at a given point per unit time. Thus, the average current I is defined as

$$I = \frac{\Delta Q}{\Delta t} \tag{17-1}$$

where ΔQ is the amount of charge that passes through a cross-section of conductor at a given point during the time interval Δt. Electric current is measured in coulombs per second; this is given a special name, the *ampere* (abbreviated amp or A), after the French physicist André Ampère (1775–1836). Thus 1 A = 1 C/s.

FIGURE 17-4

(a) Very simple electric circuit. (b) Schematic drawing of the circuit in part (a).

▫ EXAMPLE 17-1   A steady current of 2.5 A flows in a wire connected to a battery. But after 4.0 min, the current suddenly ceases because the wire is disconnected. How much charge passed through the circuit?

SOLUTION   Since the current was 2.5 A or 2.5 C/s, then in 4.0 minutes (240 s) the total charge that flowed was, from Eq. 17-1,

$$\Delta Q = I \, \Delta t$$
$$= (2.5 \text{ C/s})(240 \text{ s}) = 600 \text{ C}.$$

We saw in Chapter 15 that conductors contain many free electrons; thus, when a conducting wire is connected to the terminals of a battery as in Fig. 17-4, it is actually the negatively charged electrons that flow in the wire. When the wire is first connected, free electrons at one end of the wire are attracted into the positive terminal. At the same time, electrons leave the negative terminal of the battery and enter the wire at the other end. Thus there is a continuous flow of electrons through the wire that begins as soon as the wire is connected to *both* terminals. However, when the conventions of positive and negative charge were invented two centuries ago, it was assumed positive charge flowed in a wire. Actually, for nearly all purposes, positive charge flowing in one direction is exactly equivalent to negative charge flowing in the opposite direction, Fig. 17-5. Today, we still use the historical convention of positive current flow when discussing the direction of a current. (It also fits in well with our definition of electric potential in terms of a positive test charge.) Thus when we speak of the current flowing in a circuit we mean the direction positive charge would flow. This is sometimes referred to as *conventional current*. When we want to speak of the direction of electron flow, we will specifically state it is the electron current. In liquids and gases, both positive and negative charges (ions) can move.

FIGURE 17-5
Conventional current from + to − is equivalent to a negative (electron) current flowing from − to +.

17-3 ▌ Ohm's law; resistance and resistors

To produce an electric current in a circuit, a difference in potential is required. One way of producing a potential difference is by a battery. It was Georg Simon Ohm (1787–1854) who established experimentally that the current in a metal wire is proportional to the potential difference V applied to its ends:

$$I \propto V.$$

This is known as **Ohm's law.** If, for example, we connect a wire to a 6-V battery, the current flow will be twice what it would be if the wire were connected to a 3-V battery.

It is helpful to compare an electric current to the flow of water in a river or a pipe. If the pipe (or river) is nearly level, the flow rate is small. But if one end is somewhat higher than the other, the flow rate—or current—is much greater. The greater the difference in height, the greater the current. We saw in Chapter 16 that electric potential is analogous, in the gravitational case, to the height of a cliff; and this applies in the present case to the height through which the fluid flows. Just as an increase in height causes a greater flow of water, so a greater electric potential difference, or voltage, causes a greater current flow.

Exactly how much current flows in a wire depends not only on the voltage but also on the resistance the wire offers to the flow of electrons. The walls of a pipe, or the banks of a river and rocks in the middle, offer resistance to the flow of current. Similarly, electrons are slowed down because of interactions with the atoms of the wire. The higher this resistance, the less the current for a given voltage V. We then define resistance so that the current is inversely proportional to the resistance. When we combine this with the above proportion we have

$$I = \frac{V}{R}; \qquad (17\text{-}2)$$

R is the **resistance** of some object or device, V is the potential difference across the device, and I is the current that flows through it. This relation (Eq. 17-2) is often written

$$V = IR,$$

and is often referred to as Ohm's law. Actually, it is not a law, but rather is the *definition of resistance*. If we want to call something Ohm's law, it would be the statement that the current through a *metal conductor* is proportional to the applied voltage, $I \propto V$. But this relation does *not* apply generally for other substances such as semiconductors, vacuum tubes, transistors, and so on. Thus "Ohm's law" is not a fundamental law like Newton's laws or the laws of thermodynamics.

The unit for resistance is called the *ohm* and is abbreviated Ω (Greek capital omega). Because $R = V/I$, we see that $1.0\ \Omega$ is equivalent to $1.0\ \text{V/A}$.

☐ **EXAMPLE 17-2**   A plate on the bottom of a small tape recorder specifies that it should be connected to 6.0 V and will draw 300 mA. (*a*) What is the net resistance of the recorder? (*b*) If the voltage dropped to 5.0 V, how would the current change?

SOLUTION   (*a*) From Eq. 17-2 we have

$$R = \frac{V}{I} = \frac{6.0\ \text{V}}{0.30\ \text{A}} = 20\ \Omega.$$

(*b*) If the resistance stayed constant, the current would be approximately $I = V/R = 5.0\ \text{V}/20\ \Omega = 0.25\ \text{A}$, or a drop of 50 mA. Actually, resistance depends on temperature (Section 17-4), so this is only an approximation.

All electric devices, from heaters to light bulbs to stereo amplifiers, offer resistance to the flow of current. Generally, the connecting wires have very low resistance. In many circuits, particularly in electronic devices, *resistors* are used to control the amount of current. Resistors have resistances from less than an ohm to millions of ohms. Several kinds are shown in Fig. 17-6.

FIGURE 17-6

Some resistors. The resistance value of a given resistor is written on the exterior or may be given a color code as shown in the figure. The first two colors represent the first two digits in the value of the resistance, the third represents the power of ten that it must be multiplied by, and the fourth is the manufactured tolerance. For example, a resistor whose four colors are red, green, orange, and silver has a resistance of $25,000\Omega$ ($25\text{k}\Omega$), give or take 10 percent.

First digit
Second digit
Multiplier
Tolerance

Resistor color code

Color	Number	Multi-plier	Toler-ance (%)
Black	0	1	
Brown	1	10^1	
Red	2	10^2	
Orange	3	10^3	
Yellow	4	10^4	
Green	5	10^5	
Blue	6	10^6	
Violet	7	10^7	
Gray	8	10^8	
White	9	10^9	
Gold		10^{-1}	5%
Silver		10^{-2}	10%
No color			20%

The two main types are "wire wound" resistors, which consist of a coil of fine wire, and "composition" resistors, which are usually made of the semiconductor carbon.

When we draw a diagram of a circuit, we indicate a resistance with the symbol $-\bigwedge\!\bigwedge\!-$. Wires whose resistance is negligible, however, are shown simply as straight lines.

17-4 ■ Resistivity and superconductivity

It is found experimentally that the resistance R of a metal wire is directly proportional to its length L and inversely proportional to the cross-sectional area A. That is

$$R = \rho \frac{L}{A},\qquad(17\text{-}3)$$

where ρ, the constant of proportionality, is called the *resistivity* and depends on the material used. This relation makes sense since we expect that the resistance of a thick wire would be less than that of a thin one because a thicker wire has more area for the electrons to pass through. And you might expect the resistance to be greater if the length is greater since there would be more obstacles to electron flow.

Typical values of ρ for various materials are given in the first column of Table 17-1. (The values depend somewhat on purity, heat treatment, temperature, and other factors.) Notice that silver has the lowest resistivity and is thus the best conductor; but it is expensive. Copper is not far behind, so

TABLE 17-1

Resistivity and temperature coefficients (at 20°C)

Material	Resistivity, ρ ($\Omega \cdot$m)	Temperature Coefficient, α $(C°)^{-1}$
Conductors		
Silver	1.59×10^{-8}	0.0061
Copper	1.68×10^{-8}	0.0068
Aluminum	2.65×10^{-8}	0.00429
Tungsten	5.6×10^{-8}	0.0045
Iron	9.71×10^{-8}	0.00651
Platinum	10.6×10^{-8}	0.003927
Mercury	98×10^{-8}	0.0009
Nichrome (alloy of Ni, Fe, Cr)	100×10^{-8}	0.0004
Semiconductors[†]		
Carbon (graphite)	$(3\text{–}60) \times 10^{-5}$	-0.0005
Germanium	$(1\text{–}500) \times 10^{-3}$	-0.05
Silicon	0.1–60	-0.07
Insulators		
Glass	$10^{9}\text{–}10^{12}$	
Hard rubber	$10^{13}\text{–}10^{15}$	

[†] Values depend strongly on presence of even slight amounts of impurities.

it is clear why most wires are made of copper. Aluminum, although it has a higher resistivity, is much less dense than copper; it is thus preferable to copper in some situations, such as transmission lines, because its resistance for the same weight is less than that for copper.

▢ **EXAMPLE 17-3** Suppose you want to connect your stereo set to a remote speaker. If each wire must be 20 m long, what diameter copper wire should you use to keep the resistance less than 0.10 Ω per wire?

SOLUTION   We solve Eq. 17-3 for A and use Table 17-1:

$$A = \rho \frac{L}{R} = \frac{(1.7 \times 10^{-8} \ \Omega \cdot m)(20 \ m)}{(0.10 \ \Omega)} = 3.4 \times 10^{-6} \ m^2.$$

The cross-sectional area A of a circular wire is related to its diameter d by $A = \pi d^2/4$. The diameter must then be at least $d = \sqrt{4A/\pi} = 2.1 \times 10^{-3} \ m = 2.1$ mm.

The resistivity of a material depends somewhat on temperature. In general, the resistance of metals increases with temperature. This is not surprising, for at higher temperatures the atoms are moving more rapidly and are arranged in a less orderly fashion. So they might be expected to interfere more with the flow of electrons. If the temperature change is not too great, the resistivity of metals increases nearly linearly with temperature, That is,

$$\rho_T = \rho_0(1 + \alpha \Delta T) \tag{17-4}$$

where ρ_0 is the resistivity at some reference temperature (such as 0°C or 20°C), ρ_T is the resistivity at a temperature ΔT above the reference temperature, and α is the *temperature coefficient of resistance*. Values for α are given in Table 17-1. Note that the temperature coefficient for semiconductors can be negative; this is apparently because at higher temperatures, some of the electrons that are not normally free become free and can contribute to the current. Thus the resistance of a semiconductor can decrease with an increase in temperature, although this is not always the case.

▢ **EXAMPLE 17-4** *Resistance thermometer.* The variation in electrical resistance with temperature can be used to make precise temperature measurements. Platinum is usually used since it is relatively free from corrosive effects and has a high melting point. To be specific, suppose at 0°C the resistance of a platinum resistance thermometer is 164.2 Ω. When placed in a particular solution, the resistance is 187.4 Ω. What is the temperature of this solution?

SOLUTION   Since the resistance R is directly proportional to the resistivity ρ, we can combine Eq. 17-3 with 17-4 and write, taking 0°C as our reference temperature so $\Delta T = T$ (in °C):

$$R = R_0(1 + \alpha T).$$

Here $R_0 = \rho_0 L/A$ is the resistance of the wire at 0°C. We solve this equation for T and find

$$T = \frac{R - R_0}{\alpha R_0} = \frac{187.4 \ \Omega - 164.2 \ \Omega}{(3.927 \times 10^{-3}(C°)^{-1})(164.2 \ \Omega)} = 35.9°C.$$

399

More convenient for some applications is a *thermistor*, which consists of a metal oxide or semiconductor whose resistance also varies in a repeatable way with temperature. They can be made quite small and respond very quickly to temperature changes.

The value of α in Eq. 17-4 depends on temperature, so it is important to check the temperature range of validity of any value (say in a handbook of physical data). If the temperature range is wide, Eq. 17-4 is not adequate and terms proportional to the square and cube of the temperature are needed: $\rho_T = \rho_0[1 + \alpha\Delta T + \beta(\Delta T)^2 + \gamma(\Delta T)^3]$ where the coefficients β and γ are generally very small, but when ΔT is large their terms become significant.

At very low temperatures, the resistivity of certain metals and their compounds or alloys becomes essentially zero to within experimental measuring techniques. Materials in such a state are said to be **superconducting.** This phenomenon was first observed by H. K. Onnes (1853–1926) in 1911 when he cooled mercury below 4.2 K. He found that at this temperature the resistance of mercury suddenly dropped to zero. In general, superconductors become superconducting only below a certain *transition temperature* which is usually within a few degrees of absolute zero. Current in a ring-shaped superconducting material has been observed to flow for years in the absence of a potential difference, with no measurable decrease.

Much research has been done on superconductivity in recent years to try to understand why it occurs and to find materials that superconduct at more reasonable temperatures to reduce the cost and inconvenience of refrigeration at the required very low temperature. The first successful superconductivity theory was published by Bardeen, Cooper, and Schrieffer (the BCS theory) in 1957.

Superconductors are already being used in large magnets (we shall see in Chapter 19 that electric currents produce magnetic fields) which sharply reduce electric power needs. Energy is needed, of course, to refrigerate the superconducting material at a low temperature.[†]

17-5 ▮ Electric power

Electric energy is useful to us because it can be easily transformed into other forms of energy. Motors, whose operations we will examine in Chapter 19, transform electric energy into mechanical work.

In other devices such as electric heaters, stoves, toasters, and hair dryers, electric energy is transformed into thermal energy in a wire resistance

[†] Superconductors are being planned for use in large generating plants and for large motors where the negligible resistance will allow very large currents to be used. Also being developed are superconducting transmission cables that will reduce energy loss tremendously and will allow electric generating plants to be placed close to fuel sources (coal, gas, and so on) rather than close to population centers, thus reducing fuel transportation costs. Superconductors may also be used in high-speed ground transportation; the magnetic fields produced by superconducting magnets would be used to "levitate" vehicles over the tracks so there is essentially no friction; the levitation arises from the repulsive force between the magnet (on the train) and the eddy currents (see Section 20-6) produced in the track below.

known as a "heating element." And in an ordinary light bulb, the tiny wire filament (Fig. 17-7) becomes so hot it glows; only a few percent of the energy is transformed into visible light, and the rest, over 90 percent, into thermal energy. Light-bulb filaments and heating elements in household appliances have a resistance typically of a few ohms to a few hundred ohms.

Electric energy is transformed into thermal energy or light in such devices because the current is usually rather large, and there are many collisions between the moving electrons and the atoms of the wire. In each collision, part of the electron's kinetic energy is transferred to the atom with which it collides. As a result, the kinetic energy of the atoms increases and hence the temperature of the wire element increases. The increased thermal energy (internal energy) can be transferred as heat by conduction and convection to the air in a heater or to food in a pan, by radiation to toast in a toaster, or radiated as light.

To find the power transformed by an electric device we use the fact that the energy transformed when a charge Q moves through a potential difference V is QV (Eq. 16-1). Then the power P, which is the rate energy is transformed, is

$$P = \text{power} = \frac{\text{energy transformed}}{\text{time}} = \frac{QV}{t}.$$

FIGURE 17-7

Incandescent light bulb.

The charge that flows per second, Q/t, is simply the electric current I. Thus we have

$$P = IV. \tag{17-5}$$

This general relation gives us the power transformed by any device, where I is the current passing through it and V is the potential difference across it. It also gives the power delivered by a source such as a battery. The SI unit of electric power is the same as for any kind of power, the *watt* (1 W = 1 J/s).

The rate of energy transformation in a resistance R can be written, using Ohm's law ($V = IR$), in two other ways:

$$P = IV \tag{17-6a}$$

$$= I(IR) = I^2 R \tag{17-6b}$$

$$= \left(\frac{V}{R}\right) V = \frac{V^2}{R}. \tag{17-6c}$$

Equations 17-6b and c apply only to resistors, whereas Eq. 17-6a, $P = IV$, applies to any device.

☐   **EXAMPLE 17-5**   Calculate the resistance of a 40-W automobile headlight designed for 12 V.

SOLUTION   Since $P = 40$ W and $V = 12$ V, we use Eq. 17-6c and solve for R:

$$R = \frac{V^2}{P} = \frac{(12 \text{ V})^2}{(40 \text{ W})} = 3.6 \ \Omega.$$

This is the resistance when the bulb is burning brightly at 40 W; when the bulb is cold, the resistance is much lower.

Contact points
Outside switch
Bimetallic strip
Compressed spring
Metal strip
To electric circuit

Fuse ribbon

(a) Fuse

(b) Circuit breaker

FIGURE 17-8

(a) A fuse. When the current exceeds a certain value, the ribbon melts and the circuit opens. Then the fuse must be replaced. (b) A circuit breaker. Electric current passes through a bimetallic strip . When the current is great enough, the bimetallic strip heats sufficiently to bend so far to the left that the notch in the springloaded metal strip drops down over the end of the bimetallic strip. The circuit then opens at the contact points (one is attached to the metal strip) and the outside switch is also flipped. As the device cools down, it can be reset using the outside switch.

It is energy, not power, you must pay for on your electric bill. Since power is the *rate* energy is transformed, the total energy used by any device is simply its power consumption multiplied by the time it is on. If the power is in watts and the time is in seconds, the energy will be in joules since $1 \text{ W} = 1 \text{ J/s}$. Electric companies usually specify the energy with a much larger unit, the kilowatt-hour (kWh). One $\text{kWh} = (1000 \text{ W})(3600 \text{ s}) = 3.60 \times 10^6 \text{ J}$.

⬜ **EXAMPLE 17-6** An electric heater draws 15 A on a 120-V line. How much power does it use and how much does it cost per month (30 days) if it operates 3.0 h per day and the electric company charges $0.060 per kWh? (For simplicity, assume the current flows steadily in one direction.)

SOLUTION The power is $P = IV = (15 \text{ A})(120 \text{ V}) = 1800 \text{ W}$ or 1.8 kW. To operate it for 90 h would cost $(1.8 \text{ kW})(90 \text{ h})(\$0.060) = \$9.72$.

The electric wires that carry electricity to lights and other electric appliances have some resistance, although usually it is quite small. Nonetheless, if the current is large enough, the wires will heat up and produce thermal energy at a rate equal to I^2R where R is the wire's resistance. One possible hazard is that the current carrying wires in the wall of a building may become so hot as to start a fire. Thicker wires of course have less resistance (see Eq. 17-3) and thus can carry more current without becoming too hot. When a wire carries more current than is safe, it is said to be "overloaded." A building should, of course, be designed with wiring heavy enough for any expected load. To prevent overloading, fuses or circuit breakers are installed in circuits. They are basically switches (Fig. 17-8) that open the circuit when the current exceeds some particular value. A 20-A fuse or circuit breaker, for example, opens when the current passing through it exceeds 20 A. If a circuit repeatedly burns out a fuse or opens a circuit breaker, there are two possibilities: there may be too many devices drawing current in that circuit; or there is a fault somewhere, such as a "short." A short, or "short circuit," means that two wires have crossed (perhaps because the insulation has worn down) so the path of the current is shortened. The resistance of the circuit is then very small so the current will be very large. Short circuits should of course be remedied immediately.

Household circuits are designed with the various devices connected so that each receives the 120 V or so from the electric company (Fig. 17-9). When a fuse blows or circuit breaker opens, the total current being drawn

FIGURE 17-9

Connection of household appliances.

Switch

Light bulb 100 W

Electric heater 1800 W

Electric frying pan 1300 W

Fuse

120 V (from electric company)

should be checked first. For example, the circuit in Fig. 17-9 draws the following currents: a light bulb draws $I = P/V = 100 \text{ W}/120 \text{ V} = 0.8 \text{ A}$, a heater $1800 \text{ W}/120 \text{ V} = 15.0 \text{ A}$, and a frying pan $1300 \text{ W}/120 \text{ V} = 10.8 \text{ A}$ for a total of 26.6 A. If the circuit has a 20-A fuse, you might expect it to blow. If it has a 30-A fuse, it shouldn't blow, so a short may be the problem. (The most likely place is in the cord of one of the appliances.)

In electric circuits, heat dissipation by resistors must be considered. The physical size of a resistor is a rough indicator of the maximum permissible power it can dissipate ($= I^2 R$) without appreciable rise in temperature. Common values are $\frac{1}{4}$ W, $\frac{1}{2}$ W, and 1 W; the higher the wattage, the larger the physical size.

17-6 ▌ Alternating current

When a battery is connected to a circuit, the current flows steadily in one direction. This is called a *direct current* or dc. Electric generators at electric power plants, however, produce *alternating current* or ac.[†] An alternating current reverses direction many times per second. The electrons in a wire first move in one direction and then in the other. See Fig. 17-10. The current supplied to homes and businesses by electric companies is ac throughout virtually the entire world.

The voltage produced by an ac electric generator is sinusoidal, as we shall see in Chapter 20. The current it produces is thus sinusoidal (Fig. 17-10b). We can write the voltage as a function of time as

$$V = V_0 \sin 2\pi f t.$$

The potential V oscillates between $+V_0$ and $-V_0$; V_0 is referred to as the *peak voltage*. The frequency f is the number of complete oscillations made per second. In most areas of the United States and Canada, f is 60 Hz (cycles per second). In some countries 50 Hz is used.

From Ohm's law, if a voltage V exists across a resistance R, then the current I is

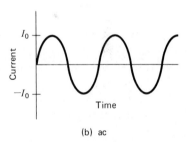

$$I = \frac{V}{R} = \frac{V_0}{R} \sin 2\pi f t = I_0 \sin 2\pi f t. \tag{17-7}$$

FIGURE 17-10

(a) Direct current. (b) Alternating current.

The quantity $I_0 = V_0/R$ is the *peak* current. The current is considered positive when the electrons flow in one direction and negative when they flow in the opposite direction. It is clear from Fig. 17-10b that an alternating current is as often positive as it is negative. Thus the average current is zero. This does not mean, however, that no power is needed or that no heat is produced in a resistor. Electrons do move back and forth in a resistor and do produce heat. Indeed, the power delivered to a resistance R at any instant is

$$P = I^2 R = I_0^2 R \sin^2 2\pi f t.$$

[†] Although it is redundant, we sometimes say "ac current" or "ac voltage" which really means "alternating current" or "alternating voltage".

Because the current is squared, we see that the power is always positive, Fig. 17-11. The quantity $\sin^2 2\pi ft$ varies between 0 and 1; and it is not too difficult to show that its average value is $\frac{1}{2}$. Thus the *average power* developed, $\bar{P}$, is

$$\bar{P} = \tfrac{1}{2}I_0^2 R.$$

Since power can also be written $P = V^2/R = (V_0^2/R)\sin^2 2\pi ft$, we also have that the average power is

$$\bar{P} = \tfrac{1}{2}\frac{V_0^2}{R}.$$

The average value of the *square* of the current or voltage is thus what is important for calculating average power: $\overline{I^2} = \tfrac{1}{2}I_0^2$ and $\overline{V^2} = \tfrac{1}{2}V_0^2$. The square root of each of these is the rms (root mean square) value of the current or voltage:

$$I_{rms} = \sqrt{\overline{I^2}} = \frac{1}{\sqrt{2}}\,I_0 = 0.707 I_0, \tag{17-8a}$$

$$V_{rms} = \sqrt{\overline{V^2}} = \frac{1}{\sqrt{2}}\,V_0 = 0.707 V_0. \tag{17-8b}$$

The rms values of V and I are sometimes called the "effective values." They are useful because they can be substituted directly into the power formulas, Eqs. 17-6, to get the average power. For example, $\bar{P} = \tfrac{1}{2}I_0^2 R = I_{rms}^2 R$. Thus a direct current whose values of I and V equal the rms values of I and V for an alternating current will produce the same power. Hence it is usually the rms value of current which is specified or measured. For example, in the United States and Canada, standard line voltage[†] is 120 V ac. The 120 V is V_{rms}; the peak voltage V_0 is

$$V_0 = \sqrt{2}V_{rms} = 170 \text{ V}.$$

In most of Europe the rms voltage is 240 V, so the peak voltage is 340 V.

EXAMPLE 17-7 Calculate the resistance and the peak current in a 1000-W hair dryer connected to a 120-V line. What happens if it is connected to a 240-V line in Britain?

SOLUTION The rms current is

$$I_{rms} = \frac{\bar{P}}{V_{rms}} = \frac{1000 \text{ W}}{120 \text{ V}} = 8.33 \text{ A}.$$

Thus $I_0 = \sqrt{2}I_{rms} = 11.8$ A. The resistance is

$$R = \frac{V_{rms}}{I_{rms}} = \frac{120 \text{ V}}{8.33 \text{ A}} = 14.4 \text{ } \Omega.$$

The resistance could equally well be calculated using peak values: $R = V_0/I_0 = 170 \text{ V}/11.8 \text{ A} = 14.4 \text{ }\Omega$. When connected to a 240-V line, the aver-

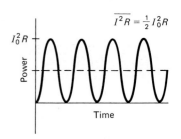

FIGURE 17-11
Power delivered to a resistor in an ac circuit.

[†] The line voltage can vary, depending on the total load; the frequency of 60 Hz, however, remains extremely steady.

age power delivered would be

$$\bar{P} = \frac{V_{\text{rms}}^2}{R} = \frac{(240 \text{ V})^2}{(14.4 \text{ }\Omega)} = 4000 \text{ W}.$$

This would undoubtedly melt the heating element or the wire coils of the motor.

This section has given a brief introduction to the simpler aspects of alternating currents. We will discuss ac circuits in more detail in Chapter 20. In Chapter 18 we will deal with the details of dc circuits only.

*17-7 ∎ The nervous system and nerve conduction

An interesting example of the flow of electric charge is our remarkable and complex nervous system, which provides us with the means for being aware of the world, for communication within the body, and to control the body's muscles. Although the detailed functioning of the nervous system is still not well understood, we do have a reasonably good understanding of how messages are transmitted within the nervous system: they are electrical signals passing along the basic element of the nervous system, the neuron.

Neurons are living cells of unusual shape (Fig. 17-12). Attached to the main cell body are several small appendages known as dendrites and a long tail called the axon. Signals are received by the dendrites and are propagated along the axon; when a signal reaches the nerve endings, it is transmitted to the next nerve or to a muscle at a connection called a synapse. (Some neurons have separate cells, called Schwann cells, wrapped around their axons; they form a layered sheath called a myelin sheath and help to insulate neurons from one another.) Neurons serve in three capacities. "Sensory neurons" carry messages from the eye, ear, skin, and other organs to the central nervous system, which consists of the brain and spinal cord. "Motor neurons" carry signals from the central nervous system to particular muscles and can signal them to contract. These two types of neuron make up the "peripheral system" as distinguished from the central nervous system. The third type of neuron is the interneuron, which transmits signals between neurons; interneurons are in the brain and spinal column, and often are connected in an incredibly complex array.

A neuron, before transmitting an electrical signal, is in the so-called "resting state." Like nearly all living cells, neurons have a net positive charge on the outer surface of the cell membrane and a negative charge on the inner surface. This was already mentioned in Section 16-10 with regard to heart muscles and the ECG. This difference in charge, or "dipole layer," means that a potential difference exists across the cell membrane. When a neuron is not transmitting a signal, this "resting potential," which is normally stated as

$$V_{\text{inside}} - V_{\text{outside}}$$

FIGURE 17-12
Neuron.

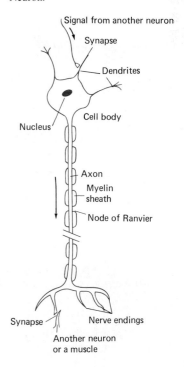

Signal from another neuron

Synapse

Dendrites

Nucleus

Cell body

Axon

Myelin sheath

Node of Ranvier

Synapse

Nerve endings

Another neuron or a muscle

is typically -60 to $-90\,mV$, depending on the type of organism. The most common ions in a cell are K^+, Na^+, and Cl^-. There are large differences in the concentrations of these ions inside and outside a cell, as indicated by the typical values given in Table 17-2. Other ions are also present, so the fluids both inside and outside the axon are electrically neutral. Because of the differences in concentration, there is a tendency for ions to diffuse across the membrane. However, in the resting state the cell membrane prevents any net flow of Na^+ (through a mechanism of "active pumping" of Na^+ out of the cell); but it does allow the flow of Cl^- ions and less so of K^+ ions, and it is these two ions that produce the dipole charge layer on the membrane. Because there is a greater concentration of K^+ inside the cell than outside, more K^+ ions tend to diffuse outward across the membrane than diffuse inward. A K^+ ion that passes through the membrane becomes attached to the outer surface of the membrane; it leaves behind an equal negative charge which lies on the inner surface of the membrane (Fig. 17-13). The fluids themselves remain neutral. Indeed, what keeps the ions on the membrane is their attraction for each other across the membrane. Independent of this process, Cl^- ions tend to diffuse *into* the cell since their concentration outside is higher. Both K^+ and Cl^- diffusion tends to charge the interior surface of the membrane negatively and the outside positively. As charge accumulates on the membrane surface, it becomes increasingly difficult for more ions to diffuse: K^+ ions trying to move outward, for example, are repelled by the positive charge already there. Equilibrium is reached when the tendency to diffuse because of the concentration difference is just balanced by the electrical potential difference across the membrane. The greater the concentration difference, the greater the potential difference across the membrane, which, as mentioned above, is in the range -60 to $-90\,mV$.

The most important aspect of a neuron is not that it has a resting potential (most cells do), but rather that it can respond to a stimulus and conduct an electrical signal along its length. A nerve can be stimulated in various ways. The stimulus could be thermal (when you touch a hot stove) or chemical (as in taste buds); it could be pressure (as on the skin or at the eardrum), or light (as in the eye); or it could be the electric stimulus of a signal coming from the brain or another neuron. In the laboratory, the stimulus is usually electrical and is applied by a tiny probe at some point on the neuron. If the stimulus exceeds some threshold, a voltage pulse will travel down the axon. This voltage pulse can be detected at a point on the axon using a voltmeter or an oscilloscope connected as in Fig. 17-14. This voltage pulse has the shape shown in Fig. 17-15, and is called an "action potential." As can be seen, the potential increases from a resting potential of about $-70\,mV$ and becomes a positive 30 or $40\,mV$. The action potential lasts for about 1 ms and travels down an axon with a speed of $30\,m/s$ to $150\,m/s$.

But what causes the action potential? Apparently, the cell membrane has the ability to alter its permeability properties. At the point where the stimulation occurs, the membrane suddenly becomes much more permeable to Na^+ than to K^+ and Cl^- ions. Thus Na^+ ions rush into the cell and the inner surface of the wall becomes positively charged, and the potential difference quickly swings positive ($\approx +30\,mV$ in Fig. 17-15). But then the membrane suddenly returns to its original characteristics: it becomes im-

TABLE 17-2

Concentrations of ions inside and outside a typical axon

	Concentration Inside axon (mol/m^3)	Concentration Outside axon (mol/m^3)
K^+	140	5
Na^+	15	140
Cl^-	9	125

FIGURE 17-13

How a dipole layer of charge forms on a cell membrane.

FIGURE 17-14

Measuring the potential difference between the inside and outside of a nerve.

FIGURE 17-15

Action potential.

permeable to Na⁺ and in fact pumps out Na⁺ ions. The diffusion of Cl⁻ and K⁺ ions again predominates and the original resting potential is restored (-70 mV in Fig. 17-15).

What causes the action potential to travel along the axon? The action potential occurs at the point of stimulation, as shown in Fig. 17-16a. The membrane momentarily is positive on the inside and negative on the outside at this point. Nearby charges are attracted toward this region as shown in part (b). The potential in these adjacent regions then drops, causing an action potential there. Thus, as the membrane returns to normal at the original point, nearby it experiences an action potential, so the action potential moves down the axon (Fig. 17-16c and d).

You may wonder if the number of ions that pass through the membrane would significantly alter the concentrations. The answer is no; and we can show this by treating the axon as a capacitor in the following example.

FIGURE 17-16

Propagation of an action potential along an axon membrane.

□ EXAMPLE 17-8 (a) Determine the capacitance of an axon 10 cm long of radius 10 μm. The thickness of the membrane is about 10^{-8} m and the dielectric constant is about 3. (b) By what factor does the concentration of Na⁺ ions in the cell change as a result of one action potential?

SOLUTION This will be an order-of-magnitude calculation.

(a) The membrane of an axon acts like a cylindrically shaped parallel-plate capacitor, with opposite charges on each side. The separation of the "plates" is the thickness of the membrane, $d \approx 10^{-8}$ m. The area A is the area of a cylinder of radius r and length l: $A = 2\pi r l \approx (6.28)(10^{-5}\text{ m}) \cdot (0.1\text{ m}) \approx 6 \times 10^{-6}$ m². From Eq. 16-7 we have

$$C = K\varepsilon_0 \frac{A}{d} \approx (3)(8.85 \times 10^{-12}\text{ C}^2/\text{N·m}^2)\frac{6 \times 10^{-6}\text{ m}^2}{10^{-8}\text{ m}} \approx 10^{-8}\text{ F}.$$

(b) Since the voltage changes from -70 mV to about $+30$ mV, the total change is about 100 mV. The amount of charge that moves is then $Q = CV \approx (10^{-8}\text{ F})(0.1\text{V}) = 10^{-9}$ C. This corresponds to $Q/e = 10^{-9}$ C/1.6×10^{-19} C $\approx 10^{10}$ ions that must flow per action potential. The volume of our cylindrical axon is $V = \pi r^2 l \approx (3)(10^{-5}\text{ m})^2(0.1\text{ m}) = 3 \times 10^{-11}$ m³;

the concentration of Na$^+$ ions inside the cell (Table 17-2) is 15 mol/m^3 = $15 \times 6.02 \times 10^{23}$ ions/m^3 $\approx 10^{25}$ ions/m^3. Thus the cell contains $10^{25} \times (3 \times 10^{-11}) \approx 3 \times 10^{14}$ Na$^+$ ions. One action potential, then, will change the concentration of Na$^+$ ions by at most 1 part in 30,000; this tiny change would not be measurable.

Thus even 1000 action potentials will not alter the concentration significantly. The sodium pump does not, therefore, have to remove Na$^+$ ions quickly after an action potential, but can operate slowly over time to maintain a relatively constant concentration.

The propagation of a nerve pulse as described here applies to an unmyelinated axon. Myelinated axons, on the other hand, are insulated from the extracellular fluid by the myelin sheath except at the nodes of Ranvier (see Fig. 17-12). Thus an action potential cannot be regenerated where there is a myelin sheath. Once such a neuron is stimulated, the pulse will still travel along the membrane, but there is resistance and the pulse becomes smaller as it moves down the axon. Nonetheless, the weakened signal can still stimulate a full-fledged action potential when it reaches a node of Ranvier. Thus the signal is repeatedly amplified at these points. Compare this to an unmyelinated neuron, in which the signal is continually amplified by repeated action potentials all along its length. This naturally requires much more energy. Development of myelinated neurons was a significant evolutionary step. For it meant reliable transmission of nerve pulses with less energy required; and the pulses travel more quickly, since ordinary conduction is faster than the repeated production of action potentials, whose speed depends on the flow of ions across the membrane.

▌ Summary

By transforming chemical energy into electric energy, an electric battery serves as a source of potential difference or *electromotive force* (emf). A simple battery consists of two electrodes made of different metals immersed in a solution or paste known as an electrolyte.

Electric current, I, refers to the rate of flow of electric charge and is measured in amperes (A): 1 A equals a flow of 1 C/s past a given point. The direction of current flow is generally taken as being that of positive charge; in a wire, it is actually negatively charged electrons that move, so they flow in a direction opposite to the direction of the "conventional current." Positive conventional current always flows from a high potential to a low potential.

Ohm's law states that the current in a good conductor is proportional to the potential difference applied to its two ends; the proportionality constant is called the *resistance* R of the material, so $V = IR$. The unit of resistance is the ohm (Ω), where $1 \, \Omega = 1$ V/A.

The resistance R of a wire is inversely proportional to its cross-sectional area A and directly proportional to its length L and to a property of the material called its resistivity: $R = \rho L/A$. The resistivity, ρ, increases with temperature for metals, but for semiconductors it may decrease. A superconductor is a material for which the electrical resistance is essentially zero.

The rate at which energy is transformed in a resistance R from electric to other forms of energy (such as heat and light) is equal to the product of current and voltage; that is, the power transformed, measured in watts, is given by $P = IV$ and can be written with the help of Ohm's law as $P = I^2R = V^2/R$. The total electric energy transformed in any device equals the product of power and the time during which the device is operated. In SI units, energy is given in joules ($1 \text{ J} = 1 \text{ W·s}$), but electric companies use a larger unit, the kilowatt-hour ($1 \text{ kWh} = 3.6 \times 10^6 \text{ J}$).

Electric current can be direct (dc), in which the current is steady in one direction; or it can be alternating (ac), in which the current reverses direction at a particular frequency, typically 60 Hz. Alternating currents are often sinusoidal in time, $I = I_0 \sin 2\pi ft$,

and are produced by an alternating voltage. The rms values of sinusoidally alternating currents and voltages are given by $I_{rms} = I_0/\sqrt{2}$ and $V_{rms} = V_0/\sqrt{2}$, where I_0 and V_0 are the peak values. The power relationship, $P = IV = I^2R = V^2/R$, is valid for the average power in alternating currents when the rms values are used.

▌Questions

1. When you turn on a water faucet, the water usually flows immediately. You don't have to wait for water to flow from the faucet valve to the spout. Why not? Is the same thing true when you connect a wire to the terminals of a battery?

2. Car batteries are often rated in ampere-hours (A·h). What does this rating mean?

3. When an electric cell is connected to a circuit, electrons flow away from the negative terminal in the circuit. But within the cell, electrons flow *to* the negative terminal. Explain.

4. Is current used up in a resistor?

5. Develop an analogy between blood circulation and an electrical circuit. Discuss what plays the role of the heart for the electric case, and so on.

6. Design a circuit in which two different switches of the type shown in Fig. 17-17 can be used to operate the same light bulb from opposite sides of a room.

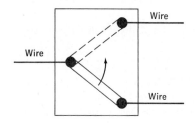

FIGURE 17-17

7. Can a copper wire and an aluminum wire of the same length have the same resistance? Explain.

8. If a large rectangular block of carbon has sides a, $2a$, $3a$, how would you connect the wires from a battery so as to obtain (a) the least resistance, (b) the greatest resistance?

9. In a car, one terminal of the battery is said to be connected to "ground." Since it is not really connected to the ground, what is meant by this expression?

10. The equation $P = V^2/R$ indicates that the power dissipated in a resistor decreases if the resistance is increased whereas the equation $P = I^2R$ implies the opposite. Is there a contradiction here? Explain.

11. What happens when a light bulb burns out?

12. Which draws more current, a 100-W light bulb or a 75-W bulb?

13. Electric power is transferred over large distances at very high voltages. Explain how the high voltage reduces power losses in the transmission lines.

14. Why is it dangerous to replace a 15-A fuse that blows repeatedly with a 25-A fuse?

15. Electric lights operated on low frequency ac (say 10 Hz) flicker noticeably. Why?

16. Suppose Franklin's original convention had been the reverse, so that electrons were considered positive. How would this affect the various results and analyses discussed in this chapter?

▌Problems

SECTIONS 17-2 AND 17-3

1. (I) A service station charges a battery using a current of 5.5 A for 6.0 h. How much charge passes through the battery?

2. (I) A current of 1.10 A flows in a wire. How many electrons are flowing past any point in the wire per second? The charge on one electron is 1.60×10^{-19} C.

3. (I) How many coulombs are there in 1 ampere-hour?

4. (I) What is the current in amperes if 1000 Na^+ ions were to flow across a cell membrane in 1.0 μs? The charge on the sodium is the same as on an electron, but positive.

5. (I) What voltage will produce 12.0 A of current through a 150 Ω resistor?

6. (I) What is the resistance of a toaster if 110 V produces a current of 4.0 A?

7. (II) Two 1.5-V batteries are connected together in series to a bulb whose resistance is 10 Ω. How many electrons leave each battery per minute?

8. (II) A bird stands on an electric transmission line carrying 1200 A. The line has 1.0×10^{-5} Ω resistance per meter and the bird's feet are 3.0 cm apart. What voltage does the bird feel?

9. (II) The *conductance* G of an object is defined as the reciprocal of the resistance $R : G = 1/R$. The unit of conductance is a mho ($= $ ohm^{-1}), which is also called the siemens (S). What is the conductance (in siemens) of an object that draws 200 mA of current at 2.0 V?

10. (III) Copper has approximately 10^{29} free electrons per cubic meter. What is the approximate average speed of electrons in a 1-mm-radius wire carrying 1.0 A?

SECTION 17-4

11. (I) What is the resistance of a 2.0-m length of copper wire 1.8 mm in diameter?

12. (I) What is the diameter of a 20-cm length of tungsten wire whose resistance is 0.10 Ω?

13. (II) A 20.0-m length of wire 1.80 mm in diameter has a resistance of 2.50 Ω. What is the resistance of a 35.0-m length of wire 3.00 mm in diameter made of the same material?

14. (II) Can a 2.0-mm-diameter copper wire have the same resistance as a tungsten wire of the same length? Give numerical details.

15. (II) A length of wire is cut in half and the two lengths are wrapped together side by side to make a thicker wire. How does the resistance of this new combination compare to the resistance of the original wire?

16. (II) A 100-W light bulb has a resistance of about 12 Ω when cold and 140 Ω when on (hot). Estimate the temperature of the filament when "on" assuming an average temperature coefficient of resistivity $\alpha = 0.0060$ (C°)$^{-1}$.

17. (II) A rectangular solid made of carbon has sides lying along the x, y, and z axes, whose lengths are 1.0 cm, 2.0 cm, and 4.0 cm, respectively. Determine the resistance for current that flows through the solid in (a) the x direction, (b) the y direction, (c) the z direction. Assume the resistivity is $\rho = 3.0 \times 10^{-5}$ Ω·m.

18. (II) A 12.0-m length of wire consists of two 6.0 m portions of equal diameter (3.0 mm), one of copper and one of aluminum, welded together. What voltage is required to produce a current of 6.0 A?

19. (III) A 1.00-Ω wire is drawn out to 3.0 times its original length. What is its resistance now?

20. (III) For some applications it is important that the value of a resistance not change with temperature. For example, suppose you made a 2.2-kΩ resistor from a carbon resistor and a nichrome wire-wound resistor connected together so the total resistance will be the sum of their separate resistances. What value should

each of these resistors have (at 0°C) so that the combination is temperature independent?

21. (III) A 38.0-Ω resistor is made from a coil of copper wire whose total mass is 11.2 g. What is the diameter of the wire and how long is it?

SECTION 17-5

22. (I) The element of an electric oven is designed to produce 3.0 kW of heat when connected to a 240-V source. What must be the resistance of the element?

23. (I) What is the current through a 6.0-W light bulb if it is connected to its proper source voltage of 12 V?

24. (I) What is the maximum power consumption of a 9.0-V transistor radio that draws a maximum of 400 mA of current?

25. (I) What is the maximum voltage that can be applied to a 200-Ω resistor rated at $\frac{1}{4}$ watt?

26. (I) An automobile starter motor draws 150 A from the 12-V battery. How much power is this?

27. (I) How many kWh does a 1300-W electric frying pan use in 15 min of operation?

28. (II) At $0.080 per kWh, what does it cost to leave a 25-W porch light on all day for a year?

29. (II) A person accidentally leaves a car with the lights on. If each of the two front lights uses 40 W and each of the two rear lights 6 W, how long will a fresh 12-V battery last if it is rated at 90 A·h? Assume the full 12 V appears across each bulb. (Hint: determine what is meant by 1 A·h.)

30. (I) What is the total amount of energy stored in a 12-V, 50-A·h car battery when it is fully charged?

31. (II) How many 100-W light bulbs, operated at 120 V, can be used without blowing a 10-A fuse?

32. (II) The heating element of a 120-V, 1800-W heater is 8.2 m long. If it is made of iron, what must its diameter be?

33. (II) (a) A particular household uses a 3.5-kW heater 2.5 h/day ("on" time), six 100-W light bulbs 5.0 h/day, a 3.3-kW electric stove element for a total of 1.2 h, and miscellaneous power amounting to 1.6 kWh/day. If electricity costs $0.070 per kWh, what will be their monthly bill (30 d)? (b) How much coal (which produces 7000 kcal/kg) must be burned by a 35-percent-efficient power plant to provide the yearly needs of this household?

34. (II) What is the efficiency of a 0.50 hp electric motor that draws 4.4 A from a 120-V line?

35. (II) A power station delivers 360 kW of power to a factory through 3.2-Ω lines. How much less power is wasted if the electricity is delivered at 40,000 V rather than 12,000 V?

36. (II) The wiring in a house must be thick enough so it doesn't become so hot as to start a fire. What diameter must a copper wire be if it is to carry a maximum current of 40 A and produce no more than 1.8 W of heat per meter of length?

37. (II) A 1200-W hair dryer is designed for 117 V. What will be the percentage change in power output if the voltage drops to 105 V? Assume no change in resistance. How would the actual change in resistivity with temperature affect your answer?

38. (III) The current in an electromagnet connected to a 240-V line is 6.60 A. At what rate must cooling water pass over the coils if the water temperature is to rise by no more than 8.00°C?

39. (III) A small immersion heater can be used in a car to heat a cup of water for coffee. If the heater can heat 200 ml of water from 5°C to 95°C in 5.0 min, approximately how much current does it draw from the 12-V battery?

40. (III) An electric heater is used to heat a room of volume 36 m³. Air is brought into the room at 5°C and is changed completely twice per hour. Heat loss through the walls amounts to approximately 500 kcal/h. If the air is to be maintained at 20°C, what minimum wattage must the heater have? (The specific heat of air is about 0.17 kcal/kg·C°.)

41. (III) An electric car makes use of storage batteries as its source of energy. Suppose that a small 650-kg postman's delivery car is to be powered by ten 70-A·h, 12-V batteries. Assume that the car is driven on the level at an average speed of 30 km/h, and the average friction force is 200 N. Assume 100 percent efficiency and neglect energy used for acceleration; note that no energy is consumed when the vehicle is stopped since the engine doesn't need to idle. (a) Determine the horsepower required; (b) after approximately how many kilometers must the batteries be recharged?

SECTION 17-6

42. (I) An ac voltage, whose peak value is 90 V, is across a 35-Ω resistor. What is the value of the rms and peak currents in the resistor?

43. (I) What is the resistance of an ordinary 60-W, 120-V_{rms} light bulb when it is on?

44. (I) Calculate the peak current in a 2.2-kΩ resistor connected to a 240-V ac source.

45. (II) The peak value of an alternating current passing through a 600-W device is 3.0 A. What is the rms voltage across it?

46. (II) Calculate the peak current passing through a 60-W light bulb connected to a 120-V ac line.

47. (II) What is the maximum instantaneous value of the power dissipated in a 100-W light bulb?

48. (II) A 10-Ω heater coil is connected to a 240-V ac line. What is the average power used? What are the maximum and minimum values of the instantaneous power?

49. (II) Suppose a current is given by the equation $I = 2.50 \sin 120t$, where I is in amperes and t in seconds. (a) What is the frequency? (b) What is the rms value of the current? (c) If this is the current through a 50.0-Ω resistor, what is the equation that describes the voltage as a function of time?

*SECTION 17-7

***50.** (I) What is the magnitude of the electric field across an axon membrane 1.0×10^{-8} m thick if the resting potential is -70 mV?

***51.** (II) A nerve is stimulated with an electric pulse. The action potential is detected at a point 3.40 cm down the axon 0.0052 s later. When the action potential is detected 7.20 cm from the point of stimulation, the time required is 0.0063 s. What is the speed of the electric pulse along the axon? (Why are two measurements needed instead of only one?)

***52.** (III) Estimate how much energy is required to transmit one action potential along the axon of Example 17-8. (Hint: one pulse is equivalent to charging and discharging the axon capacitance; see Section 16-9.) What minimum average power is required for 10^4 neurons transmitting 100 pulses per second?

***53.** (III) During the action potential, Na⁺ moves into the cell at a rate of about 3×10^{-7} mol/m²·s. How much power must be produced by the active transport system to produce this flow against a $+30$-mV potential difference? Assume that the axon is 10 cm long and 20 μm in diameter.

18 DC CIRCUITS AND INSTRUMENTS

In Chapter 17 we discussed the basic principles of current electricity. Now we will apply these principles to analyze dc circuits and to understand the operation of a number of useful instruments.[†]

When we draw a diagram of a circuit, we represent a battery by the symbol ⊣├, a capacitor by the symbol ⊣├, and a resistor by the symbol ⌇⌇⌇. Wires whose resistance is negligible compared to other resistance in the circuit are drawn simply as straight lines. For the most part in this chapter, except in Section 18-6, we will be interested in circuits operating in their steady state—that is, we won't be looking at a circuit at the moment any change is made in it, such as a battery or resistor connected or disconnected, but rather a short time later when the currents have reached their steady values.

18-1 ▌ Resistors in series and in parallel

When two or more resistors are connected end to end so the same current passes through each in turn, as in Fig. 18-1, they are said to be connected in *series*. On the other hand, if the resistors are connected so that the current from the source splits into separate branches as shown in Fig. 18-2, the resistors are said to be in *parallel;* in this case, the same potential difference exists across each. In either case, we will want to calculate the net resistance of the set of resistors. In other words, we want to find what single resistance R could replace the combination of given resistors without altering the rest of the circuit. The resistors could be like those shown in Fig. 17-6, or they could be light bulbs or other resistive electrical devices.

First we consider the series case, Fig. 18-1. The same current I passes through each resistor. If it did not, this would imply that charge was accumulating at some point in the circuit, which does not happen in the steady state. We let V represent the voltage across all three resistors; we assume all other resistance in the circuit can be ignored, and so V equals the emf of the battery. We let V_1, V_2, and V_3 be the potential differences across each of the

---

[†] Ac circuits that contain only a source of emf and resistors can be analyzed like the dc circuits in this chapter. However, ac circuits in general are more complex—for example, if they contain capacitors—and we will discuss them in Chapter 20.

FIGURE 18-1
Resistors connected in series: $R = R_1 + R_2 + R_3$.

FIGURE 18-2
Resistors connected in parallel: $1/R = 1/R_1 + 1/R_2 + 1/R_3$.

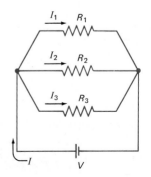

resistors R_1, R_2, and R_3. By Ohm's law, $V_1 = IR_1$, $V_2 = IR_2$, and $V_3 = IR_3$. Because of conservation of energy, the total voltage V is equal to the sum of the voltages across each resistor, so we have

$$V = V_1 + V_2 + V_3 = IR_1 + IR_2 + IR_3.$$

For the equivalent single resistor R that would draw the same current we would have

$$V = IR.$$

We equate these two expressions for V and find

$$R = R_1 + R_2 + R_3. \qquad \text{[series]} \quad (18\text{-}1)$$

This is, in fact, what we expect. When we put several resistances in series, the total resistance is the sum of the separate resistances. This applies to any number of resistances, not simply for three. Clearly, when you add more resistance to the circuit, the current will decrease. If a 12-V battery is connected to a 4-Ω resistor, the current will be 3 A. But if the 12-V battery is connected to three 4-Ω resistors in series, the total resistance is 12 Ω and the current only 1 A.

The situation is quite different for the parallel case, Fig. 18-2. Again we want to find what single resistance R is equivalent to the three in parallel. In this situation, the total current I that leaves the battery breaks into three branches. We let I_1, I_2, and I_3 be the currents through each of the resistors R_1, R_2, and R_3. Because charge is conserved, the current flowing into a junction must equal the current flowing out, so

$$I = I_1 + I_2 + I_3$$

The full voltage of the battery is applied to each resistor, so

$$I_1 = \frac{V}{R_1}, \qquad I_2 = \frac{V}{R_2}, \quad \text{and} \quad I_3 = \frac{V}{R_3}.$$

Furthermore, for the single resistor R that will draw the same current I as these three in parallel, we must have

$$I = \frac{V}{R}.$$

We now combine these equations:

$$I = I_1 + I_2 + I_3$$
$$\frac{V}{R} = \frac{V}{R_1} + \frac{V}{R_2} + \frac{V}{R_3}.$$

When we divide out the V from each term we have

$$\frac{1}{R} = \frac{1}{R_1} + \frac{1}{R_2} + \frac{1}{R_3}. \qquad \text{[parallel]} \quad (18\text{-}2)$$

For example, if three 30-Ω resistors are put in parallel, the net resistance R offered by this network is

$$\frac{1}{R} = \frac{1}{30\ \Omega} + \frac{1}{30\ \Omega} + \frac{1}{30\ \Omega} = \frac{3}{30\ \Omega} = \frac{1}{10\ \Omega},$$

An *ammeter* is used to measure current, and a *voltmeter* measures potential difference or voltage. The crucial part of each of these meters is a *galvanometer*, which works on the principle of the force between a magnetic field and a current-carrying coil of wire; we will discuss this in Chapter 19. For now we merely need to know that the deflection of the needle of a galvanometer is proportional to the current flowing through it. The *full-scale current sensitivity*, I_m, of a galvanometer is the current needed to make the needle deflect full scale, Fig. 18-14. For example, if the sensitivity I_m is 50 μA, a current of 50 μA will cause the needle to go to the end of the scale; a current of 25 μA will make it go only half way; if there is no current, the needle should be on zero and usually there is an adjustment screw to make it so.

Many meters we meet with in everyday life are galvanometers connected as ammeters or voltmeters. These include the VU meter on a tape recorder, and some of the meters on the dashboards of some cars.

A galvanometer can be used directly to measure small currents. For example, a galvanometer whose sensitivity I_m is 50 μA can measure currents from about 1 μA (currents smaller than this would be hard to read on the scale) up to 50 μA. To measure larger currents, a resistor is placed in parallel with the galvanometer. Thus an *ammeter* (represented by the symbol •—(A)—•) consists of a galvanometer (•—(G)—•) in parallel with a resistor called the *shunt resistor*. ("Shunt" is a synonym for "in parallel.") This is shown in Fig. 18-15a. The shunt resistance is R and the resistance of the galvanometer coil (which carries the current) is r. The value of R is chosen according to what full-scale deflection is wanted.

Suppose we want to design a meter to read 1.0 A at full scale using a galvanometer with a full-scale sensitivity of 50 μA and its resistance is $r = 30\ \Omega$. This means that when the total current I entering the ammeter is 1.0 A, we want the current I_G through the galvanometer to be precisely 50 μA (to give full-scale deflection). Thus, when 1.0 A flows into the meter, we want 0.999950 A $(=I_R)$ to pass through the shunt resistor R. Since the potential

FIGURE 18-14

Multimeter used as an ammeter. (Lester V. Bergman & Assoc., Inc.)

(a)

(b)

FIGURE 18-15
An ammeter is a galvanometer in parallel with a small (shunt) resistor, R. (b) A voltmeter is a galvonometer in series with a large resistor, R.

difference across the shunt is the same as across the galvanometer,

$$I_R R = I_G r,$$

then

$$R = \frac{I_G r}{I_R} = \frac{(5.0 \times 10^{-5}\text{ A})(30\ \Omega)}{(0.999950\text{ A})} = 1.5 \times 10^{-3}\ \Omega$$

or $0.0015\ \Omega$. The shunt resistor must thus have a very low resistance so that most of the current passes through it. If the current I into the meter is 0.50 A, this will produce a 25 μA current in the galvanometer and give a deflection half of full scale, as required.

A *voltmeter* (•—Ⓥ—•) also consists of a galvanometer and a resistor. But the resistor R is connected in series, Fig. 18-15b; and it is usually large. Suppose, using the same galvanometer with internal resistance $r = 30\ \Omega$ and full-scale current sensitivity of 50 μA, we want to make a voltmeter that reads from 0 to 15 V. When a potential difference of 15 V exists across the terminals of our voltmeter, we want 50 μA to be passing through it so as to give a full-scale deflection. From Ohm's law we have

$$15\text{ V} = (50\ \mu\text{A})(r + R),$$

so

$$R = \frac{15\text{ V}}{5.0 \times 10^{-5}\text{ A}} - r = 300\text{ k}\Omega.$$

Notice that $r = 30\ \Omega$ is so small compared to the value of R that it doesn't influence the calculation significantly. The scale will again be linear: if the voltage to be measured is 6.0 V, the current flowing through the voltmeter will be $(6.0\text{ V})/(3.0 \times 10^5\ \Omega) = 2.0 \times 10^{-5}\text{ A}$ or 20 μA. This will produce two-fifths of full-scale deflection, as required.

The meters discussed above are for direct current. If an alternating current passed through these meters, the needle would fluctuate wildly; or—more likely—not move at all because its inertia would not allow it to follow the rapid changes. Most *ac meters*, however, are basically like the dc meters described above except that diodes (Chapter 28) are used to allow current flow in only one direction. Thus alternating current is changed to a unidirectional current whose average value can be read by the meter. Ac meters are normally calibrated to read rms values.[†]

18-8 ▌ Use of meters and correcting for meter resistance; multimeters

Suppose you wish to determine the current I in the circuit shown in Fig. 18-16a, and the voltage V across the resistor R_1. How exactly are ammeters and voltmeters hooked up?

An ammeter, because it is to measure the current flowing in the circuit, must be inserted directly into the circuit, in series with the other elements, Fig. 18-16b.

A voltmeter, on the other hand, is connected in parallel with the circuit element across which the voltage is to be measured. It is used to measure the potential difference between two points and its two wire leads (connecting wires) are connected to the two points, as shown in Fig. 18-16c where the voltage across R_1 is being measured.

Voltmeters and ammeters can have several series or shunt resistors to offer a choice of range. *Multimeters* can measure voltage, current, and resistance. Sometimes they are called VOMs (Volt-Ohm-Meter). Meters with digital readout are called digital voltmeters (DVM) or digital multimeters (DMM). To measure resistance, the meter must contain a battery of known voltage connected to an ammeter (galvanometer with a shunt). This makes an *ohmmeter* (see problem 60). The resistor whose resistance is to be measured completes the circuit. The deflection in this case is inversely proportional to the resistance. Thus, if the resistance is small the deflection is large, and vice versa. The calibration of the scale depends on the value of the series resistor. Since an ohmmeter sends a current through the device whose resistance is to be measured, it should not be used on very delicate devices that could be damaged by the current.

The *sensitivity* of a meter is generally specified on the face. It may be given as so many ohms per volt, which indicates how many ohms of resistance there are in the meter per volt of full-scale reading. For example, if the sensitivity is 30,000 Ω/V, this means that on the 10-V scale the meter has a resistance of 300,000 Ω. The full-scale current sensitivity, I_m, discussed earlier, is just the inverse of the sensitivity in Ω/V (note that the units Ω/V = A^{-1}). For example a meter with sensitivity 30,000 Ω/V produces a full-scale deflection at 1.0 V when 30,000 Ω is in series with the galvanometer. Thus the current sensitivity is 1.0 V/3.0 × 10^4 Ω = 33 μA in this case.

It is important to know the sensitivity, for in many cases the resistance of the meter can seriously affect your results. Take the following example.

(a)

(b)

(c)

FIGURE 18-16

Measuring the current and voltage.

[†] Sometimes ac meters are calibrated to read peak-to-peak voltage (V_{pp}) or current (I_{pp})— that is, the spread from the height of a peak to the depth of a valley: $V_{pp} = 2V_0$, and $I_{pp} = 2I_0$, where V_0 and I_0 are the peak values of voltage and current (Section 17-6).

EXAMPLE 18-6 Suppose R_1 and R_2 are each 15 kΩ and are connected in series as shown in Fig. 18-16a. The battery maintains 8.0 V across them and has negligible internal resistance. A voltmeter whose sensitivity is 10,000 Ω/V is put on the 5.0-V scale. What voltage does the meter read when connected across R_1, and what error is caused by the finite resistance of the meter?

SOLUTION On the 5.0-V scale, the voltmeter has an internal resistance of (5.0 V)(10,000 Ω/V) = 50,000 Ω. When connected across R_1 as in Fig. 18-16c we have this 50 kΩ in parallel with $R_1 = 15$ kΩ; the net resistance R of these two is given by $1/R = 1/50$ kΩ $+ 1/15$ kΩ $= 13/150$ kΩ; so $R = 11.5$ kΩ. This $R = 11.5$ kΩ is in series with $R_2 = 15$ kΩ so the total resistance of the circuit is now 26.5 kΩ. Hence the current from the battery is $I = 8.0$ V/26.5 kΩ $= 0.30$ mA, and the voltage drop across R_1 and the voltmeter is $(3.0 \times 10^{-4}$ A$)(11.5 \times 10^3$ Ω$) = 3.5$ V. [That across R_2 is $(3.0 \times 10^{-4}$ A$) \cdot (15 \times 10^3$ Ω$) = 4.5$ V, for a total of 8.0 V.] If we assume the meter is precise, it will read 3.5 V. In the normal circuit, without the meter, $R_1 = R_2$ so the voltage across R_1 is half that of the battery or 4.0 V. Thus the voltmeter, because of its internal resistance, gives a low reading. In this case it is off by 0.5 V, or more than 10 percent.

This example illustrates how seriously a meter can affect a circuit and give a misleading reading. If the resistance of a voltmeter is much higher than the resistance of the circuit, however, it will have little effect and its readings can be trusted, at least to the manufactured accuracy of the meter, which for ordinary meters is typically 3 to 4 percent of full-scale deflection. An ammeter also can interfere with a circuit, but the effect is minimal if its resistance is much less than that of the circuit as a whole. For both voltmeters and ammeters, the more sensitive the galvanometer the less effect it will have. A 50,000-Ω/V meter is far better than a 1,000-Ω/V meter.

Electronic meters using transistors have sensitivities, usually specified in ohms, in the range 10^6 to 10^8 Ω. Hence they have very little effect on most circuits and their readings are reliable for nearly any circuit. Very high resistance voltmeters, called *electrometers*, may have input resistances $\gtrsim 10^{14}$ Ω.

*18-9 ∎ The potentiometer

The *potentiometer* is a device to precisely measure potential differences. It uses a "null" method, as described below, and has the further advantage of being able to measure a voltage *without drawing current*. Hence it can be used to measure an emf directly (with no voltage drop due to internal resistance). And it won't alter the currents in an operating circuit.

A diagram of a simple potentiometer is shown in Fig. 18-17. A fresh battery, termed the "working battery," is connected in series to a variable resistor R_v(⋏⋏⋏) which can be adjusted to give a convenient voltage between the points A and B. The potential difference between A and B, which we call V, produces a steady current through the precision resistor R' which has a slider (point C) to obtain an accurate variable resistance R. (This resistor R'

FIGURE 18-17
Potentiometer circuit.

427

could be a uniform wire, called a "slide wire," with a sliding contact C that moves along it.)[†] The emf to be measured, $\mathscr{E}_x$, can be placed in the circuit as shown; the slider is moved until the galvanometer indicates zero current ("null point") when the switch S is closed momentarily. In this position no current flows from the unknown source so the voltage between points A and C will be equal to $\mathscr{E}_x$, the emf of the unknown:

$$V_{AC} = \mathscr{E}_x.$$

This relationship holds only when no current flows from the unknown emf through the galvanometer, since otherwise there would be potential drops across the internal resistances of the galvanometer and the unknown source. There is a current I through R', of course, and if we let R_x represent the resistance between points A and C of R' when we are at the null point with $\mathscr{E}_x$ in the circuit, then

$$\mathscr{E}_x = IR_x.$$

A second measurement is necessary, this time with a standard cell—one whose emf $\mathscr{E}_s$ is accurately known in advance. (The standard cell should never be used to draw current, for this will affect its emf over time.) The slider is again adjusted to a new position C' so that again no current passes through the galvanometer; the resistance then between A and C' is R_s, given by

$$\mathscr{E}_s = IR_s$$

since the same current I is flowing through R'. Then

$$\frac{\mathscr{E}_x}{\mathscr{E}_s} = \frac{IR_x}{IR_s} = \frac{R_x}{R_s}$$

or

$$\mathscr{E}_x = \left(\frac{R_x}{R_s}\right)\mathscr{E}_s.$$

(If R' is a precisely uniform wire, then the ratio R_x/R_s is equal to the ratio of lengths AC/AC'.)

It should be noted that the word "potentiometer" is often used to refer simply to a variable resistance; and colloquially it is sometimes called a "pot."

*18-10 ∎ The Wheatstone bridge

The Wheatstone bridge is probably the most common type of so-called *bridge circuits* which are used to make precise measurements of resistance. Fig. 18-18 is a diagram of a Wheatstone bridge circuit. The resistance of R_1, R_2, and R_3 are accurately known. At least one of these—R_3 in our diagram—is variable; its resistance is usually read from a dial. The resistance

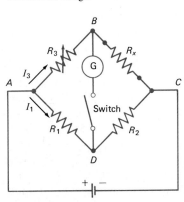

FIGURE 18-18
Wheatstone bridge.

[†] R' can also be a set of fixed precision resistors, with several taps leading out, plus a variable resistor—so that a wide range of precise values of R is available.

R_x which is to be measured (the unknown resistance) is connected so as to form the fourth arm of the bridge. When the battery is connected, current flows through each of the resistors. In operation, the variable resistor R_3 is adjusted until the points B and D are at the same potential. This is checked by momentarily closing the switch to the galvanometer which will not deflect when B and D are at the same potential. In this situation the bridge is said to be "balanced" and we have $V_{AB} = V_{AD}$, or (see Fig. 18-18)

$$I_3 R_3 = I_1 R_1.$$

I_1 is the current that passes through R_1 and also through R_2 when the bridge is balanced; I_3 is the current passing through R_3 and R_x. When the bridge is balanced the voltage across R_x equals that across R_2, so

$$I_3 R_x = I_1 R_2.$$

We now divide these two equations and find

$$\frac{I_3 R_x}{I_3 R_3} = \frac{I_1 R_2}{I_1 R_1}$$

or

$$R_x = \frac{R_2}{R_1} R_3.$$

This equation allows a precise calculation of R_x when the other three resistances are precisely known. In practice, the galvanometer is very sensitive, and so when R_3 is being adjusted the switch is closed only momentarily to check if the current is zero or not.

A Wheatstone bridge is used in many situations where resistance must be accurately measured, such as the resistance of a resistance thermometer, Example 17-4.

*18-11 ▌ Transducers and the thermocouple

A *transducer* is a device that converts one type of energy into another. A high fidelity loudspeaker is one kind of transducer—it transforms electric energy into sound energy (see Chapter 19); so is a microphone, which changes sound into an electrical signal.

Transducers are often used for measuring particular quantities. In this case they can be thought of as changing one kind of signal into another kind, usually electrical. The *resistance thermometer* and the *thermistor* mentioned in Chapter 17 are examples; basically they convert temperature into an electrical property, resistance.

We now discuss some other transducers that are useful in a variety of fields. We will meet other examples later.

A *strain gauge*, Fig. 18-19, makes use of the fact that a wire will stretch an amount proportional to the stress (or force) applied to it. When stretched, its resistance must increase since it is longer and its cross-sectional area is reduced (see Eq. 17-3). The fine wire of a strain gauge is usually bonded to a flexible backing material. When fixed tightly to a structure, the resistance

FIGURE 18-19

Wire strain gauge.

of the strain gauge changes in direct proportion to any change in stress on the structure. Since the wire must not be strained beyond its elastic limit, the change in length is generally quite small. Hence the change in resistance is quite small (less than one part per thousand) and a very sensitive Wheatstone bridge is used to measure it. Of course, strain gauges must be carefully calibrated before use. Strain gauges are used in many applications, such as by architects and engineers on models of proposed structures to determine the stress at critical points. In other applications, strain gauges have been used to determine the stress in the bones of the human skull when biting, by attaching them at various points on dry skulls.

A wire strain gauge can be attached to a membrane or diaphragm. A change in pressure against the membrane causes strain in the wire. Thus a strain gauge can be used to measure pressure. In this case it is called a *pressure transducer*. Strain gauges can be made very small (2 to 3 mm wide). When attached to a diaphragm, they can be swallowed to measure internal stomach pressure, or inserted into a vein or artery to measure blood pressure. In these and other situations there will also be variation in resistance due to temperature changes. To correct for this, an identical gauge must also be used which is subjected to the same temperature but no stress.

Another kind of pressure transducer makes use of the *piezoelectric effect*. This effect occurs in certain crystals, such as quartz, that become polarized (Section 16-8) when a mechanical force is applied and produce an emf proportional to the force. Piezoelectric crystals are used in some phonograph cartridges where they transform the movement of the stylus in a record groove into an electric signal (an emf).

The *thermocouple* is a device that produces an electrical signal when subjected to differing temperatures. Unlike the resistance thermometer (in which it is the resistance that changes with temperature), a thermocouple produces an emf. The thermocouple is based on the "thermoelectric effect." When two dissimilar metals, say iron and copper, are joined at the ends, as shown in Fig. 18-20a, it is found that an emf is produced if the two junctions are at different temperatures.[†] The magnitude of this emf depends on the temperature difference. In operation, one junction of a thermocouple is kept at a known temperature. This "reference temperature" is often 0°C. The other junction, called the "test junction," is placed where the desired temperature is to be measured. The emf is measured most accurately with a potentiometer whose terminals must be of the same metal and be kept at the same temperature so that no additional emfs are produced.

Because thermocouple junctions can be made very small, they can be used to measure temperature in small objects such as particular parts of the body and even in insects (Fig. 18-20b).

A number of different combinations of dissimilar metals are used in thermocouples. The emf as a function of temperature can be found in tables

---

[†] According to theory this is due in part to electrons in one metal occupying lower energy states than in the other, so some of them flow across the junction. This leaves one metal slightly more positive than the other, and so a *contact potential* exists between them. If the two junctions are at the same temperature, the same contact potential exists at each; they balance each other and no current will flow. However, when one of the junctions is at a higher temperature, the energy states are altered and the contact potential will be different. In this case there will be a net emf and a current will flow.

(a)

(b)

FIGURE 18-20
Thermocouple.

for each set of metals and for a particular reference temperature. Over small temperature ranges the emf is approximately proportional to the temperature differences. However, this linear relationship cannot be assumed to be true in general and tables should be consulted. Incorrect measurements can also result from heat flow along the wires which can affect the temperature of the test junction.

A *microphone* is a transducer that changes a sound wave into an electrical signal. One type of microphone transducer is the capacitor (or condenser) microphone, Fig. 18-21. The changing air pressure in a sound wave causes one plate of the capacitor C to move back and forth. We saw in Chapter 16 that the capacitance is inversely proportional to the separation of the plates. Thus a sound wave causes the capacitance to change. This in turn causes the charge Q on the plates to change ($Q = CV$) so that an electric current is generated at the same frequencies as the incoming sound wave.

18-12 ▮ Electric hazards; leakage currents

An electric shock can cause damage to the body or may even be fatal. The severity of a shock depends on the magnitude of the current, how long it acts, and through what part of the body it passes. A current passing through vital organs such as the heart or brain is especially serious for it can interfere with their operation. Current heats tissue and can cause burns. A current also stimulates nerves and muscles (whose operation, as we have seen in Sections 16-10 and 17-7, is electrical) and we feel a "shock."

Most people can "feel" a current of about 1 mA. Currents of a few mA cause pain but rarely much damage in a healthy person. However, currents above 10 mA cause severe contraction of the muscles; in this case a person may not be able to release the source of the current (say a faulty appliance or wire). Death from paralysis of the respiratory system can occur; artificial respiration, however, can often revive a victim. If a current above about 70 mA passes across the torso so that a portion passes through the heart for a second or more, the heart muscles will begin to contract irregularly and blood is not properly pumped. This condition is called "ventricular fibrillation." If it lasts for long, death results. Strangely enough, if the current is

FIGURE 18-21
Diagram of a capacitor microphone.

much larger, on the order of 1 A, the damage may be less and death less likely.[†]

The seriousness of a shock depends on the effective resistance of the body. Living tissue has quite low resistance since the fluid of cells contains ions that can conduct quite well. However, the outer layers of skin, when dry, offer much resistance. The effective resistance between two points on opposite sides of the body when the skin is dry is in the range of 10^4 to 10^6 Ω. However, when the skin is wet the resistance may be 10^3 Ω or less. A person in good contact with the ground who touches a 120-V dc line with wet hands can suffer a current

$$I = \frac{120 \text{ V}}{1000 \text{ Ω}} = 120 \text{ mA}.$$

As we saw above, this could be lethal.

Figure 18-22 shows how the circuit is completed when a person touches an electric wire. One side of a 120-V source is connected to ground by a wire connected to a buried conductor (say, a water pipe). Thus the current passes from the high-voltage wire, through the person, to the ground; it passes through the ground back to the other terminal of the source to complete the circuit. If the person in Fig. 18-22 stands on a good insulator—thick-soled shoes or a dry wood floor—there will be much more resistance in the circuit and consequently much less current will flow. However, if the person stands with bare feet on the ground, or is sitting in a bathtub, there is considerable danger. In a bathtub, not only are you wet, but the water is in contact with the drain pipe that leads to the ground. That is why it is strongly recommended not to touch anything electrical in such a situation.

FIGURE 18-22

A person receives an electric shock when the circuit is completed.

A principal danger comes from touching a bare wire whose insulation has worn off, or from a bare wire inside an appliance when you're tinkering with it. (Always unplug an electrical device before investigating its insides!) Sometimes a wire inside a device breaks or loses its insulation and comes in contact with the case. If the case is metal, it will conduct electricity. A person could then suffer a severe shock merely by touching the case. To prevent an accident, most cases are supposed to be connected to ground. Then if a "hot" wire touches the case, a short circuit to ground immediately occurs and the fuse or circuit breaker opens the circuit.

The human body acts as if it had capacitance in parallel with its resistance. A dc current can pass through the resistance, but not the capacitance. An ac current, like the changing currents discussed in Section 18-6 (more on this in Chapter 20), can exist also in the capacitive branch. Because of the additional path allowing current flow, the current for a given V_{rms} will be greater than for the same dc voltage. Thus an ac voltage is more dangerous than an equal dc voltage.

[†] Apparently, larger currents bring the entire heart to a standstill. Upon release of the current the heart returns to its normal rhythm. This may not happen when fibrillation occurs since it is often hard to stop once it starts. Fibrillation may also occur as a result of a heart attack or during heart surgery. A device known as a *defibrillator* can apply a brief high current to the heart; this causes complete heart stoppage and is often followed by resumption of normal beating.

Another danger is *leakage current*, by which we mean a current along an unintended path. Leakage currents are often capacitively coupled. For example, a wire in a lamp forms a capacitor with the metal case; charges moving in one conductor attract or repel charge in the other, so there is a current. Typical electrical codes limit leakage currents to 1 mA for any device. A 1-mA leakage current is usually harmless. It can be very dangerous, however, to a hospital patient with implanted electrodes connected to ground through the apparatus. This is because the current can pass directly through the heart as compared to the usual situation where it spreads out through the body. Although 70 mA may be needed to cause heart fibrillation (rapid irregular contractions of muscle fibers) when entering through the hands (very little of it actually passes through the heart), as little as 0.02 mA has been known to cause fibrillation when passing directly to the heart. Thus a "wired" patient is in considerable danger from leakage current even from as simple an act as touching a lamp.

*18-13 ∎ Heart pacemakers

If the heart stops, it can be made to start beating again by applying an electric stimulus through electrodes attached to the chest. The stimulus can be repeated at the normal heartbeat rate if necessary.

The heart itself contains a *pacemaker*, which sends out tiny electric pulses at a rate of 60 to 80 per minute. These are the signals that induce the start of each heartbeat. In some forms of heart disease, the pacemaker cells fail to function properly and the heart loses its beat. *Electronic pacemakers* are now commonplace for people suffering this ailment. They can be external or can be inserted beneath the skin. They produce a regular voltage pulse that starts and controls the frequency of the heartbeat. The "fixed-rate" type produces signals continuously. The "demand" type operates only when the natural pacemaker fails. The electrodes are implanted in or near the heart and the circuit usually contains a capacitor and a resistor. The charge on the capacitor increases to a certain point and then discharges; then it starts charging again. The pulsing rate depends on the values of R and C. Generally, the power source is a battery which must be replaced or recharged, depending on type. Some pacemakers obtain their energy from the heat produced by a radioactive element; the thermal energy is transformed to electricity by a thermocouple. Another type uses the heart's own contractions as its source; the critical element is a piezoelectric crystal (see Section 18-11), which produces an emf in response to the pressure of the contracting heart.

∎ Summary

When resistances are connected in series, the net resistance is the sum of the individual resistances. When connected in parallel, the reciprocal of the total resistance equals the sum of the reciprocals of the individual resistances; in a parallel connection, the net resistance is less than any of the individual resistances.

A battery behaves like a source of emf in series with an internal resistance. The emf is the potential

433

difference determined by the chemical reactions in the battery and equals the terminal voltage only when no current is drawn. When a current is drawn, the voltage at the battery's terminals is less than its emf by an amount equal to the Ir drop across the internal resistance.

When capacitors are connected in parallel, the net capacitance is the sum of the individual capacitances. When connected in series, the reciprocal of the net capacitance equals the sum of the reciprocals of the individual capacitances. When a circuit containing a resistor R in series with a capacitance C is connected to a dc source of emf, the voltage across the capacitor rises gradually in time characterized by the time constant $\tau = RC$; this is the time it takes for the voltage to reach 63 percent of its maximum value. A capacitor discharging through a resistor is characterized by the same time constant: in a time $\tau = RC$, the voltage across the capacitor drops to 37 percent of its initial value.

An ammeter, which measures current, consists of a low resistance in parallel with a galvanometer whose deflection is proportional to the current passing through it. An ammeter is placed into a circuit, in series with the other elements. A voltmeter measures potential difference and consists of a high resistance in series with a galvanometer. A voltmeter is not connected directly into a circuit but is placed in parallel with the element across which the voltage is to be measured.

Electric shocks are caused by current passing through the body. To avoid shocks, the body must not become part of a circuit by allowing different parts of the body to touch objects at different potentials; commonly, one part of the body may be touching ground and another part a high or low potential.

▮ Questions

1. Discuss the advantages and disadvantages of Christmas tree lights connected in parallel versus those connected in series.

2. If all you have is a 120-V line, would it be possible to light several 6-V lamps without burning them out? How?

3. Two light bulbs of resistance R_1 and R_2 ($> R_1$) are connected in series. Which is brighter? What if they are connected in parallel?

4. Describe carefully the difference between emf and potential difference.

5. The internal resistance of an electric cell is not actually constant. Why not?

6. Given the circuit shown in Fig. 18-23, use the words "increases," "decreases," or "stays the same" to complete the following statements:
   (a) If R_7 increases, the potential difference between A and E (assume no resistance in Ⓐ and $\mathscr{E}$) _____.
   (b) If R_7 increases, the potential difference between A and E (assume Ⓐ and $\mathscr{E}$ have resistance) _____.
   (c) If R_7 increases, the voltage drop across R_4 _____.
   (d) If R_2 decreases, the current through R_1 _____.
   (e) If R_2 decreases, the current through R_6 _____.
   (f) If R_2 decreases, the current through R_3 _____.
   (g) If R_5 increases, the voltage drop across R_2 _____.
   (h) If R_5 increases, the voltage drop across R_4 _____.
   (i) If R_2, R_5, and R_7 increase, $\mathscr{E}$ _____.

7. Why are batteries connected in series? Why in parallel? Does it matter if the batteries are nearly identical or not in either case?

8. Describe a situation in which the terminal voltage of a battery is greater than its emf.

9. The 18-V source in Fig. 18-24 is "charging" the 12-V battery. Explain how it does this.

FIGURE 18-23

FIGURE 18-24

10. Suppose that three identical capacitors are connected to a battery. Will they store more energy if connected in series or in parallel?

11. What is the main difference between a voltmeter and an ammeter?

12. Explain why an ideal ammeter would have zero resistance and an ideal voltmeter infinite resistance.

*13. What are the advantages of a potentiometer over a voltmeter? What are the advantages of a voltmeter over a potentiometer?

*14. Explain why "null" measurements allow greater accuracy than meter-reading methods.

*15. If the battery of a Wheatstone bridge wears down so that its terminal voltage drops a little, will this affect the determination of the unknown resistance?

16. Why is it more dangerous to turn on an electric appliance when you are standing outside in bare feet than when you are inside wearing shoes with thick soles?

▌Problems

SECTION 18-1

In these problems neglect the internal resistance of a battery unless the problem refers to it.

1. (I) Six 200-Ω light bulbs are connected in series. What is the total resistance of the circuit? What is their resistance if they are connected in parallel?

2. (I) Suppose you have a 500-Ω, an 800-Ω, and a 1.20 kΩ resistor. What is (a) the maximum, and (b) the minimum resistance you can obtain by combining these?

3. (II) Suppose that you have a 6.0-V battery and you wish to apply a voltage of only 1.0 V. Given an unlimited supply of 1.0 Ω resistors, how could you connect them so as to make a "voltage divider" that produced a 1.0 V output for a 6.0-V input?

4. (II) Three 100-Ω resistors can be connected together in four different ways, making combinations of series

and/or parallel circuits. What are these four ways and what is the net resistance in each case?

5. (II) What is the net resistance of the circuit connected to the battery in Fig. 18-25? Each resistance has $R = 1.0$ kΩ.

6. (II) Eight lights are connected in series across a 120-V line. (a) What is the voltage across each bulb? (b) If the current is 0.50 A, what is the resistance of each bulb and the power dissipated in each?

7. (II) Eight lights are connected in parallel to a 120-V source by two leads of total resistance 2.0 Ω. If 100 mA flows through each bulb, what is the resistance of each, and what fraction of the total power is wasted in the leads?

8. (II) Seven 8.0-W Christmas tree lights are connected in series to a 120-V source. What is the resistance of each bulb?

9. (II) What is the internal resistance of a 12-V car battery whose terminal voltage drops to 7.5 V when the starter draws 80 A?

10. (II) Two resistors when connected in series to a 120-V line use one-fourth the power that is used when they are connected in parallel. If one resistor is 2.8 kΩ, what is the resistance of the other?

11. (II) Suppose that you wish to apply a 0.25-V potential difference between two points on the body. The resistance is about 2000 Ω, and you only have a 9.0-V battery. How can you connect up one or more resistors so that you can produce the desired voltage?

12. (II) A three-way light bulb can produce 50 W, 100 W, or 150 W at 120 V. Such a bulb contains two filaments that can be connected to the 120 V individually or in parallel. Describe how the connections to the two filaments are made to give each of the three wattages; what must be the resistance of each filament?

FIGURE 18-25

12 V

13. (II) A 75-W, 120-V bulb is connected in parallel with a 40-W, 120-V bulb. What is the net resistance?

14. (II) Suppose you want to run some apparatus that is 200 m from an electric outlet. Each of the wires connecting your apparatus to the 120-V source has a resistance per unit length of 0.0055 Ω/m. If your apparatus draws 3.2 A, what will be the voltage drop across the connecting wires and what voltage will be applied to your apparatus?

15. (II) Calculate the current through each resistor in Fig. 18-25 if each resistance $R = 2.10$ kΩ. What is the potential difference between points A and B?

16. (III) Suppose that a person's body resistance is 1000 Ω. (a) What current passes through the body when the person accidentally is connected to 100 V? (b) If there is an alternative path to ground whose resistance is 20 Ω, what current passes through the person? (c) If the voltage source can produce at most 1.4 A, how much current passes through the person in case (b)?

17. (III) A 1.2-kΩ and a 1.8-kΩ resistor are connected in parallel; this combination is connected in series with a 1.9-kΩ resistor. If each resistor is rated at $\frac{1}{2}$ W, what is the maximum voltage that can be applied across the whole network?

18. (III) How many $\frac{1}{2}$-W resistors, each of the same resistance, must be used to produce an equivalent 1.2-kΩ, 5-W resistor? What is the resistance of each, and how must they be connected?

SECTION 18-2

19. (I) A battery whose emf is 6.0 V and whose internal resistance is 0.50 Ω is connected to a circuit whose net resistance is 11.6 Ω. What is the terminal voltage of the battery?

20. (I) Four 1.5-V cells are joined in series to a 9.2-Ω device. The internal resistance of each cell is 0.30 Ω. What current flows to the device?

21. (II) What is the internal resistance of a 12-V car battery whose terminal voltage drops to 7.8 V when the starter draws 70 A?

22. (II) A 1.5-V dry cell can be tested by connecting it to a low-resistance ammeter; it should be able to supply at least 30 A. What is the internal resistance of the cell in this case?

23. (II) The internal resistance of a 1.35-V mercury cell is 0.030 Ω whereas that of a 1.5-V dry cell is 0.35 Ω. Explain why three mercury cells can more effectively power a 2-W hearing aid that requires 4.0 V than can three dry cells.

24. (II) What is the current in the 8.0-Ω resistor in Fig. 18-5a.

25. (III) A battery produces 50.0 V when 5.5 A is drawn from it and 58.2 V when 1.8 A is drawn. What is the emf and internal resistance of the battery?

*SECTION 18-3

*26. (II) Determine the terminal voltage of each battery in Fig. 18-24.

*27. (II) What is the potential difference between points a and d in Fig. 18-6.

*28. (II) What is the terminal voltage of each battery in Fig. 18-6?

*29. (III) Determine the current through each of the resistors in Fig. 18-26.

*30. (III) If the 30-Ω resistor in Fig. 18-26 were shorted out (resistance = 0), what then would be the current through the 10-Ω resistor?

*31. (III) Determine the currents I_1, I_2, and I_3 in Fig. 18-27. Assume the internal resistance of each battery is $r = 1.0$ Ω. What is the terminal voltage of the 6.0-V battery?

*32. (III) What would the current I_1 be in Fig. 18-27 if the 12-Ω resistor were shorted out ($r = 1.0$ Ω)?

SECTION 18-5

33. (I) Six 1.5-μF capacitors are connected in parallel. What is the equivalent capacitance? What is their equivalent capacitance if connected in series?

34. (I) A circuit contains a 3.0-μF capacitor. However, a technician decides 4.8 μF would be better. What size

FIGURE 18-26

FIGURE 18-27

FIGURE 18-28

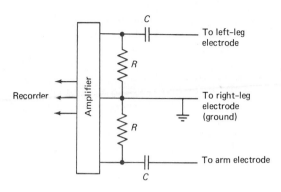

FIGURE 18-29

capacitor should be added to the circuit and how should it be connected?

35. (I) The capacitance of a portion of a circuit is to be reduced from 3600 pF to 1000 pF. What capacitance can be added to the circuit to produce this effect without removing anything from the circuit? How should the extra capacitor be connected?

36. (II) Suppose three parallel-plate capacitors, whose plates have areas A_1, A_2, and A_3 and separations d_1, d_2, and d_3, are connected in parallel. Show, using Eq. 16-6, that Eq. 18-3 is valid.

37. (II) (a) Determine the equivalent capacitance of the circuit shown in Fig. 18-28. (b) If $C_1 = C_2 = 2C_3 = 4.0$ μF, how much charge is stored on each capacitor when $V = 50$ V?

38. (II) In Fig. 18-28, let $V = 100$ V and $C_1 = C_2 = C_3 = 1200$ pF. How much energy is stored in the capacitor network?

39. (II) How much energy must a 12-V battery expend to fully charge a 0.10-μF and a 0.20-μF capacitor when they are placed (a) in parallel, (b) in series? (c) How much charge flowed from the battery in each case?

40. (II) You have three capacitors, of capacitance 2000 pF, 5000 pF, and 0.010 μF. What is the maximum and minimum capacitance that you can form from these? How do you make the connection in each case?

41. (II) A 0.20-μF and a 0.10-μF capacitor are connected in series to a 9.0-V battery. Calculate (a) the potential difference across each capacitor, and (b) the charge on each. (c) Repeat parts (a) and (b) assuming the two capacitors are in parallel.

42. (III) A 3.0-μF and a 4.0-μF capacitor are connected in series and this combination is connected in parallel with a 2.0-μF capacitor. (a) What is the net capacitance? (b) If 50 V is applied across the whole network, calculate the voltage across each capacitor.

43. (III) The variable capacitance of a radio tuner consists of six plates connected together placed alternately between six other plates, also connected together. Each plate is separated from its neighbor by 1.0 mm of air. One set of plates can move so the area of overlap varies from 1.0 cm^2 to 4.0 cm^2. (a) Are these eleven capacitors connected in series or in parallel? (b) Determine the range of capacitance values.

FIGURE 18-30

SECTION 18-6

44. (I) Electrocardiographs are often connected as shown in Fig. 18-29. The leads are said to be capacitively coupled. A time constant of 3.0 s is typical and allows rapid changes in potential to be accurately recorded. If $C = 3.0$ μF, what value must R have?

45. (II) Suppose a 4.0 μF capacitor were placed across the 6.0 Ω resistor in Fig. 18-5a. Calculate the charge on the capacitor in the steady state—that is, after the capacitor reaches its maximum charge.

46. (II) A heart pacemaker is designed to operate at 70 beats/min using a 7.0-μF capacitor. What value of resistance should be used if the pacemaker is to fire when the voltage reaches 63 percent of maximum?

47. (III) Two resistors and two uncharged capacitors are arranged as shown in Fig. 18-30. With a potential difference of 24 V across the combination, (a) what is the potential at point a with S open? (Let $V = 0$ at negative terminal of source.) (b) What is the potential at point b with the switch open? (c) When the switch is closed, what is the final potential of point b? (d) How much charge flows through the switch S when it is closed?

SECTION 18-7

48. (I) An ammeter has a sensitivity of 5000 Ω/V. What current passing through the galvanometer produces full-scale deflection?

49. (I) What is the resistance of a voltmeter on the 50-V scale if the meter sensitivity is 30,000 Ω/V?

50. (II) A galvanometer has a sensitivity of 50,000 Ω/V and internal resistance 24 Ω. How could you make this into (*a*) an ammeter reading 10 mA full scale; (*b*) a voltmeter reading 100 mV full scale?

51. (II) How could you make an ammeter with a full-scale deflection of 10 A from a galvanometer that has a 200-μA full-scale deflection when 10 mV is across it?

52. (II) A milliammeter reads 10 mA full scale. It consists of a 0.20-Ω resistor in parallel with a 30-Ω galvanometer. How can you change this ammeter to a voltmeter giving a full-scale reading of 10 V without taking the ammeter apart? What will be the sensitivity (Ω/V) of your voltmeter?

53. (II) A galvanometer has an internal resistance of 30 Ω and deflects full scale for a 60-μA current. Describe how to use this galvanometer to make (*a*) an ammeter to read currents up to 15 A; and (*b*) a voltmeter to give a full-scale deflection of 3000 V.

SECTION 18-8

54. (II) A 45-V battery of negligible internal resistance is connected to a 37-kΩ and a 22-kΩ resistor in series. What reading will a voltmeter give when used to measure the voltage across each resistor if its internal resistance is 100 kΩ? What is the percent inaccuracy due to meter resistance for each case?

55. (II) An ammeter whose internal resistance is 90 Ω reads 1.60 A when connected in a circuit containing a battery and two resistors in series whose values are 300 Ω and 500 Ω. What is the actual current when the ammeter is absent?

56. (II) A battery with $\mathscr{E} = 9.0$ V and internal resistance $r = 1.2\ \Omega$ is connected to two 5.0-kΩ resistors in series. An ammeter of internal resistance 0.50 Ω measures the current and at the same time a voltmeter with internal resistance 15 kΩ measures the voltage across one of the 5.0-kΩ resistors in the circuit. What do the ammeter and voltmeter read?

57. (III) Two 5.9-kΩ resistors are placed in series and connected to a battery. A voltmeter of sensitivity 1000 Ω/V is on the 3.0-V scale and reads 2.0 V when placed across either of the resistors. What is the emf of the battery? (Ignore the battery's internal resistance.)

58. (III) What internal resistance should the voltmeter have to be in error by less than 3 percent for the situation of Example 18-6?

59. (III) The voltage across a 120-kΩ resistor in a circuit containing additional resistance (R_2) in series with a battery (V) is measured, by a 20,000-Ω/V meter on the 100-V scale, to be 25 V. On the 30-V scale, the reading is 23 V. What is the actual voltage in the absence of the voltmeter? What is R_2?

60. (III) One type of *ohmmeter* consists of an ammeter connected to a series resistor and a battery. Fig. 18-31.

FIGURE 18-31

The scale is different from ammeters and voltmeters. Zero resistance corresponds to full-scale deflection (because maximum current flows from the battery) whereas infinite resistance corresponds to no deflection (no current flow). Suppose a 9.0-V battery is used, and the galvanometer has internal resistance of 30 Ω and deflects full scale for a current of 30 μA. What values of shunt resistance, R_{sh}, and series resistance, R_{ser}, are needed to make an ohmmeter that registers a midscale deflection (that is, half of maximum) for a resistance of 30 kΩ? (Note: an additional series resistor (variable) is also needed so the meter can be zeroed; this should be checked frequently by touching the leads together since the battery voltage can vary. Because of battery voltage variation, such ohmmeters are not precision instruments, but they are useful to obtain approximate values.)

61. (III) A 9.0-V battery (internal resistance $r = 1.0\ \Omega$) is connected to two resistors in series. A voltmeter whose internal resistance is 10.0 kΩ measures 3.0 V and 4.0 V, respectively, when connected across each of the resistors. What is the resistance of each resistor?

*SECTION 18-9

***62. (I)** A slide-wire potentiometer is balanced against a 1.0182-V standard cell when the slide wire is set at 40.2 cm out of a total length of 100.0 cm. For an unknown source, the setting is 11.9 cm. What is the emf of the unknown?

***63. (I)** The galvanometer of a potentiometer has an internal resistance of 40 Ω and can detect a current as small as 0.015 mA. What is the minimum uncertainty possible in measuring an unknown voltage?

***64. (II)** The uniform 30.00-cm-long slide wire of a potentiometer has a total resistance of 3.224 Ω. When a 1.0183-V standard cell is connected, a 22.40-cm length of wire is needed to achieve balance. When an unknown source is connected, balance is achieved only when a 9.050-Ω resistor previously connected between the working battery and point *A* in Fig. 18-17 is now included along with the slide-wire resistance, and the slide wire is set at 14.16 cm. What is the value of the unknown emf?

FIGURE 18-32

Slide-wire form of Wheatstone bridge.

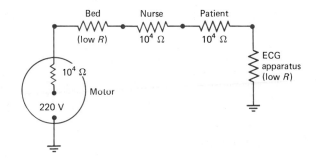

FIGURE 18-33

*65. (III) A potentiometer slide wire must be set at a 48.0-cm length to balance the emf of a cell. When a 5.0-Ω resistor is placed in parallel across the terminals of the cell, a 44.0-cm length is needed. What is the internal resistance of the cell?

*SECTION 18-10

*66. (I) A Wheatstone bridge is balanced when $R_1 = 710\ \Omega$, $R_2 = 317\ \Omega$, and $R_3 = 2.24\ k\Omega$. What is the value of the unknown resistance in the fourth arm?

*67. (I) Figure 18-32 shows a slide-wire form of Wheatstone bridge. If $R_3 = 34.5\ \Omega$ and the lengths $AC = 28.2$ cm and $CB = 71.8$ cm when the galvanometer reads zero current, what is the value of R_x? Assume the slide wire is uniform.

*68. (II) An unknown length of platinum wire 1.2 mm in diameter is placed as the unknown resistance in a Wheatstone bridge. Arms 1 and 2 have resistance of $36.0\ \Omega$ and $84.0\ \Omega$, respectively. Balance is achieved when R_3 is $4.16\ \Omega$. How long is the platinum wire?

*SECTION 18-11

*69. (I) A copper-constantan thermocouple produces emfs of about 40 μV/C°. If the reference temperature is 25°C, what must be the temperature of the test junction if the emf produced is 1.82 mV?

*70. (I) For iron-copper junctions near room temperature, the emf produced by a thermocouple is about 14 μV/C°. If a potentiometer can detect emfs as low as 0.50 μV, to what accuracy can the temperature be read?

*71. (II) The emf of a thermocouple can be written $\mathscr{E} = aT + bT^2$, where T is the temperature (°C) to be measured and the reference temperature is 0°C; a and b are constants, and for a purely linear response b would be

zero (usually it is small). (a) Determine the values for a and b if the emf is 3.62 mV at $T = 80$°C and is 6.80 mV at $T = 170$°C. (b) What is the temperature if the emf is 4.90 mV?

*72. (II) The *strain factor* K of a strain gauge is defined as the fractional change in resistance ($\Delta R/R$) divided by the fractional change in length:

$$K = \frac{\Delta R/R}{\Delta L/L},$$

and is relatively constant at a given temperature. In other words, the change in resistance ΔR is proportional to the change in length of the wire ΔL. (a) Show that this linear relationship is reasonable for small $\Delta L/L$. (b) A strain gauge with a strain factor of 1.8 is connected crosswise to a small muscle about 4.5 mm wide. The gauge is connected as the unknown arm in a Wheatstone bridge. When the muscle is relaxed, the Wheatstone bridge is balanced when $R_2/R_1 = 1.4800$ and $R_3 = 40.700\ \Omega$. When the muscle contracts, balance occurs for $R_3 = 40.736\ \Omega$. How much has the muscle widened?

SECTION 18-12

73. (II) Electricity can be a hazard in hospitals, particularly to patients who are connected to electrodes, such as an ECG. For example, suppose that the motor of a motorized bed shorts out to the bed frame, and the bed frame's connection to a ground has broken (or was not there in the first place). If a nurse touches the bed and the patient at the same time, she becomes a conductor and a complete circuit can be made through the patient to ground through the ECG apparatus. This is shown schematically in Fig. 18-33. Calculate the current through the patient.

19 MAGNETISM

Today it is clear that magnetism and electricity are closely related. This relationship was not discovered, however, until the nineteenth century. The history of magnetism begins much earlier with the ancient civilizations in Asia Minor. It was in a region of Asia Minor known as Magnesia that rocks were found that would attract each other. These rocks were called "magnets" after their place of discovery.

19-1 ▌ Magnets and magnetic fields

A magnet will attract paper clips, nails, and other objects made of iron. Any magnet, whether it is in the shape of a bar or a horseshoe, has two ends or faces, called poles; this is where the magnetic effect is strongest. If a magnet is suspended from a fine thread, it is found that one pole of the magnet will always point toward the north. It is not known for sure when this fact was discovered, but it is known that the Chinese were making use of it as an aid to navigation by the eleventh century and perhaps earlier. This is, of course, the principle of a compass. A compass needle is simply a magnet that is supported at its center of gravity so it can rotate freely. That pole of a freely suspended magnet which points toward the north is called the *north pole* of the magnet. The other pole points toward the south and is called the *south pole*.

It is a familiar fact that when two magnets are brought near one another, each exerts a force on the other. The force can be either attractive or repulsive and can be felt even when the magnets don't touch. If the north pole of one magnet is brought near the north pole of a second magnet, the force is repulsive. Similarly, if two south poles are brought close, the force is repulsive. But when a north pole is brought near a south pole, the force is attractive, Fig. 19-1. This is reminiscent of the force between electric charges; like poles repel and unlike poles attract. But do not confuse magnetic poles with electric charge. They are not the same thing.

Only iron and a few other materials such as cobalt, nickel, and gadolinium show strong magnetic effects. They are said to be *ferromagnetic* (from the Latin word *ferrum* for iron). All other materials show some slight mag-

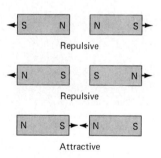

FIGURE 19-1

Like poles of a magnet repel; unlike poles attract.

FIGURE 19-2

Plotting a magnetic field line of a bar magnet.

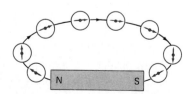

440

netic effect, but it is extremely small and can be detected only with delicate instruments. (We will look in more detail at ferromagnetism in Sections 19-3 and 19-14.)

We found it useful to speak of an electric field surrounding an electric charge. In the same way we can imagine a *magnetic field* surrounding a magnet. The force one magnet exerts on another can then be described as the interaction between one magnet and the magnetic field of the other. Just as we drew electric field lines, we can also draw magnetic field lines. They can be drawn, as for electric field lines, so that (1) the direction of the magnetic field is tangent to a line at any point, and (2) the number of lines per unit area is proportional to the magnitude of the magnetic field.

The *direction* of the magnetic field at a given point can be defined as the direction that the north pole of a compass needle would point when placed at that point. Figure 19-2 shows how one magnetic field line around a bar magnet is found using compass needles. The magnetic fields determined in this way for the field outside a horseshoe magnet and a bar magnet are shown in Fig. 19-3. Notice that because of our definition, the lines always point from the north toward the south pole of a magnet (the north pole of a magnetic compass needle is attracted to the south pole of another magnet).

We can define the **magnetic field** at any point as a vector, represented by the symbol **B**, whose direction is as defined above; the *magnitude* of **B** can be defined roughly in terms of the torque exerted on a compass needle when it makes a certain angle with the magnetic field, as in Fig. 19-4. That is, the greater the torque, the greater the magnetic field strength. We can use this (rather rough) definition for now, but a more precise definition will be given in Section 19-5. The terms "magnetic flux density" and "magnetic induction" are often used for **B** rather than our simple term "magnetic field."

The earth's magnetic field is shown in Fig. 19-5. Since the north pole of a compass needle points north, the magnetic pole which is in the geographic north is magnetically a south pole (remember that the north pole of one magnet is attracted to the south pole of a second). And the earth's magnetic pole near the geographic south pole is magnetically a north pole. The earth's magnetic poles do not, however, coincide with the geographic poles (which are on the earth's axis of rotation). The magnetic south pole, for example, is in northern Canada, about 1500 km from the geographic north pole. This must be taken into account when using a compass. The angular difference between magnetic north and true (geographical) north is called the "magnetic declination." In the United States it varies from 0° to about 25°, depending on location. Notice in Fig. 19-5 that the earth's magnetic field is not tangent

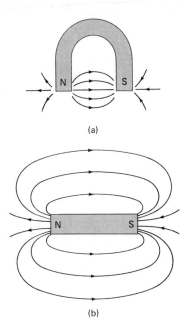

FIGURE 19-3

Magnetic field lines outside of (a) a horseshoe magnet, and (b) a bar magnet.

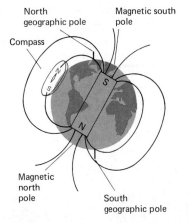

FIGURE 19-5

The earth acts like a huge magnet with its magnetic south pole near the geographic north pole.

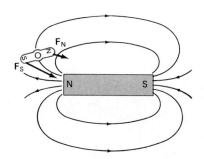

FIGURE 19-4

Forces on a compass needle that produce a torque to orient it parallel to the magnetic field lines. The torque will be zero when the needle is parallel to the magnetic field line at that point. (Only the attractive forces are shown; try drawing in the repulsive forces and show that they produce a similar torque.)

FIGURE 19-6

Deflection of a compass needle near a current-carrying wire, showing the presence and direction of the magnetic field.

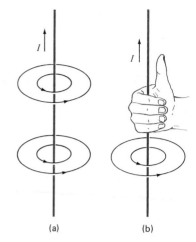

(a)   (b)

FIGURE 19-7

(a) Magnetic field lines around an electric current in a straight wire. (b) Right-hand rule for remembering the direction of the magnetic field: when the thumb points in the direction of the conventional current, the fingers wrapped around the wire point in the direction of the magnetic field.

to the earth's surface at all points. The angle **B** makes with the horizontal at any point is referred to as the angle of dip.

FIGURE 19-8

Magnetic field due to a circular loop of wire.

19-2 ▌ Electric currents produce magnetism

During the eighteenth century many natural philosophers sought to find a connection between electricity and magnetism. The first to uncover a significant connection was Hans Christian Oersted (1777–1851) in 1820. Oersted had believed for a long time in the unity of nature. Philosophically, he felt there ought to be a connection between magnetism and electricity. However, a stationary electric charge and a magnet had been shown not to have any influence on each other. But Oersted found that when a compass needle is placed near an electric wire, the needle deflects as soon as the wire is connected to a battery and a current flows.

As we have seen, a compass needle can be deflected by a magnetic field. What Oersted found was that *an electric current produces a magnetic field*. He had found a connection between electricity and magnetism.

A compass needle placed near a straight section of current-carrying wire aligns itself so it is tangent to a circle drawn around the wire, Fig. 19-6. Thus, the magnetic field lines produced by a current in a straight wire are in the form of circles with the wire at their center, Fig. 19-7a. The direction of these lines is indicated by the north pole of the compass in Fig. 19-6. There is a simple way to remember the direction of the magnetic field lines in this case. It is called a *right-hand rule:* You grasp the wire with your right hand so that your thumb points in the direction of the conventional (positive) current; then your fingers will encircle the wire in the direction of the magnetic field, Fig. 19-7b. The magnetic field lines due to a circular loop of current-carrying wire can be determined in a similar way using a compass. The result is shown in Fig. 19-8. Again the right-hand rule can be used, as shown in Fig. 19-9.

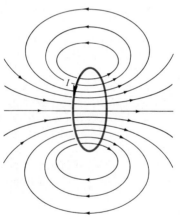

FIGURE 19-9

Right-hand rule for determining direction of magnetic field relative to current.

Magnetic field

19-3 ▌ Ferromagnetism; domains

We saw in Section 19-1 that iron (and a few other materials) can be made into strong magnets; they are said to be *ferromagnetic*. We now look more deeply into the sources of ferromagnetism.

A bar magnet, with its two opposite poles at either end, resembles an electric dipole (equal-magnitude positive and negative charges separated by a distance). Indeed, a bar magnet is sometimes referred to as a "magnetic dipole." There are opposite "poles" separated by a distance; and the magnetic field lines of a bar magnet form a pattern much like that for the electric field of an electric dipole: compare Fig. 15-20a with Fig. 19-3b. One important difference, however, is that a positive or negative electric charge can easily be isolated. But the isolation of a single magnetic pole has proved much more difficult. If a bar magnet is cut in half, you do not obtain isolated north and south poles; instead, two new magnets are produced, Fig. 19-10. If the cutting operation is repeated, more magnets are produced, each with a north and a south pole. Physicists have tried various ways to isolate a single magnetic pole, and this is an active research field today since certain theories suggest they ought to exist. But so far there is no firm experimental evidence for their existence.

Microscopic examination reveals that a magnet is actually made up of tiny regions known as *domains*, which are at most about 1 mm in length or width. Each domain behaves like a tiny magnet with a north and a south pole. In an unmagnetized piece of iron, these domains are arranged randomly as shown in Fig. 19-11a. The magnetic effects of the domains cancel each other out, so this piece of iron is not a magnet. In a magnet, the domains are preferentially aligned in one direction as shown in Fig. 19-11b (downward in this case). A magnet can be made from an unmagnetized piece of iron by placing it in a strong magnetic field. (You can make a needle magnetic, for example, by stroking it with one pole of a strong magnet.) Careful observations show in this case that the magnetization of domains may actually rotate slightly so as to be more nearly parallel to the external field; or, more commonly, the borders of domains move so that those domains whose magnetic orientation is parallel to the external field grow in size at the expense of other domains. This can be seen by comparing Figs. 19-11a and b. This explains how a magnet can pick up unmagnetized pieces of iron like paper clips or bobby pins. The magnet's field causes a slight alignment of the domains in the unmagnetized object so that the object becomes a temporary magnet with its north pole facing the south pole of the permanent magnet, and vice versa; thus attraction results. In the same way, elongated iron filings will arrange themselves in a magnetic field just as a compass needle does, and will reveal the shape of the magnetic field, Fig. 19-12.

An iron magnet can remain magnetized for a long time, and thus it is referred to as a "permanent magnet." However, if you drop a magnet on the floor or strike it with a hammer, you may jar the domains into randomness; the magnet can thus lose some or all of its magnetism. Heating a magnet too can cause a loss of magnetism; for raising the temperature increases the random thermal motion of the atoms which tends to randomize the domains.

FIGURE 19-10

If you break a magnet in half, you do not obtain isolated north and south poles; instead, two new magnets are produced, each with a north and south pole.

FIGURE 19-11

(a) Unmagnetized pieces of iron are made up of domains that are randomly arranged. (b) In a magnet, the domains are preferentially aligned in one direction. (The tip of each arrow represents the north pole of the domain.)

(a)   (b)

Above a certain temperature known as the *Curie temperature* (1043 K for iron), a magnet cannot be made at all.[†]

There is a striking similarity between the fields produced by a bar magnet and by a loop of electric current (compare Fig. 19-3b with Fig. 19-8). This suggests that the magnetic field produced by a current may have something to do with ferromagnetism, an idea proposed by Ampère in the nineteenth century. According to modern atomic theory, the atoms that make up any material can be roughly visualized as containing electrons that orbit around a central nucleus. Since the electrons are charged, they constitute an electric current and therefore produce a magnetic field. But if there is no external field, the electron orbits in different atoms are arranged so the magnetic effects due to the many orbits in all the atoms in a material effectively cancel out. However, electrons produce an additional magnetic field, almost as if they were spinning about their own axes. It is the magnetic field due to electron spin[‡] that is believed to produce ferromagnetism. In most materials, the magnetic fields due to electron spin cancel out. But in iron and other ferromagnetic materials, a complicated cooperative mechanism operates; the result is that the electrons contributing to the ferromagnetism in a domain "spin" in the same direction. Thus the tiny magnetic fields due to each of the electrons add up to give the magnetic field of a domain. And when the domains are aligned, as we have seen, a strong magnet results.

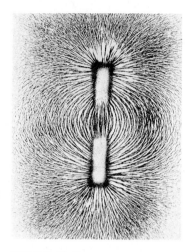

FIGURE 19-12

Iron filings line up along magnetic field lines.

It is believed possible today that *all* magnetic fields are caused by electric currents. This would explain why it has proved difficult to find a single magnetic pole. There is no way to divide up a current and obtain a single magnetic pole. Of course if an isolated pole is found, we will have to alter the idea that all magnetic fields are produced by currents.

19-4 ∎ Electromagnets and solenoids

A long coil of wire consisting of many loops of wire is called a *solenoid*. The magnetic field within a solenoid can be fairly large since it will be the sum of the fields due to the current in each loop (see Fig. 19-13). The solenoid acts like a magnet; one end can be considered the north pole and the other the south pole, depending on the direction of the current in the loops (use the right-hand rule). Since the magnetic field lines leave the north pole of a magnet, the north pole of the solenoid in Fig. 19-13 is on the right.

If a piece of iron is placed inside a solenoid, the magnetic field is increased greatly because the domains of the iron are aligned by the magnetic field of the current. The resulting magnetic field is the sum of that due to the current and that due to the iron; it can be hundreds or thousands of times that due to the current alone (see Section 19-14). This arrangement is

FIGURE 19-13

Magnetic field of a solenoid. The north pole of this solenoid, thought of as a magnet, is on the right, and the south pole is on the left.

[†] Iron, nickel, cobalt, gadolinium, and certain alloys are ferromagnetic at room temperature; several other elements and alloys have low Curie temperature and thus are ferromagnetic only at low temperatures.

[‡] The name "spin" comes from the early suggestion that the additional magnetic field arises from the electron "spinning" on its axis (as well as "orbiting" the nucleus) and this additional motion of the charge was supposed to produce the extra field. However this view of a spinning electron is completely discredited today. See Chapter 27.

called an *electromagnet*. The iron used in electromagnets acquires and loses its magnetism quite readily when the current is turned on or off; this is called "soft iron." (It is only "soft" in a magnetic sense.) Iron that holds its magnetism even when there is no externally applied field is called "hard iron." Hard iron is used in permanent magnets. Soft iron is usually used in electromagnets so that the field can be turned on and off readily. Whether iron is hard or soft depends on heat treatment and other factors.

Electromagnets find use in many practical applications, from use in motors and generators to producing large magnetic fields for research. Because the current flows continually, a great deal of waste heat (I^2R power) is often produced. Cooling coils, which are tubes carrying water, must be used to absorb the heat in bigger installations. For some applications, superconducting magnets are coming into use. The current-carrying wires are made of superconducting material (Section 17-4) kept below its transition temperature. Essentially no electric power is needed to maintain large current, which means large savings of electricity. Of course, energy is needed to keep the superconducting coils at the necessary low temperature.

Another useful device consists of a solenoid into which a rod of iron is partially inserted. This combination is also referred to as a solenoid. One use is as a doorbell (Fig. 19-14). When the circuit is closed by pushing the button, the coil effectively becomes a magnet and exerts a force on the iron rod. The rod is pulled into the coil and strikes the bell. A larger solenoid is used in the starters of cars; when you engage the starter, you are closing a circuit that not only turns the starter motor but activates a solenoid that first moves the starter into contact with the engine. Solenoids are used as switches in many other devices, such as tape recorders. They have the advantage of moving mechanical parts quickly and accurately.

FIGURE 19-14
Solenoid used as a doorbell.

19-5 ▌ Force on an electric current in a magnetic field; definition of **B**

In the Section 19-2 we saw that an electric current exerts a force on a magnet, such as a compass needle. By Newton's third law we might expect the reverse to be true as well: we should expect that *a magnet exerts a force on a current carrying wire*. This is indeed the case and this effect was also discovered by Oersted.

Let us look at the force exerted on a wire in detail. Suppose a straight wire is placed between the pole pieces of a magnet as shown in Fig. 19-15. When a current flows in the wire, a force is exerted on the wire; but this force is *not* toward one or the other poles of the magnet. Instead, the force is directed *at right angles to the magnetic field direction*. If the current is reversed in direction, the force is in the opposite direction. It is found that the direction of the force is always perpendicular to the direction of the current-carrying wire and also perpendicular to the direction of the magnetic field, **B**. This description does not completely describe the direction, however (the force could be either up or down in Fig. 19-15b and still be perpendicular to both the wire and to **B**). Experimentally the direction of the force is given by the right-hand rule illustrated in Fig. 19-15b. First you orient your right hand so that the outstretched fingers point in the direction of the

Right-hand rule

(a)   (b)

FIGURE 19-15

Force on a current-carrying wire placed in a magnetic field **B**.

(conventional) current; from this position, when you bend your fingers they should then point in the direction of the magnetic field lines; if they do not, rotate your hand and arm about the wrist until they do, remembering that straightened fingers must point along the direction of the current. When your hand is oriented in this way, then the extended thumb points in the direction of the force on the wire.

This describes the direction of the force. Now what about its magnitude? It is found experimentally that the magnitude of the force is directly proportional to the current I in the wire, to the length l of wire in the magnetic field (assumed uniform), and to the magnetic field B. The force also depends on the angle θ between the wire and the magnetic field. When the wire is perpendicular to the field lines the force is strongest; when the wire is parallel to the magnetic field lines there is no force at all. And at other angles it is proportional to $\sin \theta$ (Fig. 19-16). Thus we have

$$F \propto IlB \sin \theta.$$

Up to now we have talked of magnetic field strengths in terms of the torque exerted by the field on a compass needle (Fig. 19-4). But we have not defined the magnetic field strength precisely. In fact, the magnetic field B is defined in terms of the above proportion so that the proportionality constant is precisely 1. Thus we have

$$F = IlB \sin \theta. \qquad (19\text{-}1)$$

If the wire is perpendicular to the field ($\theta = 90°$) then the force is

$$F_{\text{max}} = IlB. \qquad [l \perp \mathbf{B}] \quad (19\text{-}2)$$

And if the wire is parallel to the field ($\theta = 0°$), the force is zero[†]

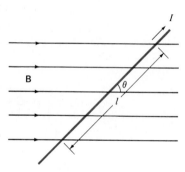

FIGURE 19-16

Current-carrying wire in a magnetic field.

[†] Both the magnitude and direction of the force **F** can be described in a single equation by using the cross-product notation we discussed in (the optional) Section 7-7:

$$\mathbf{F} = I\mathbf{l} \times \mathbf{B},$$

where l is a vector whose magnitude is the length of the wire and whose direction is along the wire (assumed straight) in the direction of the conventional (positive) current.

In our discussion we have assumed that the magnetic field is uniform. If it is not, then B in Eqs. 19-1 and 19-2 is the average field over the length l of the wire. In practical cases one can consider a wire as made up of many short segments Δl and the force on each segment is proportional to the length Δl of that segment and the magnetic field B at that segment; the total force is the sum of the individual forces.

In summary, the magnetic field vector **B** is defined as follows. The direction of **B** is the direction in which the force acting on a straight section of current-carrying wire is zero ($\theta = 0$ in Eq. 19-1). The magnitude of **B** is defined (from Eq. 19-2) as

$$B = \frac{F_{max}}{Il}$$

where F_{max} is the maximum value of the force on a straight length l of wire carrying a current I; F_{max} is obtained when the wire is perpendicular to **B**.

The SI unit for magnetic field B is the *tesla* (T). From Eq. 19-1, it is clear that $1\ T = 1\ N/A \cdot m$. An older name for the tesla is the "weber per meter squared" ($1\ Wb/m^2 = 1\ T$). Another unit commonly used to specify magnetic field is a cgs unit, the gauss (G): $1\ G = 10^{-4}\ T$. A field given in gauss should always be changed to teslas before using with other SI units. To get a "feel" for these units, we note that the magnetic field of the earth at its surface is about $\frac{1}{2}\ G$ or $0.5 \times 10^{-4}\ T$. On the other hand, strong electromagnets can produce fields on the order of a couple teslas and superconducting magnets over 10 T.

▯ **EXAMPLE 19-1** A wire carrying a 30-A current has a length $l = 12$ cm between the pole faces of a magnet at an angle $\theta = 60°$ (Fig. 19-16). The magnetic field is approximately uniform at 0.90 T. We ignore the field beyond the pole pieces. What is the force on the wire?

SOLUTION We use Eq. 19-1 and find that

$$F = IlB \sin\theta$$
$$= (30\ A)(0.12\ m)(0.90\ T)(0.866) = 2.8\ N.$$

▯ **EXAMPLE 19-2** A rectangular loop of wire hangs vertically as shown in Fig. 19-17. A magnetic field **B** is directed horizontally, perpendicular to the wire, and points out of the paper at all points as represented by the symbol ⊙. (The ⊙ is meant to resemble the tip of an arrow heading at us; if the field were pointing into the paper, we would represent this using the symbol X to resemble the tail of an arrow going away.) The magnetic field **B** is very nearly uniform along the horizontal portion of wire ab (length $l = 10.0$ cm) which is near the center of a large magnet producing the field. The top portion of the wire loop is free of the field. The loop hangs from a balance which measures a downward force $F = 3.48 \times 10^{-2}$ N when the wire carries a current $I = 0.245$ A. What is the magnitude of the magnetic field B at the center of the magnet?

SOLUTION The magnetic forces on the two vertical sections of the wire loop are in equal and opposite directions. Hence, the net magnetic force on the loop is that on the horizontal section ab whose length $l = 0.100$ m:

$$B = \frac{F}{Il} = \frac{3.48 \times 10^{-2}\ N}{(0.245\ A)(0.100\ m)} = 1.42\ T.$$

This is, incidentally, a highly precise means of determining magnetic fields.

FIGURE 19-17

Measuring a magnetic field B (Example 19-2).

19-6 ■ Force on an electric charge moving in a magnetic field

We have seen that a current-carrying wire experiences a force when placed in a magnetic field. Since a current in a wire consists of moving electric charges, we might expect that freely moving charged particles (not in a wire) would also experience a force when passing through a magnetic field. Indeed, this is the case.

From what we already know, let us calculate the force on a single moving electric charge. If n such particles of charge q pass by a given point in time t, they constitute a current $I = nq/t$. We let t be the time for a charge q to travel a distance l in a magnetic field B; then $l = vt$, where v is the velocity of the particle. Thus the force on these n particles is, by Eq. 19-1, $F = IlB \sin \theta = (nq/t)(vt)B \sin \theta = nqvB \sin \theta$. The force on *one* of the n particles is then

$$F = qvB \sin \theta. \qquad (19\text{-}3)$$

This equation gives the magnitude of the force on a particle of charge q moving with velocity v in a magnetic field of strength B; θ is the angle between **v** and **B**. The force is greatest when the particle moves perpendicular to **B** ($\theta = 90°$):

$$F_{max} = qvB. \qquad [\mathbf{v} \perp \mathbf{B}] \quad (19\text{-}4)$$

The force is *zero* if the particle moves *parallel* to the field lines ($\theta = 0°$). The *direction* of the force is perpendicular to the magnetic field **B** and to the velocity **v** of the particle. It is again given by a right-hand rule: you orient your right hand so that your outstretched fingers point along the direction of motion of the particle (**v**) and when you bend your fingers they must point along the direction of **B**; then your thumb will point in the direction of the force.[†] This is true only for *positively* charged particles, and will be "down" for the situation of Fig. 19-18. For negatively charged particles the force is in exactly the opposite direction ("up" in Fig. 19-18).

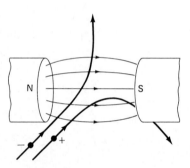

FIGURE 19-18
Force on charged particles due to a magnetic field.

□   **EXAMPLE 19-3**   A proton having a speed of 5.0×10^6 m/s in a magnetic field feels a force of 8.0×10^{-14} N toward the west when it moves vertically upward. When moving horizontally in a northerly direction, it feels zero force. What is the magnitude and direction of the magnetic field in this region?

SOLUTION   Since the proton feels no force when moving north, the field must be in a north-south direction. The right-hand rule tells us that **B** must point toward the north in order to produce a force to the west when

---

[†] Using cross-product notation (see footnote in previous section, and Section 7-7) the force can be written $\mathbf{F} = q\mathbf{v} \times \mathbf{B}$. If both an electric field **E** and a magnetic field **B** are present at a point, the force on a particle of charge q moving with velocity **v** will be

$$\mathbf{F} = q\mathbf{E} + q\mathbf{v} \times \mathbf{B}$$

(see Eq. 15-2). This relation is called the Lorentz equation.

the proton moves upward. (Your thumb points west when your right hand is oriented so that your outstretched fingers point upward and your bent fingers point north.) The magnitude of **B**, from Eq. 19-3 with $\theta = 90°$, is

$$B = \frac{F}{qv} = \frac{8.0 \times 10^{-14} \text{ N}}{(1.6 \times 10^{-19} \text{ C})(5.0 \times 10^6 \text{ m/s})} = 0.10 \text{ T}.$$

The path of a charged particle moving in a plane perpendicular to a uniform magnetic field is a circle (or the arc of a circle if the particle leaves the magnetic field region). This can be seen from Fig. 19-19. In this drawing, the magnetic field is directed *into* the paper as represented by x's. An electron at point P is moving to the right, and the force on it is therefore downward as shown. The electron is thus deflected downward. A moment later, say when it reaches point Q, the force is still perpendicular to the velocity and is in the direction shown. Since the force is always perpendicular to **v**, the magnitude of **v** does not change and the particle moves in a circular path. The force is directed toward the center of this circle at all points. Note that the electron moves clockwise in Fig. 19-19. A positive particle would feel a force in the opposite direction and would thus move in a counterclockwise circle.

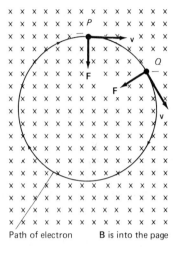

Path of electron   B is into the page

FIGURE 19-19

Force exerted by a uniform magnetic field on a moving charged particle (in this case an electron) produces a curved path.

☐   **EXAMPLE 19-4**   An electron travels at 2.0×10^7 m/s in a plane perpendicular to a 0.10-T magnetic field. Describe its path.

SOLUTION   The radius of the curved path is found using Newton's second law, $F = ma$, and the fact that the acceleration of a particle moving in a circle is $a = v^2/r$. Thus

$$F = ma$$

$$qvB = \frac{mv^2}{r}.$$

We solve for r and find:

$$r = \frac{mv}{qB}.$$

Since **F** is perpendicular to **v**, the magnitude of **v** doesn't change. From this equation we see that if **B** = constant, then r = constant and the curve must be a circle as claimed above. To get r we put in the numbers:

$$r = \frac{(9.1 \times 10^{-31} \text{ kg})(2.0 \times 10^7 \text{ m/s})}{(1.6 \times 10^{-19} \text{ C})(0.10 \text{ T})} = 1.1 \times 10^{-3} \text{ m},$$

or 1.1 mm.

*19-7 ∎ The Hall effect

When a current-carrying conductor is held firmly in a magnetic field, the field exerts a sideways force on the charges moving in the conductor. For example, if electrons move to the right in the rectangular conductor shown in Fig. 19-20a, the inward magnetic field will exert a downward force on the electrons; so they will tend to move nearer face S than face R. There will thus be a potential difference between faces R and S of the conductor. This potential difference builds up until the electric field it produces exerts a force on the moving charges that is equal and opposite to the magnetic force. This effect is called the *Hall effect* after E. H. Hall, who discovered it in 1879. The difference of potential produced is called the *Hall emf.*

Now a current of negative charges moving to the right is equivalent to positive charges moving to the left, at least for most purposes. But the Hall effect can distinguish these two. As can be seen in Fig. 19-20b, positive particles moving to the left are deflected downward, so that the bottom surface is positive relative to the top surface. This is the reverse of part (a). Indeed, the direction of the emf in the Hall effect first revealed that it is negative particles that move in most conductors. In some semiconductors, however, the Hall effect reveals that the carriers of current are positive. (More on this in Chapter 28.)

The magnitude of the Hall emf is proportional to the current and the strength of the magnetic field. The Hall effect can thus be used to measure magnetic field strengths. First the conductor, called a *Hall probe*, is calibrated with known magnetic fields. Then, for the same current, its emf output will be a measure of B. Hall probes can be made very small and are convenient and accurate to use.

The force on a particle in a magnetic field is the idea behind "electromagnetic pumping." It is used to pump metallic fluids (such as sodium) and more recently to pump blood in artificial heart machines. The basic design is shown in Fig. 19-21. An electric field is applied perpendicular to a blood vessel and to a magnetic field. The ions in the blood move vertically in the drawing as a result of the electric field. The magnetic field then exerts a sideways force on the ions that causes the fluid to flow along the blood vessel. Do positive and negative ions feel a force in the same direction?

19-8 ▮ Applications—meters, motors, loudspeakers

A number of important practical devices make use of the force that exists between a current and a magnetic field. In most of these devices, the current flows in a coil of wire.

As our first example, let us examine a *galvanometer*, the basic component of many meters, including ammeters, voltmeters, and ohmmeters (Chapter 18).

As shown in Fig. 19-22, a galvanometer consists of a loop or coil of wire suspended in the magnetic field of a permanent magnet. When current flows through the loop, which we assume is rectangular, the magnetic field exerts a force on the vertical sections of wire as shown. Notice that, by the right-hand rule (Fig. 19-15), the force on the upward current on the left is inward, whereas that on the descending current on the right is outward. These forces give rise to a net torque that tends to rotate the coil about its vertical axis. The greater the current I, the more the coil and its attached pointer will turn against the resistance of the small spring.

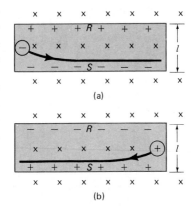

FIGURE 19-20
The Hall effect.

FIGURE 19-21
Electromagnetic pumping in a blood vessel.

FIGURE 19-22
Galvanometer.

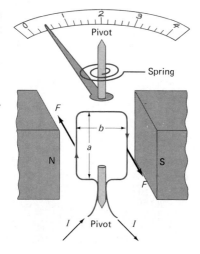

Let us calculate the magnitude of this torque. From Eq. 19-2, the force $F = IaB$, where a is the length of the vertical arm of the coil; the lever arm for each force is $b/2$, where b is the width of the coil and the "axis" is at the midpoint. The total torque is the sum of the torques due to each of the forces, so

$$\tau = IaB\frac{b}{2} + IaB\frac{b}{2} = IabB = IAB,$$

where $A = ab$ is the area of the coil. If the coil consists of N loops of wire, the current is then NI, so the torque becomes

$$\tau = NIAB. \tag{19-5}$$

This formula, derived here for a rectangular coil, is valid for any shape of flat coil. The quantity NIA is referred to as the *magnetic moment* of the coil.

This torque is opposed by the spring. By Hooke's law, the torque τ_s exerted by the spring is proportional to the angle ϕ through which it is turned. That is,

$$\tau_s = k\phi,$$

where k is the stiffness constant of the spring. Thus the coil and the attached pointer will rotate only to the point where the spring torque balances that due to the magnetic field. From Eq. 19-5 we then have $k\phi = NIAB$, or

$$\phi = \frac{NIAB}{k}. \tag{19-6}$$

Thus the deflection of the pointer, ϕ, is directly proportional to the current I flowing in the coil; this is what we want for a galvanometer.

However, there is a problem. When the loop of Fig. 19-22 is not parallel to the magnetic field, the torque will be less than that given by Eq. 19-5 since the lever arm is less. Indeed, when the loop is rotated 90° from the position shown in Fig. 19-22, the two forces act along the same line and the torque is zero. Thus the torque would depend on the angle, and ϕ would not be proportional to I as in Eq. 19-6. To solve this problem, curved pole pieces are used and the galvanometer coil is wrapped around a cylindrical iron core as shown in Fig. 19-23. The iron tends to concentrate the magnetic field lines so that B always points parallel to the face of the coil at the wire. The force is then always perpendicular to the face of the coil and the torque will not vary with angle. Thus ϕ will be proportional to I, as required.

A *chart recorder*, in which a pen graphs a signal such as an ECG on a moving roll of paper, is basically a galvanometer. The pen is attached to an arm, which is connected to the galvanometer coil. The instrument could record either voltage or current, just as any galvanometer can be connected as a voltmeter or ammeter.

An *electric motor* changes electric energy into (rotational) mechanical energy. A motor works on the same principle as a galvanometer, except that the coil is larger and is mounted on a large cylinder called the *rotor* or *armature*, Fig. 19-24. Actually, there are several coils, although only one is indicated in the figure. The armature is mounted on a shaft or axle. The permanent magnet is replaced in some motors by an electromagnet. Unlike a galvanometer, a motor must turn continuously in one direction. This presents a problem, for when the coil, which is rotating clockwise in Fig. 19-24, passes beyond the vertical position the forces would then act to return the

FIGURE 19-23
Galvanometer coil wrapped on an iron core.

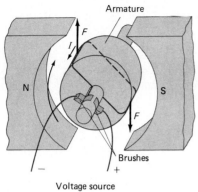

Armature

Brushes

Voltage source

FIGURE 19-24
Diagram of a simple motor.

Lead wires to armature coil

Commutator

Brushes

Battery
(dc)

FIGURE 19-25

This commutator-brush arrangement in a dc motor assures alternation of the current in the armature to keep rotation continuous. The commutators are attached to the motor shaft and turn with it while the brushes remain stationary.

FIGURE 19-26
Motor with many windings.

FIGURE 19-27
Loudspeaker.

coil back to vertical. Thus alternation of the current is necessary if a motor is to turn continuously in one direction. This can be achieved in a *DC motor* with the use of *commutators* and *brushes:* as shown in Fig. 19-25, the brushes are stationary contacts that rub against the conducting commutators mounted on the motor shaft. Every half revolution each commutator changes its connection to the other brush. Thus the current in the coil reverses every half revolution as required for continuous rotation. Most motors contain several coils, called "windings," each located in a different place on the armature, Fig. 19-26. Current flows through each coil only during a small part of a revolution, at the time when its orientation results in the maximum torque. In this way, a motor produces a much steadier torque than can be obtained from a single coil. The design of most practical motors is more complex than described here, but the general principles remain the same.

A *loudspeaker* also works on the principle that a magnet exerts a force on a current-carrying wire. The electrical output of a radio or TV set is connected to the wire leads of the speaker. The speaker leads are connected internally to a coil of wire, which is itself attached to the speaker cone, Fig. 19-27. The speaker cone is usually made of stiffened cardboard and is mounted so that it can move back and forth freely. A permanent magnet is mounted directly in line with the coil of wire. When the alternating current of an audio signal flows through the wire coil, the coil and the attached speaker cone feel a force due to the magnetic field of the magnet. As the current alternates at the frequency of the audio signal, the speaker cone moves back and forth at the same frequency, causing alternate compressions and rarefactions of the adjacent air, and sound waves are produced. A speaker thus changes electrical energy into sound energy, and the frequencies of the emitted sound waves are an accurate reproduction of the electrical input.

19-9 ∎ Discovery and properties of the electron

The electron plays a basic role in our understanding of electricity and magnetism today. But its existence was not suggested until the 1890s. We discuss it here because magnetic fields were crucial for measuring its properties.

Rigid metal frame

Coil of wire (attached to speaker cone)

Magnet

Lead-in wires

Cone

Toward the end of the nineteenth century, studies were being done on the discharge of electricity through rarefied gases. One apparatus, diagrammed in Fig. 19-28, was a glass tube fitted with electrodes and evacuated so only a small amount of gas remained inside. When a very high voltage was applied to the electrodes, a dark space seemed to extend outward from the cathode toward the opposite end of the tube; and that far end of the tube would glow. If one or more screens containing a small hole were inserted as shown, the glow was restricted to a tiny spot on the end of the tube. It seemed as though something being emitted by the cathode traveled to the opposite end of the tube; these were named *cathode rays*.

There was much discussion at the time about what these rays might be. Some scientists thought they might resemble light. But the observation that the bright spot at the end of the tube could be deflected to one side by an electric or magnetic field suggested that cathode rays could be charged particles; and the direction of the deflection was consistent with a negative charge. Furthermore, if the tube contained certain types of rarefied gas, the path of the cathode rays was made visible by a slight glow.

Estimates of the charge e of the (assumed) cathode-rays particles as well as of their charge-to-mass ratio, e/m, had been made by 1897; but in that year, J. J. Thomson (1856–1940) was able to measure e/m directly, using the apparatus shown in Fig. 19-29. Cathode rays are accelerated by a high voltage and then pass between a pair of parallel plates built into the tube; the voltage applied to the plates produces an electric field, and a pair of coils produces a magnetic field. When only the electric field is present, say with the upper plate positive, the cathode rays are deflected upward as in path a in the figure. If only a magnetic field exists, say inward in the figure, the rays are deflected downward along path c. These observations are just what is expected for a negatively charged particle. The force on the rays due to the magnetic field is

$$F = evB$$

where e is the charge and v is the velocity of the cathode rays. In the absence of an electric field, the rays are bent into a curved path, so we have, from $F = ma$,

$$evB = \frac{mv^2}{r},$$

FIGURE 19-28

Discharge tube. In some models, one of the screens is the anode (positive plate).

FIGURE 19-29

Cathode rays deflected by electric and magnetic fields.

and thus

$$\frac{e}{m} = \frac{v}{Br}.$$

The radius of curvature r can be measured and so can B. The velocity v is found by applying an electric field in addition to the magnetic field. The electric field E is adjusted so that the cathode rays are undeflected and follow path b in Fig. 19-29. In this situation, the force due to the electric field, $F = eE$, is just balanced by the force due to the magnetic field, $F = evB$. Thus we have $eE = evB$ and

$$v = \frac{E}{B}.$$

Combining this with the above equation we have

$$\frac{e}{m} = \frac{E}{B^2 r}. \tag{19-7}$$

The quantities on the right side can all be measured so that although e and m could not be determined separately, the ratio e/m could be determined. The accepted value today is $e/m = 1.76 \times 10^{11}$ C/kg. Cathode rays soon came to be called **electrons.**

It is worth noting that the "discovery" of the electron, like many others in science, is not quite so obvious as discovering gold or oil. Should the discovery of the electron be credited to the person who first saw a glow in the tube? Or to the person who first called them cathode rays? Perhaps neither one, for they had no conception of the electron as we know it today. In fact, the credit for the discovery is generally given to Thomson, but not because he was the first to see the glow in the tube. Rather it is because he believed not only that this phenomenon was due to tiny negatively charged particles and made careful measurements on them; but he also argued that these particles were constituents of atoms and not ions or atoms themselves as many thought, and he developed an electron theory of matter. His view is close to what we accept today, and this is why Thomson is credited with the "discovery." Note, however, that neither he nor anyone else ever actually saw an electron itself. We discuss this briefly, for it illustrates the fact that discovery in science is not always a clear-cut matter. In fact some philosophers of science think the word "discovery" is not always appropriate, such as in this case.

Thomson believed that an electron was not an atom, but rather a constituent, or part, of an atom. Convincing evidence for this came soon with the determination of the charge and the mass of the cathode rays. Thomson's student J. S. Townsend made the first direct (but rough) measurements of e in 1897. But it was the more refined *oil-drop experiment* of Robert A. Millikan (1868–1953) that yielded a precise value for the charge on the electron and showed that charge comes in discrete amounts. In this experiment, tiny droplets of mineral oil carrying an electric charge were allowed to fall under gravity between two parallel plates, Fig. 19-30. The electric field E between the plates was adjusted until the drop was suspended in midair; the downward pull of gravity, mg, was then just balanced by the upward force due to the electric field. Thus $qE = mg$ so the charge $q = mg/E$. The mass of the droplet

FIGURE 19-30

Millikan's oil-drop experiment.

was determined by measuring its terminal velocity in the absence of the electric field and using Stoke's equation. (See Section 9-14; the drop is too small to permit direct measurement of its radius). Sometimes the drop was charged negatively and sometimes positively, suggesting that the drop had acquired or lost electrons (presumably through friction when ejected by the atomizer). Millikan's painstaking observations and analysis presented convincing evidence that any charge was an integral multiple of a smallest charge, e, that was ascribed to the electron, and that the value of e was 1.6×10^{-19} C. (Today's precise value of e, as mentioned in Chapter 15, is $e = 1.602 \times 10^{-19}$ C.) This result, combined with the measurement of e/m, gives the mass of the electron to be $(1.6 \times 10^{-19}$ C$)/(1.76 \times 10^{11}$ C/kg$) = 9.1 \times 10^{-31}$ kg. This is less than a thousandth the mass of the smallest atom, and thus confirmed the idea that the electron is only a part of an atom. The accepted value today for the mass of the electron is $m_e = 9.11 \times 10^{-31}$ kg.

19-10 ∎ Thermionic emission and the cathode-ray tube

In the course of experiments on the electric light bulb, Thomas Edison (1847–1931) in 1883 made an interesting discovery. Into an evacuated glass bulb he inserted an electrode or plate in addition to the bulb filament. When a positive potential was applied by a battery to the plate, he found that a current would flow in the external circuit as long as the filament was hot and glowing (heated by a separate current passing through it). However, if a negative potential was applied to the plate, no current was observed to flow. When the filament was cold, no current flowed in either case. This phenomenon is known as *thermionic emission*.

Although Edison was unaware of the significance of his discovery, it eventually led to the invention of the electronic vacuum tube. Indeed, an explanation for this effect had to await the discovery of the electron over a decade later.

J. J. Thomson, in 1899, measured the value of e/m for the rays emitted in thermionic emission; he obtained the same value as that for the rays in the discharge tube produced by high voltage (Fig. 19-29), and concluded that it is electrons that are emitted in thermionic emission. Apparently what happens in thermionic emission is that electrons are being "boiled off" the filament when it is hot. When the plate is positive, the electrons are attracted to it and so a current flows. If the plate is negative, the electrons are repelled and no current flows.

We can understand how electrons might be "boiled off" a hot metal filament if we treat the electrons like molecules in a gas. This makes sense if the electrons are relatively free to move about, which is consistent with the fact that metals are good conductors. However, electrons don't readily escape from the metal. There are forces that keep them in. For example, if an electron were to escape outside the metal surface, a net positive charge would remain behind, and this would attract the electron back. To escape, an electron would have to have a certain minimum kinetic energy, just as molecules in a liquid must have a minimum KE to "evaporate" into the gaseous state.

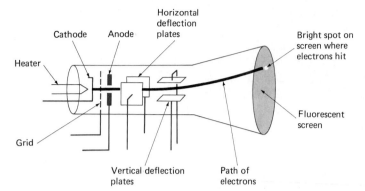

FIGURE 19-31

Cathode-ray tube. Magnetic deflection coils are often used in place of the electric deflection plates.

We saw in Chapter 10 that the average kinetic energy ($\overline{KE}$) of molecules in a gas is proportional to the absolute temperature T. We can apply this idea, but only very roughly, to free electrons in a metal as if they made up an "electron gas." Of course some electrons have more KE than average and others less. At room temperature, very few electrons would have sufficient energy to escape. At high temperatures, $\overline{KE}$ is larger and many escape; this is just like evaporation from liquids, which occurs more readily at high temperatures. Thus, significant thermionic emission occurs only at elevated temperatures. And Edison's results are readily understood.

An important vacuum tube, whose operation depends on thermionic emission, is the *cathode-ray tube* (CRT). It is the picture tube of television sets, oscilloscopes, and computer display terminals. It derives its name from the fact that inside an evacuated glass tube a beam of cathode rays (electrons) is directed to various parts of a screen to produce a "picture." A simple CRT is diagrammed in Fig. 19-31. Electrons emitted by the heated cathode are accelerated by a high voltage applied to the anode (5000–50,000 V). The electrons pass out of this "electron gun" through a small hole in the anode. The inside of the tube face is coated with a fluorescent material that glows when struck by electrons. A tiny bright spot is thus visible where the electron beam strikes the screen. Two horizontal and two vertical plates deflect the beam of electrons when a voltage is applied to them. The electrons are deflected toward whichever plate is positive. By varying the voltage on the deflection plates, the bright spot can be placed at any point on the screen.

In the picture tube of a television set, magnetic deflection coils are usually used instead of electric plates. The electron beam is made to sweep over the screen in the manner shown in Fig. 19-32. The beam is swept horizontally by the horizontal deflection plates or coils. When the horizontal deflecting field is maximum in one direction, the beam is at one edge of the screen. As the field decreases to zero, the beam moves to the center; and as it increases to a maximum in the opposite direction the beam approaches the opposite edge. When it reaches this edge, the voltage or current abruptly changes to return the beam to the opposite side of the screen. Simultaneously, the beam is deflected downward slightly by the vertical deflection plates or coils, and then another horizontal sweep is made. In the United States 525 lines constitutes a complete sweep over the entire screen. (Some European systems use almost twice that number, giving greater sharpness.) The complete picture

FIGURE 19-32

Electrons sweep across a television screen in a succession of horizontal lines.

of 525 lines is swept out in $\frac{1}{30}$ s. Actually, a single vertical sweep takes $\frac{1}{60}$ s and involves every other line; the lines in between are then swept out over the next $\frac{1}{60}$ s. We see a picture because the image is retained by the fluorescent screen and by our eyes for about $\frac{1}{20}$ s. The picture we see consists of the varied brightness of the spots on the screen. The brightness at any point is controlled by the grid (a "porous" electrode, such as a wire grid, that allows passage of electrons) which can limit the flow of electrons by means of the voltage applied to it: the more negative this voltage, the fewer electrons pass through. The voltage on the grid is determined by the video signal sent out by the station and received by the set. Accompanying this signal are signals that synchronize the grid voltage to the horizontal and vertical sweeps.

An *oscilloscope* is a device for amplifying, measuring, and visually observing an electrical signal. It provides a principle means for measuring rapidly changing signals. The signal is displayed on the screen of a CRT. In normal operation, the electron beam is swept horizontally at a uniform rate in time by the horizontal deflection plates. The signal to be displayed is applied, after amplification, to the vertical deflection plates. The visible "trace" on the screen is thus a plot of the signal voltage (vertically) versus time (horizontally).

*19-11 ∎ Mass spectrometer

A number of methods were developed in the early part of this century to measure the masses of atoms. One of the most accurate was the *mass spectrometer*[†] of Fig. 19-33. Ions are produced by heating, or by an electric current, in the source S. Those that pass through slit S_1 enter a region where there are crossed electric and magnetic fields as in Thomson's device (Fig. 19-29). Only those ions whose speed is $v = E/B$ will pass through undeflected and emerge through slit S_2. (This arrangement is called a velocity selector; see equation and discussion just before Eq. 19-7.) In this second region there is only a magnetic field B' and the ions follow a circular path. The radius of their path can be measured because the ions expose the photographic plate where they strike. Since $qvB' = mv^2/r$ and $v = E/B$, then we have

$$m = \frac{qB'r}{v} = \frac{qBB'r}{E}.$$

All the quantities on the right can be measured, and thus m can be determined. Note that for ions of the same charge, the mass of each is proportional to the radius of its path.

The masses of many atoms were measured in this way. When a pure substance was used, it was sometimes found that two or more closely spaced marks would appear on the film. For example, neon produced two marks whose radii corresponded to atoms of mass 20 and 22 atomic mass units (u). Impurities were ruled out and it was concluded that there must be two types of neon with different mass. These different forms were called *isotopes*. It was soon found that most elements are mixtures of isotopes. We shall see

FIGURE 19-33

Bainbridge mass spectrometer. The magnetic fields B and B' point out of the paper (indicated by the dots).

[†] The term *mass spectrograph* is also used.

in Chapter 29 that the difference in mass is due to different numbers of neutrons.

Mass spectrometers can be used not only to separate different elements and isotopes, but molecules as well. They are used in physics and chemistry, and in biological and biomedical laboratories.

EXAMPLE 19-5 Carbon atoms of atomic mass 12.0 u are found to be mixed with another, unknown, element. In a mass spectrometer, the carbon traverses a path of radius 22.4 cm and the unknown's path has a 26.2 cm radius. What is the unknown element? Assume they have the same charge.

SOLUTION Since mass is proportional to the radius, we have

$$\frac{m_x}{m_C} = \frac{26.2 \text{ cm}}{22.4 \text{ cm}} = 1.17$$

Thus $m_x = 1.17 \times 12.0$ u $= 14.0$ u. The other element is probably nitrogen (see the periodic table, inside the rear cover). However, it could also be an isotope of carbon or oxygen. Further physical or chemical analysis would be needed.

*19-12 ▌ Determination of magnetic field strengths; Ampère's law

The simplest magnetic field is one that is uniform—it doesn't change from one point to another. A perfectly uniform field over a large area is not easy to produce. But the field between two flat pole pieces of a magnet is nearly uniform if the area of the pole faces is large compared to their separation, Fig. 19-34. At the edges, the field "fringes" out somewhat and is not uniform. The parallel, evenly spaced field lines in the drawing indicate the field is uniform at points not too near the edge. Nonuniform magnetic fields are more common, and it is often important to know how a magnetic field varies from point to point.

We saw earlier in this chapter (Fig. 19-7) that the magnetic field due to the electric current in a long straight wire is such that the field lines are circles with the wire at the center. You might expect that the field strength at a given point would be greater if the current flowing in the wire were greater; and that the field would be less at points further from the wire. This is indeed the case. Careful experiments show that the magnetic field B at a point near the wire is directly proportional to the current I in the wire and inversely proportional to the distance r from the wire:

$$B \propto \frac{I}{r}.$$

This relation is valid as long as r, the perpendicular distance to the wire, is much less than the distance to the ends of the wire.

The proportionality constant is written, for historical reasons, as $\mu_0/2\pi$; thus

$$B = \frac{\mu_0}{2\pi} \frac{I}{r}. \qquad \text{[outside a long straight wire]} \quad (19-8)$$

FIGURE 19-34

Magnetic field between two large poles of a magnet is nearly uniform except at the edges.

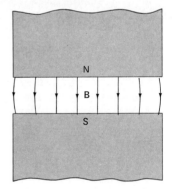

The value of the constant μ_0, which is called the "permeability of free space," is $\mu_0 = 4\pi \times 10^{-7}$ T·m/A.

EXAMPLE 19-6   A vertical electric wire in the wall of a building carries a dc current of 25 A upward. What is the magnetic field at a point 10 cm due north of this wire?

SOLUTION   According to Eq. 19-8:

$$B = \frac{(4\pi \times 10^{-7} \text{ T·m/A})(25 \text{ A})}{(2\pi)(0.10 \text{ m})} = 5.0 \times 10^{-5} \text{ T},$$

or 0.50 G. By the right-hand rule (Fig. 19-7b), the field points to the west at this point. Since this field has about the same magnitude as the earth's, a compass would not point north but in a northwesterly direction.

Equation 19-8 gives the relation between the current in a long straight wire and the magnetic field it produces. This equation is valid only for a long straight wire. The following question arises: is there a general relation between a current in a wire of any shape and the magnetic field around it? The answer is yes: the French scientist André Marie Ampère (1775–1836), proposed such a relation shortly after Oersted's discovery. Consider any (arbitrary) closed path around a current as shown in Fig. 19-35, and imagine this path as being made up of short segments each of length Δl. First, we take the product of the length of each segment times the component of **B** parallel to that segment. If we now sum all these terms, according to Ampère, the result will be equal to μ_0 times the net current I that passes through the surface enclosed by the path. This is known as **Ampère's law** and can be written mathematically as

$$\sum B_{\parallel} \Delta l = \mu_0 I. \tag{19-9}$$

The symbol $\sum$ means "the sum of" and $B_{\parallel}$ means the component of B parallel to that particular Δl. The lengths Δl are chosen so that $B_{\parallel}$ is essentially constant on each length. The sum must be made over a closed path; and I is the net current passing through the surface bounded by this closed path.

We can understand Ampère's law more easily by applying it to a simple case. We take the case of a long, straight wire carrying a current I which we have already examined and which served as an inspiration for Ampère himself. Suppose that we want to find the magnitude of B at point A, a distance r from the wire in Fig. 19-36. We know that the magnetic field lines are circles with the wire at their center. For the path to be used in Eq. 19-9 we choose a circle of radius r. (The choice of path is ours—so we choose one that will be convenient.) We choose this circular path because at any point on this path, **B** will be tangent to this circle. Thus for any short segment of the circle (Fig. 19-36), **B** will be parallel to that segment, so $B_{\parallel} = B$. Suppose that we break the circular path down into 100 segments.[†] Then Ampère's law states that

$$(B \Delta l)_1 + (B \Delta l)_2 + (B \Delta l)_3 + \cdots + (B \Delta l)_{100} = \mu_0 I.$$

[†] Actually, Ampère's law is precisely accurate when there are an infinite number of infinitesimally short segments, but this leads into calculus.

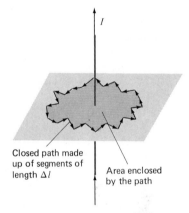

FIGURE 19-35

Arbitrary path enclosing a current, for Ampère's law. The path is broken down into segments of equal length Δl.

Closed path made up of segments of length Δl

Area enclosed by the path

FIGURE 19-36

Circular path of radius r.

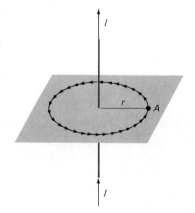

The dots represent all the terms we did not write down. Since all the segments are the same distance from the wire, we expect B to be the same at each segment. We can then factor out B from the sum:

$$B(\Delta l_1 + \Delta l_2 + \Delta l_3 + \cdots + \Delta l_{100}) = \mu_0 I.$$

The sum of the segment lengths is just the circumference of the circle, $2\pi r$. Thus we have

$$B(2\pi r) = \mu_0 I,$$

or

$$B = \frac{\mu_0 I}{2\pi r}.$$

This is just Eq. 19-8 for the field near a long, straight wire, as discussed earlier.

Ampère's law thus works for this simple case. A great many experiments indicate that Ampère's law is valid in general. However, it can be used to calculate the magnetic field mainly for simple cases. Its importance lies in the fact that it relates the magnetic field to the current in a direct and mathematically elegant way. Ampère's law is thus considered one of the basic laws of electricity and magnetism. It is valid for any situation where the currents and fields are steady and not changing in time.

We now can see why the constant in Eq. 19-8 is written $\mu_0/2\pi$; this is done so that only μ_0 appears in Eq. 19-9 (rather than, say, $2\pi k$ if we had used k in Eq. 19-8). In this way, the more fundamental equation, Ampère's law, has the simpler form.

We now use Ampère's law to calculate the magnetic field inside a long solenoid (Fig. 19-13). If the coils of the solenoid are very closely spaced, the field inside will be parallel to the axis except at the ends (Fig. 19-37). We choose the path $abcd$ shown in the figure, far from either end, for applying Ampère's law. We will consider this path as made up of four segments, the sides of the rectangle: ab, bc, cd, da. Then the left side of Eq. 19-9 becomes

$$(B_\| \Delta l)_{ab} + (B_\| \Delta l)_{bc} + (B_\| \Delta l)_{cd} + (B_\| \Delta l)_{da}.$$

Now the field outside the solenoid is so small as to be negligible compared to the field inside. Thus the first term will be zero. Furthermore, **B** is perpendicular to the segments bc and da inside the solenoid, and is nearly zero between and outside the coils. Thus these terms are zero. Therefore, the left

FIGURE 19-37

Magnetic field inside a solenoid is straight except at the ends. Dashed lines indicate the path chosen for use in Ampère's law.

side of Eq. 19-9 is simply $(B_{\parallel}\,\Delta l)_{cd} = Bl$, where B is the field inside the solenoid and l is the length cd. Now we determine the current enclosed by this loop for the right side of Eq. 19-9. If a current I flows in the wire of the solenoid, the total current enclosed by our path $abcd$ is NI, where N is the number of loops our path encircles (five in Fig. 19-37). Thus Ampère's law gives us

$$Bl = \mu_0 NI.$$

If we let $n = N/l$ be the *number of loops per unit length*, then

$$B = \mu_0 nI. \qquad [\text{solenoid}] \quad (19\text{-}10)$$

This is the magnitude of the magnetic field within a solenoid. Note that B depends only on the number of loops per unit length, n, and the current I. The field does not depend on the position within the solenoid, so B is uniform. This is strictly true only for an infinite solenoid, but it is a good approximation for real ones for points not close to the ends.

▢ **EXAMPLE 19-7** A 10-cm-long solenoid has a total of 400 turns of wire and carries a current of 2.0 A. Calculate the field inside near the center.

SOLUTION The number of turns per unit length is $n = 400/10\ \text{cm} = 4.0 \times 10^3\ \text{m}^{-1}$. Thus

$$B = \mu_0 nI = (12.57 \times 10^{-7}\ \text{T·m/A})(4.0 \times 10^3\ \text{m}^{-1})(2.0\ \text{A})$$
$$= 1.0 \times 10^{-2}\ \text{T}.$$

*19-13 ▮ Force between two parallel wires; operational definition of the ampere and the coulomb

You may have wondered how the constant μ_0 could be exactly $4\pi \times 10^{-7}$ T·m/A. With an older definition of the ampere, μ_0 was measured experimentally to be very close to this value. Today, however, μ_0 is *defined* to be exactly $4\pi \times 10^{-7}$ T·m/A. This of course could not be done if the ampere were defined independently. The ampere, the unit of current, is now defined in terms of the magnetic field B it produces using the defined value of μ_0.

Let us be more precise. Consider two long parallel conductors separated by a distance L, Fig. 19-38. They carry currents I_1 and I_2, respectively. Each current produces a magnetic field that is "felt" by the other so that each should be expected to exert a force on the other, as Ampère first pointed out. For example, the magnetic field B_1 produced by I_1 is given by Eq. 19-8; at the location of the second conductor, the magnitude of this field is

$$B_1 = \frac{\mu_0}{2\pi}\frac{I_1}{L}.$$

See Fig. 19-38b where the field due *only* to I_1 is shown. According to Eq. 19-2, the force F per unit length l on the conductor carrying current I_2 is

$$\frac{F}{l} = I_2 B_1.$$

FIGURE 19-38
(a) Two parallel conductors carrying currents I_1 and I_2. (b) Magnetic field produced by I_1.

(a)

(b)

Note that the force on I_2 is due only to the field of I_1; I_2 also produces a field, but it does not exert a force on itself. We substitute in the above formula for B_1 and find

$$\frac{F}{l} = \frac{\mu_0}{2\pi} \frac{I_1 I_2}{L}. \tag{19-11}$$

If we use the right-hand rule of Fig. 19-7b, we see that the lines of B_1 are as shown in Fig. 19-38b. Then using the right-hand rule of Fig. 19-15, we see that the force exerted on I_2 will be to the left in the figure. That is, I_1 exerts an attractive force on I_2. This is true as long as the currents are in the same direction. If I_2 is in the opposite direction, the right-hand rule indicates that the force is in the opposite direction; that is, I_1 exerts a repulsive force on I_2. Reasoning similar to that above shows that the magnetic field produced by I_2 exerts an equal but opposite force on I_1. We expect this to be true also, of course, from Newton's third law.

Thus, parallel currents in the same directions attract each other; currents in opposite directions repel. Equation 19-11 is used to define the ampere precisely. If $I_1 = I_2 = 1\,\text{A}$ exactly and the two wires are exactly $1\,\text{m}$ apart, then

$$\frac{F}{l} = \frac{(4\pi \times 10^{-7}\,\text{T·m/A})}{(2\pi)} \frac{(1\,\text{A})(1\,\text{A})}{(1\,\text{m})} = 2 \times 10^{-7}\,\text{N/m}.$$

Thus, *one ampere is defined as that current flowing in each of two long parallel conductors 1 m apart, which results in a force of exactly 2×10^{-7} N/m of length of each conductor.*

This is the precise definition of the ampere. The *coulomb* is then defined as being *exactly* one ampere-second: $1\,\text{C} = 1\,\text{A·s}$. The value of k or ε_0 in Coulomb's law (Section 15-5) is obtained from experiment.

This may seem a rather roundabout way of defining quantities. The reason behind it is the desire for *operational definitions* of quantities—that is, definitions of quantities that can actually be measured given a definite set of operations to carry out. For example, the unit of charge, the coulomb, could be defined in terms of the force between two equal charges after defining a value for ε_0 or k in Eq. 15-1. However, to carry out an actual experiment to measure the force between two charges is very difficult. For one thing, any desired amount of charge is not easily obtained precisely; and charge tends to leak from objects into the air. On the other hand, the amount of current in a wire can be varied accurately and continuously (by putting a variable resistor in a circuit). Thus the force between two current-carrying conductors is far easier to measure precisely. And this is why the ampere is defined first and then the coulomb in terms of the ampere. At the National Bureau of Standards, Washington, DC, precise measurement of current is made using circular coils of wire rather than straight lengths because it is more convenient and accurate.

Electric and magnetic field strengths are also defined operationally: the electric field in terms of the measurable force on a charge, via Eq. 15-2; and the magnetic field in terms of the force per unit length on a current-carrying wire, via Eq. 19-2.

EXAMPLE 19-8 The two wires of a 2.0-m long appliance cord are 3.0 mm apart and carry 8.0 A dc. Calculate the force between these wires.

SOLUTION Equation 19-11 gives us

$$F = \frac{(2.0 \times 10^{-7} \text{ T·m/A})(8.0 \text{ A})(2.0 \text{ m})}{(3.0 \times 10^{-3} \text{ m})} = 8.5 \times 10^{-3} \text{ N}.$$

Since the currents are in opposite directions, the force would tend to spread them apart.

*19-14 ∎ Magnetic fields in magnetic materials; hysteresis

The field of a long solenoid is directly proportional to the current. Indeed, Eq. 19-10 tells us that the field B_0 inside a solenoid is given by

$$B_0 = \mu_0 nI.$$

This is valid if there is only air inside the coil. If we put a piece of iron or other ferromagnetic material inside the solenoid, the field will be greatly increased, often by hundreds or thousands of times. This occurs because the domains in the iron become preferentially aligned by the external field. The resulting magnetic field is the sum of that due to the current and that due to the iron. It is sometimes convenient to write the total field in this case as a sum of two terms:

$$\mathbf{B} = \mathbf{B_0} + \mathbf{B_M}. \tag{19-12}$$

Here, $\mathbf{B_0}$ refers to the field due only to the current in the wire (the "external field"); it is equal to the field that would be present in the absence of a ferromagnetic material. Then $\mathbf{B_M}$ represents the additional field due to the ferromagnetic material itself; often $\mathbf{B_M} \gg \mathbf{B_0}$.

The total field inside a solenoid in such a case can also be written by replacing the constant μ_0 in Eq. 19-10 by another constant, μ, characteristic of the material inside the coil:

$$B = \mu nI; \tag{19-13}$$

μ is called the *magnetic permeability* of the material. For ferromagnetic materials μ is much greater than μ_0. For all other materials, its value is very close to μ_0.[†] The value of μ, however, is not constant for ferromagnetic materials; it depends on the value of the external field B_0, as the following experiment shows.

[†] All materials are slightly magnetic. Nonferromagnetic materials fall into two classes: *paramagnetic*, in which μ is very slightly larger than μ_0; and *diamagnetic*, in which μ is very slightly less than μ_0. Paramagnetic materials apparently contain atoms that have a net magnetic dipole moment due to orbiting electrons, and these become slightly aligned with an external field just as the galvanometer coil in Fig. 19-22 experiences a torque that tends to align it. Atoms of diamagnetic materials have no net dipole moment. However, in the presence of an external field, electrons revolving in one direction are caused to increase in speed slightly, whereas those revolving in the opposite direction are reduced in speed; the result is a slight net magnetic effect which actually opposes the external field.

FIGURE 19-39
Iron-core torus.

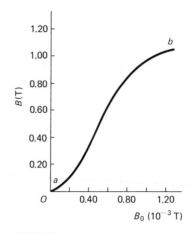

FIGURE 19-40

Total magnetic field of an iron-core torus as a function of the external field B_0.

Measurements on magnetic materials are generally done using a torus, which is essentially a long solenoid bent into the shape of a circle (Fig. 19-39), so that practically all the lines of **B** remain within the torus. Suppose the torus has an iron core that is initially unmagnetized and there is no current in the windings of the torus. Then the current I is slowly increased, and B_0 increases linearly with I. The total field B also increases, but follows the curved line shown in the graph of Fig. 19-40. (Note the different scales: $B \gg B_0$.) Initially (point a), no domains are aligned. As B_0 increases, the domains become more and more aligned until at point b, nearly all are aligned. The iron is said to be approaching *saturation*. (Point b is typically 70 percent of full saturation; the curve continues to rise very slowly, and reaches 98 percent saturation only when B_0 is increased by about a thousandfold above that at point b; the last few domains are very difficult to align.) Now suppose the external field B_0 is reduced by decreasing the current in the coils. As the current is reduced to zero, point c in Fig. 19-41, the domains do not become completely unaligned. Some permanent magnetism remains. If the current is then reversed in direction, enough domains can be turned around so $B = 0$ (point d). As the reverse current is increased further, the iron approaches saturation in the opposite direction (point e). Finally, if the current is again reduced to zero and then increased in the original direction, the total field follows the path $efgb$, again approaching saturation at point b.

Notice that the field did not pass through the origin (point a) in this cycle. The fact that the curves do not retrace themselves on the same path is called *hysteresis*. The curve $bcdefgb$ is called a *hysteresis loop*. In such a cycle, much energy is transformed to thermal energy (friction) due to realigning of the domains; it can be shown that the energy dissipated in this way is proportional to the area of the hysteresis loop.

At points c and f, the iron core is magnetized even though there is no current in the coils. These points correspond to a permanent magnet. For a permanent magnet, it is desired that ac and af be as large as possible. Materials for which this is true are said to have high *retentivity*, and may be

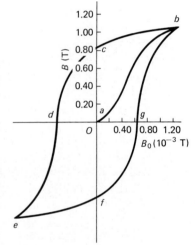

FIGURE 19-41
Hysteresis curve.

FIGURE 19-42
Hysteresis curve for soft iron.

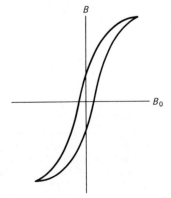

referred to as "hard." On the other hand, a hysteresis curve such as that in Fig. 19-42 occurs for so-called "soft iron" (it is soft only from a magnetic point of view). This is preferred for *electromagnets* since the field can be more readily switched off, and the field can be reversed with less loss of energy. Whether iron is "soft" or "hard" depends on how it is alloyed, heat treatment, and other factors.

A ferromagnetic material can be demagnetized—that is, made unmagnetized. This can be done by reversing the magnetizing current repeatedly while decreasing its magnitude. This results in the curve of Fig. 19-43. The heads of a tape recorder are demagnetized in this way; the alternating magnetic field acting at the heads due to a demagnetizer is strong when the demagnetizer is placed near the heads and decreases as it is moved slowly away.

FIGURE 19-43

Successive hysteresis loops during demagnetization.

▌ Summary

A magnet has two poles, north and south. The north pole is that end which points toward the north when the magnet is freely suspended. Unlike poles of two magnets attract each other, whereas like poles repel.

Permanent magnets are made up of tiny *domains*—each a tiny magnet—which are aligned. In an unmagnetized piece of iron, the domains are randomly oriented.

We can imagine that a *magnetic field* surrounds every magnet. The SI unit for *magnetic field* is the tesla (T). The force one magnet exerts on another is said to be an interaction between one magnet and the magnetic field produced by the other.

Electric currents produce magnetic fields. For example, the lines of magnetic field due to a current in a straight wire form circles around the wire and the field exerts a force on magnets placed near it.

A magnetic field exerts a force on an electric current. For a straight wire of length l carrying a current I the force has magnitude $F = IlB \sin \theta$; θ is the angle between the magnetic field of strength B and the wire;

the direction of the force is perpendicular to the wire and to the magnetic field, and is given by the right-hand rule. Similarly, a magnetic field exerts a force on a charge q moving with velocity v of magnitude $F = qvB \sin \theta$, where θ is the angle between **v** and **B**; the direction of **F** is perpendicular to **v** and to **B**. The path of a charged particle moving perpendicular to a uniform magnetic field is a circle.

The force exerted on a current-carrying wire by a magnetic field is the basis for operation of many devices, such as meters, motors, and loudspeakers.

The measurement of the charge to mass ratio (e/m) of the electron was done using magnetic and electric fields. The charge e on the electron was measured in the Millikan oil-drop experiment and then its mass was obtained from the measured value of the e/m ratio.

Electric and magnetic fields are employed in the cathode-ray tube (CRT)—which is the picture tube of TV sets, display terminals, and oscilloscopes.

▌ Questions

1. A compass needle is not always balanced parallel to the earth's surface but one end may dip downward. Explain.

2. How do you suppose the first magnets found in Magnesia were formed?

3. Why will either pole of a magnet attract an unmagnetized piece of iron?

4. Suppose you have three iron rods, two of which are magnetized but the third is not. How would you determine which two are the magnets without using any additional objects?

*5. Note that the pattern of magnetic field lines surrounding a bar magnet is similar to that of the electric field around an electric dipole. From this fact predict how the magnetic field will change with distance (*a*) when near one pole of a very long bar magnet, and (*b*) when far from a magnet as a whole.

6. Draw the magnetic field lines around a straight section of wire carrying a current horizontally to the left.

7. In what direction are the magnetic field lines surrounding a straight wire carrying a current that is moving directly toward you?

8. The magnetic field due to current in wires in your home can affect a compass. Discuss the problem in terms of currents, depending on whether they are ac or dc.

9. What kind of field or fields surround a moving electric charge?

10. Will a magnet attract any metallic object or only those made of iron? (Try it and see.)Why is this so?

11. An unmagnetized nail will not attract an unmagnetized paper clip. However, if one end of the nail is in contact with a magnet, the other end *will* attract a paper clip. Explain.

12. Another type pf magnetic switch similar to a solenoid is a *relay*. A relay is an electromagnet (the iron rod inside the coil does not move) which, when activated, attracts a piece of soft iron on a pivot. Design a relay (*a*) to make a doorbell, and (*b*) to close an electrical switch. A relay is used in the latter case when you need to switch on a circuit carrying a very large current but you do not want that large current flowing through the main switch. For example, the starter switch of a car is connected to a relay so that the large currents needed for the starter do not pass to the dashboard switch.

13. How can you make a compass without using iron or other ferromagnetic material?

14. A horseshoe magnet is held vertically with the north pole on the left and south pole on the right. A wire passing perpendicularly between the poles carries a current directly away from you. In what direction is the force on the wire?

15. Can you set a resting electron into motion with a magnetic field? With an electric field?

16. A charged particle is moving in a circle under the influence of a uniform magnetic field. If an electric field that points in the same direction as the magnetic field is turned on, describe the path the charged particle will take.

17. Each of the right-hand rules you learned in this chapter can be changed to *left-hand rules* if you are specifying the direction of movement of *negative* particles, such as electrons in a wire. Show, for each right-hand rule, that the same operations using the left hand give the same results if the direction of charge flow is for negative charges.

18. A charged particle moves in a straight line through a particular region of space. Could there be a nonzero

magnetic field in this region? If so, give two possible situations.

19. If a moving charged particle is deflected sideways in some region of space, can we conclude, for certain, that $\mathbf{B} \neq 0$ in that region?

20. In a particular region of space there is a uniform magnetic field $\mathbf{B}$; outside this region, $B = 0$. Can you inject an electron into the field so it will move in a closed circular path in the field?

21. How could you tell whether moving electrons in a certain region of space are being deflected by an electric field or by a magnetic field (or by both)?

22. A beam of electrons is directed perpendicularly toward a horizontal wire carrying a current from left to right. In what direction is the beam deflected?

23. Charged cosmic ray particles from outside the earth tend to strike the earth more frequently near the poles than at lower latitudes. Explain. (See Fig. 19-5.)

*24. In the electromagnetic pumping device shown in Fig. 19-21, does the direction of the force, and therefore of the blood flow, depend on the sign of the ions?

25. What factors determine the sensitivity of a galvanometer?

26. Bringing a magnet close to a television screen will distort the picture. Why? (This can damage a color set.)

*27. Two ions have the same mass, but one is singly ionized and the other is doubly ionized. How will their positions on the film of the mass spectrograph of Fig. 19-33 differ?

28. Two long wires carrying equal currents I are at right angles to each other, but don't quite touch. Describe the magnetic force one exerts on the other.

29. A horizontal wire carries a large current. A second wire carrying a current in the same direction is suspended below. Can the current in the upper wire hold the lower wire in suspension? Under what conditions will it be in equilibrium? Under what conditions will the equilibrium be stable?

30. A horizontal current-carrying wire, free to move, is suspended directly above a second, parallel, current-carrying wire. (*a*) In what direction is the current in the lower wire? (*b*) Can the upper wire be held in stable equilibrium due to the magnetic force of the lower wire? Explain.

*31. The presence of iron inside a solenoid can greatly increase $\mathbf{B}$. Yet without iron, greater fields could be obtained. Explain.

▌ Problems

SECTIONS 19-5 AND 19-6

1. (I) (*a*) What is the force per meter on a wire carrying a 2.50-A current when perpendicular to a 1.10-T magnetic field? (*b*) What if the angle between the wire and field is 45.0°?

2. (I) The force on a wire carrying 20.0 A is a maximum of 3.60 N when placed between the pole faces of a magnet. If the pole faces are 15.0 cm in diameter, what is the approximate strength of the magnetic field?

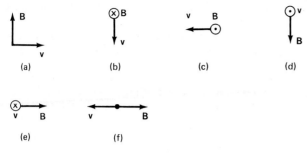

(a)   (b)   (c)   (d)

(e)   (f)

FIGURE 19-44

3. (I) How much current is flowing in a wire 2.00 m long if the force on it is 0.700 N when placed in a uniform 0.0300-T field?

4. (I) Calculate the magnetic force on a 140-m length of wire stretched between two towers carrying a 200-A current. The earth's magnetic field of 5.00×10^{-5} T makes an angle of 60.0° with the wire.

5. (I) Determine the magnitude and direction of the force on an electron traveling 3.24×10^5 m/s horizontally to the east in a vertically upward magnetic field of strength 1.50 T.

6. (I) Describe the path of a proton ($q = e$, $m = 1.67 \times 10^{-27}$ kg) that moves perpendicular to a 0.120-T magnetic field with a speed of 9.25×10^6 m/s. The field points directly toward the observer.

7. (I) Find the direction of the force on a positive charge for each diagram shown in Fig. 19-44, where **v** is the velocity of the charge and **B** is the direction of the magnetic field. ($\otimes$ means the vector points inward. $\odot$ means it points outward, toward the viewer.)

8. (I) Determine the direction of **B** for each case in Fig. 19-45 where **F** represents the force on a positively charged particle moving with velocity **v**.

9. (II) An electron experiences the greatest force as it travels 2.1×10^5 m/s in a magnetic field when it is moving southward. The force is upward and of magnitude 5.6×10^{-13} N. What is the magnitude and direction of the magnetic field?

10. (II) A proton moves in a circular path perpendicular to a 1.10-T magnetic field. The radius of its path is 4.5 cm. Calculate the energy of the proton in eV.

11. (II) A particle of charge q moves in a circular path of radius r in a uniform magnetic field B. Show that its momentum is $p = qBr$.

12. (II) An electron is accelerated through a potential difference of 5000 V. What is the strength of the magnetic field if the radius of its path is 3.4 mm?

13. (II) A doubly charged helium atom whose mass is 6.7×10^{-27} kg is accelerated by a voltage of 2800 V. (a) What will be its radius of curvature in a uniform 0.240-T field? (b) What is its period of revolution?

14. (II) A straight 2.0-mm diameter copper wire can just "float" horizontally in air because of the force of the earth's magnetic field **B** which is horizontal and of magnitude 5.0×10^{-5} T. What current does the wire carry?

15. (III) A 7.10-g bullet moves with a speed of 300 m/s perpendicular to the earth's magnetic field of 5.00×10^{-5} T. If the bullet possesses a net charge of 5.20×10^{-9} C, by what distance will it be deflected from its path due to the magnetic field after it has traveled 600 m?

16. (III) An electron enters a large solenoid at a 5.0° angle to the axis. If the field is a uniform 0.23 T, determine the radius and pitch (distance between loops) of the electron's helical path if its speed is 2.8×10^5 m/s.

SECTION 19-7

*17. (II) Show that the emf produced by the Hall effect is given by $\mathscr{E} = vBl$, where v is the speed of the charged particles in the conductor of width l (Fig. 19-20).

SECTION 19-8

18. (I) A galvanometer needle deflects full scale for a 44.0-μA current. What current will give full-scale deflection if the magnetic field weakens to 0.900 of its original value?

19. (I) If the restoring spring of a galvanometer weakens by 12.0 percent over the years, what current will give full-scale deflection if it originally required 35.0 μA?

20. (I) If the current to a motor drops by 10 percent, by what factor does the output torque change?

21. (I) A square loop of wire 12.0 cm on a side is placed with its face parallel to the magnetic field between the pole pieces of a large magnet. When 6.08 A flows in the coil, the torque on it is 0.316 N·m. What is the magnetic field strength?

22. (II) Show that the torque acting on the current loop of Fig. 19-22 is given by $\tau = NIAB \cos \theta$, where θ is the angle between the face of the coil and direction of the (uniform) magnetic field.

23. (II) A circular coil 20.0 cm in diameter and containing five loops lies flat on the ground. The earth's magnetic field at this location is 6.10×10^{-5} T; it points into the earth at an angle of 46.0° below a line pointing due

FIGURE 19-45

(a)   (b)   (c)

north. If a 15.0-A clockwise current passes through the coil (a) determine the torque on the coil; (b) which edge of the coil rises up—north, east, south, or west?

*24. (III) A square loop of aluminum wire is 25 cm on a side. It is to carry 20 A and rotate in a 2.0-T magnetic field. (a) Determine the minimum diameter of the wire so that it will not fracture from tension or shear. Assume a safety factor of 10. (See Table 8-2.) (b) What is the resistance of a single loop of this wire?

SECTION 19-9

25. (I) What is the velocity of a beam of electrons that go undeflected when passing through crossed electric and magnetic fields of magnitude 8.85×10^3 V/m and 3.67×10^{-3} T, respectively? What is the radius of the electron orbit if the electric field is turned off?

26. (I) Protons move in a circle of radius 8.10 cm in a 0.358-T magnetic field. What value of electric field could make their path straight? In what direction must it point?

27. (I) What is the value of e/m for a particle that moves in a circle of radius 8.0 mm in a 0.46-T magnetic field if a crossed 200-V/m electric field will make the path straight?

28. (II) An oil drop whose mass is determined to be 3.3×10^{-15} kg is held at rest between two large plates separated by 1.0 cm when the potential difference between them is 340 V. How many excess electrons does this drop have?

29. (III) In Millikan's oil-drop experiment, the mass of the oil drop is obtained by observing the terminal speed v_T of the freely falling drop in the absence of an electric field. Under these circumstances, the "effective" weight equals the viscous force given by Stokes's law (Section 9-14) $F = 6\pi\eta r v_T$, where η is the viscosity of air and r the radius of the drop. Also, the actual weight $mg = \frac{4}{3}\pi r^3 \rho g$ must be corrected for the buoyant force of the air; this is done by replacing ρ with $\rho - \rho_A$, where ρ is the density of the oil and ρ_A the density of air. With these preliminaries, show that the charge on the drop is given by

$$q = 18\pi \frac{d}{V} \sqrt{\frac{\eta^3 v_T^3}{2(\rho - \rho_A)g}},$$

where d is the separation of the plates (Fig. 19-30) and V is the voltage across them that just keeps the drop stationary. All of the quantities on the right side of this equation are known or can be measured. The terminal velocity v_T is determined by measuring the time it takes the drop to fall a measured distance, which is observed through a small telescope.

SECTION 19-10

30. (I) Use the ideal gas as a model to estimate the rms speed of a free electron in a metal at 300 K, and at 2500 K (the typical temperature of the cathode in a tube).

31. (II) What would be the approximate maximum deflection of the electron beam near the center of a TV screen due to the earth's 5.0×10^{-5}-T field? Assume the screen is 20 cm from the electron gun where the electrons are accelerated (a) by 2.0 kV, (b) by 30 kV. Note that in color TV sets, the beam must be directed accurately to within less than 1 mm, in order to strike the correct phosphor. Because the earth's field is significant here, mu-metal shields are used to reduce the earth's field in the CRT.

32. (III) Electrons are accelerated by 10.0 kV in a CRT. The screen is 24 cm wide and is 25 cm from the 2.8-cm-long deflection plates. Over what range must the horizontally deflecting electric field vary to sweep the beam fully across the screen?

33. (III) In a given CRT, electrons are accelerated horizontally by 25 kV. They then pass through a uniform magnetic field B for a distance of 3.5 cm which deflects them upward so they reach the top of the screen 22 cm away, 11 cm above the center. Estimate the value of B.

*SECTION 19-11

*34. (I) In a mass spectrometer, germanium atoms have radii of curvature equal to 21.0, 21.6, 21.9, 22.2 and 22.8 cm. The largest radius corresponds to an atomic mass of 76 u. What are the atomic masses of the other isotopes?

*35. (II) Suppose the electric field between the electric plates in the mass spectrometer of Fig. 19-33 is 2.18×10^4 V/m and the magnetic fields $B = B' = 0.68$ T. The source contains boron isotopes of mass numbers 10 and 11 (to get their masses, multiply by 1.67×10^{-27} kg). How far apart are the lines formed by the singly charged ions of each type on the photographic film?

*36. (II) A mass spectrometer is being used to monitor air pollutants. It is difficult, however, to separate molecules with nearly equal mass such as CO (28.0106 u) and N_2 (28.0134 u). How large a radius of curvature must a spectrometer have if these two molecules are to be separated on the film by 0.24 mm.?

*37. (II) One form of mass spectrometer accelerates ions by a voltage V before they enter a magnetic field B. The ions are assumed to start from rest. Show that the mass of an ion is $m = qB^2R^2/2V$ where R is the radius of the ions' path in the magnetic field and q is their charge.

*38. (III) The cyclotron (Fig. 19-46) is a device used to accelerate elementary particles such as protons to high speeds. Particles starting at point A with some initial velocity travel in circular orbits in the magnetic field B. The particles are accelerated to higher speeds each time they pass in the gap between the metal "dees," where there is an electric field E. (There is no electric field within the cavity of the metal dees.) The electric field changes direction each half-cycle, owing to an ac voltage $V = V_0 \sin 2\pi ft$ so that the particles are in-

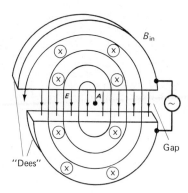

FIGURE 19-46
Cyclotron.

creased in speed at each passage through the gap. (a) Show that the frequency f of the voltage must be $f = Bq/2\pi m$, where q is the charge on the particles and m their mass. (b) Show that the kinetic energy of the particles increases by $2qV_0$ each revolution, assuming that the gap is small. (c) If the radius of the cyclotron is 2.0 m and the magnetic field strength is 0.50 T, what will be the maximum kinetic energy of accelerated protons in MeV? (d) How is a cyclotron like a swing?

*SECTION 19-12

*39. (I) If a magnetic field of no more than 1.0×10^{-3} T is to be allowed 30 cm from an electrical wire, what is the maximum current the wire can carry?

*40. (I) How strong is the magnetic field 11.0 cm from a long straight wire carrying 2.8 A?

*41. (II) What is the acceleration (in g's) of a 280-g model airplane charged to 8.0 C and traveling at 1.5 m/s as it passes within 15 cm of a wire, nearly parallel to its path, carrying a 20-A current?

*42. (II) An electron traveling 4.50×10^6 m/s is moving initially at a 45° angle to a straight wire carrying a 15.0-A current. If the electron passes within 12.0 cm of the wire, what maximum force does it feel? Assume the electron's path and the wire are in the same plane.

*43. (II) A horizontal compass is placed 15 cm due south from a straight vertical wire carrying a 30-A current downward. In what direction does the compass needle point at this location? Assume the horizontal component of the earth's field at this point is 0.45×10^{-4} T and the magnetic declination is 0°.

*44. (II) Determine the magnetic field midway between two long straight wires 10 cm apart in terms of the current I in one when the other carries 10 A. Assume these currents are (a) in the same direction, and (b) in opposite directions.

*45. (II) A compass needle in a particular location points 15°E of N outdoors. However, when it is placed 12 cm

to the east of a vertical wire inside a building it points 60°E of N. What is the magnitude and direction of the current in the wire? The earth's field there is 0.50×10^{-4} T and is horizontal.

*46. (II) A 32-cm-long solenoid, 1.2 cm in diameter, is to produce a 0.20-T magnetic field at its center. If the maximum current is 3.7 A, how many turns must the solenoid have?

*47. (II) You have 1.0 kg of copper and want to make a practical solenoid that produces the greatest possible magnetic field. Should you make your copper wire long and thin, short and fat, or what? Consider other variables, such as solenoid diameter, length, and so on.

*48. (II) A torus is a solenoid in the shape of a circle (Fig. 19-34 cannot drop abruptly to zero outside the magnet. (dashed in the figure) to determine that the magnetic field strength inside the torus is $B = \mu_0 NI/2\pi R$, where N is the total number of turns. Is the field inside a torus uniform like a solenoid's? If not, how does it vary?

*49. (II) Use Ampère's law to show that a uniform magnetic field, such as between the pole pieces of a magnet, Fig. 19-34 cannot drop abruptly to zero outside the magnet. (Hint: take as your path a rectangle with one vertical side inside the field and one vertical side completely outside the field.)

*50. (III) Suppose that a current I flows uniformly through a long cylindrical conductor of radius r_0. Use Ampère's law to show that the magnetic field inside the conductor at a distance r from the center of the conductor is

$$B = \frac{\mu_0 Ir}{2\pi r_0^2} \qquad (r < r_0).$$

Assume that the field lines are circles, just as they are outside the conductor.

*51. (III) Two long parallel wires 12.0 cm apart carry 15.0-A currents in the same direction. Determine the magnetic field strength at a point 15.0 cm from one wire and 20.0 cm from the other. (Hint: make a drawing in a plane containing the field lines and recall the rules for vector addition.)

*52. (III) The magnetic field strength B at the center of a circular coil of wire carrying a current I is

$$B = \frac{\mu_0 NI}{2r},$$

where N is the number of loops in the coil and r is its radius. Suppose that an electromagnet uses a coil 2.0 m in diameter made from square copper wire 2.0 mm on a side. The power supply produces 50 V at a maximum power output of 1.0 kW. (a) How many turns are needed to run the power supply at maximum power? (b) What is the magnetic field strength at the center of the coil? (c) If you use a greater number of turns and this same power supply (so the voltage remains at 50 V) will a greater magnetic field strength result? Explain.

*53. (I) What is the magnitude and direction of the force between two parallel wires 80 m long and 30 cm apart, each carrying 65 A in the same direction?

*54. (I) A vertical straight wire carrying a 5.0-A current exerts an attractive force per unit length of 6.0×10^{-4} N/m on a second parallel wire 8.0 cm away. What current (magnitude and direction) flows in the second wire?

*55. (II) Three long parallel wires are 12.0 cm from one another. (Looking along them, they are at three corners of an equilateral triangle.) The current in each wire is 3.0 A but that in wire A is opposite to that in wires B and C. Determine the magnetic force per unit length on each wire due to the other two.

*56. (III) A long horizontal wire carries a current of 78 A. A second wire, made of 3.5-mm-diameter copper wire and parallel to the first, but 18 cm below it, is held in suspension magnetically. (a) What is the magnitude and direction of the current in the lower wire? (b) Is it in stable equilibrium? (c) Repeat parts (a) and (b) if the second wire is suspended 20 cm above the first due to the latter's field.

*57. (II) An iron-core solenoid is 30 cm long, 1.8 cm in diameter, and has 500 turns of wire. A magnetic field of 2.2 T is produced when 48 A flows in the wire. What is the permeability μ at this high field strength?

20 ELECTROMAGNETIC INDUCTION AND FARADAY'S LAW; AC CIRCUITS

In Chapter 19 we discussed two ways in which electricity and magnetism are related: (1) an electric current produces a magnetic field; and (2) a magnetic field exerts a force on an electric current or moving electric charge. These discoveries were made in 1820–1821. Scientists then began to wonder: if electric currents produce a magnetic field, is it possible that a magnetic field can produce an electric current? Ten years later the American Joseph Henry (1797–1878) and the Englishman Michael Faraday (1791–1867) independently found that it was possible. Henry actually made the discovery first; but Faraday published his results earlier and investigated the subject in more detail. We now discuss this phenomenon and some of its world-changing applications.

20-1 ∎ Induced EMF

In his attempt to produce an electric current from a magnetic field, Faraday used the apparatus shown in Fig. 20-1. A coil of wire, X, was connected to a battery. The current that flowed through X produced a magnetic field that was intensified by the iron core. Faraday hoped that by using a strong enough battery, a steady current in X would produce a great enough magnetic field to produce a current in a second coil Y. This second circuit, Y, contained a galvanometer to detect any current but contained no battery. He met no success with steady currents. But the long-sought effect was finally observed when Faraday saw the galvanometer in circuit Y deflect strongly at the moment he closed the switch in circuit X. And the galvanometer deflected strongly in the opposite direction when he opened the switch. A *steady*

FIGURE 20-1
Faraday's experiment to induce an emf.

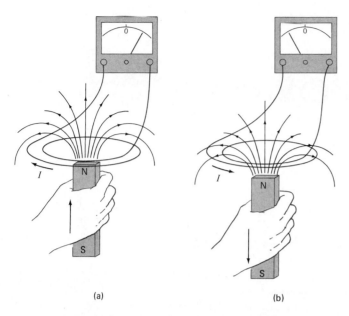

FIGURE 20-2

(a) A current is induced when a magnet is moved toward a coil. (b) The induced current is opposite when the magnet is removed. Note that the galvanometer zero is in the center of the scale and deflects left or right, depending on direction of the current.

(a)   (b)

current in X had produced *no* current in Y. Only when the current in X was starting or stopping was a current produced in Y.

Faraday concluded that although a steady magnetic field produced no current, a *changing* magnetic field can produce an electric current! Such a current is called an *induced current*. When the magnetic field through coil Y changes, a current flows as if there were a source of emf in the circuit. We therefore say that an **induced emf is produced by a changing magnetic field.**

Faraday did further experiments on **electromagnetic induction,** as this phenomenon is called. For example, Fig. 20-2 shows that if a magnet is moved quickly into a coil of wire, a current is induced in the wire. If the magnet is quickly removed, a current is induced in the opposite direction. Furthermore, if the magnet is held steady and the coil of wire is moved toward or away from the magnet, again an emf is induced and a current flows. Motion or change is required to induce an emf. It doesn't matter whether the magnet or the coil moves.

20-2 ▌ Faraday's law of induction; Lenz's law

FIGURE 20-3

Determining the flux through a flat loop of wire. This loop is square, of side l and area $A = l^2$.

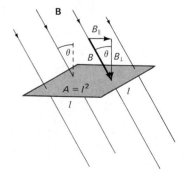

Faraday investigated quantitatively what factors influence the magnitude of the emf induced. He found first of all that it depends on time: the more rapidly the magnetic field changes, the greater the induced emf. But the emf is not simply proportional to the rate of change of the magnetic field, **B**. Rather it is proportional to the rate of change of the **magnetic flux, Φ_B,** passing through the loop of area A, which is defined as

$$\Phi_B = B_\perp A = BA \cos \theta. \qquad (20\text{-}1)$$

Here $B_\perp$ is the component of the magnetic field strength **B** perpendicular to the face of the coil; θ is the angle between **B** and a line drawn perpendicular to the face of the coil. These quantities are shown in Fig. 20-3 for a square coil of side l whose area $A = l^2$. When the face of the coil is parallel to B, $\theta = 90°$ and $\Phi_B = 0$; and when B is perpendicular to the coil, $\theta = 0°$ and

$$\Phi_B = BA \qquad\qquad [B_\perp \text{ coil face}]$$

As we saw in Chapter 19, the lines of **B** can be drawn such that the number of lines per unit area is proportional to the field strength. Then the flux Φ_B can be thought of as being proportional to the *total number of lines passing through the coil*. This is illustrated in Fig. 20-4. For $\theta = 90°$, no lines pass through the coil and $\Phi_B = 0$; Φ_B is a maximum when $\theta = 0°$. The unit of magnetic flux is the tesla-meter^2; this is called† a weber: 1 Wb = 1 T·m^2.

With this definition of the flux, we can now write down the results of Faraday's investigations: if the flux through N loops of wire changes by an amount $\Delta\Phi_B$ during a time Δt, the average induced emf during this time is

$$\mathscr{E} = -N\frac{\Delta\Phi_B}{\Delta t}. \qquad\qquad (20\text{-}2)$$

This fundamental result is known as **Faraday's law of induction.**

The minus sign in Eq. 20-2 is placed there to remind us in which direction the induced emf acts. Experiment shows that *an induced emf always gives rise to a current whose magnetic field opposes the original change in flux*. This is known as **Lenz's law.** Let us apply it to the case of relative motion between a magnet and a coil, Fig. 20-2. The changing flux induces an emf, which produces a current in the coil; and this induced current produces its own magnetic field. In Fig. 20-2a the distance between the coil and the magnet decreases; so the magnetic field, and therefore the flux, through the coil increases. The magnetic field of the magnet points upward. To oppose this upward increase, the field produced by the induced current must point *downward*. Thus Lenz's law tells us that the current must move as shown (use the right-hand rule). In Fig. 20-2b, the flux *decreases*, so the induced current produces an *upward* magnetic field that is "trying" to maintain the status quo. Thus the current must be as shown.

Let us consider what would happen if Lenz's law were not true, but was just the reverse. The induced current in this pretend situation would produce a flux in the same direction as the original change; this greater change in flux would produce an even larger current followed by a still greater change in flux, and so on. The current would continue to grow indefinitely, producing power $(= I^2 R)$ even after the original stimulus ended. This would violate the conservation of energy. Such "perpetual motion" devices do not exist. Thus Lenz's law as stated above (and not its opposite) is consistent with the law of conservation of energy.

It is important to note that an emf is induced whenever there is a change in flux. Since magnetic flux $\Phi_B = BA$, we see that an emf can be induced in two ways: (1) by a changing magnetic field B; or (2) by changing the area of the loop or its orientation θ with respect to the field. Figures 20-1 and 20-2 illustrated case 1. Examples of case 2 are illustrated in Fig. 20-5 and in the following example.

---

† Note also that 1 T = 1 Wb/m^2; this is where the old SI unit for magnetic field strength, the Wb/m^2, comes from. The magnetic field strength B, since it equals Φ_B/A, is often called the *flux density*.

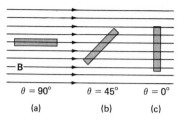

$\theta = 90°$ (a) $\theta = 45°$ (b) $\theta = 0°$ (c)

FIGURE 20-4

Magnetic flux Φ_B is proportional to the number of lines of **B** that pass through the loop.

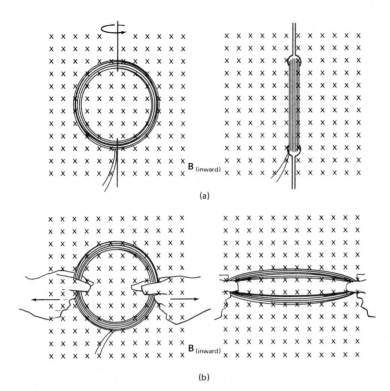

(a)

(b)

FIGURE 20-5

A current can be induced (a) by rotating a coil in a magnetic field, or (b) by changing the area of the coil. In both cases the flux through the coil is reduced.

> EXAMPLE 20-1   A square coil of side 5.0 cm contains 100 loops and is positioned perpendicular to a uniform 0.60-T magnetic field. It is quickly and uniformly pulled from the field (moving perpendicularly to **B**) to a region where B drops abruptly to zero. It takes 0.10 s for the whole coil to reach the field-free region. How much energy is dissipated in the coil if its resistance is 100 Ω?
>
> SOLUTION   The area of the coil is $A = (0.050 \text{ m})^2 = 2.5 \times 10^{-3} \text{ m}^2$. The flux is initially $\Phi_B = BA = (0.60 \text{ T})(2.5 \times 10^{-3} \text{ m}^2) = 1.5 \times 10^{-3}$ Wb. After 0.10 s the flux is zero. The rate of change of flux is constant during the 0.10 s, so the emf induced during this period is
>
> $$\mathscr{E} = -(100)\frac{(0 - 1.5 \times 10^{-3} \text{ Wb})}{(0.10 \text{ s})} = 1.5 \text{ V}.$$
>
> The current $I = \mathscr{E}/R = 1.5 \text{ V}/100 \text{ }\Omega = 15\text{mA}$. The total energy dissipated is $I^2Rt = (1.5 \times 10^{-2} \text{ A})^2(100 \text{ }\Omega)(0.10 \text{ s}) = 2.3 \times 10^{-3}$ J. From the conservation of energy principle, this is just equal to the work needed to pull the coil out of the field.

20-3 ▌ Emf induced in a moving conductor

Another way to induce an emf is shown in Fig. 20-6 and this situation helps illuminate the nature of the induced emf. Assume that a uniform magnetic field **B** is perpendicular to the area bounded by the U-shaped conductor and the movable rod resting on it. If the rod is made to move at a speed v,

FIGURE 20-6

A conducting rod is moved to the right on a U-shaped conductor in a uniform magnetic field **B** that points out of the paper.

it travels a distance $\Delta x = v\,\Delta t$ in a time Δt. Therefore, the area of the loop increases by an amount $\Delta A = lv\,\Delta t$ in a time Δt. By Faraday's law, there is an induced emf $\mathcal{E}$ given by

$$\mathcal{E} = \frac{\Delta \Phi_B}{\Delta t} = \frac{B\,\Delta A}{\Delta t} = \frac{Blv\,\Delta t}{\Delta t} = Blv, \tag{20-3}$$

where we have omitted the minus sign. This equation is valid as long as B, l, and v are mutually perpendicular. (If they are not, we use only the components of each that are mutually perpendicular.)

We can also obtain Eq. 20-3 without the use of Faraday's law. We saw in Chapter 19 that a charged particle moving perpendicular to a magnetic field B with speed v experiences a force $F = qvB$. When the rod of Fig. 20-6 moves to the right with speed v, the electrons in the rod move with this same speed. Therefore, each feels a force $F = qvB$, which acts upward in the figure. If the rod were not in contact with the U-shaped conductor, electrons would collect at the upper end of the rod, leaving the lower end positive. There must thus be an induced emf. If the rod does slide on the U-shaped conductor, the electrons will flow into it; there will thus be a clockwise (conventional) current flowing in the loop. To calculate the emf, we determine the work W needed to move a charge q from one end of the rod to the other against this potential difference: $W = \text{force} \times \text{distance} = (qvB)(l)$. The emf equals the work done per unit charge, so $\mathcal{E} = qvBl/q = Blv$, just as above.[†]

□ **EXAMPLE 20-2** An airplane travels 1000 km/h in a region where the earth's field is 5.0×10^{-5} T and is nearly vertical. What is the potential difference induced between the wing tips that are 70 m apart?

SOLUTION Since $v = 1000 \text{ km/h} = 280 \text{ m/s}$, we have $\mathcal{E} = Blv = (5.0 \times 10^{-5} \text{ T})(70 \text{ m})(280 \text{ m/s}) = 1.0 \text{ V}$. Not much to worry about.

20-4 ∎ Changing magnetic flux produces an electric field

As we just discussed, the electrons in the moving conductor of Fig. 20-6 feel a force. This implies that there must be an electric field in the conductor. Since electric field is defined as the force per unit charge, the effective field E

[†] This argument, which is basically the same as for the Hall effect, explains this one way of inducing an emf. It does not explain the general case of electromagnetic induction, however.

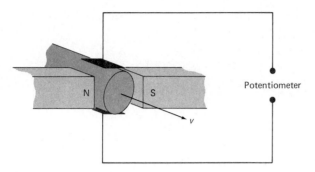

FIGURE 20-7
Measurement of blood velocity from the induced emf.

in the rod must be $E = qvB/q$, or

$$E = vB. \qquad (20\text{-}4)$$

In the situation in which a changing magnetic field (rather than a moving conductor) induces an emf (as, for example, in Fig. 20-2), there also is an induced current. And again this implies that there is an electric field in the wire. Thus we come to the important conclusion that **a changing magnetic flux produces an electric field.** This applies not only to wires and other conductors; it is actually a general result that applies to any region in space: an electric field will be produced at any point in space where there is a changing magnetic field.

    EXAMPLE 20-3 *Electromagnetic blood flow measurement.* The rate of blood flow can be measured using the apparatus shown in Fig. 20-7 since blood contains ions. Suppose that the blood vessel is 2.0 mm in diameter, the magnetic field is 0.080 T, and the measured emf is 0.10 mV. What is the flow velocity of the blood?

    SOLUTION   We solve for v in Eq. 20-3, and find that $v = \mathscr{E}/Bl = (1.0 \times 10^{-4}\ \text{V})/(0.080\ \text{T})(2.0 \times 10^{-3}\ \text{m}) = 0.63\ \text{m/s}$. (In actual practice, an alternating current is used to produce an alternating magnetic field; the induced emf is then alternating.)

20-5 ∎ Electric generators

Probably the most important practical result of Faraday's great discovery was the development of the **electric generator** or **dynamo.** A generator transforms mechanical energy into electric energy. This is just the opposite of what a motor does. Indeed, a generator is basically the inverse of a motor.[†] A simplified diagram of an ac generator is shown in Fig. 20-8; a generator consists of many coils of wire (only one is shown) wound on an armature that can rotate in a magnetic field. The axle is turned by some mechanical means and an emf is induced in the rotating coil. An electric current is thus the *output* of a generator. In Fig. 20-8 the right-hand rule tells us that, with the armature ro-

---

[†] You can, for example, actually run a car generator backward as a motor by connecting its output terminals to a battery.

FIGURE 20-8
An ac generator.

Axle turned mechanically

Armature

Slip rings

Brushes

tating counterclockwise, the (conventional) current in the wire labeled N on the armature is outward; therefore it is outward at brush N. (The brushes press against continuous slip-rings.) After one-half revolution, wire N will be where wire M is now in the drawing and the current then at brush N will be inward. Thus the current produced is alternating. Let us look at this in more detail.

In Fig. 20-9, the loop is being made to rotate clockwise in a uniform magnetic field **B**. The velocity of the two lengths ab and cd at this instant are shown. Although the sections of wire bc and da are moving, the force on electrons in these sections is toward the side of the wire, not along its length. The emf generated is then due only to the force on charges in the sections ab and cd. From the right-hand rule we see that the direction of the induced current in ab is from a toward b. And in the lower section it is from c to d; so the flow is continuous in the loop. The magnitude of the emf generated in ab is given by Eq. 20-3, except that we must take the component of the velocity perpendicular to B:

$$\mathscr{E} = Blv_\perp,$$

where l is the length of ab. From the diagram we can see that $v_\perp = v \sin \theta$, where θ is the angle the face of the loop makes with the vertical. The emf induced in cd has the same magnitude and is in the same direction. Therefore, they add and the total emf is

$$\mathscr{E} = 2NBlv \sin \theta,$$

where we have multiplied by N, the number of loops in the coil (if there is more than one). If the coil is rotating with constant angular velocity ω, then the angle $\theta = \omega t$; we also have from the angular equations (Chapter 7) that $v = \omega r = \omega(h/2)$, where h is the length of bc or ad. Thus $\mathscr{E} = 2NlB\omega(h/2) \cdot \sin \omega t$, or

$$\mathscr{E} = NAB\omega \sin \omega t, \tag{20-5}$$

where $A = lh$ is the area of the loop. (This equation holds for any shape coil, not just for a rectangle as derived.) Thus the output emf of the generator is sinusoidally alternating (Fig. 20-10 and Section 17-6). Since ω is expressed in radians per second, we can write $\omega = 2\pi f$, where f is the frequency.

Over 99 percent of the electricity used in the U.S. is produced from generators. The frequency f is 60 Hz for general use in the U.S. and Canada, although 50 Hz is used in many countries. In electric-power generating

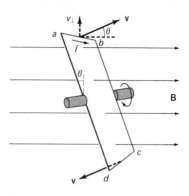

FIGURE 20-9

The emf is induced in the segments ab and cd whose velocity components perpendicular to field **B** equal $v \sin \theta$.

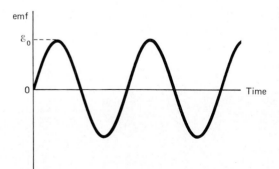

FIGURE 20-10

An ac generator produces an alternating current. The output emf $\mathscr{E} = \mathscr{E}_0 \sin \omega t$, where $\mathscr{E}_0 = NAB\omega$ (Eq. 20-5).

plants, the armature is mounted on a heavy axle connected to a turbine, which is the modern equivalent of a waterwheel. Water falling over a dam can turn the turbine at a hydroelectric plant. Most of the power generated at present in the U.S., however, is done at steam plants, where the burning of fossil fuels (coal, oil, natural gas) boils water to produce high-pressure steam that turns the turbines. Thus, a heat engine (Chapter 12) connected to a generator is the principal means of generating electric power.

EXAMPLE 20-4   The armature of a 60-Hz ac generator rotates in a 0.12-T magnetic field. If the area of the coil is 2.0×10^{-2} m², how many loops must the coil contain if the peak output is to be $\mathscr{E}_0 = 170$ V?

SOLUTION   From Eq. 20-5 we see that the maximum emf is $\mathscr{E}_0 = NAB\omega$. Since $\omega = 2\pi f = (6.28)(60 \text{ s}^{-1}) = 377 \text{ s}^{-1}$, we have

$$N = \frac{\mathscr{E}_0}{AB\omega} = \frac{170 \text{ V}}{(2.0 \times 10^{-2} \text{ m}^2)(0.12 \text{ T})(377 \text{ s}^{-1})} = 188 \text{ turns.}$$

(a)

A dc generator is much like an ac generator, except the slip rings are replaced by split-ring commutators, Fig. 20-11a, just as in a dc motor. The output of such a generator is as shown and can be smoothed out by placing a capacitor in parallel with the output. More common is the use of many armature windings, as in Fig. 20-11b, which produces a smoother output.

In the past, automobiles used dc generators. More common now, however, are ac generators or **alternators** (Fig. 20-12). The ac output is changed to dc for charging the battery with the use of diodes (Chapter 28). Most alternators differ from the generator discussed above in that the magnetic field is made to rotate within a stationary armature called a *stator*. The brushes press against a continuous ring instead of a slotted commutator and are on the input instead of the output; so large output currents are carried in solid conductors rather than through sliding-ring commutators which are subject to wear and arcing. In a car, the rotation speed can be faster so the battery can be charged even at idling speed and there will still be no problem with electrical arcing across the rings when the car travels at high speeds.

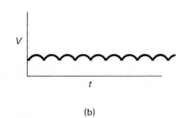

(b)

FIGURE 20-11

Dc generator (a) with one set of commutators; (b) with many sets of commutators and windings.

FIGURE 20-12

Simplified diagram of an alternator. Electromagnet current connected through continuous slip rings is preferable to output of dc generator through many sets of split commutators (Fig. 20-11b), where poor connection or arcing can occur at high speeds. Sometimes the electromagnet is replaced by a permanent magnet.

Current to produce B field

N

Slip rings

Electromagnet (rotating)

S

Emf induced in wire

+

20-6 ∎ Counter emf and torque; eddy currents

A motor turns and produces mechanical energy when a current is made to flow in it. From our description in Sec. 19-8 of a simple dc motor, you might expect that the armature would accelerate indefinitely due to the torque on it. However, as the armature of the motor turns, the magnetic flux through the coil changes and an emf is generated. This induced emf acts to oppose the motion (Lenz's law) and is called the **back,** or **counter, emf.** The greater the speed of the motor, the greater the counter emf. A motor normally turns something, but if there were no load, its speed would increase until the counter emf equaled the input voltage. In the normal situation, when there is a mechanical load, the speed of the motor is limited also by the load. The counter emf will then be less than the external voltage. The greater the load, the slower the motor rotates and the lower is the counter emf.

　EXAMPLE 20-5　The armature windings of a dc motor have a resistance of 5.0 Ω. The motor is connected to a 120-V line and when the motor reaches full speed against its normal load the counter emf is 108-V. Calculate (*a*) the current into the motor when it is just starting up; (*b*) the current when it reaches full speed.

　SOLUTION　(*a*) Initially, the motor is not turning (or turning very slowly) so there is no induced counter emf. Hence from Ohm's law the current is

$$I = \frac{V}{R} = \frac{120 \text{ V}}{5.0 \text{ Ω}} = 24 \text{ A}.$$

　(*b*) At full speed, the counter emf is a source of emf that opposes the exterior source. We represent this counter emf as a battery in the equivalent circuit shown in Fig. 20-13. In this case, Ohm's law (or Kirchhoff's rule) gives

$$120 \text{ V} - 108 \text{ V} = I(5.0 \text{ Ω}).$$

Therefore

$$I = 12 \text{ V}/5.0 \text{ Ω} = 2.4 \text{ A}.$$

FIGURE 20-13
Circuit of a motor showing induced counter emf.

This example illustrates the fact that the current is very high when a motor first starts up. This is why the lights in your house may dim when the motor of the refrigerator (or other large motor) starts up. The large initial current causes the voltage at the outlets to drop (the house wiring has resistance, so there is some voltage drop across it when large currents are drawn). If a motor is overloaded so it turns only at a slow speed, the counter emf will be reduced because of the reduced rotation speed. The current in the motor may then be sufficiently large to burn it out.

　In a generator, the situation is the reverse of that for a motor. As we saw, the mechanical turning of the armature induces an emf in the loops which is the output. If the generator is not connected to an external circuit, the emf exists at the terminals but no current flows. In this case, it takes little effort to turn the armature. But if the generator *is* connected to a device that draws current, then a current flows in the coils of the armature. Because this current-carrying coil is in a magnetic field there will be a torque exerted on it (as in a motor), and this torque opposes the motion (use the right-hand rule for the force on a wire, in Fig. 20-8). This is called a **counter torque.** The greater the load—that is, the more current that is drawn—the greater will be the counter torque. Hence the external applied torque will have to be greater to keep the generator turning. This of course makes sense from the conservation of energy principle. More mechanical-energy input is needed to produce more electrical-energy output.

Induced currents are not always confined to well-defined paths such as in wires. Consider, for example, the rotating metal wheel in Fig. 20-14a. A magnetic field is applied to a limited area as shown and points into the paper. The section of wheel in the magnetic field has an emf induced in it since the conductor is moving (carrying electrons with it). The flow of (conventional) current is upward (Fig. 20-14b) and it follows a downward return path outside the region of the magnetic field. These currents are referred to as **eddy currents** and can be present in any conductor which is moving across a magnetic field or through which the magnetic flux is changing. In Fig. 20-14, the magnetic field exerts a force on the induced currents that opposes (use the right-hand rule) the rotational motion. Eddy currents can be used in this way as a smooth braking device on, say, a rapid-transit car. In order to stop the car, an electromagnet can be turned on that applies its field either to the wheels or to the moving steel rail below. Eddy currents can dampen (reduce) the oscillation of a vibrating system. A common example is in a galvanometer where induced eddy currents keep the needle from overshooting or oscillating violently. Eddy currents, however, can be a problem. For example, eddy currents induced in the armature of a motor or generator produce heat ($P = I\mathscr{E}$) and waste energy. To reduce the eddy currents the armatures are *laminated;* that is, they are made of very thin sheets of iron that are well insulated from one another. Thus the total path length of the eddy currents is confined to each slab which increases the total resistance; hence the current is less and there is less wasted energy.

FIGURE 20-14

Production of eddy currents in a rotating wheel.

20-7 ▌ Transformers; transmission of power

A transformer is a device for increasing or decreasing an ac voltage. A transformer consists of two coils of wire known as the *primary* and *secondary* coils. The two coils can be interwoven; or they can be linked by a soft iron core (laminated to prevent eddy current losses), Fig. 20-15. Transformers are designed so that (nearly) all the flux produced by the current in the primary should pass through the secondary coil, and we assume this is true in what follows. We also assume that energy losses in the resistance of the coils and hysteresis in the iron can be ignored—a good approximation for real transformers which are typically 99 percent efficient.

When an ac voltage is applied to the primary, the changing magnetic field it produces will induce an ac voltage of the same frequency in the secondary. However, the voltage will be different according to the number of loops in each coil. From Faraday's law, the voltage or emf induced in the

FIGURE 20-15

Step-up transformer ($N_p = 4$, $N_s = 12$).

secondary is

$$V_s = N_s \frac{\Delta \Phi_B}{\Delta t},$$

where N_s is the number of turns in the secondary coil, and $\Delta \Phi_B / \Delta t$ is the rate at which the magnetic flux changes. The input primary voltage, V_p, is also related to the rate at which the flux changes[†]

$$V_p = N_p \frac{\Delta \Phi_B}{\Delta t},$$

where N_p is the number of turns in the primary coil. We divide these two equations to find

$$\frac{V_s}{V_p} = \frac{N_s}{N_p}. \tag{20-6}$$

This *transformer equation* tells how the secondary (output) voltage is related to the primary (input) voltage; V_s and V_p in Eq. 20-6 can be the rms values for both or peak values for both.

If N_s is greater than N_p, we have a *step-up transformer*. The secondary voltage is greater than the primary voltage. For example, if the secondary has twice as many turns as the primary, then the secondary voltage will be twice that of the primary. If N_s is less than N_p, we have a *step-down transformer*.

Although voltage can be increased (or decreased) with a transformer, we don't get something for nothing. Energy conservation tells us that the power output can be no greater than the power input. A well-designed transformer can be greater than 99 percent efficient so little energy is lost to heat. The power input thus essentially equals the power output; since power $P = VI$, we have

$$V_p I_p = V_s I_s$$

or

$$\frac{I_s}{I_p} = \frac{N_p}{N_s}. \tag{20-7}$$

☐   **EXAMPLE 20-6**   A transformer for a transistor radio reduces 120-V ac to 9.0-V ac. (Such a device also contains diodes to change the 9.0-V ac to dc. See Chapter 28.) The secondary contains 30 turns and the radio draws 400 mA. Calculate (*a*) the number of turns in the primary; (*b*) the current in the primary; (*c*) the power transformed.

    SOLUTION   (*a*) This is a step-down transformer, and from Eq. 20-6 we have

$$N_p = N_s \frac{V_p}{V_s} = (30)(120 \text{ V})/(9.0 \text{ V}) = 400 \text{ turns}.$$

[†] This follows from the fact that the changing flux produces a counter emf, $N_p \Delta \Phi_B / \Delta t$ in the primary that exactly balances the applied voltage V_p if the resistance of the primary can be ignored (Kirchhoff's rules).

(b) From Eq. 20-7:

$$I_p = I_s \frac{N_s}{N_p} = (0.40 \text{ A}) \left(\frac{30}{400} \right) = 0.030 \text{ A}.$$

(c) The power transformed is

$$P = I_s V_s = (9.0 \text{ V})(0.40 \text{ A}) = 3.6 \text{ W}$$

which is, assuming 100 percent efficiency, the same as the power in the primary, $P = (120 \text{ V})(0.030 \text{ A}) = 3.6 \text{ W}$.

It is important to recognize that a transformer operates only on ac. A dc current in the primary does not produce a changing flux and therefore induces no emf in the secondary. However, if a dc voltage is applied to the primary through a switch, at the instant the switch is opened or closed there will be an induced current in the secondary. For example, if the dc is turned on and off as shown in Fig. 20-16a, the voltage induced in the secondary is as shown in part (b). Notice that the secondary voltage drops to zero when the dc voltage is steady.

Transformers play an important role in the transmission of electricity. Power plants are often situated some distance from metropolitan areas; hydroelectric plants are located at a dam site and nuclear plants need much cooling water; fossil-fuel plants too are often situated far from a city because of lack of availability of land or to avoid contributing to air pollution. In any case, electricity must often be transmitted over long distances, and there is always some power loss in the transmission lines. This loss can be minimized if the power is transmitted at high voltage, as the following example shows.

FIGURE 20-16

A dc voltage turned on and off as shown in (a) produces voltage pulses in the secondary (b). Voltage scales in (a) and (b) are not necessarily the same.

EXAMPLE 20-7 An average of 120 kW of electric power is sent to a small town from a power plant 10 km away. The transmission lines have a total resistance of 0.40 Ω. Calculate the power loss if the power is transmitted at (a) 240 V and (b) 24,000 V.

SOLUTION (a) If 120 kW is sent at 240 V, the total current will be

$$I = \frac{1.2 \times 10^5 \text{ W}}{2.4 \times 10^2 \text{ V}} = 500 \text{ A}.$$

The power loss in the lines, P_L, is then

$$P_L = I^2 R = (500 \text{ A})^2 (0.40 \text{ Ω}) = 100 \text{ kW}.$$

Thus over 80 percent of all the power would be wasted as heat in the power lines!

(b) When $V = 24,000$ V

$$I = \frac{P}{V} = \frac{1.2 \times 10^5 \text{ W}}{2.4 \times 10^4 \text{ V}} = 5.0 \text{ A}.$$

The power loss is then

$$P_L = I^2 R = (5.0 \text{ A})^2 (0.40 \text{ Ω}) = 10 \text{ W},$$

which is less than 1/100 of 1 percent.

It should be clear that the greater the voltage, the less the current and thus the less power is wasted in the transmission lines. It is for this reason that power is usually transmitted at very high voltages, as high as 700 kV.

Power is generated at somewhat lower voltages than this, and the voltage in homes and factories is also much lower. The great advantage of ac, and a major reason it is in nearly universal use, is that the voltage can easily be stepped up and down by a transformer. The output voltage of an electric generating plant is stepped up prior to transmission. Upon arrival in a city, it is stepped down in stages at electric substations prior to distribution. The voltage in lines along city streets is typically 2400 V and is stepped down to 240 V or 120 V for home use by transformers.

Dc transmission has gained in popularity recently. Although changing voltage with dc is more difficult and expensive, it offers some advantages over ac. A few of these are as follows. Ac produces alternating magnetic fields which induce current in nearby wires and reduces transmitted power; this is absent in dc. Dc can be transmitted at a higher average voltage than ac since for dc the rms value equals the peak; and breakdown of insulation or of air must be avoided for the peak voltage.

*20-8 ∎ Transducers using induction: magnetic microphone, phonograph cartridge, geophone

Many microphones work on the principle of induction. In one form, a microphone is just the inverse of a loudspeaker (Section 19-8). A small coil connected to a membrane is suspended close to a small permanent magnet. The coil moves in the magnetic field when sound waves strike the membrane; the frequency of the induced emf will be just that of the impinging sound waves. In a "ribbon" microphone, a thin metal ribbon is suspended between the poles of a permanent magnet. The ribbon vibrates in response to sound waves and the emf induced in the ribbon is proportional to its velocity.

In one type of phonograph cartridge, the needle which follows the grooves of a phonograph record is connected to a tiny magnet inside the cartridge in the arm of the record player. The magnet is suspended inside a tiny coil of wire and an emf is induced in the coil according to the motion of the magnet. This tiny emf, at frequencies corresponding to the wavy shape of the record grooves, is amplified electronically before sending it to a loudspeaker. Most good-quality cartridges are of the magnetic type. Others, sometimes called "crystal pickups," make use of the piezoelectric effect (see Section 18-11).

Microphones and cartridges are two more types of transducer; each converts a mechanical signal (moving air, moving stylus) into an electrical one. Another transducer is the head of a tape recorder. Recorded tape has been magnetized at different intensities along its length corresponding to the program material. As tape passes the head, a tiny emf is induced in an induction coil due to the changing magnetic field passing the head. This signal too is amplified before going to the speaker.

In another field, geophysics, an important transducer based on electromagnetic induction is the *geophone* (sometimes referred to as a *seismometer*).

A geophone is placed in direct contact with the earth and converts the motion of the earth—whether due to an earthquake or to an explosion (such as for mineral prospecting or for detecting a bomb test)—into an electrical signal. A geophone contains a magnet and a coil of wire, one of which is fixed rigidly to the case, which moves as the earth does where it is planted; the other element is inertial and is suspended from the case by a spring. In the type shown in Fig. 20-17, the coil moves with the earth, and the relative motion of the magnet and coil produces an induced emf in the coil, which is the output of the device. In many geophones, the coil is inertial and the magnet moves with the earth.

FIGURE 20-17
A geophone.

*20-9 ∎ Inductance

Mutual inductance. If two coils of wire are placed near one another, as in Fig. 20-18, a changing current in one will induce an emf in the other. According to Faraday's law, the emf $\mathscr{E}_2$ induced in coil 2 is proportional to the rate of change of flux passing through it. Since the flux is proportional to the current flowing in coil 1, $\mathscr{E}_2$ must be proportional to the rate of change of the current in coil 1, $\Delta I_1 / \Delta t$. Thus we can write

$$\mathscr{E}_2 = -M \frac{\Delta I_1}{\Delta t}, \qquad (20\text{-}8a)$$

where the constant of proportionality, M, is called the **mutual inductance.** (The minus sign is simply a reflection of Lenz's law.) It has units of $V \cdot s/A = \Omega \cdot s$, which is called the *henry* (H): 1 H = 1 $\Omega \cdot s$. The value of M depends on whether iron is present or not, on the size of the coils, on the number of turns, and on their separation; for example, the closer the coils, the more lines of flux will pass through coil 2, so M will be greater. In some arrangements M can be calculated; more usually it is measured experimentally. If we look at the inverse situation—a changing current in coil 2 inducing an emf in coil 1—the proportionality constant, M, turns out to have the same value,

$$\mathscr{E}_1 = -M \frac{\Delta I_2}{\Delta t}. \qquad (20\text{-}8b)$$

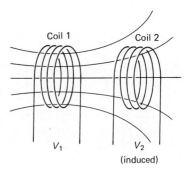

FIGURE 20-18
A changing current in one coil will induced a current in the second coil.

A transformer is an example of mutual inductance in which the coupling is maximized so that nearly all flux lines pass through both coils. However, mutual inductance has other uses as well. For example, some pacemakers, which are used to maintain blood flow in heart patients (Section 18-13), are powered externally. Power in an external coil is transmitted via mutual inductance to a second coil in the pacemaker at the heart. This has the advantage over battery-powered pacemakers that surgery is not needed to replace a battery when it wears out.

Mutual inductance can sometimes be a problem, however. Any alternating current in a circuit can induce an emf in another part of the same circuit or in a different circuit even though the conductors are not in the shape of a coil. The mutual inductance M is usually small unless multiturn coils and/or iron cores are involved. However, in situations where small signals are present, problems due to mutual inductance often arise; shielded cable, in which an inner conductor is surrounded by a cylindrical grounded conductor, is often used to reduce the problem.

Self-inductance. The concept of inductance applies also to an isolated single coil. When a changing current passes through a coil or solenoid, a changing magnetic flux is produced inside the coil, and this in turn induces an emf. This induced emf opposes the change in flux (Lenz's law); it is much like the back emf generated in a motor. For example, if the current through the coil is increasing, the increasing magnetic flux induces an emf that opposes the original current and tends to retard its increase. If the current is decreasing in the coil, the decreasing flux induces an emf in the same direction as the current, tending to maintain the original current. In either case, the induced emf $\mathscr{E}$ is proportional to the rate of change in current:

$$\mathscr{E} = -L \frac{\Delta I}{\Delta t}. \tag{20-9}$$

The constant of proportionality L is called the **self-inductance,** or simply the *inductance* of the coil. It, too, is measured in henries. The magnitude of L depends on the geometry and on the presence of an iron core.

An ac circuit always contains some inductance, but often it is quite small unless the circuit contains a coil of many turns. A coil that has significant self-inductance L is called an **inductor** or a *choke coil.* It is shown on circuit diagrams by the symbol ⌐◯◯◯◯⌐ . It can serve a useful purpose in certain circuits. Often, inductance is to be avoided in a circuit. Precision resistors are normally wire-wound and thus would have inductance as well as resistance; the inductance can be minimized by winding the wire back on itself so that the current going in the two directions cancels and little magnetic flux is produced; this is called a "noninductive winding."

If an inductor has negligible resistance, it is the inductance, or the back emf, that controls the current. If a source of alternating voltage is applied to the coil, this applied voltage will just be balanced by the induced emf of the coil given by Eq. 20-9. Thus we can see from Eq. 20-9 that if the inductance L is large, the change in the current—and therefore the current itself—will be small. The greater the inductance, the less the current. An inductance thus acts something like a resistance to impede the flow of alternating current; we use the term **impedance** for this quality of an inductor. We shall discuss impedance more fully in Sections 20-12 to 14, and we shall see that it depends not only on L but also on the frequency. Here we mention one example of its importance. The resistance of the primary in a transformer is usually quite small, perhaps less than 1 Ω. If resistance alone limited the current, tremendous currents would flow when a high voltage was applied. Indeed, a dc voltage applied to a transformer can burn it out. It is the impedance of the coil to an alternating current (or its "back" emf) that limits the current to a reasonable value.

□ **EXAMPLE 20-8** (*a*) Determine a formula for the self-inductance L of a solenoid (a long coil) containing N turns of wire in its length l and whose cross-sectional area is A. (*b*) Calculate the value of L if $N = 100$, $l = 5.0$ cm,

$A = 3.0 \text{ cm}^2$, and the solenoid is air filled. (c) Calculate L if the solenoid has an iron core with $\mu = 4000\mu_0$.

SOLUTION (a) According to Eq. 19-10, the magnetic field inside a solenoid is $B = \mu_0 nI$, where $n = N/l$. From Eqs. 20-2 and 20-9 we have $\mathscr{E} = -N(\Delta\Phi_B/\Delta t) = -L(\Delta I/\Delta t)$. Thus $L = N(\Delta\Phi_B/\Delta I)$. Since $\Phi_B = BA = \mu_0 NIA/l$, then any change in I causes a change in flux $\Delta\Phi_B = \mu_0 NA\,\Delta I/l$. Thus

$$L = N\frac{\Delta\Phi_B}{\Delta I} = \frac{\mu_0 N^2 A}{l}.$$

(b) Since $\mu_0 = 4\pi \times 10^{-7} \text{ T·m/A}$

$$L = \frac{(4\pi \times 10^{-7} \text{ T·m/A})(100)^2(3.0 \times 10^{-4} \text{ m}^2)}{(5.0 \times 10^{-2} \text{ m})} = 75 \text{ }\mu\text{H}.$$

(c) Here we replace μ_0 by $\mu = 4000\mu_0$, so L will be 4000 times larger: $L = 0.30 \text{ H} = 300 \text{ mH}$.

*20-10 ∎ Energy stored in a magnetic field

In Section 16-9 we saw that the energy stored in a capacitor is equal to $\frac{1}{2}CV^2$. By using a similar argument, we can show that the energy stored in an inductance L, carrying a current I, is

$$\text{energy} = \tfrac{1}{2}LI^2.$$

Just as the energy stored in a capacitor can be considered to reside in the electric field between its plates, so the energy if an inductor can be considered to be stored in its magnetic field.

To write the energy in terms of the magnetic field, let us use the result of the example above, that the inductance of a solenoid is $L = \mu_0 N^2 A/l$. Now the magnetic field B in a solenoid is related to the current I by $B = \mu_0 NI/l$. Thus $I = Bl/\mu_0 N$, and

$$\text{energy} = \tfrac{1}{2}LI^2 = \tfrac{1}{2}\left(\frac{\mu_0 N^2 A}{l}\right)\left(\frac{Bl}{\mu_0 N}\right)^2$$

$$= \tfrac{1}{2}\frac{B^2}{\mu_0} Al.$$

We can think of this energy as residing in the volume enclosed by the windings, which is Al. Then the energy per unit volume, or *energy density*, is

$$\text{energy density} = \tfrac{1}{2}\frac{B^2}{\mu_0}. \tag{20-10}$$

This formula, which was derived for the special case of a solenoid, can be shown to be valid for any region of space where a magnetic field exists. If a ferromagnetic material is present, μ_0 is replaced by μ. This equation is analogous to that for an electric field, $\frac{1}{2}\varepsilon_0 E^2$, Eq. 16-9.

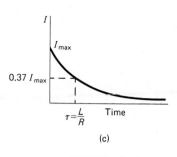

(a)

(b)

(c)

FIGURE 20-19

(a) LR circuit; (b) growth of current when connected to battery; (c) decay of current when battery is shorted out.

*20-11 ∎ LR circuit

Any inductor will have some resistance. We represent this situation by drawing the inductance L and the resistance R separately, as in Fig. 20-19a. The resistance R could also include a separate resistor connected in series. Now we ask, what happens when a dc source is connected in series to such an LR circuit. At the instant the switch connecting the battery is closed, the current starts to flow. It is, of course, opposed by the induced emf in the inductor. However, as soon as current starts to flow, there is a voltage drop across the resistance. Hence the voltage drop across the inductance is reduced and there is then less impedance to the current flow from the inductance. The current thus rises gradually as shown in Fig. 20-19b; it approaches the steady value $I_{max} = V/R$ when all the voltage drop is across the resistance. It can be shown that the time required to reach $0.63I_{max}$ is $\tau = L/R$. This is called the *time constant* of the circuit.

If the battery is suddenly shorted out, the current drops off as shown in Fig. 20-19c. The time constant is the time for the current to drop to 37 percent of the original value, and again equals L/R.

These graphs show that there is always some "reaction time" when an electromagnet, for example, is turned on or off. We also see that an LR circuit has properties similar to an RC circuit (Section 18-6). Unlike the capacitor case, however, the time constant here is *inversely* proportional to R.

*20-12 ∎ AC circuits and impedance

We have previously discussed circuits that contain combinations of resistor, capacitor, and inductor, but only when they are connected to a dc source of emf or to no source (as in the discharge of a capacitor in an RC circuit). Now we discuss these circuit elements when connected to a source of alternating emf.

First we examine, one at a time, how a resistor, a capacitor, and an inductor behave when connected to a source of alternating emf, represented by the symbol —Ⓢ—, which produces a sinusoidal voltage of frequency f; and we assume in each case that it gives rise to a current

$$I = I_0 \cos 2\pi ft$$

where t is time and I_0 is the peak current. We must remember (Section 17-6) that $V_{rms} = V_0/\sqrt{2}$ and $I_{rms} = I_0/\sqrt{2}$ (Eq. 17-8).

Resistor. When an ac source is connected to a resistor as in Fig. 20-20a, the current increases and decreases with the alternating emf according to Ohm's law: $I = V/R$. This is graphed in Fig. 20-20b, where the voltage is the solid curve and the current is the dashed curve. Because the current is zero when the voltage is zero and the current reaches a peak when the voltage does, we say that the current and voltage are *in phase.*

Energy is transformed into heat, as discussed in Section 17-6, at an average rate $P = \overline{IV} = I_{rms}^2 R = V_{rms}^2/R$.

Inductor. In Fig. 20-21a an inductor of inductance L, represented by the symbol ⎵‿‿‿‿⎵, is connected to the ac source. We ignore any resistance it might have (it is usually small). The voltage applied to the inductor will be equal to the "back" emf generated in the inductor by the changing current as given by Eq. 20-9. This is because the sum of the emfs around any closed circuit must be zero, as Kirchhoff's rule tells us. Thus

$$V - L\frac{\Delta I}{\Delta t} = 0$$

or

$$V = L\frac{\Delta I}{\Delta t},$$

where V is the sinusoidally varying voltage of the source and $L\,\Delta I/\Delta t$ is the voltage induced in the inductor. According to this equation, I is increasing most rapidly when V has its maximum value, $V = V_0$. And I will be decreasing most rapidly when $V = -V_0$. These two instants correspond to points d and b on the graph of voltage versus time in Fig. 20-21b. At points a and c, $V = 0$; the equation above tells us that $\Delta I/\Delta t = 0$ at these instants, so I is not changing; thus the curve of I versus t at these points is neither increasing nor decreasing and therefore corresponds to the maximum and minimum values of the current I. These correspond to points a and c on the graph. Going point by point in this manner, the curve of I versus t as compared to that for V versus t can be constructed, and they are shown by the dashed and solid lines respectively in Fig. 20-21b. Notice that the current reaches its peaks (and troughs) one-quarter of a cycle after the voltage does. We say that *in an inductor, the current lags the voltage by 90°.* (Remember that 360° corresponds to a full cycle, so 90° is a quarter cycle.) Alternatively, we can say that the voltage leads the current by 90°. (Since we originally chose $I = I_0 \cos 2\pi ft$, we see from the graph that V must vary as $V = -V_0 \sin 2\pi ft$.)

Because the current and voltage are out of phase by 90°, no energy is transformed in an inductor on the average; and no energy is dissipated as thermal energy. This can be seen as follows. From point c to d, the voltage is increasing from zero to its maximum. The current, however, is in the opposite direction to the voltage and is approaching zero. The average power over this interval, VI, is negative. From d to e, however, both V and I are positive so VI is positive; this contribution just balances the negative contribution of the previous quarter-cycle. Similar considerations apply to the rest of the cycle. Thus the average power transformed over one or many cycles is zero. We can see that energy from the source passes into the

(a)

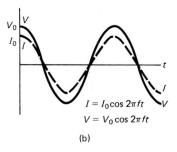

$I = I_0 \cos 2\pi ft$
$V = V_0 \cos 2\pi ft$

(b)

FIGURE 20-20

Resistor connected to ac source. Current is in phase with the voltage.

FIGURE 20-21

Inductor connected to ac source. Current lags voltage by a quarter cycle, or 90°.

(a)

$I = I_0 \cos 2\pi ft$
$V = -V_0 \sin 2\pi ft$

(b)

magnetic field of the inductor, where it is stored momentarily; then the field decreases and the energy is transferred back to the source. None is dissipated in this process. Compare this to a resistor where the current is always in the same direction as the voltage and energy is transferred out of the source and never back into it. (The product VI is never negative.) This energy is not stored, but is transformed to thermal energy in the resistor.

As mentioned in Section 20-9, the back emf of an inductor impedes the flow of an ac current. Indeed, it is found that the magnitude of the current in an inductor is directly proportional to the applied ac voltage at a given frequency. We can therefore write an equation for a pure inductor (no resistance) that is the equivalent of Ohm's law:

$$V = IX_L, \tag{20-11a}$$

where X_L is called the *inductive reactance* or *impedance* of the inductor. Normally, we use the term "reactance" to refer solely to the inductive properties. We then reserve the term "impedance" to include the total "impeding" qualities of the coil—its inductance plus any resistance it may have (more on this in the next section). In the absence of resistance, the impedance is the same as the reactance.

The quantities V and I in Eq. 20-11a can refer either to rms or to peak values. Note, however, that although this equation may relate the peak values, the peak current and voltage are not reached at the same time, so this equation is *not* valid at a particular instant, as is the case for Ohm's law; it is valid only on the average.

From the fact that $V = L\,\Delta I/\Delta t$, we see that the larger L is, the less will be the change in current ΔI during the time Δt; hence I itself will be smaller at any instant for a given frequency. Thus $X_L \propto L$. The reactance X_L also depends on the frequency; the greater the frequency, the more rapidly the magnetic flux changes in the inductor. If the emf induced by this field is to remain equal to the source emf, as it must, the magnitude of the current must then be less. Hence the greater the frequency, the greater is X_L, so $X_L \propto fL$. This is also consistent with the fact that if the frequency f is zero (so the current is dc), there is no back emf and no impedance to the flow of charge. Careful calculation (using calculus), as well as experiment, shows that the constant of proportionality is 2π. Thus

$$X_L = 2\pi f L. \tag{20-11b}$$

EXAMPLE 20-9 A coil has a resistance $R = 1.00\ \Omega$ and an inductance of 0.300 H. Determine the current in the coil if (*a*) 120 V dc is applied to it; (*b*) 120 V ac (rms) at 60.0 Hz is applied.

SOLUTION (*a*) There is no inductive impedance ($X_L = 0$ since $f = 0$), so we apply Ohm's law for the resistance:

$$I = \frac{V}{R} = \frac{120\ \text{V}}{1.00\ \Omega} = 120\ \text{A}.$$

(*b*) The inductive reactance in this case is

$$X_L = 2\pi f L = (6.28)(60.0\ \text{s}^{-1})(0.300\ \text{H}) = 113\ \Omega.$$

In comparison to this, the resistance can be ignored. Thus

$$I_{rms} = \frac{V_{rms}}{X_L} = \frac{120\ V}{113\ \Omega} = 1.06\ A.$$

(It might be tempting to say that the total impedance is $113\ \Omega + 1\ \Omega = 114\ \Omega$; this might imply that about 1 percent of the voltage drop is across the resistor, or about 1 V; and that across the inductance is 119 V. Although the 1 V_{rms} across the resistor is correct, the other statements are *not* true because of the alteration in phase in an inductor. This will be discussed in the next section.)

Capacitor. When a capacitor is connected to a battery, the capacitor plates quickly acquire equal and opposite charges; but no steady current flows in the circuit. A capacitor prevents the flow of a dc current (unless the capacitor is "leaky," in which case a leakage current would exist across the gap). However, if a capacitor is connected to an alternating source of voltage, as in Fig. 20-22a, an alternating current will flow continuously. This can happen because when the ac voltage is first turned on, charge begins to flow so that one plate acquires a negative charge and the other a positive charge. But when the voltage reverses itself, the charges flow in the opposite direction. Thus for an alternating applied voltage, an ac current is present in the circuit continuously.

Let us look at this process in more detail. First, we recall that the applied voltage must equal the voltage across the capacitor: $V = Q/C$, where C is the capacitance and Q the charge on the plates. Thus the charge Q on the plates follows the voltage; when the voltage is zero, the charge is zero; when the voltage is a maximum, the charge is a maximum. But what about the current I? At point a in Fig. 20-22b, when the voltage starts increasing, the charge on the plates is zero; thus charge flows readily toward the plates and the current is large. As the voltage approaches its maximum of V_0 (point b), the charge that has accumulated on the plates tends to prevent more charge from flowing, so the current drops to zero at point b. The charge that has accumulated now starts to flow off the plates and the current again increases (dashed curve), but in the opposite direction; it reaches a maximum (negatively) when the voltage is at point c. Thus the current follows the graph shown dashed in Fig. 20-22b. Like an inductor, the voltage and current are out of phase by 90°. But for a capacitor, the *current leads the voltage by* 90°. (Or the voltage lags the current by 90°.) This is the opposite of what happens for an inductor. (Again we have chosen $I = I_0 \cos 2\pi ft$, and we see from the graph that $V = V_0 \sin 2\pi ft$.)

Because the current and voltage are out of phase, the average power dissipated is zero just as for an inductor. Energy from the source is fed to the capacitor, where it is stored in the electric field between its plates. As the field decreases the energy returns to the source. Thus, in an ac circuit, *only a resistance will dissipate energy* to thermal energy.

A relationship between the applied voltage and the current can be written just as for an inductance:

$$V = IX_C, \tag{20-12a}$$

(a)

$$I = I_0 \cos 2\pi ft$$
$$V = V_0 \cos 2\pi ft$$

(b)

FIGURE 20-22

Capacitor connected to an ac source. Current leads voltage by a quarter cycle, or 90°.

where X_C is the *capacitive reactance* (or *impedance*) of the capacitor. This equation relates the rms or peak values for the voltage and current but is not valid at a particular instant because I and V are out of phase. X_C depends on both the capacitance C and the frequency f. The larger the capacitance, the more charge it can handle, so the less it will retard the flow of an alternating current; hence X_C will be inversely proportional to C. It is also inversely proportional to the frequency f, since, when the frequency is higher, there is less time per cycle for the charge to build up on the plates and impede the flow. Again there is a factor of 2π, and

$$X_C = \frac{1}{2\pi f C}.$$ (20-12b)

Notice that for dc conditions, $f = 0$ and X_C becomes infinite. This is as it should be, since a capacitor does not pass dc current.

It is interesting to note that the reactance of an inductor increases with frequency, but that of a capacitor decreases with frequency.

EXAMPLE 20-10   What are the peak and rms currents in the circuit of Fig. 20-22a if $C = 1.0 \ \mu F$ and $V_{rms} = 120$ V? Calculate for (a)$f = 60$ Hz, and then for (b) $f = 6.0 \times 10^5$ Hz.

SOLUTION   (a) $V_0 = \sqrt{2} V_{rms} = 170$ V. Then

$$X_C = \frac{1}{2\pi f C} = \frac{1}{(6.28)(60 \text{ s}^{-1})(1.0 \times 10^{-6} \text{ F})} = 2.7 \text{ k}\Omega.$$

Thus

$$I_0 = \frac{V_0}{X_C} = \frac{170 \text{ V}}{2.7 \times 10^3 \ \Omega} = 63 \text{ mA},$$

$$I_{rms} = \frac{V_{rms}}{X_C} = \frac{120 \text{ V}}{2.7 \times 10^3 \ \Omega} = 44 \text{ mA}.$$

(b) For $f = 6.0 \times 10^5$ Hz, X_C will be 0.27 Ω, $I_0 = 630$ A, and $I_{rms} = 440$ A. The dependence on f is dramatic.

Capacitors are used for a variety of purposes, some of which we have already described. Two other applications are illustrated in Fig. 20-23. In part (a), circuit A is said to be capacitively coupled to circuit B. The purpose of the capacitor is to prevent a dc voltage from passing from A to B but allowing an ac signal to pass relatively unimpeded. If C is sufficiently large, the ac signal will not be significantly attenuated. The capacitor in part (b) also passes ac but not dc. In this case, a dc voltage can be maintained between circuits A and B; and if the capacitance C is large enough, the capacitor offers little impedance to an ac signal leaving A. Such a signal then passes to ground instead of into B. Thus the capacitor in part (b) acts like a filter when a constant dc voltage is required; any sharp variation in voltage will pass to ground instead of into circuit B. Capacitors used in these two ways are very common in circuits.

FIGURE 20-23

Two common uses for a capacitor.

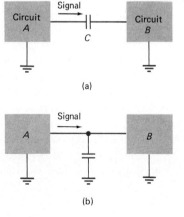

(a)

(b)

*20-13 ∎ *LRC* series AC circuit

We now examine a circuit containing all three elements in series, a resistor *R*, an inductor *L*, and a capacitor *C* (Fig. 20-24). If a given circuit contains only two of these elements, we can still use the results of this section by setting $R = 0$, $L = 0$, or $C = \infty$ (infinity) as needed. We let V_R, V_L, and V_C represent the voltage across each element at a *given instant* in time; and V_{R0}, V_{L0}, and V_{C0} represent the *maximum* (peak) values of these voltages. The voltage across each of the elements will follow the phase relations we discussed in the last section. That is, V_R will be in phase with the current, V_L will lead the current by 90°, and V_C will lag behind the current by 90°. Also, at any instant the total voltage *V* supplied by the source will be $V = V_R + V_L + V_C$. However, because the various voltages are not in phase, the rms voltages (which is what ac voltmeters usually measure) will not simply add up to give the rms voltage of the source. And V_0 will *not* equal $V_{R0} + V_{L0} + V_{C0}$. Let us now examine the circuit in detail. What we would like to find in particular is the impedance of the circuit as a whole, the rms current that flows, and the phase difference between the source voltage and the current.

First we note that the current at any instant must be the same at all points in the circuit. Thus the currents in each element are in phase with each other, although the voltages are not. We choose our origin in time, $t = 0$, so that the current *I* at any time *t* is $I = I_0 \cos 2\pi ft$.

It is convenient to analyze an *LRC* circuit using a *phasor* diagram. Arrows (like vectors) are drawn in an *xy* coordinate system. The lengths of these arrows represent the magnitudes of the peak voltages across each element:

$$V_{R0} = I_0 R, \qquad V_{L0} = I_0 X_L, \qquad \text{and} \qquad V_{C0} = I_0 X_C.$$

V_{R0} is in phase with the current and is initially drawn along the positive *x* axis. Since V_{L0} leads the current by 90°, it leads V_{R0} by 90°, so is initially drawn along the positive *y* axis. V_{C0} lags the current by 90°, so lags V_{R0} by 90°; hence V_{C0} is drawn initially along the negative *y* axis. Such a diagram is shown in Fig. 20-25a. If we let the whole diagram rotate counterclockwise at frequency *f*, we get the diagram shown in Fig. 20-25b; after a time, *t*, each arrow has rotated through an angle $2\pi ft$. Then the *projections of each arrow on the x axis represent the voltages across each element at the instant t.* For example, the projection of V_{R0} on the *x* axis is $V_{R0} \cos 2\pi ft$ (as in Fig. 20-20); and the projections of V_{L0} and V_{C0} on the *x* axis are $-V_{L0} \sin 2\pi ft$ and $V_{C0} \sin 2\pi ft$ as in Figs. 20-21b and 20-22b. Maintaining the 90° angle between each vector ensures the correct phase relations. Although these facts show the validity of a phasor diagram, what we are really interested in is how to add the voltages.

The sum of the projections of the three vectors on the *x* axis is equal to the projection of their sum. But the sum of the projections represents the instantaneous voltage across the whole circuit, *V* (equal to the source voltage). Therefore, the vector sum of these vectors will be the vector that represents the peak source voltage, V_0. This is shown in Fig. 20-26 where it is seen that V_0 makes an angle ϕ with I_0 and V_{R0}. As time passes, V_0 rotates with the other vectors, so the instantaneous voltage *V* (projection of V_0 on the *x* axis) is

$$V = V_0 \cos (2\pi ft + \phi).$$

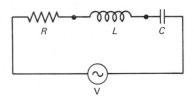

FIGURE 20-24
An *LRC* circuit.

(a)

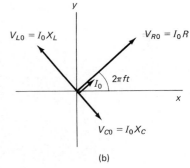

(b)

FIGURE 20-25
Phasor diagram for a series *LRC* circuit.

FIGURE 20-26
Phasor diagram for a series *LRC* circuit showing the sum vector, V_0.

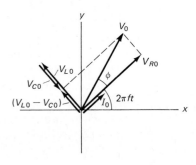

Thus we see that the voltage from the source is out of phase[†] with the current by an angle ϕ.

From this analysis we can now draw some useful conclusions. First, we determine the total impedance Z of the circuit, which is defined by the relation

$$V_{rms} = I_{rms}Z, \quad \text{or} \quad V_0 = I_0Z. \qquad (20\text{-}13)$$

From Fig. 20-26 we see, using the Pythagorean theorem (V_0 is the hypotenuse of a right triangle), that

$$V_0 = \sqrt{V_{R0}^2 + (V_{L0} - V_{C0})^2}$$
$$= \sqrt{I_0^2R^2 + (I_0X_L - I_0X_C)^2}$$
$$= I_0\sqrt{R^2 + (X_L - X_C)^2}.$$

Thus, from Eq. 20-13,

$$Z = \sqrt{R^2 + (X_L - X_C)^2} \qquad (20\text{-}14a)$$
$$= \sqrt{R^2 + (2\pi fL - 1/2\pi fC)^2}. \qquad (20\text{-}14b)$$

This gives the total impedance of the circuit. Also from Fig. 20-26 we can find the phase angle ϕ:

$$\cos\phi = \frac{V_{R0}}{V_0} = \frac{I_0R}{I_0Z},$$
$$= \frac{R}{Z}. \qquad (20\text{-}15)$$

Finally, we can determine the power dissipated in the circuit. We saw earlier that power is only dissipated by a resistance; none is dissipated by inductance or capacitance. Therefore, the average power $\bar{P} = I_{rms}^2R$. But from Eq. 20-15, $R = Z\cos\phi$. Therefore,

$$\bar{P} = I_{rms}^2 Z\cos\phi$$
$$= I_{rms}V_{rms}\cos\phi. \qquad (20\text{-}16)$$

The factor $\cos\phi$ is referred to as the *power factor* of the circuit. For a pure resistor, $\cos\phi = 1$ and $\bar{P} = I_{rms}V_{rms}$. For a capacitor or inductor, $\cos\phi = 0$ and no power is dissipated.

The test of this analysis is, of course, in experiment; and experiment is in full agreement with these results.

☐ EXAMPLE 20-11 Suppose that $R = 25.0\ \Omega$, $L = 30.0$ mH, and $C = 12.0\ \mu$F in Fig. 20-24, and that they are connected to a 90.0-V ac (rms) 500-Hz source. Calculate (*a*) the current in the circuit, (*b*) the voltmeter readings (rms) across each element, (*c*) the phase angle ϕ, and (*d*) the power dissipated in the circuit.

[†] As a check, note that *if* $R = X_C = 0$, then $\phi = 90°$, and V_0 would lead the current by 90°, as it must for an inductor alone. Similarly, if $R = L = 0$, $\phi = -90°$ and V_0 would lag the current by 90°, as it must for a capacitor alone.

$$X_L = 2\pi f L = 94.2 \ \Omega$$

$$X_C = \frac{1}{2\pi f C} = 26.5 \ \Omega$$

Then

$$Z = \sqrt{R^2 + (X_L - X_C)^2} = \sqrt{(25.0)^2 + (94.2 \ \Omega - 26.5 \ \Omega)^2}$$
$$= 72.2 \ \Omega.$$

From Eq. 20-13,

$$I_{rms} = \frac{V_{rms}}{Z} = \frac{90.0 \ V}{72.2 \ \Omega} = 1.25 \ A.$$

(b) The rms voltage across each element is

$$(V_R)_{rms} = I_{rms}R = (1.25 \ A)(25.0 \ \Omega) = 31.2 \ V$$
$$(V_L)_{rms} = I_{rms}X_L = 118 \ V$$
$$(V_C)_{rms} = I_{rms}X_C = 33.1 \ V.$$

Notice that these do *not* add up to give the source voltage, $90V_{rms}$. Indeed, the rms voltage across the inductance *exceeds* the source voltage. This is due to the fact that the different voltages are out of phase with each other. The instantaneous voltages at any time do add up, of course; this can happen because at any instant one voltage can be negative, to compensate for a large positive voltage of another. The rms voltages, however, are always positive by definition.

(c) Since $\cos \phi = R/Z = 25.0 \ \Omega/72.2 \ \Omega = 0.346$, then $\phi = 69.7°$.

(d) $\bar{P} = I_{rms}V_{rms} \cos \phi = (1.25 \ A)(90.0 \ V)(0.346) = 38.9 \ W.$

*20-14 ▮ Resonance in ac circuits; oscillators

The rms current in an *LRC* series circuit is given by (see Eqs. 20-13 and 20-14):

$$I_{rms} = \frac{V_{rms}}{Z} = \frac{V_{rms}}{\sqrt{R^2 + \left(2\pi f L - \dfrac{1}{2\pi f C}\right)^2}}. \tag{20-17}$$

Because the impedance of inductors and capacitors depends on the frequency f of the source, the current in an *LRC* circuit will depend on frequency. From Eq. 20-17 we can see that the current will be maximum at a frequency such that

$$2\pi f L - \frac{1}{2\pi f C} = 0.$$

We solve this for f, and call the solution f_0:

$$f_0 = \frac{1}{2\pi} \sqrt{\frac{1}{LC}}. \tag{20-18}$$

This is the *resonant frequency* of the circuit. At this frequency, $X_C = X_L$, so the impedance is purely resistive and $\cos \phi = 1$. A graph of I_{rms} versus f is shown in Fig. 20-27 for particular values of R, L, and C. For smaller R compared to X_L and X_C, the resonance peak will be higher and sharper.

When R is very small, we speak of an *LC circuit;* the energy in such a circuit oscillates, at frequency f_0, between the inductor and the capacitor, with some being dissipated in R (some resistance is unavoidable). To see this in more detail, consider a perfect *LC* circuit in which $R = 0$; assume at $t = 0$ that the capacitor C is charged and the switch is closed (Fig. 20-28). The capacitor immediately begins to discharge. As it does so, the current I through the inductor increases. At every instant, the potential difference across the capacitor, $V = Q/C$ (where Q is the charge on the capacitor at that instant), must equal the potential difference across the inductor, which is equal to its emf, $-L(\Delta I/\Delta t)$. At the instant when the charge on the capacitor reaches zero ($Q = 0$), the current I in the inductor has reached its maximum value, but at this instant I is not changing ($-L\,\Delta I/\Delta t = Q/C = 0$). At this moment, the magnetic field B in the inductor is also a maximum. The current next begins to decrease as the flowing charge starts to accumulate on the opposite plate of the capacitor. When the current has dropped to zero, the capacitor has again attained its maximum charge. The capacitor then begins to discharge again, with the current now flowing in the opposite direction. This process of the charge flowing back and forth from one plate of the capacitor to the other, through the inductor, continues to repeat itself. This is called an *LC oscillation* or an *electromagnetic oscillation*. Not only does the charge oscillate back and forth, but so does the energy which oscillates between being stored in the electric field of the capacitor and in the magnetic field of the inductor.

Electric resonance is used in many electronic devices. Radio and television sets, for example, use resonant circuits for tuning in a station. Many frequencies reach the circuit, but a significant current flows only for frequencies at or near the resonant frequency. Either L or C is variable so that different stations can be tuned in (more on this in Chapter 21). *LC* circuits are also used in *oscillators*, which are devices that put out an oscillating signal of particular frequency. An example is the "bias frequency" for recording on a tape recorder. Since some resistance is always present, electrical oscillators generally need a periodic input of power to compensate for the energy converted to thermal energy in the resistance.

Electrical resonance is analogous to mechanical resonance, which we discussed in Chapter 13 (see Fig. 13-9). The energy transferred to the system is a maximum at resonance whether it is electrical resonance, the oscillation of a spring, or pushing a child on a swing (Section 13-6). That this is true in the electrical case can be seen from Eq. 20-16. At resonance, $\cos \phi = 1$, and I_{rms} is a maximum. A graph of power versus frequency looks much like that for the current (Fig. 20-27).

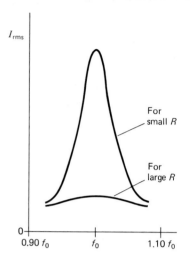

FIGURE 20-27

Current in an *LRC* circuit as a function of frequency, showing resonance peak at $f = f_0 = (\frac{1}{2\pi})\sqrt{1/LC}$.

FIGURE 20-28

A pure *LC* circuit.

An *LRC* circuit can have the elements arranged in parallel instead of in series. Resonance will occur in this case, too, but the analysis of such circuits is more involved.

▮ Summary

The *magnetic flux* passing through a loop is equal to the product of the area of the loop times the perpendicular component of the magnetic field strength. If the magnetic flux through a loop of wire changes in time, an emf is induced in the loop; the magnitude of the induced emf equals the time rate of change of the magnetic flux through the loop. This result is called *Faraday's law of induction*. The induced emf produces a current whose magnetic field opposes the original change in flux (*Lenz's law*). Faraday's law also tells us that a changing magnetic field produces an electric field; and that a straight wire of length l moving with speed v perpendicular to a magnetic field of strength B has an emf induced between its ends equal to Blv.

An electric *generator* changes mechanical energy into electrical energy. Its operation is based on Faraday's law: a coil of wire is made to rotate by mechanical means in a magnetic field, and the changing flux through the coil induces a sinusoidal current, which is the output of the generator.

A motor, which operates in the reverse of a generator, acts like a generator, in that a counter emf is induced in its rotating coil; since it opposes the input voltage, this counter emf can act to limit the current in a motor coil. Similarly, a generator acts somewhat like a motor in that a counter torque acts on its rotating coil.

A transformer, which is a device to change the magnitude of an ac voltage, consists of a primary and a secondary coil. The changing flux due to an ac voltage in the primary induces an ac voltage in the secondary. In a 100 percent efficient transformer, the ratio of output to input voltages (V_s/V_p) equals the ratio of the number of turns N_s in the secondary to the number N_p in the primary: $V_s/V_p = N_s/N_p$. The ratio of secondary to primary current is in the inverse ratio of turns: $I_s/I_p = N_p/N_s$.

▮ Questions

1. What would be the advantage, in Faraday's experiments (Fig. 20-1), of using coils with many turns?
2. What is the difference between magnetic flux and magnetic field strength? Discuss in detail.
3. Show, using Lenz's law, that the emf induced in the moving rod in Fig. 20-6 is positive at the bottom and negative at the top, so that the current flows clockwise in the circuit.
4. What direction will the current flow in Fig. 20-6 if the rod moves to the left, which decreases the area?
5. In situations where a small signal must travel over a distance, a "shielded cable" is used in which the signal wire is surrounded by an insulator and then enclosed by a cylindrical conductor. Why is a "shield" necessary?
6. What is the advantage of placing the two electric wires carrying ac close together?
7. In some early automobiles, the starter motor doubled as a generator to keep the battery charged once the car was started. Explain how this might work.

8. Explain why, exactly, the lights may dim briefly when a refrigerator motor starts up. (Hint: consider "terminal voltage.")
9. When a refrigerator starts up, the lights in older houses may dim briefly. When an electric heater is turned on, the lights may stay dimmed as long as it is on. Explain the difference.
10. Explain what is meant by the statement "A motor acts as a motor and generator at the same time." Can the same be said for a generator?
11. Use Fig. 20-8 and the right-hand rules to show why the counter torque in a generator *opposes* the motion.
12. Will an eddy current brake (Fig. 20-14) work on a copper or aluminum wheel, or must it be ferromagnetic?
13. It has been proposed that eddy currents be used to help sort solid waste for recycling. The waste is first ground into tiny pieces and iron removed with a dc magnet. The waste then is allowed to slide down an incline over permanent magnets. How will this aid in

the separation of nonferrous metals (Al, Cu, Pb, brass) from nonmetallic materials?

14. The primary of a transformer on a telephone pole has a resistance of 0.1 Ω and the input voltage is 2400 V ac. Can you estimate the current that will flow? Will it be 24,000 A? Explain.

15. A transformer designed for a 120-V ac input will often "burn out" if connected to a 120-V dc source. Explain. (*Hint:* the resistance of the primary coil is usually very low.)

16. If an aluminum sheet is held between the poles of a large bar magnet, it requires some force to pull it out of the magnetic field even though the sheet is not ferromagnetic and does not touch the pole faces. Explain.

17. A bar magnet falling inside a vertical metal tube reaches a terminal velocity even if the tube is evacuated so that there is no air resistance. Explain.

18. A metal bar, pivoted at one end, oscillates freely in the absence of a magnetic field; but in a magnetic field its oscillations are quickly damped out. Explain. (This *magnetic damping* is used in a number of practical devices.)

19. An enclosed transformer has four wire leads coming from it. How could you determine the ratio of turns on the two coils without taking the transformer apart? How would you know which wires paired with which?

20. The use of higher voltage lines in homes, say 600 V or 1200 V, would reduce energy waste. Why are they not used?

*21. Since a magnetic microphone is basically like a loudspeaker, could a loudspeaker actually serve as a micro-

phone? That is, could you speak into a loudspeaker and obtain an output signal that could be amplified? Explain. Discuss, in light of your response, how a microphone and loudspeaker differ in construction.

*22. How would you arrange two flat circular coils so that their mutual inductance was (*a*) greatest, (*b*) least (without separating them by a great distance)?

*23. If the two coils in Fig. 20-18 were connected electrically, would there still be mutual inductance?

*24. If you are given a fixed length of wire, how would you shape it to obtain the greatest self-inductance? The least?

*25. Does the emf of the battery in Fig. 20-19a affect the time needed for the *LR* circuit to reach (*a*) a given fraction of its maximum possible current, (*b*) a given value of current?

*26. How can we justify using phasor diagrams, with the potential differences as vectors, when we know that potential difference is not a vector?

*27. Can you tell whether the current in an *LRC* circuit leads or lags the applied voltage from a knowledge of the power factor, cos *ϕ*?

*28. Under what conditions is the impedance in an *LRC* circuit a minimum?

*29. An *LC* resonance circuit is often called an *oscillator* circuit. What is it that oscillates?

*30. Compare the oscillations of an *LC* circuit to the vibration of a mass *m* on a spring. What do *L* and *C* correspond to in the mechanical system?

▌Problems

SECTIONS 20-1 TO 20-4

1. (I) A 10-cm-diameter circular loop of wire is in a 0.50-T magnetic field. It is removed from the field in 0.10 s. What is the average induced emf?

2. (I) The rectangular loop shown in Fig. 20-29 is being pulled to the left, out of the magnetic field which points inward as shown. In what direction is the induced current?

3. (I) The magnetic field perpendicular to a circular loop of wire 12 cm in diameter is changed from +0.35 T to −0.15 T in 90 ms, where + means the field points

away from an observer and − toward the observer. (*a*) Calculate the induced emf. (*b*) In what direction does the induced current flow?

4. (I) The moving rod in Fig. 20-6 is 12.0 cm long and moves with a speed of 15.0 cm/s. If the magnetic field is 0.800 T, calculate (*a*) the emf developed, and (*b*) the electric field strength in the rod.

5. (I) The magnetic flux through a coil of wire containing two loops changes from −20 Wb to +15 Wb in 1.4 s. What is the emf induced in the coil?

6. (II) Suppose you are looking along a line through the centers of two circular (but separate) wire loops, one behind the other. A battery is suddenly connected to the front loop, establishing a clockwise current. (*a*) Will a current be induced in the second loop? (*b*) If so, when does this current start? (*c*) When does it stop? (*d*) In what direction is this current? (*e*) Is there a force between the two loops? (*f*) If so, in what direction?

FIGURE 20-29

FIGURE 20-30

7. (II) If the solenoid in Fig. 20-30 is being pulled away from the loop shown, in what direction is the induced current in the part of the loop closest to the viewer?

8. (II) In Fig. 20-6, the rod moves with a speed of 2.13 m/s, is 14.0 cm long, and has negligible resistance. The magnetic field is 0.215 T and the resistance of the U-shaped conductor is 22.5 Ω at a given instant. Calculate (a) the emf induced, and (b) the current flowing in the U-shaped conductor.

9. (II) A 12.0-cm-diameter circular loop of wire has a resistance of 8.15 Ω. It is initially in a 0.405-T magnetic field, with its plane perpendicular to **B**, but is removed from the field in 100 ms. Calculate the electric energy dissipated in this process.

10. (II) A 25.0-cm-diameter coil consists of 20 turns of circular copper wire 2.0 mm in diameter. A uniform magnetic field, perpendicular to the plane of the coil, changes at a rate of 6.55×10^{-3} T/s. Determine (a) the current in the loop, and (b) the rate at which thermal energy is produced.

11. (II) (a) Show that the power $P = Fv$ needed to move the conducting rod to the right in Fig. 20-6 is equal to $B^2 l^2 v^2/R$ where R is the total resistance of the circuit. (b) Show that this equals the power dissipated in the resistance, $I^2 R$.

12. (II) A *search coil for measuring B* (also called a *flip coil*) is a small coil with N turns, each of cross-sectional area A. It is connected to a so-called *ballistic galvanometer*, which is a device to measure the total charge Q that passes through it in a short time. The flip coil is placed in the magnetic field to be measured with its face perpendicular to the field. It is then quickly rotated 180°. Show that the total charge Q that flows in the induced current during this short "flip" time is proportional to the magnetic field strength B; in particular, show that B is given by

$$B = \frac{QR}{2NA}$$

where R is the total resistance of the circuit including that of the coil and the ballistic galvanometer which measures the charge Q.

13. (III) If the U-shaped conductor in Fig. 20-6 has resistivity ρ, whereas that of the moving rod is negligible, derive a formula for the current I as a function of time. Assume the rod has length l, starts at the bottom of the U at $t = 0$, and moves with uniform speed v in the magnetic field B. The cross-sectional area of the rod and all parts of the U is A.

14. (III) The magnetic field perpendicular to a single 12.0-cm-diameter circular loop of copper wire decreases uniformly from 0.350 T to zero. If the wire is 1.25 mm in diameter, how much charge moves through the coil during this operation?

SECTION 20-5

15. (I) A car generator produces 12 V when the armature turns at 600 rev/min. What will be its output at 1500 rev/min, assuming that nothing else changes?

16. (I) Calculate the peak output voltage of a simple generator whose square armature windings are 8.60 cm on a side if the armature contains 175 loops and rotates in a field of 0.300 T at a rate of 120 rev/s.

17. (I) Show that the rms output of an ac generator is $V_{rms} = NAB\omega/\sqrt{2}$.

18. (II) A 100-loop square armature coil 20 cm on a side rotates at 60 rev/s in a uniform magnetic field. If the rms output is 120 V, what is the strength of the magnetic field?

19. (II) A simple generator has a 200-loop square coil 18.0 cm on a side. How fast must it turn in a 0.550-T field to produce a 120-V peak output?

SECTION 20-6

20. (I) A motor has an armature resistance of 2.55 Ω. If it draws 4.10 A when running at full speed and connected to a 120-V line, how large is the counter emf?

21. (I) The counter emf in a motor is 90 V when operating at 1500 rpm. What would the counter emf be at 1800 rpm if the magnetic field is unchanged?

22. (II) What will be the current in the motor of Example 20-5 if the load causes it to run at half-speed?

23. (II) The magnetic field of a "shunt-wound" dc motor is produced by field coils placed in parallel to the armature coils. Suppose that the field coils have a resistance of 56.0 Ω and the armature coils 4.00 Ω. The back emf at full speed is 105 V when the motor is connected to a 115-V line. (a) Draw the equivalent circuit for the situations when the motor is just starting and when it is running full speed. (b) What is the total current drawn by the motor at start up? (c) What is the total current drawn when the motor runs at full speed?

24. (III) A small electric car overcomes a 300-N friction force when traveling 30 km/h. The electric motor is powered by ten 12-V batteries connected in series and is coupled directly to the wheels whose diameters are 40 cm. The 200 armature coils are rectangular, 20 cm by 15 cm, and rotate in a 0.40-T magnetic field. (a) How much current does the motor draw to produce the required torque? (b) What is the back emf? (c) How much power is dissipated in the coils? (d) What percent of the input power is used to drive the car?

SECTION 20-7

25. (I) A transformer changes 12 V to 18,000 V, and there are 6000 turns in the secondary. How many turns are there in the primary? Assume 100 percent efficiency.

26. (I) A transformer has 78 turns in the primary and 158 in the secondary. What kind of transformer is this and, assuming 100 percent efficiency, by what factor does it change the voltage?

27. (I) A step-up transformer increases 30 V to 120 V. What is the current in the secondary as compared to the primary? Assume 100 percent efficiency.

28. (I) Neon signs require 12 kV for their operation. To operate from a 120-V line, what must be the ratio of secondary to primary turns of the transformer? What would the voltage output be if the transformer were connected backward?

29. (I) A transformer has 1500 primary turns and 120 secondary turns. The input voltage is 120 V and the output current is 8.0 A. What is the secondary voltage and primary current?

30. (II) The output voltage of a 180-W transformer is 16.0 V and the input current is 11.0 A. (a) Is this a step-up or a step-down transformer? (b) By what factor is the voltage multiplied?

31. (II) A pair of power transmission lines each have a 0.60-Ω resistance and carry 800 A over 10 km. If the rms input voltage is 12 kV, calculate (a) the voltage at the other end, (b) the power input, (c) power loss in the lines, and (d) the power output.

32. (II) If 28 MW of power at 65 kV (rms) arrives at a town from a generator via 4.0-Ω transmission lines, calculate (a) the emf at the generator end of the lines, and (b) the fraction of the power generated that is lost in the lines.

33. (II) Show that the power loss in transmission lines, P_L, is given by $P_L = (P_T)^2 R_L / V^2$, when P_T is the power transmitted to the user, V is the delivered voltage, and R_L is the resistance of the power lines.

34. (II) If 80 kW is to be transmitted over two 0.055 Ω lines, estimate how much power is saved if the voltage is stepped up from 120 V to 1200 V and then down again, rather than simply transmitting at 120 V. Assume the transformers are each 99 percent efficient.

35. (III) Design a dc transmission line that can transmit 350 MW of electricity 600 km with only a 2 percent loss. The wires are to be made of aluminum and the voltage is 600 kV.

*SECTION 20-9

*36. (I) If the current in a 230-mH coil changes steadily from 20.0 mA to 28.0 mA in 140 ms, what is the induced emf?

*37. (I) What is the inductance L of a 1.2-m-long air-filled coil 5.7 cm in diameter containing 20,000 loops?

*38. (I) What is the inductance of a coil if it produces an emf of 6.50 V when the current in it changes from -12.0 mA to $+23.0$ mA in 11.0 ms?

*39. (II) How many turns does an air-filled coil have if it is 1.2 cm in diameter, 4.0 cm long, and its inductance is 0.25 mH. How many turns are needed if it has an iron core and $\mu = 10^3 \mu_0$?

*40. (II) A 25-V emf is induced in a 0.20-H coil by a current that rises uniformly from zero to I_0 in 1.0 ms. What is the value of I_0?

*41. (II) The wire of a tightly wound solenoid is unwound and used to make another tightly wound solenoid of half the diameter. By what factor does the inductance change?

*42. (II) A coil has 2.50-Ω resistance and 0.418-H inductance. If the current is 8.00 A and is increasing at a rate of 4.50 A/s, what is the potential difference across the coil at this moment?

*43. (II) (a) Show that the self-inductance L of a torus (Fig. 19-39) of radius R containing N loops each of radius r is

$$L \approx \frac{\mu_0 N^2 r^2}{2R}$$

if $R \gg r$. Assume the field is uniform inside the toroid; is this actually true? Is this result consistent with L for a solenoid? Should it be? (b) Calculate the inductance L of a large toroid if the diameter of the coils is 2.0 cm and the diameter of the whole ring is 50 cm. Assume the field inside the toroid is uniform. There are a total of 300 loops of wire.

*44. (II) (a) Ignoring any mutual inductance, what is the equivalent inductance of two inductors connected in series? (b) What if they are connected in parallel? (c) How does their mutual inductance (their geometrical relationship to each other) affect the results?

*45. (II) A long thin solenoid of length l and cross-sectional area A contains N_1 closely packed turns of wire. Wrapped tightly around it is an insulated coil of N_2 turns, Fig. 20-31. Assume all the flux from coil 1 (the solenoid) passes through coil 2, and calculate the mutual inductance.

*46. (III) A 32-cm-long coil with 500 loops is wound on an iron core ($\mu = 2000 \mu_0$) along with a second coil of 200 loops. The loops of each coil have a radius of 2.0 cm. If the current in the first coil drops uniformly from 4.0 A to zero in 80 ms, determine (a) the emf induced in the second coil, and (b) the mutual inductance M.

*47. (III) The potential difference across a given coil is 15.5 V at an instant when the current is 360 mA and is changing at a rate of 240 mA/s. At a later instant, the potential difference is 6.2 V while the current is 300 mA and is decreasing at a rate of 180 mA/s. Determine the inductance and resistance of the coil.

FIGURE 20-31

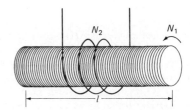

*48. (I) The magnetic field inside an air-filled solenoid 10 cm long and 2.0 cm in diameter is 0.50 T. Approximately how much energy is stored in this field?

*49. (I) How much energy is stored in 400-mH inductor at an instant when the current is 2.0 A?

*50. (II) Assuming the earth's magnetic field averages about 0.50×10^{-4} T near the surface of the earth, determine the total energy stored in this field in the first 10 km above the earth's surface.

*51. (II) Typical large values for electric and magnetic fields attained in laboratories are about 1.0×10^4 V/m and 2.0 T. (a) Determine the energy density for each field and compare. (b) What value electric field would produce the same energy density as the 2.0-T magnetic field?

*52. (II) Approximately how many time constants does it take for the potential difference across the resistor in an LR circuit (Fig. 20-19a) to drop to 1.0 percent of its original value?

*53. (II) Determine $\Delta I / \Delta t$ at $t = 0$ (when the battery is connected) for the circuit of Fig. 20-19a and show that if I continued to increase at this rate, it would reach its maximum value in one time constant.

*54. (III) Two tightly wound solenoids have the same length and circular cross-sectional area. But solenoid 1 uses wire that is half as thick as solenoid 2. (a) What is the ratio of their inductances? (b) What is the ratio of their inductive time constants (assuming no other resistance in the circuits)?

*55. (I) At what frequency will a 200-mH inductor have a reactance of 470 Ω?

*56. (I) At what frequency will a 4.10-μF capacitor have a reactance of 12.2 kΩ?

*57. (I) Plot a graph of the impedance of a 1.0 μF capacitor as a function of frequency from 10 to 1000 Hz.

*58. (I) Plot a graph of the impedance of a 2.0-mH inductor as a function of frequency from 100 to 10,000 Hz.

*59. (II) Calculate the impedance of, and rms current in, a 60-mH radio coil connected to a 400-V (rms) 3.3-kHz ac line. Ignore resistance.

*60. (II) What is the inductance of a coil whose input is 110 V at 60 Hz and the current drawn is 1.2 A?

*61. (II) (a) What is the impedance of a well-insulated 0.020-μF capacitor connected to a 1.0-kV (rms) 200-Hz line? (b) What will be the peak value of the current?

*62. (II) A capacitor is placed in parallel across a load as in Fig. 20-23b to filter out stray high-frequency signals, but to allow ordinary 60-Hz ac to pass through with little loss. Suppose that circuit B in the figure is a resistance R = 100 Ω connected to ground, and that C =

0.50 μF. What percent of the incoming current will pass through C rather than R if (a) it is 60 Hz? (b) 60,000 Hz?

*63. (II) Suppose that circuit B in Fig. 20-23a is a resistance R = 100 Ω, connected to ground, and the capacitance C = 1.0 μF. Will this capacitor act to eliminate 60 Hz ac but pass a high-frequency signal of frequency 60,000 Hz? To check this, determine the voltage drop across R for a 30-mV signal of frequency (a) 60 Hz, and (b) 60,000 Hz.

*64. (I) A 1.0-kΩ resistor and a 2.0-μF capacitor are connected in series to an ac source. Calculate the impedance of the circuit if the source frequency is (a) 60 Hz, and (b) 2.0×10^4 Hz.

*65. (I) A 10-kΩ resistor is in series with a 0.10-H inductor and an ac source. Calculate the impedance of the circuit if the source frequency is (a) 60 Hz, and (b) 2.0×10^4 Hz.

*66. (I) For a 120-V rms 60-Hz voltage, a current of 70 mA passing through the body for 1.0 s could be lethal. What must be the impedance of the body for this to occur?

*67. (II) (a) What is the rms current in an RC circuit if R = 4.7 kΩ, C = 0.10 μF, and the rms applied voltage is 120 V at 60 Hz? (b) What is the phase angle between voltage and current? (c) What is the power dissipated by the circuit? (d) What are the voltmeter readings across R and C?

*68. (II) (a) What is the rms current in an RL circuit when a 60-Hz 120-V rms ac voltage is applied, where R = 250 Ω and L = 500 mH? (b) What is the phase angle between voltage and current? (c) How much power is dissipated? (d) What are the rms voltage readings across R and L?

*69. (II) A voltage $V = 2.8 \sin 754t$ is applied to an LRC circuit. If L = 2.0 mH, R = 1.0 kΩ, and C = 3.0 μF, how much power is dissipated in the circuit?

*70. (II) A circuit contains two elements, but it is not known if they are L, R, or C. The current in this circuit when connected to a 120-V 60-Hz source is 6.4 A and lags the voltage by 30°. What are the two elements and what are their values?

*71. (II) A 50-mH inductor with a 2.0-Ω resistance is connected in series to a 200-μF capacitor and a 60-Hz 16-V (rms) source. Calculate for this circuit the (a) rms current, (b) phase angle, and (c) power dissipated.

*72. (II) What is the total impedance, phase angle, and rms current in an LRC circuit connected to a 2.0-kHz, 300-V (rms) source if L = 3.0 mH, R = 1.1 kΩ, and C = 2000 pF?

*73. (II) A 30-mH coil whose resistance is 6.0 Ω is connected to a capacitor C and a 360-Hz source voltage. If the current and voltage are to be in phase, what value must C have?

*74. (II) What is the resistance of a coil if its impedance is 25 Ω and its reactance is 20 Ω?

*75. (II) Show that for the *LRC* circuit of Fig. 20-24, if we have $I = I_0 \cos \omega t$, then $V_R = I_0 R \cos \omega t$, $V_L = I_0 \omega L \cos (\omega t + \pi/2)$, and $V_C = (I_0/\omega C) \cos (\omega t - \pi/2)$, where $\omega = 2\pi f$.

*76. (III) An inductance coil draws 1.5 A dc when connected to a 36-V battery. When connected to a 60-Hz 120-V (rms) source, the current drawn is 2.8 A (rms). Determine the inductance and resistance of the coil.

*SECTION 20-14

*77. (I) A 200-pF capacitor is connected to a 60-μH coil of resistance 2.0 Ω. What is the resonant frequency of this circuit?

*78. (I) The variable capacitor in the tuner of an AM radio has a capacitance of 1500 pF when the radio is tuned to a station at 550 kHz. (*a*) What must be the capacitance for a station at 1600 kHz? (*b*) What is the inductance (assumed constant)?

*79. (II) An *LRC* circuit has $L = 2.5$ mH and $R = 20\ \Omega$. (*a*) What value must C have to produce resonance at 40 kHz? (*b*) What will be the maximum current at resonance if the peak external voltage is 36 V?

*80. (II) A 660-pF capacitor is charged to 100 V and then quickly connected to a 75-mH inductor. Determine (*a*) the frequency of oscillation, (*b*) the peak value of the current, and (*c*) the maximum energy stored in the magnetic field of the inductor.

*81. (III) The *Q factor* of a resonant circuit is defined as the ratio of the voltage across the capacitor (or inductor) to the voltage across the resistor, at resonance. The larger the Q factor, the sharper the resonance curve will be and the sharper the tuning. (*a*) Show that the Q factor is given by the equation $Q = (1/R)\sqrt{L/C}$. (*b*) At a resonant frequency $f_0 = 1.0$ MHz, what must be the value of L and R to produce a Q factor of 1000; assume that $C = 0.010\ \mu$F.

21 ELECTROMAGNETIC WAVES

The culmination of electromagnetic theory in the nineteenth century was the prediction, and the experimental verification, that waves of electromagnetic fields could travel through space. This achievement opened a whole new world of communication—first the wireless telegraph, then radio and television. And it yielded the spectacular prediction that light is an electromagnetic wave.

The theoretical prediction of electromagnetic waves was the work of the Scottish physicist James Clerk Maxwell (1831–1879), who unified, in one magnificent theory, all the phenomena of electricity and magnetism.

21-1 ▮ Changing electric fields produce magnetic fields; Maxwell's equations

The development of electromagnetic theory in the early part of the nineteenth century by Oersted, Ampère, and others was not actually done in terms of electric and magnetic fields. The idea of the field was introduced somewhat later by Faraday, and was not generally used until Maxwell showed that all electric and magnetic phenomena could be described using only four equations involving electric and magnetic fields. These equations, known as *Maxwell's equations*, are the basic equations for all electromagnetism. They are fundamental in the sense that Newton's three laws of motion and the law of universal gravitation are for mechanics. In a sense they are even more fundamental, since they are valid even relativistically (Chapter 25) whereas Newton's laws are not. Because all of electromagnetism is contained in this set of four equations, Maxwell's equations are considered one of the great triumphs of the human mind.

Although we will not present Maxwell's equations in mathematical form since they involve calculus, we will summarize them here in words. They are: (1) a generalized form of Coulomb's law relating electric field to its sources, electric charge; (2) the same for the magnetic field, except that if there are no magnetic monopoles, magnetic field lines are continuous—they do not begin or end (as electric field lines do on charges); (3) an electric field is produced by a changing magnetic field; (4) a magnetic field is produced by an electric current or by a changing electric field.

The first part of (4), that a magnetic field is produced by an electric current, was discovered by Oersted, and the mathematical relation is given by Ampère's law. But the second part is an entirely new aspect predicted by Maxwell. Maxwell argued that if a changing magnetic field produces an electric field, as given by Faraday's law, then the reverse might be true as well: **a changing electric field will produce a magnetic field.** This was a *hypothesis* by Maxwell, based on the idea of order in the world. Indeed, the size of the effect in most cases is so small that Maxwell recognized it would be difficult to detect it experimentally.

*21-2 ∎ Maxwell's fourth equation; displacement current

Maxwell used an indirect argument to back up his idea that a changing electric field would produce a magnetic field. In a simplified fashion it goes something like this. According to Ampère's law (see Section 19-12), $\sum B_{\parallel} \Delta l = \mu_0 I$. That is, you divide any closed path you choose into short segments Δl, multiply each segment by the parallel component of the magnetic field B at that segment, and then sum all these products over the complete closed path; this sum will then equal μ_0 times the net current I that passes through the surface bounded by the path. When we applied Ampère's law (Section 19-12) to the field around a straight wire, we imagined the current as passing through the circle enclosed by our circular loop. This would be the flat surface 1 in Fig. 21-1. However, we could just as well use the sack-shaped surface 2 in the figure, since the same current I passes through it. This naturally implies that the current passing through any surface enclosed by the closed path must be the same. Thus the current flowing into the volume enclosed by surfaces 1 and 2 together equals the current that flows out of this volume. This is basically Kirchhoff's point rule, and tells us that the rate charge enters the volume equals the rate at which it leaves.

Now consider the closed path for the situation of Fig. 21-2a, where a capacitor is being discharged. Ampère's law works for surface 1, but it does not work for surface 2, since no current passes through surface 2. There is a magnetic field around the wire, so the left side of Ampère's law is not zero; yet for surface 2, the right side *is* zero, since $I = 0$. Similar things could be said for the closed path indicated in Fig. 21-2b, which surrounds the region of the electric field of the plates; there is a current passing through surface 1 but not through surface 2, and there is a magnetic field around the closed path. Actually, there is a magnetic field present only if charge is flowing to or away from the capacitor plates. In this case the electric field between the plates is changing in time. Maxwell resolved the problem of no current through surface 2 by stating that the changing electric field between the plates is *equivalent to* an electric current. He called this a **displacement current,** I_D. (The name is not especially illuminating.) An ordinary current is then called a **conduction current,** I_C. Ampère's law now becomes

$$\sum B_{\parallel} \Delta l = \mu_0 (I_C + I_D).$$

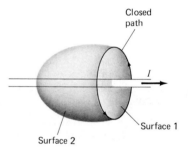

FIGURE 21-1

Ampère's law applied to two different surfaces bounded by the same closed path.

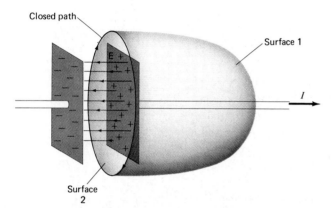

(a)

FIGURE 21-2

Capacitor discharging. No conduction current passes through surface 2 in either (a) or (b). An extra term is needed in Ampère's law.

Ampère's law will now apply even for surface 2 in Fig. 21-2a or b where the displacement current I_D refers to the changing electric field.[†]

We can write I_D in terms of the changing electric field. The charge Q on a capacitor of capacitance C is $Q = CV$, where V is the potential difference between the plates. Also recall that $V = Ed$, where d is the (small) separation of the plates and E is the electric field strength between them; we ignore any fringing of the field. Also, for a parallel-plate capacitor, $C = \varepsilon_0 A/d$, where A is the area of each plate (see Chapter 16). We combine these to obtain

$$Q = CV = \left(\varepsilon_0 \frac{A}{d} \right)(Ed) = \varepsilon_0 AE.$$

Now if the charge on the plate changes at a rate $\Delta Q/\Delta t$, the electric field changes at a proportional rate. Then, from the above expression,

$$\frac{\Delta Q}{\Delta t} = \varepsilon_0 A \frac{\Delta E}{\Delta t}.$$

Now $\Delta Q/\Delta t$ is the rate at which charge accumulates on or leaves the capacitor plates; it is therefore equal to the current flowing into or out of the

---

[†] Notice that this interpretation of the changing electric field as a displacement current fits in well with our discussion in Chapter 20 where we saw that an alternating current passes through a capacitor. It also means that Kirchhoff's point rule will be valid even at a capacitor plate, for conduction current flows into the plate, but no conduction current flows out of the plate. Instead, a "displacement current" flows out of the plate.

capacitor. If the current flow into the capacitor, $\Delta Q/\Delta t$, is set equal to the displacement current I_D between the plates, then

$$I_D = \frac{\Delta Q}{\Delta t} = \varepsilon_0 A \frac{\Delta E}{\Delta t}.$$

This can be written

$$I_D = \varepsilon_0 \frac{\Delta \Phi_E}{\Delta t}$$

where $\Phi_E = EA$ is the *electric flux*, defined in analogy to magnetic flux (Section 20-2). Then, Ampère's law becomes

$$\sum B_\parallel \Delta l = \mu_0 I_C + \mu_0 \varepsilon_0 \frac{\Delta \Phi_E}{\Delta t}. \qquad (21\text{-}1)$$

This equation embodies Maxwell's idea that a magnetic field can be caused not only by a normal electric current, but also by a changing electric field or changing electric flux. Eq. 21-1 is essentially Maxwell's fourth equation. (Actually, there is a third term on the right for the case when a magnetic field is produced by magnetized materials; but we assume in what follows that no magnets are present.)

21-3 ∎ Production of electromagnetic waves

According to Maxwell, a magnetic field will be produced in empty space if there is a changing electric field. From this, Maxwell derived another startling conclusion. If a changing magnetic field produces an electric field, the electric field can itself be changing. This changing electric field will, in turn, produce a magnetic field. The latter will be changing and so will produce a changing electric field; and so on. When Maxwell manipulated his equations, he found that the net result of these interacting changing fields was to produce a wave of electric and magnetic fields that can actually propagate (travel) through space. We now examine, in a simplified way, how such **electromagnetic waves** are produced.

Consider two conducting rods that will serve as an "antenna" (Fig. 21-3a). Suppose that these two rods are connected by a switch to the opposite terminals of a battery. As soon as the switch is closed, the upper rod

(a) (b)

FIGURE 21-3

Fields produced by charge flowing into conductors. It takes time for the **E** and **B** fields to travel outward to distant points.

quickly becomes positively charged and the lower one negatively charged. Electric field lines are formed as indicated by the lines in Fig. 21-3b. While the charges are flowing a current exists; the direction of conventional current is indicated by the arrows. A magnetic field is therefore produced; the magnetic field lines encircle the wires and therefore, in the plane of the paper, **B** points into the paper ($\otimes$) on the right and out of the paper ($\odot$) on the left. Now we ask, how far out do these fields extend? In the static case, the fields extend outward indefinitely far. However, when the switch in Fig. 21-3 is closed, the fields quickly appear nearby, but it takes time for them to reach distant points. Both electric and magnetic fields store energy, and this energy cannot be transferred to distant points at infinite speed.

This example illustrates that it takes time for electric and magnetic fields, once they are formed, to reach distant points. Now we look at a more interesting situation. Suppose that the two vertical rods are connected to an ac generator (Fig. 21-4). In part (a), the connection has just been completed. Charge starts building up and fields form just as in Fig. 21-3. The + and −

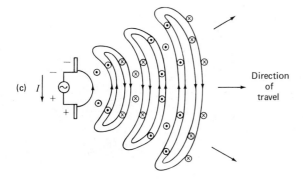

FIGURE 21-4

Sequence showing electric and magnetic fields that spread outward from oscillating charges on two conductors connected to an ac source (see the text).

signs indicate the net charge on each rod. The arrows indicate the direction of the current I. The electric field is represented by lines in the plane of the paper; and the magnetic field, according to the right-hand rule, is into ($\otimes$) or out of ($\odot$) the paper, as shown. In Fig. 21-4b, the emf of the ac generator has reached its maximum and is starting to decrease; so the current is reversed and the new magnetic field is in the opposite direction. The old fields, however, don't suddenly disappear; they are on their way to distant points. But because the new fields have changed direction, the old lines fold back to connect up to some of the new lines and form the closed loops as shown. In (c) the situation is shown a little later when the current has made several oscillations and the field lines have formed several loops (note change of scale). These field lines continue moving outward. The magnetic field lines also form closed loops; they are not shown since they are perpendicular to the paper.

The fields continue to move outward. Although the lines are shown only on the right of the source, fields also travel in other directions. (The field strengths are greatest in directions perpendicular to the oscillating charges; and they drop to zero along the direction of oscillation—above and below the antenna in Fig. 21-4.)

Several things can be noted from Fig. 21-4. First, *the electric and magnetic fields at any point are perpendicular to each other, and to the direction of motion.* Second, we can see that the fields alternate in direction (**B** is into the paper at some points and out of the paper at others; similarly for **E**). Thus the field strengths vary from a maximum in one direction to zero to a maximum in the other direction. The electric and magnetic fields are "in phase"; that is, they each are zero at the same points and reach their maximum at the same points.

If the source emf varies sinusoidally, the electric and magnetic field strengths far from the source also vary sinusoidally. This variation cannot be seen in Fig. 21-4, which shows only the field *lines.* Figure 21-5 shows the field *strengths* plotted as a function of distance. Notice that **B** and **E** are perpendicular to each other and to the direction of travel.

Electromagnetic (EM) waves are *transverse* waves and resemble other types of waves (Chapter 13). However, EM waves are waves of *fields*, not of matter as are waves on water or a rope.

We have seen in the above analysis that EM waves are produced by electric charges that are oscillating, and hence are undergoing acceleration. In fact, we can say in general that *accelerating electric charges give rise to electromagnetic waves.*

FIGURE 21-5

Electric and magnetic field strengths in an electromagnetic wave. **E** and **B** are at right angles to each other. The entire pattern moves in a direction perpendicular to both **E** and **B**.

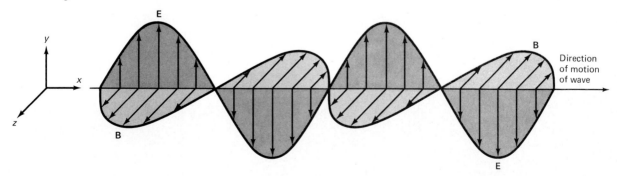

*21-4 ▮ Calculation of the speed of electromagnetic waves

Maxwell's prediction that EM waves should exist was startling. Equally startling was the speed at which they were predicted to travel. We now calculate this speed.

We shall consider a region far from the source, so that the wave fronts (the field lines in Fig. 21-4c) are essentially flat over a reasonable area instead of curved. They are then called *plane waves*, meaning that at any instant, **E** and **B** are uniform over a plane perpendicular to the direction of propagation. Let us assume that, in a particular coordinate system, the wave is traveling in the x direction with speed v; **E** is parallel to the y axis and **B** is parallel to the z axis as in Fig. 21-5.

Now let us apply Faraday's law to an imaginary rectangle $abcd$ placed in the xy plane as shown in Fig. 21-6. We assume that the side ab is in a region where the wave has not yet reached, so $E = B = 0$ there. As the wave moves, its magnetic flux through our rectangular loop changes. Instead of showing the motion of the wave in Fig. 21-6, we show instead the relative position of the rectangle (it's easier to visualize). After a short time Δt, the wave moves to the right a distance $\Delta x = v\,\Delta t$, where v is the velocity of the wave; or, equivalently, the rectangular loop moves $\Delta x = v\,\Delta t$ to the left. Rectangle $a'b'c'd'$ represents the new position of the rectangle relative to the wave. The change in magnetic flux $\Delta\Phi_B$ through the loop during the time Δt is just the flux that passes through the small rectangle $dd'c'c$. Thus $\Delta\Phi_B = B\,\Delta A = By_0\,\Delta x = By_0 v\,\Delta t$, where y_0 is the width ab or cd; $\Delta A = y_0\,\Delta x = y_0 v\,\Delta t$ is the area of $dd'c'c$; and B represents the magnetic field passing through this area (represented by the heavy arrow in Fig. 21-6). We assume Δx is small so that B is essentially constant over this small area. According to Faraday's law, then, the emf induced around the loop $abcd$ equals the rate of change of magnetic flux:

$$\mathscr{E} = \frac{\Delta\Phi_B}{\Delta t} = \frac{B\,\Delta A}{\Delta t} = \frac{By_0 v\,\Delta t}{\Delta t} = By_0 v.$$

(We omitted the minus sign for convenience.) The emf around the loop $abcd$ is the sum of the emfs in each straight section:

$$\mathscr{E} = \mathscr{E}_{ab} + \mathscr{E}_{bc} + \mathscr{E}_{cd} + \mathscr{E}_{da}.$$

Now the emf is the work done per unit charge: $\mathscr{E} = W/q = Fd/q = Ed$, where

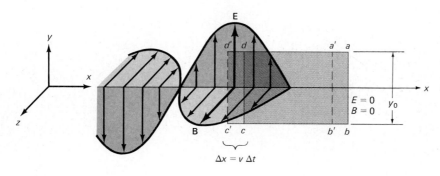

FIGURE 21-6

In a time Δt, the rectangle in the xy plane (actually stationary) moves a distance $\Delta x = v\,\Delta t$ relative to the wave (the wave moves to the right with speed v).

F is the force exerted over the distance d and the electric field E is the force per unit charge, $E = F/q$. (The result, $\mathscr{E} = Ed$, is the same as Eq. 16-2.) The term $\mathscr{E}_{ab}$ is zero since $E = 0$ in this region; and $\mathscr{E}_{bc} = \mathscr{E}_{da} = 0$, since $\mathbf{E}$ is perpendicular to the path (Fig. 21-6). Thus

$$\mathscr{E} = \mathscr{E}_{cd} = Ey_0,$$

where E is the magnitude of the electric field along cd (or $c'd'$). From Faraday's law as we wrote it above ($\mathscr{E} = By_0v$) we thus have

$$Ey_0 = By_0v$$

$$E = vB. \qquad (21\text{-}2)$$

Thus we see that at any point in space, E and B are in the ratio $E/B = v$, the velocity of the wave.

Now we consider a rectangle in the xz plane, as shown in Fig. 21-7. Again we show the rectangle moving to the left a distance $\Delta x = v\,\Delta t$ relative to the wave, although actually the rectangle is stationary and the wave is moving to the right; $abcd$ is the position initially and $a'b'c'd'$ after a time Δt. There is a changing electric flux through this loop equal to the electric field E (heavy arrow) times the increasing area $\Delta A = z_0\,\Delta x = z_0v\,\Delta t$ (where z_0 is the width $ab = cd$ of the rectangle). According to Ampère's law (Eq. 21-1 with $I_C = 0$), since there are no conduction currents, we have

$$\sum B_{\parallel}\,\Delta l = \mu_0\varepsilon_0\,\frac{\Delta\Phi_E}{\Delta t}$$

$$= \mu_0\varepsilon_0\,\frac{(E)(z_0v\,\Delta t)}{\Delta t} = \mu_0\varepsilon_0 Ez_0v.$$

The sum of $B_{\parallel}\,\Delta l$ on the sides ab, bc, and da are all zero, because either $B = 0$ or B is perpendicular to these sides. But along side cd, the contribution is Bz_0, where B is the magnetic field (heavy arrow) parallel to cd. Thus

$$Bz_0 = \mu_0\varepsilon_0 Ez_0v$$

$$B = \mu_0\varepsilon_0 vE.$$

We combine this with Eq. 21-2 and find that

$$B = \mu_0\varepsilon_0 v(vB) = \mu_0\varepsilon_0 v^2 B.$$

We cancel B on both sides and solve for v:

$$v = \frac{1}{\sqrt{\varepsilon_0\mu_0}}. \qquad (21\text{-}3)$$

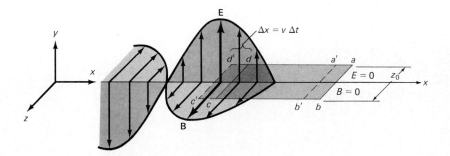

FIGURE 21-7

Rectangle in the xz plane (actually at rest) moves a distance $\Delta x = v\,\Delta t$ relative to a wave traveling to the right.

When we put in the values for ε_0 and μ_0, we find that

$$v = \frac{1}{\sqrt{\varepsilon_0 \mu_0}} = \frac{1}{\sqrt{(8.85 \times 10^{-12} \text{ C}^2/\text{N} \cdot \text{m}^2)(4\pi \times 10^{-7} \text{ N} \cdot \text{s}^2/\text{C}^2)}}$$
$$= 3.00 \times 10^8 \text{ m/s}.$$

This is a remarkable result. For this is precisely equal to the measured speed of light!

21-5 ▌ Light as an electromagnetic wave and the electromagnetic spectrum

The calculation at the end of the last section gives the result that Maxwell himself determined: that the speed of EM waves is 3.00×10^8 m/s, the same as the measured speed of light.

Light had been shown some 60 years previously to behave like a wave (we'll discuss this in Chapter 23). But nobody knew what kind of wave it was—that is, what is it that is oscillating in a light wave? Maxwell, on the basis of the calculated speed of EM waves, argued that light must be an electromagnetic wave. This soon came to be generally accepted by scientists, but not fully until after EM waves were experimentally detected. EM waves were first generated and detected experimentally by Heinrich Hertz (1857–1894) in 1887, 8 years after Maxwell's death. Hertz used a spark-gap apparatus in which charge was made to rush back and forth for a short time; this generated waves whose frequency was about 10^9 Hz. He detected them some distance away using a loop of wire in which an emf was produced when a changing magnetic field passed through. These waves were later shown to travel at the speed of light, 3.00×10^8 m/s, and to exhibit all the characteristics of light such as reflection, refraction, and interference. The only difference was that they were not visible. Hertz's experiment was a strong confirmation of Maxwell's theory.

The wavelengths of visible light were measured in the first decade of the nineteenth century, long before anyone imagined that light was an electromagnetic wave. The wavelengths were found to lie between 4.0×10^{-7} m and 7.5×10^{-7} m; or 400 nm to 750 nm (1 nm = 10^{-9} m). The frequencies of visible light rays can be found using Eq. 13-9 which we rewrite here:

$$f\lambda = c. \tag{21-4}$$

Here, c is the velocity of light, 3.00×10^8 m/s; it gets the special symbol c because of its universality for all EM waves in free space; f and λ are the frequency and wavelength of the wave. This equation tells us that the frequencies of visible light are between 4.0×10^{14} Hz and 7.5×10^{14} Hz.

But visible light is only one kind of EM wave. As we have seen, Hertz produced EM waves of much lower frequency, about 10^9 Hz. These are called *radio waves*, since frequencies in this range are used today to transmit radio and TV signals. Electromagnetic waves, or EM radiation as we sometimes call it, have been produced or detected over a wide range of frequencies. They are usually categorized as shown in Fig. 21-8. This is known as the **electromagnetic spectrum.**

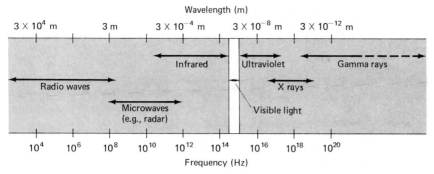

FIGURE 21-8
Electromagnetic spectrum.

Radio waves and microwaves can be produced experimentally using electronic equipment, as discussed earlier; see Fig. 21-4. Higher-frequency waves are very difficult to produce electronically. These and other types of EM waves are produced in natural processes, as emission from atoms, molecules, and nuclei (more on this later). Generally, EM waves are produced by the acceleration of electrons or other charged particles, such as electrons accelerating in the antenna of Fig. 21-4. Another example is X rays which are produced (see Chapter 27) when fast-moving electrons are rapidly decelerated upon striking a metal target. Even the visible light emitted by an ordinary incandescent light is due to electrons undergoing acceleration within the hot filament. We will meet various types of EM waves later. However, it is worth mentioning here that infrared (IR) radiation (EM waves whose frequency is just less than that of visible light) is mainly responsible for the heating effect of the sun. The sun emits not only visible light but substantial amounts of IR and UV (ultraviolet) as well. The molecules of our skin tend to "resonate" at infrared frequencies, so it is these that are preferentially absorbed and thus warm us up.

EXAMPLE 21-1   Calculate the wavelength (a) of a 60-Hz EM wave; (b) of a 1240-kHz AM radio wave.

SOLUTION   (a) Since $c = \lambda f$,

$$\lambda = \frac{c}{f} = \frac{3.0 \times 10^8 \text{ m/s}}{60 \text{ s}^{-1}} = 5.0 \times 10^6 \text{ m},$$

or 5000 km. One wavelength stretches all the way across the United States.

(b)
$$\lambda = \frac{3.0 \times 10^8 \text{ m/s}}{1.24 \times 10^6 \text{ s}^{-1}} = 240 \text{ m}.$$

Electromagnetic waves can travel along transmission lines as well as in empty space. When a source of emf is connected up to a transmission line— be it two parallel wires or a coaxial cable—the electric field within the wire is not set up immediately at all points along the wires. This is based on the same argument we used in Section 21-3 with reference to Fig. 21-4. Indeed, it can be shown that if the wires are separated by air, the electrical signal travels along the wires at the speed $c = 3.0 \times 10^8$ m/s. For example, when you flip a light switch, the light actually goes on a tiny fraction of a second later. If the wires are in a medium whose electric permittivity is ε and magnetic permeability is μ, the speed is not given by Eq. 21-3 but by $v = 1/\sqrt{\varepsilon\mu}$.

EXAMPLE 21-2   When you speak on the telephone from Los Angeles to a friend in New York 4000 km away, how long does it take your voice to travel?

SOLUTION   Since speed = distance/time, then time = distance/speed = $(4.0 \times 10^6 \text{ m})/(3.0 \times 10^8 \text{ m/s}) = 1.3 \times 10^{-2}$ s, or about $\frac{1}{100}$ s.

*21-6 ■ Energy in EM waves

Electromagnetic waves carry energy from one region of space to another. This energy is associated with the moving electric and magnetic fields. In Section 16-9 we saw that the energy density (J/m^3) stored in an electric field E is $u = \frac{1}{2}\varepsilon_0 E^2$, where u is the energy per unit volume. The energy stored in a magnetic field B, as we discussed in Section 20-10 (Eq. 20-10) is given by $u = \frac{1}{2}B^2/\mu_0$. Thus the total energy stored per unit volume in a region of space where there is an electromagnetic wave is

$$u = \tfrac{1}{2}\varepsilon_0 E^2 + \tfrac{1}{2}\frac{B^2}{\mu_0}. \tag{21-5}$$

In this equation E and B represent the electric and magnetic field strengths of the wave at any instant in a small region of space. We can write Eq. 21-5 in terms of the E field only, since from Eq. 21-3 we have $\sqrt{\varepsilon_0\mu_0} = 1/c$, and from Eq. 21-2, $B = E/c$. We insert these into Eq. 21-5 to obtain

$$u = \tfrac{1}{2}\varepsilon_0 E^2 + \tfrac{1}{2}\frac{\varepsilon_0\mu_0 E^2}{\mu_0}$$

$$= \varepsilon_0 E^2. \tag{21-6a}$$

Notice that the energy density associated with the B field is equal to that associated with the E field, so each contributes half to the total energy. We can also write the energy density in terms of the B field only, or in one term containing both: $u = \varepsilon_0 E^2 = \varepsilon_0 c^2 B^2 = \varepsilon_0 B^2/\varepsilon_0\mu_0$, or

$$u = \frac{B^2}{\mu_0}; \tag{21-6b}$$

and $u = \varepsilon_0 E^2 = \varepsilon_0 EcB = \varepsilon_0 EB/\sqrt{\varepsilon_0\mu_0}$, or

$$u = \sqrt{\frac{\varepsilon_0}{\mu_0}}\, EB. \tag{21-6c}$$

Eqs. 21-6 give the energy density in any region of space at any instant.

Now let us determine the energy that is transported per unit time per unit area perpendicular to the wave direction. This is given by a vector **S**, which is called the *Poynting vector*. The units of **S** are W/m^2. The direction of **S** is the direction in which the energy is transported, which is the direction in which the wave is moving. Let us imagine that the wave is passing through an area A perpendicular to the x axis, as shown in Fig. 21-9. In a short time Δt, the wave moves to the right a distance $\Delta x = c\,\Delta t$. The energy that has passed through A in the time Δt is the energy that now occupies the volume

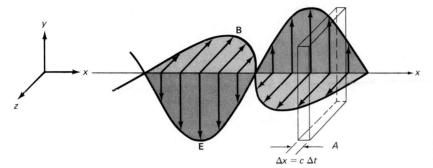

FIGURE 21-9

Electromagnetic wave carrying energy through area A.

$\Delta V = A \Delta x = Ac \Delta t$. The energy density u is $u = \varepsilon_0 E^2$, where E is the electric field in this volume at the given instant. So the energy ΔU contained in this volume is the energy density u times the volume: $\Delta U = u \Delta V = (\varepsilon_0 E^2)(Ac \Delta t)$. Therefore, the energy crossing the area A per time Δt is

$$S = \frac{\Delta U}{A \Delta t} = \varepsilon_0 c E^2.$$

Since $E = cB$ and $c = 1/\sqrt{\varepsilon_0 \mu_0}$, this can also be written

$$S = \varepsilon_0 c E^2 = \frac{cB^2}{\mu_0} = \frac{EB}{\mu_0}. \tag{21-7}$$

Equation 21-7 gives the energy transported per unit area per unit time at any *instant*. We often want to know the *average* over an extended period of time. If E and B are sinusoidal, then $\overline{E^2} = E_0^2/2$ just as for electric currents and voltages (Section 17-6), where E_0 is the *maximum* value of E. Thus we can write

$$\bar{S} = \tfrac{1}{2}\varepsilon_0 c E_0^2 = \tfrac{1}{2}\frac{c}{\mu_0} B_0^2 = \frac{E_0 B_0}{2\mu_0}, \tag{21-8}$$

where B_0 is the maximum value of B. Equation 21-7 also holds on the average (as well as instantaneously) if for E and B we use the rms values.

◻ EXAMPLE 21-3   Radiation from the sun reaches the earth (above the atmosphere) at a rate of about 1350 J/s·m². Assume that this is a single EM wave and calculate the maximum values of E and B.

SOLUTION   Since $\bar{S} = 1350$ J/s·m² $= \varepsilon_0 c E_0^2/2$, then

$$E_0 = \sqrt{\frac{2\bar{S}}{\varepsilon_0 c}} = \sqrt{\frac{2(1350 \text{ J/s·m}^2)}{(8.85 \times 10^{-12} \text{ C}^2/\text{N·m}^2)(3.0 \times 10^8 \text{ m/s})}}$$
$$= 1.01 \times 10^3 \text{ V/m}.$$

From Eq. 21-2, $B = E/c$, so

$$B_0 = \frac{E_0}{c} = \frac{1.01 \times 10^3 \text{ V/m}}{3.0 \times 10^8 \text{ m/s}} = 3.4 \times 10^{-6} \text{ T}.$$

(*B* has a small numerical value compared to E because of the different units for E and B and the way these units are defined; as we saw earlier in the section, B contributes as much energy to the wave as does E.)

Electromagnetic waves offer the possibility of transmitting information over long distances. Among the first to realize this and put it in practice was Guglielmo Marconi (1874–1937), who, in the 1890s invented and developed the wireless telegraph. With it, messages could be sent hundreds of kilometers at the speed of light without the use of wires. The first signals were merely long and short pulses that could be translated into words by a code, such as the "dots" and "dashes" of the Morse code. The next decade saw the development of vacuum tubes. Out of this early work radio and television were born. We now discuss briefly (1) how radio and TV signals are transmitted, and (2) how they are received at home.

The process by which a radio station transmits information (words and music) is outlined in Fig. 21-10. The audio (sound) information is changed into an electrical signal of the same frequencies by a transducer (microphone, phonograph cartridge, tape recorder head). This electrical signal is called an audio frequency (AF) signal, since the frequencies are in the audio range (20 to 20,000 Hz). The signal is amplified electronically and is then mixed with a radio-frequency (RF) signal. The RF frequency is determined by the values of L and C in a resonant LCR circuit (Section 20-14) and is chosen to produce a particular frequency for each station, called its *carrier frequency*. AM radio stations have carrier frequencies from about 530 to 1600 kHz. For example, "710 on your dial" means a station whose carrier frequency is 710,000 Hz. FM radio stations have much higher frequencies, between 88 and 108 MHz. The carrier frequencies for TV stations in the U.S. lie between 54 and 88 MHz for channels 2 through 6, and between 174 and 216 MHz for channels 7 through 13; UHF (ultra-high-frequency) stations have even higher carrier frequencies, between 470 and 890 MHz.

The mixing of the audio and carrier frequencies can be done in two ways. In amplitude modulation (AM) the amplitude of the audio signal is combined with that of the much higher carrier frequency, as shown in Fig. 21-11. It is called "amplitude modulation" because the amplitude of the carrier is altered ("modulate" means to change or alter). In frequency modulation (FM), the *frequencies* of the audio and carrier signals are combined. Thus the *frequency* of the carrier wave is altered by the audio signal as shown in Fig. 21-12.

The mixed signal is amplified further (since the signal contains radio frequencies, this is called an RF amplifier) and is then sent to the antenna, where the complex mixture of frequencies is sent out in the form of EM waves.

FIGURE 21-10

Block diagram of a radio transmitter.

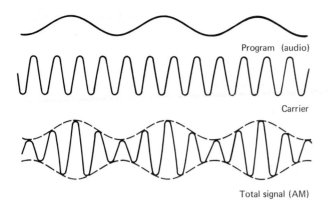

Program (audio)

Carrier

Total signal (AM)

FIGURE 21-11

In amplitude modulation (AM), the amplitudes
of the audio and carrier signals are added together.

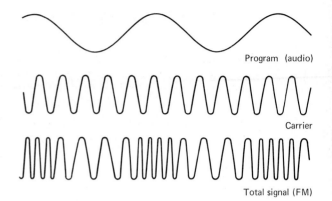

Program (audio)

Carrier

Total signal (FM)

FIGURE 21-12

In FM, the frequencies of the audio and carrier signals
are added together. This method is used by FM
radio and television.

FIGURE 21-13

Block diagram of a simple radio receiver.

A television transmitter works in a similar way, using frequency modulation, except that both audio and video (visual) signals are mixed with carrier frequencies.

Now let us look at the other end of the process, the reception of radio and TV programs at home. A simple radio receiver is diagrammed in Fig. 21-13. The EM waves sent out by all stations are received by the antenna. One kind of antenna consists of one or more conducting rods; the electric field in the EM waves exert a force on the electrons in the conductor, causing them to move back and forth at the frequencies of the waves. A second type of antenna, often found in AM radios, consists of a tubular coil of wire. This type of antenna detects the magnetic field of the wave, for the changing B field induces an emf in the coil. The signal from the antenna is very small and contains frequencies from many different stations. The receiver selects out a particular RF frequency (actually a narrow range of frequencies) corresponding to a particular station using a resonant LC circuit (Section 20-14) with a variable capacitor or inductor. A simple example is shown in Fig. 21-14. A particular station is "tuned-in" by adjusting L and C so that the resonant frequency of the circuit equals that of the station's carrier frequency. The RF signal may be amplified both before and after the tuning is done. The signal, containing both audio and carrier frequencies, next goes to the *detector* (Fig. 21-13) where "demodulation" takes place—that is, the RF carrier frequency is separated from the audio signal. The audio signal is then amplified and sent to a loudspeaker or headphones.

FIGURE 21-14

Simple tuning stage of a radio.

515

Modern receivers have more stages than those shown. Various means are used to increase the sensitivity and selectivity (ability to detect weak signals and distinguish them from other stations), and to minimize distortion of the original signal.[†]

A television receiver does similar things to both the audio and the video signals. The audio signal goes finally to the loudspeaker, and the video signal to the picture tube, a *cathode ray tube* (CRT) whose operation is discussed in Section 19-10.

> **EXAMPLE 21-4** An FM radio station transmits at 100 MHz. Calculate (a) its wavelength and (b) the value of the capacitance in the tuning circuit if $L = 0.40 \ \mu H$.
>
> **SOLUTION** (a) From Eq. 21-4, $\lambda = c/f = (3.0 \times 10^8 \text{ m/s})/(1.0 \times 10^8 \text{ s}^{-1}) = 3.0$ m. The wavelengths of other FM signals (88 to 108 MHz) are close to this. FM antennas are typically 1.5 m long, or about a half wavelength. This length is chosen so that the antenna reacts in a resonant fashion and thus is more sensitive.
>
> (b) According to Eq. 20-18, the resonant frequency is $f_0 = 1/(2\pi\sqrt{LC})$. Therefore
>
> $$C = \frac{1}{4\pi^2 f_0^2 L} = \frac{1}{4(3.14)^2(1.0 \times 10^8 \text{ s}^{-1})^2(4.0 \times 10^{-7} \text{ H})}$$
>
> $$= 6.3 \text{ pF}.$$

Of course, the capacitor is variable, so other stations can be selected.

Citizens' band (CB) receivers are similar to AM radios, except that the carrier frequencies are in the vicinity of 27 MHz. The various regions of the radio-wave spectrum are assigned by governmental agencies to various purposes. Besides those mentioned above, there are regions (or "bands") assigned for use by ships, airplanes, police, military, amateurs, satellites and space, and radar.

❚ Summary

James Clerk Maxwell synthesized an elegant theory in which all electric and magnetic phenomena could be described using four equations, now called Maxwell's equations. They are based on earlier ideas, but Maxwell added one more—that a changing electric field produces a magnetic field.

Maxwell's theory predicted that transverse *electromagnetic* (EM) *waves* would be produced by accel-

---

[†] For *FM stereo broadcasting*, two signals are carried by the carrier wave. One of these contains frequencies up to about 17,000 Hz, which includes most audio frequencies; the other signal includes the same range of frequencies, but 21,000 Hz is added to it. A stereo receiver subtracts this 21,000-Hz signal and distributes the two signals to the left and right channels. The first signal actually consists of the sum of left and right channels $(L + R)$, so mono radios detect all the sound; the second signal is the difference between left and right $(L - R)$. Hence the receiver must add and subtract the two signals to get pure left and right signals for each channel.

erating electric charges, and these waves would propagate through space at the speed of light. The oscillating electric and magnetic fields in an EM wave are perpendicular to each other and to the direction of propagation.

After EM waves were experimentally detected in the late 1800s, the idea that light is an EM wave (al-though of much higher frequency than those detected directly) became generally accepted. The *electromagnetic spectrum* includes EM waves of a wide variety of wavelengths, from microwaves and radio waves to visible light to X rays and γ rays, all of which travel through space at a speed $c = 3.0 \times 10^8$ m/s.

▌ Questions

*1. What is the direction of the displacement current in Fig. 21-2? (Note: the capacitor is discharging.)

*2. Suppose you are looking along the same direction as an electric field **E** that is increasing. Will the induced magnetic field be clockwise or counterclockwise? What if **E** points toward you and is decreasing?

3. The electric field in an EM wave traveling north oscillates in an east-west plane. Describe the direction of the magnetic field vector in this wave.

4. Is sound an electromagnetic wave? If not, what kind of wave is it?

5. Can EM waves travel through a perfect vacuum? Can sound waves?

6. How are light and sound alike? How are they different?

7. Are the wavelengths of radio and television signals longer or shorter than those detectable by the human eye?

8. What does the result of Example 21-1 tell you about the phase of an ac current that starts at a power plant as compared to its phase at a house 200 km away?

9. When you connect up two loudspeakers to a stereo amplifier, should you be sure the lead wires are equal in length so that there will not be a time lag between speakers? Explain.

10. In the electromagnetic spectrum, what type of EM wave would have a wavelength of 10^3 km? 1 km? 1 m? 1 cm? 1 mm? 1 μm?

*11. A lost person may signal by flashing a flashlight on and off using Morse code. This is actually a modulated EM wave. Is it AM or FM? What is the frequency of the carrier, approximately?

*12. Can two radio or TV stations broadcast on the same carrier frequency? Explain.

*13. If a radio transmitter has a vertical antenna, should a receiver's antenna be vertical or horizontal to obtain best reception?

*14. The carrier frequencies of FM broadcasts are much higher than for AM broadcasts. On the basis of what you learned about diffraction in earlier chapters, explain why AM signals can be detected more readily than FM signals behind low hills or buildings.

▌ Problems

*SECTION 21-2

*1. (I) Calculate the displacement current I_D between the square plates, 1.0 cm on a side, of a capacitor if the electric field is changing at a rate of 3.0×10^6 V/m·s.

*2. (II) At a given instant, a 2.8-A current flows in the wires connected to a parallel-plate capacitor. What is the rate at which the electric field is changing between the plates if the square plates are 1.00 cm on a side?

*3. (II) A 1.20-μF capacitor with parallel circular plates accumulates charge at a rate of 12.0 mC/s. What will be the magnetic field strength 15.0 cm radially outward from the center of the plates whose radii are 0.600 cm?

*4. (III) The electric field between two parallel circular capacitor plates (capacitance C) changes at a rate $\Delta E/\Delta t$. (a) If the radius of the plates is R, show that the magnetic field B a distance r from the center of the plates, if $r \leqslant R$, is $B = \frac{1}{2}\mu_0\varepsilon_0 r(\Delta E/\Delta t)$; and outside the plates, $r \geqslant R$, show that $B = (\mu_0\varepsilon_0 R^2/2r)(\Delta E/\Delta t)$. (b)

Show that the magnetic field beyond the edges of the capacitor plates is given by

$$B = \frac{\mu_0 I_D}{2\pi r}$$

where I_D is the displacement current. (c) This is the same formula for the field outside a straight wire; explain the similarity.

*SECTION 21-4

*5. (I) If the electric field in an EM wave has a peak of 1.5×10^{-4} V/m, what is the peak value of the magnetic field strength?

*6. (I) In an EM wave traveling west, the B field oscillates vertically and has a frequency of 800 kHz and an rms strength of 6.6×10^{-9} T. What is the frequency and rms strength of the electric field and what is its direction?

7. (I) What is the wavelength of a 10.85×10^9 Hz radar signal?

8. (I) What is the frequency of a microwave whose wavelength is 1.40 cm?

9. (II) Who will hear the voice of a singer first—a person in the balcony 50 m away from the stage, or a person 3000 km away at home whose ear is next to the radio? How much sooner? Assume that the microphone is a few centimeters from the singer and the temperature is 20°C.

*10. (I) The **E** field in an EM wave has a peak of 25 mV/m. What is the average rate at which this wave carries energy across unit area per unit time?

*11. (I) The magnetic field in an EM wave has an rms strength of 2.0×10^{-8} T. How much energy does this wave transport per meter squared per second?

*12. (II) How much energy is transported across a 1.0-cm^2 area per hour by an EM wave whose E field has an rms strength of 25 V/m?

*13. (II) What is the energy contained in a 1.0-cm^3 volume near the earth's surface due to radiant energy from the sun? See Example 21-3.

*14. (II) A 5.0 mW laser puts out a narrow beam 2.0 mm in diameter. What are the average (rms) values of E and B in the beam?

*15. (II) Estimate the average power output of the sun using the fact that about 1350 W/m^2 reaches the upper atmosphere of the earth.

*16. (II) A point source emits light energy uniformly in all directions at an average rate P_0 with a single frequency f. Show that the peak electric field in the wave is given by

$$E_0 = \sqrt{\frac{\mu_0 c P_0}{2\pi r^2}}.$$

*17. (II) What are E_0 and B_0 10 m from a 100-W light source? Assume the bulb emits radiation of a single frequency uniformly in all directions.

*18. (I) An FM station broadcasts at 91.5 MHz. What is the wavelength of this wave?

*19. (I) What is the wavelength of an AM station at 1500 on the dial?

*20. (I) Compare 940 on the AM dial to 94 on the FM dial.

Which has the longer wavelength, and by what factor is it larger?

*21. (I) The variable capacitor in the tuner of an AM radio has a capacitance of 1500 pF when the radio is tuned to a station at 550 kHz. What must the capacitance be for a station at the other end of the dial, 1600 kHz?

*22. (II) The oscillator of a 106 MHz FM station has an inductance of 0.30 μH. What value must the capacitance be?

*23. (II) A certain FM radio tuning circuit has a fixed capacitor $C = 900$ pF. Tuning is done by a variable inductance. What range of values must the inductance have to tune stations from 88 to 108 MHz?

*24. (II) A 100-kW radio station emits EM waves uniformly in all directions. (*a*) How much energy per second crosses a 1.0-m^2 area 100 m from the transmitting antenna? (*b*) What is the rms magnitude of the E field at this point, assuming the station is operating at full power?

*25. (II) Repeat problem 24 for a distance of 20 km from the station.

*26. (II) An amateur radio operator wishes to build a receiver that can tune a range from 14.0 MHz to 15.0 MHz. A variable capacitor has a minimum capacitance of 64 pF. (*a*) What is the required value of the inductance? (*b*) What is the maximum capacitance used on the variable capacitor?

*27. (II) A 1.40-m-long FM antenna is oriented parallel to the electric field of an EM wave. How large must the **E** field strength be to produce a 1.00-mV (rms) voltage between the ends of the antenna? What is the rate of energy transport per square meter?

*28. (II) How large an emf (rms) will be generated in an antenna that consists of a 500-loop circular coil of wire 2.0 cm in diameter if the EM wave has a frequency of 910 kHz and is transporting energy at an average rate of 1.0×10^{-4} W/m^2 at the antenna. (Hint: you can use Eq. 20-5, since it could be applied to an observer moving with the coil so that the magnetic field is oscillating with the frequency $f = \omega/2\pi$.)

*29. (II) The variable capacitance of a radio tuner consists of six plates connected together placed alternately between six other plates, also connected together. Each plate is separated from its neighbor by 1.0 mm of air. One set of plates can move so that the area of overlap varies from 1.0 cm^2 to 9.0 cm^2. (*a*) Are these capacitors connected in series or in parallel? (*b*) Determine the range of capacitance values. (*c*) What value of inductor is needed if the radio is to tune AM stations from 550 to 1600 kHz?

22 LIGHT: GEOMETRIC OPTICS

The sense of sight is extremely important to us, for it provides us with a large part of our information about the world. How do we see? What is the something called *light* that enters our eyes and causes the sensation of sight? How does light behave so that we can see the great range of phenomena that we do? The subject of light will occupy us for the next three chapters, and we will return to it in later chapters.

We see an object in one of two ways: (1) the object may be a source, such as a light bulb, a flame, or a star, in which case we see the light emitted directly from the source; or (2), more commonly, we see an object by light reflected from it. In the latter case the light may have originated from the sun, artificial lights, or some other source. An understanding of how bodies *emit* light was not achieved until the 1920s, and this will be discussed in Chapter 26. How light is *reflected* from objects was understood much earlier, and we will discuss this in Section 22-3.

22-1 ∎ The ray model of light

A great deal of evidence suggests that light travels in straight lines under a wide variety of circumstances. For example, a point source of light like the sun casts distinct shadows; and the beam of a flashlight appears to be a straight line. In fact we infer the positions of objects in our environment by assuming that light moves from the object to our eyes in straight-line paths. Our whole orientation to the physical world is based on this assumption.

This reasonable assumption has led to the **ray model** of light. This model assumes that light travels in straight-line paths called light **rays.** Actually, a ray is an idealization: it is meant to represent an infinitely narrow beam of light. When we see an object, light reaches our eyes from each point on the object. Although light rays leave each point in many different directions, normally only a small bundle of these rays can enter an observer's eye as shown in Fig. 22-1. If the person's head moves to one side, a different bundle of rays will enter the eye from each point.

We saw in Chapter 21 that light can be considered as an electromagnetic wave. Although the ray model of light does not deal with this aspect of light (we discuss the wave nature of light in Chapter 23), it has been very successful in describing many aspects of light such as reflection, refraction,

FIGURE 22-1

Light rays emerge from each single point on an object. A small bundle of rays leaving one point is shown entering a person's eye.

This bundle enters the eye

and the formation of images by mirrors and lenses. Because these explanations involve straight-line rays at various angles, this subject is referred to as *geometric optics*.

22-2 ∎ The speed of light and index of refraction

Galileo attempted to measure the speed of light by trying to measure the time required for light to travel a known distance between two hilltops. He stationed an assistant on one hilltop and himself on another and ordered the assistant to lift the cover from a lamp the instant he saw a flash from Galileo's lamp. Galileo measured the time between the flash of his lamp and when he received the light from his assistant's lamp. The time was so short that Galileo concluded it merely represented human reaction time, and that the speed of light must be extremely high.

The first successful determination that the speed of light is finite was made by the Danish astronomer Ole Roemer (1644–1710). Roemer had noted that the carefully measured period of one of Jupiter's moons (Io, with an average period of 42.5 h) varied slightly, depending on the relative motion of Earth and Jupiter. When Earth was moving away from Jupiter, the period of the moon was slightly longer, and when Earth was moving toward Jupiter, the period was slightly shorter. He attributed this variation to the extra time needed for light to travel the increasing distance to Earth when Earth is receding, or to the shorter travel time for the decreasing distance when the two planets are approaching one another. Roemer concluded that the speed of light—though great—is finite.

Since then a number of techniques have been used to measure the speed of light. Among the most important were those carried out by the American Albert A. Michelson (1852–1931). Michelson used the rotating mirror apparatus diagrammed in Fig. 22-2 for a series of high-precision experiments carried out from 1880 to the 1920s. Light from a source was directed at one face of a rotating eight-sided mirror. The reflected light traveled to a stationary mirror a large distance away and back again as shown. If the rotating mirror was turning at just the right rate the returning beam of light would reflect from one face of the mirror into a small telescope

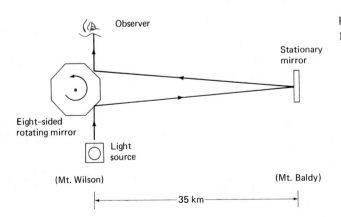

FIGURE 22-2
Michelson's speed-of-light apparatus (not to scale).

through which the observer looked. At a different speed of rotation, the beam would be deflected to one side and would not be seen by the observer. From the required speed of the rotating mirror and the known distance to the stationary mirror, the speed of light could be calculated. In the 1920s, Michelson set up the rotating mirror on the top of Mt. Wilson in southern California and the stationary mirror on Mt. Baldy (Mt. San Antonio) 35 km away. He later measured the speed of light in vacuum using a long evacuated tube.

The accepted value today for the speed of light, c, in vacuum is[†]

$$c = 2.99792458 \times 10^8 \text{ m/s}.$$

We usually round this off to

$$3.00 \times 10^8 \text{ m/s}$$

when extremely precise results are not required. In air, the speed is only slightly less. In other transparent materials such as glass and water, the speed is always less than that in vacuum. For example, in water it travels at about $\frac{3}{4}c$. The ratio of the speed of light in vacuum to the speed v in a given material is called the *index of refraction n* of that material:

$$n = \frac{c}{v}. \tag{22-1}$$

The index of refraction for various materials is given in Table 22-1. (As we shall see later, n varies somewhat with the wavelength of the light—except in vacuum—so a particular wavelength is specified, that of yellow light with wavelength $\lambda = 589$ nm.) For example, since $n = 2.42$ for diamond, the speed of light in diamond is

$$v = c/n = (3.00 \times 10^8 \text{ m/s})/2.42 = 1.24 \times 10^8 \text{ m/s}.$$

TABLE 22-1
Indices of refraction[†]

Material	$n = c/v$
Air (at STP)	1.0003
Water	1.33
Ethyl Alcohol	1.36
Glass	
Fused quartz	1.46
Crown glass	1.52
Light flint	1.58
Lucite or Plexiglas	1.51
Sodium chloride	1.53
Diamond	2.42

[†] $\lambda = 589$ nm

22-3 ∎ Reflection; image formation by a plane mirror

When light strikes the surface of an object, some of the light is reflected. The rest is either absorbed by the object (and transformed to thermal energy) or, if the object is transparent like glass or water, part of it is transmitted through. For a very shiny object such as a silvered mirror, over 95 percent of the light may be reflected.

When a narrow beam of light strikes a flat surface we define the **angle of incidence,** θ_i, to be the angle an incident ray makes with the normal to the surface ("normal" means perpendicular) and the **angle of reflection,** θ_r, to be the angle the reflected ray makes with the normal. For flat surfaces, it is found that the incident and reflected rays lie in the same plane with the normal to the surface, and that

∎ **the angle of incidence equals the angle of reflection.**

This is the **law of reflection** and is indicated in Fig. 22-3. It was known to

---

[†] As discussed in Chapter 1 (p. 8), the speed of light is now *defined* to have this value, and the length of the standard meter is defined in terms of this value of c.

the ancient Greeks, and you can confirm it yourself by shining a narrow flashlight beam at a mirror in a darkened room.

When light is incident upon a rough surface, even microscopically rough such as this page, it is reflected in many directions, Fig. 22-4. This is called *diffuse reflection*. The law of reflection still holds, however, at each small section of the surface. Because of diffuse reflection in all directions, an ordinary object can be seen from many different angles. When you move your head to the side, a different bundle of reflected rays reach your eye from each point on the object, Fig. 22-5a. But when a narrow beam of light is shone on a mirror, the light will not reach your eye unless it is placed at just the right place where the law of reflection is satisfied, as shown in Fig. 22-5b. This is what gives rise to the unusual properties of mirrors. (Galileo, using similar arguments,[†] showed that the moon must have a rough surface rather than a highly polished surface like a mirror, as some people thought.)

When you look straight in a mirror, you see what appears to be yourself as well as various objects around and behind you. Your face and the other objects look as if they are in front of you, beyond the mirror; but, of course, they are not. What you see in the mirror is an **image** of the objects.

Figure 22-6 shows how an image is formed by a plane (that is, flat) mirror. Rays from two different points on an object are shown. Rays leave each point on the object going in many directions, but only those that enclose the bundle of rays that reach the eye from the two points are shown. The diverging rays that enter the eye appear to come from behind the mirror as shown by the dashed lines. (Our eyes and brain interpret any rays that enter an eye as having traveled a straight-line path.) The point from which each bundle of rays seems to come is one point on the image. For each point on the object there is a corresponding image point. Let us concentrate on the two rays that leave the point A on the object and strike the mirror at points B and B'. The angles ADB and CDB are right angles. And angles ABD and CBD are equal because of the law of reflection. Therefore, the two triangles ABD and CDB are congruent and the length $AD = CD$. That is, the image is as far behind the mirror as the object is in front: the *image distance*, d_i (distance from mirror to image, Fig. 22-6), equals the *object distance*, d_o. From the geometry, we also see that the height of the image is the same as that of the object.

The light rays do not actually pass through the image itself. It merely *seems* like the light is coming from the image because our brains interpret any light entering our eyes as coming from in front of us. Since the rays do not actually pass through the image, a piece of white paper or film placed at the image would not detect the image. Therefore, it is called a **virtual image.** This is to distinguish it from a *real image* in which the light does pass through the image and which therefore could appear on paper or film placed at the image position. We will see that curved mirrors and lenses can form real images.

FIGURE 22-3
Law of reflection.

FIGURE 22-4

Diffuse reflection from a rough surface.

FIGURE 22-5

A beam of light from a flashlight is shined on (a) white paper; (b) a small mirror. In part (a) you can see the white light reflected at various points because of diffuse reflection. But in part (b) you see the reflected light only when your eye is placed correctly $(\theta_r = \theta_i)$.

[†] Galileo Galilei, *Dialogue Concerning the Two Chief World Systems*, trans. Stillman Drake (Berkeley: University of California Press, 1967). See Day 1.

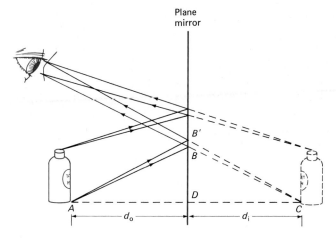

Plane mirror

FIGURE 22-6

Formation of a virtual image by a plane mirror.

EXAMPLE 22-1   A woman 1.60 m tall stands in front of a vertical plane mirror. What is the minimum height of the mirror and how high must its lower edge be above the floor if she is to be able to see her whole body? (Assume her eyes are 10 cm below the top of her head.)

SOLUTION   The situation is diagrammed in Fig. 22-7. First consider the ray from the toe, AB, which upon reflection becomes BE and enters the eye E. Since light enters the eye from point A (the toes) after reflecting at B, the mirror needs to extend no lower than B. Because the angle of reflection equals the angle of incidence, the height BD is half of the height AE. Since $AE = 1.60 \text{ m} - 0.10 \text{ m} = 1.50 \text{ m}$, $BD = 0.75 \text{ m}$. Similarly, if the woman is to see the top of her head, the top edge of the mirror only needs to reach point F, which is 5 cm below the top of her head (half of $GE = 10 \text{ cm}$). Thus $DF = 1.55 \text{ m}$ and the mirror need be only $1.55 \text{ m} - 0.75 \text{ m} = 0.80 \text{ m}$ high; and its bottom edge must be 0.75 m above the floor. In general, a mirror need be only half as tall as a person for that person to see all of himself or herself. Does this result depend on the person's distance from the mirror?

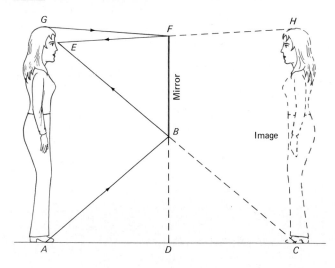

FIGURE 22-7

Seeing oneself in the mirror (Example 22-1).

These rays are the only ones that will strike the mirror, and they are essentially parallel.

FIGURE 22-8

If the object's distance is large compared to the size of the mirror, the rays are nearly parallel.

*22-4 ▌ Formation of images by spherical mirrors

Reflecting surfaces do not have to be flat. The most common *curved* mirrors are *spherical*, which means they form a section of a sphere. A spherical mirror is called *convex* if the reflection takes place on the outer surface of the spherical shape so that the center of the mirror bulges out toward the viewer; and it is called *concave* if the reflecting surface is on the inner surface of the sphere so that the center of the mirror is farther from the viewer than the edges. Concave mirrors are used as shaving or makeup mirrors, and convex mirrors are sometimes used on cars and trucks and in shops (to watch for thieves), since they take in a wide field of view.

To see how spherical mirrors form images, we first consider an object that is very far from a concave mirror; in this case the rays from each point on the object that reach the mirror will be nearly parallel, as shown in Fig. 22-8. *For an object infinitely far away* (the sun and stars approach this) *the rays would be precisely parallel.* Now consider such parallel rays falling on a concave mirror as in Fig. 22-9. The law of reflection holds for each of these rays at the point each strikes the mirror. As can be seen, they are not all brought to a single point. In order to form a sharp image, the rays must come to a point. Thus a spherical mirror will not make as sharp an image as a plane mirror will. However, if the mirror width is small compared to its radius of curvature so that the reflected rays make only a small angle upon reflection, then the rays will cross each other at nearly a single point, or *focus*, as shown in Fig. 22-10. In the case shown, the rays are parallel to the *principal axis*, which is defined as the straight line perpendicular to the curved surface at its center (line CA in the diagram). The point F, where rays parallel to the principal axis come to a focus, is called the **focal point** of the mirror; the distance between F and the center of the mirror, length FA, is

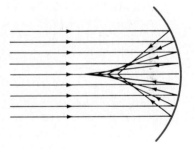

FIGURE 22-9

Parallel rays striking a concave spherical mirror do not focus at a single point.

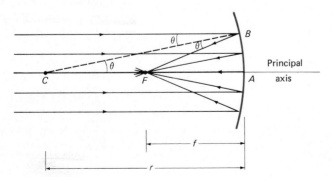

FIGURE 22-10

Rays parallel to the principal axis of a spherical mirror come to a focus at F, called the focal point, as long as the mirror is small in extent as compared to its radius of curvature, r.

called the **focal length,** f, of the mirror. Another way of defining the focal point is to say that it is the *image point for an object infinitely far away* along the principal axis. The image of the sun, for example, would be essentially at F.

Now we will show, for a mirror whose reflecting surface has a width that is small compared to the radius of curvature, that the rays do indeed meet at a common point, F, and we calculate the focal length f. We consider a ray that strikes the mirror at B in Fig. 22-10. The point C is the center of curvature of the mirror (the center of the sphere of which the mirror is a part). So the dashed line CB is equal to r, the radius of curvature, and CB is normal to the surface at B. From the law of reflection and the geometry, the three angles labeled θ are equal as indicated. The triangle CBF is isosceles because two of its angles are equal. Thus length $CF = BF$. We assume the mirror is small compared to its radius of curvature, so the angles are small and the length FB is nearly equal to length FA. In this approximation, $FA = FC$. But $FA = f$, the focal length, and $CA = 2FA = r$. Thus the focal length is half the radius of curvature:

$$f = r/2. \tag{22-2}$$

This argument only assumed that the angle θ was small; so the same result applies for all the other rays; thus all the rays pass through the same point F in this approximation of a mirror small compared to its radius of curvature.

Since it is only approximately true that the rays come to a perfect focus at F, the larger the mirror, the worse the approximation (Fig. 22-9) and the more blurred the image. This "defect" of spherical mirrors is called *spherical aberration;* we will discuss it more with regard to lenses in Chapter 24. A *parabolic* reflector, on the other hand, will reflect the rays to a perfect focus. However, because parabolic shapes are much harder to make and thus much more expensive, spherical mirrors are used for most purposes. We consider here only spherical mirrors and we will assume that they are small compared to their radius of curvature so that the image is sharp and Eq. 22-2 holds.

We saw that for an object at infinity the image is located at the focal point of a concave spherical mirror, where $f = r/2$. But where does the image lie for an object not at infinity? First consider the object shown in Fig. 22-11, which is placed at point O between F and C. Let us determine where the

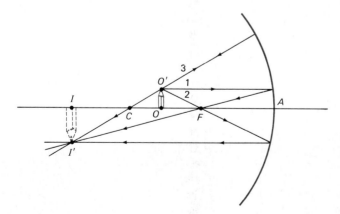

FIGURE 22-11
Rays from object at O form an image at I.

image will be for a given point O' on the object. To do this we can draw several rays and make sure these reflect from the mirror such that the reflection angle equals the incidence angle. This can involve much work and our task is simplified if we deal with three particularly simple rays. These are the rays labeled 1, 2, and 3 in the diagram. Ray 1 is drawn parallel to the axis; therefore it must pass through F after reflection (as in Fig. 22-10). Ray 2 is drawn through F; therefore it must reflect so it is parallel to the axis. Ray 3 is drawn so that it passes through C, the center of curvature, and thus is along a radius of the spherical surface; so it is perpendicular to the mirror and thus will be reflected back on itself. The point at which these rays cross is the image point I'. All other rays from the same object point will pass through this image point. To find the image point for any object point, only these three types of rays need be used. Actually, only two of these rays are needed, but the third serves as a check.

We have only shown the image point in Fig. 22-11 for a single point on the object. Other points on the object are imaged nearby so a complete image of the object is formed, as shown by the dashed outline. Because the light actually passes through the image itself, this is a **real image.** This can be compared to the virtual image formed by a plane mirror.

Image points can always be determined by drawing the three rays as described above. However, it is possible to derive an equation that gives the image distance if the object distance and radius of curvature are known. To do this we refer to Fig. 22-12. The distance of the object from the center of the mirror, called the **object distance,** is labeled d_o; and the **image distance** is labeled d_i. The height of the object OO' is called h_o and the height of the image, $I'I$, is h_i. Two rays are shown, $O'FBI'$ (same as ray 2 in the previous figure) and $O'AI'$. The ray $O'AI'$ obeys the law of reflection, of course, so the triangles $O'AO$ and $I'AI$ are similar. Therefore, we have

$$\frac{h_o}{h_i} = \frac{d_o}{d_i}.$$

For the other ray, $O'FBI'$, the triangles $O'FO$ and AFB are also similar. Therefore,

$$\frac{h_o}{h_i} = \frac{OF}{FA} = \frac{d_o - f}{f},$$

since the length $AB = h_i$ (in our approximation of a mirror small compared to its radius) and $FA = f$, the focal length of the mirror. Since the left side

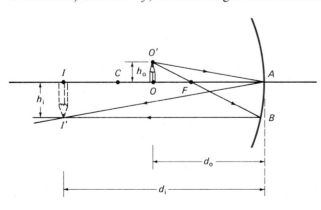

FIGURE 22-12
Diagram for deriving mirror equation.

of the two preceding expressions are the same, we equate the right sides:

$$\frac{d_o}{d_i} = \frac{d_o - f}{f}.$$

We now divide both sides by d_o and rearrange to obtain

$$\frac{1}{d_o} + \frac{1}{d_i} = \frac{1}{f}. \tag{22-3}$$

This is the equation we were seeking. It is called the **mirror equation** and relates the object and image distances to the focal length f (where $f = r/2$).

The lateral *magnification m* of a mirror is defined as the height of the image divided by the height of the object. From our first set of similar triangles above, we can write:

$$m = \frac{h_i}{h_o} = -\frac{d_i}{d_o}, \tag{22-4}$$

where the minus sign is inserted as a convention; for consistency we must be careful about the signs of all quantities in Eq. 22-4. The conventions we use are: h_o (or h_i) is considered positive for an object (or image) point above the principal axis, and negative for a point below the axis. Also d_i and d_o are positive if image and object are on the reflecting side of the mirror (as in Fig. 22-12), but if either image or object is behind the mirror the corresponding distance is negative (an example is shown in Fig. 22-13, Example 22-3). Thus the magnification (Eq. 22-4) is positive for an upright image and negative for an inverted image.

EXAMPLE 22-2 A 1.50-cm-high object is placed 20.0 cm from a concave mirror whose radius of curvature is 30.0 cm. Determine (a) the position of the image and (b) its size.

SOLUTION   (a) The focal length $f = r/2 = 15.0$ cm. Then, since $d_o = 20.0$ cm, we have from Eq. 22-3 that

$$\frac{1}{d_i} = \frac{1}{f} - \frac{1}{d_o} = \frac{1}{15.0 \text{ cm}} - \frac{1}{20.0 \text{ cm}} = 0.0167 \text{ cm}^{-1}.$$

So $d_i = 1/0.0167 \text{ cm}^{-1} = 60.0$ cm; the image is 60.0 cm from the mirror on the same side as the object. (b) From Eq. 22-4, the lateral magnification is $m = -60.0 \text{ cm}/20.0 \text{ cm} = -3.00$. Therefore the image is $(-3.00)(1.5 \text{ cm}) = -4.5$ cm high; the minus sign tells us it is inverted.

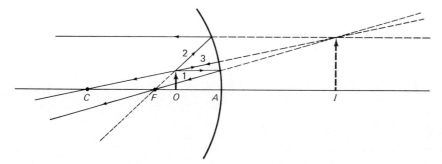

FIGURE 22-13

Object placed within the focal point F (Example 22-3). The image is *behind* the mirror and is *virtual*.

□ EXAMPLE 22-3   A 1.00-cm-high object is placed 10.0 cm from a concave mirror whose radius of curvature is 30.0 cm. (*a*) Draw a ray diagram
to locate the position of the image. (*b*) Determine the position of the image
and the magnification analytically.

SOLUTION   (*a*) Since $f = 15.0$ cm, the object is between the mirror
and the focal point. We draw the three rays as described earlier (for Fig.
22-11) and this is shown in Fig. 22-13. The rays reflected from the mirror
diverge and so never meet at a point. They appear, however, to be coming
from a point behind the mirror. The image is thus behind the mirror and
is *virtual*. (Why?) (*b*) We use Eq. 22-3 to find d_i where $d_o = 10.0$ cm:

$$\frac{1}{d_i} = \frac{1}{15.0 \text{ cm}} - \frac{1}{10.0 \text{ cm}} = \frac{2 - 3}{30.0 \text{ cm}} = -\frac{1}{30.0 \text{ cm}}.$$

Therefore $d_i = -30.0$ cm. The minus sign means the image is behind the
mirror. The lateral magnification is $m = -(-30.0 \text{ cm})/(10.0 \text{ cm}) = +3.00$.
So the image is 3.00 times larger than the object; the plus sign indicates
that the image is upright.

These examples show that a spherical mirror can produce a magnified image,
one that is larger than the object. (There is an old story—maybe fable is a
better word—that Julius Caesar spied on the British forces by setting up a
curved mirror on the coast of Gaul. Is this reasonable?)

It is useful to compare Figs. 22-11 and 22-13. We can see that if the
object is within the focal point, as in Fig. 22-13, the image is virtual, upright,
and magnified. This is how a shaving or make-up mirror is used—you must
place your head within the focal point if you are to see yourself right side
up. If the object is *beyond* the focal point as in Fig. 22-11, the image is real
and inverted. Whether the magnification is greater or less than 1.0 in this
case depends on the position of the object relative to the center of curvature,
point *C*.

The analysis used for concave mirrors can be applied to *convex* mirrors.
Even the mirror equation (Eq. 22-3) holds for a convex mirror, although the
quantities involved must be carefully defined. Fig. 22-14a shows parallel rays
falling on a convex mirror. Again spherical aberration will be present, but
we assume the mirror's size is small compared to its radius of curvature. The
reflected rays diverge, but seem to come from point *F* behind the mirror.
This is the *focal point*, and its distance from the center of the mirror is the
focal length *f*. It is easy to show that again $f = r/2$. We see that an object
at infinity produces a virtual image in a convex mirror. Indeed, no matter
where the object is placed on the reflecting side of the mirror, the image will
be virtual and erect as indicated in Fig. 22-14b. To find the image we draw
rays 1 and 3 according to the rules used before on the concave mirror.

The mirror equation, Eq. 22-3, holds for convex mirrors but the focal
length *f* must be considered negative, as must the radius of curvature. The
proof is left as a problem. It is also left as a problem to show that Eq. 22-4
for the magnification is also valid.

□ EXAMPLE 22-4   A convex rearview car mirror has a radius of curvature of 40.0 cm. Determine the location of the image and its magnification
for an object 10.0 m from the mirror.

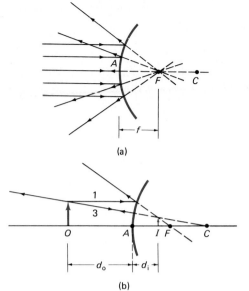

(a)

(b)

FIGURE 22-14

Convex mirror: (a) the focal point is at F, behind the mirror; (b) image I for the object at O is virtual, upright, and smaller than the object.

SOLUTION With $r = -40.0$ cm, then $f = -20.0$ cm, and the mirror equation gives:

$$\frac{1}{d_i} = \frac{1}{f} - \frac{1}{d_o} = -\frac{1}{0.200 \text{ m}} - \frac{1}{10.0 \text{ m}} = -\frac{51.0}{10.0 \text{ m}}.$$

So $d_i = 0.196$ m or 19.6 cm behind the mirror. The lateral magnification is $m = -d_i/d_o = -(-0.196 \text{ m})/(10.0 \text{ m}) = 0.0196$ or 1/51; so the upright image is reduced by a factor of 51.

We summarize the rules (the *sign conventions*) for applying Eqs. 22-3 and 22-4 to concave and convex mirrors: *When the object, image, or focal point is on the reflecting side of the mirror* (on the left in all our drawings), *the corresponding distance is considered positive. If any of these points is behind the mirror* (on the right) *the corresponding distance must be considered negative.*[†] *Object and image heights, h_o and h_i, are considered positive or negative depending on whether the point is above or below the principal axis.* It is important to be consistent with these sign conventions.

22-5 ▌ Refraction; Snell's law

When light passes from one medium into another, part of the incident light is reflected at the boundary. The remainder passes into the new medium. If a ray of light is incident at an angle to the surface (other than perpendicular),

---

[†] We have seen examples where d_i and f are negative. The object distance d_o for any material object is, of course, always positive. However, if the mirror is used in conjunction with a lens or another mirror, the image formed by the first mirror or lens becomes the object for the second mirror; it is then possible for such an "object" to be behind the second mirror, in which case, d_o would be negative. These rules are also consistent with considering the focal length of a concave mirror positive and that of a convex mirror negative.

the ray is bent as it enters the new medium. This bending is called **refraction.**
Figure 22-15a shows a ray passing from air into water. The angle θ_1 is the
angle of incidence and θ_2 is the *angle of refraction*. Notice that the ray bends
toward the normal when entering the water. This is always the case when the
ray enters a medium where the speed of light is less. If light travels from one
medium into a second where its speed is greater, the ray bends away from
the normal; this is shown in Fig. 22-15b for a ray traveling from water to air.

Refraction is responsible for a number of common optical illusions.
For example, a person standing in waist-deep water appears to have shortened legs, Fig. 22-16a. As shown in Fig. 22-16b, the rays leaving the person's
foot are bent at the surface; the observer's eye (and brain) assumes the rays
to have traveled a straight-line path, and so the feet appear to be higher
than they really are. Similarly, when you put a stick in water, it appears to
be bent.

The angle of refraction depends on the speed of light in the two media
and on the incident angle. An analytical relation between θ_1 and θ_2 was
arrived at experimentally about 1621 by Willebrord Snell (1591–1626). This
is known as **Snell's law** and is written:

$$n_1 \sin \theta_1 = n_2 \sin \theta_2; \tag{22-5}$$

θ_1 is the angle of incidence and θ_2 is the angle of refraction; n_1 and n_2 are the
respective indices of refraction in the materials (see Fig. 22-15). Snell's law[†]
is the basic *law of refraction*.

It is clear from Snell's law that if $n_2 > n_1$, then $\theta_2 < \theta_1$; that is, if light
enters a medium where n is greater (and its speed less), then the ray is bent
toward the normal. And if $n_2 < n_1$, then $\theta_2 > \theta_1$, so the ray bends away from
the normal. This is what we saw in Fig. 22-15.

[†] Snell actually was not aware that the index of refraction is related to the speed of
light in the particular medium. Only later was it found that the index of refraction could be
written as the ratio of the speed of light in vacuum to that in the given material, Eq. 22-1.
Snell's law can be derived from the wave theory of light (see next chapter, Section 23-2), and
in fact we already did so in Section 13-10 where Eq. 13-16 is just a combination of Eqs. 22-5
and 22-1.

FIGURE 22-15
Refraction.

FIGURE 22-16

Because of refraction, when a person is
standing in water, his or her legs look
shorter.

(a)

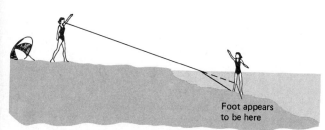

(b)

EXAMPLE 22-5   Light strikes a flat piece of glass at an incident angle of 60° as in Fig. 22-17. If the index of refraction of the glass is 1.50, (a) what is the angle of refraction θ_a in the glass; (b) what is the angle θ_b at which the ray emerges from the glass?

SOLUTION   (a) We assume the incident ray is in air so $n_1 = 1.00$ and $n_2 = 1.50$. Then, from Eq. 22-5 we have

$$\sin \theta_a = \frac{1.00}{1.50} \sin 60° = 0.577,$$

so $\theta_a = 35.2°$. (b) Since the faces of the glass are parallel, the incident angle in this case is just θ_a, so $\sin \theta_a = 0.577$. This time $n_1 = 1.50$ and $n_2 = 1.00$. Thus $\theta_b (= \theta_2)$ is

$$\sin \theta_b = \frac{1.50}{1.00} \sin \theta_a = 0.866,$$

and $\theta_b = 60.0°$. The direction of the beam is thus unchanged by passing through a plane piece of glass. It should be clear that this is true for any angle of incidence. The ray is displaced slightly to one side however. You can observe this by looking through a piece of glass (near its edge) at some object and then moving your head to the side so that you see the object directly.

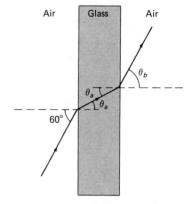

FIGURE 22-17

Light passing through a piece of glass (Example 22-5).

22-6 ■ Total internal reflection; fiber optics

When light passes from one material into a second material where the index of refraction is less (say from water into air), the light bends away from the normal, as for ray A in Fig. 22-18. At a particular incident angle, the angle of refraction will be 90°, and the refracted ray would skim the surface (ray B) in this case. The incident angle at which this occurs is called the *critical angle* θ_c; from Snell's law, θ_c is given by

$$\sin \theta_c = \frac{n_2}{n_1} \sin 90° = \frac{n_2}{n_1}. \qquad (22\text{-}6)$$

For any incident angle less than θ_c there will be a refracted ray, although part of the light will also be reflected at the boundary. However, for incident angles greater than θ_c, Snell's law would tell us that $\sin \theta_2$ is greater than 1.00. Yet the sine of an angle can never be greater than 1.00. In this case

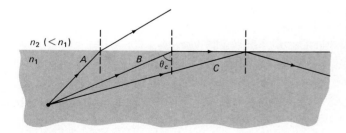

FIGURE 22-18

Since $n_2 < n_1$, light rays are totally internally reflected if $\theta > \theta_c$, as for ray C. If $\theta < \theta_c$, as for ray A, only a part of the light is reflected (this part not shown), and the rest is refracted.

there is no refracted ray at all, and *all of the light is reflected*, as for ray *C* in the diagram. This is called **total internal reflection**. But note that total internal reflection can occur only when light strikes a boundary where the medium beyond is optically less dense—that is, it has a lower index of refraction.

EXAMPLE 22-6 Describe what a person would see who looked up at the world from beneath the perfectly smooth surface of a lake or swimming pool.

SOLUTION For an air-water interface, the critical angle is given by:

$$\sin \theta_c = \frac{1.00}{1.33} = 0.750.$$

Therefore $\theta_c = 49°$. Thus the person would see the outside world compressed into a circle whose edge makes a 49° angle with the vertical. Beyond this angle the person will see reflections from the sides and bottom of the pool or lake.

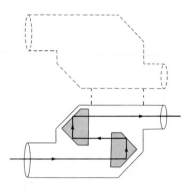

FIGURE 22-19
Prisms reflect light in binoculars.

Many optical instruments, such as binoculars, use total internal reflection within a prism to reflect light. The advantage is that very nearly 100 percent of the light is reflected, whereas even the best mirrors reflect somewhat less than 100 percent. Thus the image is brighter. For glass with $n = 1.50$, $\theta_c = 41.8°$. Therefore, 45° prisms will reflect all the light internally as shown in the binoculars of Fig. 22-19.

Total internal reflection is the principle behind *fiber optics*. Very thin glass and plastic fibers can now be made as small as a few micrometers in diameter. A bundle of such tiny fibers is called a *light pipe* since light can be transmitted along it with almost no loss. Figure 22-20 shows how light traveling down a thin fiber makes only glancing collisions with the walls so that total internal reflection occurs. Even if the light pipe is bent into a complicated shape, the critical angle won't (usually) be exceeded so light is transmitted practically undiminished to the other end. This effect is used in decorative lamps and to illuminate water streams in fountains. Light pipes can be used to illuminate difficult places to reach, such as inside the human body. They can be used to transmit telephone calls and other communication signals; the signal is a modulated light beam (variable intensity of the light beam) and is transmitted with less loss than an electrical signal in a copper wire. One sophisticated use of fiber optics, particularly in medicine, is to transmit a clear picture, Fig. 22-21. For example, a patient's stomach can be examined by inserting a light pipe through the mouth. Light is sent down one set of fibers to illuminate the stomach. The reflected light returns up another set of fibers. Light directly in front of each fiber travels up that fiber. At the opposite end, a viewer sees a series of bright and dark spots, much like a TV screen—that is, a picture of what lies at the opposite end.[†] The fibers must be optically insulated from one another, usually by a thin coating of material whose refractive index is less than that of the fiber. The fibers

FIGURE 22-20
Light reflected totally at the interior surface of a glass or transparent plastic fiber.

FIGURE 22-21
Fiber optic image.

[†] Lenses are used at each end: at the object end to bring the rays in parallel, and at the viewing end a telescope system for viewing.

must be arranged precisely parallel to one another if the picture is to be clear. The more fibers there are, and the smaller they are, the more detailed the picture. Such an "endoscope" is useful for observing the stomach or other hard to reach places for surgery or searching for lesions without surgery.

22-7 ∎ Thin lenses; ray tracing

The most important simple optical device is no doubt the thin lens. The development of optical devices using lenses dates to the sixteenth and seventeenth centuries, although the earliest record of eyeglasses dates from the late thirteenth century.[†] Today we find lenses in eyeglasses, cameras, magnifying glasses, telescopes, binoculars, microscopes, and medical instruments. A thin lens is usually circular, and its two faces are portions of a sphere. (Although cylindrical surfaces are also possible, we will concentrate on spherical.) The two faces can be concave, convex, or plane (Fig. 22-22). The importance of lenses is that they form images of objects.

Consider the rays parallel to the axis of the double convex lens which is shown in cross-section in Fig. 22-23a. We assume the lens is made of glass or transparent plastic so its index of refraction is greater than that of the air outside. From Snell's law we can see that each ray is bent toward the axis at both lens surfaces (note the dashed lines indicating the normals to each surface for the top ray). If rays parallel to the principal axis fall on a thin lens, they will be focused to a point called the **focal point,** F. This will not be precisely true for a lens with spherical surfaces. But it will be very nearly true—that is, parallel rays will be focused to a tiny region that is nearly a point— if the diameter of the lens is small compared to the radii of curvature of the two lens surfaces. This criterion is satisfied by a **thin lens,** one that is very thin compared to its diameter. It is only thin lenses that we consider here, and we assume the rays are focused to a point.

Since the rays from a distant object are essentially parallel (see Fig. 22-8), we can also say that *the focal point is the image point for an object on the principal axis at infinity.* Thus the focal point of a lens can be found by locating the point where the sun's rays (or those of some other distant object) are brought to a sharp image. The distance of the focal point from the center

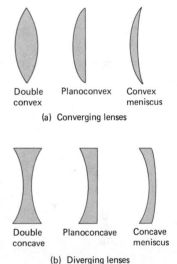

Double convex   Planoconvex   Convex meniscus

(a) Converging lenses

Double concave   Planoconcave   Concave meniscus

(b) Diverging lenses

FIGURE 22-22

Types of lenses.

[†] Rounded gem stones used as magnifiers probably date from much earlier.

(a)

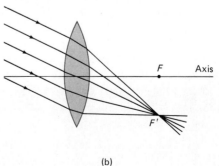

(b)

FIGURE 22-23

Parallel rays are brought to a focus by a converging lens.

of the lens is called the **focal length,** f. Since the lens is assumed thin, f can be measured from the center or an edge of the lens without substantial error. A lens can be turned around so light can pass through it from the opposite side; the focal length is the same on both sides, as we shall see later, even if the curvatures of the two lens surfaces are different. If parallel rays fall on a lens at an angle, as in Fig. 22-23b, they focus at a point F'. The plane in which all points such as F and F' fall is called the *focal plane* of the lens.

Any lens[†] which is thicker in the center than at the edges will make parallel rays converge to a point, and is called a *converging lens* (see Fig. 22-22a). Lenses which are thinner in the center than at the edges (Fig. 22-22b) are called *diverging lenses* because they make parallel light diverge, as shown in Fig. 22-24. The *focal point* F of a diverging lens is defined as that point from which refracted rays, originating from parallel incident rays, seem to emerge as shown in the figure. And the distance from F to the lens is called the focal length, just as for a converging lens.

Optometrists and ophthalmologists, instead of using the focal length, use the reciprocal of the focal length to specify the strength of eyeglass (or contact) lenses. This is called the **power,** P, of a lens:

FIGURE 22-24
Diverging lens.

$$P = \frac{1}{f}.$$

The unit for lens power is the *diopter* (D), which is an inverse meter: $1\,\text{D} = 1\,\text{m}^{-1}$. For example, a 20-cm-focal length lens has a power $P = 1/0.20\,\text{m} = 5.0\,\text{D}$. We will mainly use the focal length here, but will refer again to the power of a lens when we discuss eyeglass lenses in Chapter 24.

The most important parameter of a lens is its focal length f. For a converging lens, f is easily measured by finding the image point for the sun or other distant objects. Once f is known, the image position can be found for any object. To find the image point by drawing rays would be difficult if we had to determine all the refractive angles. Instead, we can do it very simply by making use of certain facts we already know, such as that a ray parallel to the axis of the lens passes (after refraction) through the focal point. In fact, we need consider only the three rays indicated in Fig. 22-25. These three rays, emanating from a single point on the object, are drawn as if the lens were infinitely thin (thus the sharp bend within the lens). Ray 1 is drawn parallel to the axis; therefore it is refracted by the lens so it passes through the focal point F behind the lens. Ray 2 is drawn through the focal point

[†] We are assuming the lens has index of refraction greater than that of the surrounding material, such as a glass or plastic lens in air, which is the usual situation.

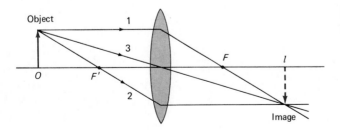

FIGURE 22-25
Finding the image by ray tracing—converging lens.

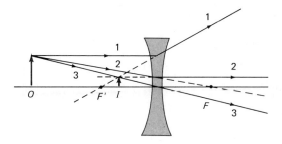

FIGURE 22-26

Finding image I by ray tracing—diverging lens.

F' on the same side of the lens as the object; it therefore emerges from the lens parallel to the axis. Ray 3 is directed toward the very center of the lens where the two surfaces are essentially parallel to each other; this ray therefore emerges from the lens at the same angle as it entered; as we saw in Example 22-5, the ray would be displaced slightly to one side, but since we assume the lens is thin, we draw ray 3 straight through as shown. Actually, any two of these rays will suffice to locate the image point, which is the point where they intersect. Drawing the third can serve as a check.

In this way we can find the image point for one point of the object (the top of the arrow in Fig. 22-25). The image points for all other points on the object can be found similarly to determine the complete image of the object. Because the rays actually pass through the image for the case shown, it is a *real image*.

By drawing the same three rays we can determine the image position for a diverging lens, as shown in Fig. 22-26. Note that ray 1 is drawn parallel to the axis, but does not pass through the focal point F behind the lens; instead it seems to come from the focal point F' in front of the lens (dashed line). Ray 2 is directed toward F and is refracted parallel by the lens. The three refracted rays seem to emerge from a point on the left of the lens. This is the image, I. Since the rays do not pass through the image, it is a *virtual image*.

22-8 ∎ The lens equation

We now derive an equation that relates the image distance to the object distance and the focal length of the lens. This will make the determination of image position quicker and more accurate than doing ray tracing. Consider the two rays shown in Fig. 22-27a for a converging lens (assumed to be very thin); h_o and h_i refer to the heights of the object and image, and d_o and d_i are their distances from the lens. The triangles $FI'I$ and FBA are similar, so

$$\frac{h_i}{h_o} = \frac{d_i - f}{f},$$

since length $AB = h_o$. Triangles OAO' and IAI' are similar. Therefore,

$$\frac{h_i}{h_o} = \frac{d_i}{d_o}.$$

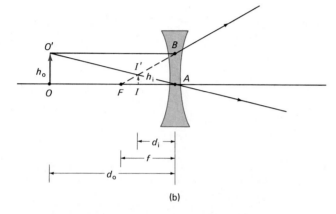

FIGURE 22-27

Deriving the lens equation for (a) converging and (b) diverging lenses.

We equate these two, divide by d_i, and rearrange to obtain

$$\frac{1}{d_o} + \frac{1}{d_i} = \frac{1}{f}. \qquad (22\text{-}7)$$

This is called the **lens equation.** It relates the image distance d_i to the object distance d_o and the focal length f. It is the most useful equation in geometric optics. (Interestingly, it is exactly the same as the mirror equation, Eq 22-3). Note that if the object is at infinity, then $1/d_o = 0$, so $d_i = f$. Thus the focal length is the image distance for an object at infinity, as mentioned earlier.

From the rays in Fig. 21-27b, we can derive the lens equation for a diverging lens. Triangles IAI' and OAO' are similar; and triangles IFI' and AFB are similar. Thus

$$\frac{h_i}{h_o} = \frac{d_i}{d_o} \quad \text{and} \quad \frac{h_i}{h_o} = \frac{f - d_i}{f},$$

When these are equated and simplified, we obtain

$$\frac{1}{d_o} - \frac{1}{d_i} = -\frac{1}{f}.$$

This equation becomes the same as Eq. 22-7 if we make f and d_i negative; that is, we take f to be negative for a diverging lens, and d_i negative when the image is on the same side of the lens as the light comes from. Thus Eq. 22-7 will be valid for both converging and diverging lenses, and for *all* situations, if we use the following conventions[†]:

1. The focal length is positive for converging lenses and negative for diverging lenses.

2. The object distance is positive if it is on the side of the lens from which the light is coming (this is normally the case, although when lenses are used in combination, it might not be so); otherwise, it is negative.

3. The image distance is positive if it is on the opposite side of the lens from where the light is coming; if it is on the same side, d_i is negative. Equivalently, the image distance is positive for a real image and negative for a virtual image.

4. Object and image heights, h_o and h_i, are positive for points above the axis, and negative for points below the axis.

The lateral *magnification m* of a lens is defined as the ratio of the image height to object height, $m = h_i/h_o$. From Fig. 22-27 and the conventions just stated, we have

$$m = \frac{h_i}{h_o} = -\frac{d_i}{d_o}. \tag{22-8}$$

The minus sign has been inserted, in accord with convention 4 above, so that for an upright image the magnification is positive, and for an inverted image m is negative.

From convention 1, above, it follows that the power of a converging lens, in diopters, is positive, whereas the power of a diverging lens is negative. A converging lens is sometimes referred to as a *positive lens*, and a diverging lens as a *negative lens*.

EXAMPLE 22-7   What is the position and size of the image of a 22.4-cm-high giant insect placed 1.50 m from a +50.0-mm-focal-length camera lens?

SOLUTION   The camera lens is converging, with $f = +5.00$ cm, so Eq. 22-7 gives

$$\frac{1}{d_i} = \frac{1}{f} - \frac{1}{d_o} = \frac{1}{5.00 \text{ cm}} - \frac{1}{150 \text{ cm}} = \frac{30.0 - 1.0}{150 \text{ cm}},$$

and $d_i = 150$ cm$/29.0 = 5.17$ cm or 51.7 mm behind the lens. The magnification $m = -d_i/d_o = -5.17$ cm$/150$ cm $= -0.0345$, so $h_i = (-0.0345) \cdot (22.4$ cm$) = -0.773$ cm. The image is 7.73 mm high and is inverted as in Fig. 22-27a. Notice that the image is 1.7 mm farther from the lens than would be the image for an object at infinity. This is an example of the fact that when focusing a camera lens, the closer the object is to the camera, the farther the lens must be from the film.

EXAMPLE 22-8   An object is placed 10 cm from a 15-cm-focal-length converging lens. Determine the image position and size (*a*) analytically, (*b*) using a ray diagram.

[†] These conventions are the same for mirrors, except for 3. See Section 22-4.

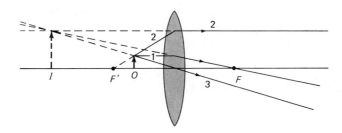

FIGURE 22-28

An object placed within the focal point of a converging lens produces a virtual image (Example 22-8).

SOLUTION   (a) Since $f = 15$ cm and $d_o = 10$ cm,

$$\frac{1}{d_i} = \frac{1}{15 \text{ cm}} - \frac{1}{10 \text{ cm}} = -\frac{1}{30 \text{ cm}}$$

so $d_i = -30$ cm. Since d_i is negative, the image must be virtual and on the same side of the lens as the object. The magnification $m = -(-30 \text{ cm})/(10 \text{ cm}) = 3.0$; so the image is three times as large as the object and is upright. (b) The ray diagram is shown in Fig. 22-28 and confirms the result in part (a).

It is a general rule that when an object is placed between a converging lens and its focal point, the image is virtual.

▢   **EXAMPLE 22-9**   Where must a small insect be placed if a 25-cm-focal-length diverging lens is to form a virtual image 20 cm in front of the lens?

SOLUTION   Since $f = -25$ cm and $d_i = -20$ cm, then

$$\frac{1}{d_o} = -\frac{1}{25 \text{ cm}} + \frac{1}{20 \text{ cm}} = \frac{1}{100 \text{ cm}}.$$

So the object must be 100 cm in front of the lens. The ray diagram is basically that of Fig. 22-27b.

▢   **EXAMPLE 22-10**   To measure the focal length of a diverging lens, a converging lens is placed in contact with it as in Fig. 22-29. The sun's rays are focused by this combination at a point 28.5 cm on the other side of the lenses as shown. If the converging lens has a focal length f_C of 16.0 cm, what is the focal length f_D of the diverging lens?

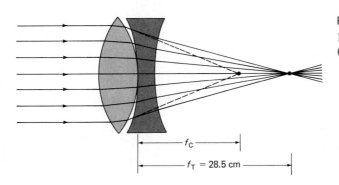

FIGURE 22-29

Determining the focal length of a diverging lens (Example 22-10).

SOLUTION   Let $f_T = 28.5$ cm refer to the focal length of the total combination. If the diverging lens were absent, the converging lens would form the image at its focal point—that is, at a distance $f_C = 16.0$ cm behind it (dashed lines in Fig. 22-29). When the diverging lens is placed next to the converging lens (we assume both lenses are thin and the space between them is negligible) we treat the image formed by the first lens as the *object* for the second (diverging) lens. Since this object lies to the right of the diverging lens, this is one of the unusual situations where d_o is negative. Thus, for the diverging lens, the object is virtual and $d_o = -16.0$ cm; and it forms the image a distance $d_i = 28.5$ cm away (this was given). Thus

$$\frac{1}{f_D} = \frac{1}{d_o} + \frac{1}{d_i} = \frac{1}{-16.0 \text{ cm}} + \frac{1}{28.5 \text{ cm}} = -0.0274 \text{ cm}^{-1}.$$

So $f_D = -1/(0.0274 \text{ cm}^{-1}) = -36.5$ cm. Note that the converging lens must have a focal length whose magnitude is less than that of the diverging lens if this technique is to work.

This example is our first illustration of how to deal with lenses used in combination. In general, when light passes through several lenses, the image formed by one lens becomes the object for the next lens. The total magnification will be the product of the separate magnification of each lens. We will see more examples in Chapter 24, where we discuss optical instruments such as telescopes and microscopes.

*22-9 ∎ The lens-maker's equation

In this optional section we will prove that parallel rays are brought to a focus at a *single* point for a thin lens. At the same time, we will also derive an equation that relates the focal length of a lens to the radii of curvature of its two surfaces, which is known as the lens-maker's equation.

In Fig. 22-30 a ray parallel to the axis of a lens is refracted at the front surface of the lens at point A_1 and is refracted at the back surface at point

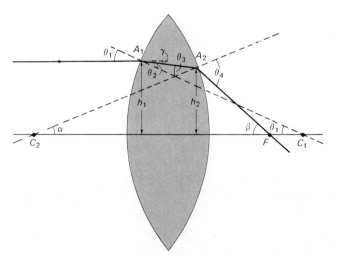

FIGURE 22-30

Diagram of ray passing through a lens for derivation of the lens-maker's equation.

A_2; it then passes through point F, which we call the focal point for this ray. Point A_1 is a height h_1 above the axis, and point A_2 is height h_2 above the axis. C_1 and C_2 are the centers of curvature of the two lens surfaces, so the length $C_1A_1 = R_1$, the radius of curvature of the front surface; and $C_2A_2 = R_2$ is the radius of the second surface. The thickness of the lens has been grossly exaggerated so the various angles would be clear. But we will assume that the lens is actually very thin and that angles between the rays and the axis are small. In this approximation $h_1 \approx h_2$, and the sines and tangents of all the angles will be equal to the angles themselves in radians; for example, $\sin \theta_1 \approx \tan \theta_1 \approx \theta_1$.

To this approximation, then, Snell's law tells us that

$$\theta_1 = n\theta_2$$
$$\theta_4 = n\theta_3$$

where n is the index of refraction of the glass, and we assume that the lens is surrounded by air ($n = 1$). Notice also that

$$\theta_1 = \sin \theta_1 = \frac{h_1}{R_1}$$

$$\alpha = \frac{h_2}{R_2}$$

$$\beta = \frac{h_2}{f}.$$

The last follows because the distance from F to the lens (assumed very thin) is f. From the diagram the angle γ is defined as

$$\gamma = \theta_1 - \theta_2.$$

A careful examination of the diagram shows also that

$$\alpha = \theta_3 - \gamma.$$

This can be seen by drawing a horizontal line to the left from point A_2, which divides the angle θ_3 into two parts. The upper part equals γ and the lower part equals α. (The opposite angles between an oblique line and two parallel lines are equal.) Thus $\theta_3 = \gamma + \alpha$. Finally, by drawing a horizontal line to the right from point A_2, we divide θ_4 into two parts. The upper part is α and the lower is β. Thus

$$\theta_4 = \alpha + \beta.$$

We now combine all these equations:

$$\alpha = \theta_3 - \gamma = \frac{\theta_4}{n} - (\theta_1 - \theta_2) = \frac{\alpha}{n} + \frac{\beta}{n} - \theta_1 + \theta_2,$$

or

$$\frac{h_2}{R_2} = \frac{h_2}{nR_2} + \frac{h_2}{nf} - \frac{h_1}{R_1} + \frac{h_1}{nR_1}.$$

Since the lens is thin, $h_1 \approx h_2$ and we can cancel this from all the numerators. We then multiply through by n and rearrange to find that

$$\frac{1}{f} = (n - 1)\left(\frac{1}{R_1} + \frac{1}{R_2}\right). \tag{22-9}$$

This is the **lens-maker's equation.** It relates the focal length of a lens to the radii of curvature of its two surfaces and its index of refraction. Notice that f does not depend on h_1 or h_2. Thus the position of the point F does not depend on where the ray strikes the lens. Hence all rays parallel to the axis of a thin lens will pass through the same point F, which we wished to prove.

In our derivation both surfaces are convex and R_1 and R_2 are considered positive.[†] Equation 22-9 also works for lenses with one or both surfaces concave; but for a concave surface, the radius must be considered *negative*.

Notice in Eq. 22-9 that the equation is symmetrical in R_1 and R_2. Thus, if a lens is turned around so that light impinges on the other surface, the focal length is the same even if the two lens surfaces are different.

EXAMPLE 22-11   A concave meniscus lens (Fig. 22-22a) is made from glass with $n = 1.50$. The radius of curvature of the convex surface is 22.4 cm and that of the concave surface is 46.2 cm. (*a*) What is the focal length? (*b*) Where will it focus an object 2.00 m away?

SOLUTION $R_1 = 22.4$ cm and $R_2 = -46.2$ cm; the latter is negative since it refers to the concave surface. Then

$$\frac{1}{f} = (1.50 - 1.00)\left(\frac{1}{22.4 \text{ cm}} - \frac{1}{46.2 \text{ cm}}\right) = 0.0114 \text{ cm}^{-1}.$$

So $f = 89$ cm and is converging. Notice that if we turn the lens around so that $R_1 = -46.2$ cm and $R_2 = +22.4$ cm we get the same result. (*b*) From the lens equation

$$\frac{1}{d_i} = \frac{1}{f} - \frac{1}{d_o} = \frac{1}{0.89 \text{ m}} - \frac{1}{2.00 \text{ m}} = 0.62 \text{ m}^{-1}$$

so $d_i = 1.60$ m.

EXAMPLE 22-12   A Lucite planoconcave lens has one flat surface and the other has $R = -18.4$ cm. What is the focal length?

SOLUTION   From Table 22-1, n for Lucite is 1.51. A plane surface has infinite radius of curvature; if we call this R_1, then $1/R_1 = 0$. Therefore,

$$\frac{1}{f} = (1.51 - 1.00)\left(-\frac{1}{18.4 \text{ cm}}\right).$$

So $f = (-18.4 \text{ cm})/0.51 = -36.0$ cm, and the lens is diverging.

▌ Summary

Light appears to travel in straight-line paths, called *rays*, at a speed v that depends on the *index of refraction*, n, of the material: $v = c/n$, where c is the speed of light in vacuum.

When light reflects from a flat surface, the angle of reflection equals the angle of incidence. This *law of reflection* explains why mirrors can form *images*. In a plane mirror, the image is virtual, upright, the same size as the object, and as far behind the mirror as the object is in front.

[†] Some books use a different convention—for example, R_1 and R_2 are considered positive if their centers of curvature are to the right of the lens, in which case minus signs appear in their equivalent of Eq. 22-9.

When light passes from one transparent medium into another, the rays bend or refract. The law of *refraction* (*Snell's law*) states that $n_1 \sin \theta_1 = n_2 \sin \theta_2$, where n_1 and θ_1 are the index of refraction and angle with the normal to the surface for the incident ray, and n_2 and θ_2 are for the refracted ray.

When light rays reach the boundary of a material where the index of refraction decreases, the rays will be *totally internally reflected* if the incident angle, θ, is such that Snell's law would predict $\sin \theta_2 > 1$; this occurs if θ_1 exceeds the critical angle θ_c given by $\sin \theta_c = n_2/n_1$.

A lens uses refraction to produce a real or virtual image. Parallel rays of light are focused to a point, called the *focal point*, by a converging lens. The distance of the focal point from the lens is called the *focal*

length f of the lens. After parallel rays pass through a diverging lens, they appear to diverge from a point, its focal point; and the corresponding focal length is considered negative. The power of a lens, equal to $1/f$, is given in diopters, which are units of inverse meters (m^{-1}). The position and size of the image formed by a lens of a given object can be found by ray tracing. Algebraically the relation between image and object distances, d_i and d_o, and the focal length is given by the *lens equation*: $1/d_o + 1/d_i = 1/f$. The ratio of image height to object height, which equals the magnification, is equal to $-d_i/d_o$.

When using the various equations of geometrical optics, it is important to remember the sign conventions for all quantities involved.

▌Questions

1. Give arguments to show why the moon must have a rough surface rather than a polished mirrorlike surface.

2. When you look at yourself in a tall plane mirror, you see the same amount of your body whether you are close to the mirror or far away. (Try it and see.) Use ray diagrams to show why this should be true.

3. When you look at the moon's reflection from a ripply sea it appears elongated. Explain.

4. Although a plane mirror reverses left and right, it doesn't reverse up and down. Explain.

*5. Archimedes is said to have burned the whole Roman fleet in the harbor of Syracuse by focusing the rays of the sun with a huge spherical mirror. Is this reasonable?

*6. Show with diagrams that the magnification of a concave mirror is less than 1 if the object is beyond the center of curvature C and is greater than 1 if it is within this point.

*7. If a concave mirror produces a real image, is the image necessarily inverted?

*8. When you use a concave mirror, you cannot see an inverted image of yourself unless you place your head beyond the center of curvature C. Yet you can see an inverted image of another object placed between C and F as in Fig. 22-11. Explain. (Hint: you can see a real image only if your eye is behind the image, so that it can be formed.)

*9. Using the rules for the three rays discussed with reference to Fig. 22-11, draw ray 2 for Fig. 22-14b.

*10. What is the focal length of a plane mirror?

*11. Does the mirror equation, Eq. 22-3, hold for a plane mirror? Explain.

*12. What is the magnification of a plane mirror?

13. How might you determine the speed of light in a solid, rectangular, transparent object?

14. When a wide beam of parallel light enters water at an angle, the beam broadens. Explain.

15. What is the angle of refraction when a light ray meets the boundary between two materials perpendicularly?

16. When you look down into a swimming pool or a lake, are you likely to underestimate or overestimate its depth? Explain. How does the apparent depth vary with the viewing angle? (Use ray diagrams.)

17. Draw a ray diagram to show why a stick looks bent when part of it is under water.

18. How are you able to "see" a round drop of water on a table even though the water is transparent and colorless?

19. Can a light ray traveling in air be totally reflected when it strikes a smooth water surface if the incident angle is right?

20. When you look up at an object in air from beneath the water in a pool, does the object appear to be the same size as when you see it directly in air? Explain.

21. Where must the film be placed if a camera lens is to make a sharp image of an object very far away?

22. Can a diverging lens form a real image under any circumstances? Explain.

23. Why must the converging lens have a shorter focal length than the diverging lens if the latter's focal length is to be determined by combining them as in Example 22-10?

24. Show that a real image formed by a thin lens is always inverted, whereas a virtual image is always upright if the object is real.

25. Light rays are reversible. Is this consistent with the lens equation?

26. Can real images be projected on a screen? Can virtual images? Can either be photographed? Discuss carefully.

27. A thin converging lens is moved closer to a nearby object. Does the real image formed change (a) in position, (b) in size? If yes, describe how.

*28. An unsymmetrical lens (say planoconvex) forms an image of a nearby object. Does the image point change if the lens is turned around?

*29. Does the focal length of a lens depend on the material surrounding it? What about the focal length of a spherical mirror? Explain.

*30. The thicker a double convex lens is in the center as compared to its edges, the shorter its focal length for a given lens diameter. Explain.

*31. A lens is made of a material with an index of refraction $n = 1.30$. In air it is a converging lens. Will it still be a converging lens if placed in water? Explain, using a ray diagram.

*32. Compare the mirror equation with the lens equation. Discuss similarities and differences, and especially compare the sign conventions for the quantities involved.

▌ Problems

SECTION 22-2

1. (I) What is the speed of light in (a) ethyl alcohol, (b) lucite?

2. (I) The speed of light in ice is 2.29×10^8 m/s. What is the index of refraction of ice?

3. (I) Light is emitted from an ordinary light bulb filament in wave-train bursts of about 10^{-8} s in duration. What is the length in space of such wave trains?

4. (I) How long does it take light to reach us from the sun, 1.49×10^8 km away?

5. (I) Our nearest star (other than the sun) is 4.2 light years away. That is, it takes 4.2 years for the light to reach earth. How far away is it in meters?

6. (II) If $n = 1.00030 \pm 0.000010$ for air, what error will this introduce in the value for the speed of light in air?

7. (II) What is the minimum angular speed at which Michelson's eight-sided mirror would have had to rotate in order that light would be reflected into an observer's eye by succeeding mirror faces (Fig. 22-2)?

SECTION 22-3

8. (I) Suppose that you want to take a photograph of yourself as you look at your image in a mirror 2.2 m away. For what distance should the camera lens be focused?

9. (II) Stand up two plane mirrors so they form a right angle as in Fig. 22-31. When you look into this double

mirror, you see yourself as others see you, instead of reversed as in a single mirror. Make a ray diagram to show how this occurs.

10. (II) A person whose eyes are 1.48 m above the floor stands 2.40 m in front of a vertical plane mirror whose bottom edge is 40 cm above the floor. What is the horizontal distance to the base of the wall supporting the mirror of the nearest point on the floor that can be seen reflected in the mirror?

11. (II) Suppose you are 60 cm from a plane mirror. What area of the mirror is used to reflect the rays entering one eye from a point on the tip of your nose if your pupil diameter is 5.5 mm?

12. (III) Show that if two plane mirrors meet at an angle ϕ, a single ray reflected (successively) from both mirrors is deflected through an angle of 2ϕ independent of the incident angle. Assume $\phi < 90°$ and that only two reflections, one from each mirror, take place.

13. (III) Suppose a third mirror is placed beneath the two shown in Fig. 22-31, so that all three are perpendicular to each other. (a) Show that for such a "corner reflector," any incident ray will return in its original direction after three reflections. (b) What happens if it makes only two reflections?

*SECTION 22-4

*14. (I) What is the radius of a concave reflecting surface that brings parallel light to a focus 18.9 cm in front of it?

*15. (I) How far from a concave mirror (radius 24.0 cm) must an object be placed if its image is to be at infinity?

*16. (II) You try to look at yourself in a silvered ball of diameter 56.0 cm when you are 2.30 m away. Where is your image? Is it real or virtual? Can you see yourself clearly?

*17. (II) Show, using a ray diagram, that the magnification m of a convex mirror is $m = -d_i/d_o$ just as for a concave mirror. (Hint: consider a ray from the top of the object that reflects at the center of the mirror.)

*18. (II) A dentist wants a small mirror that, when 2.10 cm from a tooth, will produce a $5.5 \times$ upright image. What kind of mirror must be used and what must its radius of curvature be?

FIGURE 22-31

*19. (II) Use ray diagrams to show that the mirror equation, Eq. 22-3, is valid for a convex mirror as long as f is considered negative.

*20. (II) Use the mirror equation to show that the magnitude of the magnification of a concave mirror is less than 1 if the object is beyond the center of curvature C ($d_o > r$) and is greater than 1 if the object is within C ($d_o < r$).

*21. (II) A 2.40-cm-tall object is placed 22.0 cm from a spherical mirror. It produces a virtual image 3.20 cm high. (a) What type of mirror is being used? (b) Where is the image located? (c) What is the radius of curvature of the mirror?

*22. (II) The magnification of a convex mirror is $0.35\times$ for objects 4.0 m away. What is the focal length of this mirror?

*23. (II) (a) A plane mirror can be considered a limiting case of a spherical mirror; specify what this limit is. (b) Determine an equation that relates the image and object distances in this limit of a plane mirror. (c) Determine the magnification of a plane mirror in this same limit. (d) Are your results in parts (b) and (c) consistent with the discussion of Section 22-3 on plane mirrors?

*24. (III) What is the radius of a concave mirror that gives a $1.6\times$ magnification of a face 30 cm from it?

*25. (III) A 1.58-m-tall person stands 4.50 m from a convex mirror and notices that he looks only half as tall as he does in a plane mirror placed at the same distance. What is the radius of curvature of the convex mirror? (Assume that $\sin\theta \approx \theta$.)

SECTION 22-5

26. (I) A flashlight beam strikes the surface of a pane of glass ($n = 1.50$) at a $45°$ angle. What is the angle of refraction?

27. (I) A diver shines a flashlight upward from beneath the water at a $28.0°$ angle to the vertical. At what angle does the light leave the water?

28. (I) Rays of the sun are seen to make a $15.0°$ angle to the vertical beneath the water. At what angle above the horizon is the sun?

29. (II) Light is incident on an equilateral crown glass prism at a $45.0°$ angle to one face, Fig. 22-32. Calculate the angle at which light emerges from the opposite face. Assume that $n = 1.56$.

30. (II) A bright light is 2.60 m below the surface of a swimming pool and 1.48 m from one edge of the pool. At what angle does the light leave the water at the edge of the pool? Assume that the water reaches the top edge of the pool.

31. (II) Prove in general that for a light beam incident on a uniform layer of transparent material, as in Fig. 22-17, the direction of the emerging beam is parallel to the incident beam, independent of the incident angle θ and the thickness of the material.

32. (II) An aquarium filled with water has flat glass sides whose index of refraction is 1.58. A beam of light from outside the aquarium strikes the glass at a $48.9°$ angle to the perpendicular. What is the angle of this light ray when it enters the glass and then the water? What would it be if the ray entered the water directly?

33. (III) When light passes through a prism, the angle that the refracted ray makes relative to the incident ray is called the deviation angle δ, Fig. 22-33. This angle is a minimum when the ray passes through the prism symmetrically—that is, for a prism shaped like an isosceles triangle, when the ray inside the prism is parallel to the base. Show that this minimum deviation angle, δ_m, is related to the prism's index of refraction n by

$$n = \frac{\sin\frac{1}{2}(\phi + \delta_m)}{\sin\phi/2}$$

where ϕ is the apex angle.

34. (III) A light ray is incident on a flat piece of glass as in Fig. 22-17. Show that if the incident angle θ is small, the ray is displaced a distance $d = t\theta(n - 1)/n$ where t is the thickness of the glass and θ is in radians.

SECTION 22-6

35. (I) What is the critical angle for a diamond–water interface?

36. (I) The critical angle for a certain liquid-air surface is $57°$. What is the index of refraction of the liquid?

37. (II) A beam of light is emitted 6.0 cm beneath the surface of a liquid and strikes the surface 4.0 cm from the point directly above the source. If total internal reflection occurs, what can you say about the index of refraction of the liquid?

38. (II) The end faces of a cylindrical glass rod ($n = 1.54$) are perpendicular to the sides. Show that a light ray entering an end face at any angle will be totally internally reflected inside the rod when it strikes the sides. Assume the rod is in air. What if it were in water?

39. (II) If the apex angle of a prism is $\phi = 70°$ (see Fig. 22-32) what is the minimum incident angle for a ray if it is to emerge from the opposite side, if $n = 1.58$?

40. (III) (a) What is the minimum index of refraction for a glass or plastic prism to be used in binoculars (Fig. 22-19) so that total internal reflection occurs at $45°$? (b) Will binoculars work if its prisms are immersed in water? Assume $n = 1.50$. (c) What minimum n is needed if the prisms are immersed in water?

SECTIONS 22-7 AND 22-8

41. (I) A sharp image is located 60.0 mm behind a 50.0-mm-focal length converging lens. Calculate the object distance.

FIGURE 22-32   FIGURE 22-33

42. (I) A leaf is placed 77.0 cm in front of a -98.0-mm-focal-length lens. Where is the image? Is it real or virtual?

43. (I) A certain lens focuses an object 22.5 cm away as an image 33.0 cm on the other side of the lens. What type of lens is it and what is its focal length? Is the image real or virtual?

44. (I) (a) An object 24.0 cm in front of a certain lens is imaged 5.10 cm in front of that lens (on the same side as the object). What type of lens is this and what is its focal length? Is the image real or virtual? (b) What if the image were located, instead, 35.0 cm in front of the lens?

45. (I) (a) What is the power of a 17.0-cm-focal-length lens? (b) What is the focal length of a -8.5-diopter lens? Are these lenses converging or diverging?

46. (II) A -5.0-diopter lens is held 12.0 cm from an ant 1.0 mm high. What is the position, type, and height of the image?

47. (II) (a) How far from a 50.0-mm-focal-length lens must an object be placed if its image is to be magnified $3.00 \times$ and be real? (b) What if the image is to be virtual and magnified $3.00 \times$?

48. (II) Repeat problem 47 for a -50.0-mm-focal-length lens.

49. (II) How large is the image of the sun on the film used in a camera with a 50-mm-focal-length lens? The sun's diameter is 1.4×10^6 km and it is 1.5×10^8 km away.

50. (II) A 36.5-cm-focal length converging lens is 25.0 cm behind a diverging lens. Parallel light strikes the diverging lens; after passing through the converging lens, the light is again parallel. What is the focal length of the diverging lens?

51. (II) (a) A 2.70-cm high insect is 2.20 m from a 135-mm-focal-length lens. Where is the image, how high is it, and what type is it? (b) What if $f = -135$ mm?

52. (II) In a slide or movie projector, the film acts as the object whose image is projected on a screen. If a 100-mm-focal-length lens is to project an image on a screen 4.0 m away, how far from the lens should the slide be? If the slide is 36 mm wide, how wide will the picture be on the screen?

53. (II) A 35-mm slide (picture size is actually 24 by 36 mm) is to be projected on a screen 1.00 by 1.50 m placed 6.0 m from the projector. What focal-length lens should be used if the image is to cover the screen?

54. (II) Show analytically that the image formed by a converging lens is real and inverted if the object is beyond the focal point ($d_o > f$), and is virtual and upright if the object is within the focal point ($d_o < f$). Describe the image if the object is a virtual image (formed by another lens) for which $-d_o > f$, and for which $0 < -d_o < f$.

55. (II) A diverging lens with $f = -32.5$ cm is placed 11.0 cm behind a converging lens with $f = 14.5$ cm. Where will an object at infinity be focused?

56. (II) Two 32.0-cm-focal-length converging lenses are placed 21.5 cm apart. An object is placed 55.0 cm in front of one. Where will the final image formed by the second lens be located? What is the total magnification?

57. (III) A diverging lens is placed next to a converging lens of focal length f_c as in Fig. 22-29. If f_T represents the focal length of the combination, show that the focal length of the diverging lens, f_D, is given by

$$\frac{1}{f_D} = \frac{1}{f_T} - \frac{1}{f_C}.$$

58. (III) (a) Show that if two lenses of focal lengths f_1 and f_2, respectively, are placed next to each other, the focal length of the combination f_T is given by $f_T = f_1 f_2/(f_1 + f_2)$. (b) Show that the power P of the combinations of two lenses is the sum of their separate powers, $P = P_1 + P_2$.

59. (III) How far apart are an object and an image formed by a 65-cm-focal-length converging lens if the image is $3.50 \times$ larger than the object and is real?

60. (III) Show that the lens equation can be written in the *Newtonian form*:

$$xx' = f^2$$

where x is the distance of the object from the focal point on the front side of the lens, and x' is the distance of the image to the focal point on the other side of the lens.

61. (III) A bright object is placed on one side of a converging lens of focal length f and a white screen for viewing the image is on the opposite side. The distance $d_T = d_i + d_o$ between the object and the screen is kept fixed, but the lens can be moved. Show that (a) if $d_T > 4f$, there will be *two* positions where the lens can be placed and a sharp image will be produced on the screen, (b) if $d_T < 4f$, there will be no lens position where a sharp image is formed. (c) Determine the distance between the two lens positions in part (a), and the ratio of the image sizes.

*SECTION 22-9

*62. (I) A double concave lens has surface radii of 20.6 cm and 10.2 cm. What is the focal length if $n = 1.58$?

*63. (I) Both surfaces of a double convex lens have radii of 28.0 cm. If the focal length is 26.2 cm, what is the index of refraction of the lens material?

*64. (I) A planoconvex lens is to have a focal length of 28.0 cm. If made from fused quartz, what must be the radius of curvature of the convex surface?

*65. (I) A glass ($n = 1.50$) planoconcave lens has a focal length of -26.6 cm. What is the radius of the concave surface?

*66. (II) An object is placed 120 cm from a glass lens ($n = 1.56$) with one concave surface of radius 18.0 cm and one convex surface of radius 13.5 cm. Where is the final image? What is the magnification?

*67. (III) A glass lens ($n = 1.55$) in air has a power of $+3.0$ diopters. What would its power be if submerged in water?

*68. (III) A lens whose index of refraction is n is submerged in a material of index of refraction n' ($n' \neq 1$). Derive the equivalent of Eqs. 22-7, 8 and 9 for this lens.

23

THE *WAVE* NATURE OF LIGHT

That light carries energy is obvious to anyone who has focused the sun's rays with a magnifying glass on a piece of paper and burned a hole in it. But how does light travel and in what form is this energy carried? In our discussion of waves in Chapter 13, we noted that energy can be carried from place to place in basically two ways: by particles or by waves. In the first case, material bodies or particles can carry energy, such as a thrown baseball or rushing water. In the second case, water waves and sound waves, for example, can carry energy over long distances even though mass itself does not travel these distances. In view of this, what can we say about the nature of light: does light travel as a stream of particles away from its source; or does it travel in the form of waves that spread outward from the source? Historically, this question has turned out to be a difficult one. For one thing, light does not reveal itself in any obvious way as being made up of tiny particles nor do we see tiny light waves passing by as we do water waves. The evidence seemed to favor first one side and then the other until about 1830 when most physicists had accepted the wave theory. By the end of the nineteenth century, light was considered to be an *electromagnetic wave* (Chapter 21). Although modifications had to be made in the twentieth century (Chapter 26), the wave theory of light has proved very successful. We now investigate the evidence for the wave theory and how it has explained a wide range of phenomena.

23-1 ▮ Waves versus particles; Huygens' principle and diffraction

The Dutch scientist Christian Huygens (1629–1695), a contemporary of Newton, proposed a wave theory of light that had much merit. Still useful today is a technique he developed for predicting the future position of a wave front when an earlier position is known. This is known as **Huygens' principle** and can be stated as follows: *every point on a wave front can be considered as a source of tiny wavelets that spread out in the forward direction at the speed of the wave itself; the new wave front is the envelope of all the wavelets* (that is, *the tangent to all of them*).

FIGURE 23-1

Huygens' principle used to determine wave front *CD* when *AB* is given.

As a simple example of the use of Huygens' principle, consider the wave front *AB* in Fig. 23-1 which is traveling away from a source *S*. We assume the medium is *isotropic*—that is, the speed *v* of the waves is the same in all directions. To find the wave front a short time *t* after it is at *AB*, tiny circles are drawn whose radius $r = vt$. The centers of these tiny circles are on the original wave front *AB* and the circles represent Huygens' (imaginary) wavelets. The tangent to all these wavelets, the line *CD*, is the new position of the wave front.

Huygens' principle is particularly useful when waves impinge on an obstacle and the wave fronts are partially interrupted. Huygens' principle predicts that waves bend in behind an obstacle as shown in Fig. 23-2. This is just what water waves do as we saw in Chapter 13 (Figs. 13-29 and 13-30). The bending of waves behind obstacles into the "shadow region" is known as **diffraction.** Since diffraction occurs for waves, but not for particles, it can serve as one means for distinguishing the nature of light.

Does light exhibit diffraction? In the mid-seventeenth century, a Jesuit priest, Francesco Grimaldi (1618–1663), had observed that when sunlight entered a darkened room through a tiny hole in a screen, the spot on the opposite wall was larger than would be expected from geometric rays; he also observed that the border of the image was not clear but was surrounded by colored fringes. Grimaldi attributed this to the diffraction of light. Newton, who favored a particle theory, was aware of Grimaldi's result. He felt that Grimaldi's result was due to the interaction of light corpuscles ("little bodies") with the edges of the hole. If light were a wave, he said, the light waves should bend more than that observed. Newton's argument seems reasonable. Yet, as we saw in Chapter 13, diffraction is large only when the size of the obstacle or the hole is on the order of the wavelength of the wave (Fig. 13-30). Newton did not guess that the wavelengths of visible light might be incredibly tiny, and thus diffraction effects would be very small. (Indeed this is why geometric optics using rays is so successful—normal openings and obstacles are much larger than the wavelength of the light, and so relatively little diffraction or bending occurs.)

23-2 ∎ Huygens' principle and the law of refraction

The laws of reflection and refraction were well known in Newton's time. The law of reflection could not distinguish the two theories. For when waves reflect from an obstacle, the angle of incidence equals the angle of reflection (Fig. 13-21); the same is true of particles—think of a tennis ball without spin striking a flat surface.

The law of refraction is another matter. Consider light entering a medium where it is bent toward the normal as when it travels from air into water. As shown in Fig. 23-3, this effect can be constructed using Huygens' principle if we assume the speed of light is less in the second medium ($v_2 < v_1$). That is, in time *t*, the point *B* on wave front *AB* goes a distance $v_1 t$ to reach point *D*; point *A*, on the other hand, travels a distance $v_2 t$ to

FIGURE 23-2

Huygen's principle is consistent with diffraction (a) around the edge of an obstacle, (b) through a large hole, (c) through a small hole whose size is on the order of the wavelength of the wave.

(a)

(b)

(c)

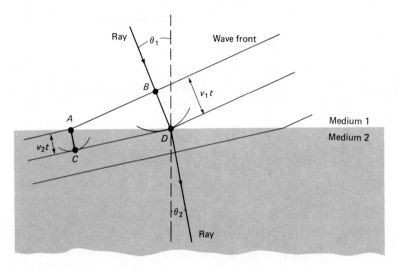

FIGURE 23-3

Refraction explained, using Huygens' principle.

reach point C. Huygens' principle is applied to points A and B to obtain the curved wavelets shown at C and D. The wave front is tangent to these two wavelets so the new wave front is the line CD. Hence the rays (which are perpendicular to the wave fronts) bend toward the normal if $v_2 < v_1$ as drawn.[†] Newton's corpuscle theory predicted the opposite result: if the path of light corpuscles entering a new medium changes direction, it must be because the medium exerts a force on the corpuscles at the boundary. This force was assumed to act perpendicular to the boundary and thus would affect only the perpendicular component of the corpuscles' velocity. When light enters a medium such as water where it is bent toward the normal, the force must accelerate the corpuscles so the perpendicular component of velocity is increased; only in this way will the refracted angle be less than the incident. In Newton's theory, then, the speed of light would be greater in the second medium $(v_2 > v_1)$. Thus the wave theory predicts that the speed of light in water (say) is less than in air and Newton's corpuscle theory predicts the reverse. An experiment to actually measure the speed of light in water and confirm the wave theory prediction was not done until 1850 (by the French physicist Jean Foucault) and by then the wave theory was fully accepted as we shall see in the next section.

It is easy to show that Snell's law of refraction follows directly from Huygens' principle, given that the speed of light v in any medium is related to the speed in a vacuum, c, and the index of refraction, n, by Eq. 22-1: $v = c/n$. From the Huygens' construction of Fig. 23-3, angle ADC is equal to θ_2 and angle BAD is equal to θ_1. Then for the two triangles that have the common side AD, we have

$$\sin \theta_1 = \frac{v_1 t}{AD}, \qquad \sin \theta_2 = \frac{v_2 t}{AD}.$$

We divide these two equations and obtain:

$$\frac{\sin \theta_1}{\sin \theta_2} = \frac{v_1}{v_2},$$

[†] This is basically the same as the discussion around Fig. 13-25.

or, since $v_1 = c/n_1$ and $v_2 = c/n_2$,

$$n_1 \sin \theta_1 = n_2 \sin \theta_2$$

which is Snell's law of refraction. (The law of reflection can be derived from Huygens' principle in a similar way, and this is left as a problem at the end of the chapter.)

When a light wave travels from one medium to another, its frequency does not change but its wavelength does. This can be seen from Fig. 23-3, where we assume each of the lines representing a wave front corresponds to a crest (peak) of the wave. Then

$$\frac{\lambda_2}{\lambda_1} = \frac{v_2 t}{v_1 t} = \frac{v_2}{v_1} = \frac{n_1}{n_2}.$$

If medium 1 is a vacuum (or air), so $n_1 = 1$, $v_1 = c$, and we call λ_1 simply λ, then the wavelength in another medium of index of refraction n ($= n_2$) will be

$$\lambda_n = \frac{\lambda}{n}. \tag{23-1}$$

This result is consistent with the frequency f being unchanged since $c = f\lambda$; combining this with $v = f\lambda_n$ in a medium where $v = c/n$ gives $\lambda_n = v/f = c/nf = f\lambda/nf = \lambda/n$.

23-3 ∎ Interference—Young's double-slit experiment

In 1801, the Englishman Thomas Young (1773–1829) obtained convincing evidence for the wave nature of light and was even able to measure the wavelength. Figure 23-4a shows a schematic diagram of Young's famous double-slit experiment. Light from a single source (Young used the sun) falls on a screen containing two closely spaced slits S_1 and S_2. If light consists of tiny particles, we might expect to see two bright lines on a screen placed behind the slits as in (b). But Young observed instead a series of bright lines as in (c). Young was able to explain this result as a *wave-interference* phenomenon. To see this, imagine plane waves of light of a single wavelength (called "monochromatic") falling on the two slits as shown in Fig. 23-5. Because of diffraction, the waves leaving the two small slits spread out as

FIGURE 23-4

(a) Young's double-slit experiment. (b) If light consists of particles, we would expect to see two bright lines on the screen behind the slits. (c) Young observed many lines.

(a) (b) (c)

shown. This is equivalent to the interference pattern produced when two rocks are thrown into a lake (Fig. 13-27), or when sound from two loud-speakers interferes (Fig. 14-10).

To see how an interference pattern is produced on the screen, we make use of Fig. 23-6. Waves of wavelength λ are shown entering the slits S_1 and S_2 which are a distance d apart. The waves spread out in all directions after passing through the slits, but they are shown only for three different angles θ. In (a) the waves reaching the center of the screen are shown ($\theta = 0$); the waves from the two slits travel the same distance, so they are in phase: constructive interference occurs and there is a bright spot at the center of the screen. There will also be constructive interference when the paths of the two rays differ by one wavelength (or any whole number of wavelengths) as shown in (b). But if one ray travels an extra distance of one-half a wavelength (or $\frac{3}{2}\lambda$, $\frac{5}{2}\lambda$, and so on) the two waves are exactly out of phase when they reach the screen, and so destructive interference occurs and the screen is dark, Fig. 23-6c. Thus there will be a series of bright and dark lines (or *fringes*).

To determine exactly where the bright lines fall, first note that Fig. 23-6 is somewhat exaggerated; in real situations, the distance d between the slits is very small compared to the distance l to the screen. The rays from each slit for each case will therefore be essentially parallel and θ is the angle they make with the horizontal. From the shaded right triangle in Fig. 23-6b, the extra distance traveled by the lower ray is $d \sin \theta$. **Constructive interference** will occur on the screen when $d \sin \theta$ equals a whole number of wavelengths.

$$d \sin \theta = m\lambda, \qquad m = 0, 1, 2, \ldots \quad \begin{bmatrix} \text{constructive} \\ \text{interference} \end{bmatrix} \quad (23\text{-}2a)$$

The value of m is called the *order* of the interference fringe. **Destructive interference** occurs when the extra distance $d \sin \theta$ is $\frac{1}{2}$, $\frac{3}{2}$, and so on, wavelengths:

$$d \sin \theta = (m + \tfrac{1}{2})\lambda, \qquad m = 0, 1, 2, \ldots \quad \begin{bmatrix} \text{destructive} \\ \text{interference} \end{bmatrix} \quad (23\text{-}2b)$$

The intensity of the bright fringes is greatest for the center one ($m = 0$) and decreases for higher orders, as shown in Fig. 23-7.

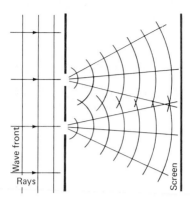

FIGURE 23-5

If light is a wave, light passing through one of two slits should interfere with light passing through the other slit.

FIGURE 23-6

How the wave theory explains the pattern of lines seen in the double-slit experiment.

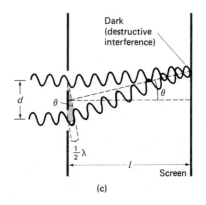

EXAMPLE 23-1 A screen containing two slits 0.100 mm apart is 1.20 m from the viewing screen. Light of wavelength $\lambda = 500$ nm falls on the slits from a distant source. Approximately how far apart will the bright interference fringes be on the screen?

SOLUTION Since $d = 0.100$ mm $= 1.00 \times 10^{-4}$ m and $l = 1.20$ m, the first-order fringe ($m = 1$) occurs at an angle θ given by:

$$\sin \theta_1 = \frac{m\lambda}{d} = \frac{(1)(500 \times 10^{-9}\text{ m})}{1.00 \times 10^{-4}\text{ m}} = 5.00 \times 10^{-3}.$$

This is a very small angle, so we can take $\sin \theta = \theta$. The first-order fringe will occur a distance x_1 above the center of the screen (see Fig. 23-8) given by $x_1/l = \theta_1$, so

$$x_1 = l\theta_1 = (1.20\text{ m})(5.00 \times 10^{-3}) = 6.00\text{ mm}.$$

The second fringe ($m = 2$) will occur at

$$x_2 = l\theta_2 = l\frac{2\lambda}{d} = 12.0\text{ mm}$$

above the center, and so on. Thus the fringes are 6.00 mm apart.

From Eqs. 23-2 we can see that, except for the zeroth order fringe at the center, the position of the fringes depends on wavelength. Consequently, when white light falls on the two slits, as Young found in his experiments, the central fringe is white but the first (and higher) order fringes contain a spectrum of colors like a rainbow; θ was found to be smallest for violet light and largest for red. By measuring the position of these fringes, Young became the first to determine the wavelengths of visible light (using Eqs. 23-2). In doing so, he showed that what distinguishes different colors physically is their wavelength.

EXAMPLE 23-2 White light passes through two slits 0.50 mm apart and an interference pattern is observed on a screen 2.5 m away. The first-order fringe resembles a rainbow with violet and red light at either end. The violet light falls about 2.0 mm and the red 3.5 mm from the center of the central white fringe. Estimate the wavelengths of the violet and red lights.

FIGURE 23-7

Interference fringes produced by double-slit experiment detected by photographic film placed on the viewing screen. The arrow marks the central fringe. (From F. W. Sears, M. W. Zemansky, and Hugh D. Young, *College Physics*, 4th ed., © 1974, Addison-Wesley Publishing Co., Inc., Reading Mass, Figure 42-3, Reprinted with permission.) (b) Intensity of light in the interference pattern.

(a)

(b)

FIGURE 23-8

For small angles, the interference fringes occur at distances $x = \theta l$ above the center ($m = 0$) fringe.

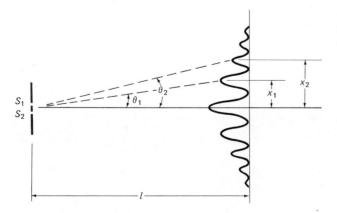

SOLUTION   We use Eq. 23-2a with $m = 1$ and $\sin \theta = \theta$. Then for violet light, $x = 2.0$ mm, so

$$\lambda = \frac{d\theta}{m} = \frac{d}{m}\frac{x}{l} = \frac{(5.0 \times 10^{-4}\text{ m})}{1}\frac{(2.0 \times 10^{-3}\text{ m})}{(2.5\text{ m})} = 4.0 \times 10^{-7}\text{ m}$$

or 400 nm. For red light, $x = 3.5$ mm so λ is about 700 nm.

The two slits in Fig. 23-6 act as if they were sources of radiation. They are called **coherent sources** because the waves leaving them bear the same phase relationship to each other at all times (because ultimately the waves come from a single source to the left of the two slits in Fig. 23-6). An interference pattern is observed only when the sources are coherent. If two tiny light bulbs replaced the two slits (or separate light bulbs illuminated each slit) an interference pattern would not be seen; the light emitted by one light bulb would have a random phase with respect to the second bulb, and the screen would be more or less uniformly illuminated. Two such sources are called **incoherent sources.**

23-4 ▌ The visible spectrum and dispersion

The two most obvious properties of light are readily describable in terms of the wave theory of light: intensity (or brightness) and color. The *intensity* of light is related to the square of the amplitude of the wave, just as for any wave (see Section 13-9, or Eqs. 21-7 and 21-8). The *color* of the light is related

FIGURE 23-9

The spectrum of visible light, showing the range of wavelengths for the various colors.

to the wavelength or frequency of the light. Visible light—that to which our eyes are sensitive—falls in the wavelength range of about 400 nm to 750 nm.[†] This is known as the *visible spectrum*, and within it lie the different colors from violet to red as shown in Fig. 23-9. Light with wavelength shorter than 400 nm is called ultraviolet (UV) and that with wavelength greater than 750 nm is called infrared (IR).[‡] Although human eyes are not sensitive to UV or IR, some types of photographic film do respond to them.

It is a familiar fact that a prism separates white light into a rainbow of colors. This is due to the fact that the index of refraction of a material depends on the wavelength. This is shown for several materials in Fig. 23-10. White light is a mixture of all visible wavelengths; and when incident on a prism as in Fig. 23-11, the different wavelengths are bent to varying degrees. Since the index of refraction is greater for the shorter wavelengths, violet light is bent the most and red the least as indicated. This spreading of white light into the full spectrum is called **dispersion.**

FIGURE 23-10

Index of refraction as a function of wavelength.

† Sometimes the Angstrom (Å) unit is used when referring to light: $1\text{ Å} = 1 \times 10^{-10}$ m; then visible light falls in the wavelength range of 4000 Å to 7500 Å.

‡ The complete electromagnetic spectrum is illustrated in Fig. 21-8.

Rainbows are a spectacular example of dispersion (by water in this case). You can see rainbows when you look at falling water with the sun at your back. Figure 23-12 shows how red and violet rays are bent by spherical water droplets and are reflected off the back surface. Red is bent the least and so reaches the observer's eyes from droplets higher in the sky. Thus the top of the rainbow is red.

Diamonds achieve their brilliance from a combination of dispersion and total internal reflection. Since diamonds have a very high index of refraction of about 2.4, the critical angle for total internal reflection is only 25°. Incident light therefore strikes many of the internal surfaces before it strikes one at less than 25° and emerges. After many such reflections the light has traveled far enough that the colors have become sufficiently separated to be seen individually and brilliantly by the eye after leaving the crystal.

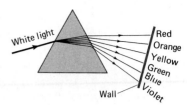

FIGURE 23-11

White light dispersed by a prism into the visible spectrum.

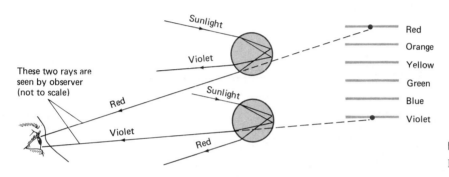

FIGURE 23-12

Formation of a rainbow.

23-5 ▌ Diffraction by a single slit or disc

Young's double-slit experiment put the wave theory of light on a firm footing. But full acceptance came only with studies on diffraction more than a decade later.

We have already discussed diffraction briefly with regard to water waves (Section 13-10) as well as for light (Section 23-1) and we have seen that it refers to the spreading or bending of waves around edges. Now we look at diffraction in more detail.

A part of the history of the wave theory of light belongs to Augustin Fresnel (1788–1827) who in 1819 presented to the French Academy a wave theory of light that predicted and explained interference and diffraction effects. Almost immediately Siméon Poisson (1781–1840) pointed out a counter-intuitive inference: that according to Fresnel's wave theory, if light from a point source were to fall on a solid disk, then light diffracted around the edges should constructively interfere at the center of the shadow (Fig. 23-13). That prediction seemed very unlikely. But when the experiment was actually carried out by François Arago, the bright spot was seen at the very center of the shadow! This was strong evidence for the wave theory.

Figure 23-14a is a photograph of the shadow cast by a coin using a (nearly) point source of light (a laser in this case). The bright spot is clearly present at the center. Note that there also are bright and dark fringes beyond

FIGURE 23-13

If a light is a wave, a bright spot will appear at the center of the shadow of a solid disk illuminated by a point source of monochromatic light.

(a)

(b)

(c)

FIGURE 23-14

Diffraction pattern of (a) a penny (courtesy of P. M. Rinard, from *Am. J. Phys.* 44, 1976, p. 70), (b) a razor blade (from F. W. Sears, *Optics,* © 1949, Addison-Wesley Publishing Co., Inc., Reading, Mass., Figure 9-8; reprinted with permission), (c) a single slit, illuminated by a nearly point source of monochromatic light (from M. Cagnet, M. Françon, and J. C. Thrierr, *Atlas of Optical Phenomena,* Springer-Verlag, Berlin, 1962).

the shadow. These resemble the interference fringes of a double slit. Indeed, they are due to interference of waves diffracted around different parts of the disk, and the whole is referred to as a *diffraction pattern*. A diffraction pattern exists around any sharp object illuminated by a point source, as shown in Fig. 23-14b and c. We are not always aware of them because most sources of light in everyday life are not points; so light from different parts of the source washes out the pattern.

   To see how a diffraction pattern arises, we will analyze the important case of monochromatic light passing through a narrow slit. We will assume that parallel rays (or plane waves) of light fall on the slit of width D as shown in Fig. 23-15. If the viewing screen is infinitely far away, or a lens is placed behind the slit to focus parallel rays on the screen, the diffraction pattern is called *Fraunhofer diffraction*. If the screen is close and no lenses are used, it is called *Fresnel diffraction*. The analysis in the latter case is rather involved, so we consider only the case of Fraunhofer diffraction. As we know from studying water waves and from Huygens' principle, the waves passing through the slit spread out in all directions. We will now examine how the waves passing through different parts of the slit interfere with each other.

   Since the screen is assumed to be very far away, the rays heading for any point are essentially parallel. First we consider rays that pass straight

FIGURE 23-15

Analysis of diffraction patterns formed by light passing through a narrow slit.

(a) $\theta = 0$
Bright

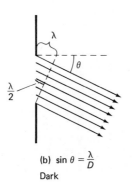

(b) $\sin \theta = \dfrac{\lambda}{D}$
Dark

(c) $\sin \theta = \dfrac{3\lambda}{2D}$
Bright

(d) $\sin \theta = \dfrac{2\lambda}{D}$
Dark

through as in Fig. 23-15a. They are all in phase so there will be a central bright spot on the screen. In part (b), we consider rays moving at an angle θ such that the ray from the top of the slit travels exactly one wavelength farther than the ray from the bottom edge. The ray passing through the very center of the slit will travel one half wavelength farther than the ray at the bottom of the slit. These two rays will be exactly out of phase with one another and so will destructively interfere. Similarly, a ray slightly above the bottom one will cancel a ray that is the same distance above the central one. Indeed, each ray passing through the lower half of the slit will cancel with a corresponding ray passing through the upper half. Thus, all the rays destructively interfere in pairs, and so no light will reach the viewing screen at this angle. The angle θ at which this occurs can be seen from the diagram to occur when $\lambda = D \sin \theta$, so

$$\sin \theta = \lambda/D. \qquad \text{[first minimum]} \quad (23\text{-}3)$$

The light intensity is a maximum at $\theta = 0°$ and decreases to a minimum (intensity = zero) at the angle θ given by Eq. 23-3.

Now consider a larger angle θ such that the top ray travels $\frac{3}{2}\lambda$ farther than the bottom ray as in Fig. 23-15c. In this case, the rays from the bottom third of the slit will cancel in pairs with those in the middle third since they will be $\lambda/2$ out of phase. However, light from the top third of the slit will still reach the screen, so there will be a bright spot, but not nearly as bright as the central spot at $\theta = 0°$. For an even larger angle θ such that the top ray travels 2λ farther than the bottom ray, Fig. 23-15d, rays from the bottom quarter of the slit will cancel with those in the quarter just above it since the path lengths differ by $\lambda/2$. And the rays through the quarter of the slit just above center will cancel with those through the top quarter. At this angle there will again be a minimum of zero intensity in the diffraction pattern. A plot of the intensity as a function of angle is shown in Fig. 23-16. This corresponds well with the photo of Fig. 23-14c. Notice that minima (zero intensity) occur at

$$D \sin \theta = m\lambda, \qquad m = 1, 2, 3, \ldots.$$

but not at $m = 0$ where there is the strongest maximum. Between the minima, smaller brightness maxima occur.

FIGURE 23-16

Intensity in diffraction pattern of a single slit as a function of $\sin \theta$.

FIGURE 23-17
Example 23-3.

☐ **EXAMPLE 23-3** Light of wavelength 750 nm passes through a slit 1.0×10^{-3} mm wide. How wide is the central maximum (a) in degrees, and (b) in centimeters, on a screen 20 cm away?

SOLUTION (a) The first minimum occurs at

$$\sin \theta = \frac{\lambda}{D} = \frac{7.5 \times 10^{-7} \text{ m}}{1 \times 10^{-6} \text{ m}} = 0.75.$$

So $\theta = 49°$. This is the angle between the center and the first minimum, Fig. 23-17. The angle subtended by the whole central maximum, between the minima above and below the center, is twice this or 98°. (b) The width of the central maximum is $2x$ where $\tan \theta = x/20$ cm. So $2x = 2(20 \text{ cm}) \cdot (\tan 49°) = 46$ cm; a large width of the screen will be illuminated. It will not be terribly bright, normally, since the amount of light that passes through such a small slit will be small and it is spread over a large area.

From Eq. 23-3 we can see that the smaller the aperture D, the larger the central diffraction maximum. This is consistent with our earlier study of waves in Chapter 13.

23-6 ▌ Diffraction grating

A large number of equally spaced parallel slits is called a **diffraction grating,** although the term "interference grating" might be as appropriate. Gratings are often made by ruling very fine lines on glass with a diamond tip. The untouched spaces between the lines serve as the slits. Photographic transparencies of an original grating serve as inexpensive gratings. Photographic reduction can be used to make very fine gratings. Gratings containing 10,000 lines per centimeter are common today, and are very useful for precise measurements of wavelengths. A diffraction grating containing slits is called a *transmission grating. Reflection gratings* are also possible; they can be made by ruling fine lines on a metallic or glass surface from which light is reflected and examined. The analysis is basically the same as for a transmission grating, and we consider only the latter here.

The analysis of a diffraction grating is much like that of Young's double-slit experiment. We assume parallel rays of light are incident on the grating as shown in Fig. 23-18. We also assume that the slits are narrow enough so that diffraction by each of them spreads light over a very wide angle on a distant screen behind the grating, and interference can occur with light from all the other slits. Light rays that pass through each slit without deviation ($\theta = 0°$) interfere constructively to produce a bright spot at the center of the screen. At an angle θ such that rays from adjacent slits travel an extra distance of $\Delta l = m\lambda$ where m is an integer, again constructive interference occurs. Thus, if d is the distance between slits,

$$\sin \theta = \frac{m\lambda}{d}, \qquad m = 0, 1, 2, \ldots \qquad \text{[principal maxima]} \quad (23\text{-}4)$$

is the criterion to have a brightness maximum. This is the same equation as for the double-slit situation, and again m is called the *order* of the pattern.

There is an important difference between a double-slit and a multiple-slit pattern, however. The bright maxima are much *sharper* and *narrower* for a grating. Why this happens can be seen as follows. Suppose that the angle θ is increased just slightly beyond that required for a maximum. In the case of only two slits, the two waves will be only slightly out of phase, so nearly full constructive interference occurs. This means the maxima are wide (see Fig. 23-7). For a grating, the waves from two adjacent slits will also not be significantly out of phase. But waves from one slit and those from a second one a few hundred slits away may be exactly out of phase; all or nearly all the light will cancel in pairs in this way. For example, suppose the angle θ is different from its first-order maximum so that the extra path length for a pair of adjacent slits is not exactly λ but rather 1.0010λ. The wave through one slit and another one 500 slits below will be out of phase by 1.5000λ, or exactly $1\frac{1}{2}$ wavelengths, so the two will cancel. A pair of slits, one below each of these, will also cancel. That is, the light from slit 1 cancels with that from slit 501; light from slit 2 cancels with that from slit 502, and so on.

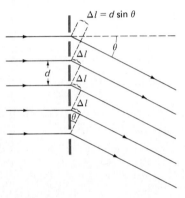

FIGURE 23-18

Diffraction grating.

FIGURE 23-19

Intensity as a function of viewing angle θ (or position on screen) for (a) two slits, (b) six slits. For a diffraction grating, the number of slits is very large ($\sim 10^4$) and the peaks are narrower still.

(a)

(b)

Atomic hydrogen

Sodium

Helium

Neon

Mercury

Molecular hydrogen

Solar absorption spectrum

FIGURE 23-22   Line spectra for the elements indicated and spectrum of the sun, showing absorption lines.   (Courtesy of Bausch and Lomb Incorporated)

| $m = 2$ | | $m = 2$ $m = 1$ $m = 1$ | | $m = 0$ | | $m = 1$ | $m = 1$ $m = 2$ | | $m = 2$ |

700
nm

400 700
nm nm

400
nm

Both λ

400
nm

700 400
nm nm

700
nm

(a)

| $m = 2$ | $m = 1$ | $m = 0$ | $m = 1$ | $m = 2$ |

Rainbow
(fainter)

Rainbow

White

Rainbow

Rainbow
(fainter)

(b)

FIGURE 23-20

Spectra produced by a grating: (a) two wavelengths, 400 nm and 700 nm; (b) white light. The second order will be dimmer than the first order, normally; if grating spacing is small enough, second order may be missing. (Higher orders not shown.)

Thus even for a tiny angle[†] corresponding to an extra path length of $\frac{1}{1000}\lambda$, there is much destructive interference, and so the maximum must be very narrow. The more lines there are in a grating the sharper will be the peaks (see Fig. 23-19). Because a grating produces much sharper (and brighter) lines than two slits alone, it is a far more precise device for measuring wavelengths.

Suppose the light striking a diffraction grating is not monochromatic, but rather consists of two or more distinct frequencies. Then for all orders other than $m = 0$, each wavelength will produce a maximum at a different angle (Fig. 23-20a). If white light strikes a grating, the central ($m = 0$) maximum will be a sharp white peak; but for all other orders, there will be a distinct spectrum of colors spread out over a certain angular width, Fig. 23-20b. Because a diffraction grating spreads out light into its component wavelengths, the resulting pattern is called a *spectrum*.

EXAMPLE 23-4 Calculate the first- and second-order angles for light of wavelength 400 nm and 700 nm if the grating contains 10,000 lines/cm.

SOLUTION Since the grating contains 10^4 lines/cm $= 10^6$ lines/m, the separation between slits is $d = 1/10^6$ m$^{-1} = 1.0 \times 10^{-6}$ m. In first order ($m = 1$), the angles are

$$\sin \theta_{400} = \frac{m\lambda}{d} = \frac{1(4.0 \times 10^{-7}\text{ m})}{1.0 \times 10^{-6}\text{ m}} = 0.400$$

$$\sin \theta_{700} = 0.700,$$

so $\theta_{400} = 23.6°$ and $\theta_{700} = 44.0°$. In second order

$$\sin \theta_{400} = \frac{2(4.0 \times 10^{-7}\text{ m})}{1.0 \times 10^{-6}\text{ m}} = 0.800$$

$$\sin \theta_{700} = 1.40,$$

so $\theta_{400} = 53.0°$ but the second order does not exist for $\lambda = 700$ nm since $\sin \theta$ cannot exceed 1. No higher orders will appear.

[†] Depending on the number of slits, there may or may not be complete cancellation for such an angle, so there will be very tiny peaks between the main maxima (see Fig. 23-19b), but they are usually much too small to be seen.

EXAMPLE 23-5 White light containing wavelengths from 400 nm to 750 nm strikes a grating containing 4000 lines/cm. Show that the violet of the third-order spectrum overlaps the red of the second order.

SOLUTION   The grating spacing is $d = 1/(4000\ \text{cm}^{-1}) = 2.50 \times 10^{-6}$ m. The violet of the third order occurs at an angle θ given by

$$\sin \theta = \frac{(3)(4.00 \times 10^{-7}\ \text{m})}{(2.50 \times 10^{-6}\ \text{m})} = 0.480.$$

Red in second order occurs at

$$\sin \theta = \frac{(2)(7.50 \times 10^{-7}\ \text{m})}{(2.50 \times 10^{-6}\ \text{m})} = 0.600,$$

which is clearly a greater angle.

The diffraction grating is the essential component of a spectroscope, a device for precise measurement of wavelengths, and we discuss it next.

23-7 ∎ The spectroscope and spectroscopy

A **spectroscope**, Fig. 23-21, is a device to measure wavelengths accurately using a diffraction grating (discussed in Section 23-6), or a prism, to separate different wavelengths of light. Light from a source passes through a narrow slit S in the collimator. The slit is at the focal point of the lens L, so parallel light falls on the grating. The movable telescope can bring the rays to a focus. Nothing will be seen in the viewing telescope unless it is positioned at an angle θ that corresponds to a diffraction peak (first order is usually used) of a wavelength emitted by the source. The angle θ can be measured[†] to very high accuracy, so the wavelength of a line can be determined to high accuracy using Eq. 23-4:

$$\sin \theta = \frac{m\lambda}{d},$$

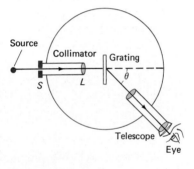

FIGURE 23-21
Spectroscope.

where m is an integer representing the order, and d is the distance between grating lines. (The line you see in a spectroscope corresponding to each wavelength is actually an image of the slit S; so the narrower the slit, the narrower—but dimmer—is the line, and the more precise its measurement. If the light contains a continuous range of wavelengths, you will then see a continuous spectrum in the spectroscope.)

In some spectroscopes, a reflection grating or a prism is used. A prism, because of dispersion (Section 23-4), bends light of different wavelengths into different angles. A prism has the disadvantage that it produces less sharp lines and is less able to separate closely spaced lines. But it has the advantage of deflecting more light (and so is more useful for dim sources) than a normal diffraction grating, since with the latter most of the light passes straight through to the central peak. However, reflection gratings can now be made

---

[†] The angle θ for a given wavelength is usually measured on both sides of center because the grating cannot always be aligned precisely; the average of the two values is then taken.

that have shaped grooves, so that a large portion of the light is reflected at an appropriate angle to give a strong first-order peak.

If the spectrum of a source is recorded (say on film) rather than viewed by the eye, the device is called a *spectrometer*, as compared to a *spectroscope*, although these terms are sometimes used interchangeably. Devices which can also measure the intensity of light of a given wavelength are called *spectrophotometers*.

An important use of any of these devices is for the identification of atoms or molecules. When a gas is heated or a large electric current is passed through it, the gas emits a characteristic **line spectrum.** That is, only certain wavelengths of light are emitted, and these are different for different elements and compounds.[†] Figure 23-22 (color plate)[‡] shows the line spectra for a number of elements in the gas state. Line spectra occur only for gases at high temperatures and low pressure. The light from heated solids, such as a light bulb filament, and even from a dense gaseous object such as the sun, produces a *continuous spectrum* including a wide range of wavelengths.

As can be seen in Fig. 23-22, the sun's "continuous spectrum" contains a number of *dark* lines (only the most prominent are shown), called *absorption lines*. Atoms and molecules absorb light at the same wavelengths at which they emit light. The sun's absorption lines are due to absorption by atoms and molecules in the cooler outer atmosphere of the sun, as well as by atoms and molecules in the earth's atmosphere. A careful analysis of all these thousands of lines reveals that at least two-thirds of all elements are present in the sun's atmosphere. The presence of elements in the atmosphere of other planets, in interstellar space, and in stars is also determined by spectroscopy.

Spectroscopy is useful for determining the presence of certain types of molecules in laboratory specimens where chemical analysis would be difficult. For example, biological DNA and different types of protein absorb light in particular regions of the spectrum (such as in the UV). The material to be examined, which is often in solution, is placed in a monochromatic light beam whose wavelength is chosen by placement of a prism or diffraction grating. The amount of absorption, as compared to a standard solution without the specimen, can reveal not only the presence of a particular type of molecule but also its concentration.

Light emission and absorption also occur outside the visible part of the spectrum, such as in the UV and IR regions. Since glass absorbs light in these regions, reflection gratings and mirrors (in place of lenses) are used. Special types of film, or photocell detectors, are used for detection.

23-8 ∎ Interference by thin films

Interference of light gives rise to many everyday phenomena such as the bright colors reflected from soap bubbles and from thin oil films on water. In these and other cases, the colors are a result of constructive interference

---

[†] Why atoms and molecules emit line spectra was a great mystery for many years and played a central role in the development of modern quantum theory as we shall see in Chapter 26.

[‡] See the page following page 556.

between light reflected from the two surfaces of the thin film. To see how this happens, consider a smooth surface of water on top of which is a thin uniform layer of another substance, Fig. 23-23. Part of the incident light (say from the sun or street lights) is reflected at A on the top surface, and part of that transmitted is reflected at B on the lower surface. The part reflected at the lower surface must travel the extra distance ABC. If the distance ABC is equal to one or a whole number of wavelengths, the two waves will interfere constructively and the light will be bright.[†] But if ABC equals $\frac{1}{2}\lambda$, $\frac{3}{2}\lambda$, and so on, the two waves will be out of phase and destructive interference occurs. The wavelength λ is that in the film (see Eq. 23-1).

When white light falls on such a film, the path ABC will equal λ (or $m\lambda$ with $m =$ an integer) for only one wavelength at a given viewing angle. This color will be seen as very bright. For light viewed at a slightly different angle, the path ABC will be longer or shorter and a different color will undergo constructive interference. Thus, for an extended (nonpoint) source emitting white light, a series of bright colors will be seen next to one another. Variations in thickness of the film will also alter the length ABC and therefore affect the color of light that is most strongly reflected.

When a curved glass surface is placed in contact with a flat glass surface, Fig. 23-24 a series of concentric rings is seen when illuminated from above by monochromatic light. These are called *Newton's rings*[‡] and they are due to interference between rays reflected by the top and bottom surfaces of the *air gap* between the two pieces of glass. Because this gap (which is equivalent to a thin film) increases in width from the central contact point out to the edges, the extra path length for the lower ray (equal to BCD) varies; where it equals $0, \frac{1}{2}\lambda, \lambda, \frac{3}{2}\lambda, 2\lambda$, and so on, it corresponds to constructive and destructive interference; and this gives rise to the series of bright and dark lines seen in Fig. 23-24b.

Note that the point of contact of the two glass surfaces (A in Fig. 23-24a) is dark in Fig. 23-24b. Since the path difference is zero here, we expect the rays reflected from each surface to be in phase and this point to

---

[†] As is discussed shortly, this is true if the refractive index of the film is less than that of the water.

[‡] Although Newton gave an elaborate description of them, they had been first observed and described by his contemporary Robert Hooke. Newton did not realize their significance in support of a wave theory of light.

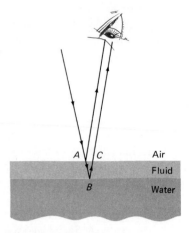

FIGURE 23-23

Light reflected from upper and lower surfaces of a thin film of fluid lying on water.

(a)

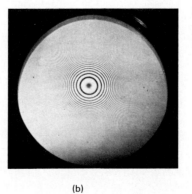

(b)

FIGURE 23-24

Newton's rings. (Part b courtesy of Bausch & Lomb Incorporated.)

be bright. But it is dark, which tells us the two rays must be completely out of phase; this can happen only because one of the waves undergoes a change in phase of 180° upon reflection, corresponding to $\frac{1}{2}$ cycle or $\frac{1}{2}\lambda$. Indeed, this and other experiments reveal that *a beam of light reflected by a material whose index of refraction is greater than that in which it is traveling changes phase by $\frac{1}{2}$ cycle.* If the index is less than that of the material in which it is traveling, no phase change occurs. (This corresponds to the reflection of a wave traveling along a rope when it reaches the end; as we saw in Fig. 13-20, if the end is tied down, the rope changes phase and the pulse flips over; but if the end is free, no phase change occurs; see also Fig. 13-22.) Thus the ray reflected by the curved surface above the air gap in Fig. 23-24a undergoes no change in phase. That reflected at the lower surface, where the beam in air strikes the glass, undergoes a $\frac{1}{2}\lambda$ phase change. Thus the two rays reflected at the point of contact A of the two glass surfaces (where the air gap approaches zero thickness) will be $\frac{1}{2}\lambda$ out of phase, and a dark spot occurs. Other dark bands will occur when the path difference BCD in Fig. 23-24a is equal to an integral number of wavelengths. Bright bands will occur when the path difference is $\frac{1}{2}\lambda$, $\frac{3}{2}\lambda$, and so on, since the phase change at one surface effectively adds another $\frac{1}{2}\lambda$.

EXAMPLE 23-6 A very fine wire 7.35×10^{-3} mm in diameter is placed between two flat glass plates as in Fig. 23-25a. Light whose wavelength in air is 600 nm falls (and is viewed) perpendicularly to the plates, and a series of bright and dark bands is seen, Fig. 23-25b. How many light and dark bands will there be in this case? Will the area next to the wire be bright or dark?

SOLUTION The thin film is the wedge of air between the two glass plates. Because of the phase change at the lower surface, there will be a dark band when the path difference is 0, λ, 2λ, 3λ, and so on. Since the light rays are perpendicular to the plates, the extra path length equals $2t$ where t is the thickness of the air gap at any point; so dark bands occur where

$$2t = m\lambda, \qquad m = 0, 1, 2, \dots.$$

Bright bands occur when $2t = (m + \frac{1}{2})\lambda$, where m is an integer. At the position of the wire, $t = 7.35 \times 10^{-6}$ m. At this point there will be $(2)(7.35 \times 10^{-6} \text{ m})/(6.00 \times 10^{-7} \text{ m}) = 24.5$ wavelengths. Since this is a "half integer," the area next to the wire will be bright. There will be a total of 25 dark lines along the plates, corresponding to path lengths of 0λ, 1λ, 2λ, 3λ, ... 24λ, including the one at the point of contact A ($m = 0$). Between them there will be 24 bright lines plus the one at the end, or 25. The bright and dark bands will be straight only if the glass plates are extremely smooth. If they are not, the pattern is uneven, as in Fig. 23-25c. This is a very precise way of testing a glass surface for smoothness.

If the wedge between the two glass plates of Example 23-6 is filled with some transparent substance other than air—say water—the pattern shifts. This is because the wavelength of the light changes. In a material where the index of refraction is n, the wavelength is $\lambda_n = \lambda/n$ where λ is the wavelength

FIGURE 23-25

(a) Light rays reflected from upper and lower surfaces of a thin wedge of air interfere to produce bright and dark bands. (b) Pattern observed when glass plates are optically flat (courtesy of The Van Keuren Co.); (c) pattern when plates are not so flat (courtesy of Bausch & Lomb Incorporated).

(a)

(b)

(c)

in a vacuum (see Eq. 23-1). For instance, if the thin wedge of Example 23-6 were filled with water, $\lambda_n = 600 \text{ nm}/1.33 = 450 \text{ nm}$; and instead of 25 dark lines there would be 33.

When white light (rather than monochromatic light) is incident on the thin wedge of Figs. 23-24a or 23-25a a colorful series of fringes is seen. This is because constructive interference occurs in the reflected light at different locations along the wedge for different wavelengths. Such a difference in thickness is part of the reason bright colors appear when light is reflected from a soap bubble or a thin layer of oil on a puddle or lake; which wavelengths appear brightest also depends on the viewing angle as we saw earlier.

EXAMPLE 23-7   A soap bubble appears green ($\lambda = 540 \text{ nm}$) at its point nearest the viewer. What is its minimum thickness? Assume $n = 1.35$.

SOLUTION   The light is reflected perpendicularly from the point on a spherical surface nearest the viewer. Therefore the path difference is $2t$ where t is the thickness of the soap film. Light reflected from the outer surface undergoes a $\frac{1}{2}\lambda$ phase change whereas that on the inner surface does not. Therefore, green light is bright when the minimum path difference equals $\frac{1}{2}\lambda$. Thus $2t = \lambda/2n$ so $t = 540 \text{ nm}/(4)(1.35) = 100 \text{ nm}$.

An important application of thin-film interference is in the coating of glass to make it "nonreflecting," particularly for lenses. Glass reflects about 4 percent of the light incident upon it. Good-quality cameras, microscopes, and other optical devices may contain six to ten thin lenses. Reflection from all these surfaces can reduce the light level considerably and multiple reflections produce a background haze that reduces the quality of the image. A very thin coating on the lens surfaces can reduce these problems considerably. The amount of reflection at a boundary depends on the difference of index of refraction between the two materials. Ideally, the coating material should have an index of refraction which is the geometric mean of those for air and glass, so that the amount of reflection at each surface is about equal; then destructive interference can occur nearly completely for one particular wavelength depending on the thickness of the coating. Nearby wavelengths will at least partially destructively interfere, but it is clear that a single coating cannot eliminate reflections for all wavelengths. Nonetheless, a single coating can reduce total reflection from 4 percent to 1 percent of the incident light. Often the coating is designed to eliminate the center of the spectrum (around 550 nm); the extremes of the spectrum—red and violet—will not be reduced as much. Since a mixture of red and violet produces purple, the light seen reflected from such coated lenses is purple. Lenses containing two or three separate coatings can more effectively reduce a wider range of reflecting wavelengths.

EXAMPLE 23-8   What is the thickness of an optical coating of MgF_2 whose index of refraction is $n = 1.38$ and is designed to eliminate wavelengths centered at 550 nm when incident normally on glass for which $n = 1.50$?

SOLUTION   Light reflected at both the front and back surfaces of the coating change phase. Thus, for normal incidence, the thickness of the coating to produce destructive interference must be a half integral number of wavelengths. That is $2t = (m + \frac{1}{2})\lambda$ where m is an integer. The minimum thickness ($m = 0$) is usually chosen because destructive interference will then occur over the widest angle. Then $t = \lambda_n/4 = \lambda/4n = (550 \text{ nm})/(4)(1.38) = 99.6 \text{ nm}$.

*23-9 ∎ Michelson interferometer

Interference by thin films is the basis of the *Michelson interferometer* (Fig. 23-26),[†] invented by the American Albert A. Michelson (Section 22-2). Monochromatic light from a single point on an extended source is shown striking a half-silvered mirror M_s (called the *beam splitter*). Half of the beam passes through to a fixed mirror M_2, where it is reflected back. The other half is reflected by M_s up to a mirror M_1 which is movable (by a fine-thread screw), where it is also reflected back. Upon its return, part of beam 1 passes through M_s and reaches the eye; and part of beam 2, on its return, is reflected by M_s into the eye. (A compensator plate C of transparent glass, usually cut from the same plate as M_s, is placed in the path of beam 2 so both beams pass through the same thickness of glass, to within a fraction of a wavelength.) If the two path lengths are identical, the two coherent beams entering the eye constructively interfere and brightness will be seen. If the movable mirror is moved a distance $\lambda/4$, one beam will travel an extra distance equal to $\lambda/2$ (because it travels back and forth over the distance $\lambda/4$); then the two beams will destructively interfere and darkness will be seen. As M_1 is moved farther, brightness will recur (when the path difference is λ), then darkness, and so on.

The eye sees the image of M_2 very close to the position of M_1; so in effect, there is a thin layer of air (a "thin film") between this image and M_1. If M_1 is tilted slightly, then the air gap is wedge shaped as in Fig. 23-25,

[†] There are other types of interferometer but Michelson's is the best known.

FIGURE 23-26

Michelson interferometer.

and a series of bright and dark lines are seen; when mirror M_1 is then moved the bright and dark lines move to the left or right. When the path difference is many wavelengths, the observer no longer sees a nearly uniform brightness or darkness when the mirrors M_1 and M_2 are precisely aligned. Because the path difference is different for different angles of view, one sees a series of rings, much like the Newton ring pattern of Fig. 23-24b. When mirror M_1 is moved, dark and bright spots seem to emerge from (or disappear into) the center of the pattern.

Very precise length measurements can be made with an interferometer. The motion of mirror M_1 by only $\frac{1}{4}\lambda$ produces a clear difference between brightness and darkness. For $\lambda = 400$ nm, this means a precision of 100 nm or 10^{-4} mm! By observing the sideways motion of the fringes when the mirrors are not precisely aligned, even greater precision can be obtained. By counting the number of fringes, or fractions thereof, extremely precise length measurements can be made.

Michelson saw that the interferometer could be used to determine the length of the standard meter in terms of the wavelength of a particular light. In 1960, that standard was chosen to be a particular orange-red line in the spectrum of krypton-86 (krypton atoms with atomic mass 86). Careful repeated measurements of the old standard meter (the distance between two marks on a platinum–iridium bar kept in Paris) were made to establish 1 meter as being 1,650,763.73 wavelengths of this light, which was *defined* to be the meter. In 1983, the meter was redefined in terms of the speed of light (Section 1-3).

23-10 ∎ Polarization

An important and useful property of light is that it can be *polarized*. To see what this means, let us examine waves traveling on a rope. A rope may be set into vibration in a vertical plane as in Fig. 23-27a or in a horizontal plane as in part (b). In either case the wave is said to be **plane-polarized**— that is, the oscillations are in a plane.

If we now place an obstacle containing a vertical slit in the path of the wave, Fig. 23-28, a vertically polarized wave passes through, but a horizontally polarized wave will not. If a horizontal slit were used, the vertically polarized wave would be stopped. If both types of slit are used, neither wave could pass through. Note that polarization can exist *only* for *transverse waves*, and not for longitudinal waves. The latter vibrate only along the direction of motion and neither orientation of slit would stop them.

That light can be polarized was recognized only in the nineteenth century. However, even in Newton's time a phenomenon that depends on polarization was already known—namely that certain types of crystal, such as iceland spar, refract light into two rays; such crystals are called *doubly refracting* (Section 23-12). Certain other crystals, such as tourmaline, would not transmit one or the other of these two rays, depending on the orientation of the crystal. Today we recognize that the two rays in a doubly refracting crystal are plane-polarized in mutually perpendicular directions, and that the

(a)

(b)

FIGURE 23-27
Transverse waves on a rope polarized (a) in a vertical plane and (b) in a horizontal plane.

tourmaline acts as a "slit" to eliminate one or the other if properly oriented. However, it was not until after Young's and Fresnel's work in the early 1800s that this phenomenon was recognized as evidence that light is a transverse wave. A half century later, Maxwell's theory of light as electromagnetic (EM)

FIGURE 23-28
Vertically polarized wave passes through a vertical slit, but a horizontally polarized wave will not.

(a)

(b)

waves was fully consistent with the facts of polarization since an EM wave is a transverse wave. The direction of polarization in a plane-polarized EM wave is taken as the direction of the electric field vector (Fig. 21-5).

Light is not necessarily polarized. It can be *unpolarized*, which means that the source has vibrations in many planes at once, Fig. 23-29. An ordinary incandescent light bulb emits unpolarized light, as does the sun.

Plane-polarized light can be obtained from unpolarized light using certain crystals such as tourmaline. Or, more commonly today, we can use a *Polaroid sheet*. (Polaroid materials were invented in 1929 by Edwin Land.) A Polaroid sheet consists of complicated long molecules arranged with their axes parallel. Such a Polaroid acts as a series of parallel slits to allow one orientation of polarization to pass through nearly undiminished (this direction is called the *axis* of the Polaroid) whereas a perpendicular polarization is absorbed almost completely.[†] If a beam of plane-polarized light strikes a Polaroid whose axis is at an angle θ to the incident polarization direction, the beam will emerge plane-polarized parallel to the Polaroid axis and its amplitude will be reduced by $\cos \theta$, Fig. 23-30. Thus a Polaroid passes only that component of polarization (the electric field vector, **E**) that is parallel to its axis. Since the intensity of a light beam is proportional to the square of the amplitude, we see that the intensity of a plane-polarized beam transmitted by a polarizer is

$$I = I_0 \cos^2 \theta, \qquad (23\text{-}5)$$

where θ is the angle between the polarizer axis and the plane of polarization of the incoming wave, and I_0 is the incoming intensity.[‡]

A Polaroid can be used as a *polarizer* to produce plane-polarized light from unpolarized light, since only the component of light parallel to the axis

FIGURE 23-29
Vibration of the electric field vector in unpolarized light.

[†] How this occurs can be explained at the molecular level. An electric field **E** that oscillates parallel to the long molecules can set electrons into motion along the molecules, thus doing work on them and transferring energy; hence, if **E** is parallel to the "slits," it gets absorbed. An electric field **E** perpendicular to the long molecules (and thus perpendicular to the "slits") does not have this possibility of doing work and so passes through freely. Thus Fig. 23-28 does not apply for the **E** field of an EM wave (although it would for the **B** field). When we speak of the *axis* of a Polaroid, we mean the direction for which **E** is passed, so a Polaroid axis is perpendicular to the long molecules and the "slits" between them.

[†] Equation 23-5 is often referred to as "Malus' law," after Etienne Malus, a contemporary of Fresnel.

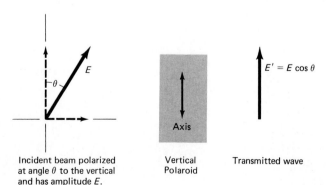

FIGURE 23-30

Vertical Polaroid transmits only the vertical component of a wave incident upon it.

Incident beam polarized at angle θ to the vertical and has amplitude E.

Vertical Polaroid

Transmitted wave

$E' = E \cos \theta$

is transmitted. A Polaroid can also be used as an *analyzer* to determine (1) if light is polarized and (2) what is the plane of polarization. A Polaroid acting as an analyzer will pass the same amount of light independent of the orientation of its axis if the light is unpolarized; try rotating one lens of a pair of Polaroid sunglasses while looking through it at a light bulb. If the light is polarized, however, when you rotate the Polaroid the transmitted light will be a maximum when the plane of polarization is parallel to the Polaroid's axis, and a minimum when perpendicular to it. If you do this while looking at the sky, preferably at right angles to the sun's direction, you will see that skylight is polarized. (Direct sunlight is unpolarized, but don't look directly at the sun, even through a polarizer, for damage to the eye may occur.) If the light transmitted by an analyzer Polaroid falls to zero at one orientation, then the light is 100 percent plane-polarized. If it merely reaches a minimum, the light is *partially polarized.*

Unpolarized light consists of light with random directions of polarization (electric field vector). Each of these polarization directions can be resolved into components along two mutually perpendicular directions. Thus, an unpolarized beam can be thought of as two plane-polarized beams of equal magnitude perpendicular to one another. When two Polaroids are *crossed*—that is, their axes are perpendicular to one another—unpolarized light can be entirely stopped (or nearly so—Polaroids are not quite perfect). As shown in Fig. 23-31, unpolarized light is made plane-polarized by the first Polaroid (the polarizer). The second Polaroid, the analyzer, then eliminates this component since its axis is perpendicular to the first. You can try this with Polaroid sunglasses. It should be clear that Polaroid sunglasses eliminate 50 percent of unpolarized light because of their polarizing property; they absorb even more because they are colored.

EXAMPLE 23-9   Unpolarized light passes through two Polaroids; the axis of one is vertical and that of the other is at 60° to the vertical. What is the orientation and intensity of the transmitted light?

SOLUTION   The first Polaroid eliminates half the light so the intensity is reduced by half: $I_1 = \frac{1}{2}I_0$. The light reaching the second polarizer is vertically polarized and so is reduced in intensity (Eq. 23-5) by

$$I_2 = I_1(\cos 60°)^2 = \tfrac{1}{4}I_1.$$

Thus $I_2 = \frac{1}{8}I_0$. The transmitted light has an intensity one-eighth that of the original and is plane-polarized at a 60° angle to the vertical.

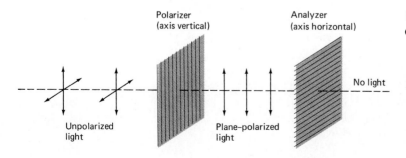

Polarizer
(axis vertical)

Analyzer
(axis horizontal)

No light

Unpolarized
light

Plane-polarized
light

FIGURE 23-31

Crossed Polaroids completely eliminate light.

FIGURE 23-32

Light reflected from a nonmetallic surface, such as the smooth surface of water in this lake, is partially polarized parallel to the surface.

Another means of producing polarized light from unpolarized light is by reflection. When light strikes a nonmetallic surface at any angle other than perpendicular, the reflected beam is polarized preferentially in the plane parallel to the surface, Fig. 23-32. In other words, the component with polarization in the plane perpendicular to the surface is preferentially transmitted or absorbed. You can check this fact by rotating Polaroid sunglasses while looking through them at a flat surface of a lake or road. Since most outdoor surfaces are horizontal, Polaroid sunglasses are made with their axis vertical to eliminate the stronger component, and thus reduce glare. This is well known by fishermen who wear Polaroids to eliminate reflected glare from the surface of a lake or stream and thus see beneath the water more clearly.

The amount of polarization in the reflected beam depends on the angle; it varies from no polarization at normal incidence to 100 percent polarization at an angle known as the *polarizing angle*, θ_p.[†] This angle is related to the index of refraction of the two materials on either side of the boundary by the equation

$$\tan \theta_p = \frac{n_2}{n_1}, \tag{23-6a}$$

where n_1 is the index of refraction of the material in which the beam is traveling, and n_2 is that of the medium beyond the reflecting boundary. If the beam is traveling in air, $n_1 = 1$ and

$$\tan \theta_p = n. \tag{23-6b}$$

The polarizing angle θ_p is also called *Brewster's angle*, and Eqs. 23-6 *Brewster's law*, after the Scottish physicist David Brewster (1781–1868) who worked it out experimentally in 1812. Equations 23-6 can be derived from the electromagnetic wave theory of light. It is interesting that at Brewster's angle, the reflected and transmitted rays make a 90° angle to each other; that is,

---

[†] Although all of one component is transmitted into a transparent medium at this angle, some of the other component is transmitted as well, so the transmitted beam is *not* 100 percent polarized.

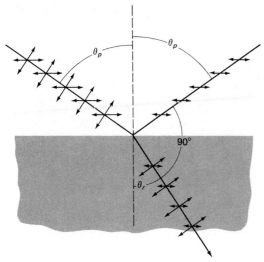

FIGURE 23-33

At θ_p the reflected light is plane-polarized parallel to the surface, and $\theta_p + \theta_r = 90°$, where θ_r is the refraction angle.

$\theta_p + \theta_r = 90°$, Fig. 23-33. This can be seen by substituting Eq. 23-6a, $n_2 = n_1 \tan \theta_p = n_1 \sin \theta_p / \cos \theta_p$, into Snell's law, $n_1 \sin \theta_p = n_2 \sin \theta_r$, and get $\cos \theta_p = \sin \theta_r$, which can only hold if $\theta_p = 90° - \theta_r$.

EXAMPLE 23-10 (a) At what incident angle is sunlight reflected from a lake plane-polarized? (b) What is the refraction angle?

SOLUTION (a) We use Eq. 23-6b with $n = 1.33$ so $\tan \theta_p = 1.33$ and $\theta_p = 53.1°$.
  (b) $\theta_r = 90.0° - \theta_p = 36.9°$.

*23-11 ▌ Optical activity

When a beam of plane-polarized light is passed through crystals and solutions, it is found that the plane of polarization is rotated through an angle. For example, Fig. 23-34 shows light passing through a polarizer and then through a sugar solution. The analyzer Polaroid behind the solution does not cut out all the light when placed at 90° to the polarizer. If the analyzer is rotated through an angle ϕ, however, there is no transmitted light; this indicates that the plane of polarization has been rotated through an angle ϕ by the intervening substance. Such substances are said to be *optically active*.

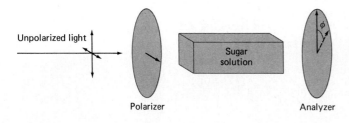

FIGURE 23-34

Sugar solution rotates the plane of polarization of incident light so that it is no longer horizontal but at an angle ϕ. The analyzer thus extinguishes the light when rotated ϕ from the vertical (crossed) position.

Unpolarized light

Sugar solution

Polarizer

Analyzer

Optical activity is due to asymmetry of the molecules, which may have a spiral shape as, for example, some proteins do. Substances which rotate the polarization plane to the right as viewed along the direction of the beam as in Fig. 23-34 are called *dextrorotatory* (also called *right-handed* or *positive*). Those that rotate it to the left are called *levorotatory* (*left-handed*, or *negative*). The common sugar, dextrose or D-glucose, is dextrorotatory. Most naturally occurring amino acids and proteins are levorotatory.

The angle of rotation ϕ depends on the path length l (meters) through the substance and on the concentration c (kg/m^3) if it is in solution. For dilute solutions, this is a linear relationship and ϕ (in radians) is given by

$$\phi = \alpha l c.$$

The constant α is a property of the substance and is called the specific rotation or the *specific optical rotatory power* (it depends on the temperature and the wavelength of light used). Some typical values are given in Table 23-1.

Since ϕ is proportional to the concentration, optical activity is a standard method to measure concentrations of solutions such as for sugars. It is also helpful in determining the three-dimensional shape of large molecules, such as proteins, or their change in shape when conditions are changed. For example, a protein in an α-helix shape will have a large (negative) value for α, but may alter its shape when the pH or temperature is changed, and this will change the value of α. Much information on the shape and properties of molecules has been obtained in this way.

Glass and plastics become optically active when put under stress; the plane of polarization is rotated most where the stress is greatest. Models of bones or machine parts made of plastic can be observed between crossed Polaroids to determine where the points of greatest stress lie. This is called "optical stress analysis."

TABLE 23-1
Specific optical rotatory power, α[†]

Substance	α(rad · m^2/kg)[‡]
Alanine	$+ 3.14 \times 10^{-4}$
Leucine	$- 19.2 \times 10^{-4}$
Cysteine	$- 28.8 \times 10^{-4}$
Insulin	$- 59 \times 10^{-4}$
Collagen	$-500 \text{ to } -700 \times 10^{-4}$

[†] Substances in water solution at 20°C for $\lambda = 589$ nm.
[‡] + is dextrorotatory; − is levorotatory.

*23-12 ∎ Double refraction: birefringence and dichroism

In many transparent materials, the speed of light is the same in all directions. Such materials are said to be *isotropic*. In some crystals and solutions, however, the speed of light is different in different directions, and these are referred to as being *anisotropic*. Such substances are also said to be *doubly refracting* or *birefringent* because of the unusual phenomenon they give rise to.

There is one particular direction in a birefringent crystal, such as calcite, known as its *optic axis* (this is not a single line, but a direction in the crystal). If ordinary light enters a crystal along its optic axis nothing abnormal is noted. But if unpolarized light falls at an angle to the optic axis, as in Fig. 23-35, a very unusual thing happens: there are two refracted rays. In the case shown, the incident ray is normal to the surface and the optic axis lies in the plane shown. One refracted ray, called the *ordinary ray* (*o*) passes straight through in a normal way. But the other ray, called the *extraordinary ray* (*e*), is refracted at an angle.

As can be seen, Snell's law does not hold in this case for the *e* ray. It does hold, however, for the *o* ray.

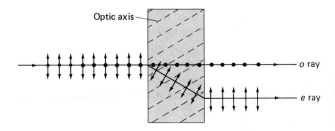

FIGURE 23-35

Unpolarized light, incident normally on a birefringent crystal such as calcite, is separated into two refracted beams. The dashed lines indicate the direction of the optic axis.

The *e* ray and *o* ray are found to be plane-polarized in mutually perpendicular directions. This is indicated in Fig. 23-35 by the dots on the *o* ray indicating the vibrations are perpendicular to the paper; and the polarization of the *e* ray is indicated by the short arrows.

The phenomenon of double refraction can be explained if we make the following assumption: that the speed of light is different depending on the orientation of the polarization vector with respect to the crystal's optic axis. As can be seen in Fig. 23-35, the polarization of the *o* ray is perpendicular to the optic axis; so its speed will be the same in all directions, as long as its polarization remains perpendicular to the optic axis. The *e* ray, on the other hand, has components of polarization both parallel and perpendicular to the optic axis, and it therefore travels with a different speed in different directions; it travels with the same speed as the *o* ray if its polarization vector is perpendicular to the optic axis; at other angles it has a higher speed (or, in certain crystals, a lower speed) which reaches a maximum (or minimum) when its polarization is parallel to the optic axis. (In what follows, we assume a crystal in which this speed is higher.) Thus the index of refraction n_o for the ordinary ray is the same for all directions whereas that for the extraordinary ray, n_e, depends on angle; usually n_e is specified for the *e* ray traveling perpendicular to the optic axis where its polarization direction is parallel to the optic axis. These principal values for n_o and n_e are given in Table 23-2 for several crystals.

Now we use the above assumptions, plus Huygens' principle, to explain double refraction. In Fig. 23-36a we examine the ordinary wave. Huygens' wavelets are shown for two points on the wavefront just as it has entered the crystal. Since these wavelets travel at the same speed in all directions,

TABLE 23-2

Principal indices of refraction for some doubly refracting crystals ($\lambda = 589$ nm)

Crystal	n_o	n_e
Ice	1.309	1.313
Quartz	1.544	1.553
Calcite	1.658	1.486
Dolomite	1.681	1.500

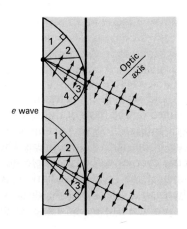

FIGURE 23-36

Huygens' principle used to explain double refraction.

The formula $\sin \theta = m\lambda/d$ for constructive interference also holds for a *diffraction grating*, which consists of many parallel slits or lines, each separated from its neighbors by a distance d. The peaks of constructive interference are much brighter and sharper for a diffraction grating than for the simple two-slit apparatus. A diffraction grating (or a prism) is used in a *spectroscope* to separate different colors or to observe *line spectra*, since for a given order m, θ depends on λ; precise determination of wavelength can thus be done with a spectroscope by careful measurement of θ.

Diffraction refers to the fact that light, like other waves, bends around objects it passes and spreads out after passing through narrow slits. This bending gives rise to a diffraction pattern due to interference between rays of light that travel different distances. Light passing through a very narrow slit of width D will produce a pattern with a bright central maximum of half-width θ given by $\sin \theta = \lambda/D$, flanked by fainter lines to either side.

Light reflected from the front and rear surfaces of a thin film of transparent material can interfere. Such thin-film interference has many practical applications, such as lens coatings and Newton's rings.

In *unpolarized light* the electric field vectors vibrate at all angles. If the electric vector vibrates only in one plane the light is said to be *plane-polarized*. Light can also be partially polarized. When an unpolarized light beam passes through a Polaroid sheet or a doubly refracting crystal, the emerging beams are plane-polarized. That a light beam is polarized is determined experimentally by the fact that when it passes through a Polaroid, the intensity varies as the Polaroid is rotated. Thus a Polaroid can act as polarizer or as an analyzer. Light can also be partially or fully polarized by reflection; if light traveling in air is reflected from a medium of index of refraction n, it will be completely plane-polarized if the incident angle θ_p is given by $\tan \theta_p = n$. The fact that light can be polarized shows that it must be a transverse wave.

▌ Questions

1. Does Huygens' principle apply to sound waves? To water waves?

2. What is the evidence that light is energy?

3. Why is light sometimes described as rays and sometimes as waves?

4. Two rays of light from the same source destructively interfere if their path lengths differ by how much?

5. If Young's double-slit experiment were submerged in water, how would the fringe pattern be changed?

6. Why was the observation of the double-slit interference pattern more convincing evidence for the wave theory of light than the observation of diffraction?

7. Compare a double-slit experiment for sound waves to that for light waves. Discuss the similarities and differences.

8. How would you produce "beats" with light waves, a well-known phenomenon with sound waves? How might you detect light "beats"?

9. Why doesn't the light from the two headlights of a distant car produce an interference pattern?

10. When white light passes through a flat piece of window glass, it is not broken down into colors as it is by a prism. Explain.

11. For both converging and diverging lenses, discuss how the focal length for red light differs from that for violet light.

12. We can hear sounds around corners, but we cannot see around corners; yet both sound and light are waves. Explain the difference.

13. Hold one hand close to your eye and focus on a distant light source through a narrow slit between two fingers. (Adjust your fingers to get best pattern.) Describe the pattern you see. Is this Fresnel or Fraunhofer diffraction?

14. A rectangular slit is twice as high as it is wide. Will the light spread more in the horizontal or in the vertical plane? Describe the pattern.

15. What is the difference in the interference patterns formed by two slits 10^{-4} cm apart and by a diffraction grating containing 10^4 lines/cm?

16. Explain why there are tiny peaks between the main peaks produced by a diffraction grating illuminated with monochromatic light. Why are the peaks so tiny?

17. Why is a diffraction grating preferable to a prism for use in a spectroscope?

18. Why are interference fringes noticeable only for a *thin* film like a soap bubble and not for a thick piece of glass, say?

19. Why are Newton's rings (Fig. 23-24) closer together farther from the center?

20. Some coated lenses appear greenish yellow when seen by reflected light. What wavelengths do you suppose they are designed to eliminate completely?

*21. How could a Michelson interferometer be used to measure the index of refraction of air?

22. What does polarization tell us about the nature of light?

23. What is the difference between a polarizer and an analyzer?

24. How can you tell if a pair of sunglasses is polarizing or not?

25. Sunlight will not pass through two Polaroids whose axes are at right angles. What happens if a third Polar-

oid, with axis at 45° to each of the other two, is placed between them?

*26. For which of the materials in Table 23-2 is the speed of the extraordinary ray less when its polarization vector is parallel to the optic axis than when it is perpendicular?

*27. What would be the color of the sky if the earth had no atmosphere?

*28. If the earth's atmosphere were 50 times denser than it is, would sunlight still be white, or would it be some other color?

▌Problems

SECTION 23-2

1. (II) Derive the law of reflection—namely, that the angle of incidence equals the angle of reflection from a flat surface—using Huygens' principle for waves.

SECTION 23-3

2. (I) The second-order fringe when 700 nm light falls on two slits is observed at a 15° angle to the initial beam direction. How far apart are the slits?

3. (I) Monochromatic light falling on two slits 0.026 mm apart produces the fourth-order fringe at a 6.4° angle. What is the wavelength of the light used?

4. (II) Monochromatic light falls on two slits 0.020 mm apart. The fringes on a screen 3.00 m away are 9.2 cm apart. What is the wavelength of the light?

5. (II) A parallel beam of 600 nm light falls on two small slits 5.0×10^{-2} mm apart. How far apart are the fringes on a screen 5.0 m away?

6. (II) Light of wavelength 680 nm falls on two slits and produces an interference pattern in which the fourth-order fringe is 28 mm from the central fringe on a screen 1.0 m away. What is the separation of the two slits?

7. (II) Television and radio waves can reflect from nearby mountains or from airplanes, and the reflections can interfere with the direct signal from the station. (a) Determine what kind of interference will occur when 75-MHz television signals arrive at a receiver directly from a distant station, and reflected from an airplane 118 m directly above the receiver. (Assume no change in phase of the signal upon reflection.) (b) What kind of interference will occur when the plane has moved 22 m farther toward the station?

8. (II) The shortest-wavelength visible light falls on two slits 2.80×10^{-2} mm apart. The slits are immersed in water, as is a viewing screen 18.5 cm away. How far apart are the fringes on the screen?

9. (II) If 520-nm and 660-nm light passes through two slits 0.50 mm apart, how far apart are the second-

order fringes for these two wavelengths on a screen 1.5 m away?

10. (II) Suppose a thin piece of glass were placed in front of the lower slit in Fig. 23-6 so that the two waves enter the slits 180° out of phase. Describe in detail the interference pattern on the screen.

11. (III) A screen contains two slits a distance d apart. If light of wavelength λ strikes it at an angle θ_i to the normal, determine the angle θ_m at which the mth-order maximum occurs.

SECTION 23-4

12. (I) By what percent, approximately, does the speed of red light (700 nm) exceed that of violet light (400 nm) in silicate flint glass? (See Fig. 23-10.)

13. (II) A light beam strikes a piece of glass at a 50.00° incident angle. The beam contains two wavelengths, 500 nm and 700 nm, for which the index of refraction of the glass is 1.4810 and 1.4742, respectively. What is the angle between the two refracted beams?

14. (II) A parallel beam of light containing two wavelengths, $\lambda_1 = 400$ nm and $\lambda_2 = 650$ nm, strikes a piece of silicate flint glass at an incident angle of 60°. Calculate the angle between the two color beams inside the glass (see Fig. 23-10).

15. (III) A double convex lens whose radii of curvature are both 12.0 cm is made of crown glass. Find the distance between the focal points for 400-nm and 700-nm light. (Hint: use the lens-maker's equation, Eq. 22-9, and Fig. 23-10.)

SECTION 23-5

16. (I) If 680-nm light falls on a slit 3.10×10^{-2} mm wide, what is the angular width of the central diffraction peak?

17. (I) Monochromatic light falls on a slit 2.2×10^{-3} mm wide. If the angle between the first dark fringes on either side of the central maximum is 30°, what is the wavelength of the light used?

18. (II) How wide is the central diffraction peak on a screen 3.50 m behind a 0.0112 mm-wide slit illuminated by 400-nm light?

19. (II) When 450 nm light falls on a slit, the central diffraction peak on a screen 2.0 m away is 6.0 cm wide. Calculate the slit width.

20. (II) For a given wavelength λ, what is the maximum slit width for which there will be no diffraction minima?

21. (II) How wide is a slit if it diffracts 590-nm light so that its central diffraction peak is 2.0 cm wide on a screen 1.80 m away?

SECTIONS 23-6 AND 23-7

22. (I) The first-order line of 550-nm light falling on a diffraction grating is observed at a 12° angle. How far apart are the slits?

23. (I) At what angle will 710-nm light produce a third-order maximum when falling on a grating whose slits are 1.70×10^{-3} cm apart?

24. (II) How many lines per centimeter does a grating have if the third-order occurs at a 22.0° angle for 650-nm light?

25. (II) A grating has 5000 lines/cm. How many spectral orders can be seen when it is illuminated by white light?

26. (II) White light containing wavelengths from 400 to 700 nm falls on a grating with 8000 lines/cm. How wide is the first-order spectrum on a screen 2.70 m away?

27. (II) How many lines per centimeter must a grating have if there is to be no second-order spectrum for any visible wavelength?

28. (II) Show that the second- and third-order spectra of white light produced by a diffraction grating always overlap.

29. (II) The wings of a certain beetle have a series of parallel lines across it. When 600-nm light is reflected from the wing, the wing appears bright when viewed at an angle of 49°. How far apart are the lines? The light is incident normally.

30. (II) When yellow sodium light, $\lambda = 589$ nm, falls on a diffraction grating, its first-order peak on a screen 20 cm away falls 2.12 cm from the central peak. Another source produces a line 3.61 cm from the central peak. What is its wavelength?

31. (II) Two first order spectrum lines are measured by a 10,000 line/cm spectroscope at angles, on each side of center, of $+26°38'$, $+41°02'$ and $-26°18'$, $-40°27'$. What are the wavelengths?

32. (III) Suppose the angles measured in problem 31 were produced when the spectrometer (but not the source) was submerged in water. What then would be the wavelength?

SECTION 23-8

33. (I) How far apart are the dark fringes in Example 23-6 if the glass plates are each 10 cm long?

34. (II) A lens appears greenish yellow ($\lambda = 570$ nm is strongest) when white light reflects from it. What

minimum thickness of coating ($n = 1.25$) do you think is used on such a (glass) lens, and why?

35. (II) A total of 33 bright and 33 dark Newton's rings (not counting the dark spot at the center) are observed when 450-nm light falls normally on a planoconvex lens resting on a flat glass surface (Fig. 23-24). How much thicker is the center than the edges?

36. (II) A fine metal foil separates one end of two pieces of optically flat glass as in Fig. 23-25. When light of wavelength 550 nm is incident normally, 37 dark lines are observed (one at each end). How thick is the foil?

37. (II) If a soap bubble is 120 nm thick, what color will appear at the center when illuminated normally by white light? Assume that $n = 1.34$.

38. (II) How thick (minimum) should the air layer be between two flat glass surfaces if the glass is to appear bright when 540-nm light is incident normally? What if the glass is to appear dark?

39. (II) A thin film of alcohol ($n = 1.36$) lies on a flat glass plate ($n = 1.58$). When monochromatic light, whose wavelength can be changed, is incident normally, the reflected light is a minimum for $\lambda = 520$ nm and a maximum for $\lambda = 650$ nm. What is the thickness of the film?

40. (II) When a Newton's ring apparatus (Fig. 23-24) is immersed in a liquid, the diameter of the eighth dark ring decreases from 2.92 cm to 2.48 cm. What is the refractive index of the liquid?

41. (III) Suppose you viewed the light *transmitted* through a thin film on a flat piece of glass. Draw a diagram, similar to Fig. 23-23, and describe the conditions required for maxima and minima; consider all possible values of index of refraction. Discuss the relative size of the minima compared to the maxima and to zero.

42. (III) Show that the radius r of the mth dark Newton's ring, as viewed from directly above (Fig. 23-24), is given by $r = \sqrt{m\lambda R}$ where R is the radius of curvature of the curved glass surface and λ is the wavelength of light used. Assume that the thickness of the air gap is much less than R at all points and that $r \ll R$.

43. (III) A planoconvex Lucite lens 4.0 cm in diameter is placed on a flat piece of glass as in Fig. 23-24. When 520-nm light is incident normally, 68 bright rings are observed, the last one right at the edge. What is the radius of curvature of the lens surface, and what is the focal length of the lens?

*SECTION 23-9

*44. (II) What is the wavelength of the light entering an interferometer if 950 fringes are counted when the movable mirror moves 0.276 mm?

*45. (II) A micrometer is connected to the movable mirror of an interferometer. When the micrometer bears on a thin metal foil, the net number of fringes that move, compared to the empty micrometer, is 218. What is the thickness of the foil? The wavelength of light used is 606 nm.

*46. (II) How far must the mirror M_1 in a Michelson interferometer be moved if 1000 fringes of 510-nm light are to pass by a reference line?

*47. (III) One of the beams of an interferometer passes through a small glass container containing a cavity 1.20 cm wide. When a gas is allowed to slowly fill the container, a total of 188 dark fringes are counted to move past a reference line. The light used has a wavelength of 546 nm. Calculate the index of refraction of the gas, assuming that the interferometer is in vacuum.

*48. (III) The yellow sodium D lines have wavelengths of 589.0 and 589.6 nm. When they are used to illuminate a Michelson interferometer, it is noted that the interference fringes disappear and reappear periodically as the mirror M_1 is moved. Why does this happen? How far must the mirror move between one disappearance and the next?

SECTION 23-10

49. (I) What is Brewster's angle for an air-glass ($n = 1.50$) surface?

50. (I) What is Brewster's angle for a piece of glass ($n = 1.56$) submerged in water?

51. (I) Two polarizers are at 45° to one another. Unpolarized light falls on them. What fraction of the light intensity is transmitted?

52. (II) Two Polaroids are aligned so that the light passing through them is a maximum. At what angle should one of them be placed so that the intensity is reduced by half?

53. (II) At what angle above the horizon is the sun when light reflecting off a smooth lake is polarized most strongly?

54. (II) At what angle should the axes of two Polaroids be placed so as to reduce the intensity of the incident unpolarized light to (a) $\frac{1}{3}$, (b) $\frac{1}{10}$?

55. (II) Unpolarized light passes through five successive Polaroid sheets each of whose axis makes a 45° angle with the previous one. What is the intensity of the transmitted beam?

56. (II) Two polarizers are oriented at 44.0° to one another. Light polarized at a 22.0° angle to each polarizer passes through both. What reduction in intensity takes place?

57. (II) Unpolarized light falls on two polarizer sheets whose axes are at right angles. (a) What fraction of the incident light intensity is transmitted? (b) What fraction is transmitted if a third polarizer is placed between the first two so that its axis makes a 45° angle with each of their axes? (c) What if the third polarizer is in front of the other two?

58. (II) What would Brewster's angle be for reflections off the surface of water for light coming from beneath the surface? Compare to the angle for total internal reflection, and to Brewster's angle from above the surface.

*SECTION 23-11

*59. (II) Calculate the concentration of an insulin solution if the plane of polarization is rotated 18.2° after passing through 11.0 cm of the solution.

*60. (II) A 1.1×10^3 kg/m^3 solution of the sugar fructose 8.0 cm thick produces a rotation in the plane of polarization of incident light by $-117°$. What is the optical rotatory power of fructose?

*SECTION 23-12

*61. (III) A region of a birefringent biological specimen is 1.75×10^{-3} cm thick with indices of refraction 1.322 and 1.331 for light whose wavelength in air is 590 nm. What phase difference arises between the two rays after passing through this region? That is, by what fraction of a wavelength are the two beams out of step when they exit?

24 OPTICAL INSTRUMENTS

In our discussion of the behavior of light in the two previous chapters, we also described a few instruments such as the spectroscope and the Michelson interferometer. In this chapter we will discuss some additional instruments. Most of these use lenses, such as the camera, telescope, microscope, and the human eye. To describe their operation, we will use ray diagrams. However, we will see that some aspects of their operation depend on the wave nature of light.

24-1 ▌ The camera

The basic elements of a camera are a lens, a light-tight box, a shutter to let light pass through the lens only briefly, and a sensitized plate or piece of film (Fig. 24-1). When the shutter is opened, light from external objects in the field of view are focused by the lens as an image on the film. The film contains light-sensitive chemicals which undergo change when light strikes them. In the development process chemical reactions cause the changed areas to turn black so that the image is recorded on the film.[†] You can see an image yourself by removing the camera back and viewing through a piece of tissue or wax paper placed at the position of the film with the shutter open.

There are three main adjustments on good-quality cameras: shutter speed, f-stop, and focusing, and we now discuss them.

Shutter speed. This refers to how long the shutter is open and the film exposed. It may vary from a second or more ("time exposures") to $\frac{1}{1000}$ s or less. To avoid blurring from camera movement, speeds faster than $\frac{1}{50}$ s are normally used; if the object is moving, faster shutter speeds are needed to "stop" the action.

f-stop. The amount of light reaching the film must be carefully controlled to avoid *underexposure* (too little light for any but the brightest objects to show up) or *overexposure* (too much light, so that all bright objects look the same, with a consequent lack of contrast and a "washed-out"

FIGURE 24-1
A simple camera.

Viewfinder

Lens

Iris diaphragm or "stop"

Shutter

Film

[†] This is called a *negative*, since the black areas correspond to bright objects and vice versa. The same process occurs during printing to produce a black-and-white "positive" picture from the negative. Color film makes use of three dyes corresponding to the primary colors.

appearance). To control the exposure, a "stop" or iris diaphragm whose opening is of variable diameter is placed behind the lens (Fig. 24-1). The size of the opening is varied to compensate for bright or dark days, the sensitivity of the film used, and for different shutter speeds. The size of the opening is specified by the *f-stop*, defined as

$$f\text{-stop} = \frac{f}{D},$$

where f is the focal length of the lens and D the diameter of the opening. For example, a 50-mm-focal-length lens set at $f/2$ has an opening $D = 25$ mm; when set at $f/8$, the opening is only $6\frac{1}{4}$ mm. The faster the shutter speed (or the darker the day), the greater the opening must be to get a proper exposure. This corresponds to a smaller f-stop number. The smallest f-number of a lens (largest opening) is referred to as the *speed* of the lens. It is common to find $f/2.0$ lenses today, and even some as fast as $f/1.0$. Fast lenses are expensive to make and require many elements in order to reduce the defects present in simple thin lenses (Section 24-6). The advantage of a fast lens is that it allows pictures to be taken under poorly lighted conditions.

Lenses normally stop down to $f/16$, $f/22$, or $f/32$. Although the lens opening can usually be varied continuously, there are nearly always markings for specific lens openings: the standard f-stop markings are 1.0, 1.4, 2.0, 2.8, 4.0, 5.6, 8, 11, 16, 22, and 32. Notice that each of these corresponds to a diameter of about $\sqrt{2} = 1.4$ times smaller. Since the amount of light reaching the film is proportional to the area of the opening (and therefore proportional to the diameter squared), we see that each standard f-stop corresponds to a factor of 2 in light intensity reaching the film.

Focusing. Focusing is the operation of placing the film at the correct position for the sharpest image. The image distance is a minimum for objects at infinity (the symbol ∞ is used for infinity) and is equal to the focal length. For closer objects the image distance is greater than the focal length, as can be seen from the lens equation, $1/f = 1/d_o + d_i$. To focus on nearby objects, the lens must therefore be moved away from the film, and this is usually done by turning a ring on the lens.

If the lens is focused on a nearby object, a sharp image of it will be formed. But the rays from an object far away will be out of focus—they will form a circle on the film as shown (exaggerated) in Fig. 24-2; the distant object will thus produce an image consisting of overlapping circles and will be blurred. These circles are called *circles of confusion*. If you want to have near and distant objects sharp at the same time, you can try setting the lens focus at an intermediate position. Neither near nor distant objects will then be perfectly sharp, but the circles of confusion may be small enough that the blurriness is not noticeable.

Rays from nearby object

Rays from distant object

"Circle of confusion" for distant object (exaggerated)

FIGURE 24-2

When the lens is positioned to focus on a nearby object, points on a distant object produce circles and are therefore blurred. (The effect is greatly exaggerated).

For a given distance setting, there is a range of distances over which the circles of confusion will be small enough that the images will be reasonably sharp. This is called the *depth of field*. For a particular circle of confusion diameter (typically taken to be 0.03 mm for 35-mm cameras), the depth of field depends on the lens opening. If the lens opening is smaller, the circles of confusion will be smaller, since only rays through the central part of the lens are accepted, and these form a smaller circle of confusion (Fig. 24-2). Hence at smaller lens openings, the depth of field is greater.

Other factors also affect the sharpness of the image, such as the graininess of the film, diffraction, and lens aberrations relating to the quality of the lens itself. (Lens quality and diffraction effects will be discussed in Sections 24-6 and 24-7).

Camera lenses are categorized into normal, telephoto, and wide angle, according to focal length and film size. A *normal* lens is one that covers the film with a field of view that corresponds approximately to that of normal vision. A normal lens for 35-mm film has a focal length in the vicinity of 50 mm. A *telephoto* lens, as its name implies, acts like a telescope to magnify images. They have longer focal lengths than a normal lens: as we saw in Chapter 22 (Eq. 22-8), the height of the image for a given object distance is proportional to the image distance, and the image distance will be greater for a lens with longer focal length. For distant objects, the image height is very nearly proportional to the focal length (can you prove this?). Thus a 200-mm telephoto lens for use with a 35-mm camera gives a $4\times$ magnification over the normal 50-mm lens. A *wide-angle* lens has a shorter focal length than normal. A wider field of view is included and objects appear smaller. A *zoom lens* is one whose focal length can be changed so that you seem to zoom up to, or away from, the subject as you change the focal length.

Two types of viewing systems are common in cameras today. In a *range-finder* camera, you view through a small window just above the lens as in Fig. 24-1. The name derives from the fact that a dual-mirror system is usually used to determine the object distance for focusing. In a *single-lens reflex* camera (SLR), you view directly through the lens with the use of prisms and mirrors (Fig. 24-3). A mirror hangs at a 45° angle behind the lens and flips up out of the way just before you take a picture. SLRs have the great advantage that you can see exactly what you will get on film.

FIGURE 24-3

SLR, showing how light is viewed through the lens with the help of a movable mirror and a prism. (Courtesy of Olympus Camera Corporation.)

□   EXAMPLE 24-1   How far must a 50-mm-focal-length camera lens be moved from its infinity setting in order to sharply focus an object 3.0 m away?

SOLUTION   When focused at infinity, the lens is 50 mm from the film. When focused at $d_o = 3.0$ m, the image distance is given by the lens equation,

$$\frac{1}{d_i} = \frac{1}{f} - \frac{1}{d_o} = \frac{1}{50 \text{ mm}} - \frac{1}{3000 \text{ mm}}.$$

We solve for d_i and find $d_i = 50.8$ mm, so the lens moves 0.8 mm.

□   EXAMPLE 24-2   A light meter reads that a lens setting of $f/8$ is correct for a shutter speed of $\frac{1}{250}$ s under certain conditions. What would be the correct lens opening for a shutter speed of $\frac{1}{500}$ s?

SOLUTION   The amount of light entering the lens is proportional to the area of the lens opening. Since the exposure is reduced by half (from $\frac{1}{250}$ s to $\frac{1}{500}$ s), the area must be doubled. Since the area is proportional to D^2, where D is the diameter of the opening, then D must be increased by $\sqrt{2} = 1.41$. Thus the f stop must be $8.0/1.41 = f/5.6$.

24-2 ∎ The human eye; corrective lenses

The human eye resembles a camera in its basic structure (Fig. 24-4). The eye is an enclosed volume into which light passes through a lens. A diaphragm, called the *iris* (the colored part of your eye), adjusts automatically to control the amount of light entering the eye. The hole in the iris through which light passes (the *pupil*) is black because no light is reflected from it (it's a hole), and very little light is reflected back out from the interior of the eye. The *retina*, which plays the role of the film in a camera, is on the curved rear surface. It consists of a complex array of nerves and receptors known as rods and cones which act as transducers that change light energy into electrical signals that travel along the nerves. The reconstruction of the image from all these tiny receptors is done mainly in the brain, although some analysis is apparently done in the complex interconnected nerve network at the retina itself. At the center of the retina is a small area called the *fovea*, about 0.25 mm in diameter, where the cones are very closely packed and the sharpest image and best color discrimination are found.

Unlike a camera, the eye contains no shutter. The equivalent operation is carried out by the nervous system, which analyzes the signals to form images at the rate of about 30 per second. This can be compared to motion picture or television cameras, which operate by taking a series of still pictures at a rate of 24 (movies) or 30 (television) per second. The rapid projection of these on a screen or television set gives the appearance of motion.

The lens of the eye does little of the bending of the light rays. Most of the refraction is done at the front surface of the *cornea*, which also acts as

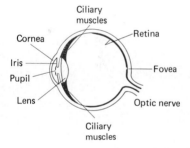

FIGURE 24-4

Diagram of a human eye.

a protective covering. The lens acts as a fine adjustment for focusing at different distances. This is accomplished by the ciliary muscles (Fig. 24-4) which change the curvature of the lens so that its focal length is changed. To focus on a distant object, the muscles are relaxed and the lens is thin, Fig. 24-5a. To focus on a nearby object, the muscles contract causing the center of the lens to be thicker, Fig. 24-5b, thus shortening the focal length. This adjustment is called *accommodation*.

The closest distance at which the eye can focus clearly is called the *near point* of the eye. For young adults it is typically 25 cm, although younger children can often focus on objects as close as 10 cm; as people grow older, the ability to accommodate is reduced and the near point increases. A given person's *far point* is the farthest distance at which an object can be seen clearly. For some purposes it is useful to speak of a *normal eye* (a sort of average over the population), which is defined as one having a near point of 25 cm and a far point of infinity.

The "normal" eye is more of an ideal than a commonplace. A large part of the population have eyes that do not accommodate within the normal range of 25 cm to infinity, or have some other defect. Two common defects are nearsightedness and farsightedness. Both can be corrected to a large extent with lenses—either eyeglasses or contact lenses.

Nearsightedness or *myopia* refers to an eye that can focus only on nearby objects; the far point is not infinity but some shorter distance, so distant objects are not seen clearly. It is usually caused by an eyeball that is too long, although sometimes it is the curvature of the cornea that is too great. In either case, images of distant objects are focused in front of the retina. A diverging lens, because it causes parallel rays to diverge, allows the rays to be focused at the retina (Fig. 24-6a) and thus corrects this defect.

Farsightedness or *hyperopia* refers to an eye which cannot focus on nearby objects. Although distant objects are usually seen clearly, the near point is somewhat greater than the "normal" 25 cm which makes reading difficult. This defect is caused by an eyeball that is too short, or (less often) by a cornea that is not sufficiently curved. It is corrected by a converging lens, Fig. 24-6b. Similar to hyperopia is *presbyopia* which refers to the less-

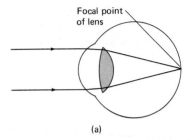

Focal point of lens

(a)

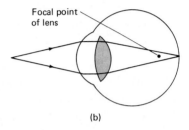

Focal point of lens

(b)

FIGURE 24-5

Accommodation by eye: (a) lens relaxed, focused on infinity; (b) lens thickened, focused on nearby object.

(a) Nearsighted eye

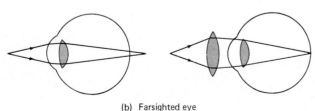

(b) Farsighted eye

FIGURE 24-6

(a) A nearsighted eye, which cannot focus clearly on distant objects, can be corrected by use of a diverging lens. (b) A farsighted eye, which cannot focus clearly on nearby objects, can be corrected by use of a converging lens.

ening ability of the eye to accommodate as one ages, and the near point moves out. Converging lenses also compensate for this.

Astigmatism is usually caused by an out-of-round cornea or lens so that point objects are focused as short lines. It is as if the cornea were spherical with a cylindrical section superimposed. As shown in Fig. 24-7, a cylindrical lens focuses a point into a line parallel to its axis. An astigmatic eye focuses rays in a vertical plane, say, at a shorter distance than it does for rays in a horizontal plane. Astigmatism is corrected with the use of a compensating cylindrical lens. Lenses for eyes that are nearsighted or far-sighted as well as astigmatic are ground with superimposed spherical and cylindrical surfaces, so that the radius of curvature of the correcting lens is different in different planes. Astigmatism is tested for by looking with one eye at a pattern like that in Fig. 24-8. An astigmatic eye focuses rays in a vertical plane, say, at a shorter distance than it does in a horizontal plane. Sharply focused lines appear dark, whereas those that are spread out slightly appear dimmer or gray.

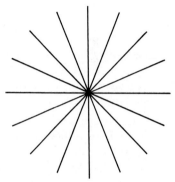

FIGURE 24-7
A cylindrical lens forms a line image of a point object because it is converging in one plane only.

EXAMPLE 24-3   A particular farsighted person has a near point of 100 cm. Reading glasses must have what lens power so that this person can read a newspaper at a distance of 25 cm? Assume the lens is very close to the eye.

SOLUTION   When the object is placed 25 cm from the lens, we want the image to be 100 cm away on the *same* side of the lens, and so it will be virtual, Fig. 24-9. Thus $d_o = 25$ cm, $d_i = -100$ cm and

$$\frac{1}{f} = \frac{1}{25 \text{ cm}} + \frac{1}{-100 \text{ cm}} = \frac{1}{33 \text{ cm}}.$$

So $f = 33$ cm $= 0.33$ m. The power P of the lens is $P = 1/f = +3.0$ D. The plus sign indicates it is a converging lens.

FIGURE 24-8
Test for astigmatism.

EXAMPLE 24-4   A nearsighted eye has near and far points of 12 cm and 17 cm, respectively. What lens power is needed for this person to see distant objects clearly, and what then will be the near point? Assume that each lens in 2.0 cm from the eye.

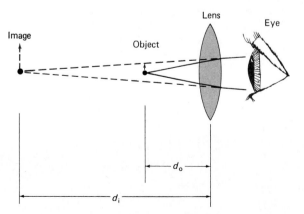

FIGURE 24-9
Lens of a reading glass (Example 24-3).

SOLUTION   The lens must image distant objects ($d_o = \infty$) so they are 17 cm from the eye, or 15 cm in front of the lens ($d_i = -15$ cm):

$$\frac{1}{f} = -\frac{1}{15 \text{ cm}} + \frac{1}{\infty} = -\frac{1}{15 \text{ cm}}.$$

So $f = -15$ cm $= -0.15$ m or $P = 1/f = -6.7$ D; the minus sign indicates it must be a diverging lens. For the near point, the image must be 12 cm from the eye or 10 cm from the lens, so $d_i = -0.10$ m and

$$\frac{1}{d_o} = \frac{1}{f} - \frac{1}{d_i} = -\frac{1}{0.15 \text{ m}} + \frac{1}{0.10 \text{ m}} = \frac{1}{0.30 \text{ m}}.$$

So $d_o = 30$ cm which means the near point when the person is wearing glasses is 30 cm in front of the lens.

Contact lenses could be used to correct the eye in Example 24-4. Since contacts are placed directly on the cornea, we would not subtract out the 2.0 cm. That is, for distant objects $d_i = -17$ cm, so $P = 1/f = -5.9$ diopters. Thus we see that a contact lens and an eyeglass lens will require slightly different focal lengths for the same eye because of their different placement relative to the eye.

24-3 ▌ The magnifying glass

Much of the remainder of this chapter will deal with optical devices that are used to produce magnified images of objects. We first discuss the *simple magnifier* or *magnifying glass* which is simply a converging lens.

How large an object appears, and how much detail we can see on it, depends on the size of the image it makes on the retina. This, in turn, depends on the angle subtended by the object at the eye. For example, a penny held 30 cm from the eye looks twice as high as one held 60 cm away, because the angle it subtends is twice as great (Fig. 24-10). When we want to examine detail on an object, we bring it up close to our eyes so that it subtends a greater angle. However, our eyes can accommodate only up to a point (the near point), and we will assume a standard distance of 25 cm as the near point in what follows.

A magnifying glass allows us to place the object closer to our eye so that it subtends a greater angle. As shown in Fig. 24-11a, the object is placed at the focal point or just within it. Then the converging lens produces a virtual image, which must be at least 25 cm from the eye if the eye is to focus on it. If the eye is relaxed, the image will be at infinity, and in this case the object is exactly at the focal point. (You make this slight adjustment yourself when you "focus" on the object by moving the lens.)

A comparison of part (a) of Fig. 24-11 with part (b), in which the same object is viewed at the near point with the unaided eye, reveals that the angle the object subtends at the eye is much larger when the magnifier is used. The **angular magnification** or **magnifying power**, M, of the lens is defined as the ratio of the angle subtended by an object when using the lens to that

FIGURE 24-10

When the same object is viewed at a shorter distance, the image on the retina is greater; so the object appears larger and more detail can be seen. The angle θ the object subtends in (a) is greater than in (b).

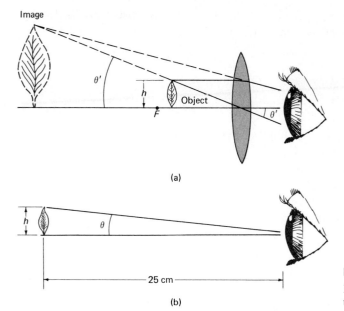

Image

θ'

h

Object

F

θ'

(a)

h

θ

25 cm

(b)

FIGURE 24-11

Leaf viewed (a) through a magnifying glass and (b) with the unaided eye focused at its near point.

subtended using the unaided eye without the lens at a distance of 25 cm:

$$M = \frac{\theta'}{\theta}$$

where θ and θ' are shown in Fig. 24-11. This can be written in terms of the focal length f of the lens as follows. Suppose first that the image in Fig. 24-11a is at the near point N of the eye: $d_i = -N$, where $N = 25$ cm for the normal eye. Then the object distance d_o is given by

$$\frac{1}{d_o} = \frac{1}{f} - \frac{1}{d_i} = \frac{1}{f} + \frac{1}{N}$$

or $d_o = Nf/(f + N)$. (We see from this equation that $d_o < f$, as shown in Fig. 24-11a, since $N/(f + N)$ must be less than 1.) Let h be the height of an object, and we assume h is small so the angles θ and θ' are approximately equal to their sines and tangents; then $\theta' = h/d_o = (f + N)h/(Nf)$ and $\theta = h/N$. Thus

$$M = \frac{\theta'}{\theta} = \frac{(f + N)h}{Nf}\left(\frac{N}{h}\right)$$

or

$$M = 1 + \frac{N}{f}. \qquad \begin{bmatrix} \text{eye focused at near point, } N, \\ \text{which} = 25 \text{ cm for normal eye} \end{bmatrix} \quad (24\text{-}1a)$$

If the eye is relaxed when using the magnifying glass, the image is then at infinity, and the object is then precisely at the focal point. In this case $\theta' = h/f$ so

$$M = \frac{\theta'}{\theta} = \left(\frac{h}{f}\right)\left(\frac{N}{h}\right) = \frac{N}{f}. \qquad \begin{bmatrix} \text{eye focused at } \infty \, ; N = \\ 25 \text{ cm for normal eye} \end{bmatrix} \quad (24\text{-}1b)$$

It is clear that the magnification is slightly greater when the eye is focused at its near point than when relaxed. And the shorter the focal length of the lens, the greater the magnification.

☐ EXAMPLE 24-5 An 8-cm-focal-length converging lens is used as a magnifying glass by a person with normal eyes. Calculate (a) the maximum magnification; (b) the magnification when the eye is relaxed.

SOLUTION (a) The maximum magnification is obtained when the eye is focused at its near point:

$$M = 1 + \frac{N}{f} = 1 + \frac{25}{8} \approx 4 \times .$$

(b) With the eye focused at infinity, $M = 25$ cm/8 cm $\approx 3 \times$.

24-4 ∎ Telescopes

A telescope is used to magnify objects that are very far away; in most cases the object can be considered to be at infinity.

Galileo, although he did not invent it,[†] developed the telescope into a usable and important instrument; he was the first to train it on the heavens where he made a number of world-shaking discoveries (the moons of Jupiter,

---

[†] Galileo built his first telescope in 1609 after having heard of such an instrument existing in Holland. The first telescopes magnified only 3 to 4 times, but Galileo soon made a 30-power instrument. The first Dutch telescope seems to date from about 1604, but there is a reference suggesting it may have been copied from an Italian one from as early as 1590. Kepler (see Chapter 4) gave a ray description (1611) of both the Keplerian and Galilean telescopes (that is, those with two lenses). The former is named for him because he first described it, although he did not build it.

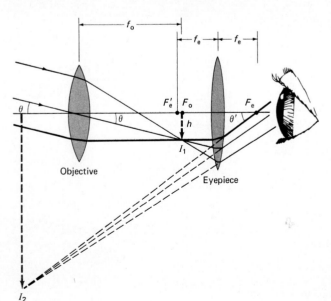

FIGURE 24-12

Astronomical telescope (refracting).

the phases of Venus, sunspots, the structure of the moon's surface, that the Milky Way is made up of a huge number of individual stars, among others).

Several types of astronomical telescope exist. The common *refracting* type, sometimes called "Keplerian," contains two converging lens located at opposite ends of a long tube, Fig. 24-12. The lens closest to the object is called the *objective lens* and forms a real image I_1 of the object at its focal point F_o (or near it if the object is not at infinity). Although this image, I_1, is smaller than the original object, it subtends a greater angle and is very close to the second lens, called the *eyepiece*, which acts as a magnifier. That is, the eyepiece magnifies the image produced by the objective to produce a second, greatly magnified image, I_2, which is virtual. If the viewing eye is relaxed, the eyepiece is adjusted so the image I_2 is at infinity; then the real image I_1 is at the focal point F'_e of the eyepiece, and the distance between the lenses is $f_o + f_e$ for an object at infinity.

To find the total magnification of this telescope, we note that the angle an object subtends as viewed by the unaided eye is just the angle θ subtended at the telescope objective. From Fig. 24-12 we can see that $\theta \approx h/f_o$ where h is the height of the image I_1 and we assume θ is small so $\tan \theta \approx \theta$. Note too that the darkest of the three rays drawn in the figure is parallel to the axis before it strikes the eyepiece and therefore passes through the focal point F_e. Thus $\theta' \approx h/f_e$ and the total magnifying power (angular magnification) of this telescope is

$$M = \frac{\theta'}{\theta} = -\frac{f_o}{f_e}, \qquad (24-2)$$

where we have inserted a minus sign to indicate that the image is inverted. To achieve a large magnification, the objective lens should have a long focal length and the eyepiece a short one.

For an astronomical telescope to produce bright images of distant stars, the objective lens must be large to allow in as much light as possible. The construction and grinding of large lenses is very difficult. Therefore, the largest telescopes are *reflecting* telescopes using a curved mirror as the objective, Fig. 24-13, since a mirror has only one surface to be ground and can be supported along its entire surface. (A large lens, supported at its edges, would sag under its own weight.)[†] Normally, the eyepiece lens or mirror (see Fig. 24-13) is removed so that the real image formed by the objective can be recorded on film.

A *terrestrial telescope* (for use in viewing objects on earth), unlike its astronomical counterpart, must provide an upright image. Two designs are shown in Fig. 24-14. The Galilean type shown in part (a), which Galileo used for his great astronomical discoveries, has a diverging lens as eyepiece which intercepts the converging rays from the objective lens before they reach a focus, and acts to form a virtual upright image. This design is often used in opera glasses. The tube is reasonably short, but the field of view is small. The second type, shown in Fig. 24-14b, is often called a *spyglass* and makes use of a third lens ("field lens") which acts to make the image upright as

---

[†] Another advantage of mirrors is that they exhibit no chromatic aberration since the light doesn't pass through them; and they can be ground in a parabolic shape to correct for spherical aberration (see Section 24-6). The reflecting telescope was first proposed by Newton.

FIGURE 24-13

A concave mirror can be used as the objective of an astronomical telescope. Either a lens or a mirror can be used as the eyepiece. Arrangement (a) is called the Newtonian focus and (b) the Cassegrainian focus. Other arrangements are also possible

Concave mirror (objective)

Eyepiece (lens)

(a)

Eyepiece (mirror)

(b)

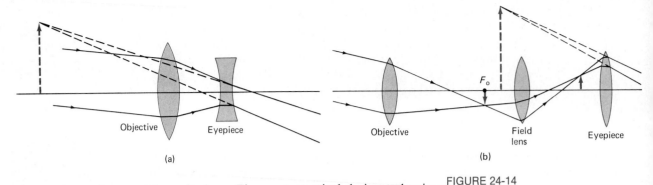

Objective   Eyepiece

(a)

F_o Objective   Field lens   Eyepiece

(b)

FIGURE 24-14

Terrestrial telescopes that produce an upright image: (a) Galilean; (b) spyglass or field-lens type.

shown. A spyglass must be quite long. The most practical design today is the *prism binocular* which was shown in Fig. 22-19. The objective and eyepiece are converging lenses. The prisms reflect the rays by total internal reflection and shorten the physical size of the device, and they also act to produce an upright image; one prism reinverts the image in the vertical plane, the other in the horizontal plane.

EXAMPLE 24-6   A Galilean telescope has an objective lens whose focal length is 28 cm and an eyepiece with focal length -8.0 cm. What is the magnification?

SOLUTION $M = -f_o/f_e = -(28 \text{ cm})/(-8.0 \text{ cm}) = 3.5 \times$.

24-5 ▮ Compound microscope

The compound microscope, like the telescope, has both objective and eyepiece (or ocular) lenses, Fig. 24-15. The design is different from that for a telescope since a microscope is used to view objects that are very close, so the object distance is very small. The object is placed just beyond the objective's focal point as shown in Fig. 24-15a. The image I_1 formed by the objective lens is real, quite far from the lens, and much enlarged. This image is magnified by the eyepiece into a very large virtual image, I_2, seen by the eye.

The overall magnification of a microscope is the product of the magnifications produced by the two lenses. The image I_1 formed by the objective is a factor M_o greater than the object itself; from Fig. 24-15 and Eq. 22-8 for the lateral magnification of a simple lens, we have

$$M_o = \frac{d_i}{d_o} = \frac{l - f_e}{d_o},$$

where l is the distance between the lenses (equal to the length of the barrel), and we ignored the minus sign in Eq. 22-8 which only tells us that the image is inverted. The eyepiece acts like a simple magnifier; if we assume that the eye is relaxed, its angular magnification M_e is (from Eq. 24-1b).

$$M_e = \frac{N}{f_e},$$

where the near point $N = 25$ cm for the normal eye. Since the eyepiece

(a)

(b)

FIGURE 24-15

Compound microscope: (a) ray diagram; (b) actual microscope.

enlarges the imaged formed by the objective, the overall angular magnification M is

$$M = M_e M_o = \left(\frac{N}{f_e}\right)\left(\frac{l - f_e}{d_o}\right) \quad (24\text{-}3a)$$

$$\approx \frac{Nl}{f_e f_o}. \quad (24\text{-}3b)$$

The approximation, Eq. 24-3b, is accurate when f_e and f_o are small compared to l, so $l - f_e \approx l$ and $d_o \approx f_o$ (Fig. 24-15). This is a good approximation for large magnifications, since these are obtained when f_o and f_e are very small (they are in the denominator of Eq. 24-3b). In order to make lenses of very short focal length, which can be done best for the objective, compound lenses involving several elements must be used to avoid serious abberations, as discussed in the next section.

☐ EXAMPLE 24-7 A compound microscope consists of a $10 \times$ eyepiece and a $50 \times$ objective 18.0 cm apart. Determine (a) the overall magnification, (b) the focal length of each lens, and (c) the position of the object when the final image is in focus with the eye relaxed. Assume a normal eye so $N = 25$ cm.

SOLUTION (a) The overall magnification is $10 \times 50 = 500 \times$. (b) The eyepiece focal length is $f_e = N/M_e = 25$ cm$/10 = 2.5$ cm. It is easier to next find d_o (part c) before we find f_o since we can use the equation for M_o above: solving for d_o, we find $d_o = (l - f_e)/M_o = 15.5$ cm$/50 = 0.31$ cm. Then, from the lens equation with $d_i = l - f_e = 15.5$ cm (see Fig. 24-15a),

$$\frac{1}{f_o} = \frac{1}{d_o} + \frac{1}{d_i} = \frac{1}{0.31 \text{ cm}} + \frac{1}{15.5 \text{ cm}} = \frac{51}{15.5 \text{ cm}};$$

so $f_o = 0.30$ cm. (c) We just calculated $d_o = 0.31$ cm, which is very close to f_o.

Opaque objects are generally illuminated by a source placed above them. More commonly, the objects to be viewed are transparent, such as cells or tissue. In this case, light is normally passed through the object from a source beneath the microscope stage (see Fig. 24-15b). The illumination system must be carefully designed if maximum sharpness and contrast are to be achieved. Usually, a *condenser* is employed, which is a set of two or three lenses, although inexpensive condensers may be a single lens or mirror. The purpose of the condenser is to gather light over a wide angle from the source, and to "condense" it down to a narrow beam that will illuminate the object strongly and uniformly. A number of different designs are employed. The source is often placed in the focal plane of the condenser so that light from each point on the source is parallel when it passes through the object.

Some specialized microscope types are described in Section 24-10.

24-6 ▮ Lens aberrations

In Sections 22-8 and 9 we developed a theory of image formation by a thin lens. We found, for example, that all rays from each point on an object are brought to a single point as the image. This, and other results, were based on approximations such as that all rays make small angles with one another and we can use $\sin \theta \approx \theta$. Because of these approximations, we expect deviations from the simple theory and these are referred to as *lens aberrations*. There are several types of aberration; we will briefly discuss each of them separately, but all may be present at one time.

First consider a point object on the axis of a lens. Rays from this point that pass through the outer regions of the lens are brought to a focus at a different point than those that pass through the center of the lens; this is called *spherical aberration*, and is shown exaggerated in Fig. 24-16. Consequently, the image seen on a piece of film (say) will not be a point but a tiny circular patch of light. If the film is placed at the point C as indicated, the circle will have its smallest diameter, which is referred to as the *circle of least confusion*. Spherical aberration is present whenever spherical surfaces are used. It can be corrected by using nonspherical lens surfaces, but to grind such lenses is very expensive. It can be minimized with spherical surfaces by choosing the curvatures so that equal amounts of bending occur at each lens surface; a lens can only be designed like this for one particular object distance. Spherical aberration is usually corrected (by which we mean reduced greatly) by the use of several lenses in combination.

For object points off the lens axis, additional aberrations occur. Rays passing through different parts of the lens cause spreading of the image that is noncircular. We won't go into the details but merely point out that there are two effects: *coma* (because the image is comet-shaped rather than a circle) and *off-axis astigmatism.*[†] Furthermore, the image points for objects off the axis but at the same distance from the lens do not fall on a flat plane

FIGURE 24-16

Spherical aberration (exaggerated). Circle of least confusion obtained at C.

[†] Although the effect is the same as for astigmatism in the eye (Section 24-2) the cause is different. Off-axis astigmatism is no problem in the eye because objects are clearly seen only at the fovea which is on the lens axis.

but on a curved surface—that is, the focal plane is not flat. (We expect this since the points on a flat plane, such as the film in a camera, are not equidistant from the lens.) This aberration is known as *curvature of field* and is obviously a problem in cameras and other devices where the film is placed in a flat plane. In the eye, however, the retina is curved, which compensates for this effect. Another aberration, known as *distortion*, is a result of variation of magnification at different distances from the lens axis. Thus a straight line object not passing through the axis may form a curved image; a square grid of lines may be distorted to produce "pincushion distortion" or "barrel distortion," Fig. 24-17; the latter is common in extreme wide-angle lenses.

All the above aberrations occur for monochromatic light and hence are referred to as *monochromatic aberrations*. If the light is not monochromatic, there will also be *chromatic aberration*. This aberration arises because of dispersion—the variation of index of refraction of transparent materials with wavelength (Section 23-4). For example, blue light is bent more than red light by glass; so if white light is incident on a lens, the different colors are focused at different points, Fig. 24-18, and there will be colored fringes in the image. Chromatic aberration can be eliminated for any two colors (and reduced greatly for all others) by the use of two lenses made of different materials with different indices of refraction and dispersion. Normally one lens is converging and the other diverging and they are often cemented together, Fig. 24-19. This is called an *achromatic doublet* (or "color corrected" lens).

It is not possible to fully correct all aberrations. Combining two or more lenses together can reduce them. High quality lenses used in cameras, microscopes and other devices are *compound lenses* consisting of many simple lenses (referred to as *elements*). A typical high quality camera lens may contain six to eight (or more) elements.

For simplicity we will normally indicate lenses in diagrams as if they were simple lenses; but it must be remembered that good quality lenses are compound.

The human eye is also subject to aberrations. In the course of evolution, however, these have been minimized. Spherical aberration, for example, has been largely corrected since (1) the cornea is less curved at the edges than at the center, and (2) the lens is less dense at the edges than at the center. Both effects cause rays at the outer edges to be bent less strongly, and thus help to reduce spherical aberration. Chromatic aberration is partially compensated for by the fact that the lens absorbs the shorter wavelengths appreciably and the retina is less sensitive to the blue and violet wavelengths. This is just the region of the spectrum where dispersion—and thus chromatic aberration—is greatest (Fig. 23-10).

24-7 ∎ Limits of resolution; the Rayleigh criterion

The ability of a lens to produce distinct images of two point objects very close together is called the **resolution** of the lens. The closer the two images can be and still be seen as distinct (rather than overlapping blobs), the

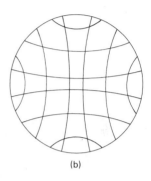

FIGURE 24-17
Distortion. Lenses may image a square grid of perpendicular lines to produce (a) barrel distortion or (b) pincushion distortion.

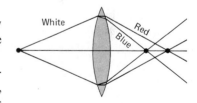

FIGURE 24-18
Chromatic aberration. Different colors are focused at different points.

FIGURE 24-19
Achromatic doublet.

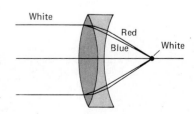

higher the resolution. The resolution of a camera lens, for example, is often specified as so many lines per millimeter.[†] This is determined by photographing a standard set of parallel lines on fine grain film. The minimum spacing of lines distinguishable on film using the lens gives the resolution.

There are two principle factors that limit the resolution of a lens. The first is lens aberrations. As we saw, because of spherical and other aberrations, a point object is not a point on the image but a tiny blob. Careful design of compound lenses can reduce aberrations significantly, but they cannot be eliminated entirely. The second factor that limits resolution is *diffraction*, which cannot be corrected for because it is a natural result of the wave nature of light. We discuss it now.

In Section 23-5 we saw that because light travels as a wave, light from a point source passing through a slit is spread out into a diffraction pattern (Figs. 23-14 and 23-16). A lens, because it has edges, acts like a slit. When a lens forms the image of a point object, the image is actually a tiny diffraction pattern. Thus an image would be blurred even if aberrations were absent.

In the analysis that follows, we assume that the lens is free of aberrations, so that we can focus our attention on diffraction effects and how much they limit the resolution of a lens. In Fig. 23-16 we saw that the diffraction pattern produced by light passing through a rectangular slit has a central maximum in which most of the light falls. This central peak falls to a minimum on either side of its center at an angle $\theta \approx \sin \theta = \lambda/D$ (this is Eq. 23-3) where D is the diameter of the slit, λ is the wavelength of light used, and we assume θ is small. There are also low-intensity fringes beyond. For a lens, or any circular hole, the image of a point object will consist of a *circular* central peak (called the *diffraction spot* or *Airy disk*) surrounded by faint circular fringes, as in Fig. 24-20a. The central maximum has an angular half width given by

$$\theta = \frac{1.22\lambda}{D}.$$

This differs from that for a slit (Eq. 23-3) by the factor 1.22. This factor comes from the fact that the width of a circular hole is not uniform (like a rectangular slit) but varies from its diameter D to zero. A careful analysis shows that the "average" width is $D/1.22$; hence we get the equation above rather than Eq. 23-3. The intensity of light in the diffraction pattern of a circular opening is shown in Fig. 24-21. Any image is thus a complex diffraction pattern. For most purposes we need consider only the central spot since the concentric rings are so much dimmer.

If two point objects are very close, the diffraction patterns of their images will overlap as shown in Fig. 24-20b. As the objects are moved closer, a point is reached where you can't tell if there are two overlapping images or a single image. Where this happens may be judged differently by different observers. However, a generally accepted criterion is one proposed by Lord Rayleigh (1842–1919). This **Rayleigh criterion** states that *two images are just resolvable when the center of the diffraction disk of one is directly over the first minimum in the diffraction pattern of the other*. This is shown

---

[†] This may be specified at the center of the field of view as well as at the edges, where it is usually less because of off-axis aberrations.

(a)

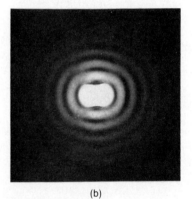

(b)

FIGURE 24-20

Photographs of images (greatly magnified) formed by a lens: (a) a single point object; (b) two point objects barely resolved. (From M. Cagnet, M. Françon, and J. C. Thrierr, *Atlas of Optical Phenomena*, Springer-Verlag, Berlin, 1962.)

FIGURE 24-21

Intensity of light across the diffraction pattern of a circular hole.

in Fig. 24-22. Since the first minimum is at an angle $\theta = 1.22\lambda/D$ from the central maximum, Fig. 24-22 shows us that two objects can be considered just resolvable if they are separated by this angle θ:

$$\theta = \frac{1.22\lambda}{D}. \qquad (24\text{-}4)$$

This is the limit on resolution set by the wave nature of light due to diffraction.

24-8 ∎ Resolution of telescopes and microscopes

It might be thought that a microscope or telescope could be designed to produce any desired magnification depending on the choice of focal lengths. But this is not possible, because of diffraction. An increase in magnification above a certain point merely results in magnification of the diffraction pattern. This would be highly misleading since we might think we are seeing details of an object when we are really seeing details of the diffraction pattern. To examine this we apply the Rayleigh criterion: two objects (or two nearby points on one object) are just resolvable if they are separated by an angle θ (Fig. 24-22) given by Eq. 24-4

$$\theta = \frac{1.22\lambda}{D}.$$

This is valid for either a microscope or a telescope, where D is the diameter of the objective lens. For a telescope, the resolution is specified by stating θ as given by this equation.[†]

For a microscope, it is more convenient to specify the actual distance s between two points that are just barely resolvable, Fig. 24-22. Since objects are normally placed near the focal point of the objective, $\theta = s/f$ or $s = f\theta$. If we combine this with Eq. 24-4, we obtain for the *resolving power* (RP)

$$\text{RP} = s = f\theta = \frac{1.22\lambda f}{D}. \qquad (24\text{-}5)$$

This distance s is called the resolving power of the lens because it is the minimum separation of two object points that can just be resolved. The above equation is often written in terms of the angle of acceptance, α, of the objective lens as defined in Fig. 24-23. The derivation is long and we only quote the result;

$$\text{RP} = s = \frac{1.22\lambda}{2\sin\alpha} = \frac{0.61\lambda}{\sin\alpha}.$$

The resolving power can be increased by placing a drop of oil that encloses the object and the front surface of the objective. This is called an *oil-immersion objective*. In the oil, the wavelength of the light is reduced to λ/n

[†] Telescopes with large-diameter objectives are usually limited not by diffraction but by other effects such as turbulence in the atmosphere. The resolution of a high-quality microscope, on the other hand, is normally limited by diffraction since microscope objectives are complex compound lenses containing many elements of small diameter (since f is small).

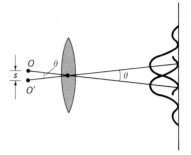

FIGURE 24-22

The *Rayleigh criterion*. Two images are just resolvable when the center of the diffraction peak of either is directly over the first minimum in the diffraction pattern of the other. The two point objects O and O' subtend an angle θ at the lens; one ray only is drawn for each point to indicate the center of the diffraction pattern of its image.

FIGURE 24-23

Objective lens of a microscope, showing the angle of acceptance, α.

where n is the oil's index of refraction. Thus the resolving power becomes

$$RP = \frac{0.61\lambda}{n \sin \alpha}. \tag{24-6}$$

The oil typically has $n \approx 1.5$ although n may be as great as 1.8. Thus oil immersion increases the resolution by 50 percent or more.

The quantity ($n \sin \alpha$) is called the *numerical aperture* (NA) of the lens:

$$NA = n \sin \alpha. \tag{24-7}$$

It is usually specified on the objective lens housing along with the magnification. The larger the value of the NA, the finer the resolving power.

□ **EXAMPLE 24-8** What is the theoretical minimum angular separation of two stars that can just be resolved by the 200-inch telescope on Mt. Palomar? Assume the stars emit white light of average wavelength 550 nm.

SOLUTION Since $D = 200$ in $= 5.1$ m, we have from Eq. 24-4 that $\theta = 1.22\lambda/D = 1.22(5.50 \times 10^{-7}$ m$)/5.1$ m $= 1.3 \times 10^{-7}$ rad or 0.75×10^{-5} deg. This is the limit set by diffraction. In actual fact, the resolution is not this good because of aberrations and, more importantly, turbulence in the atmosphere. In fact, large-diameter objectives are not justified by increased resolution, but by their greater light-gathering ability—they allow more light in, so dimmer objects can be seen.

□ **EXAMPLE 24-9** Determine the NA and RP of the best oil-immersion microscopes where the index of refraction of the oil is $n = 1.8$ and $\sin \alpha \approx 0.90$. Assume that $\lambda = 550$ nm.

SOLUTION The NA $= n \sin \alpha = 1.6$. The resolving power is RP $= 0.61\lambda/$NA $= (0.61)(5.50 \times 10^{-7}$ m$)/(1.6) \approx 2 \times 10^{-7}$ m $= 200$ nm. This is the best resolution that a visible-light microscope can attain.

Diffraction sets an ultimate limit on the detail that can be seen on any object. In Eq. 24-5 we note that the focal length of a lens cannot be made less than (approximately) the radius of the lens, and even that is very difficult—see the lens-maker's equation (Eq. 22-9); in this best case[†], Eq. 24-5 gives, with $f \approx D/2$,

$$RP \approx \frac{\lambda}{2}. \tag{24-8}$$

Thus we can say, to within a factor of 2 or so, that

■ **it is not possible to resolve detail of objects smaller than the wavelength of the radiation being used.**

This is an important and useful rule of thumb.

[†] The same result can be obtained from Eq. 24-6 since $\sin \alpha$ can never exceed 1 and typically is 0.6 to 0.9 at most. With oil immersion, Eq. 24-6 gives, at best, RP $\approx \lambda/3$ which corresponds to the result of Example 24-9.

Compound lenses are now designed so well that the actual limit on resolution is often set by diffraction—that is, by the wavelength of the light used. To obtain greater detail one must use radiation of shorter wavelength. The use of UV radiation can increase the resolution by a factor of perhaps 2. Far more important, however, was the discovery in the early twentieth century that electrons have wave properties (Chapter 26) and that their wavelengths can be very small; this fact is used in the electron microscope (Section 26-6), which can magnify 100 to 1000 times more than a visible-light microscope because of the much shorter wavelengths. X rays too have very short wavelengths and are often used to study objects in great detail using special techniques (Section 24-11).

24-9 ∎ Resolution of the human eye and useful magnification

The resolution of the human eye is limited by several factors, all of roughly the same order of magnitude. The resolution is best at the fovea, where the cone spacing is smallest, about 3 μm ($= 3000$ nm). The diameter of the pupil varies from about 0.1 cm to about 0.8 cm; so for $\lambda = 550$ nm (where the eye's sensitivity is greatest), the diffraction limit is about $\theta \approx 1.22\lambda/D \approx 8 \times 10^{-5}$ rad to 6×10^{-4} rad. Since the eye is about 2 cm long, this corresponds to a resolving power of $s \approx (8 \times 10^{-5}$ rad$)(2 \times 10^{-2}$ m$) \approx 2$ μm at best, to about 15 μm at worst (pupil small). Spherical and chromatic aberration also limit the resolution to about 10 μm. The net result is that the eye can resolve objects whose angular separation is about 5×10^{-4} rad at best. This corresponds to objects separated by 1 cm at a distance of about 20 m.

The typical near point of a human eye is about 25 cm. At this distance, the eye can just resolve objects that are (25 cm)(5×10^{-4} rad) $\approx 10^{-4}$ m apart. Since the best light microscopes can resolve objects no smaller than about 200 nm (see Example 24-9), the useful magnification is limited to about

$$\frac{10^{-4} \text{ m}}{200 \times 10^{-9} \text{ m}} = 500 \times .$$

In practice, magnifications of about 1000× are often used to minimize eyestrain. Any greater magnification would simply make visible the diffraction pattern produced by the microscope objective.

*24-10 ∎ Specialty microscopes and contrast

All the resolving power a microscope can attain will be useless if the object to be seen cannot be distinguished from the background. The difference in brightness between the image of an object and the image of the surroundings is called *contrast*. Achieving high contrast is an important problem in microscopy. The problem arises in biology, for example, because cells consist largely of water and are almost uniformly transparent to light. We now discuss two special types of microscope that can increase contrast: the interference and phase-contrast microscopes.

Interference microscope. The interference microscope makes use of the wave properties of light in a direct way. It is one of the most effective means to increase contrast in a transparent object. To see how it works, let us consider a transparent object—perhaps a bacterium—in a water solution (Fig. 24-24). Light enters uniformly from the left and is coherent (meaning in phase) at all points such as a and b. If the object is as transparent as the water solution, the beam leaving at d will be as bright as that at c. There will be no contrast and the object will not be seen. However, if the object's refractive index is slightly different from that of the surrounding medium, the wavelength within the object will be altered as shown. Hence the waves at points c and d will differ in phase, if not in amplitude. This appears at first to be of no help, since the eye responds only to differences in amplitude or brightness, and does not detect this difference in phase. What the interference microscope does is to change this difference in phase into a difference of amplitude. It does this by superimposing the light that passes through the sample onto a reference beam that does not pass through the object, so that they interfere. One way of doing this is shown in Fig. 24-25. Light from a source is split into two equal beams by a half-silvered mirror, MS_1. One beam passes through the object and the second (comparison beam) passes through an identical system without the object. The two meet again and are superposed by the half-silvered mirror MS_2 before entering the eyepiece and the eye. The path length (and amplitude) of the comparison beam is adjustable. It can be adjusted, for example, so that the background is dark; that is, full destructive interference occurs. Light passing through the object (beam bd in Fig. 24-24) will also interfere with the comparison beam; but because of its different phase, the interference will not be completely destructive. Thus it will appear brighter than the background. Where the object varies in thickness, the phase difference between beams ac and bd in Fig. 24-24 will be different; and this will affect the amount of interference. Hence variation in the thickness of the object will appear as variations in brightness in the image. As an example, suppose that the object is a bacterium 1.0 μm thick and has a refractive index of 1.35; then if yellow light ($\lambda = 550$ nm in air) is used, there will be (1.0 μm)/(550 nm/1.35) = 2.46 wavelengths in the bacterium and (1.0 μm)/(550 nm/1.33) = 2.42 wavelengths in the water. Thus the two waves will be out of phase by 0.04 wavelengths, or $14°$ ($= 0.04 \times 360°$).

Phase-contrast microscope. The phase-contrast microscope also makes use of interference and differences in phase to produce a high-contrast image. Although it has certain limitations, it is far simpler to construct and operate than an interference microscope. To describe the operation of a phase-contrast microscope in detail, we would have to discuss the diffraction theory of image formation—how the diffraction pattern produced by each point on the object contributes to the final image. This is quite complicated, so we give only a simplified description: Figure 24-26 is a simplified diagram of a phase-contrast microscope. The object to be viewed is illumi-

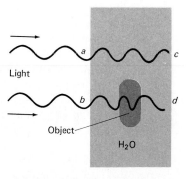

FIGURE 24-24

Object—perhaps a bacterium—in a water solution.

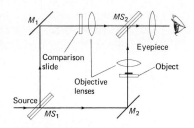

FIGURE 24-25

Diagram of an interference microscope.

nated from below as usual; to be specific we assume that rays from each point on the source are made parallel by a set of condensing lenses (not shown). However, a plate with a ring-shaped hole is placed above the source, so the light can only pass through this annular ring. Light that is not deviated by the object (beam is shown shaded) is brought into focus by the objective lens in the *source image plane;* if the source is effectively at infinity (because of the condensing lenses), as is assumed here, the source image plane is at the focal point of the lens. Light that strikes the object, on the other hand, is diffracted or scattered by the object. Each point on the object then serves as a source for rays diverging from that point (dashed lines in the figure). These rays are brought to a focus in the *object image plane,* which is behind the source image plane (because the object is so close to the lens). The undeviated light from the source diverges, meanwhile, from its image plane and provides a broad bright background at the object image plane. The object is transparent, however, and the image will not be seen clearly since there will be little contrast. Contrast is achieved by inserting a circular glass *phase plate* at the source image plane. The phase plate has a groove as shown (or a raised portion) in the shape of a ring. This ring is positioned so that all the undeviated rays pass through it. Most of the rays deviated by the object, on the other hand, do not pass through this ring (see Fig. 24-26). Because the rays deviated by the object travel through a different thickness of glass than the undeviated source rays, the two can be out of phase and can interfere destructively at the object image plane. Thus the image of the object will contrast sharply with the background. Actually, because only a small fraction of the light is deviated by the object, the background light will be much stronger and so the contrast will not be great. To compensate for this, the grooved ring on the phase plate is darkened to absorb a good part of the undeviated light so that its intensity is more nearly equal to that of the deviated light. Then nearly complete destructive interference can occur at particular points and the contrast will be very high. The chief limitation of the phase-contrast microscope is that images tend to have "halos" around them as a result of diffraction from the phase plate opening. Because of this artifact, care must be taken in the interpretation of images.

24-11 ▌ X rays and X-ray diffraction

In 1895, W. C. Roentgen (1845–1923) discovered that when electrons were accelerated by a high voltage in a vacuum tube and allowed to strike a glass (or metal) surface inside the tube, fluorescent minerals some distance away would glow, and film would become exposed. Roentgen attributed these effects to a new type of radiation (different from cathode rays). They were given the name X rays after the algebraic symbol *x* meaning an unknown quantity. He soon found that X rays penetrated through some materials better than through others, and within a few weeks he presented the first X-ray photograph (of his wife's hand). The production of X rays today is done in a tube (Fig. 24-27) similar to Roentgen's, using voltages of typically 30 to 150 kV.

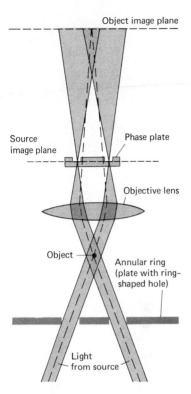

FIGURE 24-26

Phase-contrast microscope. Light beam from source that is undeviated is shown shaded for clarity. Rays deviated by object, and which form the image of the object, are shown dashed.

FIGURE 24-27

X-ray tube. Electrons emitted by a heated filament in a vacuum tube are accelerated by high voltage. When they strike the surface of the anode, the "target", X rays are emitted.

Investigations into the nature of X rays indicated they were not charged particles (such as electrons) since they could not be deflected by electric or magnetic fields. It was suggested that they might be a form of invisible light. However, they showed no diffraction or interference effects using ordinary gratings. Of course, if their wavelengths were much smaller than the typical grating spacing of 10^{-6} m $(= 10^3$ nm) no effects would be expected. Around 1912, it was suggested by Max von Laue (1879–1960) that if the atoms in a crystal were arranged in a regular array (see Fig. 10-1a), a theory generally held by scientists though not then fully tested, such a crystal might serve as a diffraction grating for very short wavelengths on the order of the spacing between atoms, estimated to be about 10^{-10} m $(= 10^{-1}$ nm). Experiments soon showed that X rays scattered from a crystal did indeed show the peaks and valleys of a diffraction pattern (Fig. 24-28). Thus it was shown, in a single blow, that X rays have a wave nature and that atoms are arranged in a regular way in crystals. Today X rays are recognized as electromagnetic radiation with wavelengths in the range of about 10^{-2} nm to 10 nm, the range readily produced in an X-ray tube.

We saw in Sections 24-7 and 24-8 that light of shorter wavelength provides greater resolution when we are examining an object microscopically. Since X rays have much shorter wavelengths than visible light, they should in principle offer much greater resolution. However, there seems to be no effective material to use as lenses for the very short wavelengths of X rays. Instead the clever but complicated technique of *X-ray diffraction* (or *crystallography*) has proved very effective for examining the microscopic world of atoms and molecules. In a simple crystal such as NaCl, the atoms are arranged in an orderly cubical fashion, Fig. 24-29, with atoms spaced a distance d apart. Suppose that a beam of X rays is incident on the crystal at an angle θ to the surface, and that the two rays shown are reflected from two subsequent planes of atoms as shown. The two rays will constructively interfere if the extra distance ray I travels is a whole number of wavelengths farther than what ray II travels. This extra distance is $2d \sin \theta$. Therefore, constructive interference will occur when

$$m\lambda = 2d \sin \theta, \qquad m = 1, 2, 3, \ldots \qquad (24\text{-}9)$$

where m can be any integer. (Notice that θ is *not* the angle with respect to the normal to the surface.) This is called the *Bragg equation* after W. L. Bragg (1890–1971), who derived it and who, together with his father W. H. Bragg (1862–1942), developed the theory and technique of X-ray diffraction by crystals in 1912–13. Thus, if the X-ray wavelength is known and the angle θ at which constructive interference occurs is measured, d can be obtained. This is the basis for X-ray crystallography.

Actual X-ray diffraction patterns are quite complicated. First of all, a crystal is a three-dimensional object, and X rays can be diffracted from different planes at different angles within the crystal, as shown in Fig. 24-30. Although the analysis is complex, a great deal can be learned about any substance that can be put in crystalline form. If the substance is not a single crystal but a mixture of many tiny crystals—as in a metal or a powder—then instead of a series of spots, as in Fig. 24-28, a series of circles is obtained, Fig. 24-31; each circle corresponds to diffraction of a certain order m (Eq. 24-9), from a particular set of parallel planes.

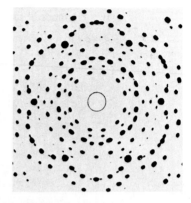

FIGURE 24-28

Diffraction pattern produced by X rays scattered from a beryl crystal. (Courtesy of Eastman Kodak.)

FIGURE 24-29

X-ray diffraction by a crystal.

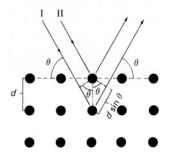

FIGURE 24-30

There are many possible planes existing within a crystal from which X rays can be diffracted.

(a)

(b)

FIGURE 24-31

(a) Diffraction of X rays from a poly-crystalline substance produces a set of circular rings as in (b) which is for polycrystalline sodium acetoacetate. (Part b Lester V. Bergman & Assoc., Inc.)

X-ray diffraction has been very useful in determining the structure of biologically important molecules. Often it is possible to make a crystal of such molecules. The analysis is complex, and it is usually necessary to make various guesses of the structure of the molecule. Predictions of the diffraction patterns for each guessed structure can then be compared to that actually obtained. For larger molecules, such as proteins and nucleic acids, an important innovation has been the "heavy atom technique." Since very large atoms scatter X rays more strongly than the ordinary C, N, O, and H atoms of biological molecules, heavy atoms can be used as "markers." The heavy atoms are chemically added to particular spots on the molecule (say, a protein)—hopefully without disturbing its structure significantly. Analysis of the *changes* in the resulting diffraction pattern gives helpful information.

Even when a good crystal cannot be obtained, if the molecule under study has a regularly repeating shape (such as many proteins and DNA have), X-ray diffraction can reveal it. In a sense, each molecule is then like a single crystal and a sample is a collection of such tiny crystals. Indeed, it was with the help of X-ray diffraction that the α structure of proteins (Section 15-10) was proposed. And in 1953 J. D. Watson and F. H. C. Crick worked out the double-helix structure of DNA with the aid of X-ray diffraction photographs.

Around 1960, the first detailed structure of a protein molecule was elucidated with the aid of X-ray diffraction; this was for myoglobin, a relative of the important constituent of blood, hemoglobin. Soon the structure of hemoglobin itself was worked out and since then the structures of a great many molecules have been determined with the help of X rays.

*24-12 ∎ X-ray imaging and computerized tomography (CAT scanning)

For a conventional medical (or dental) X-ray photograph, the X rays emerging from the tube (Fig. 24-27, Section 24-11) pass through the body and are detected on photographic film or a fluorescent screen, Fig. 24-32a. The rays travel in very nearly straight lines through the body with minimal deviation since at X-ray wavelengths there is little diffraction or refraction. There is absorption (and scattering), however; and the difference in absorption by different structures in the body is what gives rise to the image produced by the transmitted rays. The less the absorption, the greater the transmission and the darker the film. The image is, in a sense, a "shadow"

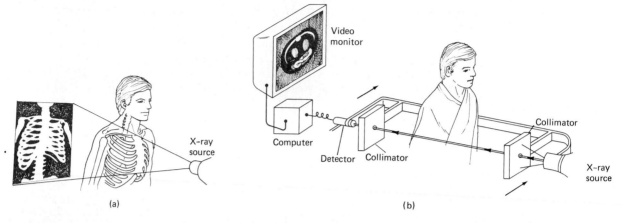

(a) (b)

of what the rays have passed through. (The X-ray image is *not* produced by focusing rays with lenses as is the case for the instruments discussed earlier in this chapter.)

Within months of Roentgen's 1895 discovery of X rays, they had already become a powerful tool for medical diagnosis, and have remained so to this day. Although many technical advances have been made over the years, the basic principles have not changed significantly. At least not until the 1970s when a resolutionary new technique called *computerized tomography* (CT) was developed.

In conventional X-ray images, the entire thickness of the body is projected onto the film; structures overlap and in many cases are difficult to distinguish. A tomographic image, on the other hand, is an image of a *slice* through the body. (The word *tomography* comes from the Greek: *tomos* = slice, *graph* = picture.) Structures and lesions previously impossible to visualize can now be seen with remarkable clarity. The principle behind CT is shown in Fig. 24-32b: a thin collimated beam of X rays (to "collimate" means to "make straight") passes through the body to a detector which measures the transmitted intensity; measurements are made at a large number of points as the source and detector are moved past the body together. The apparatus is then rotated slightly about the body axis and again scanned; this is repeated at (say) 1° intervals for 180°. The intensity of the transmitted beam for the many points of each scan, and for each angle, are sent to a computer which reconstructs the image of the slice (more on this in a moment). Note that the imaged slice is perpendicular to the long axis of the body. For this reason, CT is sometimes called *computerized axial tomography* (CAT), although the abbreviation CAT, as in CAT scan, can also stand for *computer-assisted tomography*.

The use of a single detector as in Fig. 24-32b requires a few minutes for the many scans needed to form a complete image. Much faster scanners use a fan beam, Fig. 24-33, in which beams through the entire body are detected simultaneously by many detectors; the source and detectors are then rotated about the patient on an apparatus called a gantry. At each of the hundreds of angular positions of the apparatus, several hundred detectors can measure the intensity of transmitted rays simultaneously, so an image requires only a few seconds.

FIGURE 24-32

(a) Conventional X-ray imaging, which is essentially shadowing. (b) Tomographic imaging: the X-ray source and detector move together horizontally, the transmitted intensity being measured at a large number of points; then the source-detector assembly is rotated slightly (say 1°) and another scan is made. This is repeated for perhaps 180°. The computer reconstructs the image of the slice and it is presented on a TV monitor (cathode ray tube).

FIGURE 24-33

Fan-beam scanner. Rays through the entire body are measured simultaneously at each angle. The source and detector rotate to take measurements at different angles. In another type of fan-beam scanner, there are detectors around the entire 360° of circle and they remain fixed as the source moves.

But how is the image formed? We can think of the slice to be imaged as being divided into many tiny picture elements (or *pixels*), which could be squares, as was done to make the image shown in Fig. 24-34. For CT, the width of each pixel is chosen according to the width of the detectors and/or the width of the X-ray beams; this sets the resolution of the image, which is typically about 2 mm. An X-ray detector measures the intensity of the transmitted beam after it has passed through the body; subtracting this value from the intensity of the beam at the source, we get the total absorption. Note that only the total absorption (called "a projection") along each beam line can be measured (the sum of the absorptions that take place for each of the pixels in a line). To form an image, we need to determine how much radiation is absorbed at each pixel; we can then assign a "grayness value" to each pixel according to how much radiation was absorbed. The image, then, is made up of tiny spots (pixels) of varying shades of gray, as is a black and white television picture. Often the amount of absorption is color coded; the colors in the resulting image have nothing to do, however, with the actual color of the object.

Finally, we must discuss how this image is formed even though all we can measure is the total absorption along each beam line in the slice. It can be done only by using the many beam scans made at a great many different angles. Suppose the image is to be an array of 100×100 elements for a total of 10^4 pixels. If we have 100 detectors and measure the absorption projections at 100 different angles, then we get 10^4 pieces of information; from this information an image can be reconstructed, but not precisely. If more angles are measured, the reconstruction of the image can be done more accurately.

There are a number of mathematical reconstruction techniques, all of which are complicated and require the use of a computer. To suggest how it is done, we consider a very simple case using the so-called "iterative" technique ("to iterate" is from the Latin "to repeat"). Although this technique is less used now than the more direct "Fourier transform" and "back projection" techniques, it is the simplest to explain. Suppose our sample slice is divided into the simple 2×2 pixels as shown in Fig. 24-35. The number in each pixel represents the amount of absorption by the material in that area (say, in tenths of a percent): that is, 4 represents twice as much absorption as 2. But we cannot directly measure these values. All we can measure are the projections—the total absorption along each beam line—and these are shown in the diagram as the sum of the absorptions for the pixels along each line at four different angles. We are given these projections (at the tip of each arrow), and we want now to work back from them to see how close we can get to the true absorption value for each pixel. We start with each pixel being assigned a zero value, Fig. 24-36a. In the iterative technique we use the projections to estimate the absorption value in each square, and repeat for each angle. The angle 1 projections are 7 and 13; we divide each of these equally between their two squares: each square in the left column gets $3\frac{1}{2}$ (half of 7), and each square in the right column gets $6\frac{1}{2}$ (half of 13); see Fig. 24-36b. Next we use the projections at angle 2; we calculate the difference between the measured projections at angle 2 (6 and 14) and the projections based on the previous estimate (top row: $3\frac{1}{2} + 6\frac{1}{2} = 10$; same

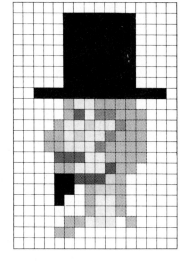

FIGURE 24-34

Example of an image made up of many tiny squares. This one has rather poor resolution.

FIGURE 24-35

Simple 2×2 image showing true absorption values and measured projections.

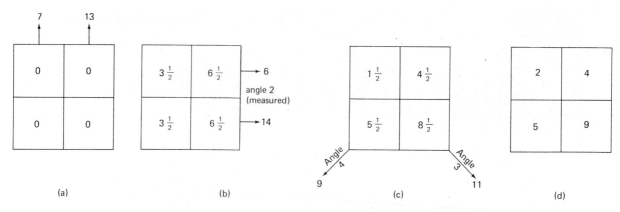

FIGURE 24-36

Reconstructing the image using projections in an iterative procedure.

for bottom row); and we distribute this difference equally to the squares in that row. For the top row we have

$$3\tfrac{1}{2} + \frac{6-10}{2} = 1\tfrac{1}{2} \quad \text{and} \quad 6\tfrac{1}{2} + \frac{6-10}{2} = 4\tfrac{1}{2};$$

and for the bottom row

$$3\tfrac{1}{2} + \frac{14-10}{2} = 5\tfrac{1}{2} \quad \text{and} \quad 6\tfrac{1}{2} + \frac{14-10}{2} = 8\tfrac{1}{2}.$$

These values are inserted as shown in Fig. 24-36c. Next, the projection at angle 3 gives

$$\text{(upper left) } 1\tfrac{1}{2} + \frac{11-10}{2} = 2 \quad \text{and} \quad \text{(lower right) } 8\tfrac{1}{2} + \frac{11-10}{2} = 9$$

and that for angle 4 gives

$$\text{(lower left) } 5\tfrac{1}{2} + \frac{9-10}{2} = 5 \quad \text{and} \quad \text{(upper right) } 4\tfrac{1}{2} + \frac{9-10}{2} = 4.$$

The result, shown in Fig. 24-36d, corresponds exactly to the true values. (Note that in real situations the true values are not known, which is why these computer techniques are required.) To obtain these numbers precisely, we used six pieces of information (two each at angles 1 and 2, one each at angles 3 and 4). For the much larger number of pixels used for actual images, exact values are generally not attained. The calculation is considered sufficiently precise when the difference between calculated and measured projections is sufficiently small. The above example illustrates the "convergence" of the process: the first iteration (b to c in Fig. 24-36) changed the values by 2, the last iteration (c to d) by only $\tfrac{1}{2}$.

Figure 24-37 illustrates what actual CT images look like. It is generally agreed that CT scanning has revolutionized some areas of medicine by providing much less invasive, and/or more accurate, diagnosis.

Computerized tomography can also be applied to ultrasound imaging (Section 14-10) and to emissions from radioisotopes and nuclear magnetic resonance (Sections 30-8 and 30-9).

FIGURE 24-37
Two CT images, with different resolutions, each showing a cross section of a brain. The bright area is an unusual growth (a tumor). (Photograph by Martin M. Rotker, Taurus Photos, Science Photo Library International, New York.)

▌ Summary

A *camera* lens forms an image on film by allowing light in through a shutter. The lens is focused by moving it relative to the film, and its *f*-stop (or lens opening) must be adjusted for the brightness of the scene and the chosen shutter speed. The *f*-stop is defined as the ratio of the focal length to the diameter of the lens opening.

The human *eye* also adjusts for the available light—by opening and closing the iris. It focuses not by moving the lens but by adjusting the shape of the lens to vary its focal length. The image is formed on the retina, which contains an array of receptors known as rods and cones. Diverging eyeglass or contact lenses are used to correct the defect of a nearsighted eye, which cannot focus well on distant objects; converging lenses are used to correct for defects in which the eye cannot focus on close objects.

A *simple magnifier* is a converging lens that forms a virtual image of an object placed at (or within) the focal point. The *angular magnification*, when viewed by a relaxed normal eye, is $M = N/f$, where f is the focal length of the lens and N is the near point of the eye (25 cm for a "normal" eye).

An astronomical *telescope* consists of an objective lens or mirror and an eyepiece that magnifies the real image formed by the objective. The magnification is equal to the ratio of the objective and eyepiece focal lengths, and the image is inverted. A terrestrial telescope uses extra lenses, prisms, or a diverging lens as eyepiece, so that the final image is right side up.

A compound *microscope* also uses objective and eyepiece lenses, and the final image is also inverted. The total magnification is the product of the magnifications of the two lenses and is approximately $(N/f_e)(l/f_o)$ where l is the distance between the lenses, N is the nearpoint of the eye, and f_o and f_e are the focal lengths of objective and eyepiece.

Microscopes, telescopes, and other optical instruments are limited in the formation of sharp images by *lens aberrations*. These include spherical aberration, in which rays passing through the edge of a lens are not focused at the same point as those that pass near the center; and chromatic aberration, in which different colors are focused at different points. Compound lenses, consisting of several elements, can largely correct for aberrations.

The wave nature of light also limits the sharpness or *resolution* of images. Because of diffraction, it is not possible to discern details smaller than the wavelength of the radiation being used. This limits the useful magnification of a light microscope to about $1000\times$.

X rays are a form of electromagnetic radiation of very short wavelength. They are produced when high speed electrons, accelerated by high voltage in an evacuated tube, strike a glass or metal target.

Questions

1. Why is the depth of field greater, and the image sharper, when a camera lens is "stopped down" to a larger f number?

2. Why must a camera lens be moved farther from the film to focus on a closer object?

3. A "pinhole" camera uses a tiny pinhole instead of a lens. Show, using ray diagrams, how reasonably sharp images can be formed using such a pinhole camera. In particular, consider two point objects 2.0 cm apart that are 1.0 m from a 1-mm-diameter pinhole. Show that on a piece of film 5 cm behind the pinhole each object produces a tiny, easily resolvable spot.

4. Why are bifocals needed mainly by older persons and not generally by younger people?

5. Explain why swimmers with good eyes see distant objects as blurry when they are under water. Use a diagram and also show why goggles correct this problem.

6. Will a nearsighted person who wears corrective lenses be able to see clearly underwater when wearing glasses? Use a diagram to show why or why not.

7. The human eye is much like a camera—yet when a camera shutter is left open and the camera moved, the image will be blurred; but when you move your head with your eyes open, you still see clearly. Explain.

8. Reading glasses use converging lenses. A simple magnifier is also a converging lens. Are reading glasses therefore magnifiers? Discuss the similarities and differences between converging lenses as used for these two different purposes.

9. Is the image formed on the retina of the human eye upright or inverted? Discuss the implications of this for our perception of objects.

10. Complete the ray diagram of Fig. 24-14a by showing the intermediate image and the focal points.

11. Inexpensive microscopes sold for children's use usually produce images that are colored at the edges. Why?

12. Spherical aberration in a thin lens is minimized if rays are bent equally by the two surfaces. If a plano-convex lens is used to form a real image of an object at infinity, which surface should face the object? Use ray diagrams to show why.

13. Which aberrations present in a simple lens are not present (or are greatly reduced) in the human eye?

14. Atoms have diameters of about 10^{-8} cm. Can visible light be used to "see" an atom? Why or why not?

15. Does diffraction limit the resolution of images formed by (a) spherical mirrors, (b) plane mirrors?

16. Do diffraction effects occur for virtual as well as real images?

17. What are the advantages (give at least two) for the use of large reflecting mirrors in astronomical telescopes.

18. Which color of visible light would give the best resolution in a microscope?

Problems

SECTION 24-1

1. (I) A 135-mm-focal-length lens has f stops ranging from $f/3.5$ to $f/32$. What is the corresponding range of lens diameters?

2. (I) A television camera lens has a 14-cm focal length and a lens diameter of 6.0 cm. What is its f number?

3. (I) A properly exposed photograph is taken at $f/11$ and $\frac{1}{125}$ s. What lens opening would be required if the shutter speed were $\frac{1}{1000}$ s?

4. (I) A light meter reports that a camera setting of $\frac{1}{100}$ s at $f/11$ will give a correct exposure. But the photographer wishes to use $f/16$ to increase the depth of field. What should the shutter speed be?

5. (II) Suppose that a correct exposure is $\frac{1}{250}$ s at $f/11$. Under the same conditions, what exposure time would be needed for a pinhole camera if the pinhole diameter is 1.0 mm and the film is 5.0 cm from the hole?

6. (II) A 135-mm-focal-length lens can be adjusted so that it is 135 to 140 mm from the film. For what range of object distances can it be adjusted?

7. (II) A nature photographer wishes to photograph a 4.5-m-high elephant from a distance of 100 m. What focal-length lens should be used if the image is to fill the 24-mm height of the film?

8. (II) Show that for objects very far away (assume infinity), the magnification of a camera lens is proportional to its focal length.

SECTION 24-2

9. (I) A human eyeball is about 2.0 cm long and the pupil has a maximum diameter of about 5.0 mm. What is the "speed" of this lens?

10. (II) A person's left eye is corrected by a -5.5-diopter lens, 2.0 cm from the eye. (a) Is this person near- or far-sighted? (b) What is this person's far point without glasses?

11. (II) Reading glasses of what power are needed for a person whose near point is 130 cm so that he can read at 25 cm? Assume a lens-eye distance of 2.0 cm.

12. (II) A man's left eye can see objects clearly only if they are between 18 and 35 cm away. (a) What power

of contact lens is required so that objects far away are sharp? (b) What, then, will be his near point?

13. (II) About how much longer is the near-sighted eye in Example 24-4 than 2.0 cm of a normal eye?

14. (II) A woman can see clearly with her right eye only when objects are between 36 and 250 cm away. Prescription bifocals should have what powers so that she can see distant objects clearly (upper part) and be able to read a book 25 cm away (lower part)? Assume that the glasses will be 2.0 cm from the eye.

15. (II) One lens of a nearsighted person's eyeglasses has a focal length of −20.0 cm and the lens is 1.6 cm from the eye. If the person switches to contact lenses that are placed directly on the eye, what should be the focal length of the corresponding contact lens?

16. (II) What is the focal length of the eye lens system when viewing an object (a) at infinity? (b) 20 cm from the eye? Assume that the lens-retina distance is 2.0 cm.

17. (III) A 50-year-old man uses +3.0 diopter lenses to be able to read a newspaper 25 cm away. Ten years later, he finds that he must hold the paper 45 cm away to see clearly with the same lenses. What power lenses does he need now? (Distances are measured from the lens.)

SECTION 24-3

18. (I) What is the focal length of a magnifying glass of 2.5 × magnification for a relaxed normal eye?

19. (I) A magnifier is rated at 3.0× for a normal eye focusing on an image at its near point. (a) What is its focal length? (b) What will it be if the 3.0 × refers to a relaxed eye?

20. (II) A 6.0-cm-focal-length lens is used as a simple magnifier. To obtain maximum magnification where must the object be placed for a normal eye?

21. (II) A child has a near point of 10 cm. What is the maximum magnification the child can obtain using an 8.5-cm-focal-length magnifier? Compare to that for a normal eye.

22. (II) A 2.60-mm-wide stone is viewed with an 8.50-cm-focal-length lens. A normal eye views the image at its near point. Calculate (a) the angular magnification (b) the width of the image, (c) the object distance from the lens.

23. (II) A small insect is placed 3.80 cm from a +4.00-cm-focal-length lens. Calculate (a) the position of the image, (b) the angular magnification.

24. (II) What is the magnifying power of a +22-diopter lens? Assume a relaxed normal eye.

25. (II) A magnifying glass is rated at 3.5× for a normal eye that is relaxed. What would be the magnification for a relaxed eye whose near point is (a) 40 cm, (b) 15 cm?

SECTION 24-4

26. (I) A 10× binocular has 3.0-cm-focal-length eyepieces. What is the focal length of the objective lenses?

27. (I) What is the magnification of an astronomical telescope whose objective lens has a focal length of 50 cm and whose eyepiece has a focal length of 3.1 cm? What is the overall length of the telescope when adjusted for a relaxed eye?

28. (II) An astronomical telescope has an objective with focal length 88 cm and a +48-D eyepiece. What is the total magnification?

29. (II) What is the magnifying power of an astronomical telescope using a reflecting mirror whose radius of curvature is 4.50 m and an eyepiece whose focal length is 2.6 cm?

30. (II) A lawyer lost in the mountains tries to make a telescope using the lenses from his reading glasses. They have powers of +1.5 and +6.0 D, respectively. (a) What maximum magnification telescope is possible? (b) Which lens should be used as the eyepiece?

31. (II) A Galilean telescope adjusted for a relaxed eye is 33 cm long. If the objective lens has a focal length of 36 cm, what is the magnification?

32. (III) A 50× astronomical telescope is adjusted for a relaxed eye when the two lenses are 84 cm apart. What is the focal length of each lens?

33. (III) A 6.0× pair of binoculars has an objective focal length of 24 cm. If the binoculars are focused on an object 4.0 m away (from the objective), what is the magnification? (The 6.0 × refers to objects at infinity; Eq. 24-2 holds only for objects at infinity and not for nearby ones.)

SECTION 24-5

34. (I) A 900× microscope uses a 0.40-cm-focal-length objective lens. If the tube length is 16.8 cm, what is the focal length of the eyepiece? Assume a normal eye and that the final image is at infinity.

35. (II) The eyepiece of a compound microscope has a focal length of 2.60 cm and the objective has $f = 0.800$ cm. If an object is placed 0.850 cm from the objective lens, calculate (a) the distance between the lenses when the microscope is adjusted for a relaxed eye, and (b) the total magnification.

36. (II) A microscope has a 15× eyepiece and a 40× objective 17.5 cm apart. Calculate (a) the total magnification, (b) the focal length of each lens, and (c) where the object must be for a normal relaxed eye to see it in focus.

37. (II) A microscope has a 2.0-cm-focal-length eyepiece and 1.0-cm objective. Calculate (a) the position of the object if the distance between the lenses is 18.0 cm, and (b) the total magnification assuming a relaxed normal eye.

38. (II) Repeat problem 37 assuming that the final image is located 25 cm from the eyepiece (near point of a normal eye).

SECTION 24-6

39. (II) An achromatic lens is made of two very thin lenses placed in contact which have focal lengths $f_1 = -24$ cm and $f_2 = +15$ cm. (a) Is the combination converging or diverging? (b) What is the net focal length?

40. (I) What is the angle of acceptance α of a microscope oil-immersion objective and its resolving power if $n = 1.60$ and the NA $= 1.35$? Use $\lambda = 500$ nm.

41. (I) What is the angular resolution limit set by diffraction for the 100-in Mt. Wilson telescope ($\lambda = 500$ nm)?

42. (II) A microscope objective is immersed in oil ($n = 1.60$) and accepts light scattered from the object up to $60°$ on either side of vertical. (*a*) What is the numerical aperture? (*b*) What is the approximate resolution of the microscope for 550-nm light?

43. (II) Two stars 10 light-years away are barely resolved by a 100-in (mirror diameter) telescope. How far apart are the stars? Assume $\lambda = 500$ nm.

44. (II) A certain sea organism has a pattern of dots on its surface with an average spacing of 0.40 μm. If the specimen is viewed using 550-nm light, what minimum value must the numerical aperture be in order that dots be resolved?

45. (III) What minimum magnification would be required to see the dots on the organism of problem 44?

46. (III) Suppose that you wish to construct a telescope that can resolve features 10 km across on the moon, 384,000 km away. You have a 2.0-m-focal-length objective lens whose diameter is 10 cm. What focal-length eyepiece is needed if your eye can resolve objects 0.10 mm apart at a distance of 25 cm? What is the resolution limit set by the size of the objective lens (that is, by diffraction)? Use $\lambda = 500$ nm.

47. (III) (*a*) Can a human eye distinguish the two headlights, 1.8 m apart, on a truck 10 km away? Consider only diffraction and assume an eye diameter of 5.0 mm and a wavelength of 500 nm. (*b*) What is the maximum distance at which they can be resolved?

*SECTION 24-10

*48. (II) Show that the phase difference (in radians) between the two waves ac and bd in Fig. 24-24 is $\delta = (2\pi/\lambda)(n_2 - n_1)t$, where n_2 and n_1 are the refractive

7	2	3
4	5	8
1	6	3

FIGURE 24-38

indices of the object and the medium, t the thickness of the object, and λ the wavelength of light used.

SECTION 24-11

49. (II) X rays of wavelength 0.128 nm fall on a crystal whose atoms, lying in planes, are spaced 0.300 nm apart. At what angle must the X rays be directed if the first diffraction maximum is to be observed?

50. (II) First-order Bragg diffraction is observed at $16.2°$ from a crystal with spacing between atoms of 0.24 nm. (*a*) At what angle will second order be observed? (*b*) What is the wavelength of the X rays?

51. (III) If X-ray diffraction peaks corresponding to the first three orders ($m = 1$, 2, and 3) are measured, can both the X-ray wavelength λ and lattice spacing d be determined? Prove your answer.

*SECTION 24-12

*52. (II) (*a*) Suppose in a conventional X-ray image that the X-ray beam consisted of parallel rays; what would be the magnification of the image. (*b*) Suppose, instead, the X rays came from a point source (as in Fig. 24-32a) that is 15 cm in front of a human body 25 cm thick, and film is pressed against the person's back. Determine and discuss the range of magnifications that result.

*53. (III) Suppose a simple 3×3 image has the absorption values shown in Fig. 24-38. (*a*) If the projections are measured along vertical, horizontal, and diagonal directions (as was done in Fig. 24-35), what are their values? (*b*) What values for each square do you get if you reconstruct the image as was done in the text?

25 RELATIVITY

Physics at the end of the nineteenth century looked back on a period of great progress. The theories developed over the preceding three centuries had been very successful in explaining a wide range of natural phenomena. Newtonian mechanics beautifully explained the motion of objects on earth and in the heavens; furthermore, it formed the basis for successful treatments of fluids, wave motion, and sound. Kinetic theory, on the other hand, explained the behavior of gases and other materials. And Maxwell's theory of electromagnetism not only brought together and explained electric and magnetic phenomena, but it predicted the existence of EM waves that would behave in every way just like light—so light came to be thought of as an electromagnetic wave. Indeed, it seemed that the natural world, as seen through the eyes of physicists, was very well explained; a few puzzles remained, and it was felt that these would soon be explained using already known principles.

But it did not turn out so simply. Instead, these puzzles were only to be solved by the introduction, in the early part of the twentieth century, of two revolutionary new theories that changed our whole conception of nature: the *theory of relativity* and *quantum theory*.

Physics as it was known at the end of the nineteenth century (what we've covered up to now in this book) is referred to as *classical physics*. The new physics that grew out of the great revolution at the turn of the twentieth century is now called *modern physics*. In this chapter we present the special theory of relativity, which was first proposed by Albert Einstein in 1905. In the following chapters we discuss the equally momentous quantum theory and then nuclear and elementary particle physics.

25-1 ∎ Galilean–Newtonian relativity

Einstein's special theory of relativity deals with how we observe events, particularly how objects and events are observed from different frames of reference.[†] This had, of course, already been discussed by Galileo and Newton, and we first briefly discuss these earlier ideas.

[†] A reference frame is a set of coordinate axes fixed to some body (or group of bodies) such as the earth, a train, the moon, and so on. See Section 1-2.

(a)
Reference frame = car

(b)
Reference frame = earth

FIGURE 25-1

A coin is dropped by a person in a moving car. (a) In the reference frame of the car, the coin falls straight down. (b) In a reference frame fixed on the earth, the coin follows a curved (parabolic) path. The upper views show the moment of the coin's release, and the lower views a short time later.

We will mainly deal with so-called **inertial reference frames**. An inertial reference frame (as mentioned in Section 4-4), is one in which Newton's first law, the law of inertia, is valid. That is, if an object experiences no net force due to other bodies, the object either remains at rest or remains in motion with constant velocity in a straight line. Rotating or otherwise accelerating frames of reference are noninertial frames,[†] and we won't be concerned with them here. The earth is not quite an inertial frame (it rotates). But it is close enough that for most purposes we can consider it an inertial frame.

A reference frame that moves with constant velocity with respect to an inertial frame is itself also an inertial frame.

Both Galileo and Newton were keenly aware of what we now call the *relativity principle: that the basic laws of physics are the same in all inertial reference frames.* You may have recognized its validity in everyday life: for example, objects move in the same way in a smoothly moving (constant velocity) train or airplane as they do on earth. (This assumes no vibrations or rocking—for they would make the reference frame noninertial.) When you walk, drink a cup of soup, play Ping-Pong, or drop a pencil on the floor, the bodies move just as they do on earth. Suppose you are in a car traveling rapidly along at constant velocity. If you release a coin from above your head inside the car, how will it fall? It falls straight downward and hits the floor directly below the point of release, Fig. 25-1a. (If you drop the object out the car's window, this won't happen because the moving air drags the object backward.) This is just how objects fall on the earth—straight down—and thus our experiment in the moving car is in accord with the relativity principle.

Note in this example, however, that to an observer on the earth the object follows a curved path, Fig. 25-1b. The actual path followed by the object is different as viewed from different frames of reference. This does not violate the relativity principle, because this principle states that the *laws* of physics are the same in all inertial frames. The same law of gravity, and the same laws of motion, apply in both reference frames. The difference in Figs. 25-1a and b is that in the earth's frame of reference, the coin has an initial velocity (equal to that of the car); the laws of physics therefore predict it will

[†] On a rotating platform, for example, (say a merry-go-round) an object at rest starts moving outward even though no force is exerted on it. This is therefore not an inertial frame.

follow a parabolic path like any projectile. In the car's reference frame, there is no initial velocity and the laws of physics predict it will fall straight down. The laws are the same in both reference frames although the specific paths are different.[†]

Galilean–Newtonian relativity involves certain unprovable assumptions that make sense from everyday experience. It is assumed that the lengths of objects are the same in one reference frame as in another, and that time passes at the same rate in different reference frames. In classical mechanics, then, space and time are considered to be *absolute:* their measurement doesn't change from one reference frame to another. The mass of an object, as well as all forces, are assumed to be unchanged by a change in inertial reference frame.

The position of an object is, of course, different when specified in different reference frames, and so is velocity. For example, a person may walk inside a bus toward the front with a speed of 5 km/h; but if the bus moves 40 km/h with respect to the earth, the person's speed is then 45 km/h with respect to the earth. The acceleration of a body, however, is the same in any inertial reference frame according to classical mechanics. This is because the change in velocity, and the time, will be the same. For example, the person in the bus may accelerate from 0 to 5 km/h in 1.0 seconds, so $a = 5$ km/h/s in the reference frame of the bus. With respect to the earth, the acceleration is $(45 \text{ km/h} - 40 \text{ km/h})/(1.0 \text{ s}) = 5$ km/h/s, which is the same.

Since neither F, m, nor a changes from one inertial frame to another, then Newton's second law, $F = ma$ does not change. Thus it satisfies the relativity principle. It is easily shown that the other laws of mechanics also satisfy the relativity principle.

That the laws of mechanics are the same in all inertial reference frames implies that no one inertial frame is special in any sense; we express this important conclusion by saying that *all inertial reference frames are equivalent* for the description of mechanical phenomena. No one inertial reference frame is any better than another. A reference frame fixed to a car or an aircraft traveling at constant velocity is as good as one fixed on the earth. When you travel smoothly at constant velocity in a car or airplane, it is just as valid to say you are at rest and the earth is moving as it is to say the reverse. There is no experiment one can do to tell which frame is "really" at rest and which is moving. Thus there is no way to single out one particular reference frame as being at absolute rest.

The situation changed somewhat in the last half of the nineteenth century. When Maxwell presented his comprehensive and very successful theory of electromagnetism (Chapter 21), he showed that light can be considered as an electromagnetic wave. Maxwell's equations predicted that the velocity of light c would be 3.00×10^8 m/s; and this is just what is measured within experimental error. The question then arose: in what reference frame does light have precisely the value predicted by Maxwell's theory? For it was assumed that light would have a different speed in different frames of reference. For example, if observers were traveling on a rocket ship at a speed of 1.0×10^8 m/s toward a source of light, we might expect them to measure the speed of

[†] Galileo, in his great book *Dialogues on the Two Chief Systems of the World*, described a similar experiment and predicted the same results. Galileo's example involved a sailor dropping a knife from the top of the mast of a sailing vessel. If the vessel moves at constant speed, where will the knife hit the deck (ignoring earth's rotation and air resistance)?

the light reaching them to be 3.0×10^8 m/s $+ 1.0 \times 10^8$ m/s $= 4.0 \times 10^8$ m/s. But Maxwell's equations have no provision for relative velocity. They merely predicted the speed of light to be $c = 3.0 \times 10^8$ m/s. This seemed to imply there must be some special reference frame where c would have this value.

We discussed in Chapters 13 and 14 that waves travel on water and along ropes or strings, and sound waves travel in air and other materials. Since nineteenth-century physicists viewed the material world in terms of the laws of mechanics, it was natural for them to assume that light too must travel in some *medium;* they called this transparent medium the *ether* and assumed it permeated all space.[†] It was therefore presumed that the velocity of light given by Maxwell's equations must be with respect to this ether.

However, it appeared that Maxwell's equations did *not* satisfy the relativity principle. They were not the same in all inertial reference frames. They were simplest in the frame where $c = 3.00 \times 10^8$ m/s; that is, in a reference frame at rest in the ether. In any other reference frame, extra terms would have to be added to take into account the relative velocity. Thus, although most of the laws of physics obeyed the relativity principle, the laws of electricity and magnetism apparently did not. Instead, they seemed to single out one reference frame that was better than any other—a reference frame that could be considered absolutely at rest.

Scientists soon set out to determine the speed of the earth relative to this absolute frame whatever it might be. A number of clever experiments were designed to do this. The most direct were performed by A. A. Michelson and E. W. Morley. The details of their experiment are discussed in the next section. Briefly, what they did was measure the difference in the speed of light in different directions. They expected to find a difference depending on the orientation of their apparatus with respect to the ether. For just as a boat has different speeds relative to the land when it moves upstream, downstream, or across the stream, so too light would be expected to have different speeds depending on the velocity of the ether past the earth.

Strange as it may seem, they detected no difference at all. This was a great puzzle. A number of explanations were put forth over a period of years; but they led to contradictions or were otherwise not generally accepted.

Then in 1905, Albert Einstein proposed a radical new theory that reconciled these many problems in a simple way. But at the same time, as we shall soon see, it completely changed our ideas of space and time.

*25-2 ∎ The Michelson–Morley experiment

The Michelson–Morley experiment was designed to measure the speed of the *ether*—the medium in which light was assumed to travel—with respect to the earth. The experimenters thus hoped to find an absolute reference frame, one that could be considered to be at rest.

[†] The medium for light waves could not be air, since light travels from the sun to earth through nearly empty space. Therefore another medium was postulated, the ether. The ether was not only transparent but, because of difficulty in detecting it, was assumed to have zero density.

One of the possibilities nineteenth-century scientists considered was that the ether is fixed relative to the sun, for even Newton had taken the sun as the center of the universe. If this were the case (there was no guarantee, of course), the earth's speed of about 3×10^4 m/s in its orbit around the sun would produce a change of 1 part in 10^4 in the speed of light (3.0×10^8 m/s). Direct measurement of the speed of light to this accuracy was not possible. But A. A. Michelson, later with the help of E. W. Morley, was able to use his interferometer (Section 23-9) to measure the difference in the speed of light in different directions to this accuracy. This famous experiment is based on the principle shown in Fig. 25-2. Part (a) is a simplified diagram of the Michelson interferometer, and for simplicity it is assumed that the ether "wind" is moving with speed v to the right. (Alternately, the earth is assumed to move to the left with respect to the ether at speed v.)

Whether constructive or destructive interference occurs at the center of the interference pattern depends on the relative phases of the two beams after they have traveled their separate paths. To examine this we will use an analogy of a boat traveling up and down, and across, a river whose current moves with speed v as shown in Fig. 25-2b. In still water the boat can travel with speed c (not the speed of light in this case). The light from the source is split into two beams by the half-silvered mirror M_s; one beam travels to mirror M_1 and the other to mirror M_2. The beams are reflected by M_1 and M_2 and are joined again after passing through M_s. The now superposed beams interfere with one another and the resultant is viewed by the observer's eye as an interference pattern (discussed in Section 23-9).

First we consider beam 2 in Fig. 25-2a which travels parallel to the "ether wind." In its journey from M_s to M_2, we expect the light to travel with speed $c + v$, just as a boat traveling downstream (see Fig. 25-2b) acquires the speed of the river current. Since the beam travels a distance l_2, the time it takes to go from M_s to M_2 is $t = l_2/(c + v)$. To make the return trip from M_2 to M_s, the light is moving against the ether wind (like the boat going upstream) so its relative speed is expected to be $c - v$; the time for the return trip is $l_2/(c - v)$. The total time required for beam 2 to travel from

FIGURE 25-2

The Michelson–Morley experiment. (a) Michelson interferometer. (b) Boat analogy: boat 1 goes across the stream and back; boat 2 goes downstream and back upstream. (c) Calculation of velocity of boat (or light beam) traveling perpendicular to the current (or either wind).

(a)

(b)

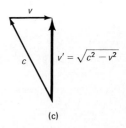

(c)

M_s to M_2 and back to M_s is

$$t_2 = \frac{l_2}{c+v} + \frac{l_2}{c-v}$$

$$= \frac{2l_2}{c(1 - v^2/c^2)}.$$

The second line was obtained from the first by finding the common denominator and factoring out c^2 in the denominator.

Now let us consider beam 1, which travels crosswise to the ether wind. Here the boat analogy (part b) is especially helpful. The boat wants to get from wharf A to wharf B directly across the stream. If it heads directly across, the stream's current will drag it downstream. To reach wharf B, the boat must head at an angle upstream. The precise angle depends on the magnitudes of c and v, but is of no interest to us in itself. Part c of Fig. 25-2 shows how to calculate the velocity v' of the boat relative to earth as it crosses the stream. Since c, v, and v' form a right triangle, we have that $v' = \sqrt{c^2 - v^2}$. The boat has the same velocity when it returns. If we now apply these principles to light beam 1 in Fig. 25-2a, we see that the beam travels with a speed $\sqrt{c^2 - v^2}$ in going from M_s to M_1 and back again. The total distance traveled is $2l_1$, so the time required for beam 1 to make the round trip is $2l_1/\sqrt{c^2 - v^2}$ or,

$$t_1 = \frac{2l_1}{c\sqrt{1 - v^2/c^2}}.$$

Notice that the denominator in this equation for t_1 involves a square root, whereas that for t_2 (above) does not.

If $l_1 = l_2 = l$, we see that beam 1 will lag behind beam 2 by an amount

$$\Delta t = t_2 - t_1 = \frac{2l}{c}\left(\frac{1}{1 - v^2/c^2} - \frac{1}{\sqrt{1 - v^2/c^2}}\right).$$

If $v = 0$, then $\Delta t = 0$ and the two beams will return in phase since they were initially in phase. But if $v \neq 0$, then $\Delta t \neq 0$, and the two beams will return out of phase. If this change of phase from the condition $v = 0$ to that for $v = v$ could be measured, then v could be determined. But the earth cannot be stopped. Furthermore, it is not possible to independently assume $l_1 = l_2$.

Michelson and Morley realized that they could detect the difference in phase (assuming that $v \neq 0$) if they rotated their apparatus by 90°, for then the interference pattern between the two beams should change. In the rotated position, beam 1 would now move parallel to the ether and beam 2 perpendicular to it. Thus the roles could be reversed and in the rotated positions, the times (designated by primes) would be

$$t_1' = \frac{2l_1}{c(1 - v^2/c^2)} \quad \text{and} \quad t_2' = \frac{2l_2}{c\sqrt{1 - v^2/c^2}}.$$

The time lag between the two beams in the nonrotated position (unprimed) would be

$$\Delta t = t_2 - t_1 = \frac{2l_2}{c(1 - v^2/c^2)} - \frac{2l_1}{c\sqrt{1 - v^2/c^2}}.$$

In the rotated position the time difference would be

$$\Delta t' = t'_2 - t'_1 = \frac{2l_2}{c\sqrt{1 - v^2/c^2}} - \frac{2l_1}{c(1 - v^2/c^2)}.$$

When the rotation is made, the fringes of the interference pattern (Section 23-9) will shift an amount determined by the difference in these:

$$\Delta t - \Delta t' = \frac{2}{c}(l_1 + l_2)\left(\frac{1}{1 - v^2/c^2} - \frac{1}{\sqrt{1 - v^2/c^2}}\right).$$

This can be considerably simplified if we assume that $v/c \ll 1$. For in this case we can use the binomial expansion[†] so that

$$\frac{1}{1 - v^2/c^2} \approx 1 + \frac{v^2}{c^2} \quad \text{and} \quad \frac{1}{\sqrt{1 - v^2/c^2}} \approx 1 + \frac{1}{2}\frac{v^2}{c^2}.$$

Then

$$\Delta t - \Delta t' \approx \frac{2}{c}(l_1 + l_2)\left(1 + \frac{v^2}{c^2} - 1 - \frac{1}{2}\frac{v^2}{c^2}\right)$$

$$\approx (l_1 + l_2)\frac{v^2}{c^3}.$$

Now we take $v = 3.0 \times 10^4$ m/s, the speed of the earth in its orbit around the sun. In Michelson and Morley's experiments, the arms l_1 and l_2 were about 11 m long. The time difference would then be about $(22 \text{ m}) \times (3.0 \times 10^4 \text{ m/s})^2/(3.0 \times 10^8 \text{ m/s})^3 \approx 7.0 \times 10^{-16}$ s. For visible light of wavelength $\lambda = 5.5 \times 10^{-7}$ m, say, the frequency would be $f = c/\lambda = (3.0 \times 10^8 \text{ m/s})/(5.5 \times 10^{-7} \text{ m}) = 5.5 \times 10^{14}$ Hz, which means that wave crests pass by a point every $1/(5.5 \times 10^{14} \text{ Hz}) = 1.8 \times 10^{-15}$ s. Thus, with a time difference of 7.0×10^{-16} s, Michelson and Morley should have noted a movement in the interference pattern of $(7.0 \times 10^{-16} \text{ s})/(1.8 \times 10^{-15} \text{ s}) = 0.4$ fringe. They could easily have detected this, since their apparatus was capable of observing a fringe shift as small as 0.01 fringe.

But they found *no significant fringe shift whatever*! They set their apparatus at various orientations. They made observations day and night, so that they would be at various orientations with respect to the sun (due to the earth's rotation). They tried at different seasons of the year (the earth at different locations due to its orbit around the sun). Never did they observe a significant fringe shift.

This "null" result was one of the great puzzles of physics at the end of the nineteenth century. To explain it was a difficult challenge. One possibility was that the ether is not at rest with respect to the sun or other stars, but instead is at rest with respect to the earth; in this case v would be zero and no fringe shift would be expected. But this implies the earth is somehow a preferred object; only with respect to the earth would the speed of light be c as predicted by Maxwell's equations. This is tantamount to assuming the

---

[†] The binomial expansion states that $(1 \pm x)^n = 1 \pm nx + [n(n - 1)/2]x^2 + \cdots$. In our case we have, therefore, $(1 - x)^{-1} \approx 1 + x$, and $(1 - x)^{-1/2} \approx 1 + \frac{1}{2}x$, where only the first term is kept, since $x = v^2/c^2$ is assumed to be small.

earth is the central body of the universe, an ancient idea that had been rejected centuries earlier. Another possibility was that the ether was dragged along by the earth and other bodies, so that its speed at the earth's surface would be zero. However, experiments in high-flying balloons, where at least some ether movement might be detected, also gave a null result. Both possibilities were negated by the observation of "aberration of starlight." Over an observation period of a year, the stars appear to move in tiny ellipses; this is explained, if the earth is moving relative to the ether, by assuming a telescope must be pointed not exactly at the star to be observed but rather at an angle as explained in Fig. 25-3. If the ether were at rest with respect to the earth, you could point directly at the star and no "aberration angle" would be present. But the aberration angle is there, so the ether could not be at rest with respect to the earth.

Another theory to explain the null result was put forth independently by G. F. Fitzgerald and H. A. Lorentz (in the 1890s) in which they proposed that any length (including the arm of an interferometer) contracts by a factor $\sqrt{1 - v^2/c^2}$ in the direction of motion through the ether. According to Lorentz, this could be due to the ether affecting the forces between the molecules of a substance which were assumed to be electrical in nature. This theory, as useful as it was, was eventually replaced by the far more comprehensive theory proposed by Albert Einstein in 1905—the special theory of relativity.

25-3 ▋ Postulates of the special theory of relativity

The problems that existed at the turn of the century with regard to electromagnetic theory and Newtonian mechanics were beautifully resolved by Einstein's introduction of the theory of relativity in 1905. Einstein, however, was apparently not influenced directly by the null result of the Michelson–Morley experiment. He had studied and admired the theoretical work of Lorentz. What motivated Einstein were certain questions regarding electromagnetic theory and light waves. For example, he asked himself: "What would I see if I rode a light beam?" The answer was that instead of a traveling electromagnetic wave, he would see electric and magnetic fields at rest whose magnitude changed in space, but did not change in time. Such fields, he realized, had never been detected and indeed were not consistent with Maxwell's electromagnetic theory. He argued, therefore, that it was unreasonable to think that the speed of light relative to any observer could be reduced to zero, or in fact reduced at all; this became the second postulate of his theory of relativity.

Einstein concluded that the inconsistencies he found in electromagnetic theory were due to the assumption that an absolute space exists. In his famous 1905 paper, he proposed doing away completely with the idea of the ether and the accompanying assumption of an absolute reference frame at rest. This proposal was embodied in two postulates. The first postulate was an extension of the Newtonian relativity principle to include not

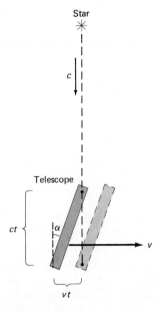

FIGURE 25-3

Aberration of starlight. If a telescope moves with the earth at speed v with respect to the ether, the telescope must be aimed slightly away from the star so that light entering the telescope will travel down the tube (since the tube moves to the right with speed v), rather than hitting the sides. Thus an overhead star will appear to be at an angle $\alpha = v/c$ to the vertical and, as the earth revolves, will seem to trace a circular path.

only the laws of mechanics but also those of the rest of physics, including electricity and magnetism:

■ *First postulate* (*the relativity principle*): **The laws of physics have the same form in all inertial reference frames.**

The second postulate is consistent with the first:

■ *Second postulate* (*constancy of the speed of light*): **Light propagates through empty space with a definite speed c independent of the speed of the source or observer.**

These two postulates form the foundation of Einstein's **special theory of relativity.** It is called "special" to distinguish it from his later "general theory of relativity," which deals with noninertial (accelerating) reference frames. The special theory, which is what we will discuss here, deals only with inertial frames.

The second postulate seems to be the hardest to accept, for it violates commonsense notions. First of all, we have to think of light traveling through empty space. Giving up the ether is not too hard, however, for after all it could never be detected. But the second postulate also tells us that the speed of light in vacuum is always the same, 3.00×10^8 m/s, no matter what the speed of the observer or the source. Thus, a person traveling toward or away from a source of light will measure the same speed for that light as someone at rest with respect to the source. This conflicts with our everyday notions, for we would expect to have to add in the velocity of the observer. Part of the problem is that in our everyday experience we do not measure velocities anywhere near the speed of light; thus we can't expect our everyday experience to be helpful when dealing with such a high velocity. On the other hand, the Michelson—Morley experiment is fully consistent with the second postulate[†].

The beauty of Einstein's proposal is evident. For by doing away with the idea of an absolute reference frame, it was possible to reconcile Maxwell's electromagnetic theory with mechanics. The speed of light predicted by Maxwell's equations *is* the speed of light in vacuum in *any* reference frame.

Einstein's theory required giving up commonsense intuitive notions of space and time, and in the following sections we will examine some strange but interesting consequences of Einstein's theory. Our arguments for the most part will be simple ones. We will use a technique that Einstein himself did: we will imagine very simple experimental situations in which little mathematics is needed. In this way we can see many of the consequences of relativity theory without getting involved in detailed calculations. Einstein called these "gedanken" experiments, which is German for "thought" experiments. In Section 25-13 we will look more deeply into the mathematics of relativity.

[†] The Michelson–Morley experiment can also be considered as evidence for the first postulate, for it was intended to measure the motion of the earth relative to an absolute reference frame; its failure to do so implies the absence of any such preferred frame.

One of the important consequences of the theory of relativity is that we can no longer regard time as an absolute quantity. No one doubts that time flows onward and never turns back. But, as we shall see in this section and the next, the time between two events, and even whether two events are simultaneous, depends on the observer's reference frame.

Two events are said to occur simultaneously if they occur at exactly the same time. But how do we tell if two events occur precisely at the same time? If they occur at the same point in space—such as two apples falling on your head at the same time—it is easy. But if the two events occur at widely separated places, it is more difficult to know since we have to take into account the time it takes for the light to reach us. Because light travels at finite speed, a person who sees two events must calculate back to find out when they actually occurred. For example, if two events are *observed* to occur at the same time, but one actually took place farther from the observer than the other, then the former must have occurred earlier, and the two events were not simultaneous.

To avoid making calculations, we will now make use of a simple thought experiment. We assume an observer, called O, is located exactly halfway between points A and B where two events occur, Fig. 25-4. The two events may be lightning that strikes the points A and B, as shown, or any other type of events. For brief events like lightning, only short pulses of light will travel outward from A and B and reach O. O "sees" the events when the pulses of light reach point O. If the two pulses reach O at the same time, then the two events had to be simultaneous. This is because the two light pulses travel at the same speed; and since the distance OA equals OB, the time for the light to travel from A to O and B to O must be the same. Observer O can then definitely state the two events occurred simultaneously. On the other hand, if O sees the light from one event before that from the other, then it is certain the former event occurred first.

The question we really want to examine is this: if two events are simultaneous to an observer in one reference frame, are they also simultaneous to another observer moving with respect to the first? Let us call the observers O_1 and O_2 and we assume they are fixed in reference frames 1 and 2 that move with speed v relative to one another. These two reference frames can be thought of as train cars (Fig. 25-5). O_2 says that O_1 is moving to the right with speed v, as in (a); and O_1 says O_2 is moving to the left, as in (b). Both viewpoints are legitimate according to the relativity principle.

FIGURE 25-4

A moment after lightning strikes points A and B, the pulses of light are traveling toward O, but O "sees" the lightning only when the light reaches him or her.

Light coming from the two events at A and B

(a)

(b)

FIGURE 25-5

Observers O_1 and O_2, on two different trains (two different reference frames), are moving with relative velocity v. O_2 says that O_1 is moving to the right (a); O_1 says that O_2 is moving to the left (b). Both viewpoints are legitimate—it all depends on your reference frame.

(There is, of course, no third point of view which will tell us which one is "really" moving.)

Now suppose two events occur that are seen by both observers. Let us assume again that the two events are the striking of lightning and that the lightning marks both trains where it struck: at A_1 and B_1 on O_1's train, and at A_2 and B_2 on O_2's train. For simplicity, we assume O_1 happens to be exactly halfway between A_1 and B_1, and that O_2 is halfway between A_2 and B_2. We must now put ourselves in one reference frame or the other. Let us put ourselves in O_2's reference frame, so we see O_1 moving to the right with speed v. Let us also assume that the two events occur *simultaneously* in O_2's frame, and just at the instant when O_1 and O_2 are opposite each other, Fig. 25-6a. A short time later, Fig. 25-6b, the light reaches O_2 from A_2 and B_2 at the same time (we assumed this). Since O_2 knows (or measures) the distances O_2A_2 and O_2B_2 as equal, O_2 knows the two events are simultaneous.

But what does observer O_1 see? From our (O_2) reference frame, we can see that O_1 moves to the right during the time the light is traveling to O_1 from A_1 and B_1. As can be seen in Fig. 25-6b, the light from B_1 has already passed O_1, whereas that from A_1 has not yet reached O_1. Therefore, it is clear that O_1 sees the light from B_1 before that from A_1. Now O_1's frame is as good as O_2's. Light travels at the same speed for O_1 as for O_2, and this speed is the same from A_1 to O_1 as from B_1 to O_1. Since the distance O_1A_1 equals O_1B_1, observer O_1 must conclude that the event at B_1 occurred before the event at A_1.

We thus find that *two events which are simultaneous to one observer are not necessarily simultaneous to a second observer.*

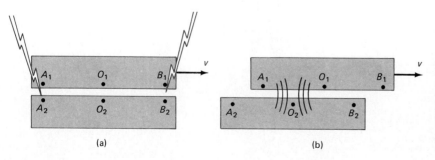

(a)

(b)

FIGURE 25-6

Thought experiment on simultaneity. To observer O_2, the reference frame of O_1 is moving to the right. In (a), one lightning bolt strikes the two reference frames at A_1 and A_2, and a second lightning bolt strikes at B_1 and B_2. According to observer O_2, the two bolts of lightning strike simultaneously. (b) A moment later the light from the two events reaches O_2 at the same time (simultaneously). But in O_1's reference frame, the light from B_1 has already reached O_1, whereas the light from A_1 has not yet reached O_1. So in O_1's reference frame the event at B_1 must have preceded the event at A_1. Time is not absolute!

It may be tempting to ask: "Which observer is right, O_1 or O_2?" The answer, according to relativity, is that they are *both* right. There is no "best" reference frame we can choose to determine which observer is right. Both frames are equally good. We can only conclude that simultaneity is not an absolute concept, but is relative. We are not aware of it in everyday life, however, since the effect is noticeable only when the relative speed of the two reference frames is very large (near c), or the distances involved are very large.

Because of the principle of relativity, the argument we gave for the thought experiment of Fig. 25-6 can be done from O_1's reference frame as well. In this case, O_1 will be at rest and will see event B_1 occur before A_1. But O_1 will recognize (by drawing a diagram equivalent to Fig. 25-6) that O_2, who is moving with speed v to the left, will see the two events as simultaneous. The analysis is left as an exercise.

25-5 ∎ Time dilation

The fact that two events simultaneous to one observer may not be simultaneous to a second observer suggests that time itself is not absolute; could it be that time passes differently in one reference frame than in another? This is, indeed, just what Einstein's theory of relativity predicts, as the following thought-experiment shows.

(a)

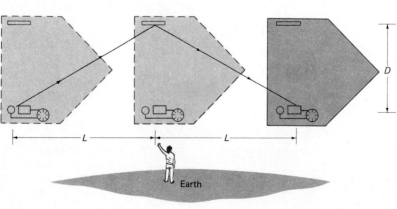

(b)

FIGURE 25-7

Time dilation can be shown by a thought experiment: the time it takes for light to travel over and back on a spaceship is longer for the earth observer (b) than for the observer on the space-ship (a).

Figure 25-7 shows a spaceship traveling past earth at high speed. The point of view of an observer on the spaceship is shown in part (a), and that of an observer on earth in part (b). Both observers have accurate clocks. The person on the spaceship (a) flashes a light and measures the time it takes the light to travel across the spaceship and return after reflecting from a mirror. The light travels a distance $2D$ at speed c so the time required, which we call Δt_0, is

$$\Delta t_0 = \frac{2D}{c}.$$

The observer on earth, Fig. 25-7b, observes the same process. But to this observer, the spaceship is moving; so the light travels the diagonal path shown in going across the spaceship, reflecting off the mirror, and returning to the sender. Although the light travels at the same speed to this observer (the second postulate), it travels a greater distance. Hence the time required, as measured by the earth observer, will be greater than that measured by the observer on the spaceship. The time interval, Δt, as observed by the earth observer can be calculated as follows. In the time Δt, the spaceship travels a distance $2L = v\,\Delta t$ where v is the speed of the spaceship (Fig. 25-7b). Thus, the light travels a total distance on its diagonal path of $2\sqrt{D^2 + L^2}$ and therefore

$$c = \frac{2\sqrt{D^2 + L^2}}{\Delta t} = \frac{2\sqrt{D^2 + v^2(\Delta t)^2/4}}{\Delta t}.$$

We square both sides and solve for Δt to find

$$c^2 = \frac{4D^2}{(\Delta t)^2} + v^2,$$

$$\Delta t = \frac{2D}{c\sqrt{1 - v^2/c^2}}.$$

We combine this with the formula above for Δt_0 and find:

$$\Delta t = \frac{\Delta t_0}{\sqrt{1 - v^2/c^2}}. \qquad (25\text{-}1)$$

Since $\sqrt{1 - v^2/c^2}$ is always less than 1, we see that $\Delta t > \Delta t_0$. That is, the time between the two events (the sending of the light, and its reception at the receiver) is *greater* for the earth observer than for the traveling observer. This is a general result of the theory of relativity, and is known as **time dilation.** Stated simply, the time dilation effect says that *moving clocks are measured to run slowly.* However, we should not think that the clocks are somehow at fault. To the contrary, we assume the clocks are good ones. Time is actually measured to pass more slowly in any moving reference frame as compared to your own. This remarkable result is an inevitable outcome of the two postulates of the theory of relativity.

The concept of time dilation may be hard to accept, for it violates our commonsense understanding. We can see from Eq. 25-1 that the time-dilation effect is negligible unless v is reasonably close to c. The speeds we experience in everyday life are very much smaller than c, so it is little wonder we haven't noticed time dilation as a real effect. Experiments have been done to test the time-dilation effect, and have confirmed Einstein's predictions. In 1971, for example, extremely precise atomic clocks were flown around the world in jet planes. Since the speed of the planes (10^3 km/h) is much less than c, the clocks had to be accurate to nanoseconds (10^{-9} s) in order to detect the time-dilation effect. They were this accurate and confirmed Eq. 25-1 to within experimental error. Time dilation had been confirmed decades earlier, however, by observation on "elementary particles" (see Chapter 31) which have very small masses (typically 10^{-30} to 10^{-27} kg) and so require little energy to be accelerated to speeds close to c. Many of these elementary particles are not stable and decay after a time into simpler particles. One example is the muon whose mean lifetime is 2.2 μs when at rest. Careful experiments showed that when a muon is traveling at high speeds, its lifetime increases just as predicted by the time-dilation formula.

□ EXAMPLE 25-1   What will be the mean lifetime of a muon as measured in the laboratory if it is traveling $0.60c = 1.8 \times 10^8$ m/s with respect to the laboratory? Its mean life at rest is 2.2×10^{-6} s.

SOLUTION   If an observer were to move along with the muon (the muon would be at rest to this observer) the muon would have a mean life of 2.2×10^{-6} s. To an observer in the lab, the muon lives longer because of time dilation. From Eq. 25-1 with $v = 0.60c$, we have

$$\Delta t = \frac{\Delta t_0}{\sqrt{1 - v^2/c^2}} = \frac{2.2 \times 10^{-6} \text{ s}}{\sqrt{1 - \frac{0.36c^2}{c^2}}} = \frac{2.2 \times 10^{-6} \text{ s}}{\sqrt{0.64}} = 2.8 \times 10^{-6} \text{ s.}$$

We need to make a comment about the use of Eq. 25-1 and the meaning of Δt and Δt_0. The equation is true only when Δt_0 represents the time interval between the two events in a reference frame where the two events occur at *the same point in space* (as in Fig. 25-7a where the two events are the light flash being sent and being received). This time interval, Δt_0, is often called the **proper time.** Then Δt in Eq. 25-1 represents the time interval between the two events as measured in a reference frame moving with speed v with respect to the first. In Example 25-1 above, Δt_0 (and not Δt) was set equal to 2.2×10^{-6} s because it is only in the rest frame of the muon that the two events ("birth" and "decay") occur at the same point in space.

Time dilation has aroused interesting speculation about space travel. Under the old time regime, to reach a star 100 light-years away would not be possible by ordinary mortals (1 light-year is the distance light can travel in 1 year = 3.0×10^8 m/s $\times 3.15 \times 10^7$ s = 9.5×10^{15} m). Even if a spaceship could travel at close to the speed of light, it would take over 100 years to reach such a star. But time dilation tells us that the time involved would be less for an astronaut. In a spaceship traveling at $v = 0.999c$, the time for such a trip would be only about $\Delta t_0 = \Delta t \sqrt{1 - v^2/c^2} = (100 \text{ yr})\sqrt{1 - (0.999)^2} = 4.5$ yr. Thus a person could make such a trip. Time dilation allows such a trip, but the enormous practical problems of achieving such speeds will not be overcome in the foreseeable future.

Notice, in this example, that whereas 100 years would pass on earth, only 4.5 years would pass for the astronaut on the trip. Is it just the clocks that would slow down for the astronaut? The answer is no. All processes, including life processes, run more slowly for the astronaut according to the earth observer. But to the astronaut, time would pass in a normal way. The astronaut would experience 4.5 years of normal sleeping, eating, reading, and so on. And people on earth would experience 100 years of ordinary activity.

25-6 ▌ Length contraction

Not only time intervals are different in different reference frames; space intervals—lengths and distances—are different as well, according to the special theory of relativity, and we illustrate this with a thought experiment.

An observer on Earth watches a spacecraft traveling at speed v from Earth to, say, Neptune, Fig. 25-8a. The distance between the planets, as measured by an Earth observer, is L_0. The time required for the trip, measured from Earth, is $\Delta t = L_0/v$. In Fig. 25-8b we see the point of view of an observer on the spacecraft. In this frame of reference, the spaceship is at rest; Earth and Neptune move with speed v. (We assume v is much greater than the relative speed of Neptune and Earth, so the latter can be ignored.) But the time between the departure of Earth and arrival of Neptune is less for the spacecraft observer than for the earth observer, because of time dilation. From Eq. 25-1, the time for the trip as viewed by the spacecraft is $\Delta t_0 = \Delta t \sqrt{1 - v^2/c^2}$. Since the spacecraft observer measures the same speed but less time between these two events, he must also measure the distance as less. If we let L be the distance between the planets as viewed by the spacecraft observer, then $L = v \Delta t_0$. We have already seen that $\Delta t_0 = \Delta t \sqrt{1 - v^2/c^2}$ and $\Delta t = L_0/v$, so we have $L = v \Delta t_0 = v \Delta t \sqrt{1 - v^2/c^2} = L_0 \sqrt{1 - v^2/c^2}$. That is

$$L = L_0 \sqrt{1 - v^2/c^2}. \tag{25-2}$$

This is a general result of the special theory of relativity and applies to lengths of objects as well as to distance. The result can be stated most simply in words as: *the length of an object is measured to be shorter when it is*

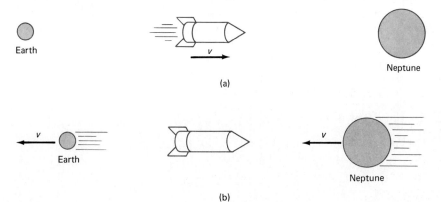

(a)

(b)

FIGURE 25-8

(a) A spaceship traveling at very high speed from Earth to Neptune, as seen from Earth's frame of reference. (b) As viewed by an observer on the spaceship, Earth and Neptune are moving at the very high velocity v. Note in (b) that each planet does not look shortened since at high speeds we see the trailing edge (as in Fig. 25-9), and the net effect is to leave its appearance as a circle.

moving than when it is at rest. This is called **length contraction.** The length L_0 in Eq. 25-2 is called the **proper length.** It is the length of the object (or distance between two points) as measured by an observer at rest with respect to it. Equation 25-2 gives the length that will be measured when the object travels past an observer at speed v. It is important to note, however, that the length contraction occurs *only along the direction of motion*. For example, the moving spaceship in Fig. 25-8a is shortened in length but its height is the same as when it is at rest.

Length contraction, like time dilation, is not noticeable in everyday life because the factor $\sqrt{1 - v^2/c^2}$ in Eq. 25-2 differs from 1.00 significantly only when v is very large.

☐ **EXAMPLE 25-2** A spaceship passes the earth at speed $v = 0.80c$. Describe the changes in length of a meter stick as it is slowly rotated from vertical to horizontal by a person inside, as viewed (*a*) by another person in the spaceship, (*b*) by an observer on earth.

SOLUTION (*a*) The meter stick looks 1.0 m long in all orientations since it is at rest. (*b*) The meter stick varies in length from 1.0 m (vertical) to $L = (1.0 \text{ m}) \times \sqrt{1 - (0.80)^2} = 0.60$ m long in the horizontal direction (assuming that is the direction of motion).

Equation 25-2 tells us what the length of an object will be *measured* to be when traveling at speed v. The *appearance* of the object is another matter. Suppose, for example, you are traveling to the left past a tall building at speed $v = 0.85c$. This is equivalent to the building moving past you to the right at speed v. The building will look narrower (and the same height), but you will also be able to see the side of the building even if you are directly in front of it. This is shown in Fig. 25-9b—part (a) shows the building at rest. The fact that you see the side is not really a relativistic effect, but is due to the finite speed of light. To see how this occurs, we look at Fig. 25-9c which is a top view of the building, looking down. At the instant shown, the observer O is directly in front of the building. Light from points A and B reach O at the same time. If the building were at rest, light from point C could

(a)

(b)

(c)

FIGURE 25-9

Building seen (a) at rest, (b) moving at high speed. (c) Diagram explains why the side of the building is seen (see the text).

never reach O. But the building is moving at very high speed and does "get out of the way" so that light from C can reach O. Indeed, at the instant shown, light from point C when it was at an earlier location (C' on the diagram) can reach O because the building has moved. In order to reach the observer at the same time as light from A and B, light from C had to leave at an earlier time since it must travel a greater distance. Thus it is light from C' that reaches the observer at the same time as light from A and B. This, then, is how an observer might see both the front and side of an object at the same time even when directly in front of it.[†] It can be shown, by the same reasoning, that spherical objects will actually still have a circular outline even at high speeds. That is why the planets in Fig. 25-8b are drawn round rather than contracted.

25-7 ∎ The twin paradox

Not long after Einstein proposed the special theory of relativity, an apparent paradox was pointed out. According to this "twin paradox," suppose one of a pair of 20-year-old twins takes off in a spaceship traveling at very high speed to a distant star and back again, while the other twin remains on earth. According to the earth twin, the traveling twin will age less. Whereas 20 years might pass for the earth twin, perhaps only 1 year (depending on the spacecraft's speed) would pass for the traveler. Thus, when the traveler returns, the earthbound twin could expect to be 40 years old whereas his twin would be only 21.

This is the viewpoint of the twin on the earth. But what about the traveling twin? Since "everything is relative," all inertial reference frames are equally good. Won't the traveling twin make all the claims the earth twin does, only in reverse? Can't the astronaut twin claim that since the earth is moving away at high speed, time passes more slowly on earth and the twin on earth will age less? This is the opposite of what the earth twin predicts. They cannot both be right, for after all the spacecraft returns to earth and a direct comparison of ages and clocks can be made.

The resolution lies in the fact that the starting and ending points are fixed relative to the earth twin. The traveling twin must accelerate at the beginning and end of the trip and also when turning around out in space. We assume these acceleration periods occupy only a tiny portion of the total time. But while traveling at constant velocity on both outward and homeward journeys, the traveling twin measures the distance he must travel as shorter than what his earthbound twin measures. Why? Because of length contraction, and it is the earth observer who measures the proper length. Since the traveling twin measures a shorter travel distance, and both twins measure the same relative speed, this confirms that the traveling twin measures a shorter time and thus returns to earth having aged less than the

---

[†] It would be an error to think that the building in Fig. 25-9b would look rotated. This is not correct since in that case side A would look shorter than side B. In fact, if the observer is directly in front these sides appear equal in height. Thus the building looks contracted in its front face, but we also see the side, as described above.

earthbound twin. Even when the acceleration periods are considered, Einstein's general theory of relativity, which deals with accelerating reference frames, confirms this result. The ultimate judge, of course, is experiment, and the 1971 experiment of precise clocks sent around the world in jet planes (discussed in Section 25-5) confirms that less time passes for the traveler.

25-8 ∎ Four-dimensional space–time

Let us imagine a person is on a train moving at a very high speed, say 0.65c, Fig. 25-10. This person begins a meal at 7:00 and finishes at 7:15, according to a clock on the train. The two events, beginning and ending the meal, take place at the same point on the train. So the proper time between these two events is 15 min. To observers on earth, the meal will take longer—20 min according to Eq. 25-1. Let us assume that the meal was served on a 20-cm-diameter plate. To observers on the earth, the plate is only 15 cm wide (length contraction). Thus to observers on the earth the food looks smaller but lasts longer.

In a sense these two effects, time dilation and length contraction, balance each other. When viewed from the earth, what the food seems to lose in size it gains in length of time it lasts. Space, or length, is exchanged for time.

Considerations like this led to the idea of *four-dimensional space–time:* space takes up three dimensions and time is a fourth dimension. Space and time are intimately connected. Just as when we squeeze a balloon we make one dimension larger and another smaller, so when we examine objects and events from different reference frames a certain amount of space is exchanged for time, or vice versa.

(a)   (b)

FIGURE 25-10

According to an accurate clock on a fast-moving train, a person begins dinner at 7:00 (a) and finishes at 7:15 (b). At the beginning of the meal, observers on the earth set their watches to correspond with the clock on the train. These observers measure the eating time as 20 min.

Although the idea of four dimensions may seem strange, it refers to the fact that any object or event is specified by four quantities—three to describe where in space, and one to describe when in time. The really unusual aspect of four-dimensional space–time is that space and time can intermix: a little of one can be exchanged for a little of the other when the reference frame is changed.

It is difficult for most of us to understand the idea of four-dimensional space–time. Somehow we feel, just as physicists did before the advent of relativity, that space and time are completely separate entities. Yet we have found in our thought experiments that they are not completely separate. Our difficulty in accepting this is reminiscent of the situation in the seventeenth century at the time of Galileo and Newton. Before Galileo, the vertical direction, that in which objects fall, was considered to be distinctly different from the two horizontal dimensions. Galileo showed that the vertical dimension differs only in that it happens to be the direction in which gravity acts. Otherwise, all three dimensions are equivalent, a fact that we all accept today. Now we are asked to accept one more dimension, time, which we had previously thought of as being somehow different. This is not to say that there is no distinction whatsoever between space and time. What relativity has shown is that space and time determinations are not independent of one another.

25-9 ∎ Mass increase

The three basic mechanical quantities are length, time intervals, and mass. The first two have been shown to be relative—their value depends on the reference frame from which they are measured. We might expect that mass, too, is a relative quantity. Indeed, Einstein showed that *the mass of an object increases as its speed increases* according to the formula

$$m = \frac{m_0}{\sqrt{1 - v^2/c^2}}. \tag{25-3}$$

In this **mass-increase** formula, m_0 is the *rest mass* of the object—the mass it has as measured in a reference frame in which it is at rest; and m is the mass it will be measured to have in a reference frame in which it moves at speed v (see Fig. 25-11). Einstein's derivation of Eq. 25-3 was based on the assumption that the law of conservation of momentum is valid.

Relativistic mass increase has been tested countless times on tiny elementary particles (such as muons) and the mass has been found to increase in accord with Eq. 25-3.

▢ EXAMPLE 25-3 An electron ($m_0 = 9.1 \times 10^{-31}$ kg) enters a magnetic field $B = 1.5$ T perpendicularly to the field lines with a speed $v = 0.50c = 1.50 \times 10^8$ m/s. What will be the radius of curvature r of its path?

SOLUTION We use $F = ma$ and obtain $evB = mv^2/r$ (see Chapter 19), where $e = 1.6 \times 10^{-19}$ C. Thus

$$r = \frac{mv}{eB} = \frac{m_0 v}{\sqrt{1 - v^2/c^2}\, eB} = 6.6 \times 10^{-4} \text{ m.}$$

FIGURE 25-11

Increase in mass of an object as its velocity increases.

25-10 ∎ The ultimate speed

A basic result of the special theory of relativity is that no object can equal or exceed the speed of light. That the speed of light is a natural speed limit in the universe can be seen from any one of Eqs. 25-1, 25-2, and 25-3. It is perhaps easiest to see it from Eq. 25-3, the mass-increase formula, $m = m_0/\sqrt{1 - v^2/c^2}$. As an object is accelerated to greater and greater speeds, its mass becomes larger and larger. Indeed, if v were to equal c, the denominator in this equation would be zero and the mass m would become infinite. To accelerate an object up to $v = c$ would thus require infinite energy, and so is not possible. The fact that light itself travels at the speed c implies that its rest mass (if such a thing makes sense) must be zero; of course, light is never at rest.

If v were to exceed c, then the factor $\sqrt{1 - v^2/c^2}$ would be the square root of a negative number, which is imaginary: so lengths, time intervals, and mass would not be real. We conclude, then, that ordinary objects cannot equal or exceed the speed of light. However, as was pointed out in the late 1960s, Einstein's equations do not rule out the possibility that objects exist whose speed is *always* greater than c. If such particles exist (the name "tachyon"—meaning "fast"—was proposed) the rest mass m_0 would have to be imaginary; in this way the mass m would be the ratio of two imaginary numbers for $v > c$, which is real. For such hypothetical particles, c would be a *lower* limit on their speed. In spite of extensive searches for tachyons, none has been found. It seems that the speed of light *is* the ultimate speed in the universe.

25-11 ∎ $E = mc^2$; mass and energy

When a steady net force is applied to an object of rest mass m_0, the object increases in speed. Since the force is acting through a distance, work is done on the object and its energy increases. As the speed of the object approaches c, the speed cannot increase indefinitely since it cannot exceed c. On the other hand, the mass of the object increases with increasing speed. That is, the work done on an object not only increases its speed but also contributes to increasing its *mass*. Normally, the work done on an object increases its energy. This new twist from the theory of relativity leads to the idea that mass is a form of energy, a crucial part of Einstein's theory of relativity.

To find the mathematical relationship between mass and energy, Einstein assumed that the work-energy theorem (Chapter 5) is still valid in relativity. That is, the net work done on a particle is equal to its change in kinetic energy (KE). Using this theorem, Einstein showed that at high speeds the formula KE $= \frac{1}{2}mv^2$ is not correct. You might think that using Eq. 25-3 for m would give KE $= \frac{1}{2}m_0v^2/\sqrt{1 - v^2/c^2}$, but this formula, too, is wrong. Instead, Einstein showed that the kinetic energy of a particle is given by

$$\text{KE} = mc^2 - m_0c^2, \tag{25-4}$$

where m is the mass of the particle traveling at speed v and m_0 is its rest mass.

But what does the second term in Eq. 25-4—the m_0c^2—mean? Consistent with the idea that mass is a form of energy, Einstein called m_0c^2 the *rest energy* of the object. We can rearrange Eq. 25-4 to get $mc^2 = m_0c^2 + $ KE. We call mc^2 the *total energy E* of the particle (assuming no potential energy), and we see that the total energy equals the rest energy plus the kinetic energy:

$$E = mc^2 \qquad\qquad\qquad (25\text{-}5a)$$

or

$$E = m_0c^2 + \text{KE}. \qquad\qquad (25\text{-}5b)$$

Here we have Einstein's famous formula $E = mc^2$.

For a particle at rest in a given reference frame, its total energy is $E_0 = m_0c^2$ which we have called its rest energy. This formula mathematically relates the concepts of energy and mass. But if this idea is to have any meaning from a practical point of view, then mass ought to be convertible to energy and vice versa. That is, if mass is just one form of energy then it should be convertible to other forms of energy just as other types of energy are interconvertible. Einstein suggested that this might be possible, and indeed this has been experimentally confirmed countless times and forms the basis for many important processes.

The interconversion of mass and energy is most easily detected in nuclear and elementary particle physics. For example, the neutral pion (π^0) of rest mass 2.4×10^{-28} kg is observed to decay into pure electromagnetic radiation (photons); the π^0 completely disappears in the process. The amount of electromagnetic energy produced is found to be exactly equal to that predicted by Einstein's formula, $E = m_0c^2$. The reverse process is also commonly observed in the laboratory: electromagnetic radiation under certain conditions can be converted into material particles such as electrons. On a larger scale, the energy produced in nuclear power plants is a result of the loss in mass of the uranium fuel as it undergoes the process called fission (Chapter 30). Even the radiant energy we receive from the sun is an instance of $E = mc^2$; the sun's mass is continually decreasing as it radiates energy outward.

The relation $E = mc^2$ is now believed to apply to all processes, although the changes are often too small to measure. That is, when the energy of a system changes by an amount ΔE, the mass of the system changes by an amount Δm given by

$$\Delta E = (\Delta m)(c^2).$$

In a chemical reaction where heat is gained or lost, the masses of the reactants and the products will be different. Even when water is heated on a stove, the mass of the water increases very slightly. This example is easy to see because as heat is added, the temperature and therefore the average speed of the molecules increases; and Eq. 25-3 tells us that the mass also increases.

EXAMPLE 25-4   A π^0 meson ($m_0 = 2.4 \times 10^{-28}$ kg) travels at a speed $v = 0.80c = 2.4 \times 10^8$ m/s. What is its kinetic energy? Compare to a classical calculation.

$$m = \frac{m_0}{\sqrt{1 - v^2/c^2}} = \frac{2.4 \times 10^{-28} \text{ kg}}{\sqrt{1 - (0.80)^2}} = 4.0 \times 10^{-28} \text{ kg.}$$

Thus its KE is

$$\text{KE} = (m - m_0)c^2 = (4.0 \times 10^{-28} \text{ kg} - 2.4 \times 10^{-28} \text{ kg})(3.0 \times 10^8 \text{ m/s})^2$$
$$= 1.4 \times 10^{-11} \text{ J.}$$

Notice that the units of mc^2 are kg·m^2/s^2 which is the joule. A classical calculation would give $\text{KE} = \frac{1}{2}m_0v^2 = \frac{1}{2}(2.4 \times 10^{-28} \text{ kg})(2.4 \times 10^8 \text{ m/s})^2 = 6.9 \times 10^{-12}$ J, but this is not a correct result.

EXAMPLE 25-5   How much energy would be released if the π^0 meson in the above example is transformed by decay completely into electromagnetic radiation?

SOLUTION   The rest energy of the π^0 is

$$E_0 = m_0c^2 = (2.4 \times 10^{-28} \text{ kg})(3.0 \times 10^8 \text{ m/s})^2 = 2.2 \times 10^{-11} \text{ J.}$$

This is how much energy would be released if the π^0 decayed at rest. If it has $\text{KE} = 1.4 \times 10^{-11}$ J, the total energy released would be 3.6×10^{-11} J.

EXAMPLE 25-6   The energy required or released in nuclear reactions and decays comes from a change in mass between the initial and final particles. In one type of radioactive decay (Chapter 29) an atom of uranium ($m = 232.03714$ u) decays to an atom of thorium ($m = 228.02873$ u) plus an atom of helium ($m = 4.00260$ u) where the masses given are in atomic mass units (1 u $= 1.6606 \times 10^{-27}$ kg). Calculate the energy released in this decay.

SOLUTION   The initial mass is 232.03714 u and after the decay it is 228.02873 u + 4.00260 u = 232.03133 u, so there is a decrease in mass of 0.00581 u. This mass, which equals (0.00581 u)(1.66 $\times 10^{-27}$ kg) $= 9.64 \times 10^{-30}$ kg, is changed into energy. By $E = mc^2$ we have

$$E = (9.64 \times 10^{-30} \text{ kg})(3.0 \times 10^8 \text{ m/s})^2 = 8.68 \times 10^{-13} \text{ J.}$$

Since 1 MeV $= 1.60 \times 10^{-13}$ J, the energy released is 5.4 MeV.

Equation 25-4 for the kinetic energy can be written in terms of the speed v of the object with the help of Eq. 25-3:

$$\text{KE} = m_0c^2\left(\frac{1}{\sqrt{1 - v^2/c^2}} - 1\right). \tag{25-6}$$

At low speeds, $v \ll c$, we can expand the square root in Eq. 25-6 using the binomial expansion (see the footnote in Section 25-2). Then we get

$$\text{KE} \approx m_0c^2\left(1 + \frac{1}{2}\frac{v^2}{c^2} + \cdots - 1\right)$$
$$\approx \frac{1}{2}m_0v^2,$$

where the dots in the first expression represent very small terms in the expansion which we have neglected since we assumed that $v \ll c$. Thus we see that at low speeds, the relativistic form for kinetic energy reduces to the classical form $\text{KE} = \frac{1}{2}m_0 v^2$. This is, of course, what must be. Relativity would not be a valuable theory if it did not predict accurate results at low speed as well as at high. Indeed, the other equations of special relativity also reduce to their classical equivalents at ordinary speeds: length contraction, time dilation, and mass increase all disappear for $v \ll c$ since $\sqrt{1 - v^2/c^2} \approx 1$.

A useful relation between the total energy E of a particle and its momentum p can also be derived. Since $E = mc^2$ and $p = mv$, where $m = m_0/\sqrt{1 - v^2/c^2}$, we have

$$E^2 = m^2 c^4 = m^2 c^2(c^2 + v^2 - v^2)$$
$$= m^2 c^2 v^2 + m^2 c^2(c^2 - v^2)$$
$$= p^2 c^2 + \frac{m_0^2 c^4(1 - v^2/c^2)}{1 - v^2/c^2}$$

or

$$E^2 = p^2 c^2 + m_0^2 c^4. \qquad (25\text{-}7)$$

Thus the total energy can be written in terms of the momentum p or in terms of the kinetic energy (Eq. 25-5).

25-12 ▌ Relativistic addition of velocities

Consider a rocket ship that travels away from the earth with speed v, and assume that this rocket has fired off a second rocket that travels at speed u' with respect to the first (Fig. 25-12). We might expect that the speed u of rocket 2 with respect to earth is $u = v + u'$, which in the case shown in the figure is $u = 0.60c + 0.60c = 1.20c$. But, as discussed in Section 25-10, no object can travel faster than the speed of light in any reference frame. Indeed, Einstein showed that since length and time are different in different reference frames, the old addition-of-velocities formula is no longer valid. Instead, the correct formula is

$$u = \frac{v + u'}{1 + vu'/c^2} \qquad (25\text{-}8)$$

for a motion along a straight line. We derive this formula in the next section. If u' is in the opposite direction from v, then u' must have a minus sign and $u = (v - u')/(1 - vu'/c^2)$.

▢ EXAMPLE 25-7   Calculate the speed of rocket 2 in Fig. 25-12 with respect to earth.

SOLUTION   Rocket 2 moves with speed $u' = 0.60c$ with respect to rocket 1. Rocket 1 has speed $v = 0.60c$ with respect to earth. The speed of rocket 2 with respect to earth is therefore

$$u = \frac{0.60c + 0.60c}{1 + \dfrac{(0.60c)(0.60c)}{c^2}} = \frac{1.20c}{1.36} = 0.88c.$$

FIGURE 25-12

Rocket 2 is fired from rocket 1 with speed $u' = 0.60c$. What is the speed of rocket 2 with respect to the earth?

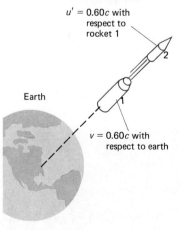

$u' = 0.60c$ with respect to rocket 1

Earth

$v = 0.60c$ with respect to earth

Notice that Eq. 25-8 reduces to the classical form for velocities small compared to the speed of light since $1 + vu'/c^2 \approx 1$ for v and $u' \ll c$. Thus $u \approx v + u'$.

Let us test our formula in one more case, that of the speed of light. Suppose that rocket 1 in Fig. 25-12 sends out a beam of light so that $u' = c$. Equation 25-8 tells us that the speed of this light with respect to earth is

$$u = \frac{0.60c + c}{1 + \dfrac{(0.60)(c)}{c^2}} = \frac{1.60c}{1.60} = c,$$

as it must if the second postulate of relativity is to hold.

*25-13 ▌ Galilean and Lorentz transformations

We now examine in detail the mathematics of relating quantities in one inertial reference frame to the equivalent quantities in another. In particular, we will see how positions and velocities *transform* (that is, change) from one frame to the other.

We begin with the classical or Galilean viewpoint. Consider two reference frames S and S' which are each characterized by a set of coordinate axes, Fig. 25-13. The axes x and y (z is not shown) refer to S and x' and y' to S'. The x' and x axes overlap one another, and we assume that frame S' moves to the right (in the x direction) at speed v with respect to S. And for simplicity let us assume the origins O and O' of the two reference frames are superimposed at time $t = 0$.

Now consider an event that occurs at some point P (Fig. 25-13) represented by the coordinates x', y', z' in reference frame S' at the time t'. What will be the coordinates of P in S? Since S and S' overlap precisely initially, after a time t, S' will have moved a distance vt'. Therefore, at time t', $x = x' + vt'$. The y and z coordinates, on the other hand, are not altered by motion along the x axis; thus $y = y'$ and $z = z'$. Finally, since time is assumed to be absolute in Galilean–Newtonian physics, clocks in the two frames will agree with each other; so $t = t'$. We summarize these in the following

FIGURE 25-13

Inertial reference frame S' moves to the right at speed v with respect to frame S.

Galilean transformation equations:

$$x = x' + vt'$$
$$y = y'$$
$$z = z'$$
$$t = t'.$$

(25-9)

These equations give the coordinate of an event in the S frame when those in the S' frame are known. If those in the S system are known, then the S' coordinates are obtained from

$$x' = x - vt, \qquad y' = y, \qquad z' = z, \qquad t' = t.$$

These four equations are the "inverse" transformation and are very easily obtained from Eqs. 25-9. Notice that the effect is merely to exchange primed and unprimed quantities and replace v by $-v$. This makes sense because from the S' frame, S moves to the left (negative x direction) with speed v.

Now suppose that the point P in Fig. 25-13 represents an object that is moving. Let the components of its velocity vector in S' be u'_x, u'_y, and u'_z (we use u to distinguish it from the relative velocity of the two frames, v). Now $u'_x = \Delta x'/\Delta t'$, $u'_y = \Delta y'/\Delta t'$, and $u'_z = \Delta z'/\Delta t'$, by the definition of velocity. For example, if at time t'_1 the particle is at x'_1 and a short time later, t'_2, it is at x'_2, then

$$u'_x = \frac{x'_2 - x'_1}{t'_2 - t'_1} = \frac{\Delta x'}{\Delta t'}.$$

Now the velocity of P as seen from S will have components u_x, u_y, and u_z. We can show how these are related to the velocity components in S' by using Eqs. 25-9. For example,

$$u_x = \frac{\Delta x}{\Delta t} = \frac{x_2 - x_1}{t_2 - t_1} = \frac{(x'_2 + vt'_2) - (x'_1 + vt'_1)}{t'_2 - t'_1}$$
$$= \frac{(x'_2 - x'_1) + v(t'_2 - t'_1)}{t'_2 - t'_1}$$
$$= \frac{\Delta x'}{\Delta t'} + v = u'_x + v.$$

For the other components, $u'_y = u_y$ and $u'_z = u_z$, so we have

$$u_x = u'_x + v$$
$$u_y = u'_y$$
$$u_z = u'_z.$$

(25-10)

These are known as the *Galilean velocity transformation* equations. We see that the y and z components of velocity are unchanged, but the x components differ by v. This is just what we have used before when dealing with relative velocity. For example, if S' is a train and S the earth, and the train moves with speed v with respect to earth, a person walking toward the front of the train with speed u'_x will have speed $u_x = u'_x + v$ with respect to the earth.

The Galilean transformations, Eqs. 25-9 and 25-10 are valid only when the velocities involved are much less than c. We can see, for example, that the first of Eqs. 25-10 will not work for the speed of light; for light traveling in S' with speed $u'_x = c$ will have speed $c + v$ in S, whereas the theory of relativity insists it must be c in S. Clearly, then, a new set of transformation equations are needed to deal with relativistic velocities.

We will derive the required equations in a simple way, again looking at Fig. 25-13. We assume the transformation is linear and of the form

$$x = \gamma(x' + vt'), \qquad y = y', \qquad z = z'.$$

That is, we modify the first of Eqs. 25-9 by multiplying by a factor γ which is yet to be determined; but we assume the y and z equations are unchanged since there is no length contraction in these directions. We won't assume a form for t, but will derive it. The inverse equations must have the same form with v replaced by $-v$. (The principle of relativity demands it, since S' moving to the right with respect to S is equivalent to S moving to the left with respect to S'.) Therefore

$$x' = \gamma(x - vt).$$

Now if a light pulse leaves the common origin of S and S' at time $t = t' = 0$, after a time t it will have traveled a distance $x = ct$ or $x' = ct'$ along the x axis. Therefore, from the equations for x and x' above,

$$ct = \gamma(ct' + vt') = \gamma(c + v)t',$$

$$ct' = \gamma(ct - vt) = \gamma(c - v)t.$$

We substitute t' from the second equation into the first and find $ct = \gamma(c + v)\gamma(c - v)(t/c) = \gamma^2(c^2 - v^2)t/c$. We cancel out the t on each side and solve for γ to find

$$\gamma = \frac{1}{\sqrt{1 - v^2/c^2}}.$$

Now that we have found γ, we need only find the relation between t and t'. To do so, we combine $x' = \gamma(x - vt)$ with $x = \gamma(x' + vt')$:

$$x' = \gamma(x - vt) = \gamma(\gamma[x' + vt'] - vt).$$

We solve for t and find $t = \gamma(t' + vx'/c^2)$. In summary

$$x = \frac{1}{\sqrt{1 - v^2/c^2}}(x' + vt')$$

$$y = y'$$

$$z = z' \qquad\qquad (25\text{-}11)$$

$$t = \frac{1}{\sqrt{1 - v^2/c^2}}(t' + vx'/c^2).$$

These are called the *Lorentz transformation equations*. They were first proposed, in a slightly different form, by Lorentz in 1904 to explain the null result of the Michelson–Morley experiment and to make Maxwell's equations take the same form in all inertial systems. A year later Einstein derived them independently based on his theory of relativity. Notice that not only is the x equation modified as compared to the Galilean transformation, but so is the t equation; indeed, we see directly in this last equation how the space and time coordinates mix.

The relativistically correct velocity equations are readily obtained. For example (we let $\gamma = 1/\sqrt{1 - v^2/c^2}$),

$$u_x = \frac{\Delta x}{\Delta t} = \frac{\gamma(\Delta x' + v\,\Delta t')}{\gamma(\Delta t' + v\,\Delta x'/c^2)} = \frac{(\Delta x'/\Delta t') + v}{1 + (v/c^2)(\Delta x'/\Delta t')}$$

$$= \frac{u'_x + v}{1 + vu_x/c^2}.$$

The others are obtained in the same way and we collect them here:

$$u_x = \frac{u'_x + v}{1 + vu'_x/c^2}$$

$$u_y = \frac{u'_y\sqrt{1 - v^2/c^2}}{1 + vu'_x/c^2} \tag{25-12}$$

$$u_z = \frac{u'_z\sqrt{1 - v^2/c^2}}{1 + vu'_x/c^2}.$$

The first of these equations is just Eq. 25-8 of the previous section. As we saw there, velocities do not add in our commonsense (Galilean) way, because of the denominator $(1 + vu'_x/c^2)$. We see, too, that the y and z components of velocity are also altered and that they depend on the x' component of velocity.

☐ **EXAMPLE 25-8** Derive the length contraction formula, Eq. 25-2, from the Lorentz transformation equations.

SOLUTION Let an object of length L_0 be at rest on the x axis in S. The coordinates of its two end points are x_1 and x_2, so that $x_2 - x_1 = L_0$. At any instant in S', the end points will be at x'_1 and x'_2 as given by the Lorentz transformation equations. The length measured in S' is $L = x'_2 - x'_1$. An observer in S' measures this length by measuring x'_2 and x'_1 at the same time (in the S' frame), so $t'_2 = t'_1$. Thus, from the first of Eqs. 25-11,

$$L_0 = x_2 - x_1 = \frac{1}{\sqrt{1 - v^2/c^2}}(x'_2 + vt'_2 - x'_1 - vt'_1).$$

Since $t'_2 = t'_1$, we have

$$L_0 = \frac{1}{\sqrt{1 - v^2/c^2}}(x'_2 - x'_1) = \frac{L}{\sqrt{1 - v^2/c^2}},$$

or

$$L = L_0\sqrt{1 - v^2/c^2}$$

which is Eq. 25-2.

SOLUTION   The time Δt_0 between two events that occur at the same place ($x'_2 = x'_1$) in S' is measured to be $\Delta t_0 = t'_2 - t'_1$. Since $x'_2 = x'_1$, then from the last of Eqs. 25-11 the time Δt between the events as measured in S is

$$\Delta t = t_2 - t_1 = \frac{1}{\sqrt{1 - v^2/c^2}} \left(t'_2 + \frac{vx'_2}{c^2} - t'_1 - \frac{vx'_1}{c^2} \right)$$

$$= \frac{1}{\sqrt{1 - v^2/c^2}} (t'_2 - t'_1) = \frac{\Delta t_0}{\sqrt{1 - v^2/c^2}}$$

which is Eq. 25-1. Notice that we chose S' to be the frame where the two events occur at the same place, so that $x'_1 = x'_2$ and these terms would cancel out.

25-14 ∎ The impact of special relativity

A great many experiments have been performed to test the predictions of the special theory of relativity. Within experimental error, no contradictions have been found. The vast majority of scientists have therefore accepted relativity as an accurate description of nature.

At speeds much less than the speed of light, the relativistic formulas reduce to the old classical ones, as we have discussed. We would, of course, hope—or rather, insist—that this be true since Newtonian mechanics works so well for objects moving with speeds $v \ll c$. This insistence that a more general theory (such as relativity) give the same results as a more restricted theory (such as classical mechanics which works for $v \ll c$) is called the **correspondence principle.** The two theories must correspond where their realms of validity overlap. Relativity thus does not contradict classical mechanics. Rather, it is a more general theory, of which classical mechanics is now considered to be a special case.

The importance of relativity is not simply that it gives more accurate results, especially at very high speeds. Much more than that, it has changed the way we view the world. The concepts of space and time are now seen to be relative, and intertwined with one another, whereas before they were considered absolute and separate. Even our concepts of matter and energy have changed: either can be converted to the other. The impact of relativity extends far beyond physics. It has influenced the other sciences, and even the world of art and literature; it has, indeed, entered the general culture.

From a practical point of view, we do not have much opportunity in our daily lives to use the mathematics of relativity. For example, the factor $\sqrt{1 - v^2/c^2}$ which appears in many relativistic formulas has a value of 0.995 when $v = 0.1c$. Thus, for speeds even as high as $0.1c = 3 \times 10^7$ m/s, the factor $\sqrt{1 - v^2/c^2}$ in relativistic formulas gives a numerical correction of less than 1 percent. For speeds less than $0.1c$, or unless mass and energy get interchanged, we thus don't need to use the more complicated relativistic formulas, and can use the simpler classical formulas.

▌Summary

An *inertial reference frame* is one in which Newton's law of inertia holds. Inertial reference frames can move at constant velocity relative to each other; accelerating reference frames are noninertial.

The special theory of relativity is based on two principles: the *relativity principle*, which states that the laws of physics are the same in all inertial reference frames, and the principle of the *constancy of the speed of light*, which states that the speed of light in empty space has the same value in all inertial reference frames.

One consequence of relativity theory is that two events that are simultaneous in one reference frame may not be simultaneous in another. Other effects are *time dilation:* moving clocks are measured to run slowly; *length contraction:* the length of an object is measured to be shorter when it is moving than when it is at rest; *mass increase:* the mass of a body increases with speed; and velocity addition must be done in a special way. All these effects are significant only at high speeds, close to the speed of light, which itself is the ultimate speed in the universe.

The theory of relativity has changed our notions of space and time, and of mass and energy. Space and time are seen to be intimately connected, with time being the fourth dimension in addition to space's three dimensions. Mass and energy are interconvertible; the equation

$$E = mc^2$$

tells how much energy E is needed to create a mass m, or vice versa. The law of conservation of energy must include mass as a form of energy.

▌Questions

1. Give some examples of noninertial reference frames.

2. A woman stands on top of a moving railroad car. She throws a heavy ball straight up (it seems to her). Ignoring air resistance, will the ball land on the car or behind it?

*3. Discuss a Michelson–Morley type of experiment done with sound waves. Under what conditions would you expect or not expect a null result?

4. Does the earth really go around the sun? Or is it also valid to say that the sun goes around the earth? Discuss in view of the first principle of relativity (that there is no best reference frame).

5. If you were on a spaceship traveling at $0.5c$ away from a star, at what speed would the starlight pass you?

6. Will two events that occur at the same place and same time for one observer be simultaneous to a second observer moving with respect to the first?

7. (*a*) Explain why two events will be simultaneous to each of two observers moving with respect to each other if the two events occur at the same point in each reference frame. (*b*) Under what other conditions will two events be simultaneous to each observer?

8. Analyze the thought experiment of Section 25-4 from O_1's point of view. (Make a diagram analogous to Fig. 25-6.)

9. The time-dilation effect says that "moving clocks run slowly." Actually, this effect has nothing to do with motion affecting the functioning of clocks. What then does it deal with?

10. Does time dilation mean that time actually passes more slowly in moving reference frames or that it only *seems* to pass more slowly?

11. Today's subways are said to age people prematurely. Suppose that in the future, subway trains which traveled very close to the speed of light could be designed. How do you think this would affect the aging process?

12. A young-looking woman astronaut has just arrived home from a long trip. She rushes up to an old gray-haired man and in the ensuing conversation refers to him as her son. How is this possible?

13. If you were traveling away from earth at a speed of $0.5c$, would you notice a change in your heartbeat? Would your mass, height, or waistline change? What would observers on earth using telescopes say about these things?

14. The predictions of the theory of relativity seem to contradict some of our everyday notions about the world and therefore don't seem to make sense. Examine these notions in detail and determine if any measurable contradiction exists.

15. Discuss how our everyday lives would be different if the speed of light were only 25 m/s.

16. Do mass increase, time dilation, and length contraction occur at ordinary speeds, say 90 km/h?

17. Suppose the speed of light were infinite. What would happen to the relativistic predictions of length contraction, time dilation, and mass increase?

18. Explain how the length-contraction and time-dilation formulas might be used to indicate that c is the limiting speed in the universe.

19. Consider an object of mass m to which is applied a constant force for an indefinite period of time. Discuss how its velocity and mass change with time.

20. A white-hot iron bar is cooled to room temperature. Does its mass change?

21. Does the equation $E = mc^2$ conflict with the conservation of energy principle? Explain.

22. Does $E = mc^2$ apply to particles that travel at the speed of light? Does it apply only to them?

23. An electron is limited to travel at speeds less than c. Does this put an upper limit on the momentum of an electron? If so, what is this upper limit?

24. If mass is a form of energy, does this mean that a spring has more mass when compressed than when relaxed?

25. It is not correct to say that "matter can neither be created nor destroyed." What must we say instead?

▮ Problems

SECTIONS 25-5 TO 25-7

1. (I) A beam of a certain type of elementary particle travels at a speed of 2.85×10^8 m/s. At this speed, the average lifetime is measured to be 2.50×10^{-8} s. What is the particle's lifetime at rest?

2. (I) What is the speed of a beam of pions if their average lifetime is measured to be 3.5×10^{-8} s? At rest, their lifetime is 2.6×10^{-8} s.

3. (I) A spaceship passes you at a speed of $0.80c$. You measure its length to be 90 m. How long would it be when at rest?

4. (I) You are sitting in your sports car when a very fast sports car passes you at a speed of $0.18c$. A person in the car says his car is 6.00 m long and yours is 6.15 m long. What do you measure for these two lengths?

5. (I) Suppose you decide to travel to a star 85 light-years away. How fast would you have to travel so the distance would be only 20 light-years?

6. (I) If you were to travel to a star 50 light-years from earth at a speed of 2.0×10^8 m/s, what would you measure this distance to be?

7. (II) A certain star is 36 light-years away. How long would it take a spacecraft traveling $0.98c$ to reach that star from earth as measured by observers: (a) on earth; (b) on the spacecraft? (c) What is the distance traveled according to observers on the spacecraft? (d) What will the spacecraft occupants compute their speed to be from the results of (b) and (c)?

8. (II) A friend of yours travels by you in her fast sports car at a speed of $0.760c$. It appears to be 5.80 m long and 1.45 m high. (a) What will be its length and height at rest? (b) How many seconds did you see elapse on your friend's watch when 20.0 s passed on yours? (c) How fast did you appear to be traveling to your friend? (d) How many seconds did she see elapse on your watch when she saw 20.0 s pass on hers?

9. (II) The nearest star to earth is Alpha Centauri, 4.0 light-years away. (a) At what constant velocity must a spacecraft travel from earth if it is to reach the star in 3.0 years, as measured by travelers on the spacecraft? (b) How long does the trip take according to earth observers?

10. (III) How fast must a pion be traveling to travel 12.0 m before it decays? The average lifetime, at rest, is 2.6×10^{-8} s.

SECTION 25-9

11. (I) What is the mass of a proton traveling at $v = 0.85c$?

12. (I) At what speed will an object's mass be twice its rest mass?

13. (II) At what speed v will the mass of an object be 1 percent greater than its rest mass?

14. (II) Escape velocity from the earth is 40,000 km/h. What would be the percent increase in mass of a 3.8×10^5-kg spacecraft traveling at that speed?

15. (II) (a) What is the speed of an electron whose mass is 10,000 times its rest mass? Such speeds are reached in the Stanford Linear Accelerator, SLAC. (b) If the electrons travel in the lab through a tube 3.0 km long (as at SLAC), how long is this tube in the electrons' reference frame?

16. (II) Derive a formula showing how the density of an object changes with speed v.

SECTION 25-11

17. (I) How much energy can be obtained from conversion of 1.0 mg of mass? How much mass could this energy raise to a height of 100 m?

18. (I) What is the kinetic energy of an electron whose mass is 5.0 times its rest mass?

19. (I) A certain chemical reaction requires 2.56×10^4 J of energy input for it to go. What is the increase in mass of the products over the reactants?

20. (II) Calculate the rest energy of an electron in joules and in MeV (1 MeV = 1.60×10^{-13} J).

21. (II) Suppose a spacecraft of rest mass 20,000 kg is accelerated to $0.25c$. (a) How much kinetic energy would it have? (b) If you used the classical formula for KE, by what percentage would you be in error?

22. (II) (a) By how much does the mass of the earth increase each year as a result of the sunlight reaching it? (b) How much mass does the sun lose per year? (Radiation from the sun reaches the earth at a rate of about 1400 W/m^2 of area perpendicular to the energy flow.)

23. (II) Calculate the rest energy of a proton in MeV (1 eV = 1.60 × 10^{-19} J).

24. (II) What is the mass and speed of an electron that has been accelerated by a voltage of 200 kV?

25. (II) Calculate the kinetic energy and momentum of a proton traveling 9.2 × 10^7 m/s.

26. (II) What is the momentum of a 500-MeV proton (that is, one with KE = 500 MeV)?

27. (II) What is the speed of a proton accelerated by a potential difference of 250 MV?

28. (II) What is the speed of an electron whose KE is 1.00 MeV?

29. (II) Calculate the mass of a proton (m_0 = 1.67 × 10^{-27} kg) whose kinetic energy is half its total energy. How fast is it traveling?

30. (II) Calculate the kinetic energy and momentum of a proton (m_0 = 1.67 × 10^{-27} kg) traveling 8.3 × 10^7 m/s. By what percentages would your calculations have been in error if you had used classical formulas?

31. (II) What is the speed and momentum of a proton (m_0 = 1.67 × 10^{-27} kg) whose kinetic energy is half its rest energy.

32. (II) An electron (m_0 = 9.11 × 10^{-31} kg) is accelerated from rest to speed v by a conservative force. In this process its potential energy decreases by 4.20 × 10^{-14} J. Determine the electron's speed, v.

33. (II) How many grams of matter would have to be totally destroyed to run a 100-W light bulb for 1 year?

34. (II) How much energy would be required to break a helium nucleus into its constituents, two protons and two neutrons? The rest masses of proton (including an electron), neutron, and helium are 1.00783 u, 1.00867 u, and 4.00260 u respectively. (This is called the *total binding energy* of the $^4_2$He nucleus.)

35. (II) Make a graph of the kinetic energy versus momentum for (a) a particle of nonzero rest mass. (b) a particle with zero rest mass.

36. (III) Show that the kinetic energy of a particle of rest mass m_0 is related to its momentum p by the equation $p = \sqrt{(\text{KE})^2 + 2(\text{KE})(m_0 c^2)}/c$.

37. (III) A pi meson of rest mass m_π decays at rest into a muon (rest mass m_μ) and a neutrino of zero rest mass. Show that the kinetic energy of the muon is $\text{KE}_\mu = (m_\pi - m_\mu)^2 c^2/2m_\pi$.

SECTION 25-12

38. (I) A person on a rocket traveling at 0.50c (with respect to the earth) observes a meteor come from behind and pass her at a speed she measures as 0.50c. How fast is the meteor moving with respect to the earth?

39. (II) Two spaceships leave the earth in opposite directions, each with a speed of 0.50c with respect to the earth. (a) What is the velocity of spaceship 1 relative to spaceship 2? (b) What is the velocity of spaceship 2 relative to spaceship 1?

40. (II) A spaceship leaves earth traveling 0.68c. A second spaceship leaves the first at a speed of 0.86c with respect to the first. Calculate the speed of the second ship with respect to earth if it is fired (a) in the same direction the first spaceship is already moving, (b) directly backward toward earth.

*SECTION 25-13

*41. (I) Suppose in Fig. 25-13 that the origins of S and S' overlap at $t = t' = 0$ and that S' moves at speed $v = 30$ m/s with respect to S. In S', a person is resting at a point whose coordinates are $x' = 15$ m, $y' = 10$ m, and $z' = 0$. Calculate this person's coordinates in $S(x, y, z)$ at (a) $t = 5.0$ s, and (b) $t = 10.0$ s. Use the Galilean transformation.

*42. (I) Repeat problem 41 using the Lorentz transformation and a relative speed $v = 1.0 × 10^8$ m/s, but calculate at time t' equals (a) 5.0 μs and (b) 10.0 μs.

*43. (II) A spaceship traveling 0.60c away from earth fires a module with a speed of 0.80c at right angles to its own direction of travel (as seen by the first spaceship). What is the speed of this second spaceship and its direction of travel (relative to the first spaceship's direction) as seen by an observer on earth?

*44. (II) In problem 41, suppose that the person moves with a velocity whose components are $u'_x = u'_y = 20$ m/s. What will be his velocity with respect to S? (Give magnitude and direction.)

*45. (II) In problem 42, suppose that the person moves with a velocity (in a rocket) whose components are $u'_x = u'_y = 1.2 × 10^8$ m/s. What will be the person's velocity (magnitude and direction) with respect to S?

*46. (III) In the old West, a marshal riding on a train traveling 50 m/s sees a duel between two men standing on the earth 50 m apart parallel to the train. The marshal's instruments indicate that in his reference frame the two men fired simultaneously. (a) Which of the two men, the first one the train passes (A) or the second (B) should be arrested for firing the first shot? That is, in the gunfighters' frame of reference, who fired first? (b) How much earlier did he fire? (c) Who was struck first?

*47. (III) If a particle moves in the xy plane of system S (Fig. 25-13) in a direction that makes an angle θ with the x axis, show that it makes an angle θ' in S' given by $\tan \theta' = (\sin \theta)\sqrt{1 - v^2/c^2}/(\cos \theta - v/u)$.

26 EARLY QUANTUM THEORY AND MODELS OF THE ATOM

The second aspect of the revolution that shook the world of physics in the early part of the twentieth century (the first half was Einstein's theory of relativity) is the quantum theory. Unlike the special theory of relativity—whose basic tenets were put forth mainly by one person in a single year—the revolution of quantum theory required almost three decades to unfold, and many scientists contributed to its development. It began in 1900 with Planck's quantum hypothesis and culminated in the mid-1920s with the theory of quantum mechanics of Schrödinger and Heisenberg which has been so effective in explaining the structure of matter.

26-1 ▮ Planck's quantum hypothesis

One of the observations that was unexplained at the end of the nineteenth century was the spectrum of light emitted by hot objects. We saw in Chapter 11 that all objects emit radiation whose total intensity is proportional to the fourth power of the Kelvin temperature (T^4). At normal temperatures, we are not aware of this electromagnetic radiation because of its low intensity. At higher temperatures, there is sufficient infrared radiation that we can feel heat if we are close to the object. At still higher temperatures (on the order of 1000 K) objects actually glow, such as a red-hot electric stove burner or element in a toaster. At temperatures above 2000 K objects glow with a yellow or whitish color, such as white-hot iron and the filament of a light bulb. As the temperature increases, the electromagnetic radiation emitted by bodies is strongest at higher and higher frequencies.

The spectrum of light emitted by a hot dense object is shown in Fig. 26-1 for an idealized *blackbody;* such a body would absorb all the radiation falling on it and the radiation it emits, called *blackbody radiation*, is the easiest to deal with. As can be seen, the spectrum contains a continuous range of frequencies; such a continuous spectrum is emitted by any heated solid or liquid, and even by dense gases. The 6000-K curve in Fig. 26-1, corresponding to the temperature of the sun, peaks in the visible part of the spectrum. For lower temperatures the total radiation drops considerably and the peak occurs at higher wavelengths; hence the blue end of the visible spectrum (and the UV) is relatively weaker. (This is why objects glow with

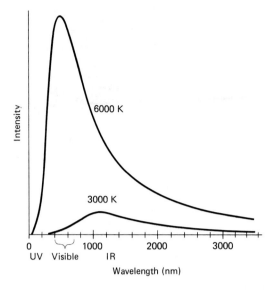

FIGURE 26-1

Spectrum of frequencies emitted by a blackbody at two different temperatures.

a red color at not so high temperatures.) It is found that the wavelength at the peak of the spectrum, λ_p, is related to the Kelvin temperature T by:

$$\lambda_p T = 2.90 \times 10^{-3} \text{ m} \cdot \text{K}.$$

This is known as *Wien's displacement law* and for the sun's temperature gives us $\lambda_p = (2.90 \times 10^{-3} \text{ m} \cdot \text{K})/(6.0 \times 10^3 \text{ K}) = 500$ nm, which is in the visible region of the spectrum.

A major problem facing scientists in the 1890s was to explain black-body radiation. Maxwell's electromagnetic theory had predicted that oscillating electric charges produce electromagnetic waves, and the radiation emitted by a hot object could be due to the oscillations of electric charges in the molecules of the material. Although this would explain where the radiation came from, it did not correctly predict the observed spectrum of emitted light. Two important theoretical curves based on classical ideas were those proposed by Wien (1896) and by Rayleigh (1900); the latter was slightly modified later by Jeans and since then has been known as the Rayleigh–Jeans theory. As experimental data came in, it became clear that neither Wien's nor the Rayleigh–Jeans formulations were in accord with experiment: Wien's was accurate at short wavelength but deviated from experiment at longer wavelengths whereas the reverse was true for the Rayleigh–Jeans theory (see Fig. 26-2).

The break came in late 1900 when Max Planck (1858–1947) proposed an empirical formula that nicely fit the data. He then sought a theoretical basis for the formula and within 2 months found that he could obtain the formula by making a new and radical (though not so recognized at the time) assumption: that the energy distributed among the molecular oscillators is not continuous but instead consists of a finite number of very small discrete amounts, each related to the frequency of oscillation by

$$E_{\min} = hf.$$

Here h is a constant, now called *Planck's constant*, whose value was esti-

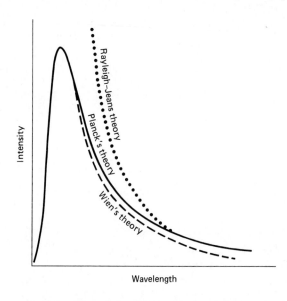

FIGURE 26-2

Comparison of the Wien and the Rayleigh–Jeans
theories to that of Planck which closely follows
experiment.

mated by Planck by fitting his formula for the blackbody radiation curve
to experiment. The value accepted today is

$$h = 6.626 \times 10^{-34} \text{ J} \cdot \text{s}.$$

Planck's assumption suggests that the energy of any molecular vibration
could be only some whole number multiple of hf:

$$E = nhf, \qquad n = 1, 2, 3, \ldots. \tag{26-1}$$

This is often called *Planck's quantum hypothesis* ("quantum" means "fixed
amount") although little attention was brought to this point at the time. In
fact, it appears that Planck considered it more as a mathematical device to
get the "right answer" rather than as a discovery comparable to those of
Newton. Planck himself continued to seek a classical explanation for the
introduction of h. The recognition that this was an important and radical
innovation did not come until later, after about 1905 when others, particularly Einstein, entered the field.

The quantum hypothesis, Eq. 26-1, as we accept it today, says that the
energy can be $E = hf$, or $2hf$, or $3hf$, and so on, but there cannot be vibrations whose energy lies between these values. That is, energy would not be
a continuous quantity as had been believed for centuries; rather it is *quantized*—it exists only in discrete amounts. The smallest amount of energy
possible (hf) is called the *quantum of energy*. Another way of expressing the
quantum hypothesis is that not just any amplitude of vibration is possible.
The possible values for the amplitude are related to the frequency f.

26-2 ■ Photon theory of light and the photoelectric effect

In 1905, the same year as he introduced the special theory of relativity,
Einstein made a bold extension of the quantum idea by proposing a new
theory of light. Planck's work suggested that the vibrational energy of mole-

cules in a radiating object is quantized with energy $E = nhf$. Einstein reasoned that if the energy of the molecular oscillators is thus quantized, then to conserve energy, the light ought to be emitted in packets or quanta, each with an energy

$$E = hf. \qquad (26\text{-}2)$$

Again h is Planck's constant. Since all light ultimately comes from a radiating source, this suggests that perhaps *light is transmitted as tiny particles,* or **photons,** as they are now called, rather than as waves. This, too, was a radical departure from classical ideas. Einstein proposed a simple test of the quantum theory of light: quantitative measurements on the photoelectric effect.

The **photoelectric effect** refers to the fact that when light shines on a metal surface, electrons are emitted from the surface. (The photoelectric effect occurs in other materials but is most easily observed with metals.) This effect can be observed using the apparatus shown in Fig. 26-3. A metal plate P along with a smaller electrode C are placed inside an evacuated glass tube, called a *photocell.* The two electrodes are connected to an ammeter and a source of emf as shown. When the photocell is in the dark, the ammeter reads zero. But when light of sufficiently high frequency is shone on the plate, the ammeter indicates a current flowing in the circuit. To explain completion of the circuit we can imagine electrons flowing across the tube from the plate to the "collector" C as shown in the diagram.

That electrons should be emitted when light shines on a metal is consistent with the electromagnetic (EM) wave theory of light, since the electric field of an EM wave could exert a force on electrons in the metal and thrust some of them out. Einstein pointed out, however, that the wave theory and the photon theory of light give very different predictions on the details of this effect. For example, one thing that can be measured with the apparatus of Fig. 26-3 is the maximum kinetic energy ($\mathrm{KE_{max}}$) of the emitted electrons. This can be done by using a variable voltage source and reversing the terminals so that electrode C is negative and P is positive. The electrons emitted from P will be repelled by the negative electrode, but if this reverse voltage is small enough, the fastest electrons will still reach C and there will be a current in the circuit. If the reversed voltage is increased, a point is reached where the current reaches zero—no electrons have sufficient kinetic energy to reach C. This is called the "stopping potential" V_0, and from its measurement $\mathrm{KE_{max}}$ can be determined using conservation of energy (loss of KE = gain in PE):

$$\mathrm{KE_{max}} = eV_0.$$

Now let us examine the details of the photoelectric effect in view of the wave theory versus Einstein's particle theory. First the wave theory, assuming monochromatic light. The two important properties of a light wave are its intensity and its frequency (or wavelength). When these two quantities are varied, the wave theory makes the following predictions.

1. If the light intensity is increased, the number of electrons ejected and their maximum KE should be increased because the higher intensity means a greater electric field amplitude, and the greater electric field should thrust electrons out with higher speed.

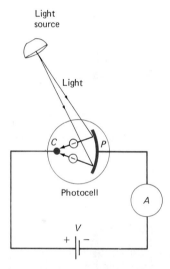

FIGURE 26-3
Photoelectric effect.

2. The frequency of the light should not affect the KE of the ejected electrons. Only the intensity should affect KE_{max}.

The photon theory makes completely different predictions. First we note that in a monochromatic beam, all photons have the same energy ($= hf$). Increasing the intensity of the light beam means increasing the number of photons in the beam but does not affect the energy of each as long as the frequency is not changed. According to Einstein's theory, an electron is ejected from the metal by a collision with a single photon. In the process, all the photon energy is transferred to the electron and the photon ceases to exist. Since electrons are held in the metal by attractive forces, some minimum energy W_0 (called the *work function*, which is on the order of a few electron volts for most metals) is required just to get an electron out of the surface. If the frequency f of the incoming light is so low that hf is less than W_0, then the photons will not have enough energy to eject any electrons at all. If $hf > W_0$, then electrons will be ejected and energy will be conserved in the process. That is, the input energy (of the photon), hf, will equal the outgoing KE of the electron plus the energy required to get it out of the metal, W:

$$hf = \text{KE} + W. \tag{26-3a}$$

For the least tightly held electrons, W is the work function W_0, and KE in this equation becomes KE_{max}:

$$hf = \text{KE}_{max} + W_0. \tag{26-3b}$$

Many electrons will require more energy than the bare minimum (W_0) to get out of the metal, and thus the KE of such electrons will be less than the maximum.

From these considerations, the photon theory makes the following predictions.

1. An increase in intensity of the light beam means more photons are incident, so more electrons will be ejected; but since the energy of each photon is not changed, the maximum KE of electrons is not changed.

2. If the frequency of the light is increased, the maximum KE of the electrons increases linearly, according to Eq. 26-3b. That is,

$$\text{KE}_{max} = hf - W_0.$$

This is plotted in Fig. 26-4.

3. If the frequency f is less than the "cutoff" frequency f_0, where $hf_0 = W_0$, no electrons will be ejected at all.

These predictions of the photon theory are clearly very different from the predictions of the wave theory. In 1913–1914 careful experiments were carried out by R. A. Millikan. The results were fully in agreement with Einstein's photon theory.

A number of other experiments were carried out in the early twentieth century which also supported the photon theory. One of these was the **Compton effect** (1923), named after its discoverer, A. H. Compton (1892–1962). Compton scattered short-wavelength light (actually X rays) from various materials. He found the scattered light had a slightly lower frequency than

FIGURE 26-4

Photoelectric effect: maximum kinetic energy of ejected electrons increases linearly with frequency of incident light. No electrons are emitted if $f < f_0$.

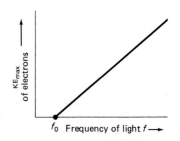

did the incident light, indicating a loss of energy. This, he showed, could be explained on the basis of the photon theory as incident photons colliding with electrons of the material, Fig. 26-5; he applied the laws of conservation of energy and momentum to such collisions and found that the predicted energies of scattered photons was in accord with experimental results. Thus the photon theory of light rests on a firm experimental foundation.

> **EXAMPLE 26-1** Calculate the energy of a photon of blue light, $\lambda = 450$ nm.
>
> SOLUTION   Since $f = c/\lambda$, we have
>
> $$E = hf = \frac{hc}{\lambda} = \frac{(6.63 \times 10^{-34} \text{ J·s})(3.0 \times 10^8 \text{ m/s})}{(4.5 \times 10^{-7} \text{ m})} = 4.4 \times 10^{-19} \text{ J,}$$
>
> or $(4.4 \times 10^{-19} \text{ J})/(1.6 \times 10^{-19} \text{ J/eV}) = 2.7$ eV.

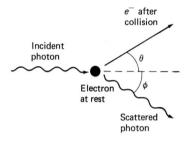

FIGURE 26-5
Compton effect.

> **EXAMPLE 26-2** What is the maximum kinetic energy and speed of an electron ejected from a sodium surface whose work function is $W_0 = 2.28$ eV when illuminated by light of wavelength (a) 410 nm; (b) 550 nm?
>
> SOLUTION   (a) For $\lambda = 410$ nm, $hf = hc/\lambda = 4.85 \times 10^{-19}$ J or 3.03 eV. From Eq. 26-3b, $\text{KE}_{\text{max}} = 3.03$ eV $- 2.28$ eV $= 0.75$ eV or 1.2×10^{-19} J. Since $\text{KE} = \frac{1}{2}mv^2$ where $m = 9.1 \times 10^{-31}$ kg,
>
> $$v = \sqrt{2 \text{ KE}/m} = 5.1 \times 10^5 \text{ m/s.}$$
>
> Notice that we used the nonrelativistic equation for KE. If v had turned out to be more than about $0.1c$, our calculation would have been inaccurate by more than a percent or so, and we would probably prefer to redo it using the relativistic form (Eq. 25-6).
>
> (b) For $\lambda = 550$ nm, $hf = 3.60 \times 10^{-19}$ J $= 2.25$ eV. Since this photon energy is less than the work function, no electrons are ejected.

The photoelectric effect, besides playing an important historical role in confirming the photon theory of light, also has many practical applications. Burglar alarms and automatic door openers often make use of the photocell circuit of Fig. 26-3. When a person interrupts the beam of light, the sudden drop in current in the circuit activates a switch—often a solenoid—to operate a bell or open the door. Many smoke detectors use the photoelectric effect to detect tiny amounts of smoke that interrupt the flow of light and so alter the electric current. Photographic light-meters use this circuit as well. For many applications today, the vacuum-tube photocell of Fig. 26-3 has been replaced by a semiconductor device known as a *photodiode*. In these semiconductors, the absorption of a photon liberates a bound electron, which changes the conductivity of the material, so the current through a photodiode is altered.

The photon theory of light is useful in biology and medicine; an example is the following.

EXAMPLE 26-3   In *photosynthesis*, which is the process by which pigments such as chlorophyll in plants use the energy of sunlight to change CO_2 to useful carbohydrate, it is believed that about nine photons are needed to change one molecule of CO_2 to carbohydrate and O_2. Assuming light of wavelength $\lambda = 670$ mm (chlorophyll absorbs most strongly in the range 650 nm to 700 nm), how efficient is the photosynthetic process? The reverse chemical reaction has a heat of combustion of 4.9 eV/molecule of CO_2.

SOLUTION   The energy of nine photons, each of energy $hf = hc/\lambda$ is $(9)(6.6 \times 10^{-34}\ \text{J·s})(3.0 \times 10^8\ \text{m/s})/(6.7 \times 10^{-7}\ \text{m}) = 2.7 \times 10^{-18}$ J or 17 eV. Thus the process is (4.9 eV/17 eV) = 29 percent efficient.

26-3 ▌ Photon interactions; pair production

The photon is truly a relativistic particle—it travels at the speed of light. Thus we must use relativistic formulas for dealing with its mass, energy, and momentum. The mass m of any particle is given by $m = m_0/\sqrt{1 - v^2/c^2}$. Since $v = c$ for a photon, the denominator is zero. So the rest mass, m_0, of a photon must also be zero, or its energy $E = mc^2$ would be infinite. Of course a photon is never at rest. The momentum of a photon, from Eq. 25-7 with $m_0 = 0$, is

$$p = \frac{E}{c}.$$

Since $E = hf$, the momentum of a photon is related to its wavelength by

$$p = \frac{hf}{c} = \frac{h}{\lambda}. \tag{26-4}$$

When a photon passes through matter, it interacts with the atoms and electrons. There are four important types of interactions that a photon can undergo. First, the photon can be scattered off an electron (or a nucleus) and in the process lose some energy; this is the Compton effect (Fig. 26-5). But notice that the photon is not slowed down. It still travels with speed c, but its frequency will be less. A second type of interaction is the photoelectric effect; a photon may knock an electron out of an atom and in the process itself disappear. The third process is similar: the photon may knock an atomic electron to a higher-energy state in the atom if its energy is not sufficient to knock it out altogether. In this process the photon also disappears, and all its energy is given to the atom; such an atom is then said to be in an excited state, and we shall discuss this more later. Finally, a photon can actually create matter. The most common process is the production of an electron and a positron, Fig. 26-6. (A positron has the same mass as an electron, but the opposite charge, $+e$.)[†] This is called **pair production** and the

† Positrons do not last long in nature because, when they collide with an electron, they annihilate each other and produce two or more photons. This pair annihilation process is also an example of $E = mc^2$.

FIGURE 26-6

Pair production: a photon disappears and produces an electron and a positron.

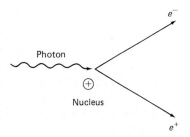

photon disappears in the process. This is an example of rest mass being created from pure energy, and it occurs in accord with Einstein's equation $E = mc^2$. Notice that a photon cannot create a single electron since electric charge would not then be conserved.

□ **EXAMPLE 26-4** What is the minimum energy of a photon, and its wavelength, that can produce an electron-positron pair?

SOLUTION Because $E = mc^2$, the photon must have energy $E = 2(9.1 \times 10^{-31} \text{ kg})(3.0 \times 10^8 \text{ m/s})^2 = 1.64 \times 10^{-13}$ J or 1.02 MeV. A photon with less energy cannot undergo pair production. Since $E = hf = hc/\lambda$, the wavelength of a 1.02-MeV photon is

$$\lambda = \frac{hc}{E} = \frac{(6.6 \times 10^{-34} \text{ J·s})(3.0 \times 10^8 \text{ m/s})}{(1.64 \times 10^{-13} \text{ J})} = 1.2 \times 10^{-12} \text{ m},$$

which is 0.0012 nm. Thus the wavelength must be very short. Such photons are in the gamma-ray (or very short X-ray) region of the electromagnetic spectrum.

Pair production cannot occur in empty space for energy and momentum could not simultaneously be conserved. In Example 26-4, for instance, energy is conserved, but the electron–positron pair has no momentum to carry away the initial momentum of the photon. Indeed, it can be shown that at any energy an additional massive object, such as an atomic nucleus (see Fig. 26-6), must take part in the interaction to carry off some of the momentum.

26-4 ∎ Wave–particle duality; the principle of complementarity

The photoelectric effect, the Compton effect, and other experiments (see, for example, Section 27-9 on X rays) have placed the particle theory of light on a firm experimental basis. But what about the classic experiments of Young and others (Chapter 23) on interference and diffraction which showed that the wave theory of light also rests on a firm experimental basis?

We seem to be in a dilemma. Some experiments indicate that light behaves like a wave; and others indicate that it behaves like a stream of particles. These two theories seem to be incompatible but both have been shown to have validity. Physicists have finally come to the conclusion that this duality of light must be accepted as a fact of life. It is referred to as the **wave-particle duality.** Apparently, light is a more complex phenomenon than just a simple wave or a simple beam of particles.

To clarify the situation, the great Danish physicist Niels Bohr (1885–1962) proposed his famous **principle of complementarity.** It states that to understand any given experiment, we must use either the wave or the photon theory, but not both. Yet we must be aware of both the wave and particle aspects of light if we are to have a full understanding of light. Therefore these two aspects of light complement one another.

It is not possible to "visualize" this duality. We cannot picture a combination of wave and particle. Instead, we must recognize that the two aspects of light are different "faces" that light shows to experimenters.

Part of the difficulty stems from how we think. Visual pictures (or models) in our minds are based on what we see in the everyday world. We apply the concepts of waves and particles to light because in the macroscopic world we see that energy is transferred from place to place by these two methods. We cannot see directly whether light is a wave or particle—so we do indirect experiments. To explain the experiments we apply the models of waves or of particles to the nature of light. But these are abstractions of the human mind. When we try to conceive of what light really "is," we insist on a visual picture. Yet there is no reason why light should conform to these models (or visual images) taken from the macroscopic world. The "true" nature of light—if that means anything—is not possible to visualize. The best we can do is recognize that our knowledge is limited to the indirect experiments, and that in terms of everyday language and images, light reveals both wave and particle properties.

It is worth noting that Einstein's equation $E = hf$ itself links the particle and wave properties of a light beam. In this equation E refers to the energy of a particle; and on the other side of the equation we have the frequency f of the corresponding wave.

26-5 ❚ Wave nature of matter

In 1923 Louis de Broglie (1892–) extended the idea of the wave-particle duality. He sensed deeply the symmetry in nature and argued that if light sometimes behaves like a wave and sometimes like a particle, then perhaps those things in nature thought to be particles—such as electrons and other material objects—might also have wave properties. De Broglie proposed that the wavelength of a material particle would be related to its momentum in the same way as for a photon,[†] Eq. 26-4, $p = h/\lambda$. That is, for a particle of mass m traveling with speed v, the wavelength λ is given by

$$\lambda = \frac{h}{mv}. \qquad (26\text{-}5)$$

This is sometimes called the *de Broglie wavelength* of a particle.

> EXAMPLE 26-5 Calculate the de Broglie wavelength of a 0.20-kg ball moving with a speed of 15 m/s.
>
> SOLUTION $\lambda = h/mv = (6.6 \times 10^{-34} \text{ J·s})/(0.20 \text{ kg})(15 \text{ m/s})$
> $= 2.2 \times 10^{-34} \text{ m}.$

This is an incredibly small wavelength. Even if the speed were extremely small, say 10^{-4} m/s, the wavelength would be about 10^{-29} m. Indeed, the

---

[†] de Broglie chose this formula (rather than, say, $E = hf$, which is not consistent with $p = h/\lambda$ for a particle with rest mass), because it allowed him to explain, or give a reason for, Bohr's model of the atom. This will be discussed in Section 26-10.

wavelength of any ordinary object is much too small to be measured and detected. The problem is that the properties of waves, such as interference and diffraction, are significant only when the size of objects or slits is not much larger than the wavelength. And there are no known objects or slits to diffract waves only 10^{-30} m long, so the wave properties of ordinary objects go undetected.

But tiny elementary particles, such as electrons are another matter. Since the mass m appears in the denominator in Eq. 26-5, a very small mass should give a much larger wavelength.

EXAMPLE 26-6 Determine the wavelength of an electron that has been accelerated through a potential difference of 100 V.

SOLUTION We assume that the speed of the electron will be much less than c so we use nonrelativistic mechanics. (If this assumption were to come out wrong, we would have to recalculate using relativistic formulas.) The gain in KE will equal the loss in PE, so $\frac{1}{2}mv^2 = eV$ and

$$v = \sqrt{2\,eV/m} = \sqrt{(2)(1.6 \times 10^{-19}\ \text{C})(100\ \text{V})/(9.1 \times 10^{-31}\ \text{kg})}$$
$$= 5.9 \times 10^6\ \text{m/s}.$$

Then

$$\lambda = \frac{h}{mv} = \frac{(6.6 \times 10^{-34}\ \text{J·s})}{(9.1 \times 10^{-31}\ \text{kg})(5.9 \times 10^6\ \text{m/s})} = 1.2 \times 10^{-10}\ \text{m}$$

or 0.12 nm.

From this example, we see that electrons can have wavelengths on the order of 10^{-10} m. Although small, this wavelength can be detected: the spacing of atoms in a crystal is on the order of 10^{-10} m and the orderly array of atoms in a crystal could be used as a type of diffraction grating, as was done earlier for X rays (see Section 24-11). C. J. Davisson and L. H. Germer performed the crucial experiment; they scattered electrons from the surface of a metal crystal and, in early 1927, observed that the electrons came off in regular peaks. When they interpreted these peaks as a diffraction pattern, the wavelength of the diffracted electron wave was found to be just that predicted by de Broglie, Eq. 26-5. In the same year, G. P. Thomson (son of J. J. Thomson), using a different experimental arrangement, also detected diffraction of electrons (see Fig. 26-7). Later experiments showed that protons, neutrons, and other particles also have wave properties.

Thus the wave–particle duality applies to material objects as well as to light. The principle of complementarity applies to matter as well. That is, we must be aware of both the particle and wave aspects in order to have an understanding of matter, including electrons. But again we must recognize that a visual picture of a "wave–particle" is not possible.

We might ask ourselves: "What is an electron?" The early experiments of J. J. Thomson (see Section 19-9) indicated a glow in a tube that moved when a magnetic field was applied. The results of these and other experiments were best interpreted as being caused by tiny negatively charged particles which we now call electrons. No one, however, has actually seen an electron directly. The drawings we sometimes make of electrons as tiny spheres with

FIGURE 26-7

Diffraction patterns of electrons scattered from aluminum foil, as recorded on film. (Courtesy of Education Development Center, Newton, MA 02160.)

a negative charge on them are merely convenient pictures (now recognized to be inaccurate). Again we must rely on experimental results, some of which are best interpreted using the particle model and others using the wave model. These again are mere pictures that we use to extrapolate from the macroscopic world to the tiny microscopic world of the atom. And there is no reason to expect that these models somehow reflect the reality of an electron. We thus use a wave or a particle model (whichever works best in a situation) so that we can talk about what is happening. But we shouldn't be led to believe that an electron *is* a wave or a particle. Instead we could say that an electron is the set of its properties that we can measure. Bertrand Russell said it well when he wrote that an electron is "a logical construction."

*26-6 ∎ Electron microscopes

The idea that electrons have wave properties led to the development of the *electron microscope*, which can produce images of much greater magnification than a light microscope. Figures 26-8 and 26-9 are diagrams of two types, the *transmission electron microscope*, which produces a two dimensional image, and the more recently developed *scanning electron miscroscope*, which produces images with a three-dimensional quality. In each design, the objective and eyepiece lenses are actually magnetic fields that exert forces on the electrons to bring them to a focus; the fields are produced by carefully designed current-carrying coils of wire. Photographs using each type are shown in Fig. 26-10.

Electrons accelerated by voltages on the order of 10^5 V have wavelengths on the order of 0.004 nm. The maximum resolution obtainable would be on this order (see Sections 24-7 and 24-8) but in practice aberrations in

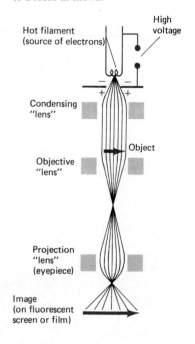

FIGURE 26-8

Transmission electron microscope. The squares represent magnetic field coils for the "magnetic lenses," which bend the electron paths and bring them to a focus as shown.

FIGURE 26-9

Scanning electron microscope. Scanning coils move an electron beam back and forth across the specimen. Secondary electrons produced when the beam strikes the specimen are collected and modulate intensity of the beam in the CRT to produce a picture.

FIGURE 26-10

Micrographs of the green alga *Tetraspora* sp.: (a) Made by the transmission electron microscope; (b) made by the scanning electron microscope. (Courtesy of J.D. Pickett-Heaps)

the magnetic lenses limit the resolution in transmission electron microscopes to at best about 0.2 to 0.5 nm. This is still 10^3 times finer than that attainable with a light microscope, and corresponds to a useful magnification of about a million. Such magnifications are difficult to attain, and more common magnifications are 10^4 to 10^5. The maximum resolution attainable with a scanning electron microscope is somewhat less, about 5 to 10 nm at best.

26-7 ∎ Early models of the atom

The idea that matter is made up of atoms was accepted by most scientists by 1900. With the discovery of the electron in the 1890s, scientists began to think of the atom itself as having a structure and that electrons were part of that structure. We now trace, in the remainder of this chapter and the next, the development of our modern understanding of the atom, and of the quantum theory with which it is intertwined.[†]

A typical model of the atom in the 1890s visualized the atom as a homogeneous sphere of positive charge inside of which there were the negatively charged electrons, a little like plums in a pudding, Fig. 26-11. J. J. Thomson, soon after his discovery of the electron in 1897, argued that the electrons in this model should be moving.

Around 1911, Ernest Rutherford (1871–1937) and his colleagues performed experiments whose results contradicted Thomson's model of the atom. In these experiments a beam of positively charged "alpha (α) particles" was directed at a thin sheet of metal foil such as gold, Fig. 26-12a. (These

FIGURE 26-11

Plum-pudding model of the atom.

Positively charged material

[†] Some readers may say: "Tell us the facts as we know them today, and don't bother us with the historical background and its outmoded theories." Such an approach would not only ignore the creative aspect of science and thus give a false impression of how science develops, but it would not really be possible to understand today's view of the atom and the quantum without discussing the concepts that led to it.

newly discovered α particles were emitted by certain radioactive materials and were soon shown to be ionized helium atoms—see Chapter 29.) It was expected from Thomson's model that the alpha particles would not be deflected significantly since electrons are so much lighter than alpha particles, and the alpha particles didn't approach any massive concentration of positive charge to strongly repel them. The experimental results completely contradicted these predictions. It was found that most of the alpha particles passed through the foil unaffected, as if the foil were mostly empty space. And of those deflected, a few were deflected at very large angles—some even in nearly the direction from which they had come. This could happen, Rutherford reasoned, only if the positively charged alpha particles were being repelled by a massive positive charge concentrated in a very small region of space (see Fig. 26-12b). He theorized that the atom must consist of a tiny but massive positively charged nucleus, containing over 99.9 percent of the mass of the atom, surrounded by electrons some distance away. The electrons would be moving in orbits about the nucleus—much like the planets move around the sun—because if they were at rest they would fall into the nucleus due to electrical attraction, Fig. 26-13. Rutherford's experiments suggested that the nucleus must have a radius of about 10^{-15} to 10^{-14} m. From kinetic theory, and especially Einstein's analysis of Brownian movement (see Section 10-1), the radius of atoms was estimated to be about 10^{-10} m. Thus the electrons would seem to be at a distance from the nucleus of about 10,000 to 100,000 times the radius of the nucleus itself; so an atom would be mostly empty space.

Rutherford's "planetary" model of the atom was a major step toward how we view the atom today. It was not, however, a complete model and presented some major problems as we shall see.

26-8 ∎ Atomic spectra: key to the structure of the atom

At the beginning of this chapter we saw that heated solids (as well as liquids and dense gases) emit light with a continuous spectrum of wavelengths. This radiation is assumed to be due to oscillations of atoms and molecules which are largely governed by the interaction of each atom or molecule with its neighbors.

Rarefied gases can also be excited to emit light. This is done by intense heating or more commonly by applying a high voltage to a "discharge tube" containing the gas at low pressure, Fig. 26-14. The radiation from excited gases had been observed early in the nineteenth century, and it was found that the spectrum was not continuous, but *discrete*. Since excited gases emit light of only certain wavelengths, when this light is analyzed through the slit of a spectroscope or spectrometer, a *line spectrum* is seen rather than a continuous spectrum. The line spectrum emitted by a number of elements in the visible region was shown in Fig. 23-22 (color plate). The *emission spectrum* is characteristic of the material and can serve as a type of "fingerprint" for identification of the gas. We also saw (Chapter 23) that if a continuous spectrum passes through a gas, dark lines are observed in the spectrum cor-

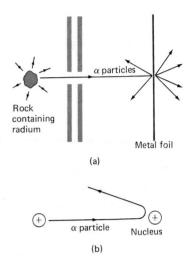

FIGURE 26-12

(a) Experimental setup for Rutherford's experiment: α particles emitted by radium strike metallic foil and some rebound backward; (b) backward rebound of α particles explained as repulsion from heavy positively charged nucleus.

FIGURE 26-13

Rutherford's model of the atom, in which electrons orbit a tiny positive nucleus (not to scale). The atom is visualized as mostly empty space.

responding to lines normally emitted by the gas. This is called an *absorption spectrum*, and it became clear that gases absorb light at the same frequencies at which they emit. With the use of film sensitive to ultraviolet and to infrared light, it was found that gases emit and absorb discrete frequencies in these regions as well as in the visible.

For our purposes here, the importance of the line spectra is that they are emitted (or absorbed) by gases that are not dense. In such thin gases, the atoms are far apart on the average and hence the light emitted or absorbed must be by individual atoms rather than through interactions between atoms as in a solid, liquid, or dense gas. Thus the line spectra serve as a key to the structure of the atom: any theory of atomic structure must be able to explain why atoms emit light only of discrete wavelengths, and it should be able to predict what these frequencies are.

Hydrogen is the simplest atom—it has only one electron orbiting its nucleus. It also has the simplest spectrum. The spectrum of most atoms shows little apparent regularity; but the spacing between lines in the hydrogen spectrum decreases in a regular way (Fig. 26-15). Indeed, in 1885, J. J. Balmer (1825–1898) showed that the four visible lines in the hydrogen spectrum (measured to be 656 nm, 486 nm, 434 nm, and 410 nm) would fit the following formula

$$\frac{1}{\lambda} = R\left(\frac{1}{2^2} - \frac{1}{n^2}\right), \qquad n = 3, 4, \dots \qquad (26\text{-}6)$$

where n takes on the values 3, 4, 5, 6 for the four lines, and R, called the *Rydberg constant*, has the value $R = 1.097 \times 10^7 \text{ m}^{-1}$. Later it was found that this **Balmer series** of lines extended into the UV region, ending at $\lambda = 365$ nm, as shown in Fig. 26-15. Balmer's formula, Eq. 26-6, also worked for these lines with higher integer values of n. The lines near 365 nm became too close together to distinguish but the limit of the series at 365 nm corresponds to $n = \infty$ (so $1/n^2 = 0$ in the formula).

Later experiments on hydrogen showed that there were other series of lines in the UV and IR and these had a pattern just like the Balmer series, but at different wavelengths. Each of these series was found to fit a formula resembling Eq. 26-6 but with the $1/2^2$ replaced by $1/1^2$, $1/3^2$, $1/4^2$, and so on. For example, the so-called *Lyman series* contains lines with wavelengths from 91 nm to 122 nm and fits the formula

$$\frac{1}{\lambda} = R\left(\frac{1}{1^2} - \frac{1}{n^2}\right), \qquad n = 2, 3, \dots.$$

And the wavelengths of the *Paschen series* fit

$$\frac{1}{\lambda} = R\left(\frac{1}{3^2} - \frac{1}{n^2}\right), \qquad n = 4, 5, \dots.$$

The Rutherford model, as it stood, was unable to explain why atoms emit line spectra. It had other difficulties as well. According to the Rutherford model, electrons orbit the nucleus and since their paths are curved the electrons are accelerating. Hence they should give off light like any other accelerating electric charge (Chapter 21). Then, since energy is conserved, the electron's own energy must decrease to compensate. Hence electrons would

FIGURE 26-14
Gas discharge tube.

FIGURE 26-15
Balmer series of lines for hydrogen.

be expected to spiral into the nucleus. As they spiraled inward, their frequency would increase gradually and so too would the frequency of the light emitted. Thus the two main difficulties of the Rutherford model are these: (1) it predicts that light of a continuous range of frequencies will be emitted, whereas experiment shows line spectra; (2) it predicts that atoms are unstable—electrons quickly spiral into the nucleus—but we know that atoms in general are stable, since the matter around us is stable.

Clearly Rutherford's model was not sufficient. Some sort of modification was needed, and it was Niels Bohr who provided it by adding an essential idea—the quantum hypothesis.

↗ 26-9 ∎ The Bohr model

Bohr had studied in Rutherford's laboratory for several months in 1912 and was convinced that Rutherford's planetary model of the atom had validity. But in order to make it work he felt that the newly developing quantum theory would somehow have to be incorporated in it. The work of Planck and Einstein had shown that in heated solids, the energy of oscillating electric charges must change discontinuously—from one discrete energy state to another with the emission of a quantum of light. Perhaps, Bohr argued, the electrons in an atom also cannot lose energy continuously but must do so in quantum "jumps." In working out his theory during the next year, Bohr postulated that electrons move about the nucleus in circular orbits, but that only certain orbits are allowed. He further postulated that an electron in each orbit would have a definite energy and would move in the orbit *without radiating energy* (even though this violated classical ideas). He thus called the possible orbits **stationary states.** Light is emitted, he hypothesized, only when an electron jumps from one stationary state to another of lower energy. When such a jump occurs, a single photon of light would be emitted whose energy, since energy is conserved, is given by

$$hf = E_u - E_l \tag{26-7}$$

where E_u refers to the energy of the upper state and E_l the energy of the lower state.

He next set out to determine what energies these orbits would have, since then the spectrum of light emitted could be predicted from Eq. 26-7. When he became aware of the Balmer formula in early 1913, he had the key he was looking for. Bohr quickly found that his theory would be in accord with the Balmer formula if he assumed the electron's angular momentum L was equal to an integer n times $h/2\pi$. As we saw in Chapter 7, angular momentum is given by $L = I\omega$ where I is the moment of inertia and ω is the angular velocity; for a single particle of mass m moving in a circle of radius r with speed v, $I = mr^2$ and $\omega = v/r$; hence $L = I\omega = (mr^2)(v/r) = mvr$. Bohr's *quantum condition* then is

$$L = mvr_n = n\frac{h}{2\pi}, \qquad n = 1, 2, 3, \ldots . \tag{26-8}$$

Here n is an integer and r_n is the radius of the nth possible orbit. The allowed orbits are numbered 1, 2, 3, . . . , according to the value of n, which is called the **quantum number** of the orbit.

Equation 26-8 did not have a firm theoretical foundation. Bohr had searched for some "quantum condition," and such tries as $E = hf$ (where E represents the energy of the electron in an orbit) did not give results in accord with experiment. Bohr's reason for using Eq. 26-8 was simply that it worked; and we now look at how.

An electron in a circular orbit of radius r_n would have a centripetal acceleration v^2/r_n produced by the electrical force of attraction between the negative electron and the positive nucleus according to Coulomb's law,

$$F = k \frac{(Ze)(e)}{r^2},$$

Fig. 26-16. The charge on the electron is $q_1 = -e$, and that on the nucleus is $q_2 = +Ze$, where Z is the number of positive charges[†] (or protons). For the hydrogen atom, $Z = +1$.

From Newton's second law, $F = ma$, we substitute $a = v^2/r_n$ and Coulomb's law for F and obtain

$$k \frac{Ze^2}{r_n^2} = \frac{mv^2}{r_n}.$$

We solve this for r_n and substitute for v from Eq. 26-8 (which says $v = nh/2\pi m r_n$):

$$r_n = \frac{kZe^2}{mv^2} = \frac{kZe^2 4\pi^2 m r_n^2}{n^2 h^2}.$$

We solve for r_n (it appears on both sides, so we cancel one of them) and find

$$r_n = \frac{n^2 h^2}{4\pi^2 m k Z e^2}. \tag{26-9}$$

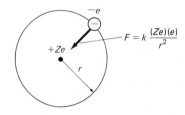

FIGURE 26-16

Electric force (Coulomb's law) keeps negative electron in orbit around a positively charged nucleus.

This equation gives the radii of the possible orbits. The smallest orbit is for $n = 1$ and for hydrogen ($Z = 1$) has the value

$$r_1 = \frac{(1)^2 (6.626 \times 10^{-34}\,\text{J·s})^2}{4(3.14)^2 (9.11 \times 10^{-31}\,\text{kg})(9.00 \times 10^9\,\text{N·m}^2/\text{C}^2)(1)(1.602 \times 10^{-19}\,\text{C})^2}$$

$$r_1 = 0.529 \times 10^{-10}\,m. \tag{26-10}$$

The radius of the smallest orbit in hydrogen, r_1, is sometimes called the *Bohr radius*. From Eq. 26-9 we can see that the radii of the larger orbits[‡] increase

---

[†] We include Z in our derivation so we can treat other single-electron ("hydrogenlike") atoms such as the ions He$^+$ ($Z = 2$) and Li^{2+} ($Z = 3$). Helium in the neutral state has two electrons; if one electron is missing the remaining He$^+$ ion consists of one electron revolving around a nucleus of charge $+2e$; similarly, doubly ionized lithium, Li^{2+}, also has a single electron, and in this case $Z = 3$.

[‡] Be careful not to believe that these well-defined orbits actually exist. The Bohr model is only a model, not reality; and, as we shall see in the next chapter, the idea of electron orbits has been rejected. Instead, today electrons are better thought of as forming "clouds," as discussed in Chapter 27.

as n^2, so

$$r_2 = 4r_1 = 2.11 \times 10^{-10} \text{ m}$$
$$r_3 = 9r_1 = 4.74 \times 10^{-10} \text{ m},$$

and so on. The first four are shown in Fig. 26-17. Notice that, according to Bohr's model, an electron can exist only in the orbits given by Eq. 26-9; it cannot be in between.

In each of its possible orbits, the electron would have a definite energy, as the following calculation shows. The total energy equals the sum of the kinetic and potential energies. The potential energy of the electron is given by $\text{PE} = qV = -eV$, where V is the potential due to a point charge $+Ze$ as given by Eq. 16-3: $V = kQ/r = kZe/r$. So

$$\text{PE} = -eV = -k\frac{Ze^2}{r}.$$

The total energy E_n for an electron in the nth orbit of radius r_n is the sum of its kinetic and potential energies:

$$E_n = \tfrac{1}{2}mv^2 - \frac{kZe^2}{r_n}.$$

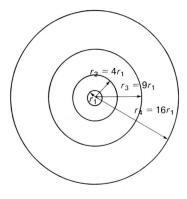

FIGURE 26-17

Possible orbits in the Bohr model of hydrogen; $r_1 = 0.529 \times 10^{-10}$ m.

When we substitute v from Eq. 26-8 and r_n from Eq. 26-9 into this equation, we obtain

$$E_n = -\frac{2\pi^2 Z^2 e^4 m k^2}{h^2}\frac{1}{n^2}, \qquad n = 1, 2, 3, \ldots. \qquad (26\text{-}11)$$

For hydrogen ($Z = 1$), the lowest energy level has $n = 1$, and when numbers are substituted into Eq. 26-11, we find

$$E_1 = -2.17 \times 10^{-18} \text{ J} = -13.6 \text{ eV},$$

where we have converted joules to electron volts, as is customary in atomic physics. Since n^2 appears in the denominator of Eq. 26-11, the energies of the larger orbits are given by

$$E_n = \frac{-13.6 \text{ eV}}{n^2}.$$

For example,

$$E_2 = \frac{-13.6 \text{ eV}}{4} = -3.40 \text{ eV},$$

$$E_3 = \frac{-13.6 \text{ eV}}{9} = -1.51 \text{ eV},$$

we see from Eq. 26-11 that not only are the orbit radii quantized, but so is the energy. Notice that although the energy for the larger orbits has a smaller numerical value, all the energies are less than zero. Thus -3.4 eV is a greater energy than -13.6 eV. Hence the orbit closest to the nucleus (r_1) has the lowest energy. The reason the energies have negative values has to do with the way we defined the zero for potential energy; for two point charges, $\text{PE} = kq_1q_2/r$ corresponds to zero PE when the two are infinitely far apart.

Thus, an electron without KE that is free from the atom would have $E = 0$, corresponding to $n = \infty$. Electrons that are part of an atom would have $E < 0$.[†] Also, it is clear that energy would be required to remove an electron from an atom—this is called the **binding energy** or **ionization energy**—so if $E = 0$ for a free electron, an electron bound to an atom must have $E < 0$. That is, energy must be added to bring its energy up, from a negative value, to zero. The ionization energy for hydrogen has been measured to be 13.6 eV, and this corresponds precisely to removing an electron from the lowest state, $E_1 = -13.6$ eV, up to $E = 0$ where it will be free.

It is useful to show the various possible energy values as horizontal lines on an energy-level diagram.[‡] This is shown for hydrogen in Fig. 26-18. The quantum number n that labels the orbit radii also labels the energy levels. The lowest *energy level* or *energy state* has energy E_1, and is called the **ground state.** The higher states, E_2, E_3, and so on, are called **excited states.** The electron in a hydrogen atom can be in any one of these levels according to Bohr theory. But it could never be in between, say at -9.0 eV. At room temperature, nearly all H atoms will be in the ground state. At higher temperatures, or during an electric discharge when there are many collisions between free electrons and atoms, many electrons can be in excited states. Once

[†] This is not unlike the case of the gravitational potential energy of a stone, PE = mgh; if we measure h from the top of a table, say, then when the stone is on the floor below the table, it will have negative potential energy.

[‡] Note that an electron can have any energy above $E = 0$, for there it is free; thus there is a continuum of energy states above $E = 0$, as indicated in the energy-level diagram of Fig. 26-18.

FIGURE 26-18

Energy-level diagram for the hydrogen atom, showing origin of spectral lines for the Lyman, Balmer, and Paschen series.

in an excited state, an electron can jump down to a lower state, and give off a photon in the process. This is, according to the Bohr model, the origin of the emission spectra of excited gases. The vertical arrows in Fig. 26-18 represent the transitions or jumps that correspond to the various observed spectral lines. For example, an electron jumping from the level $n = 3$ to $n = 2$ would give rise to the 656-nm line in the Balmer series, and the jump from $n = 4$ to $n = 2$ would give rise to the 486-nm line (see Fig. 26-15). We can predict wavelengths of the spectral lines emitted by combining Eq. 26-7 with Eq. 26-11. Since $hf = hc/\lambda$, we have

$$\frac{1}{\lambda} = \frac{hf}{hc} = \frac{1}{hc} (E_n - E_{n'})$$

or

$$\frac{1}{\lambda} = \frac{2\pi^2 Z^2 e^4 m k^2}{h^3 c} \left(\frac{1}{n'^2} - \frac{1}{n^2} \right), \qquad (26\text{-}12)$$

where n refers to the upper state and n' to the lower state. This theoretical formula has the same form as the experimental Balmer formula, Eq. 26-6, with $n' = 2$. Thus we see that the Balmer series of lines corresponds to transitions or "jumps" that bring the electron down to the second energy level. Similarly, $n' = 1$ corresponds to the Lyman series and $n' = 3$ to the Paschen series (see Fig. 26-18). When the constant $(2\pi^2 Z^2 e^4 m k^2 / h^3 c)$ in Eq. 26-12 is evaluated with $Z = 1$, it is found to have the measured value of the Rydberg constant, $R = 1.0974 \times 10^7 \text{ m}^{-1}$, in accord with experiment.

The great success of Bohr's theory is that it explains why atoms emit line spectra and accurately predicts, for hydrogen, the wavelengths of emitted light. The Bohr theory also explains absorption spectra: photons of just the right wavelength can knock an electron from one energy level to a higher one. To conserve energy, the photon must have just the right energy. This explains why a continuous spectrum passing through a gas will have dark (absorption) lines at the same frequencies as the emission lines.

The Bohr theory also ensures the stability of atoms. It establishes stability by fiat: the ground state is the lowest state for an electron and there is no lower energy level to which it can go and emit more energy. Finally, as we saw above, the Bohr theory accurately predicts the ionization energy of 13.6 eV for hydrogen.

EXAMPLE 26-7 Use Fig. 26-18 to determine the wavelength of the first Lyman line, the transition from $n = 2$ to $n = 1$.

SOLUTION In this case, $hf = E_2 - E_1 = 13.6 \text{ eV} - 3.4 \text{ eV} = 10.2 \text{ eV} = 1.63 \times 10^{-18} \text{ J}$. Since $\lambda = c/f$, we have

$$\lambda = \frac{c}{f} = \frac{hc}{E_2 - E_1} = \frac{(6.63 \times 10^{-34} \text{ J·s})(3.00 \times 10^8 \text{ m/s})}{1.63 \times 10^{-18} \text{ J}}$$

$$= 1.22 \times 10^{-7} \text{ m}$$

or 122 nm, which is in the UV.

EXAMPLE 26-8   Determine the wavelength of light emitted when a hydrogen atom makes a transition from the $n = 6$ to the $n = 2$ energy level of the Bohr model.

SOLUTION   We can use Eq. 26-12 or its equivalent, Eq. 26-6, with $R = 1.097 \times 10^7 \text{ m}^{-1}$. Thus

$$\frac{1}{\lambda} = (1.097 \times 10^7 \text{ m}^{-1})(\tfrac{1}{4} - \tfrac{1}{36}) = 2.44 \times 10^6 \text{ m}^{-1}.$$

So $\lambda = 4.10 \times 10^{-7}$ m or 410 nm. This is the fourth line in the Balmer series, Fig. 26-15, and is violet in color.

EXAMPLE 26-9   Use the Bohr model to determine the ionization energy of the ion He^+ which has a single electron. Also calculate the minimum wavelength a photon must have to cause ionization.

SOLUTION   We want to determine the energy required to lift the electron from its ground state to the free state, $E = 0$. The ground state energy of He^+ is given by Eq. 26-11 with $n = 1$ and $Z = 2$. Since all the symbols in Eq. 26-11 are the same as for the calculation for hydrogen, except Z is 2 instead of 1, we see that E_1 will be $Z^2 = 2^2 = 4$ times larger than for hydrogen. That is

$$E_1 = 4(-13.6 \text{ eV}) = -54.4 \text{ eV}.$$

Thus, to ionize the He^+ ion should require 54.4 eV, and this agrees with experiment. The minimum wavelength photon that can cause ionization will have energy $hf = 54.4$ eV and wavelength $\lambda = c/f = hc/hf = (6.63 \times 10^{-34} \text{ J·s}) \times (3.00 \times 10^8 \text{ m/s})/(54.4 \text{ eV})(1.60 \times 10^{-19} \text{ J/eV}) = 22.8$ nm.   If the atom absorbed a photon of greater energy (wavelength shorter than 22.8 nm), the atom could still be ionized and the freed electron would have kinetic energy of its own.

In the last example, we saw that E_1 for the He^+ ion is four times lower than that for hydrogen. Indeed, the energy-level diagram for He^+ looks just like that for hydrogen, Fig. 26-18, except that the numerical values for each energy level are four times larger. It is important to note, however, that we are talking here about the He^+ *ion*. Normal (neutral) helium has two electrons and its spectrum is entirely different.

We should note that Bohr made some radical assumptions that were at variance with classical ideas. He assumed that electrons in fixed orbits do not radiate light even though they are accelerating, and he assumed that angular momentum is quantized. Furthermore, he was not able to say how an electron moved when it made a transition from one energy level to another. On the other hand, there is no real reason to expect that in the tiny world of the atom electrons would behave as ordinary-sized objects do. Nonetheless, he felt that where quantum theory overlaps with the macroscopic world, it should predict classical results. This is the *correspondence principle* already mentioned in regard to relativity (Section 25-14). This principle does seem to work for Bohr's theory of the hydrogen atom. The orbit sizes and energies are quite different for $n = 1$ and $n = 2$, say. But orbits with $n = 100,000,000$ and $100,000,001$ would be very close in size and energy (see Fig. 26-18). Indeed, jumps between such large orbits, which would approach everyday sizes, would be imperceptible; such orbits would thus appear to be continuous, which is what we expect in the everyday world.

26-10 ▮ de Broglie's hypothesis

Bohr's theory was largely of an *ad hoc* nature. Assumptions were made so that theory would agree with experiment. But Bohr could give no reason why the orbits were quantized. Finally, 10 years later, a reason was found by de Broglie.

We saw in Section 26-5 that in 1923, Louis de Broglie proposed that material particles, such as electrons, have a wave nature; and that this hypothesis was confirmed by experiment several years later.

One of de Broglie's original arguments in favour of the wave nature of electrons was that it provided an explanation for Bohr's theory of the hydrogen atom. According to de Broglie, a particle of mass m moving with speed v would have a wavelength (Eq. 26-5) of

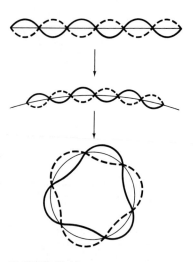

FIGURE 26-19

An ordinary standing wave compared to a circular standing wave.

$$\lambda = \frac{h}{mv}.$$

Each electron orbit in an atom, he proposed, is actually a standing wave. As we saw in Chapter 13, when a violin or guitar string is plucked, a vast number of wavelengths are excited. But only certain ones—those that have nodes at the ends—are sustained. These are the *resonant* modes of the string. All other wavelengths interfere with themselves upon reflection and their amplitudes quickly drop to zero. Since electrons move in circles, according to Bohr's theory, de Broglie argued that the electron wave must be a *circular* standing wave that closes on itself, Fig. 26-19. If the wavelength of a wave does not close on itself, as in Fig. 26-20, destructive interference takes place as the wave travels around the loop and it quickly dies out. Thus, the only waves that persist are those for which the circumference of the circular orbit contains a whole number of wavelengths, Fig. 26-21. The circumference of a Bohr orbit of radius r_n is $2\pi r_n$ so we must have

FIGURE 26-20

When a wave does not close (and hence interferes with itself), it rapidly dies out.

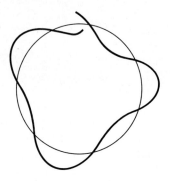

$$2\pi r_n = n\lambda, \qquad n = 1, 2, 3, \ldots.$$

When we substitute $\lambda = h/mv$, we get

$$2\pi r_n = \frac{nh}{mv},$$

or

$$mvr_n = \frac{nh}{2\pi}.$$

This is just the *quantum condition* proposed by Bohr on an *ad hoc* basis, Eq. 26-8. And it is from this equation that the discrete orbits and energy levels were derived. Thus we have an explanation for the quantized orbits and en-

$n = 2$ $\qquad n = 3$ $\qquad n = 5$

FIGURE 26-21

Standing circular waves for two, three, and five wavelengths on the circumference; n, the number of wavelengths, is also the quantum number.

ergy states in the Bohr model: they are due to the wave nature of the electron and the fact that only resonant "standing" waves can persist. This implies that the *wave–particle duality* is at the root of atomic structure.

It should be noted in viewing the circular electron waves of Fig. 26-21 that the electron is not to be thought of as following the oscillating wave pattern. In the Bohr model of hydrogen, the electron, considered as a particle, moves in a circle. The circular wave, on the other hand, represents the *amplitude* of the electron "matter wave," and in Fig. 26-21 the wave amplitude is shown superimposed on the circular path of the particle orbit for convenience.

Bohr's theory worked well for hydrogen and for one-electron ions. It did not prove as successful for multielectron atoms. We will discuss this and other problems with the Bohr theory in the next chapter, and we will see how a new and radical theory, quantum mechanics, finally solved the problem of atomic structure and gave us a very different view of the atom: the idea of electrons in well-defined orbits was replaced with a picture of electron "clouds." And this new (and now generally accepted) theory of quantum mechanics has given us a wholly different view of the basic mechanisms underlying physical processes.

▮ Summary

Quantum theory has its origins in the *quantum hypothesis* that molecular oscillations are quantized: their energy E can only be integer (n) multiples of hf, where h is Planck's constant and f is the natural frequency of oscillation: $E = nhf$. This hypothesis explained the spectrum of radiation emitted by (black) bodies at high temperature.

Einstein proposed that for some purposes light might be pictured as being emitted and absorbed as quanta (particles) we now call *photons*, each with energy

$$E = hf.$$

He proposed the photoelectric effect as a test for the photon theory of light. In this effect, the photon theory says that each incident photon can strike an electron in a material and eject it if it has sufficient energy; the maximum energy of ejected electrons is then linearly related to the frequency of the incident light. The photon theory is also supported by the Compton effect and the observation of electron-positron *pair production*.

The wave–particle duality refers to the idea that light and matter (such as electrons) have both wave and particle properties. The wavelength of a material object is given by

$$\lambda = \frac{h}{mv},$$

where mv is the momentum of the object. The *principle*

of complementarity states that we must be aware of both the particle and wave properties of light and of matter for a complete understanding of them.

Early models of the atom include Thomson's modified plum-pudding model and Rutherford's planetary (or nuclear) model. Rutherford's model, which was created to explain the back-scattering of alpha particles from thin metal foils, assumes that an atom consists of a massive positively charged nucleus surrounded (at a relatively great distance) by electrons.

To explain the line spectra emitted by atoms, as well as their stability, Bohr proposed a theory which postulated (1) that electrons bound in an atom can only occupy orbits for which the angular momentum is quantized, which results in discrete values for the radius and energy; (2) that an electron in such a *stationary state* emits no radiation; (3) that, if an electron jumps to a lower state, it emits a photon whose energy equals the difference in energy between the two states; (4) that the angular momentum L of atomic electrons is quantized by the rule $L = nh/2\pi$, where n is an integer called a *quantum number*. The $n = 1$ state in hydrogen is the *ground state*, which has an energy $E_1 = -13.6$ eV; higher values of n correspond to *excited states* and their energies are $E_n = -13.6$ eV/n^2. Atoms are excited to these higher states by collisions with other atoms or electrons or by absorption of a photon of just the right frequency.

De Broglie's hypothesis that electrons (and other matter) have a wavelength $\lambda = h/mv$ gave an explanation for Bohr's quantized orbitals by bringing in the wave–particle duality: the orbits correspond to circular standing waves in which the circumference of the orbit equals a whole number of wavelengths.

▌ Questions

1. What can be said about the relative temperature of whitish-yellow, reddish, and bluish stars?
2. If energy is radiated by all objects, why can't we see them in the dark? (See also Chapter 11.)
3. Does a light bulb at a temperature of 2500 K produce as white a light as the sun at 6000 K? Explain.
4. Why do jewellers often examine diamonds in daylight rather than with indoor light?
5. "Orthochromatic" film is not sensitive to red light. Give an explanation based on the photon theory of light.
6. If the threshold wavelength in the photoelectric effect increases when the emitting metal is changed, what can you say about the work functions of the two metals?
7. Explain why the existence of a cutoff frequency in the photoelectric effect more strongly favors a particle theory rather than a wave theory of light.
8. UV light causes sunburn, whereas visible light does not. Explain.
9. If an X-ray photon is scattered by an electron, does its wavelength change? If so, does it increase or decrease?
10. Explain how the photoelectric circuit of Fig. 26-3 could be used in (a) a burglar alarm, (b) a smoke detector, (c) a photographic light meter, (d) a spectrophotometer (see Section 23-7), (e) a film sound track.
11. Why do we say that light has wave properties? Why do we say that light has particle properties?
12. Why do we say that electrons have wave properties? Why do we say that electrons have particle properties?

13. What is the difference between a photon and an electron? Be specific.
14. If an electron and a proton travel at the same speed, which has the shorter wavelength?
15. In Rutherford's planetary model of the atom, what keeps the electrons from flying off into space?
16. Which of the following can emit a line spectrum: (a) gases, (b) liquids, (c) solids. Which can emit a continuous spectrum?
17. Why doesn't the O_2 gas in the air around us give off light?
18. How can you tell if there is oxygen on the sun?
19. When a wide spectrum of light passes through hydrogen gas at room temperature, absorption lines are observed that correspond only to the Lyman series. Why don't we observe the other series?
20. Explain how the closely spaced energy levels for hydrogen near the top of Fig. 26-18 correspond to the closely spaced spectral lines at the top of Fig. 26-15.
21. Discuss the differences between Rutherford's and Bohr's theory of the atom.
22. In a helium atom, which contains two electrons, do you think that on the average the electrons are closer to the nucleus or farther away than in a hydrogen atom? Why?
23. How can the spectrum of hydrogen contain so many lines when hydrogen contains only one electron?
24. Why was the Balmer series of lines from hydrogen the first to be observed and analyzed rather than, say, the Lyman or Paschen series?

▌Problems

SECTION 26-1

1. (I) An HCl molecule vibrates with a natural frequency of 8.1×10^{13} Hz. What is the difference in energy (in joules and electron volts) between possible values of the oscillation energy?

2. (I) (a) At what temperature will the peak of a blackbody spectrum be at 1.0 nm? (b) What is the wavelength at the peak of a blackbody spectrum if the body is at a temperature of 800 K?

3. (II) A child's swing has a natural frequency of 0.40 Hz. (a) What is the separation between possible energy values (in joules)? (b) If the swing reaches a height of 30 cm above its lowest point and has a mass of 20 kg (including the child), what is the value of the quantum number n? (c) What is the fractional change in energy between levels whose quantum numbers are n (as just calculated) and $n + 1$? Would quantization be measurable in this case?

SECTION 26-2

4. (I) What is the energy of a photon of wavelength (a) 400 nm, and (b) 700 nm?

5. (I) Calculate the wavelength of a 4.6-eV photon in nanometers.

6. (I) What is the energy of photons (in eV) emitted by a 100-MHz FM radio station?

7. (I) What minimum frequency of light is needed to eject electrons from a metal whose work function is 3.9×10^{-19} J?

8. (II) What is the longest wavelength of light that will emit electrons from a metal whose work function is 2.3 eV?

9. (II) Barium has a work function of 2.48 eV. What is the maximum kinetic energy of electrons if the metal is illuminated by light of wavelength 480 nm? What is their speed?

10. (II) What is the maximum kinetic energy of electrons ejected from barium ($W_0 = 2.48$ eV) when illuminated by white light, $\lambda = 400$ to 700 nm?

11. (II) When UV light of wavelength 280 nm falls on a metal surface, the maximum kinetic energy of emitted electrons is 0.85 eV. What is the work function of the metal?

12. (II) Show that the energy E (in electron volts) of a photon whose wavelength is λ (meters) is given by

$$E = 1.24 \times 10^{-6}/\lambda.$$

13. (II) Sunlight reaching the earth has an intensity of 1300 W/m². How many photons per square meter per second does this represent? Take the average wavelength to be 550 nm.

14. (II) The threshold wavelength for emission of electrons from a given surface is 380 nm. What will be the maximum kinetic energy of ejected electrons when the wavelength is changed to (a) 480 nm, (b) 280 nm?

15. (II) A certain type of film is sensitive only to light whose wavelength is less than 640 nm. What is the energy (kcal/mol) needed for the chemical reaction to occur which causes the film to change?

16. (II) If a 100-W light bulb emits 3.0 percent of the input energy as visible light (average wavelength 550 nm) uniformly in all directions, approximately how many photons per second of visible light will strike the pupil (4.0 mm diameter) of the eye of an observer 10 km away?

17. (II) When 230-nm light falls on a metal, the current through a photoelectric circuit (Fig. 26-3) is brought to zero at a reverse voltage of 1.64 V. What is the work function of the metal?

18. (III) In the Compton effect, a 0.100-nm photon strikes a free electron in a head-on collision and knocks it into the forward direction. The rebounding photon recoils directly backwards. Use conservation of energy and momentum to determine (a) the kinetic energy of the electron, and (b) the wavelength of the recoiling photon. Assume the electron's kinetic energy is given by the nonrelativistic formula. (Note: see Eq. 26-4.)

SECTION 26-3

19. (I) How much total kinetic energy will an electron–positron pair have if produced by a 3.6-MeV photon?

20. (II) What is the momentum and the effective mass of a 0.10-nm X-ray photon?

21. (II) What is the minimum photon energy needed to produce a μ^+–μ^- pair? The mass of each μ is 207 times the mass of the electron. What is the wavelength of such a photon?

22. (II) An electron-positron pair, each with a kinetic energy of 350 keV, is produced from a photon. What was the energy and wavelength of the photon?

SECTION 26-5

23. (I) Calculate the wavelength of a 0.10-kg ball traveling at 0.20 m/s.

24. (I) What is the wavelength of a neutron ($m_0 = 1.67 \times 10^{-27}$ kg) traveling 1.0×10^4 m/s?

25. (II) By what potential difference must a proton ($m_0 = 1.67 \times 10^{-27}$ kg) be accelerated to have a wavelength $\lambda = 0.0011$ nm?

26. (II) What is the wavelength of an electron of energy (a) 5.0 eV, and (b) 50 eV?

27. (II) Show that if an electron and a proton have the same kinetic energy, the proton has the shorter wavelength.

28. (II) Calculate the ratio of KE of an electron to that of a proton whose wavelengths are equal. Assume that the speeds are much less than the speed of light.

29. (II) What is the wavelength of an N_2 molecule in the air at room temperature? (Hint: see Chapter 10.)

30. (III) Show that the wavelength of a particle of mass m_0 with kinetic energy KE is given by the relativistic formula $\lambda = hc/\sqrt{(\text{KE})^2 + 2m_0c^2(\text{KE})}$.

31. (III) What is the kinetic energy and wavelength of a "thermal" neutron (one that is in equilibrium at room temperature—see Chapter 10).

32. (III) Show that for a particle of rest mass m_0, if $\lambda = h/mv$ then it cannot be true that $E = hf$ where E is (a) kinetic energy, or (b) KE plus rest mass energy, and $v = f\lambda$ is the speed of the particle.

*SECTION 26-6

*33. (II) What voltage is needed to produce electron wavelengths of 0.10 nm? (Assume that electrons are nonrelativistic.)

*34. (III) Electrons are accelerated by 1200 V in an electron microscope. To achieve a resolution of 4.0 nm, what numerical aperture is required?

*35. (III) What is the theoretical limit of resolution for an electron microscope whose electrons are accelerated through 50 kV? (Relativistic formulas should be used.)

SECTION 26-7

36. (II) In certain of Rutherford's experiments, the α particles (mass $= 6.68 \times 10^{-27}$ kg) had a kinetic energy of 4.8 MeV. How close could they get to a gold nucleus (charge $= +79e$)? Ignore the recoil motion of the nucleus.

SECTION 26-9

37. (I) How much energy is needed to ionize a hydrogen atom in the $n = 3$ state?

38. (I) For what maximum kinetic energy is a collision between an electron and a hydrogen atom in its ground state definitely elastic?

39. (I) The third longest wavelength in the Paschen series in hydrogen (Fig. 26-18) corresponds to what transition?

40. (I) (a) Determine the wavelength of the second Balmer line ($n = 4$ to $n = 2$ transition) using Fig. 26-18. Determine likewise (b) the wavelength of the first Lyman line and (c) the wavelength of the fourth Balmer line.

41. (I) Calculate the ionization energy of doubly ionized lithium, Li^{2+}, which has $Z = 3$.

42. (II) What is the longest wavelength light capable of ionizing a hydrogen atom in the ground state?

43. (II) At low temperatures, nearly all the atoms in hydrogen gas will be in the ground state. What minimum frequency photon is needed if the photoelectric effect is to be observed?

44. (II) Construct the energy-level diagram for the He^+ ion (see Fig. 26-18).

45. (II) Construct the energy-level diagram for doubly ionized lithium, Li^{2+}.

46. (II) By what fraction does the mass of an H atom decrease when it makes an $n = 2$ to $n = 1$ transition?

47. (II) What is the potential energy and kinetic energy of an electron in the ground state of the hydrogen atom?

48. (II) An excited hydrogen atom could, in principle, have a radius of 1.0 mm. What would be the value of n for a Bohr orbit of this size? What would its energy be?

49. (II) Calculate the ratio of the gravitational to electric force for the electron in a hydrogen atom. Can the gravitational force be safely ignored?

50. (II) Are nonrelativistic formulas justified in the Bohr atom? To check, calculate the electron's velocity, v, in terms of c for the ground state of hydrogen.

51. (II) Suppose an electron was bound to a proton, as in the hydrogen atom, by the gravitational force rather than by the electric force. What would be the radius, and energy, of the first Bohr orbit?

52. (II) Electrons accelerated by a potential difference of 12.3 V pass through a gas of hydrogen atoms at room temperature. What wavelengths of light will be emitted?

53. (II) Atoms can be formed in which a muon (mass $=$ 207 times the mass of an electron) replaces one of the electrons in an atom. Calculate, using Bohr theory, the energy of the photon emitted when a muon makes a transition from $n = 2$ to $n = 1$ in a muonic $^{208}_{82}Pb$ atom (lead whose nucleus has a mass 208 times the proton mass and charge $+82e$).

54. (III) *Correspondence principle*: Show that for large values of n, the difference in radius Δr between two adjacent orbits (with quantum numbers n and $n - 1$) is given by

$$\Delta r = r_n - r_{n-1} \approx \frac{2r_n}{n}$$

so $\Delta r/r_n \to 0$ as $n \to \infty$ in accordance with the correspondence principle. [Note that we can check the correspondence principle by either considering large values of n ($n \to \infty$) or by letting $h \to 0$. Are these equivalent?]

SECTION 26-10

55. (III) Suppose a particle of mass m is confined to a one-dimensional box of width L. According to quantum theory, the particle's wave (with $\lambda = h/mv$) is a standing wave with nodes at the edges of the box. (a) Show the possible modes of vibration on a diagram. (b) Show that the kinetic energy of the particle has quantized energies given by KE $= n^2h^2/8mL^2$ where n is an integer. (c) Calculate the ground-state energy ($n = 1$) for an electron confined to a box of width 0.50×10^{-10} m. (d) What is the ground-state energy for a baseball ($m = 140$ g) in a box 0.50 m wide? (e) An electron confined to a box has a ground-state energy of 10 eV. What is the width of the box?

27 QUANTUM MECHANICS OF *ATOMS*

Bohr's model of the atom gave us a first (though rough) picture of what an atom is like. It proposed an explanation for why there should be emission and absorption of light by atoms at discrete wavelengths, as well as the stability of atoms. The wavelengths of the line spectra and ionization energy for hydrogen and one-electron ions are calculated to be in excellent agreement with experiment. But the Bohr theory had important limitations. It was not able to predict the line spectra for more complex atoms—not even for the neutral helium atom, which has only two electrons. Nor could it explain why the emission lines, when viewed with great precision, actually consist of two or more very closely spaced lines (referred to as *fine structure*). The Bohr theory also didn't explain why some spectral lines were brighter than others. And it couldn't explain the bonding of atoms in molecules or in solids and liquids.

From a theoretical point of view, too, the Bohr theory was not really satisfactory. For it was a strange mixture of classical and quantum ideas. And the wave–particle duality was still not really resolved.

We mention these limitations of the Bohr theory not to disparage it—for it was a landmark in the history of science. Rather, we had to mention them to show why, in the early 1920s, it became increasingly evident that a new, more comprehensive theory was needed. It was not long in coming. Less than 2 years after de Broglie gave us his matter-wave hypothesis, Erwin Schrödinger (1887–1961) and Werner Heisenberg (1901–1976) independently developed a new comprehensive theory. Their separate approaches were quite different but were soon shown to be fully compatible.

27-1 ▌ Quantum mechanics—a new theory

The new theory, called **quantum mechanics,** unifies the wave–particle duality into a single consistent theory. As a theory, quantum mechanics has been extremely successful. It has successfully dealt with the spectra emitted by complex atoms, even the fine details. It explains the relative brightness of spectral lines and how atoms form molecules. It is also a much more general theory that covers all quantum phenomena from blackbody radiation to atoms and molecules. It has explained a wide range of natural phenomena and from its predictions many new practical devices have become possible. Indeed, it has been so successful that it is accepted today by nearly all physicists as the fundamental theory underlying physical processes.

Quantum mechanics deals mainly with the microscopic world of atoms and light; but in our everyday macroscopic world we do perceive light and we accept that ordinary objects are made up of atoms. This new theory must therefore also account for the verified results of classical physics; that is, when it is applied to macroscopic phenomena, quantum mechanics must be able to produce the old classical laws. This, the *principle of correspondence* (already mentioned in Section 26-9), is met fully by quantum mechanics. This doesn't mean we throw away classical theories such as Newton's laws. In the everyday world the latter are far easier to apply and give an accurate description. But when we deal with high speeds, close to the speed of light, we must use the theory of relativity; and when we deal with the tiny world of the atom, we use quantum mechanics.

Although we won't go into the detailed mathematics of quantum mechanics, we will discuss the main ideas and how they involve the wave and particle properties of matter to explain atomic structure and other applications.

*27-2 ∎ The wave function and its interpretation; the double-slit experiment

The important properties of any wave are its wavelength, frequency, and amplitude. For an electromagnetic wave, the wavelength determines whether the light is visible or not, and if so, what color it is. We also have seen that the wavelength (or frequency) is a measure of the energy of the corresponding photon ($E = hf$). The amplitude of an electromagnetic wave at any point is the strength of the electric (or magnetic) field at that point, and is related to the intensity of the wave (the brightness of the light).

For material particles such as electrons, quantum mechanics relates the wavelength to momentum according to de Broglie's formula, $\lambda = h/mv$. But what about the amplitude of a matter wave? In quantum mechanics the amplitude of, say, an electron wave is called the **wave function** and is given the symbol Ψ (the Greek letter psi). Thus Ψ represents the amplitude, as a function of time and position, of a new kind of field which we might call a "matter" field or a matter wave.

To calculate the wave function Ψ in a given situation (say, for an electron in an atom) is one of the basic tasks of quantum mechanics. Indeed, the development of an equation to do so was Schrödinger's great contribution. The *Schrödinger wave equation*, as it is called, is a differential equation (so we will not deal with it here) and is considered to be the basic equation for the description of nonrelativistic material particles. Since our treatment of quantum mechanics is necessarily brief (whole books are needed for the subject), the actual form of the Schrödinger equation, and how it is solved, will not concern us here. What will be useful are the solutions obtained and the meaning of the wave function, Ψ, itself. One way to interpret Ψ is simply as the amplitude at any point in space and time of a "matter wave", just as

E (the electric field vector) represents the amplitude of an electromagnetic wave. Another interpretation is possible, however, based on the wave–particle duality. To understand this, we make an analogy with light.

We saw in Chapter 13 that the intensity I of any wave is proportional to the square of the amplitude. This holds true for light waves as well, as we saw in Chapter 21; that is,

$$I \propto E^2$$

where E is the electric field strength. From the *particle* point of view, the intensity of a light beam is proportional to the number of photons, N, that pass through a given area per unit time. The more photons (of a given wavelength) there are, the greater the intensity. Thus

$$I \propto E^2 \propto N.$$

This proportion can be turned around so we have

$$N \propto E^2.$$

That is, the number of photons (striking a page of this book, say) is proportional to the square of the electric field strength.

If the light beam is very weak, only a few photons will be involved. Indeed, it is possible to "build up" a photograph in a camera using very weak light so the effect of individual photons can be seen. If we are dealing with only one photon, the relationship above ($N \propto E^2$) can be interpreted in a slightly different way. At any point the square of the electric field strength, E^2, is a measure of the *probability* that a photon will be at that location. At points where E^2 is large, there is a high probability the photon will be there; where E^2 is small, the probability is low.

We can interpret matter waves in the same way. The wave function Ψ may vary in magnitude from point to point in space and time. If Ψ describes a collection of many electrons, then Ψ^2 at any point will be proportional to the number of electrons expected to be found at that point. When dealing with small numbers of electrons we can't make very exact predictions, so Ψ^2 takes on the character of a probability. If Ψ, which depends on time and position, represents a single electron (say in an atom) then Ψ^2 is interpreted as follows: *Ψ^2 at a certain point in space and time represents the probability of finding the electron at the given position and time.*

To understand this better, let's examine the familiar double-slit experiment, and consider it both for light and for electrons.

Consider two slits whose size and separation are on the order of the wavelength of whatever we direct at them, either light or electrons, Fig. 27-1. We know very well what would happen in this case for light, since this is just Young's double-slit experiment: an interference pattern would be seen on the screen behind. If light were replaced by electrons with wavelength comparable to the slit size, they too would produce an interference pattern. (Recall the Davisson–Germer experiment, Fig. 26-7.) In the case of light, the pattern would be visible to the eye or could be recorded on film. For electrons, a fluorescent screen could be used (it glows where an electron strikes).

Now, if we reduced the flow of electrons (or photons) so that they passed through the slits one at a time, we would see a flash each time one struck the screen. At first, the flashes would seem random. Indeed, there is

FIGURE 27-1

Parallel beam of light or electrons falls on two slits whose sizes are comparable to the wavelength. An inteference pattern is observed.

Light or electrons

Intensity on screen

no way to predict just where any one electron would hit the screen. If we let the experiment run for a long time, however, and kept track of where each electron hit the screen, we would soon see a pattern emerging—namely the interference pattern predicted by the wave theory. Thus, although we could not predict where a given electron would strike the screen, we could predict probabilities. (The same can be said for photons.) The probability, as mentioned before, is proportional to Ψ^2. Where Ψ is zero, we would get a minimum in the interference pattern. And where Ψ is a maximum, we would get a peak in the interference pattern.

Since the interference pattern would occur even when electrons (or photons) passed through the slits one at a time, it is clear that the interference pattern would not arise from the interaction of one electron with another. It is as if the electron passed through both slits at the same time! This is possible, for remember that an electron is not precisely a particle. It is as much a wave as it is a particle, and a wave could certainly travel through both slits at once. But what would happen if we covered one of the slits so we knew that the electron passed through the other one, and after a time we covered the second slit so the electron had to have passed through the first? The result would be that no interference pattern would be seen. We would see instead two bright areas on the screen behind each slit. This confirms our idea that if both slits were open, each electron would pass through both, as if it were a wave. Yet each would make a tiny spot on the screen as if it were a particle.

The main point of this discussion is this. If we treat electrons (and other matter) as if they were waves, then Ψ represents the wave amplitude. If we treat them as particles, then we must treat them on a *probabilistic* basis. The square of the wave function, Ψ^2, is the probability of finding a given electron at a given point. We cannot predict—or even follow—the path of a single electron precisely through space and time.

27-3 ▌ The Heisenberg uncertainty principle

Whenever a measurement is made, some uncertainty or error is always involved. For example, you cannot make an absolutely exact measurement of the length of a table. Even with a measuring stick that has markings 1 mm apart there will be an inaccuracy of about $\frac{1}{2}$ mm or so. More precise instruments will produce more precise measurements; but there is always some uncertainty involved in a measurement no matter how good the measuring device. We expect that by using more precise instruments the uncertainty in a measurement can be made indefinitely small.

But according to quantum mechanics there is actually a limit to the accuracy of certain measurements. This limit is not a restriction on how well instruments can be made; rather, it is inherent in nature. It is the result of two factors: the wave–particle duality; and the unavoidable interaction between the thing observed and the observing instrument. Let us look at this in more detail.

To make a measurement on an object without somehow disturbing it, at least a little, is not possible. Consider trying to locate a Ping-Pong ball in a completely dark room. You grope about trying to find its position; and just when you touch it with your finger, it bounces away. Whenever we measure the position of an object, whether it's a Ping-Pong ball or an electron, we always touch it with something else which gives us the information about its position. To locate a lost Ping-Pong ball in a dark room you could probe about with your hand or a stick; or you could shine a light and detect the light reflecting off the ball. When you search with your hand or a stick, you find the ball's position when you touch it. But when you touch the ball you unavoidably bump it, and give it some momentum; thus you won't know its *future* position. The same would be true, but to a much lesser extent, if you observe the Ping-Pong ball using light: in order to "see" the ball, at least one photon must scatter from it, and the reflected photon must enter your eye or some other detector. When a photon strikes an ordinary-sized object it does not appreciably alter the motion or position of the object. But when a photon strikes a very tiny object like an electron, it can transfer much of its momentum to the object and thus greatly change the object's motion and position in an unpredictable way. The mere act of measuring the position of an object at one time makes our knowledge of its future position inaccurate.

Now let us see where the wave–particle duality comes in. Imagine a thought experiment in which we are trying to measure the position of an object (say an electron) with photons, although the arguments would be similar if we were using an electron microscope. As we saw in Chapter 24, objects can be seen to an accuracy no greater than the wavelength of the radiation used. If we want an accurate position measurement, we must use a short wavelength. But a short wavelength corresponds to high frequency and high energy (since $E = hf$); and the more energy the photons have, the more momentum they can give the object when they strike it. If photons of longer wavelength, and correspondingly lower energy are used, the object's motion when struck by the photons will not be affected as much; but its position will be less accurately known. Thus the act of observing produces a significant uncertainty in either the *position* or the *momentum* of the electron. This is the essence of the *uncertainty principle* first enunciated by Heisenberg in 1927.

Quantitatively, we can make an approximate calculation of the magnitude of this effect. If we use light of wavelength λ, the position can be measured at best to an accuracy of about λ. That is, the uncertainty in the position measurement, Δx, is approximately

$$\Delta x \approx \lambda.$$

Suppose that the object can be detected by a single photon. The photon has a momentum $p = h/\lambda$; and when it strikes our object it will give some or all of this momentum to the object. Therefore, the final momentum of our object will be uncertain in the amount

$$\Delta p \approx \frac{h}{\lambda}$$

since we can't tell beforehand how much momentum will be transferred. The product of these uncertainties is

$$(\Delta x)(\Delta p) \approx h.$$

Of course, the uncertainties could be worse than this, depending on the apparatus and the number of photons needed for detection. In Heisenberg's more careful calculation, he found that at the very best

$$(\Delta x)(\Delta p) \gtrsim \frac{h}{2\pi}. \tag{27-1}$$

This is a mathematical statement of **Heisenberg's uncertainty principle.** It tells us that we cannot measure both the position and momentum of an object precisely at the same time. The more accurately we try to measure the position, so that Δx is small, the greater will be the uncertainty in momentum, Δp. If we try to measure the momentum very precisely, then the uncertainty in the position becomes large. The uncertainty principle does not forbid single exact measurements, however. For example, in principle we could measure the position of an object exactly. But then its momentum would be completely unknown. Thus, although we might know the position of the object exactly at one instant, we could have no idea at all where it would be a moment later.

Another useful form of the uncertainty principle relates energy and time, and we examine this as follows. The object to be detected has an uncertainty in position $\Delta x \approx \lambda$. Now the photon (or electron) used to detect it travels with speed $v (= c$ for a photon); and it takes a time $\Delta x/v \approx \lambda/v$ to pass through the distance of uncertainty. Hence, the measured time when our object is at a given position is uncertain by about

$$\Delta t \approx \frac{\lambda}{v}.$$

Since the photon can transfer some or all of its energy $(= hf = hv/\lambda)$ to our object, the uncertainty in energy of our object as a result is

$$\Delta E \approx \frac{hv}{\lambda}.$$

The product of these two uncertainties is

$$(\Delta E)(\Delta t) \approx h.$$

Heisenberg's more careful calculation gives

$$(\Delta E)(\Delta t) \gtrsim \frac{h}{2\pi}. \tag{27-2}$$

This form of the uncertainty principle tells us that the energy of an object can be uncertain, or may even be nonconserved, by an amount ΔE for a time $\Delta t \approx h/(2\pi \Delta E)$.

We have been discussing the position and velocity of an electron as if it were a particle. But it isn't a particle. Indeed, we have the uncertainty principle because an electron—and matter in general—is not purely particulate. What the uncertainty principle really tells us is that if we insist on thinking of the electron as a particle, then there are certain limitations on this simplified view—namely that the position and velocity cannot both be known precisely at the same time; and that the energy can be uncertain (or nonconserved) in the amount ΔE for a time $\Delta t \approx h/(2\pi \Delta E)$.

Because Planck's constant, h, is so small, the uncertainties expressed in the uncertainty principle are usually negligible on the macroscopic level. But at the level of the atom, the uncertainties are significant. Because ordinary sized objects are made up of atoms containing nuclei and electrons, the uncertainty principle is relevant to our understanding of all of nature. The uncertainty principle expresses, perhaps most clearly, the probabilistic nature of quantum mechanics; it thus is often used as a basis for philosophic discussion.

27-4 ▌ Philosophic implications; probability versus determinism

The classical Newtonian view of the world is a deterministic one (see Section 4-10). One of its basic ideas is that once the position and velocity of an object are known at a particular time, its future position can be predicted if the forces on it are known. For example, if a stone is thrown a number of times with the same initial velocity and angle, and the forces on it remain the same, the path of the projectile will always be the same. If the forces are known (gravity and air resistance, if any) the stone's path, and where it will travel, can be precisely predicted. This mechanistic view implies that the future unfolding of the universe, assumed to be made up of particulate bodies, is completely determined.

This classical deterministic view of the physical world has been radically altered by quantum mechanics. As we saw in the analysis of the double-slit experiment (Section 27-2) electrons all prepared in the same way will not all end up in the same place; according to quantum mechanics, certain probabilities exist that an electron will arrive at different points. This is very different from the classical view, in which the path of a particle is precisely predictable from the initial position and velocity and the forces exerted on it. According to quantum mechanics, the position and velocity of an object cannot even be known accurately at the same time. This is expressed in the uncertainty principle, and is a result of the fact that basic entities, such as electrons, are not considered simply as particles since they have wave properties as well. Quantum mechanics only allows us to calculate the probability that, say, an electron (when thought of as a particle) will be observed at various different places. Quantum mechanics says there is some inherent unpredictability in nature.

Since matter is considered to be made up of atoms, even ordinary-sized objects are expected to be governed by probability, rather than by strict determinism. For example, there is a finite (but very small) probability that when you throw a stone its path will suddenly curve upward instead of following the downward-curved parabola of normal projectile motion. Quantum mechanics predicts with very high probability that ordinary objects will behave just as the classical laws of physics predict; but these predictions are probabilities, not certainties. The reason that macroscopic objects behave in accordance with classical laws with very high probability is due to the large number of molecules involved: when large numbers of objects are present in a statistical situation, deviations from the average are neligible. It is the average configuration of vast numbers of molecules that follows the so-called fixed laws of classical physics with such high probability, and gives rise to an apparent "determinism." Deviations from classical laws are readily observed when small numbers of molecules are dealt with. We can say, then, that although there are no precise deterministic laws in quantum mechanics, there are statistical laws based on probability.

It is important to note that there is a difference in the probability imposed by quantum mechanics and that used in the nineteenth century to understand thermodynamics and the behavior of gases in terms of molecules (Chapters 10 and 12). In thermodynamics, probability is used because there are far too many molecules to be kept track of. But the molecules were still assumed to move and interact in a deterministic way according to Newton's laws. Probability in quantum mechanics is quite different; it is seen as *inherent* in nature, and not as a limitation on our abilities to calculate.

Although a few physicists have not given up the deterministic view of nature and have refused to accept quantum mechanics as a complete theory—one was Einstein—nonetheless, the vast majority of physicists do accept quantum mechanics and the probabilistic view of nature. This view, which as presented here is the generally accepted one, is called the *Copenhagen interpretation* of quantum mechanics in honor of Niels Bohr's home, since it was largely developed there through discussions between Bohr and other prominent physicists.

Because electrons are not simply particles, they cannot be thought of as following particular paths in space and time. This suggest that a description of matter in space and time may not be completely correct. This deep and far-reaching conclusion has been a lively topic of discussion among philosophers. Perhaps the most important and influential philosopher of quantum mechanics was Bohr. He argued that a space–time description of actual atoms and electrons is not possible; but a description of experiments on atoms or electrons must be given in terms of space and time and other concepts familiar to ordinary experience, such as waves and particles. Yet we must not let our *descriptions* of experiments lead us into believing that atoms or electrons themselves actually exist in space and time as particles. This distinction between our interpretations of experiments and what is "really" happening in nature is crucial.

27-5 ∎ Quantum mechanical view of atoms

At the beginning of this chapter, we discussed the limitations of the Bohr theory of atomic structure. Now we examine the quantum mechanical theory of atoms, which is far more complete than the old Bohr theory. It retains certain aspects of the older theory, such as that electrons in an atom exist only in discrete states of definite energy; and that a photon of light is emitted (or absorbed) when an electron makes a transition from one state to another. But quantum mechanics is not merely an extension of the Bohr theory. It is a much deeper theory, and has provided us with a very different view of the atom. According to quantum mechanics, electrons do not exist in well-defined circular orbits as in the Bohr theory. Rather, the electron (because of its wave nature) is spread out in space, as a "cloud" of negative charge. The size and shape of the electron cloud can be calculated for a given state of an atom. For the ground state in the hydrogen atom the electron cloud is spherically symmetric as shown in Fig. 27-2. The electron cloud roughly indicates the "size" of an atom; but just as a cloud may not have a distinct border, atoms do not have a precise boundary or a well-defined size. Not all electron clouds have a spherical shape as we shall see later in this chapter.

The electron cloud can be interpreted using either the particle or the wave viewpoint. Remember that by a particle we mean something that is localized in space—it has a definite position at any given instant. But a wave is spread out in space. Thus the electron cloud, spread out in space as in Fig. 27-2, is a result of the wave nature of electrons. Electron clouds can also be interpreted as *probability distributions* for a particle. If you were to make 500 different measurements of the position of an electron (considering it as a particle) the majority of the results would show the electron at points where the probability is high (dark area in Fig. 27-2); only occasionally would the electron be found where the probability is low. We cannot predict the path an electron will follow. As we saw in Section 27-3, after one measurement of its position we cannot predict exactly where it will be at a later time. We can only calculate the probability that it will be found at different points. This is clearly different from classical Newtonian physics. Indeed, as Bohr later pointed out, it is meaningless even to ask how an electron gets from one state to another when the atom emits a photon of light.

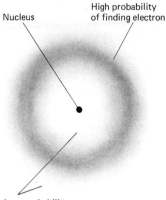

Nucleus

High probability of finding electron

Low probability of finding electron

FIGURE 27-2

Electron cloud or "probability distribution" for the ground state of the hydrogen atom. The cloud is densest—indicating the highest probability—at a distance from the nucleus of 0.53×10^{-10} m, which is just what the Bohr model predicts for the radius of the first orbit. But unlike the Bohr model, quantum mechanics tells us the electron can be within or beyond that distance at any given time.

27-6 ▮ Quantum mechanics of the hydrogen atom; quantum numbers

We now look more closely at what quantum mechanics tells us about the hydrogen atom. Much of what we say here also applies to more complex atoms, which are discussed in the next section.

Quantum mechanics predicts exactly the same energy levels (Fig. 26-18) for the hydrogen atom as does the Bohr theory. That is,

$$E_n = -\frac{13.6 \text{ eV}}{n^2} \qquad n = 1, 2, 3, \dots$$

where n is an integer. In the simple Bohr theory, there was only one quantum number, n. In quantum mechanics, it turns out that four different quantum numbers are needed to specify each state in the atom.

The quantum number n from Bohr theory is retained in quantum mechanics and is called the **principal quantum number.** It can have any integer value from 1 to ∞. The total energy of a state in the hydrogen atom depends on n, as we saw above.

The **orbital quantum number,** l, is related to the angular momentum of the electron; l can take on integer values from 0 to $(n - 1)$. For the ground state, $n = 1$, l can only be zero; but for $n = 3$, say, l can be 0, 1, or 2. The actual magnitude of the angular momentum L is related to the quantum number l by

$$L = \sqrt{l(l+1)}\,\frac{h}{2\pi}.$$

The value of l does not affect the total energy in the hydrogen atom;[†] only n does. But in atoms with two or more electrons, the energy does depend on l as well as n, as we shall see in the next section.

The **magnetic quantum number,** m_l, is related to the direction of the electron's angular momentum and it can take on integer values ranging from $-l$ to $+l$; for example, if $l = 2$, then m_l can be -2, -1, 0, $+1$, or $+2$. Since angular momentum is a vector, it is not surprising that both its magnitude and its direction would be quantized. For $l = 2$, the five different directions allowed can be represented by the diagram of Fig. 27-3. In quantum mechanics the direction of the angular momentum is usually specified by giving its component along the z axis. Then L_z is related to m_l by the equation

$$L_z = m_l \frac{h}{2\pi}.$$

The name for m_l derives not from theory (which relates it to L_z) but from experiment: it was found that when a gas discharge tube was placed in a magnetic field, the spectral lines were split into several very closely spaced lines. (This is called the Zeeman effect.) This implies that the energy levels must be split (Fig. 27-4), and thus the energy of a state depends not only on n but also on m_l when a magnetic field is applied—hence the name "magnetic quantum number." (Why the energy should depend on the direction of **L** can be seen from a semiclassical view of a moving electron as an electric current which interacts with the magnetic field—see Chapter 19.)

Finally, there is the **spin quantum number,** m_s, and it can have only two values, $m_s = +\frac{1}{2}$ and $m_s = -\frac{1}{2}$. The existence of this quantum number did not come out of Schrödinger's original theory, as did n, l, and m_l. Instead, a subsequent modification due to P. A. M. Dirac showed that it was present as a relativistic effect. The fact that m_s was needed, however, first came from experiment. A careful study of the spectral lines of hydrogen showed that each actually consisted of two (or more) very close lines. This was called **fine structure.** It was at first hypothesized that this tiny splitting of energy levels might be due to angular momentum associated with a spinning of the electron; that is, the electron might spin on its axis as well as orbit the nucleus just as the earth spins on its axis as it orbits the sun. The interaction between the tiny current of the spinning electron could then interact with the magnetic field due to the orbiting charge and cause the small observed splitting of energy levels. (So the energy depends slightly on m_l and m_s.) This picture of the electron as spinning is, however, wholly discredited today. We cannot even view an electron as a localized object, much less a spinning one. What is important is that the electron can have two different states due to some intrinsic property and (perhaps unfortunately) we still call this property

[†] Not quite precisely true due to normally very small interactions within the atom; see discussion of *fine structure* later in this section.

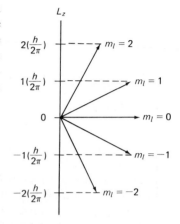

L_z

$2(\frac{h}{2\pi})$ —————➚ $m_l = 2$

$1(\frac{h}{2\pi})$ —————➙ $m_l = 1$

0 ————➙ $m_l = 0$

$-1(\frac{h}{2\pi})$ —————➘ $m_l = -1$

$-2(\frac{h}{2\pi})$ —————➘ $m_l = -2$

FIGURE 27-3

Quantization of angular momentum direction for $l = 2$.

FIGURE 27-4

When a magnetic field is applied, an $n = 3$, $l = 2$ energy level is split into five separate levels, corresponding to the five values of m_l (2, 1, 0, -1, -2). An $n = 2$, $l = 1$ level is split into three levels ($m_l = 1$, 0, -1). Transitions can occur between levels (not all are shown), with photons being given off with several frequencies, which are only slightly different.

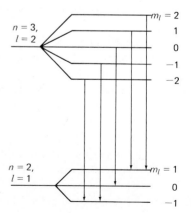

$n = 3$, $l = 2$

$m_l = 2$
1
0
-1
-2

$n = 2$, $l = 1$

$m_l = 1$
0
-1

TABLE 27-1

Quantum numbers for an electron

Name	Symbol	Possible Values
Principal	n	$1, 2, 3, \ldots, \infty$.
Orbital	l	For a given n: l can be $0, 1, 2, \ldots, n-1$.
Magnetic	m_l	For given n and l: m_l can be $l, l-1, \ldots, 0, \ldots, -l$.
Spin	m_s	For each set of n, l, and m_l: m_s can be $+\frac{1}{2}$ or $-\frac{1}{2}$.

"spin." The two possible values of $m_s (+\frac{1}{2}$ and $-\frac{1}{2})$ are often said to be "spin up" and "spin down," referring to the two possible directions of the spin angular momentum.

The possible values of the four quantum numbers for an electron in the hydrogen atom are summarized in Table 27-1.

EXAMPLE 27-1 How many different states are possible for an electron whose principal quantum number is $n = 3$?

SOLUTION For $n = 3$, l can have the values $l = 2, 1, 0$. For $l = 2$, m_l can be $2, 1, 0, -1, -2$ which is five different possibilities; for each of these, m_s can be either up or down $(+\frac{1}{2}$ or $-\frac{1}{2})$; so for $l = 2$ there are $2 \times 5 = 10$ states. For $l = 1$, m_l can be $1, 0, -1$, and since m_s can be $+\frac{1}{2}$ or $-\frac{1}{2}$ for each of these, we have 6 more possible states. Finally, for $l = 0$, m_l can only be 0, and there are only two states corresponding to $m_s = +\frac{1}{2}$ and $-\frac{1}{2}$. The total number of states is $10 + 6 + 2 = 18$, as detailed in the following table:

n	l	m_l	m_s	n	l	m_l	m_s
3	2	2	$\frac{1}{2}$	3	2	-2	$-\frac{1}{2}$
3	2	2	$-\frac{1}{2}$	3	1	1	$\frac{1}{2}$
3	2	1	$\frac{1}{2}$	3	1	1	$-\frac{1}{2}$
3	2	1	$-\frac{1}{2}$	3	1	0	$\frac{1}{2}$
3	2	0	$\frac{1}{2}$	3	1	0	$-\frac{1}{2}$
3	2	0	$-\frac{1}{2}$	3	1	-1	$\frac{1}{2}$
3	2	-1	$\frac{1}{2}$	3	1	-1	$-\frac{1}{2}$
3	2	-1	$-\frac{1}{2}$	3	0	0	$\frac{1}{2}$
3	2	-2	$\frac{1}{2}$	3	0	0	$-\frac{1}{2}$

Although l and m_l do not significantly affect the energy levels in hydrogen, they do affect the electron probability distribution in space. For $n = 1$, l and m_l can only be zero and the electron distribution is as shown in Fig. 27-2. For $n = 2$, l can be 0 or 1. The distribution for $n = 2$, $l = 0$ is shown in Fig. 27-5a, and it is seen to differ from that for the ground state, although it is still spherically symmetric. For $n = 2$, $l = 1$, the distribution is not spherically symmetric, but is more barbell-shaped as shown in Fig. 27-5b. The distribution shown is for one value of m_l, say $m_l = 1$; for $m_l = 0$ and $m_l = -1$, the distribution is also barbell-shaped but is symmetric around the other axes.

Although the spatial distributions of the electron can be calculated for

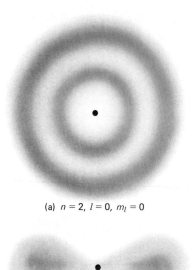

(a) $n = 2,\ l = 0,\ m_l = 0$

(b) $n = 2,\ l = 1,\ m_l = 1$

FIGURE 27-5
Electron cloud (or probability distribution) for $n = 2$ states in hydrogen.

the various states, it is difficult to measure them experimentally. Indeed, most of the experimental information about the atom has come from a careful examination of the emission spectra under various conditions.

27-7 ∎ Complex atoms; the exclusion principle

We have discussed the hydrogen atom in detail because it is the simplest to deal with. Now we briefly discuss more complex atoms, those that contain more than one electron, and whose energy levels can be determined experimentally from an analysis of the emission spectra. The number of electrons in a neutral atom is called its *atomic number*, Z; Z is also the number of positive charges (protons) in the nucleus, and determines what kind of atom it is. That is, Z determines most of the properties that distinguish one kind of atom from another.

Although modifications to the Bohr theory had been attempted in order to deal with complex atoms, the development of quantum mechanics in the years after 1925 proved far more successful. The mathematics becomes very difficult, however, since in multielectron atoms, each electron is not only attracted to the nucleus but is repelled by the other electrons.

The simplest approach has been to treat each electron in an atom as occupying a particular state characterized by the quantum numbers n, l, m_l, and m_s. But to understand the possible arrangements of electrons in an atom, a new principle was needed. It was introduced by Wolfgang Pauli (1900–1958) and is called the **Pauli exclusion principle.** It states:

∎ **No two electrons in an atom can occupy the same quantum state.**

That is, no two electrons can have exactly the same set of the quantum numbers n, l, m_l, and m_s. The Pauli exclusion principle[†] forms the basis not only for understanding complex atoms, but also for understanding molecules and bonding, and other phenomena as well.

Let us now look at the structure of some of the simpler atoms when they are in the ground state. After hydrogen, the next simplest atom is helium with two electrons. Both electrons can have $n = 1$, since one can have spin up ($m_s = +\frac{1}{2}$) and the other spin down ($m_s = -\frac{1}{2}$), thus satisfying the exclusion principle. Of course, since $n = 1$, l and m_l must be zero (Table 27-1). Thus the two electrons have the quantum numbers indicated in the table in the margin.

Lithium has three electrons, two of which can have $n = 1$. But the third cannot have $n = 1$ without violating the exclusion principle. Hence the third electron must have $n = 2$. Since it happens that the $n = 2$, $l = 0$ level has a lower energy than $n = 2$, $l = 1$, the electrons in the ground state have the quantum numbers indicated in the table in the margin. Of course, the quantum numbers of the third electron could also be $(3, 1, -1, \frac{1}{2})$, say; but the atom in this case would be in an excited state since it would have greater energy. It would not be long before it jumped to the ground state with the emission of a photon. At room temperature, unless energy is put in (as in a discharge tube), the vast majority of atoms are in the ground state.

We can continue in this way to describe the quantum numbers of each electron in the ground state of larger and larger atoms. That for sodium, with its eleven electrons, is shown in the table in the margin.

The ground-state configuration for all atoms is given in the *periodic table* which is displayed inside the rear cover of this book, and discussed in the next section.

*27-8 ∎ The periodic table of elements

A century ago, Dmitri Mendeleev (1834–1907) arranged the then known elements into what we now call the *periodic table* of the elements. The atoms were arranged according to increasing mass, but also so that elements with similar chemical properties would fall in the same column. Today's version is shown inside the rear cover. Each square contains the atomic number Z, the symbol for the element, and the atomic mass (in atomic mass units). Finally, in the lower left corner the configuration of the ground state of the atom is given. This requires some explanation. Electrons with the same value of n are referred to as being in the same *shell*. Electrons with $n = 1$ are in one shell (called the K shell); those with $n = 2$ are in a second shell (the L shell), and so on. Electrons with the same value of l are referred to as being in the same *subshell*. Letters are often used to specify the value of l as indicated in Table 27-2. That is, $l = 0$ is the s subshell; $l = 1$ is the p subshell; $l = 2$ is the d subshell; beginning with $l = 3$, the letters follow the alphabet,

[†] It applies to identical particles whose spin quantum number is a half-integer ($\frac{1}{2}$, $\frac{3}{2}$, and so on), including electrons, protons, and neutrons, but not to particles with integer spin (0, 1, 2, and so on), such as the photon and π meson.

Helium, $Z = 2$

n	l	m_l	m_s
1	0	0	$\frac{1}{2}$
1	0	0	$-\frac{1}{2}$

Lithium, $Z = 3$

n	l	m_l	m_s
1	0	0	$\frac{1}{2}$
1	0	0	$-\frac{1}{2}$
2	0	0	$\frac{1}{2}$

Sodium, $Z = 11$

n	l	m_l	m_s
1	0	0	$\frac{1}{2}$
1	0	0	$-\frac{1}{2}$
2	0	0	$\frac{1}{2}$
2	0	0	$-\frac{1}{2}$
2	1	1	$\frac{1}{2}$
2	1	1	$-\frac{1}{2}$
2	1	0	$\frac{1}{2}$
2	1	0	$-\frac{1}{2}$
2	1	-1	$\frac{1}{2}$
2	1	-1	$-\frac{1}{2}$
3	0	0	$\frac{1}{2}$

TABLE 27-2
Values of l

Value of l	Letter symbol	Maximum number of electrons in subshell
0	s	2
1	p	6
2	d	10
3	f	14
4	g	18
5	h	22
⋮	⋮	⋮

f, g, h, i, and so on. (The first letters s, p, d, and f were originally abbreviations of "sharp," "principal," "diffuse," and "fundamental," and were experimental terms referring to the spectra.) The Pauli exclusion principle limits the number of electrons possible in each shell and subshell. Since for any value of l there are $2l + 1$ different m_l values (m_l can be any integer from 1 to l, from -1 to $-l$, or zero), and two different m_s values, there can be at most $2(2l + 1)$ electrons in any l subshell. For example, for $l = 2$, five m_l values are possible (2, 1, 0, -1, -2), and for each of these m_s can be $+\frac{1}{2}$ or $-\frac{1}{2}$, for a total of $2(5) = 10$ states. Table 27-2 lists the maximum number of electrons that can be in each subshell.

Since the energy levels depend almost entirely on the values of n and l, it is customary to specify the electron configuration simply by giving the n value and the appropriate letter for l; and the number of electrons in each subshell is given as a superscript. The ground state configuration of sodium, for example, is written as $1s^2 2s^2 2p^6 3s^1$. In the periodic table this is simplified by specifying the configuration only of the outermost electrons and any other nonfilled subshells. The grouping of atoms in the periodic table is according to increasing mass; there is also a regularity according to chemical properties. And although this is discussed in chemistry textbooks, we discuss it briefly here since it is a result of quantum mechanics.

All the noble gases (in the last column of the table) have completely filled shells or subshells. That is, their outermost subshell is completely full, the electron distribution is spherically symmetric, and there is no opportunity for additional electrons. This is the reason they are nonreactive (more on this when we discuss molecules and bonding in Chapter 28). Column seven contains the "halogens," which lack one electron from a filled shell. Because of the shapes of the orbits (see Section 28-1), an additional electron can be accepted from another atom, and hence these are quite reactive with a valence of -1 (meaning that when such an electron is accepted, the resulting ion has a net charge of $-1e$). At the left of the table, column I contains the alkali metals, all of which have a single outer s electron. This electron spends most of its time outside the inner closed shells and subshells which shield it from most of the nuclear charge; indeed, it is relatively far from the nucleus and is attracted to it by a net charge of only $+1e$, because of the shielding effect of the other electrons. Hence this outer electron is easily removed and can spend much of its time around another atom, forming a molecule; this is why the alkali metals have a valence of $+1$. The other columns of the table can be treated similarly.

The presence of the transition elements in the center of the table, as well as the lanthanides (rare earths) and actinides below, is a result of incomplete inner shells. For the lowest Z elements, the subshells are filled in a simple order: first $1s$, then $2s$, followed by $2p$, $3s$, and $3p$. You might expect that $3d$ ($n = 3$, $l = 2$) would be filled next, but it isn't. Instead, the $4s$ level actually has a slightly lower energy than the $3d$, so it fills first (K and Ca). Only then does the $3d$ shell start to fill up, beginning with Sc. (The $4s$ and $3d$ levels are close, so some elements have only one $4s$ electron, such as Cr.) Most of the chemical properties are governed by the relatively loosely held $4s$ electrons and hence these "transition elements" mainly have valences of $+1$ or $+2$. A similar effect is responsible for the rare earths, which are shown at the bottom of the periodic table for convenience; all have very similar chemical properties because their outer $6s$ or $7s$ electrons control their chemical properties, whereas the different numbers of electrons in the unfilled inner shells have little effect.

27-9 ▌ X-ray spectra and atomic number

The line spectra of atoms in the visible, UV, and IR regions of the EM spectrum are mainly due to transitions between states of the outer electrons. For these electrons, much of the charge of the nucleus is shielded from them by the negative charge on the inner electrons. But the innermost electrons in the $n = 1$ shell see the full charge of the nucleus. Since the energy of a level is proportional to Z^2 (see Eq. 26-11), for an atom with $Z = 50$, we would expect wavelengths about $50^2 = 2500$ times shorter than those found in the Lyman series of hydrogen (around 100 nm), or 10^{-2} to 10^{-1} nm. These are in the X-ray region of the spectrum.

As we saw in Section 24-11, X rays are produced when electrons accelerated by a high voltage strike the metal target inside the X-ray tube. If we look at the spectrum of wavelengths emitted by an X-ray tube we see that the spectrum consists of two parts: a continuous spectrum with a cutoff at some λ_0 which depends only on the voltage across the tube; and peaks which are superimposed on top. A typical example is shown in Fig. 27-6. The smooth curve and the cutoff wavelength λ_0 move to the left as the voltage across the tube increases. But the peaks (labeled K_α and K_β in Fig. 27-6) remain at the same wavelength when the voltage is changed, but are located at different places when different target materials are used. This suggests that the peaks are characteristic of the material used. Indeed, we can explain them by imagining that the electrons accelerated by the high voltage of the tube can reach sufficient energies that when they collide with the atoms of the target, they can knock out one of the very tightly held inner electrons. These *characteristic X rays* (the peaks in Fig. 27-6) are photons emitted when an electron in an upper state drops down to fill the vacated lower state.

Measurement of the characteristic X-ray spectra has allowed a determination of the inner energy levels of atoms. It has also allowed the determination of Z values for many atoms, since (as we have seen) the wavelength of the shortest X rays emitted will be proportional to Z^2. Actually, for an electron jumping from say the $n = 2$ to the $n = 1$ level, the wavelength is inversely proportional to $(Z - 1)^2$ because the nucleus is shielded by the one electron that still remains in the $1s$ level. In 1914, H. G. J. Moseley found that a plot of $\sqrt{1/\lambda}$ versus Z produced a straight line, Fig. 27-7. The Z values of a number of elements were determined by fitting them to such a "Moseley plot." The work of Moseley put the concept of atomic number on a firm experimental basis.

☐ EXAMPLE 27-2   Estimate the wavelength for an $n = 2$ to $n = 1$ transition in molybdenum ($Z = 42$).

SOLUTION   We use the Bohr formula, Eq. 26-12, with Z^2 replaced by $(Z - 1)^2 = (41)^2$. Or, more simply, we can use the result of Example 26-7

FIGURE 27-6

Spectrum of X rays emitted from a molybdenum target in an X-ray tube operated at 50 kV.

FIGURE 27-7

Plot of $\sqrt{1/\lambda}$ versus Z for K_α X-ray lines.

for the $n = 2$ to $n = 1$ transition in hydrogen ($Z = 1$). Since $\lambda \propto 1/(Z - 1)^2$, we will have

$$\lambda = (1.22 \times 10^{-7} \text{ m})/(41)^2 = 0.073 \text{ nm}.$$

This is close to the measured value (Fig. 27-6) of 0.071 nm.

Now we briefly analyze the continuous part of an X-ray spectrum (Fig. 27-6) based on the photon theory of light. When electrons strike the target, they collide with atoms of the material and give up most of their energy as heat (about 99 percent, so X-ray tubes must be cooled). However, electrons can also give up energy by emitting a photon of light. As shown in Fig. 27-8, an electron can be decelerated by the positive nucleus of an atom; but an accelerating charge can emit radiation (Chapter 21), and in this case it is called *bremsstrahlung* (German for "braking radiation"). Because energy is conserved, the energy of the emitted photon, hf, must equal the loss of kinetic energy of the electron, $\Delta \text{KE} = \text{KE} - \text{KE}'$, so

$$hf = \Delta \text{KE}.$$

An electron may lose all or a part of its energy in such a collision. The continuous X-ray spectrum (Fig. 27-6) is explained as being due to such bremsstrahlung collisions in which varying amounts of energy are lost by the electrons. The shortest wavelength X ray produced (the highest frequency) must be due to an electron that gives up all its kinetic energy to one photon in a single collision. Since the initial kinetic energy of an electron is equal to the energy given it by the accelerating voltage, V, then $\text{KE} = eV$; in a single collision in which the electron is brought to rest we have

$$hf_0 = eV$$

or

$$\lambda_0 = \frac{hc}{eV} \tag{27-3}$$

where $\lambda_0 = c/f_0$ is the cutoff wavelength, Fig. 27-6. This prediction for λ_0 corresponds precisely with that observed experimentally. This result is further evidence that X rays are a form of light[†] and that the photon theory of light is valid.

FIGURE 27-8
Bremsstrahlung photon produced by an electron that is decelerated by a positive nucleus.

□ **EXAMPLE 27-3** What is the shortest wavelength X-ray photon emitted in an X-ray tube subjected to 50 kV?

SOLUTION From Eq. 27-3,

$$\lambda_0 = \frac{(6.6 \times 10^{-34} \text{ J·s})(3.0 \times 10^8 \text{ m/s})}{(1.6 \times 10^{-19} \text{ C})(5.0 \times 10^4 \text{ V})} = 2.5 \times 10^{-11} \text{ m}$$

or 0.025 nm. This agrees well with experiment, Fig. 27-6.

[†] If X rays were not photons but rather neutral particles with rest mass m_0, Eq. 27-3 would not hold.

*27-10 ∎ Fluorescence and phosphorescence

When an atom is excited from one energy state to a higher one by the absorption of a photon, it may return to the lower level in a series of two (or more) jumps if there is an energy level in between (Fig. 27-9). The photons emitted will consequently have lower energy and frequency than the absorbed photon. This is called *fluorescence*; common fluorescent rocks and paints can emit visible light after absorbing UV light.

The wavelength for which fluorescence will occur depends on the energy levels of the particular atoms. Because the frequencies are different for different substances, and because many substances fluoresce readily, fluorescence is a powerful tool for identification of compounds. It is also used for assaying—determining how much of a substance is present—and for following substances along a natural pathway as in plants and animals. For detection of a given compound, the stimulating light must be monochromatic, and solvents or other materials present must not fluoresce in the same region of the spectrum. Often the observation of fluorescent light being emitted is sufficient; in other cases, spectrometers are used to measure the wavelengths and intensities of the light.

Fluorescent light bulbs work in a two-step process. The applied voltage accelerates electrons that strike atoms of the gas in the tube and cause them to be excited. When the excited atoms jump down to their normal levels, they emit UV photons which strike a fluorescent coating on the inside of the tube. The light we see is a result of this material fluorescing in response to the UV light striking it.

Materials such as those used for luminous watch dials are said to be *phosphorescent*. In a phosphorescent substance, atoms can be excited by absorption of a photon to an energy level said to be metastable. When an atom is raised to a normal excited state, it drops back down within about 10^{-8} s. *Metastable* states last much longer—even a few seconds or longer. In a collection of such atoms, many of the atoms will descend to the lower state fairly soon, but many will remain in the excited state for over an hour. Hence light will be emitted even after long periods. When you put your watch dial close to a bright lamp, it excites many atoms to metastable states, and you can see the glow a long time after.

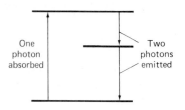

FIGURE 27-9
Fluorescence.

*27-11 ∎ Lasers and holography

A laser is a device that can produce a very narrow intense beam of monochromatic coherent light. The emitted beam is a nearly perfect plane wave. An ordinary light source, on the other hand, emits light in all directions (so the intensity decreases rapidly with distance), and the emitted light is incoherent. It is incoherent (the different parts of the beam are not in phase with each other) since the excited atoms which emit the light act independently so each photon emitted can be considered as a short wave train, typically 30 cm long and lasting 10^{-8} s; these wave trains bear no phase relation to one another.

The action of a laser is based on quantum theory. We have seen that a photon can be absorbed by an atom if (and only if) its energy hf corresponds to the energy difference between an occupied energy level of the atom and an available excited state, Fig. 27-10a. This is, in a sense, a resonant situation. If the atom is already in the excited state, it may of course jump spontaneously to the lower state with the emission of a photon. However, if a photon with this same energy strikes the excited atom, it can stimulate the atom to make the transition sooner to the lower state, Fig. 27-10b. This is called **stimulated emission,** and it can be seen that not only do we still have the original photon, but also a second one of the same frequency as a result of the atom's transition. And these two photons are exactly *in phase*, and moving in the same direction. This is how coherent light is produced in a laser; hence the name "laser," which is an acronym for **l**ight **a**mplification by **s**timulated **e**mission of **r**adiation.

Normally, most atoms are in the lower state, so incident photons will mostly be absorbed. In order to obtain the coherent light from stimulated emission, two conditions must be satisfied. First, the atoms must be excited to the higher state. That is, an *inverted population* is needed, one in which more atoms are in the upper state than in the lower one, so that *emission* of photons will dominate over absorption. And secondly, the higher state must be a *metastable state*—a state in which the electrons remain longer than usual so that the transition to the lower state occurs by stimulated emission rather than spontaneously. How these conditions are achieved for different lasers will be discussed shortly. For now, we assume that the atoms have been excited to an upper state. Figure 27-11 is a schematic diagram of a laser: the "lasing" material is placed in a long narrow tube at the ends of which are two mirrors, one of which is partially transparent (perhaps 1 or 2 percent). Some of the excited atoms drop down fairly soon after being excited. One of these is the atom shown on the far left in Fig. 27-11. If the emitted photon strikes another atom in the excited state, it stimulates this atom to emit a photon of the same frequency and in phase with it. These two photons then move on to strike other atoms causing more stimulated emission. As the process continues, the number of photons multiplies. When the photons strike the end mirrors, most are reflected back and as they move in the opposite direction, they continue to stimulate more atoms to emit photons. As the photons move back and forth between the mirrors, a small percentage pass through the partially transparent mirror at one end. These photons make up the narrow coherent external laser beam.

Inside the tube, some spontaneously emitted photons will not be emitted parallel to the axis, and these will merely go out the side of the tube and not contribute to the main beam. Thus the beam can be very narrow. In a well-designed laser, the spreading of the beam is limited only by diffraction, so the

FIGURE 27-10
(a) Absorption of a photon.
(b) Stimulated emission.

Mirror — Partially transparent mirror

FIGURE 27-11
Laser diagram, showing excited atoms stimulated to emit light.

angular spread is $\approx \lambda/D$ (see Eq. 23-3) where D is the diameter of the end mirror; the diffraction spreading can be incredibly small. The light energy, instead of spreading out in space as it does for an ordinary light source, is directed in a pencil-thin beam.

The excitation of the atoms in a laser can be done in several ways to produce the necessary inverted population. In a ruby laser, the lasing material is a ruby rod consisting of Al_2O_3 with a small percentage of aluminum (Al) atoms replaced by chromium (Cr) atoms. The Cr atoms are the ones involved in lasing. The atoms are excited by strong flashes of light of wavelength 550 nm, which corresponds to a photon energy of 2.2 eV. As shown in Fig. 27-12, the atoms are excited from state E_0 to state E_2. This process is called *optical pumping*. The atoms quickly decay either back to E_0 or to the intermediate state E_1 which is metastable with a lifetime of about 3×10^{-3} s (compared to 10^{-8} s for ordinary levels). With strong pumping action, more atoms can be forced into the E_1 state than are in the E_0 state. Thus we have the inverted population needed for lasing. As soon as a few atoms in the E_1 state jump down to E_0, they emit photons which produce stimulated emission of the other atoms and the lasing action begins. A ruby laser thus emits a beam whose photons have energy 1.8 eV and a wavelength of 694.3 nm (or "ruby-red" light).

In a helium-neon (He-Ne) laser, the lasing material is a gas, a mixture of about 15 percent He and 85 percent Ne. This combination works as a lasing material because of certain compatible properties of the two gases. In this laser, the atoms are excited by applying a high voltage to the tube so that an electric discharge takes place within the gas. In the process, some of the He atoms are raised to the metastable state E_1 shown in Fig. 27-13 which corresponds to a jump of 20.61 eV. Now Ne atoms have an excited state that is almost exactly the same energy above the ground state, 20.66 eV. The He atoms do not quickly return to the ground state by spontaneous emission, but instead often give their excess energy to a Ne atom when they collide; in such a collision, the He drops to the ground state and the Ne atom is excited to the state E_3' (the prime refers to neon states). The slight difference in energy (0.05 eV) is supplied by the kinetic energy of the moving molecules. In this manner, the E_3' state in Ne—which is metastable—becomes more populated than the E_2' level. This inverted population between E_3' and E_2' is what is needed for lasing.

FIGURE 27-12

Energy levels of chromium in ruby crystal. Photons of energy 2.2 eV "pump" atoms from E_0 to E_2, which then decay to metastable state E_1. Lasing action occurs by stimulated emission of photons in transition from E_1 to E_0.

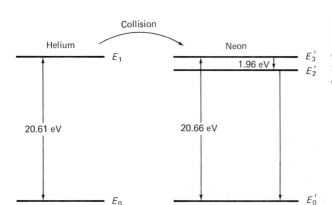

FIGURE 27-13

Energy levels for He and Ne. He is excited in electric discharge to the E_1 state. This energy is transferred to the E_3' level of Ne by collision. E_3' is metastable and decays to E_2' by stimulated emission.

Another type of laser makes use of a *pn* junction in which the transitions occur between the bottom of the conduction band and the upper part of the valence band (Section 28-3).

The excitation of the atoms in a laser can be done continuously or in pulses. In a *pulsed laser*, the atoms are excited by periodic inputs of energy; the multiplication of photons continues until all the atoms have been stimulated to jump down to the lower state, and the process is repeated with each pulse. In a *continuous laser*, the energy input is continuous so that as atoms are stimulated to jump down to the lower level, they are soon excited back up to the upper level. In either case, of course, the laser is not a source of energy. Energy must be put in, and the laser converts a part of this into an intense narrow beam. Thus a laser, like an electronic amplifier, uses energy to amplify a certain type of signal.

The unique feature of light from a laser is, as mentioned before, that it is a coherent narrow beam of a single frequency (or several distinct frequencies). Because of this feature, the laser has found many applications. Lasers are a useful surgical tool. The narrow intense beam can be used to destroy tissue in a localized area; or, because of the heat produced, a laser beam can be used to "weld" broken tissue, such as a detached retina. Tiny organelles within a living cell have been destroyed by researchers using lasers to study how the absence of that organelle affects the behavior of the cell. The finely focused beam of a laser has been used to destroy cancerous and precancerous cells; at the same time the heat seals off capillaries and lymph vessels, thus "cauterizing" the wound in the process to prevent spread of the disease. The intense heat produced in a small area by a laser beam is also used for welding and machining metals and for drilling tiny holes in hard materials. The beam of a laser is narrow in itself (typically a few mm). But because the beam is coherent, monochromatic, and essentially parallel and narrow, lenses can be used to focus the light into incredibly small areas without the usual aberration problems; the limit thus becomes diffraction, and the energy crossing unit area per unit time can be very large. The precise straightness of a laser beam is also useful to surveyors for lining up equipment precisely, especially in inaccessible places.

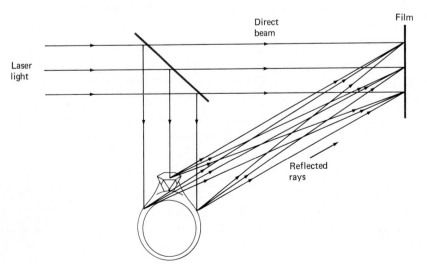

FIGURE 27-14
Making a hologram.

One of the most interesting applications of laser light is the production of three-dimensional images called **holograms.** In an ordinary photograph, the film simply records the intensity of light reaching it at each point; when the photograph or transparency is viewed, light reflecting from it or passing through it gives us a two-dimensional picture. In holography, the images are formed by interference, without lenses. When a laser hologram is made on film, a broadened laser beam is split into two parts by a half-silvered mirror, Fig. 27-14. One part goes directly to the film; the rest passes to the object to be photographed, from which it is reflected to the film. Light from every point on the object reaches each point on the film, and the interference of the two beams allows the film to record both the intensity and relative phase of the light at each point. After the film is developed, it is placed again in a laser beam and a three-dimensional image of the object is seen. You can walk around such an image and see it from different sides as if it were the original object. Yet, if you try to touch it with your hand, there will be nothing material there.

The details of how the image is formed are quite complicated. But we can get the basic idea by considering one single point on the object. In Fig. 27-15a the rays OA and OB have reflected from one point on our object. The rays CA and DB come directly from the source and interfere with OA and OB at points A and B on the film. A set of interference fringes is produced as shown in Fig. 27-15b. The spacing between the fringes changes from top to bottom as shown. Why this happens is explained in Fig. 27-16. Thus the hologram of a single point object would have the pattern shown in Fig. 27-15b. The film in this case looks like a diffraction grating with variable spacing. Hence, when coherent laser light is passed back through the developed film, the diffracted rays in the first order maxima occur at slightly different angles because the spacing changes (remember Eq. 23-4, $\sin \theta = m\lambda/d$; so where the spacing d is greater, the angle θ is less). Hence, the rays diffracted upward (in first order) seem to diverge from a single point, Fig. 27-17. This is a virtual image of the original object, which can be seen with the eye. Rays diffracted in first order *downward* converge to make a real image, which can be seen and also photographed. (Note that the straight-through undiffracted rays are of no interest.) Of course real objects consist of many points, so a hologram will be a complex interference pattern which,

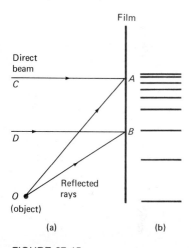

FIGURE 27-15

Light from point O on the object interferes with light of direct beam (rays CA and DB).

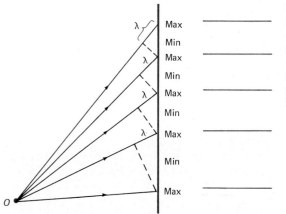

FIGURE 27-16

Each of the rays shown leaving point O is one wavelength shorter than the one above it. If the top ray is in phase with the direct beam (not shown), which has the same phase at all points on the screen, all the rays shown produce constructive interference. From this diagram it can be seen that the fringe spacing increases toward the bottom.

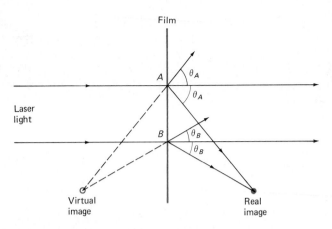

FIGURE 27-17

Laser beam strikes film that is like a diffraction grating of variable spacing. Rays corresponding to the first diffraction maxima are shown emerging. The angle $\theta_A > \theta_B$ because the spacing at B is greater than at A ($\sin \theta = \lambda/d$). Hence real and virtual images of the point are reproduced as shown.

when laser light is incident on it, will reproduce an image of the object. Each image point will be at the correct (three-dimensional) position with respect to other points, so the image accurately represents the original object; and it can be viewed from different angles as if viewing the original object. Holograms can be made in which a viewer can walk entirely around the image (360°) and see all sides of it.

▌ Summary

In 1925, Schrödinger and Heisenberg separately worked out a new theory, *quantum mechanics*, which is now considered to be the basic theory at the atomic level. It is a statistical theory rather than a deterministic one.

An important aspect of quantum mechanics is the Heisenberg *uncertainty principle*. It results from the wave–particle duality and the unavoidable interaction between the observed object and the observer. One form of the uncertainty principle states that the position x and momentum p of an object cannot both be measured precisely at the same time; the products of the uncertainties, $(\Delta x)(\Delta p)$, can be no less than $h/2\pi$:

$$(\Delta p)(\Delta x) \gtrsim \frac{h}{2\pi}.$$

Another form states that the energy can be uncertain, or nonconserved, by an amount ΔE for a time $\Delta t \approx h/(2\pi \Delta E)$.

According to quantum mechanics, the electrons in an atom do not have well-defined orbits but instead exist as a "cloud." Electron clouds can be interpreted as an electron wave spread out in space, or as a probability distribution for electrons as particles.

According to quantum mechanics, the state of an electron in an atom is specified by four quantum numbers: n, l, m_l, and m_s. The principal quantum number, n, can take on any integer value (1, 2, 3, . . .) and corresponds to the quantum number of the old Bohr theory; l can take on values from 0 up to $n - 1$; m_l can take on integer values from $-l$ to $+l$; and m_s can be $+\frac{1}{2}$ or $-\frac{1}{2}$. The energy levels in the hydrogen atom depend on n, whereas in other atoms they depend on n and l; when an external magnetic field is applied, the spectral lines are split (the Zeeman effect), indicating that the energy depends also on m_l in this case. Even in the absence of a magnetic field, precise measurements of spectral lines show a tiny splitting of the lines called fine structure, whose explanation is that the energy depends very slightly on m_l and m_s.

The arrangement of electrons in multielectron atoms is governed by the *Pauli exclusion principle*, which states that no two electrons can occupy the same quantum state—that is, they cannot have the same set of quantum numbers n, l, m_l, and m_s. Electrons, as a result, are grouped into shells (according to the value of n) and subshells (according to l).

X rays, which are a form of electromagnetic radiation of very short wavelength, are produced when high-speed electrons strike a target. The spectrum of X rays so produced consists of two parts: a continuous spectrum produced when the electrons are decelerated by nuclei of the target; and peaks representing photons emitted by atoms of the target after being excited by collision with the high speed electrons.

▌ Questions

*1. Compare a matter wave Ψ to (a) a wave on a string, (b) an EM wave. Discuss similarities and differences.

2. Explain why Bohr's theory of the atom is not compatible with quantum mechanics, particularly the uncertainty principle.

3. Explain why it is that the more massive an object is the easier it becomes to predict its future position.

4. In view of the uncertainty principle, why does a baseball seem to have a well-defined position and speed whereas an electron does not?

5. Discuss whether something analogous to the uncertainty principle operates when taking a public opinion survey. That is, do we alter what we are trying to measure when we take such a survey?

6. A cold thermometer is placed in a hot bowl of soup. Will the temperature reading of the thermometer be the same as the temperature of the hot soup before the measurement was made?

7. If Planck's constant were much larger than it is, how would this affect our everyday life?

8. In what ways is Newtonian mechanics contradicted by quantum mechanics?

9. Discuss the differences between Bohr's view of the atom and the quantum-mechanical view.

10. The 589-nm yellow line in sodium is actually two very closely spaced lines. It is due to an "internal" Zeeman effect. Can you explain this? (*Hint*: put yourself in the reference frame of the electron.)

11. Which of the following electron configurations are forbidden? (a) $1s^22s^22p^63s^3$; (b) $1s^22s^22p^43s^24p^2$; (c) $1s^22s2p^83s^1$.

12. Give the complete electron configuration for a uranium atom.

*13. In what column of the periodic table would you expect to find the atom with each of the following configurations? (a) $1s^22s^22p^63s^2$; (b) $1s^22s^22p^63s^23p^6$; (c) $1s^22s^2\,2p^63s^23p^64s^1$; (d) $1s^22s^22p^5$.

*14. The ionization energy for neon ($Z = 10$) is 21.6 eV and that for sodium ($Z = 11$) is 5.1 eV. Explain the large difference.

*15. Why do chlorine and iodine exhibit similar properties?

*16. Explain why potassium and sodium exhibit similar properties.

*17. Why are the chemical properties of the rare earths so similar?

18. Why do we not expect perfect agreement between measured values of X-ray lines and those calculated using Bohr theory, as in Example 27-2?

*19. Certain dyes and other materials fluoresce by emitting visible light when UV light falls on them. Can infrared light produce fluorescence?

*20. Compare spontaneous emission to stimulated emission.

*21. How does laser light differ from ordinary light? How is it the same?

*22. Example how a 0.0005-W laser beam, photographed at a distance, can seem much stronger than a 1000-W street lamp.

▌ Problems

SECTION 27-3

1. (I) A proton is traveling with a speed of $(8.880 \pm 0.012) \times 10^5$ m/s. With what maximum accuracy can its position be ascertained?

2. (I) An electron remains in an excited state of an atom for typically 10^{-8} s. What is the minimum uncertainty in the energy of the state (in eV)? What is this uncertainty, in percent, of the first excited state in hydrogen?

3. (I) If an electron's position can be measured to an accuracy of 1.6×10^{-8} m, how accurately can its velocity be known?

4. (II) Estimate the least energy of a neutron contained in a typical nucleus of radius 10^{-15} m.

5. (II) A 12-g bullet leaves a rifle at a speed of 480 m/s. (a) What is the wavelength of this bullet? (b) If the position of the bullet is known to an accuracy of 0.65 cm (radius of the barrel), what is the minimum uncertainty in its momentum? (c) If the accuracy of the bullet were determined only by the uncertainty principle (an unreasonable assumption), by how much might the bullet miss a pinpoint target 300 m away?

6. (II) An electron and a 150-g baseball are each traveling 220 m/s measured to an accuracy of 0.065 percent. Calculate and compare the uncertainty in position of each.

7. (II) Use the uncertainty principle to show that if an

electron were present in the nucleus ($r \approx 10^{-15}$ m), its kinetic energy would be hundreds of MeV. (Since such electron energies are not observed, we conclude that electrons are not present in the nucleus.)

8. (III) The uncertainty principle can be stated in terms of angular qualities as follows:

$$\Delta L \, \Delta\phi \gtrsim \frac{h}{2\pi}.$$

Here, L stands for angular momentum along a given axis, and ϕ for the angular position measured in a plane perpendicular to that axis. (a) Make a plausibility argument for this relation. (b) Electrons in atoms have well-defined, quantized, values of angular momentum, with no uncertainty. What does this say about the uncertainty in angular position and the concept of electron orbits?

9. (III) Use the uncertainty principle to estimate the position uncertainty for the electron in the ground state of the hydrogen atom. (Hint: determine the momentum using the Bohr model of Section 26-9.) How does this compare to the Bohr radius?

*10. (III) In a double-slit experiment on electrons (or photons), suppose that we use indicators to determine which slit each electron went through (Section 27-2). These indicators must tell us the y coordinate to within $a/2$ where a is the distance between slits. Use the uncertainty principle to show that the interference pattern will be destroyed. (Note: first show that the angle θ between maxima and minima of the interference pattern is given by $\lambda/2a$, Fig. 27-18.)

11. (III) How accurately can the position of a 1.50-keV electron be measured assuming its energy is known to 1.00 percent?

SECTIONS 27-6 AND 27-7

12. (I) How many different states are possible for an electron whose principal quantum number is $n = 4$?

13. (I) List the quantum numbers for each electron in the ground state of carbon ($Z = 6$).

14. (I) How many electrons can be in the $n = 6$, $l = 3$ subshell?

15. (I) For $n = 6$, what values can l have?

FIGURE 27-18

16. (I) For $n = 5$, $l = 3$, what are the possible values of m_l and m_s?

17. (I) List the quantum numbers for each electron in the ground state of aluminum ($Z = 13$).

18. (II) Calculate the magnitude of the angular momentum of an electron in the $n = 3$, $l = 2$ state of hydrogen.

19. (II) Using the Bohr formula for the radius of an electron orbit, estimate the average distance from the nucleus for an electron in the innermost ($n = 1$) orbit in uranium ($Z = 92$). Approximately how much energy would be required to remove this innermost electron?

20. (II) Estimate the binding energy of the third electron in lithium using Bohr theory. (Hint: this electron has $n = 2$ and "sees" a net charge of approximately $+1e$.) The measured value is 5.36 eV.

21. (II) The ionization (binding) energy of the outermost electron in boron is 8.26 eV. (a) Determine the "effective charge," Z_{eff}, seen by this electron. (Hint: use the Bohr model.) (b) Estimate the average orbital radius.

22. (II) Show that the maximum number of electrons allowed in any subshell is equal to $2(2l + 1)$ where l is the angular momentum quantum number of the subshell.

SECTION 27-9

23. (I) What are the shortest-wavelength X rays emitted by electrons striking the face of a 30-kV TV picture tube? What are the longest wavelengths?

24. (I) If the shortest-wavelength bremsstrahlung X rays emitted from an X-ray tube have $\lambda = 0.019$ nm, what is the voltage across the tube?

25. (II) Use the result of Example 27-2 to estimate the X-ray wavelength emitted when a Ti ($Z = 22$) atom jumps from $n = 2$ to $n = 1$.

26. (II) Estimate the wavelength for an $n = 2$ to $n = 1$ transition in iron ($Z = 26$).

27. (II) Use Bohr theory to estimate the wavelength for an $n = 3$ to $n = 1$ transition in molybdenum. The measured value is 0.063 nm. Why do we not expect perfect agreement?

28. (II) An X-ray tube operates at 100 kV with a current of 20 mA and nearly all the electron energy goes into heat. If the specific heat capacity of the 78-g plate is 0.11 kcal/kg·C°, what will be the temperature rise per minute if no cooling water is used?

*SECTION 27-11

*29. (II) A laser used to weld detached retinas puts out 25-ms-long pulses of 640-nm light which average 0.50-W output during a pulse. How much energy can be deposited per pulse and how many photons does each pulse contain?

*30. (III) Estimate the angular spread of a laser beam due to diffraction if the beam emerges through a 5.0-mm-diameter mirror. Assume that $\lambda = 694$ nm. How large would be the diameter of this beam if it struck a satellite 1000 km above the earth?

28 MOLECULES AND SOLIDS

Since its development in the 1920s, quantum mechanics has had a profound influence on our lives, both intellectually and technologically. Even the way we view the world has changed, as we saw in Chapter 27. In the present chapter we will discuss how quantum mechanics has given us an understanding of the structure of molecules and matter in bulk, as well as a number of important applications including semiconductor devices, lasers, and applications to biology. Our discussions will necessarily be qualitative for the most part.

*28-1 ▌ Bonding in molecules

The study of molecules and their formation is considered both physics and chemistry. One of the great successes of quantum mechanics was to give scientists, at last, an understanding of the nature of chemical bonds. Since it is based in physics, and because this understanding is so important in many fields, we discuss it here.

By a molecule we mean a group of two or more atoms that are strongly held together so that it functions like a single unit. When atoms make such an attachment, we say that a chemical *bond* has been formed. There are two main types of strong chemical bond: covalent and ionic. Many bonds are actually intermediate between these two types.

Covalent bonds. To understand how covalent bonds are formed, we take the simplest case, the bond that holds two hydrogen atoms together to form the hydrogen molecule H_2. The mechanism is basically the same for other covalent bonds. As two H atoms approach each other, the electron clouds begin to overlap, and the electrons from each atom can "orbit" both nuclei. (This is sometimes called "sharing" electrons.) If both electrons are in the ground state ($n = 1$) of their respective atoms, two possibilities can occur: they can have their spins parallel (both up or both down), or they can be opposite ($m_s = +\frac{1}{2}$ for one, $m_s = -\frac{1}{2}$ for the other). We shall now see that a bond is formed only when the spins are opposite. First suppose that the spins are the same; the two electrons cannot both be in the lowest energy state, for then they would have identical quantum numbers. The exclusion principle tells us that the two electrons can then not come close together. That is, since no two electrons can occupy the same quantum state, if they have the same quantum numbers, they must be different in some other way—namely, by being in different places in space. When the two atoms approach, the electrons will stay away from each other as shown by the probability distributions of Fig. 28-1. The positively charged nuclei then repel each other, and no bond is formed. If the spins are opposite, the two electrons are automatically in different quantum states and then they can come close together spatially. In this case, the probability distribution looks like Fig. 28-2. As can be seen, the electrons spend much of their time between the two nuclei. The two positively charged nuclei are attracted to this negatively charged electron cloud between them and it is this attraction that holds the two atoms together to form a molecule. This is a *covalent bond*.

The probability distributions of Figs. 28-1 and 28-2 can perhaps be better understood on the basis of waves. What the exclusion principle requires is that when the spins are the same, there is destructive interference of the electron waves in the region between the two atoms. But when the spins are opposite, constructive interference occurs in the region between the two atoms, resulting in a large amount of negative charge there. Thus a covalent bond can be said to be the result of constructive interference of electron waves in the space between the two atoms, and the electrostatic attraction of the two positive nuclei for the negative charge concentration between them.

Why a bond is made can also be understood from the energy point of view. When the two H atoms approach close to one another, if the spins of their electrons are opposite, the electrons can occupy the same space as discussed above. This means that each electron can now move about in the space of two atoms instead of in the volume of only one. Because each electron now occupies more space, it is less well localized. From the uncertainty principle, with Δx increased, the momentum can be less. Another way of seeing this is using de Broglie waves (Section 26-10). Because each electron has a larger "orbit," its wavelength λ can be longer; and its momentum $p = h/\lambda$ (Eq. 26-5) can be less. With less momentum, each electron has less energy when the two atoms combine than when they are separate. That is, the molecule has less energy than the two separate atoms, and so is more stable. It takes an energy input to break the H_2 molecule into two separate H atoms, so the H_2 molecule stays together (unless given some energy input). This is what we mean by a *bond*. The energy required to break a bond is called the *bond energy* or *binding energy*.

Ionic bonds. An ionic bond is, in a sense, an extreme case of the covalent bond. Instead of the electrons being shared equally, they are shared unequally. For example, in sodium chloride (NaCl) the outer electron of the sodium spends nearly all its time around the chlorine (Fig. 28-3). The chlorine atom has a net negative charge as a result of the extra electron, whereas the sodium is positive. The electrostatic attraction between these two charged atoms is what holds them together and forms the bond. This is called an *ionic bond* because it is the attraction between the two ions (Na^+ and Cl^-) that holds them together. But to understand the ionic bond, we

FIGURE 28-1

Electron probability distribution (electron cloud) for two H atoms when their spins are the same.

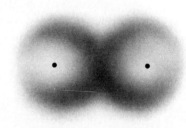

FIGURE 28-2

Electron probability distribution (cloud) around two H atoms when their spins are opposite. In this case a bond is formed because the positive nuclei are attracted to the concentration of negative charge between them. This is a hydrogen molecule, H_2.

FIGURE 28-3

Probability distribution for last electron of Na in NaCl.

must understand why the extra electron from the sodium spends so much of its time around the chlorine. After all, the chlorine is neutral; why should it attract another electron? The answer lies in the probability distributions of the two neutral atoms. Sodium contains 11 electrons, 10 of which are in spherically symmetric closed shells (Fig. 28-4). The last electron spends most of its time beyond these closed shells. Because the closed shells have a total charge of $-10e$ and the nucleus has $+11e$, the outermost electron in sodium feels a net attraction due to $+1e$. It is thus not held terribly strongly. On the other hand, 12 of chlorine's 17 electrons form closed shells or subshells (corresponding to $1s^2 2s^2 2p^6 3s^2$). These 12 form a spherically symmetric shield around the nucleus. The other five electrons are in $3p$ orbits, which are barbell-shaped. Two of these $3p$ electrons are in the dumbbell-shaped orbit (or cloud) shown in Fig. 28-5. Two more are in an orbit perpendicular to the page (not shown). The fifth is in an orbit perpendicular to these and is shown only in faint outline. The exclusion principle allows one more electron to be in this orbit (it will have spin opposite to that of the electron already there). If an extra electron—say from a Na atom—happens to be in the vicinity, it can be in this orbit, say at point x in Fig. 28-5. It can experience an attraction due to as much as $+5e$ because the $+17e$ of the nucleus is shielded at this point mainly by the 12 inner electrons. Clearly, the outer electron of a Na atom will be more strongly attracted by the chlorine than by the $+1e$ of its own atom.[†] This is why we get the charge distribution of Fig. 28-3, and thus an ionic bond.

Partial ionic character of covalent bonds. A pure covalent bond in which the electrons are shared equally occurs mainly for symmetrical molecules like H_2, O_2, and Cl_2. When the atoms involved are different from each other,

---

[†] Coulomb's law also involves distance between the two charges, but that is not greatly different in the two cases.

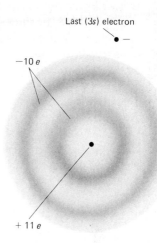

Last (*3s*) electron

$-10\,e$

$+11\,e$

FIGURE 28-4

In a neutral sodium atom, the 10 inner electrons shield the nucleus; so the single outer electron is attracted by a net charge of $+1e$.

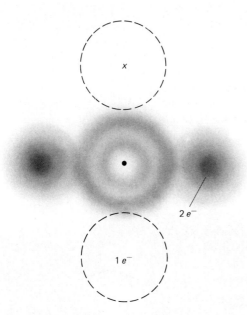

x

$2\,e^-$

$1\,e^-$

FIGURE 28-5

Neutral chlorine atom. The $+17e$ of the nucleus is shielded by the 12 electrons in the inner shells and subshells. Five electrons are in dumbbell-shaped clouds ($l = 1$). An extra electron at x will be attracted by a net charge that can be as much as $+5e$.

it is usual to find that the shared electrons are found more on one atom than the other. The extreme case is an ionic bond, but in intermediate cases the covalent bond is then said to have a *partial ionic character*. The molecules themselves are *polar*—that is, one part (or parts) of the molecule has a net positive charge and other parts a negative charge. An example is the water molecule, H_2O (Fig. 28-6). The shared electrons gather around the oxygen atom more than around the two hydrogens. The reason is similar to that causing ionic bonds. Oxygen has eight electrons ($1s^2 2s^2 2p^4$), of which four form a spherically symmetric core and the other four are in barbell-shaped orbits. One possible barbell-shaped orbit is empty, so electrons from hydrogen atoms can be attracted by a net charge of $+4e$. They are also attracted by the H nuclei, so they actually orbit the H atoms as well as the O atom. The net effect is that there is a net positive charge on each H atom (less than $+1e$), because the electron spends only part of its orbit there. And there is a net negative charge on the O atom.

*28-2 ∎ Weak bonds

Once a bond is made, energy must normally be put in to break the bond and separate the atoms. As mentioned in the previous section, this energy is called the *bond energy* or *binding energy*. The binding energy for covalent and ionic bonds is typically 2 to 5 eV. These bonds, which hold atoms together to form molecules, are often called *strong bonds* to distinguish them from so-called *weak bonds*. The term "weak bond" as we use it here refers to an attachment between molecules because of simple electrostatic attraction—such as *between* polar molecules (and not *within* a polar molecule, which is a strong bond). The strength of the attachment is much less than for the strong bonds; binding energies are typically in the range 0.04 to 0.3 eV—hence their name "weak bonds."

Weak bonds are generally the result of attraction between dipoles. For example Fig. 28-7 shows two molecules that have a permanent dipole moment attracting one another. Besides such *dipole–dipole bonds*, there can also be *dipole–induced dipole* bonds, in which a polar molecule with a permanent dipole moment can induce a dipole moment in an otherwise electrically balanced (nonpolar) molecule, just as a single charge can induce a separation of charge in a nearby object (see Fig. 15-5). There can even be an attraction between nonpolar molecules: even though a molecule may not have a permanent dipole moment on the average, we can think of its electrons as moving about so that at any instant there may be a separation of charge which can induce a dipole moment in a nearby molecule so a weak attraction occurs. All these weak bonds are referred to as *van der Waals bonds*.

When one of the atoms in a dipole–dipole bond is hydrogen, it is often called a *hydrogen bond*. A hydrogen bond is generally the strongest of the weak bonds; this is because the hydrogen atom is the smallest atom and thus can be approached more closely. Hydrogen bonds also have a partial "covalent" character; that is, electrons between the two dipoles may be shared to a small extent.

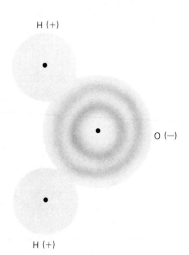

FIGURE 28-6
The water molecule is polar.

FIGURE 28-7
The C^+—O^- and H^+—N^- dipoles attract each other. (These dipoles may be part of thymine and adenine molecules, respectively.)

Weak bonds are important in liquids and solids when strong bonds are absent (see Section 28-3), and they are also very important for understanding the activities of a cell in molecular biology. The average kinetic energy of molecules in a cell is around 0.04 eV, about the magnitude of weak bonds. This means that a weak bond can readily be broken just by a molecular collision. Hence weak bonds are not very permanent—they are, instead, brief attachments. But because of this, they play an important role in the cell, as we saw in Chapter 15. On the other hand, strong bonds—those that hold molecules together—are almost never broken simply by molecular collision. Thus they are relatively permanent. They can be broken by chemical action (the making of even stronger bonds), and this usually happens in the cell with the aid of an enzyme.

*28-3 ▌ Bonding in solids

The molecules of a solid are held together in a number of ways. The most common are by covalent bonding (such as between the carbon atoms of diamond) or ionic bonding (as in a NaCl crystal). Often the bonds are partially covalent and partially ionic.

Another type of bond occurs in metals. Metals have relatively free outer electrons, and present **metallic bond** theories indicate that in a metallic solid, these outer electrons roam rather freely among all the metal atoms which, without their outer electrons, act like positive ions. The electrostatic attraction between the metal ions and the negative electron "gas" (as it is called) is what is believed to hold the solid together. This theory nicely accounts for the shininess of smooth metal surfaces: the free electrons can vibrate at any frequency; so when light of almost any frequency falls on a metal, the electrons can vibrate in response and re-emit light of that same frequency. Hence the reflected light can have mainly the same frequencies as the incident light. Compare this to ordinary materials that have a distinct color—the electrons exist only in certain energy states and thus can resonate only at certain frequencies.

The atoms or molecules of some materials, such as the inert gases, can only form *weak bonds* with each other. As we just saw, weak bonds have very low binding energies and would not be expected to hold atoms together as a liquid or solid at room temperature. Such materials condense only at very low temperatures, where the atomic kinetic energy is small and the bonds can then hold the atoms together.

*28-4 ▌ Potential-energy diagrams

It is useful to analyze the interaction between two objects—say, between two atoms or molecules—with the use of a potential-energy diagram, which is simply a plot of the potential energy versus distance.

Let us first examine the simple case of one point charge, q_1, in the presence of a second point charge, q_2. As we saw in Chapter 16, the potential energy of charge q_2 in the presence of q_1 a distance r away is

$$\text{PE} = k \frac{q_1 q_2}{r},$$

where $k = 9.0 \times 10^9 \text{ N·m}^2/\text{C}^2$.

If the two charges are the same sign, the PE is positive for all values of r; a graph of PE versus r in this case is shown in Fig. 28-8a. The force is repulsive (the charges have the same sign) and we see that the curve rises as r decreases.

If the two charges are of opposite sign, the PE will be negative because the product $q_1 q_2$ will be negative. The force is attractive in this case and $\text{PE} = k(q_1 q_2/r)$ versus r looks like Fig. 28-8b. The PE in this case becomes more *negative* as r decreases. Keep in mind this fact: for a repulsive force, the PE increases as r decreases; and for an attractive force, the PE decreases as r decreases.

Now let us look at the potential-energy diagram for the formation of a covalent bond, such as for the hydrogen molecule. The potential energy of one H atom in the presence of the other is plotted in Fig. 28-9. The PE decreases as the atoms approach, because the electrons concentrate between the two nuclei (Fig. 28-2), so attraction occurs. However, at very short distances, the electrons would be "squeezed out"—there is no room for them between the two nuclei. Without the electrons between them, each nucleus would feel a repulsive force[†] due to the other, so the curve rises as r decreases further. There is an optimum separation of the atoms, r_0 in Fig. 28-9, where the energy is lowest. This is the point of greatest stability for the hydrogen molecule, and r_0 is the average separation of atoms in the H_2 molecule. The depth of this "well" is the binding energy, as shown. This is how much energy must be put into the system to separate the two atoms to infinity, where the PE = 0.

[†] In larger atoms, say oxygen or nitrogen, repulsion also occurs at short distances, because the closed inner electron shells begin to overlap and the exclusion principle forbids their coming too close.

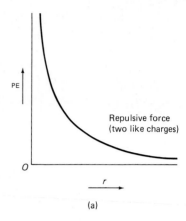

Repulsive force
(two like charges)

(a)

Attractive force
(unlike charges)

(b)

FIGURE 28-8

Potential energy as a function of separation for two point charges of (a) like sign and (b) opposite sign.

FIGURE 28-9

Potential-energy diagram for H_2 molecule; r is separation of the two H atoms.

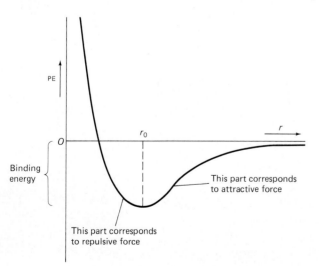

Binding energy

This part corresponds to attractive force

This part corresponds to repulsive force

The potential-energy diagram for ionic bonds is similar in shape. In NaCl, for example, the Na^+ and Cl^- ions attract each other at distances above some r_0, but at smaller distances the overlapping of inner electron shells gives rise to repulsion. The two atoms thus are most stable at some intermediate separation r_0.

*28-5 ▌ Bonds and activation energy

For some bonds, the potential-energy curve has the shape shown in Fig. 28-10. There is still an optimum distance r_0 at which the molecule is stable. But when the atoms approach from a large distance, the force is initially repulsive rather than attractive. The atoms thus do not interact spontaneously. Instead, some additional energy must be injected into the system to get it over the "hump" (or barrier) in the potential-energy diagram. This required energy is called the *activation energy*.

The curve of Fig. 28-10 is far more common than that of Fig. 28-9. The activation energy often reflects a need to break other bonds, before the one under discussion can be made. For example, to make water from O_2 and H_2, the H_2 and O_2 molecules must first be broken into H and O atoms; this is what the activation energy represents. Then the H and O atoms can combine to form H_2O with the release of a great deal more energy than was put in initially. The initial activation energy can be provided by applying an electric spark to a mixture of H_2 and O_2, breaking a few of these molecules into H and O atoms; the resulting explosive release of energy when these atoms combine to form H_2O quickly provides the activation energy needed for further reactions, so further H_2 and O_2 molecules are broken up and recombined to form H_2O.

In a living cell, many chemical reactions are occurring that involve the making and breaking of strong bonds. The activation energies are often on the order of several eV. Such energies are not easy to overcome in the cell.

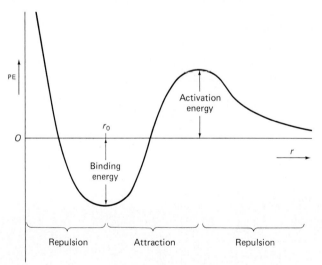

FIGURE 28-10

Potential-energy diagram for a bond requiring an activation energy.

This is where enzymes come in. They act as catalysts, which means that they act to lower the activation energy so that reactions can occur that otherwise would not. Enzymes act by distorting the bonding electrons so that the initial bonds are easily broken.

Sometimes the potential energy of a bond looks like that of Fig. 28-11. In this case the energy of the bonded molecule, at a separation r_0, is greater than when there is no bond ($r = \infty$). That is, an energy *input* is required to make the bond. (Hence the binding energy is negative.) And there is energy release when the bond is broken. Such a bond is stable only because there is the barrier of the activation energy. This type of bond is very important in biological cells, for it is by this means that energy can be stored efficiently in certain molecules, particularly ATP (adenosine triphosphate). The bond that connects the last phosphate group (designated Ⓟ in Fig. 28-11) to the rest of the molecule (ADP, meaning adenosine diphosphate, since it contains only two phosphates) is of the form shown in Fig. 28-11. Energy is actually stored in this bond.[†] When the bond is broken (ATP → ADP + Ⓟ) energy is released and this energy can be used to make other chemical reactions "go."

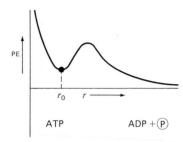

FIGURE 28-11
Potential-energy diagram for formation of ATP from ADP and phosphate (Ⓟ).

*28-6 ▌ Molecular spectra

When atoms combine to form molecules, the energy levels of the outer electrons are altered because they now interact with each other. Additional energy levels also become possible because the atoms can vibrate with respect to each other, and the molecule as a whole can rotate. The energy levels for both vibrational and rotational motion are quantized. However, they are very close together, particularly for the rotational levels (typically 10^{-3} eV apart). Each atomic energy level thus becomes a set of closely spaced levels corresponding to the vibrational and rotational motions (Fig. 28-12). Transitions from one level to another appear as many very closely spaced lines. In fact, the lines are not always distinguishable, and such spectra are called *band spectra*. Each type of molecule has its own characteristic spectrum, which can be used for identification and for determination of structure.

When atoms or molecules are in the solid (or liquid) state, their outer electrons also overlap. Hence their energy levels are changed somewhat, and because of the interaction, the energy levels are spread out into *energy bands* as in Fig. 28-12. Because there are so many atoms present, the energy levels

[†] The second phosphate is also connected by such a "high-energy" bond, as it is sometimes called.

FIGURE 28-12
The individual energy levels of an isolated atom become bands of closely spaced levels in molecules as well as in solids and liquids.

in such a band are extremely close together. They thus seem continuous, which is why the spectrum of heated solids, such as red-hot iron or a light bulb filament, appears continuous. The energy bands in semiconductors play an important role for transistors and diodes as we discuss in the next four sections.

*28-7 ∎ Semiconductors

Quantum mechanics has been a great tool for understanding the structure of solids. This active field of research today is called *solid-state physics* (or *condensed-matter physics*, so as to include liquids as well). We now examine, very briefly, one aspect of this field—semiconductors—and some of its applications in modern electronics. First we look at some properties of the most commonly used semiconductors, germanium and silicon.

An atom of silicon or germanium has four outer electrons which act to hold the atoms in the regular lattice structure of the crystal, Fig. 28-13a. Germanium and silicon acquire useful properties for use in electronics only when a tiny amount of impurity is introduced into the crystal structure (about 1 part in 10^6). This is called "doping" the semiconductor. Two kinds of semiconductors can be made, depending on the type of impurity used. If the impurity is a material whose atoms have five outer electrons, such as arsenic, we have the situation shown in Fig. 28-13b. Only four of arsenic's electrons fit into the crystal structure. The fifth does not fit in and can move relatively freely, much like the electrons in a conductor. Because of this small number of extra electrons, a doped semiconductor becomes slightly conducting. An arsenic-doped silicon crystal is called an *n-type semiconductor* because electrons (*n*egative charge) carry the electric current.

In a *p-type semiconductor*, a small amount of impurity with three outer electrons—such as gallium—is added to the semiconductor. As shown in Fig. 28-14a, there is a "hole" in the lattice structure next to a gallium atom since it has only three outer electrons. Electrons from nearby silicon atoms can jump into this hole and fill it. But this leaves a "hole" where that electron had previously been, Fig. 28-14b. The vast majority of atoms are silicon, so the hole is almost always next to a silicon atom. Since silicon atoms require four outer electrons to be neutral, this means there is a net positive charge

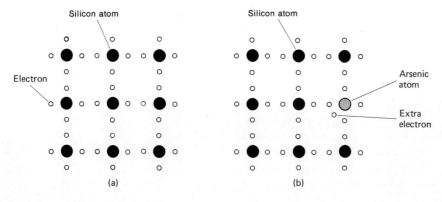

(a)   (b)

FIGURE 28-13

A silicon crystal. (a) Four (outer) electrons surround each silicon atom. (b) Silicon crystal doped with a few arsenic atoms; the extra electron doesn't fit into the crystal lattice and so is free to move about. This is an *n*-type semiconductor.

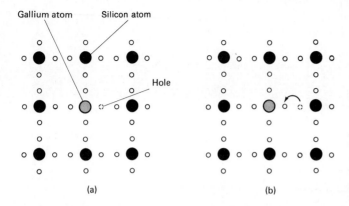

Gallium atom   Silicon atom

Hole

(a)   (b)

FIGURE 28-14

A *p*-type semiconductor, gallium-doped silicon.
(a) Gallium has only three outer electrons, so there is
an empty spot, or hole, in the structure. (b) Electrons
from silicon atoms can jump into the hole and fill
it. As a result, the hole moves to a new location (to
the right in this figure).

charge at a hole. Whenever an electron moves to fill a hole, the positive hole
is then at the previous position of that electron. Another electron can then
fill this hole, and the hole thus moves to a new location; and so on. This
type of semiconductor is called *p-type* because it is the *positive* holes that
seem to carry the electric current. Note, however, that *p*-type and *n*-type
semiconductors have *no net charge* on them.

*28-8 ∎ Electron band theory

Semiconductor behavior can be explained at the atomic level using the so-
called *electron band theory* of solids which is based on quantum mechanics.
Electrons in a single atom can occupy only certain energy levels, as we have
seen. The lowest possible energy level is called the ground state and higher
ones are called excited states. In a solid, these energy levels spread out into
wide "bands" because of the interaction among the atoms, as was described
in Section 28-6. The outer electrons can be considered to be in either of
two bands:[†] the lower *valence band*, which corresponds to the ground state;
or the upper *conduction band*. No electron can have an energy in the "forbidden" energy gap between the two bands. Normally the electrons reside
in the valence band where they are held rather tightly to individual atoms.
In an insulator, the valence band is full, the conduction band is essentially
empty, and the forbidden gap between them is fairly large; so there are no
free states for the electrons to move into, unless given sufficient energy (say
by collision with another electron) to jump up to the higher conduction
band. Once there, an electron would no longer be bound to a particular
atom but could move about freely in the lattice. If the gap between the bands
is large, the material will be an insulator since very few electrons would have
sufficient energy to reach the conduction band; see Fig. 28-15a. In a good
conductor, on the other hand, there is no gap (Fig. 28-15b); the two bands
may overlap, or there is simply one band that is not filled and the electrons are free to move easily to other states; they can thus move about freely
and carry an electric current. In a pure (or *intrinsic*) semiconductor such as
carbon, germanium or silicon, the forbidden energy gap between valence and

---

[†] Additional bands, of lower energy, normally have all states occupied.

(a) Insulator   (b) Conductor   (c) Semiconductor

FIGURE 28-15
Energy bands in solids, showing (idealized) only bands for outer electrons. The bands for other electrons are of lower energy.

conduction bands is small, Fig. 28-15c. A small percentage of electrons can have enough energy to jump the gap, so there will be a very slight amount of conduction. (At room temperature, the average KE of electrons is on the order of $kT \approx \frac{1}{40}$ eV.) If the temperature is raised, more electrons will have enough energy to jump the gap; this effect can often more than offset the effects of more frequent collisions due to increased disorder at higher temperature, so that the resistivity of semiconductors can *decrease* with temperature (see Table 17-1).

In a doped semiconductor, the impurity provides additional energy states between the bands as shown in Fig. 28-16. In an *n*-type semiconductor, the impurity energy level lies just below the conduction band. Electrons in this energy level can easily acquire sufficient energy to reach the conduction band. Since this level supplies electrons to the conduction band, it is called a *donor* level. In *p*-type semiconductors, the impurity energy level is just above the valence band. It is called an *acceptor* level because electrons from the valence band can easily jump into it. A positive hole is left behind; and as other electrons move into this hole, the hole moves about as discussed earlier.

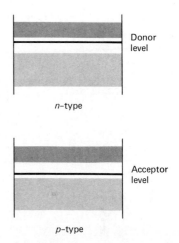

n-type

p-type

FIGURE 28-16
Impurity levels in doped semiconductors.

*28-9 ∎ Semiconductor diodes; rectification

Semiconductor diodes and transistors are essential components of modern electronic devices. The miniaturization achieved today allows many thousands of diodes, transistors, resistors, and so on, to be placed on a single *chip* only a centimeter on a side. We now discuss, briefly and simply, the operation of diodes and transistors.

When an *n*-type semiconductor is joined to a *p*-type, a *p-n junction diode* is formed. Separately, the two semiconductors are electrically neutral. When joined, a few electrons near the junction diffuse from the *n*-type into the *p*-type semiconductor, where they fill a few of the holes. The *n*-type is left with a positive charge, and the *p*-type acquires a net negative charge. Thus a potential difference is established, with the *n* side positive relative to the *p* side, and this prevents further diffusion of electrons.

If a battery is connected to a diode with the positive terminal to the *p* side and the negative terminal to the *n* side as in Fig. 28-17a, such a voltage opposes the internal potential difference and the diode is said to be *forward biased*. If the voltage is great enough (about 0.3 V for Ge, 0.6 V for Si at room temperature) a current will flow. The positive holes in the *p*-type semiconductor are repelled by the positive terminal of the battery and the electrons

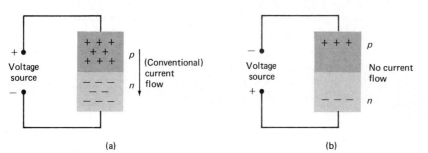

FIGURE 28-17

Schematic diagram showing how a semiconductor diode operates. Current flows when the voltage is connected in forward bias, as in (a), but not when in reverse bias, as in (b).

in the *n*-type are repelled by the negative terminal of the battery. The holes and electrons meet at the junction, and the electrons cross over and fill the holes. A current is flowing. Meanwhile, the positive terminal of the battery is continually pulling electrons off the *p* end, forming new holes, and electrons are being supplied by the negative terminal at the *n* end. Consequently a large current flows through the diode.

When the diode is *reverse biased*, as in Fig. 28-17b, the holes in the *p* end are attracted to the battery's negative terminal and the electrons in the *n* end are attracted to the positive terminal. The current carriers do not meet near the junction and, ideally, no current flows.

A graph of current versus voltage for a typical diode is shown in Fig. 28-18. As can be seen, a real diode does allow a small amount of reverse current to flow. For most practical purposes, this is negligible.[†]

EXAMPLE 28-1 The diode whose current–voltage characteristics are shown in Fig. 28-18 is connected in series with a 4.0-V battery and a resistor. If a current of 10 mA is to pass through the diode, what resistance must the resistor have?

SOLUTION From Fig. 28-18 we see that the voltage drop across the diode is about 0.7 V when the current is 10 mA. Therefore, the voltage drop across the resistor is $4.0 \text{ V} - 0.7 \text{ V} = 3.3 \text{ V}$, so $R = V/I = (3.3 \text{ V})/(1.0 \times 10^{-2} \text{ A}) = 330 \ \Omega$.

[†] Reverse current in Ge at room temperature is typically a few μA; for Si, a few pA. The reverse current increases rapidly with temperature, however, and may render a diode ineffective above 200°C.

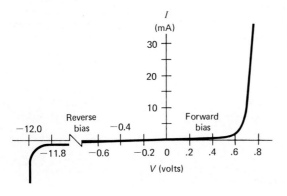

FIGURE 28-18

Current through a diode as a function of applied voltage.

If the voltage across a diode connected in reverse bias is increased greatly, a point is reached where breakdown occurs. The electric field across the junction becomes so large that ionization of atoms results. The electrons thus pulled off their atoms contribute to a larger and larger current as breakdown continues. The voltage remains constant over a wide range of currents. This is shown on the far left in Fig. 28-18. This property of diodes can be used to accurately regulate a voltage supply. A diode designed for this purpose is called a *zener diode*. When placed across the output of an unregulated power supply, a zener diode can maintain the voltage at its own breakdown voltage as long as the supply voltage is always above this point. Zener diodes can be obtained corresponding to voltages of a few volts to hundreds of volts.

A diode is called a *nonlinear device* since the current is not proportional to the voltage; that is, a graph of current versus voltage (Fig. 28-18) is not a straight line as it is for a resistor (which *is* linear). Transistors are also *nonlinear* devices.

Since a *p-n* junction diode allows current to flow only in one direction, it can serve as a *rectifier*—to change ac into dc. A simple rectifier circuit is shown in Fig. 28-19a where the arrow inside the symbol for a diode indicates the direction in which a diode conducts conventional (+) current. The ac source applies a voltage across the diode alternately positive and negative. Only during half of each cycle will a current pass through the diode; so only then is there a current through the resistor R. Hence the voltage V_{ab} across R as a function of time looks like that shown in the graph in Fig. 28-19b. This *half-wave rectification* is not exactly dc, but it is unidirectional. More useful is a *full–wave rectifier* circuit which uses two diodes (or sometimes four) as shown in Fig. 28-20a. At any given instant, either one diode or the other will conduct current to the right. Therefore the output across the load resistor R will be as shown in Fig. 28-20b; actually this is the voltage if the capacitor C were not in the circuit. The capacitor tends to store charge, and thus helps to smooth out the current as shown in Fig. 28-20c.

Rectifier circuits are important because most line voltage is ac, and most electronic devices require a dc voltage for their operation. Hence, diodes are found in nearly all electronic devices including radio and TV sets, calculators, and computers.

ac source

(a)

V_{ab}

Time

(b)

FIGURE 28-19

(a) Simple rectifier circuit, using a semiconductor diode. (b) Output voltage across R as a function of time.

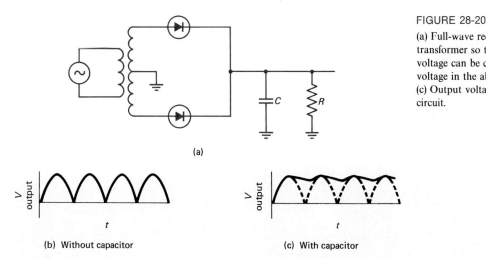

(a)

(b) Without capacitor

(c) With capacitor

FIGURE 28-20

(a) Full-wave rectifier circuit (including a transformer so the magnitude of the voltage can be changed). (b) Output voltage in the absence of capacitor C. (c) Output voltage with capacitor in the circuit.

FIGURE 28-21

(a) Schematic diagram of *npn* and *pnp* transistors.
(b) Symbols for *npn* and *pnp* transistors.

*28-10 ▮ Transistors; amplifiers and integrated circuits

A simple *junction transistor* consists of a crystal of one type of semiconductor sandwiched between two crystals of the opposite type. Both *pnp* and *npn* transistors are made and they are shown schematically in Fig. 28-21a. The three semiconductors are given the names collector, base, and emitter. The symbols for *npn* and *pnp* transistors are shown in Fig. 28-21b. The arrow is always placed on the emitter and indicates the direction of (conventional) current flow in normal operation.

The operation of a transistor can be analyzed as follows. Consider an *npn* transistor connected as shown in Fig. 28-22. A voltage V_{CE} is maintained between the collector and emitter by the battery $\mathcal{E}_C$. The voltage applied to the base is called the *base bias voltage*, V_{BE}; if it is positive, electrons in the emitter are attracted into the base. Since the base region is very thin (perhaps 1 μm), most of the electrons flow right across into the collector, which is maintained at a positive voltage. A large current, I_C, flows between collector and emitter and a much smaller current, I_B, through the base. A small variation in the base voltage due to an input signal causes a large change in the collector current and therefore a large change in the voltage drop across the

FIGURE 28-22

An *npn* transistor used as an amplifier.

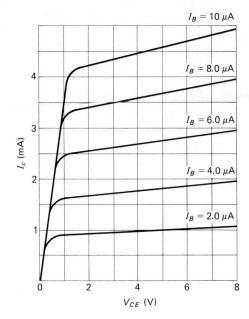

$I_B = 10\ \mu A$

$I_B = 8.0\ \mu A$

$I_B = 6.0\ \mu A$

$I_B = 4.0\ \mu A$

$I_B = 2.0\ \mu A$

I_c (mA)

V_{CE} (V)

FIGURE 28-23

Typical transistor characteristics. Collector current I_C is plotted against collector voltage V_{CE} for various values of the base current I_B.

output resistor R_C. Hence a transistor can amplify a small signal into a larger one. In fact transistors are the basic elements in modern electronic amplifiers of all sorts.

A *pnp* transistor operates in the same fashion, except that holes move instead of electrons. The collector voltage is negative, and so is the base voltage in normal operation.

The operation of a transistor is commonly examined using graphs called *characteristic curves*. For example, it is common to plot the collector current I_C versus collector voltage V_{CE} for different values of the base current I_B. Such a graph is shown in Fig. 28-23 for a typical *npn* transistor. As can be seen, a few microamps in the base corresponds to milliamps in the collector. If the base current is zero, there is no collector current. The greater the base current (caused by a greater base voltage), the greater the collector current will be. Normally, a small ac signal is to be amplified. Such a signal, when added to the base bias voltage, causes the voltage and current at the base to vary. Thus, what is important for amplification is the *change* in collector current for a given *change* in base current. *Gain* is defined as the ratio output/input, and to see how to calculate it from a graph such as Fig. 28-23, we take an example.

◻   EXAMPLE 28-2   Refer to Fig. 28-23. Suppose that the base bias voltage is such that the base current is 6.0 μA when there is no input signal. Suppose also that the collector voltage, V_{CE}, is kept at 5.0 V. Then from the graph we see that the collector current is 2.8 mA. Now let an ac input signal, whose voltage and current we represent by v and i, be applied to the base. Suppose that i alternates from 2.0 μA to $-2.0\ \mu A$. Calculate the current gain and voltage gain assuming that $R_B = 5.0\ k\Omega$ and $R_C = 1.0\ k\Omega$ (in Fig. 28-22).

SOLUTION   The total base current oscillates between 4.0 μA and 8.0 μA (6.0 $\pm$ 2.0 μA). From Fig. 28-23 we see that if V_{CE} remains at 5.0 V, the collector current varies from 1.8 mA to 3.8 mA. That is, I_C varies ± 1.0 mA from its original value of 2.8 mA. Thus a 2.0-μA alternating current in the base causes a 1.0-mA alternating current in the collector. The current gain β is given by

$$\beta = \frac{\text{output ac current}}{\text{input ac current}} = \frac{1.0 \times 10^{-3}\,\text{A}}{2.0 \times 10^{-6}\,\text{A}} = 500.$$

The voltage, too, will be amplified. The input voltage v of the signal is related to the current i by $v = iR_B$. (The transistor offers little resistance, since the base–emitter junction is like a forward-biased diode.) $R_B = 5.0\,\text{k}\Omega$ and $R_C = 1.0\,\text{k}\Omega$; then the peak ac input voltage is $v = (2.0 \times 10^{-6}\,\text{A})(5.0 \times 10^3\,\Omega) = 10$ mV; the peak ac output voltage is $(1.0 \times 10^{-3}\,\text{A})(1.0 \times 10^3\,\Omega) = 1.0$ V. Thus the voltage gain is $(1.0\,\text{V})/(10 \times 10^{-3}\,\text{V}) = 100$.

We assumed in this example that V_{CE} does not change. In more typical situations, a battery or other dc source ($\mathscr{E}_C$ in Fig. 28-22) maintains a constant voltage across a resistor R_C and the transistor in series; so the voltage V_{CE} varies, depending on the current I_C that passes through R_C. This is dealt with in the problems at the end of this chapter.

In Fig. 28-22 two batteries are shown: $\mathscr{E}_C$ supplies the collector voltage and $\mathscr{E}_B$ the base bias voltage. In practice, only one source is usually used, and the base bias voltage is obtained using a resistance voltage divider as in Fig. 28-24. Transistors can be connected in many other ways as well, and many new and innovative uses have been found for them.

Transistors were a great advance in miniaturization of electronic circuits. Although individual transistors are very small compared to vacuum tubes, they are huge compared to *integrated circuits* or "chips." Tiny amounts of impurities can be placed at particular locations within a single silicon crystal. These can be arranged to form diodes, transistors, and resistors (which are simply undoped semiconductors). Capacitors and inductors can also be formed, although they are often connected separately. A tiny chip, only 1 cm on a side, may contain thousands of transistors and other circuit elements.

FIGURE 28-24

Typical transistor circuit involving an *npn* transistor. $\mathscr{E}_C = +12$ V and $\mathscr{E}_B$ is determined by the resistors.

❚ Summary

Quantum mechanics explains the bonding together of atoms to form molecules. In a *covalent bond*, the electron clouds of two or more atoms overlap because of constructive interference between the electron waves; the positive nuclei are attracted to this concentration of negative charge between them, which forms the bond. An *ionic bond* is an extreme case of a covalent bond in which one or more electrons from one atom spend much more time around the other atom than around their own; the atoms then act as oppositely charged ions which attract each other— which is the bond. These *strong bonds* hold molecules together, as well as hold atoms and molecules together in solids. Also important are *weak bonds*, which are generally dipole attractions between molecules. In metals, the electrostatic force between the free electrons and the positive ions is believed to be the mechanism for the *metallic bond*.

The energy states in molecules, as well as in liquids and solids, are often very closely spaced groups of levels, giving rise to *band spectra*.

Semiconductor devices make use of two types of doped semiconductors: *n*-type, which contain impurities with extra electrons that can move rather freely; and *p*-type which contain impurities with fewer than the normal number of electrons so that positively charged "holes" can move about. A semiconductor diode consists of a *pn*-junction and allows current to flow in one direction only; it can be used as a rectifier to change ac to dc.

Common transistors consist of three semiconductor sections, either as *pnp* or *npn*. Transistors can amplify electrical signals and find many other uses. An integrated circuit consists of a tiny semiconductor crystal or "chip" on which many transistors, diodes, resistors, and other circuit elements have been constructed using careful placement of impurities.

▌Questions

*1. What type of bond would you expect for (*a*) the N_2 molecule, (*b*) the HCl molecule, (*c*) Fe atoms in a solid?

*2. Describe how the molecule $CaCl_2$ could be formed.

*3. Does the H_2 molecule have a permanent dipole moment? Does O_2? Does H_2O? Explain.

*4. Although the molecule H_3 is not stable, the ion H_3^+ is. Explain, using the Pauli exclusion principle.

*5. If conduction electrons are free to roam about in a metal, why don't they leave the metal entirely?

*6. A silicon semiconductor is doped with phosphorus. Will these atoms be donors or acceptors? What type of semiconductor will this be?

*7. Can a diode be used to amplify a signal?

*8. Figure 28-25 shows a "bridge-type" full-wave rectifier. Explain how the current is rectified and how current flows during each half cycle.

*9. Compare the resistance of a *pn* junction diode when connected in forward bias to when connected in reverse bias.

*10. Explain how a transistor could be used as a switch.

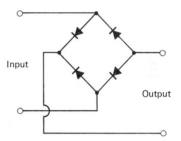

FIGURE 28-25

*11. If $\mathscr{E}_C$ were reversed in Fig. 28-22 how would the amplification be altered?

*12. Describe how a *pnp* transistor can operate as an amplifier.

*13. In a transistor, the base–emitter junction and the base–collector junction are basically diodes. Are these junctions reverse-biased or forward-biased in the application shown in Fig. 28-22?

*14. What purpose does the capacitor in Fig. 28-24 serve?

▌Problems

*SECTIONS 28-1 TO 28-4

*1. (II) Estimate the binding energy of a KCl molecule by calculating the electrostatic potential energy when the K^+ and Cl^- ions are at their stable separation of 0.28 nm. Assume each has a charge of magnitude $1.0e$.

*2. (II) The measured binding energy of KCl is 4.43 eV. From the result of problem 1 estimate the contribution to the binding energy of the repelling electron clouds at the equilibrium distance $r_0 = 0.28$ nm.

*3. (II) Estimate the binding energy of the H_2 molecule; assume the two H nuclei are 0.8 nm apart and the two electrons spend 50 percent of their time midway between them.

*4. (III) Estimate the binding energy of the H_2 molecule by calculating the difference in KE of the electrons between separate atoms and the molecule using the uncertainty principle; take Δx for the electrons in the separated atoms to be the radius of the first Bohr orbit, 0.053 nm, and for the molecule take Δx to be the separation of the nuclei, 0.074 nm.

*SECTION 28-9

*5. (I) A silicon diode whose current–voltage characteristics are given in Fig. 28-18 is connected in series with a battery and a 200-Ω resistor. What battery voltage is needed to produce a 10-mA current?

FIGURE 28-26

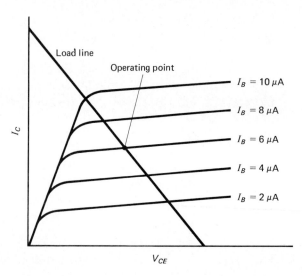

FIGURE 28-27

"Load-Line" curve superimposed on the characteristic curves of the transistor of Fig. 28-23, showing the "Operating Point" for a base bias current of 6.0 μA. The numerical values are to be filled in when doing Problem 18.

*6. (II) An ac voltage of 240 V rms is to be rectified. Estimate the average current in the output resistor R (15 kΩ) for (a) a half-wave rectifier (Fig. 28-19), and (b) a full-wave rectifier (Fig. 28-20) without capacitor.

*7. (III) A 240 V rms 60-Hz voltage is to be rectified with a full-wave rectifier as in Fig. 28-20, where $R = 15$ kΩ and $C = 30$ μF. (a) What will be the average current (approximately)? (b) What if $C = 0.10$ μF? (Hint: see Section 18-6.)

*8. (III) A full-wave rectifier (Fig. 28-20) uses two diodes to rectify a 74-V rms ac voltage. If $R = 15$ kΩ and $C = 30$ μF, what will be the approximate percent variation in the output voltage? This is called *ripple voltage* (see Fig. 28-20c). (Hint: see Section 18-6 and assume that the discharge of the capacitor is approximately linear.)

*9. (III) Suppose that the diode of Fig. 28-18 is connected in series to a 100-Ω resistor and a 2.0-V battery. What current flows in the circuit? (Hint: draw a line on Fig. 28-18 representing the current in the resistor as a function of the voltage across the diode; the intersection of this line with the characteristic curve will give the answer.)

*10. (III) A zener diode voltage regulator is shown in Fig. 28-26. Suppose that $R = 2.0$ kΩ and that the diode breakdown voltage is 130 V; the diode is rated at a maximum current of 100 mA. (a) If $R_{load} = 15$ kΩ, over what range of supply voltages will the circuit maintain the voltage at 130 V? (b) If the supply voltage is 200 V, over what range of load resistance will the voltage be regulated?

*11. (III) A silicon diode passes significant current only if the forward bias voltage exceeds about 0.6 V. Estimate the average current in the output resistor R of (a) a half-wave rectifier (Fig. 28-19), and (b) a full-wave rectifier (Fig. 28-20) without a capacitor. Assume that $R = 200$ Ω in each case and that the ac voltage is 10.0 V rms in each case.

*SECTION 28-10

*12. (I) What collector current results if the collector voltage on the transistor of Fig. 28-23 is 5.0 V and the base current is 5.0 μA?

*13. (I) What base current will produce a collector current of 3.5 mA for a collector voltage of 6.0 V for the transistor of Fig. 28-23?

*14. (II) Suppose that the current gain of the transistor in Fig. 28-22 is $\beta = 100$. If $R_C = 3.0$ kΩ, calculate the output voltage for a time-varying input current of 2.0 μA.

*15. (II) If the current gain of the transistor amplifier in Fig. 28-22 is $\beta = 60$, what value must R_C have if a 1.0 μA base current is to produce an output voltage of 0.40 V?

*16. (II) A transistor, whose current gain $\beta = 80$, is connected as in Fig. 28-22 with $R_B = 2.0$ kΩ and $R_C = 6.5$ kΩ. Calculate (a) the voltage gain, and (b) the power amplification.

*17. (II) An amplifier has a voltage gain of 90 and a 10-kΩ load (output) resistance. What is the peak output current through the load resistor if the input voltage is an ac signal with a peak of 0.080 V?

*18. (III) Suppose that the transistor represented by the curves in Fig. 28-23 is connected as shown in Fig. 28-22 with $R_C = 1.0$ kΩ, $R_B = 4.5$ kΩ, and $\mathscr{E}_C = 5.0$ V. (a) Show that $V_{CE} = 5.0$ V $- I_C (10^3 \ \Omega)$. (b) Draw a straight line representing this equation on the graph of Fig. 28-23. (Hint: note that $V_{CE} = 5.0$ V when $I_C = 0$ and that $I_C = 5$ mA when $V_{CE} = 0$.) This is called the *load line*; its intersection with the curve corresponding to the bias base current I_B is called the *operating point* (see Fig. 28-27). (c) I_C and V_{CE} must satisfy this equation and therefore must lie on the line just drawn as well as on the proper characteristic curves. Use this fact to calculate the variation in I_C when I_B varies from 4.0 μA to 8.0 μA. That is, assume that the base current, when no signal is present, is 6.0 μA; and an alternating signal current of ±2.0 μA is applied. (d) Calculate the current gain and (e) the voltage gain.

29 NUCLEAR PHYSICS AND RADIOACTIVITY

In the early part of the twentieth century Rutherford's experiments led to the idea that at the center of an atom there is a tiny but massive nucleus. At the same time that the quantum theory was being developed and scientists were attempting to understand the structure of the atom and its electrons, investigations into the nucleus itself had also begun. In this chapter and the next, we take a brief look at *nuclear physics.*

29-1 ∎ Structure of the nucleus

An important question to physicists in the early part of this century was whether the nucleus had a structure, and what this structure might be. It turns out that the nucleus is a complicated entity and is not fully understood even today. However, by the early 1930s a model of the nucleus had been developed that is still useful. According to this model, a nucleus is considered as an aggregate of two types of particles: protons and neutrons. (Of course, we must remember that these "particles" also have wave properties, but for ease of visualization and language we often refer to them simply as "particles.") A **proton** is the nucleus of the simplest atom, hydrogen. It has a positive charge ($= +e = +1.6 \times 10^{-19}$ C) and a mass

$$m_p = 1.6726 \times 10^{-27} \text{ kg.}$$

The **neutron,** whose existence was ascertained only in 1932 by the Englishman James Chadwick (1891–1974), is electrically neutral ($q = 0$), as its name implies; its mass, which is almost identical to that of the proton, is

$$m_n = 1.6750 \times 10^{-27} \text{ kg.}$$

These two constituents of a nucleus, neutrons and protons, are referred to collectively as **nucleons.**

Although the hydrogen nucleus consists of a single proton alone, the nuclei of other elements consist of both neutrons and protons. The different types of nuclei are often referred to as **nuclides.** The number of protons in a nucleus (or nuclide) is called the **atomic number** and is designated by the symbol Z. The total number of nucleons, neutrons plus protons, is designated by the symbol A and is called the **atomic mass number.** This name is used since the mass of a nucleus is very closely A times the mass of one nucleon. A nuclide with 7 protons and 8 neutrons thus has $Z = 7$ and $A = 15$. The *neutron number N* is $N = A - Z$.

To specify a given nuclide, we need give only A and Z. A special symbol is commonly used which takes the form

$$_Z^A X$$

where X is the chemical symbol for the element (see the periodic table inside of rear cover), A is the atomic mass number, and Z is the atomic number. For example, $_7^{15}N$ means a nitrogen nucleus containing 7 protons and 8 neutrons for a total of 15 nucleons. In a neutral atom, the number of electrons orbiting the nucleus is equal to the atomic number Z (since the charge on an electron has the same magnitude but opposite sign to that of a proton). The main properties of an atom are determined by the number of electrons. Hence Z determines what kind of atom it is: carbon, oxygen, gold, or whatever. It is redundant to specify both the symbol of a nucleus and its atomic number Z as described above. If the nucleus is nitrogen, for example, we know immediately that $Z = 7$. The subscript Z is thus sometimes dropped and $_7^{15}N$ is then written simply ^{15}N; in words we say "nitrogen fifteen."

For a particular type of atom (say carbon), nuclei are found to contain different numbers of neutrons, although they all have the same number of protons. For example carbon nuclei always have 6 protons but they may have 5, 6, 7, 8, 9, or 10 neutrons. Nuclei that contain the same number of protons but different numbers of neutrons are called **isotopes.** Thus $_6^{11}C$, $_6^{12}C$, $_6^{13}C$, $_6^{14}C$, $_6^{15}C$, and $_6^{16}C$ are all isotopes of carbon. Of course, the isotopes of a given element are not all equally common. For example, 98.9 percent of naturally occurring carbon (on earth) is the isotope $_6^{12}C$ and about 1.1 percent is $_6^{13}C$. These percentages are referred to as the **natural abundances.**[†] Many isotopes that do not occur naturally can be produced in the laboratory by means of nuclear reactions (more on this later). Indeed, all elements beyond uranium ($Z > 92$) do not occur naturally and are only produced artificially.

The approximate size of nuclei was determined originally by Rutherford by scattering of charged particles. Of course, we cannot speak about a definite size for nuclei because of the wave–particle duality: their spatial extent must remain somewhat fuzzy. Nonetheless a rough "size" can be measured, and the most accurate recent measurements have been done by scattering of high-speed electrons off nuclei. It is found that nuclei have a roughly spherical shape with a radius that increases with A according to the approximate formula

$$r \approx (1.2 \times 10^{-15} \text{ m})(A^{1/3}). \qquad (29\text{-}1)$$

Since the volume of a sphere is $V = \frac{4}{3}\pi r^3$, we see that the volume is proportional to the number of nucleons, $V \propto A$. This is what we would expect if nucleons were like impenetrable billiard balls: if you double the number of balls, you double the total volume. Hence, all nuclei have nearly the same density.

[†] The mass values for the elements as given in the periodic table (inside rear cover) are averaged over the natural abundances.

TABLE 29-1

Rest masses in kilograms, unified atomic mass units, and MeV/c²

Object	Mass		
	kg	u	MeV/c^2
Electron	9.1095×10^{-31}	0.00054858	0.51100
Proton	1.67265×10^{-27}	1.007276	938.28
$_1^1$H atom	1.67356×10^{-27}	1.007825	938.79
Neutron	1.67500×10^{-27}	1.008665	939.57

The masses of nuclei can be determined, in one method, by measuring the radius of curvature of fast-moving nuclei in a magnetic field using a mass spectrometer, as discussed in Section 19-11. Indeed, as mentioned there, the existence of different isotopes of the same element was discovered using this device. Nuclear masses are specified in *unified atomic mass units* (u). On this scale, a neutral $_6^{12}$C atom is given the precise value 12.000000 u. A neutron then has a measured mass of 1.008665 u, a proton 1.007276 u, and a neutral hydrogen atom, $_1^1$H (proton plus electron) 1.007825 u. The masses of many nuclides are given in Appendix D. (It should be noted that the masses in this table, as is customary, are for the *neutral atom*, and not for a bare nucleus.)

Masses are often specified using the electron-volt energy unit. This can be done because mass and energy are related, and the precise relationship is given by Einstein's equation $E = mc^2$ (Chapter 25). Since the mass of a neutral $_1^1$H atom is 1.6736×10^{-27} kg or 1.007825 u, then 1.0000 u = $(1.0000/1.007825) \cdot (1.6736 \times 10^{-27}$ kg$) = 1.6606 \times 10^{-27}$ kg; this is equivalent to an energy $E = mc^2 = (1.6606 \times 10^{-27}$ kg$)(2.998 \times 10^8$ m/s$)^2/(1.602 \times 10^{-19}$ J/eV$) = 931.5$ MeV. Thus

$$1 \text{ u} = 1.6606 \times 10^{-27} \text{ kg} = 931.5 \text{ MeV}/c^2.$$

The rest masses of some of the basic particles are given in Table 29-1.

29-2 ∎ Binding energy and nuclear forces

The total mass of a stable nucleus is always less than the sum of the masses of its constituent protons and neutrons as the following example shows.

EXAMPLE 29-1   Compare the mass of a $_2^4$He nucleus to that of its constituent nucleons.

SOLUTION   The mass of a neutral $_2^4$He atom, from Appendix D, is 4.002603 u. The mass of two neutrons and two protons (including the two electrons) is

$$\begin{aligned} 2m_n &= 2.017330 \text{ u} \\ 2m(_1^1\text{H}) &= \underline{2.015650 \text{ u}} \\ &\quad\; 4.032980 \text{ u}. \end{aligned}$$

(We almost always deal with masses of neutral atoms—that is, nuclei with Z electrons—since this is how masses are measured. We must therefore be sure to balance out the electrons when we compare masses, which is why we used the mass of ${}^1_1\text{H}$ in this example rather than that of the proton alone.)

Thus the mass of ${}^4_2\text{He}$ is measured to be $4.032980 \text{ u} - 4.002603 \text{ u} = 0.030377 \text{ u}$ less than the masses of its constituents. How can this be? Where has this mass gone?

It has, in fact, gone into energy of another kind (such as radiation, kinetic energy, and so on). The mass (or energy) difference in the case of ${}^4_2\text{He}$, is $(0.030377 \text{ u})(931.5 \text{ MeV/u}) = 28.30 \text{ MeV}$. This difference is referred to as the **total binding energy** of the nucleus. The total binding energy represents the amount of energy that must be put into a nucleus in order to break it apart into its constituent protons and neutrons. If the mass of, say, a ${}^4_2\text{He}$ nucleus were exactly equal to the mass of two neutrons plus two protons, the nucleus could fall apart without any input of energy. To be stable the mass *must* be less than that of its constituents. Note that the binding energy is not something a nucleus has—it is energy it "lacks" relative to the total mass of its separate constituents.

This situation can be compared to the binding energy of electrons in an atom. We saw in Chapter 26 that the binding energy of the one electron in the hydrogen atom, for example, is 13.6 eV. The mass of a ${}^1_1\text{H}$ atom is less than that of a single proton plus a single electron by 13.6 eV. Compared to the total mass (938 MeV) this is incredibly small (1 part in 10^8), and for practical purposes the mass difference can be ignored. The binding energies of nuclei are on the order of 10^6 times greater than the binding energies of electrons in atoms and are therefore far more important.

The **average binding energy per nucleon** is defined as the total binding energy of a nucleus divided by A, the total number of nucleons. For ${}^4_2\text{He}$ it is $28.3 \text{ MeV}/4 = 7.1 \text{ MeV}$. Figure 29-1 shows the average binding energy per nucleon as a function of A for stable nuclei. The curve rises as A increases and reaches a plateau at about 8 MeV per nucleon above $A \approx 15$. Beyond about $A = 60$, the curve decreases slowly, indicating that larger nuclei are held together a little less tightly than those in the middle of the periodic table. (We will see later that these facts allow the release of nuclear energy in the processes of fission and fusion.)

We can analyze nuclei not only from the point of view of energy but also from the point of view of the forces that hold them together. We would not expect a collection of protons and neutrons to come together spontaneously since protons are all positively charged and thus exert repulsive forces on each other. Indeed, the question arises as to how a nucleus stays together at all in view of the fact that the electric force between protons would tend to break it apart. Since stable nuclei *do* stay together, it is clear that another force must be acting. Because this new force is stronger than the electric force (which in turn is much stronger than gravity at the nuclear level) it is called the **strong nuclear force**. The strong nuclear force is an attractive force that acts between all nucleons—protons and neutrons alike. Thus protons attract each other via the nuclear force at the same time they

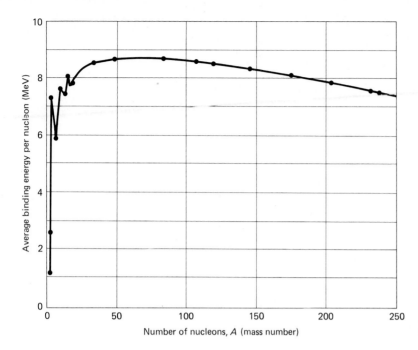

FIGURE 29-1

Average binding energy per nucleon as a function of mass number A for stable nuclei.

repel each other via the electric force. Neutrons, since they are electrically neutral, only attract other neutrons or protons via the nuclear force.

The nuclear force turns out to be far more complicated than the gravitational and electromagnetic forces. A precise mathematical description is not yet possible. Nonetheless, a great deal of work has been done to try to understand the nuclear force. One important aspect of the strong nuclear force is that it is a *short-range* force: it acts only over a very short distance. It is very strong between two nucleons if they are less than about 10^{-15} m apart; but it is essentially zero if they are separated by a distance greater than this. Compare this to electric and gravitational forces, which can act over great distances and are therefore called *long-range* forces. The strong nuclear force has some strange quirks. For example, if a nuclide contains too many or too few neutrons relative to the number of protons, the binding of the nucleons is reduced; nuclides that are too unbalanced in this regard are unstable. As shown in Fig. 29-2, stable nuclei tend to have the same number of protons as neutrons ($N = Z$) up to about $A \approx 30$ or 40. Beyond this, stable nuclei contain more neutrons than protons. This reflects the fact that as Z increases, the electrical repulsion increases, so a greater number of neutrons—which exert only the attractive nuclear force—are required to maintain stability. For very large Z, no number of neutrons can overcome the greatly increased electric repulsion; indeed, there are no completely stable nuclides above $Z = 82$.

What we mean by a stable nucleus is one that stays together indefinitely. What then is an unstable nucleus? It is one that comes apart; and this results in radioactive decay. Before we discuss the important subject of radioactivity (the next section), we note that there is a second type of nuclear force that is much weaker than the strong nuclear force. It is called the **weak nuclear force,** and we are aware of its existence only because it shows

FIGURE 29-2

Number of neutrons versus number of protons for stable nuclides, which are represented by dots.

itself in certain types of radioactive decay. These two nuclear forces, the strong and the weak, together with the gravitational and electromagnetic forces, comprise the four known types of force in nature (more on this in Chapter 31).

29-3 ∎ Radioactivity

Nuclear physics had its beginnings in 1896. In that year, Henri Becquerel (1852–1908) made an important discovery: in his studies of phosphorescence, he found that a certain mineral (which happened to contain uranium) would darken a photographic plate even when the plate was wrapped to exclude light. It was clear that the mineral emitted some new kind of radiation which, unlike X rays, occurred without any external stimulus. This new phenomenon eventually came to be called **radioactivity.**

Soon after Becquerel's discovery, Marie Curie (1867–1934) and her husband, Pierre Curie (1859–1906), isolated two previously unknown elements which were very highly radioactive. These were named polonium and radium. Other radioactive elements were soon discovered as well. The radioactivity was found in every case to be unaffected by the strongest physical and chemical treatments including strong heating or cooling and the action of strong chemical reagents. It soon became clear that the source of radioactivity must be deep within the atom, that it must emanate from the nucleus. And it became apparent that radioactivity is the result of the *disintegration* or *decay* of an unstable nucleus. Certain isotopes are not stable under the action of the nuclear force, and they decay with the emission of some type of radiation or "rays."

Many unstable isotopes occur in nature, and such radioactivity is called "natural radioactivity." Other unstable isotopes can be produced in the laboratory by nuclear reactions (Section 30-1); these are said to be produced "artificially" and they are said to have "artificial radioactivity."

Rutherford and others began studying the nature of the rays emitted in radioactivity about 1898. They found that the rays could be classified into three distinct types according to their penetrating power. One type of radiation could barely penetrate a piece of paper. The second type could pass through as much as 3 mm of aluminum. The third was extremely penetrating: it could pass through several centimeters of lead and still be detected on the other side. They named these three types of radiation alpha (α), beta (β), and gamma (γ), respectively, after the first three letters of the Greek alphabet.

Each type of ray was found to have a different charge and hence is bent differently in a magnetic field, Fig. 29-3; α rays are positively charged, β rays are negatively charged, and γ rays are neutral. It was soon found that all three types of radiation consisted of familiar kinds of particles. Gamma rays are very high energy photons whose energy is even higher than that of X rays. Beta rays are electrons, identical to those that orbit the nucleus (but they are created within the nucleus itself). Alpha rays (or α particles) are simply the nuclei of helium atoms, ^4_2He; that is, an α ray consists of two protons and two neutrons bound together.

FIGURE 29-3

Alpha and beta rays are bent in opposite directions by a magnetic field, whereas gamma rays are not bent at all.

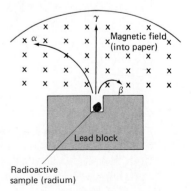

We now discuss each of these three types of radioactivity, or decay, in more detail.

29-4 ▌ Alpha decay

When a nucleus emits an α particle ($^4_2$He), it is clear that the remaining nucleus will be different from the original: for it has lost two protons and two neutrons. Radium 226 ($^{226}_{88}$Ra), for example, is an α emitter. It decays to a nucleus with $Z = 88 - 2 = 86$ and $A = 226 - 4 = 222$. The nucleus with $Z = 86$ is radon (Rn). Thus the radium decays to radon with the emission of an α particle. This is written

$$^{226}_{88}\text{Ra} \rightarrow {}^{222}_{86}\text{Rn} + {}^4_2\text{He}.$$

It is clear that when α decay occurs, a new element is formed. The *daughter* nucleus ($^{222}_{86}$Rn in this case) is different from the *parent* nucleus ($^{226}_{88}$Ra in this case). This changing of one element into another is called **transmutation.**

Alpha decay occurs because the strong nuclear force is unable to hold very large nuclei together. Because the nuclear force is a short-range force, it acts only between neighboring nucleons. But the electric force can act clear across the nucleus. For very large nuclei, the large Z means the repulsive electric force becomes very large (Coulomb's law) and acts between all protons; the strong nuclear force, since it acts only between neighboring nucleons, is overpowered and is unable to hold the nucleus together.

We can express the instability in terms of energy (or mass): the mass of the parent nucleus is greater than the mass of the daughter nucleus plus the mass of the α particle. The mass difference appears as kinetic energy, which is carried away mainly by the α particle. If the parent has *less* mass than the daughter plus the α particle, spontaneous decay could not occur, for the conservation of energy law would be violated.

▢ **EXAMPLE 29-2** Calculate the kinetic energy of the α particle emitted when a $^{232}_{92}$U nucleus (mass = 232.03714 u) decays to $^{228}_{90}$Th (228.02873 u).

SOLUTION Since the mass of a $^4_2$He nucleus is 4.002603 u the total mass in the final state is 228.02873 u + 4.002603 u = 232.03133 u. The mass lost when $^{232}_{92}$U nucleus decays is 232.03714 u − 232.03133 u = 0.00581 u. This mass appears as kinetic energy; since 1 u = 931.5 MeV, the KE released in (0.00581 u) · (931.5 MeV/u) ≈ 5.4 MeV. (Using conservation of momentum, it can be shown that the α particle in this decay has a KE of about 5.3 MeV. Thus the daughter nucleus—which recoils in the opposite direction from the emitted α particle—has about 0.1 MeV of kinetic energy.)

Why, you may wonder, do nuclei emit this combination of four nucleons called an α particle? Why not just four nucleons, or even one? The answer lies in the fact that the α particle is very strongly bound and its mass is so much less than four separate nucleons. As we saw in Example 29-1, two protons and two neutrons separately have a total mass of about 4.03298 u. A $^{228}_{90}$Th nucleus plus four nucleons has a total mass of 232.06171 u which is greater than the mass of the parent nucleus. Such a decay could not occur because it would violate conservation of energy. Similarly, it is almost always true that the emission of a single nucleon is energetically not possible.

29-5 ∎ Beta decay

Transmutation of elements also occurs when a nucleus decays by β decay—that is, with the emission of an electron or β particle. The nucleus $^{14}_{6}C$, for example, decays as follows:

$$^{14}_{6}C \rightarrow \, ^{14}_{7}N + \, ^{0}_{-1}e.$$

(The symbol $^{0}_{-1}e$ stands for an electron whose charge corresponds to $Z = -1$ and, since it is not a nucleon and has very small mass, has $A = 0$.) No nucleons are lost when an electron is emitted, and the total number of nucleons, A, is the same in the daughter as in the parent. But because an electron has been emitted, the charge on the daughter is different from the parent. The parent had $Z = +6$. In the decay, the nucleus loses a charge of -1, so the nucleus remaining behind (from charge conservation) must have an extra $+$ charge for a total of 7. So the daughter has $Z = 7$, which is a nitrogen nucleus.

It must be carefully noted that the electron emitted in β decay is *not* an orbital electron. Instead, the electron is created *within the nucleus itself*. It is as if one of the neutrons changes to a proton and in the process (to conserve charge) throws off an electron. Indeed, free neutrons actually do decay in this fashion: $n \rightarrow p + e^{-}$. Because of their origin in the nucleus, the electrons emitted in β decay are often referred to as "β particles," rather than as electrons, to remind us of their origin; they are, nonetheless, indistinguishable from orbital electrons.

☐ EXAMPLE 29-3 How much energy is released when $^{14}_{6}C$ decays to $^{14}_{7}N$ by β emission? Use Appendix D.

SOLUTION Because the masses given in Appendix D are those of the neutral atom, we have to keep track of the electrons involved. Assume the parent nucleus has six orbiting electrons so it is neutral and its mass is 14.003242 u. The daughter, in this decay $^{14}_{7}N$, is not neutral however since it has the same six electrons circling it but the nucleus has a charge of $+7e$. However, the mass of this daughter with its six electrons, plus the mass of the emitted electron (which makes a total of seven electrons), is just the mass of a neutral nitrogen atom. That is, the total mass in the final state is

(mass of $^{14}_{7}N$ nucleus + 6 electrons) + (mass of 1 electron)

and this is equal to

mass of neutral $^{14}_{7}N$ (includes 7 electrons)

which, from Appendix D has a mass of 14.003074 u. Hence the mass after decay is 14.003074 u, whereas before decay it is 14.003242 u; so the mass difference is 0.000168 u which corresponds to 0.156 MeV or 156 keV.

According to this example, we would expect the emitted electron to have a kinetic energy of 156 keV. (The daughter nucleus, because its mass is very much larger than that of the electron, recoils with very low velocity and hence gets very little of the kinetic energy.) Indeed, very careful measurements indicate that a few emitted β particles do have kinetic energy close to this calculated value; but the vast majority of emitted electrons have somewhat less energy. In fact, the energy of the emitted electron can be anywhere from zero up to the maximum value as calculated above. This was found to be true for any β decay. It was as if the law of conservation of energy was being violated! Careful experiments also indicated that linear momentum and angular momentum also did not seem to be conserved. Physicists were troubled at the prospect of having to give up these laws, which had worked so well in all previous situations. In 1930, Wolfgang Pauli proposed an alternate solution: perhaps a new particle that was very difficult to detect was emitted during β decay in addition to the electron. This hypothesized particle could be carrying off the energy, momentum, and angular momentum required to maintain the conservation laws. This new particle was named the **neutrino**—meaning "little neutral one"—by the great Italian physicist Enrico Fermi (1901–1954) who in 1933 worked out a detailed theory of β decay. (It was Fermi, in this theory, who postulated the existence of the fourth force in nature, which we call the weak nuclear force.) The neutrino has zero charge and seems to have zero rest mass, although there are recent suggestions that it may have a very tiny rest mass. If its rest mass is zero, it is much like a photon in that it is neutral and travels at the speed of light; but it is far more difficult to detect. In 1956, complex experiments produced further evidence for the existence of the neutrino; but by then most physicists had already accepted its existence.

The symbol for the neutrino is the Greek letter nu (v). The correct way of writing the decay of $^{14}_{6}C$ is then

$$^{14}_{6}C \rightarrow \, ^{14}_{7}N + \, _{-1}^{0}e + \bar{v}.$$

The bar ($^-$) over the neutrino symbol is to indicate that it is an "antineutrino." (More on antiparticles later; why this is called an antineutrino rather than simply a neutrino need not concern us now; it is discussed in Chapter 31).

Many isotopes decay by electron emission. They are always isotopes that have too many neutrons compared to the number of protons. That is, they are isotopes that lie above the stable isotopes plotted in Fig. 29-2. But what about unstable isotopes that have too few neutrons compared to their number of protons—those that fall below the stable isotopes of Fig. 29-2? These, it turns out, decay by emitting a **positron** instead of an electron. A positron (sometimes called an e^+ or β^+ particle) has the same mass as the electron, but it has a positive charge of $+1e$. Because it is so like an electron, except for its charge, it is called the *antiparticle*[†] to the electron. An example of a β^+ decay is that of $^{19}_{10}Ne$:

$$^{19}_{10}Ne \rightarrow \, ^{19}_{9}F + \, _{1}^{0}e + v$$

where $_{1}^{0}e$ stands for a positron (e^+). (Note that the v emitted here is a neu-

[†] Discussed in Chapter 31. Briefly, an antiparticle has the same mass as its corresponding particle, but opposite charge.

trino, whereas that emitted in β^- decay is called an antineutrino. Thus an antielectron (= positron) is emitted with a neutrino, whereas an antineutrino is emitted with an electron; this is discussed in Chapter 31).

Besides β^- and β^+ emission, there is a third related process. This is *electron capture* and occurs when a nucleus absorbs one of its orbiting electrons. An example is ${}^{7}_{4}\text{Be}$, which as a result becomes ${}^{7}_{3}\text{Li}$: the process is written

$$ {}^{7}_{4}\text{Be} + {}^{0}_{-1}e \rightarrow {}^{7}_{3}\text{Li} + v. $$

Usually it is an electron in the innermost (K) shell that is captured, and then it is called "K-capture." The electron disappears in the process and a proton in the nucleus becomes a neutron; a neutrino is emitted as a result. This process is inferred experimentally by detection of emitted X rays (due to electrons jumping down to fill the empty state) of just the proper wavelength.

In β decay it is the weak nuclear force that plays the crucial role. The neutrino is unique in that it interacts with matter only via the weak force, which is why it is so hard to detect.

29-6 ∎ Gamma decay

Gamma rays are photons having very high energy. And the decay of a nucleus by emission of a γ ray is much like emission of photons by excited atoms. Like an atom, a nucleus itself can be in an excited state. When it jumps down to a lower energy state, or to the ground state, it emits a photon. The possible energy levels of a nucleus are much farther apart in energy than those of an atom: on the order of keV or MeV, as compared to a few eV for electrons in an atom. Hence the emitted photons have energies that can range from a few keV to several MeV. For a given decay, the γ ray always has the same energy. Since a γ ray carries no charge, there is no change in the element as a result of a γ decay.

How does a nucleus get into an excited state? It may occur because of a violent collision with another particle; or more commonly the nucleus remaining after a previous radioactive decay may be in an excited state. A typical example is shown in the energy-level diagram of Fig. 29-4. ${}^{12}_{5}\text{B}$ can decay by β decay directly to the ground state of ${}^{12}_{6}\text{C}$; or it can go by β decay to an excited state of ${}^{12}_{6}\text{C}$ which then decays by emission of a 4.4 MeV γ ray to the ground state.[†]

In some cases a nucleus may remain in an excited state for some time before it emits a γ ray as in Fig. 29-4. The nucleus is then said to be in a *metastable state* and is called an *isomer*.

(What, you may wonder, is the difference between a γ ray and an X ray: they both are electromagnetic radiation (photons) and their range of energies overlap to some extent. The difference is not intrinsic. We use the term X

FIGURE 29-4

Energy-level diagram, showing how ${}^{12}_{5}\text{B}$ can decay to the ground state of ${}^{12}_{6}\text{C}$ by β decay (total energy released = 13.4 MeV) or can β-decay to an excited state of ${}^{12}_{6}\text{C}$ (indicated by *), which subsequently decays to its ground state by emitting a 4.4-MeV γ ray.

[†] An excited nucleus can sometimes return to the ground state by another process known as *internal conversion*. In this process, the excited nucleus interacts with one of the oribital electrons and ejects this electron from the atom with the same KE that an emitted γ ray would have had.

TABLE 29-2

The three types of radioactive decay[†]

α decay
$$N(A, Z) \rightarrow N(A - 4, Z - 2) + {}_2^4\text{He}$$

β decay
$$\beta^- : N(A, Z) \rightarrow N(A, Z + 1) + {}_{-1}^0 e + \bar{\nu}$$
$$\beta^+ : N(A, Z) \rightarrow N(A, Z - 1) + {}_1^0 e + \nu$$
Electron capture: $N(A, Z) + {}_{-1}^0 e \rightarrow N(A, Z - 1) + \nu$

γ decay
$$N^*(A, Z) \rightarrow N(A, Z) + \gamma$$

[†] $N(A, Z)$ means a nucleus with atomic number Z and mass number A; * means excited state of a nucleus.

ray if the photon is produced by an electron-atom interaction, and a γ ray if the photon is produced in a nuclear process.)

29-7 ∎ Conservation of nucleon number and other conservation laws

In all three types of radioactive decay the classical conservation laws hold. Energy, linear momentum, angular momentum, and electric charge are all conserved; these quantities are the same before the decay as after. But a new conservation law is also revealed. This is the **law of conservation of nucleon number.** According to this law, the total number of nucleons (A) remains constant in any process, although one type can change into the other type (protons into neutrons or vice versa). This law holds in all three types of decay. See Table 29-2 which gives a summary of α, β, γ decay.

29-8 ∎ Half-life and rate of decay

A macroscopic sample of any radioactive isotope consists of a vast number of radioactive nuclei. These nuclei do not all decay at one time. Rather, they decay one by one over a period of time. This is a random process: we can't predict exactly when a given nucleus will decay. But we can determine, on a probabilistic basis, approximately how many nuclei in a sample will decay over a given time period.

The number of decays ΔN that occur in a very short time interval Δt is found to be proportional to Δt and to the total number N of radioactive nuclei present:

$$\Delta N = -\lambda N \, \Delta t. \tag{29-2}$$

In this equation, λ is a constant of proportionality called the **decay constant,** which is different for different isotopes. The greater λ is, the greater the rate of decay and the more radioactive that isotope is said to be. The quantity ΔN in Eq. 29-2 represents the number of decays that occur in the short time interval Δt; we call this ΔN (meaning "change in N") because each decay that occurs corresponds to a decrease by one in the number N of nuclei present. That is, radioactive decay is a "one-shot" process, Fig. 29-5; once a particular

(a)

(b)

(c)

Legend

⬤ ${}_6^{14}\text{C}$ atom (parent)

◯ ${}_7^{14}\text{N}$ atom (daughter)

FIGURE 29-5

Radioactive nuclei decay one by one. Hence the number of parent nuclei in a sample is continually decreasing. When a ${}_6^{14}\text{C}$ nucleus emits the electron, it becomes a ${}_7^{14}\text{N}$ nucleus.

parent nucleus decays into its daughter, it cannot do it again. The minus sign in Eq. 29-2 is needed to indicate that N is decreasing.

The number of parent nuclei in a sample decreases in time as shown in Fig. 29-6a. Since the number of decays per second, $\Delta N/\Delta t$ (called the *activity* of the sample), is proportional to N (see Eq. 29-2), it, too, decreases in time with the same time constant (Fig. 29-6b). Both of these curves are examples of "exponential" curves.[†]

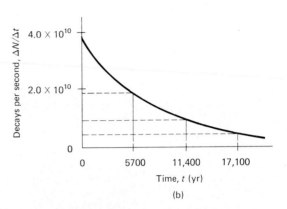

(a)   (b)

The rate of decay of any isotope is often specified by giving its "half-life" rather than the decay constant. The **half-life** of an isotope is defined as the time it takes for half the original amount of isotope in a given sample to decay. For example, the half-life of $^{14}_{6}C$ is about 5700 years (more precisely 5730 years). If at some time a piece of petrified wood contains, say, 1.00×10^{22} $^{14}_{6}C$ nuclei, then some 5700 years later it will contain only 0.50×10^{22} of these nuclei. After another 5700 years it will contain 0.25×10^{22} nuclei, and so on. This is shown in Fig. 29-6a. Since the rate of decay $\Delta N/\Delta t$ is proportional to N, it, too, decreases by a factor of 2 every half-life (Fig. 29-6b).

The half-lives of known radioactive isotopes vary from about 10^{-22} s to 10^{28} s (about 10^{21} yr). The half-lives of many isotopes are given in Appendix D. It should be clear that the half-life (which we designate $T_{\frac{1}{2}}$) bears an inverse relationship to the decay constant. The longer the half-life of an isotope, the more slowly it decays, and hence λ is smaller; and very active isotopes (large λ) have very short half-lives. The precise relationship between half-life and decay constant is (calculus is needed to prove it)

$$T_{\frac{1}{2}} = \frac{0.693}{\lambda}. \tag{29-3}$$

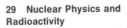

FIGURE 29-6

(a) The number N of parent nuclei in a given sample of $^{14}_{6}C$ decreases exponentially. (b) The number of decays per second also decreases exponentially. The half-life of $^{14}_{6}C$ is about 5700 yr, which means that the number of parent nuclei, N, and the rate of decay, $\Delta N/\Delta t$, decrease by half every 5700 yr.

☐ EXAMPLE 29-4 The isotope $^{14}_{6}C$ has a half-life of 5730 yr. If at some time a sample contains 1.00×10^{22} carbon-14 nuclei, what is the activity of the sample?

[†] The N versus t curve has the form $N = N_0 e^{-\lambda t}$, where N_0 is the number of parent nuclei present at time $t = 0$. The curve for rate of decay is written $\Delta N/\Delta t = (\Delta N/\Delta t)_0 e^{-\lambda t}$, where $(\Delta N/\Delta t)_0$ is the rate of decay at $t = 0$. The symbol e is a number called the "natural exponential" and has the value $e = 1 + \frac{1}{1} + \frac{1}{1 \cdot 2} + \frac{1}{1 \cdot 2 \cdot 3} + \cdots = 2.718 \ldots$. Why we use e instead of 2 or 10, say, to represent this exponential arises from calculus, and we will not deal with it here.

and obtain

$$\lambda = \frac{0.693}{T_{\frac{1}{2}}} = \frac{0.693}{(5730 \text{ yr})(3.15 \times 10^7 \text{ s/yr})} = 3.84 \times 10^{-12} \text{ s}^{-1},$$

since there are $(60)(60)(24)(365) = 3.15 \times 10^7$ s in a year. From Eq. 29-2, the activity or rate of decay (we ignore the minus sign) is

$$\frac{\Delta N}{\Delta t} = \lambda N = (3.84 \times 10^{-12} \text{ s}^{-1})(1.00 \times 10^{22})$$

$$= 3.84 \times 10^{10} \text{ decays/s}.$$

Note that the graph of Fig. 29-6b starts at this value, corresponding to the original value of $N = 1.0 \times 10^{22}$ nuclei in Fig. 29-6a.

EXAMPLE 29-5   A laboratory has 1.49 μg of pure $^{13}_{7}$N, which has a half-life of 10.0 min (600 s). (a) How many nuclei are present initially? (b) What is the activity initially? (c) What is the activity after 1.0 h? (d) After approximately how long will the activity drop to less than one per second?

SOLUTION   (a) Since the atomic mass is 13.0, then 13.0 g will contain 6.02×10^{23} nuclei (Avogadro's number). Since we have only 1.49×10^{-6} g, the number of nuclei, N, that we have is given by the ratio

$$\frac{N}{6.02 \times 10^{23}} = \frac{1.49 \times 10^{-6} \text{ g}}{13.0 \text{ g}},$$

so $N = 6.90 \times 10^{16}$ nuclei. (b) From Eq. 29-3, $\lambda = (0.693)/(600 \text{ s}) = 1.16 \times 10^{-3} \text{ s}^{-1}$. Then

$$\frac{\Delta N}{\Delta t} = \lambda N = 8.00 \times 10^{13} \text{ decays/s}.$$

(c) Since the half-life is 10.0 min, the decay rate decreases by half every 10.0 min. We can make the following table of activity after given periods of time:

Time (min)	Activity (decays/s)
0	8.00×10^{13}
10	4.00×10^{13}
20	2.00×10^{13}
30	1.00×10^{13}
40	0.500×10^{13}
50	0.250×10^{13}
60	0.125×10^{13}

Thus after 1.0 h, the activity is 1.25×10^{12} decays/s. A simpler way of seeing this is to note that 60 min is 6 half-lives, so the activity will decrease to $(\frac{1}{2})(\frac{1}{2})(\frac{1}{2})(\frac{1}{2})(\frac{1}{2})(\frac{1}{2}) = (\frac{1}{2})^6 = \frac{1}{64}$ of its original value, or $(8.00 \times 10^{13})/(64) =$

1.25×10^{12} per second. (*d*) After each hour, the activity drops by a factor of 64, so our table is extended to read as follows:

Time (h)	Activity
0	8.00×10^{13}
1	1.25×10^{12}
2	1.95×10^{10}
3	3.05×10^8
4	4.77×10^6
5	7.45×10^4
6	1.16×10^3
7	$1.8 \times 10^1 = 18$
8	<1

So the activity drops to less than one per second within 8 h.

FIGURE 29-7

Decay series begining with $^{238}_{92}$U. Nuclei in the series are specified by a dot representing A and Z values. Half-lives are given in seconds (s), minutes (min), hours (h), days (d), or years (yr). Note that a horizontal arrow represents β decay (A does not change), whereas a diagonal line represents α decay (A changes by 4, Z changes by 2).

*29-9 ■ Decay series

It is often the case that one radioactive isotope decays to another isotope that is also radioactive. Sometimes this daughter decays to a further isotope which also is radioactive. Such successive decays are said to be a *decay series*. An important example is illustrated in Fig. 29-7. As can be seen, $^{238}_{92}U$ decays by α emission to $^{234}_{90}Th$, which in turn decays by β decay to $^{234}_{91}Pa$. The series continues as shown, with several possible branches near the bottom. For example, $^{218}_{84}Po$ can decay either by α decay to $^{214}_{82}Pb$ or by β decay to $^{218}_{85}At$. The series ends at the stable lead isotope $^{206}_{82}Pb$. Other radioactive series also exist.

Because of such decay series, certain radioactive elements are found in nature that otherwise would not be. For when the solar system acquired its present form about 5 billion years ago, it is believed nearly all nuclides were formed (by the fusion process, Section 30-3). Many isotopes with short half-lives decayed quickly and no longer exist in nature today. But long-lived isotopes, such as $^{238}_{92}U$ with a half-life of 4.5×10^9 yr, still do exist in nature today; indeed, about half of the original $^{238}_{92}U$ still remains (assuming that the origin of the solar system was about 5×10^9 yr ago). We might expect, however, that radium ($^{226}_{88}Ra$), with a half-life of 1600 yr, would long since have disappeared from the earth. Indeed, the original $^{226}_{88}Ra$ nuclei must by now have all decayed. However, because $^{238}_{92}U$ decays (in several steps) to $^{226}_{88}Ra$, the supply of $^{226}_{88}Ra$ is continually replenished, which is why it still exists on earth today. The same can be said for many other radioactive nuclides.

*29-10 ■ Stability and tunneling

We have seen that radioactive decay occurs only when the mass of the parent nucleus is greater than the sum of the masses of the daughter nucleus and all particles emitted. For example, $^{238}_{92}U$ decays to $^{234}_{90}Th$, because the mass of $^{238}_{92}U$ is greater than the mass of the $^{234}_{90}Th$ plus the mass of the α particle. Since systems tend to go in the direction that reduces their internal or potential energy (a ball rolls downhill, a positive charge moves toward a negative charge), you may wonder why an unstable nucleus doesn't fall apart immediately. In other words, why do $^{238}_{92}U$ ($T_{\frac{1}{2}} = 4.5 \times 10^9$ yr) and other isotopes have such long half-lives?

The answer has to do with quantum theory and the nature of the forces involved. One way to view the situation is with the aid of a potential-energy diagram, as in Fig. 29-8. Here we consider the particular case of the decay $^{238}_{92}U \rightarrow ^{234}_{90}Th + ^4_2He$. The curved line represents the potential energy, including rest mass. Actually, it is simplest to imagine the α particle as a separate entity in the $^{238}_{92}U$ nucleus; then the curve of Fig. 29-8 can represent the potential energy of the α particle when it is within the uranium nucleus (A on the diagram), and when it is free of the uranium nucleus (C). In order to get to point C, the α particle has to get by the barrier shown. In classical

FIGURE 29-8

Potential-energy diagram for the decay $^{238}_{92}U \rightarrow ^{234}_{90}Th + ^4_2He$.

$M(^{238}_{92}U)$ $[M(^{234}_{90}Th) + M(^4_2He)]$

physics, this could only be done by putting in an energy equal to the height of the barrier above point A. (A similar situation was discussed in Section 28-5, where the required input energy was called the "activation energy.") Nuclei decay spontaneously, however, without any input of energy. How, then, does the α particle get from point A to point C? It actually passes through the barrier in a process known as **tunneling.** Classically, this could not happen, because an α particle at point B (within the barrier) would be violating the conservation-of-energy principle. The uncertainty principle, however, tells us that energy conservation can be violated by an amount ΔE for a length of time Δt given by

$$(\Delta E)(\Delta t) \approx \frac{h}{2\pi}.$$

We saw in Section 27-3 that this is a result of the wave–particle duality. Thus quantum mechanics allows conservation of energy to be violated for brief periods that may be long enough for an α particle to "tunnel" through the barrier. The higher and wider the barrier, the less time the α particle has to escape and the less likely it is to do so. It is therefore the height and width of this barrier that controls the rate of decay and half-life of an isotope.

*29-11 ▌ Radioactive dating

Radioactive decay has many interesting applications. One is the technique of *radioactive dating* in which the age of ancient materials can be determined.

The age of any object made from once living matter, such as wood, can be determined using the natural radioactivity of $^{14}_{6}C$. All living plants absorb carbon dioxide (CO_2) from the air, utilizing the carbon and expelling oxygen. The vast majority of these carbon atoms are $^{12}_{6}C$, but a small fraction, about 1.3×10^{-12}, is the radioactive isotope $^{14}_{6}C$. The ratio of $^{14}_{6}C$ to $^{12}_{6}C$ in the atmosphere has remained roughly constant over many thousands of years, in spite of the fact that $^{14}_{6}C$ decays with a half-life of about 5730 yr. This is because neutrons in the cosmic radiation that impinges on the earth from outer space collide with atoms of the atmosphere. In particular, collisions with nitrogen nuclei produce the following nuclear transformation: $n + {}^{14}_{7}N \rightarrow {}^{14}_{6}C + p$. That is, a neutron strikes and is absorbed by a $^{14}_{7}N$ nucleus, and a proton is knocked out in the process. The remaining nucleus is $^{14}_{6}C$. This continual production of $^{14}_{6}C$. in the atmosphere roughly balances the loss of $^{14}_{6}C$ by radioactive decay. As long as a plant or tree is alive, it continually uses the carbon from carbon dioxide in the air to build new tissue and to replace old. Animals eat plants, so they too are continually receiving a fresh supply of carbon for their tissues. Organisms cannot distinguish[†] $^{14}_{6}C$ from $^{12}_{6}C$, and since the ratio of $^{14}_{6}C$ to $^{12}_{6}C$ in the atmosphere remains nearly constant, the ratio of the two isotopes within the living organism remains nearly constant as well. But when an organism dies, carbon dioxide is no longer absorbed and utilized; and because the $^{14}_{6}C$ decays radio-

---

[†] Organisms operate almost exclusively via chemical reactions—which involve only the outer orbital electrons of the atom; extra neutrons in the nucleus have almost no effect.

actively, the ratio of $^{14}_{6}C$ to $^{12}_{6}C$ in a dead organism decreases in time. Since the half-life of $^{14}_{6}C$ is about 5730 yr, the $^{14}_{6}C/^{12}_{6}C$ ratio decreases by half every 5730 yr. If, for example, the $^{14}_{6}C/^{12}_{6}C$ ratio of an ancient wooden tool is half of what it is in living trees, then the object must have been made from a tree that was felled about 5730 yr ago. Actually, corrections must be made for the fact that the $^{14}_{6}C/^{12}_{6}C$ ratio in the atmosphere has not remained precisely constant over time. The determination of what this ratio has been over the centuries has required using techniques such as comparing the expected ratio to actual ratio for objects whose age is known, such as very old trees whose annual rings can be counted.

Carbon dating is useful only for determining the age of objects less than about 40,000 yr old. The amount of $^{14}_{6}C$ remaining in older objects is usually too small to measure accurately. However, radioactive isotopes with longer half-lives can be used in certain circumstances to obtain the age of older objects. For example, the decay of $^{238}_{92}U$, because of its long half-life of 4.5×10^9 years, is useful to determine the age of rocks on a geologic time scale. When molten material solidifies into rock. the uranium present in the material becomes fixed in its position and the daughter nuclei that result from the decay of uranium will also be fixed in that position. Thus by measuring the amount of $^{238}_{92}U$ remaining in the material relative to the amount of daughter nuclei, a scientist can determine the time when the rock solidified.

Radioactive dating methods using $^{238}_{92}U$ and other isotopes have shown the age of the oldest earth rocks to be about 4×10^9 yr. The age of rocks in which the oldest fossilized organisms are embedded indicates that life appeared at least 3 billion years ago. The earliest fossilized remains of mammals are found in rocks 200 million years old, and the first human-like creatures seem to have appeared about 2 million years ago. Radioactive dating has been indispensable for the reconstruction of earth's history and the evolution of its biological organisms.

*29-12 ▮ Detection of radiation

Individual particles such as electrons, protons, α particles, neutrons, and γ rays are not detected directly by our senses. Consequently, a variety of instruments have been developed to detect them.

One of the most common is the *Geiger counter*. As shown in Fig. 29-9, it consists of a cylindrical metal tube filled with a certain type of gas. A long wire runs down the center and is kept at a high positive voltage ($\sim 10^3$ V) with respect to the outer cylinder. The voltage is just slightly less than that required to ionize the gas atoms. When a charged particle enters through a thin "window" at one end of the tube, it ionizes a few atoms of the gas. The freed electrons are attracted toward the positive wire and as they are accelerated they strike and ionize additional atoms. An "avalanche" of electrons is quickly produced, and when it reaches the wire anode it produces a voltage pulse. The pulse, after being amplified, can be sent to an electronic counter, which keeps track of how many particles have been detected. Or the pulses can be sent to a loudspeaker and each detection of a particle is heard as a "click."

FIGURE 29-9

Diagram of a Geiger counter.

A *scintillation counter* makes use of a solid, liquid, or gas known as a *scintillator* or *phosphor*. The atoms of a scintillator are easily excited when struck by an incoming particle and emit visible light when they return to their ground states. Typical scintillators are crystals of NaI and certain plastics. A solid scintillator is wrapped with opaque material to keep it light-tight—except for one face, which is cemented to a photomultiplier tube. The *photomultiplier (PM) tube* converts the energy of the scintillator-emitted photon(s) into an electric signal. A PM tube is a vacuum tube containing several electrodes (typically 8 to 14), called dynodes, which are maintained at successively higher voltages as shown in Fig. 29-10. At its top surface is a photoelectric surface, called the photocathode, whose work function is low enough that electrons are easily released when struck by a photon from the scintillator. Such an electron is accelerated toward the first dynode. When it strikes the first dynode, the electron has acquired sufficient kinetic energy so that it can eject two to five more electrons. These, in turn, are accelerated to the second dynode, and a multiplication process begins. The number of electrons striking the last dynode may be 10^6 or more. Thus the passage of a particle through the scintillator results in an electric signal at the output of the PM tube that can be sent to an electronic counter just as for a Geiger tube. Because a scintillator crystal is much more dense than the gas of a Geiger counter, it is a more efficient detector—especially for γ rays, which interact less with matter than do β rays.

In tracer work and other biological experiments (Section 30-7) liquid scintillators are often used. Radioactive samples taken at different times or from different parts of an organism are placed directly in small bottles containing the liquid scintillator. This is particularly convenient for detection of β rays from ^3_1H and $^{14}_6\text{C}$, which have very low energies and have difficulty passing through the outer covering of a crystal scintillator or Geiger tube. A PM tube is still used to produce the electric signal.

A recent development is the *semiconductor detector*. It consists of a reversed biased *pn* junction diode. A particle passing through the junction causes ionization, and the released charges produce a short electrical pulse that can be counted just as for Geiger and scintillation counters.

The three devices discussed so far are used for counting the number of particles (or decays of a radioactive isotope). Other devices have been developed that allow the track of a charged particle to be seen. The simplest is the *photographic emulsion*. A particle passing through a layer of photographic emulsion ionizes the atoms along its path. This results in a chemical change at these points; and when the emulsion is developed, the particle's path is revealed.

In a *cloud chamber*, a gas is cooled to a temperature slightly below its usual condensation point. (It is said to be "supercooled.") The gas molecules begin to condense on any ionized molecules present. Thus the ions produced when a charged particle passes through serve as centers on which tiny bubbles form (Fig. 29-11). Light scatters more from these bubbles than from the gas background, so a photo of the cloud chamber at the right moment shows the track of the particle.

The *bubble chamber*, invented in 1952 by D. A. Glaser (1926–   ), makes use of a superheated liquid. The liquid is kept close to its boiling point and the bubbles characteristic of boiling form around the ions pro-

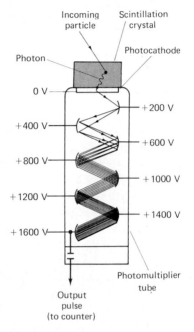

FIGURE 29-10

Scintillation counter with a photomultiplier tube.

FIGURE 29-11

In a cloud or bubble chamber, bubbles are formed by the passage of a charged particle.

(a)

(b)

FIGURE 29-12

(a) Bubble chamber; (b) particle tracks in bubble chamber. (Courtesy of Fermilab.)

duced by the passage of a charged particle (Fig. 29-12). A photograph of the chamber thus reveals the path of particles that recently passed through. Because the bubble chamber uses a liquid—often liquid hydrogen—the density of atoms is much greater than in a cloud chamber; hence it is a much more efficient device for observing particles and their interactions with the nuclei of the liquid. Usually, a magnetic field is applied across the chamber and the momentum of the moving particles can be determined from the radius of curvature of their paths.

A *spark chamber* consists of a set of closely spaced parallel plates; alternate plates are grounded and the ones in between are kept at very high voltage. When a charged particle passes through, the ions produced in the gas between the plates avalanche and a large current results which produces a visible spark. Thus the path of the particle is made visual. In a high-voltage *streamer chamber* there are only two plates, widely spaced, and the electric discharge closely follows the particle's path, which is seen as a glow or streamer. Spark and streamer chambers have the advantage that they recycle more quickly than bubble chambers (so more particles or events can be observed in a given time), although they can handle fewer tracks per photograph.

▌ Summary

Nuclear physics is the study of atomic nuclei. Nuclei contain *protons* and *neutrons*, which are collectively known as *nucleons*. The total number of nucleons, A, is the *atomic mass number*. The number of protons, Z, is the *atomic number*. The number of neutrons equals $A–Z$. *Isotopes* are nuclei with the same Z, but with different numbers of neutrons. For an element X, an isotope of given Z and A is represented by ^A_Z X.

The nuclear radius is proportional to $A^{1/3}$, indicating that all nuclei have about the same density. Nuclear masses are specified in atomic mass units (u), where the mass of $^{12}_6$C is defined as exactly 12.000000 u, or in terms of their energy equivalent (because $E = mc^2$), where 1 u $= 931.5$ MeV/c^2.

The mass of a nucleus is less than the sum of the masses of its constituent nucleons. The difference in mass (times c^2) is the *total binding energy;* it represents the energy needed to break the nucleus into its constituent nucleons. The *binding energy per nucleon* averages about 8 MeV per nucleon, and is lowest for very light and very heavy nuclei.

Unstable nuclei undergo *radioactive decay;* they change into other nuclei with the emission of an α, β, or γ particle. An α particle is a $^4_2$He nucleus; a β particle is an electron or positron; and a γ ray is a high-energy photon. In β decay, a *neutrino* is also emitted. The transformation of the parent into the daughter nucleus is called *transmutation* of the elements. Radioactive decay occurs spontaneously only when the rest mass of the products is less than the mass of the parent nucleus; the loss in mass appears as kinetic energy of the products.

Nuclei are held together by the *strong nuclear force;* the *weak nuclear force* makes itself apparent in β-decay; these two forces, plus the gravitational and electromagnetic forces, are the four known types of force. Electric charge, linear and angular momentum, mass-energy, and *nucleon number* are *conserved* in all decays.

Radioactive decay is a statistical process. The number that decay (ΔN) in a time Δt is proportional to the number N of parent nuclei present: $\Delta N = -\lambda N \, \Delta t$; the proportionality constant, λ, is called the *decay constant* and is characteristic of the given nucleus. The *half-life*, $T_{\frac{1}{2}}$, is the time required for half the nuclei of a radioactive sample to decay. It is related to the decay constant by $T_{\frac{1}{2}} = 0.693/\lambda$.

▌ Questions

1. How do we know there is such a thing as the strong nuclear force?

2. What do different isotopes of a given element have in common? How are they different?

3. What are the elements represented by the X in the following: (a) $^{232}_{92}X$; (b) $^{18}_{7}X$; (c) $^{1}_{1}X$; (d) $^{82}_{38}X$; (e) $^{247}_{97}X$?

4. How many protons and how many neutrons do each of the isotopes in question 3 have?

5. Why are the atomic masses of many elements (see the periodic table) not close to whole numbers?

6. What is the experimental evidence in favor of radioactivity being a nuclear process?

7. The isotope $^{64}_{29}$Cu is unusual in that it can decay by γ, β^-, and β^+ emission. What is the resulting nuclide for each case?

8. A $^{238}_{92}$U nucleus decays to a nucleus containing how many neutrons?

9. Describe, in as many ways as possible, the differences between α, β, and γ rays.

10. What element is formed by the radioactive decay of (a) $^{24}_{11}$Na(β^-); (b) $^{22}_{11}$Na(β^+); (c) $^{210}_{84}$Po(α)?

11. What element is formed by the decay of (a) $^{32}_{15}$P(β^-); (b) $^{35}_{16}$S(β^-); (c) $^{211}_{83}$Bi(α)?

12. Fill in the missing particle or nucleus:
   (a) $^{45}_{20}$Ca $\rightarrow$? $+ e^- + \bar{v}$
   (b) $^{58}_{29}$Cu $\rightarrow$? $+ \gamma$
   (c) $^{46}_{24}$Cr $\rightarrow$ $^{46}_{23}$V $+$?
   (d) $^{234}_{94}$Pu $\rightarrow$? $+ \alpha$
   (e) $^{239}_{93}$Np $\rightarrow$ $^{239}_{92}$U $+$?

13. Immediately after a $^{238}_{92}$U nucleus decays to $^{234}_{90}$Th $+$ $^4_2$He, the daughter thorium nucleus still has 92 electrons circling it. Since thorium normally holds only 90 electrons, what do you suppose happens to the two extra ones?

14. Do isotopes that undergo electron capture generally lie above or below the line of stability in Fig. 29-2? ·

15. Why are many artificially produced radioactive isotopes rare in nature?

16. Can hydrogen or deuterium emit an α particle?

*17. Describe how the potential energy curve for an α particle in an α-emitting nucleus differs from that for a stable nucleus.

*18. Can $^{14}_6$C dating be used to measure the age of stone walls and tablets of ancient civilizations?

*19. Explain the absence of β^+ emitters in the radioactive decay series of Fig. 29-7.

▌ Problems

SECTION 29-1

1. (I) What is the rest energy of an α particle in MeV/c^2?

2. (I) A pi meson has a mass of 139 MeV/c^2. What is this in atomic mass units?

3. (I) (a) What is the approximate radius of a $^{64}_{29}$Cu nucleus? (b) Approximately what is the value of A for a nucleus whose radius is 3.6×10^{-15} m?

4. (II) How much energy must an α particle have to just "touch" the surface of a $^{238}_{92}$U nucleus?

5. (II) (a) Determine the density of nuclear matter in kg/m^3. (b) What would be the radius of the earth if it had its actual mass but had the density of nuclei? (c) What would be the radius of a $^{238}_{92}U$ nucleus if it had the density of the earth?

6. (II) What stable nucleus (approximately) has half the radius of a uranium nucleus? (Hint: find A and use Appendix D to get Z.)

SECTION 29-2

7. (I) Estimate the total binding energy for $^{40}_{20}Ca$, using Fig. 29-1.

8. (I) Use Fig. 29-1 to estimate the total binding energy of $^{238}_{92}U$.

9. (II) Use Appendix D and calculate the binding energy of ^2_1H.

10. (II) Calculate the total binding energy and the binding energy per nucleon for ^6_3Li. Use Appendix D.

11. (II) Calculate the binding energy of the last neutron in a $^{12}_6C$ nucleus. (Hint: compare the mass of $^{12}_6C$ and $^{11}_6C + ^1_0n$; use Appendix D.)

12. (II) Calculate the binding energy per nucleon for a $^{12}_6C$ nucleus.

13. (II) (a) Show that the nucleus ^8_4Be (mass = 8.005308 u) is unstable to decay into two α particles. (b) Is $^{12}_6C$ stable against decay into three α particles? Show why or why not.

SECTIONS 29-3 TO 29-7

14. (I) $^{60}_{27}Co$ in an excited state emits a 1.33-MeV γ ray as it jumps to the ground state. What is the mass of the excited cobalt atom?

15. (II) Show that the decay $^{11}_6C \rightarrow ^{10}_5B + p$ is not possible because energy would not be conserved.

16. (II) A $^{232}_{92}U$ nucleus emits an α particle with $KE = 5.32$ MeV. What is the final nucleus and what is the approximate atomic mass (in u) of the final atom?

17. (II) When $^{23}_{10}Ne$ (mass = 22.9945 u) decays to $^{23}_{11}Na$ (mass = 22.9898 u), what is the maximum kinetic energy of the emitted electron? What is its minimum energy? What is the energy of the neutrino in each case?

18. (II) The nuclide $^{32}_{15}P$ decays by emitting an electron whose maximum kinetic energy can be 1.71 MeV. (a) What is the daughter nucleus? (b) What is its atomic mass (in u)?

19. (II) The isotope $^{218}_{84}Po$ can decay by either α or β^- emission. What is the energy release in each case? The mass of $^{218}_{84}Po$ is 218.008969 u.

20. (II) What is the energy of the α particle emitted in the decay $^{210}_{84}Po \rightarrow ^{206}_{82}Pb + \alpha$?

21. (II) How much energy is released in electron capture by beryllium: $^7_4Be + {}_{-1}^0e \rightarrow ^7_3Li + \nu$?

22. (III) The nuclide $^{191}_{76}Os$ decays with β^- energy of 0.14 MeV accompanied by γ rays of energy 0.042 MeV and 0.129 MeV. (a) What is the daughter nucleus? (b) Draw an energy-level diagram showing the ground states of the parent and daughter and excited states of the daughter. To which of the daughter states does β decay of $^{191}_{76}Os$ occur?

23. (III) Show that when a nucleus decays by β^+ decay, the total energy released is equal to $(M_P - M_D - 2m_e)c^2$, where M_P and M_D are the masses of the parent and daughter atoms (neutral) and m_e is the mass of an electron or positron.

24. (III) Use the result of problem 23 to determine the maximum kinetic energy of β^+ particles released when $^{11}_6C$ decays to $^{11}_5B$. What is the maximum energy the neutrino can have? What is its minimum energy?

25. (III) In α decay of, say, a $^{226}_{88}Ra$ nucleus, show that the nucleus carries away a fraction $1/(1 + A_D/4)$ of the total energy available, where A_D is the mass number of the daughter nucleus. (Hint: use conservation of momentum as well as conservation of energy.) Approximately what percentage of the energy available is thus carried off by the α particle in the case cited?

SECTIONS 29-8 TO 29-11

26. (I) A radioactive material registers 1280 counts per minute on a Geiger counter at one time, and 6 h later registers 320 counts per minute. What is its half-life?

27. (I) In a series of decays, the nuclide $^{235}_{92}U$ becomes $^{207}_{82}Pb$. How many α and β^- particles are emitted in this series?

28. (I) What is the decay constant of $^{238}_{92}U$ whose half-life is 4.5×10^9 yr?

29. (I) The decay constant of a given nucleus is 6.2×10^{-5} s^{-1}. What is its half-life?

30. (I) What is the activity of a sample of $^{14}_6C$ that contains 2.0×10^{20} nuclei?

31. (I) How many nuclei of $^{238}_{92}U$ remain in a rock if the activity registers 1.8×10^4 decays per second?

32. (I) What fraction of a sample of $^{68}_{32}Ge$, whose half-life is about 9 months, will remain after 4.5 yr?

33. (II) $^{124}_{55}Cs$ has a half-life of 30.8 s. (a) If we have 7.5 μg initially, how many nuclei are present? (b) How many are present 2.0 min later? (c) What is the activity at this time? (d) After how much time will the activity drop to less than about 1 per second?

34. (II) Calculate the activity of a pure 3.5-μg sample of $^{32}_{15}P$ ($T_{\frac{1}{2}} = 1.24 \times 10^6$ s).

35. (II) The activity of a sample of $^{35}_{16}S$ ($T_{\frac{1}{2}} = 7.5 \times 10^6$ s) is 5.2×10^6 decays per second. What is the mass of sample present?

36. (II) A sample of $^{233}_{92}U$ ($T_{\frac{1}{2}} = 1.6 \times 10^5$ yr) contains 2.0×10^{18} nuclei. (a) What is the decay constant? (b) Approximately how many disintegrations will occur per minute?

*37. (II) Use Fig. 29-7 and calculate the relative decay rates for α and β decay of $^{218}_{84}Po$.

38. (II) A 50-g sample of pure carbon contains 1.3 parts in 10^{12} (atoms) of $^{14}_{6}C$. How many disintegrations occur per second?

39. (II) The $^{3}_{1}H$ isotope of hydrogen, which is called *tritium* (because it contains three nucleons), has a half-life of 12.33 yr. It can be used to measure the age of objects up to about 100 yr. It is produced in the upper atmosphere by cosmic rays and is brought to earth by rain. As an application, determine approximately the age of a bottle of wine whose $^{3}_{1}H$ radiation is about $\frac{1}{10}$ that present in new wine.

40. (II) The rubidium isotope $^{87}_{37}Rb$, a β emitter with a half-life of 4.9×10^{10} yr, is used to determine the age of rocks and fossils. Rocks containing fossils of early animals contain a ratio of $^{87}_{38}Sr$ to $^{87}_{37}Rb$ of 0.010. Assuming that there was no $^{87}_{38}Sr$ present when the rocks were formed, calculate the age of these fossils. (Hint: use Eq. 29-2.)

41. (III) An ancient club is found that contains 240 g of carbon and has an activity of 5.0 decays per second. Determine its age using the fact that in living trees the ratio of $^{14}C/^{12}C$ atoms is about 1.3×10^{-12}.

30 NUCLEAR ENERGY; EFFECTS AND USES OF RADIATION

We continue our study of nuclear physics in this chapter. We begin with a discussion of nuclear reactions, after which we examine the huge energy releasing processes of fission and fusion. The remainder of the chapter deals with the effects of nuclear radiation when it passes through matter, particularly biological matter, and how radiation is used medically for therapy and diagnosis, including recently developed imaging techniques.

30-1 ∎ Nuclear reactions and the transmutation of elements

When a nucleus undergoes α or β decay, the daughter nucleus is that of a different element from the parent. The transformation of one element into another, called *transmutation*, also occurs by means of nuclear reactions. A **nuclear reaction** is said to occur when a given nucleus is struck by another nucleus, or by a simpler particle such as a γ ray or neutron, so that an interaction takes place. Ernest Rutherford was the first to report seeing a nuclear reaction. In 1919 he observed that some of the α particles passing through nitrogen gas were absorbed and protons emitted. He concluded that nitrogen nuclei had been transformed into oxygen nuclei via the reaction

$$\mathstrut^4_2\text{He} + \mathstrut^{14}_7\text{N} \to \mathstrut^{17}_8\text{O} +$$

where $^4_2$Hc is an α particle and $^1_1$H is a proton.

Since then, a great many nuclear reactions have been observed. Indeed, many of the radioactive isotopes used in the laboratory are made by means of nuclear reactions. Nuclear reactions can be made to occur in the laboratory, but they also occur regularly in nature. In Chapter 29 we saw an example of this: $^{14}_6$C is continually being made in the atmosphere via the reaction $n + \mathstrut^{14}_7\text{N} \to \mathstrut^{14}_6\text{C} + p$.

Nuclear reactions are sometimes written in a shortened form; for example, the reaction $n + \mathstrut^{14}_7\text{N} \to \mathstrut^{14}_6\text{C} + p$ is written $^{14}_7\text{N}(n, p)^{14}_6\text{C}$. The symbols outside the parentheses on the left and right represent the initial and final nuclei, respectively. The symbols inside the parentheses represent the bombarding particle (first) and the emitted small particle (second).

In any nuclear reaction, the total electric charge is conserved, and the total number of nucleons is conserved. These conservation laws are often useful, as the following example shows.

☐ EXAMPLE 30-1   A neutron is observed to strike an $^{16}_{8}O$ nucleus and a deuteron is given off. (A *deuteron*, or *deuterium*, is the isotope of hydrogen containing one proton and one neutron, $^{2}_{1}H$.) What is the nucleus that results?

SOLUTION   We have the reaction $^{1}_{0}n + {}^{16}_{8}O \rightarrow ? + {}^{2}_{1}H$. The total number of nucleons initially is $16 + 1 = 17$, and the total charge is $8 + 0 = 8$; the same totals apply to the right side of the reaction. Hence the product nucleus must have $Z = 7$ and $A = 15$. From the periodic table, we find that it is nitrogen that has $Z = 7$, so the nucleus produced is $^{15}_{7}N$. The reaction can be written $^{16}_{8}O(n, d)^{15}_{7}N$ where d represents deuterium, $^{2}_{1}H$.

Energy (as well as momentum) is conserved in nuclear reactions. This fact can be used to determine whether a given reaction can occur or not. For example, if the total mass of the products is less than the total mass of the initial particles, then energy will be released by the reaction—it will appear as kinetic energy of the outgoing particles. But if the total mass of the products is greater than the total mass of the initial reactants, the reaction requires energy; the reaction will not occur unless the bombarding particle has sufficient kinetic energy. This is demonstrated in the next example.

☐ EXAMPLE 30-2   Can the reaction $^{13}_{6}C(p, n)^{13}_{7}N$ occur when $^{13}_{6}C$ is bombarded by 2.0 MeV protons?

SOLUTION   We look up the masses of the nuclei in Appendix D. The total masses before and after the reaction are:

Before	After
$m(^{13}_{6}C) = 13.003355$	$m(^{13}_{7}N) = 13.005739$
$m(^{1}_{1}H) = \underline{\ 1.007825}$	$m(^{1}_{0}n) = \underline{\ 1.008665}$
14.011180	14.014404

(We must use the mass of the $^{1}_{1}H$ atom rather than that of the bare proton because the masses of $^{13}_{6}C$ and $^{13}_{7}N$ include the electrons and we must include an equal number of electrons on each side of the equation since none are created or destroyed.) The products have an excess mass of $0.003224 \text{ u} \times 931.5 \text{ MeV/u} = 3.00 \text{ MeV}$. This reaction requires energy, and the 2.0 MeV protons do not have enough to make it go. Hence this reaction won't occur. The proton would have to have somewhat more than 3.00 MeV of KE; 3.00 MeV would be enough to conserve energy, but a proton of this energy would produce the $^{13}_{7}N$ and $^{1}_{0}n$ with no KE and hence no momentum; since the incident proton has momentum, conservation of momentum would be violated. A more complicated calculation shows that to conserve both energy and momentum the minimum proton energy required (called the *threshold energy*) is 3.23 MeV in this case. See problem 7.

The artificial transmutation of elements took a great leap forward in the 1930s when Enrico Fermi realized that neutrons would be the most effective projectiles for causing nuclear reactions and in particular for producing new elements. Because neutrons have no net electric charge, they are not repelled by positively charged nuclei as protons or alpha particles are. Hence the probability of a neutron reaching the nucleus and causing a reaction is much greater than for charged projectiles,[†] particularly at low energies. Between 1934 and 1936, Fermi and his co-workers in Rome produced many previously unknown isotopes by bombarding different elements with neutrons. Fermi realized that if the heaviest known element, uranium, were bombarded with neutrons, it might be possible to produce new elements whose atomic numbers were greater than that of uranium. After several years of hard work, it was suspected that two new elements had been produced, neptunium ($Z = 93$) and plutonium ($Z = 94$). The full confirmation that such "transuranic" elements could be produced came several years later at the University of California, Berkeley. The reactions are shown in Fig. 30-1.

It was soon shown that what Fermi actually had observed when he bombarded uranium was an even stranger process—one that was destined to play an extraordinary role in the world at large.

30-2 ∎ Nuclear fission; nuclear reactors

In 1938, the German scientists Otto Hahn and Fritz Strassmann made an amazing discovery. Following up on Fermi's work, they found that uranium bombarded by neutrons sometimes produced smaller nuclei which were roughly half the size of the original uranium nucleus. Lise Meitner and Otto Frisch, two refugees from Nazi Germany working in Scandinavia, quickly realized what had happened: the uranium nucleus, after absorbing a neutron, actually had split into two roughly equal pieces. This was startling, for until then the known nuclear reactions involved knocking out only a tiny fragment (for example, n, p, or α) from a nucleus.

This new phenomenon was named **nuclear fission** because of its resemblance to biological fission (cell division). It occurs much more readily for $^{235}_{92}U$ than for the more common $^{238}_{92}U$. The process can be visualized by imagining the uranium nucleus to be like a liquid drop. According to this **liquid-drop model,** the neutron absorbed by the $^{235}_{92}U$ nucleus gives the nucleus extra internal energy (like heating a drop of water). This intermediate state, or **compound nucleus,** is $^{236}_{92}U$ (because of the absorbed neutron). The extra energy of this nucleus—it is in an excited state—appears as increased motion of the individual nucleons which causes the nucleus to take on abnormal elongated shapes, Fig. 30-2. When the nucleus elongates into the shape shown in Fig. 30-2c, the attraction of the two ends via the short-range nuclear force is greatly weakened by the excess separation distance, but the electric repulsive force is weakened only slightly and becomes predominant; so the nucleus splits in two. The two resulting nuclei, N_1 and N_2, are called

[†] That is, positively charged particles. Electrons rarely cause nuclear reactions because they do not partake of the strong nuclear force.

FIGURE 30-1

Neptunium and plutonium are produced in this series of reactions, after bombardment of $^{238}_{92}U$ by neutrons.

(a) $n + {}^{238}_{92}U \rightarrow {}^{239}_{92}U$

Neutron captured by $^{238}_{92}U$

(b) $^{239}_{92}U \rightarrow {}^{239}_{93}Np + e^- + \bar{\nu}$

$^{239}_{92}U$ decays by β decay to neptunium 239.

(c) $^{239}_{93}Np \rightarrow {}^{239}_{94}Pu + e^- + \bar{\nu}$

$^{239}_{93}Np$ itself decays by β decay to produce plutonium 239.

FIGURE 30-2

Fission of a $^{235}_{92}U$ nucleus after capture of a neutron.

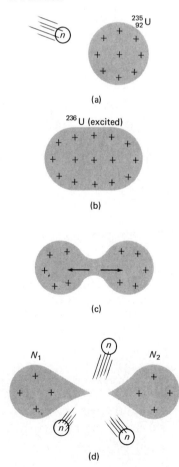

(a)

(b)

(c)

(d)

fission fragments, and in the process a number of neutrons (typically two or three) are also given off. The reaction can be written

$$_0^1n + {}_{92}^{235}U \rightarrow {}_{92}^{236}U \rightarrow N_1 + N_2 + \text{neutrons}. \tag{30-1}$$

The compound nucleus, ${}_{92}^{236}U$, exists for less than 10^{-12} s, so the process occurs very quickly. The two fission fragments have roughly half the mass of the uranium, although rarely are they exactly equal in mass. A typical fission reaction is

$$_0^1n + {}_{92}^{235}U \rightarrow {}_{56}^{141}Ba + {}_{36}^{92}Kr + 3{}_0^1n, \tag{30-2}$$

although many others also occur.

A tremendous amount of energy is released in a fission reaction because the mass of ${}_{92}^{235}U$ is considerably greater than that of the fission fragments. This can be seen from the binding-energy-per-nucleon curve of Fig. 29-1; the binding energy per nucleon for uranium is about 7.6 MeV/nucleon, but for fission fragments which have intermediate mass (in the center portion of the graph, $Z \approx 100$), the average binding energy per nucleon is about 8.5 MeV/nucleon. Since the fission fragments are more tightly bound, they have less mass. The difference in mass (or energy) between the original uranium nucleus and the fission fragments is about $8.5 - 7.6 = 0.9$ MeV per nucleon. Since there are 236 nucleons involved in each fission, the total energy released per fission is

$$(0.9 \text{ MeV/nucleon})(236 \text{ nucleons}) \approx 200 \text{ MeV}.$$

This is an enormous amount of energy on the nuclear scale. At a practical level, the energy from one fission is, of course, tiny. However, a great deal of energy at the macroscopic level would be available if many such fissions could occur at once. A number of physicists, including Fermi, recognized that the neutrons released in each fission (Eq. 30-1 or 30-2) could be used to create a **chain reaction:** one neutron initially causes one fission of a uranium nucleus; the two or three neutrons released can go on to cause additional fissions, so the process multiplies as shown schematically in Fig. 30-3. If a *self-sustaining chain reaction* was actually possible in practice, the enormous energy available in fission could be released. Fermi and his co-workers (at

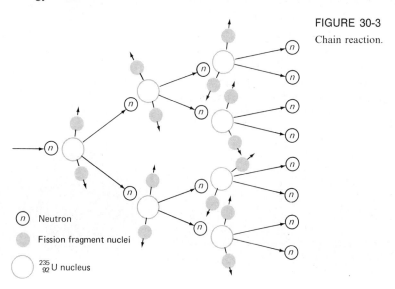

FIGURE 30-3

Chain reaction.

Ⓝ Neutron

⬤ Fission fragment nuclei

◯ ${}_{92}^{235}U$ nucleus

the University of Chicago) showed it was possible by constructing the first **nuclear reactor** in 1942.

Several problems have to be overcome in any nuclear reactor. First, the probability that a $^{235}_{92}U$ nucleus will absorb a neutron is large only for slow neutrons, but the neutrons emitted during a fission, and which are needed to sustain a chain reaction, are moving very fast. A substance known as a **moderator** must be used to slow down the neutrons. The most effective moderator will consist of atoms whose mass is as close as possible to that of the neutrons. (To see why this is true, recall from Chapter 6 that a billiard ball striking an equal mass at rest can itself be stopped in one collision; but a billiard ball striking a heavy object bounces off with nearly the same speed it had). The best moderator would thus contain $^{1}_{1}H$ atoms; unfortunately, $^{1}_{1}H$ tends to absorb neutrons; but *deuterium*, $^{2}_{1}H$, does not absorb many neutrons and is thus an almost ideal moderator. Either $^{1}_{1}H$ or $^{2}_{1}H$ can be used in the form of water; in the latter case, it is *heavy water*, where the hydrogen atoms have been replaced by deuterium. Another common moderator is *graphite*, which consists of $^{12}_{6}C$ atoms.

A second problem is that the neutrons produced in one fission may be absorbed and produce other nuclear reactions with other nuclei in the reactor, rather than produce further fissions. In a "light water" reactor the $^{1}_{1}H$ nuclei absorb neutrons, as does $^{238}_{92}U$ to form $^{239}_{92}U$ in the reaction $n + {}^{238}_{92}U \rightarrow {}^{239}_{92}U + \gamma$. Naturally occurring uranium contains 99.3 percent $^{238}_{92}U$ and only 0.7 percent fissionable $^{235}_{92}U$. To increase the probability of fission of $^{235}_{92}U$ nuclei, natural uranium is often **enriched**[†] to increase the percentage of $^{235}_{92}U$ using processes such as diffusion or centrifugation.

The third problem is that some neutrons will escape through the surface of the reactor core before they cause further fissions (Fig. 30-4). Thus the mass of fuel must be sufficiently large for a self-sustaining chain reaction to take place. The minimum mass of uranium needed is called the **critical mass.** The value of the critical mass depends on the moderator, the fuel ($^{239}_{94}Pu$ may be used[‡] instead of $^{235}_{92}U$), and how much the fuel is enriched, if at all. Typical values are on the order of a few kilograms (that is, not grams nor thousands of kilograms).

To have a self-sustaining chain reaction, it is clear that on the average at least one neutron produced in each fission must go on to produce another fission. The average number of neutrons per each fission that do go on to produce further fissions is called the *multiplication factor*, f. For a self-sustaining chain reaction we must have $f \geqslant 1$. If $f < 1$, the reactor is "subcritical"; if $f > 1$, it is "supercritical." Reactors are equipped with movable *control rods* (usually of cadmium or boron), whose function is to absorb neutrons and maintain the reactor[§] at just barely "critical," $f = 1$.

[†] Enrichment is not usually necessary for reactors using heavy water as moderator.

[‡] $^{238}_{92}U$ will fission, but only with fast neutrons ($^{238}_{92}U$ is more stable than $^{235}_{92}U$). The probability of absorbing a fast neutron and producing a fission is too low to produce a self-sustaining chain reaction.

[§] The release of neutrons and subsequent fissions caused by them occurs so quickly that manipulation of the control rods to maintain $f = 1$ would not be possible if it weren't for the fortunate fact that a small percentage of neutrons released are "delayed." They come from the decay of neutron-rich fission fragments (or their daughters) having lifetimes on the order of seconds—sufficient to allow enough reaction time to maintain $f = 1$.

FIGURE 30-4

If the amount of uranium exceeds the critical mass, as in (b), a sustained chain reaction is possible. If the mass is less than critical, as in (a), most neutrons escape before additional fissions occur, and the chain reaction is not sustained.

(a)

(b)

Nuclear reactors have been built for use in research and to produce electric power. Fission produces many neutrons and a "research reactor" is basically an intense source of neutrons. These neutrons can be used as projectiles in nuclear reactions to produce nuclides not found in nature, including isotopes used as tracers and for medical therapy. A "power reactor" is used to produce electricity. The energy released in the fission process appears as heat, which is used to boil water and produce steam to drive a turbine connected to an electric generator. Figure 30-5 shows a typical design. The *core* of a nuclear reactor consists of the fuel and a moderator (water in most U.S. commercial reactors). The fuel is usually uranium enriched so that it contains 2 to 4 percent $^{235}_{92}U$. Water or other liquid (such as liquid sodium) is allowed to flow through the core. The thermal energy it absorbs is used to produce steam in the heat exchanger, so the fissionable fuel acts as the heat input for a heat engine (see Sections 12-5 and 12-10).

Many problems are associated with nuclear power plants. There is, of course, the usual thermal pollution associated with any heat engine (Section 12-10). But probably the most serious problems are associated with the radioactive fission fragments produced in the reactor, plus radioactive nuclides produced by neutrons interacting with the structural parts of the reactor. Fission fragments, like their uranium or plutonium parents, have about 50 percent more neutrons than protons. Nuclei with atomic number in the typical range for fission fragments ($Z \approx 30$ to 60) are stable only if they have more nearly equal numbers of protons and neutrons (see Fig. 29-2). Hence the highly neutron-rich fission fragments are very unstable and decay radioactively. The accidental release of highly radioactive fission fragments into the atmosphere poses a serious threat to human health (see Section 30-4). Also a serious problem is the disposal of the spent fuel, which contains highly radioactive fission fragments. Leakage of the radioactive wastes is possible and has in fact occurred. Indeed, a satisfactory method of disposal is still not at hand. Finally, the lifetime of nuclear power plants is limited to

FIGURE 30-5

Nuclear reactor. The heat generated by the fission process in the fuel rods is carried off by hot water or liquid sodium and is used to boil water to steam in the heat exchanger. The steam drives a turbine to generate electricity and is then cooled in the condenser.

some 30 years, due to buildup of radioactivity and the fact that the structural materials themselves are weakened by the intense conditions inside. Decommissioning of a power plant will be a major undertaking, and could take on a number of forms. The cost of any method of decommissioning a large plant will probably be extremely large.

So-called *breeder reactors* have the same problems, but were proposed as a solution to the problem of limited supplies of fissionable uranium. A breeder reactor is one in which some of the neutrons produced in the fission of $^{235}_{92}U$ are absorbed by $^{238}_{92}U$, and $^{238}_{94}Pu$ is produced via the set of reactions shown in Fig. 30-1. $^{239}_{94}Pu$ is fissionable with slow neutrons, so after separation it can be used as a fuel in a nuclear reactor. Thus a breeder reactor "breeds" new fuel[†] ($^{239}_{94}Pu$) from otherwise useless $^{238}_{92}U$. Since natural uranium is 99.3 percent $^{238}_{92}U$, this means that the supply of fissionable fuel could be increased by more than a factor of 100. But breeder reactors present additional problems. First, plutonium has a very long half-life of 24,000 years, and is a highly toxic substance that poses a serious danger to health. Also important is the fact that plutonium produced can readily be used in a bomb. Thus the use of a breeder reactor, even more than a conventional uranium reactor, presents the danger of nuclear proliferation: even poor nations might be able to produce nuclear bombs. And the possibility of theft of fuel by terrorists who could produce a bomb further increases the probability of nuclear holocaust.

It is clear that nuclear power presents many risks. Other large-scale energy-conversion methods, such as conventional coal-burning steam plants, also present health and environmental hazards, some of which were discussed in Chapter 12. The solution to the world's needs for energy is not only technological, but economic and political as well.

The first use of fission was not, however, to produce electric power. Instead it was first used as a fission bomb (the "atomic bomb"). In early 1940, with Europe already at war, Hitler banned the sale of uranium from the Czech mines he had recently taken over. Research into the fission process suddenly was enshrouded in secrecy by Germany and by the western powers. Physicists in the United States were alarmed. A group of them approached Einstein—a man whose name was a household word—to write a letter to President Roosevelt about the possibilities of using nuclear fission for a bomb far more powerful than any previously known, and inform him that Germany might already have begun development of such a bomb. Roosevelt responded by authorizing the program known as the Manhattan Project, to see if a bomb could be built. Work began in earnest after Fermi's demonstration in 1942 that a sustained chain reaction was possible. A new secret laboratory was developed on an isolated mesa in New Mexico known as Los Alamos. Under the direction of J. Robert Oppenheimer (1904–1967), it became the home of famous scientists from all over Europe and the U.S.; the roster of names of those who worked there during the next couple of years is like a who's who of science.

One of the problems was how to build a bomb that was subcritical during transport but which could be made supercritical (to produce a chain

---

[†] A breeder reactor does *not* produce more fuel than it uses, for a nearly endless supply, as popular articles often imply.

reaction) at just the right moment. This was achieved by using two pieces of uranium each of which was less than the critical mass but when placed together became greater than the critical mass. The two masses would be kept separate until the moment of detonation arrived; then a special kind of gun would force the two pieces together very quickly, a chain reaction of explosive proportions would occur, and a tremendous amount of energy would be released very suddenly. The first fission bomb was tested in the New Mexico desert in July, 1945. It was successful. In early August, a fission bomb using uranium was dropped on Hiroshima and a second, using plutonium, was dropped on Nagasaki. World War II ended shortly thereafter, but the dropping of these bombs aroused much controversy.

Scientists were later criticized for working on the bomb, and many regretted having done so. Critical judgment, however, should perhaps be placed in the context of the times. In the early 1940s, Hitler's armies were overrunning Europe, and it was believed that German scientists were trying to build a bomb (they never did). Japan had entered the war and the world seemed in danger of being overwhelmed by oppressive powerful regimes. The scientists at Los Alamos were aware they were developing a dangerous weapon. But they had to make a choice. Stopping Hitler and his allies was uppermost in their minds.

Besides its great destructive power, a fission bomb produces many highly radioactive fission fragments, as does a nuclear reactor as already discussed. When a fission bomb explodes, these radioactive isotopes are released into the atmosphere and are known as *radioactive fallout*.

Testing of nuclear bombs in the atmosphere after World War II was a cause of concern, for the movement of air masses spread the fallout all over the globe. Radioactive fallout eventually settles to the earth, particularly in rainfall, and is absorbed by plants and grasses and enters the food chain. This is a far more serious problem than the same radioactivity on the exterior of our bodies, since α and β particles are largely absorbed by clothing and the outer (dead) layer of skin. But once inside our bodies via food, the isotopes are in direct contact with living cells. One particularly dangerous radioactive isotope is $^{90}_{38}Sr$, which is chemically much like calcium and becomes concentrated in bone, where it causes bone cancer and destruction of bone marrow. The 1963 treaty signed by over 100 nations that bans nuclear weapons testing in the atmosphere was motivated because of the hazards of fallout.

30-3 ∎ Fusion

The mass of every stable nucleus is less than the sum of the masses of its constituent protons and neutrons. For example, the mass of the helium isotope ^4_2He is less than the mass of two protons plus the mass of two neutrons. Thus, if two protons and two neutrons were to come together to form a helium nucleus there would be a loss of mass. This mass loss is manifested in the release of a large amount of energy. The process of building up nuclei by bringing together individual protons and neutrons, or building larger nuclei by combining small nuclei, is called **nuclear fusion.** A glance at

Fig. 29-1 shows how small nuclei can combine to form larger ones with the release of energy: it is because the binding energy per nucleon is smaller for light nuclei than it is for those of intermediate size ($A = 50$ to 100). It is believed that all the elements in the universe were originally formed through the process of fusion. Today, fusion is continually taking place within the stars, including our sun, producing the intense light energy they emit.

> **EXAMPLE 30-3**   One of the simplest fusion reactions involves the production of deuterium, $_1^2H$, from a neutron and a proton: $_1^1H + _0^1n \rightarrow _1^2H + \gamma$. How much energy is released in this reaction?
>
> SOLUTION   From Appendix D the initial rest mass is 1.007825 u $+ 1.008665$ u $= 2.016490$ u and after the reaction the mass is that of the $_1^2H$, namely 2.014102 u. The energy released is thus $(0.002388$ u$)(931.5$ MeV/u$) = 2.22$ MeV, and is carried off by the $_1^2H$ nucleus and the γ ray.

The possibility of utilizing the energy released in fusion to make a power reactor is very attractive, but so far a successful reactor has not been achieved. The fusion reactions most likely to succeed in a reactor involve the isotopes of hydrogen, $_1^2H$ (deuterium) and $_1^3H$ (tritium), and are as follows, with the energy released given in parentheses:

$$_1^2H + _1^2H \rightarrow _1^3H + _1^1H \qquad (4.03 \text{ MeV}) \qquad (30\text{-}3a)$$

$$_1^2H + _1^2H \rightarrow _2^3He + _0^1n \qquad (3.27 \text{ MeV}) \qquad (30\text{-}3b)$$

$$_1^2H + _1^3H \rightarrow _2^4He + _0^1n \qquad (17.59 \text{ MeV}) \qquad (30\text{-}3c)$$

$$_1^3H + _1^3H \rightarrow _2^4He + 2_0^1n \qquad (11.33 \text{ MeV}). \qquad (30\text{-}3d)$$

The energy released in fusion reactions is greater for a given mass of fuel than in fission. Furthermore, fusion presents less of a radioactive-waste problem. For fuel, a fusion reactor could use deuterium, which is very plentiful in the water of the oceans (the natural abundance of $_1^2H$ is 0.015 percent, or about 1 g of deuterium per 60 L of water).

Unfortunately, considerable difficulties still exist for making a usable fusion reactor. The problems are associated with the fact that all nuclei have a positive charge and thus repel each other. However, if they can be brought close enough together so that the short-range attractive nuclear force can come into play, the latter can pull the nuclei together and fusion will occur. In order for the nuclei to get close enough together they must have very high speeds. Since high speed of atoms corresponds to high temperature, very high temperatures are required for fusion to occur; hence fusion devices are often referred to as *thermonuclear devices*. The sun and other stars are very hot, many millions of degrees, so the nuclei are moving fast enough for fusion to take place, and the energy released keeps the temperature high so that further fusion reactions can occur. The sun and the stars represent self-sustaining thermonuclear reactors, but on earth such high temperatures are not easily attained in a controlled manner.

It was realized after World War II that the temperature produced within a fission (or "atomic") bomb was close to 10^8 K; this suggested that a fission bomb could be used to ignite a fusion bomb (popularly known as a thermonuclear or hydrogen bomb) to release the vast energy of fusion. The uncontrollable release of fusion energy in an H-bomb was fairly easy to obtain. But to realize usable energy from fusion at a slow and controlled rate turned out to be very difficult. The high temperatures needed can now be produced by an infusion of concentrated energy, such as from a high-powered laser. The real difficulty is to contain the nuclei long enough for sufficient reactions to occur that a usable amount of energy is obtained. At the temperatures needed for fusion (approximately 10^8 K) the atoms are ionized, and this collection of nuclei and electrons is referred to as a **plasma.** Ordinary materials vaporize at a few thousand degrees at best, and hence could not be used to contain a high temperature plasma. One technique is to confine a plasma with a magnetic field; unfortunately, all attempted configurations of magnetic field confinement have developed "leaks" and the charged particles leak out before sufficient fusion takes place. One promising configuration of magnetic fields is the so-called Tokamak, and confinement times are being increased. The plasma must also be sufficiently dense so that a significant number of interactions takes place during the confinement time. In 1957, J. D. Lawson showed that for the output energy of a fusion reactor to exceed the input energy, the product of the ion density n (number/cm^3) and confinement time τ must satisfy (approximately)

$$n\tau \gtrsim 3 \times 10^{14} \text{ s/cm}^3.$$

This is known as the *Lawson criterion*, and over the years it has been more closely approached.

Another containment technique is to form solid pellets of fuel which are quickly heated by intense laser or electron beams. This *inertial confinement* technique too shows promise, but controlled fusion remains elusive.

30-4 ∎ Passage of radiation through matter; radiation damage

When we speak of *radiation*, we include α, β, γ, and X rays, as well as protons, neutrons, and other particles such as pions (see Chapter 31). Because charged particles can ionize the atoms or molecules of any material they pass through, they are sometimes referred to as *ionizing radiation*. And because radiation produces ionization, it can cause considerable damage to materials, particularly to biological tissue.

Charged particles, such as α and β rays and protons, cause ionization because of the electric force. That is, when they pass through a material they can attract or repel electrons strongly enough to remove them from the atoms of the material. Since the α and β rays emitted by radioactive substances have energies on the order of 1 MeV (more precisely, from 10^4 to 10^7 eV), whereas ionization of atoms and molecules requires from about 10 to 50 eV, it is clear that a single α or β particle can cause thousands of ionizations.

Neutral particles also give rise to ionization when they pass through materials. For example, X-ray and γ-ray photons can ionize atoms by knocking out electrons by means of the photoelectric and Compton effects. Furthermore, if a γ ray has sufficient energy (greater than 1.02 MeV), it can undergo pair production: an electron and a positron are produced (Section 26-3). The charged particles produced in all three of these processes themselves can go on to produce further ionization. Neutrons, on the other hand, interact with matter mainly by collisions with nuclei, with which they interact strongly. Often the nucleus is broken apart by such a collision, which alters the molecule of which it was a part. And the fragments produced can move on to cause ionization.

Radiation passing through matter can cause considerable damage. In metals and other structural materials, the strength of the material can be weakened if the radiation is very intense. This is a considerable problem in nuclear reactor power plants, and for space vehicles that must pass through areas of intense cosmic radiation.

Very important is the radiation damage produced in biological organisms. The damage is due primarily to the ionization produced in cells. Several related processes can occur. Ions or radicals are produced which are highly reactive and take part in chemical reactions that interfere with the normal operation of the cell. An important example is the production of the (neutral) free radicals H and OH from water, which is the major constituent of cells; these free radicals are very reactive and cause chemical reactions that may break chemical bonds in vital molecules such as proteins.

All forms of radiation can ionize atoms by knocking out electrons. If these are bonding electrons, the molecule may break apart; or its structure may be altered so that it does not perform its normal function or performs a harmful function. In the case of proteins, the loss of one molecule is not serious if there are other copies of that particular one in the cell, and additional ones can be made from its gene. However, large doses of radiation may damage too many molecules so that new copies cannot be made quickly enough, and the cell dies. Damage to the DNA is more serious, since a cell may have only one copy. Each alteration in the DNA affects a gene and can alter the molecule it codes for, so that needed proteins or other materials cannot be made at all. Again the cell may die. The death of a single cell is not normally a problem, since the body can replace it with a new one. (There are exceptions, such as neurons, which are *not* replaceable, so their loss is serious.) But if many cells die, the organism may not be able to recover. On the other hand, a cell may survive but be defective. It may go on dividing and produce many more defective cells to the detriment of the whole organism. Thus radiation can cause cancer—the rapid production of defective cells.

Radiation damage to biological organisms is often separated into categories according to its location in the body: "somatic" and "genetic." Somatic damage refers to that in any part of the body except the reproductive organs. Somatic damage can seriously affect that particular organism, causing cancer and at high doses radiation sickness (characterized by nausea, fatigue, loss of body hair, and other symptoms) or even death. Genetic damage refers to damage to the reproductive apparatus and so affects an individual's offspring. Damage to the genes results in mutation, the majority of which are harmful; and mutations, if they occur in the reproductive organs, are transmitted to future generations. Radiation, including that from diagnostic use of X rays, is commonplace, and its effect on the future of the human race is a cause of great concern.

The passage through matter (including the human body) of charged particles, such as α and β rays, as well as γ and X rays, can ionize atoms and molecules and cause considerable damage. Radiation can also be used to treat certain diseases, particularly cancer. It is therefore important to be able to quantify the amount, or *dose*, of radiation that passes through a material. This is the subject of *dosimetry*. We now discuss the more important ways of measuring radiation.

The strength of a source can be specified at a given time by stating the *source activity*, or how many disintegrations occur per second. The traditional unit is the *curie* (Ci), defined as

$$1 \text{ Ci} = 3.70 \times 10^{10} \text{ disintegrations per second.}$$

Although the curie is still in common use, the proper SI unit for source activity is the becquerel (Bq), defined as

$$1 \text{ Bq} = 1 \text{ disintegration/s.}$$

Of course commercial suppliers of *radionuclides* (radioactive nuclides) specify the activity at a given time; since the activity decreases in time, particularly for short-lived isotopes, it is important to take this into account.

The source activity ($\Delta N/\Delta t$) is related to the half-life, $T_{\frac{1}{2}}$, by (see Section 29-8):

$$\frac{\Delta N}{\Delta t} = \lambda N = \frac{0.693}{T_{\frac{1}{2}}} N.$$

◻   EXAMPLE 30-4   In a certain experiment, 0.016 μCi of $^{32}_{15}$P is injected into a medium containing a culture of bacteria. After 1 h, the cells are washed and a detector that is 70 percent efficient (counts 70 percent of emitted β rays) records 720 counts per minute from all the cells. What percentage of the original $^{32}_{15}$P was taken up by the cells?

SOLUTION   The total number of disintegrations per second originally was $(0.016 \times 10^{-6})(3.7 \times 10^{10}) = 590$. The counter could be expected to count 70 percent of this or 410 per second. Since it counted $720/60 = 12$ per second, then $12/410 = 0.029$ or 2.9 percent was incorporated into the cells.

Another type of measurement, in contrast to the activity of the source as just discussed, is the exposure or *absorbed dose*—that is, the effect the radiation has on the absorbing material. The earliest unit of dosage was the *roentgen* (R) which was defined in terms of the amount of ionization produced by the radiation. Today 1 R is defined as the amount of X or γ radiation that deposits 0.878×10^{-2} J of energy per kilogram of air.

The roentgen was largely superseded by another unit of absorbed dose applicable to any type of radiation, the rad: *1 rad is that amount of radiation which deposits 1.00×10^{-2} J/kg of any absorbing material.* (This is quite close to the roentgen for X and γ rays.) The proper SI unit for absorbed dose is the gray (Gy):

$$1 \text{ Gy} = 1 \text{ J/kg} = 100 \text{ rad},$$

and is slowly coming into use. The absorbed dose depends not only on the strength of a given radiation beam (number of particles per second) and the energy per particle, but also on the type of material that is absorbing the radiation. Since bone, for example, is denser than flesh and absorbs more of the radiation normally used, the same beam passing through a human body deposits a greater dose (in rads or grays) in bone than in flesh.

The gray and the rad are physical units of dose—the energy deposited per unit mass of material. They are, however, not the most meaningful units for measuring the biological damage produced by radiation. This is because equal doses of different types of radiation cause differing amounts of damage. For example, 1 rad of α radiation does 10 to 20 times the amount of damage as does 1 rad of β or γ rays. This is largely due to the fact that α rays (and other heavy particles such as protons and neutrons) move much more slowly than equal energy β and γ rays because of their greater mass. Hence ionizing collisions occur closer together so more irreparable damage is done. The *relative biological effectiveness* (RBE) or *quality factor* (QF) of a given type of radiation is defined as the number of rads of X or γ radiation that produces the same biological damage as 1 rad of the given radiation. The QF for several types of radiation is given in Table 30-1. The numbers are approximate since they depend somewhat on the energy of the particles and on the type of damage that is used as the criterion.

The product of the dose in rads and the QF gives a unit known as the *rem* (which stands for *rad equivalent man*):

$$\text{effective dose (in rem)} = \text{dose (in rad)} \times \text{QF}$$

This unit is being replaced by the SI unit for "effective dose," the sievert (Sv):

$$\text{effective dose (Sv)} = \text{dose (Gy)} \times \text{QF}.$$

By this definition, 1 rem of any type of radiation does approximately the same amount of biological damage. For example, 50 rem of fast neutrons does the same damage as 50 rem of γ rays. But note that 50 rem of fast neutrons is only 5 rads whereas 50 rem of γ rays is 50 rads.

We are constantly exposed to low-level radiation from natural sources: cosmic rays, natural radioactivity in rocks and soil, and naturally occurring radioactive isotopes that occur in our food such as $^{40}_{19}\text{K}$. The natural radioactive background supplies about 0.13 rem per year per person on the average. From medical X rays the average person receives about 0.07 rem per year. The U.S. government specifies the recommended upper limit of allowed radiation for an individual in the general populace at about 0.5 rem per year, exclusive of natural sources. However, since even low doses of radiation are believed by many scientists to increase the chances of cancer or genetic defects, the altitude today is to keep the radiation dose as low as possible.

People who work around radiation—in hospitals, in power plants, in research—often are subjected to much higher doses than 0.5 rem/yr. The upper limit for such occupational exposures has been set somewhat higher, on the order of 5 rem/yr whole-body dose (presumably because such people know what they are getting into).

TABLE 30-1
Quality factor (QF) of different kinds of radiation

Type	QF
X and γ rays	≈ 1
β (electrons)	≈ 1
Fast protons	1
Slow neutrons	≈ 3
Fast neutrons	Up to 10
α particles and heavy ions	Up to 20

Large doses of radiation can cause reddening of the skin, drop in white-blood-cell count, and a large number of unpleasant symptoms such as nausea, fatigue, and loss of body hair. Such effects are sometimes referred to as radiation sickness. Large doses can also be fatal, although the time span of the dose is important. A short dose of 1000 rem is nearly always fatal. A 400-rem dose in a short period of time is fatal in 50 percent of the cases. However, the body possesses remarkable repair processes so that a 400-rem dose spread over several weeks is not usually fatal. It will, nonetheless, cause considerable damage to the body.

☐ EXAMPLE 30-5 What whole-body dose is received by a 70-kg patient exposed to a 1000-Ci $^{60}_{27}$Co source if 2.0 percent of the γ rays reach the patient? $^{60}_{27}$Co emits γ rays of energy 1.33 MeV and 1.17 MeV in equal amounts. Approximately 50 percent of the γ rays interact in the body and deposit all their energy. (The rest pass through.)

SOLUTION The average γ-ray energy is 1.25 MeV so the total energy passing through the body is $(1000 \text{ Ci})(3.7 \times 10^{10} \text{ decays/Ci·s}) \cdot (1.25 \text{ MeV})(0.020) = 9.3 \times 10^{11}$ MeV/s. (The factor 0.020 represents the 2.0 percent that reach the body.) Only half this energy is deposited in the body, so the dose rate is $(\frac{1}{2})(9.3 \times 10^{11} \text{ MeV/s})/(1.6 \times 10^{-13} \text{ J/MeV}) = 7.4 \times 10^{-2}$ J/s. Since 1 Gy = 1 J/kg, the whole-body dose rate is $(7.4 \times 10^{-2} \text{ J/s})/(1 \text{ J/kg·Gy})(70 \text{ kg}) = 1.1 \times 10^{-3}$ Gy/s.

*30-6 ▌ Radiation therapy

The applications of radioactivity and radiation to human beings and other organisms is a vast field that has filled many books. This subject is often called **radiation biology,** and when applied in medicine it is called **nuclear medicine.** Nuclear medicine involves two basic aspects: (1) radiation *therapy*—the treatment of disease—which we discuss in this section; and (2) the *diagnosis* of disease, which we discuss in the following sections of this chapter.

Radiation can cause cancer. It can also be used to treat it. Rapidly growing cancer cells are especially susceptible to destruction by radiation. Nonetheless, large doses are needed to kill the cancer cells, and some of the surrounding normal cells are inevitably killed as well. It is for this reason that cancer patients receiving radiation therapy often suffer side effects characteristic of radiation sickness. To minimize the destruction of normal cells, a narrow beam of γ or X rays is often used when the cancerous tumor is well localized. The beam is directed at the tumor, and the source (or body) is rotated so that the beam passes through various parts of the body to keep the dose at any one place as low as possible—except at the tumor and its immediate surroundings, where the beam passes at all times. The radiation may be from a radioactive source such as $^{60}_{27}$Co; or it may be from an X-ray machine that produces photons in the range 200 keV to 5 MeV. Protons, neutrons, electrons, and pions, which are produced in particle accelerators (Section 31-2), are also being used in cancer therapy.

In some cases a tiny radioactive source may be inserted directly inside a tumor; this will eventually kill the majority of the cells. A similar technique is used to treat cancer of the thyroid with the radioactive isotope $^{131}_{53}I$. The thyroid gland tends to concentrate any iodine present in the bloodstream; so when $^{131}_{53}I$ is injected into the blood, it becomes concentrated in the thyroid, particularly in any area where abnormal growth is taking place. The intense radioactivity emitted can then destroy the defective cells.

Although radiation can increase the lifespan of many patients, it is not completely effective. It is not always possible to kill all the diseased cancer cells, so a recurrence of the disease is possible. Many cases, especially when the cancerous cells are not well localized in one area, are difficult to treat at all without seriously damaging the rest of the organism.

Another application of radiation is for sterilizing bandages, surgical equipment, and even packaged foods since bacteria and viruses can be killed or deactivated by large doses of radiation.

*30-7 ∎ Tracers and imaging in research and medicine

Radioactive isotopes are commonly used in biological and medical research as *tracers*. A given compound is artificially synthesized using a radioactive isotope such as $^{14}_{6}C$ or $^{3}_{1}H$. Such "tagged" molecules can then be traced as they move through an organism or as they undergo chemical reaction. The presence of these tagged molecules (or parts of them, if they undergo chemical change) can be detected by a Geiger or scintillation counter, which detects emitted radiation (see Section 29-12). The details of how food molecules are digested, and to what parts of the body they are diverted, can be traced in this way. Radioactive tracers have been used to determine how amino acids and other essential compounds are synthesized by organisms. The permeability of cell walls to various molecules and ions can be determined using radioactive isotopes: the tagged molecule or ion is injected into the extracellular fluid and the radioactivity present inside and outside the cells is measured as a function of time.

In many cases, the cells or chemicals involved are removed from the organism and placed in the Geiger or scintillation counter. To follow a process in time, different sets of organisms are treated for different periods of time before the cells or chemicals are extracted. (See Example 30-4.)

In a technique known as *autoradiography*, the position of the radioactive isotopes is detected on film. For example, the distribution of carbohydrates produced in the leaves of plants from absorbed CO_2 can be observed by keeping the plant in an atmosphere where the carbon atom in the CO_2 is $^{14}_{6}C$. After a time, a leaf is placed firmly on a photographic plate and the emitted radiation darkens the film most strongly where the isotope is most strongly concentrated (Fig. 30-6). Autoradiography using labeled nucleotides (that form DNA) has revealed much about the details of DNA replication (Fig. 30-7).

For medical diagnosis, the radionuclide most often used today is $^{99m}_{43}Tc$, a long-lived excited state of technicium-99 (the "m" in the symbol

FIGURE 30-6

Autoradiograph of a mature leaf of the squash plant *Cucurbita melopepo* exposed for 30 s to $^{14}CO_2$. The photosynthetic (green) tissue has become radioactive; the nonphotosynthetic tissue of the veins is free of ^{14}C and therefore does not blacken the x-ray sheet. This technique is very useful in following patterns of nutrient transport in plants. (Courtesy of R. Turgeon.)

stands for "metastable" state). It is formed when $^{99}_{42}$Mo decays. The great usefulness of $^{99m}_{43}$Tc derives from its convenient half-life of 6 h (short, but not too short) and the fact that it can combine with a large variety of compounds. The compound to be labeled with the radionuclide is so chosen because it concentrates in the organ or region of the anatomy to be studied. Detectors outside the body then record, or image, the distribution of the radioactively labeled compound. The detection can be done by a single detector (Fig. 30-8) which is moved across the body, measuring the intensity of radioactivity at a large number of points. The image represents the relative intensity of radioactivity at each point. The relative radioactivity is a diagnostic tool; for example, high or low radioactivity may represent overactivity or underactivity of an organ or part of an organ, or in another case may represent a lesion or tumor. More complex *gamma cameras* make use of many detectors which simultaneously record the radioactivity at a variety of points. The measured intensities can be displayed on a CRT (TV monitor or oscilloscope screen), and allow "dynamic" studies (that is, images that change in time) to be performed.

*30-8 ▌ Emission Tomography

The images formed using the standard techniques of nuclear medicine, as briefly discussed in the previous section, are produced from radioactive tracer sources within the *volume* of the body. It is also possible to image the radioactive emissions in a single plane or slice through the body using the computed tomography techniques discussed in Section 24-12. A basic gamma camera is moved around the patient to measure the radioactive intensity from the tracer at many points and angles; the data are processed in much the same way as for X-ray CT scans (Section 24-12). This is referred to as *single photon emission tomography* (SPET).

Another recently developed technique is *positron emission tomography* (PET), which makes use of positron emitters such as $^{11}_{6}$C, $^{13}_{7}$N, $^{15}_{8}$O, and $^{18}_{9}$F. When such a nuclide β-decays, the emitted positron travels at most a few millimeters before it collides with a normal electron. In this collision, the

FIGURE 30-7

An autoradiogram of a fiber of chromosomal DNA isolated from the higher plant *Arabidopsis thaliana*. The dashed arrays of silver grains show the Y-shaped growing point of replicating DNA. (Courtesy of J. Van't Hof.)

FIGURE 30-8

Collimated gamma ray detector for scanning (moving) over patient. The collimator is necessary to select γ rays that come in a straight line from the patient. Without the collimator, γ rays from all parts of the body could strike the scintillator producing a very poor image.

Photomultiplier tube

Scintillator crystal

Lead collimator

Collimating hole

Patient

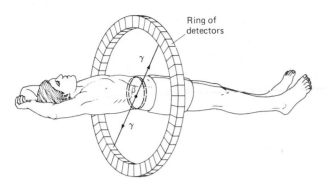

FIGURE 30-9

Positron emission tomography system showing a ring of detectors to detect the two annihilation γ rays $(e^+ e^- \rightarrow 2\gamma)$ emitted at $180°$ to each other.

positron and electron are annihilated, producing two γ rays $(e^+ e^- \rightarrow 2\gamma)$ each having 510 keV. The two γ rays fly off in opposite directions $(180° \pm 0.25°)$ since they must have almost exactly equal and opposite momenta to conserve momentum (the momenta of the e^+ and e^- are essentially zero compared to the momenta of the γ rays). Because the photons travel along the same line in opposite directions, their detection in coincidence by rings of detectors surrounding the patient (Fig. 30-9) readily establishes the line along which the emission took place. If the difference in time of arrival of the two photons could be determined accurately, the actual position of the emitting nuclide on that line could be calculated; present-day electronics can measure times to at best ± 300 ps, so at the γ ray's speed $(c = 3 \times 10^8 \text{ m/s})$, the actual position could be determined only to about $(3 \times 10^8 \text{ m/s}) \cdot (300 \times 10^{-12} \text{ s}) \approx 10$ cm. Although there may be future potential for time-of-flight measurements to determine position, today computed tomography techniques are used instead, similar to those for X-ray CT, which can reconstruct PET images with a resolution on the order of 1 cm. One big advantage of PET is that no collimators are needed (as for detection of a single photon—see Fig. 30-8). Thus fewer photons are "wasted" and lower doses can be administered to the patient with PET.

Both PET and single-photon-emission systems can give images related to biochemistry, metabolism, and function. This is to be compared to X-ray CT scans (Section 24-12) whose images reflect shape and structure—that is, the anatomy of the imaged region.

*30-9 ∎ Nuclear magnetic resonance (NMR) and NMR imaging

Nuclear magnetic resonance (NMR) is a phenomenon which soon after its discovery in 1946 became a powerful research tool in a variety of fields from physics to chemistry and biochemistry. Presently it is also being developed into an important medical imaging technique. We first briefly discuss the phenomenon, and then we will look at its applications.

We saw in Chapter 19 (Section 19-8) that a loop of wire carrying an electric current has a torque on it when placed in a magnetic field; the loop is said to have a magnetic moment, and the torque on the loop tends to align it so the direction of its magnetic moment (taken to be perpendicular to the plane of the coil) is parallel to the field, **B**. (The coil's plane is perpendicular to **B**.) When the coil is so aligned, it has its lowest potential energy. Then in Chapter 27 (Section 27-6) we saw that atomic electrons in their orbital motions act as tiny current loops, and furthermore that they have an intrinsic angular momentum called *spin*, which also acts like the magnetic moment of a current loop. When atoms are placed in a magnetic field, atomic energy levels split into several closely spaced levels (see Fig. 27-4).

Nuclei, too, exhibit these magnetic properties. Many nuclides have magnetic moments, but we will examine only the simplest, the hydrogen (H) nucleus, since it is the one most used even for medical imaging. The $^1_1$H nucleus consists of a single proton. Its spin angular momentum (and its magnetic moment), like that of the electron, can take on only two values; when placed in a magnetic field we call these "spin up" (parallel to the field) and "spin down" (antiparallel to the field), as suggested in Fig. 30-10. When a magnetic field is present, the energy of the nucleus splits into two levels as shown in Fig. 30-11, with the spin-up (parallel to field) having the lower energy. (This is like the Zeeman effect for atomic levels—see Fig. 27-4.) The difference in energy, ΔE, between these two levels is proportioned to the total magnetic field, B_T, at the nucleus:

$$\Delta E = kB_T$$

where k is a proportionality constant that is different for different nuclides.

In a standard nuclear magnetic resonance (NMR) setup, the sample to be examined is placed in a static magnetic field. A radiofrequency (RF) pulse of electromagnetic radiation (that is, photons) is applied to the sample. If the frequency, f, of this pulse corresponds precisely to the energy difference between the two energy levels (Fig. 30-11), so that

$$hf = \Delta E = kB_T, \tag{30-4}$$

then the photons of the RF beam will be absorbed, exciting many of the nuclei from the lower state to the upper state. (This is a resonance phenomenon since there is significant absorption only if f is very near $f = kB_T/h$; hence the name "nuclear magnetic resonance.") For $^1_1$H, the frequency is 42.57 MHz for a field $B_T = 1.0$ T. (For $^{31}_{15}$P, for example, it is $f = 17.24$ MHz.) This is the frequency for free H nuclei. If the H atoms are bound in a molecule, the total magnetic field B_T at the H nuclei will be the sum of the external applied field plus the magnetic field due to electrons and nuclei of neighboring atoms. Since f is proportional to B_T, the value of f will be slightly different for the bound H atoms than free atoms for the same external field. This change in frequency, which can be measured, is called the "chemical shift." A great deal has been learned about the structure of molecules and bonds using this NMR technique.

FIGURE 30-10
Schematic picture of electron represented in magnetic field **B** (pointing upward) with its two possible states of spin, up and down.

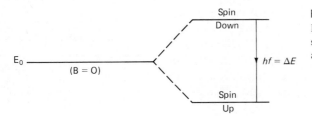

FIGURE 30-11
Energy E_0 in the absence of a magnetic field splits into two levels in the presence of a magnetic field.

Magnets

RF coils

FIGURE 30-12
Typical NMR imaging setup.

Now let us look at the new and still developing technique of medical imaging using NMR. Because hydrogen is the commonest element in the human body and gives the strongest NMR signals, it is the element most often used. The experimental apparatus is shown in Fig. 30-12. The large coils set up the static magnetic field, and the RF coils produce the RF pulse of electromagnetic waves (photons) that causes the nuclei to jump from the lower state to the upper one (Fig. 30-11). These same coils (or another coil) can detect the absorption of energy, or the emitted radiation (also of frequency $f = \Delta E/h$) when the nuclei jump back down to the lower state.

The formation of a two-dimensional or three-dimensional image can be done using the techniques of computed tomography similar to those for X rays as discussed in Section 24-12. The simplest thing to measure for creating an image is the intensity of absorbed and/or reemitted radiation from many different points of the body, and this would be a measure of the density of H atoms at each point. But how do we determine from what part of the body a given photon comes? One technique is to give the static magnetic field a gradient; that is, instead of applying a uniform magnetic field, B_T, the field is made to vary with position across the width of the sample (or patient). Since the frequency absorbed by the H nuclei is proportional to B_T (Eq. 30-4), only one plane within the body will have the proper value of B_T to absorb photons of a particular frequency f; by varying f, absorption by different planes can be measured. Alternately, if the field gradient is applied *after* the pulse, the frequency of the emitted photons will be a measure of where they were emitted. See Fig. 30-13. If a magnetic field gradient in one direction is applied during excitation (absorption of photons) and photons of a single frequency are transmitted, only H nuclei in one thin slice will be excited; by applying a gradient in a different direction, perpendicular to the first, during reemission, the frequency f of the reemitted radiation will represent depth in that slice. Other ways of varying the magnetic field throughout the volume of the body can be used in order to correlate NMR frequency with position.

FIGURE 30-13

A static field that is stronger at the bottom than the top. The frequency of absorbed, or emitted, light is proportional to B in NMR.

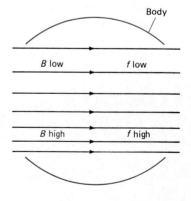

Body

B low f low

B high f high

TABLE 30-2

Technique	Where discussed in this book	Resolution
Conventional X ray	Section 24-12	$\frac{1}{2}$ mm
CT scan, X-ray	Section 24-12	$\frac{1}{2}$ mm
Nuclear Medicine	Section 30-7	1 cm
SPET (single photon)	Section 30-8	1 cm
PET (positron emission)	Section 30-8	1 cm
NMR	Section 30-9	2 mm
Ultrasound	Section 14-10	2 mm

A reconstructed image based on the density of H atoms (that is, the intensity of absorbed or emitted radiation) is not very interesting. Much more useful are images based on the rate at which the nuclei decay back to the ground state. The nuclei, after being excited to the upper state of Fig. 30-11, descend to the ground state one by one, just as in radioactive decay; for a collection of many excited nuclei, the decay rate is exponential, just as for radioactive decay (Fig. 29-6), except that instead of one decay constant, there are two of them; they are usually expressed as "relaxation times," T_1 and T_2. These relaxation times can be determined for each point (picture element) of the image. Images using this information can have very good contrast, providing a great deal of information even in soft tissue.

This NMR technique (sometimes called "spin-echo") is producing images of great diagnostic value, both in the imaging of structure (anatomy) as well as in the functional and biochemical aspects. An NMR image is shown in Fig. 30-14.

NMR imaging is considered to be noninvasive. We can calculate the energy of the photons involved: as mentioned earlier, in a 1.0 T magnetic field, $f = 42.57$ MHz for $^1_1$H. This corresponds to an energy of $hf = (6.6 \times 10^{-34} \text{ J·s})(43 \times 10^6 \text{ Hz}) \approx 3 \times 10^{-26}$ J or about 10^{-7} eV. Since molecular bonds are on the order of 1 eV, it is clear that the RF photons can cause little cellular disruption; this should be compared to X or γ rays whose energies are 10^4 to 10^6 eV and thus do cause significant damage. The static magnetic fields, though often large (~ 0.1 to 1 T), are believed to be harmless (except for people wearing heart pacemakers), although not a great deal of research has been done in this area, so caution is necessary.

NMR imaging has the great advantage over X-ray CT scanning of being less invasive. However, it does not quite reach, as yet, the best resolution of standard CT scanning. Table 30-2 lists the recently developed techniques we have discussed for imaging the interior of the body, along with the optimum resolution attainable today. Of course resolution is only one factor that must be considered; it must be remembered that the different imaging techniques provide different types of information, useful for different types of diagnosis.

FIGURE 30-14

NMR image of a vertical section through the head, revealing the sinuses. (Reproduced with permission, © 1984, Howard Sochurek.)

▌ Summary

A *nuclear reaction* occurs when two nuclei collide and two or more other nuclei (or particles) are produced. In this process, as in radioactivity, *transmutation* (change) of elements occurs.

In *fission* a heavy nucleus such as uranium splits into two intermediate-sized nuclei after being struck by a neutron. $^{235}_{92}U$ is fissionable by slow neutrons, whereas some fissionable nuclei require fast neutrons. Much energy is released in fission because the binding energy per nucleon is lower for heavy nuclei than it is for intermediate-sized nuclei, so the mass of a heavy nucleus is greater than the total mass of its fission products. The fission process releases neutrons so that a *chain reaction* is possible. The *critical mass* is the minimum mass of fuel needed to sustain a chain reaction. In a *nuclear reactor* or nuclear bomb, a moderator is needed to slow down the released neutrons.

The *fusion* process, in which small nuclei combine to form larger ones, also releases energy. It has not yet been possible to build a fusion reactor for power generation, because of the difficulty in containing the fuel long enough at the high temperature required.

Radiation can cause damage to biological tissue, and it is important to quantify amounts of radiation; this is the subject of *dosimetry*. The curie (Ci) and the becquerel (Bq) are units that measure the activity or rate of decay of a sample: 1 Ci = 3.70×10^{10} disintegrations per second whereas 1 Bq = 1 disintegration/s. The rad measures the amount of energy deposited per unit mass of absorbing material: 1 rad is the amount of radiation that deposits 10^{-2} J/kg of material. The SI unit of absorbed dose in the gray: 1 Gy = 1 J/kg. The *rem* = rad × QF where QF is the "quality factor" of a given type of radiation; 1 rem of any type of radiation does approximately the same amount of biological damage. The average dose received per person per year is about 0.20 rem. The SI unit for effective dose is the sievert: 1 sv = 10^2 rem.

❚ Questions

1. Fill in the missing particles or nuclei: (*a*) $^{137}_{56}Ba(n, \gamma)$?; (*b*) $^{137}_{56}Ba(n, ?)$ $^{137}_{55}Cs$; (*c*) $^{2}_{1}H(d, ?)$ $^{4}_{2}He$; (*d*) $^{197}_{79}Au(\alpha, d)$?.

2. The isotope $^{32}_{15}P$ is produced by an (*n*, *p*) reaction. What must be the target nucleus?

3. When $^{22}_{11}Na$ is bombarded by deuterons ($^{2}_{1}H$) an α particle is emitted. What is the resulting nuclide?

4. Why are neutrons such good projectiles for producing nuclear reactions?

5. A proton strikes a $^{20}_{10}Ne$ nucleus, and an α particle is observed to come out. What is the residual nucleus? Write down the reaction equation.

6. Are fission fragments β^+ or β^- emitters?

7. If $^{235}_{92}U$ released only 1.5 neutrons per fission on the average, would a chain reaction be possible? What would be different?

8. $^{235}_{92}U$ releases an average of 2.5 neutrons per fission compared to 2.7 for $^{239}_{94}Pu$. Pure samples of which of these two nuclei do you think would have the smaller critical mass?

9. The energy from nuclear fission appears in the form of thermal energy—but the thermal energy of what?

10. Why would a porous block of uranium be more likely to explode if kept under water rather than in air?

11. A reactor that uses highly enriched uranium can use ordinary water (instead of heavy water) as a moderator and still have a self-sustaining chain reaction. Explain.

12. Why must the fission process release neutrons if it is to be useful?

13. Discuss the relative merits and disadvantages, including pollution and safety, of power generation by fossil fuels, nuclear fission, and nuclear fusion.

14. What is the reason for the "secondary system" in Fig. 30-5? That is, why is the water heated by the fuel in a nuclear reactor not used directly to drive the turbines?

15. Discuss how the course of history might have been changed if, during World War II, scientists had refused to work on developing a nuclear bomb. Do you think it would have been possible to delay the building of a bomb indefinitely?

16. Research in molecular biology is moving toward the ability to perform genetic manipulations on human beings. The moral implications of future discoveries along these lines has led to a warning that this may be the molecular biologists' "Hiroshima." Discuss.

17. Does $E = mc^2$ apply in (*a*) fission, (*b*) fusion, (*c*) nuclear reactions?

18. Light energy emitted by the sun and stars comes from the fusion process. What conditions in the interior of stars makes this possible?

19. What is the basic difference between fission and fusion?

20. Why is the recommended maximum radiation dose higher for women beyond the child-bearing age than for younger women?

*21. How might radioactive tracers be used to find a leak in a pipe?

*22. How might radioactive tracers be used to measure (*a*) engine wear, (*b*) tire wear?

▌Problems

SECTION 30-1

1. (I) Determine whether the reaction $^2_1H(d, n)^3_2He$ requires a threshold energy.

2. (I) Is the reaction $^{238}_{92}U (n, \gamma)^{239}_{92}U$ possible with slow neutrons? Explain.

3. (II) Does the reaction $^7_3Li(p, \alpha)^4_2He$ require energy or does it release energy? How much energy?

4. (II) (a) Can the reaction $^{24}_{12}Mg(n, d)^{23}_{11}Na$ occur if the bombarding particles have 10.00 MeV of KE? (b) If so, how much energy is released?

5. (II) (a) Can the reaction $^7_3Li(p, \alpha)^4_2He$ occur if the incident proton has KE = 500 keV? (b) If so, what is the total kinetic energy of the products?

6. (II) In the reaction $^{14}_7N(\alpha, p)^{17}_8O$, the incident α particles have 7.68 MeV of kinetic energy. (a) Can this reaction occur? (b) If so, what is the total kinetic energy of the products? The mass of $^{17}_8O$ is 16.999131 u.

7. (III) Use conservation of energy and momentum to show that a bombarding proton must have an energy of 3.23 MeV in order to make the reaction $^{13}_6C(p, n)^{13}_7N$ occur. (See Example 30-2.)

8. (III) Show, using the laws of conservation of energy and momentum, that for a nuclear reaction requiring energy the minimum kinetic energy of the bombarding particle (the *threshold energy*) is equal to $[Qm_{pr}/(m_{pr} - m_b)]$, where Q is the energy required (difference in total mass between products and reactants), m_b is the rest mass of the bombarding particle and m_{pr} the total rest mass of the products. Assume the target nucleus is at rest before an interaction takes place.

SECTION 30-2

9. (I) Calculate the energy released in the fission reaction $^1_0n + ^{235}_{92}U \rightarrow ^{88}_{38}Sr + ^{136}_{54}Xe + 12^1_0n$. Use Appendix D, and assume the initial KE of the neutron is very small.

10. (I) What is the energy released in the fission reaction of Eq. 30-2? (The masses of $^{141}_{56}Ba$ and $^{92}_{36}Kr$ are 140.9141 u and 91.9250 u, respectively.)

11. (I) How many fissions take place per second in a 25-MW reactor? Assume that 200 MeV is released per fission.

12. (II) One means of enriching uranium is by diffusion of the gas UF_6. Calculate the ratio of the speeds of molecules of this gas containing $^{235}_{92}U$ and $^{238}_{92}U$, on which this process depends.

13. (II) Suppose that the average power consumption, day and night, of an average house is 300 W. What mass of $^{235}_{92}U$ would have to undergo fission to supply the electrical needs of such a house for a year? (Assume 200 MeV is released per fission.)

14. (II) What mass of $^{235}_{92}U$ is required to operate a 500-MW reactor for 1 yr?

15. (II) What was the mass loss of $^{235}_{92}U$ actually fissioned in the first atomic bomb, whose energy was the equivalent of about 20 kilotons of TNT (1 kiloton of TNT releases 5×10^{12} J)?

16. (III) Suppose that the neutron multiplication factor is 1.0004. If the average time between successive fissions in a chain of reactions is 1.0 ms, by what factor will the reaction rate increase in 1.0 s?

17. (III) Consider a system of nuclear power plants that produce 4000 MW. (a) What total mass of $^{235}_{92}U$ fuel would be required to operate these plants for 1 yr, assuming that 200 MeV is released per fission? (b) Typically 6 percent of the $^{235}_{92}U$ nuclei that fission is converted to $^{90}_{38}Sr$, a β^- emitter with a half-life of 29 yr. What is the total radioactivity of the $^{90}_{38}Sr$, in curies, after its production? (Neglect the fact that some of it decays during the 1-yr period.)

SECTION 30-3

18. (I) What is the average kinetic energy of protons at the center of a star where the temperature is 10^7 K?

19. (II) If an average house requires 300 W of electric power on average, how much deuterium would have to be used in a year to supply these electrical needs? Assume the reaction of Eq. 30-3b.

20. (II) Show that the energy released in the fusion reaction $^2_1H + ^3_1H \rightarrow ^4_2He + n$ is 17.59 MeV.

21. (II) Show that the energy released when two deuterium nuclei fuse to form ^3_2He with the release of a neutron is 3.27 MeV.

22. (II) Calculate the energy release per gram of fuel for the reaction of Eq. 30-3b. Compare to the energy release per gram of uranium in the fission process.

23. (II) Calculate the energy release per gram of fuel for the reactions of Eqs. 30-3a and 30-3c. Compare to the energy release per gram of uranium in the fission process.

24. (II) In the so-called carbon cycle that occurs in the sun, ^4_2He is built from four protons starting with $^{12}_6C$. First $^{12}_6C$ absorbs a proton to form nucleus X_1. X_1 decays by β^+ emission to X_2. X_2 absorbs a proton to become X_3, which itself absorbs a proton to become X_4. X_4 decays to X_5 by β^+ decay and X_5 reacts via $X_5(p, \alpha)X_6$. (a) Determine the intermediate nuclei and write out each step in detail, and show that X_6 is again $^{12}_6C$, which is thus not used up in the process. (b) Determine the energy release for each step.

25. (III) How much energy (J) is contained in 1.00 kg of water if the natural deuterium is used in the fusion reaction of Eq. 30-3a? Compare this to the energy obtained from the burning of 1.00 kg of gasoline, about 5×10^7 J.

26. (I) A 0.018-μCi sample of $^{32}_{15}$P is injected into an animal for tracer studies. If a Geiger counter intercepts 20 percent of the emitted β particles and is 90 percent efficient in counting them, what will be the counting rate?

27. (I) An average adult body has about 0.10 μCi of $^{40}_{19}$K, which comes from food. How many decays occur per second?

28. (I) A dose of 500 rem of γ rays in a short period would be lethal to about half the people subjected to it. How many rads is this?

29. (I) Fifty rads of α-particle radiation is equivalent to how many rads of X rays, in terms of biological damage?

30. (I) How many rads of slow neutrons will do as much biological damage as 50 rad of fast neutrons?

31. (I) How much energy is deposited in the body of a 70-kg adult exposed to a 50-rad dose?

32. (II) A 1.0-mCi source of $^{32}_{15}$P (in NaHPO$_4$), a β^- emitter, is implanted in an organ where it is to administer 5000 rad. The half-life of $^{32}_{15}$P is 14.3 days and 1 mCi delivers about 1 rad/min. Approximately how long should the source remain implanted?

33. (II) About 35 eV is required to produce one ion pair in air. Calculate the number of ion pairs produced per cm^3 when air absorbs 1 R (roentgen) of X rays.

34. (II) $^{57}_{27}$Co emits 122-keV γ rays. If a 70-kg person swallowed 2.0 μCi of $^{57}_{27}$Co, what would be the dose rate (rad/day) averaged over the whole body? Assume that 50 percent of the γ-ray energy is deposited in the body. (Hint: determine the rate of energy deposited in the body and use the definition of the rad.)

35. (II) What is the mass of a 1.0-μCi $^{14}_{6}$C source?

36. (II) In a certain town the average yearly background radiation consists of 25 mrad of X and γ rays plus 3.0 mrad of particles having a QF of 10. How many rems will a person receive per year on the average?

37. (II) A shielded γ-ray source yields a dose rate of 0.050 rad/h at a distance of 1.0 m for an average-sized person. If workers are allowed a maximum dose rate of 5.0 rem/yr, how close to the source may they operate assuming a 40-h work week? Assume that the intensity of radiation falls off as the square of the distance. (It actually falls off more rapidly than $1/r^2$ because of absorption in the air, so the answer above will give a better-than-permissible value.)

31

ELEMENTARY PARTICLES

In this, the final chapter, we discuss the exciting subject of *elementary particle* physics, which represents the human endeavor to understand the basic building blocks of all matter.

In the years after World War II, it was found that if the incoming particle in a nuclear reaction has sufficient energy, new types of particles can be produced. In order to produce high-energy particles, physicists have constructed various types of particle accelerators. Most commonly they accelerate protons or electrons, although heavy ions can also be accelerated depending on the design. These high-energy accelerators have been used to probe the nucleus more deeply, to produce and study new particles, and to give us information about the basic forces and constituents of nature.

31-1 ■ High-energy projectiles

Particles accelerated to high speeds are projectiles that can probe the interior of nuclei they strike. An important factor is that faster moving projectiles can reveal more detail. The wavelength of projectile particles is given by de Broglie's wavelength formula

$$\lambda = \frac{h}{mv},$$

from which we see that the greater the momentum of the bombarding particle, the shorter the wavelength and the more detail that can be obtained; as discussed in Chapter 24, resolution of details is limited by the wavelength, so we see one reason why particle accelerators of higher and higher energy have been built in recent years.

□   EXAMPLE 31-1   To explore the distribution of charge within nuclei, very-high-energy electrons are used. (Electrons are used rather than protons because they do not partake in the strong nuclear force, so only the electric force is involved.) Experiments at the Stanford linear accelerator using electrons with KE of 1.3 GeV obtained the charge distribution for the bismuth nucleus shown in Fig. 31-1. What is the expected resolution—that is, the size of the smallest details that can be detected?

FIGURE 31-1

Distribution of electric charge for bismuth nucleus, determined using 1.3-GeV electrons.

Distance from center of nucleus (fm)

SOLUTION   1.3 GeV = 1300 MeV which is about 2500 times the mass of the electron. We are clearly dealing with relativistic speeds here, and it is easily shown that the speed of the electron is nearly $c = 3.0 \times 10^8$ m/s. (From Eq. 25-4, $\text{KE} = mc^2 - m_0c^2 \approx mc^2$; and from Eq. 25-7, $E^2 = p^2c^2 + m_0^2c^4 \approx p^2c^2$ since m_0c^2 is small compared to pc; hence $p^2c^2 = m^2v^2c^2 \approx m^2c^4$, so $v \approx c$.) Therefore

$$\lambda = \frac{h}{mv} \approx \frac{h}{mc} = \frac{hc}{mc^2}$$

where $mc^2 = 1.3$ GeV. Hence

$$\lambda = \frac{(6.6 \times 10^{-34} \text{ J·s})(3.0 \times 10^8 \text{ m/s})}{(1.3 \times 10^9 \text{ eV})(1.6 \times 10^{-19} \text{ J/eV})} = 0.96 \times 10^{-15} \text{ m},$$

or 0.96 fm (1 fm = 10^{-15} m = 1 femtometer or 1 fermi—in honor of Enrico Fermi). This is somewhat less than the size of nuclei. Notice in Fig. 31-1 that nuclei have a nearly uniform charge distribution on the interior, but their boundary is not distinct. Studies with even higher-energy electrons bombarding protons and neutrons show structural details that may indicate nucleons consist of still smaller particles.

High-energy particles produced in accelerators are used not only to investigate nuclear structure, but also to produce many new types of particles, as we shall soon see. But first we discuss the operation of several types of particle accelerators.

31-2 ∎ Particle accelerators

Van de Graaff accelerator.   A Van de Graaff generator, invented in 1931, is diagrammed in Fig. 31-2. A nearly spherical large hollow conductor is supported by an insulating column above the base. The sphere is charged to a high potential by a nonconducting moving belt in the following way. A voltage of typically 50,000 V is applied to a pointed conductor A, which "sprays" positive charge onto the moving belt. (Rather, electrons are pulled off the belt onto the electrode A.) The belt carries the positive charge into the interior of the sphere, where it is "wiped off" the belt at B and races to the outer surface of the spherical conductor—remember that charge collects on the outer surface of any conductor, since the charges repel each other and try to get as far from each other as possible. As more and more charge is brought upward, the sphere becomes more highly charged and reaches greater voltages. The process requires energy, of course, since the upward-moving charged belt is repelled by the charged sphere; the energy is supplied by the motor driving the belt. Very high potential differences can be obtained in this way, often giving the accelerated particles as much as 30 MeV of KE when two or more generators are used in tandem. Connected to the Van de Graaff generator is an evacuated tube which serves as the particle accelerator. A source of H or He ions (p or α) is located inside the

GENERATOR

FIGURE 31-2

Van de Graaff generator and accelerator.

tube and the large positive voltage repels them so they are accelerated toward the grounded target at the far end of the tube.

The cyclotron.   The cyclotron was developed in 1930 by E. O. Lawrence (1901–1958), at the University of California at Berkeley. It uses a magnetic field to maintain the ions—usually protons—in nearly circular paths (Chapter 19). The protons move within two D-shaped cavities, as shown in Fig. 31-3. Each time they pass into the gap between the "dees," a voltage is applied that accelerates them. This increases their speed and also increases the radius of curvature of their path. After many revolutions, the protons acquire high kinetic energy and reach the outer edge of the cyclotron. They then either strike a target placed inside the cyclotron, or leave the cyclotron with the help of a carefully placed "bending magnet" and are directed to an external target. The voltage applied to the dees to produce the acceleration must be alternating. When the protons are moving to the right across the gap in Fig. 31-3, the right dee must be negative and the left one positive. A half-cycle later, the protons are moving to the left, so the left dee must be negative in order to accelerate them. The frequency of the applied voltage must be equal to that of the circulating protons, and this can be determined as follows. When the ions are circulating within the dees, the net force F on each is simply that due to the magnetic field B, so $F = qvB$, where q is the charge on the ion and v its speed at a given moment (Eq. 19-4). (We ignore the electric force due to the applied voltage since it acts only when the protons are between the dees.) Since the protons move in circles, the acceleration is centripetal and equals v^2/r, where r is the radius of the ion's path at a given moment. We use Newton's second law, $F = ma$, and

FIGURE 31-3

Diagram of a cyclotron. The magnetic field, applied by a large electromagnet, points into the paper. A is the ion source. The field lines shown are for the electric field in the gap.

find that

$$F = ma$$

$$qvB = \frac{mv^2}{r}$$

or

$$v = \frac{qBr}{m}.$$

The time required for a complete revolution is the period T and is equal to

$$T = \frac{\text{distance}}{\text{speed}} = \frac{2\pi r}{qBr/m} = \frac{2\pi m}{qB}.$$

Hence the frequency of revolution f is

$$f = \frac{1}{T} = \frac{qB}{2\pi m}. \qquad (31\text{-}1)$$

This is known as the *cyclotron frequency*.

EXAMPLE 31-2   A small cyclotron of maximum radius $R = 0.25$ m accelerates protons in a 1.7-T magnetic field. Calculate (*a*) what frequency is needed for the applied alternating voltage, and (*b*) the kinetic energy of protons when they leave the cyclotron.

SOLUTION   (*a*) From Eq. 31-1

$$f = \frac{qB}{2\pi m} = \frac{(1.6 \times 10^{-19}\ \text{C})(1.7\ \text{T})}{(6.28)(1.67 \times 10^{-27}\ \text{kg})} = 2.6 \times 10^7\ \text{Hz}.$$

(*b*) The protons leave the cyclotron at $r = R = 0.25$ m. Then, since $v = qBr/m$,

$$\text{KE} = \tfrac{1}{2}mv^2 = \tfrac{1}{2}m\frac{q^2 B^2 R^2}{m^2} = \frac{q^2 B^2 R^2}{2m}$$

$$= \frac{(1.6 \times 10^{-19}\ \text{C})^2(1.7\ \text{T})^2(0.25\ \text{m})^2}{(2)(1.67 \times 10^{-27}\ \text{kg})} = 1.4 \times 10^{-12}\ \text{J},$$

or 8.8 MeV. Note that the magnitude of the voltage applied to the dees does not affect the final energy; but the higher this voltage, the fewer revolutions are required to bring the protons to full energy.

An interesting aspect of the cyclotron is that the frequency of the applied voltage, as given by Eq. 31-1, does not depend on the radius r. That is, the frequency does not have to be changed as the ions start from the source and are accelerated to paths of larger and larger radii. Unfortunately, this is only true at nonrelativistic energies. For at higher speeds, the mass of the ions will increase according to Einstein's formula, $m = m_0/\sqrt{1 - v^2/c^2}$, where m_0 is the rest mass. This is already a problem for the 34-MeV protons in the example above, for the KE is about 3 percent of the mass. As can be seen from Eq. 31-1, as the mass increases, the frequency of the applied voltage must be reduced. To achieve large energies, complex electronics is needed to decrease the frequency as a packet of protons increases in speed and reaches larger orbits. Such a modified cyclotron is called a *synchrocyclotron*.

Synchrotron.   Another way to deal with the increase in mass with speed is to increase the magnetic field *B* as the particles speed up. Such a device is called a *synchrotron*. The largest circular accelerators today fall into this category, and they are enormous. The Fermi National Accelerator Laboratory (Fermilab) at Batavia, Illinois, has a radius of 1.0 km and that at CERN (European Center for Nuclear Research) in Geneva, Switzerland, is 1.1 km in radius. They can accelerate protons to 500 GeV and that at Fermilab is expected to soon reach 1000 GeV using superconducting magnets. These large synchrotrons do not use enormous magnets 1 km in radius. Instead, a narrow ring of magnets is used (see Fig. 31-4) which are all placed at the same radius from the center of the circle. The magnets are interrupted by gaps where high voltage accelerates the ions. After the ions are injected, they must then be maintained in motion in a circle of constant radius. This is done by giving them considerable energy initially in a much smaller accelerator, and then slowly increasing the magnetic field as they speed up in the large synchrotron.

(a)

One problem of any accelerator is that accelerating electric charges radiate electromagnetic energy (see Chapter 21). Since ions or electrons are accelerated in any accelerator, we can expect considerable energy to be lost by radiation. The effect increases with speed and is especially important in circular machines where centripetal acceleration is present. It is mainly important in synchrotrons and hence is called *synchrotron radiation*. Synchrotron radiation can actually be useful, however. Since it consists of photons of a wide range of frequencies (typically UV and X-ray range), it can be used in experiments—or for treatment—where intense beams of photons are needed.

(b)

Linear accelerators.   A Van de Graaff is essentially a linear accelerator since the ions move in a linear path. But the name *linear accelerator* is usually reserved for a more complex arrangement in which particles are accelerated many, many times along a straight-line path. Figure 31-5 is a diagram of a simple "linac." The ions pass through a series of tubular conductors. The voltage applied to the tubes must be alternating so that when the ions reach a gap, the tube in front of them is negative and the one they just left is positive. This assures that they are accelerated at each gap. As the ions increase in speed, they cover more distance in the same amount of time. Consequently, the tubes must be longer the farther they are from the source. The Heavy Ion Linear Accelerator (Hilac) at Berkeley, California, can accel-

FIGURE 31-4

(a) Aerial view of Fermilab at Batavia, Illinois; the accelerator is a circular ring 2.0 km in diameter. (Courtesy of Fermilab.) (b) Interior photograph, showing magnets that keep ions moving in a circle. (Courtesy of Fermilab.)

FIGURE 31-5

Simple linear accelerator.

erate ions from hydrogen through uranium, imparting to them as much as 8.5 MeV per nucleon. Linear accelerators are of particular importance for accelerating electrons. Because of their small mass, electrons reach high speeds very quickly. (Indeed, an electron linac such as the one shown in Fig. 31-5 would have tubes nearly equal in length, since the electrons would be traveling close to $c = 3.0 \times 10^8$ m/s for almost the entire distance.) The largest electron linear accelerator is that at Stanford, called SLAC (Stanford Linear Accelerator Center). It is over 3 km (2 miles) long and can accelerate electrons to over 30 GeV.

Colliding beams. The usual arrangement in *high-energy-physics* experiments (as this field in which high-energy accelerators are used is called) is to allow the beam of particles from an accelerator to strike a stationary target. An important way to increase the energy of a collision is the recent development of *colliding beams*—that is, the target particles are moving as well as the projectile particles. This can be accomplished with only one accelerator through the use of *storage rings*. The accelerator accelerates one type of particle (say electrons or protons) to a maximum energy and then magnets are used to steer these particles into one circular storage ring where the particles can continue to circulate for many hours. It then accelerates a second type of particle (say positrons, or it could be the same as the first type, such as protons), and these are sent to a second storage ring. The two storage rings overlap at several places and the two beams, circulating in opposite directions, collide head on with each other at these points of intersection. Storage rings for colliding-beam experiments are in use at a number of facilities around the world and have played an important role in recent advances in elementary particle physics.

31-3 ∎ Beginnings of elementary particle physics—the Yukawa particle

By the mid 1930s, it was recognized that all atoms can be considered to be made up of neutrons, protons, and electrons. The basic constituents of the universe were no longer considered to be atoms but rather the proton, neutron, and electron. Besides these three *elementary particles*, as they could be called, there were several others also known: the positron (a positive electron), the neutrino, and the γ particle (or photon), for a total of six elementary particles.

Looking back, things seemed fairly simple in 1935. But in the decades that followed, hundreds of other subnuclear particles were discovered. The properties and interactions of these particles, and which ones should be considered as fundamental or "elementary," became the substance of research in the field of **elementary particle physics.**

Elementary particle physics, as it exists today, can be said to have begun in 1935 when the Japanese physicist Hideki Yukawa (1907–1981) predicted the existence of a new particle that would in some way mediate the strong nuclear force. To understand Yukawa's idea, we first look at the electromagnetic force. When we first discussed electricity we saw that the electric force acts over a distance, without contact. To better perceive how a force can act over a distance, we saw that Faraday introduced the idea of a *field*. The force that one charged particle exerts on a second can be said to be due to the electric field set up by the first. Similarly, the magnetic field can be said to carry the magnetic force. Later (Chapter 21) we saw that electromagnetic fields can travel through space as waves. Finally, in Chapter 26 we saw that electromagnetic radiation (light) can be considered as either a wave or as a collection of particles called photons. Because of this wave—particle duality, it is possible to imagine that the electromagnetic force between charged particles is due (1) to the EM field set up by one and felt by the other, or (2) to an exchange of photons or γ-particles between them. It is (2) that we want to concentrate on here, and an example of how an exchange of particles could give rise to a force is illustrated in the crude picture of Fig. 31-6. In part (a), two children start throwing pillows at each other; each catch results in the child being moved backward by the impulse. This is the equivalent of a repulsive force. On the other hand, if the two children exchange pillows by grabbing them out of the other person's hand, they will be pulled toward each other, as when an attractive force acts. For the electromagnetic force between two charged particles, it is photons that are exchanged between the two particles that give rise to the force. A simple diagram describing this is shown in Fig. 31-7. Such a diagram is called a *Feynman diagram* and the theory on which it is based, *quantum electro-dynamics* (QED). The case shown is the simplest in which a single photon is exchanged. One of the charged particles emits the photon and recoils somewhat as a result; and the second particle absorbs the photon. In any such collision or *interaction*, energy and momentum are transferred from one charged particle to the other, and it is carried by the photon. Because the photon is absorbed by the second particle very shortly after it is emitted by the first, it is not observable, and it is referred to as being a *virtual* photon, as compared to one that is free and can be detected by instruments.

Now to Yukawa's prediction. In analogy to photon exchange to mediate the electromagnetic force, Yukawa argued that there ought to be a particle that mediates the strong nuclear force—the force that holds nucleons together in the nucleus. Just as the photon is called the quantum of the electromagnetic field or force, so the Yukawa particle would represent the quantum of the strong nuclear force.

Yukawa predicted that this new particle would have a mass intermediate between that of the electron and the proton. Hence it was called a *meson*, meaning "in the middle," and Fig. 31-8 is a Feynman diagram of meson exchange. We can make a rough approximation of the mass of the meson as follows. Suppose the proton on the left in Fig. 31-8 is at rest. For it to emit a meson would require energy (to make the mass) which would have to come from nowhere; such a process would violate conservation of energy. But the uncertainty principle allows nonconservation of energy of an amount ΔE if it occurs only for a time Δt given by $(\Delta E)(\Delta t) \approx h/2\pi$. We set ΔE equal to the energy needed to create the mass m of the meson: $\Delta E = mc^2$. Now conservation of energy is violated only as long as the meson exists, which is the time Δt required for the meson to pass from one nucleon to the other. If we assume the meson travels at relativistic speed, close to

(a) Repulsive force (children throwing pillows)

(b) Attractive force (children grabbing pillows from each other's hand)

FIGURE 31-6

Forces equivalent to particle exchange. (a) Repulsive force (children throwing pillows). (b) Attractive force (children grabbing pillows from each other's hand).

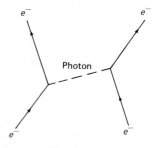

FIGURE 31-7

Feynman diagram, showing how a photon acts as carrier of the electromagnetic force between two electrons.

FIGURE 31-8

Meson exchange when a proton and neutron interact via the strong nuclear force.

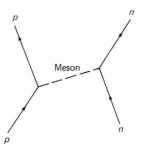

the speed of light c, then Δt will be at most about $\Delta t = d/c$, where d is the maximum distance that can separate the interacting nucleons. Thus we have

$$\Delta E \, \Delta t \approx \frac{h}{2\pi}$$

$$mc^2 \left(\frac{d}{c}\right) \approx \frac{h}{2\pi}$$

or

$$mc^2 \approx \frac{hc}{2\pi d}. \tag{31-2}$$

The range of the strong nuclear force is about $d \approx 1.5 \times 10^{-15}$ m (this is the maximum distance away it can be felt) so

$$mc^2 \approx \frac{(6.6 \times 10^{-34} \text{ J·s})(3.0 \times 10^8 \text{ m/s})}{(6.28)(1.5 \times 10^{-15} \text{ m})} \approx 2.2 \times 10^{-11} \text{ J} = 130 \text{ MeV}.$$

The mass of the predicted meson is thus very roughly 130 MeV/c^2 or about 250 times the electron mass of 0.51 MeV/c^2. (Note, incidentally, that since the electromagnetic force has infinite range ($d = \infty$), Eq. 31-2 tells us that the exchanged particle for the electromagnetic force, the photon, will have zero rest mass.)

Just as photons can be observed as free particles, as well as acting in an exchange, so it was expected that mesons might be observed directly. Such a meson was searched for in the cosmic radiation that enters the earth's atmosphere from the sun and other sources in the universe. In 1937 a new particle was discovered whose mass was 106 MeV (207 times the electron mass). This is quite close to the mass predicted. But it turned out that this new particle, called the *muon*, did not interact strongly with matter. It could hardly mediate the strong nuclear force if it didn't interact via the strong nuclear force. Thus the muon, which can have either a positive or a negative charge and seems to be nothing more than a very massive electron, is not the Yukawa particle.

The particle predicted by Yukawa was finally found in 1947. It is called the "π" or pi meson, or simply the *pion*. It comes in three charge states, $+$, $-$, or 0. The π^+ and π^- have mass of 139.6 MeV/c^2 and the π^0 a mass of 135.0 MeV/c^2. All three interact strongly with matter. Soon after their discovery in cosmic rays, pions were produced in the laboratory using a particle accelerator. Reactions observed included

$$p + p \rightarrow p + p + \pi^0$$
$$p + p \rightarrow p + n + \pi^+. \tag{31-3}$$

The incident proton from the accelerator must have sufficient energy to produce the additional mass of the pion. A number of other mesons were discovered in subsequent years which were also considered to mediate the strong nuclear force. (The recent theory of quantum chromodynamics, involving quarks, has replaced mesons with gluons as exchange particles for the strong force; see Section 31-10.)

TABLE 31-1
The four forces in nature

Type	Relative Strength (approx.)	Field Particle
Strong nuclear	1	Mesons/gluons[†]
Electromagnetic	10^{-2}	Photon
Weak nuclear	10^{-13}	$W^{\pm}$ and Z^0
Gravitational	10^{-40}	Graviton (?)

[†] See Section 31-10.

So far we have discussed the particles that mediate the electromagnetic and strong nuclear forces. But there are four known types of force—or interaction—in nature. What about the other two: the weak nuclear force, and gravity? Theorists believe that these are also mediated by particles. The particles presumed to transmit the weak force are referred to as the W^+, W^-, and Z^0. In 1983, after extensive searches, the long-awaited discovery of the W and Z^0 particles was announced by Carlo Rubbia,[†] the scientist behind the building of the proton-antiproton collider at the very high energy accelerator at CERN that produced the results, and leader of a large group of scientists (over 100) who worked on the project. The quantum of the gravitational force, called the *graviton*, has not yet been identified. A comparison of the four forces is given in Table 31-1, where they are listed according to their (approximate) relative strengths. Notice that although gravity may be the most obvious force in daily life (because of the huge mass of the earth), on a nuclear scale it is much the weakest of the four forces and its effect at the nuclear level can nearly always be ignored.

31-4 ■ Particles and antiparticles

The positron, as we have seen, is basically a positive electron. That is, many of its properties are the same as for the electron, such as mass, but it has the opposite charge. The positron is said to be the **antiparticle** to the electron. After the positron was discovered in 1932, it was predicted that other particles also ought to have antiparticles. In 1955 the antiparticle to the proton was found, the *antiproton* ($\bar{p}$), see Fig. 31-9. (The bar over the p is used to indicate antiparticle.) A large amount of energy was needed to produce this massive particle (mass = proton's mass). Its discovery (by E. Segré and O. Chamberlain) was made only after the completion of the large accelerator (the Bevatron) at the University of California at Berkeley. Soon after, the antineutron ($\bar{n}$) was found. Most other particles also have antiparticles. But the photon, the π^0, and a few other particles do not have distinct antiparticles—or we say that they are their own antiparticles.

Antiparticles are produced in nuclear reactions when there is sufficient energy available, and they do not live very long in the presence of matter. For example, when a positron encounters an electron, the two annihilate each other. The energy of their vanished mass, plus any kinetic energy they possessed, is converted to the energy of γ rays or of other particles. Annihilation occurs for all other particle–antiparticle pairs.

[†] For this work, Rubbia was awarded the 1984 Nobel Prize, along with Simon van der Meer.

FIGURE 31-9
Liquid-hydrogen bubble-chamber photograph of an antiproton ($\bar{p}$) colliding with a proton, producing a hyperon pair ($\bar{p} + p \rightarrow \Xi^- + \bar{\Xi}^-$) which subsequently decays into other particles. The drawing indicates the assignment of a particle to each track. Neutral-particle paths are shown by dashed lines since neutral particles produce no bubbles. (Courtesy of Brookhaven National Laboratory.)

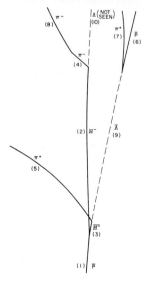

31-5 ▌ Particle interactions and conservation laws

One of the important uses of high-energy accelerators is to study the interactions of elementary particles with each other. As a means of ordering this subnuclear world, the conservation laws are indispensable. The laws of conservation of energy, of momentum, of angular momentum, and of electric charge are found to hold precisely in all particle interactions. (Although the uncertainty principle allows, for example, nonconservation of energy for times $\Delta t \approx h/2\pi\,\Delta E$, as we have seen, the times involved are much too short for experimental observation and violation has never been detected.

A study of particle interactions has revealed a number of new conservation laws, some of which we now discuss. These new conservation laws (just like the old ones) are ordering principles: they help to explain why some reactions occur and others do not. For example, the following reaction has never been found to occur:

$$p + n \nrightarrow p + p + \bar{p}$$

even though charge, energy, and so on, are conserved ($\bar{p}$ means an antiproton and $\nrightarrow$ means the reaction does not occur). To understand why it doesn't occur, physicists hypothesized a new conservation law, the conservation of **baryon number.** (Baryon number is the same as nucleon number—which we saw earlier is conserved in nuclear reactions—but is applied to other particles in addition to nucleons.) An important addition to this law is the proposal that whereas all nucleons have baryon number $B = +1$, all antinucleons (antiprotons, antineutrons) have $B = -1$. The reaction above does not conserve baryon number since on the left side we have $B = (+1) + (+1) = +2$ and on the right $B = (+1) + (+1) + (-1) = +1$. On the other hand, the following reaction does conserve B and *does* occur if the incoming proton has sufficient energy:

$$p + n \rightarrow \quad p + n + \bar{p} + p$$
$$B = +1 + 1 = +1 + 1 - 1 + 1.$$

As indicated, $B = +2$ on both sides of this equation. From these and other reactions, the conservation of baryon number has been established as a basic law of physics.

Other useful "number" laws are associated with weak interactions, mainly decays. In ordinary β decay an electron or positron is emitted along with a neutrino or antineutrino; in a similar type of decay a muon is emitted instead of an electron. The neutrino that accompanies an emitted electron (v_e) is found to be different from the neutrino that accompanies an emitted muon (v_μ). Each of these neutrinos has an antiparticle: $\bar{v}_e$ and $\bar{v}_\mu$. In ordinary β decay we have, for example

$$n \rightarrow p + e^- + \bar{v}_e$$

but never $n \rightarrow p + e^- + \bar{v}_\mu$ or $n \rightarrow p + e^- + \bar{v}_e + v_e$. To explain why these do not occur, the concept of electron lepton number, L_e, was invented. If the electron (e^-) and the electron neutrino (v_e) are given $L_e = +1$, and e^+ and $\bar{v}_e$ are given $L_e = -1$, whereas all other particles have $L_e = 0$, then all observed decays conserve L_e. For example, in $n \rightarrow p + e^- + \bar{v}_e$, $L_e = 0$ initially, and $L_e = 0 + (+1) + (-1) = 0$ after the decay. Decays that do not conserve L_e but would obey the other conservation laws are not observed to occur; hence it is believed that L_e is conserved in all interactions.

In a decay involving muons, such as

$$\pi^+ \rightarrow \mu^+ + v_\mu,$$

a second quantum number, muon lepton number (L_μ) is conserved. The μ^- and v_μ are assigned $L_\mu = +1$ and μ^+ and $\bar{v}_\mu$ have $L_\mu = -1$, whereas other particles have $L_\mu = 0$. It is believed that L_μ is also conserved in all interactions or decays. Similar assignments can be made for a third lepton number, L_τ, associated with the newly discovered τ lepton and its neutrino, v_τ.

Keep in mind that antiparticles have not only opposite electric charge from their particles, but also opposite B, L_e, L_μ, and L_τ.

EXAMPLE 31–3 Which of the following decay schemes is possible for muon decay: (a) $\mu^- \rightarrow e^- + \bar{v}_e$; (b) $\mu^- \rightarrow e^- + \bar{v}_e + v_\mu$; (c) $\mu^- \rightarrow e^- + v_e$? All of these particles have $L_\tau = 0$.

SOLUTION A μ^- has $L_\mu = +1$, and $L_e = 0$. This is the initial state, and the final state (after decay) must also have $L_\mu = +1$, $L_e = 0$. In (a), the final state has $L_\mu = 0 + 0 = 0$, and $L_e = +1 - 1 = 0$; L_μ would not be conserved and indeed this decay is not observed to occur. The final state of (b) has $L_\mu = 0 + 0 + 1 = +1$ and $L_e = +1 - 1 + 0 = 0$, so both L_μ and L_e are conserved. This is in fact the most common decay mode of the μ^-. Finally (c) does not occur because L_e (= +2 in final state) is not conserved.

Recent theoretical work suggests that the conservation laws of baryon and lepton numbers may be only approximate, rather than exact, as we shall discuss in Section 31-11.

31-6 ▌ Particle classification

In the decades following the discovery of the π meson in the late 1940s, a great many other subnuclear particles were discovered; today they number in the hundreds. Much theoretical and experimental work has been done to try to understand this multitude of particles. One important aid to understanding is to arrange the particles in categories according to their properties. One way of doing this is according to their interactions. Since not all particles take part in all four of the forces known in nature, this fact is useful as a classification scheme. Table 31-2 lists many of the stable (and "long-lived") particles classified in this way along with many of their properties. The *photon* takes part only in the electromagnetic force, and it is in a class by itself. The **leptons** are those particles that do not interact via the strong force but do interact via the weak nuclear force (as well as the much weaker gravitational force); those that carry electric charge also interact via the electromagnetic force. The four well-known leptons are the electron, the muon, and two types of neutrino: the electron neutrino, v_e, and the muon

TABLE 31-2

Elementary particles (stable or "long-lived")[†]

Category	Particle Name	Symbol	Antiparticle	Rest Mass (MeV/c^2)	B	L_e	L_μ	L_τ	S	Lifetime (s)	Principal Decay Modes
Photon	Photon	γ	Self	0	0	0	0	0	0	Stable	
Leptons	Electron	e^-	e^+	0.511	0	+1	0	0	0	Stable	
	Neutrino (e)	ν_e	$\bar{\nu}_e$	0 (?)	0	+1	0	0	0	Stable	
	Muon	μ^-	μ^+	105.7	0	0	+1	0	0	2.20×10^{-6}	$e^-\bar{\nu}_e\nu_\mu$
	Neutrino (μ)	ν_μ	$\bar{\nu}_\mu$	0 (?)	0	0	+1	0	0	Stable	
	Tau	τ^-	τ^+	1784.	0	0	0	−1	0	$<4 \times 10^{-13}$	$\mu^-\bar{\nu}_\mu\nu_\tau, e^-\bar{\nu}_e\nu_\tau$, hadrons
	Neutrino (τ)	ν_τ	$\bar{\nu}_\tau$	0 (?)	0	0	0	−1	0	Stable	
Hadrons											
Mesons	Pion	π^+	π^-	139.6	0	0	0	0	0	2.60×10^{-8}	$\mu^+\nu_\mu$
		π^0	Self	135.0	0	0	0	0	0	0.83×10^{-16}	2γ
	Kaon	K^+	K^-	493.7	0	0	0	0	+1	1.24×10^{-8}	$\mu^+\nu_\mu, \pi^+\pi^0$
		K_S^0	$\overline{K_S^0}$	497.7	0	0	0	0	+1	0.89×10^{-10}	$\pi^+\pi^-, 2\pi^0$
		K_L^0	$\overline{K_L^0}$	497.7	0	0	0	0	+1	5.2×10^{-8}	$\pi^\pm e^\mp \overset{(-)}{\nu}_e$ $\pi^\pm \mu^\mp \overset{(-)}{\nu}_\mu$ $3\pi^0$
	Eta	η^0	Self	548.8	0	0	0	0	0	$<10^{-18}$	$2\gamma, 3\pi$
Baryons	Proton	p	$\bar{p}$	938.3	+1	0	0	0	0	Stable[‡]	
	Neutron	n	n	939.6	+1	0	0	0	0	920	$pe^-\bar{\nu}_e$
	Lambda	Λ^0	$\bar{\Lambda}^0$	1115.6	+1	0	0	0	−1	2.6×10^{-10}	$p\pi^-, n\pi^0$
	Sigma	Σ^+	$\bar{\Sigma}^-$	1189.4	+1	0	0	0	−1	0.80×10^{-10}	$p\pi^0, n\pi^+$
		Σ^0	$\bar{\Sigma}^0$	1192.5	+1	0	0	0	−1	6×10^{-20}	$\Lambda^0\gamma$
		Σ^-	$\bar{\Sigma}^+$	1197.3	+1	0	0	0	−1	1.5×10^{-10}	$n\pi^-$
	Xi	Ξ^0	$\bar{\Xi}^0$	1315	+1	0	0	0	−2	2.9×10^{-10}	$\Lambda^0\pi^0$
		Ξ^-	Ξ^+	1321	+1	0	0	0	−2	1.64×10^{-10}	$\Lambda^0\pi^-$
	Omega	Ω^-	Ω^+	1672	+1	0	0	0	−3	0.82×10^{-10}	$\Xi^0\pi^0, \Lambda^0K^-$

[†] See also Table 31-4.

[‡] $>2 \times 10^{32}$ yr.

neutrino, ν_μ. They each have antiparticles as indicated in Table 31-2. The recent evidence for another lepton, the τ, and its neutrino, ν_τ, brings the total number of leptons today to six.

The third category of particle is the **hadron.** Hadrons are those particles that can interact via the strong nuclear force. Hence they are said to be **strongly interacting particles.** They also interact via the other forces, but the strong force predominates at short distances. The hadrons include nucleons, pions, and a large number of other particles. They are divided into two subgroups:[†] **baryons,** which are those particles that have baryon number +1

[†] Originally, particles were divided according to their mass into leptons (meaning light particles), baryons (meaning "heavy"), and those of intermediate mass, the mesons (meaning "middle"). The newer classification according to their interactions is not always consistent with this. The muon, for example, is now called a lepton (it doesn't interact strongly) although it was once called the mu meson because of its mass. Other exceptions are the ψ particles which are very heavy but have $B = 0$, and so are classified as mesons (Section 31-9), and the τ which is a lepton although it is heavier than some baryons.

(or −1 in the case of their antiparticles); and **mesons,** which have baryon number = 0.

Notice that the baryons Λ, Σ, Ξ, and Ω all decay to lighter-mass baryons, and eventually to a proton or neutron. All these processes conserve baryon number. Since there is no lighter particle than the proton with $B = +1$, if baryon number is strictly conserved, the proton itself cannot decay and is stable (but see Section 31-11).

31-7 ∎ Particle stability and resonances

The particles listed in Table 31-2 are those that are either stable or are rather long-lived (lifetime $\gtrsim 10^{-19}$ s). The lifetime of an unstable particle depends on which force is most active in causing the decay. When we say the strong nuclear force is stronger than the electromagnetic, we mean that two particles will interact more quickly and more frequently if this force is acting. When a stronger force influences a decay, that decay occurs more quickly. Decays caused by the weak force have lifetimes of 10^{-10} s or longer. Particles that decay via the electromagnetic force have much shorter life-times, typically about 10^{-16} to 10^{-19} s. The unstable particles listed in Table 31-2 decay either via the weak or the electromagnetic interaction; those that involve a γ (photon) are electromagnetic, and the others shown decay via the weak interaction (note the lifetimes).

A great many particles have been found that can decay via the strong interaction, and these are not listed in Table 31-2. Such particles decay into other strongly interacting particles (say, n, p, π, but not involving γ, e, and so on) and their lifetimes are very short, typically 10^{-23} s. In fact their life-times are so short that they do not go far enough to be detected before decaying. Their decay products can be detected, however, and it is from them that the existence of such short-lived particles is inferred. To see how this is done, let us consider the first such particle discovered (by Fermi, Fig. 31-10). Fermi used a beam of π^+ directed through a hydrogen target (protons) with varying amounts of energy. A graph of the number of inter-actions (π^+ scattered) versus the pion's kinetic energy is shown in Fig. 31-11. The large peak around 200 MeV was much higher than expected and cer-

FIGURE 31-10
Enrico Fermi (1901–1954).
(The Bettmann Archive.)

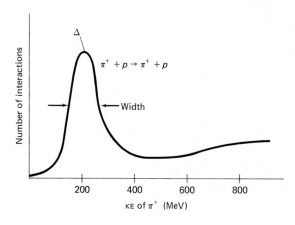

FIGURE 31-11
Number of π^+ being scattered by a proton target as a function of the incident π^+ kinetic energy.

tainly much higher than the number of interactions at neighboring energies. This led Fermi to conclude that the π^+ and proton combined momentarily to form a short-lived particle before coming apart again, or at least they resonated back and forth for a short time. Indeed, the large peak in Fig. 31-11 resembles a resonance curve (see Figs. 13-9 and 20-27) and this new "particle"—now called the Δ—is referred to as a **resonance.** Hundreds of other resonances have been found in a similar way. Many resonances are regarded as excited states of other particles such as the nucleon.

The width of the resonances—in Fig. 31-11 the width of the Δ peak is about 100 MeV—is an interesting application of the uncertainty principle. If a particle lives only 10^{-23} s, then its mass (that is, its rest energy) will be uncertain by an amount $\Delta E \approx h/2\pi \, \Delta t \approx (6.6 \times 10^{-34} \text{ J·s})/(6)(10^{-23} \text{ s}) \approx 10^{-11}$ J ≈ 100 MeV, which is what is observed. Actually, the lifetimes of $\approx 10^{-23}$ s for such resonances are inferred by the reverse process: from the measured width being ≈ 100 MeV.

31-8 ∎ Strange particles

In the early 1950s, certain of the newly found particles, namely the K, Λ, and Σ, were found to behave rather strangely in two ways. First, they were always produced in pairs. For example, the reaction

$$\pi^- + p \to K^0 + \Lambda^0$$

occurred with high probability, but the reaction $\pi^- + p \to K^0 + n$ was never observed to occur; this seemed strange since no known conservation law would have been violated and there was plenty of energy available. The second feature of these *strange particles* (as they came to be called) was that although they were clearly produced via the strong interaction (that is, at a high rate), they did not decay at a rate characteristic of the strong interaction; and they decayed only into strongly interacting particles (for example $K \to 2\pi$, $\Sigma^+ \to p + \pi^0$). Instead of lifetimes of 10^{-23} s as expected for strongly interacting particles, strange particles have lifetimes of 10^{-10} to 10^{-8} s which are characteristic of the weak interaction.

To explain these observations, a new quantum number, **strangeness,** and a new conservation law, *conservation of strangeness*, were introduced. By assigning the strangeness numbers (S) indicated in Table 31-2, the production of strange particles in pairs was readily explained; antiparticles were assigned opposite strangeness from their particles: one of each pair was assigned $S = +1$ and the other $S = -1$ (see Table 31-2). For example, in the reaction $\pi^- + p \to K^0 + \Lambda^0$, the initial state has strangeness $S = 0 + 0 = 0$, and the final state has $S = +1 - 1 = 0$, so strangeness is conserved. But for $\pi^- + p \nrightarrow K^0 + n$, the initial state has $S = 0$ and the final state has $S = +1 + 0 = +1$, so strangeness would not be conserved; and the reaction isn't observed.

To explain the decay of strange particles, it is assumed that strangeness is conserved in the strong interaction but is *not* conserved in the weak interaction. Thus, although the strange particles were forbidden by strangeness conservation to decay to lower mass nonstrange particles via the strong interaction, they could undergo such decay by means of the weak interaction. This would occur much more slowly, of course, which accounts for their longer lifetimes of 10^{-10} to 10^{-8} s.

The conservation of strangeness was the first example of a "partially conserved" quantity. In this case, the quantity strangeness is conserved by strong interactions but not by weak.

31-9 ∎ Quarks

Nearly all observed particles fall into one of two groups: leptons or hadrons. The principal difference between these two groups is that the hadrons interact via the strong interaction whereas the leptons do not. Another important difference that physicists had to deal with in the 1960s was that there were only four known leptons (e^-, μ^-, ν_e, ν_μ) but there are well over a hundred hadrons.

The leptons are considered to be truly elementary particles since they do not seem to break down into smaller entities, do not show any internal structure, and have no measurable size. (Attempts to determine the size of leptons have put an upper limit of about 10^{-18} m.)

The hadrons, on the other hand, are more complex. Experiments indicate they do have an internal structure. And the fact that there are so many of them suggests that they can't all be elementary. To deal with this problem, M. Gell-Mann and G. Zweig in 1963 independently proposed that none of the hadrons so far observed is elementary. Instead, they proposed that the hadrons are made up of combinations of three, more fundamental,

TABLE 31-3
Properties of quarks and antiquarks

				Quarks				
Name	**Symbol**	**Spin**	**Charge**	**Baryon Number**	**Strangeness**	**Charm**	**Bottomness**	**Topness**
Up	u	$\frac{1}{2}$	$+\frac{2}{3}e$	$\frac{1}{3}$	0	0	0	0
Down	d	$\frac{1}{2}$	$-\frac{1}{3}e$	$\frac{1}{3}$	0	0	0	0
Strange	s	$\frac{1}{2}$	$-\frac{1}{3}e$	$\frac{1}{3}$	-1	0	0	0
Charmed	c	$\frac{1}{2}$	$+\frac{2}{3}e$	$\frac{1}{3}$	0	$+1$	0	0
Bottom	b	$\frac{1}{2}$	$-\frac{1}{3}e$	$\frac{1}{3}$	0	0	$+1$	0
Top	t	$\frac{1}{2}$	$+\frac{2}{3}e$	$\frac{1}{3}$	0	0	0	$+1$

				Antiquarks				
Name	**Symbol**	**Spin**	**Charge**	**Baryon Number**	**Strangeness**	**Charm**	**Bottomness**	**Topness**
Up	$\bar{u}$	$\frac{1}{2}$	$-\frac{2}{3}e$	$-\frac{1}{3}$	0	0	0	0
Down	$\bar{d}$	$\frac{1}{2}$	$+\frac{1}{3}e$	$-\frac{1}{3}$	0	0	0	0
Strange	$\bar{s}$	$\frac{1}{2}$	$+\frac{1}{3}e$	$-\frac{1}{3}$	$+1$	0	0	0
Charmed	$\bar{c}$	$\frac{1}{2}$	$-\frac{2}{3}e$	$-\frac{1}{3}$	0	-1	0	0
Bottom	$\bar{b}$	$\frac{1}{2}$	$+\frac{1}{3}e$	$-\frac{1}{3}$	0	0	-1	0
Top	$\bar{t}$	$\frac{1}{2}$	$-\frac{2}{3}e$	$-\frac{1}{3}$	0	0	0	-1

pointlike entities called **quarks.**[†] Quarks, then, would be considered truly elementary particles, like leptons. The three quarks were labeled *u*, *d*, *s*, and given the names *up*, *down*, and *sideways* (or, more commonly now, *strange*). They were assumed to have fractional charge ($\frac{1}{3}$ or $\frac{2}{3}$ the charge on the electron—that is, less than the previously thought smallest charge); other properties of quarks and antiquarks are indicated in Table 31-3. All hadrons known at the time could be constructed in theory from these three types of quark. Mesons would consist of a quark–antiquark pair. For example, a π^+ meson is considered a $u\bar{d}$ pair (note that for the $u\bar{d}$ pair, $Q = \frac{2}{3}e + \frac{1}{3}e = +1e$, $B = \frac{1}{3} - \frac{1}{3} = 0$, $S = 0 + 0 = 0$, as it must for a π^+). On the other hand a $K^+ = u\bar{s}$ with $Q = +1$, $B = 0$, $S = +1$. Baryons, on the other hand, would consist of three quarks; for example a neutron is $n = ddu$ whereas an antiproton is $\bar{p} = \bar{u}\bar{u}\bar{d}$.

Soon after the quark theory was proposed, physicists began looking for these fractionally charged particles. Although there is indirect experimental evidence in favor of their existence, direct detection of them remains elusive. Indeed, there are suggestions that quarks are so tightly bound together that they may not ever exist in the free state.

In 1964, several physicists proposed that there ought to be a fourth quark. Their argument was based on the expectation that there exists a deep symmetry in nature, including a connection between quarks and leptons. If there are four leptons (as was thought in the 1960s) then symmetry in nature would suggest there should also be four quarks. The fourth quark was said to be *charmed;* its charge would be $+\frac{2}{3}e$ and it would have another property to distinguish it from the other three quarks. This new property, or quantum number, was called **charm** (see Table 31-3). Charm was assumed to be like strangeness: it would be conserved in strong and electromagnetic interactions, but would not be conserved by the weak. The new charmed quark would have charm $C = +1$ and its antiquark $C = -1$.

Experimentally, however, there seemed to be no need for a charmed quark. Before 1974, all known hadrons could be explained as combinations of the three original quarks. In fact, hadrons corresponding to all three quark, and quark–antiquark, combinations had been found. Furthermore the Ω^- baryon had been predicted by the three-quark theory ($\Omega^- = sss$) and was discovered soon after, Fig. 31-12. But in 1974, a new heavy meson was discovered simultaneously by two different groups of experimenters. This new meson, called the J/ψ (often simply the ψ), did not fit the old three-quark scheme. The J/ψ, whose mass is 3100 MeV/c^2, far higher than other known mesons, could also not be an excited state of a smaller-mass meson (into which it would decay) because its lifetime would have to be about 10^{-23} s. In fact, its lifetime was found to be 1000 times greater than this, about 10^{-20} s. It soon became clear that the existence of the J/ψ could be accounted for on the basis of the charmed quark: a J/ψ would be a combination of a charmed quark and its antiquark ($c\bar{c}$). The charm of the J/ψ itself is zero ($C = +1 - 1 = 0$), so it can decay strongly into hadrons (such as several pions, $p\bar{p}$, $\Lambda\bar{\Lambda}$, etc), which is observed. The question of why it

---

[†] Gell-Mann chose the word from the phrase "Three quarks for Muster Mark" in James Joyce's *Finnegan's Wake*.

FIGURE 31-12

Liquid-hydrogen bubble-chamber photograph showing the production of a negatively charged omega baryon (Ω^-) by the interaction of a negative *K meson* (K^-) with a proton (a hydrogen nucleus in the bubble chamber). (Courtesy of Brookhaven National Laboratory.)

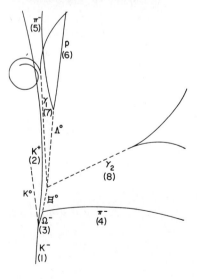

lives 1000 times longer than other strongly decaying particles is theorized as being due to the fact that the c and $\bar{c}$ quarks that make it up must each be converted into noncharmed quarks that become the hadrons seen in the decay. This process would inhibit the rate at which the decay could occur, so the charmed-quark model yielded a useful explanation.

Soon after the J/ψ was discovered, a similar meson, called the ψ', was also found whose mass is about 3685 MeV/c^2. Many other ψ-like mesons were also found, and these are all believed to be bound states of a $c\bar{c}$ pair.

Although the J/ψ meson and its relatives have no net charm themselves, it stands to reason that in the decay of one of the larger-mass $c\bar{c}$ combinations there ought to be mesons that do have charm. That is, the c and $\bar{c}$ quarks ought to appear in separate particles and lend a charm to them of $+1$ or -1. In such a decay process, other quarks would also be produced. For example the decay

$$\psi \rightarrow D^+ + D^-$$

might be written as

$$\psi = c\bar{c} \rightarrow c\bar{c}d\bar{d} \rightarrow c\bar{d} + \bar{c}d.$$

The $d\bar{d}$ quarks in the third step are produced from energy; since the d and $\bar{d}$ quarks have opposite quantum numbers and charge, no conservation laws are violated—only energy is required to produce their mass. The $c\bar{d}$ combination which has $Q = +1$, $B = 0$, $S = 0$, $C = +1$, has been dubbed the D^+ meson, and $\bar{c}d$ is its antiparticle, D^-. This meson, along with its neutral sister, D^0, was searched for and believed found in 1977 with a mass around 1870 MeV/c^2. More recent experiments indicate that charmed baryons also exist. Some of the new particles are listed in Table 31-4.

Another important recent development is the experimental evidence for the tau (τ) lepton, with a mass of 1784 MeV/c^2. This lepton, like the electron and muon, presumably has a neutrino associated with it. Thus the family of leptons is at present believed to have six members. This would upset the balance between leptons and quarks, the presumed basic building

TABLE 31-4

Partial list of hadrons associated with charm and bottomness ($L_e = L_\mu = L_\tau = 0$)

Category	Particle	Anti-particle	Rest Mass (MeV/c^2)	Baryon Number	Strange-ness	Charm	Bottom-ness	Lifetime(s)	Principal Decay Modes
Mesons	D^+	D^-	1869	0	0	+1	0	9×10^{-13}	K + others, e + others
	D^0	$\overline{D^0}$	1865	0	0	+1	0	5×10^{-13}	K + others
	F^+	F^-	2021	0	+1	+1	0	$\sim 2 \times 10^{-13}$	$n\pi$, $n3\pi$
	J/ψ (3100)	Self	3097	0	0	0	0	$\sim 1 \times 10^{-20}$	Hadrons
	ψ' (3685)	Self	3686	0	0	0	0	$\sim 3 \times 10^{-20}$	$J/\psi\,\pi\,\pi$, hadrons
	Υ	Self	9460	0	0	0	0	$\sim 1 \times 10^{-20}$	$\mu^+\mu^-$, e^+e^-
	B^+	B^-	5271	0	0	0	-1	$\sim 10^{-13}$	$\overline{D^0}\pi^+$
	B^0	$\overline{B^0}$	5274	0	0	0	-1	$\sim 10^{-13}$	$\overline{D^0}\pi^+\pi^-$
Baryon	Λ_C^+	Λ_C^-	2280	+1	0	+1	0	1×10^{-13}	Hadrons

blocks of matter, unless two new quarks also exist. Indeed, theoretical physicists have postulated the existence of a fifth and sixth quark. These have been named *top* and *bottom* quarks, since they resemble the "up" and "down" quarks. (Some physicists prefer the names *truth* and *beauty* for these t and b quarks.) The names apply also to the new properties (quantum numbers) that distinguish the new quarks from the old quarks. (These are included in Table 31-3.) Indeed, another new meson, Υ, has been detected at about 9400 MeV/c^2, which is considered to be a $b\bar{b}$ combination. And in 1983, the B meson was observed, which has "bottomness" (consisting of only one b quark plus a non-b quark). Evidence for the top (t) quark first appeared in 1984 from Rubbia's group at CERN; its mass was judged to be in the range 30-50 GeV/c^2 (as compared to about 5.2 GeV/c^2 for the b quark) and was detected in the decay of the W particle.

31-10 ∎ The "standard model": quantum chromodynamics (QCD) and the electroweak theory

Not long after the quark theory was proposed, it was suggested that quarks have another property (or quality) called **color.** The distinction between the six quarks [u, d, s, c, b, t] was referred to as **flavor.** According to theory, each of the flavors of quark can have three colors, usually designated red, green, and blue. [Note that the names "color" and "flavor" have nothing to do with our senses, but are purely whimsical—as are other names (like charm) in this new field.] The antiquarks are colored antired, antigreen, and antiblue. Baryons are made up of three quarks, one of each color; mesons consist of a quark–antiquark pair of a particular color and its anticolor; thus baryons and mesons are white or colorless.

Originally, the idea of quark color was proposed to preserve the Pauli exclusion principle (Section 27-7) which applies to particles of spin $\frac{1}{2}$ (or any half-integral spin, like $\frac{3}{2}$, $\frac{5}{2}$, and so on), such as electrons and nucleons. Since quarks have spin $\frac{1}{2}$, they ought to obey the exclusion principle; yet for three particular baryons (*uuu, ddd, sss*) all three quarks would have the same quantum numbers, and at least two of them have their spin in the same direction (since there are only two choices, spin up [$m_S = +\frac{1}{2}$] or spin down [$m_S = -\frac{1}{2}$]). This would seem to violate the exclusion principle, but if quarks have an additional quantum number (color), which could be different for each quark, it would serve to distinguish them and the exclusion principle would hold. Although quark color, and the resulting increase in number of quarks (three-fold), was thus originally an *ad hoc* idea, it also served to bring the theory into better agreement with experiment, such as predicting the correct lifetime of the π^0 meson. The idea of color soon became, in addition, a central feature of the theory as determining the force binding quarks together in a hadron. Each quark is assumed to carry a *color charge*, analogous to electric charge, and the strong force between quarks is often referred to as the **color force.** This new theory of the strong force is called **quantum chromodynamics** (*chrome* = color in Greek), or QCD, to indicate that the force acts between color charges (and not between, say, electric charges). The strong force between hadrons[†] is considered to be a force between the quarks that make them up, as suggested in Fig. 31-13. The particles that transmit the force (analogous to photons for the EM force) are called **gluons** (a play on "glue"). There are eight gluons, according to the theory, all massless, and six of them have color charge.[‡] Thus gluons have replaced mesons (Table 31-1) as the particles mediating the strong (color) force.

The weak force, as we have seen, is thought to be mediated by the W^+, W^-, and Z^0 particles. It acts between the "weak charge" that each particle has. Each elementary particle thus has electric charge, weak charge, color charge, and gravitational mass, although one or more of these could be zero. For example, all leptons have color charge of zero, so they do not interact via the strong force.

To summarize, the latest theories consider the truly elementary particles to be the photon, leptons, quarks, gluons, and $W^\pm$ and Z^0. The photon and the leptons are observed in experiments, and finally so were the W^+, W^-, and Z^0. But so far only combinations of quarks (baryons, mesons) have been observed, and it seems likely that free quarks are unobservable. On the other hand, some physicists believe that leptons and quarks are not fundamental, but are composites of still more fundamental objects. Only the future can tell.

One important aspect of new theoretical work is the attempt to find a unified basis for the different forces in nature. This was a long-held hope of Einstein, which he was never able to fulfill. A so-called "gauge" theory that unified the weak and electromagnetic interactions was put forward in the 1960s by Weinberg, Glashow, and Salam. In this **electroweak theory,** the weak and electromagnetic forces are seen as two different manifestations of a single, more fundamental, electroweak interaction. The electroweak theory has had many successes, including the prediction of the $W^\pm$ particles, as intermediates for the weak force, having masses of $81 \pm 2 \, \text{GeV}/c^2$ in excellent agreement with the measured (1983) values of $83 \pm 3 \, \text{GeV}/c^2$ (and similar accuracy for the Z^0 particle). The electroweak theory and QCD for the strong interaction are often referred to today as the "standard model."

31-11 ▌ Grand unified theories

With the success of the unified electroweak theory, attempts have recently been made also to incorporate it and QCD for the strong (color) force into a so-called **grand unified theory** (GUT). One type of such a grand unified theory of the electromagnetic, weak, and strong forces has been worked out

[†] The strong force between hadrons appears feeble, however, in comparison to the force directly between quarks within a hadron.

[‡] Compare to the EM interaction where the photon has no electric charge. Because gluons have color charge, they could attract each other and form composite particles (photons cannot). Such "glueballs" are being searched for and may have been observed.

FIGURE 31-13

(a) Force between two quarks, holding them together as a proton, say, is carried by a gluon which in this case involves a change in color. (b) Strong interaction $\pi^- p \to \pi^0 n$ with the exchange of a charged π meson (+ or − depending on whether it is considered moving to the left or to the right). (c) Quark representation of the same interaction $\pi^- p \to \pi^0 n$; the intermediate pion ($d\bar{u}$ or $\bar{d}u$) can be interpreted as the annihilation of a $u\bar{u}$ pair with production of a $d\bar{d}$ pair. The wavy lines between quarks represent gluon exchanges holding the hadrons together.

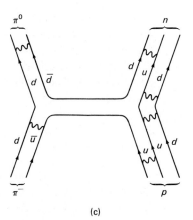

in which there is only one class of particle—leptons and quarks belonging to the same family and able to change freely from one type to the other—and the three forces are different aspects of a single underlying force. The unity is predicted to occur, however, only on a scale of less than about 10^{-31} m. If two elementary particles (leptons or quarks) approach each other to within this *unification scale*, the apparently fundamental distinction between them would not exist at this level, and a quark could readily change to a lepton, or vice versa. Baryon and lepton numbers would not be conserved. Leptons and quarks would belong to a single family. The weak, electromagnetic, and strong (color) force would blend to a force of a single strength.

How could a lepton become a quark, or vice versa? The theory predicts the existence of particles that can be exchanged between a quark and a lepton that allows one to change into the other, just as the charged pion exchanged between the π^- and p in Fig. 31-13b allows the proton to become a neutron. The mass of the new exchange particles, consistent with the uncertainty principle as applied earlier in this chapter (see Eq. 31-2 with $d \approx 10^{-31}$ m), would be about 10^{15} GeV/c^2, or 10^{15} times the proton mass. With such an incredibly large rest energy, there is little hope of producing them in the laboratory. It is also this huge mass which keeps baryon and lepton numbers conserved in observed reactions, since the likelihood of producing such a massive particle, even as a virtual exchange particle, is extremely small at even the highest laboratory energies.

What happens between the unification distance of 10^{-31} m and more normal (larger) distances is referred to as *symmetry breaking*. As an analogy, consider an atom in a crystal. Deep within the atom, there is much symmetry—in the innermost regions the electron cloud is spherically symmetric; no particular direction is singled out. Further out, this symmetry breaks down—the electron clouds are distributed preferentially along the lines joining the atoms in the crystal. In a similar way, at 10^{-31} m the force between elementary particles appears as one—it is symmetrical and does not single out one type of "charge" over another. But at larger distances, that symmetry is broken and we see three distinct forces. (In the "standard model" of electroweak interactions, Section 31-10, the symmetry breaking between the electromagnetic and the weak interactions occurs at about 10^{-18} m.)

Since unification occurs at such tiny distances and huge energies, the theory is difficult to test experimentally. But it is not completely impossible. One possibly testable prediction is the basis for the idea, suggested at the end of Section 31-5, that the proton might decay (via, for example, $p \rightarrow \pi^0 e^+$) and violate conservation of baryon number. This could happen if two quarks approached to within 10^{-31} m of each other; but it is very unlikely at normal temperature and energy, so the decay of a proton can only be an unlikely process. In the simplest form of GUT, the theoretical estimate of the proton lifetime is $\approx 10^{31}$ yr, and this has just come within the realm of testability. Proton decays have still not been seen and experiments in 1984 put the lower limit on the proton lifetime to be 2×10^{32} yr, almost an order of magnitude greater than this prediction. This may seem a disappointment, but on the other hand, it presents a challenge; indeed more complex GUTs are not affected by this result.

Another interesting prediction of unified theories relates to cosmology. It is thought that during the first 10^{-40} s after the theorized big bang that created the universe, the temperature was so extremely high that particles had energies corresponding to the unification scale; then baryon number would not have been conserved. This could account for the observed predominance of matter ($B > 0$) over antimatter ($B < 0$) in the universe.

This last example is interesting, for it illustrates a deep connection between investigations at either end of the size scale: theories about the tiniest objects (elementary particles) have a strong bearing on the understanding of the universe as a whole.

Even more ambitious than grand unified theories are attempts to incorporate also gravity, and thus unify all four forces in nature into a single theory. The world of elementary particles is opening new vistas. What happens in the near future is bound to be exciting.

▌ Summary

Particle accelerators are used to accelerate charged particles such as electrons and protons to very high energy. High-energy particles have short wavelength and so can be used to probe in great detail; high kinetic energy also allows the creation of new particles through collision (via $E = mc^2$). Van de Graaff and linear accelerators use high voltage to accelerate particles along a line. Cyclotrons and synchrotrons use a magnetic field to keep the particles in a circular path and they are accelerated at intervals by high voltage.

Just as the electromagnetic force can be said to be due to an exchange of photons, the strong nuclear force is thought to be carried by *mesons* that have rest mass, or, according to more recent theory, by massless *gluons*.

An *antiparticle* has the same mass as a particle but opposite charge. Certain other properties may also be opposite: for example, the antiproton has *baryon number* (nucleon number) opposite to that for the proton. In all nuclear and particle reactions, the following conservation laws hold: momentum, mass-energy, angular momentum, electric charge, baryon number, and the three lepton numbers. Certain particles have a property, called *strangeness*, which is conserved by the strong force but not by the weak force.

Particles can be classified as *leptons* and *hadrons*, plus the photon. Leptons participate in the weak and electromagnetic interactions. Hadrons participate in the strong interaction as well. The hadrons can be classified as *mesons*, with baryon number zero, and *baryons* with nonzero baryon number.

All particles, except for the photon, electron, neutrinos, and (so far at least) proton, decay with measurable half-lives varying from 10^{-23} s to 10^3 s. The half-life depends on which force is predominant in the decay. Weak decays have half-lives greater than about 10^{-10} s. Electromagnetic decays have half-lives on the order of 10^{-16} to 10^{-19} s. The shortest-lived particles, called *resonances*, decay via the strong interaction and live typically for only about 10^{-23} s.

The latest theories of elementary-particle physics postulate the existence of *quarks* as the basic building blocks of the hadrons. Initially, three quarks were proposed. Recent evidence suggests that a fourth, *charmed*, quark is needed and also a fifth and sixth. It is expected that there are the same number of quarks as leptons, and that quarks and leptons are the truly elementary particles. Quarks are said to have *color*, and, according to *quantum chromodynamics*, the strong color force acts between their color charges and is transmitted by *gluons*. A unified theory of forces suggests that at very short distances (10^{-31} m) and very high energy, the weak, electromagnetic, and strong forces appear as a single force, and the fundamental difference between quarks and leptons disappears.

▌ Questions

1. What limits the maximum energy attainable for protons in an ordinary cyclotron? How is this overcome in a synchrotron?

2. A proton in a synchrotron has a speed of $0.99c$. What must be done to increase its energy?

3. Give a reaction between two nucleons that could produce a π^-, similar to Eq. 31-3.

4. Draw a Feynman diagram for $n + p \rightarrow n + p + \pi^0$.

5. Draw a Feynman diagram for $p\bar{p}$ annihilation with the production of two pions.

6. Draw a Feynman diagram for the photoelectric effect.

7. What would an "antiatom," made up of antiparticles to constituents of normal atoms, consist of? What might happen if *antimatter*, made of such antiatoms, came in contact with our normal world of matter?

8. Why is it that a neutron decays via the weak interaction even though the neutron and one of its decay products (proton) are strongly interacting?

9. Which of the four interactions (strong, electromagnetic, weak, gravitational) does an electron take part in? A neutrino? A proton?

10. Check that charge and baryon number are conserved in each of the decays in Table 31-2.

11. Which of the particle decays in Table 31-2 occur via the electromagnetic interaction?

12. Which of the particle decays in Table 31-2 occur by the weak interaction?

13. Which of the following decays are possible? For those that are forbidden, explain which laws are violated.
    (a) $\Xi^0 \rightarrow \Sigma^+ + \pi^-$
    (b) $\Omega^- \rightarrow \Sigma^0 + \pi^- + \nu$
    (c) $\Sigma \rightarrow \Lambda + \gamma + \gamma$

14. Which of the following reactions are possible, and by what interaction could they occur? For those forbidden, explain why.
    (a) $\pi^- p \rightarrow K^+ \Sigma^-$
    (b) $\pi^+ p \rightarrow K^+ \Sigma^+$
    (c) $\pi^- p \rightarrow \Lambda^0 K^0 \pi^0$
    (d) $\pi^+ p \rightarrow \Sigma^0 \pi^0$
    (e) $\pi^- p \rightarrow p e^- \bar{\nu}_e$
    (f) $\pi^- p \rightarrow K^0 p \pi^0$
    (g) $K^- p \rightarrow \Lambda^0 \pi^0$
    (h) $K^+ n \rightarrow \Sigma^+ \pi^0 \gamma$
    (i) $K^+ \rightarrow \pi^0 \pi^0 \pi^+$
    (j) $\pi^+ \rightarrow e^+ \nu_e$

15. What are the quark combinations that can form a (a) neutron, (b) antineutron, (c) Λ^0, (d) $\overline{\Xi^0}$?

16. What particles do the following quark combinations produce? (a) uud, (b) $\bar{u}\bar{u}\bar{s}$, (c) $\bar{u}s$, (d) $d\bar{u}$, (e) $\bar{c}s$.

17. What is the quark combination needed to produce a D^0 meson ($Q = B = S = 0, C = +1$)?

18. The F^+ meson has $S = C = +1$, $B = 0$. What quark combination would produce it?

19. Draw a possible quark Feynman diagram (see Fig. 31-13c) for the reaction $K^- p \rightarrow K^- p$.

▌ Problems

SECTIONS 31-1 AND 31-2

1. (I) What is the total energy of a proton whose kinetic energy is 15.0 GeV?

2. (I) Calculate the wavelength of 20-GeV electrons.

3. (I) What strength of magnetic field is used in a cyclotron in which protons make 1.9×10^7 revolutions per second?

4. (I) If α particles are accelerated by the cyclotron of Example 31-2, what must be the frequency of voltage applied to the dees?

5. (I) If the cyclotron of Example 31-2 accelerated α particles, what maximum energy could they attain? What would their speed be? Repeat for deuterons (^2_1H).

6. (I) What is the time for one complete revolution for a very high energy proton in the 2.0-km-diameter Fermilab accelerator?

7. (I) The voltage across the dees of a cyclotron is 50 kV.

How many revolutions do protons make to reach a kinetic energy of 10 MeV?

8. (I) What is the total energy of a proton whose kinetic energy is 25 GeV? What is its wavelength?

9. (II) What is the wavelength, and maximum resolving power attainable, using 400-GeV protons at Fermilab?

10. (II) Protons are injected into the 1.0-km-radius Fermilab synchrotron with an energy of 8.0 GeV. If they are accelerated by 2.5 MV each revolution, how far do they travel and approximately how long does it take for them to reach 400 GeV?

11. (III) What magnetic field intensity is needed at the 1.0-km-radius Fermilab synchrotron for 400-GeV protons? Use the relativistic mass.

12. (III) Show that the energy of a particle (charge e) in a synchrotron, in the relativistic limit ($v \approx c$), is given by E (in eV) $= Brc$, where B is magnetic field strength and r the radius of the orbit (SI units).

13. (I) Two protons are heading toward each other with equal speeds. What minimum energy must each have if a π^0 meson is to be created in the process? (See Table 31-2.)

14. (I) How much energy is released in the decay $\pi^+ \rightarrow \mu^+ + \nu_\mu$?

15. (I) About how much energy is released when a Λ^0 decays? (See Table 31-2.)

16. (I) (a) How much energy is released when an electron and a positron annihilate each other? (b) How much energy is released when a proton and an antiproton annihilate each other?

17. (I) How much energy is required to produce a neutron–antineutron pair?

18. (II) What are the wavelengths of the two photons when a proton–antiproton pair at rest annihilate?

19. (II) What would be the wavelengths of the two photons produced when an electron–positron pair, each with 300 keV of KE, annihilate?

20. (II) In the rare decay $\pi^+ \rightarrow e^+ + \nu_e$, what is the kinetic energy of the positron?

21. (II) What minimum kinetic energy must a neutron and proton each have if they are traveling at the same speed toward each other, collide, and produce a K^+K^- pair in addition to themselves? (See Table 31-2.)

22. (II) The mass of a π^0 can be measured by observing the reaction $\pi^- + p \rightarrow \pi^0 + n$ at very low incident π^- kinetic energy (assume it is zero). The neutron is observed to be emitted with a KE of 0.60 MeV. Use conservation of energy and momentum to determine the π^0 mass.

23. (III) Could a π^+ meson be produced if a 100-MeV proton struck a proton at rest? What minimum KE must it have?

SECTIONS 31-7 TO 31-11

24. (I) The measured width of the ψ meson is 63 keV. Estimate its lifetime.

25. (I) The measured width of the ψ' meson is 215 keV. Estimate its lifetime.

26. (I) The newly discovered B^- meson is presumed to be a $b\bar{u}$ quark combination. (a) Show that this is consistent for all quantum numbers. (b) What are the quark combinations for B^+, B^0, $\overline{B^0}$?

27. (II) (a) Show that the so-called unification distance of 10^{-31} m in recent unified theory is equivalent to an energy of about 10^{15} GeV. Use either the uncertainty principle or de Broglie's wavelength formula and explain how they apply. (b) Calculate what temperature this corresponds to.

28. (II) Symmetry breaking occurs in the electroweak theory at about 10^{-18} m. Show that this corresponds to an energy which is on the order of the mass of the $W^\pm$.

29. (II) Draw possible Feynman diagrams using quarks (as in Fig. 31-13c) for the reactions (a) $pn \rightarrow pn$, (b) $\bar{p}p \rightarrow \pi^+\pi^-$.

APPENDIX A
MATHEMATICAL REVIEW

A-1 ■ Relationships, proportionality, and equations

One of the important aspects of physics is the search for relationships between different quantities—that is, determining how one quantity affects another. For example, how does temperature affect the air pressure in a tire? Or how does the net force on an object affect its acceleration? Sometimes a given quantity is affected by two or more quantities; for instance, the acceleration of an object is related to both its mass and the applied force. If it is suspected that a relationship exists between two or more quantities, one can try to determine the precise nature of this relationship. This is done by varying one of the quantities and measuring how the other varies as a result. If it is likely that a particular quantity will be affected by more than one factor or quantity, only one of the latter is varied at a time, while the others are held constant.[†]

As a simple example, the ancients found that if one circle has twice the diameter of a second circle, the first also has twice the circumference. If the diameter is three times as large, the circumference is also three times as large. In other words, an increase in the diameter results in a proportional increase in the circumference. We say that the circumference is directly proportional to the diameter. This can be written in symbols as $C \propto D$, where "$\propto$" means "is proportional to" and C and D refer to the circumference and diameter of a circle, respectively. The next step is to change this proportionality to an equation, which will make it possible to link the two quantities numerically. This merely entails inserting a proportionality constant, which in many cases is determined by measurement. (In some cases it can be chosen arbitrarily, as in Section 3-4). The ancients found that the ratio of the circumference to the diameter of any circle was 3.1416 (to keep only the first few decimal places). This number is designated by the Greek letter π. It is the constant of proportionality for the relationship $C \propto D$, and to obtain an equation we insert it into the proportion and change the $\propto$ to $=$. Thus $C = \pi D$.

[†] When one quantity affects another, we often use the expression "is a function of" to indicate this dependence; for example, we say that the pressure in a tire is a function of the temperature.

Other kinds of proportionality occur as well. For example, the area of a circle is proportional to the *square* of its radius. That is, if the radius is doubled, the area becomes four times as large; and so on. In this case we can write $A \propto r^2$, where A stands for the area and r for the radius of the circle.

Sometimes two quantities are related in such a way that an increase in one leads to a proportional decrease in the other. This is called *inverse proportion*. For example, the time required to travel a given distance is inversely proportional to the speed of travel. The greater the speed, the less time it takes. We can write this inverse proportion as: time $\propto$ 1/speed. The larger the denominator of a fraction, the smaller is the fraction as a whole. For example, $\frac{1}{4}$ is smaller than $\frac{1}{2}$. Thus if the speed is doubled, the time is halved, which is what we want to express by this inverse proportionality relationship.

Whatever kind of proportion is found to hold, it can be changed to an equality by insertion of the proper proportionality constant. Quantitative statements or predictions about the physical world can then be made with the equation.

A-2 ▮ Accuracy and significant figures

Accurate measurements are an important part of physics. But no measurement is absolutely precise; there is an uncertainty associated with every measurement. Uncertainty arises from different sources; among the most important, other than blunders, are the limited accuracy of every measuring instrument and the inability to read an instrument beyond some fraction of the smallest division shown. For example, if you were to use a centimeter ruler to measure the width of a spice box, the result could be claimed to be accurate to about 0.1 cm, the smallest division on the ruler (although half of this value might be a valid claim as well). This is because it is difficult for the observer to interpolate between the smallest divisions, and the ruler itself has probably not been manufactured to an accuracy much better than this.

When giving the result of a measurement, it is good practice to state the precision, or **estimated uncertainty,** in the measurement. For example, the width of a board would be written as 23.2 ± 0.1 cm. The ± 0.1 cm ("plus or minus 0.1 cm") represents the estimated uncertainty in the measurement, so that the actual width most likely lies between 23.1 and 23.3 cm. The *percent uncertainty* is simply the ratio of the uncertainty to the measured value, multiplied by 100; for example, if the measurement is 23.2 and the uncertainty about 0.1 cm, the percent uncertainty is

$$\frac{0.1}{23.2} \times 100 = 0.4\%.$$

Often the uncertainty in a measured value is not specified explicitly; in this case, it is generally accepted that the uncertainty is approximately one or two units in the last digit specified. Although this is not as precise as actually specifying the uncertainty, it is often adequate. For example if a length is given as 23.2 cm, the uncertainty is assumed to be about 0.1 cm (or perhaps 0.2 cm). It is important in this case that you not write 23.20 cm, for this implies an uncertainty of 0.01 cm; it assumes that the length is probably between 23.19 cm and 23.21 cm when actually you only believe it is between 23.1 and 23.3 cm.

The number of reliably known digits in a number is called the number of **significant figures.** Thus there are four significant figures in the number 23.21 and two in the number 0.062 cm.

When making measurements, or when doing calculations, one should never keep more digits in the final answer than the number of significant figures in the least accurate factor in the calculation. For example, to calculate the area of a rectangle 11.3 by 6.8 cm, the result of multiplication would be 76.84 cm^2, but this answer is clearly not accurate to 0.01 cm^2, since (using the outer limits of the assumed uncertainty) the result could be between $11.2 \times 6.7 = 75.04$ cm^2 and $11.4 \times 6.9 = 78.46$ cm^2. At best, we can quote the answer as 77 cm^2, which implies an uncertainty of about 1 or 2 cm^2. As a general rule, *the final result of a multiplication or division should have only as many digits as the least significant number used in the calculation.* In our example, 6.8 cm was the least significant number and it had two significant figures; thus the result 76.84 cm^2 is rounded off to 77 cm^2. (However, it is usually good practice to keep an extra significant figure throughout a calculation, if possible, even though it isn't quoted in the final result.)

Another way to indicate significant figures is shown in Section A-4.

A-3 ▌ Exponents

When we write 10^4 we mean that you multiply 10 times itself four times: $10^4 = 10 \times 10 \times 10 \times 10 = 10,000$. The superscript 4 is called an *exponent*, and 10 is said to be raised to the fourth power. Any number or symbol can be raised to a power; special names are used when the exponent is 2 (a^2 is "a squared") or 3 (a^3 is "a cubed"). For any other power we say a^n is "a to the nth power." If the exponent is 1, it is usually dropped: $a^1 = a$, since no multiplication is involved.

The rules for multiplying numbers expressed as powers are as follows:

$$(a^n)(a^m) = a^{n+m}. \qquad \text{(A-1)}$$

That is, the exponents are added. To see why, consider the result of the multiplication of 3^3 by 3^4:

$$(3^3)(3^4) = (3)(3)(3) \times (3)(3)(3)(3) = (3)^7.$$

Here the sum of the exponents is $3 + 4 = 7$, so rule A-1 works. Notice that this rule works only if the base numbers (a in Eq. A-1) are the same. Thus we *cannot* use the rule of summing exponents for $(6^3)(5^2)$; these numbers would have to be written out. However, if the base numbers are different but the exponents are the same, we can write a second rule:

$$(a^n)(b^n) = (ab)^n. \qquad \text{(A-2)}$$

For example, $(5^3)(6^3) = (30)^3$ since

$$(5)(5)(5)(6)(6)(6) = (30)(30)(30).$$

The third rule involves a power raised to another power: $(a^3)^2$ means $(a^3)(a^3)$, which is equal to $a^{3+3} = a^6$. The general rule is then

$$(a^n)^m = a^{nm}. \qquad \text{(A-3)}$$

In this case, the exponents are multiplied.

Negative exponents are used for reciprocals. Thus

$$\frac{1}{a} = a^{-1}, \qquad \frac{1}{a^3} = a^{-3},$$

and so on. The reason for using negative exponents is that it allows us to use the multiplication rules given above. For example, $(a^5)(a^{-3})$ means

$$\frac{(a)(a)(a)(a)(a)}{(a)(a)(a)} = a^2.$$

Rule A-1 gives us the same result:

$$(a^5)(a^{-3}) = a^{5-3} = a^2.$$

What does an exponent of zero mean? That is, what is a^0? Any number raised to the zeroth power is defined as being equal to 1:

$$a^0 = 1.$$

This definition is used because it follows from the rules for adding exponents. For example,

$$a^3 a^{-3} = a^{3-3} = a^0 = 1.$$

But *does* $a^3 a^{-3}$ actually equal 1? Yes, because

$$a^3 a^{-3} = \frac{a^3}{a^3} = 1.$$

Fractional exponents are used to represent *roots*. For example, $a^{1/2}$ means the square root of a; that is $a^{1/2} = \sqrt{a}$. Similarly, $a^{1/3}$ means the cube root of a, and so on. The fourth root of a means that if you multiply the fourth root of a by itself four times, you again get a:

$$(a^{1/4})^4 = a.$$

This is consistent with rule A-3 since $(a^{1/4})^4 = a^{4/4} = a^1 = a$.

A-4 ▮ Powers of 10, or exponential notation

Writing out very large and very small numbers such as the distance of Neptune from the sun, 4,500,000,000 km, or the diameter of a typical atom, 0.00000001 cm, is inconvenient and prone to error. It also leaves in question the number of significant figures. (How many of the zeros are significant in the number 4,500,000,000 km?) We therefore make use of the "powers of 10," or exponential notation. The distance from Neptune to the sun is then expressed as 4.50×10^9 km (assuming that the value is significant to three digits) and the diameter of an atom 1.0×10^{-8} cm. This way of writing numbers is based on the use of exponents, where a^n signifies a multiplied by itself n times. For example, $10^4 = 10 \times 10 \times 10 \times 10 = 10{,}000$. Thus $4.50 \times 10^9 = 4.50 \times 1{,}000{,}000{,}000 = 4{,}500{,}000{,}000$. Notice that the exponent (9 in this case) is just the number of places the decimal point is moved to the right to obtain the fully written out number (4.500,000,000.)

When two numbers are multiplied (or divided), you first multiply (divide) the simple parts and then the powers of 10. Thus 2.0×10^3 multiplied by 5.5×10^4 equals $(2.0 \times 5.5) \times (10^3 \times 10^4) = 11 \times 10^7$, where we have used the rule for adding exponents (Section A-3). Similarly, 8.2×10^5 divided by 2.0×10^2 equals

$$\frac{8.2 \times 10^5}{2.0 \times 10^2} = \frac{8.2}{2.0} \times \frac{10^5}{10^2} = 4.1 \times 10^3.$$

For numbers less than 1, say 0.01, the exponent or power of 10 is written with a negative sign: $0.01 = 1/100 = 1/10^2 = 1 \times 10^{-2}$. Similarly, $0.002 = 2 \times 10^{-3}$. The decimal point has again been moved the number of places expressed in the exponent. The negative exponent allows calculations to come out correctly. Thus $0.020 \times 3600 = 72$; in exponential notation $(2.0 \times 10^{-2}) \times (3.6 \times 10^3) = 7.2 \times 10^1 = 72$.

Notice also that $10^1 \times 10^{-1} = 10 \times 0.1 = 1$, and by the law of exponents, $10^1 \times 10^{-1} = 10^0$. Therefore, $10^0 = 1$.

When writing a number in exponential notation, it is usual to make the simple number be between 1 and 10. Thus it is conventional to write 4.5×10^9 rather than 45×10^8, although they are the same number.[†] This notation also allows the number of **significant figures** to be clearly expressed. We write 4.50×10^9 if this value is accurate to three significant figures, but 4.5×10^9 if it is accurate to only two.

A-5 ▌ Algebra

Physical relationships between quantities can be represented as equations involving symbols (usually letters of the alphabet) that represent the quantities. The manipulation of such equations is the field of algebra, and is used a great deal in physics. An equation involves an equals sign, which tells us that the quantities on either side of the equals sign have the same value. Examples of equations are

$$3 + 8 = 11$$
$$2x + 7 = 15$$
$$a^2 b + c = 6.$$

The first equation involves only numbers, so is called an arithmetic equation. The other two equations are algebraic since they involve symbols. In the third equation, the quantity $a^2 b$ means the product of a times a times b: $a^2 b = a \times a \times b$.

[†] Another convention used, particularly with computers, is that the simple number be between 0.1 and 1. Thus we would write 4,500,000,000 as 0.450×10^{10}.

Often we wish to solve for one (or more) symbols, and we treat it as an *unknown*. For example, in the equation $2x + 7 = 15$, x is the unknown; this equation is true, however, only when $x = 4$. Determining what value (or values) the unknown(s) can have to satisfy the equation(s) is called *solving the equation*. To solve an equation, the following rule can be used:

> *An equation will remain true if any operation performed on one side is also performed on the other side:* for example, (*a*) addition or subtraction of a number or symbol; (*b*) multiplication or division by a number or symbol; (*c*) raising each side of the equation to the same power, or taking the same root (such as square root).

EXAMPLE A-1   Solve for x in the equation

$$2x + 7 = 15.$$

SOLUTION   We first subtract 7 from both sides:

$$2x + 7 - 7 = 15 - 7$$
$$2x = 8.$$

Then we divide both sides by 2 to get

$$\frac{2x}{2} = \frac{8}{2}$$
$$x = 4,$$

and this solves the equation.

EXAMPLE A-2   (*a*) Solve the equation

$$a^2 b + c = 24$$

for the unknown a in terms of b and c. (*b*) Solve for a assuming that $b = 2$ and $c = 6$.

SOLUTION   (*a*) We are trying to solve for a, so we first subtract c from both sides

$$a^2 b = 24 - c,$$

then divide by b:

$$a^2 = \frac{24 - c}{b},$$

and finally take square roots,

$$a = \sqrt{\frac{24 - c}{b}}.$$

(*b*) If we are given that $b = 2$ and $c = 6$, then

$$a = \sqrt{\frac{24 - 6}{2}} = 3.$$

To check a solution, we put it back into the original equation (this is really a check that we did all the manipulations correctly). In the equation

$$a^2b + c = 24$$

we put in $a = 3$, $b = 2$, $c = 6$ and find

$$(3)^2(2) + (6) \stackrel{?}{=} 24$$
$$24 = 24,$$

which checks.

Two or more unknowns

If we have two or more unknowns, one equation is not sufficient to find them. In general, if there are n unknowns, n independent equations are needed. For example, if there are two unknowns, we need two equations. If the unknowns are called x and y, a typical procedure is to solve one equation for x in terms of y, and substitute this into the second equation.

EXAMPLE A-3   Solve the following pair of equations for x and y:

$$3x - 2y = 19$$
$$x + 4y = -3.$$

SOLUTION   We solve the second equation for x in terms of y by subtracting $4y$ from both sides:

$$x = -3 - 4y.$$

We substitute this expression for x into the first equation, and simplify

$$3(-3 - 4y) - 2y = 19$$
$$-9 - 12y - 2y = 19 \qquad \text{(carried out the multiplication by 3)}$$
$$-14y = 28 \qquad \text{(added 9 to both sides)}$$
$$y = -2 \qquad \text{(divided both sides by } -14).$$

Now that we know $y = -2$, we substitute this into the expression for x:

$$x = -3 - 4y$$
$$= -3 - 4(-2) = -3 + 8 = 5.$$

Our solution is $x = 5$, $y = -2$. We check this solution by putting these values back into the original equations:

$$3x - 2y \stackrel{?}{=} 19$$
$$3(5) - 2(-2) \stackrel{?}{=} 19$$
$$15 + 4 \stackrel{?}{=} 19$$
$$19 = 19 \qquad \text{(it checks)}$$

and

$$x + 4y \stackrel{?}{=} -3$$
$$5 + 4(-2) \stackrel{?}{=} -3$$
$$-3 = -3 \qquad \text{(it checks).}$$

Other methods for solving two or more equations, such as the method of determinants, can be found in an algebra textbook.

The quadratic formula

We sometimes encounter equations that involve an unknown, say x, that appears not only to the first power, but squared as well. Such a *quadratic equation* can be written in the form

$$ax^2 + bx + c = 0.$$

The quantities a, b, and c are typically given numbers or constants.[†] The general solutions to such an equation are given by the *quadratic formula*:

$$x = \frac{-b \pm \sqrt{b^2 - 4ac}}{2a}.$$

The $\pm$ sign indicates that there are two solutions for x: one where the plus sign is used, the other where the minus sign is used.[‡]

> ☐   **EXAMPLE A-4**   Find the solutions for x in the equation
>
> $$3x^2 - 5x = 2.$$
>
> **SOLUTION**   First we write this equation in the standard form $ax^2 + bx + c = 0$ by subtracting 2 from both sides:
>
> $$3x^2 - 5x - 2 = 0.$$
>
> In this case, a, b, and c in the standard formula take the values $a = 3$, $b = -5$, and $c = -2$. The two solutions for x are
>
> $$x = \frac{+5 + \sqrt{25 - (4)(3)(-2)}}{(2)(3)} = \frac{5 + 7}{6} = 2$$
>
> and
>
> $$x = \frac{+5 - \sqrt{25 - (4)(3)(-2)}}{(2)(3)} = \frac{5 - 7}{6} = -\frac{1}{3}.$$

In this example the two solutions are $x = 2$ and $x = -\frac{1}{3}$. In physics problems, however, only one of these solutions may correspond to a real-life solution; in this case the other solution is discarded. In other cases both solutions may correspond to physical reality.

Notice, incidentally, that b^2 must be greater than $4ac$, so that $\sqrt{b^2 - 4ac}$ yields a real number. If $(b^2 - 4ac)$ is less than zero (negative), there is no real solution. The square root of a negative number is called *imaginary*.

[†] Or one or more of them could be variables, in which case additional equations are needed.

[‡] A second-order equation—one in which the highest power of x is 2—has two solutions; a third-order equation—involving x^3—has three solutions, and so on.

A-6 ■ Plane geometry

FIGURE A-1

We review here a number of theorems involving angles and triangles that are useful in physics.

1. *Equal angles.* Two angles are equal if any of the following conditions are true:
   (*a*) They are vertical angles (Fig. A-1); *or*
   (*b*) the left side of one is parallel to the left side of the other, and the right side of one is parallel to the right side of the other (the left and right sides are as seen from the vertex, where the two sides meet; Fig. A-2); *or*
   (*c*) the left side of one is perpendicular to the left side of the other, and the right sides are likewise perpendicular (Fig. A-3).

2. *The sum of the angles* in any plane triangle is 180°.

FIGURE A-2

3. *Similar triangles.* Two triangles are said to be similar if all three of their angles are equal (in Fig. A-4, $\theta_1 = \phi_1, \theta_2 = \phi_2, \theta_3 = \phi_3$). Similar triangles thus have the same basic shape but may be different sizes and have different orientations. Two useful theorems about similar triangles are:
   (*a*) Two triangles are similar if any two of their angles are equal. (This follows because the third angles must also be equal since their sum is 180°.)
   (*b*) The ratio of corresponding sides of two similar triangles are equal. That is (Fig. A-4)

$$\frac{a_1}{b_1} = \frac{a_2}{b_2} = \frac{a_3}{b_3}.$$

FIGURE A-3

4. *Congruent triangles.* Two triangles are congruent if one can be placed precisely on top of the other. That is, they are similar triangles and they have the same size. Two triangles are congruent if any of the following holds:
   (*a*) The three corresponding sides are equal.
   (*b*) Two sides and the enclosed angle are equal ("side-angle-side").
   (*c*) Two angles and the enclosed side are equal ("angle-side-angle").

5. *Right triangles.* A right triangle has one angle that is 90° (a *right angle*); that is, the two sides that meet at the right angle are perpendicular. The two other (acute) angles in the right triangle add up to 90°.

6. *Pythagorean theorem.* In any right triangle, the square of the length of the hypotenuse (the side opposite the right angle) is equal to the sum of the squares of the lengths of the other two sides. From Fig. A-5,

$$c^2 = a^2 + b^2.$$

FIGURE A-4

A-7 ■ Logarithms

FIGURE A-5

Logarithms are defined in the following way:

$$\text{if} \quad y = A^x, \quad \text{then} \quad x = \log_A y.$$

That is, the logarithm of a number y to the base A is that number which, as the exponent of A, gives back the number y. For *common logarithms*, the base is 10, so

$$\text{if} \quad y = 10^x, \quad \text{then} \quad x = \log y.$$

The subscript 10 on $\log_{10}$ is usually omitted when dealing with common logs. Another base sometimes used is the natural number $e = 2.718\ldots$[†] Such logarithms are called *natural logarithms* and are written ln. Thus

$$\text{if}\quad y = e^x,\qquad \text{then}\quad x = \ln y.$$

For any number y, the two types of logarithm are related by

$$\ln y = 2.3026 \log y.$$

Some simple rules for logarithms are as follows:

$$\log (ab) = \log a + \log b. \tag{A-4}$$

This is true because if $a = 10^n$ and $b = 10^m$, then $ab = 10^{n+m}$. From the definition of logarithm, $\log a = n$, $\log b = m$, and $\log (ab) = n + m$; hence $\log (ab) = n + m = \log a + \log b$. In a similar way, we can show that

$$\log \left(\frac{a}{b}\right) = \log a - \log b \tag{A-5}$$

and

$$\log a^n = n \log a. \tag{A-6}$$

These three rules apply not only to common logs but to natural or any other kind of logarithm.

Logs were once used as a technique for simplifying certain types of calculation. Because of the advent of electronic calculators and computers, they are not often used any more for this purpose. However, logs do appear in certain physical equations, so it is helpful to know how to deal with them. If you do not have a calculator that calculates logs, you can easily use a *log table*, such as the small one (Table A-1) below. The number N is given to two digits (some tables give N to three or more digits); the first digit is in the vertical column to the left, the second digit is in the horizontal row across the top. For example, the table tells us that $\log 1.0 = 0.000$, $\log 1.1 = 0.041$, and $\log 4.1 = 0.613$; note that the table does not include the decimal point—

[†] The natural number e can be written as an infinite series:

$$e = 1 + \frac{1}{1} + \frac{1}{1 \cdot 2} + \frac{1}{1 \cdot 2 \cdot 3} + \frac{1}{1 \cdot 2 \cdot 3 \cdot 4} + \cdots.$$

TABLE A-1
Short table of common logarithms

N	0.0	0.1	0.2	0.3	0.4	0.5	0.6	0.7	0.8	0.9
1	000	041	079	114	146	176	204	230	255	279
2	301	322	342	362	380	398	415	431	447	462
3	477	491	505	519	531	544	556	568	580	591
4	602	613	624	633	643	653	663	672	681	690
5	699	708	716	724	732	740	748	756	763	771
6	778	785	792	799	806	813	820	826	833	839
7	845	851	857	863	869	875	881	887	892	898
8	903	908	914	919	924	929	935	940	944	949
9	954	959	964	968	973	978	982	987	991	996

it is understood. The table gives logs for numbers between 1.0 and 9.9; for larger or smaller numbers we use rule A-4:

$$\log (ab) = \log a + \log b.$$

For example,

$$\log (380) = \log (3.8 \times 10^2) = \log (3.8) + \log (10^2).$$

From the table, $\log 3.8 = 0.580$; and from rule A-6, $\log (10^2) = 2 \log (10) = 2$ since $\log (10) = 1$. [This follows from the definition of the logarithm: if $10 = 10^1$, then $1 = \log (10)$.] Thus

$$\log (380) = \log (3.8) + \log (10^2)$$
$$= 0.580 + 2$$
$$= 2.580.$$

Similarly,

$$\log (0.081) = \log (8.1) + \log (10^{-2})$$
$$= 0.908 - 2 = -1.092.$$

Sometimes we need to do the reverse process: find the number N whose log is, say, 2.670. This is called "taking the antilogarithm." To do so, we separate our number 2.670 into two parts, making the separation at the decimal point:

$$\log N = 2.670 = 2 + 0.670$$
$$= \log 10^2 + 0.670.$$

We now look in the table to see what number has its log equal to 0.670; none does, so we must *interpolate*: we see that $\log 4.6 = 0.663$ and $\log 4.7 = 0.672$. So the number we want is between 4.6 and 4.7, and closer to the latter by 7/9. Approximately we can say that $\log 4.68 = 0.670$. Thus

$$\log N = 2 + 0.670$$
$$= \log (10^2) + \log (4.68) = \log (4.68 \times 10^2),$$

so $N = 4.68 \times 10^2 = 468$. If the given logarithm is negative, say, -2.180, we proceed as follows:

$$\log N = -2.180 = -3 + 0.820$$
$$= \log 10^{-3} + \log 6.6 = \log 6.6 \times 10^{-3},$$

so $N = 6.6 \times 10^{-3}$. Notice that what we did was to add to our given logarithm the next largest integer (3 in this case) so that we have an integer, plus a decimal number between 0 and 1.0 whose antilogarithm can be looked up in the table.

APPENDIX B
ORDER OF MAGNITUDE: RAPID ESTIMATING

We are sometimes interested only in an approximate value for a quantity. This might be because an accurate calculation would take more time than it is worth or would require additional data that are not available. In other cases, we may want to make a rough estimate in order to check an accurate calculation made on a calculator to make sure that no blunders were made when entering the numbers. Also, the correct power of 10 may be lost on a calculator or slide rule, and a rough estimate can be used to obtain it.

In general, a rough estimate is made by rounding off all numbers to one significant figure and its power of 10, and after the calculation is made, again only one significant figure is kept. Such an estimate is called an *order-of-magnitude* estimate and can be assumed to be accurate within a factor of 10, and usually better; in fact, the phrase "order of magnitude" is sometimes used to refer simply to the power of 10.

As an example, suppose that a person wants to find out how much water there is in a particular lake, which is roughly circular, about 1 km across, and has an average depth of 10 m. To find the volume, we simply multiply the average depth of the lake times its surface area (as if it were a cylinder). We approximate the surface to be a circle, so the area is πr^2, which is approximately $3 \times (5 \times 10^2 \text{ m})^2 \approx 8 \times 10^5 \text{ m}^2$, where the radius r is 500 m and π was rounded off to 3 ($\approx$ means "approximately equal to"). Then the volume is about $(8 \times 10^5 \text{ m}^2) \times (10 \text{ m}) = 8 \times 10^6 \text{ m}^3$, which is on the order of 10^7 m^3. Because of all the estimates that went into this calculation, the order-of-magnitude estimate (10^7) is probably better to quote than the 8×10^6 figure.

APPENDIX C
DIMENSIONAL ANALYSIS

When we speak of the *dimensions* of a quantity, we are referring to the type of units that must be used. The dimensions of area, for example, are always length squared (abbreviated $[L^2]$, using square brackets) and the units can be square meters, square feet, and so on. Velocity, on the other hand, can be measured in units of km/h, m/s, and mi/h, but the dimensions are always a length $[L]$ divided by a time $[T]$, that is, $[L/T]$. The formula for a quantity may be different in different cases, but the dimensions remain the same. For example, the area of a triangle of base b and height h is $A = \frac{1}{2}bh$, whereas the area of a circle of radius r is $A = \pi r^2$. The formulas are different in the two cases, but the dimensions in both cases are the same: $[L^2]$.

When we specify the dimensions of a quantity, we usually do so in terms of base quantities, not derived quantities (see Section 1-3). The four common base quantities are length $[L]$, time $[T]$, mass $[M]$, and electric current $[I]$. (There are also three other, less used, basic quantities—see Section 1-3.) Thus, force, which by Newton's second law has the same units as mass $[M]$ times acceleration $[L/T^2]$, has dimensions of $[ML/T^2]$.

Dimensions can be used as a help in working out relationships, and such a procedure is referred to as *dimensional analysis*. One useful technique is the use of dimensions to check a relationship for correctness. Two simple rules apply here. First, we can add or subtract quantities only if they have the same dimensions (we do not add centimeters and pounds); second, the quantities on each side of an equals sign must have the same dimensions.

For example, suppose that you derived the equation $v = v_0 + \frac{1}{2}at^2$, where v is the speed of an object after a time t, when it starts with an initial speed v_0 and undergoes an acceleration a. Let us do a dimensional check to see if this equation is correct. We write a dimensional equation as follows, remembering that the dimensions of speed are $[L/T]$ and of acceleration are $[L/T^2]$:

$$\left[\frac{L}{T}\right] \stackrel{?}{=} \left[\frac{L}{T}\right] + \left[\frac{L}{T^2}\right][T^2]$$

$$\stackrel{?}{=} \left[\frac{L}{T}\right] + [L].$$

The dimensions are incorrect: on the right side, we have the sum of two quantities whose dimensions are not the same. Thus we conclude that an error was made in the derivation of the original equation.

If such a dimensional check does come out correct, it does not prove that the equation is correct; for example, a dimensionless numerical factor (such as $\frac{1}{2}$ or 2π) could be wrong. Thus a dimensional check can only tell you when a relationship is wrong; it cannot tell you if it is completely right.

Another use of dimensional analysis is for a quick check on an equation you are not sure about. For example, suppose that you cannot remember whether the equation for the period T of a mass m on the end of a spring with spring constant k is $T = 2\pi\sqrt{k/m}$ or is $T = 2\pi\sqrt{m/k}$. A dimensional check (with the dimensions of k being $[M/T^2]$, since from Hooke's law $k = \text{force/distance}$) shows that the latter is correct:

$$[T] = \sqrt{\frac{[M]}{[M/T^2]}},$$

whereas the former is not,

$$[T] \neq \sqrt{\frac{[M/T^2]}{[M]}}.$$

Finally, an important use of dimensional analysis, but one with which much care must be taken, is to obtain the *form* of an equation. That is, we may want to determine how one quantity depends on others. To take a concrete example, let us try to find an expression for the period T of a simple pendulum. First we try to figure out what T could depend on, and make a list of these variables. It might depend on its length l, on the mass m of the bob, on the angle of swing θ, and on the acceleration due to gravity, g. It might also depend on air resistance (we would use the viscosity of air), the gravitational pull of the moon, and so on; but everyday experience suggests that the earth's gravity is the major force involved, so we ignore the others. So let us assume that T is a function of l, m, θ, and g, and that each of these factors is present to some power:

$$T = Cl^w m^x \theta^y g^z.$$

C is a dimensionless constant, and w, x, y, and z are exponents we want to solve for. We now write down the dimensional equation for this relationship:

$$[T] = [L]^w [M]^x [L/T^2]^z;$$

because θ has no dimensions (a radian is a length over a length—see Eq. 7-1), it does not appear. We simplify and obtain

$$[T] = [L]^{w+z} [M]^x [T]^{-2z}.$$

To have dimensional consistency, we must have

$$1 = -2z$$
$$0 = w + z$$
$$0 = x.$$

We solve these equations and find that $z = -\frac{1}{2}$, $w = \frac{1}{2}$, and $x = 0$. Thus our desired equation must be

$$T = C\sqrt{l/g}f(\theta), \qquad \text{(C-1)}$$

where $f(\theta)$ is some function of θ that we cannot determine using this technique. Nor can we determine in this way the dimensionless constant C. (Of course, to obtain C and f, we would have to do an analysis such as that in Chapter 13 using Newton's laws, which reveals that $C = 2\pi$ and $f \approx 1$ for small θ.) But look what we *have* found, using only dimensional consistency. We obtained the form of the expression that relates the period of a simple pendulum to the major variables of the situation, l and g (see Eq. 13-8) and saw that it indeed does not depend on the mass m.

How did we do it? And how useful is this technique? Basically, we had to use our intuition as to what variables were important and which were not. This is not always easy, and often requires a lot of insight. As to usefulness, the final result in our example could have been obtained from Newton's laws, as in Chapter 13. But in many physical situations, such a derivation from other laws cannot be done. In those situations dimensional analysis can be a powerful tool.

In the end, any expression derived by the use of dimensional analysis (or by any other means, for that matter) must be checked against experiment. For example, in our derivation of Eq. C-1, we can compare the periods of two pendula of different lengths, l_1 and l_2, whose amplitudes (θ) are the same. For, using Eq. C-1, we would have

$$\frac{T_1}{T_2} = \frac{C\sqrt{l_1/g}f(\theta)}{C\sqrt{l_2/g}f(\theta)} = \sqrt{\frac{l_1}{l_2}}.$$

Because C and $f(\theta)$ are the same for both pendula, they cancel out, so we can experimentally determine if the ratio of the periods varies as the ratio of the square roots of the lengths. This comparison to experiment checks our deviation, at least in part. C and $f(\theta)$ could be determined by further experiment.

APPENDIX D
SELECTED ISOTOPES†

(1) Atomic Number Z	(2) Element	(3) Symbol	(4) Mass Number, A	(5) Atomic Mass‡	(6) Percent Abundance, Or Decay Mode If Radioactive	(7) Half-life (If Radioactive)
0	(Neutron)	n	1	1.008665	β^-	10.6 min
1	Hydrogen	H	1	1.007825	99.985	
	Deuterium	D	2	2.014102	0.015	
	Tritium	T	3	3.016049	β^-	12.33 yr
2	Helium	He	3	3.016029	0.00014	
			4	4.002603	≈ 100	
3	Lithium	Li	6	6.015123	7.5	
			7	7.016005	92.5	
4	Beryllium	Be	7	7.016930	EC, γ	53.3 days
			9	9.012183	100	
5	Boron	B	10	10.012938	19.8	
			11	11.009305	80.2	
6	Carbon	C	11	11.011433	β^+, EC	20.4 min
			12	12.000000	98.89	
			13	13.003355	1.11	
			14	14.003242	β^-	5730 yr
7	Nitrogen	N	13	13.005739	β^+	9.96 min
			14	14.003074	99.63	
			15	15.000109	0.37	
8	Oxygen	O	15	15.003065	β^+, EC	122 s
			16	15.994915	99.76	
			18	17.999159	0.204	
9	Fluorine	F	19	18.998403	100	
10	Neon	Ne	20	19.992439	90.51	
			22	21.991384	9.22	
11	Sodium	Na	22	21.994435	β^+, EC, γ	2.602 yr
			23	22.989770	100	
			24	23.990964	β^-, γ	15.0 h
12	Magnesium	Mg	24	23.985045	78.99	
13	Aluminum	Al	27	26.981541	100	
14	Silicon	Si	28	27.976928	92.23	
			31	30.975364	β^-, γ	2.62 h

† Data are taken from *Chart of the Nuclides*, 12th ed., 1977, and from C. M. Lederer and V. S. Shirley, eds., *Table of Isotopes*, 7th ed., John Wiley & Sons, Inc., New York, 1978.
‡ The masses given in columns (5) are those for the neutral atom, including the Z electrons.

(1) Atomic Number Z	(2) Element	(3) Symbol	(4) Mass Number, A	(5) Atomic Mass[‡]	(6) Percent Abundance, Or Decay Mode If Radioactive	(7) Half-life (If Radioactive)
15	Phosphorus	P	31	30.973763	100	
			32	31.973908	β^-	14.28 days
16	Sulfur	S	32	31.972072	95.0	
			35	34.969033	β^-	87.4 days
17	Chlorine	Cl	35	34.968853	75.77	
			37	36.965903	24.23	
18	Argon	Ar	40	39.962383	99.60	
19	Potassium	K	39	38.963708	93.26	
			40	39.964000	β^-, EC, γ, β^+	1.28×10^9 yr
20	Calcium	Ca	40	39.962591	96.94	
21	Scandium	Sc	45	44.955914	100	
22	Titanium	Ti	48	47.947947	73.7	
23	Vanadium	V	51	50.943963	99.75	
24	Chromium	Cr	52	51.940510	83.79	
25	Manganese	Mn	55	54.938046	100	
26	Iron	Fe	56	55.934939	91.8	
27	Cobalt	Co	59	58.933198	100	
			60	59.933820	β^-, γ	5.271 yr
28	Nickel	Ni	58	57.935347	68.3	
			60	59.930789	26.1	
29	Copper	Cu	63	62.929599	69.2	
			65	64.927792	30.8	
30	Zinc	Zn	64	63.929145	48.6	
			66	65.926035	27.9	
31	Gallium	Ga	69	68.925581	60.1	
32	Germanium	Ge	72	71.922080	27.4	
			74	73.921179	36.5	
33	Arsenic	As	75	74.921596	100	
34	Selenium	Se	80	79.916521	49.8	
35	Bromine	Br	79	78.918336	50.69	
36	Krypton	Kr	84	83.911506	57.0	
37	Rubidium	Rb	85	84.911800	72.17	
38	Strontium	Sr	86	85.909273	9.8	
			88	87.905625	82.6	
			90	89.907746	β^-	28.8 yr
39	Yttrium	Y	89	88.905856	100	
40	Zirconium	Zr	90	89.904708	51.5	
41	Niobium	Nb	93	92.906378	100	
42	Molybdenum	Mo	98	97.905405	24.1	
43	Technecium	Tc	98	97.907210	β^-, γ	4.2×10^6 yr
44	Ruthenium	Ru	102	101.904348	31.6	
45	Rhodium	Rh	103	102.90550	100	
46	Palladium	Pd	106	105.90348	27.3	
47	Silver	Ag	107	106.905095	51.83	
			109	108.904754	48.17	
48	Cadmium	Cd	114	113.903361	28.7	
49	Indium	In	115	114.90388	95.7; β^-	5.1×10^{14} yr
50	Tin	Sn	120	119.902199	32.4	
51	Antimony	Sb	121	120.903824	57.3	

(1) Atomic Number Z	(2) Element	(3) Symbol	(4) Mass Number, A	(5) Atomic Mass‡	(6) Percent Abundance, Or Decay Mode If Radioactive	(7) Half-life (If Radioactive)
52	Tellurium	Te	130	129.90623	34.5; β^-	2×10^{21} yr
53	Iodine	I	127	126.904477	100	
			131	130.906118	β^-, γ	8.04 days
54	Xenon	Xe	132	131.90415	26.9	
			136	135.90722	8.9	
55	Cesium	Cs	133	132.90543	100	
56	Barium	Ba	137	136.90582	11.2	
			138	137.90524	71.7	
57	Lanthanum	La	139	138.90636	99.911	
58	Cerium	Ce	140	139.90544	88.5	
59	Praseodymium	Pr	141	140.90766	100	
60	Neodymium	Nd	142	141.90773	27.2	
61	Promethium	Pm	145	144.91275	EC, α, γ	17.7 yr
62	Samarium	Sm	152	151.91974	26.6	
63	Europium	Eu	153	152.92124	52.1	
64	Gadolinium	Gd	158	157.92411	24.8	
65	Terbium	Tb	159	158.92535	100	
66	Dysprosium	Dy	164	163.92918	28.1	
67	Holmium	Ho	165	164.93033	100	
68	Erbium	Er	166	165.93031	33.4	
69	Thulium	Tm	169	168.93423	100	
70	Ytterbium	Yb	174	173.93887	31.6	
71	Lutecium	Lu	175	174.94079	97.39	
72	Hafnium	Hf	180	179.94656	35.2	
73	Tantalum	Ta	181	180.94801	99.988	
74	Tungsten (wolfram)	W	184	183.95095	30.7	
75	Rhenium	Re	187	186.95577	62.60; β^-	4×10^{10} yr
76	Osmium	Os	191	190.96094	β^-, γ	15.4 days
			192	191.96149	41.0	
77	Iridium	Ir	191	190.96060	37.3	
			193	192.96294	62.7	
78	Platinum	Pt	195	194.96479	33.8	
79	Gold	Au	197	196.96656	100	
80	Mercury	Hg	202	201.97063	29.8	
81	Thallium	Tl	205	204.97441	70.5	
82	Lead	Pb	206	205.97446	24.1	
			207	206.97589	22.1	
			208	207.97664	52.3	
			210	209.98418	α, β^-, γ	22.3 yr
			211	210.98874	β^-, γ	36.1 min
			212	211.99188	β^-, γ	10.64 h
			214	213.99980	β^-, γ	26.8 min
83	Bismuth	Bi	209	208.98039	100	
			211	210.98726	α, β^-, γ	2.15 min
84	Polonium	Po	210	209.98286	α, γ	138.38 days
			214	213.99519	α, γ	164 μs

(1) Atomic Number Z	(2) Element	(3) Symbol	(4) Mass Number, A	(5) Atomic Mass‡	(6) Percent Abundance, Or Decay Mode If Radioactive	(7) Half-life (If Radioactive)
85	Astatine	At	218	218.00870	α, β^-	≈ 2 s
86	Radon	Rn	222	222.017574	α, γ	3.8235 days
87	Francium	Fr	223	223.019734	α, β^-, γ	21.8 min
88	Radium	Ra	226	226.025406	α, γ	1.60×10^3 yr
89	Actinium	Ac	227	227.027751	α, β^-, γ	21.773 yr
90	Thorium	Th	228	228.02873	α, γ	1.9131 yr
			232	232.038054	100; α, γ	1.41×10^{10} yr
91	Protactinium	Pa	231	231.035881	α, γ	3.28×10^4 yr
92	Uranium	U	232	232.03714	α, γ	72 yr
			233	233.039629	α, γ	1.592×10^5 yr
			235	235.043925	0.72; α, γ	7.038×10^8 yr
			236	236.045563	α, γ	2.342×10^7 yr
			238	238.050786	99.275; α, γ	4.468×10^9 yr
			239	239.054291	β^-, γ	23.5 min
93	Neptunium	Np	239	239.052932	β^-, γ	2.35 days
94	Plutonium	Pu	239	239.052158	α, γ	2.41×10^4 yr
95	Americium	Am	243	243.061374	α, γ	7.37×10^3 yr
96	Curium	Cm	245	245.065487	α, γ	8.5×10^3 yr
97	Berkelium	Bk	247	247.07003	α, γ	1.4×10^3 yr
98	Californium	Cf	249	249.074849	α, γ	351 yr
99	Einsteinium	Es	254	254.08802	α, γ, β^-	276 days
100	Fermium	Fm	253	253.08518	EC, α, γ	3.0 days
101	Mendelevium	Md	255	255.0911	EC, α	27 min
102	Nobelium	No	255	255.0933	EC, α	3.1 min
103	Lawrencium	Lr	257	257.0998	α	≈ 35 s
104	Rutherfordium (?)	Rf	261	261.1087	α	1.1 min
105	Hahnium (?)	Ha	262	262.1138	α	0.7 min
106			263	263.1184	α	0.9 s

ANSWERS TO ODD-NUMBERED PROBLEMS

CHAPTER 1

1 109 km/h
3 7.8 h
5 (a) 89 km/h, (b) 25 m/s, (c) 81 ft/s
7 (a) 11.4 mi/h, (b) 0
9 910 km/h
11 3800 cars/h
13 5.6 s
15 -8.6 m/s^2, -0.88 g
19 1.8 m/s^2
21 32 m/s
23 (a) 84 m, (b) 53 m
27 Stop
31 13.2 g
33 (a) 31.1 m, (b) 5.04 s
35 7.0 m/s
37 27.9 m
39 5.83 s
41 6
45 (a) 3.22 m/s, (b) 1.51 s, 2.17 s, (c) it passes this point on the way up and on the way down.
47 1.9 m
49 (a) 0.28 m/s, (b) 1.1 m/s, (c) 0.28 m/s, (d) 1.6 m/s, (e) -1.0 m/s
51 (a) 50 s, (b) 87 s to 106 s, (c) same as (b), and almost constant 0 to 40 s, 65 s to 75 s, (d) 70 s
55 (a) graph; 2.0 m/s, 2.0 m/s, 2.0 m/s, (b) -2.0 m/s for each, (c) 3.0 m/s for each

CHAPTER 2

1 103 km, 33° S of W
5 (a) 2.8 in $+ x$ direction, (b) 9.8, $+ x$, (c) 9.8, $-x$
7 (a) parallel, (b) perpendicular, (c) $V_2 = 0$
9 64.6, 53.1°
11 (a) 62.6, $-31.1°$; (b) 77.5, 71.9°
13 (7, 4, 5), 9.5

15 (25 m, 18 m, 36 m), 47 m
17 (-2245 m, 4186 m, 2150 m), 5214 m
19 (a) 800 km/h, 600 km/h, (b) 1600 km, 1200 km
21 $v_t/\tan \theta$
23 3.8 min
25 1.41 m/s
27 1.16 m/s
29 (a) 89 m, (b) 111 s, (c) 26°
31 150 s, 3750 m; 23.1 s, 577 m
33 7.4 s
35 (a) $Dv/(v^2 - u^2)$, (b) $D/\sqrt{v^2 - u^2}$
37 12 m
41 2.6 s
43 22 m
45 0.89 s, 0.97 m
49 6 times
53 $\theta = \tan^{-1}(gt/v_0)$

CHAPTER 3

1 850 N
3 9.90 kg
5 4200 N
7 (a) 6.90 N, (b) 120 N, (c) 610 N, (d) 0
9 4.2 m/s^2, down
11 807 N
13 4.373×10^4 N, 3.957×10^4 N
15 5.95 m/s^2
17 2.2×10^4 N, yes
19 (a) 9.9 m/s, (b) 4.0×10^3 N
21 SW
23 110 N, 0
25 98 N
27 (a) 424 N, (b) 11 m/s
29 10 kg
31 1.5 m
33 (b) 37½ m
35 23.4 N, 46.7 N
37 90.7 N, 0.613
39 -8.0 m/s^2
41 3980 N
43 16 m/s

45 (a) 1.05 m/s^2, (b) 420 N, (c) 280 N
47 143 N
49 1.36 m/s^2

CHAPTER 4

1 (a) 0.261 m/s^2, (b) 6.53 N
3 5.95×10^{-3} m/s^2, 3.56×10^{22} N
5 22.6 m/s, yes
7 5.6 m/s
9 0.68
13 (a) 0.162 N, (b) 6.73 N
15 (a) inside of outer edge, (b) 1.5×10^3 rev/day
17 yes, 6.0×10^3 N down the slope
21 759 N
23 3.8 m/s^2
25 2.45 m/s^2
27 (a) 4.0 kg, 4.0 kg, (b) 39 N, 8.0 N
29 (a) 9.79 m/s^2, (b) 4.35 m/s^2
31 2.64×10^6 m
33 $g_p r^2/G$
35 3.3×10^{-8} N toward center of square
37 2.53 m/s^2 up
39 2.0×10^{27} kg
41 (a) 1.96 m/s^2, down, (b) 510 N up, (c) 118 N, down
43 (a) 18 N toward moon, (b) 220 N away
45 (b) 9.6×10^{26} kg
47 84.5 min
49 225 days
51 2.93×10^4 m/s, 3.03×10^4 m/s

CHAPTER 5

1 2.9×10^3 J
3 9.4×10^6 J
5 (a) 9.5×10^4 J, (b) 2.2×10^5 J
7 32 J
9 (a) 0, (b) -9.2×10^2 J, (c) 5.6×10^3 J

11 5.4×10^3 J
13 192 J
15 2.5×10^{-7} J
17 3.86×10^5 J
19 180 N
21 2.1 m/s, 4.2 m/s
23 (a) 1.8×10^5 J, (b) 22 m/s, (c) 3.2 m
25 70 J
27 (a) 7.0×10^5 J, (b) 7.0×10^5 J, (c) yes
29 3.3 m, no
31 6.62 m/s
33 21 m/s
35 1.1×10^3 J
37 (a) 20 m/s, (b) 1.9 N
39 (a) $\sqrt{2\,gL}$, (b) $\sqrt{gL}$
41 24.3 m/s, 10.1 m/s, 18.9 m/s
43 (a) 7.64×10^4 J, (b) 717 N
45 99 $Mg/2h$
47 5.0×10^2 N
49 0.54%
51 40 hp
53 5.4×10^6 J
55 0.95 m/s
57 375 W (0.50 hp)
59 (a) 54.2 m/s, (b) 9.9×10^5 W

CHAPTER 6

1 0.270 kg · m/s
3 6.00×10^7 N
5 17.3 m/s
7 3.6×10^2 m/s
9 1.09×10^4 m/s
13 131 kg
15 130 N, no
17 $\sqrt{2}\ mv$
19 1.98×10^3 J
21 (b) 0.72 N · s, (c) 2.25 g
23 2.70 m/s, 10.8 m/s
25 (a) 0.480 kg, (b) 0.75
29 (a) 4.2 m/s, 4.2 m/s, (b) 1.8 m along incline
31 (b) 1.13 m/s, $-32.0°$
33 $+x$ direction, 4.80 m/s; 3.00 m/s
35 (a) 2.5×10^{-16} km/s, (b) 1.67×10^{-17}, (c) 0.188 J
37 1800 J (larger), 2700 J (smaller)
39 141°
43 4.84×10^{-11}m
45 9.4%
47 (a) 2.0 m SE of center, (b) 1.44m at 2.3° S of E
49 (a) ($\frac{7}{3}$) D, (b) 5 D

CHAPTER 7

1 (a) 0.524, (b) 1.57, (c) 7.33
3 6.8 km
5 2.1×10^2 rad/s
7 -8.4 rad/s^2

9 R_2/R_1
11 (a) 1.99×10^{-7} rad/s, (b) 7.27×10^{-5} rad/s
13 $L/2$, $L/2$
15 (a) -56 rad/s^2, (b) 3.0×10^2 rev
17 0.63 rad/s^2
21 14.3 m
23 (a) $a_T = \alpha r$, $a_c = r\alpha^2 t^2$, (b) $\tan^{-1}$ $(1/4\pi N)$
25 106 N · m
27 0.126 kg · m^2
29 (a) 2.99×10^{-3} kg · m^2, (b) 0.108 N · m
31 3.61 kg
33 1.60×10^3 rev
35 $g(m_2 - m_1)/(m_1 + m_2 + I/R_0^2)$
37 (a) 3.34 rad/s^2, (b) 7.01 m/s^2, (c) 352 m/s^2, (d) 2.57×10^3 N, (e) 1.14°
39 2.19×10^4 J
41 2.09×10^5 J
43 (b) 2.2×10^3 J, (c) 24 min
45 2.95 m/s
47 1.89 kg · m^2/s
49 0.38 rev/s
51 8.2×10^{-6}
53 0.40 rad/s; no change (if jump off radially)
55 (a) $\omega_w I_w/I_p$, (b) $\omega_w I_w/2I_p$, (c) $-\omega_w I_w/I_p$, (d) 0

CHAPTER 8

1 1.44 N
3 5.42 kg
5 2.27 N
7 -2.94×10^3 N, 1.47×10^4 N
9 1.47×10^3 N down; 1.96×10^3 N up
11 6.47 kg
13 196 N; 196 N, 30°
15 2.12×10^3 N, no
17 (a) 176 N, 32 N, (b) 0.26
19 (a) tip, (b) 50 cm
21 30 kg
23 6.14 kg
25 2.1×10^2 N
27 207 N, 1.76×10^3 N
29 960 N; 330 N (to the left), 1510 N (down)
31 17.5
33 $2\pi\ r_2/P$
35 608 N
37 yes ; 2.5 m more (at top)
39 (a) 9.8×10^4 N/m^2, (b) 1.96×10^{-6}
41 1.7 mm (right); 0.96 mm (right, if same size)
43 0.15%, 0.13%
45 9.0×10^7 N/m^2; 900 atm
47 1.7×10^{-2} J
49 5.1×10^4 N
51 0.20 cm^2
53 0.77 cm
55 9 on each side, 1.25 m apart
57 2.68 m
59 12.0 m

CHAPTER 9

1 2.7×10^{11} kg
3 0.89
5 7.4×10^7 N/m^2, 730 atm
7 1.65×10^4 N/m^2
9 2.4×10^5 N, 2.4×10^5 N
11 1.37×10^{-2} m^2 (= 137 cm^2)
13 0.051 atm
15 1.9×10^9 N
17 5.8×10^3 kg
21 5×10^{18} kg
23 (a) 0.34 kg, (b) 1.5×10^4 N
25 1.068
27 5.45×10^3 kg/m^3
29 0.83
31 12 kg
33 0.95×10^3 kg/m^3
35 (a) 810 kg/m^3, (b) $P_{sol}(w_{air} - w_{liq})/w_{air}$
37 0.107
41 4.0×10^{-2} N/m
43 No
45 2.3×10^{-3} N down
47 3
51 1.01×10^5 N
53 2.4 atm
55 1.1×10^{-3} m^3/s (1.1 L/s)
57 5.9×10^5 N
59 (a) 102 m/s, (b) 0.934 atm
61 0.13
63 3.7×10^{-3} Pa · s
65 14.2 cm
67 1.6×10^{-4} m^3/s
69 $P_G = 2.07$ atm
71 4.6 years

CHAPTER 10

1 3.5 kg
3 75
5 $-40°$
7 (a) low, (b) 0.024%
9 13.48×10^3 kg/m^3
11 72.4 cm^3
15 1.58 mL
17 (a) more of the cube is submerged, (b) 0.44%
19 (a) 3.78 cm, (b) (approx.) $(\beta_{Hg} - \beta_{gl})$ $\Delta T\, V_{bulb}/\pi r_{tube}^2$
21 (a) 9.4×10^7 N/m^2, (b) no, (c) 9.4×10^6 N/m^2, yes
23 (a) 310 K, (b) 300 K, (c) 77K
25 (a) 4.27×10^3 K, 1.50×10^7 K, (b) 6%, 2×10^{-3}%
27 0.2
29 1.43 kg/m^3
31 4.15 kg
33 (a) 774 kg, (b) 65 kg leaves
35 0.080 (8.0%)
37 11.9L, no!
39 2.92 cm^3

41 2.69×10^{25} molecules/m³
43 2.6×10^2 molecules/cm³
45 1800 N
47 6.1×10^3 m/s
49 899°C
51 3.7×10^2 m/s
53 $v_{rms} \approx 300$ m/s, $P \approx 5 \times 10^{-22}$ atm
57 1.0043
59 (a) solid or vapor, (b) $-56.6° \leq T \leq$ 31°C, 5.11 atm $\leq P \leq 73$ atm
61 vapor
63 $92\frac{1}{2}$°C
65 2.06×10^3 N/m²
67 5.5 kg
69 120°C
71 29%
73 (a) 0.5×10^{-10} mol/s, (b) 0.6 s

CHAPTER 11

1 1.7×10^5 J
3 1.67×10^6 J
5 170 kg/h
7 7.8×10^3 kg
9 1.1×10^2 kg
11 0.0576 kg
13 (a) $C = Mc$, (b) 1 kcal/C°, (c) 35 kcal/C°
15 1.14 C°
17 9.3 C°
19 0.56 kcal/kg · C°
21 6.7 min
23 3.07 cm; it would melt
25 9.6×10^2 kcal
27 (a) 5.8 h, (b) 44 h
29 1.7 g/min
31 3.0 g
33 16 kW
35 0.75 mm
37 23 W, $\approx 1/10$
39 (a) 1.3×10^{-2} kcal/s, (b) 2.2 kcal/s
43 (a) 1.1×10^5 J/s, (b) 1.6×10^8 J, (c) $157
45 $3\frac{1}{4}$ days
47 about $4\frac{1}{2}$ g/h

CHAPTER 12

3 (a) 1.69×10^5 J, (b) 2.08×10^6 J
5 (a) 0, (b) -850 J, (c) fall
7 (a) 248 J, (b) 248 J
9 -5.5×10^5 J
11 115 W
13 21.5%
15 562 °C
17 2.15×10^{10} J/h
19 687°C
21 67 kg
23 11.3 km³/day, 56.4 km²
25 0.31 kcal/K
27 -2.9 kcal/K

29 0.256 cal/K · s
31 330 J/K
33 (a) 5/16 (b) 1/64
35 208 m²
37 (a) 7.8×10^6 kWh, (b) 0.39 MW
39 3.3×10^{13} J/day

CHAPTER 13

1 35 N/m
3 1.2 Hz
5 8.2×10^2 N/m; 3.2 Hz
7 (a) 1.5 m/s, (b) 0.83 m/s, (c) 0.40 J, (d) $x = 0.12 \cos 4\pi t$
9 1.05 kg
11 (a) 0.42 m, (b) 1.18 Hz, (c) 4.8 J, (d) 4.1 J, 0.70 J
13 0.286 s
15 5.5 m/s
19 416 m/s
21 $A_1 = \sqrt{10}\, A_2$
27 $\rho_{H_2O} g A$
29 0.248 m
31 (a) $-20°$, (b) 6.2°, (c) 20°
33 $\sqrt{2gL(1 - \cos \theta_0)}$
35 $\frac{1}{3}$
37 3.3 m/s
39 545 m, 188 m; 3.41 m, 2.78 m
41 less dense; $\sqrt{2} \times$
43 1.7×10^2 Hz
45 1.1×10^3 m
47 $\sqrt{2}$
49 (a) 1:4, (b) 1:2
53 8.1 km/s
55 9.0°
57 (a) $\sin^{-1} (v_i/v_r)$, (b) $\geq 52.4°$
59 440 Hz, 880 Hz, 1320 Hz, 1760 Hz
61 0.526 m
63 0.85 m/s
65 308 Hz

CHAPTER 14

1 (a) 5.76 mm, (b) 1.37 mm
3 70.5 m
5 40 dB
7 1.6×10^6
9 92 dB
11 1.6×10^{-9} J/s
13 25 W
15 (a) higher, $2 \times$, (b) 4:1
17 8.92×10^{-5} m
19 45 dB
21 (a) 130 dB, (b) 95 dB, (c) 53 dB
23 about 40 dB
25 25 N
27 6.6 cm
29 (a) 107 Hz, 322 Hz, 536 Hz, 750Hz; (b) 214 Hz, 429 Hz, 643 Hz, 858 Hz
31 3400 Hz, yes

33 (a) 280 m/s, 40 N, (b) 0.188 m, (c) 880 Hz; 1320 Hz
35 635 Hz
37 31 N
39 0.5 Hz
41 4.4 Hz
43 (a) (132 ± 0.5)Hz, no; (b) 0.76%
45 (a) 343 Hz, (b) 1029 Hz, 1715 Hz
47 (a) 11.9s⁻¹, (b) 86.8 m
49 (a) 849 Hz, (b) 951 Hz
51 ± 133 Hz
53 26 Hz
55 0.205 m/s
57 1.7×10^2 m/s
61 0.595; 0.25

CHAPTER 15

1 6.25×10^{14}
3 6.0×10^{-3} N
5 68 N
7 5.08 m
9 8.8 N, 34 N, 25 N
11 9.14×10^4 N toward center
13 2.1×10^{-10} m
15 $+0.402\, Q_0$ placed between the given charges, $0.366\, l$ from $-Q_0$
17 6.34 μC, 843.6 μC; -6.25 μC, 856.25 μC
19 5.0×10^3 N/C in direction opposite to force
21 813 N/C, up
23 1.02×10^{-7} N/C, up
25 A: 2.6×10^6 N/C, up; B: 6.7×10^6 N/C, at 56° above $+x$ axis. Yes.
27 4.90×10^7 N/C, away from center
29 6.78×10^5 C, negative
31 7.9×10^{-4} N/C
33 1.4×10^7
35 (a) 5×10^{-10} N, (b) 8×10^{-10}N, (c) 6×10^{-5} N

CHAPTER 16

1 7.5 cm
3 48 V
5 1.28×10^{-16} J
7 4.0 kV, B
9 -350 V
11 3.53×10^{-2} eV
13 3.4°
15 7.2 V
17 3.0×10^{-2} J
19 (a) 6.96×10^{-3} m beyond -2.0 μC charge; (b) 1.25×10^{-3} m from -2.0 μC charge toward the 30 μC charge
21 (a) 2.0×10^5 V; (b) 1.3×10^6 V/m, 27° N of E
23 (a) 4.3×10^{-2} V, (b) 3.1×10^{-2} V, (c) -3.1×10^{-2} V
27 3.87 μF

29 35 pF
31 7.5×10^3 N/C
33 0.22 μC
35 $Q_1 = Q_0C_1/(C_1 + C_2)$, $Q_2 = Q_0C_2/(C_1 + C_2)$; $V_1 = V_2 = Q_0/(C_1 + C_2)$
37 164 V; 660 μC, 1640 μC
39 1.0×10^{-7} J/m^3
41 Doubles

CHAPTER 17

1 1.2×10^5 C
3 3600 C
5 1800 V
7 1.1×10^{20}
9 0.10 S
11 1.3×10^{-2} Ω
13 1.58 Ω
15 ¼ of original
17 (a) 3.8×10^{-4} Ω, (b) 1.5×10^{-3} Ω, (c) 6.0×10^{-3} Ω
19 9.0 × greater
21 0.173 mm, 53.4 m
23 0.50 A
25 7.1 V
27 0.33 kWh
29 11.7 h
31 12
33 (a) $36.35, (b) 2.2×10^3 kg/y
35 2.6 kW less
37 19.5%; less decrease
39 21 A
41 (a) 2.2 hp, (b) 150 km
43 240 Ω
45 2.8×10^2 V
47 200 W
49 (a) 19.1 Hz, (b) 1.77 A, (c) $V = 125 \sin 120t$
51 35 m/s
53 5×10^{-9} W

CHAPTER 18

1 1200 Ω; 33.3 Ω
3 One possibility: six in series, tap off of one.
5 1⅝ Ω
7 1.18×10^3 Ω; 0.0133
9 0.056 Ω
11 one possibility: 70 kΩ in series with body and battery
13 125 Ω
15 3.52 mA, 2.20 mA, 1.32 mA, 0.88 mA, 0.44 mA; 1.85V
17 42.5 V
19 5.75 V
21 0.060 Ω

23 Higher terminal voltage (4.00 vs. 3.98), and less power wasted (0.02 W vs 0.3 W)
25 62.2 V; 2.2 Ω
27 25.7 V
29 0.19 A, 0.21 A, 0.33 A, 0.024 A, 0.36 A
31 0.77 A, 0.71 A, 0.055 A; 5.94 V
33 9.0 μF; 0.25 μF
35 1385 pF in series
37 (a) $C_1 + C_2C_3/(C_2 + C_3)$; (b) 200 μC, 67 μC, 67 μC
39 (a) 4.3×10^{-5} J, (b) 9.6×10^{-6} J, (c) 3.6 μC, 0.80 μC
41 (a) 3.0 V, 6.0 V, (b) 0.60 μC, 0.60 μC, (c) 9.0V, 9.0V; 1.8 μC, 0.90 μC
43 (a) parallel, (b) 9.7 pF to 39 pF
45 11.4 μC
47 (a) 8 V, (b) 16 V, (c) 8 V, (d) −5.8 μC
49 1.5×10^6 Ω
51 1.0×10^{-3} Ω in parallel
53 (a) 1.2×10^{-4} Ω in parallel; (b) 50 MΩ in series
55 1.78 A
57 7.94 V
59 26 V, 220 kΩ
61 5.0 kΩ, 6.7 kΩ
63 0.60 mV
65 0.45 Ω
67 13.6 Ω
69 $70\frac{1}{2}°$
71 (a) a = 4.99×10^{-5} V/C°, b = -5.83×10^{-8} V/(C°)^2; (b) 113°
73 7 mA

CHAPTER 19

1 (a) 2.75 N, (b) 1.94 N
3 11.7 A
5 7.78×10^{-14} N, north
7 (a) out, (b) right, (c) up, (d) right, (e)down, (f) zero
9 17 T, west
13 (a) 4.5 cm, (b) 5.5×10^{-7} s^{-1}
15 2.20×10^{-8} m
19 30.8 μA
21 3.61 T
23 (a) 9.98×10^{-5} N · m, (b) north
25 2.41×10^6 m/s; 3.73 mm
27 1.2×10^5 C/kg
31 (a) 6.6 mm, (b) 1.7 mm
33 7.6×10^{-3} T
35 0.98 mm
39 1500 A
41 1.2×10^{-4} g
43 42° W of N
45 24 A, down
51 3.33×10^{-5} T
53 0.225 N attraction
55 2.6×10^{-5} N/m (A), 1.5×10^{-5} N/m (B and C)
57 2.8×10^{-5} T · m/A

CHAPTER 20

1 39 mV
3 (a) 63 mV, (b) clockwise
5 50 V
7 down
9 2.57×10^{-5} J
13 $BlvA/\rho(l + 2vt)$
15 30 V
19 5.36 rev/s
21 108 V
23 (b) 30.8 A, (c) 4.55 A
25 4
27 $I_2 = \frac{1}{4}I_1$
29 9.6 V; 0.64 A
31 (a) 11 kV, (b) 9.6 MW, (c) 0.77 MW, (d) 8.8 MW
35 Wires of cross-section A = 1.6×10^{-3}m^2 (d = 4.6 cm)
37 1.07 H
39 265, 8
41 ½
43 (b) 23 μH
45 $M = \mu_o N_1 N_2 A/l$
47 31 Ω, 18 H
49 0.80 J
51 (a) 4.4×10^{-4} J/m^3(**E**); 1.6×10^6 J/m^3 (**B**); (b) 6.0×10^8 V/m
53 V/L
55 374 Hz
59 1.24 kΩ, 0.32 A
61 (a) 40 kΩ, (b) 36 mA
63 (a) 1.1 mV, (b) 30 mV
65 (a) 10 kΩ, (b) 16 kΩ
67 (a) 4.5 mA, (b)−80°, (c) 0.094 W, (d) 21 V, 118 V
69 3.3 mW
71 (a) 2.7 A, (b) 70°, (c) 15 W
73 6.5 μF
77 1.45 MHz
79 (a) 6.3×10^{-9} F, (b) 1.8 A
81 2.5 μH, 1.6×10^{-2} Ω

CHAPTER 21

1 2.7×10^{-9} A
3 1.60×10^{-8} T
5 5.0×10^{-13} T
7 2.763 cm
9 At radio, 0.14 s
11 9.6×10^{-2} W/m^2
13 4.5×10^{-12} J
15 3.82×10^{26} W
17 7.8 V/m, 2.6×10^{-8} T
19 200 m
21 177 pF
23 3.63×10^{-9} H to 2.41×10^{-9} H
25 (a) 2.0×10^{-5} W/m^2, (b) 8.7×10^{-2} V/m
27 7.14×10^{-4} V/m; 1.35×10^{-9} W/m^2

CHAPTER 22

1 (a) 2.20×10^8 m/s, (b) 1.99×10^8 m/s
3 3m
5 4.0×10^{16} m
7 3.4×10^3 rad/s
11 5.9 mm^2
15 12.0 cm
21 (a) concave, (b) 29.3 cm behind mirror, (c) 177 cm
23 (a) $r \to \infty$, (b) $d_i = -d_0$, (c) 1, (d) yes
25 4.50 m
27 38.6° to the vertical
29 58.3°
35 33.3°
37 n > 1.8
39 54°
41 30.0 cm
43 converging; 13.4 cm; real
45 (a) 5.88 D, (b) −11.8 cm; (a) is converging (b) is diverging
47 (a) 66.7 mm, (b) 33.3 mm
49 0.47 cm
51 (a) 144 mm behind lens, real, inverted, 1.77 mm high; (b) 127 mm in front of lens, virtual, upright, 1.56 mm high
53 14.1 cm
55 3.92 cm behind diverging lens
59 3.76 m
61 (c) $\sqrt{d_T^2 - 4fd_T}$; (d) $[(d_T + \sqrt{d_T^2 - 4fd_T})/(d_T - \sqrt{d_T^2 - 4fd_T})]^2$
63 1.53
65 −13.3 cm
67 0.90 D

CHAPTER 23

3 725 nm
5 6.0 cm
7 (a) destructive, (b) destructive
9 0.84 mm
11 $d \sin \theta_m - d \sin \theta_i = \pm m \lambda$ where $m = 0$ for $\theta_m = \theta_i$
13 0.160°
15 0.53 cm
17 569 nm
19 0.030 mm
21 0.106 mm
23 7.20°
25 On each side of center, two full spectra, and part of another.
27 12,500 cm^{-1}
29 795 nm
31 445.7 nm, 652.6 nm
33 4.00 mm
35 7.42 μm
37 red-orange (643 nm)
39 478 nm
43 11.4 cm; 22.3 cm

45 6.61×10^{-5} m
47 1.0043
49 56.3°
51 $\frac{1}{4}$
53 36.9°
55 $\frac{1}{32}$ of original
57 (a) 0, (b) $\frac{1}{8}$, (c) 0
59 4.9×10^2 kg/m^3
61 $\Delta\lambda/\lambda = 0.27$

CHAPTER 24

1 39 mm, 4.2 mm
3 $f/4$
5 $^1/_{12}$ s
7 530 mm
9 $f/4$
11 3.6 D
13 0.27 cm
15 −21.6 cm
17 4.8 D
19 (a) 12.5 cm, (b) 8.3 cm
21 2.2×, 3.9×
23 (a) −76.0 cm, (b) 6.58×
25 (a) 5.6×, (b) 2.1×
27 16×; 53 cm
29 87×
31 12×
33 6.4×
35 (a) 16.2 cm, (b) 154×
37 (a) 1.07 cm, (b) 1.90×
39 (a) converging, (b) 40 cm
41 2.4×10^{-7} rad
43 2.3×10^{10} m
45 250×
47 (a) yes, (b) 15 km
49 12.3°
53 (b) 4.3 4.0 2.8
           5.3 4.8 6.0
           1.4 3.3 4.3

CHAPTER 25

1 7.81×10^{-9} s
3 150 m
5 $0.97c$
7 (a) 36.7y, (b) 7.3y, (c) 7.2 l-y, (d) $0.98c$
9 (a) $0.80c$, (b) 5.0 y
11 $1.9u = 3.2 \times 10^{-27}$ kg
13 $0.140 \, c$
15 (a) $(1 - 0.5 \times 10^{-8}) \, c$, (b) 30 cm
17 9.0×10^{13} J ; 9.2×10^{10} kg
19 2.84×10^{-13} kg
21 (a) 5.90×10^{19} J; (b) 4.7%
23 938 MeV
25 7.6×10^{-12} J(48 MeV); 1.6×10^{-19} kg · m/s
27 1.84×10^8 m/s
29 $m = 2m_0 = 3.34 \times 10^{-27}$ kg; $0.87c$

31 2.24×10^8 m/s; 5.61×10^{-19} kg · m/s
33 3.5×10^{-5} g
39 (a) $0.80 \, c$, (b) $-0.80 \, c$
41 (a) (165, 10, 0) m, (b) (315, 10, 0) m
43 0.87c, 47°
45 2.2×10^8 m/s, 27°

CHAPTER 26

1 0.34 eV
3 (a) 2.7×10^{-34} J, (b) 2.2×10^{35}, (c) 4.5×10^{-36}, no
5 270 nm
7 5.9×10^{14} Hz
9 0.110 eV; 1.96×10^5 m/s
11 3.59 eV
13 3.6×10^{21} photons/s · m^2
15 45 kcal/mol
17 3.76 eV
19 2.58 MeV
21 212 MeV, 5.86×10^{-15} m
23 3.3×10^{-32} m
25 680 V
29 2.8×10^{-11} m
31 0.038 eV, 0.147 nm
33 150 V
35 3.3×10^{-12} m
37 1.51 eV
39 $n = 6 \to n = 3$
41 122 eV
43 3.28×10^{15} Hz
47 −27.2 eV, 13.6 eV
49 4.4×10^{-40}, yes
51 1.20×10^{29} m, -2.64×10^{-78} eV
53 14.2 MeV
55 (c) 1.5×10^2 eV, (d) 9.8×10^{-48} eV, (e) 0.19 nm

CHAPTER 27

1 5.3×10^{-11} m
3 $\pm 7.2 \times 10^3$ m/s
5 (a) 1.2×10^{-34} m, (b) 1.6×10^{-32} kg · m/s, (c) 8.5×10^{-31} m
9 5.3×10^{-11} m ($\approx$Bohr radius)
11 1.01 nm
13 $(1, 0, 0, \frac{1}{4})$, $(1, 0, 0, -\frac{1}{2})$, $(2, 0, 0, \frac{1}{2})$, $(2, 0, 0, -\frac{1}{2})$, $(2, 1, 0, \frac{1}{2})$, $(2, 1, 0, -\frac{1}{2})$, but not unique
15 0, 1, 2, 3, 4, 5
19 5.8×10^{-13} m; 115 keV
21 (a) $1.56 \, e$, (b) 1.36×10^{-10} m
23 4.14×10^{-11} m; no upper limit
25 0.278 nm
27 0.061 nm
29 1.3×10^{-2} J, 4.0×10^{16} photons

CHAPTER 28

1 5.1 eV
3 5.4 eV
5 2.7 V
7 (a) 23 mA, (b) 16 mA
9 12 mA
11 (a) 24 mA, (b) 47 mA
13 7.5 μA
15 6.7 kΩ
17 0.72 mA

CHAPTER 29

1 3727 MeV/c^2
3 (a) 4.8×10^{-15} m, (b) 27
5 (a) 2.3×10^{17} kg/m^3, (b) 180 m, (c) 2.6×10^{-10} m
7 350 MeV
9 2.224 MeV
11 18.72 MeV
13 (b) yes
17 4.38 MeV, 0; 0, 4.38 MeV
19 6.12 MeV; 0.25 MeV

21 0.862 MeV
25 (b) 98.3%
27 7α, 4 β$^-$
29 3.1 h
31 3.7×10^{21}
33 (a) 3.6×10^{16}, (b) 2.3×10^{15}, (c) 5.2×10^{13} s^{-1}, (d)24 min
35 3.3×10^{-12} kg
37 $\lambda_\beta = 60 \ \lambda_\alpha$
39 41 y
41 2.1×10^4y

CHAPTER 30

1 No
3 Releases 17.35 MeV
5 (a) yes, (b) 17.85 MeV
9 126.5 MeV
11 7.8×10^{17} s^{-1}
13 1.15×10^{-4} kg
15 1.11 g
17 (a) 1.54×10^3 kg, (b) 4.8×10^6 Ci
19 0.12 g
23 9.70×10^{10} J/g, 3.39×10^{11} J/g; 8.2×10^{10} J/g

25 3.23×10^9 J; 65 times larger
27 3.7×10^3 s^{-1}
29 1000 rad
31 35 J
33 2.0×10^9
35 2.2×10^{-10} kg
37 4.6 m

CHAPTER 31

1 15.9 GeV
3 1.25 T
5 34 MeV, 4.1×10^7 m/s; 17 MeV, 4.1×10^7 m/s
7 100
9 3.1×10^{-18} m
11 1.34 T
13 67.5 MeV
15 37.7 MeV, 42.3 MeV
17 1.87915 GeV
19 1.53×10^{-12} m
21 493.4 MeV, 494.0 MeV
23 No; 289 MeV
25 3.1×10^{-21} s
27 (b) 10^{28} K

INDEX

The abbreviation *defn* means only the definition of the term is cited; *fn* after a reference means it is found in a footnote; *p* means it is found in a problem or question.

A-scan, 347
Aberrations:
　astigmatism, 583, 590
　chromatic, 591
　of eye, 591
　of lenses, 590–91
　spherical, 525, 528, 590–91
Aberration of starlight, 614
Absolute temperature scale, 226
Absolute zero, 225–26
Absorbed dose, 738
Absorption coefficient, 329
Absorption lines, 559, 651
Absorption spectra, 651, 656
Ac circuits, 487–96
　resonance in, 494–96
Acceleration, 13–26, 50–51, 69–75
　angular, 139–41
　average, 13
　centripetal, 70–75
　of gravity, 20–21, 39–42, 81–82
　instantaneous, 13–14
　linear, 13–26, 139–40
　related to force, 50–53
　tangential, 62
　uniform, 15–19
　uniform rotational, 141
　distinguished from velocity, 14, 22, 71
Accelerators (*see* Particle accelerators)
Acceptor level, 697
Accommodation of eye, 582
Achromatic lens, 591
Acoustics (*see* Sound)
Action at a distance, 361
Action potential, 406–8
Action-reaction (Newton's third law), 53–54
Activation energy, 235, 693–94, 720
Activity of nuclear radiation, 716, 738
Addition of velocities:
　classical, 35–39
　relativistic, 629–30

Adhesion, 198–99
Adiabatic process, 273
Air columns, vibration of, 333–36
Air conditioners, 279
Airfoil, 205
Airplane wing, 205
Airy disc, 592
Algebra, review of, 777–80
Alpha decay, 710–12, 715
　and tunneling, 719–20
Alpha particles, 710–11
Alternating current (ac), 403–5, 487–96
　meters for, 425–26
Alternators, 478
Ammeter, 424–27
Ampere (unit), 395, 462
　operational definition of, 462
Ampère, André, 395, 444, 459
Ampère's law, 459–61, 503–5
Amplification, electronic, 700–702
Amplifiers, 700–702
Amplitude:
　displacement, 329, 336
　pressure, 329, 336
　of vibration, 295
　of wave, 306–7, 326, 328–29
Amplitude modulation (AM), 514–15
AM radio, 514–16
Aneroid barometer, 190
Aneroid gauge, 189–90
Angle:
　Brewster's, 568
　of contact, 198
　critical, 531
　of incidence, 312, 314, 521, 530
　phase, 493
　radian measure of, 138
　of reflection, 312–13, 521
　of refraction, 314, 530
Angstrom (unit), 552 *fn*
Angular acceleration, 139–41
Angular magnification, 584–85
Angular momentum, 149–51, 153–54
　law of conservation of, 150–51, 153–54
　quantized in atoms, 652, 671–72
　spin, 672–73
　vector nature of, 151–54
Angular quantities, 137–41

Angular velocity, 138–41
Anode, 394
Antielectron (*see* Positron)
Antimatter, 758
Antineutrino, 713, 758
Antineutron, 758
Antinodes, 318, 334–36
Antiparticles, 713, 758
Antiproton, 758, 759
Apparent weightlessness, 83–84
Approximate calculations, 65, 784
Arago, F., 553
Arches, 175–76
Archimedes, 193–94
Archimedes' principle, 193–96
Area under a curve or graph, 25–26
Aristotle, 1, 48–49
Armature, 451, 476
Astigmatism:
　of eye, 583
　of spherical lens, 590
Astronomical telescope, 586–87
Atmosphere (unit of pressure), 187
Atmosphere, scattering of light by, 572–73
Atmospheric pressure, 187–88, 190
Atomic bomb, 733–34
Atomic mass, 218
Atomic mass number, 705–6
Atomic mass unit, 218, 707
Atomic number, 674, 705–6
　determined from x-ray analysis, 677–78
Atomic spectra, 650–52, 655–56
　fine structure of, 663, 672
Atomic structure, 354–55, 649–78
　Bohr model of, 652–59
　of complex atoms, 674–78
　of hydrogen, 652–59, 663, 670–74
　planetary model of, 354–55, 650–52
　plum pudding model of, 649
　quantum mechanics of, 670–79
　shells and subshells of, 675–77
Atomic theory, 217–19
　(*see also* Atomic structure, Atoms, Kinetic theory)
Atoms, 217–19, 354–55
　complex, 674–78
　electric charge in, 354–55
　energy levels in, 655–57, 671–81

799

PERIODIC TABLE OF THE ELEMENTS[†]

Transition elements

Key:
- Symbol — **Cl** 17 — Atomic number
- Atomic mass‡ — 35.453
- Electron configuration — $3p^5$

Main group elements

Group I	Group II	Group III	Group IV	Group V	Group VI	Group VII	Group 0
H 1 — 1.0079 — $1s^1$							**He** 2 — 4.0026 — $1s^2$
Li 3 — 6.94 — $2s^1$	**Be** 4 — 9.01218 — $2s^2$	**B** 5 — 10.81 — $2p^1$	**C** 6 — 12.011 — $2p^2$	**N** 7 — 14.0067 — $2p^3$	**O** 8 — 15.9994 — $2p^4$	**F** 9 — 18.9984 — $2p^5$	**Ne** 10 — 20.17 — $2p^6$
Na 11 — 22.9898 — $3s^1$	**Mg** 12 — 24.305 — $3s^2$	**Al** 13 — 26.9815 — $3p^1$	**Si** 14 — 28.0855 — $3p^2$	**P** 15 — 30.9738 — $3p^3$	**S** 16 — 32.06 — $3p^4$	**Cl** 17 — 35.453 — $3p^5$	**Ar** 18 — 39.948 — $3p^6$
K 19 — 39.0983 — $4s^1$	**Ca** 20 — 40.08 — $4s^2$	**Ga** 31 — 69.735 — $4p^1$	**Ge** 32 — 72.59 — $4p^2$	**As** 33 — 74.9216 — $4p^3$	**Se** 34 — 78.96 — $4p^4$	**Br** 35 — 79.904 — $4p^5$	**Kr** 36 — 83.80 — $4p^6$
Rb 37 — 85.467 — $5s^1$	**Sr** 38 — 87.62 — $5s^2$	**In** 49 — 114.82 — $5p^1$	**Sn** 50 — 118.69 — $5p^2$	**Sb** 51 — 121.75 — $5p^3$	**Te** 52 — 127.60 — $5p^4$	**I** 53 — 126.90 — $5p^5$	**Xe** 54 — 131.30 — $5p^6$
Cs 55 — 132.905 — $6s^1$	**Ba** 56 — 137.33 — $6s^2$	**Tl** 81 — 204.37 — $6p^1$	**Pb** 82 — 207.2 — $6p^2$	**Bi** 83 — 208.980 — $6p^3$	**Po** 84 — (209) — $6p^4$	**At** 85 — (210) — $6p^5$	**Rn** 86 — (222) — $6p^6$
Fr 87 — (223) — $7s^1$	**Ra** 88 — 226.025 — $7s^2$						

Transition elements

3	4	5	6	7	8	9	10	11	12
Sc 21 — 44.9559 — $3d^14s^2$	**Ti** 22 — 47.90 — $3d^24s^2$	**V** 23 — 50.9415 — $3d^34s^2$	**Cr** 24 — 51.996 — $3d^54s^1$	**Mn** 25 — 54.9380 — $3d^54s^2$	**Fe** 26 — 55.847 — $3d^64s^2$	**Co** 27 — 58.9332 — $3d^74s^2$	**Ni** 28 — 58.71 — $3d^84s^2$	**Cu** 29 — 63.546 — $3d^{10}4s^1$	**Zn** 30 — 65.38 — $3d^{10}4s^2$
Y 39 — 88.9059 — $4d^15s^2$	**Zr** 40 — 91.22 — $4d^25s^2$	**Nb** 41 — 92.9064 — $4d^45s^1$	**Mo** 42 — 95.94 — $4d^55s^1$	**Tc** 43 — 98.9062 — $4d^55s^2$	**Ru** 44 — 101.07 — $4d^75s^1$	**Rh** 45 — 102.906 — $4d^85s^1$	**Pd** 46 — 106.4 — $4d^{10}5s^0$	**Ag** 47 — 107.868 — $4d^{10}5s^1$	**Cd** 48 — 112.41 — $4d^{10}5s^1$
57–71[§]	**Hf** 72 — 178.49 — $5d^26s^2$	**Ta** 73 — 180.947 — $5d^36s^2$	**W** 74 — 183.85 — $5d^46s^2$	**Re** 75 — 186.207 — $5d^56s^2$	**Os** 76 — 190.2 — $5d^66s^2$	**Ir** 77 — 192.22 — $5d^76s^2$	**Pt** 78 — 195.09 — $5d^96s^1$	**Au** 79 — 196.966 — $5d^{10}6s^1$	**Hg** 80 — 200.59 — $5d^{10}6s^2$
89–103[‖]	**Rf** 104 — (261) — $6d^27s^2$	**Ha** 105 — (260) — $6d^37s^2$	106 — (263)	107 — (262)	108 — (265)	109 — (266)			

§Lanthanide series

La 57 139.906 $5d^16s^2$	**Ce** 58 140.12 $5d^14f^16s^2$	**Pr** 59 140.908 $4f^36s^2$	**Nd** 60 144.24 $4f^46s^2$	**Pm** 61 (145) $4f^56s^2$	**Sm** 62 150.4 $4f^66s^2$	**Eu** 63 151.96 $4f^76s^2$	**Gd** 64 157.25 $5d^14f^76s^2$	**Tb** 65 158.925 $5d^14f^86s^2$	**Dy** 66 162.50 $4f^{10}6s^2$	**Ho** 67 164.930 $4f^{11}6s^2$	**Er** 68 167.26 $4f^{12}6s^2$	**Tm** 69 168.934 $4f^{13}6s^2$	**Yb** 70 173.04 $4f^{14}6s^2$	**Lu** 71 174.967 $5d^14f^{14}6s^2$

‖Actinide series

Ac 89 (227) $6d^17s^2$	**Th** 90 232.038 $6d^27s^2$	**Pa** 91 231.036 $5f^26d^17s^2$	**U** 92 238.029 $5f^36d^17s^2$	**Np** 93 237.048 $5f^36d^17s^2$	**Pu** 94 (244) $5f^66d^07s^2$	**Am** 95 (243) $5f^76d^07s^2$	**Cm** 96 (247) $5f^76d^17s^2$	**Bk** 97 (247) $5f^86d^17s^2$	**Cf** 98 (251) $5f^{10}6d^07s^2$	**Es** 99 (253) $5f^{11}6d^07s^2$	**Fm** 100 (257) $5f^{12}6d^07s^2$	**Md** 101 (258) $5f^{13}6d^07s^2$	**No** 102 (259) $6d^07s^2$	**Lr** 103 (260) $7s^27p^1$

† Atomic mass values averaged over isotopes in percentages they occur on earth's surface.
‡ For many unstable elements, mass number of the most stable known isotope is given in parentheses.